Endogenous Toxins

Edited by
Peter J. O'Brien and
W. Robert Bruce

Further Reading

Geacintov, N. E., Broyde, S. (eds.)

The Chemical Biology of DNA Damage

2010
ISBN: 978-3-527-32295-4

Külpmann, W. R. (ed.)

Clinical Toxicological Analysis
Procedures, Results, Interpretation

2009
ISBN: 978-3-527-31890-2

Knasmüller, S., DeMarini, D. M., Johnson, I., Gerhäuser, C. (eds.)

Chemoprevention of Cancer and DNA Damage by Dietary Factors

2008
ISBN: 978-3-527-32058-5

Meyers, R. A. (ed.)

Cancer
From Mechanisms to Therapeutic Approaches

2007
ISBN: 978-3-527-31768-4

Dübel, S. (ed.)

Handbook of Therapeutic Antibodies

2007
ISBN: 978-3-527-31453-9

Brigelius-Flohé, R., Joost, H.-G. (eds.)

Nutritional Genomics
Impact on Health and Disease

2006
ISBN: 978-3-527-31294-8

Endogenous Toxins

Diet, Genetics, Disease and Treatment

Edited by
Peter J. O'Brien and W. Robert Bruce

VOLUME I

WILEY-VCH Verlag GmbH & Co. KGaA

The Editors

Prof. Peter J. O'Brien
University of Toronto
Faculty of Pharmacy, Room 1004
College Street 144
Toronto, ON M5S 3M2
Canada

Prof. W. Robert Bruce
University of Toronto
Fitz Gerald Building, Room 342
College Street 150
Toronto, ON M5S 3E2
Canada

All books published by Wiley-VCH are carefully produced. Nevertheless, authors, editors, and publisher do not warrant the information contained in these books, including this book, to be free of errors. Readers are advised to keep in mind that statements, data, illustrations, procedural details or other items may inadvertently be inaccurate.

Library of Congress Card No.:
applied for

British Library Cataloguing-in-Publication Data
A catalogue record for this book is available from the British Library.

Bibliographic information published by the Deutsche Nationalbibliothek
The Deutsche Nationalbibliothek lists this publication in the Deutsche Nationalbibliografie; detailed bibliographic data are available on the Internet at <http://dnb.d-nb.de>.

© 2010 WILEY-VCH Verlag GmbH & Co. KGaA, Weinheim

All rights reserved (including those of translation into other languages). No part of this book may be reproduced in any form – by photoprinting, microfilm, or any other means – nor transmitted or translated into a machine language without written permission from the publishers. Registered names, trademarks, etc. used in this book, even when not specifically marked as such, are not to be considered unprotected by law.

Composition   Laserwords Private Limited, Chennai, India
Printing   Strauss GmbH, Mörlenbach
Bookbinding   Litges & Dopf GmbH, Heppenheim
Cover Design   Adam Design, Weinheim

Printed in the Federal Republic of Germany
Printed on acid-free paper

ISBN: 978-3-527-32363-0

The Authors

Contents

VOLUME I

Preface *XXXI*

List of Contributors *XXXIII*

Abbreviations *XLIII*

Part One Endogenous Toxins Associated with Excessive Sugar, Fat, Meat, or Alcohol Consumption *1*

Sub-Part Chemistry and Biochemistry *3*

1 **Endogenous DNA Damage** *5*
Erin G. Prestwich and Peter C. Dedon
1.1 Introduction *5*
1.2 Oxidatively Damaged DNA *8*
1.2.1 Base Oxidation *9*
1.2.2 Tandem Lesions: Multiple Products Derived from a Single Oxidation Event *12*
1.2.3 Base Nitration *13*
1.2.4 Halogenation *14*
1.2.5 2-Deoxyribose Oxidation *14*
1.3 DNA Alkylation by Endogenous Electrophiles *15*
1.3.1 Lipid Peroxidation-Derived Adducts *17*
1.3.2 Malondialdehyde-Like Adducts *21*
1.3.3 DNA Glycation *22*
1.3.4 Estrogen-Derived DNA Adducts *24*
1.4 DNA and RNA Deamination *25*
1.4.1 Hydrolytic Deamination *26*
1.4.2 Nitrosative Chemistry and Inflammation *26*
1.4.3 Enzymatic Deamination *27*

Endogenous Toxins. Diet, Genetics, Disease and Treatment.
Edited by Peter J. O'Brien and W. Robert Bruce
Copyright © 2010 WILEY-VCH Verlag GmbH & Co. KGaA, Weinheim
ISBN: 978-3-527-32363-0

1.5	Summary 28
	References 28

2 Modification of Cysteine Residues in Protein by Endogenous Oxidants and Electrophiles 43

Norma Frizzell and John W. Baynes

2.1	Introduction 43
2.2	Autoxidation of Cysteine 44
2.3	Glutathione 44
2.3.1	Glutathionylation 45
2.3.2	Glutathione as an Electrophile Trap 45
2.4	Lipoxidation 46
2.4.1	Glyceraldehyde-3-Phosphate Dehydrogenase (GAPDH) 47
2.4.2	Albumin 48
2.4.3	Actin 48
2.4.4	Quantitative Analysis 49
2.5	Maillard Reactions of Cysteine 50
2.5.1	Transglycosylation 50
2.5.2	Carboxyalkylation 51
2.6	Succination 53
2.6.1	Discovery and Metabolic Origin of S-(2-Succinyl)cysteine (2SC) 53
2.6.2	2SC in Skeletal Muscle in Diabetes 55
2.6.3	2SC in Adipose Tissue in Diabetes 55
2.6.4	Overview on Succination of Protein 57
2.7	Conclusion 57
	Acknowledgment 58
	References 58

3 Endogenous Macromolecule Radicals 65

Arno G. Siraki and Marilyn Ehrenshaft

3.1	Introduction 65
3.2	What is a Free Radical? 66
3.2.1	Superoxide Anion 67
3.2.2	Singlet Oxygen 68
3.2.3	Hydrogen Peroxide 68
3.2.4	Peroxynitrite and Carbonate Radicals 69
3.2.5	Nitrogen-Based Radicals 70
3.2.6	Hypochlorous Acid 71
3.2.7	Thiyl Radicals 72
3.2.8	Lipid Peroxidation 73
3.2.9	Summary 73
3.3	Free Radicals as Initiators of Macromolecule Damage 74
3.4	Macromolecule Radical Formation Induced by Small Molecule Free Radicals 78

3.5	Examples and Potential Application of Immuno-Spin Trapping (IST) for the Detection of Macromolecule Radicals *80*	
3.5.1	Superoxide-Induced Macromolecule Radicals *80*	
3.5.2	Singlet Oxygen–Induced Macromolecule Radicals *81*	
3.5.3	Hydroxyl Radical–Induced Macromolecule Radicals *82*	
3.5.4	Peroxynitrite, Nitrogen Dioxide Radicals, and Nitric Oxide *83*	
3.5.5	Carbonate Radicals *84*	
3.5.6	HOCl-Induced Macromolecule Radicals *85*	
3.5.7	Thiyl Radical–Induced Macromolecule Radicals *86*	
3.5.8	Myeloperoxidase Protein Radical Detection *87*	
3.5.9	Heme Protein Radicals–A Unique Exception of Substrate-Induced Protein Radical Formation *88*	
3.5.10	Example of *In vivo* Detection of Macromolecule Radicals Using IST *89*	
3.5.11	Future Directions *91*	
3.6	Immunological Detection of Oxidized Tryptophan Residues in Proteins *92*	
	References *92*	
4	**Alcohol-Derived Bioadducts** *103*	
	Geoffrey M. Thiele and Lynell W. Klassen	
4.1	Introduction *103*	
4.2	The Generation of Reactive Metabolites *104*	
4.3	Ethanol Metabolites React with Proteins *106*	
4.4	Acetaldehyde Adducts *107*	
4.4.1	Formation and Types of Adducts *107*	
4.4.2	Biological Detection and/or Significance *109*	
4.5	Malondialdehyde *111*	
4.5.1	Formation and Types of Adducts *111*	
4.5.2	Biological Detection and/or Significance *113*	
4.6	Malondialdehyde–Acetaldehyde (MAA) Adducts *114*	
4.6.1	Formation and Types of Adducts *114*	
4.6.2	Biological Detection and/or Significance *114*	
4.7	4-Hydroxyalkenals *116*	
4.7.1	Formation and Types of Adducts *116*	
4.7.2	Biological Detection and/or Significance *116*	
4.8	Free Radical–Derived Adducts *117*	
4.8.1	Formation and Detection of Adducts *117*	
4.8.2	Biological Detection and/or Significance *117*	
4.9	5-Deoxy-D-xylulose-1-phosphate (DXP) *118*	
4.10	Acetaldehyde–Glucose Amadori Product Adduct *119*	
4.11	Immune Responses to Proteins Modified with Alcohol Metabolites *119*	
4.12	The Role of Alcohol-Derived Bioadducts in Alcohol-Related Tissue Injury *121*

Contents

4.13	Future Directions *123*	
	Acknowledgments *124*	
	References *124*	

5 **Iron from Meat Produces Endogenous Procarcinogenic Peroxides** *133*
Denis E. Corpet, Françoise Guéraud, and Peter J. O'Brien
5.1 Introduction *133*
5.2 Toxic Effects of Iron: Molecular Mechanisms *134*
5.2.1 Fe-/Cu-Catalyzed Fenton Chemistry *135*
5.2.2 Fe-/Cu-Catalyzed Oxidation of Proteins *135*
5.2.3 Fe-/Cu-Catalyzed Protein Carbonylation *136*
5.2.4 Fe-/Cu-Catalyzed Advanced Glycation End Products (AGEs) Formation *136*
5.2.5 Fe-/Cu-Catalyzed Lipid Peroxidation and Advanced Lipoxidation End Products (ALEs) *138*
5.2.6 Fe-/Cu-Catalyzed Oxidation of Nucleic Acids *138*
5.2.7 Role of Lysosomal Fe/Cu in Oxidative Stress Cytotoxicity *139*
5.2.8 Role of ROS and Iron in Cellular Redox Signaling that Could Cause Cell Transformation *140*
5.3 Procarcinogenic Effects of Iron: *In vitro* Studies *140*
5.3.1 Iron in Food *140*
5.3.2 Iron in Cell Culture *141*
5.4 Procarcinogenic Effects of Iron in the Gut *141*
5.4.1 Free Radicals in Gut Lumen *141*
5.4.2 Hemin Effect in the Colon of Rats *141*
5.4.3 Iron-Induced Lipid Peroxidation Products *142*
5.4.4 Effect of Heme Iron on Experimental Carcinogenesis *142*
5.4.5 Carcinogenesis Prevention *142*
5.5 Procarcinogenic Effects of Iron Inside the Body *143*
5.5.1 Iron Concentration is Tightly Limited in Body Fluids *143*
5.5.2 Iron Overload Causes Severe Diseases *143*
5.5.3 Colorectal Cancer and Iron Overload *144*
5.6 General Conclusion *144*
References *144*

Sub-Part Molecular Toxicology Mechanisms of Dietary Endogenous Toxins *151*

6 **Short Chain Sugars as Endogenous Toxins** *153*
Ludmil T. Benov
6.1 Definition and Properties *153*
6.2 Short Chain Sugars and Reactive Oxygen Species *154*
6.3 Dicarbonyls *156*
6.4 Direct Reactions with Biomolecules *156*
6.5 Short Chain Sugars and Their Sources *157*
6.5.1 Trioses and Triosephosphates *157*

6.6	Glyceraldehyde 3-phosphate Dehydrogenase	159
6.7	Triosephosphate Isomerase	159
6.7.1	Glycolaldehyde	159
6.7.2	Short Chain Sugars in Diabetes and Aging	160
6.7.2.1	Diabetes	160
6.7.2.2	Aging, Glycation, and Dietary Restriction	161
	Acknowledgment	162
	References	162
7	**Fructose-Derived Endogenous Toxins**	**173**
	Peter J. O'Brien, Cynthia Y. Feng, Owen Lee, Q. Dong, Rhea Mehta,	
	Jeff Bruce, and W. Robert Bruce	
7.1	Introduction	174
7.2	Recent Increased Consumption of Fructose	174
7.3	Health Concerns Associated with High Chronic Consumption of Fructose	176
7.3.1	Sugar and Sugar Intolerances	176
7.3.2	Fructose and Obesity – Nonalcoholic Liver Disease (NAFLD)	177
7.4	Sugars as a Source of Endogenous Reactive Carbonyl Formation and AGEs	178
7.4.1	Fructose versus Glucose Metabolism by Glycolysis to Carbonyl Metabolites	178
7.4.2	Fructose Oxidation to Endogenous Genotoxic Carbonyl Products and Protein Advanced Glycation End Products (AGEs)	179
7.4.3	Rodent Models for Fatty Liver (Steatosis) and NASH (Steatohepatitis)	183
7.4.4	The Two-Hit Hypothesis for NASH Hepatotoxicity	184
7.4.4.1	Introduction	184
7.4.4.2	The First Hit: Steatosis Mechanisms and Fructose	184
7.4.4.3	The Second Hit: Hepatocyte Inflammation Oxidative Stress Model with Fructose	185
7.4.4.4	The Second Hit: ROS Formation Mechanisms	187
7.4.4.5	Fructose-Induced *In Vivo* Liver Toxicity Studies in Rats	187
7.5	Rat Hepatocyte Studies on Endogenous Toxins Formed by Fructose Metabolism and/or Oxidation	188
7.5.1	Introduction	188
7.5.2	Glyceraldehyde (GA)	188
7.5.2.1	GA Enzymic Formation	188
7.5.2.2	GA Detoxication Metabolizing Enzymes	188
7.5.2.3	GA Toxicity Mechanisms	189
7.5.2.4	GA Oxidative Stress Enhanced Toxicity: Inflammation Model	189
7.5.3	Dihydroxyacetone (DHA)	190
7.5.3.1	DHA Enzymic or Nonenzymic Formation	190
7.5.3.2	DHA Enzymic Metabolism or Oxidation	190
7.5.3.3	DHA Toxicity Mechanisms	190

7.5.3.4	DHA Oxidative Stress Enhanced Toxicity: Inflammation Model	190
7.5.4	Glycolaldehyde (GOA)	190
7.5.4.1	GOA Enzymic Formation	190
7.5.4.2	GOA Detoxication Metabolizing Enzymes	191
7.5.4.3	GOA Toxicity Mechanisms	192
7.5.4.4	GOA Oxidative Stress Enhanced Toxicity: Inflammation Model	192
7.5.5	Glyoxal (G) and Methylglyoxal (MG)	193
7.5.5.1	G and MG Enzymic and Nonenzymic Formation	193
7.5.5.2	Gloxal and MG Detoxification Metabolizing Enzymes	193
7.5.5.3	Glyoxal and MG Toxicity Mechanisms	196
7.5.5.4	Glyoxal and MG Oxidative Stress Enhanced Cytotoxicity: Inflammation Model	197
7.5.6	Glycolate as a Peroxisomal ROS Source	198
7.5.6.1	Glycolate (GL) Enzymic Formation	198
7.5.6.2	GL Detoxification Metabolizing Enzymes	198
7.5.6.3	GL Toxicity	198
7.5.6.4	GL Oxidative Stress Enhanced Toxicity: Inflammation Model	198
7.5.7	Glyoxylate as a Carbonyl Toxin	198
7.5.7.1	Glyoxylate (Gx) Enzymic Formation	198
7.5.7.2	Gx Detoxication Metabolizing Enzymes	199
7.5.7.3	Gx Toxicity Mechanisms	199
7.5.7.4	Glyoxylate Oxidative Stress Enhanced Cytotoxicity: Inflammation Model	200
7.5.8	Oxalic Acid	200
7.5.8.1	Oxalic Acid Enzymic Formation	200
7.5.8.2	Oxalic Acid Detoxication Metabolizing Enzymes and Toxicogenetic Basis	200
7.5.8.3	Oxalic Acid Toxicity Mechanisms: Hepatic Endogenous Metabolite and Kidney Toxin	201
7.5.8.4	Oxalic Acid Oxidative Stress Enhanced Toxicity: Inflammation Model	201
7.6	Cancer Risk and Genotoxicity of Fructose or Carbonyl Metabolites	202
7.7	Disease Prevention by Fruits and Vegetables versus Fructose Concern	203
7.8	Conclusions	203
	References	204
8	**Glyceraldehyde-Related Reaction Products**	**213**
	Teruyuki Usui, Hirohito Watanabe, and Fumitaka Hayase	
8.1	Maillard Reaction	213
8.2	Maillard-Related Diseases	214
8.3	Glyceraldehyde-Modified Protein	215
8.3.1	Cytotoxicity for HL-60 Cells	216
8.3.2	Cytotoxicity for Caco-2 Cells	217

8.3.3	Cytotoxicity for PC12 Cells	218
8.4	Glyceraldehyde-Derived AGEs	219
8.5	Cytotoxicity and Oxidative Stress	220
	References	223

9 Estrogens as Endogenous Toxins 227

Jason Matthews

9.1	Introduction	228
9.2	Estrogen Synthesis	229
9.3	Receptor-Mediated Estrogen Signaling	229
9.3.1	Estrogen Receptor α and β	230
9.3.2	ER-Mediated Mechanism of Cell Proliferation	232
9.3.3	Rapid Nongenomic Effects of Estrogens	232
9.4	DNA Damage Induced by Estrogen	233
9.5	Oxidative Metabolism of Estrogen	233
9.5.1	Formation of Catechol Estrogens	234
9.5.2	4-Hydroxylated Oxidation of Estrogens by CYP1B1	234
9.5.3	Direct DNA Damage by Estrogen Metabolites	236
9.5.4	Indirect DNA Damage and ROS Generation	239
9.6	Role for Estrogen Receptor in ROS Generation	240
9.6.1	ER in the Mitochondria	241
9.7	Treatment of Estrogen-Dependent Diseases	241
9.8	Conclusions	242
	References	243

10 Reactive Oxygen Species, Hypohalites, and Reactive Nitrogen Species in Liver Pathophysiology 249

Hartmut Jaeschke

10.1	Introduction	249
10.2	Reactive Oxygen Species	250
10.2.1	Nature of Reactive Oxygen Species	250
10.2.2	Sources of ROS and Their Pathophysiological Relevance	250
10.2.3	Mechanisms of ROS-Mediated Cell Death	253
10.3	Reactive Nitrogen Species	253
10.3.1	Sources of Peroxynitrite	253
10.3.2	Peroxynitrite and Acetaminophen Hepatotoxicity	254
10.4	Hypohalites	256
10.4.1	Sources of Hypochlorous Acid	256
10.4.2	Hypochlorite and Host Defense Functions	256
10.4.3	Hypochlorite in Liver Pathophysiology	256
10.5	Summary	259
	References	259

Part Two Genetics: Endogenous Toxins Associated with Inborn Errors of Metabolism *267*

11 Oxalate and Primary Hyperoxaluria *269*
Christopher J. Danpure
11.1 Oxalate as an Endogenous Toxin *269*
11.2 Hereditary Oxalate Overproduction – The Primary Hyperoxalurias *270*
11.2.1 Clinical Characteristics *270*
11.2.2 Enzyme and Metabolic Defects *271*
11.2.3 Crystal Structures of AGT and GRHPR *273*
11.2.4 Genes, Mutations, and Polymorphisms *274*
11.2.5 Genotype–Phenotype Correlations *275*
11.2.5.1 Pro11Leu *275*
11.2.5.2 Gly170Arg *277*
11.2.5.3 Ile244Thr *278*
11.2.5.4 Gly41Arg *278*
11.2.5.5 Gly82Glu *279*
11.2.5.6 Ser205Pro *279*
11.2.5.7 Other Mutations *279*
11.2.6 Diagnosis *280*
11.2.7 Treatment *280*
11.3 Cytotoxicity of Oxalate, Calcium Oxalate, and Related Metabolites *283*
11.3.1 Indirect Glycolate Toxicity and Screening for Chemotherapeutic Agents *283*
11.4 Conclusions *284*
References *284*

12 Pathophysiology of Endogenous Toxins and Their Relation to Inborn Errors of Metabolism and Drug-Mediated Toxicities *291*
Vangala Subrahmanyam
12.1 Introduction *292*
12.2 Pathophysiology of Hepatobiliary System *293*
12.2.1 Arginine Vasopressin and Bile Flow *294*
12.3 Inborn Errors of Bile Acid/Salt Transporter Deficiencies *295*
12.3.1 Progressive Familial Cholestasis Type 1 (PFIC1) and Benign Recurrent Intrahepatic Cholestasis (BRIC) *295*
12.3.2 Progressive Familial Cholestasis Types 2 and 3 (PFIC2 and PFIC3) *298*
12.3.3 Dubin-Johnson Syndrome *299*
12.3.4 Sitosterolemia *299*
12.4 Drugs and Cholestatic Liver Disease *299*
12.4.1 Bosentan Hepatotoxicity – Role of Inhibition of Bile Acid/Salt Transporters *300*

12.4.2	Troglitazone Hepatotoxicity-Role of Bile Salt Export Pump (BSEP/Bsep) *301*	
12.4.3	Cholestatic Hepatotoxicity Associated with Ethinyl Estradiol: Role of MRP2 *302*	
12.5	Phospholipases and Phospholipidosis *302*	
12.5.1	Lysosomal PhospholipaseA2 Deficiency and Phospholipidosis *303*	
12.5.2	Aminoglycoside Antibiotic-Induced Phospholipidosis and Nephrotoxicity *304*	
12.6	Mechanisms of Toxicity Associated with Accumulated Metabolites during Inborn Errors of Metabolism *305*	
12.7	Pathophysiology of Phenylketonuria and Hyperphenylalaninemia *305*	
12.7.1	Oxidative Stress, Mitochondrial Damage, and Antioxidant Status in Phenylketonuria *306*	
12.7.2	Glucose-Induced Oxidative Stress Mediated Toxicity: Implications in Diabetes *307*	
12.7.3	Drugs-Induced Hyperglycemia *308*	
12.7.4	Antipsychotic-Induced Hyperglycemia and Diabetes *309*	
12.8	Conclusions *309*	
	References *309*	
13	**Mechanisms of Toxicity in Fatty Acid Oxidation Disorders** *317*	
	J. Daniel Sharer	
13.1	Introduction *317*	
13.2	Mitochondrial β-Oxidation *319*	
13.2.1	Mobilization and Activation of Long Chain Fatty Acids *319*	
13.2.2	Carnitine-Mediated Transport *321*	
13.2.2.1	Plasma Membrane Carnitine Transporter *322*	
13.2.2.2	Carnitine Palmitoyl Transferase Type 1 (CPT 1) *322*	
13.2.2.3	Carnitine Acylcarnitine Translocase (CACT) *323*	
13.2.2.4	Carnitine Palmitoyl Transferase Type 2 (CPT 2) *324*	
13.2.3	β-Oxidation Enzymes *324*	
13.2.3.1	Very Long Chain Acyl-CoA Dehydrogenase (VLCAD) *325*	
13.2.3.2	Long Chain Acyl-CoA Dehydrogenase (LCAD) *325*	
13.2.3.3	Medium Chain Acyl-CoA Dehydrogenase (MCAD) *326*	
13.2.3.4	Short Chain Acyl-CoA Dehydrogenase (SCAD) *326*	
13.2.3.5	3-Hydroxy Acyl CoA Dehydrogenases *326*	
13.2.3.6	Mitochondrial β-Ketothiolase *327*	
13.2.3.7	Mitochondrial Trifunctional Protein (MTP) *328*	
13.2.3.8	Electron Transfer Flavoprotein (ETF)/Electron Transfer Flavoprotein Dehydrogenase (ETFDH) *328*	
13.2.4	Mechanisms of Toxicity Associated with Disorders of Mitochondrial β-Oxidation *328*	
13.2.4.1	Energy Depletion *329*	
13.2.4.2	Accumulation of Toxic Compounds *331*	

13.2.4.3	Sequestration of Important Biological Compounds 334
13.3	Peroxisomal Fatty Acid Oxidation 335
13.3.1	β-Oxidation 335
13.3.1.1	Acyl-CoA Oxidase 336
13.3.1.2	Bifunctional Protein 337
13.3.1.3	ABCD1 337
13.3.2	α-Oxidation 338
13.3.2.1	Phytanoyl-CoA Hydroxylase 338
13.3.2.2	2-Methyl-acyl-CoA Racemase (AMACR) 338
13.3.3	Mechanisms of Toxicity Associated with Disorders of Peroxisomal FAO 339
13.3.3.1	Peroxisomal β-Oxidation Defects 339
13.3.3.2	Peroxisomal α-Oxidation Defects 340
13.4	Conclusions 341
	Acknowledgments 342
	References 342

14	**Homocysteine as an Endogenous Toxin in Cardiovascular Disease** 349
	Sana Basseri, Jennifer Caldwell, Shantanu Sengupta, Arun Kumar, and Richard C. Austin
14.1	Cardiovascular Disease Risk 349
14.2	Overview of Atherosclerosis 349
14.2.1	Lesion Development and Disease Progression 350
14.2.2	Contribution of Atherosclerosis to Cardiovascular Disease 351
14.3	Homocysteine Metabolism 351
14.4	Hyperhomocysteinemia and CVD 351
14.5	Genetic Causes of Hyperhomocysteinemia 353
14.6	Recent Studies 354
14.6.1	Mouse Models of Hyperhomocysteinemia and Atherosclerosis 355
14.6.2	Animal Study Findings 355
14.6.3	Clinical Trials Examining the Effects of Folate and B Vitamin Supplementation on Cardiovascular Disease 356
14.6.3.1	Folate 356
14.6.3.2	Vitamin B_{12} 357
14.6.3.3	Vitamin B_6 357
14.6.3.4	Riboflavin 358
14.6.4	Effect of Lifestyle on tHcy Levels 358
14.7	Potential Mechanisms through which Hyperhomocysteinemia Contributes to Atherogenesis 359
14.7.1	Protein-S-Homocysteinylation 359
14.7.2	Protein-N-Homocysteinylation 361
14.7.3	Nitric Oxide Production and Oxidative Stress 362
14.7.4	Smooth Muscle Cell Proliferation 363
14.7.5	Endothelial Cell Proliferation 363
14.7.6	ER Stress 364

14.7.7	Vascular Matrix Studies	364
14.7.8	Findings from Yeast Studies	365
14.7.9	Inflammation	365
14.8	Conclusions	366
	References	367

15 Uric Acid Alterations in Cardiometabolic Disorders and Gout 379
Renato Ippolito, Ferruccio Galletti, and Pasquale Strazzullo

15.1 Introduction 379
15.2 Dual Properties of Uric Acid as Antioxidant or Pro-Oxidant Molecule 380
15.3 Effects of Uric Acid on the Arterial Wall Properties and Endothelial Function 381
15.4 Causes of Hyperuricemia and Hypouricemia 381
15.5 Hyperuricemia, Gout, and Their Treatment 382
15.6 Uric Acid and Cardiovascular Disease 385
    References 387

16 Genetic Defects in Iron and Copper Trafficking 395
Douglas M. Templeton

16.1 Introduction and Terminology 395
16.2 Iron 396
16.2.1 Physiology 396
16.2.2 Hereditary Hemochromatoses 401
16.2.3 African Siderosis 402
16.2.4 Friedreich Ataxia 402
16.2.5 X-Linked Sideroblastic Anemias 403
16.2.6 Hypotransferrinemia 404
16.2.7 Hereditary Hyperferritinemia Cataract Syndrome 404
16.2.8 Hypochromic Microcytic Anemia with Iron Overload 405
16.2.9 Aceruloplasminemia 405
16.3 Copper 406
16.3.1 Physiology 406
16.3.2 ATP7A/B 407
16.3.3 Other Copper Toxicoses 409
16.3.4 Amyotrophic Lateral Sclerosis (ALS) 409
16.3.5 Other Copper-Related Genes 409
16.4 Summary 410
    References 411

17 Polyglutamine Neuropathies: Animal Models to Molecular Mechanisms 419
Kelvin Hui and Jeffrey Henderson

17.1 Neurobiology of Polyglutamine Diseases 419
17.1.1 Huntington's Disease 420

17.1.2	Kennedy's Disease (SBMA)	421
17.1.3	DRPLA/Smith's Disease	422
17.1.4	Spinocerebellar Ataxia 1 (SCA1)	422
17.1.5	SCA2	423
17.1.6	SCA3/Machado-Joseph Disease	423
17.1.7	SCA6	424
17.1.8	SCA7	424
17.1.9	SCA17	425
17.2	Animal Models of Polyglutamine Diseases	425
17.2.1	Worm and Fly	426
17.2.2	Murine Models	427
17.2.3	Critical Issues in Model Dependence	427
17.3	Mechanisms of Polyglutamine-Induced Neural Injury	429
17.3.1	Conformation of Polyglutamine Sequences	429
17.3.2	Polyglutamine Aggregation	430
17.3.3	Polyglutamine-Based Disruption of Protein Function	432
17.3.4	Toxic Gain of Function in Pathogenic PolyQ Proteins	433
17.3.5	Disruption of Ubiquitin-Mediated Proteasomal Degradation and Induction of ER Stress	434
17.3.6	CAG Repeats and RNA-Mediated Toxicity	436
17.3.7	Questions Remaining	436
17.4	Future Perspectives	437
	Acknowledgments	439
	References	439

VOLUME II

Abbreviations XXXI

Part Three Examples of Endogenous Toxins Associated with Acquired Diseases or Animal Disease Models 449

18	**Alcohol-Induced Hepatic Injury**	**451**
	Emanuele Albano	
18.1	Introduction	451
18.2	Ethanol Metabolism and Toxicity	453
18.2.1	Ethanol Metabolism in the Liver	453
18.2.2	Ethanol-Induced Oxidative Stress	456
18.2.3	Alcohol Effects on Methionine Metabolism	459
18.3	Mechanisms of Alcohol Hepatotoxicity	460
18.3.1	Alcoholic Steatosis	460
18.3.2	Alcoholic Steatohepatitis	464
18.3.2.1	The Mechanism Promoting Inflammatory Reactions in ALD	464

18.3.2.2	Role of Immune Reactions in Alcohol-Induced Hepatic Inflammation *466*	
18.3.2.3	Mechanisms in Alcohol-Mediated Hepatocyte Killing *468*	
18.3.2.4	Alcohol and the Formation of Mallory Bodies *469*	
18.3.3	Alcoholic Cirrhosis *469*	
18.3.4	Alcohol and Liver Cancer *472*	
18.4	Conclusions *474*	
	References *475*	
19	**Ethanol-Induced Endotoxemia and Tissue Injury** *485*	
	Radhakrishna K. Rao	
19.1	Alcoholic Endotoxemia *486*	
19.1.1	Endotoxemia in Patients with ALD *486*	
19.1.2	Endotoxemia in Experimental Models of ALD *488*	
19.1.3	Gender Differences in Endotoxemia *489*	
19.1.4	Bacterial Translocation *489*	
19.2	Causes of Alcoholic Endotoxemia *490*	
19.2.1	Delayed Endotoxin Clearance *490*	
19.2.2	Bacterial Overgrowth *491*	
19.2.3	Increased Gut Permeability to Endotoxins *491*	
19.3	Influence of Alcoholic Endotoxemia on Different Organs *493*	
19.3.1	Liver *493*	
19.3.2	Pancreas *495*	
19.3.3	Lung *495*	
19.4	Mechanism of Tissue Damage by Alcoholic Endotoxemia *496*	
19.4.1	Cellular Targets *496*	
19.4.2	Receptors and Signaling *497*	
19.5	Factors that Ameliorate Alcoholic Endotoxemia and Tissue Damage *498*	
19.5.1	Synthetic Drugs *498*	
19.5.2	Dietary Components *499*	
19.5.3	Plant Extracts *499*	
19.5.4	Probiotics *500*	
19.6	Summary and Perspectives *500*	
	Acknowledgments *501*	
	References *501*	
20	**Gut Microbiota, Diet, Endotoxemia, and Diseases** *511*	
	Patrice D. Cani and Nathalie M. Delzenne	
20.1	Introduction *511*	
20.2	Gut Microbiota and Energy Homeostasis *512*	
20.2.1	Gut Microbiota and Energy Harvest *512*	
20.2.1.1	Gut Microbiota and Adipose Tissue Development *513*	
20.2.1.2	Lipogenesis *513*	
20.2.1.3	Specific SCFA Receptors *514*	

20.3	Energy Harvest, Obesity, and Metabolic Disorders: Paradoxes?	514
20.4	Role of the Gut Microbiota in the Inflammatory Associated with Obesity	515
20.4.1	Metabolic Endotoxemia and Metabolic Disorders	516
20.4.2	Metabolic Endotoxemia and Nutritional Intervention	517
20.4.3	Metabolic Endotoxemia, Gut Microbiota, and Fatty Liver Diseases	518
20.4.4	Selective Changes in Gut Microbiota and NASH	518
20.5	Metabolic Endotoxemia and High-Fat Feeding: Human Evidence	519
20.6	Conclusion	520
	References	520

21	**Nutrient-Derived Endogenous Toxins in the Pathogenesis of Type 2 Diabetes at the β-Cell Level**	**525**
	Christine Tang, Andrei I. Oprescu, and Adria Giacca	
21.1	Introduction	525
21.2	Acute Effect of Glucose and FFA on Insulin Secretion	526
21.3	Insulin Secretory Abnormalities in Type 2 Diabetic Patients	527
21.4	β-Cell Glucotoxicity	528
21.4.1	Chronic Effect of Glucose on β-Cell Function and Mass	528
21.4.2	Level of Impairment of β-Cell Function by Chronic Glucose Exposure	529
21.4.2.1	Insulin Gene Transcription	529
21.4.2.2	Insulin Biosynthesis	530
21.4.2.3	ATP Production	531
21.4.2.4	Late Stages of Insulin Secretion	531
21.4.2.5	β-Cell Mass	532
21.5	β-Cell Lipotoxicity	532
21.5.1	Chronic Effect of Free Fatty Acids on β-Cell Function and Mass	532
21.5.2	Level of Impairment of β-Cell Function by Chronic Free Fatty Acids Exposure	533
21.5.2.1	Insulin Gene Transcription	533
21.5.2.2	Insulin Biosynthesis	533
21.5.2.3	ATP Production	534
21.5.2.4	Late Stages of Insulin Secretion	534
21.5.2.5	β-Cell Mass	535
21.6	Glucolipotoxicity	535
21.7	Oxidative Stress as an Endogenous Toxin	536
21.7.1	Reactive Oxygen Species as a Derived Toxin from Chronic Glucose and Fat Exposure	536
21.7.2	Sites of Reactive Oxygen Species Generation	537
21.7.3	Oxidative Stress, Type 2 Diabetes, and β-Cell Dysfunction	537
21.8	Oxidative Stress, β-Cell Glucotoxicity, and Lipotoxicity	538
21.8.1	Evidence for a Role of Oxidative Stress in β-Cell Glucotoxicity and Lipotoxicity	538

21.8.2	Sites of Oxidative Stress–Induced Impairment of β-Cell Function by Glucotoxicity and Lipotoxicity 540	
21.8.2.1	Glucose Oxidation and Uncoupling Protein 2 540	
21.8.2.2	Insulin Gene Transcription and Insulin Biosynthesis 540	
21.8.3	Downstream Signaling Mechanisms of Oxidative Stress–Induced Impairment of β-cell Function by Glucotoxicity and Lipotoxicity 541	
21.8.3.1	JNK 541	
21.8.3.2	NFκB 541	
21.8.3.3	β-Cell Insulin Resistance 542	
21.8.3.4	Endoplasmic Reticulum (ER) Stress 542	
21.9	Conclusion 542	
	References 543	
22	**Endogenous Toxins and Susceptibility or Resistance to Diabetic Complications** **557**	
	Paul J. Beisswenger	
22.1	Introduction and Background 557	
22.1.1	Elevated Glucose, Pathways of Glycation/Oxidation, and Diabetic Complications 557	
22.2	Synthetic Pathways for Glycation Products 558	
22.2.1	Amadori Products 558	
22.2.2	Methylglyoxal (MG) 558	
22.2.3	3-Deoxyglucosone (3DG) 559	
22.2.4	Glyoxal 559	
22.2.5	Synthetic Pathways for Oxidation Products 559	
22.2.6	Formation of Advanced Glycation End-Products (AGEs) 560	
22.3	Enzymatic Deglycation Pathways 560	
22.3.1	Deglycation Systems for Amadori Products 561	
22.3.2	Methylglyoxal (MG) Detoxification 561	
22.3.3	Detoxification Pathways for 3-Deoxyglucosone (3DG) 562	
22.3.4	Endogenous Binders of α-Dicarbonyls as Protective Mechanisms 562	
22.3.5	The Removal of AGEs 562	
22.3.6	Pathways that Lead to Removal of Oxidative Products 563	
22.4	Susceptibility to Diabetic Complications Varies Widely among Individuals 563	
22.4.1	Slow and Rapid Nonenzymatic Glycation and Propensity to Diabetic Complications 564	
22.4.2	Impaired Deglycation of Amadori Products and Diabetic Complications 564	
22.5	Overproduction of α-Dicarbonyls is Characteristic of Individuals Who are Prone to Diabetic Complications 566	
22.5.1	Impaired Degradation of Methylglyoxal and Diabetic Complications 567	
22.5.2	3DG Detoxification and Diabetic Complications 568	

22.6	Oxidative Stress and Propensity to Diabetic Nephropathy 569
22.6.1	Nitrosative Stress and Diabetic Complications 570
22.6.2	Protective Upregulation of Antioxidant Enzymes and Diabetic Complications 570
22.7	Conclusions 571
	References 571

23 Serum Advanced Glycation End Products Associated with NASH and Other Liver Diseases 577

Hideyuki Hyogo, Sho-ichi Yamagisti, and Susumu Tazuma

23.1	Introduction 577
23.2	Formation Pathways of AGEs 578
23.3	AGEs' Association in the Liver 579
23.4	Circulating AGEs and Liver Disease 581
23.4.1	Circulating AGEs in Liver Cirrhosis 581
23.4.2	Circulating AGEs in NASH 581
23.5	Possible Molecular Mechanisms by which the AGEs–RAGE System is Involved in Liver Disease 584
23.5.1	AGEs in Hepatic Sinusoidal Endothelial Cells 584
23.5.2	AGEs in Hepatic Stellate Cells 584
23.5.3	AGEs in Hepatocytes 585
23.5.4	RAGE Involvement in Liver Diseases 588
23.6	Conclusions 589
	Acknowledgment 589
	References 589

24 Oxidative Stress in the Pathogenesis of Hepatitis C 595

Tom S. Chan and Marc Bilodeau

24.1	Hepatitis C 595
24.1.1	Incidence and Economical Burden of Hepatitis C 596
24.1.2	The Hepatitis C Virus 597
24.1.3	Conventional Treatment of Hepatitis C 598
24.2	The Molecular Basis of Oxidative Stress in Hepatitis C 599
24.2.1	Oxidative Stress in HCV 599
24.2.2	ROS Induced by Nonparenchymal Cells of the Liver (Kupffer Cells, Lymphocytes, Stellate Cells) 601
24.2.2.1	Innate and Adaptive Immunity in HCV: the Snowball Effect 601
24.2.2.2	Phagocyte NADPH Oxidase (NOX2) 602
24.2.2.3	Nonphagocyte NOX 603
24.2.3	Pro-oxidant Components of HCV 604
24.2.3.1	Oxidative Stress Induced by HCV Core Protein 605
24.2.3.2	Oxidative Stress Induced by HCV NS5A Protein 606
24.3	The Emerging Role of Antioxidants in the Treatment of HCV 608
24.4	Summary and Conclusions 610
24.4.1	Summary 610

24.4.2	Conclusions 610	
	References 610	
25	**Oxidized Low Density Lipoprotein Cytotoxicity and Vascular Disease** 619	
	Steven P. Gieseg, Elizabeth Crone, and Zunika Amit	
25.1	Introduction to Vascular Disease 620	
25.2	oxLDL Formation 622	
25.3	Toxicity of oxLDL 625	
25.4	Types of oxLDL 626	
25.5	Cell-Mediated Oxidation and Atherosclerotic Plaques 628	
25.6	Endogenous Antioxidants 630	
25.7	Mechanism of oxLDL Cytotoxicity 632	
25.8	Oxysterol and the Mitochondria 633	
25.9	Oxysterols and Calcium 634	
25.10	NADPH Oxidase and Apoptosis versus Necrosis 635	
25.11	The Future 636	
	References 637	
26	**Oxidative Stress in Breast Cancer Carcinogenesis** 647	
	Lisa J. Martin and Norman Boyd	
26.1	Introduction 647	
26.2	Markers of Oxidative Stress 648	
26.2.1	MDA 649	
26.2.2	Isoprostanes 649	
26.2.3	Biological Sampling 650	
26.3	Oxidative Stress and Breast Cancer 650	
26.4	Oxidative Stress and Breast Cancer Risk Factors 651	
26.5	Mammographic Density (MD) 652	
26.5.1	Mammographic Density and Breast Cancer Risk 653	
26.5.2	Mammographic Density and Breast Tissue Composition 654	
26.5.3	Age, Other Risk Factors, and Mammographic Density as a Marker of Susceptibility 654	
26.5.4	Genetic Factors 655	
26.6	Association of Hormones, Mitogens, and Mutagens with Mammographic Density 655	
26.6.1	Levels of Blood Hormones and Growth Factors (Mitogens) 658	
26.6.2	Urinary MDA and Mammographic Density 659	
26.6.3	Urinary Isoprostane and MD 660	
26.6.4	Comparison of MDA and Isoprostanes as Markers of Lipid Peroxidation 661	
26.7	Potential Mechanisms for the Association of Mitogens and MDA with Mammographic Density and Breast Cancer Risk 662	
26.8	Summary 663	
	References 664	

27		**Lifestyle, Endogenous Toxins, and Colorectal Cancer Risk** 673
		Gail McKeown-Eyssen, Jeff Bruce, Owen Lee, Peter J. O'Brien, and W. Robert Bruce
	27.1	Introduction 673
	27.2	Lifestyle Risk Factors for CRC 674
	27.3	Oxidative Stress Relates to Lifestyle and CRC Risk 676
	27.4	Energy Excess Relates Lifestyle and CRC Risk 681
	27.5	Interaction of Toxicity of Energy Excess and Oxidative Stress 683
	27.6	Future Research 684
		References 687
28		**Dopamine-Derived Neurotoxicity and Parkinson's Disease** 695
		Jose Luis Labandeira-Garcia
	28.1	The Neurotransmitter Dopamine 695
	28.2	Dopaminergic Alterations – Parkinson's Disease 696
	28.3	Levodopa Therapy for PD and Progress of Dopaminergic Neuron Degeneration 697
	28.4	Dopamine Toxicity and Aging 698
	28.5	Dopamine Toxicity and Animal Models of Parkinsonism 698
	28.6	Mechanisms of Dopamine Neurotoxicity – Vesicular and Cytosolic Dopamine 700
	28.7	Neuromelanin Formation and Dopamine Toxicity 701
	28.8	Interaction between Dopamine Toxicity and Protein Aggregation 702
	28.9	Receptor-Dependent Dopamine Toxicity, Dopamine-Mediated Excitotoxicity, and Relevance for Huntington's Disease 703
	28.10	Dopamine and Neuroprotection 704
	28.11	Conclusion 705
		Acknowledgments 706
		References 706
29		**Dopamine Catabolism and Parkinson's Disease: Role of a Reactive Aldehyde Intermediate** 715
		Jonathan A. Doorn
	29.1	Dopamine Biosynthesis and Catabolism 715
	29.2	DOPAL as a Metabolite of DA: Importance to Pharmacology and Toxicology 718
	29.3	DOPAL Metabolism 719
	29.4	Mechanisms for Elevation in Levels of DOPAL 720
	29.5	DOPAL Toxicity 722
	29.6	DOPAL Reactivity with Proteins 723
	29.7	Relevance of DOPAL to PD: Summary and "Big Picture" 725
		Acknowledgments 726
		References 726

30	**Tetrahydropapaveroline, an Endogenous Dicatechol Isoquinoline Neurotoxin** *733*
	Young-Joon Surh and Hyun-Jung Kim
30.1	Introduction *733*
30.2	Biosynthesis of THP *734*
30.2.1	Nonenzymatic versus Enantioselective Formation *734*
30.2.2	Effects of Alcohol *734*
30.3	Neurotoxic Potential of THP *735*
30.3.1	Parkinsonism *736*
30.3.1.1	Inhibition of Dopamine Biosynthesis *736*
30.3.1.2	Inhibition of Dopamine Uptake through the Dopamine Transporter *736*
30.3.1.3	Inhibition of Mitochondrial Respiration *737*
30.3.1.4	Inhibition of Serotonin Production *738*
30.3.2	Implications for L-DOPA Paradox *738*
30.3.3	Alcohol Dependence *739*
30.4	Biochemical Mechanisms Underlying THP-Induced Neurotoxicity *740*
30.4.1	Redox Cycling *740*
30.4.2	Oxidative Cell Death *740*
30.4.3	DNA Damage *743*
30.5	Cellular Protection against THP-Induced Injuries *743*
30.6	Concluding Remarks *744*
	References *745*
31	**Chemically Induced Autoimmunity** *747*
	Michael Schiraldi and Marc Monestier
31.1	Introduction *747*
31.1.1	Autoimmunity *748*
31.2	Human Disease *749*
31.2.1	Drug-Induced Lupus *749*
31.2.2	Drug-Induced Immune Hemolytic Anemia *751*
31.2.3	Primary Biliary Cirrhosis *752*
31.2.4	D-Penicillamine-Induced Myasthenia Gravis *752*
31.2.5	Mercury and Systemic Autoimmune Disease *753*
31.2.6	Hydrocarbon Exposure and Goodpasture's Syndrome *754*
31.2.7	Pesticide Exposure and Autoimmunity *755*
31.2.8	Eosinophilic Pneumonia *755*
31.3	Animal Models of Chemically Induced Autoimmunity *756*
31.3.1	Heavy Metals *756*
31.3.2	Pristane *758*
31.3.3	Phthalate-Induced Autoimmunity *759*
31.3.4	D-Penicillamine-Induced Autoimmunity *760*
31.4	Conclusion *760*
	References *761*

32	**Endogenous Toxins Associated with Life Expectancy and Aging**	769

Victoria Ayala, Jordi Boada, José Serrano, Manuel Portero-Otín, and Reinald Pamplona

32.1	Introduction 769	
32.2	Metabolism(s) 771	
32.3	The Rate of Generation of Damage Induced by Mitochondrial Free Radicals and Life Span 772	
32.4	Structural Components that are Highly Resistant to Oxidative Stress and Life Span 774	
32.4.1	The First Antioxidant Defense Line 774	
32.4.2	Resistance to Oxidative Damage and Life Span 775	
32.5	Oxidative Stress, Aging, and Dietary Restriction 778	
32.5.1	Caloric Restriction 778	
32.5.2	Protein Restriction 779	
32.5.3	Methionine Restriction 781	
32.6	Nutritional Considerations on Methionine: A Public Health Matter 783	
	Acknowledgments 783	
	References 783	

Part Four Therapeutics Proposed for Decreasing Endogenous Toxins 787

33	**Therapeutic Potential for Decreasing the Endogenous Toxin Homocysteine: Clinical Trials**	789

Wolfgang Herrmann and Rima Obeid

33.1	Homocysteine Metabolism 789
33.2	Physiological Determinants of Plasma Homocysteine Level 791
33.3	Causes of Hyperhomocysteinemia 791
33.4	The Role of Vitamin B_{12} (Cobalamin) 792
33.4.1	Cobalamin Metabolism 792
33.4.2	Sources of Cobalamin 794
33.4.3	Absorption, Excretion, and Homeostasis of Cobalamin 794
33.4.4	Cobalamin Deficiency 795
33.5	Folic Acid, Folate (Vitamin B_9) 795
33.5.1	Folate Metabolism 795
33.5.2	Biological Roles of Folate 796
33.5.3	Consequences of Folate Deficiency 797
33.5.4	Cobalamin Deficiency Causes Folate Trap 798
33.6	Vitamin B_6 798
33.6.1	Function 799
33.6.2	Vitamin B_6 Deficiency 800
33.6.2.1	Vitamin B_6 Deficiency, Hyperhomocysteinemia, and Cardiovascular Diseases 800
33.6.2.2	Vitamin B_6 Deficiency and Cognitive as well as Immune Dysfunction 801

33.6.2.3	Disease Treatment *802*	
33.7	Homocysteine (Hcy) as an Endogenous Toxin – Human and Experimental Studies *802*	
33.7.1	Homocysteine Toxicity in the CNS *802*	
33.7.1.1	Mechanisms of Homocysteine Toxicity in the Central Nervous System *803*	
33.7.1.2	Homocysteine in Dementia and Cognitive Decline *807*	
33.7.1.3	Treatment with B Vitamin Supplements *808*	
33.7.1.4	Homocysteine in Patients with Parkinson Disease *810*	
33.7.1.5	Homocysteine in Cerebrovascular Diseases *810*	
33.7.1.6	Reduction of Stroke Risk – Primary and Secondary Prevention Studies *811*	
33.7.2	Homocysteine in the Endothelial System *812*	
33.7.2.1	Homocysteine-Lowering Treatment in Reduction of Cardiovascular Risk *813*	
33.7.2.2	Can Treating HHcy Provide Protection against Cardiovascular Diseases? *813*	
33.7.3	Homocysteine Toxicity in Alcoholic Liver Disease *815*	
33.7.3.1	Fatty Liver and HHcy are Mutual Findings in Chronic Alcoholism *815*	
33.7.4	Homocysteine and Bone Health *816*	
33.7.4.1	Bone Health and Intervention Studies with B Vitamins *817*	
33.7.5	Homocysteine Toxicity in Patients with Renal Diseases *818*	
33.7.5.1	Role of Hyperhomocysteinemia as Vascular Risk Factor in Patients with ESRD *818*	
33.7.5.2	Renal Function and Hyperhomocysteinemia *818*	
33.7.5.3	Reduced Remethylation Pathway in ESRD *819*	
33.7.5.4	Cellular Uptake of Vitamin B_{12} in Patients with Chronic Renal Failure *820*	
33.7.5.5	Response of Homocysteine to Vitamin Treatment in Dialysis Patients *820*	
33.7.5.6	Effect of Homocysteine-Lowering Treatment on Transmethylation *821*	
	References *821*	
34	**Prevention of Oxidative Stress–Induced Diseases by Natural Dietary Compounds: The Mechanism of Actions** *841*	
	Tin Oo Khor, Ka-Lung Cheung, Avantika Barve, Harold L. Newmark, and Ah-Ng Tony Kong	
34.1	Introduction *841*	
34.2	Cytoprotective Effect of Dietary Compounds: Antioxidant Effects *843*	
34.2.1	Phenolic Dietary Compounds *843*	
34.2.1.1	EGCG *844*	
34.2.1.2	Curcumin *846*	
34.2.1.3	Dibenzoylmethane *847*	

34.2.1.4	Resveratrol 847
34.2.1.5	Genistein 848
34.2.2	Isothiocyanates 848
34.2.3	Garlic Organosulfur Compounds (OSCs) 850
34.2.4	Vitamin E 850
34.3	Conclusion Remarks 851
34.4	Tribute to Professor Harold L. Newmark 851
	Acknowledgments 852
	References 852
35	**Genotoxicity of Endogenous Estrogens** 859
	James S. Wright
35.1	Introduction 859
35.2	Estrogens and Women's Health 862
35.3	Causes of Estrogen Genotoxicity 862
35.3.1	Significance of ERα and ERβ: the Estrogen Receptors 862
35.3.2	Quinone Formation is Tumorigenic 863
35.3.3	Experimental Support for the Quinone Hypothesis 865
35.3.4	Factors Affecting Estrogen Concentrations 869
35.4	Enzymatic Reactions that Affect Estrogen Genotoxicity 870
35.4.1	Reactions that Increase Toxicity 870
35.4.1.1	Catechol Formation from E2 870
35.4.1.2	P450 and the Redox Cycle 871
35.4.1.3	Sulfatase Creates More Free Estrogen 871
35.4.1.4	Aromatase Creates More E2 via E1 871
35.4.2	Reactions that Reduce Toxicity 872
35.4.2.1	COMT Deactivates Catechol 872
35.4.2.2	NQO1 Deactivates the Quinone 873
35.4.2.3	GSH, Ascorbate, and Antioxidant Intervention Experiments 873
35.4.2.4	Sulfotransferase Creates Bound Estrogen 874
35.5	Conclusions 874
	Acknowledgment 876
	References 876
36	**Design of Nutritional Interventions for the Control of Cellular Oxidation** 881
	Elizabeth P. Ryan and Henry J. Thompson
36.1	Overview 881
36.2	Antioxidant–Chronic Disease Conundrum Terms 883
36.3	The Design of an Intervention 884
36.3.1	The Oxidizing Species 884
36.3.2	The Macromolecular Target 885
36.3.3	The Target Tissue 888
36.3.4	The Pathogenesis of the Disease 890
36.3.5	The Antioxidants 891

36.4	Designing the Next Generation of Nutritional Interventions 894	
36.4.1	Define Individuals Who are at Increased Risk for Cancer 894	
36.4.1.1	Risk for Oxidative Stress–Induced Mutations 895	
36.4.1.2	Risk for Altered Signal Transduction 895	
36.4.2	Identify Candidate Mechanisms that Accounts for the Abnormal Levels of Oxidative Products 896	
36.4.3	Define a Nutritional Intervention that is Tailored to Correct the Defect 896	
36.5	Overall Principles 897	
36.6	The Broader Perspective 897	
36.7	Summary 898	
	Acknowledgments 900	
	References 900	

Appendix Questions for Discussion 907

1. What are "endogenous toxins?" 907
2. Why consider "alcohol drinking" in a review of endogenous toxins? 907
3. What endogenous toxins are involved in the development of type 2 diabetes? 908
4. What endogenous sugar/fatty acid toxins are involved in the development of nonalcoholic steatohepatitis (NASH)? 909
5. Inflammation and oxidative stress are involved in many disease processes. Are these chronic diseases a direct result of oxidative stress or are they a result of even more reactive endogenous toxins formed with the interaction of oxidative stress with labile metabolites? 910
6. Endogenous toxins resulting from inborn errors of metabolism appear to be involved in an increasing number of diseases. Can dietary modification (supplements or deletions) reduce the levels of these toxins? 911
7. Do differences in endogenous toxins explain the differences in chronic disease incidence with time and across populations in the world? 912

References 913

Index 915

Preface

Welcome to this, the first book on *Endogenous Toxins*!

The idea for this "conference in a book" came to us when we examined the likely importance of endogenous toxins in the origins of epithelial cancers and liver disease. We noted that a wide range of studies in different disciplines showed that toxicities originating within the body could contribute to the development of chronic disease. However, there was no book on this subject and there appeared to be little communication between various researchers comparing endogenous toxins assessed in one field of study with those in related fields. We imagined that a book focusing on endogenous toxins would encourage a wider appreciation of the importance of these toxins in various diseases and in the aging processes. Furthermore, it could encourage the development of novel therapies for decreasing endogenous toxins. We thought that this could be to our common benefit. A further understanding of the work of others in the field would identify overlapping interests that could be exploited by any one of us. Such a volume could thus be helpful to epidemiologists, by identifying new hypotheses relating lifestyle factors with endogenous toxins and chronic disease; helpful to chemists and biochemists, by identifying likely important areas of investigation; and helpful to investigators in all disciplines, by providing a broader perspective of the problem and approaches taken by others in this complex field.

Accordingly, we assembled a group of authors who had made significant contributions to the study of endogenous toxins and asked each to contribute a short chapter reviewing their particular field of interest for others unfamiliar with it. The chapters have been arranged in *Endogenous Toxins* in four parts or "sessions".

Part One is concerned with chemistry and biochemistry, and the formation and reactivity of endogenous toxins, particularly those associated with diet; Part Two with the association of increased endogenous toxin levels with inborn errors of metabolism; Part Three with examples of endogenous toxins that appear to be associated with disease; and Part Four with therapeutics that have been proposed for decreasing endogenous toxins. We think that together their contributions encompass the major part of the field of endogenous toxins.

Let the authors speak. We will meet again afterward to begin a discussion which we expect will be continued...

Toronto, August 2009

Peter J. O'Brien and
W. Robert Bruce

List of Contributors

Emanuele Albano
University of East Piedmont
"A. Avogadro"
Department of Medical Sciences
Via Solaroli 17
28100 Novara
Italy

Zunika Amit
University of Canterbury
Free Radical Biochemistry
Laboratory
School of Biological Sciences
Private Bag 4800
Christchurch 8140
New Zealand

Richard C. Austin
McMaster University
Department of Medicine
St. Joseph's Healthcare
711 Concession Street
Hamilton
Ontario, L8V 1C3
Canada

Victoria Ayala
University of Lleida-IRBLLEIDA
Department of Experimental
Medicine
c/Montserrat Roig 2
25008 Lleida
Spain

Avantika Barve
Rutgers University
Department of Pharmaceutics
160 Frelinghuysen Road
Piscataway, NJ, 08854
USA

Sana Basseri
McMaster University
Department of Medicine
St. Joseph's Healthcare
711 Concession Street
Hamilton
Ontario, L8V 1C3
Canada

John W. Baynes
University of South Carolina
Department of Exercise Science
Public Health Research Center
927 Assembly St.
Columbia, SC 29208
USA

Paul J. Beisswenger
Dartmouth Medical School
Section of Endocrinology
Diabetes and Metabolism,
Department of Medicine
Remsen 311, HB 7515
Hanover, NH, 03755
USA

Endogenous Toxins. Diet, Genetics, Disease and Treatment.
Edited by Peter J. O'Brien and W. Robert Bruce
Copyright © 2010 WILEY-VCH Verlag GmbH & Co. KGaA, Weinheim
ISBN: 978-3-527-32363-0

Ludmil T. Benov
Kuwait University
Department of Biochemistry
Faculty of Medicine
P.O. Box 24923
Safat, 13110
Kuwait

Marc Bilodeau
Centre Hospitalier de l'Université
de Montréal (CHUM)
Centre de Recherche Hôpital
St-Luc, 264, René-Lévesque Est
Montréal, H2X 1P1, QC
Canada

Jordi Boada
University of Lleida-IRBLLEIDA
Department of Experimental
Medicine
c/Montserrat Roig 2
25008 Lleida
Spain

Norman Boyd
Campbell Family Institute for
Breast Cancer Research
Ontario Cancer Institute
610 University Ave.
Toronto, Ontario M5G 2M9
Canada

Jeff Bruce
University of Toronto
Graduate Department of
Pharmaceutical Sciences
Faculty of Pharmacy
144 College Street
Toronto, Ontario M5S 3M2
Canada

W. Robert Bruce
University of Toronto
Department of Nutritional
Sciences
Faculty of Medicine
Toronto, Ontario M5S 2E3
Canada

Jennifer Caldwell
McMaster University
Department of Medicine
St. Joseph's Healthcare
711 Concession Street
Hamilton
Ontario, L8V 1C3
Canada

Patrice D. Cani
Université catholique de Louvain
Department of Pharmaceutical
Sciences
Louvain Drug Research Institute
Unit of Pharmacokinetics
Metabolism
Nutrition and Toxicology
73 Avenue Mounier
Brussels 1200
Belgium

Tom S. Chan
Centre Hospitalier de l'Université
de Montréal (CHUM)
Centre de Recherche Hôpital
St-Luc
264 René-Lévesque Est
Montréal, H2X 1P1, QC
Canada

Ka-Lung Cheung
Rutgers University
Department of Pharmaceutics
160 Frelinghuysen Road,
Piscataway, NJ 08854
USA

Denis E. Corpet
Université de Toulouse
ENVT, UMR1089
23 Chemin des Capelles
31076 Toulouse
France

Elizabeth Crone
University of Canterbury
Free Radical Biochemistry
Laboratory
School of Biological Sciences
Private Bag 4800
Christchurch 8140
New Zealand

Christopher J. Danpure
University College London
Department of Cell and
Developmental Biology
Division of Biosciences
Gower Street
London WC1E 6BT
UK

Peter C. Dedon
Massachusetts Institute of
Technology
Department of Biological
Engineering
77 Massachusetts Avenue
Cambridge, MA 02139
USA
and
Massachusetts Institute of
Technology
Center for Environmental Health
Sciences
77 Massachusetts Avenue
Cambridge, MA 02139
USA

Nathalie M. Delzenne
Université catholique de Louvain
Department of Pharmaceutical
Sciences
Louvain Drug Research Institute
Unit of Pharmacokinetics
Metabolism
Nutrition and Toxicology
73 Avenue Mounier
1200 Brussels
Belgium

Jonathan A. Doorn
The University of Iowa
Division of Medicinal and Natural
Products Chemistry
College of Pharmacy
115 S. Grand Avenue
Iowa City, IA 52242-1112
USA

Marilyn Ehrenshaft
National Institute of
Environmental Health Sciences
National Institutes of Health
Laboratory of Pharmacology
111 TW Alexander Dr. RTP
Morrisville, NC 27709
USA

Cynthia Y. Feng
University of Toronto
Graduate Department of
Pharmaceutical Sciences
Faculty of Pharmacy
144 College Street
Toronto, Ontario M5S 3M2
Canada

Norma Frizzell
University of South Carolina
Department of Exercise Science
Public Health Research Center
927 Assembly St.
Columbia, SC 29208
USA

Ferruccio Galletti
University of Naples
Department of Clinical and
Experimental Medicine
Federico II Medical School
Via S. Pansini 5
80131 Naples
Italy

Adria Giacca
University of Toronto
Department of Physiology
Faculty of Medicine
1 King's College Circle
Medical Sciences Building
Toronto, Ontario M5S 1A8
Canada
and
University of Toronto
Institute of Medical Science
Faculty of Medicine
1 King's College Circle
Medical Sciences Building
Toronto, Ontario M5S 1A8
Canada
and
University of Toronto
Department of Medicine, 1 King's
College Circle
Medical Sciences Building
Toronto, Ontario M5S 1A8
Canada

Steven P. Gieseg
University of Canterbury
Free Radical Biochemistry
Laboratory
School of Biological Sciences
Private Bag 4800
Christchurch 8140
New Zealand

Françoise Guéraud
INRA
UMR-Xénobiotiques
180 ch. Tournefeuille
31027 Toulouse cedex 3
France

Fumitaka Hayase
Meiji University
Department of Agricultural
Chemistry
1-1-1 Higashi-mita Tama-ku
Kawasaki, Kanagawa 214–8571
Japan

Jeffrey Henderson
University of Toronto
Faculty of Pharmacy
27 King's College Circle
Toronto, Ontario M5S 1A1
Canada

Wolfgang Herrmann
University Hospital of the
Saarland
Department of Clinical Chemistry
and Laboratory Medicine
Gebäude 57
66424 Homburg
Germany

Kelvin Hui
University of Toronto
Faculty of Pharmacy
27 King's College Circle
Toronto, Ontario M5S 1A1
Canada

Hideyuki Hyogo
Hiroshima University
Department of Medicine and
Molecular Science
Graduate School of Biomedical
Sciences
1-2-3 Kasumi Minami-ku
Hiroshima, 734-8551
Japan

Renato Ippolito
University of Naples
Department of Clinical and
Experimental Medicine
Federico II Medical School
Via S. Pansini 5
80131 Naples
Italy

Hartmut Jaeschke
University of Kansas Medical
Center
Pharmacology Toxicology and
Therapeutics
3901 Rainbow Blvd.
Kansas City, KS 66160
USA

Tin Oo Khor
Rutgers University
Department of Pharmaceutics
160 Frelinghuysen Road
Piscataway, NJ 08854
USA

Hyun-Jung Kim
Chung-Ang University
College of Pharmacy
221 Heukseok-dong
Dongjak-gu
156-756 Seoul
South Korea

Lynell W. Klassen
University of Nebraska Medical
Center
Department of Internal Medicine
Section of Rheumatology/
Immunology
983025 Nebraska Medical Center
Omaha, NE 68198-3025
USA
and
Department of Veterans Affairs
Omaha VA Medical Center
4101 Woolworth Avenue
Omaha, NE 68105
USA

Ah-Ng Tony Kong
Rutgers University
Department of Pharmaceutics
160 Frelinghuysen Road
Piscataway, NJ 08854
USA

Jose Luis Labandeira-Garcia
University of Santiago de
Compostela
Department of Morphological
Sciences
Laboratory of Neuroanatomy and
Experimental Neurology
Faculty of Medicine
15782 Santiago de Compostela
Spain
and
Hospitales Universitarios Virgen
del Rocío
Networking Research Center on
Neurodegenerative Diseases
(CIBERNED)
c/Manuel Siurot
41013 Seville
Spain

Owen Lee
University of Toronto
Graduate Department of
Pharmaceutical Sciences
Faculty of Pharmacy
144 College Street
Toronto, Ontario M5S 3M2
Canada

Lisa J. Martin
Campbell Family Insitute for
Breast Cancer Research
Ontario Cancer Institute
610 University Avenue
Toronto, Ontario M5G 2M9
Canada

Jason Matthews
University of Toronto
Department of Pharmacology and
Toxicology
Medical Sciences Building
1 King's College Circle
Toronto, Ontario M55/A8
Canada

Gail McKeown-Eyssen
University of Toronto
Dalla Lana School of
Public Health
155 College Street
Toronto, Ontario M5T 3M7
Canada

Rhea Mehta
University of Toronto
Graduate Department of
Pharmaceutical Sciences
Faculty of Pharmacy
144 College Street
Toronto, Ontario M5S 3M2
Canada

Marc Monestier
Temple University School of
Medicine
Department of Microbiology and
Immunology
3400 North Broad Street
Philadelphia, PA 19140
USA

Harold L. Newmark
Rutgers University
Department of Chemical Biology
160 Frelinghuysen Road
Piscataway, NJ 08854
USA

Rima Obeid
University Hospital of the
Saarland
Department of Clinical Chemistry
and Laboratory Medicine
Gebäude 57
66424 Homburg
Germany

Peter J. O'Brien
University of Toronto
Graduate Department of
Pharmaceutical Sciences
Faculty of Pharmacy
144 College Street
Toronto, Ontario M5S 3M2
Canada

Andrei I. Oprescu
University of Toronto
Institute of Medical Science
Faculty of Medicine
1 King's College Circle
Medical Sciences Building
Toronto, Ontario M5S 1A8
Canada

Reinald Pamplona
University of Lleida-IRBLLEIDA
Department of Experimental
Medicine
c/Montserrat Roig 2
25008 Lleida
Spain

Manuel Portero-Otín
University of Lleida-IRBLLEIDA
Department of Experimental
Medicine
c/Montserrat Roig 2
25008 Lleida
Spain

Erin G. Prestwich
Massachusetts Institute of
Technology
Department of Biological
Engineering
77 Massachusetts Avenue
Cambridge, MA 02139
USA

Radhakrishna K. Rao
University of Tennessee Health
Science Center
Department of Physiology
894 Union Avenue
Memphis TN 38163
USA

Elizabeth P. Ryan
Colorado State University
Department of Clinical Sciences
Department of Horticulture
Cancer Prevention Laboratory
1173 Campus Delivery
Fort Collins, CO 80523
USA

Michael Schiraldi
Temple University School of
Medicine
Department of Microbiology and
Immunology
3400 North Broad Street
Philadelphia, PA 19140
USA

Shantanu Sengupta
Proteomics and Structural Biology
Division
Institute of Genomics and
Integrative Biology
Mail Road
Delhi, 110007
India

José Serrano
University of Lleida-IRBLLEIDA
Department of Experimental
Medicine
c/Montserrat Roig 2
25008 Lleida
Spain

J. Daniel Sharer
University of Alabama
Department of Genetics
720 20th Street
South Birmingham, AL 35294
USA

Arno G. Siraki
NIEHS-NIH
Laboratory of Pharmacology and
Chemistry
111 Alexander Drive
Research Triangle Park, NC 27709
USA

Pasquale Strazzullo
University of Naples
Department of Clinical and
Experimental Medicine
Federico II Medical School
Via S. Pansini 5
80131 Naples
Italy

Vangala Subrahmanyam
Sai Advantium Pharma Ltd
Plot 1, Bldg 2
Chrysalis Enclave International
Biotech Park, Phase 2 Hinjewadi
Pune, 411 057
Maharashtra
India

Young-Joon Surh
Seoul National University
College of Pharmacy
56-1 Sillim-9-dong
Gwanak-gu
Seoul 151-742
Seoul
South Korea

Christine Tang
University of Toronto
Department of Physiology
Faculty of Medicine
1 King's College Circle
Medical Sciences Building
Toronto, Ontario M5S 1A8
Canada

Susumu Tazuma
Hiroshima University
Department of General Medicine
Division of Clinical
Pharmacotherapeutics
Graduate School of Biomedical
Sciences
1-2-3 Kasumi Minami-ku
Hiroshima, 734-8551
Japan

Douglas M. Templeton
University of Toronto
Department of Laboratory
Medicine and Pathobiology
1 King's College Circle
Toronto, ON M5S 1A8
Canada

Geoffrey M. Thiele
University of Nebraska Medical
Center
Department of Internal Medicine
Section of Rheumatology/
Immunology
983025 Nebraska Medical Center
Omaha, NE 68198-3025
USA
and
Department of Veterans Affairs
Omaha VA Medical Center
4101 Woolworth Avenue
Omaha, NE 68105
USA
and University of Nebraska
Medical Center
Department of Pathology and
Microbiology
983025 Nebraska Medical Center
Omaha, NE 68198-3135
USA

Henry J. Thompson
Colorado State University
Cancer Prevention Laboratory
1173 Campus Delivery
Fort Collins, CO 80523
USA

Teruyuki Usui
Meiji University
Department of Agricultural Chemistry
1-1-1 Higashi-mita Tama-ku
Kawasaki Kanagawa
Japan

Hirohito Watanabe
Meiji University
Department of Life Science
1-1-1 Higashi-mita Tama-ku
Kawasaki, Kanagawa, 214 8571
Japan

James S. Wright
Carleton University
Department of Chemistry
1125 Colonel By Drive
Ottawam
Ontario K1S 5B6
Canada

Sho-ichi Yamagishi
Kurume University School of Medicine
Department of Pathophysiology and Therapeutics of Diabetic Vascular Complications
67 Asahi-machi
Kurume, 830-0011
Japan

Abbreviations

Aβ	β-amyloid
AA	aromatic amine
AADC	aromatic acid decarboxylase
AAPH	2,2′-azobis(amidinopropane) dihydrochloride
ABC	ATP-binding cassette (transporter)
ABCA1	ATP-binding cassette transporter (member 1)
ABCB7	ATP-binding cassette, sub-family B, member 7
AC	acetyl-CoA carboxylase
ACAT	acetyl-CoA acetyltransferase
ACC	acetyl-CoA carboxylase
ACDH	acyl-Coenzyme A dehydrogenase
ACF	aberrant crypt foci
ACh	acetylcholine
ACOX2	product of the acyl-CoA oxidase family, branched-chain acyl-CoA oxidase
ACR	acrolein
AD	Alzheimer's disease
ADAR	adenosine deaminase
ADAT	adenosine deaminases that act on tRNA
ADH	alcohol dehydrogenase
AdoCbl	adenosylcobalamin
ADP	adenosine diphosphate
AF1	activation function 1
AF2	activation function 2
AFLP	acute fatty liver of pregnancy
2-AG	2-arachidonylglycerol
AGE	advanced glycated endproduct
AGEs-BSA	advanced glycation end products-modified bovine serum albumin
AGT	alanine:glyoxylate transaminase
AHR	aryl hydrocarbon receptor
AHRE	aryl hydrocarbon receptor responsive element
AICAR	5-aminoimidazole-4-carboxamide ribonucleotide

Endogenous Toxins. Diet, Genetics, Disease and Treatment.
Edited by Peter J. O'Brien and W. Robert Bruce
Copyright © 2010 WILEY-VCH Verlag GmbH & Co. KGaA, Weinheim
ISBN: 978-3-527-32363-0

AID	activation-induced cytidine deaminase
AIF	apoptosis-inducing factor
aiPLA2	lysosomal phospholipase A2
Akt	members of protein kinase family, involved in insulin resistance
ALA	aminolevulinic acid
ALAS2	δ-aminolevulinic acid synthase 2
ALD	alcoholic liver disease
ALDH	aldehyde dehydrogenase
ALDP	product of the gene defective in ALD, a membrane transporter of ABC family
ALE	advanced lipoxidation end product
ALS	amyotrophic lateral sclerosis
ALT	alanine transaminase
AMA	antimitochondrial autoantibodies
AMACR	2-methyl-acyl-CoA racemase
AMP	adenylate monophosphate
AMPA	a compound that is a specific agonist for the AMPA receptor
AMPK	AMP-activated kinase
AMPK	AMP-dependent protein kinase
ANA	antinuclear antibodies
ANG	a gene thats product is a potent mediator of new blood vessel formation
ANT	adenine nucleotide translocase
AOM	azoxymethane
AP	apurinic and apyrimidinic sites on DNA
APAP	acetaminophen
Apc	a gene whose dysfunction results in Adenomatous Polyposis Coli
apoB	apolipoprotein B
APOBEC-1	apolipoprotein B mRNA editing catalytic subunit 1
APP	amyloid precursor protein
AR	aldehyde/aldose reductase
ARE	antioxidant response element
ARNT	aryl hydrocarbon receptor nuclear translocator
ASH	alcoholic steatohepatitis
ASK-1	apoptosis signaling kinase-1
ATAC	Arimidex tamoxifen, alone or in combination
ATF4	activating transcription factor 4
ATM1	a mitochondrial inner membrane ABC transporter
ATOX1	encodes a copper chaperone that binds and transports cytosolic copper
ATP	adenosine triphosphate
AV	arginine vasopressin

AXH	a domain of ataxin-1
BAC	bacterial artificial chromosome
BAD	Bcl-2 associated death promotor (apoptotic)
Bak	(Bcl-2 homologous antagonist/killer) a pro-apoptotic gene of the Bcl-2 family
BALB	inbred strain of mouse
BAP	bone alkaline phosphatase
Bax	a pro-apoptotic protein than inactivates Bcl-2
BBB	blood-brain barrier
BCFA	branched-chain fatty acid
Bcl-2	mitochodrial protooncogene encoding a growth factor
BCOX	peroxisomal branched chain acyl-CoA oxidase
BDNF	brain-derived neurotrophic factor
BH4	tetrahydrobiopterin
BHMT	betaine homocysteine methyltransferase
Bid	a cytosolic pro-apoptotic Bcl-2, translocates to mitochondria, activates Bax
BiP	binding immunoglobulin protein
BMD	bone mineral density
BMI	body mass index
BN	brown Norway
BP	bifunctional protein
BRIC	benign recurrent intrahepatic cholestasis
BSA	bovine serum albumin
BSEP	bile salt exporter protein
C/EBP-α	CCAAT/enhancer binding protein-α
c-JUN	in combination with c-Fos, forms the AP-1 early response transcription factor
C-K	cathepsin K
CACNL1A4	mutations in this Cacnl1a4 calcium channel gene is associated with seizures
CACT	carnitine acylcarnitine translocase
CaOx	calcium oxalate
CAT	catalase
Cbl	cobalamin
CBP	CREB-binding protein
CBS	cystathionine-β-synthase
CCER	childhood cerebral
CCS	copper chaperone for superoxide dismutase
CD14	cluster of differentiation 14 cell surface marker protein involved in LPS binding
CD36	cluster of differentiation 36 cell surface marker protein binding many ligands
CDNB	1-chloro-2,4-dinitrobenzene
CE	catechol estrogen

CEBPβ	CCAAT/enhancer-binding protein beta
CEC	S-(carboxyethyl)-cysteine
CEL	carboxyethyl lysine
CGF	connective tissue growth factor
CGL	cystathionine γ-lyase
CHOP	C/EBP homologous protein
ChREBP	carbohydrate responsive element binding protein
CI	chemical ionization, confidence interval
CIA	chemically induced autoimmunity
Cic	Capicua
β-CIT SPECT	[(123)I]beta-CIT (2beta-carbomethoxy-3beta-(4-iodophenyl)tropane) single photon emission computed tomograph
cJNK	c-Jun-terminal kinase
CL	citrate lyase
CLDN1	claudin 1 human gene
CMA	carboxymethylarginine
CMC	S-(carboxymethyl)-cysteine
CML	carboxymethyl lysine
CNS	central nervous system
CoA	coenzyme A
COMMD1	copper metabolism (Murr1) domain containing 1
COMT	catechol O-methyl transferase
COPT1, 2	copper transporter 1, 2
COX	cyclooxygenase
Cp	ceruloplasmin
cPLA2	cytosolic phospholipase A2
CPT	carnitine palmitoyltransferase
CR	caloric restriction
CRC	colorectal cancer
CREB	cAMP responsive element binding protein
CRP	C-reactive protein
CRX	cone-rod homeobox, a nuclear transcription factor in the retina
CS	citrate synthase
CT	chlorotyrosine
CTGF	connective tissue growth factor
CTLA-4	cytotoxic T lymphocyte associated antigen-4
CTR1, 2	copper transporter analogous to COPT
CUP1	copper ion binding, with overexpression protecting from copper excess
CUTC	cutC copper transporter homolog (E. coli)
CVD	cardiovascular disease

CXC	"X" for two N-terminal cysteines of chemokines separated by one amino acid
CYP	cytochrome P450 isoenzyme
CYP2E1	cytochrome mixed-function oxidase system involved in the metabolism of xenobiotics
CYS	cystathionine
DA	dopamine
DAG	diacylglycerol
DAO	D-amino acid oxidase
DAT	dopamine transporter
DBA	inbred strain of mouse
DBD	DNA-binding domain
DBP	D-bifunctional protein
DC	dendritic cell
DCF	dichlorofluorescin
DCFH	dichlorofluorescein
Dcytb	duodenal cytochrome b
3DF	3-deoxyfructose
3DG	3-deoxyglucosone
3-DGA	3-deoxy 2-keto gluconic acid
DHA	dihydroxyacetone
DHA	docosahexaenoic acid
DHAP	dihydroxyacetone phosphate
DHCA	dihydroxycholestanoic acid
DHEAS	dehydroepiandrosterone-sulfate
DHF	dihydrofolate
DHFR	dihydrofolate reductase
DHN-MA	1,4-dihydroxynonane mercapturic acid
DIA	D-penicillamine-induced autoimmunity
DIIHA	drug-induced immune hemolytic anemia
DIL	drug-induced lupus
DM	dystrophia myotonica
DMEM	Dulbecco's modified Eagle's medium
DMG	dimethylglycine
DMPO	5,5-dimethyl-1-pyrroline N-oxide
DMSO	dimethyl sulfoxide
DMT	divalent metal transporter
DN	diabetic nephropathy
7,8-DNP	7,8-dihydroneopterin
DNPH	dinitrophenylhydrazine
dNTP	deoxynucleoside triphosphates
DODE	9,12-dioxo-10(E)-dodecenoic acid
DOLD	3-deoxyglucosone-derived lysine dimer
DOPAC	3,4-dihydroxyphenylacetic acid
DOPAL	3,4-dihydroxyphenylacetaldehyde

DOPET	3,4-dihydroxyphenylethanol
DPD	deoxypyridinoline
DPPC	dipalmitoylphosphatidylcholine
DR	diet restricted
DRPLA	dentatorubral pallidoluysian atrophy
DSPC	disaturated phosphatidylcholine
DT	diaphorosea flavoprotein that reversibly catalyzes the oxidation of NADH or NADPH
dTMP	thymidylate
dUMP	deoxyuridine monophosphate
DXP	5-deoxy-d-xylulose-1-phosphate
DZA	deazaadenosine
E1	estrone
E2	17β-estradiol
E2(E1)-3,4-Q	E2-3,4-quinone
E2(E1)-3,4-SQ	E2(E1)-3,4-semiquinone
E217G	estradiol-17beta-D glucuronide
E3	estriol
EAA	excitatory amino acid
EAE	experimental autoimmune encephalomyelitis
EBP	element (or enhancer) binding protein
EC	endothelial cell
ECD	electrochemical detection
ECM	extracellular matrix
EDE	4,5-epoxy-2(E)-decenal
EE	ethinyl estradiol
EEG	electroencephalogram
EG	ethylene glycol
EGF	epidermal growth factor
EGFR	epidermal growth factor receptor
EGP	early glycation product
Egr-1	early growth response-1
EH	epoxide hydrolase
EIF4	eukaryotic translation initiation factor
ELUMO	energy of the lowest unoccupied molecular orbital
EMA	ethylmalonic acid
e-NOS	constitutively expressed NOS
EP	eosinophilic pneumonia
EPR	electron paramagnetic resonance
EpRE	electrophile-responsive element
ER	endoplasmic reticulum
ER	estrogen receptor
ERα	estrogen receptor α
ERβ	estrogen receptor β
ERAD	endoplasmic reticulum–associated degradation

ERE	estrogen response element
ERK	extracellular signal-regulated kinase (MAPK)
ERKO	estrogen receptor α knockout
ERO1	endoplasmic reticulum oxidoreduction 1
ERT	enzyme replacement therapy
ES	embryonic stem
ESR	electron spin resonance
ESRD	end stage renal disease
EST	expressed sequence tag
ET	endothelin
ETC	electron transport chain
ETF	electron transfer flavoprotein
ETFDH	electron transfer flavoprotein dehydrogenase
EULAR	European League Against Rheumatism
F3P	fructose-3-phosphate
FABP	fatty acid binding protein
FAD	flavin adenine dinucleotide (oxidized)
$FADH_2$	flavin adenine dinucleotide (reduced)
FAO	fatty acid oxidation
FAOD	fatty acid oxidation disorder
FAPy-G	2,6-diamino-5-formamidopyrimidine
FAS	fatty acid synthase, fetal alcohol syndrome
FDA	Food and Drug Administration, U.S.A.
FEEL-1, -2	endocytic receptors for advanced glycation end products
FFA	free fatty acid
FGF	fibroblast growth factor
FIAF	fasting-induced adipose factor
FICZ	6-formoindolo[3,2-b]carbazole
FL	fructose lysine
FL3P	fructoselysine-3-phosphate
FMN	flavin mononucleotide
FN3K	3-phosphokinase
FN3K	fructosamine-3-kinase
FN3KRP	fructosamine-3-kinase–related protein
FOXO	forkhead box O transcription factors related to insulin signaling pathway
FRL	free radical leak
FXN	frataxin
FXTAS	fragile X tremor/ataxia syndrome
G	glyoxal
GA	glyceraldehyde
GA3P	glyceraldehyde 3-phosphate
GABA	gamma-aminobutyric acid
GADD	growth arrest and DNA damage induced genes
GAP	glyceraldehyde 3-phosphate

GAPDH	glyceraldehyde phosphate dehydrogenase
GBM	glomerular basement membrane
GC/MS	gas chromatography with mass spectrometry
GE	glucose to ethanolamine
GFP	green flourescent protein as a reporter protein
GFR	glomerular filtration rate
GK	the Goto and Kakisaki strain of rats, a spontaneous model of T2DM
GL	glycolate
GLA-BSA	glyceraldehyde-modified bovine serum albumin
GLA-CAS	glyceraldehyde-modified casein
GLAP	glyceraldehyde-derived pyridinium compound
GLUT	glucose transporter
GM-CSF	granulocyte/macrophage colony stimulating factor
GMP	guanylate monophosphate
GO	glycolate oxidase
GOA	glycolaldehyde
GOLD	a glyoxal-derived AGE
GOT	aspartate aminotransferase (AST)
GPR	G-protein coupled receptor
Gpx	glutathione peroxidase
GR	glutathione reductase
GREACE	study on atorvastatin and coronary-heart-disease evaluation
GRHPR	glyoxylate reductase/hydroxypyruvate reductase
GRP	glucose-regulated protein, endoplasmic reticulum chaperone
GS	glutathionyl radical
GS	Goodpasture's syndrome
GSH	glutathione
GSH-Px	glutathione peroxidase
GSIS	glucose-stimulated insulin secretion
GSK3	glycogen synthase kinase, inactivates the enzyme
GSSG	oxidised glutathione
GST	glutathione S-transferase
GTP	guanosine-5'-triphosphate
GX	glyoxylate
GXA	glyoxylic acid
3HAA	3-hydroxyanthranilic acid
HAMP	hepcidin antimicrobial peptide
HAS	human serum albumin
HAT	histone acetyl transferase
HbA1ach	acetaldehyde-induced hemoglobin fraction
HBP1	HMG-box transcription factor 1 protein
HCB	hexachlorobenzene

HCC	hepatocellular carcinoma
HCV	hepatitis C virus
Hcy	homocysteine
HD	Huntington's disease
HDAC	histone deacetylase
HDL	Huntington disease-like
HDL	high-density lipoprotein
HEL	hen egg lysozyme
HELLP	hemolysis with elevated liver enzyme activity and low platelet
HEP	high-ethanol-preferring
HEPES	buffering agent
HER	hydroxyethyl free radical
Herp	homocysteine-responsive endoplasmic reticulum protein
HFCS	high-fructose corn syrup
HFE	major histocompatibility complex-encoded gene
HFE2	hemochromatosis type II
HFE2B	hemochromatosis designated type IIB
HFE3	hemochromatosis type III
HFE4	hemochromatosis type IV
HFI	hereditary fructose intolerance
HGI	hemoglobin glycosylation index
HgIA	mercury-induced autoimmunity
Hh	Hedgehog
HHcy	hyperhomocysteinemia
HI	hydroimidazolone
HIF	hypoxia inducible factor
HIV	human immunodeficiency virus
HJV	hemojuvelin
HMB-1	high mobility group box 1
HMDM	human monocyte–derived macrophage
HMEC	human mammary epithelial cell
HMG	3-hydroxy-3-methylglutaryl-
HMGB	contains a HMG-box domain
HMP	hypochlorous acid–modified protein
HMW	high molecular weight
4-HNE	4-hydroxynonenal
4HNE	4-hydroxy-2-nonenal
HNE	hydroxynonenal
hnRNP	heterogenous nuclear ribonucleoprotein
HO-1	heme oxygenase-1
holoHC	holohaptocorrin
holoTC	holotranscobalamin
HOMA-IR	homeostasis model of insulin resistance

HOPE	a randomized control study of folate, vitamin B6 and B12	
HPETE	hydroperoxy eicosatetraenoic acid	
HPNE	4-hydroperoxy-2(E)-nonenal	
HPODE	13-hydroperoxy-octadecadienoic acid	
HPR	hydroxypyruvate reductase	
HRT	hormone replacement therapy	
HSC	hepatic stellae cells	
17β-HSD-1	17β-hydroxysteroid dehydrogenase type 1	
Htt	huntingtin protein	
Huh-7	hepatocyte-derived cells	
HVA	homovanilic acid	
IAPP	islet amyloid polypeptide	
IARC	International Agency for Research on Cancer	
IBD	inflammatory bowel disease	
ICAM	intercellular adhesion molecule	
ICTP	telopeptide of human collagen I	
IDO	indoleamine 2,3-dioxygenase	
IEM	inborn errors of metabolism	
IFABP	intestinal fatty acid binding protein	
IGF	insulin-like growth factor	
IGF-1	insulin-like growth factor 1	
IGFPB	insulin-like growth factor binding protein	
IHA	immune hemolytic anemia	
IkB	inhibitor of KappaB	
IKK	IKK, inhibitor of KappaB Kinase	
IL	interleukin	
IL-1α	interleukin 1α	
IMM	inner mitochondrial membrane	
INF-γ	interferon-γ	
INOS	inducible nitric oxide synthase	
iNOS	inducible nitric oxide synthetase	
IP	intra-peritoneal	
iPLA2	independent phospholipase A2	
iPS	induce potential stem	
IR	insulin resistance	
IRE	iron-responsive element	
IRP	iron regulatory protein	
IRS	insulin receptor substrate	
IST	immuno-spin trapping	
JNK	c-jun N-terminal kinase	
JNK	N-terminal jun kinase	
JNK/AP1	c-Jun N-terminal kinase/activator protein-1	
K/O	knockout	
KAR2	karyogamy 2	
KCI	Krebs cycle intermediate	

KEAP	human kelch-like ECH-associated protein
KHA	ketohydroxyglutarate aldolase
LA	lipoic acid
LAD	liver alcohol disease (ALD)
LBD	ligand-binding domain
LBP	lipopolysaccharide-binding protein
LC-CoA	long chain acyl CoA
LC-MS	liquid chromatography/mass spectroscopy
LC-MS/MS	liquid chromatography/tandem mass spectroscopy
LCAD	long chain acyl-Coenzyme A dehydrogenase
LCFA-CoA	long chain fatty-CoA
LCHAD	long chain 3-hydroxyacyl-Coenzyme A dehydrogenase
LDH	lactate dehydrogenase
LDL	low density lipoprotein
LDLR	low-density lipoprotein receptor
L-DOPA	L-3,4-dihydroxyphenylalanine
LEW	Lewis rat strain
LFA	lymphocyte function antigen
LIFE	Losartan Intervention For Endpoint reduction in hypertension
LLNA	large, neutral amino acid
LM3α	rabbit alcohol-induced cytochrome enzyme (CYP 2E1)
LNAA	large, neutral amino acid
LNAAT	large neutral amino acid transporter
LPL	lipoprotein lipase
LPLA2	lysosomal phospholipase A2
LPO	lipid peroxidation
LPS	lipopolysaccharide
MAA	malondialdehyde-acetaldehyde
MAA	malonyldialdehyde-acetaldehyde adduct
MAA-Alb	malondialdehyde-acetaldehyde adducts of albumin
MAD	multiple acyl-CoA dehydrogenase
MAO	monoamine oxidase
MAP	mitogen activated protein
MAPK	mitogen-activated protein kinase
MAT	methionine adenosyltransferase
MBG	mean blood glucose
2MBCFA	2-methyl branched-chain fatty acid
MCAD	medium chain acyl-Coenzyme A dehydrogenase
MCAD	mitochondrial medium chain acyl CoA dehydrogenase
MCP	monocyte chemoattractant protein
M-CSF	monocyte colony-stimulating factor
MD	mammographic density
MDA	malondialdehyde
MDR	multidrug resistance

MEC	Ministry of Education
MeCbl	methylcobalamin
MEOS	microsomal ethanol oxidizing system
METH	methamphetamine
5-methylTHF	5-methyltetrahydrofolate
MetR	methionine restriction
MetSO	methionine sulfoxide
2-MF	2-methylene-3(2H)-furanone
5-MF	5-methylene-5(2H)-furanone
MG	methylglyoxal
MG	myasthenia gravis
MG-H1	a methylglyoxal-derived AGE
MGO	methylglyoxal
MGS	methylglyoxal synthase
MHC	major histocompatibility complex
MIP-1	macrophage inflammatory protein 1
MitROS	mitochondrial reactive oxygen species
mLDL	minimally oxidized low density lipoprotein
MMA	methylmalonic acid
MMP	matrix metalloprotease
MnSOD	manganese superoxide dismutase
MOLD	methylglyoxal-derived lysine dimer
MOPS	buffering agent
MPDP	1-methyl-4-phenyl-2,3-dihydropyridinium
MPO	myeloperoxidase
MPP	1-methyl-4-phenyl-pyridinium ion
MPT	membrane permeability transition
MPTP	1-methyl-4-phenyl-1,2,3,6-tetrahydropyridine
MRI	magnetic resonance imagining
MRP	multidrug resistance associated protein
MS	mass spectrometry
MS	metabolic syndrome
MS	methionine synthase
M/SCHAD	medium/short chain 3-hydroxyacyl-CoA dehydrogenase
MSF-1	macrophage stimulating factor 1
mtDNA	mitochondrial DNA
MTHFR	5,10-methylenetetrahydrofolate reductase
mTOR	mammalian target of rifamycin, a serine/threonine protein kinase
MTP	mitochondrial trifunctional protein
MTS	mitochondrial targeting sequence
MTT	an agent used to evaluate mitochondrial reducing activity
MUFA	monounsaturated fatty acyl

MURR	a multifunctional protein involved in copper homeostasis
MW	molecular weight
NAC	N-acetylcysteine
NAD	nicotinamide adenine dinucleotide
NADH	nicotinamide adenine dinucleotide
NADPH	nicotinamide adenine dinucleotide phosphate
NAFLD	nonalcoholic fatty liver disease
NALD	*neonatal adrenoleukodystrophy*
NAPQI	N-acetyl-p-benzoquinone imine
NARSAD	national Alliance for Research on Schizophrenia and Depression
NASH	nonalcoholic steatohepatitis
NBT	nitroblue tetrazolium
NCE	new chemical entities
nDNA	nuclear DNA
Neh	a domain in Nrf2
NF-κB	nuclear factor κB
NFK	N-formylkynurenine
Nfr2	nuclear factor erythroid-2-related factor
NFT	neurofibrillary tangle
NGF	nerve growth factor
NHANES	National Health and Nutrition Examination Survey
NHE	4-hydroxy-2(E)-nonenal
NIH	National Institutes of Health
NK	natural killer cell
NKT	natural killer T cells
NMDA	a synthetic compound agonist at the NMDA receptor
NO	nitric oxide
NOC	N-nitroso compound
NOS	nitric oxide synthase
NOX	NADPH oxidase
NQO	NAD(P)H-dependent quinone oxireductase
Nrf-2	a redox-sensitive transcription factor
NS	not statistically significant
NSAID	nonsteroidal anti-inflammatory drug
3-NT	3-nitrotyrosine
NTCP	sodium taurocholate cotransporting polypeptide
NZB/W	New Zealand Black X White mouse cross F1
OCTN	an organic cation transporter
OGG1	8-oxoguanine DNA glycosylase protein
OGTT	oral glucose tolerance test
6-OHDA	6-hydroxydopamine
8-OHdG	8-hydroxy-2-deoxyguanosine
OHdG	hydroxydeoxyguanosine

4-OHE1(E2)-1-N7Gua	4-hydroxyestradiol-1(α,β)-N7-guanine
4-OHE2	4-hydroxylated E2
2-OHE2	2-hydroxylated E2
2-OHE2-6-N3Ade	2-hydroxyestradiol-6(α,β)-N3-adenine
OMIM	online mendelian inheritance in man
oMLT	oral methionine-loading test
ONE	4-oxononenal
2-OMe E2	2-methoxy E2
OPTIMA	Oxford Project to Investigate Memory and Ageing
OR	odds ratio
OS	oxidative stress
Ox	oxazolone
8-oxo-dG	8-oxo-7,8-dihydro-2'-deoxyguanosine
8-oxo-G	7,8-dihydro-8-oxoguanine
8-oxodG	8-hydroxy-2-deoxyguanosine
8-oxodG	8-hydroxyguanosine
oxLDL	oxidized low density lipoprotein
OxS	oxidative stress
PAA	3-(phenylamino)-alanine
PAF-AH	platelet-activating factor acetylhydrolase
PAH	phenylalanine hydroxylase
PAP	3-phenylamino-1,2-propanediol
2D-PAGE	two-dimensional polyacrylamide gel electrophoresis
PBC	primary biliary cirrhosis
PBD	peroxisome biogenesis disorder
PC	mouse teratoma-derived cell line
PCD	programmed cell death
PCH	phytanoyl-CoA hydroxylase
PD	Parkinson's disease
PDGF	platelet-derived growth factor
PDH	pyruvate dehydrogenase
PDTC	pyrrolidine dithiocarbamate
PEG	polyethylene glycol
PEMT	phosphatidylcholine methyl transferase
PERK	protein kinase-like endoplasmic reticulum kinase
PEX	a group of genes identified as important for peroxisomal synthesis
PFIC	progressive familial cholestasis
PFK	phosphofructokinase
PH	primary hyperoxaluria
PH1	primary hyperoxaluria type 1
PH2	primary hyperoxaluria type 2
PHD	prolyl-4-hydroxylase
PHF	hyperphosphorylation of tau protein
PHS	prostaglandin H synthase

PI3K	phosphoinositide 3-kinase
PIH	pregnancy-induced hypertension
PKB	protein kinase B
PKC	protein kinase C
PKU	phenylketonuria
PL	phospholipid
PL	pyridoxal
PLA1	phospholipase A1
PLA2	phospholipase A2
PLB	phospholipase B
PLC	phospholipase C
PLD	phospholipase D
PLP	pyridoxal 5'-phosphate
PM	pyridoxamine
PME	protein phosphatase methylesterase
PMN	polymorphonuclear leukocytes
PMP	pyridoxamine 5'-phosphate
PN	pyridoxine
polyQ	polyglutamine
PP2A	protein phosphatase 2A
PPAR	peroxisome-proliferator activator receptor
PPAR-α	peroxisome proliferator-activated factor-α
PPAR-γ	peroxisome proliferator-activated receptor γ
PPMT	protein phosphatase methyl transferase
PR	protein restriction
pRB	retinoblastoma protein
Prdx	peroxiredoxin
PS	photosensitizer
PS	presenilin
PTS	peroxisomal targeting sequence
PUFA	polyunsaturated fatty acids
Px	pancreatectomized
QBP1	polyQ disrupting peptide
RA	rheumatoid arthritis
Rac-1	Ras-related C3 botulinum toxin substrate 1
RAE-1	retinoic acid early inducible gene 1
Raf	a serine/threonine kinase frequently mutated in various cancers
RAGE	receptor for advanced glycation endproduct
RBC	red blood cells
RBM17	a splicing factor (SPF45)
RCO	reactive carbonyl compound
RCS	reactive carbonyl species
RERE	arginine-glutamic acid rich nuclear protein (atrophin-2)
RIG-1	a viral helicase

RNS	reactive nitrogen species
ROS	reactive oxygen species
RSNO	nitrosothiol
RT-PCR	reverse transcription polymerase chain reaction
RTP	endoplasmic reticulum stress-response gene
RXR	retinoid X receptor
S-AH	S-adenosylhomocysteine
S-AM	S-adenosylmethionine
SAH	S-adenosylhomocysteine
SAM	S-adenosylmethionine
SAMe	S-adenosylmethionine
SB-203580	an inhibitor of MAP kinase reactivating kinase
SBMA	spinal bulbar muscular atrophy
2SC	S-(2-Succinyl)cysteine
SCA	spinocerebellar ataxia
SCAD	short chain acyl-Coenzyme A dehydrogenase
SCAP	SREBP cleavage activation protein
SCFA	short chain carboxylic acid
SCO1, 2	genes and proteins required for the assembly of cytochrome oxidase
SCOX	straight-chain acyl-CoA oxidase
SCR	scavenger receptor
SD	standard deviation
SDS-PAGE	sodium dodecyl sulfate polyacrylamide gel electrophoresis
SEC	sinusoidal liver endothelial cell
SER	smooth endoplasmic reticulum
SERM	selective estrogen receptor modulator
SGLT	sodium-dependent glucose cotransport
SH	sulfhydryl
SHBG	sex hormone binding globulin
siRNA	small inhibitory RNA
SIRT1	sirtuin 1
SJL	inbred mouse strain
SKH-1	hairless mouse strain
SLC22A5	a membrane transporter gene associated with primary carnitine deficiency
SLE	systemic lupus erythematosus
SMC	smooth muscle cell
SMRT	transcriptional corepressor
SNc	substantia nigra compacta
SNP	single nucleotide polymorphism
SOC	store-operated Ca^{2+} entry channel
SOD	superoxide dismutase
SPF45	a splicing factor (RBM17)

sPLA2	secreted phospholipase A2
SPT	serine:pyruvate aminotransferase
SPT	serine:pyruvate transaminase
SQ	o-semiquinone
SR	scavenger receptor
SR-A	scavenger receptors on the surface of macrophages
sRAGE	soluble receptor for advanced glycation end product
SRB	steroid receptor binding protein
Srx	sulfiredoxin
SSAO	semicarbazide-sensitive Cu-amine oxidase
STAGA	a complex chromatin-acetylating transcription coactivator
STAT	signal transducer and activator of transcription
SUA	serum uric acid
t-PA	tissue plasminogen activator
T2DM	type 2 diabetes mellitus
TAGE	toxic advanced glycation end product
TAR	total antioxidant reactivity
TBA	thiobarbituric acid
TBARS	thiobarbituric acid reactive substance
TBP	TATA binding protein
TC	transcobalamin
TCA	the tricarboxylic acid cycle
TCDD	2,3,7,8-tetrachlorodibenzo-p-dioxin
TCEP	tris(2-carboxyethyl)phosphine
TCR	T cell receptor
TEMPOL	4-hydroxy-2,2,6,6-tetramethylpiperidinyloxy
TER	transepithelial electrical resistance
TF	tissue factor
TfR	transferrin receptor
TFTC	a complex chromatin-acetylating transcription coactivator
TGF	transforming growth factor
TGF-β	transforming growth factor β
TGF-1	transforming growth factor 1
TGN	trans-Golgi network
TH	tyrosine hydroxylase
THCA	trihydroxycholestanoic acid
tHcy	total homocysteine
THIQ	tetrahydroisoquinoline
THP	tetrahydropapaveroline
TIA	transient ischemic attack
TLR	Toll-like receptor
TMX	tetramethylmurexide, a dye used with Mitotracker
TNF	tumor necrosis factor

TNF-α	tumor necrosis factor α
TOM20	a component of the TOM (translocase of outer membrane) complex
TOS	toxic oil syndrome
TPMET	transplasma membrane electron transport
TRAMP	transgenic adenocarcinoma of mouse prostate
TRAP5b	tartrate resistant acid phosphatase 5b
Treg	regulatory T cell
Treg	regulatory T
Trpc-1	TRPC1 transient receptor potential cation channel, subfamily C, member 1
TUNEL	terminal transferase-mediated dUTP nick end labeling
UA	uric acid
UCP	uncoupling protein
UCP2	uncoupling protein 2
UL	tolerable upper intake level
UPR	unfolded protein response
UPS	ubiquitin-mediated proteasomal degradation system
UVB	ultraviolet radiation 290 to 320 nm
UVRR	ultraviolet resonance Raman
VCAM	vascular endothelial adhesion molecule
VCP	valosin-containing protein
VD	voltage-dependent
VEGF	vascular endothelial growth factors
VLCAD	very long chain acyl-Coenzyme A dehydrogenase
VLCFA	very long chain fatty acid
VLDL	very low density lipoprotein
VMAT	vesicular monoamine transporter
VSMC	vascular smooth muscle cell
VTA	ventral tegmental area
vWF	von Willebrand factor
WCRF/AICR	World Cancer Research Fund/American Institute for Cancer Research
Wnt-1	wingless-type MMTV integration site family, member 1
XALD	X-linked adrenoleukodystrophy
XO	xanthine oxidase
XO	xanthine oxidoreductase
YAC	yeast artificial chromosome
ZDF	Zucker diabetic fatty
ZO	zonula occludens

Part One
Endogenous Toxins Associated with Excessive Sugar, Fat, Meat, or Alcohol Consumption

Chemistry and Biochemistry

1
Endogenous DNA Damage
Erin G. Prestwich and Peter C. Dedon

With the advent of ultrasensitive analytical technology, there has been a major shift in thinking about pathological damage to cellular molecules caused by endogenous processes. There is now ample evidence for a cause-and-effect linkage between the toxicities originating within the body and the development of many human diseases, which is perhaps best illustrated in the epidemiological link between chronic inflammation and cancer. In this paradigm, activated immune cells generate reactive oxygen and nitrogen species that damage DNA and RNA directly, or by generating alkylating agents in reactions with other biomolecules. There are also numerous fundamental physiological chemicals and associated chemistries that can cause nucleic acid damage in cells, including alkylating agents such as S-adenosyl methionine and α, β-unsaturated dicarbonyl products of carbohydrate metabolism and lipid peroxidation, and reactive oxygen species generated by mitochondria and fatty acid metabolism in the peroxisome. These chemistries can damage both nucleobases and the sugar–phosphate backbone of DNA through oxidation, halogenation, nitration, alkylation, and deamination, which can lead to mutation and cancer directly or by cytotoxicity-induced cell turnover. This chapter thus focuses on recent studies of the sources and products of DNA and RNA damage caused by endogenous processes, with an emphasis on adducts not addressed in other review articles, and on oxidative stress and inflammation.

1.1
Introduction

The past several decades have witnessed a major shift in thinking about pathological damage to cellular molecules, particularly DNA, with a growing appreciation for endogenous processes in addition to exogenous sources. There is now ample evidence for a cause-and-effect linkage between the toxicities originating within the body and the development of many human diseases, which is perhaps best illustrated in the epidemiological link between chronic inflammation and cancer. The many causes

Endogenous Toxins. Diet, Genetics, Disease and Treatment.
Edited by Peter J. O'Brien and W. Robert Bruce
Copyright © 2010 WILEY-VCH Verlag GmbH & Co. KGaA, Weinheim
ISBN: 978-3-527-32363-0

of inflammation, from infection to diet, lead to immune cell-mediated generation of cytokines and reactive oxygen and nitrogen species that cause damage to cellular biomolecules [1]. Given the absolute requirement for mutations in carcinogenesis, research has focused on the mutagenicity and cytotoxicity of DNA damage and epigenetic phenomena in the cancer-inflammation link, under the broader theme of chemical carcinogenesis [2]. This has led to intense study of the mechanisms and products of endogenous DNA damage, and consequently there are numerous published reviews on the subject [1–12]. Two recent reviews by Nair et al. and by De Bont and van Larebeke address many of the endogenous base lesions, with a thorough compilation of estimates of the quantities of lesions in mammalian cells and tissues [3, 9], though many of these numbers are highly suspect in light of the challenges posed by artifacts in the measurement of endogenous DNA damage [13, 14]. This chapter thus focuses on recent studies of the sources and products of DNA and RNA damage caused by endogenous processes, with an emphasis on adducts not addressed in other review articles and on oxidative stress and inflammation.

Historically, DNA damage was originally viewed as a problem associated with exposures to environmental chemical and physical agents, such as foodborne nitrosamines, polycyclic aromatic hydrocarbons, and ionizing radiation. While there was considerable evidence for endogenous damage to DNA involving "spontaneous" (i.e., simple hydrolysis) deamination and depurination of DNA bases [15–17], there was little appreciation for most of the endogenously formed DNA and RNA damage products due mainly to a failure to detect them as a result of the limited sensitivity and specificity of available analytical technologies. The historical context for the study of endogenous DNA damage and its relationship to the evolution of analytical technology has been elegantly summarized by Prof. Lawrence Marnett in a review article published in 2000 [5]. With the emergence of highly sensitive mass spectrometry technologies, we now have analytical tools capable of detecting the rarest chemical products of endogenous damage in relatively small samples of cells and tissue, to the point of quantifying a single damaged molecule among trillions of undamaged targets. This has led to an appreciation for the ubiquity of endogenous DNA and RNA damage and the importance of DNA and RNA repair processes as "housekeeping" activities in addition to their roles in stress response.

As has been reviewed extensively [1, 2, 7, 9, 18–21], there are numerous fundamental endogenous (i.e., physiological) chemicals and chemistries that can cause DNA and RNA damage in cells, including endogenous alkylating agents such as S-adenosylmethionine [10, 22]; dicarbonyl products of carbohydrate metabolism [21, 23]; and reactive oxygen and nitrogen species generated by immune cells, mitochondria, and fatty acid metabolism in peroxisomes [1, 18, 24]. Though recently challenged as incorrect [25], perhaps the most widely recognized paradigm for endogenous damage involves oxidative stress arising from "leakage" of superoxide from electron transport systems in mitochondria or $O_2^{\bullet-}$ generation by xanthine and NADPH oxidases. With the subsequent reduction of $O_2^{\bullet-}$ to hydrogen peroxide by superoxide dismutases, the two reactive oxygen species can participate in the redox reactions of copper- and iron-catalyzed Fenton chemistry to generate highly oxidizing hydroxyl radicals. The reduction potential of a hydroxyl radical (2.31 V vs. NHE; [26]) is such that this species is capable of oxidizing all types of cellular molecules, including lipids, proteins, carbohydrates, and nucleic acids, as well as small metabolites. The direct damage of biomolecules can lead to pathological dysfunction on the road to cell death. However, many of the products of biomolecular damage are toxic electrophiles capable of causing damage by reaction with endogenous nucleophiles such as DNA, RNA, and proteins. For example, the peroxidation of polyunsaturated fatty acids generates a host of electrophilic products that react to form DNA and RNA base adducts, such as the well-studied etheno adducts discussed later in this chapter and elsewhere in this book.

While the importance of endogenous DNA damage is becoming increasingly clear, the definition of what constitutes "endogenous" is becoming blurred as we learn more about the complex interplay between exogenous factors such as infectious agents, life style, and diet, and the chemical biology of endogenous oxidative stress, inflammation, and other DNA-damaging processes in cells and tissues. This point is perhaps best illustrated by chronic inflammation. There is now overwhelming epidemiological evidence for a cause-and-effect relationship between chronic inflammation and increased risk of human disease, with a growing appreciation for exogenous causes of chronic inflammation. Perhaps the strongest relationship exists with cancer [18, 27–29], such as the association of cancers of the colon [30, 31], stomach [32, 33], bladder [34, 35], and liver [36] with inflammatory bowel disease and infections with *Helicobacter pylori*, *Schistosoma haematobium*, and hepatitis virus, respectively.

The basis for this relationship lies in the infiltration of immune cells into tissues at sites of inflammation as part of the response to infection or tissue injury, with secretion of a battery of chemically reactive oxygen and nitrogen species by activated macrophages and neutrophils [37, 38] (Figure 1.1). The chemical mediators of inflammation shown in Figure 1.1 span a wide range of reactions, including nitrosation, nitration, oxidation, and halogenation. Activated macrophages generate large quantities of nitric oxide (NO) and $O_2^{\bullet-}$ [1, 18, 27]. Although NO at low levels (nanomolar) is a physiological regulator of the cardiovascular, nervous, and immune systems [39–45], the high NO concentration (≤ 1 μM) [46–48] at sites of inflammation leads to reactions with oxygen and $O_2^{\bullet-}$ to generate nitrous anhydride (N_2O_3; Figure 1.1), a potent nitrosating agent capable of deaminating proteins and DNA bases, and peroxynitrite ($ONOO^-$), respectively. The protonated form of $ONOO^-$ rapidly ($t_{1/2} \sim 1$ second) undergoes homolysis to yield hydroxyl radical and the weak oxidant, $NO_2^{\bullet}$, while further reaction of $ONOO^-$ with CO_2 in tissues results in the formation of nitrosoperoxycarbonate ($ONOOCO_2^-$), which also undergoes homolytic scission ($t_{1/2} \sim 50$ milliseconds) to form carbonate radical anion and $NO_2^{\bullet}$. The neutrophil contribution to chemical mediators of inflammation arises from myeloperoxidase-mediated generation of hypochlorous acid (HOCl), a potent oxidizing and halogenating agent, and conversion of nitrite to $NO_2^{\bullet}$ via a nitryl chloride (NO_2Cl) intermediate [49–52].

As a paradigm for the array of endogenous processes that lead to DNA damage, the reactive chemical mediators of inflammation are capable of damaging nucleic

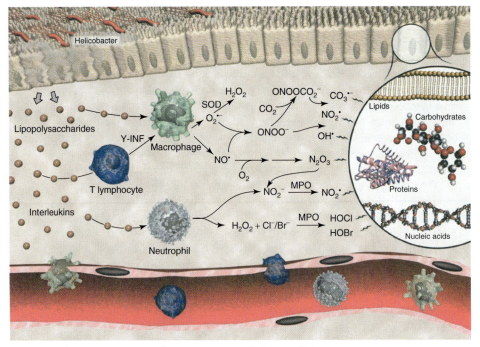

Fig. 1.1 Generation of reactive oxygen and nitrogen during inflammation. Illustration by Jeff Dixon.

acids by two routes. One involves direct reaction with DNA and RNA, such as oxidation, halogenation, and nitrosative deamination. Alternatively, the reactive chemicals can cause nucleic acid damage indirectly by formation of DNA adducts with electrophiles generated from other reactions with polyunsaturated fatty acids, proteins, carbohydrates, small molecule metabolites, and even nucleic acids themselves. The following sections of this chapter address the specific chemistries of DNA and RNA damage arising in endogenous reactions.

1.2
Oxidatively Damaged DNA

Depending upon their inherent redox energetics, endogenous oxidants can react with either the nucleobase or sugar moieties in DNA and RNA, producing a variety of damage products covered in the following sections. Although both ribose and 2-deoxyribose are susceptible to hydrogen atom abstraction by strong oxidants (e.g., hydroxyl radical), only the better-defined chemistry of 2-deoxyribose oxidation will be covered in this chapter. The chemistry of one-electron oxidation of nucleobases is further complicated by sequence context effects on the spectrum of damage products and on charge migration phenomena that dictate the final location of the damage.

1.2.1
Base Oxidation

Three types of oxidation reactions are performed on DNA and RNA bases: one-electron removal, nucleophilic addition, and bond insertion. The latter is performed by singlet oxygen, which will not be covered in this review but has been reviewed extensively [53, 54].

Figure 1.2 shows the variety of guanine (G) damage products arising from one-electron oxidation and hydroxyl radical attack, with specific pathways involved in hydroxyl radical reactions with G. Most of the products are common to both pathways and to many types of oxidants as a result of the multistep nature of the damage chemistry. However, some of the products shown are agent specific, such as the nitro-containing products that are unique to $ONOO^-$ and $ONOOCO_2^-$ as a result of the formation of $NO_2^{\bullet}$ (Figure 1.7) [1]. Regarding one-electron oxidations, guanine (G) has the lowest reduction potential of the five canonical nucleobases (1.29 V vs. NHE for the G neutral radical, $G(-H)^{\bullet}$; 1.58 V vs. NHE for G radical cation, $G^{\bullet+}$; [55]) and is thus the most readily oxidized. "Classical" one-electron oxidants such as photoactivated riboflavin oxidize G to form $G^{\bullet+}$, with the resulting electron hole migrating through the π-stack of B-DNA in competition with trapping to form stable products [56–59]. Recent evidence for sequence-specific variation in the ionization potential (IP) of G [60–62] explains the existence of damage "hot spots" at sites containing multiple adjacent Gs (e.g., GG, GGG), with the electron hole migrating to sites with the lowest G IPs where stable damage products finally form. However, other one-electron oxidants, notably $ONOO^-$ and $ONOOCO_2^-$, paradoxically select the least oxidizable Gs (i.e., those with the highest IP) as preferential targets [63], which complicates the predictive power of charge transfer models of DNA oxidation.

Of particular importance to the development of biomarkers of oxidative DNA damage, the low reduction potential of 8-oxo-G (0.74 V vs. NHE; [64]) makes it significantly more susceptible to further oxidation to more stable products such as Sp (Figure 1.2). This relative instability demands great care to avoid artifacts of oxidation during DNA isolation and processing steps involved in the quantification of 8-oxo-G. Further, the ready oxidation of 8-oxo-G suggests that the other more stable G oxidation products would serve as more abundant and possibly better biomarkers of DNA oxidation.

As opposed to the predominance of $G^{\bullet+}$ as the initial product of one-electron oxidation of G, hydroxyl radical reacts with DNA bases mainly by nucleophilic attack, with formation of 8-OH-$G^{\bullet}$ and 4-OH-$G^{\bullet}$ radicals in the case of G. Reduction of 8-OH-$G^{\bullet}$ can produce a hemi-aminal that opens the imidazole ring to form 2,6-diamino-5-formamidopyrimidine (FAPy-G), while a second one-electron oxidation of 8-OH-$G^{\bullet}$ results in the formation of 7,8-dihydro-8-oxoguanine (8-oxo-G) (Figure 1.2) [54]. FAPy-G can also be formed by the ring opening of 8-oxo-G. The loss of a water molecule from 4-OH-$G^{\bullet}$ at physiological pH results in the neutral G radical, $G(-H)^{\bullet}$ [65], which is converted to a ring-opened imidazolone that eventually reacts with water to form the stable oxazolone (Ox) [7, 66–68] (Figure 1.2). The

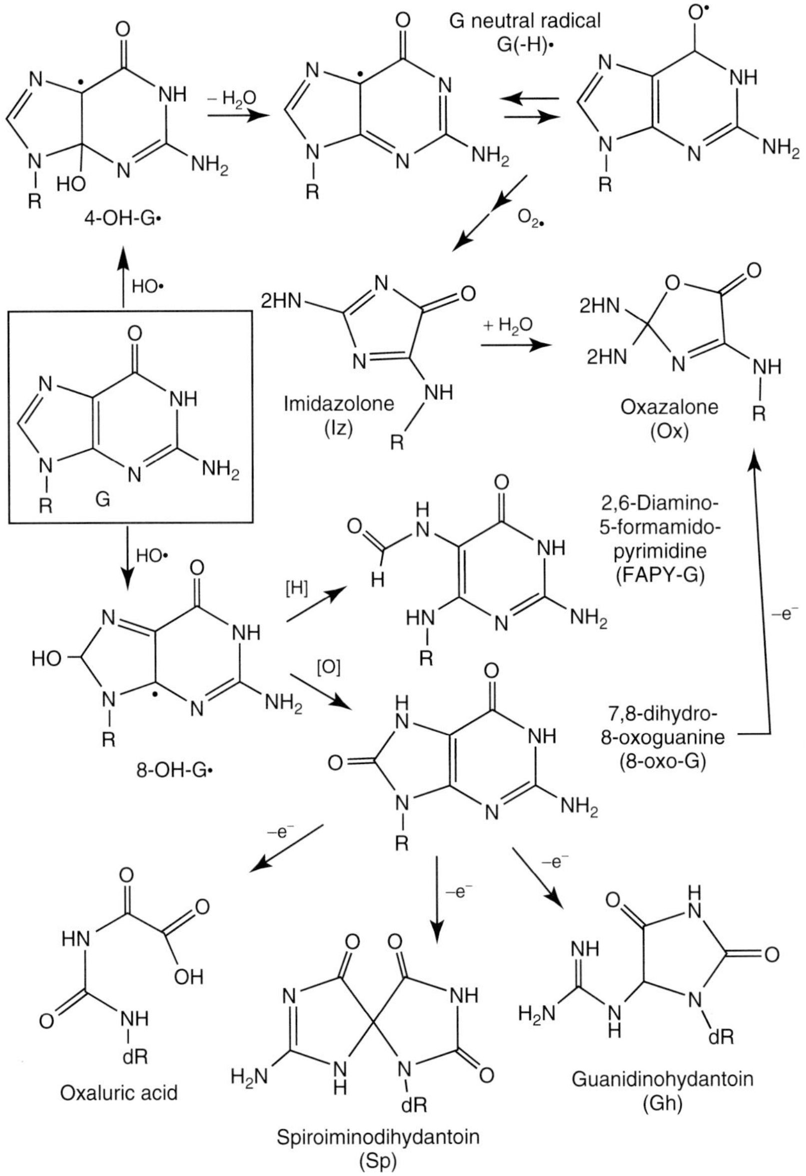

Fig. 1.2 Hydroxyl radical and one-electron oxidation of guanine.

complicating factor here is the balance between the fates of 8-OH-G•, 4-OH-G•, and G(−H)•. A bias toward 8-OH-G• would lead to a predominance of stable damage products (i.e., 8-oxo-G and Fapy-G) formed at the initial site of G oxidation, while higher proportions of 4-OH-G• and its dehydration product G(−H)• could lead to migration of the radical to neighboring G bases.

Unlike one-electron oxidation, attack by hydroxyl radical is less discriminate on the basis of either the identity of the nucleobase or its sequence location, with high levels of adenine (A), thymine (T), and cytosine (C) oxidation. Though less frequent than stable products formed at G, the oxidation of A by hydroxyl radical is similar to that of guanine, with addition at the C-4 and C-8 and formation of 8-oxo-A and FAPy-A products (Figure 1.3).

The reaction of T with hydroxyl radical is complicated by the possibility of both nucleophilic addition to the C5–C6 double bond or hydrogen atom abstraction from the C5-methyl group (Figure 1.4) [7, 67, 69]. In all cases, the resulting radical species can react with molecular oxygen to form peroxyl radicals that degrade to a variety of products (Figure 1.4). Peroxyl radicals at the C5 or C6 and at the allyl position can be reduced to their respective hydroperoxides. The allyl hydroperoxide can then be converted to 5-hydroxymethyluracil or 5-formyluracil (Figure 1.4), while both C5 and C6 hydroperoxides can be reduced to yield thymine glycol. Alternatively, the C6 hydroperoxide (i.e., 5-hydroperoxy-6-hydroxy-5,6-dihydrothymidine) can be converted to 5-hydroxy-5-methylhydantoin (Figure 1.4).

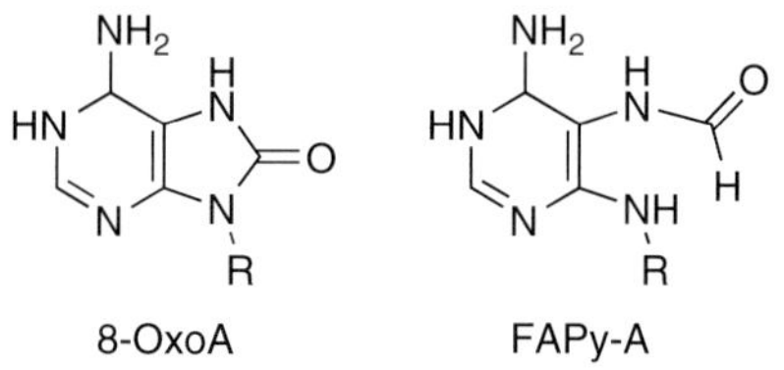

Fig. 1.3 Adenine oxidation products.

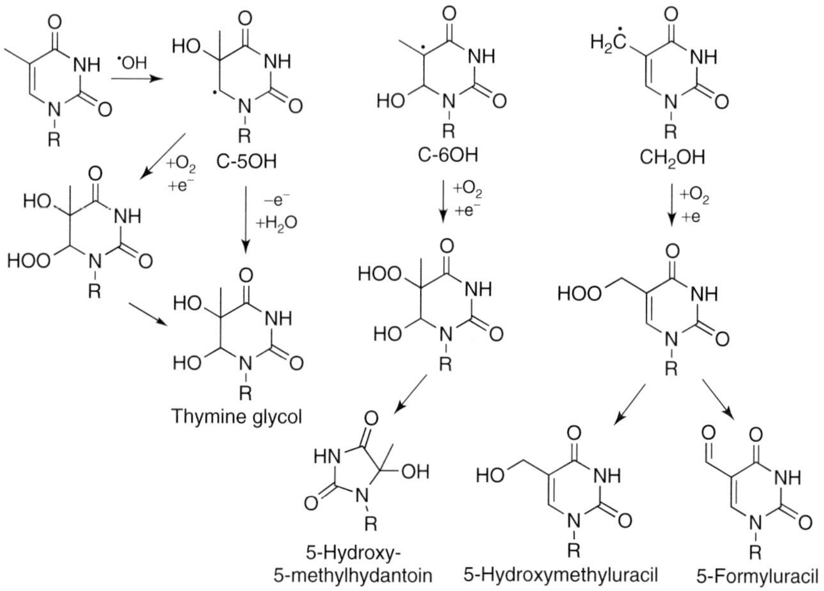

Fig. 1.4 Hydroxyl radical oxidation of thymine.

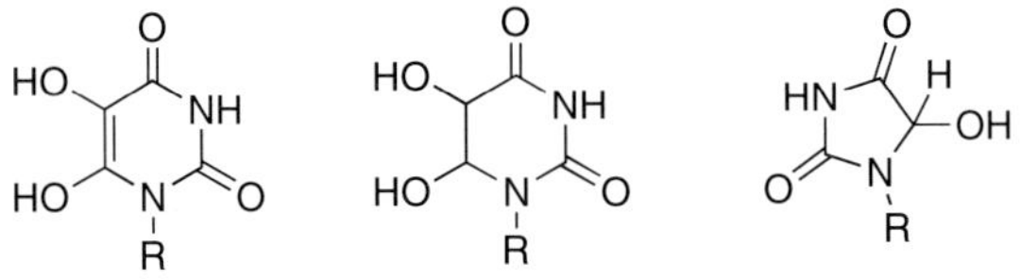

5,6-Hydroxyuridine   **Uridine glycol**   **5-Hydroxyhydantoin**

Fig. 1.5 Hydroxyl radical oxidation of cytosine.

The chemistry of C reaction with hydroxyl radicals is made more complicated by the fact that the primary products of the addition of hydroxyl radicals to the C4–C5 double bond are unstable toward deamination of the exocyclic amine [7, 67]. Following addition of hydroxyl radical to C5, the reactions of the resulting C6 radical are analogous to those with T: formation of a peroxyl radical with molecular oxygen and reduction to form cytosine glycol (Figure 1.5). However, oxidation of the pyrimidine ring system in cytosine glycol leads to deamination of the exocyclic amine to form uridine glycol, which can subsequently lose a water molecule to produce 5-hydroxyuridine (Figure 1.5). The reactions subsequent to hydroxyl radical attack at the C6 position can alternatively partition to form 5-hydroxyhydantoin.

1.2.2
Tandem Lesions: Multiple Products Derived from a Single Oxidation Event

Recent studies have revealed a wide range of DNA lesions in which a single oxidation event leads to multiple local damage products. While often associated with DNA damage produced by ionizing radiation, the so-called tandem lesions were originally described by Box and coworkers in the intrastrand cross-links between adjacent purine and pyrimidine bases, as illustrated in Figure 1.6 ([70]; reviewed more recently in [8]). These observations have been expanded and mechanistically characterized mainly by the Box, Cadet, and Wang groups to reveal that a single C, T, or U radical species is sufficient to cause the formation of intrastrand cross-link lesions [8, 14, 70]. Similarly, Cadet and Box have demonstrated tandem lesions comprising 8-oxo-G adjacent to a T- or C-derived formylamine residue (Figure 1.6).

DNA lesions comprising both base and 2-deoxyribose oxidation events have also been described. For example, the electrophiles generated by 2-deoxyribose oxidation, as described shortly, can react with local DNA bases to produce adducts, with the end result of a strand break and base lesion derived from a single oxidation event [6]. Similarly, intranucleotide cross-links can form between the base and 2-deoxyribose moieties to form cyclic nucleotide lesions, as illustrated in Figure 1.6 [8, 54, 71, 72]. As originally described by the groups of Cadet and Dizdaroglu, the diastereomeric pairs of 5′,8-*cyclo*-dA and -dG are likely derived from an initial 5′-oxidation of 2-deoxyribose in DNA, with the 5′-radical attacking the C8 position of G. The stereochemistry changes on the basis of the context. In nucleosides and single-stranded DNA, the R stereochemistry predominates, while in double-stranded DNA the S isomers dominate [73–75] (Figure 1.6). Recent studies demonstrated that the second step in the formation of cyclopurine

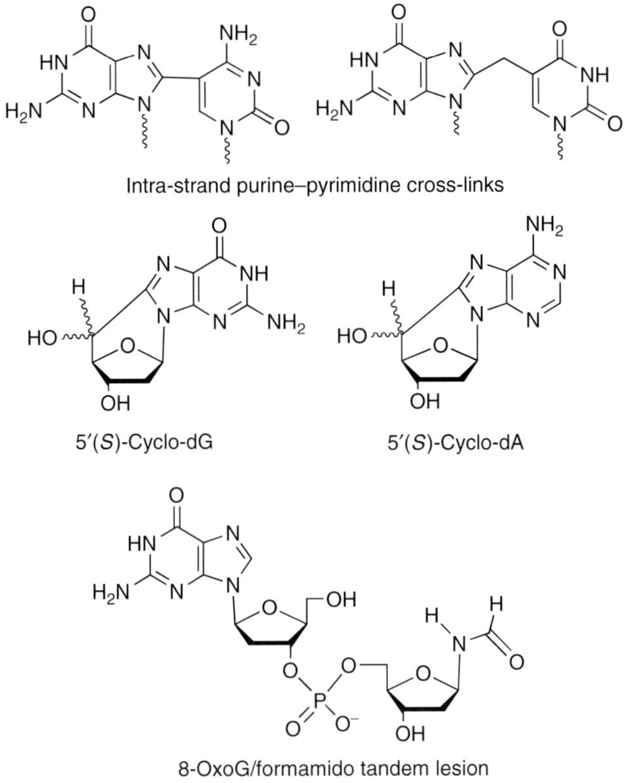

Intra-strand purine–pyrimidine cross-links

5′(S)-Cyclo-dG

5′(S)-Cyclo-dA

8-OxoG/formamido tandem lesion

Fig. 1.6 Tandem DNA lesions.

adducts is relatively slow; therefore, the initial radical can react with intracellular components such as molecular oxygen, potentially inhibiting adduct formation [72, 74]. These lesions have been enhanced in mammalian cellular DNA *in vivo* under conditions of oxidative stress [76, 77].

1.2.3
Base Nitration

The oxidation chemistry associated with reactive nitrogen species receives a significant contribution of nucleobase nitration, mediated primarily by peroxynitrite (ONOO$^-$) and its CO_2 conjugate, nitrosoperoxycarbonate (ONOOCO$_2^-$). As shown in Figure 1.7, G oxidation by these agents results in the formation of several nitrated species unique to one-electron base oxidation reactions, including 8-nitro-G, which is unstable toward depurination, and 5-guanidino-4-nitroimidazole [1]. Evidence points to nitration products of dA as well, but they have not been fully characterized [78].

1.2.4
Halogenation

The final class of oxidatively induced DNA and RNA base lesions, the halogenation products, appears to be unique to granulocytes such as neutrophils by way of myeloperoxidase-generated hypohalous acids. The reaction of HOCl with DNA leads to the formation of 5-chloro-C, and 8-chloro-G and -A (Figure 1.7) [79–81]. Additionally, HOCl can oxidize proteins, carbohydrates, and polyunsaturated fatty acids to generate adduct-forming electrophiles [82]. Given the apparent strong association between chloro-tyrosine levels and cardiovascular disease [83], it is possible that similar granulocyte-mediated chemistry with DNA and RNA will yield useful biomarkers of inflammation.

1.2.5
2-Deoxyribose Oxidation

Although studies of base damage have dominated the literature, there is growing evidence that oxidation of 2-deoxyribose in DNA by endogenous processes plays an important role in the genetic toxicology of oxidative stress beyond the formation of simple "strand breaks." A recent review of the chemical biology of 2-deoxyribose oxidation covers the topic thoroughly [6], so this portion of the chapter on endogenous DNA damage will only survey the product spectrum. As discussed shortly, several of the electrophilic 2-deoxyribose oxidation products lead to the formation of protein–DNA cross-links, and protein and DNA adducts [6]. They also contribute to complex DNA lesions caused by ionizing radiation and hydroxyl radicals, with closely opposed strand breaks and oxidized abasic sites [84, 85].

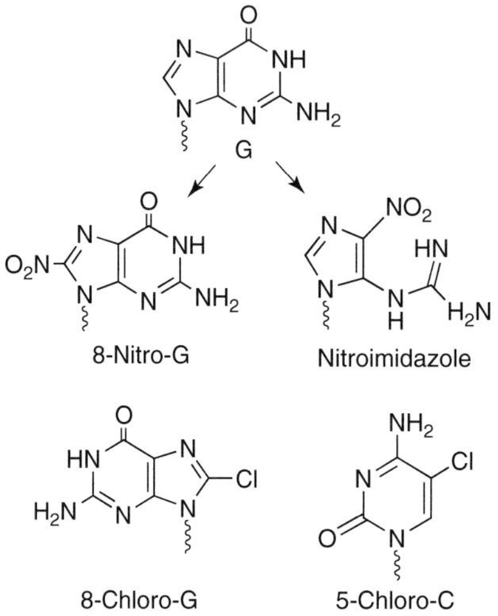

Fig. 1.7 Base nitration and halogenation.

Oxidation of each of the five positions in 2-deoxyribose in DNA occurs with hydroxyl radical and other strong oxidants with an initial hydrogen atom abstraction to form a carbon-centered radical that adds molecular oxygen to form a peroxyl radical [6]. Although the product spectra for 2-deoxyribose oxidation has been studied under both aerobic and anaerobic conditions, the more biologically relevant spectrum of products formed under aerobic conditions is shown in Figure 1.8 [6]. The degradation of the peroxyl radical results in the formation of a variety of electrophilic and genotoxic products that differ for each position (Figure 1.8). *1′-chemistry*: Oxidation of the 1′-carbon produces a 2′-deoxyribonolactone abasic site that can undergo β/δ eliminations to form 5-methylene-5(2H)-furanone (5-MF). *2′-chemistry*: Oxidation of the 2′-position during γ-irradiation generates a d-erythrose abasic site. *3′-chemistry*: Oxidation of the 3′-position has recently been shown to partition along two pathways to form a strand break with 3′-phosphoglycolaldehyde, 5′-phosphate, and base propenoic acid residues; or a 3′-oxo-nucleoside residue that undergoes β/δ eliminations to release 2-methylene-3(2H)-furanone (2-MF). *4′-chemistry*: Recent studies have revealed novel facets of 4′-oxidation, with partitioning along three pathways. In all cases, one pathway leads to a 2-deoxypentos-4-ulose abasic site. The other pathways all entail the formation of a strand break with a 3′-phosphoglycolate residue. *5′-chemistry*: 5′-Oxidation also partitions to form two sets of products – a 3′-formylphosphate-ended fragment and a 2-phosphoryl-1,4-dioxo-2-butane residue that undergoes β elimination to form a *trans*-butenedialdehyde species; or a nucleoside-5′-aldehyde residue that can undergo β/δ eliminations to produce furfural.

A second source of reactive electrophiles derived from 2-deoxyribose involves repair of native abasic sites arising from base excision repair and hydrolysis [86]. As illustrated in Figure 1.9, bi- and monofunctional DNA glycosylases release damaged bases leaving an abasic site that undergoes further reactions with AP lyase and endonuclease to release several α, β-unsaturated aldehydes.

The generation of the variety of reactive electrophiles in oxidation of 2-deoxyribose in DNA has several biological consequences. One involves formation of DNA adducts, which is best illustrated with the base propenals derived from 4′-oxidation [6]. Base propenals are structural analogs of β-hydroxyacrolein, the enol tautomer of malondialdehyde from lipid peroxidation, and react with G to form the M_1dG pyrimidopurinone adduct (Figure 1.12) at rates >100 times faster than malondialdehyde [87]. Recent studies in *Escherichia coli* demonstrated that base propenals represent a major source of M_1dG in cells subjected to oxidative stress [88].

1.3
DNA Alkylation by Endogenous Electrophiles

A second major source of endogenous DNA damage involves alkylation of DNA bases by endogenous electrophiles. This review will not address the simple

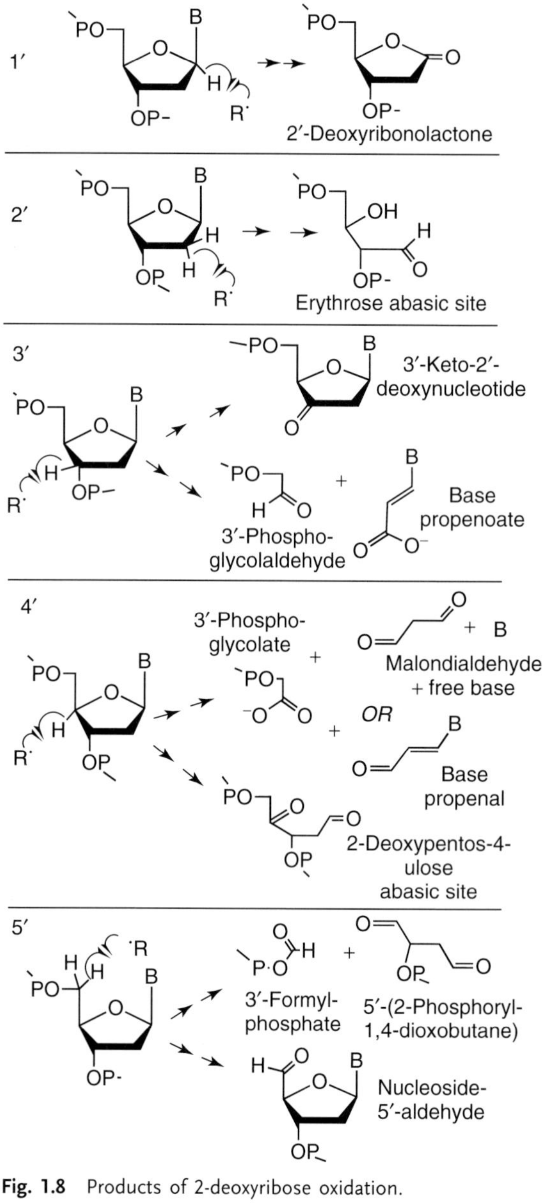

Fig. 1.8 Products of 2-deoxyribose oxidation.

alkylating agents, such as S-adenosylmethionine, and anticancer agents such as the nitrogen mustards. Instead, the review focuses on recent advances in endogenous DNA adducts derived from oxidation of lipids, DNA, carbohydrates, and proteins, as well as estrogen metabolism.

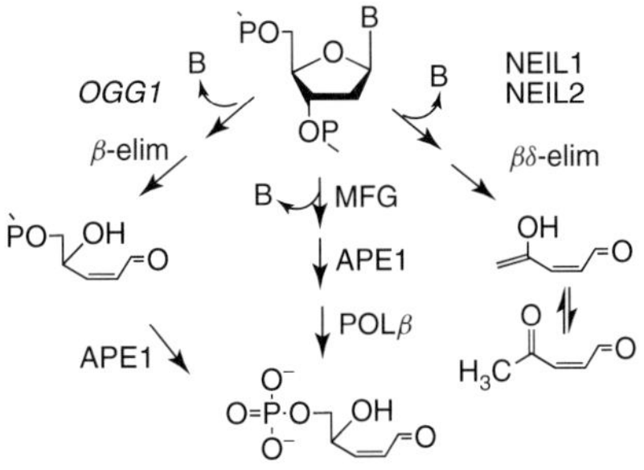

Fig. 1.9 Formation of electrophilic 2-deoxyribose remnants from processing of abasic sites during base excision repair (APE1, AP-endonuclease 1; OGG1, 8-oxo-G glycosylase; NEIL1/2, human DNA glycosylases; MFG, monofunctional glycosylase; POLβ, DNA polymerase β. Adapted from [86].

1.3.1
Lipid Peroxidation-Derived Adducts

Lipid peroxidation represents a rich source of DNA adduct-forming electrophiles as a result of the presence of a variety of different fatty acid substrates and the complexity of the peroxidation chemistry. There are numerous reviews addressing lipid peroxidation-derived adducts, most notably by Helmut Bartsch and the late Jagadeesan Nair [9, 89] and the most recent review by Ian Blair [90], which address the full spectrum of DNA adducts derived from the diversity of lipid peroxidation products and the methods to quantify them in biological systems.

The generation of DNA-reactive electrophiles from polyunsaturated fatty acids begins with hydrogen atom abstraction from the methylene groups located between cis double bonds that define polyunsaturated fatty acids (Figure 1.10). The resulting carbon-centered radical equilibrates at either end of the double bond before adding molecular oxygen to form a peroxyl radical, which is reduced to a hydroperoxide in the chain termination event. These lipid hydroperoxides are relatively stable in the absence of metals and can be reduced by glutathione (GSH) peroxidases to fatty acid alcohols. However, in the presence of redox active metals and other reductants, the hydroperoxides undergo multistep reactions leading to the formation of a variety of α, β-unsaturated carbonyl species that react as bifunctional alkylating agents with DNA bases [7, 9, 89, 90]. Some major lipid peroxidation products are shown in Figure 1.10 and Figure 1.11, all of which react with endogenous nucleophiles to form GSH, protein, and DNA adducts [90].

Perhaps the best-studied DNA adducts arising from lipid peroxidation products are the unsubstituted etheno adducts of A, G, and C shown in Figure 1.10. Elevated levels of these lesions have been found in conditions of oxidative stress in human and mouse atherosclerotic tissues [91, 92], mouse models of chronic inflammation [93], urine of thalassemia patients [94], and urine of patients with chronic hepatitis, cirrhosis, or hepatocellular carcinoma [9]. Although there is no definitive proof for the identity of the lipid peroxidation product that serves as a source for the

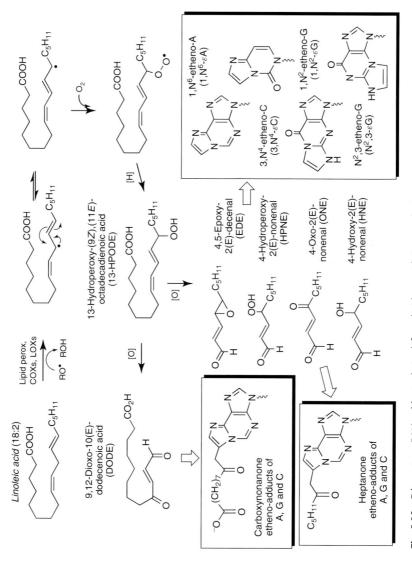

Fig. 1.10 Etheno-type DNA adducts derived from lipid peroxidation products.

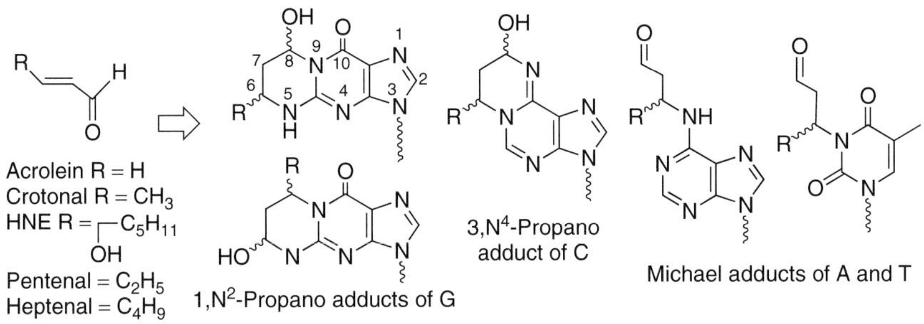

Fig. 1.11 Propano-type DNA adducts derived from lipid peroxidation products.

unsubstituted etheno adducts, studies of adduct chemistry *in vitro* appear to require species with multiple oxidation sites such as EDE or HPNE, with HPNE exhibiting significantly higher reactivity (studies summarized in [90]).

The more recently discovered substituted etheno adducts have been the subject of significant study. Although there are many possible structural permutations due to the variety of polyunsaturated fatty acids and potential peroxidation products, the Blair group has identified two types of substituted etheno adducts *in vitro* and *in vivo*: heptanone and carboxynonanone adducts (Figure 1.10) [90, 95]. The reaction of (13S)-HPODE or its degradation product, ONE, with dG, dC, and dA leads to the formation of the respective heptanone adducts, HεdG, Hε-dC, and HεdA [90, 96–98]. The proposed mechanism for the formation of these adducts involves initial formation of an ethano adduct by nucleophilic addition of N-2 of dG to the C-1 aldehyde of ONE followed by the attack of the N1 at C2 of the resulting α,β-unsaturated ketone [90, 96–98]. By similar mechanisms, (13S)-HPODE-derived DODE gives rise to the analogous carboxynonanone DNA adducts [99], with reactions of (13S)-HPODE in the presence of dG giving rise to equimolar mixtures of the heptanone and carboxynonanone adducts. As a demonstration of the potential complexity of the etheno adduct spectrum, Blair and coworkers identified a carboxypentanone adduct of G in reactions of 5-lipoxygenase-derived hydroperoxy eicosatetraenoic acid (HPETE) with dG [100]. The lipid peroxidation products arising from reactions of cyclooxygenase and lipoxygenase with polyunsaturated fatty acids are stereospecific in nature, which suggests that the various stereoisomers of the lipid peroxidation products and their DNA adducts may be useful as biomarkers capable of distinguishing enzymatic versus chemically derived lipid damage products [90, 95].

While the α,β-unsaturated carbonyl-containing lipid peroxidation products can react to form the aromatic etheno adducts following an additional oxidation event (e.g., epoxidation), the parent "enal" species can react to form simpler Michael adducts of DNA such as the propano adducts shown in Figure 1.11. These DNA lesions are derived from HNE, heptenal, pentenal, crotonaldehyde, and acrolein. Studies performed *in vitro* with nucleosides and nucleic acids reveal the potential for conjugation of multiple copies of the electrophiles with the nucleobases (e.g., paraldol dimers of crotonaldehyde observed by Hecht and coworkers [101, 102]). However, these complex structures, while chemically interesting, may have limited biological relevance given the high concentrations of electrophile and prolonged exposure required for the addition reactions. Building on previous reviews of the propano adducts [103], this review will focus only on the monoadducts of the various species.

The mechanism of propano adduct formation common to most of these molecules involves an initial Michael addition of the exocyclic amino group of G, A, or C to the double bond of the electrophile followed by nucleophilic attack of the heterocyclic nitrogen of the base at the carbonyl carbon. There is a strong regioselectivity for the reactions. HNE, crotonaldehyde, and other substituted enals (e.g., pentenal, heptenal) all react to form the 8-hydroxy-propano adducts with G in DNA (Figure 1.11), while acrolein has been observed to form both the 8- and 6-hydroxy-propano adducts, the latter by Michael addition starting at the heterocyclic nitrogen (e.g., N^1 of G) [104–109]. However, acrolein has been shown to favor the 8-hydroxy-propano-dG adduct over the 6-hydroxy-form [105], while crotonaldehyde reacts to favor a predominance of the trans configuration of the hydroxyl and methyl groups of the 8-hydroxy form (Figure 1.11). Similar adducts form with A and C, though these are less well explored, while T forms the simple Michael adduct via its heterocyclic nitrogen (Figure 1.11) [105, 109–114]. With implications for the biological effects of the propano adduct structure in general, it has been observed that all types of propano adducts undergo ring opening to aldehydic forms in DNA, with subsequent formation of cross-links with proteins and neighboring DNA bases [115–125].

The biological consequences of the propano adducts have been the subject of recent intense study in cells, rodents, and humans. In keeping with a major caveat of the DNA adduct world, the major adduct is not always the most toxic: the 8-hydroxy-1,N^2-propano-dG formed by acrolein is far less mutagenic than the minor 6-hydroxy form [126–129]. On the other hand, the propano-G adduct of HNE, the major adduct formed even under peroxidizing conditions *in vitro* [107], was found to be poorly repaired relative to HNE-derived etheno adducts in cells exposed to HNE [107]. The crotonaldehyde adducts may inhibit DNA synthesis and induce miscoding in human cells [110]. The 6S/8S diastereomer of 8-hydroxy-6-methyl-1,N^2-propano-dG causes miscoding more frequently than 6R/8R and leads predominantly to G to T transversions [130–132]. Interestingly, the formation of adducts by crotonaldehyde, and

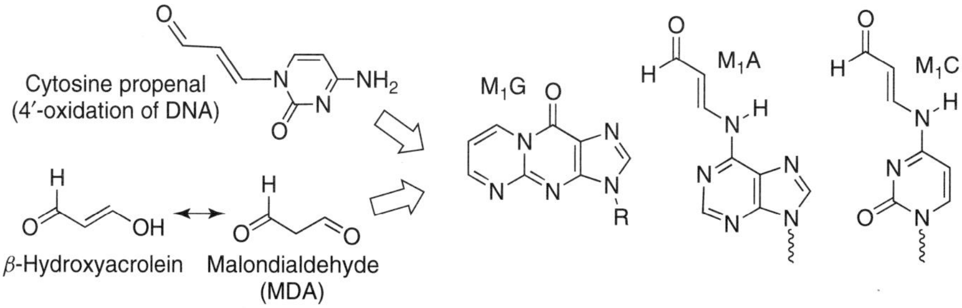

Fig. 1.12 Formation of M_1G from base propenals and malondialdehyde.

probably the entire class of enal compounds, is enhanced by the presence of histone proteins [133, 134], with evidence for amine-catalyzed conversion of acetaldehyde to crotonaldehyde [135]. With the advent of ultrasensitive analytical methods to quantify the propano adducts in cells and tissues [9], including LC/MS–MS [136] and ^{32}P-postlabeling [137, 138], the lesions have been detected in a variety of rodent and human tissues [9, 103, 110, 136, 138–140].

1.3.2
Malondialdehyde-Like Adducts

The well-known lipid peroxidation product, malondialdehyde, has been the subject of study in the genetic toxicology community for two decades since the discovery that it reacts with DNA bases to form a variety of adducts, some of which are illustrated in Figure 1.12, the most well characterized of which is the pyrimidopurinone adduct of G, M_1G. As such, the so-called malondialdehyde adducts have been discussed in numerous review articles [5, 6, 9, 141–146]. M_1A and M_1C are formed by the addition of the exocyclic amino group of the nucleobase to the aldehydic carbon followed by the elimination of water (Figure 1.12). A similar mechanism is presumed to take place in the formation of M_1G; however, there is an elimination of two molecules of water. When base paired in DNA or under basic conditions, M_1dG undergoes hydrolytic ring opening to N^2-oxopropenyl-dG [147–152], a phenomenon that has been exploited to develop an analytical method for quantifying M_1G [153–155]. Further, the exocyclic ring of M_1G is susceptible to direct nucleophilic attack with subsequent ring opening [148], which suggests that the alkyl adduct has the potential to migrate to neighboring protein or nucleic acid nucleophiles.

While the M_1G adduct was originally described as a product of DNA reaction with malondialdehyde *in vitro*, recent observations suggest a more complicated chemical basis for its formation *in vivo* [88, 155]. As discussed earlier, base propenals arising from 4'-oxidation of DNA (Figure 1.8) are structural analogs of the DNA-reactive β-hydroxyacrolein enol tautomer of malondialdehyde (Figure 1.12) and have been shown to be significantly more reactive in forming M_1G in DNA as a result of the presence of the nucleobase as a better leaving group than hydroxide [87]. More recent studies in *E. coli* revealed that base propenals, and not malondialdehyde, were the major source of M_1G in cells subjected to oxidative stress [88]. In light of the potential "mobility" of M_1G [147–152] and the potential for transfer of the oxopropenyl group to and from DNA via N^ε-oxopropenyllysine adducts in histone proteins [156], it will be difficult to precisely define the source of M_1G adducts *in vivo*.

1.3.3
DNA Glycation

With bifunctional reactivity similar to α, β-unsaturated carbonyl compounds, α-oxoaldehydes arising from glycolysis and lipid peroxidation represent yet another source of endogenous DNA alkylation. Among the best-studied reactive α-oxoaldehydes are glyoxal, methylglyoxal, and 3-deoxyglucosone. These electrophiles are capable of forming stable adducts with DNA and proteins to form the so-called advanced glycation end products that are involved in the pathophysiology of diabetes, vascular disease, arthritis, and neurodegeneration, as well as age-related pathologies [21, 23]. Again, the abundance of review articles covering glycation [157–162] motivates consideration of only the most recent advances in the chemical biology of DNA adduct formation.

The dicarbonyl species react with DNA to form adducts that are inherently unstable relative to other types of DNA adducts, which has made it difficult to assess the formation of such adducts *in vivo*. Nonetheless, several adducts in two different classes have been identified, as shown in Figure 1.13. The imidazopurinone adducts formed with glyoxal and G, with the hydroxyl groups mainly in the trans configuration [163], were originally identified four decades ago [164] and later characterized by Kasai and coworkers [165]. They have been quantified *in vitro* using both a ^{32}P-postlabeling [163, 166] and by mass spectrometric methods [167, 168] in purified DNA treated with glyoxal [163, 166] and in tissues from rats treated with 2-hydroxyethyl-containing nitrosamines [163, 166]. Contrary to M_1G, the imidazopurinone glyoxal adducts were shown to be more stable in duplex DNA than in nucleoside form [165]. In addition to the imidazopurinone adducts of G, Pluskota-Karwatka *et al.* identified an acyclic N^2-carboxymethyl adduct of G that proved to be more stable than the imidazopurinone adduct [132].

As shown in Figure 1.13, glyoxal adducts have also been identified with dC and dA in DNA. Kasai and coworkers identified a C5-hydroxyacetyl adduct of dC [165] while Olsen *et al.* identified an acyclic N6-hydroxyacetyl adduct of A [169] (Figure 1.13). Kasai also observed that dC was prone to deamination to 2′-deoxyuridine (dU) by glyoxal, probably by means of a transient imidazo adduct with glyoxal [165]. This is analogous to the deamination of dC by hydroxyl radical adduct formation. Lunec and coworkers have developed a potentially useful antibody against the putative 5-hydroxyacetyl-C adduct of glyoxal, with significant selectivity for this adduct among the several adducts arising in glyoxal-treated DNA [170–172]. The antibody was used to show the presence of the 5-hydroxyacetyl-C adduct *in vivo* [170].

One potentially important feature of DNA reactions with glyoxal entails the formation of intrastrand cross-links of dC, dA, and dG with dG. This was originally described by Kasai and coworkers [165] and later characterized by Brock *et al.* [173]. This behavior may explain the comet assay evidence for DNA cross-link formation by glyoxal [174].

Finally, while there are numerous sources of glyoxal, including lipid, protein, and carbohydrate oxidation, another source was recently added: DNA oxidation. Kasai and coworkers originally demonstrated that glyoxal and its

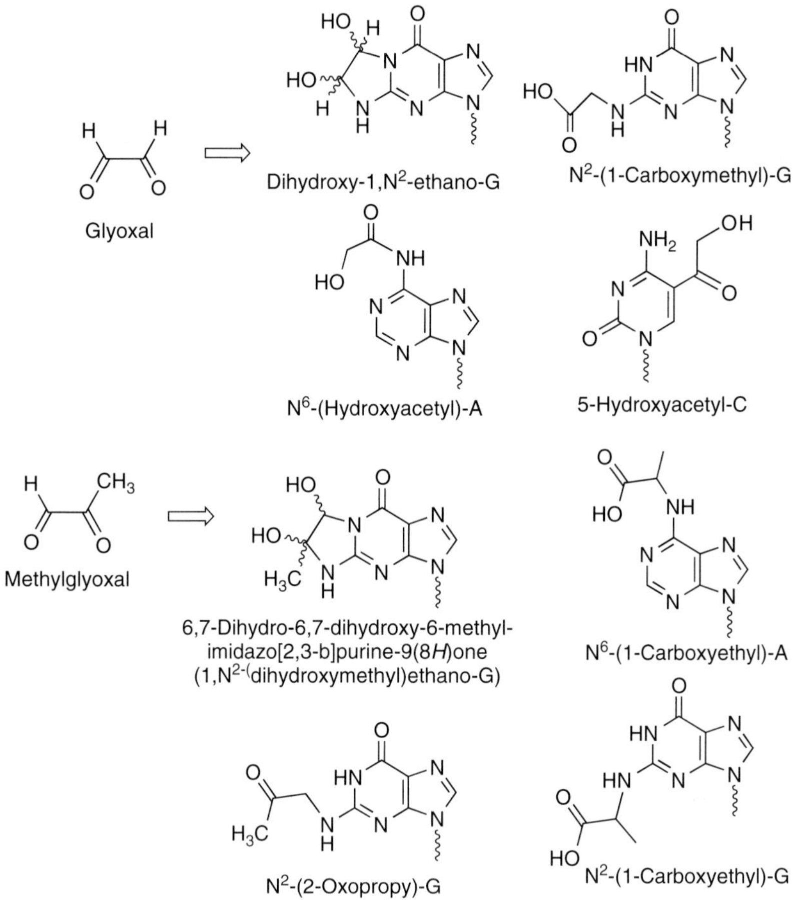

1.3 DNA Alkylation by Endogenous Electrophiles | 23

Fig. 1.13 Glycation adducts of DNA.

1,N^2-imidazo adducts of dG were generated in DNA subjected to oxidation by the Fe(II)–EDTA complex [175]. We later observed that the glyoxal was derived from the 3'-phosphoglycolaldehyde residue arising from 3'-oxidation of 2-deoxyribose in DNA (Figure 1.8), by a mechanism involving a radical-independent phosphate–phosphonate rearrangement that led to oxidation of the glycolaldehyde to glyoxal [176].

With regard to methylglyoxal, the situation with endogenous DNA adducts is similar to that with glyoxal but more complicated. Under biologically relevant conditions of electrophile and nucleophile ratios, the adduct structures include the imidazo-type cyclic adducts, with the associated diastereomeric forms, and both 1-carboxyethyl [177] and 2-oxopropyl adducts of the exocyclic amine [178] (Figure 1.13). However, the 2-oxopropyl adduct species required prolonged incubation at elevated temperatures and the structural characterization was minimal [178]. An additional controversy surrounds the claim by Casida and coworkers

[179] that the originally proposed imidazopurinone structure for the methylglyoxal adduct of dG defined by Vaca et al. [180] is incorrect. Using a more rigorous set of analytical methods, Casida and coworkers proposed an acyclic N2,7-bis(1-hydroxy-2-oxopropyl)-dG species as the correct structure for the putative imidazopurinone [179]. However, the adduct syntheses in both cases were performed with a 15- to 20-fold molar excess of methylglyoxal [179, 180], which would favor such multiple adduction reactions. Thornalley and coworkers more recently performed a synthesis with dG and methylglyoxal at a one-to-one molar ratio and obtained the expected cyclic imidazopurinone adduct in 62% yield, with structural confirmation by an even more rigorous set of analyses [181]. This controversy illustrates the challenges posed by reactive dicarbonyl-containing species capable of polymerization and other multimeric reactions with endogenous nucleophiles. It is highly unlikely that electrophile concentrations *in vivo* reach a level capable of sustaining multiple adduction reactions with DNA. Further, studies comparing different methylglyoxal concentrations revealed that the bis adducts were not detectable at low (10 µM) concentrations approaching biological relevance [182].

Several groups have developed analytical methods to quantify the methylglyoxal adducts of DNA [183–185]. Wang and coworkers have performed perhaps the most credible quantitative study of DNA glycation in their development and application of an isotope-dilution, electrospray-ionization, HPLC-coupled tandem mass spectrometric method to quantify the N^2-(1-carboxyethyl)-G adduct of methylglyoxal [185]. They observed a background (i.e., uninduced) level of one lesion per 10^7 nucleotides, with cells exposed to either methylglyoxal or glucose showing dose-dependent increases in the adducts [185]. Using a similar approach, Bidom et al. observed time-dependent increases in the level of the N^2-(1-carboxyethyl)-G adduct in cultured cells [183].

1.3.4
Estrogen-Derived DNA Adducts

There are striking parallels between established mechanisms and paradigms for DNA damage by exogenous and endogenous agents. For example, polycyclic aromatic hydrocarbons, such as benzo[a]pyrene, are metabolized by cytochrome P450s to form electrophilic species capable of DNA adduct formation and radical generation by quinone redox chemistry. An endogenous parallel for this phenomenon has emerged in studies of endogenous DNA damage caused by estrogen metabolism, as recently reviewed by Bolton [186]. The established carcinogenicity of endogenous and therapeutic estrogens has been proposed to involve stimulation of cell proliferation with the consequential mutagenesis of DNA replication. However, estrogens have also been shown to be metabolized to catechols by P450 with further oxidation to *o*-quinones that can cause damage by DNA alkylation or oxidation by redox cycling. The DNA damage associated with endogenous human estrogen quinones involves N3-adenine or N7-guanine DNA adducts that readily undergo depurination to form abasic sites [186]. It is not too great a stretch of imagination to predict analogous DNA damage chemistry arising from other metabolites of endogenous species.

1.4 DNA and RNA Deamination

Perhaps one of the most complicated forms of endogenous damage to nucleic acids involves the removal of the exocyclic amines of G, A, and C in processes generally referred to as *deamination*. As shown in Figure 1.14, the deaminated counterparts of these three canonical nucleobases include hypoxanthine (2-deoxyinosine/dI and inosine/rI as nucleosides) derived from A; uracil (2-deoxyuridine/dU, uridine/rU) from C; and xanthine (2-deoxyxanthosine/dX, xanthosine/rX) and oxanine (2-deoxyoxanosine/dO, oxanosine/rO) derived from G. The complexity of nucleobase deamination derives from the many mechanisms leading to their presence in DNA, RNA, and the nucleotide pool, which include simple hydrolysis, nistrosative deamination caused by chemical mediators of inflammation and environmental exposure to nitrite, enzymatically mediated deamination, and misincorporation of purine nucleotide biosynthetic intermediates. As with other sections of this chapter, there are numerous review articles covering the various aspects of DNA and RNA deamination [1, 187–195], therefore only the most recent publications will be considered here.

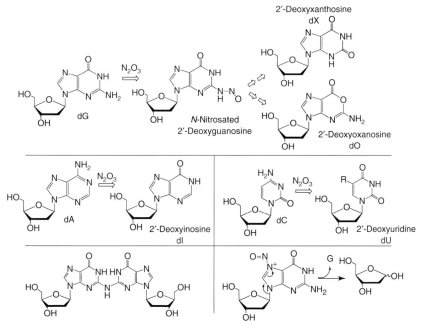

Fig. 1.14 Products of nucleobase deamination.

1.4.1
Hydrolytic Deamination

Among the simplest endogenous mechanisms is hydrolytic deamination that occurs in any aqueous environment [190, 196, 197]. The rate of hydrolytic deamination of DNA occurs in the order 5-methyl-dC > dC > dA > dG [198, 199], with a half-life for dC ranging from 10^2 to 10^3 years for single-stranded DNA and 10^4 to 10^5 years in double-stranded DNA [200–202]. Methylation of the 5-position of dC increases the rate of deamination by up to 20-fold [201, 202], with the high rate deamination of 5'-methyl-dC proposed to account for the high frequency of C → T mutations at CpG sites [203]. However, it must be kept in mind that CpG motifs are also hot spots for reactions with many genotoxic agents [204].

1.4.2
Nitrosative Chemistry and Inflammation

The deamination of DNA nucleobases *in vivo* is mediated mainly by the nitrosative chemistry of nitrous anhydride (N_2O_3), the autoxidative product of nitric oxide ($NO^\bullet$) [1, 48]. As illustrated in Figure 1.14, nitrosative deamination leads to the universal xanthine, hypoxanthine, uracil, and thymine, the latter arising from 5-methyl-C, as well as several novel derivatives including abasic sites, oxanine, and inter- or intrastrand G-G cross-links [1].

Oxanine presents a unique problem as one of the two deamination products arising from G. It has been observed to form in purified DNA exposed to nitrite under acidic conditions [205, 206], but it has not been detected by LC–MS or LC–MS/MS under biologically relevant conditions in purified DNA and cells exposed to NO and O_2 *in vitro* [207, 208], or in tissues from a mouse model of NO overproduction [93]. To explain this discrepancy, Glaser and coworkers have proposed a model [209] in which, under conditions of neutral pH, the initially formed diazonium ion at N2 of G cannot undergo reactions leading to O due to the conformational restriction of double-stranded DNA and catalytic interference from the base-paired C. The model adequately accounts for most, if not all, of the observed deamination products under different conditions and predicts that significant levels of O should form from G in nucleosides, nucleotides, and single-stranded DNA under conditions of nitrosative stress [209].

With respect to the other base deamination products (X, I, and U), the cellular environment provides an approximately 4-fold protective effect against nitrosative deamination [207, 208], with significant elevations of X, I, and U only when the cells are exposed to toxic concentrations of NO and associated N_2O_3 [208]. Similar results were obtained in animal models of inflammation [93, 210]. It is possible that the modest increases in the steady-state levels of DNA deamination products result from limited exposure of nuclear DNA to nitrosating species or from a balance between the rates of formation and repair of nucleobase deamination lesions in DNA.

1.4.3
Enzymatic Deamination

An emerging body of evidence points to enzymatically mediated DNA and RNA deamination as a source of endogenous damage that plays a role in human disease. In terms of DNA, perhaps the most notable example of an enzymatic deamination mechanism involves the dC deaminase activity termed *activation-induced cytidine deaminase* (*AID*) that functions in three facets of immunoglobulin diversification: class switch recombination, somatic hypermutation, and gene conversion (reviewed in [188, 189, 191, 192, 203, 211–214]). These diverse mechanisms probably require a variety of different DNA repair pathways and ancillary proteins in addition to AID [214]. For example, one mechanism involves AID-mediated conversion of dC to dU followed, in the case of somatic hypermutation, by mutagenic responses to dU, including uracil N-glycosylase-mediated formation of abasic sites followed by error-prone mismatch repair [215, 216] or polymerase bypass of unrepaired dU [216, 217]. In the case of class switch recombination, the conversion of dC to dU by AID initiates controlled, sequence-specific recombination [214].

With regard to RNA deamination, there are at least three recognized adenosine deaminases (ADARs) that act on unspliced transcripts at rA in an exon that is base paired with an adjacent intron, with the resulting rI being read as rG during translation [189, 218–220]. Currently known genes affected by A-to-I editing include the calcium-gated glutamate receptor [221–223], the 5-hydroxytryptamine receptor [224, 225], the potassium channel KCNA1 [226, 227], and several identified by comparative genomics [228, 229]. One interesting consequence of ADAR sequence searches was the discovery of a family of ADAR-related adenosine deaminases that act on tRNA (ADAT or Tad) [230]. RNA editing also involves conversion of rC to rU in several transcripts [231–233]. For example, intestinal apolipoprotein B (apoB) [234] is edited by apolipoprotein B mRNA editing catalytic subunit 1 (APOBEC-1), a member of the apobec family of cytidine deaminases and a relative of AID protein discussed earlier [231].

The point of discussing this RNA biology is to highlight the potential consequences of RNA damage caused by endogenous processes. RNA editing by both A-to-I and C-to-U mechanisms is certain to be widespread, so nitrosative deamination even at relatively low levels could alter gene expression. Indeed, RNA damage can disrupt protein synthesis and cause errors in protein translation and impact neurodegenerative diseases [235]. The damaged RNA may also have effects beyond the transcript itself in that proteins termed *vigilins* bind to promiscuously A-to-I-edited transcripts and target them, or their degradation products, to heterochromatic regions of the nucleus [236]. It is possible that RNA damaged by other mechanisms may have functional consequences beyond those expected for interfering with translation, such as microRNA or RNAi function.

1.5
Summary

The spectrum of endogenous lesions is rapidly expanding as analytical technology increases in sensitivity and specificity. While the past decades have witnessed the cataloguing of most of the probable (and improbable) DNA and RNA damage products in studies performed with isolated DNA, the next decades will establish the biological relevance of these lesions. Quantitative dose–response studies with different stressors will provide new insights into the chemistry and kinetics of formation and repair of endogenous lesions and a better understanding of their role in various states of pathology and disease. As illustrated by the use of DNA deamination for generation of immunoglobulin diversity, recent evidence points to the potential for endogenous DNA damage to play a role in the normal physiology of signaling pathways. One such piece of evidence involves the recent observation of site-specific DNA oxidation and repair during the estrogen receptor–dependent transcription in human cells [237]. This model entails transcriptional enhancement by localized demethylation of histone proteins. The hydrogen peroxide produced by the demethylation of histone H3 is proposed to cause local DNA damage, the repair of which triggers chromatin and DNA conformational changes that are essential for estrogen-induced transcription [237]. It is thus possible that controlled application of endogenous DNA damage may play a role in gene expression, with the potential for interference by other uncontrolled forms of endogenous DNA damage. This and other phenomena suggest that the next several decades will reveal significant opportunities for moving the study of endogenous nucleic acid damage to the chemistry–biology interface.

References

1 Dedon, P.C. and Tannenbaum, S.R. (2004) Reactive nitrogen species in the chemical biology of inflammation. *Arch Biochem Biophys*, **423**, 12–22.

2 Loeb, L.A. and Harris, C.C. (2008) Advances in chemical carcinogenesis: a historical review and prospective. *Cancer Res*, **68**, 7 6863–6872.

3 De Bont, R. and van Larebeke, N. (2004) Endogenous DNA damage in humans: a review of quantitative data. *Mutagenesis*, **19**, 169–185.

4 Epe, B. (2002) Role of endogenous oxidative DNA damage in carcinogenesis: What can we learn from repair-deficient mice? *Biol Chem*, **383**, 467–475.

5 Marnett, L.J. (2000) Oxyradicals and DNA damage. *Carcinogenesis*, **21**, 361–370.

6 Dedon, P.C. (2008) The chemical toxicology of 2-deoxyribose oxidation in DNA. *Chem Res Toxicol*, **21**, 206–219.

7 Pluskota-Karwatka, D. (2008) Modifications of nucleosides by endogenous mutagens-DNA adducts arising from cellular processes. *Bioorg Chem*, **36**, 198–213.

8 Wang, Y. (2008) Bulky DNA lesions induced by reactive oxygen species. *Chem Res Toxicol*, **21**, 276–281.

9 Nair, U., Bartsch, H., and Nair, J. (2007) Lipid peroxidation-induced DNA damage in cancer-prone inflammatory diseases: A review of published adduct types and levels in humans. *Free Radic Biol Med*, **43**, 1109–1120.

10 Sturla, S.J. (2007) DNA adduct profiles: chemical approaches to addressing the biological impact of

DNA damage from small molecules. *Curr Opin Chem Biol*, **11**, 293–299.

11 Cooke, M.S., Olinski, R. and Loft, S. (2008) Measurement and meaning of oxidatively modified DNA lesions in urine. *Cancer Epidemiol Biomarkers Prev*, **17**, 3–14.

12 Evans, M.D., Dizdaroglu, M., and Cooke, M.S. (2004) Oxidative DNA damage and disease: Induction, repair and significance. *Mutat Res*, **567**, 1–61.

13 Son, J., Pang, B., McFaline, J.L., Taghizadeh, K., and Dedon, P.C. (2008) Surveying the damage: the challenges of developing nucleic acid biomarkers of inflammation. *Mol Biosyst*, **4**, 902–908.

14 Cadet, J., Douki, T., Gasparutto, D., and Ravanat, J.L. (2003) Oxidative damage to DNA: formation, measurement and biochemical features. *Mutat Res*, **531**, 5–23.

15 Lindahl, T. (1993) Instability and decay of the primary structure of DNA. *Nature*, **362**, 709–714.

16 Nakamura, J. and Swenberg, J.A. (1999) Endogenous apurinic/apyrimidinic sites in genomic DNA of mammalian tissues. *Cancer Res*, **59**, 2522–2526.

17 Atamna, H., Cheung, I., and Ames, B.N. (2000) A method for detecting abasic sites in living cells: age-dependent changes in base excision repair. *Proc Natl Acad Sci U S A*, **97**, 686–691.

18 Ohshima, H., Tatemichi, M., and Sawa, T. (2003) Chemical basis of inflammation-induced carcinogenesis. *Arch Biochem Biophys*, **417**, 3–11.

19 Klaunig, J.E. and Kamendulis, L.M. (2004) The role of oxidative stress in carcinogenesis. *Annu Rev Pharmacol Toxicol*, **44**, 239–267.

20 O'Brien, P.J., Siraki, A.G., and Shangari, N. (2005) Aldehyde sources, metabolism, molecular toxicity mechanisms, and possible effects on human health. *Crit Rev Toxicol*, **35**, 609–662.

21 Thornalley, P.J. (2007) Endogenous α-oxoaldehydes and formation of protein and nucleotide advanced glycation endproducts in tissue damage. *Novartis Found Symp*, **285**, 229–243; discussion 243–6.

22 Sedgwick, B., Bates, P.A., Paik, J., Jacobs, S.C., and Lindahl, T. (2007) Repair of alkylated DNA: recent advances. *DNA Repair (Amst)*, **6**, 429–442.

23 Thornalley, P.J. (2008) Protein and nucleotide damage by glyoxal and methylglyoxal in physiological systems–role in ageing and disease. *Drug Metab Drug Interact*, **23**, 125–150.

24 Ames, B.N., Shigenaga, M.K., and Hagen, T.M. (1993) Oxidants, antioxidants, and the degenerative diseases of aging. *Proc Natl Acad Sci U S A*, **90**, 7915–7922.

25 Hoffmann, S., Spitkovsky, D., Radicella, J.P., Epe, B., and Wiesner, R.J. (2004) Reactive oxygen species derived from the mitochondrial respiratory chain are not responsible for the basal levels of oxidative base modifications observed in nuclear DNA of mammalian cells. *Free Radic Biol Med*, **36**, 765–773.

26 Buettner, G.R. (1993) The pecking order of free radicals and antioxidants: lipid peroxidation, α-tocopherol, and ascorbate. *Arch Biochem Biophys*, **300**, 535–543.

27 Ohshima, H. (2003) Genetic and epigenetic damage induced by reactive nitrogen species: implications in carcinogenesis. *Toxicol Lett*, **140–141**, 99–104.

28 Balkwill, F. and Mantovani, A. (2001) Inflammation and cancer: back to virchow? *Lancet*, **357**, 539–545.

29 Shacter, E. and Weitzman, S.A. (2002) Chronic inflammation and cancer. *Oncology (Williston Park)*, **16**, 217–226.

30 Levin, B. (1992) Ulcerative colitis and colon cancer: biology and surveillance. *J Cell Biochem Suppl*, **16G**, 47–50.

31 Farrell, R.J. and Peppercorn, M.A. (2002) Ulcerative colitis. *Lancet*, **359**, 331–340.

32 Asaka, M., Takeda, H., Sugiyama, T., and Kato, M. (1997) What role does helicobacter pylori play in gastric cancer? *Gastroenterology*, **113**, S56–S60.

33 Ebert, M.P., Yu, J., Sung, J.J., and Malfertheiner, P. (2000) Molecular alterations in gastric cancer: the role of helicobacter pylori. *Eur J Gastroenterol Hepatol*, **12**, 795–798.

34 Badawi, A.F., Mostafa, M.H., Probert, A., and O'Connor, P.J. (1995) Role of schistosomiasis in human bladder cancer: evidence of association, aetiological factors, and basic mechanisms of carcinogenesis. *Eur J Cancer Prev*, **4**, 45–59.

35 Mostafa, M.H., Sheweita, S.A., and O'Connor, P.J. (1999) Relationship between schistosomiasis and bladder cancer. *Clin Microbiol Rev*, **12**, 97–111.

36 Groopman, J.D. and Kensler, T.W. (2005) Role of metabolism and viruses in aflatoxin-induced liver cancer. *Toxicol Appl Pharmacol*, **206**, 131–137.

37 Nathan, C. and Shiloh, M.U. (2000) Reactive oxygen and nitrogen intermediates in the relationship between mammalian hosts and microbial pathogens. *Proc Natl Acad Sci U S A*, **97**, 8841–8848.

38 Nathan, C.F. (1987) Secretory products of macrophages. *J Clin Invest*, **79**, 319–326.

39 Bredt, D. and Snyder, S. (1994) Nitric oxide: a physiologic messenger molecule. *Neuron*, **63**, 175–195.

40 Gross, S. and Wolin, M. (1995) Nitric oxide: pathophysiological mechanisms. *Annu Rev Physiol*, **57**, 737–769.

41 Lancaster, J. (1992) Nitric oxide in cells. *Am Sci*, **80**, 248–259.

42 MacMicking, J., Xie, Q., and Nathan, C. (1997) Nitric oxide and macrophage function. *Annu Rev Immunol*, **15**, 323–350.

43 Moncada, S., Palmer, R.M., and Higgs, E.A. (1991) Nitric oxide: physiology, pathophysiology, and pharmacology. *Pharmacol Rev*, **43**, 109–142.

44 Nathan, C. (1992) Nitric oxide as a secretory product of mammalian cells. *FASEB J*, **6**, 3051–3064.

45 Tamir, S. and Tannenbaum, S. (1996) The role of nitric oxide (NO) in the carcinogenic process. *Biochem Biophys Acta*, **1288**, F31–F36.

46 Miwa, M., Stuehr, D.J., Marletta, M.A., Wishnok, J.S., and Tannenbaum, S.R. (1987) Nitrosation of amines by stimulated macrophages. *Carcinogenesis*, **8**, 955–958.

47 Stuehr, D.J. and Marletta, M.A. (1987) Synthesis of nitrite and nitrate in murine macrophage cell lines. *Cancer Res*, **47**, 5590–5594.

48 Lewis, R.S., Tamir, S., Tannenbaum, S.R., and Deen, W.M. (1995) Kinetic analysis of the fate of nitric oxide synthesized by macrophages in vitro. *J Biol Chem*, **270**, 29350–29355.

49 van der Vliet, A., Eiserich, J.P., Halliwell, B., and Cross, C.E. (1997) Formation of reactive nitrogen species during peroxidase-catalyzed oxidation of nitrite. A potential additional mechanism of nitric oxide-dependent toxicity. *J Biol Chem*, **272**, 7617–7625.

50 Hazen, S.L., Zhang, R., Shen, Z., Wu, W., Podrez, E.A., MacPherson, J.C., Schmitt, D., Mitra, S.N., Mukhopadhyay, C., Chen, Y., Cohen, P.A., Hoff, H.F., and Abu-Soud, H.M. (1999) Formation of nitric oxide-derived oxidants by myeloperoxidase in monocytes: pathways for monocyte-mediated protein nitration and lipid peroxidation in vivo. *Circ Res*, **85**, 950–958.

51 Wu, W., Chen, Y., and Hazen, S.L. (1999) Eosinophil peroxidase nitrates protein tyrosyl residues. Implications for oxidative damage by nitrating intermediates in eosinophilic inflammatory disorders. *J Biol Chem*, **274**, 25933–25944.

52 Eiserich, J.P., Hristova, M., Cross, C.E., Jones, A.D., Freeman, B.A., Halliwell, B., and van der Vliet, A. (1998) Formation of nitric oxide-derived inflammatory oxidants by myeloperoxidase in neutrophils. *Nature*, **391**, 393–397.

53 Cadet, J., Douki, T., Pouget, J.P., and Ravanat, J.L. (2000) Singlet oxygen DNA damage products: formation and measurement. *Methods Enzymol*, **319**, 143–153.

54 Cadet, J., Douki, T., and Ravanat, J.L. (2008) Oxidatively generated damage to the guanine moiety of DNA: mechanistic aspects and formation in cells. *Acc Chem Res*, **41**, 1075–1083.

55 Steenken, S. and Jovanovic, S.V. (1997) How easily oxidizable is DNA? One-electron reduction potentials of adenosine and guanosine radicals in aqueous solution. *J Am Chem Soc*, **119**, 617–618.

56 Henderson, P.T., Jones, D., Hampikian, G., Kan, Y., and Schuster, G.B. (1999) Long-distance charge transport in duplex DNA: the phonon-assisted polaron-like hopping mechanism. *Proc Natl Acad Sci U S A*, **96**, 8353–8358.

57 Giese, B. (2002) Long-distance electron transfer through DNA. *Annu Rev Biochem*, **71**, 51–70.

58 Nunez, M.E. and Barton, J.K. (2000) Probing DNA charge transport with metallointercalators. *Curr Opin Chem Biol*, **4**, 199–206.

59 Schuster, G.B. (2000) Long-range charge transfer in DNA: transient structural distortions control the distance dependence. *Acc Chem Res*, **33**, 253–260.

60 Saito, I., Nakamura, T., Nakatani, K., Yoshioka, Y., Yamaguchi, K., and Sugiyama, H. (1998) Mapping of the hot spots for DNA damage by one-electron oxidation: efficacy of GG doublets and GGG triplets as a trap in long-range hole migration. *J Am Chem Soc*, **120**, 12686–12687.

61 Sugiyama, H. and Saito, I. (1996) Theoretical studies of GG-specific photocleavage of DNA via electron transfer: Significant lowering of ionization potential and 5′-localization of homo of stacked GG bases in B-form DNA. *J Am Chem Soc*, **118**, 7063–7068.

62 Senthilkumar, K., Grozema, F.C., Guerra, C.F., Bickelhaupt, F.M., and Siebbeles, L.D.A. (2003) Mapping the sites of selective oxidation of guanines in DNA. *J Am Chem Soc*, **125**, 13658–13659.

63 Margolin, Y., Cloutier, J.F., Shafirovich, V., Geacintov, N.E., and Dedon, P.C. (2006) Paradoxical hotspots for guanine oxidation by a chemical mediator of inflammation. *Nat Chem Biol*, **2**, 365–366.

64 Steenken, S., Jovanovic, S.V., Bietti, M., and Bernhard, K. (2000) The trap depth (in DNA) of 8-oxo-7,8-dihydro-2′deoxyguanosine as derived from electron-transfer equilibria in aqueous solution. *J Am Chem Soc*, **122**, 2373–2374.

65 Candeias, L.P. and Steenken, S. (2000) Reaction of HO with guanine derivatives in aqueous solution: formation of two different redox-active oh-adduct radicals and their unimolecular transformation reactions. Properties of G(-H). *Chemistry*, **6**, 475–484.

66 Cadet, J., Berger, M., Buchko, G.W., Joshi, P.C., Raoul, S., and Ravanat, J.-L. (1994) 2,2-diamino-4-[(3,5-di-o-acetyl-2-deoxy-β-d-erythro-pentofuranosyl)amino]-5-(2H)-oxazolone: a novel and predominant radical oxidation product of 3′,5′-di-O-acetyl-2′-deoxyguanosine. *J Am Chem Soc*, **116**, 7403–7404.

67 Burrows, C. and Muller, J. (1998) Oxidative nucleobase modifications leading to strand scission. *Chem Rev*, **98**, 1109–1151.

68 Misiaszek, R., Crean, C., Joffe, A., Geacintov, N.E., and Shafirovich, V. (2004) Oxidative DNA damage associated with combination of guanine and superoxide radicals and repair mechanisms via radical trapping. *J Biol Chem*, **279**, 32106–32115.

69 Breen, A.P. and Murphy, J.A. (1995) Reactions of oxyl radicals with DNA. *Free Radic Biol Med*, **18**, 1033–1077.

70 Box, H.C., Budzinski, E.E., Dawidzik, J.B., Gobey, J.S., and Freund, H.G. (1997) Free radical-induced tandem

base damage in DNA oligomers. *Free Radic Biol Med*, **23**, 1021–1030.

71 Dizdaroglu, M., Jaruga, P., Birincioglu, M., and Rodriguez, H. (2002) Free radical-induced damage to DNA: mechanisms and measurement. *Free Radic Biol Med*, **32**, 1102–1115.

72 Brooks, P.J. (2008) The 8,5′-cyclopurine-2′-deoxynucleosides: candidate neurodegenerative DNA lesions in xeroderma pigmentosum, and unique probes of transcription and nucleotide excision repair. *DNA Repair (Amst)*, **7**, 1168–1179.

73 Dirksen, M.L., Blakely, W.F., Holwitt, E., and Dizdaroglu, M. (1988) Effect of DNA conformation on the hydroxyl radical-induced formation of 8,5′-cyclopurine 2′-deoxyribonucleoside residues in DNA. *Int J Radiat Biol*, **54**, 195–204.

74 Chatgilialoglu, C., Bazzanini, R., Jimenez, L.B., and Miranda, M.A. (2007) (5′s)- and (5′r)-5′,8-cyclo-2′-deoxyguanosine: Mechanistic insights on the 2′-deoxyguanosin-5′-yl radical cyclization. *Chem Res Toxicol*, **20**, 1820–1824.

75 Jaruga, P., Birincioglu, M., Rodriguez, H., and Dizdaroglu, M. (2002) Mass spectrometric assays for the tandem lesion 8,5′-cyclo-2′-deoxyguanosine in mammalian DNA. *Biochemistry*, **41**, 3703–3711.

76 Randerath, K., Zhou, G.D., Somers, R.L., Robbins, J.H., and Brooks, P.J. (2001) A 32P-postlabeling assay for the oxidative DNA lesion 8,5′-cyclo-2′-deoxyadenosine in mammalian tissues: Evidence that four type II i-compounds are dinucleotides containing the lesion in the 3′ nucleotide. *J Biol Chem*, **276**, 36051–36057.

77 Zhou, G.D., Randerath, K., Donnelly, K.C., and Jaiswal, A.K. (2004) Effects of nqo1 deficiency on levels of cyclopurines and other oxidative DNA lesions in liver and kidney of young mice. *Int J Cancer*, **112**, 877–883.

78 Sodum, R.S. and Fiala, E.S. (2001) Analysis of peroxynitrite reactions with guanine, xanthine, and adenine nucleosides by high-pressure liquid chromatography with electrochemical detection: C8-nitration and -oxidation. *Chem Res Toxicol*, **14**, 438–450.

79 Badouard, C., Masuda, M., Nishino, H., Cadet, J., Favier, A., and Ravanat, J.L. (2005) Detection of chlorinated DNA and RNA nucleosides by HPLC coupled to tandem mass spectrometry as potential biomarkers of inflammation. *J Chromatogr B Analyt Technol Biomed Life Sci*, **827**, 26–31.

80 Masuda, M., Suzuki, T., Friesen, M.D., Ravanat, J.L., Cadet, J., Pignatelli, B., Nishino, H., and Ohshima, H. (2001) Chlorination of guanosine and other nucleosides by hypochlorous acid and myeloperoxidase of activated human neutrophils. Catalysis by nicotine and trimethylamine. *J Biol Chem*, **276**, 40486–40496.

81 Shen, Z., Wu, W., and Hazen, S.L. (2000) Activated leukocytes oxidatively damage DNA, RNA, and the nucleotide pool through halide-dependent formation of hydroxyl radical. *Biochemistry*, **39**, 5474–5482.

82 Anderson, M.M., Hazen, S.L., Hsu, F.F., and Heinecke, J.W. (1997) Human neutrophils employ the myeloperoxidase-hydrogen peroxide-chloride system to convert hydroxy-amino acids into glycolaldehyde, 2-hydroxypropanal, and acrolein. A mechanism for the generation of highly reactive α-hydroxy and α, β-unsaturated aldehydes by phagocytes at sites of inflammation. *J Clin Invest*, **99**, 424–432.

83 Hazen, S.L. and Heinecke, J.W. (1997) 3-chlorotyrosine, a specific marker of myeloperoxidase-catalyzed oxidation, is markedly elevated in low density lipoprotein isolated from human atherosclerotic intima. *J Clin Invest*, **99**, 2075–2081.

84 Povirk, L.F. and Goldberg, I.H. (1985) Endonuclease-resistant

apyrimidinic sites formed by neo-carzinostatin at cytosine residues in DNA: evidence for a possible role in mutagenesis. *Proc Natl Acad Sci U S A*, **82**, 3182–3186.

85 Weinfeld, M., Rasouli-Nia, A., Chaudhry, M.A., and Britten, R.A. (2001) Response of base excision repair enzymes to complex DNA lesions. *Radiat Res*, **156**, 584–589.

86 Hazra, T.K., Das, A., Das, S., Choudhury, S., Kow, Y.W., and Roy, R. (2007) Oxidative DNA damage repair in mammalian cells: a new perspective. *DNA Repair (Amst)*, **6**, 470–480.

87 Dedon, P.C., Plastaras, J.P., Rouzer, C.A., and Marnett, L.J. (1998) Indirect mutagenesis by oxidative DNA damage: formation of the pyrimidopurinone adduct of deoxyguanosine by base propenal. *Proc Natl Acad Sci U S A*, **95**, 11113–11116.

88 Zhou, X., Taghizadeh, K., and Dedon, P.C. (2005) Chemical and biological evidence for base propenals as the major source of the endogenous M1dG adduct in cellular DNA. *J Biol Chem*, **280**, 25377–25382.

89 Bartsch, H. and Nair, J. (2004) Oxidative stress and lipid peroxidation-derived DNA-lesions in inflammation driven carcinogenesis. *Cancer Detect Prev*, **28**, 385–391.

90 Blair, I.A. (2008) DNA adducts with lipid peroxidation products. *J Biol Chem*, **283**, 15545–15549.

91 Nair, J., De Flora, S., Izzotti, A., and Bartsch, H. (2007) Lipid peroxidation-derived etheno-DNA adducts in human atherosclerotic lesions. *Mutat Res*, **621**, 95–105.

92 Godschalk, R.W., Albrecht, C., Curfs, D.M., Schins, R.P., Bartsch, H., van Schooten, F.J., and Nair, J. (2007) Decreased levels of lipid peroxidation-induced DNA damage in the onset of atherogenesis in apolipoprotein e deficient mice. *Mutat Res*, **621**, 87–94.

93 Pang, B., Zhou, X., Yu, H.-B., Dong, M., Taghizadeh, K., Wishnok, J.S., Tannenbaum, S.R., and Dedon, P.C. (2007) Lipid peroxidation dominates the chemistry of DNA adduct formation in a mouse model of inflammation. *Carcinogenesis*, **28**, 1807–1813.

94 Meerang, M., Nair, J., Sirankapracha, P., Thephinlap, C., Srichairatanakool, S., Fucharoen, S., and Bartsch, H. (2008) Increased urinary 1,N6-ethenodeoxyadenosine and 3,n4-ethenodeoxycytidine excretion in thalassemia patients: Markers for lipid peroxidation-induced DNA damage. *Free Radic Biol Med*, **44**, 1863–1868.

95 Williams, M.V., Lee, S.H., Pollack, M., and Blair, I.A. (2006) Endogenous lipid hydroperoxide-mediated DNA-adduct formation in min mice. *J Biol Chem*, **281**, 10127–10133.

96 Rindgen, D., Lee, S.H., Nakajima, M., and Blair, I.A. (2000) Formation of a substituted 1,n(6)-etheno-2′-deoxyadenosine adduct by lipid hydroperoxide-mediated generation of 4-oxo-2-nonenal. *Chem Res Toxicol*, **13**, 846–852.

97 Lee, S.H., Arora, J.A., Oe, T., and Blair, I.A. (2005) 4-hydroperoxy-2-nonenal-induced formation of 1,N2-etheno-2′-deoxyguanosine adducts. *Chem Res Toxicol*, **18**, 780–786.

98 Pollack, M., Oe, T., Lee, S.H., Silva Elipe, M.V., Arison, B.H., and Blair, I.A. (2003) Characterization of 2′-deoxycytidine adducts derived from 4-oxo-2-nonenal, a novel lipid peroxidation product. *Chem Res Toxicol*, **16**, 893–900.

99 Lee, S.H., Silva Elipe, M.V., Arora, J.S., and Blair, I.A. (2005) Dioxododecenoic acid: a lipid hydroperoxide-derived bifunctional electrophile responsible for etheno DNA adduct formation. *Chem Res Toxicol*, **18**, 566–578.

100 Jian, W., Lee, S.H., Arora, J.S., Silva Elipe, M.V., and Blair, I.A. (2005) Unexpected formation of etheno-2′-deoxyguanosine adducts from 5(s)-hydroperoxyeicosatetraenoic

acid: Evidence for a bis-hydroperoxide intermediate. *Chem Res Toxicol*, **18**, 599–610.

101 Wang, M., McIntee, E.J., Cheng, G., Shi, Y., Villalta, P.W. and Hecht, S.S. (2000) Identification of paraldol-deoxyguanosine adducts in DNA reacted with crotonaldehyde. *Chem Res Toxicol*, **13**, 1065–1074.

102 Wang, M., McIntee, E.J., Cheng, G., Shi, Y., Villalta, P.W. and Hecht, S.S. (2001) A schiff base is a major DNA adduct of crotonaldehyde. *Chem Res Toxicol*, **14**, 423–430.

103 Chung, F.L., Nath, R.G., Nagao, M., Nishikawa, A., Zhou, G.D., and Randerath, K. (1999) Endogenous formation and significance of 1,N2-propanodeoxyguanosine adducts. *Mutat Res*, **424**, 71–81.

104 Winter, C.K., Segall, H.J., and Haddon, W.F. (1986) Formation of cyclic adducts of deoxyguanosine with the aldehydes trans-4-hydroxy-2-hexenal and trans-4-hydroxy-2-nonenal in vitro. *Cancer Res*, **46**, 5682–5686.

105 Chung, F.L., Young, R., and Hecht, S.S. (1984) Formation of cyclic 1,N2-propanodeoxyguanosine adducts in DNA upon reaction with acrolein or crotonaldehyde. *Cancer Res*, **44**, 990–995.

106 Chung, F.L. and Hecht, S.S. (1983) Formation of cyclic 1,N2–adducts by reaction of deoxyguanosine with α-acetoxy-N-nitrosopyrrolidine, 4-(carbethoxynitrosamino)butanal, or crotonaldehyde. *Cancer Res*, **43**, 1230–1235.

107 Douki, T., Odin, F., Caillat, S., Favier, A., and Cadet, J. (2004) Predominance of the 1,N2-propano 2′-deoxyguanosine adduct among 4-hydroxy-2-nonenal-induced DNA lesions. *Free Radic Biol Med*, **37**, 62–70.

108 Douki, T. and Ames, B.N. (1994) An hplc-ec assay for 1,N2-propano adducts of 2′-deoxyguanosine with 4-hydroxynonenal and other α,β-unsaturated aldehydes. *Chem Res Toxicol*, **7**, 511–518.

109 Pan, J. and Chung, F.L. (2002) Formation of cyclic deoxyguanosine adducts from omega-3 and omega-6 polyunsaturated fatty acids under oxidative conditions. *Chem Res Toxicol*, **15**, 367–372.

110 Zhang, S., Villalta, P.W., Wang, M., and Hecht, S.S. (2007) Detection and quantitation of acrolein-derived 1,N2-propanodeoxyguanosine adducts in human lung by liquid chromatography-electrospray ionization-tandem mass spectrometry. *Chem Res Toxicol*, **20**, 565–571.

111 Pawlowicz, A.J., Munter, T., Klika, K.D., and Kronberg, L. (2006) Reaction of acrolein with 2′-deoxyadenosine and 9-ethyladenine–formation of cyclic adducts. *Bioorg Chem*, **34**, 39–48.

112 Pawlowicz, A.J., Munter, T., Zhao, Y., and Kronberg, L. (2006) Formation of acrolein adducts with 2′-deoxyadenosine in calf thymus DNA. *Chem Res Toxicol*, **19**, 571–576.

113 Pawlowicz, A.J., Klika, K.D., and Kronberg, L. (2007) The structural identification and conformational analysis of the products from the reaction of acrolein with 2′deoxycytidine, 1-methylcytosine and calf thymus DNA. *Eur J Org Chem*, 1429–1437.

114 Pawlowicz, A.J. and Kronberg, L. (2008) Characterization of adducts formed in reactions of acrolein with thymidine and calf thymus DNA. *Chem Biodivers*, **5**, 177–188.

115 Cho, Y.J., Kim, H.Y., Huang, H., Slutsky, A., Minko, I.G., Wang, H., Nechev, L.V., Kozekov, I.D., Kozekova, A., Tamura, P., Jacob, J., Voehler, M., Harris, T.M., Lloyd, R.S., Rizzo, C.J., and Stone, M.P. (2005) Spectroscopic characterization of interstrand carbinolamine cross-links formed in the 5′-CpG-3′ sequence by the acrolein-derived γ-OH-1,N2-propano-2′-deoxyguanosine DNA adduct. *J Am Chem Soc*, **127**, 17686–17696.

116 Cho, Y.J., Kozekov, I.D., Harris, T.M., Rizzo, C.J., and Stone, M.P. (2007) Stereochemistry modulates

the stability of reduced interstrand cross-links arising from R- and S-α-CH3-γ-OH-1,N2-propano-2′-deoxyguanosine in the 5′-CpG-3′ DNA sequence. *Biochemistry*, **46**, 2608–2621.

117 Cho, Y.J., Wang, H., Kozekov, I.D., Kozekova, A., Kurtz, A.J., Jacob, J., Voehler, M., Smith, J., Harris, T.M., Rizzo, C.J., Lloyd, R.S., and Stone, M.P. (2006) Orientation of the crotonaldehyde-derived N2-[3-oxo-1(S)-methyl-propyl]-dG DNA adduct hinders interstrand cross-link formation in the 5′-CpG-3′ sequence. *Chem Res Toxicol*, **19**, 1019–1029.

118 Cho, Y.J., Wang, H., Kozekov, I.D., Kurtz, A.J., Jacob, J., Voehler, M., Smith, J., Harris, T.M., Lloyd, R.S., Rizzo, C.J., and Stone, M.P. (2006) Stereospecific formation of interstrand carbinolamine DNA cross-links by crotonaldehyde- and acetaldehyde-derived α-ch3-γ-oh-1,N2-propano-2′-deoxyguanosine adducts in the 5′-cpg-3′ sequence. *Chem Res Toxicol*, **19**, 195–208.

119 Huang, H., Wang, H., Lloyd, R.S., Rizzo, C.J., and Stone, M.P. (2008) Conformational interconversion of the trans-4-hydroxynonenal-derived (6S,8R,11S) 1,n(2)-deoxyguanosine adduct when mismatched with deoxyadenosine in DNA. *Chem Res Toxicol*, **22**, 187–200.

120 Huang, H., Wang, H., Qi, N., Kozekova, A., Rizzo, C.J., and Stone, M.P. (2008) Rearrangement of the (6S,8R,11S) and (6R,8S,11R) exocyclic 1,N2-deoxyguanosine adducts of trans-4-hydroxynonenal to N2-deoxyguanosine cyclic hemiacetal adducts when placed complementary to cytosine in duplex DNA. *J Am Chem Soc*, **130**, 10898–10906.

121 Huang, H., Wang, H., Qi, N., Lloyd, R.S., Rizzo, C.J., and Stone, M.P. (2008) The stereochemistry of trans-4-hydroxynonenal-derived exocyclic 1,N2-2′-deoxyguanosine adducts modulates formation of interstrand cross-links in the 5′-CpG-3′ sequence. *Biochemistry*, **47**, 11457–11472.

122 Kozekov, I.D., Nechev, L.V., Moseley, M.S., Harris, C.M., Rizzo, C.J., Stone, M.P., and Harris, T.M. (2003) DNA interchain cross-links formed by acrolein and crotonaldehyde. *J Am Chem Soc*, **125**, 50–61.

123 Stone, M.P., Cho, Y.J., Huang, H., Kim, H.Y., Kozekov, I.D., Kozekova, A., Wang, H., Minko, I.G., Lloyd, R.S., Harris, T.M., and Rizzo, C.J. (2008) Interstrand DNA cross-links induced by α, β-unsaturated aldehydes derived from lipid peroxidation and environmental sources. *Acc Chem Res*, **41**, 793–804.

124 Minko, I.G., Kozekov, I.D., Kozekova, A., Harris, T.M., Rizzo, C.J., and Lloyd, R.S. (2008) Mutagenic potential of DNA-peptide crosslinks mediated by acrolein-derived DNA adducts. *Mutat Res*, **637**, 161–172.

125 Sanchez, A.M., Kozekov, I.D., Harris, T.M., and Lloyd, R.S. (2005) Formation of inter- and intrastrand imine type DNA-DNA cross-links through secondary reactions of aldehydic adducts. *Chem Res Toxicol*, **18**, 1683–1690.

126 Yang, I.Y., Chan, G., Miller, H., Huang, Y., Torres, M.C., Johnson, F., and Moriya, M. (2002) Mutagenesis by acrolein-derived propanodeoxyguanosine adducts in human cells. *Biochemistry*, **41**, 13826–13832.

127 Yang, I.Y., Hossain, M., Miller, H., Khullar, S., Johnson, F., Grollman, A., and Moriya, M. (2001) Responses to the major acrolein-derived deoxyguanosine adduct in escherichia coli. *J Biol Chem*, **276**, 9071–9076.

128 Yang, I.Y., Johnson, F., Grollman, A.P., and Moriya, M. (2002) Genotoxic mechanism for the major acrolein-derived deoxyguanosine adduct in human cells. *Chem Res Toxicol*, **15**, 160–164.

129 VanderVeen, L.A., Hashim, M.F., Nechev, L.V., Harris, T.M., Harris, C.M., and Marnett, L.J. (2001) Evaluation of the mutagenic potential of

the principal DNA adduct of acrolein. *J Biol Chem*, **276**, 9066–9070.

130 Eder, E. and Hoffman, C. (1992) Identification and characterization of deoxyguanosine-crotonaldehyde adducts. Formation of 7,8 cyclic adducts and 1,N2,7,8 bis-cyclic adducts. *Chem Res Toxicol*, **5**, 802–808.

131 Stein, S., Lao, Y., Yang, I.Y., Hecht, S.S., and Moriya, M. (2006) Genotoxicity of acetaldehyde- and crotonaldehyde-induced 1,N2-propanodeoxyguanosine DNA adducts in human cells. *Mutat Res*, **608**, 1–7.

132 Pluskota-Karwatka, D., Pawlowicz, A.J., Tomas, M., and Kronberg, L. (2008) Formation of adducts in the reaction of glyoxal with 2′-deoxyguanosine and with calf thymus DNA. *Bioorg Chem*, **36**, 57–64.

133 Inagaki, S., Esaka, Y., Goto, M., Deyashiki, Y., and Sako, M. (2004) LC-MS study on the formation of cyclic 1,N2-propano guanine adduct in the reactions of DNA with acetaldehyde in the presence of histone. *Biol Pharm Bull*, **27**, 273–276.

134 Sako, M., Inagaki, S., Esaka, Y., and Deyashiki, Y. (2003) Histones accelerate the cyclic 1,N2-propanoguanine adduct-formation of DNA by the primary metabolite of alcohol and carcinogenic crotonaldehyde. *Bioorg Med Chem Lett*, **13**, 3497–3498.

135 Theruvathu, J.A., Jaruga, P., Nath, R.G., Dizdaroglu, M., and Brooks, P.J. (2005) Polyamines stimulate the formation of mutagenic 1,N2-propanodeoxyguanosine adducts from acetaldehyde. *Nucleic Acids Res*, **33**, 3513–3520.

136 Zhang, S., Villalta, P.W., Wang, M., and Hecht, S.S. (2006) Analysis of crotonaldehyde- and acetaldehyde-derived 1,N(2)-propanodeoxyguanosine adducts in DNA from human tissues using liquid chromatography electrospray ionization tandem mass spectrometry. *Chem Res Toxicol*, **19**, 1386–1392.

137 Pan, J., Davis, W., Trushin, N., Amin, S., Nath, R.G., Salem, N. Jr, and Chung, F.L. (2006) A solid-phase extraction/high-performance liquid chromatography-based (32)P-postlabeling method for detection of cyclic 1,N(2)-propanodeoxyguanosine adducts derived from enals. *Anal Biochem*, **348**, 15–23.

138 Budiawan and Eder, E. (2000) Detection of 1,N(2)-propanodeoxyguanosine adducts in DNA of Fischer 344 rats by an adapted (32)P-post-labeling technique after per os application of crotonaldehyde. *Carcinogenesis*, **21**, 1191–1196.

139 Chung, F.L., Pan, J., Choudhury, S., Roy, R., Hu, W., and Tang, M.S. (2003) Formation of trans-4-hydroxy-2-nonenal- and other enal-derived cyclic DNA adducts from omega-3 and omega-6 polyunsaturated fatty acids and their roles in DNA repair and human p53 gene mutation. *Mutat Res*, **531**, 25–36.

140 Chung, F.L., Zhang, L., Ocando, J.E., and Nath, R.G. (1999) Role of 1,N2-propanodeoxyguanosine adducts as endogenous DNA lesions in rodents and humans. *IARC Sci Publ*, 45–54.

141 Blair, I.A. (2001) Lipid hydroperoxide-mediated DNA damage. *Exp Gerontol*, **36**, 1473–1481.

142 Lee, S.H. and Blair, I.A. (2001) Oxidative DNA damage and cardiovascular disease. *Trends Cardiovasc Med*, **11**, 148–155.

143 Marnett, L.J. (2002) Oxy radicals, lipid peroxidation and DNA damage. *Toxicology*, **181–182**, 219–222.

144 Poirier, M.C. (1997) DNA adducts as exposure biomarkers and indicators of cancer risk. *Environ Health Perspect*, **105** (Suppl 4), 907–912.

145 Marnett, L.J. (1994) DNA adducts of α,β-unsaturated aldehydes and dicarbonyl compounds. *IARC Sci Publ*, **125**, 151–163.

146 Marnett, L.J. and Burcham, P.C. (1993) Endogenous DNA adducts: potential and paradox. *Chem Res Toxicol*, **6**, 771–785.

147 Mao, H., Schnetz-Boutaud, N.C., Weisenseel, J.P., Marnett, L.J., and Stone, M.P. (1999) Duplex DNA catalyzes the chemical rearrangement of a malondialdehyde deoxyguanosine adduct. *Proc Natl Acad Sci U S A*, **96**, 6615–6620.

148 Schnetz-Boutaud, N., Daniels, J.S., Hashim, M.F., Scholl, P., Burrus, T., and Marnett, L.J. (2000) Pyrimido[1,2-α]purin-10(3H)-one: a reactive electrophile in the genome. *Chem Res Toxicol*, **13**, 967–970.

149 Schnetz-Boutaud, N.C., Saleh, S., Marnett, L.J., and Stone, M.P. (2001) Structure of the malondialdehyde deoxyguanosine adduct M1G when placed opposite a two-base deletion in the (CpG)3 frameshift hotspot of the salmonella typhimurium hisd3052 gene. *Adv Exp Med Biol*, **500**, 513–516.

150 Wang, Y., Schnetz-Boutaud, N.C., Saleh, S., Marnett, L.J., and Stone, M.P. (2007) Bulge migration of the malondialdehyde opdg DNA adduct when placed opposite a two-base deletion in the (CpG)3 frameshift hotspot of the salmonella typhimurium hisd3052 gene. *Chem Res Toxicol*, **20**, 1200–1210.

151 Riggins, J.N., Pratt, D.A., Voehler, M., Daniels, J.S., and Marnett, L.J. (2004) Kinetics and mechanism of the general-acid-catalyzed ring-closure of the malondialdehyde-DNA adduct, N2-(3-oxo-1-propenyl)deoxyguanosine (N2OPdG-), to 3-(2′-deoxy-β-d-erythro-pentofuranosyl)pyrimido[1,2-α]purin-10(3H)-one (M1dG). *J Am Chem Soc*, **126**, 10571–10581.

152 Riggins, J.N., Daniels, J.S., Rouzer, C.A., and Marnett, L.J. (2004) Kinetic and thermodynamic analysis of the hydrolytic ring-opening of the malondialdehyde-deoxyguanosine adduct, 3-(2′-deoxy-β-d-erythro-pentofuranosyl)-pyrimido[1,2-α]purin-10(3H)-one. *J Am Chem Soc*, **126**, 8237–8243.

153 Jeong, Y.C., Nakamura, J., Upton, P.B., and Swenberg, J.A. (2005) Pyrimido[1,2-a]-purin-10(3H)-one, M1G, is less prone to artifact than base oxidation. *Nucleic Acids Res*, **33**, 6426–6434.

154 Jeong, Y.C., Sangaiah, R., Nakamura, J., Pachkowski, B.F., Ranasinghe, A., Gold, A., Ball, L.M., and Swenberg, J.A. (2005) Analysis of M1G-dR in DNA by aldehyde reactive probe labeling and liquid chromatography tandem mass spectrometry. *Chem Res Toxicol*, **18**, 51–60.

155 Jeong, Y.C. and Swenberg, J.A. (2005) Formation of M1G-dR from endogenous and exogenous ros-inducing chemicals. *Free Radic Biol Med*, **39**, 1021–1029.

156 Plastaras, J.P., Riggins, J.N., Otteneder, M., and Marnett, L.J. (2000) Reactivity and mutagenicity of endogenous DNA oxopropenylating agents: Base propenals, malondialdehyde, and N(epsilon)-oxopropenyllysine. *Chem Res Toxicol*, **13**, 1235–1242.

157 Stitt, A.W. and Curtis, T.M. (2005) Advanced glycation and retinal pathology during diabetes. *Pharmacol Rep*, **57** (Suppl), 156–168.

158 Munch, G., Schinzel, R., Loske, C., Wong, A., Durany, N., Li, J.J., Vlassara, H., Smith, M.A., Perry, G., and Riederer, P. (1998) Alzheimer's disease–synergistic effects of glucose deficit, oxidative stress and advanced glycation endproducts. *J Neural Transm*, **105**, 439–461.

159 Sensi, M., Pricci, F., Andreani, D., and Di Mario, U. (1991) Advanced nonenzymatic glycation endproducts (age): their relevance to aging and the pathogenesis of late diabetic complications. *Diabetes Res*, **16**, 1–9.

160 Thornalley, P.J. (1998) Glutathione-dependent detoxification of α-oxoaldehydes by the glyoxalase system: involvement in disease mechanisms and antiproliferative activity of glyoxalase i inhibitors. *Chem Biol Interact*, **111–112**, 137–151.

161 Thornalley, P.J. (2002) Glycation in diabetic neuropathy: characteristics, consequences, causes, and therapeutic options. *Int Rev Neurobiol*, **50**, 37–57.

162 Thornalley, P.J. and Stern, A. (1984) The production of free radicals during the autoxidation of monosaccharides by buffer ions. *Carbohydr Res*, **134**, 191–204.

163 Loeppky, R.N., Cui, W., Goelzer, P., Park, M., and Ye, Q. (1999) Glyoxal-guanine DNA adducts: detection, stability and formation in vivo from nitrosamines. *IARC Sci Publ*, 155–168.

164 Shapiro, R. and Hachmann, J. (1966) The reaction of guanine derivatives with 1,2-dicarbonyl compounds. *Biochemistry*, **5**, 2799–2807.

165 Kasai, H., Iwamoto-Tanaka, N., and Fukada, S. (1998) DNA modifications by the mutagen glyoxal: adduction to G and C, deamination of C and GC and GA cross-linking. *Carcinogenesis*, **19**, 1459–1465.

166 Loeppky, R.N., Ye, Q., Goelzer, P., and Chen, Y. (2002) DNA adducts from n-nitrosodiethanolamine and related β-oxidized nitrosamines in vivo: (32)p-postlabeling methods for glyoxal- and o(6)-hydroxyethyldeoxyguanosine adducts. *Chem Res Toxicol*, **15**, 470–482.

167 Dennehy, M.K. and Loeppky, R.N. (2005) Mass spectrometric methodology for the determination of glyoxaldeoxyguanosine and O6-hydroxyethyldeoxyguanosine DNA adducts produced by nitrosamine bident carcinogens. *Chem Res Toxicol*, **18**, 556–565.

168 Olsen, R., Ovrebo, S., Thorud, S., Lundanes, E., Thomassen, Y., Greibrokk, T., and Molander, P. (2008) Sensitive determination of a glyoxal-DNA adduct biomarker candidate by column switching capillary liquid chromatography electrospray ionization mass spectrometry. *Analyst*, **133**, 802–809.

169 Olsen, R., Molander, P., Ovrebo, S., Ellingsen, D.G., Thorud, S., Thomassen, Y., Lundanes, E., Greibrokk, T., Backman, J., Sjoholm, R., and Kronberg, L. (2005) Reaction of glyoxal with 2'-deoxyguanosine, 2'-deoxyadenosine, 2'-deoxycytidine, cytidine, thymidine, and calf thymus DNA: identification of DNA adducts. *Chem Res Toxicol*, **18**, 730–739.

170 Cooke, M.S., Mistry, N., Ahmad, J., Waller, H., Langford, L., Bevan, R.J., Evans, M.D., Jones, G.D., Herbert, K.E., Griffiths, H.R., and Lunec, J. (2003) Deoxycytidine glyoxal: Lesion induction and evidence of repair following vitamin C supplementation in vivo. *Free Radic Biol Med*, **34**, 218–225.

171 Mistry, N., Evans, M.D., Griffiths, H.R., Kasai, H., Herbert, K.E., and Lunec, J. (1999) Immunochemical detection of glyoxal DNA damage. *Free Radic Biol Med*, **26**, 1267–1273.

172 Mistry, N., Podmore, I., Cooke, M., Butler, P., Griffiths, H., Herbert, K., and Lunec, J. (2003) Novel monoclonal antibody recognition of oxidative DNA damage adduct, deoxycytidine-glyoxal. *Lab Invest*, **83**, 241–250.

173 Brock, A.K., Kozekov, I.D., Rizzo, C.J., and Harris, T.M. (2004) Coupling products of nucleosides with the glyoxal adduct of deoxyguanosine. *Chem Res Toxicol*, **17**, 1047–1056.

174 Kuchenmeister, F., Schmezer, P., and Engelhardt, G. (1998) Genotoxic bifunctional aldehydes produce specific images in the comet assay. *Mutat Res*, **419**, 69–78.

175 Murata-Kamiya, N., Kamiya, H., Iwamoto, N., and Kasai, H. (1995) Formation of a mutagen, glyoxal, from DNA treated with oxygen free radicals. *Carcinogenesis*, **16**, 2251–2253.

176 Awada, M. and Dedon, P.C. (2001) Formation of the 1,N2-glyoxal adduct of deoxyguanosine by phosphoglycolaldehyde, a product of 3'-deoxyribose oxidation in DNA. *Chem Res Toxicol*, **14**, 1247–1253.

177 Papoulis, A., al-Abed, Y., and Bucala, R. (1995) Identification of

N2-(1-carboxyethyl)guanine (CEG) as a guanine advanced glycosylation end product. *Biochemistry*, **34**, 648–655.

178 Li, Y., Cohenford, M.A., Dutta, U., and Dain, J.A. (2008) The structural modification of DNA nucleosides by nonenzymatic glycation: an in vitro study based on the reactions of glyoxal and methylglyoxal with 2′-deoxyguanosine. *Anal Bioanal Chem*, **390**, 679–688.

179 Schneider, M., Quistad, G.B., and Casida, J.E. (1998) N2,7-bis(1-hydroxy-2-oxopropyl)-2′-deoxyguanosine: identical noncyclic adducts with 1,3-dichloropropene epoxides and methylglyoxal. *Chem Res Toxicol*, **11**, 1536–1542.

180 Vaca, C.E., Fang, J.L., Conradi, M., and Hou, S.M. (1994) Development of a 32P-postlabelling method for the analysis of 2′-deoxyguanosine-3′-monophosphate and DNA adducts of methylglyoxal. *Carcinogenesis*, **15**, 1887–1894.

181 Fleming, T., Rabbani, N., and Thornalley, P.J. (2008) Preparation of nucleotide advanced glycation endproducts–imidazopurinone adducts formed by glycation of deoxyguanosine with glyoxal and methylglyoxal. *Ann NY Acad Sci*, **1126**, 280–282.

182 Frischmann, M., Bidmon, C., Angerer, J., and Pischetsrieder, M. (2005) Identification of DNA adducts of methylglyoxal. *Chem Res Toxicol*, **18**, 1586–1592.

183 Bidmon, C., Frischmann, M., and Pischetsrieder, M. (2007) Analysis of DNA-bound advanced glycation end-products by LC and mass spectrometry. *J Chromatogr B Analyt Technol Biomed Life Sci*, **855**, 51–58.

184 Breyer, V., Frischmann, M., Bidmon, C., Schemm, A., Schiebel, K., and Pischetsrieder, M. (2008) Analysis and biological relevance of advanced glycation end-products of DNA in eukaryotic cells. *FEBS J*, **275**, 914–925.

185 Yuan, B., Cao, H., Jiang, Y., Hong, H., and Wang, Y. (2008) Efficient and accurate bypass of N2-(1-carboxyethyl)-2′-deoxyguanosine by DinB DNA polymerase in vitro and in vivo. *Proc Natl Acad Sci U S A*, **105**, 8679–8684.

186 Bolton, J.L. and Thatcher, G.R. (2008) Potential mechanisms of estrogen quinone carcinogenesis. *Chem Res Toxicol*, **21**, 93–101.

187 Kow, Y.W. (2002) Repair of deaminated bases in DNA. *Free Radic Biol Med*, **33**, 886–893.

188 Visnes, T., Doseth, B., Pettersen, H.S., Hagen, L., Sousa, M.M., Akbari, M., Otterlei, M., Kavli, B., Slupphaug, G., and Krokan, H.E. (2008) Review. Uracil in DNA and its processing by different DNA glycosylases. *Philos Trans R Soc Lond B Biol Sci*, **364**, 563–568.

189 Anant, S. and Davidson, N.O. (2003) Hydrolytic nucleoside and nucleotide deamination, and genetic instability: a possible link between RNA-editing enzymes and cancer? *Trends Mol Med*, **9**, 147–152.

190 Barnes, D.E. and Lindahl, T. (2004) Repair and genetic consequences of endogenous DNA base damage in mammalian cells. *Annu Rev Genet*, **38**, 445–476.

191 Pham, P., Bransteitter, R., and Goodman, M.F. (2005) Reward versus risk: DNA cytidine deaminases triggering immunity and disease. *Biochemistry*, **44**, 2703–2715.

192 Chelico, L., Pham, P., and Goodman, M.F. (2008) Stochastic properties of processive cytidine DNA deaminases aid and apobec3g. *Philos Trans R Soc Lond B Biol Sci*, **364**, 583–593.

193 Goodman, J.E., Hofseth, L.J., Hussain, S.P., and Harris, C.C. (2004) Nitric oxide and p53 in cancer-prone chronic inflammation and oxyradical overload disease. *Environ Mol Mutagen*, **44**, 3–9.

194 Cristalli, G., Costanzi, S., Lambertucci, C., Lupidi, G., Vittori, S., Volpini, R., and Camaioni, E. (2001) Adenosine deaminase: functional implications and different classes of inhibitors. *Med Res Rev*, **21**, 105–128.

195 Yonekura, S.I., Nakamura, N., Yonei, S., and Zhang-Akiyama, Q.M. (2008) Generation, biological consequences and repair mechanisms of cytosine deamination in DNA. *J Radiat Res (Tokyo)*, **50**, 19–26.

196 Holliday, R. and Grigg, G.W. (1993) DNA methylation and mutation. *Mutat Res*, **285**, 61–67.

197 Lutsenko, E. and Bhagwat, A.S. (1999) Principal causes of hot spots for cytosine to thymine mutations at sites of cytosine methylation in growing cells. A model, its experimental support and implications. *Mutat Res*, **437**, 11–20.

198 Lindahl, T. and Nyberg, B. (1974) Heat-induced deamination of cytosine residues in deoxyribonucleic acid. *Biochemistry*, **13**, 3405–3410.

199 Shapiro, R. and Klein, R.S. (1966) The deamination of cytidine and cytosine by acidic buffer solutions. Mutagenic implications. *Biochemistry*, **5**, 2358–2362.

200 Frederico, L.A., Kunkel, T.A., and Shaw, B.R. (1990) A sensitive genetic assay for the detection of cytosine deamination: determination of rate constants and the activation energy. *Biochemistry*, **29**, 2532–2537.

201 Zhang, X. and Mathews, C.K. (1994) Effect of DNA cytosine methylation upon deamination-induced mutagenesis in a natural target sequence in duplex DNA. *J Biol Chem*, **269**, 7066–7069.

202 Shen, J.-C., Rideout, W.M. III, and Jones, P.A. (1994) The rate of hydrolytic deamination of 5-methylcytosine in double-stranded DNA. *Nucleic Acids Res*, **22**, 972–976.

203 Krokan, H.E., Drablos, F., and Slupphaug, G. (2002) Uracil in DNA–occurrence, consequences and repair. *Oncogene*, **21**, 8935–8948.

204 Pfeifer, G.P. (2006) Mutagenesis at methylated cpg sequences. *Curr Top Microbiol Immunol*, **301**, 259–281.

205 Suzuki, T., Yamaoka, R., Nishi, M., Ide, H., and Makino, K. (1996) Isolation and characterization of a novel product, 2′-deoxyoxanosine, from 2′-deoxyguanosine, oligodeoxynucleotide, and calf thymus DNA treated with nitrous acid and nitric oxide. *J Am Chem Soc*, **118**, 2515–2516.

206 Suzuki, T., Kanaori, K., Tajima, K., and Makino, K. (1997) Mechanism and intermediate for formation of 2′-deoxyoxanosine. *Nucleic Acids Symp Ser*, **37**, 313–314.

207 Dong, M., Wang, C., Deen, W.M., and Dedon, P.C. (2003) Absence of 2′-deoxyoxanosine and presence of abasic sites in DNA exposed to nitric oxide at controlled physiological concentrations. *Chem Res Toxicol*, **16**, 1044–1055.

208 Dong, M. and Dedon, P.C. (2006) Relatively small increases in the steady-state levels of nucleobase deamination products in DNA from human TK6 cells exposed to toxic levels of nitric oxide. *Chem Res Toxicol*, **19**, 50–57.

209 Glaser, R., Wu, H., and Lewis, M. (2005) Cytosine catalysis of nitrosative guanine deamination and interstrand cross-link formation. *J Am Chem Soc*, **127**, 7346–7358.

210 Lim, K.S., Huang, S.H., Jenner, A., Wang, H., Tang, S.Y., and Halliwell, B. (2006) Potential artifacts in the measurement of DNA deamination. *Free Radic Biol Med*, **40**, 1939–1948.

211 Reynaud, C.A., Aoufouchi, S., Faili, A., and Weill, J.C. (2003) What role for aid: Mutator, or assembler of the immunoglobulin mutasome? *Nat Immunol*, **4**, 631–638.

212 Longacre, A. and Storb, U. (2000) A novel cytidine deaminase affects antibody diversity. *Cell*, **102**, 541–544.

213 Harris, R.S., Sheehy, A.M., Craig, H.M., Malim, M.H., and Neuberger, M.S. (2003) DNA deamination: not just a trigger for antibody diversification but also a mechanism for defense against retroviruses. *Nat Immunol*, **4**, 641–643.

214 Chaudhuri, J. and Alt, F.W. (2004) Class-switch recombination: interplay of transcription, DNA deamination

and DNA repair. *Nat Rev Immunol*, **4**, 541–552.
215 Petersen-Mahrt, S.K., Harris, R.S., and Neuberger, M.S. (2002) Aid mutates E. Coli suggesting a DNA deamination mechanism for antibody diversification. *Nature*, **418**, 99–103.
216 Rada, C., Di Noia, J.M., and Neuberger, M.S. (2004) Mismatch recognition and uracil excision provide complementary paths to both ig switching and the a/t-focused phase of somatic mutation. *Mol Cell*, **16**, 163–171.
217 Di Noia, J. and Neuberger, M.S. (2002) Altering the pathway of immunoglobulin hypermutation by inhibiting uracil-DNA glycosylase. *Nature*, **419**, 43–48.
218 Gerber, A.P. and Keller, W. (2001) RNA editing by base deamination: more enzymes, more targets, new mysteries. *Trends Biochem Sci*, **26**, 376–384.
219 Gott, J.M. and Emeson, R.B. (2000) Functions and mechanisms of RNA editing. *Annu Rev Genet*, **34**, 499–531.
220 Maas, S. and Rich, A. (2000) Changing genetic information through RNA editing. *Bioessays*, **22**, 790–802.
221 Sommer, B., Kohler, M., Sprengel, R., and Seeburg, P.H. (1991) RNA editing in brain controls a determinant of ion flow in glutamate-gated channels. *Cell*, **67**, 11–19.
222 Higuchi, M., Single, F.N., Kohler, M., Sommer, B., Sprengel, R., and Seeburg, P.H. (1993) RNA editing of AMPA receptor subunit Glur-B: A base-paired intron-exon structure determines position and efficiency. *Cell*, **75**, 1361–1370.
223 Seeburg, P.H., Higuchi, M., and Sprengel, R. (1998) RNA editing of brain glutamate receptor channels: mechanism and physiology. *Brain Res Brain Res Rev*, **26**, 217–229.
224 Burns, C.M., Chu, H., Rueter, S.M., Hutchinson, L.K., Canton, H., Sanders-Bush, E., and Emeson, R.B. (1997) Regulation of serotonin-2c receptor g-protein coupling by RNA editing. *Nature*, **387**, 303–308.
225 Sanders-Bush, E., Fentress, H., and Hazelwood, L. (2003) Serotonin 5-HT2 receptors: Molecular and genomic diversity. *Mol Interv*, **3**, 319–330.
226 Bhalla, T., Rosenthal, J.J., Holmgren, M., and Reenan, R. (2004) Control of human potassium channel inactivation by editing of a small mRNA hairpin. *Nat Struct Mol Biol*, **11**, 950–956.
227 Hoopengardner, B., Bhalla, T., Staber, C., and Reenan, R. (2003) Nervous system targets of RNA editing identified by comparative genomics. *Science*, **301**, 832–836.
228 Levanon, E.Y., Eisenberg, E., Yelin, R., Nemzer, S., Hallegger, M., Shemesh, R., Fligelman, Z.Y., Shoshan, A., Pollock, S.R., Sztybel, D., Olshansky, M., Rechavi, G., and Jantsch, M.F. (2004) Systematic identification of abundant a-to-i editing sites in the human transcriptome. *Nat Biotechnol*, **22**, 1001–1005.
229 Levanon, E.Y., Hallegger, M., Kinar, Y., Shemesh, R., Djinovic-Carugo, K., Rechavi, G., Jantsch, M.F., and Eisenberg, E. (2005) Evolutionarily conserved human targets of adenosine to inosine RNA editing. *Nucleic Acids Res*, **33**, 1162–1168.
230 Keegan, L.P., Leroy, A., Sproul, D., and O'Connell, M.A. (2004) Adenosine deaminases acting on RNA (ADARs): RNA-editing enzymes. *Genome Biol*, **5**, 209.
231 Jarmuz, A., Chester, A., Bayliss, J., Gisbourne, J., Dunham, I., Scott, J., and Navaratnam, N. (2002) An anthropoid-specific locus of orphan C to U RNA-editing enzymes on chromosome 22. *Genomics*, **79**, 285–296.
232 Blanc, V. and Davidson, N.O. (2003) C-to-U RNA editing: mechanisms leading to genetic diversity. *J Biol Chem*, **278**, 1395–1398.
233 Anant, S., Blanc, V., and Davidson, N.O. (2003) Molecular regulation, evolutionary, and functional adaptations associated with C to U editing of mammalian apolipoprotein B

mRNA. *Prog Nucleic Acid Res Mol Biol*, **75**, 1–41.

234 Anant, S. and Davidson, N.O. (2001) Molecular mechanisms of apolipoprotein B mRNA editing. *Curr Opin Lipidol*, **12**, 159–165.

235 Kong, Q., Shan, X., Chang, Y., Tashiro, H., and Lin, C.L. (2008) RNA oxidation: a contributing factor or an epiphenomenon in the process of neurodegeneration. *Free Radic Res*, 1–5.

236 Wang, Q., Zhang, Z., Blackwell, K., and Carmichael, G.G. (2005) Vigilins bind to promiscuously A-to-I-edited RNAs and are involved in the formation of heterochromatin. *Curr Biol*, **15**, 384–391.

237 Perillo, B., Ombra, M.N., Bertoni, A., Cuozzo, C., Sacchetti, S., Sasso, A., Chiariotti, L., Malorni, A., Abbondanza, C., and Avvedimento, E.V. (2008) DNA oxidation as triggered by h3k9me2 demethylation drives estrogen-induced gene expression. *Science*, **319**, 202–206.

2
Modification of Cysteine Residues in Protein by Endogenous Oxidants and Electrophiles

Norma Frizzell and John W. Baynes

Cysteine is among the most readily oxidized and reactive amino acid in cellular proteins. Glutathione, present in millimolar concentrations in cells, plays an essential role in maintaining protein cysteine residues in the reduced state. Despite this protection, in addition to glutathionylation, a wide range of products are formed from cysteine, resulting from enzymatic and spontaneous oxidation, nitrosation, alkylation, and acylation reactions. This chapter presents an overview on nonenzymatic oxidation and alkylation reactions of cysteine by reactive oxygen species and endogenous electrophiles, with emphasis on products of lipoxidation and glycoxidation reactions and the modification of proteins by fumarate. We end with a discussion of the role that these reactions and products have in regulatory biology.

2.1
Introduction

Cysteine is among the most oxidizable and reactive amino acid in proteins. In physiological buffers, cysteine, either free or peptide form, autoxidizes spontaneously to cystine; the reaction is catalyzed by trace concentrations of transition metal ions (Fe, Cu) and is inhibited by chelators, such as EDTA. The rate of oxidation of cysteine is accelerated under alkaline and inhibited under acidic solutions, pointing to the thiolate ion as the substrate for oxidation. The cysteine thiol group has a nominal $pK_a \approx 8$, but low-pK_a cysteine residues are common elements in the active or binding sites of enzymes, transporters, signal transducers, transcription factors, and cytoskeletal proteins. *In vivo*, these cysteine residues exist primarily as thiolate anions. The thiolate ion is the reactive nucleophilic form of cysteine and is highly susceptible to oxidative insults, either autoxidation (oxidation by reactive oxygen species (ROS)) or reaction with endogenous and exogenous electrophiles; the latter reaction forms a thioether. After a brief discussion of cysteine autoxidation and the protective role of glutathione (GSH), this chapter focuses on the modification of cysteine during lipoxidation and glycoxidation reactions and reaction with the mitochondrial intermediate, fumarate, during mitochondrial stress. The chapter ends with an overview on the role of these reactions and products in cell signaling and regulatory biology.

Endogenous Toxins. Diet, Genetics, Disease and Treatment.
Edited by Peter J. O'Brien and W. Robert Bruce
Copyright © 2010 WILEY-VCH Verlag GmbH & Co. KGaA, Weinheim
ISBN: 978-3-527-32363-0

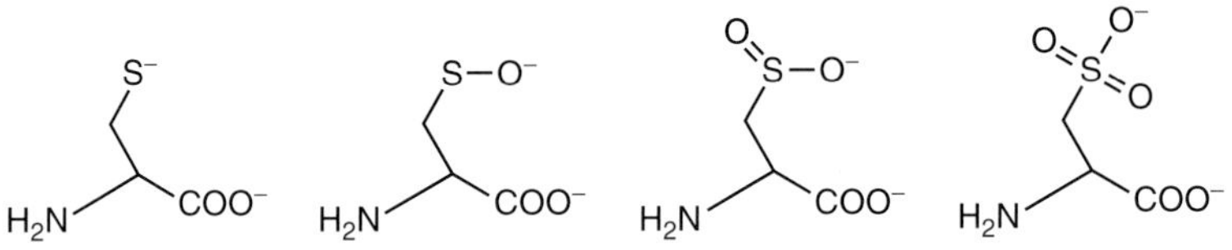

Fig. 2.1 Products of autoxidation of cysteine.

Structures shown: Cysteine (thiolate), Sulfenic acid, Sulfinic acid, Sulfonic acid (cysteic acid).

2.2
Autoxidation of Cysteine

The major products of autoxidation of cysteine by ROS are the sulfenic (RSOH), sulfinic (RSO$_2$H), and sulfonic (RSO$_3$H) acids (Figure 2.1); the sulfonic acid is known as *cysteic acid*. Oxidation to the sulfenic acid, which occurs spontaneously under air, is reversible by thiol compounds *in vitro* and by glutathionylation (below) *in vivo*. Stronger oxidizing conditions and reagents, for example, H$_2$O$_2$ and ONOO$^-$, yield sulfinic acid, which is reducible by ATP- and GSH-dependent sulfiredoxins (Srx). In yeast, there is evidence that the disulfide and sulfinic acid forms of a peroxiredoxin, resulting from different levels of oxidative stress (OxS), induce distinct antioxidant responses; with more severe OxS, sulfinic acid form induces expression of Srx (reviewed in [1]). The sulfonic acid, cysteic acid, is the irreversible end product of cysteine oxidation; it can be recovered from proteins by acid hydrolysis. Cysteic acid is commonly measured as a surrogate for the total cysteine + cystine content of proteins following oxidation with performic acid. Cysteine is also subject to oxidation by reactive nitrogen species (RNS), yielding a nitrosothiol (RSNO). The chemistry and regulatory biology of nitric oxide, peroxynitrite, and nitrosation of proteins are discussed in later chapters in this monograph.

2.3
Glutathione

GSH, present at ∼5 mM concentration in cells, plays an essential role in maintaining protein cysteine residues in the reduced state. GSH itself is maintained in the reduced state by GSH reductase, using the coenzyme NADPH, produced by dehydrogenases in the pentose phosphate pathway. Inadequate production of GSH leads to intra- and intermolecular disulfide cross-linking of proteins. Heinz bodies, consisting of disulfide cross-linked hemoglobin molecules, precipitate in the cytoplasm of red cells in glucose-6-phosphate dehydrogenase deficiency [2].

While GSH serves as a shield for protection of thiol groups in protein, maintaining them in the thiol or thiolate state, its cysteine residue, with a pK_a ≈ 9.2 [3], is relatively unreactive at physiological pH. In contrast, low-pK_a cysteine residues in proteins are much more reactive and are prime targets for oxidation by ROS and by electrophilic toxins and carcinogens. All of these reactions yield an oxidized cysteine residue – an autoxidation product (sulfenic, sulfinic, or sulfonic acid), a disulfide cross-link (cystine), or a thioether adduct. To the best of our knowledge, the formation of sulfonic acid (cysteic acid) and cysteine thioethers is biologically irreversible – the damage is removed only by turnover of the protein.

2.3.1
Glutathionylation

Glutathionylation, the formation of a mixed disulfide with GSH, is probably the most common chemical modification of thiol groups in the cell. Glutathionylation protects vital sulfhydryl groups in proteins from further oxidation during OxS. Because of the high concentration of GSH and the ratio of GSH : GSSG (oxidized glutathione) (~100 : 1) in cells, it is unlikely that glutathionylation of proteins is an equilibrium process. For most proteins with multiple sulfhydryl groups, only one or two of the several cysteine residues will form GSH adducts during OxS [4]. The specificity of glutathionylation is affected by the pK_a of the cysteine residue and by steric constraints; however, glutathionylation of proteins is also catalyzed by glutathione-S-transferases (GSTs). The most likely pathway for glutathionylation involves the formation of sulfenic acid during exposure of a protein to ROS. Sulfenic acid is reduced by GSH, either spontaneously or by GST catalysis, to form H_2O and a glutathionylated protein. *In vivo*, the reversal of glutathionylation by GSH is catalyzed by glutaredoxins, regenerating the native, reduced protein or enzyme. Dalle-Donne *et al.* [5] have proposed alternative routes to glutathionylation, including reduction of thiyl radicals and nitroso adducts on protein, but the significance of these alternate routes is uncertain.

The complementary stability and reversibility of glutathionylation of proteins makes it possible both to quantify total glutathionylation of cellular proteins and to identify specific glutathionylated proteins [6]. Following removal of free GSH by gel permeation chromatography, total glutathionylation can be estimated by reducing the protein, for example, with tris(2-carboxyethyl)phosphine (TCEP), followed by quantitation of free GSH by HPLC. Prior reduction of the protein with an irreversible thiol reagent, such as N-ethylmaleimide, and addition of chelators to buffers are recommended to limit artifactual glutathionylation during preparation of proteins. Identification of specific glutathionylated proteins may be accomplished by 2D-PAGE, followed by Western blotting with anti-GSH antibody, then in-gel digestion and analysis by mass spectrometry. Immunoprecipitation of glutathionylated proteins should also be useful for their isolation and identification.

2.3.2
Glutathione as an Electrophile Trap

In addition to its role in protecting protein cysteine residues from autoxidation, GSH also shields proteins from modification by endogenous electrophiles during OxS. Thus, 4-hydroxynonenal (HNE), an electrophilic lipid peroxidation (LPO) product discussed in the next section (Figure 2.2), is found in urine as N-acetylcysteine

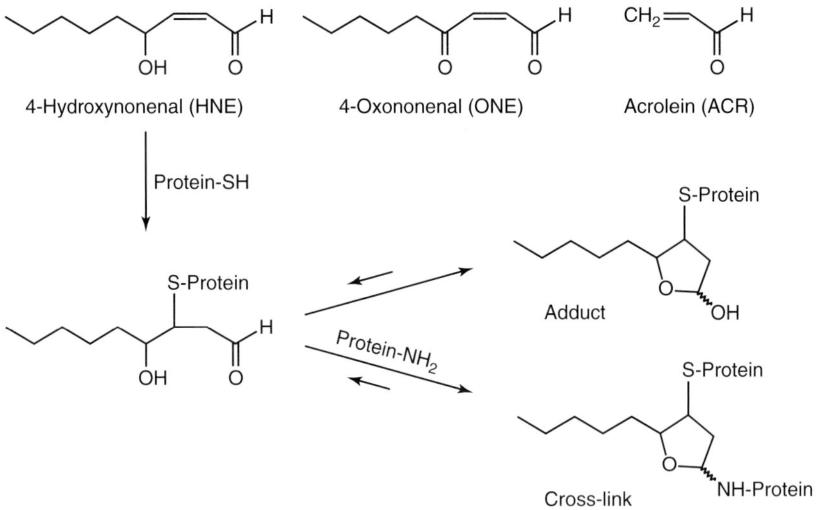

Fig. 2.2 Characteristic electrophilic products of autoxidation of fatty acids. The HNE–Cys adduct is shown in its open-chain aldehyde form and in the cyclic hemiacetal conformation. The aldehyde conformer may form intra- and intermolecular cross-links through Schiff base linkages with lysine residues.

derivatives, N-acetylcysteine-nonanol and N-acetylcysteine-nonenoic acid and its lactone [7]. These N-acetylcysteine derivatives, known as mercapturic acids, are derived in large part from reaction of HNE with GSH by both spontaneous and GST-catalyzed reactions, followed by oxidation or reduction of the HNE–GSH adduct and proteolytic degradation of the GSH. Mercapturic acids may also be derived from proteolytic degradation of HNE-modified proteins. The mercapturic acids of HNE and other lipid-derived electrophiles, such as acrolein (ACR) (Figure 2.2), increase in urine of animals exposed to OxS [8, 9]. Thus, GSH protects against oxidative modification of cysteine residues in protein, both by ROS and by reactive electrophiles. When these defenses are evaded or overwhelmed, oxidative modification of thiol groups during OxS activates antioxidant defenses.

2.4
Lipoxidation

Polyunsaturated fatty acids (PUFA) are among the most readily oxidized compounds in biological systems. Because LPO initiates radical chain reactions and electrons are relatively mobile along the fatty acid carbon chain, a wide range of reactive, unsaturated, conjugated hydroperoxy, epoxy, and carbonyl compounds are produced on oxidation of even a homogeneous lipid – Spiteller has listed over 50 compounds produced on oxidation of linoleic acid alone [10]. The most studied among these LPO products are malondialdehyde (MDA), ACR, HNE, and

4-oxononenal (ONE). These compounds are formed during OxS and inflammation, react with proteins and DNA, and induce a wide range of biological responses. Although MDA reacts with lysine residues and cross-links proteins, it does not appear to form stable adducts with thiol compounds [11], so it will not be considered further. ACR, HNE, and ONE react with lysine, histidine, and cysteine residues in protein, leading to an array of chemical modifications and producing cross-links that are part of a group of compounds known as advanced lipoxidation end products (ALEs) [12]. These ALEs have a residual aldehyde functional group, therefore they contribute to the formation of protein carbonyls (carbonylation) (Figure 2.2) [13], a common marker of OxS, and also mediate secondary cross-linking reactions. There are excellent current reviews on the chemistry and reactivity of ACR [14], HNE, and ONE [15], and a recent volume of *Redox Reports* [16] is dedicated to research on the chemistry and biochemistry of HNE. Rather than recapitulate the content of these papers, we will focus on the reaction of HNE with cysteine residues in protein as a model for reaction of endogenous electrophiles with cysteine residues in protein. To the best of our knowledge, ACR and ONE, although they are less and more reactive than HNE, respectively, have similar specificity in modification of proteins and yield similar biological responses.

2.4.1
Glyceraldehyde-3-Phosphate Dehydrogenase (GAPDH)

In their original studies on the reaction of HNE with protein, Uchida and Stadtman showed that glyceraldehyde-3-phosphate dehydrogenase (GAPDH), which has an active site cysteine residue, was inhibited by HNE in a time- and concentration-dependent fashion [17, 18]. The enzyme was incubated with high concentrations of HNE (≥ 10 mol HNE/mol GAPDH subunit), and adducts to Cys, His, and Lys residues were readily detected by amino acid analysis. Cys and His were the most reactive, consistent with their greater nucleophilicity, compared to lysine. Protein carbonyls, detected by reaction with dinitrophenylhydrazine (DNPH), also increased, indicating a Michael addition reaction (Figure 2.2). However, only a fraction of the modified HNE–amino acids could be recovered by NaBH$_4$ reduction, acid hydrolysis, and amino acid analysis of the protein. SDS-PAGE analysis of the NaBH$_4$-reduced protein indicated that aldehyde functional groups of HNE were involved in mostly intramolecular cross-linking reactions through formation of Schiff bases (aldimines) with neighboring lysine residues (Figure 2.2). These cross-links are relatively stable, but would react with DNPH in assays for protein carbonyls. One of the surprising results of analysis of HNE-modified GAPDH was that the most reactive cysteines were located on the surface of the protein, that is, not the low-pK_a cysteine in the active site. The most reactive histidine residue was also located prominently on the surface of the protein. These results suggest that HNE and also ONE, because of their high reactivity, are relatively nonselective in modifying amino acids in protein.

2.4.2
Albumin

As with GAPDH, HNE reacts with albumin to yield Cys, His, and Lys adducts. Two comprehensive studies on modification of human serum albumin by HNE [19, 20] differed significantly in experimental design (ratio of HNE : protein), instrumental methods, and data analysis, and not surprisingly led to some differences in identification of the most reactive amino acids in the protein. However, in both studies, Cys-34 was identified as the most reactive amino acid in the protein, followed by low-pK_a histidine and lysine residues in hydrophobic (fatty acid) binding sites in the protein. In subsequent work, Aldini *et al.* also detected HNE (and ACR) adducts to Cys-34 when HSA was incubated with mildly oxidized low density lipoproteins (LDL) [21] or when HNE was added to whole plasma [22]. Although Cys-34 in HSA is the major thiol compound in plasma and is a potent trap for HNE (and presumably for ACR), there are no reports on the detection of the HNE (or ACR) adducts to Cys-34 on HSA isolated from human plasma – despite the fact that both HNE and ACR are measurable at low micromolar concentrations in plasma and increase in plasma of patients with diseases involving renal failure or OxS (reviewed in 23). Because of the high reactivity of these compounds, it is possible that the "free" compounds measured in plasma may be present on protein as aldimine adducts, which would be released during the assay procedure. Alternatively, HNE may be formed in plasma by LPO reactions during sample preparation.

2.4.3
Actin

The cytoskeletal protein actin has also been studied in some detail. Cys-374 is the primary site of reaction with HNE [24]. This residue is both surface exposed and has a low pK_a; four other cysteine residues in actin, which are less exposed or nucleophilic, are not modified by HNE. Histidine also reacts, but only at high ratios of HNE : protein, after modification of Cys-374. Cys-374 is reactive in both globular (G) and fibrous (F) actin, and modification of this cysteine residue by HNE does not appear to affect polymerization of the protein. In contrast, more extensive modification of the protein – which is unlikely to occur *in vivo* even at pathological concentrations of HNE – leads to modification of His residues, which then interferes with the polymerization reaction *in vitro* [25]. In studies with isolated rat hearts, Eaton *et al.* [26] detected a >50-fold increase in HNE-modified proteins, based on immunohistochemistry, in the heart following ischemic-reperfusion injury. Most of the HNE was localized in sarcolemma membranes, but significant staining was detected on longitudinal striations, consistent with modification of actin. In rats treated with the oxidant stressor ferric nitriloacetate, Ozeki *et al.* [27] also identified actin as a target of HNE modification in the kidney; the identification of HNE–actin was confirmed by immunoprecipitation with anti-actin antibody, followed by Western blot analysis with anti-HNE antibody. HNE–actin is also elevated in regions of the brain of subjects with mild cognitive impairment, an early stage in the development of Alzheimer's disease [28].

2.4.4
Quantitative Analysis

Although several proteins have been used in model studies on the reaction of HNE with protein, the actual extent or site specificity of modification of any of these proteins by HNE in response to OxS *in vivo* is unknown. Requena et al. [29] reported a GC/MS assay for HNE–Lys in proteins and were able to detect this adduct at approximately 1 mmol/mol Lys in oxidized LDL. However, HNE–Lys was not detectable in native LDL or in skin collagen or lens proteins from diabetic or aged subjects. Evidence for increased modification of proteins by HNE during OxS is based exclusively on sensitive immunohistochemistry or Western blotting techniques, but these assays do not distinguish between the Cys, His, and Lys adducts. In the absence of chemical data to support the immunological observations, there is no information on the relative extent of HNE modification of Cys, His, and Lys residues in proteins *in vivo*. There are several reasons for this lack of information. First, much of the HNE that is produced *in vivo* is detoxified by redox enzymes or by reaction with GSH, limiting the extent of reaction with proteins. Secondly, because of its reactivity, the extent of modification of a specific protein by HNE would be limited by its reaction with a wide range of ambient proteins. Thirdly, HNE reacts with Cys, His, and Lys residues in protein, so that three different products are formed. Fourthly, because of its reactivity, HNE may react at multiple sites on proteins, limiting the detection of specific HNE-peptides by mass spectrometric techniques; enrichment of modified peptides by immunoaffinity techniques may address this problem, but there are no published reports on the success of this approach to date. Finally, HNE adducts to protein continue to react to form secondary adducts and cross-links, so that a range of products may be formed, limiting the detection of a characteristic product. Sayre et al. [15] have described several furan and pyrrole cross-links that are formed following the Michael addition of HNE or ONE to lysine residues in protein.

Because of the lack of quantitative information on the extent of reaction of HNE with specific proteins *in vivo*, it is difficult to argue that lipoxidative modification of proteins by HNE (or other LPO products) has a significant, site-directed effect on the function of any specific protein in response to OxS or disease. However, recent studies on Alzheimer's and prion diseases emphasize that HNE is only part of the story. At sites of neuronal damage, the increase in HNE adducts, measured by immunohistochemistry, is accompanied by accumulation of ACR adducts [30] and other oxidation products, including the advanced glycation end products (AGE), N^ε–(carboxymethyl)lysine (CML), and pentosidine, and the protein oxidation product, nitrotyrosine [31, 32]. The impact of this broad lipoxidative damage, as well as total oxidative damage to proteins, can probably be best analogized as death by a thousand needles – chemical modification of a wide range of proteins by an array of reactive electrophiles – rather than by a dagger or stake aimed at a specific protein target in the cell. Indeed, to date, there are no published results on quantitative analysis of HNE modifications of specific cellular proteins by chemical or instrumental (nonimmunological) methods. However, using LC/MS/MS, Orioli et al. [33] have recently detected a 20- to 30-fold increase in HNE–His peptides in the urine of Zucker obese rats. The methodology described in this study should be applicable for quantitation and comparison of the extents of HNE modification of Cys, His, and Lys residues in specific immunoprecipitated proteins or whole tissues in response to OxS. At this point, however, there is no evidence that lipoxidative damage inactivates a significant fraction of any specific protein during OxS or in disease.

2.5
Maillard Reactions of Cysteine

2.5.1
Transglycosylation

Maillard or nonenzymatic browning reactions between reducing sugars and protein yield a wide range of products, including both adducts and cross-links in protein [34]. Glycation, the first step in this reaction, involves formation of a glycosylamine adduct to a lysine or amino-terminal residue in protein, which dehydrates to a Schiff base, then rearranges to a metastable Amadori adduct on protein (Figure 2.3). Glucose (and other sugars) acts as a weak electrophile in the glycation reaction. Szwergold et al. [35, 36] have proposed that low molecular weight thiol compounds, primarily GSH, have an important role in limiting intracellular glycation of proteins. Mechanistically, GSH is proposed to participate in a transglycation reaction by which GSH discharges the Schiff base adduct on a lysine residue, yielding a thiohemiacetal, which is in equilibrium with the cyclic S-glycoside (Figure 2.3). The transglycosylation product either hydrolyzes slowly or eventually appears in urine as a cyclic thiazolidine. In support of their hypothesis, the authors employ NMR spectroscopy to show that N-acetylcysteine discharges the glycosylamine adduct of glucose to ethanolamine (GE) and that incubation of GE with free cysteine yields the stable thiazolidine adduct. Evidence for the biological significance of this pathway is based on the observations that depletion of GSH leads to increased glycation of hemoglobin in red cells incubated with glucose *in vitro* and that the glucose–cysteine thiazolidine, measured by GC/MS, is increased in urine of diabetic subjects. By this mechanism, GSH might also participate in removal of other carbonyl adducts formed on cysteine residues in proteins, including thiohemiacetals of glucose, glycolytic intermediates, and dicarbonyl compounds (below). Conversely, glucose, as an endogenous electrophile – admittedly a weak one, but present at high concentration in plasma and in some cells – might also react reversibly with thiol groups in proteins, affecting the activity of sulfhydryl enzymes in diabetes.

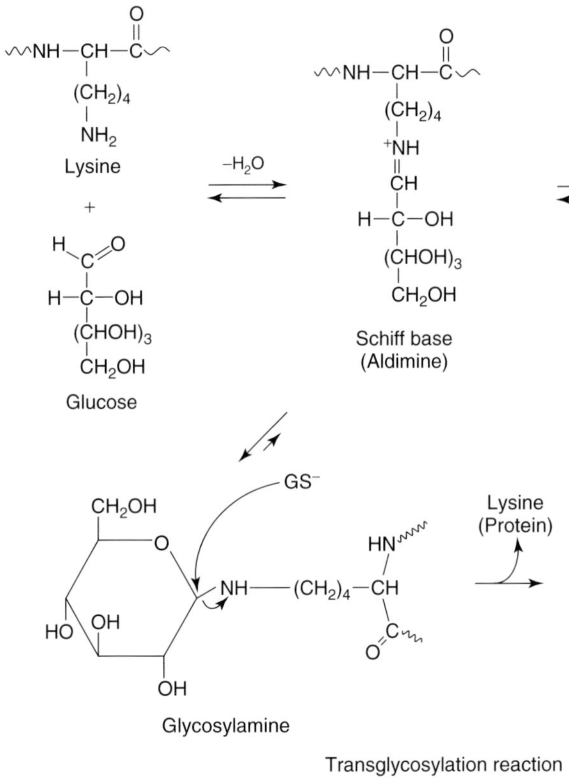

Fig. 2.3 Chemistry of glycation and transglycosylation by GSH. By this mechanism, the concentration of the Schiff base, the precursor of the Amadori adduct, is decreased by transglycosylation with GSH, thereby decreasing the rate of glycation (formation of the Amadori adduct) of protein. GS⁻ = thiolate anion of glutathione.

2.5.2
Carboxyalkylation

The Amadori adducts on protein undergo amine-catalyzed degradation reactions, yielding a variety of reactive carbonyl and dicarbonyl compounds, including 1- and 3-deoxyglucosones, shorter-chain aldoses and ketoses such as glyceraldehyde and dihydroxyacetone, and dicarbonyl compounds such as glyoxal (GO) and methylglyoxal (MGO). These reactive carbonyl compounds (RCOs) react with proteins, forming adducts and cross-links, known as *advanced glycation end products*. AGE accumulate in tissue proteins with age, and accelerated accumulation of AGE as a result of hyperglycemia is implicated in the pathogenesis of diabetic complications [37]. AGE also accumulate in tissues in renal disease, even in the absence of hyperglycemia, because of decreased renal clearance and detoxification of RCO, a condition known as *carbonyl stress* [38].

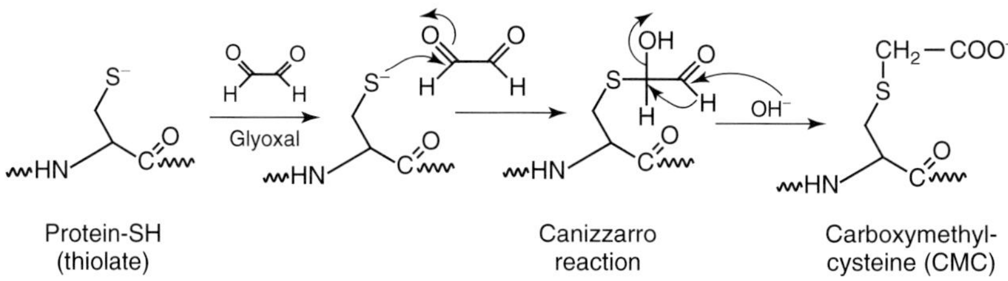

Fig. 2.4 Pathway for formation of carboxymethylcysteine (CMC) by reaction of glyoxal with cysteine residues in protein. Carboxyethylcysteine (CEC) is formed by an analogous reaction involving methylglyoxal.

While most studies on the Maillard reaction and AGEs *in vivo* have focused on the modification of lysine and arginine residues, cysteine is an obvious target for modification by electrophilic RCOs. Reaction of cysteine with dicarbonyl compounds is normally considered a reversible reaction, yielding a thiohemiacetal or thiohemiketal. The GSH hemiacetal is an intermediate in detoxification of dicarbonyl compounds, such as GO and MGO, by the glyoxalase pathway. However, stable, irreversible products of reaction of dicarbonyl compounds with cysteine, S-(carboxymethyl)-cysteine (CMC) and S-(carboxyethyl)-cysteine (CEC), have been measured in tissue proteins [39, 40]. These compounds may be formed by a Cannizzaro reaction (Figure 2.4) [39] or by an amine-catalyzed rearrangement [41] of the thiol-carbonyl adduct. CMC is present at concentrations comparable to CML, a characteristic and prominent AGE in tissue proteins [42]. Like CML, CMC increases with age in human skin collagen and in muscle protein, including myofibrillar (intracellular) proteins, of diabetic rats [39]. The increase in CMC correlates with the increase in CML in both aging and diabetes [39, 40]. CMC and CEC are also increased in plasma proteins of persons with type 1 and type 2 diabetes, and their concentrations correlate with that of hemoglobin A_{1c}, a biomarker of glycemia, and urinary albumin:creatinine ratio, a biomarker of nephropathy [40].

The biological significance of the thiol AGEs is uncertain. Zeng and Davies have suggested that CMC may not be present *per se* in tissue protein, but may be formed on hydrolysis of cysteine–lysine cross-links in protein [41]. This interesting proposal implicates modification of cysteine residues by dicarbonyl compounds in the formation of intra- or intermolecular cross-links in proteins, analogous to the cross-linking of proteins by HNE. In support of this proposal, Zeng and Davies demonstrate that several low molecular weight thiols (cysteine, cysteamine, GSH) are potent inhibitors of both glycation and cross-linking of proteins by GO and MGO [43] – although this inhibition may be attributed to the antioxidant and/or chelating activity of thiol compounds, rather than their reaction with dicarbonyl intermediates. In other studies, Zeng et al. [44] demonstrated inactivation of several cathepsins and papain (cysteine proteases) by GO and MGO *in vitro*. Inactivation of papain by GO was paralleled by the loss of the active site cysteine residue and a corresponding increase in the CMC content of the enzyme. Although Arg is generally considered the primary site of reaction of dicarbonyl compounds with proteins, modification of cysteine by dicarbonyl compounds *in vivo* may also contribute to inactivation of thiol enzymes, transporters, and other proteins in aging and chronic disease. Specific proteins modified by CMC and CEC have not been identified either *in vivo* or in cell culture, but progress in this direction is likely now that an antibody has been developed, which is specific for CMC, without cross-reaction with CML or N^ε-(carboxyethyl)lysine (CEL) [45]. Needless to say, identification of target proteins and measurement of the extent of modification of these proteins are essential for assessing the biological significance of carboxyalkylation of cysteine residues in proteins.

2.6
Succination

2.6.1
Discovery and Metabolic Origin of S-(2-Succinyl)cysteine (2SC)

2SC was identified by Alderson *et al.* [46] (2SC) as a nonenzymatic, irreversible chemical modification of proteins. It is formed by a Michael addition reaction of the Krebs cycle intermediate (KCI), fumarate, with cysteine residues in proteins (Figure 2.5). The term *succination* was introduced to distinguish formation of 2SC, an acid-stable thioether, from succinylation, in which a hydrolyzable ester, thioester, or amide bond is formed on protein. 2SC was originally detected on extracellular proteins (albumin, plasma proteins, and skin collagen), but is now known to be widely distributed in cellular proteins. Increases in 2SC in muscle [46] and adipose [47] tissue proteins were detected in animal models of both type 1 (rat) [46, 48] and type 2 (mouse) diabetes [47], and also during differentiation of fibroblasts to adipocytes in high (30 mM) glucose medium [49] and in c2c12 myocytes grown in high glucose medium (N. Frizzell, unpublished). In both muscle and adipose tissue in diabetes, and in adipocyte cell culture, the increase in succination was correlated with an increase in tissue or cellular fumarate concentration. While

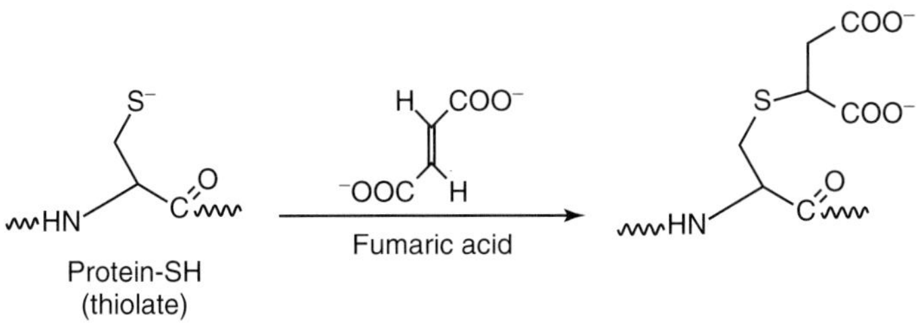

Fig. 2.5 Formation of 2-(S-succinyl)cysteine (2SC) by reaction of fumarate with cysteine residues in protein.

other KCIs probably increase in concert with fumarate, fumarate appears to be unique in its reactivity with proteins. Levels of 2SC in muscle and adipose tissue of diabetic rats are significantly higher than the concentration of the AGE/ALEs, CML, or CMC. Given the lower cysteine content of proteins, compared to lysine, the fractional carboxyalkylation of cysteine is much greater than that of lysine. In addition, considering the more prominent role of cysteine in the active sites of numerous enzymes and proteins, modification of cysteine is likely to have a much greater impact on metabolism in diabetes.

The increase in fumarate and 2SC is probably secondary to *flooding* of the mitochondrion with excess substrate, glucose and lipids, in diabetes. The pressure from the pathological excess of fuels in diabetes, without the increase in energy expenditure, would cause a buildup of KCIs and intramitochondrial NADH, leading to hyperpolarization of the inner mitochondrial membrane (IMM). The increase in fumarate and 2SC in muscle and adipose tissue in diabetes by substrate flooding may be exacerbated by true hypoxia as a result of microvascular disease in diabetes or from chronic pseudohypoxia, characterized by an increase in NADH during normoxia in diabetes [50]. Although fumarate (and other KCIs) increases severalfold in muscle during strenuous exercise [51], when lack of oxygen (hypoxia) causes a buildup of reduced redox coenzymes, the reaction of fumarate with protein is slow, so that the transient increase in muscle fumarate during hypoxia is unlikely to lead to a significant increase in 2SC during exercise.

While the contribution of (pseudo)hypoxia to accumulation of fumarate in muscle is uncertain, there is evidence of true hypoxia in adipose tissue in both obesity [52, 53] and type 2 diabetes [54], as well as recent evidence that growth of adipocytes under hypoxic conditions causes insulin resistance, OxS, and a proinflammatory state [55]. Adipocytes grown in high glucose medium also have increased levels of biomarkers of OxS and inflammation, and agents that cause hyperpolarization of the IMM also increase biomarker production, even at normal glucose concentration [55]. Mitochondrial superoxide production is also decreased by agents that depolarize the IMM, such as electron transport inhibitors and chemical uncouplers, and by overexpression of uncoupling proteins [56]. The increase in 2SC that develops in adipocytes grown in high glucose medium and in adipose tissue of diabetic animals suggests that 2SC may be useful as a general biomarker of OxS and mitochondrial stress and glucotoxicity.

Fumarate formed in the mitochondrion would leak downhill along its concentration gradient into the cytoplasm through passive, anion transporters in the IMM [57]. Fumarate transporters in other membranes [58] would mediate its transport between cellular compartments and eventually its excretion from the cell. Indeed, when adipocytes are grown in high (30 mM) glucose medium, the concentration of fumarate in the medium increases by $\sim$10-fold [59]. The increase in fumarate concentration in tissues in diabetes is mirrored by less pronounced but significant increases in fumarate and its precursor, succinate, in urine of diabetic rats [60]. Similar increases in fumarate are observed in urine of obese (insulin resistant) Zucker rats [61], suggesting that urinary 2SC may be an early indicator of mitochondrial stress in prediabetes.

2.6.2
2SC in Skeletal Muscle in Diabetes

In studies on aortic endothelial cells in culture, Brownlee et al. [62, 63] showed that inactivation of GAPDH during OxS led to intracellular accumulation of triose phosphates and their precursors (hexose phosphates) and the AGE/ALE precursor MGO. In a unifying hypothesis on the origin of diabetic complications [64], Brownlee proposed that these increases in metabolite concentration activated major metabolic pathways of hyperglycemic damage to endothelial cells, including the polyol, hexosamine, and protein kinase C pathways and formation of MGO–AGEs. Although poly-ADP-ribosylation was proposed as the mechanism by which ROS production led to inactivation of GAPDH [65], the hyperpolarization of the IMM and increased ROS production in endothelial cells grown in high glucose medium suggested that, as in adipocytes, an increase in succination of protein might be involved in the inactivation of GAPDH in endothelial and other cells during hyperglycemia in diabetes.

GAPDH has two reactive cysteine residues per subunit, one of which is in the active site; the other is also a nucleophilic cysteine residue, and modification of this cysteine by HNE inactivates the enzyme [18]. Blatnik et al. [48] showed that incubation of GAPDH with fumarate led to inactivation of the enzyme by approximately equal modification of these two cysteine residues. They also showed that, in the streptozotocin-induced (type 1) diabetic rat, the increase in 2SC in total muscle protein correlated with the loss in specific activity of GAPDH. Analysis of GAPDH immunoprecipitated from control and diabetic muscle indicated that the loss in enzymatic activity of GAPDH in diabetic muscle was consistent with the extent of succination of the protein, measured by HPLC–ESI and MALDI-TOF mass spectrometry. Semiquantitative analysis by both of these techniques [65] indicated that up to 20% of GAPDH in muscle was inactivated by succination in this animal model of severe diabetes. The increase in succination of skeletal muscle protein in diabetes is consistent with evidence of muscle mitochondrial flooding in diabetes [66, 67] and suggests that the resultant increase in succination of protein in muscle (and other tissues) may underlie metabolic imbalances implicated in the development of diabetic complications.

2.6.3
2SC in Adipose Tissue in Diabetes

Nagai et al. [49] analyzed the succinated proteome of adipocytes grown in high glucose medium. Western blot analysis, using polyclonal anti-2SC antibody, revealed only a single succinated protein in fibroblasts, while over 60 succinated proteins were identified by 2D-PAGE analysis of adipocytes. Adiponectin was identified as one of the major succinated proteins based on 1D- and 2D-PAGE and Western blot analysis of the immunoprecipitated protein. Thirteen other proteins were identified by MALDI-TOF analysis of tryptic peptides from spots extracted from the 2D-PAGE. These included cytoskeletal proteins (actins (2 isoforms), annexins

(2 isoforms), vimentin and tubulin) tropomyosin, heat-shock- and chaperone proteins (HSP 90β and HSP 70 protein 5/BiP), regulatory proteins (prolyl-4-hydroxylase (PHD)/protein disulfide isomerase and protein 14-3-3γ), a protease (cathepsin B), and a major fatty acid binding protein (FABP4), in addition to GAPDH and adiponectin. Each of the identified proteins has a free cysteine residue, and in six cases a peptide with the molecular mass of the succinated peptide was detected in adipocyte extracts by MALDI-TOF/TOF mass spectrometry.

Nagai *et al.* [49] suggested how inactivation of several proteins by succination might be involved in adipogenesis or the pathological changes in diabetes. Thus, succination of cytoskeletal proteins might weaken the cytoskeleton, permitting changes in cellular morphology to accommodate the increase in lipid volume during adipogenesis. PHD catalyzes the hydroxylation of hypoxia inducible factor-1α (HIF), which is the rate limiting step in turnover of this transcription factor. Inhibition of PHD by succination would enhance the HIF-dependent induction of glycolytic enzymes and angiogenic factors (e.g., VEGF) in response to adipose tissue hypoxia [68, 69]. Interestingly, PHD is inhibited allosterically by fumarate [70], suggesting that irreversible succination may complement chronic allosteric inhibition of the enzyme. Protein 14-3-3γ modulates signal transduction by binding to phosphorylated proteins [71]. Its binding to phosphorylated insulin receptor substrate 1 (IRS-1) is implicated in development of insulin resistance, so that succination of this protein might limit the development of insulin resistance. In contrast to these potential positive effects of succination, inhibition of heat shock/chaperone proteins and protein disulfide isomerase may contribute to the development of endoplasmic reticulum stress in diabetes [72], exacerbating the disease process.

Adiponectin, the predominant adipokine among about 25 proteins secreted by adipocytes, was examined in detail. This $\sim$30 kDa glycoprotein associates intracellularly with trimeric, hexameric, and other high molecular weight (HMW) complexes. It acts primarily on receptors in muscle where the trimeric form enhances fatty acid oxidation, and in the liver where HMW species promote the inhibition of gluconeogenesis [73]. Levels of both total and HMW adiponectin are decreased in plasma of type 2 diabetic patients, and hypoadiponectinemia is implicated in the development of insulin resistance [74].

Adiponectin contains two conserved cysteine residues, one of these near the N-terminus in a collagen-like domain in the protein. This cysteine is critical for the oligomerization of adiponectin from the trimeric species to the HMW isoforms as mutation of this cysteine to serine prevents the formation of species greater than the trimer [75]. This cysteine is a target for succination by fumarate *in vitro*, and 2SC was detected only in the monomeric form of adiponectin. Densitometric analysis of 2D-PAGE Western blots indicates that 5–10% of adiponectin in adipocytes and in adipose tissue of db/db mice is succinated [47]. The fact that succinated adiponectin was not detectable in either the cell growth medium or in plasma argues that succination may inhibit adiponectin assembly and secretion, contributing to the decrease in plasma adiponectin concentration and HMW isoforms of adiponectin in obesity and diabetes.

2.6.4
Overview on Succination of Protein

Fumarate is not usually considered an endogenous electrophile, but it clearly reacts with cysteine residues in proteins. 2SC, formed by reaction of fumarate with protein, appears to be a biomarker of mitochondrial flooding, resulting from excess substrate, possibly exacerbated by (pseudo)hypoxia. Notably, in contrast to the lack of information on the extent and site specificity of AGE and ALE formation on proteins in response to OxS, both the extent and specificity of modification of proteins *in vivo* by fumarate have been measured for GAPDH [48, 65] and adiponectin [47].

The overall significance of succination in regulatory biology or pathology is still uncertain, but it is interesting to note the overlap between succinated and glutathionylated proteins. Thus, 5 of the 11 proteins identified as succinated in adipocytes grown in high glucose medium [49] were also members of a set of 42 proteins identified as targets of glutathionylation *in vitro* or in cell culture systems [76], suggesting that the two processes may be related. Townsend has commented that "the actual number of cellular S-glutathionylated proteins is not large, relative to the proteome" [76], which is also consistent with the limited number of succinated proteins detected in adipocytes in cell culture [49] and in adipose tissue of diabetic rats [47]. Whether these same proteins are preferentially modified by reactive electrophiles, such as HNE, is unknown. Further characterization of the 2SC proteome in various tissues and quantitative analysis of the extent of protein succination should lead to a broader perspective on the role of succination in the regulation of metabolism and signal transduction in diabetes.

2.7
Conclusion

Cysteine residues in protein are subject to a wide range of modifications by ROS and endogenous electrophiles of both enzymatic and nonenzymatic origin. These electrophiles increase in tissues during OxS and inflammation, stimulate the antioxidant response element (ARE) [77], and have broad effects on cell signaling cascades and metabolism [78]. Progress in understanding the regulatory role of cysteine modifications will require identification of critical target proteins, analysis of the site specificity of modification and the resultant impact on protein structure and function, and careful quantification of the extent of protein modification under physiological and pathological conditions. These are challenging tasks, because of the large variety of electrophiles produced, the range of products formed, and the low intracellular concentration of many of the target proteins involved in signal transduction. With the resolution and sensitivity available from modern liquid chromatography–mass spectrometry systems, we anticipate rapid progress in understanding the effects of electrophilic modification of protein cysteine residues on signal transduction and regulatory biology.

Acknowledgment

Research in the authors' laboratory was supported by NIH Research Grant DK19971.

References

1 Jacob, C., Knight, I., and Winyard, P.G. (2006) Aspects of the biological redox chemistry of cysteine: from simple redox responses to sophisticated signaling pathways. *Biol Chem*, 387, 1385–1397.

2 Ronquist, G. and Theodorsson, E. (2007) Inherited, non-spherocytic hemolysis due to deficiency of glucose-6-phosphate dehydrogenase. *Scand J Clin Lab Invest*, 67, 105–111.

3 Jung, G., Breitmeier, E., and Voelter, W. (1972) Dissociation constants of glutathione. *Eur J Biochem*, 24, 438–445.

4 Hamnell-Pamment, Y., Lind, C., Palmberg, C., Bergman, T., and Cotgreave, I.A. (2005) Determination of the site-specificity of S-glutathionylated cellular protein. *Biochem Biophys Res Commun*, 332, 362–369.

5 Dalle-Donne, I., Rossi, R., Giustarini, D., Colombo, R., and Milzani, A. (2007) S-glutathionylation in protein redox regulation. *Free Radic Biol Med*, 43, 883–898.

6 Della-Donne, I., Milzani, A., Gagliano, N., Colombo, R., Giustarini, D., and Rossi, R. (2008) Molecular mechanisms and potential clinical significance of S-glutathionylation. *Antioxid Redox Signal*, 10, 445–473.

7 Kuiper, H.C., Miranda, C.L., Sowell, J.D., and Stevens, J.F. (2008) Mercapturic acid conjugates of 4-hydroxy-2-nonenal and 4-oxo-2-nonenal metabolites are in vivo markers of oxidative stress. *J Biol Chem*, 283, 17131–17138.

8 Gueraud, F., Peiro, G., Bernard, H., Alary, J., Creminon, C., Debrauwer, L., Rathahao, E., Drumare, M.F., Canlet, C., Wal, J.M., and Bories, G. (2006) Enzyme immunoassay for a urinary metabolite of 4-hydroxynonenal as a marker of lipid peroxidation. *Free Radic Biol Med*, 40, 54–62.

9 Carmella, S.G., Chen, M., Zhang, Y., Zhang, S., Hatsukami, D.K., and Hecht, S.S. (2007) Quantitation of acrolein-derived (3-hydroxypropyl)mercapturic acid in human urine by liquid chromatography – atmospheric pressure chemical ionization tandem mass spectrometry: effects of cigarette smoking. *Chem Res Toxicol*, 20, 986–990.

10 Spiteller, G. (1998) Linoleic acid peroxidation – the dominant lipid peroxidation process in low density lipoprotein – and its relationship to chronic diseases. *Chem Phys Lipids*, 95, 105–162.

11 Esterbauer, H., Schaur, R.J., and Zollner, H. (1991) Chemistry and biochemistry of 4-hydroxynonenal, malondialdehyde and related aldehydes. *Free Radic Biol Med*, 11, 81–128.

12 Baynes, J.W. and Thorpe, S.R. (2000) Glycoxidation and lipoxidation in atherogenesis. *Free Radic Biol Med*, 28, 1708–1716.

13 Liu, Q., Raina, A.K., Smith, M.S., Sayre, L.M., and Perry, G. (2003) Hydroxynonenal, toxic carbonyls, and

Alzheimer disease. *Mol Aspects Med*, **24**, 305–313.
14. Stevens, J.F. and Maier, C.S. (2008) Acrolein: sources, metabolism, and biomolecular interactions relevant to human health and disease. *Mol Nutr Food Res*, **52**, 7–25.
15. Sayre, L.M., Lin, D., Yuan, Q., Zhu, X., and Tang, X. (2006) Protein adducts generated from products of lipid oxidation: focus on HNE and ONE. *Drug Metab Rev*, **38**, 651–675.
16. Grune, T. (ed.) (2007) HNE (4-hydroxynonenal) and related aldehydes: analysis, effects, detoxification. *Redox Rep*, **12** (1-2).
17. Uchida, K. and Stadtman, E.R. (1993) Covalent attachment of 4-hydroxynonenal to glyceraldehyde-3-phosphate dehydrogenase. *J Biol Chem*, **268**, 6386–6393.
18. Ishii, T., Tatsuda, E., Kumazawa, S., Nakayama, T., and Uchida, K. (2003) Molecular basis of enzyme inactivation by an endogenous electrophile 4-hydroxy-2-nonenal: identification of modification sites in glyceraldehyde-3-phosphate dehydrogenase. *Biochemistry*, **42**, 3474–3480.
19. Aldini, G., Gamberoni, L., Orioli, M., Beretta, G., Regazzoni, L., Facino, R.M., and Carini, M. (2006) Mass spectrometric characterization of covalent modification of human serum albumin by 4-hydroxy-trans-2-nonenal. *J Mass Spectrom*, **41**, 1149–1161.
20. Szapacs, M.E., Riggins, J.N., Zimmerman, L.J., and Liebler, D.C. (2006) Covalent adduction of human serum albumin by 4-hydroxy-2-nonenal: kinetic analysis of competing alkylation reactions. *Biochemistry*, **45**, 10521–10528.
21. Aldini, G., Regazzoni, L., Orioli, M., Rimondi, I., Facino, M.F., and Carini, M. (2008) A tandem MS precursor-ion scan approach to identify variable covalent modification of albumin Cys34: a new tool for studying vascular carbonylation. *J Mass Spectrom*, **43**, 1470–1481.
22. Aldini, G., Vistoli, G., Regazzoni, L., Gamberoni, L., Facino, R.M., Yamaguchi, S., Uchida, K., and Carini, M. (2008) Albumin is the main nucleophilic target of human plasma: a protective role against pro-atherogenic electrophilic reactive carbonyls species? *Chem Res Toxicol*, **21**, 824–835.
23. Aldini, G., Orioli, M., and Carini, M. (2007) α, β-Unsaturated aldehydes adducts to actin and albumin as potential biomarkers of carbonylation damage. *Redox Rep*, **12**, 20–25.
24. Aldini, G., Dalle-Donne, I., Vistoli, G., Facino, R.M., and Carini, M. (2005) Covalent modification of actin by 4-hydroxy-trans-2-nonenal (HNE): LC-ESI-MS/MS evidence for Cys374 Michael adduction. *J Mass Spectrom*, **40**, 946–954.
25. Dalle-Donne, I., Carini, M., Vistoli, G., Gamberoni, L., Giustarini, D., Colombo, R., Facino, R.M., Rossi, R., Milzani, A., and Aldini, G. (2007) Actin Cys374 as a nucleophilic target of α, β-unsaturated aldehydes. *Free Radic Biol Med*, **42**, 583–598.
26. Eaton, P., Li, J.M., Hearse, D.J., and Shattock, M.J. (1999) Formation of 4-hydroxy-2-nonenal-modified proteins in ischemic rat heart. *Am J Physiol*, **276**, H935–H943.
27. Ozeki, M., Miyagawa-Hayashino, A., Akatsuka, S., Shirase, T., Lee, W.H., Uchida, K., and Toyokuni, S. (2005) Susceptibility of actin to modification by 4-hydroxy-2-nonenal. *J Chromatogr B*, **827**, 119–126.
28. Reed, T., Perluigi, M., Sultana, R., Pierce, W.M., Klein, J.B., Turner, D.M., Coccia, R., Markesbery, W.R., and Butterfield, D.A. (2008) Redox proteomic identification of 4-hydroxy-2-nonenal-modified proteins in amnestic mild cognitive impairment∼: insight into the role of lipid peroxidation in the progression and pathogenesis of Alzheimer's disease. *Neurobiol Dis*, **30**, 107–120.
29. Requena, J.R., Fu, M.X., Ahmed, M.U., Jenkins, A.J., Lyons, T.J., Baynes, J.W., and Thorpe, S.R. (1997) Quantification of malondialdehyde

and 4-hydroxynonenal adducts to lysine residues in native and oxidized low-density lipoprotein. *Biochem J*, **322**, 317–325.

30 Calingasan, N.Y., Uchida, K., and Gibson, G.E.J. (1999) Protein bound acrolein∼: a novel marker of oxidative stress in Alzheimer's disease. *Neurochemistry*, **72**, 751–756.

31 Pamplona, R., Dalfo, E., Ayala, V., Bellmunt, M.J., Prat, J., Ferrer, I., and Portero-Otin, M. (2005) Proteins in human brain cortex are modified by oxidation, glycoxidation, and lipoxidation: effects of Alzheimer disease and identification of lipoxidation targets. *J Biol Chem*, **280**, 21522–21530.

32 Pamplona, R., Naudi, A., Gavin, R., Pastrana, M.A., Sajnani, G., Ilieva, E.V., Del Rio, J.A., Portero-Otin, M., Ferrer, I., and Requena, J.R. (2008) Increased oxidation, glycoxidation, and lipoxidation of brain proteins in prion disease. *Free Radic Biol Med*, **45**, 1159–1166.

33 Orioli, M., Aldini, G., Benfatto, M.C., Facino, R.M., and Carini, M. (2007) HNE Michael adducts to histidine and histidine-containing peptides as biomarkers of lipid-derive carbonyl stress in urines: LC-MS/MS profiling in Zucker obese rats. *Anal Chem*, **79**, 9174–9184.

34 Thorpe, S.R. and Baynes, J.W. (2003) Maillard reaction products in tissue proteins: new products and new perspectives. *Amino Acids*, **25**, 275–281.

35 Szwergold, B.S., Howell, S.K., and Beisswenger, P.J. (2005) Transglycation – a potential new mechanism for deglycation of Schiff bases. *Ann N Y Acad Sci*, **1043**, 845–864.

36 Szwergold, B.S. (2006) α-Thiolamines such as cysteine and cysteamine act as effective transglycating agents due to formation of irreversible thiazolidine derivatives. *Med Hypotheses*, **66**, 698–707.

37 Goh, S.Y. and Cooper, M.E. (2008) The role of advanced glycation end products in progressions and complications of diabetes. *J Clin Endocrinol Metab*, **93**, 1143–1152.

38 Miyata, T., Sugiyama, S., Saito, A., and Kurokawa, K. (2001) Reactive carbonyl compounds related uremic toxicity ("carbonyl stress"). *Kidney Int Suppl*, **78**, S25–S31.

39 Alt, N., Carson, J.A., Alderson, N.L., Wang, Y., Nagai, R., Henle, T., Thorpe, S.R., and Baynes, J.W. (2004) Chemical modification of muscle protein in diabetes. *Arch Biochem Biophys*, **425**, 200–206.

40 Mostafa, A.A., Randell, E.W., Vasdev, S.C., Gill, V.D., Han, Y., Gadag, V., Raouf, A.A., and El Said, H. (2007) Plasma protein advanced glycation end products, carboxymethyl cysteine, and carboxyethyl cysteine, are elevated and related to nephropathy in patients with diabetes. *Mol Cell Biochem*, **302**, 35–42.

41 Zeng, J. and Davies, M.J. (2005) Evidence for formation of adducts and S-(carboxymethyl)cysteine on reaction of α-dicarbonyl compounds with thiol groups on amino acids, peptides, and proteins. *Chem Res Toxicol*, **18**, 1232–1241.

42 Thorpe, S.R. and Baynes, J.W. (2002) CML: a brief history, in *The Maillard Reaction in Food, Chemistry and Medical Science: Update for the Postgenomic Era* (eds S. Horiuchi N. Taniguchi F. Hayase T. Kurata, and T. Osawa), Elsevier, Amsterdam, pp. 91–99.

43 Zeng, J. and Davies, M.J. (2006) Protein and low molecular mass thiols as targets and inhibitors of glycation reactions. *Chem Res Toxicol*, **19**, 1668–1676.

44 Zeng, J., Dunlop, R.A., Rodgers, K.J., and Davies, M.J. (2006) Evidence for inactivation of cysteine proteases by reactive carbonyls via glycation of reactive site thiols. *Biochem J*, **398**, 197–206.

45 Mera, K., Nagai, M., Brock, J.W.C., Fujiwara, Y., Murata, T., Maruyama, T., Baynes, J.W., Otagiri, M., and Nagai, R. (2008) Glutaraldehyde is an effective cross-linker for production of antibodies against advanced glycation end-products. *J Immunol Methods*, **334**, 82–90.

46 Alderson, N.A., Wang, Y., Alt, N., Blatnik, M., Frizzell, N., Nagai, R., Walla, M.D., Carson, J.A., Thorpe, S.R., and Baynes, J.W. (2005) S-(2-succinyl)cysteine: a novel chemical modification of tissue proteins by a Krebs cycle intermediate. *Arch Biochem Biophys*, **450**, 1–8.

47 Frizzell, N., Rajesh, M., Jepson, M.J., Nagai, R., Carson, J.A., Thorpe, S.R., and Baynes, J.W., (2007) Succination of thiol groups in adipose tissue proteins in diabetes: succination inhibits polymerization and secretion of adiponectin. *J. Biol. Chem*, July 10, Epub ahead of print.

48 Blatnik, M., Frizzell, N., Thorpe, S.R., and Baynes, J.W. (2008) Inactivation of glyceraldehyde-3-phosphate dehydrogenase by fumarate in diabetes. *Diabetes*, **57**, 41–49.

49 Nagai, R., Brock, J.W.C., Blatnik, M., Baatz, J.E., Bethard, J., Walla, M.D., Thorpe, S.R., Baynes, J.W., and Frizzell, N. (2007) Succination of protein thiols during adipocyte maturation: a biomarker of mitochondrial stress. *J Biol Chem*, **282**, 34219–34228.

50 Williamson, J.R., Chang, K., Frangos, M., Hasan, K.S., Ido, Y., Kawamura, T., Nyengaard, J.R., van den Enden, M., Kilo, C., and Tilton, R.G. (1993) Hyperglycemic pseudohypoxia and diabetic complications. *Diabetes*, **42**, 801–803.

51 Gibala, M.J., Tarnopolsky, M.A., and Graham, T.E. (1997) Tricarboxylic acid cycle intermediates in human muscle at rest and during prolonged exercise. *Am J Physiol Endocrinol Metab*, **35**, E239–E244.

52 Hosogai, N., Fukuhara, A., Oshima, K., Miyata, Y., Tanaka, S., Segawa, K., Furukawa, S., Tochino, Y., Komuro, R., Matsuda, M., and Shimomura, I. (2007) Adipose tissue hypoxia in obesity and its impact on adipocytokines dysregulation. *Diabetes*, **56**, 901–911.

53 Trayhurn, P., Wang, B., and Wood, I.S. (2008) Hypoxia in adipose tissue: a basis for the dysregulation of tissue function in obesity. *Br J Nutr*, **100**, 227–237.

54 Yin, J., Gao, Z., He, Q., Zhou, D., Guo, Z.K., and Ye, J. (2009) Role of hypoxia in obesity-induced disorders of glucose and lipid metabolism in adipose tissue. *Am J Physiol Endocrinol Metab*, **296**, E333–E342.

55 Regazzetti, C., Peraldi, P., Gremeaux, T., Najem-Lendom, R., Ben-Sahra, I., Cormont, M., Bost, F., Le Marchand-Brustel, Y., Tanti, J.F., and Giorgetti-Peraldi, S. (2009) Hypoxia decreases insulin signaling pathways in adipocytes. *Diabetes*, **58**, 95–103.

56 Lin, Y., Berg, A.H., Iyengar, P., Lam, T.K.T., Giacca, A., Combs, T.P., Barzilai, N., Rhodes, C.J., Fantus, I.G., Brownlee, M., and Scherer, P.E. (2005) The hyperglycemia-induced inflammatory response in adipocytes: the role of reactive oxygen species. *J Biol Chem*, **280**, 4617–4626.

57 Passarella, S., Atlante, A., Barile, M., and Quagliariello, E. (1987) Anion transport in rat brain mitochondria: fumarate uptake via the dicarboxylate carrier. *Neurochem Res*, **12**, 255–264.

58 Finder, D.R. and Hardin, C.D. (1999) Transport and metabolism of exogenous fumarate and 3-phosphoglycerate in vascular smooth muscle. *Mol Cell Biochem*, **195**, 113–121.

59 Lai, R.K. and Goldman, P. (1992) Organic acid profiling in adipocyte differentiation of 3T3-F442A cells: increased production of Krebs cycle acid metabolites. *Metabolism*, **41**, 545–547.

60 Zhang, S., Gowda, G.A.N., Asiago, V., Shanaiah, N., Barbas, C., and Raftery, D. (2008) Correlative and quantitative 1H NMR-based metabolomics reveals specific metabolic pathway disturbances in diabetic rats. *Anal Biochem*, **383**, 76–84.

61 McDevitt, J., Wilson, S., Her, G.R., Stobiecki, M., and Goldman, P. (1990) Urinary organic acid profiles in fatty Zucker rats: indication

of impaired oxidation of butyrate and hexanoate. *Metabolism*, **39**, 1012–1020.
62 Nishikawa, T., Edelstein, D., Du, X.L., Yamagishi, S., Matsumura, T., Kaneda, Y., Yorek, M.A., Beebe, D., Oates, P.J., Hammes, H.P., Giardino, I., and Brownlee, M. (2000) Normalizing mitochondrial superoxide production blocks three pathways of hyperglycaemic damage. *Nature*, **404**, 787–790.
63 Du, X., Matsumura, T., Edelstein, D., Rossetti, L., Zsengeller, Z., Szabo, C., and Brownlee, M. (2003) Inhibition of GAPDH activity by poly(ADP-ribose) polymerase activates three major pathways of hyperglycemic damage in endothelial cells. *J Clin Invest*, **112**, 1049–1057.
64 Brownlee, M. (2005) The pathobiology of diabetic complications: a unifying mechanism. *Diabetes*, **54**, 1615–1625.
65 Blatnik, M., Thorpe, S.R., and Baynes, J.W. (2008) Succination of proteins by fumarate: mechanism of inactivation of glyceraldehyde-3-phosphate dehydrogenase in diabetes. *Ann N Y Acad Sci*, **1126**, 272–275.
66 Abdul-Ghani, M.A. and DeFronzo, R.A. (2008) Mitochondrial dysfunction, insulin resistance, and type 2 diabetes mellitus. *Curr Diab Rep*, **8**, 173–178.
67 Hojlund, K., Mogensen, M., Sahlin, K., and Beck-Nielsen, H. (2008) Mitochondrial dysfunction in type 2 diabetes and obesity. *Endocrinol Metab Clin North Am*, **37**, 713–731.
68 Wenger, R.H. (2002) Cellular adaptation to hypoxia: O2-sensing protein hydroxylases, hypoxia-inducible transcription factors, and O2-regulated gene expression. *FASEB J*, **16**, 1151–1162.
69 Aragones, J., Schneider, M., VanGeyte, K., and Carmeliet, P. (2008) Deficiency or inhibition of oxygen sensor Phd1 induces hypoxia intolerance by reprogramming basal metabolism. *Nat Genet*, **40**, 170–180.
70 Koivunen, P., Hirsila, M., Remes, A.M., Hassinen, I.E., Kivirikko, K.I., and Myllyharju, J. (2007) Inhibition of HIF hydroxylases by citric acid cycle intermediates: possible links between cell metabolism and stabilization of HIF. *J Biol Chem*, **282**, 4524–4532.
71 Xiang, X., Yuan, M., Song, Y., Ruderman, N., Wen, R., and Luo, Z. (2002) 14-3-3 facilitates insulin-stimulated trafficking of insulin receptor substrate-1. *Mol Endocrinol*, **16**, 552–562.
72 Ozcan, U., Cao, Q., Yilmaz, E., Lee, A.H., Iwakoshi, N.N., Ozdelen, E., Tuncman, G., Gorgun, C., Glimcher, L.H., and Hotamisligil, G.S. (2004) Endoplasmic reticulum stress links obesity, insulin action, and type 2 diabetes. *Science*, **306**, 457–461.
73 Pajvani, U.V., Hawkins, M., Combs, T.P., Rajala, M.W., Doebber, T., Berger, J.P., Wagner, J.A., Wu, M., Knopps, A., Xiang, A.H., Utzschneider, K.M., Kahn, S.E., Olefsky, J.M., and Scherer, P.E. (2004) Complex distribution, not absolute amount of adiponectin, correlates with thiazolidinedione-mediated improvement in insulin sensitivity. *J Biol Chem*, **279**, 12152–12162.
74 Lu, J.Y., Huang, K.C., Chang, L.C., Huang, Y.S., Chi, Y.C., Su, T.C., Chen, C.L., and Yang, W.S. (2008) Adiponectin: a biomarker of obesity-induced insulin resistance in adipose tissue and beyond. *J Biomed Sci*, **15**, 565–576.
75 Pajvani, U.B., Du, X., Combs, T.P., Berg, A.H., Rajala, M.W., Schulthess, T., Engel, J., Brownlee, M., and Scherer, P.E. (2003) Structure-function studies of the adipocyte-secreted hormone Acrp30/adiponectin. Implications for metabolic regulation and bioactivity. *J Biol Chem*, **278**, 9073–9085.
76 Townsend, D.M. (2007) S-Glutathionylation: Indicator of

cell stress and regulator of the unfolded protein response. *Mol Interv*, **7**, 313–324.

77 Zhang, D.D. (2006) Mechanistic studies of the Nr2-Keap1 signaling pathway. *Drug Metab Rev*, **38**, 769–789.

78 Forman, H.J., Fukuto, J.M., Miller, T., Zhang, H., Rinna, A., and Levy, S. (2008) The chemistry of cell signaling by reactive oxygen and nitrogen species and 4-hydroxynonenal. *Arch Biochem Biophys*, **477**, 183–195.

3
Endogenous Macromolecule Radicals

Arno G. Siraki and Marilyn Ehrenshaft

The generation and presence of free radicals within a biological milieu can lead to the formation of endogenous macromolecule radicals on DNA, lipids, and proteins. All of these reactions can have significant cellular consequences, but because proteins comprise the majority of macromolecules in the cell, they also represent the predominant target of oxidative and radical-mediated events. Here we review the chemistry leading to the production of primary free radicals and the electron transfer events between the primary radical to macromolecules to produce the secondary radical. We focus on the detection of secondary macromolecule radical products on protein and DNA using immuno-spin trapping, and discuss this approach with examples.

3.1
Introduction

Endogenous formation of macromolecule radicals is a subject that presents its own unique challenges. One of the most fundamental challenges is what should be considered background based on technical limitations and what the "true" background radical levels are. In this chapter, we briefly introduce primary free radicals (molecules that are precursors to macromolecule radicals) and then discuss electron transfer between primary free radicals and macromolecule targets (secondary free radicals). We highlight the latest techniques in detection of macromolecule (protein and DNA) free radicals, known as *immuno-spin trapping* (IST) developed by Ron Mason and coworkers, and discuss its applications. Although protein radical formation is the key component of the catalytic mechanism of many enzymes (see [1] for a comprehensive review), they will not be discussed at length here. Instead, we focus on macromolecule radical formation that is not part of the catalytic mechanism of enzyme which could have toxic consequences. We discuss the direct and indirect roles of small molecule free radicals in the generation of macromolecule free radicals (see Figure 3.1). This will provide a general introduction to endogenous free radicals, but more attention to reactive

Endogenous Toxins. Diet, Genetics, Disease and Treatment.
Edited by Peter J. O'Brien and W. Robert Bruce
Copyright © 2010 WILEY-VCH Verlag GmbH & Co. KGaA, Weinheim
ISBN: 978-3-527-32363-0

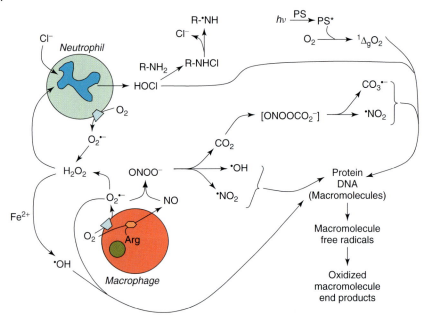

Scheme 3.1 A brief picture of some sources of endogenous free radicals that can lead to macromolecule free radicals. This illustration highlights the potential events during infection, where neutrophils and macrophages are activated by a foreign pathogen, resulting in enzyme activation. Not all sources of free radicals are shown here. (PS, photosensitizer; other abbreviations in text.)

oxygen species, hypohalides (e.g., hypochlorous acid), and nitric oxide is described by Hartmut Jaeschke in Chapter 10. Furthermore, low density lipoprotein may be considered a macromolecule due to its high molecular weight; however, it is not discussed here and the reader should refer to Chapter 25 by Steven Gieseg for details.

3.2
What is a Free Radical?

A free radical is an atom or group of atoms (molecule) with one or more unpaired electrons [2]. The free radical may be formed by the loss or gain of an electron. A free radical is usually reactive, and has a tendency to abstract an electron (or sometimes donate its electron) in order to achieve stability. The first organic free radical (triphenylmethyl radical or trityl) was synthesized by Moses Gomberg in 1900 [3]. Although met with scepticism for decades, this was a landmark breakthrough that would begin research that would eventually lead to exploring the role of free radical in health and disease. Oxygen itself is technically a free radical (or a diradical) because it contains two unpaired electrons, one in the π_x^*2p, the other electron in the π_y^*2p orbital. However, each unpaired electron has parallel spins, which does not make it a relatively reactive species. It is sometimes symbolized as $^3O_2$ to designate that it is in the triplet ground state.

3.2.1
Superoxide Anion

The most common example of the latter is the one-electron reduction (gain of an electron) of oxygen to superoxide ($O_2^{\bullet-}$). There are many sources of endogenous $O_2^{\bullet-}$, which include mitochondrial respiration and turnover of enzymes (e.g. xanthine oxidase, glycolate oxidase). Interestingly, when these normal processes are impaired, superoxide formation will increase. Mitochondria produce a flux of $O_2^{\bullet-}$ normally, but this is significantly increased when mitochondrial respiration is inhibited [4]. Also, impairment of nitric oxide synthase (by the absence of cofactors) can also produce the same result [5]. Interestingly, in both of these cases the electrons in the catalytic process are diverted to oxygen instead of the intended substrate. The cytochrome P450 enzymes have also been shown to produce $O_2^{\bullet-}$ during the oxidation of various substrates [6], and xanthine oxidase catalytically forms $O_2^{\bullet-}$ [7].

During infection, however, $O_2^{\bullet-}$ is produced by NADPH oxidase, a membrane-bound enzyme on neutrophils and macrophages. Superoxide is produced to aid in the bactericidal killing of invading pathogens. Thus, one cannot state that $O_2^{\bullet-}$ formation (or the formation of any of the free radicals discussed here) is categorically a deleterious process. In fact, individuals lacking NADPH oxidase (chronic granulomatous disease) require careful monitoring and treatment to prevent widespread infection [8]. Having said this, $O_2^{\bullet-}$ formation by NADPH oxidase has been implicated in neurotoxicity [9] and alcohol-induced liver disease [10], for example. The formation of $O_2^{\bullet-}$ can be generalized as follows:

$$O_2 + e^- \rightarrow O_2^{\bullet-}$$
$$O_2^{\bullet-} + H^+ \rightleftharpoons (HO_2^{\bullet}) \quad (pK_a = 4.88)$$

The perhydroxyl radical (protonated form of superoxide) is sometimes used but infrequently because the superoxide anion dominates at physiological pH. Superoxide anion formation is kept in check by the presence of superoxide dismutase (SOD), an enzyme that catalyzes the conversion of $O_2^{\bullet-}$ to hydrogen peroxide (H_2O_2). Although the dismutation of superoxide occurs chemically, its order of magnitude is more efficient when catalyzed enzymatically:

$$2O_2^{\bullet-} + 2H^+ \rightarrow H_2O_2 + O_2$$

The rate constant for this reaction catalyzed by SOD is 2×10^9 M^{-1} s^{-1} [11], and the nonenzymatic rate constant is 8×10^5 M^{-1} s^{-1} [12]. As shown in this reaction, H_2O_2 is formed. The latter is not a free radical; however, it has the potential to form highly reactive free radicals.

3.2.2
Singlet Oxygen

An electronically excited configuration of oxygen called *singlet oxygen* ($^1\Delta_g O_2$) has both electrons with opposite spin in the π_x^*2p orbital [13]. This can result when O_2 absorbs energy released from pigments, dyes, or photosensitizers (PSs) (e.g., rose bengal, porphyrins, St. John's Wort, etc.) after they absorb energy from light. This electronic configuration makes $^1\Delta_g O_2$ significantly more reactive than oxygen in the ground state.

$$\text{Rose Bengal} + h\nu \rightarrow [\text{Rose Bengal}]^* + O_2 \rightarrow {}^1\Delta_g O_2$$

It is expected that the toxicology of $^1\Delta g O_2$ would occur at sites that are exposed to light (eye and skin). A role for $^1\Delta g O_2$ has been proposed as a cause for age-related macular degeneration [14, 15]. Although not a free radical, $^1\Delta_g O_2$ reactivity is such that it can lead to free radical products. Singlet oxygen can add across double bonds forming endoperoxides (or dioxetanes), which break down to carbonyl compounds. It reacts with cholesterol, proteins, vitamins, and lipids, and is approximately 1500-fold more reactive toward linoleic acid than oxygen [16, 17]. For example, its reaction with histidine produces a dioxetane intermediate which leads to aspartic acid formation. The rate constants for the reaction of $^1\Delta g O_2$ with histidine, tryptophan, tyrosine, cysteine, and methionine are on the order of 10^7–10^8 M^{-1} s^{-1} [16]. It can also damage DNA, resulting in 8-hydroxyguanine (or 8-hydroxydeoxyguanosine) oxidation and strand breaks [18]. The rate constant for this reaction, however, is on the order of 10^6 M^{-1} s^{-1}, which is less than it is for the amino acids suggesting that protein may be more significant targets for $^1\Delta_g O_2$ attack.

3.2.3
Hydrogen Peroxide

In addition to the dismutation of $O_2^{\bullet-}$ discussed above, H_2O_2 is also a normal product of many endogenous reactions. Hydrogen peroxide has been experimented with in animals and plants since the nineteenth century. In fact, it was postulated as a remedy to treat diabetes because of its ability to oxidize glucose [19]. Many enzymes produce H_2O_2 as a by-product. For example, catecholamine catabolism by monoamine oxidase produces hydrogen peroxide through the following reaction:

$$RNH_2 + O_2 + H_2O \rightarrow RC{=}O + NH_3 + H_2O_2 \text{ (general)}$$

It should be noted that although H_2O_2 is indicated as the product, $O_2^{\bullet-}$ may be its precursor because the mechanism of monoamine oxidase has been suggested to be through single electron transfer [20] and a tyrosyl radical has been detected during enzyme catalysis as an intermediate [21]. Once formed, H_2O_2 can diffuse throughout the cell. Similar to superoxide anion, it also has detoxifying enzymes present in cells. Catalase was discovered first [22], followed 60 years later by glutathione (GSH) peroxidase [23]. Both enzymes catalyze the reduction of H_2O_2.

$$2H_2O_2 \rightarrow 2H_2O + O_2 \text{ (catalase)}$$

In contrast to $O_2^{\bullet-}$, however, H_2O_2 will reduce very slowly nonenzymatically, which can be increased in the present of light. There is ample evidence that H_2O_2 carries out cell signaling (see [24] for a recent review), which is not within the scope of this chapter. Although not a free radical, H_2O_2 is capable of oxidation. Acute high concentrations have shown that H_2O_2 can activate caspases and oxidize peroxiredoxins to sulfinic acid through thiol oxidation [25]. The potential harm induced by H_2O_2 is unleashed in the presence of free transition metals (typically iron intracellularly) to form hydroxyl radical ($HO^{\bullet}$):

$$Fe^{2+} + H_2O_2 \rightarrow Fe^{3+} + HO^{\bullet} + HO^{-}$$

Iron is used in the above reaction as an example; copper(II), chromium(V), cobalt(I), and vanadium(IV) can all substitute as catalysts to generate hydroxyl radicals [26]. Cerium, a lanthanide series metal, has also recently been shown to catalyze $HO^{\bullet}$ formation from H_2O_2 [27]. Hydroxyl radical is also generated upon exposure to γ-ray, X-ray, and UV irradiation causing cleavage of the water molecule:

$$H_2O + h\nu \rightarrow HO^{\bullet} + H^{\bullet}$$

The $HO^{\bullet}$ is the most reactive radical known, and will react with other molecules at diffusion limited rate.

SOD was introduced above as an enzyme that detoxifies $O_2^{\bullet-}$. However, it also carries out a reverse reaction, where H_2O_2 is consumed and causes inactivation of the enzyme. This "peroxidase activity" was observed by Hodgson and Fridovich [28], but was explained years later by the formation of $HO^{\bullet}$ that was formed by the enzyme [29].

$$\text{SOD-Cu}^{1+} + O_2^{\bullet-} + 2H^+ \rightarrow \text{SOD-Cu}^{2+} + H_2O_2 \text{ (SOD ``forward'' reaction)}$$
$$\text{SOD-Cu}^{1+} + H_2O_2 \rightarrow \text{SOD-Cu}^{2+} - {}^{\bullet}OH + HO^{-} \text{ (SOD ``peroxidase activity'')}$$

(modified from [30]).

This reaction causes specific oxidation of His-61 of SOD [31] as the $HO^{\bullet}$ is localized to the active site copper, which causes inactivation of the enzyme. However, the peroxidase activity of the enzyme is enhanced by bicarbonate (discussed later).

3.2.4
Peroxynitrite and Carbonate Radicals

Peroxynitrite ($ONOO^-$, in equilibrium with peroxynitrous acid, ONOOH) is a potential generator of free radicals. $ONOO^-$ is formed by the diffusion limited reaction (rate constant = 6.7×10^9 M^{-1} s^{-1} [32]) of superoxide with nitric oxide (NO$^\bullet$):

$$O_2^{\bullet-} + NO^\bullet \rightarrow ONOO^-$$
$$ONOO^- + H^+ \rightleftharpoons ONOOH$$

Nitric oxide is formed by nitric oxide synthase and was discovered as the endothelium-derived relaxing factor, a key player involved in vasodilation [33]. Once formed, ONOOH (pK_a = 7.5) undergoes spontaneous degradation (half-life ~1.6 ms) to form HO$^\bullet$ and nitrogen dioxide radicals ($^\bullet$NO$_2$) [34, 35]. The reaction of ONOOH with bicarbonate (in equilibrium with CO$_2$) leads to the formation of carbonate radicals (CO$_3^{\bullet-}$) [36]. This likely occurs through a nitrosoperoxycarbonate intermediate [37], and the formation of HCO$_3^\bullet$ was proven by direct electron spin resonance spectroscopy [38].

$$ONOOH \rightarrow HO^\bullet + {}^\bullet NO_2$$
$$ONOO^- + CO_2 \rightarrow [ONOOCO_2^-] \rightarrow CO_3^{\bullet-} + {}^\bullet NO_2$$

The CO$_3^{\bullet-}$ was first generated with pulse radiolysis and work was done to determine the reactivity of this radical with biological targets [39]. A great deal of interest arose about the formation of endogenous CO$_3^{\bullet-}$ as a product of the peroxidase activity of SOD, especially as it relates to familial amyotrophic lateral sclerosis (Lou Gherrig's disease). The familial disease encodes for a mutation in SOD. However, there appeared to be no difference in wild type and mutant SOD in their production of free radicals [40]. In addition, there may be a role for CO$_3^{\bullet-}$ in other neurodegenerative diseases [41]. The role of CO$_3^{\bullet-}$ in SOD peroxidase activity is as follows:

$$SOD\text{-}Cu^{2+} - {}^\bullet OH + HCO_3^- \rightarrow SOD\text{-}Cu^{2+} + H_2O + CO_3^{\bullet-}$$
$$CO_3^{\bullet-} + H^+ \leftrightarrow HCO_3^\bullet$$

The pK_a of HCO$_3^\bullet$ has been reported as 9.3–9.6 [42, 43]; however, there are some conflicts in the literature [44, 45]. The CO$_3^{\bullet-}$ will be mostly in the anionic form at physiological pH (if considered to be a base). The HO$^\bullet$ formed in the active site of SOD is too reactive to attack distal targets. Although the CO$_3^{\bullet-}$ has a relatively high oxidation potential, it is less reactive than the HO$^\bullet$ and is capable of attacking distal macromolecules.

3.2.5
Nitrogen-Based Radicals

The nitrogen dioxide radical ($^\bullet$NO$_2$) that forms from ONOOH scission may participate in protein nitration or oxidation reactions; however, this has not yet been dissociated from NO nitration of protein thiols. Because there is complex interconversion and radical–radical recombination with reactive nitrogen species, there is generally a pool of reactive nitrogen species that may carry out protein nitration, nitrosation, or oxidation. The reaction of $^\bullet NO_2$ with $NO^\bullet$ produces N_2O_3 (nitrous anhydride), which is able to nitrosate small thiols with a rate constant on the order of $10^7\,M^{-1}\,s^{-1}$ [46]. Overall, the nitrogen-based radicals or reactive nitrogen species that are capable of protein nitration are $NO^\bullet$, NO-heme, N_2O_3, N_2O_4. $^\bullet NO_2$ may mediate macromolecule nitration as well as oxidize macromolecules. They may also add to lipids undergoing lipid peroxidation leading to lipid–protein adducts. Peroxidases are also capable of oxidizing nitrite to $^\bullet NO_2$, which can recycle the pool of reactive nitrogen species.

3.2.6
Hypochlorous Acid

A portion of the bactericidal activity of neutrophils is attributed to their ability to form hypochlorous acid (HOCl), which is essentially bleach. Neutrophils possess a protein, first named verdoperoxidase for its green color [47], that is unique in its ability to catalyze HOCl formation

$$H_2O_2 + Cl^- \rightarrow H_2O + HOCl + {}^-OCl + H^+$$

As is clear from the reaction, the presence of H_2O_2 is required as a reactant. The major source of it is likely derived from the formation of $O_2^{\bullet-}$ from NADPH oxidase, which will dismutate to H_2O_2 as discussed above. Therefore, the *in vivo* partnership of NADPH oxidase and myeloperoxidase results in the formation of HOCl. Chloride (Cl^-) is abundant in plasma (100 mM), although it is not a good substrate. Other halides and pseudohalides can also be substrates [48]. HOCl is in equilibrium with hypochlorite ion (^-OCl), with a $pK_a = 7.54$ [49]. HOCl can react with Cl^- to produce molecular chlorine, Cl_2.

HOCl, Cl_2, and a chlorinated myeloperoxidase intermediate have been proposed to be the chlorinating (and oxidizing) agent(s) [50, 51]. Recent detailed work on deducing the chlorinating mechanism has been performed by Obinger's group, whose findings show that the target molecule may dictate which intermediate will chlorinate the target (either HOCl or the chlorinated myeloperoxidase intermediate) [52]. The link between macromolecule chlorination and radical formation is due to the general instability of chloramines that break down to aminyl radicals [53]:

$$R\text{-}CH_2NH_2 + HOCl \rightarrow H_2O + R\text{-}CH_2NHCl \text{ (unstable chloramine)}$$
$$R\text{-}CH_2NHCl \rightarrow R\text{-}CH_2N^\bullet H + Cl^-$$
$$[R\text{-}^\bullet CH_2\text{-}NH] \rightarrow R\text{-}CH=NH + H^+ + H_2O+ \rightarrow R\text{-}CH=O + NH_3$$

(It should be noted that there is also a two electron (nonradical) pathway that also will result in an aldehyde and ammonia without radical intermediates [54]). The aminyl radicals eventually form aldehyde products in analogy with monoamine oxidase–catalyzed reactions. This degradation reaction will take longer with taurine (a relatively stable chloramine), which is abundant in neutrophils (the very cell that produces the chlorinating agents). The high concentration of taurine in neutrophils is believed to act as a sink for chlorinating intermediates to prevent host damage [55]. However, some have speculated that taurine–chloramine may transport chlorinating potential to carry out specific oxidation at distal sites [56, 57].

Finally, it has been proposed that superoxide and Fe(II) may react with HOCl to produce HO$^\bullet$ [58, 59]:

$$O_2^{\bullet-} + HOCl \rightarrow HO^\bullet + O_2 + Cl^-$$
$$Fe(II) + HOCl \rightarrow HO^\bullet + Fe(III) + HCl$$

3.2.7
Thiyl Radicals

Although a ubiquitous antioxidant, GSH radicals have the potential to damage proteins by binding to cysteine residues, thereby forming mixed disulfides. GSH oxidation can take place through interaction with free radicals. However, these reactions are not restricted to GSH, as other endogenous thiols (especially cysteine) can be more reactive. The reactions shown here will generalize for thiyl radical formation:

$$RSH + {}^\bullet OH \rightarrow RS^\bullet + H_2O$$
$$RS^\bullet + RS^- \rightarrow RSSR^{\bullet-}$$
$$RSSR^{\bullet-} + O_2 \rightarrow; RSSR + O_2^{\bullet-}$$

These reactions produce $O_2^{\bullet-}$; however, RS$^\bullet$ itself has significant oxidizing potential. The glutathionyl radical (GS$^\bullet$) can oxidize lipids, peptides [60–63], and thymine [64]. This reaction, however, can go both ways as follows:

$$\text{(repair)} \quad GS^\bullet + \text{macromolecule-H} \leftrightarrow \text{macromolecule}^\bullet + GSH \quad \text{(damage)}$$

In this instance, although the macromolecule may be repaired by RSH, the RS$^\bullet$ product may react with oxygen as well as other targets.

$$GS^\bullet + O_2 \rightarrow GSOO^\bullet$$
$$GSOO^\bullet + GSH \rightarrow GSOOH + GS^\bullet$$

Although GSH is an essential antioxidant, the reactivity of some small molecule free radicals may be too fast for excess (exogenous) GSH to afford any additional protection. Since proteins constitute the majority of the dry weight of cells, they will likely be the first targets of the most reactive radicals (discussed later).

3.2.8
Lipid Peroxidation

Lipids are targets of small molecule free radicals but can also interact with macromolecules to induce damage. Lipid peroxides have been studied since the mid twentieth century, with publications describing increases in lipid peroxides occurring in Vitamin E deficient rats [65]. Lipid peroxidation consists of three stages: initiation, propagation, and termination. This reaction sequence has stood the test of time ever since it was proposed [66, 67]:

$$RH + \text{one-electron oxidant} \rightarrow R^\bullet + H+ \quad \text{(initiation)}$$
$$R^\bullet + O_2 \rightarrow ROO^\bullet \quad \text{(propagation)}$$
$$ROO^\bullet + RH \rightarrow ROOH + R^\bullet$$
$$R^\bullet + R^\bullet \rightarrow 2R \quad \text{(termination)}$$
$$R^\bullet + ROO^\bullet \rightarrow ROOR$$
$$R^\bullet \text{ or } ROO^\bullet + \text{antioxidant} - H \rightarrow RH \text{ or } ROOH$$

The antioxidant is considered to be tocopherol because of its ability to partition into lipid. This chain reaction has recently been suggested to include an alterative step where cross chain hydroperoxide dimers break down to increase the pool of radicals to continue propagation [68].

Because radical–radical recombination is generally a kinetically favorable reaction, there has been interest in the reaction of NO and its related redox intermediates with lipid-derived radicals [69, 70]. Since radical–radical recombination results in termination, the reaction of NO with lipid-derived radicals has been regarded as an antioxidant mechanism [71], and recent work shows important anti-inflammatory effects mediated by NO–lipid molecules [69, 72]; also reviewed in [73].

Lipid peroxides or aldehyde products of lipid peroxidation can react with protein forming lipid–protein adducts [74]. Protein radicals or ferryl-heme react with lipid (linoleic acid) [75, 76] and protein has been shown to protect lipid membranes from oxidation by thiyl radicals [60]. In addition, protein oxidation preceded lipid oxidation in U937 cells exposed to peroxyl radicals [77]. To our knowledge, the formation of protein or DNA radicals from lipid radicals has not been shown, although it is conceivable.

3.2.9
Summary

We have introduced key endogenous free radicals that have the potential to form endogenous macromolecule radicals or oxidize macromolecules. Rate constants were referred to above as an indication of the second-order reactivity between a radical and a target. A more general way to appreciate the reactivity of free radicals is to compare their redox potentials. A partial list of redox potential is shown below in Table 3.1. The more positive the redox potential is (shown in millivolts), the

Tab. 3.1 Redox potentials of endogenous small molecule free radicals

Redox couple	One-electron reduction potential (mV)
$HO^\bullet, H^+/H_2O$	+2300
$CO_3^{\bullet-}, H^+/HCO_3^-$	+1780[a]
$ONOO^-, +2H^+/{^\bullet}NO_2 + H_2O$	+1680[b]
$RO^\bullet, H^+/ROH$	+1600
$ROO^\bullet, H^+/ROOH$	+1000
$^\bullet NO_2/NO_2^-$	+990[c]
$O2^{\bullet-}, 2H^+/H_2O_2$	+940
$GS^\bullet/GS^-$	+920
$Cysteine^\bullet, +H^+/Cysteine$	+920–1100

From [78].
[a][79].
[b][80].
[c][81].

more oxidizing is the radical. As it was previously discussed, hydroxyl radical ranks at the top of the list (most reactive, therefore, highest redox potential).

3.3
Free Radicals as Initiators of Macromolecule Damage

Although this chapter is dedicated to the formation of macromolecule free radicals, the introduction to small molecule free radicals was necessary since they can be the source for generating macromolecule radicals. The tendency of these molecules is to restore their electron pair at extremely variable rates. The $HO^\bullet$ is the most reactive radical known, and will react with other molecules at diffusion limited rate. There have been examples of "specific" hydroxyl radical scavengers in the literature [82, 83], and in fact the addition of millimolar of certain chemicals *in vitro* appears to prevent $HO^\bullet$ radical-induced toxicity [84]. However, what is likely occurring is the addition of an additional exogenous target for $HO^\bullet$ attack, rather than a specific target. Having said this, there are still differences in rate constants for $HO^\bullet$ attack on different protein sites. Hydroxyl radical reacts fastest with tryptophan, tyrosine, histidine, and cysteine with a rate constant of $\sim 10^{10}$ M^{-1} s^{-1}. The least reactive amino acids toward $HO^\bullet$ are alanine, aspartate, asparagine, and glycine, which react with rate constants on the order of 10^7 M^{-1} s^{-1}. The protein backbone reacts at 10^8 M^{-1} s^{-1} [85, 86]. Less reactive radicals will exaggerate these differences. Therefore, although $HO^\bullet$ will react with any given molecule at a diffusion limited rate, there still exist differences in rate constants, which could be explained by the redox potential of the target molecule. For example, the rate constant for $HO^\bullet$ reactivity with glycine is 1.7×10^7 M^{-1} s^{-1} and for tyrosine it is 1.3×10^{10} M^{-1} s^{-1}

Tab. 3.2 Rate constants for the reaction of hydroxyl radical with selected biomolecules

Biomolecules	One-electron redox potential (mV)[a]
Glycine	1220[b]
Tyrosine	930[b]
Adenine	750[c]; 1320[d]
Cytosine	1440[d]
Guanine	620[c]; 1040[d]
Thymine	780[c]; 1290[d]

[a][87].
[b]pH 7.
[c]pH 13.
[d][88].

(Table 3.2), indicating that the latter reacts 1000-fold faster. This may be explained by the redox potential of these two amino acids: it is more favorable to oxidize tyrosine (redox potential 0.93 V) than glycine (redox potential = 1.22 V).

Rate constants for the reaction with DNA bases are much closer. The rate constant for the reaction of HO$^\bullet$ with DNA bases are on the same order (10^9). Perhaps the same argument of redox potential can be made for why 8-oxo-deoxyguanosine is used as a marker for oxidative DNA damage; as guanine has a lower oxidation potential than the other bases.

Amino acids in protein, as well as in DNA bases, have been characterized to form stable end products after free radical reactions. Table 3.3 shows a compilation of stable (and some unstable) end products that are formed from free radical reactions with amino acids as well as nucleic acids. It appears that amino acids that do not have an easily oxidizable side chain form fever end products or have not been shown to form modified products in proteins. Chains have not been shown to be oxidized in intact proteins or peptides. The more labile sites appear to be detected more readily. This may relate to the rate constant for reaction with HO$^\bullet$ as discussed above. The amino acids with the highest rate constants are shown in bold (cysteine, tryptophan, and tyrosine), and appear to have the most easily oxidizable side chains. It is difficult to categorically state this as a rule, because the rate constants shown were performed with free amino acids; peptides would be more informative because the amino and carboxylate groups would be in amide linkages, allowing "isolation" of the side chain. There is competition for the site of attack in free amino acids between the side chain and amino or carboxylate groups. In addition, the extremely high reactivity of HO$^\bullet$ also makes it difficult to distinguish what targets form end products more easily than others.

Tab. 3.3 Oxidized end products of amino and nucleic acids formed via free radicals

Amino acid	Example of free radical intermediates/mechanism	Rate constant for reaction with HO• [85] ($M^{-1} s^{-1}$)	Oxidized end product	References[a]
Alanine	HO• attack	7.7×10^7	Pyruvate[b], propionate[b], acetaldehyde[b], ethylamine[b]	[89]
Arginine	HO• attack (addition, followed by loss of guanidine)	3.5×10^9	Glutamic semialdehyde, urea, 5-hydroxy-2-aminovaleric acid	[89–94]
Asparagine	H_2O_2 oxidation	4.9×10^7	Aspartate formation (deamination)	[95]
Aspartate	HO• attack on α carbon of side chain carboxyl	7.5×10^7	Oxalic acid	[96]
Cysteine	R–S• formed via one-electron (HO•) oxidation; ONOO⁻ attack forming oxyacids, R–S•, RSO•, RSSR•⁻; $^1\Delta_g O_2$	3.4×10^{10}	Cystine, mixed disulfides, Cys-OH, Cys-OOH, Cys-O₂H (oxyacids); RSNO/RSNO₂	[96–100]
Glutamine	HO• attack	5.4×10^8	4-Hydroxyglutamic acid	[94]
Glutamate	HO• attack on γ carbon (beside carboxyl)	1.6×10^8	Oxalic acid	[96]
Glycine	HO•; HO•/CO₂•⁻ attack	1.7×10^7	Glyoxalic acid (deamination)[h]; aminomalonic acid	[89, 94]
Histidine	N- or C-centered radical via HO• attack; ONOO⁻ attack; $^1\Delta_g O_2$	5.0×10^9	2-Oxohistidine; asparagine or aspartate; (histidinyl radical)	[93,100–103]
Isoleucine	HO• attack	1.8×10^9	Hydroxyisoleucine	[94]
Leucine	HO• attack	1.7×10^9	α-Ketoisocaproate[b], isovaleraldehyde[b], and isovalerate[h]; 3- and 4-hydroxyleucine; 5-hydroxyleucine	[93, 94, 96, 104]
Lysine	HO• attack (addition, followed by deamination)	3.5×10^8	Aminoadipic semialdehyde; 3-, 4-, 5-hydroxylysine	[90, 91, 104]
Methionine	-CH₂-S•-CH₃ or -CH₂-S•-(OH)CH₃ (HO• attack); demethylation; ONOO⁻ attack; $^1\Delta_g O_2$	8.3×10^9	Sulfoxides, sulfones;	[96, 100, 105–108]

Tab. 3.3 Continued

Amino acid	Example of free radical intermediates/mechanism	Rate constant for reaction with HO• [85] ($M^{-1} s^{-1}$)	Oxidized end product	References[a]
Phenylalanine	HO• addition; ONOO⁻ attack	6.5×10^9	Dihydroxyphenylalanine derivatives; nitrophenylalanine, nitrotyrosine, dityrosine	[77, 109, 110]
Proline	HO• attack (addition, followed by ring opening)	4.8×10^8	Glutamic semialdehyde	[90, 91]
Serine	HO• attack	3.2×10^8	Glyoxal[b]	[89]
Threonine	HO• attack	5.1×10^8	γ-Methyl hydroxythreonine	[111]
Tryptophan	Indole radical (N- or C-centered) via singlet oxygen or other 1e oxidant; HO• attack (ring hydroxylation); ONOO⁻ attack; $^1\Delta_g O_2$	1.3×10^{10}	N-Formylkynurenine, kynurenine; hydroxylated derivatives; nitrotryptophan	[100, 112–116]
Tyrosine	HO• addition; phenoxyl radical and/or aromatic carbon-centered radical; ONOO⁻ attack; $^1\Delta_g O_2$	1.3×10^{10}	Dihydroxyphenylalanine; dityrosine; nitrotyrosine, chlorotyrosine	[89, 98, 100, 109, 117, 118]
Valine	HO• attack	7.6×10^8	3-Hydroxyvaline[b]	[96, 104]

DNA base				
Guanine/ guanosine	Guanine radical via $CO_3^{•-}$; •OH addition; $^1\Delta_g O_2$; HOCl + UV light (chloramine degradation)	9.2×10^9	8-oxo-Guanosine, spiroiminodihydantoin; thymine-guanine cross-link; 2,2,4-triamino-5(2H)-oxazolone; [N-centered radical][c]	[119–123]
Cytidine/ cytosine	•OH addition; HOCl + UV light (chloramine degradation)	5.6×10^9	Cytosine glycol, 5-hydroxycytosine, 5-hydroxyuracil, 6-hydroxy-5,6-dihydrouracil; [N- and C-centered radical]	[123, 124]

continued overleaf

Tab. 3.3 Continued

Amino acid	Example of free radical intermediates/mechanism	Rate constant for reaction with HO• [85] (M^{-1} s^{-1})	Oxidized end product	References[a]
Adenine/ adenosine	HOCl + UV light (chloramine degradation)	6.1×10^9	8-oxo-Adenosine, 2-hydroxyadenine, [N- and C-centered radical]	[123, 124]
Thymine	•OH addition; HOCl + UV light (chloramine degradation)	6.4×10^9	Thymine glycol, 5,6-dihydrothymine, 6-hydroxy-5,6-dihydrothymine, 5-(hydroxymethyl)-uracil [C-centered radical]	[123, 124]

[a] References indicate only where oxidized end products have been reported, but not necessarily free radical intermediates.
[b] Only for the free amino acids (not in protein). Note that free amino acids such as leucine have been shown to be oxidized by Fenton systems [125], but these appear to be deamination reactions (not the side chain), which would not likely occur in proteins unless the site of damage is the protein backbone (–CONH–).
[c] Unstable products shown in square brackets.

3.4
Macromolecule Radical Formation Induced by Small Molecule Free Radicals

Macromolecule free radicals have been detected by various techniques and different systems. Their endogenous detection, if this is considered *in vivo*, is limited by the technical difficulty in detecting macromolecule free radicals *in vivo*. Where appropriate, we will indicate where *in vivo* detection has been observed. In other cases, endogenous enzymes that have been shown to form free radicals will be described. Specifically, attention will be drawn to where IST has been applied.

The issues to be addressed first are (i) why macromolecules, and (ii) which macromolecules are targets for oxidative attack by free radicals. In order to address why macromolecules, it is an issue of rate constant and concentration. The rate constant for the reaction of HO• with albumin = 8×10^{10} M^{-1} s^{-1}; DNA = 8×10^8 M^{-1} s^{-1}; linoleic acid (lipid) = 8×10^9 M^{-1} s^{-1} [85]. As mentioned above, HO• is not a particularly selective oxidant leads us to the next point–concentration. The rate equation for a chemical reaction is shown in its simplest form:

$$r = k[BM]^n[R^\bullet]^m$$

where [BM] is the concentration of a biomolecule (protein or DNA), [R•] is the concentration of a given radical, n and m are the reaction order (which are typically second order for free radical reactions), and k is the rate constant. It is a rudimentary

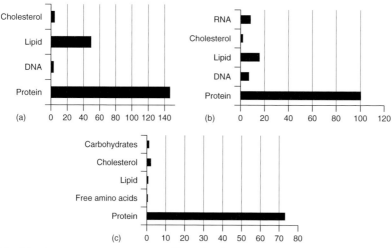

Fig. 3.1 Approximate concentration of biomolecules in (a) liver (g kg^{-1} wet weight), (b) leukocytes (g/10^{12} cells), and (c) plasma (g l^{-1}). From [86] with permission.

concept that concentration determines the rate of reaction. A useful illustration to this end is the following figure (Figure 3.1) that shows the relative amounts of macromolecules and other molecules in different cells and tissues.

As is evident from this figure, proteins appear to represent the overwhelming majority of macromolecules. As such, they will be the first target of oxidative attack. Even if we assume that GSH is present at 10 mM, this converts to 3.1 g l^{-1}, which is less than the protein concentration, although we still need to keep in mind the rate constant for a reaction rate as well.

Various techniques are used to detect free radicals. These are summarized in Table 3.4. The union between immunology and electron spin resonance spin trapping is the latest advance in the detection of macromolecules free radicals in biology (IST). There are considerable applications that have emerged as a result, which are pointed out in the following sections where applicable. It must be noted that the combination of the techniques listed in Table 3.4, just like any aspect of scientific research, provides the most information possible and allows for solid conclusions to be derived. No one technique can answer every question.

The IST approach is shown in Figure 3.2. DMPO (5,5-dimethyl-1-pyrroline N-oxide) is a spin trap, a molecule that reacts with a free radical to form a spin adduct. The macromolecule free radical (shown as •P) is usually short-lived while the spin adduct has a longer half-life. The spin adduct, however, will also lose its paramagnetic character once reduced or oxidized, which makes it undetectable by electron spin resonance. To extend the detection window for DMPO spin adducts, antiserum was made using an analog of the stable, oxidized, nitrone adduct. And, while the oxidized nitrone adduct cannot be detected by electron spin resonance because there is no unpaired electron, it can now be detected in cells and tissues using immunological methods.

Tab. 3.4 Advantages and disadvantages of techniques for the detection of free radicals

Technique	Advantages	Disadvantages
Inhibition by free radical scavengers (antioxidants)	Simple	Identity of radical must be deduced; little or no structural information
Enzyme inhibition	Simple; specific with specific enzyme inhibitors	Only the enzyme generating or detoxifying the radical can provide indirect evidence for the presence of a given radical/proradical
Pulse radiolysis	Optical detection	Radical may not be identified
Mass spectrometry	Sensitivity; provides exact molecular weight	Lacks specificity
Electron spin resonance Direct	Provides structural data	Radicals are typically short-lived or require high concentrations; not feasible with protein
Spin trapping	Provides structural data and radical stabilization	Some structural data may be lost, and high concentrations may also be needed
Immuno-spin trapping (anti-DMPO) (IST)	Molecular weight of radical known (protein); most sensitive technique; allows for versatile immunological assays	Not all radicals are trapped by DMPO; require high concentrations; DMPO will react with initiating radicals.

Selected descriptions from [126, 127].

The seminal publication that validated this method used myoglobin as a model system [128]. Myoglobin treated with H_2O_2 produces protein radicals on the surface of the protein, and in the presence of DMPO these radicals can be trapped by the latter. Detweiler *et al.* showed that the spin trapped radicals thus produced were initially detectable by ESR (Figure 3.3a). This ESR spectrum, however, was only detectable for a few minutes while an ELISA assay using the anti-DMPO antibody that was carried out after the spectrum decayed could still detect the DMPO adducts (Figure 3.3c). These EPR-silent adducts could also be detected by mass spectrometry (Figure 3.3b).

3.5
Examples and Potential Application of Immuno-Spin Trapping (IST) for the Detection of Macromolecule Radicals

3.5.1
Superoxide-Induced Macromolecule Radicals

A study on mitochondrial $O_2^{\bullet-}$ generation by isolated NADH dehydrogenase showed that not only did this enzyme produce $O_2^{\bullet-}$ but also reacted with it [129].

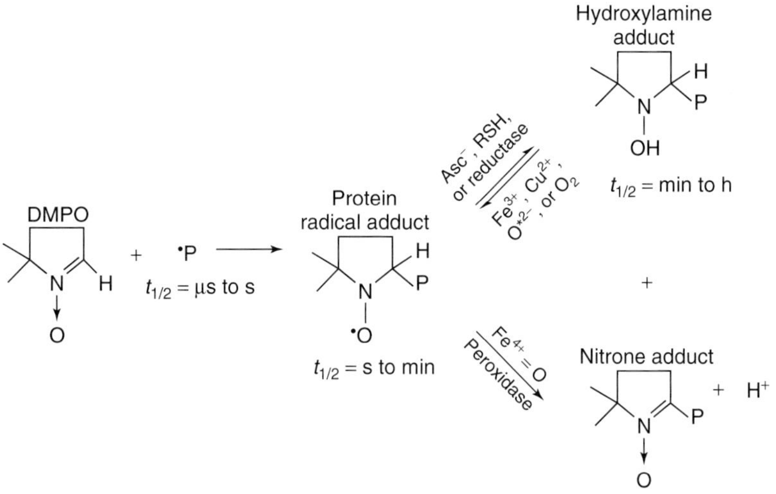

Fig. 3.2 The basis of IST. From [127] with permission.

The formation of $O_2^{•-}$ required the presence of the prosthetic FMN group as well as an intact FMN-binding domain and resulted in inactivation of the enzyme. IST showed that a DMPO adduct was formed on a 51 kDa subunit and proteolytic digests of this subunit and analysis by LC/MS/MS indicated that Cys206 and Tyr177 were the sites of radical formation.

Spinal cord mitochondria from mice expressing a transgenic SOD mutant (G93A) showed impaired respiration that was restored when DMPO was added [130]. IST showed that the mutant SOD possessed anti-DMPO antibody recognition. Nitrotyrosine was also detected, suggesting the presence of NO while mitochondria from the transgenic mutants had increased $O_2^{•-}$ formation. Either of these oxidizing species, or both, could lead to radical formation on the mutant enzyme.

3.5.2
Singlet Oxygen–Induced Macromolecule Radicals

As discussed previously, there are many different compounds, called *photosensitizers*, that absorb light and transfer their energy to oxygen forming $^1\Delta_gO_2$. Ketoprofen is a nonsteroidal anti-inflammatory drug also a photosensitizer. Upon exposure to light, ketoprofen achieves an excited state, which transfers energy to O_2 [131]. The resulting $^1\Delta_gO_2$ can damage other macromolecules that are present. Ketoprofen exposed to UVA light in the presence of hemoglobin resulted in the formation of hemoglobin radicals in red blood cells, which were detected using IST [132]. In addition, work using IST and confocal microscopy determined the site of $^1\Delta_gO_2$ generation by rose bengal in keratinocytes. This study showed that the intracellular compartmentalization of rose bengal coincided with the site of protein radical formation [133]. The application of confocal microscopy to intracellular protein

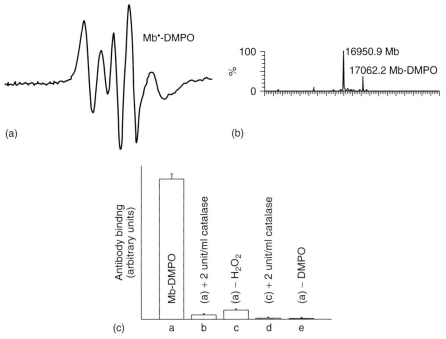

Fig. 3.3 (a) Electron spin resonance spectrum of 50 μM myoglobin, 50 μM H_2O_2, and 10 mM DMPO. From [128] with permission. (b) Mass spectrometry spectrum of a sample from Figure 3.4a, showing the presence of the DMPO adduct. (c) IST (anti-DMPO recognition) of 1 μM myoglobin, 1 μM H_2O_2, and 10 mM DMPO.

radical detection is a valuable tool for the visualization of cellular macromolecule radicals; first performed in mouse hepatocytes [134] (discussed later).

3.5.3
Hydroxyl Radical–Induced Macromolecule Radicals

A copper-catalyzed $HO^{\bullet}$-generating system was used to study the damage to human serum albumin by IST [135]. In addition, immunoassays were performed for protein carbonyl formation (using anti-dinitrophenylhydrazine) and DMPO was found to inhibit protein carbonyl formation. This suggests that protein radical formation preceded protein carbonyl formation.

IST application to DNA radical detection is also of significant interest. As shown in Table 3.3 above, $HO^{\bullet}$ attack of DNA results in 8-oxo-deoxyguanosine formation. A study was performed to compare IST with anti-8-oxo-deoxyguanosine antibody detection of DNA damage in a copper-catalyzed $HO^{\bullet}$-generating system. The results of this study using isolated DNA and rat hepatocyte nuclei treated with $Cu(II)/H_2O_2$ found that IST was more sensitive than the 8-oxo-deoxyguanosine antibody in both [136].

3.5.4
Peroxynitrite, Nitrogen Dioxide Radicals, and Nitric Oxide

The reaction rate constant for ONOO$^-$ with human serum albumin (9.7 × 10^3 M^{-1} s^{-1}) was derived and compared with that of human serum albumin with the single cysteine residue blocked. It was concluded that this residue was a major target in this protein [137]. In general, the reaction of ONOO$^-$ with proteins results in protein inactivation. A compilation of the proteins inactivated by ONOO$^-$ showed that tryptophan, tyrosine, cysteine, histidine, and methionine became modified (oxidized or nitrated) after treatment with ONOO$^-$ [138]. The authors wish to remind the reader that ONOO$^-$ decomposition results in reactive radical products (HO$^\bullet$, CO$_3^{\bullet-}$, and $^\bullet$NO$_2$), which are all capable of radical generation; as such, it is challenging to determine whether a single oxidant is the primary oxidant.

Nitrated DNA bases appear to form by reacting with ONOO$^-$. In a study that used purine derivates, it was found that once a base is nitrated (e.g., 8-nitroguanosine), a subsequent reaction with ONOO$^-$ will produce an oxidized base (8-oxoguanosine) [139]. Nitrite oxidation by myeloperoxidase/H$_2$O$_2$ has been shown to result in 8-nitroguanine or 8-nitrodeoxyguanosine [140]. By analogy, $^\bullet$NO$_2$ formed enzymatically can also react with tyrosine residues of protein to produce

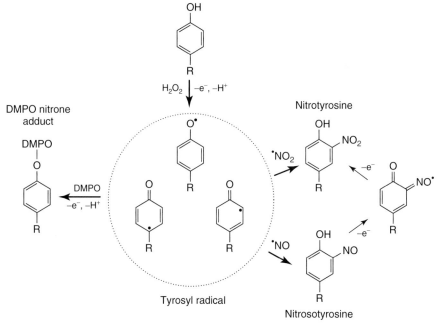

Fig. 3.4 Fate of tyrosine in sperm whale myoglobin treated with H$_2$O$_2$ and NO/$^\bullet$NO$_2$. Treatment of myoglobin with H$_2$O$_2$ generates the protein radical which can recombine with NO/$^\bullet$NO$_2$ to form nitrotyrosine. From [142] with permission.

3-nitrotyrosine. Although these nitrated products are not radicals, they have potential to form radicals. In the case of 8-nitroguanosine, reaction with $ONOO^-$ can produce radical intermediates on DNA, which rearrange to 8-oxoguanosine. 3-Nitrotyrosine has been proposed to undergo one-electron reduction analogous to many xenobiotics and drugs to a nitro anion radical capable of redox cycling and induction of oxidative stress. The reaction of $ONOO^-$ with tyrosine produces 3-nitrotyrosine and dityrosine; both species are produced through radical intermediates [98]. To the author's knowledge, IST has not been performed in this system.

The reaction of $ONOO^-$ with hemoglobin results in the displacement of O_2 from the heme and coordination of $ONOO^-$ in the active site. The –O–O– bond breaks through homolysis, resulting in nitrite formation, methemoglobin, and to a lesser extent ferryl-heme and $^\bullet NO_2$. IST analysis of the products of the reaction of peroxynitrite with hemoglobin reveals the formation of protein radicals [141].

Another study investigated the reactions of NO and $^\bullet NO_2$ with the tyrosyl radical formed by the treatment of sperm whale myoglobin with H_2O_2 [142]. IST showed that addition of either NO or $^\bullet NO_2$ could attenuate DMPO adduct formation, and in the absence of DMPO myoglobin was nitrated via the formation of nitrotyrosine. Interestingly, nitrotyrosine formation was abolished in the presence of DMPO, suggesting that tyrosine radical formation precedes nitrotyrosine formation. The mechanism proposed is shown in Figure 3.4.

3.5.5
Carbonate Radicals

$CO_3^{\bullet-}$ oxidizes deoxyguanosine (free or in an oligonucleotide) to form radicals leading to intrastrand DNA cross-links [120]. $CO_3^{\bullet-}$ attack on 5'-d(CCTAC**G**CTACC) results in the formation of spiroiminodihydantoin and a guanine–thymine cross-link [143]. In the presence of $^\bullet NO_2$, a nitro adduct can be formed resulting in 8-nitroguanosine adducts [144], which was reported to follow $CO_3^{\bullet-}$ oxidation of guanosine to a guanosine radical, with radical–radical recombination between it and $^\bullet NO_2$. To date, IST has not been applied for the detection of DNA radicals formed by $CO_3^{\bullet-}$.

In another study, the peroxidase activity of SOD showed that the addition of H_2O_2 to SOD resulted in protein radicals, which was detected using IST [145]. In addition, the inclusion of bicarbonate in the reaction enhanced protein radical formation, implying that $CO_3^{\bullet-}$ was formed and reacted with the protein. These results support the concept that bicarbonate oxidation by the active site bound $HO^\bullet$ results in $CO_3^{\bullet-}$ that can oxidize distal sites (in this case, sites on the same protein removed from the active site). These findings were recently extended to show that the presence of bicarbonate resulted in protein radical formation, but prevented protein fragmentation. This suggests that the active site $HO^\bullet$ plays a role in fragmentation, and that the $CO_3^{\bullet-}$ has a different site of attack [146].

3.5.6
HOCl-Induced Macromolecule Radicals

Treatment of cytochrome c with HOCl was shown to result in aggregation and protein radical formation [147]. Derivatization of the amino acid side chains (lysine was succinylated and tyrosine was iodinated) was used to determine that radicals were formed on lysine and to a lesser extent on tyrosine residues. Figure 3.5 uses IST to show the extent of attenuation of free radical and protein aggregation when either lysine or tyrosine is derivatized.

These results suggest that HOCl chlorinated the lysine residues to chloramines, which then degraded to nitrogen centered radicals. It was proposed that tyrosine oxidation could occur via electron transfer from the tyrosine to the lysine residue or via a ferryl-heme (analogous to H_2O_2).

Another study that may relate to the formation of ferryl-heme as a generator of protein radicals used confocal microscopy to visualize radicals formed within cells [134]. A hypothesis proposed in this article was that inactivation of catalase by HOCl (reported first by [148]) could be a mechanism to increase the susceptibility of an invading pathogen to the damaging effects of HOCl. This study used IST to show that HOCl reaction with catalase resulted in protein aggregation and the production of radicals (Figure 3.6). Investigation of the cellular localization of the catalase radicals showed that they appeared in the peroxisomes, as expected (Figure 3.7). Interestingly, in catalase knockout mice, the protein radicals were localized to the cell membrane, clearly indicating specificity for HOCl-induced damage. Catalase was proposed to form a ferryl-heme species when treated with

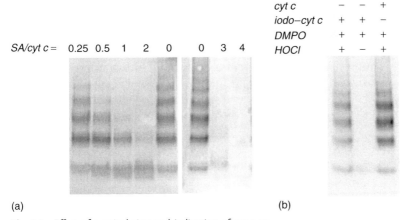

(a) (b)

Fig. 3.5 Effect of succinylation and iodination of cyt c on HOCl-induced protein radicals and protein aggregates detected by Western blot using anti-DMPO nitrone adduct polyclonal antibody. (a) Increasing succinylation decreased protein radical formation (anti-DMPO detection) and (b) iodination of cytochrome c decreased protein radical formation. Used with permission from [147].

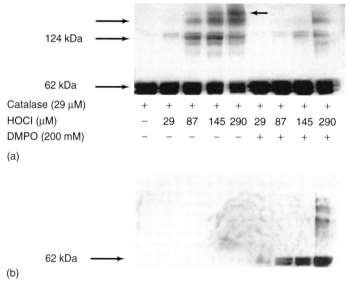

Fig. 3.6 SDS-PAGE analysis of incubations of catalase (29 μM in heme) exposed to different HOCl concentrations (29–290 μM) in the absence or in the presence of DMPO (200 mM) at pH 6.9. (a) Gel was stained for protein using Coomassie brilliant blue dye. (b) Same as (a), but the autoradiograph picture of catalase Western blot, using anti-DMPO as the primary antibody, is shown. From [134] with permission.

HOCl, which would then cause electron transfer and formation of a tyrosyl radical that was trapped by DMPO. Although this is not the chloramine pathway to radical formation as discussed previously, it represents an interesting effect of HOCl.

HOCl has also been shown to induce DNA damage (in isolated DNA or bases) and chlorinated bases have been proposed as markers for HOCl-induced DNA damage [149]. Analogous to protein damage, chloramines generated by the reaction of HOCl with the amine group of DNA bases results in products that will eventually decay to radicals [123, 150]. It would be interesting if these findings could be carried out *in vitro* to determine the conditions for generating DNA radicals within intact cells.

3.5.7
Thiyl Radical–Induced Macromolecule Radicals

A lactoperoxidase protein radical was detected in the presence of GSH using IST [151]. This finding has been attributed in part to the GS$^\bullet$, which was shown to be formed by lactoperoxidase using electron spin resonance. SOD appeared to increase, and catalase to decrease protein radical formation, suggesting that H_2O_2 has a role in generation of the protein free radical. Indeed, H_2O_2 has been shown to induce self-peroxidation of lactoperoxidase resulting in a protein radical and lactoperoxidase cross-linking through tyrosine-289 [152]. Although a GS$^\bullet$ may be capable of oxidizing amino acid residues, the effect of H_2O_2 cannot be ignored.

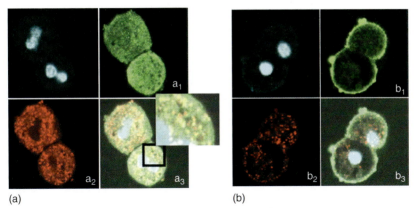

Fig. 3.7 Confocal microscopy images of the colocalization of catalase (red stain) and protein-DMPO adducts (green stain) obtained by treating mouse hepatocytes (2.5 × 10^6 cells/ml) with HOCl. (a) Cells were treated with three pulses of HOCl (20 µM, 30-minute intervals) in the presence of DMPO: (a_1) anti-DMPO, (a_2) anti-catalase, and (a_3) the overlay of (a_1) and (a_2). (b) The same as in (a) but using hepatocytes from catalase knockout mice. Note that in (b_3) only the membrane shows anti-DMPO binding with little or no colocalization with anti-catalase. Cells in (b_2) show very little catalase expression. From [134] with permission.

Recall that thiyl radicals can react with a parent thiol forming disulfide radical anion and superoxide, and the disproportionation of superoxide forms H_2O_2:

$$RSSR^{\bullet-} + O_2 \rightarrow RSSR + O_2^{\bullet-}$$

$$2O_2^{\bullet-} + 2H^+ \rightarrow H_2O_2 + O_2$$

$$\text{Lactoperoxidase} + H_2O_2 \rightarrow \text{[lactoperoxidase compound I]}$$

$$\rightarrow \text{lactoperoxidase}^{\bullet} \text{ AND/OR}$$

$$GS^{\bullet} + \text{lactoperoxidase} \rightarrow \text{lactoperoxidase}^{\bullet} + GSH$$

Therefore, the lactoperoxidase radical formation in this system may be indirectly related to thiyl radicals since the latter generates H_2O_2, although the $GS^{\bullet}$ may also have a role.

3.5.8
Myeloperoxidase Protein Radical Detection

Recent studies with human myeloperoxidase, which is found in neutrophils, showed that metabolism of aromatic amine (AA) drugs could induce the formation of a protein free radical on this enzyme [153, 154]. In contrast to the self-peroxidation reaction of lactoperoxidase which required only H_2O_2 to form a protein free radical, with myeloperoxidase an exogenous substrate was required:

Myeloperoxidase + H_2O_2 → [myeloperoxidase compound I] + H_2O

[Myeloperoxidase compound I] + AA → [myeloperoxidase compound II] + AA radical

AA radical + myeloperoxidase → myeloperoxidase radical + AA

In promyelocytic leukemia cells (HL-60), the presence of an AA and H_2O_2 resulted in the intracellular formation and detection of a myeloperoxidase protein free radical, suggesting that this reaction could occur endogenously. The role of a myeloperoxidase protein free radical is currently unknown since it was found that this reaction did not cause enzyme inactivation.

3.5.9
Heme Protein Radicals – A Unique Exception of Substrate-Induced Protein Radical Formation

IST has also shown and confirmed that heme enzymes which reacted with H_2O_2 typically produced protein radicals. This has been known for some time using electron spin resonance; however, the protein concentrations were high as is required by this technique. IST, on the other hand, has allowed for the investigation of proteins that are difficult to obtain in high concentration, such as thyroid peroxidase (see below), which is a reflection of the sensitivity of this technique.

In vitro experiments with hemoglobin treated with H_2O_2 showed that protein radicals could be detected using IST [155]. In this work, the amino acids suspected to be involved in radical formation were tyrosine and cysteine; subsequent work showed that the DMPO was bound to tyrosine-42, tyrosine-24, cysteine-93, and histidine-20 [156]. Blocking the tyrosine residue resulted in abolishing the detection of protein radicals, whereas cysteine blocking was not as significant in preventing protein radical detection. The treatment of hemoglobin with H_2O_2 results in the formation of ferryl-heme and a protein radical – there is no direct attack by a radical species, but rather the transfer of electrons from the protein surface (summarized in Figure 3.8). The treatment of red blood cells with H_2O_2 also shows a dose-dependent increase in hemoglobin radical formation (Figure 3.9).

Another interesting example of H_2O_2-driven radical formation of protein is the case of thyroid peroxidase [157]. The endogenous substrate for thyroid peroxidase is I^- (a reason for the iodination of table salt). As with other peroxidases, H_2O_2 is a required cosubstrate which combined with I^- results in tyrosine iodination in thyroglobulin. However, in the absence of I^- but in the presence of H_2O_2, it was shown using IST that thyroid peroxidase formed a protein radical (Figures 3.10 and 3.11). When I^- was added back to the reaction, there was a dose-dependent inhibition of protein radical formation. Derivatization of tyrosine, tryptophan, and cysteine residues showed that protein radical formation was attenuated only when the tyrosine residues were blocked. The authors speculated that the formation of

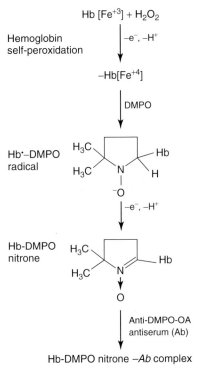

Fig. 3.8 Sequence of events when hemoglobin is treated with H_2O_2 and analyzed by IST. From [153] with permission.

oxidized thyroid peroxidase could be immunogenic and hypothesized that iodine deficiency could induce this effect *in vivo*.

Human neuroglobin is a recently discovered member of the globin family and similar to myoglobin and hemoglobin, it can bind O_2, NO, and CO. As its name suggests, it is present in the brains of various mammals. A unique feature of neuroglobin is that it is hexacoordinated without an external ligand (the sixth ligand is from histidine). A mutant enzyme was produced (H64V) and studies showed that reaction of the mutant (but not wild-type enzyme) with H_2O_2 resulted in protein aggregation (via tyrosine-88) and protein radical formation as detected by IST [158]. This is yet another example of H_2O_2-induced heme protein radical formation.

3.5.10
Example of *In vivo* Detection of Macromolecule Radicals Using IST

Recent work in a mouse sepsis model has shown protein radical formation in the spleen [159]. Carboxypeptidase B1-DMPO adducts were identified in spleens of LPS-treated mice. Interestingly, in sham-treated, control mice a significant amount of carboxypeptidase B1-DMPO adducts was detected, suggesting that this

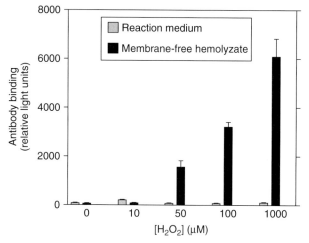

Fig. 3.9 IST detection of hemoglobin-derived nitrone adducts in red blood cells exposed to H_2O_2. From [153] with permission.

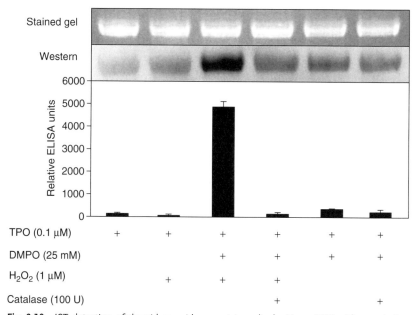

Fig. 3.10 IST detection of thyroid peroxidase protein radicals. From [155] with permission.

protein forms free radicals under normal physiological conditions. The enzymes that played a role in protein radical formation were xanthine oxidase (xanthine oxidase knockout mice showed reduced protein radical formation) and nitric oxide synthase. Since xanthine oxidase forms $O_2^{\bullet -}$, combination with NO would result

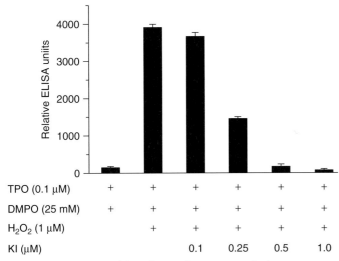

Fig. 3.11 Attenuation of thyroid peroxidase protein radicals by iodide as detected by IST. From [155] with permission.

in $ONOO^-$, which may have a role as a mediator of radical formation (nitrotyrosine was also detected).

3.5.11
Future Directions

Detection of macromolecule free radicals in living cells, tissues, and organisms is a continuing challenge. Although the IST technique developed by Mason and coworkers is a considerable advance, more research is needed to explore the biological significance of macromolecule radicals. One of the key hurdles is the identification of the site of radical formation. The success of this characterization depends on the protein in question, and a method has been developed to address this problem [160]. Protein and DNA oxidation is, in general, a toxic event. Macromolecule radicals will form stable end products. DNA radicals rapidly form 8-oxo(hydroxy)-dG, and protein oxidation can lead to aldehydes. With DNA, the toxic outcome of mutagenesis is possible, but the consequences of protein oxidation can only be dealt with on a case-by-case basis because of the unique function of each protein. In many, but not all cases, protein radical formation results in protein inactivation. Protein oxidation generally results in protein catabolism via ubiquitination and proteasomal activation; however, there may be signaling events or radial transfer reactions that mediate the toxic outcome. The future research in this area should focus on the biological events triggered by macromolecule radicals, a basic template for which is shown in Figure 3.12.

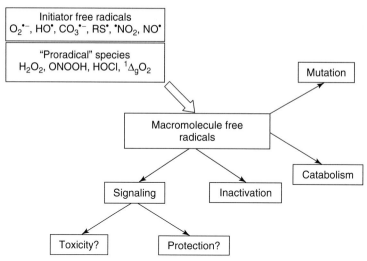

Fig. 3.12 Initiators of macromolecule free radical formation and its general consequences.

3.6
Immunological Detection of Oxidized Tryptophan Residues in Proteins

A newly developed antiserum is aimed at detecting oxidatively modified tryptophan residues in proteins in a manner comparable to that of IST [161]. Tryptophan, whether as a free amino acid or as a residue within a peptide or a protein, can react with free radicals, $^1\Delta_g O_2$, or ozone to form a derivative, N-formylkynurenine, with a cleaved indole ring. Initial antiserum validation experiments have shown that the new anti-N-formylkynurenine antiserum is sensitive enough to detect N-formylkynurenine in proteins with as few as one or two tryptophan residues and in mixtures of proteins and in cells as depicted in the following reaction.

(A synthetic hapten was used for production of antiserum to detect N-Formylkynurenine)

References

1 Stubbe, J. and van der Donk, W.A. (1998) Protein radicals in enzyme catalysis. *Chem Rev*, **98**, 705–762.

2 Koppenol, W.H. (1990) What is in a name? Rules for radicals. *Free Radic Biol Med*, **9**, 225–227.

3 Gomberg, M. (1900) An Instance of Trivalent Carbon: Triphenylmethyl, pp. 757–771.
4 Turrens, J.F. (2003) Mitochondrial formation of reactive oxygen species. *J Physiol*, **552**, 335–344.
5 Vasquez-Vivar, J., Kalyanaraman, B., Martasek, P., Hogg, N., Masters, B.S., Karoui, H., Tordo, P., and Pritchard, K.A. Jr (1998) Superoxide generation by endothelial nitric oxide synthase: the influence of cofactors. *Proc Natl Acad Sci U S A*, **95**, 9220–9225.
6 Dybing, E., Nelson, S.D., and Mitchell, J.R. (1976) Oxidation of α-methyldopa and other catechols by cytochrome P 450 generated superoxide anion: possible mechanism of methyldopa hepatitis. *Mol Pharmacol*, **12**, 911–920.
7 Fridovich, I. (1970) Quantitative aspects of the production of superoxide anion radical by milk xanthine oxidase. *J Biol Chem*, **245**, 4053–4057.
8 Roos, D. (1994) The genetic basis of chronic granulomatous disease. *Immunol Rev*, **138**, 121–157.
9 Qin, L., Liu, Y., Wang, T., Wei, S.J., Block, M.L., Wilson, B., Liu, B., and Hong, J.S. (2004) NADPH oxidase mediates lipopolysaccharide-induced neurotoxicity and proinflammatory gene expression in activated microglia. *J Biol Chem*, **279**, 1415–1421.
10 Kono, H., Rusyn, I., Yin, M., Gabele, E., Yamashina, S., Dikalova, A., Kadiiska, M.B., Connor, H.D., Mason, R.P., Segal, B.H., Bradford, B.U., Holland, S.M., and Thurman, R.G. (2000) NADPH oxidase-derived free radicals are key oxidants in alcohol-induced liver disease. *J Clin Invest*, **106**, 867–872.
11 Michel, E., Nauser, T., Sutter, B., Bounds, P.L., and Koppenol, W.H. (2005) Kinetics properties of Cu,Zn-superoxide dismutase as a function of metal content. *Arch Biochem Biophys*, **439**, 234–240.
12 Plonka, A., Mayer, J., Metodiewa, D., Gebicki, J.L., Zgirski, A., and Grabska, M. (1986) Superoxide radical dismutation by copper proteins. *J Radioanal Nucl Chem Articles*, **101**, 221–225.
13 Halliwell, B. and Gutteridge, J.M. (1990) Role of free radicals and catalytic metal ions in human disease: an overview. *Methods Enzymol*, **186**, 1–85.
14 Beatty, S., Koh, H.H., Phil, M., Henson, D., and Boulton, M. (2000) The role of oxidative stress in the pathogenesis of age-related macular degeneration. *Surv Ophthalmol*, **45**, 115–134.
15 Schey, K.L., Patat, S., Chignell, C.F., Datillo, M., Wang, R.H., and Roberts, J.E. (2000) Photooxidation of lens α-crystallin by hypericin (active ingredient in St. John's Wort). *Photochem Photobiol*, **72**, 200–203.
16 Choe, E. and Min, D.B. (2005) Chemistry and reactions of reactive oxygen species in foods. *J Food Sci*, **70**.
17 Morgan, P.E., Pattison, D.I., Hawkins, C.L., and Davies, M.J. (2008) Separation, detection, and quantification of hydroperoxides formed at side-chain and backbone sites on amino acids, peptides, and proteins. *Free Radic Biol Med*, **45**, 1279–1289.
18 Devasagayam, T.P.A., Steenken, S., Obendorf, M.S.W., Schulz, W.A., and Sies, H. (1991) Formation of 8-hydroxy(deoxy)guanosine and generation of strand breaks at guanine residues in DNA by singlet oxygen. *Biochemistry*, **30**, 6283–6289.
19 Day, J. (1868) Peroxide of hydrogen as a remedy in diabetes. *Lancet*, **91**, 45–46.
20 Silverman, R.B., Hoffman, S.J., and Catus, W.B. (1980) A mechanism for mitochondrial monoamine oxidase catalyzed amine oxidation. *J Am Chem Soc*, **102**, 7126–7128.
21 Rigby, S.E., Hynson, R.M., Ramsay, R.R., Munro, A.W., and Scrutton, N.S. (2005) A stable tyrosyl radical in monoamine oxidase A. *J Biol Chem*, **280**, 4627–4631.

22 May, D.W. (1901) Catalase, a new enzyme of general occurrence. *Science*, **14**, 815–816.

23 Mills, G.C. (1960) Glutathione peroxidase and the destruction of hydrogen peroxide in animal tissues. *Arch Biochem Biophys*, **86**, 1–5.

24 Veal, E.A., Day, A.M., and Morgan, B.A. (2007) Hydrogen peroxide sensing and signaling. *Mol Cell*, **26**, 1–14.

25 Cox, A.G., Pullar, J.M., Hughes, G., Ledgerwood, E.C., and Hampton, M.B. (2008) Oxidation of mitochondrial peroxiredoxin 3 during the initiation of receptor-mediated apoptosis. *Free Radic Biol Med*, **44**, 1001–1009.

26 Valko, M., Morris, H., and Cronin, M.T. (2005) Metals, toxicity and oxidative stress. *Curr Med Chem*, **12**, 1161–1208.

27 Heckert, E.G., Seal, S., and Self, W.T. (2008) Fenton-like reaction catalyzed by the rare earth inner transition metal cerium. *Environ Sci Technol*, **42**, 5014–5019.

28 Hodgson, E.K. and Fridovich, I. (1975) The interaction of bovine erythrocyte superoxide dismutase with hydrogen peroxide: inactivation of the enzyme. *Biochemistry*, **14**, 5294–5299.

29 Yim, M.B., Chock, P.B., and Stadtman, E.R. (1990) Copper, zinc superoxide dismutase catalyzes hydroxyl radical production from hydrogen peroxide. *Proc Natl Acad Sci U S A*, **87**, 5006–5010.

30 Goss, S.P., Singh, R.J., and Kalyanaraman, B. (1999) Bicarbonate enhances the peroxidase activity of Cu,Zn-superoxide dismutase. Role of carbonate anion radical. *J Biol Chem*, **274**, 28233–28239.

31 Uchida, K. and Kawakishi, S. (1994) Identification of oxidized histidine generated at the active site of Cu,Zn-superoxide dismutase exposed to H2O2. Selective generation of 2-oxo-histidine at the histidine 118. *J Biol Chem*, **269**, 2405–2410.

32 Huie, R.E. and Padmaja, S. (1993) The reaction of no with superoxide. *Free Radic Res Commun*, **18**, 195–199.

33 Moncada, S., Palmer, R.M., and Higgs, E.A. (1991) Nitric oxide: physiology, pathophysiology, and pharmacology. *Pharmacol Rev*, **43**, 109–142.

34 Beckman, J.S., Beckman, T.W., Chen, J., Marshall, P.A., and Freeman, B.A. (1990) Apparent hydroxyl radical production by peroxynitrite: implications for endothelial injury from nitric oxide and superoxide. *Proc Natl Acad Sci U S A*, **87**, 1620–1624.

35 Augusto, O., Gatti, R.M., and Radi, R. (1994) Spin-trapping studies of peroxynitrite decomposition and of 3-morpholinosydnonimine N-ethylcarbamide autooxidation: direct evidence for metal-independent formation of free radical intermediates. *Arch Biochem Biophys*, **310**, 118–125.

36 Radi, R., Cosgrove, T.P., Beckman, J.S., and Freeman, B.A. (1993) Peroxynitrite-induced luminol chemiluminescence. *Biochem J*, **290** (Pt 1), 51–57.

37 Denicola, A., Freeman, B.A., Trujillo, M., and Radi, R. (1996) Peroxynitrite reaction with carbon dioxide/bicarbonate: kinetics and influence on peroxynitrite-mediated oxidations. *Arch Biochem Biophys*, **333**, 49–58.

38 Bonini, M.G., Radi, R., Ferrer-Sueta, G., Ferreira, A.M.D.C., and Augusto, O. (1999) Direct EPR detection of the carbonate radical anion produced from peroxynitrite and carbon dioxide. *J Biol Chem*, **274**, 10802–10806.

39 Adams, G.E., Aldrich, J.E., Bisby, R.H., Cundall, R.B., Redpath, J.L., and Willson, R.L. (1972) Selective free radical reactions with proteins and enzymes: reactions of inorganic radical anions with amino acids. *Radiat Res*, **49**, 278–289.

40 Singh, R.J., Karoui, H., Gunther, M.R., Beckman, J.S., Mason, R.P.,

and Kalyanaraman, B. (1998) Re-examination of the mechanism of hydroxyl radical adducts formed from the reaction between familial amyotrophic lateral sclerosis-associated Cu,Zn superoxide dismutase mutants and H2O2. *Proc Natl Acad Sci U S A*, **95**, 6675–6680.

41 Zhang, H., Andrekopoulos, C., Joseph, J., Crow, J., and Kalyanaraman, B. (2004) The carbonate radical anion-induced covalent aggregation of human copper, zinc superoxide dismutase, and α-synuclein: intermediacy of tryptophan- and tyrosine-derived oxidation products. *Free Radic Biol Med*, **36**, 1355–1365.

42 Chen, S.N. and Hoffman, M.Z. (1974) Reactivity of the carbonate radical in aqueous solution. Tryptophan and its derivatives. *J Phys Chem*, **78**, 2099–2102.

43 Wu, G., Katsumura, Y., Muroya, Y., Lin, M., and Morioka, T. (2002) Temperature Dependence of Carbonate Radical in NaHCO3 and Na2CO3 Solutions: Is the Radical a Single Anion? pp. 2430–2437.

44 Eriksen, T.E., Lind, J., and Merenyi, G. (1985) On the acid-base equilibrium of the carbonate radical. *Radiat Phys Chem*, **26**, 197–199.

45 Armstrong, D.A., Waltz, W.L., and Rauk, A. (2006) Carbonate radical anion–thermochemistry. *Can J Chem*, **84**, 1614–1619.

46 Keshive, M., Singh, S., Wishnok, J.S., Tannenbaum, S.R., and Deen, W.M. (1996) Kinetics of S-nitrosation of thiols in nitric oxide solutions. *Chem Res Toxicol*, **9**, 988–993.

47 Agner, K. (1941) Verdoperoxidase: a ferment isolated from leukocytes. *Acta Physiol Scand*, **2**, 1–62.

48 Dunford, H.B. (1999) *Heme Peroxidases*, John Wiley & Sons, Inc., New York.

49 Carrell Morris, J. (1966) The acid ionization constant of HOCl from 5 to 35°. *J Phys Chem*, **70**, 3798–3805.

50 Harrison, J.E. and Schultz, J. (1976) Studies on the chlorinating activity of myeloperoxidase. *J Biol Chem*, **251**, 1371–1374.

51 Marquez, L.A. and Dunford, H.B. (1994) Chlorination of taurine by myeloperoxidase. Kinetic evidence for an enzyme-bound intermediate. *J Biol Chem*, **269**, 7950–7956.

52 Ramos, D.R., Garcia, M.V., Canle, L.M., Santaballa, J.A., Furtmuller, P.G., and Obinger, C. (2008) Myeloperoxidase-catalyzed chlorination: the quest for the active species. *J Inorg Biochem*, **102**, 1300–1311.

53 Hawkins, C.L. and Davies, M.J. (1998) Hypochlorite-induced damage to proteins: formation of nitrogen-centred radicals from lysine residues and their role in protein fragmentation. *Biochem J*, **332** (Pt 3), 617–625.

54 Antelo, J.M., Arce, F., and Parajó, M. (1996) Kinetic study of the decomposition of N-chloramines. *J Phys Org Chem*, **9**, 447–454.

55 Marcinkiewicz, J., Grabowska, A., Bereta, J., Bryniarski, K., and Nowak, B. (1998) Taurine chloramine down-regulates the generation of murine neutrophil inflammatory mediators. *Immunopharmacology*, **40**, 27–38.

56 Midwinter, R.G., Cheah, F.C., Moskovitz, J., Vissers, M.C., and Winterbourn, C.C. (2006) IkappaB is a sensitive target for oxidation by cell-permeable chloramines: inhibition of NF-kappaB activity by glycine chloramine through methionine oxidation. *Biochem J*, **396**, 71–78.

57 Peskin, A.V. and Winterbourn, C.C. (2006) Taurine chloramine is more selective than hypochlorous acid at targeting critical cysteines and inactivating creatine kinase and glyceraldehyde-3-phosphate dehydrogenase. *Free Radic Biol Med*, **40**, 45–53.

58 Candeias, L.P., Patel, K.B., Stratford, M.R., and Wardman, P. (1993) Free hydroxyl radicals are formed on reaction between the neutrophil-derived species superoxide anion and hypochlorous acid. *FEBS Lett*, **333**, 151–153.

59 Candeias, L.P., Stratford, M.R., and Wardman, P. (1994) Formation of hydroxyl radicals on reaction of hypochlorous acid with ferrocyanide, a model iron(II) complex. *Free Radic Res*, **20**, 241–249.

60 Tweeddale, H.J., Kondo, M., and Gebicki, J.M. (2007) Proteins protect lipid membranes from oxidation by thiyl radicals. *Arch Biochem Biophys*, **459**, 151–158.

61 Nauser, T., Pelling, J., and Schoneich, C. (2004) Thiyl radical reaction with amino acid side chains: rate constants for hydrogen transfer and relevance for posttranslational protein modification. *Chem Res Toxicol*, **17**, 1323–1328.

62 Nauser, T. and Schöneich, C. (2003) Thiyl radicals abstract hydrogen atoms from the αC-H bonds in model peptides: absolute rate constants and effect of amino acid structure. *J Am Chem Soc*, **125**, 2042–2043.

63 Schoneich, C. (2008) Mechanisms of protein damage induced by cysteine thiyl radical formation, *Chem Res Toxicol*, **21**, 1175–1179.

64 Nauser, T. and Schoneich, C. (2003) Thiyl radical reaction with thymine: absolute rate constant for hydrogen abstraction and comparison to benzylic C-H bonds. *Chem Res Toxicol*, **16**, 1056–1061.

65 Dam, H. and Granados, H. (1945) Role of unsaturated fatty acids in changes of adipose and dental tissues in vitamin E deficiency. *Science*, **102**, 327–328.

66 Bolland, J.L. (1946) Kinetic studies in the chemistry of rubber and related materials. I. The thermal oxidation of ethyl linoleate. *Proc R Soc Lond A Math Phys Sci*, **186**, 218–236.

67 Bateman, L. and Gee, G. (1948) A kinetic investigation of the photochemical oxidation of certain non-conjugated olefins. *Proc R Soc Lond A Math Phys Sci*, **195**, 376–391.

68 Schneider, C., Porter, N.A., and Brash, A.R. (2008) Routes to 4-hydroxynonenal: fundamental issues in the mechanisms of lipid peroxidation. *J Biol Chem*, **283**, 15539–15543.

69 Cui, T., Schopfer, F.J., Zhang, J., Chen, K., Ichikawa, T., Baker, P.R.S., Batthyany, C., Chacko, B.K., Feng, X., Patel, R.P., Agarwal, A., Freeman, B.A., and Chen, Y.E. (2006) Nitrated fatty acids: endogenous anti-inflammatory signaling mediators. *J Biol Chem*, **281**, 35686–35698.

70 O'Donnell, V.B., Eiserich, J.P., Chumley, P.H., Jablonsky, M.J., Krishna, N.R., Kirk, M., Barnes, S., Darley-Usmar, V.M., and Freeman, B.A. (1999) Nitration of unsaturated fatty acids by nitric oxide-derived reactive nitrogen species peroxynitrite, nitrous acid, nitrogen dioxide, and nitronium ion. *Chem Res Toxicol*, **12**, 83–92.

71 Rubbo, H., Parthasarathy, S., Barnes, S., Kirk, M., Kalyanaraman, B., and Freeman, B.A. (1995) Nitric oxide inhibition of lipoxygenase-dependent liposome and low-density lipoprotein oxidation: termination of radical chain propagation reactions and formation of nitrogen-containing oxidized lipid derivatives. *Arch Biochem Biophys*, **324**, 15–25.

72 Ichikawa, T., Zhang, J., Chen, K., Liu, Y., Schopfer, F.J., Baker, P.R., Freeman, B.A., Chen, Y.E., and Cui, T. (2008) Nitroalkenes suppress lipopolysaccharide-induced signal transducer and activator of transcription signaling in macrophages: a critical role of mitogen-activated protein kinase phosphatase 1. *Endocrinology*, **149**, 4086–4094.

73 Kalyanaraman, B. (2004) Nitrated lipids: a class of cell-signaling molecules. *Proc Natl Acad Sci U S A*, **101**, 11527–11528.

74 Trostchansky, A. and Rubbo, H. (2007) Lipid nitration and formation of lipid-protein adducts: biological insights. *Amino Acids*, **32**, 517–522.

75 Østdal, H., Davies, M.J., and Andersen, H.J. (2002) Reaction between protein radicals and other biomolecules. *Free Radic Biol Med*, **33**, 201–209.

76 Rao, S.I., Wilks, A., Hamberg, M., and Ortiz de Montellano, P.R. (1994) The lipoxygenase activity of myoglobin. Oxidation of linoleic acid by the ferryl oxygen rather than protein radical. *J Biol Chem*, **269**, 7210–7216.

77 Gieseg, S., Duggan, S., and Gebicki, J.M. (2000) Peroxidation of proteins before lipids in U937 cells exposed to peroxyl radicals. *Biochem J*, **350** (Pt 1), 215–218.

78 Buettner, G.R. (1993) The pecking order of free radicals and antioxidants: lipid peroxidation, α-tocopherol, and ascorbate. *Arch Biochem Biophys*, **300**, 535–543.

79 Huie, R.E., Clifton, C.L., and Neta, P. (1991) Electron-transfer reaction rates and equilibria of the carbonate and sulfate radical anions. *Radiat Phys Chem*, **38**, 477–481.

80 Merenyi, G. and Lind, J. (1997) Thermodynamics of peroxynitrite and its CO_2 adduct. *Chem Res Toxicol*, **10**, 1216–1220.

81 Augusto, O., Bonini, M.G., Amanso, A.M., Linares, E., Santos, C.C., and De Menezes, S.L. (2002) Nitrogen dioxide and carbonate radical anion: two emerging radicals in biology. *Free Radic Biol Med*, **32**, 841–859.

82 Cameron, N.E., Tuck, Z., McCabe, L., and Cotter, M.A. (2001) Effects of the hydroxyl radical scavenger, dimethylthiourea, on peripheral nerve tissue perfusion, conduction velocity and nociception in experimental diabetes. *Diabetologia*, **44**, 1161–1169.

83 Desesso, J.M., Scialli, A.R., and Goeringer, G.C. (1994) D-mannitol, a specific hydroxyl free radical scavenger, reduces the developmental toxicity of hydroxyurea in rabbits. *Teratology*, **49**, 248–259.

84 Shertzer, H.G., Bannenberg, G.L., and Moldéus, P. (1992) Evaluation of iron binding and peroxide-mediated toxicity in rat hepatocytes. *Biochem Pharmacol*, **44**, 1367–1373.

85 Buxton, G.V., Greenstock, C.L., Helman, W.P., and Ross, A.B. (1988) Critical review of rate constants for reactions of hydrated electrons, hydrogen atoms and hydroxyl radicals (.OH/.O-) in aqueous solution. *J Phys Chem Ref Data*, **17**, 513–886.

86 Davies, M.J. (2005) The oxidative environment and protein damage. *Biochim Biophys Acta*, **1703**, 93–109.

87 Wardman, P. (1989) Reduction potentials of one-electron couples involving free radicals in aqueous solution. *J Phys Chem Ref Data*, **18**, 1637–1755.

88 Faraggi, M., Broitman, F., Trent, J.B., and Klapper, M.H. (1996) One-electron oxidation reactions of some purine and pyrimidine bases in aqueous solutions. Electrochemical and pulse radiolysis studies. *J Phys Chem*, **100**, 14751–14761.

89 Collinson, E. and Swallow, A.J. (1956) The radiation chemistry of organic substances. *Chem Rev*, **56**, 471–568.

90 Requena, Js.R., Chao, C.-C., Levine, R.L., and Stadtman, E.R. (2001) Glutamic and aminoadipic semialdehydes are the main carbonyl products of metal-catalyzed oxidation of proteins. *Proc Natl Acad Sci U S A*, **98**, 69–74.

91 Amici, A., Levine, R.L., Tsai, L., and Stadtman, E.R. (1989) Conversion of amino acid residues in proteins and amino acid homopolymers to carbonyl derivatives by metal-catalyzed oxidation reactions. *J Biol Chem*, **264**, 3341–3346.

92 Climent, I. and Levine, R.L. (1991) Oxidation of the active site of glutamine synthetase: conversion of arginine-344 to γ-glutamyl semialdehyde. *Arch Biochem Biophys*, **289**, 371–375.

93 Stadtman, E.R. (1990) Metal ion-catalyzed oxidation of proteins: biochemical mechanism and biological consequences. *Free Radic Biol Med*, **9**, 315–325.

94 Davies, M.J., Fu, S., Wang, H., and Dean, R.T. (1999) Stable markers of oxidant damage to proteins and their application in the study of human disease. *Free Radic Biol Med*, **27**, 1151–1163.

95 Roberts, C.R., Roughley, P.J., and Mort, J.S. (1989) Degradation of human proteoglycan aggregate induced by hydrogen peroxide. Protein fragmentation, amino acid modification and hyaluronic acid cleavage. *Biochem J*, **259**, 805–811.

96 Garrison, W.M. (1987) Reaction mechanisms in the radiolysis of peptides, polypeptides, and proteins. *Chem Rev*, **87**, 381–398.

97 Radi, R., Beckman, J.S., Bush, K.M., and Freeman, B.A. (1991) Peroxynitrite oxidation of sulfhydryls. The cytotoxic potential of superoxide and nitric oxide. *J Biol Chem*, **266**, 4244–4250.

98 Van Der Vliet, A., Eiserich, J.P., O'Neill, C.A., Halliwell, B., and Cross, C.E. (1995) Tyrosine modification by reactive nitrogen species: a closer look. *Arch Biochem Biophys*, **319**, 341–349.

99 Bonini, M.G. and Augusto, O. (2001) Carbon dioxide stimulates the production of thiyl, sulfinyl, and disulfide radical anion from thiol oxidation by peroxynitrite. *J Biol Chem*, **276**, 9749–9754.

100 Michaeli, A. and Feitelson, J. (1994) Reactivity of singlet oxygen toward amino acids and peptides. *Photochem Photobiol*, **59**, 284–289.

101 Uchida, K. and Kawakishi, S. (1993) 2-Oxo-histidine as a novel biological marker for oxidatively modified proteins. *FEBS Lett*, **332**, 208–210.

102 Farber, J.M. and Levine, R.L. (1986) Sequence of a peptide susceptible to mixed-function oxidation. Probable cation binding site in glutamine synthetase. *J Biol Chem*, **261**, 4574–4578.

103 Alvarez, B., Demicheli, V., Duran, R., Trujillo, M., Cervenansky, C., Freeman, B.A., and Radi, R. (2004) Inactivation of human Cu,Zn superoxide dismutase by peroxynitrite and formation of histidinyl radical. *Free Radic Biol Med*, **37**, 813–822.

104 Morin, B., Bubb, W.A., Davies, M.J., Dean, R.T., and Fu, S. (1998) 3-Hydroxylysine, a potential marker for studying radical-induced protein oxidation. *Chem Res Toxicol*, **11**, 1265–1273.

105 Burk, G.K. and Schoffa, G. (1969) ESR measurement of short-lived free radicals obtained by addition of OH to DL-methionine. *Int J Protein Res*, **1**, 113–116.

106 Yang, S.F. (1970) Sulfoxide formation from methionine or its sulfide analogs during aerobic oxidation of sulfite. *Biochemistry*, **9**, 5008–5014.

107 Schoneich, C. and Bobrowski, K. (1993) Intramolecular hydrogen transfer as the key step in the dissociation of hydroxyl radical adducts of (alkylthio)ethanol derivatives. *J Am Chem Soc*, **115**, 6538–6547.

108 Pryor, W.A., Jin, X., and Squadrito, G.L. (1994) One- and two-electron oxidations of methionine by peroxynitrite. *Proc Natl Acad Sci U S A*, **91**, 11173–11177.

109 Gieseg, S.P., Simpson, J.A., Charlton, T.S., Duncan, M.W., and Dean, R.T. (1993) Protein-bound 3,4-dihydroxyphenylalanine is a major reductant formed during hydroxyl radical damage to proteins. *Biochemistry*, **32**, 4780–4786.

110 van der Vliet, A., O'Neill, C.A., Halliwell, B., Cross, C.E., and Kaur, H. (1994) Aromatic hydroxylation and nitration of phenylalanine and tyrosine by peroxynitrite. Evidence for hydroxyl radical production from peroxynitrite. *FEBS Lett*, **339**, 89–92.

111 Nukuna, B.N., Goshe, M.B., and Anderson, V.E. (2001) Sites of hydroxyl radical reaction with amino acids identified by 2H NMR detection of induced 1H/2H exchange. *J Am Chem Soc*, **123**, 1208–1214.

112 Galzigna, L., Previero, A., Reggiani, A., and Coletti, M.A. (1964) Inattivazione della gramicidina dovuta alla conversione triptofano → N'-Formilchinurenina. *Experientia*, **20**, 669–670.

113 Pirie, A. (1971) Formation of N'-formylkynurenine in proteins from lens and other sources by exposure to sunlight. *Biochem J*, **125**, 203–208.

114 Fujimori, E. (1982) Crosslinking and photoreaction of ozone-oxidized calf-lens α-crystallin. *Invest Ophthalmol Vis Sci*, **22**, 402–405.

115 Armstrong, R.C. and Swallow, A.J. (1969) Pulse- and gamma-radiolysis of aqueous solutions of tryptophan. *Radiat Res*, **40**, 563–579.

116 Alvarez, B., Rubbo, H., Kirk, M., Barnes, S., Freeman, B.A., and Radi, R. (1996) Peroxynitrite-dependent tryptophan nitration. *Chem Res Toxicol*, **9**, 390–396.

117 Fletcher, G.L. and Okada, S. (1961) Radiation-induced formation of dihydroxyphenylalanine from tyrosine and tyrosine-containing peptides in aqueous solution. *Radiat Res*, **15**, 349–354.

118 Kettle, A.J. (1996) Neutrophils convert tyrosyl residues in albumin to chlorotyrosine. *FEBS Lett*, **379**, 103–106.

119 Joffe, A., Geacintov, N.E., and Shafirovich, V. (2003) DNA lesions derived from the site selective oxidation of guanine by carbonate radical anions. *Chem Res Toxicol*, **16**, 1528–1538.

120 Crean, C., Lee, Y.L., Yun, B.H., Geacintov, N.E., and Shafirovich, V. (2008) Oxidation of Guanine by Carbonate Radicals Derived from Photolysis of Carbonatotetramminecobalt(III) Complexes and the pH Dependence of Intrastrand DNA Cross-Links Mediated by Guanine Radical Reactions, pp. 1985–1991.

121 Cadet, J., Berger, M., Buchko, G.W., Joshi, P.C., Raoul, S., and Ravanat, J.-L. (1994) 2,2-Diamino-4-[(3,5-di-O-acetyl-2-deoxy-β-D-erythro-pentofuranosyl)amino]-5-(2H)-oxazolone: a novel and predominant radical oxidation product of 3′,5′-Di-O-acetyl-2′-deoxyguanosine. *J Am Chem Soc*, **116**, 7403–7404.

122 Ravanat, J.-L., Di Mascio, P., Martinez, G.R., Medeiros, M.H.G., and Cadet, J. (2000) Singlet oxygen induces oxidation of cellular DNA. *J Biol Chem*, **275**, 40601–40604.

123 Hawkins, C.L. and Davies, M.J. (2001) Hypochlorite-induced damage to nucleosides: formation of chloramines and nitrogen-centered radicals. *Chem Res Toxicol*, **14**, 1071–1081.

124 Bjelland, S. and Seeberg, E. (2003) Mutagenicity, toxicity and repair of DNA base damage induced by oxidation. *Mutat Res*, **531**, 37–80.

125 Stadtman, E.R. (1993) Oxidation of free amino acids and amino acid residues in proteins by radiolysis and by metal-catalyzed reactions. *Annu Rev Biochem*, **62**, 797–821.

126 Aust, S.D., Chignell, C.F., Bray, T.M., Kalyanaraman, B., and Mason, R.P. (1993) Free radicals in toxicology. *Toxicol Appl Pharmacol*, **120**, 168–178.

127 Mason, R.P. (2004) Using anti-5,5-dimethyl-1-pyrroline N-oxide (anti-DMPO) to detect protein radicals in time and space with immuno-spin trapping. *Free Radic Biol Med*, **36**, 1214–1223.

128 Detweiler, C.D., Deterding, L.J., Tomer, K.B., Chignell, C.F., Germolec, D., and Mason, R.P. (2002) Immunological identification of the heart myoglobin radical formed by hydrogen peroxide. *Free Radic Biol Med*, **33**, 364–369.

129 Chen, Y.R., Chen, C.L., Zhang, L., Green-Church, K.B., and Zweier, J.L. (2005) Superoxide generation from mitochondrial NADH dehydrogenase induces self-inactivation with specific protein radical formation. *J Biol Chem*, **280**, 37339–37348.

130 Cassina, P., Cassina, A., Pehar, M., Castellanos, R., Gandelman, M., de Leon, A., Robinson, K.M., Mason, R.P., Beckman, J.S., Barbeito, L., and Radi, R. (2008) Mitochondrial dysfunction in SOD1G93A-bearing astrocytes promotes motor neuron degeneration: prevention by mitochondrial-targeted antioxidants. *J Neurosci*, **28**, 4115–4122.

131 Nakajima, A., Tahara, M., Yoshimura, Y., and Nakazawa, H. (2005) Determination of free radicals generated from light exposed

ketoprofen. *J Photochem Photobiol A Chem*, **174**, 89–97.

132 He, Y.Y., Ramirez, D.C., Detweiler, C.D., Mason, R.P., and Chignell, C.F. (2003) UVA-ketoprofen-induced hemoglobin radicals detected by immuno-spin trapping. *Photochem Photobiol*, **77**, 585–591.

133 He, Y.Y., Council, S.E., Feng, L., Bonini, M.G., and Chignell, C.F. (2008) Spatial distribution of protein damage by singlet oxygen in keratinocytes. *Photochem Photobiol*, **84**, 69–74.

134 Bonini, M.G., Siraki, A.G., Atanassov, B.S., and Mason, R.P. (2007) Immunolocalization of hypochlorite-induced, catalase-bound free radical formation in mouse hepatocytes. *Free Radic Biol Med*, **42**, 530–540.

135 Ramirez, D.C., Mejiba, S.E.G., and Mason, R.P. (2005) Copper-catalyzed protein oxidation and its modulation by carbon dioxide: enhancement of protein radicals in cells. *J Biol Chem*, **280**, 27402–27411.

136 Ramirez, D.C., Mejiba, S.E., and Mason, R.P. (2006) Immuno-spin trapping of DNA radicals. *Nat Methods*, **3**, 123–127.

137 Alvarez, B., Ferrer-Sueta, G., Freeman, B.A., and Radi, R. (1999) Kinetics of peroxynitrite reaction with amino acids and human serum albumin. *J Biol Chem*, **274**, 842–848.

138 Alvarez, B. and Radi, R. (2003) Peroxynitrite reactivity with amino acids and proteins. *Amino Acids*, **25**, 295–311.

139 Lee, J.M., Niles, J.C., Wishnok, J.S., and Tannenbaum, S.R. (2002) Peroxynitrite reacts with 8-nitropurines to yield 8-oxopurines. *Chem Res Toxicol*, **15**, 7–14.

140 Byun, J., Henderson, J.P., Mueller, D.M., and Heinecke, J.W. (1999) 8-Nitro-2'-deoxyguanosine, a specific marker of oxidation by reactive nitrogen species, is generated by the myeloperoxidase-hydrogen peroxide-nitrite system of activated human phagocytes. *Biochemistry*, **38**, 2590–2600.

141 Romero, N., Radi, R., Linares, E., Augusto, O., Detweiler, C.D., Mason, R.P., and Denicola, A. (2003) Reaction of human hemoglobin with peroxynitrite. Isomerization to nitrate and secondary formation of protein radicals. *J Biol Chem*, **278**, 44049–44057.

142 Nakai, K. and Mason, R.P. (2005) Immunochemical detection of nitric oxide and nitrogen dioxide trapping of the tyrosyl radical and the resulting nitrotyrosine in sperm whale myoglobin. *Free Radic Biol Med*, **39**, 1050–1058.

143 Crean, C., Uvaydov, Y., Geacintov, N.E., and Shafirovich, V. (2008) Oxidation of single-stranded oligonucleotides by carbonate radical anions: generating intrastrand cross-links between guanine and thymine bases separated by cytosines. *Nucleic Acids Res*, **36**, 742–755.

144 Shafirovich, V., Mock, S., Kolbanovskiy, A., and Geacintov, N.E. (2002) Photochemically catalyzed generation of site-specific 8-nitroguanine adducts in DNA by the reaction of long-lived neutral guanine radicals with nitrogen dioxide. *Chem Res Toxicol*, **15**, 591–597.

145 Ramirez, D.C., Gomez Mejiba, S.E., and Mason, R.P. (2005) Mechanism of hydrogen peroxide-induced Cu,Zn-superoxide dismutase-centered radical formation as explored by immuno-spin trapping: the role of copper- and carbonate radical anion-mediated oxidations. *Free Radic Biol Med*, **38**, 201–214.

146 Ramirez, D.C., Gomez-Mejiba, S.E., Corbett, J.T., Deterding, L.J., Tomer, K.B., and Mason, R. (2008) Cu, Zn-superoxide dismutase-driven free radical modifications: copper- and carbonate radical anion-initiated protein radical chemistry. *Biochem J*.

147 Chen, Y.R., Chen, C.L., Liu, X., Li, H., Zweier, J.L., and Mason, R.P. (2004) Involvement of protein radical, protein aggregation, and effects on NO metabolism in the hypochlorite-mediated oxidation of

148. Mashino, T. and Fridovich, I. (1988) Reactions of hypochlorite with catalase. *Biochim Biophys Acta*, **956**, 63–69.

149. Whiteman, M., Jenner, A., and Halliwell, B. (1999) 8-Chloroadenine: a novel product formed from hypochlorous acid-induced damage to calf thymus DNA. *Biomarkers*, **4**, 303–310.

150. Hawkins, C.L. and Davies, M.J. (2002) Hypochlorite-induced damage to DNA, RNA, and polynucleotides: formation of chloramines and nitrogen-centered radicals. *Chem Res Toxicol*, **15**, 83–92.

151. Bonini, M.G., Siraki, A.G., Bhattacharjee, S., and Mason, R.P. (2007) Glutathione-induced radical formation on lactoperoxidase does not correlate with the enzyme's peroxidase activity. *Free Radic Biol Med*, **42**, 985–992.

152. Lardinois, O.M., Medzihradszky, K.F., and Ortiz de Montellano, P.R. (1999) Spin trapping and protein cross-linking of the lactoperoxidase protein radical. *J Biol Chem*, **274**, 35441–35448.

153. Siraki, A.G., Bonini, M.G. Jiang, J. Ehrenshaft, M., and Mason, R.P. (2007) Aminoglutethimide-induced protein free radical formation on myeloperoxidase: a potential mechanism of agranulocytosis. *Chem Res Toxicol*, **20**, 1038–1045.

154. Siraki, A.G., Deterding, L.J., Bonini, M.G. Jiang, J. Ehrenshaft, M., Tomer, K.B., Mason, R.P. (2008) Procainamide, but not N-acetylprocainamide, induces protein free radical formation on myeloperoxidase: a potential mechanism of agranulocytosis. *Chem Res Toxicol*, **21**, 1143–1153.

155. Ramirez, D.C., Chen, Y.-R., and Mason, R.P. (2003) Immunochemical detection of hemoglobin-derived radicals formed by reaction with hydrogen peroxide: involvement of a protein-tyrosyl radical. *Free Radic Biol Med*, **34**, 830–839.

156. Deterding, L.J., Ramirez, D.C., Dubin, J.R., Mason, R.P., and Tomer, K.B. (2004) Identification of free radicals on hemoglobin from its self-peroxidation using mass spectrometry and immuno-spin trapping: observation of a histidinyl radical. *J Biol Chem*, **279**, 11600–11607.

157. Ehrenshaft, M. and Mason, R.P. (2006) Protein radical formation on thyroid peroxidase during turnover as detected by immuno-spin trapping. *Free Radic Biol Med*, **41**, 422–430.

158. Lardinois, O.M., Tomer, K.B., Mason, R.P., and Deterding, L.J. (2008) Identification of protein radicals formed in the human neuroglobin-H_2O_2 reaction using immuno-spin trapping and mass spectrometry. *Biochemistry*, **47**, 10440–10448.

159. Chatterjee, S., Ehrenshaft, M., Bhattacharjee, S., Deterding, L.J., Bonini, M.G., Corbett, J.T., Kadiiska, M.B., Tomer, K.B., and Mason, R.P. (2009) Immuno-spin-trapping of a post-translational carboxypeptidase B1 radical formed by a dual role of xanthine oxidase and endothelial nitric oxide synthase in acute septic mice. *Free Radic Biol Med*, **46** (4), 454–461.

160. Lardinois, O.M., Detweiler, C.D., Tomer, K.B., Mason, R.P., and Deterding, L.J. (2008) Identifying the site of spin trapping in proteins by a combination of liquid chromatography, ELISA, and off-line tandem mass spectrometry. *Free Radic Biol Med*, **44**, 893–906.

161. Ehrenshaft, M., de Silva, S.O., Perdivara, I., Bilski, P., Sik, R.H., Chignell, C.F., Tomer, K.B., and Mason, R.P. (2009) Immunological detection of N-formylkynurenine in oxidized proteins. *Free Radic Biol Med*, **46** (9), 1260–1266.

4
Alcohol-Derived Bioadducts
Geoffrey M. Thiele and Lynell W. Klassen

Chronic ethanol abuse commonly leads to serious damage of a variety of tissues, especially the liver, muscle, and brain. Through many years of investigation, it is now believed that direct toxicity of ethanol is not the major factor resulting in tissue damage. Rather, it is the metabolites of ethanol that have become more interesting, and have resulted in major changes in the direction of research efforts. These metabolites have been shown to covalently modify (adduct) proteins, and in many cases have been detected in humans and laboratory animals as a result of the chronic consumption of ethanol. More importantly, it has been shown that the adduction of proteins with these metabolites may play a significant role in the development and/or progression of tissue damage associated with alcohol abuse including alcoholic liver disease, alcoholic myopathy, and alcoholic cardiomyopathy. Thus, it is the purpose of this chapter to review how the metabolites of ethanol are capable of adducting proteins; and additionally, to examine which of these adducts may be relevant in the development and/or progression of the alcohol induced tissue damage.

4.1
Introduction

Ethanol is normally encountered as the fermentation product of carbohydrates by yeasts (fermented drinks) and bacteria (gut). The extensive abuse of ethanol has many deleterious social, physical, and economic consequences. Chronic ethanol abuse commonly leads to serious damage of a variety of tissues, especially liver, muscle, and brain.

Ethanol-induced tissue damage may occur through direct and indirect toxicity. Yet, despite many years of research, the molecular mechanisms involved in the toxicity by ethanol and its metabolites remain to be elucidated. From these studies, it is now believed that direct toxicity of ethanol is not a major factor of tissue damage in individuals chronically consuming ethanol. Therefore, the role(s) that the metabolites play have become more interesting (indirect toxicity) and have

Endogenous Toxins. Diet, Genetics, Disease and Treatment.
Edited by Peter J. O'Brien and W. Robert Bruce
Copyright © 2010 WILEY-VCH Verlag GmbH & Co. KGaA, Weinheim
ISBN: 978-3-527-32363-0

resulted in major changes in the direction of research efforts. These metabolites have been shown to covalently modify (adduct) proteins and in many cases have been detected in humans and laboratory animals as a result of the consumption of ethanol. More importantly, it has been shown that the adduction of proteins with these metabolites may play a significant role in the development and/or progression of tissue damage associated with alcohol abuse, including alcoholic liver disease [1], alcoholic myopathy [2, 3], and alcoholic cardiomyopathy [4, 5]. Thus, it is the purpose of this chapter to review how the metabolites of ethanol modify various proteins that are relevant to the indirect toxicity that has been attributed to them.

4.2
The Generation of Reactive Metabolites

Following the consumption of ethanol, a small amount is oxidized by the stomach (first-pass metabolism), and some is most likely eliminated unmetabolized by the kidneys and lungs. Most of the ethanol (~90%) is oxidized by the liver following absorption from the gastrointestinal tract and recirculation via the hepatic portal vein. The liver is normally able to metabolize ~10–15 g ethanol/hour [6], and the majority of this is performed by hepatocytes. These cells contain three systems capable of oxidizing ethanol; the alcohol dehydrogenase (ADH) family of enzymes in the cytosol; catalase found in peroxisomes; and the microsomal ethanol oxidizing system (MEOS) in the endoplasmic reticulum.

Cytosolic ADH is an enzyme, which catalyzes the oxidation of ethanol to acetaldehyde and is the major pathway for the elimination of ethanol (Figure 4.1). It is especially important at low blood ethanol concentrations, and in evolutionary terms is presumed to have developed in response to the production of ethanol by bacterial fermentation in the gut. During ADH-mediated oxidation of ethanol, hydrogen is abstracted from the ethanol and transferred to the enzyme cofactor nicotinamide adenine dinucleotide (NAD^+), converting it to its reduced form NADH. During prolonged ethanol oxidation NADH produced by ADH action alters the redox state ($NADH/NAD^+$) of the liver, thus altering carbohydrate, fat, and protein metabolism [7].

Studies on human liver ADH have revealed a complex series of isoenzymes that arise from the association of different subunits to produce zinc-containing, active dimeric enzymes [6, 8, 9]. Subunits appear to originate from seven gene loci, ADH1 to ADH7 [10], and their frequency of appearance is related to ethnicity. Since these isoenzymes are efficient under different conditions, their expression alters the ability of an individual to clear ethanol from their system.

The peroxisomal enzyme catalase system has been shown to oxidize ethanol *in vitro* in the presence of a H_2O_2 generating system [11]. *In vivo*, H_2O_2 is normally generated by the β-oxidation of fatty acids such as octanoate, palmitate, and oleate. However, the peroxisomal enzymes do not metabolize short chain fatty acids such

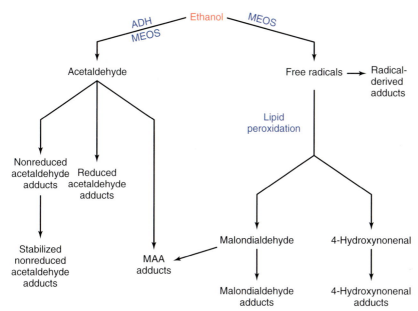

Fig. 4.1 The cells of the liver contain three systems capable of oxidizing ethanol: the alcohol dehydrogenase (ADH) family of enzymes in the cytosol; catalase found in peroxisomes; and the microsomal ethanol oxidizing system (MEOS) in the endoplasmic reticulum. Cytosolic ADH is an enzyme which catalyzes the oxidation of ethanol to acetaldehyde and is the major pathway for the elimination of ethanol. Additionally, an NADPH-dependent microsomal ethanol oxidising system (MEOS) has been isolated from hepatic tissue and shown to contain cytochrome P4502E1 which is extremely important in consistently heavy drinking individuals. However, it appears that the peroxisomal enzyme catalase system may play an important role in ethanol metabolism only in the absence of ADH activity and is a minor player. The oxidation of ethanol, mainly by the liver, produces a series of metabolites that can readily react with macromolecules to produce both stable and unstable covalent adducts. Much work has been done on the chemical nature of adducts produced by the reaction of ethanol metabolites with proteins. However, the majority of studies are confined to *in vitro* investigations in which metabolites are reacted with model proteins or peptides. The chemical characterization of adducts formed *in vivo*, in animals fed ethanol or in samples from alcoholics, has not yet been systematically carried out, largely due to an incomplete understanding of the chemistry behind adduct formation, the lack of consistent animals models, and the small amount of these adducts in the liver tissues. As shown in this figure, the oxidation of ethanol can lead to a number of different protein adducts, which is highly dependent upon which oxidation pathway is initiated, whether lipid peroxidation develops in response to free radicals, the redox potential, and many other factors that are still under investigation.

as octanoate [12] and β-oxidation is inhibited by NADH produced by the action of ADH. Thus, peroxisomal production of H_2O_2 during ethanol oxidation is so low that the concentration of H_2O_2 in hepatocytes becomes negligible. Therefore, it appears that this system may play an important role in ethanol metabolism only in the absence of ADH activity [13, 14].

Following chronic ethanol consumption there is an adaptive increase in the rate of ethanol oxidation, which cannot be explained in terms of ADH or catalase activity. This led to proposals that an inducible system may be involved in ethanol metabolism. Eventually, an NADPH-dependent microsomal (another name for the SER fraction) ethanol oxidizing system (MEOS) was isolated from hepatic tissue and shown to contain cytochrome P450, NADPH-cytochrome P450 reductase, and lipids [15]. The major cytochrome associated with this system, which metabolized ethanol at much higher rates than other P450 isoforms, was isolated from many different animals – LM3a from rabbit [16], P450j from rats [17], and P-450j-like from humans [18] – and was eventually designated CYP2E1 under a new nomenclature system [19].

The CYP2E1-containing MEOS is an inducible system that is under complex control. For a more comprehensive review of 30 years of research on MEOS, see [20]. MEOS has a low affinity for ethanol and becomes the major oxidizing system only when blood alcohol levels are consistently high over a period of several weeks. On prolonged drinking, this system becomes induced and in a consistently heavy drinking person accounts for up to 70% of ethanol metabolism.

Both ADH and MEOS oxidize ethanol to acetaldehyde (Figure 4.1). Acetaldehyde is a highly reactive species, which could lead to cellular damage if a significant amount were to accumulate. To prevent this, acetaldehyde is oxidized to acetic acid by a family of aldehyde dehydrogenase enzymes that are present in the mitochondria and the cytoplasm [8, 21, 22]. Aldolase can catalyze the condensation of acetaldehyde and the glycolytic intermediate dihydroxyacetone phosphate (DHAP) to produce the reactive sugar 5-deoxylulose-1-phosphate [23], though its production has not yet been demonstrated in a living organism.

MEOS, as well as producing acetaldehyde, also produces a series of free radicals such as α-hydroxyethyl, hydroxyl, and other radicals, which can lead to cellular damage by reaction with proteins and unsaturated fatty acids in biological membranes. The attack on membranes by free radicals leads to lipid peroxidation and generates the reactive aldehydes, malondialdehyde (MDA) and 4-hydroxynonenal, by the cleavage of lipid peroxides.

4.3
Ethanol Metabolites React with Proteins

The oxidation of ethanol, mainly by the liver, produces a series of metabolites that can readily react with macromolecules to produce both stable and unstable covalent adducts. Many researchers have shown that these modified proteins are immunogenic, leading to the generation of antibodies that react to the modifications as well as with the carrier protein [24–27]. Thus, it has been proposed that an aberrant immune response against modified proteins may play a role in the etiology of tissue damage in ethanol abuse.

Much work has been done on the chemical nature of adducts produced by the reaction of ethanol metabolites with proteins. However, the majority of studies is confined to *in vitro* investigations in which metabolites are reacted with model proteins or peptides. The chemical characterization of adducts formed *in vivo*, in animals fed ethanol or in samples from alcoholics, has not yet been systematically carried out, largely due to an incomplete understanding of the chemistry behind adduct formation, the lack of consistent animal models, and the small amount of these adducts in the liver tissues.

4.4 Acetaldehyde Adducts

4.4.1 Formation and Types of Adducts

A much more intensive effort has been made to understand the reactivity of acetaldehyde with macromolecules than with many of the other metabolites. This is most likely due to the fact that all enzymatic metabolism of ethanol leads to acetaldehyde production such that a rough molar equivalency between the ethanol consumed and the acetaldehyde produced is reached. Thus, acetaldehyde is likely to reach the highest concentration of any of the metabolites. A study using hepatocytes in culture has shown that concentrations as high as 1 mM may be achieved during chronic ethanol oxidation [28].

Acetaldehyde is able to react with a variety of nucleophilic groups due to the electrophilic nature of the carbonyl carbon [29]. The major protein nucleophilic groups that react with acetaldehyde at physiological pH and temperature appear to be the ε-amino group of internal lysine residues and the α-amino group of the N-terminal amino acid (Figures 4.2 and 4.3). The initial adduct is almost certainly a Schiff base formed by the carbonyl group reacting with protein amino groups [30, 31]. Other adducts may also be formed with thiol, hydroxyl, and imidazole groups. It is likely that unstable Schiff base adducts are the precursors of more stable adducts [32] with stabilization occurring after further reactions or rearrangements. One such stabilization event (Figure 4.2) occurs when a Schiff base forms on α-amino groups of the N-terminal amino acid followed by cyclization to produce a 2-methyl-imidazolidin-4-one derivative [33, 34]. A study has shown that this type of adduct is likely to be formed and to accumulate *in vivo* [33]. Potentially important reactions that may stabilize Schiff bases formed on ε-amino groups *in vivo* involve addition across the double bond, either by reduction or nucleophilic addition by a thiol group (Figure 4.3). Initially it was suggested that the reduction scenario, which results in the formation of N-ethyl lysine, was likely to predominate during chronic ethanol oxidation. Conflicting results have emerged from two laboratories on whether N-ethyl lysine is formed in the liver of ethanol-fed rats [35, 36]. This has become even more confusing as immunization with proteins modified under nonreducing conditions results in antibodies to N-ethyl lysine, the reduced adduct [37]. These data strongly suggest that during the processing and presentation of the nonreduced acetaldehyde adduct there must be a conversion of this relatively

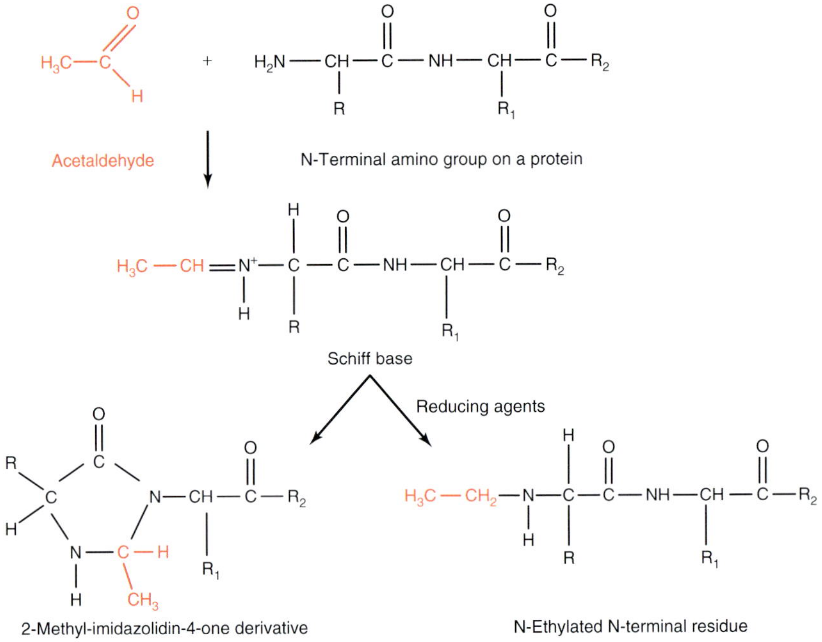

Fig. 4.2 Acetaldehyde is able to react with a variety of nucleophilic groups due to the electrophilic nature of the carbonyl carbon. The major protein nucleophilic groups that react with acetaldehyde at physiological pH and temperature appear to be the ε-amino group of internal lysine residues and the α-amino group of the N-terminal amino acid. The initial adduct is almost certainly a Schiff base formed by the carbonyl group reacting with protein amino groups. Other adducts may also be formed with thiol, hydroxyl, and imidazole groups. It is likely that unstable Schiff base adducts are the precursors of more stable adducts with stabilization occurring after further reactions or rearrangements. One such stabilization event occurs when a Schiff base forms on α-amino groups of the N-terminal amino acid followed by cyclization to produce a 2-methyl-imidazolidin-4-one derivative.

unstable adduct to the stable reduced adduct. However, no mechanism for this conversion has been elucidated. Regardless, it may explain the presence of the antibody to the reduced acetaldehyde adduct under conditions where this adduct should never be formed.

Adducts formed *in vitro* in the absence of strong reducing agents are both chemically [31] and immunologically [35] different from the ethylated amino acid derivatives generated in their presence. Polylysine and proteins lacking free thiol groups have been shown to form considerable amounts of stable adducts with acetaldehyde [31] in the absence of reducing agents.

The reaction of acetaldehyde with peptides containing thiol groups and with free cysteine at physiological pH and temperature has also been studied. NMR analysis confirmed that the product with free cysteine or with peptides with an N-terminal cysteine residue is the expected thiazolidine derivative [33]. This reaction is rapid

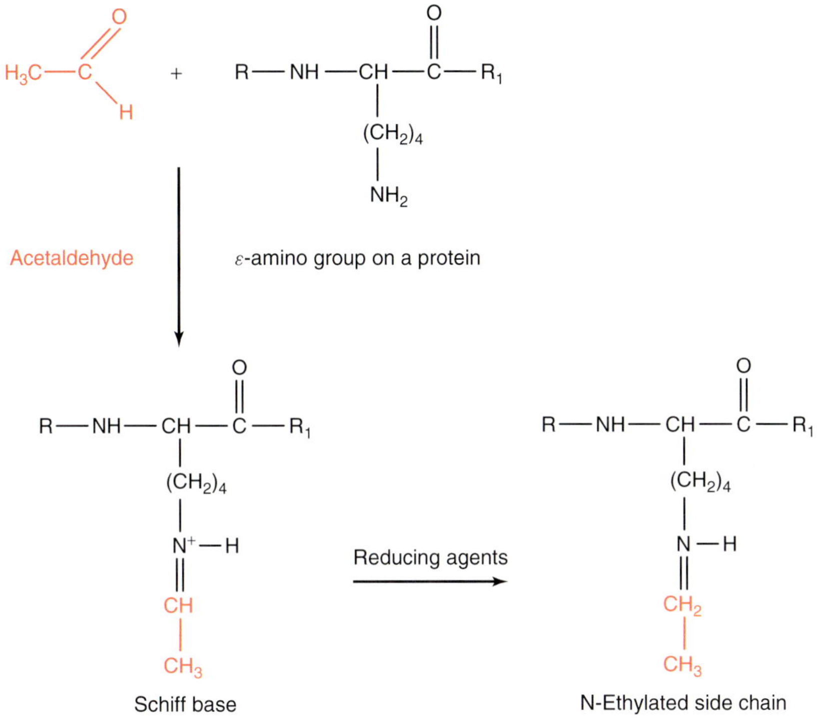

Fig. 4.3 Potentially important reactions that may stabilize Schiff bases formed on ε-amino groups *in vivo* involve addition across the double bond, either by reduction or nucleophilic addition by a thiol group. Initially, it was suggested that the reduction scenario, which results in the formation of N-ethyl lysine, was likely to predominate during chronic ethanol oxidation. Many studies have suggested that antibodies to N-ethyl lysine are found in the tissues of humans and animals chronically consuming ethanol. However, no mechanism for this conversion has been elucidated, as no agent that may induce the requisite level of reduction has been found in the liver during this highly oxidative reactive state.

at pH 7.4 and 37 °C and therefore likely to occur *in vivo*. However, no adduct formation has been detected by NMR when a peptide with an internal cysteine residue was reacted with a large excess of acetaldehyde, suggesting that the expected hemimercaptal is not formed to any extent under these conditions [33].

4.4.2
Biological Detection and/or Significance

Much evidence exists to demonstrate that acetaldehyde-modified proteins are generated during the metabolism of ethanol within the liver. Initial studies, using cell-free liver homogenates [30] and liver slices [38], showed that acetaldehyde generated during ethanol oxidation reacted with hepatic proteins to produce acetaldehyde–protein adducts. It was also shown that unstable adducts were formed and then stabilized during more prolonged incubation to produce stable products. These data were confirmed using polyclonal antisera raised against proteins modified by acetaldehyde *in vitro* in Western and immunohistochemical analysis of liver samples from rodents and humans consuming alcohol.

However, while these studies showed that acetaldehyde had reacted with hepatic protein(s), the number and cellular location of the protein targets was very different in each study. Behrens et al. [39] found that cytochrome P4502E1 (an enzyme which is part of MEOS and oxidizes ethanol to acetaldehyde) in the endoplasmic reticulum was modified, whereas Lin et al. [40] could only detect a 37-kDa cytosolic protein (later identified as Δ^4-3-ketosteroid-5β reductase) [41]. In contrast, Worrall et al. [42] and Yokoyama et al. [43] detected multiple modified cytosolic proteins, including the 37 kDa one seen by Lin and colleagues. Furthermore, Nicholls et al. [44], using ELISAs, demonstrated the presence of modified proteins in hepatic cytosol, membrane, and mitochondrial fractions from ethanol-fed rats.

Niemela and coworkers [45] used immunohistochemistry to demonstrate modified proteins in the cytosol of hepatocytes from human alcoholics. Further studies have shown that acetaldehyde-modified epitopes are located in the cytoplasm and are more intense in the centrilobular regions of the liver [46, 47], which are the areas associated with the highest levels of ethanol oxidation. Studies using rodents have been useful in the investigation of adduct formation as it has been shown that antibodies from human alcoholics react with cytosolic proteins from ethanol fed but not control rats, suggesting that similar, if not identical, immunogenic epitopes are being generated in alcoholics [48].

Acetaldehyde can also reversibly bind to red cells and plasma proteins and travel to extrahepatic tissues [49, 50]. The concentration of circulating proteins modified by acetaldehyde has been measured in two studies, both of which used ELISA technology. Lin and coworkers [51] were able to demonstrate larger amounts of acetaldehyde-modified serum proteins in samples from alcoholics than from nondrinking controls. Interestingly, in this study, a clearer differentiation between alcoholic and controls was seen if an antiserum raised against acetaldehyde-modified myoglobin rather than modified hemocyanin was used. The reasons for this discrepancy are not clear, but it is possible that the myoglobin acts as a carrier protein to make adducts more representative of those carried on serum proteins than does hemocyanin. In a more recent study, it was shown that alcoholics had statistically significant higher levels of acetaldehyde-modified albumin than heavy drinkers or control social drinkers [52]. Furthermore, the heavy drinkers had higher levels than the controls. These data suggest that the level of acetaldehyde-modified albumin is related to the amount of alcohol consumed. Albumin was utilized in these studies as it was presumed that it would be a representative protein that could be modified by physiological levels of acetaldehyde since it is synthesized within the liver and then secreted into the circulation. Thus, it has the opportunity to be modified during its synthesis and secretion, as well as when in the bloodstream. In rats fed ethanol, it has been shown that albumin and

IgG (a protein of extrahepatic origin) are both modified by acetaldehyde suggesting that modification also takes place [36].

Recent studies utilizing rats fed a liquid diet containing ethanol have shown that skeletal muscle (soleus and plantaris) contained unreduced, possibly imidazolidinone, adducts [2, 3] and cardiac muscle contained unreduced acetaldehyde and malondialdehyde-acetaldehyde (MAA) adducts in 2 : 1 ratio[5, 36].

However, most of the effort on the measurement of proteins modified *in vivo* has centered on the modification of hemoglobin by acetaldehyde. Early studies on the reactivity of acetaldehyde with purified proteins *in vitro* showed that acetaldehyde readily reacts with numerous sites on the globin monomers, particularly at the amino terminus of the ε-chain [34]. The initial studies on the isolation and characterization of acetaldehyde-modified hemoglobin from the blood of alcoholics were unsuccessful [53]. However, Niemela and coworkers [54] used an ELISA to detect acetaldehyde-modified hemoglobin in the blood of volunteers after acute ethanol dosing over 8 hours. Using a similar ELISA, it was shown that acetaldehyde-modified hemoglobin levels were higher in women who subsequently delivered children with fetal alcohol syndrome than those who delivered normal children [55]. Similarly, Niemela and Israel [56] and Lin and colleagues [57] showed that alcoholics had higher levels of acetaldehyde-modified hemoglobin than control patients.

The development of sensitive cation exchange liquid chromatographic methods [58–60] has enabled this modified hemoglobin, often termed HbA_{1ach}, to be studied in more detail. Analysis of modified hemoglobin from alcoholics has revealed that the likely sites of modification are the N-termini of the α- and β-globin chains [61], as well as unspecified internal residues [62]. Chromatographic techniques with higher resolution are now suggesting that acetaldehyde-modified hemoglobin exists as at least three species, each of which appears to be related to alcohol consumption [63–65]. However, as yet the modifications leading to the three forms of hemoglobin have not been characterized.

4.5
Malondialdehyde

4.5.1
Formation and Types of Adducts

MDA is formed in abnormally large amounts during ethanol metabolism, as a result of the breakdown of lipid peroxides formed on unsaturated fatty acids. It is also formed in high levels during oxidative stress induced by other agents and its reactions have been studied in considerable detail. The principal form of MDA, which exists in aqueous solution at neutral pH is the enolate anion which is of low reactivity. Thus, under physiological conditions, MDA is not a highly reactive compound [66]. However, the reactivity increases at lower pH as β-hydroxyacrolein becomes the predominant form, and can undergo a Michael type 1,4-addition in a

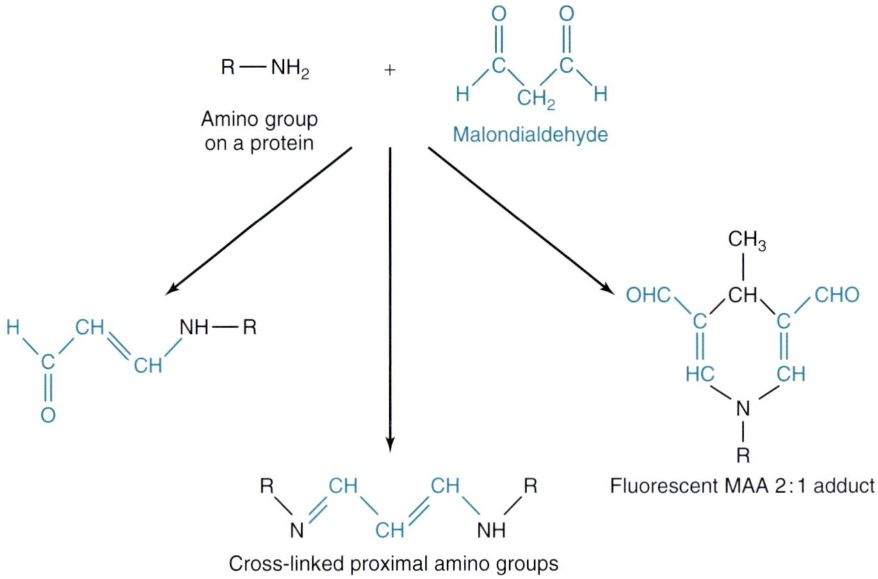

Fig. 4.4 Malondialdehyde is formed in abnormally large amounts during ethanol metabolism, as a result of the breakdown of lipid peroxides formed on unsaturated fatty acids. It is also formed in high levels during oxidative stress induced by other agents and its reactions have been studied in considerable detail. The principal form of malondialdehyde which exists in aqueous solution at neutral pH is the enolate anion which is of low reactivity. Thus, under physiological conditions, malondialdehyde is not a highly reactive compound. However, the reactivity increases at lower pH as β-hydroxyacrolein becomes the predominant form, and can undergo a Michael type 1,4-addition in a similar manner to other α,β-unsaturated aldehydes (i.e.,-4-hydroxy-2-nonenal). This type of reaction is favored at low pH by resonance stabilization and results in a β-substituted acrolein derivative as an end product (1 : 1 adduct). This compound contains a double CC bond which can further react with another amino group at low pH to give a 1 : 2 adduct. Proteins are much more reactive under physiological conditions than are free amino acids. For example, reaction with polylysine led to three types of modified α-amino groups. About one-fifth was found to be the unstable amino-propenal, approximately 1% to be a dihydropyridine derivative (also forms when malondialdehyde reacts in concert with acetaldehyde to form MAA), and the majority to be stable cross-linked structures based on amino-imino-propenal derivatives.

similar manner to other α,β-unsaturated aldehydes (i.e.,-4-hydroxy-2-nonenal). This type of reaction is favored at low pH by resonance stabilization and a β-substituted acrolein derivative is obtained as an end product (1 : 1 adduct) [67]. This compound contains a double CC bond which can further react with another amino group at low pH to give a 1 : 2 adduct (Figure 4.4).

A study of the reactivity of MDA with amino acids at pH 4.2 showed that histidine, tyrosine, tryptophan, and arginine reacted exclusively via their α-amino groups to give a 1 : 1 adduct [68]. The reaction of cysteine with MDA leads to a product that contains two cysteine residues and three molecules of MDA. Neutral pH does not appear to favor the reaction with amino groups since reactivity with cysteine (amino and thiol groups) but not glycine (only an amino group) could be demonstrated [66].

Proteins are much more reactive under physiological conditions than are free amino acids. It has been suggested that proteins may present amino acids in a much more favorable environment such that greater reactivity occurs. For example, reaction with polylysine led to three types of modified ε-amino groups [69]. About one-fifth was found to be the unstable amino-propenal, approximately 1% to be a dihydropyridine derivative (also forms when MDA reacts in concert with acetaldehyde to form MAA), and the majority to be stable cross-linked structures based on amino-imino-propen derivatives. Similar results were obtained when bovine serum albumin was reacted with MDA under physiological conditions. MDA was able to modify about 40% of the total amino groups available. The amino-propenal adduct was found on 11 of 26 modified ε-amino groups and the rest were amino-imino-propen cross-linked structures. Under these conditions, about 1 in 12 protein molecules carried a single dihydropyridine derivative. Other studies have shown that histidine, tyrosine, arginine, and methionine residues may also be modified to some extent [70].

4.5.2
Biological Detection and/or Significance

Elevated levels of MDA have been detected in humans drinking about 100 g of ethanol per day [71]. The increase in MDA production appears to continue for several weeks after cessation of drinking [72]. MDA-modified proteins have been detected *in vivo* in a wide range of conditions. For example, hepatic cytosol and plasma from rats with hepatic iron overload contained elevated levels of MDA [73]. In animal models of alcoholic liver disease, several researchers have shown that 4-hydroxynonenal- and MDA-modified proteins are associated with areas of inflammation and necrosis [74–76]. Furthermore, Ohhira and coworkers [77] have recently shown that 4-hydroxynonenal-modified proteins are associated with iron deposits in hepatocytes, and appear to be a marker of oxidative stress–induced damage.

MDA-modified proteins have been detected in humans and many different animal models *in vivo* by a number of different methods. For example, French *et al.* [78] showed that MDA-modified proteins are present in ethanol-fed rat livers using fluorometric methods, whereas Niemela *et al.* [76] found these adducts in livers of ethanol-fed rats and minipigs and human alcoholics using immunohistochemistry. More importantly, the proteins that have been modified with MDA have begun to be identified, and this should help in determining the functional consequences of this modification. For example, apoB 100 adducted with MDA has been identified in atheromas from hyperlipidemic Watanabe rabbits by immunohistochemistry [79, 80] and by Western blotting [81]. MDA-modified collagen that colocalized with the areas of focal necrosis [82] has been detected by immunohistochemical analysis of livers from ethanol-fed rats. Additionally, cytosol and plasma from rats with hepatic iron overload contained MDA-protein adducts [73].

Probably the best characterized MDA adducts are the MDA-modified plasma low density lipoproteins (LDLs) that have been shown to be slightly increased in patients with chronic stable angina pectoris, and greatly increased in other atherosclerotic diseases such as acute myocardial infarction and carotid atherosclerosis [83]. The LDLs (apoB and apoE) have been shown to be modified with MDA following oxidative stress and results in an increase in atherogenicity, most likely due to the interaction of these modified LDLs with scavenger receptors on macrophages [84]. As outlined for acetaldehyde-modification of proteins, it is presumed that MDA modification may also alter the function of the protein, and result in the production of an immune response against modified proteins. This has been shown by many investigators who have immunized rabbits with MDA-modified LDL, rabbit serum albumin or bovine serum albumin to produce antibodies that react with other proteins that have been modified with MDA [81, 82, 85].

4.6
Malondialdehyde–Acetaldehyde (MAA) Adducts

4.6.1
Formation and Types of Adducts

Tuma and coworkers have recently shown that the binding of acetaldehyde to proteins is markedly increased in the presence of MDA [86]. Incubation of proteins with acetaldehyde and MDA led to the generation of two distinct adducts, one of which was fluorescent (Figure 4.5). These hybrid adducts were termed *MalondiAldehyde-Acetaldehyde* adducts by Tuma and coworkers [86]. They have been identified as 4-methyl-1,4-dihydropyridine-3,5-dicarbaldehyde (MAA 2 : 1, fluorescent and derived from two molecules of MDA and one molecule of acetaldehyde) and 2-formyl-3-(alkylamino)butanal (MAA 1 : 1, nonfluorescent and derived from one molecule of acetaldehyde and one molecule of MDA) derivatives of protein amino groups [87]. There is now evidence to suggest that MAA 1 : 1 adducts react with MDA-derived Schiff bases to form the MAA 2 : 1 adduct [88]. The MAA 2 : 1 adduct is unique in that it makes carrier proteins immunogenic in the absence of adjuvant [26] and may therefore play an important role in the etiology of tissue damage associated with alcohol abuse. In contrast, the 1 : 1 product has been suggested to have a potential correlation with the consumption of ethanol as it would require more acetaldehyde than MDA in order to form.

4.6.2
Biological Detection and/or Significance

Polyclonal antibodies have been generated against the fluorescent adduct (2 : 1 product) and have been used to show that these adducts are formed in the liver of

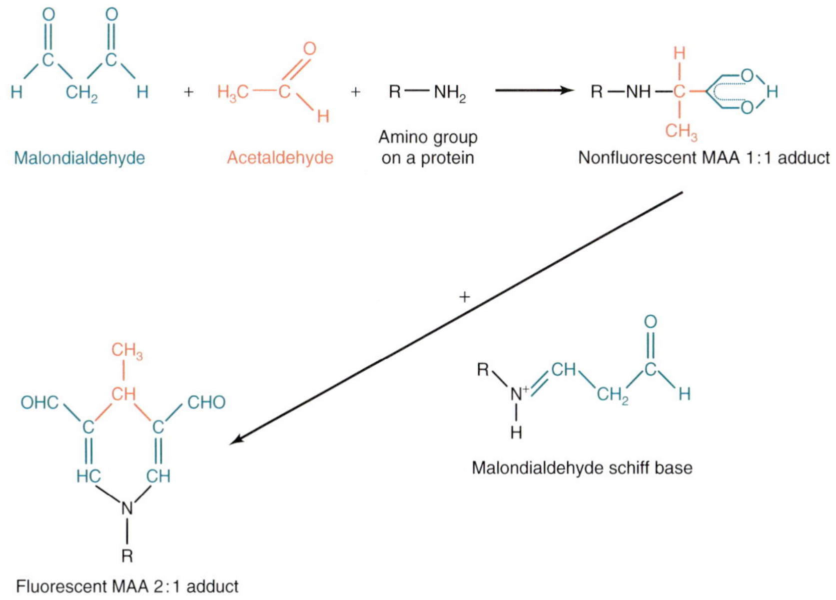

Fig. 4.5 It has been recently shown that the binding of acetaldehyde to proteins is markedly increased in the presence of malondialdehyde. Incubation of proteins with acetaldehyde and malondialdehyde led to the generation of two distinct adducts, one of which was fluorescent. These hybrid adducts were termed *MalondiAldehyde-Acetaldehyde (MAA) adducts*. They have been identified as 4-methyl-1,4-dihydropyridine-3,5-dicarbaldehyde (MAA 2 : 1, fluorescent and derived from two molecules of malondialdehyde and one molecule of acetaldehyde) and 2-formyl-3-(alkylamino)butanal (MAA 1 : 1, nonfluorescent and derived from one molecule of acetaldehyde and one molecule of malondialdehyde) derivatives of protein amino groups. There is now evidence to suggest that MAA 1 : 1 adducts react with malondialdehyde-derived Schiff bases to form the MAA 2 : 1 adduct [88]. The MAA 2 : 1 adduct is presumed to play an important role in the etiology of tissue damage associated with alcohol abuse. In contrast, the 1 : 1 product has been suggested to have a potential correlation with the consumption of ethanol as it would require more acetaldehyde than malondialdehyde in order to form.

ethanol-fed rats [36, 86, 89]. The fluorescent adduct is unique amongst the adducts so far discovered in that it makes carrier proteins immunogenic in the absence of adjuvant [26] and may therefore play an important role in the etiology of tissue damage associated with alcohol abuse. There is also indirect evidence that the MAA 1 : 1 adduct is also formed *in vitro* and *in vivo* [88]. Studies *in vitro* have shown that the MAA 2 : 1 adduct appears to be formed by an MAA 1 : 1 adduct reacting with an MDA Schiff base such that the 1 : 1 adduct is transferred onto the Schiff base. This hypothesis was tested using liver from ethanol-fed rats. When ethanol-fed rat liver was perfused with MDA, a higher concentration of MAA 2 : 1 adducts was seen compared to livers perfused with vehicle alone. This suggested that MAA 1 : 1 adducts were present in the liver of ethanol-fed rats, and were converted to the 2 : 1 adduct after perfusion with MDA [88].

4.7
4-Hydroxyalkenals

4.7.1
Formation and Types of Adducts

4-Hydroxyalkenals such as 4-hydroxy-2-nonenal have three main functional groups, that is, the aldehyde group, the CC double bond, and the hydroxy group, which can participate alone or in sequence in reactions with other molecules. As such the reactivity of 4-hydroxy-2-nonenal is much more complicated than that of the other metabolites [66]. The addition of 4-hydroxyalkenals to cells or tissues results in a rapid loss of thiol groups suggesting that they are the initial targets of these molecules. A saturated aldehyde bound to the thiol-containing molecule via a thio-ether linkage at carbon atom 3 is the initial product formed. This can then undergo intramolecular rearrangement to produce a five-membered cyclic hemiacetal derivative. If an excess of thiol groups is present, the initial conjugate can react with another thiol to produce a thiazolidine derivative [66].

4-Hydroxy-2-nonenal can react with various amino acid residues in proteins. When reacted with human apolipoprotein B (M_r 500 000), 5 mM hydroxynonenal modified 2 thiol groups and 45 lysine, 23 serine, 7 histidine, and 51 tyrosine residues [90]. The binding to the lysine residues was reversible, indicating that it was probably due to the formation of Schiff bases. Amino groups can also react with hydroxynonenal via nucleophilic Michael addition of the amino group to the CC double bond, which can then be stabilized as a cyclic hemiacetal by loss of water. Schiff bases are believed to be a minor product in reactions with amino groups, but they can be stabilized by loss of water to generate a pyrrole derivative.

4.7.2
Biological Detection and/or Significance

The process of lipid peroxidation not only results in the perturbation of cellular membrane lipids but is also associated with the production of aldehydic products such as MDA and HNE. These aldehydes are documented to mediate a number of cellular toxicities [66]. Since it has been shown that the potentially toxic intermediates compete for the same routes of enzymatic detoxification, this competition may represent one mechanism whereby these aldehydes could accumulate and potentiate ethanol-induced hepatocellular injury. Certainly, HNE–protein adducts have been found to localize in the areas of hepatic injury in experimental models of ALD in the rat [73]. The *in situ* localization of these aldehyde–protein adducts in the injured liver is significant, because acetaldehyde [91, 92], as well as products of lipid peroxidation, can directly stimulate collagen production and collagen gene expression in cell cultures of fibroblasts and stellate cells [93, 94].

4.8
Free Radical–Derived Adducts

4.8.1
Formation and Detection of Adducts

Acetaldehyde produced by the action of ADH or CYP2E1 on ethanol in hepatocytes is normally further oxidized by aldehyde dehydrogenase to produce acetate. This is normally considered to be detoxification as the potent electrophile acetaldehyde is converted to relatively innocuous acetate. However, there are other metabolic fates available for ethanol and acetaldehyde. Metabolism of ethanol by CYP2E1 becomes more pronounced after chronic alcohol consumption [20] and can generate free radicals as well as acetaldehyde [95]. Radicals such as the superoxide anion [96, 97] and the 1-hydroxyethyl radical have been shown to be produced both *in vitro* and *in vivo* [36, 71, 98–103]. Furthermore, the enymatic oxidation of acetaldehyde to acetate can also lead to the generation of free radicals [104–107].

Enzymes and subcellular fractions that have been shown to oxidize acetaldehyde to free radical intermediates include xanthine oxidase [106], aldehyde oxidase [107], mitochondria [108], and microsomes [105]. In the majority of these studies, free radical production was inferred from indirect measurements, and acetaldehyde-derived free radicals were not identified. Exceptions were the spin-trapping studies that demonstrated acetyl radical formation by xanthine oxidase [104] and rat liver and brain microsomes [105]. These systems are known producers of reactive oxygen species, such as the superoxide anion radical and hydrogen peroxide, which in the presence of transition metal ions, produces the highly damaging hydroxyl radical. Consequently, it is assumed that acetyl radical production comes from hydroxyl radical attack on acetaldehyde. However, a more recent study using spin trapping and electron paramagnetic resonance has now shown that xanthine oxidase and microsomes generate both the acetyl and methyl radicals *in vitro*, though only the latter could be detected *in vivo* in rats [109].

4.8.2
Biological Detection and/or Significance

At present there is no clear evidence of covalent adducts formed by methyl or acetyl radicals in the livers of animals fed ethanol or of human alcoholics, but there is a strong body of evidence to show that 1-hydroxyethyl radicals generated during ethanol oxidation react with cellular macromolecules such as proteins to form adducts in both animals and humans. Initially, the evidence for adduct formation was indirect and was inferred by the generation of antibodies reactive to rat liver microsomes incubated *in vitro* with ethanol or albumin modified by chemically derived hydroxyethyl radicals [102]. These antibodies could recognize the radical-derived epitopes independently of the carrier protein. Similar antibodies were also seen in rats chronically fed ethanol [36, 102, 110]. Similar studies on alcoholics have shown that they generate high titers of IgG and IgA reactive to hydroxyethyl radical–derived epitopes, whereas patients with nonalcoholic liver disease and healthy controls had low or nonexistent titers [99]. Similar antibodies were also observed in heavy drinking individuals without clinical signs of liver damage. To determine the likely source of the radicals, the induction of CYP2E1 was measured by determining the rate of oxidation of chloroxazone. This study showed that 78% of alcoholics had highly elevated rates of oxidation suggesting induction of CYP2E1, but the remainder had normal rates indicating no CYP2E1 induction [111]. Furthermore, individuals without CYP2E1 induction did not generate IgG reactive to hydroxyethyl radical–modified proteins, whereas induced individuals did. This study also showed that CYP2E1 induction was lower in alcoholics with apparently undamaged livers when compared with those with alcoholic liver disease. This study confirmed that hydroxyethyl radicals are produced in humans as a result of chronic ethanol oxidation, and the level of CYP2E1 activity greatly influences their production and the magnitude of the immune response against radical-modified proteins.

Hydroxyl radicals are reactive species that can react with proteins and DNA [112, 113]. Western analysis of microsomal proteins from human liver microsomes incubated with ethanol showed that at least four proteins with apparent molecular weights of 78, 60, 52, and 40 kDa carried hydroxyethyl radical–derived modifications [99]. The 52 kDa protein was identified as the likely source of the radicals, CYP2E1. Inhibitors of CYP2E1 activity such as diallylsulfide and phenylethylisothiocyanate greatly decrease the modification of CYP2E1 and decrease the titer of antiadduct IgG in rats fed ethanol [99]. In further studies, Western analysis of plasma membrane proteins from ethanol-treated hepatocytes revealed that sera from patients with alcoholic liver disease recognized three main proteins, one of which was also recognized by anti-hydroxyethyl radical–derived adduct antiserum and anti-CYP2E1 antibodies [114]. It was also observed that isolated rat hepatocytes, exposed *in vitro* to ethanol, were killed by antibody-dependent cell-mediated cytotoxicity upon addition of sera from patients with alcoholic liver disease and mononuclear cells from normal controls [115].

4.9
5-Deoxy-D-xylulose-1-phosphate (DXP)

Formation of 5-deoxy-d-xylulose-1-phosphate (DXP) from acetaldehyde and fructose-1,6-bisphosphate was first demonstrated by Meyerhof, Lohmann, and Schuster in 1936 using muscle extracts to catalyze the reaction [116]. The glycolytic enzymes aldolase and triose isomerase were believed to be involved. These enzymes are found in virtually every type of cell suggesting that most cells may be able to synthesize DXP if acetaldehyde is available. Evidence for the formation of DXP in the liver came from studies by Hoberman who showed that extracts of livers perfused *in situ* with [^{3}H]acetaldehyde contained a small amount of a highly acidic, phosphate-containing, nonvolatile radiolabeled compound. Hoberman speculated that this compound was DXP, formed by the action of aldolase on acetaldehyde and DHAP, and confirmed this by chromatographic comparison with chemically synthesized DXP. Hoberman further characterized the synthesis of DXP using erythrocytes, cells that actively undergo glycolysis to provide DHAP in a system uncomplicated by the regulatory influences of mitochondrial respiration [23]. Later studies showed that DXP could react with hemoglobin to form an adduct that was stable to repeated precipitation by trichloroacetic acid [117]. This adduct is likely to be formed by a series of reactions similar to those of nonenzymic glycation. Initially, the DXP probably forms a Schiff base on an amino group which then undergoes the Amadori rearrangement to finally yield a ketoamine adduct. The product contains a hydroxyketone group that can react with another amino group in a similar manner and could cause cross-linking if the second amino group was on a different peptide. Apart from an *in vitro* study showing that DXP reacts with hemoglobin to produce covalent modifications, there is no evidence to show that similar adducts are formed *in vivo*.

4.10
Acetaldehyde–Glucose Amadori Product Adduct

The nonenzymatic addition of glucose to proteins such as hemoglobin has been shown to occur under conditions such as diabetes, where the prevailing glucose concentration is high. In this interaction, an open chain form of the sugar reacts with an amino group to form a Schiff base. This can then undergo the Amadori rearrangement to yield a product that can undergo dehydration to yield a diketone which is the precursor to a chemically diverse group of adducts termed *advanced glycation end product*s or AGEs. AGEs, through a series of chemical rearrangements, lead to the formation of heterocyclic structures with a characteristic yellow-brown color and the ability to cross-link proximal amino groups. The progressive accumulation of AGEs in long-lived connective tissue and vascular wall components account in part for the increased cardiovascular disease associated with diabetes. One study has shown that acetaldehyde can interfere in the conversion of the glucose Amadori product into AGEs by forming an acetaldehyde–glucose Amadori product adduct that is incapable of further rearrangement into AGEs. At present, there is little evidence on the formation of these adducts *in vivo*. However, one study, in which diabetic rat were fed ethanol in a liquid diet, showed that hemoglobin–AGE formation was reduced by 52% when compared to diabetic controls [118]. However, there is no evidence for the formation of these acetaldehyde–glucose Amadori product adducts in the liver.

4.11
Immune Responses to Proteins Modified with Alcohol Metabolites

There is now a fairly extensive literature to show that proteins modified by reactive metabolites act as neoantigens to elicit an immune response against the modification (adduct; hapten) and/or against the carrier protein. However, although research has largely been limited to the detection of antibodies reactive with adducts, there are reports of cellular immune responses against proteins modified with the metabolites of ethanol [119, 120].

Many studies have now shown that rodents fed ethanol and human alcoholics generate antibodies reactive to proteins modified by ethanol metabolites *in vitro*. In 1986, Israel and coworkers [121] were the first to demonstrate the production of anti-acetaldehyde antibodies by animals (mice) chronically oxidizing ethanol. In a later study, the magnitude of this response was shown to be related to the time that the animals (rats) were fed ethanol [27] and most likely the cumulative ethanol dose. A further study revealed that antibodies reactive to at least two different classes of modification were generated in the ethanol-fed animals [44]. Similar studies have demonstrated the generation of antibodies against proteins modified by α-hydroxyethyl radicals [36, 110], MDA [36], 4-hydroxynonenal [122], and MAA [26].

There is also a considerable body of evidence relating to the production of antibodies against epitopes derived from ethanol metabolites in humans. Initial studies were carried out on antibodies reactive to acetaldehyde-derived modifications. Thus, several groups were able to detect antibodies reactive against acetaldehyde-modified proteins before they could demonstrate the presence of the modified proteins *per se* [25, 27, 123]. These studies used ELISAs to measure reactivity of antibodies in plasma or serum with proteins modified by acetaldehyde *in vitro*. The conditions used to modify the proteins for use in the assays varied widely, making it likely that different modifications were being made to the proteins, and therefore different types of immunoreactivity were being detected in each study. In experiments with rodents, there was a clear distinction between animals fed ethanol and their ethanol-naïve controls by which any animal with a high immunoreactivity against acetaldehyde-modified proteins could be definitively identified as an "alcoholic." This was not so in the human studies as high immunoreactivity was observed in some social drinkers (<50 g ethanol per week for males and <30 g per week for females) and nondrinkers [124]. However, more individuals with high immunoreactivity were observed in the alcoholic groups in all of the studies [25, 123, 124].

At present most interest on antibody reactivity has focused on their interaction with modified proteins. It should, however, be noted that antibodies generated against modified proteins will also react with modified lipids. Studies by Trudell and coworkers [125, 126] have shown that antibodies raised against acetaldehyde-modified albumin will also react with acetaldehyde-modified dioleoyl phosphatidylethanolamine, presumably because the modification in both cases is of a similar nature. Thus, while it is possible that modified lipids are at least in part responsible for the generation of the antiadduct antibodies *in vivo*, it is much easier to use modified proteins to detect and measure them.

Research efforts to date have primarily focused on the reactivity of antibodies with the chemically derived modifications rather than looking at the reactivity with the carrier protein. Once a protein is modified, the immune system generates antibodies against the carrier protein *per se* and the chemical modification. In an early study [124], it was noted that immunoreactivity with unmodified proteins was higher in alcoholics than in patients with nonalcoholic liver disease or social drinkers. There is also a potential for antibodies to be raised against an epitope that consists of part of the protein and the modification. There is some evidence to suggest that this actually occurs. Koskinas and coworkers [127] demonstrated that antibodies from 70% of patients with alcoholic hepatitis reacted with a 200-kDa cytosolic protein when it was modified *in vitro* by acetaldehyde and then treated with a reducing agent. Similar antibodies could be seen only in 25% of controls and 35% of patients with nonalcoholic liver disease. The antibodies detected in this study must have a degree of carrier protein specificity since they only detected the 200-kDa protein even when other modified proteins were present. However, they did not recognize the 200-kDa protein when it was unmodified, suggesting that the epitope they reacted with was in large part due to the process of modification.

When examining antibody responses to proteins modified as a result of alcohol metabolism, the main research effort has been on those modified by acetaldehyde. The effect of alcohol consumption on the plasma concentrations of MDA and antibodies directed against MDA adducts on proteins has been investigated in supposedly healthy men who consumed only low or moderate amounts of alcohol, and in alcoholic patients without severe liver disease [72]. Plasma concentrations of MDA and antibodies reactive to MDA-modified proteins were higher in the alcoholics when compared to controls. In the alcoholic group, three weeks after cessation of drinking, plasma concentrations of MDA and the titer of antibodies were decreased. In another study, higher titers of antibodies reactive to MDA-modified protein were observed in alcoholics, but the elevation was not statistically significant when compared to controls [128].

In a recent study of alcoholics, heavy drinkers, and social drinking controls, significantly higher IgG immunoreactivity to MAA 2 : 1 adduct was observed [129]. Alcoholics with hepatitis or cirrhosis expressed the highest immunoreactivity, whereas patients with nonalcoholic liver disease had much lower levels. These data suggest that IgG immunoreactivity to MAA 2 : 1 adducts may be related to alcoholic liver damage.

4.12
The Role of Alcohol-Derived Bioadducts in Alcohol-Related Tissue Injury

A major consequence of chronic long-term alcohol abuse is tissue injury including alcohol-related tissue injury. It has been proposed that adduct formation is important in alcoholic liver disease because of two predominant mechanisms: (i) covalent modification of cellular macromolecules by reactive compounds or metabolites resulting in the loss of cell function and (ii) the initiation of an immune response against macromolecules covalently modified by reactive metabolites.

A close correlation between the extent of covalent binding of some drugs and their toxicity has been established [130]. A recent refinement of the general hypothesis is that covalent modification of specific proteins leads to a loss of function resulting in cell death [131], and ultimately tissue damage.

An example of the "loss of function" mechanism relating to alcohol abuse is the impairment of tubulin polymerization into microtubules. Modification of ∼5% of α-tubulin monomers by acetaldehyde *in vitro* leads to the complete inhibition of polymerization [132], which is likely to account for the observed defective secretion of proteins from the livers of alcoholics, leading to increased intracellular protein levels and to osmotic ballooning [133]. Another potential example of loss of function is the impaired synthesis of glutathione, resulting from the reaction of acetaldehyde with cysteine to form the stable thiazolidine [134]. Glutathione plays a major protective role against reactive oxygen species in cells. Depletion of cellular glutathione levels is presumed to be one of the factors leading to cellular damage in alcoholic liver disease. Kera and coworkers [135] suggested that this reaction may also be important by disrupting the γ-glutamyl cycle for the uptake of amino acids by cells.

Many investigators have reported that the covalent modification of reactive metabolites to proteins makes them immunogenic. These reactive metabolites are too small to be immunogenic on their own, but when covalently bound to protein can result in antibody response to the metabolite, the protein, and/or the combination. Relevant to the "neoantigen" mechanism, both rats fed ethanol [42] and human alcoholics [45] have been shown to produce modified hepatic proteins and to generate antibodies reactive to these modifications [48]. As outlined above, these reactions have been well characterized, with almost all immune responses related to the production of antibodies. Indeed, only two groups have reported a T cell response to acetaldehyde-modified proteins [119, 120].

It was recently shown by Thiele *et al.* [26] that modification of soluble proteins with the MAA adduct could induce antibody responses to the MAA epitope and the unmodified protein epitopes on the soluble protein. In the absence of any adjuvants, it was possible to show that smaller immunizing doses favored the production of antibodies to the unmodified parts of the soluble protein, whereas larger doses favored the production of antibodies to the MAA epitope. These data were important as it was shown that an alcohol-related metabolite can induce an antibody response in the absence of adjuvants, and suggested a mechanism by which antibody to the MAA adduct or its carrier may be generated *in vivo*.

Classically, if there is an antibody response then there must be a T cell response. Using the highly characterized immunogen, hen egg lysozyme (HEL), to investigate how MAA modification stimulates antibody and T cell responses in the absence of adjuvants, Willis *et al.* [136] was able to show that the MAA modification of proteins appears to alter the immunogenicity of proteins as measured by antibody production and T cell proliferative responses. Additionally, it appeared that these responses were most likely mediated by scavenger receptors that recognize proteins adducted with MAA. Thus, these data strongly suggested a mechanism by which proteins modified with oxidative products associated with chronic ethanol consumption may alter immune responses that may play an active role in the development and/or progression of alcoholic liver disease.

These studies were performed with exogenous proteins, but preliminary data from our laboratories suggested that the modification of self proteins with MAA induces antibody and T cell responses to the self protein. Experiments were designed to determine whether proteins in mouse liver cytosol modified with MAA are capable of inducing immune responses, and play some role in the pathophysiology of ALD. Mice immunized with MAA-modified mouse liver cytosol showed a two- to threefold increase in serum GOT levels over the control group, with liver histology confirming the presence of significant damage. Antibodies to the MAA epitope were observed in the serum of these animals by ELISA, but the most interesting observation was the antibody response to unmodified mouse liver cytosol. Importantly, mice immunized with nothing or mouse liver cytosol showed no antibody responses to these antigens by ELISA or Western Blot analysis, and no increase in liver enzymes or alterations in liver histology. Thus, these data strongly suggested that MAA modification may be a mechanism by which immune responses may be initiated to the self-carrier protein resulting in an autoimmune reaction, and the involvement of the immune system in the development and/or progression of ALD.

In other studies with MAA-modified albumins, it was shown that receptor-mediated endocytosis was impaired following chronic alcohol consumption in male Wistar rats [137]. This impairment was seen for both *in situ* perfused livers and isolated sinusoidal liver endothelial cells (SECs). These studies indicated that MAA-Alb is degraded by SECs, and chronic alcohol consumption impairs the degradation of MAA-Alb. Additionally, this impairment was shown to be due to a defect in the internalization rather than binding or degradation of the modified protein. It was also shown that MAA-modified proteins can cause the secretion of cytokines/chemokines by SECs, Kupffer cells, and Stellate cells [138, 139]. Thus, the prolonged exposure of these adducts to these various cell types may have a unique role in inducing inflammatory responses.

Additionally, attempts have been made to correlate the presence of these immune responses with alcoholic liver disease. However, it has been difficult to generate significant liver damage in rodents using ethanol alone. Thus, until a reliable animal model of alcoholic liver disease can be developed, these correlations will be difficult to prove.

4.13
Future Directions

The measurement of both ethanol metabolite–modified proteins and their associated antibodies has shown promise in elucidating their roles in the etiology of alcohol-related tissue injury. However, there is one major problem that has so far clouded previous results and is holding up future work. That problem is the chemical nature of the modifications formed *in vivo* and *in vitro*. Without an understanding of the nature of the adducts formed during modification procedures used to prepare modified proteins *in vitro*, it is impossible to know what is being measured by the antibodies generated from their use. The situation is no better if the modified protein is used as the coating antigen for an ELISA as the immunoreactivity measured is then largely undefined. To rectify this situation, the authors and others are undertaking studies to chemically define the adducts produced under various conditions [33, 86, 134, 140]. Production of defined adducts has allowed the generation of some antibodies with known reactivity, enabling measurement of specific adducts to be carried out [36, 129]. Such antibodies so far generated include those with reactivity to N-ethylated amino groups [140] or those with reactivity against the MAA 2 : 1 adduct [86].

Reagents with known chemical structure or immunoreactivity would greatly advance this area of research and increase our level of understanding of the role of alcohol-derived bioadducts in the development and/or progression of ALD.

Acknowledgments

The authors wish to thank all members of the Experimental Immunology Laboratory and the Liver Study Unit at the Omaha VA Medical Center for their invaluable support and insights over the past 20 years. Additionally, we would like to acknowledge the contributions of our collaborators Simon Worrall and Emanuele Albano without whom some of our recent studies could not have been completed.

References

1 Klassen, L.W., Tuma, D., and Sorrell, M.F. (1995) Immune mechanisms of alcohol-induced liver disease. *Hepatology*, **22**, 355–357.
2 Preedy, V.R., Adachi, J., Ueno, Y., Ahmed, S., Mantle, D., Mullatti, N., Rajendram, R., and Peters, T.J. (2001) Alcoholic skeletal muscle myopathy: definitions, features, contribution of neuropathy, impact and diagnosis. *Eur J Neurol*, **8**, 677–687.
3 Worrall, S., Niemela, O., Parkkila, S., Peters, T.J., and Preedy, V.R. (2001) Protein adducts in type i and type ii fibre predominant muscles of the ethanol-fed rat: preferential localisation in the sarcolemmal and subsarcolemmal region. *Eur J Clin Invest*, **31**, 723–730.
4 Niemela, O., Parkkila, S., Worrall, S., Emery, P.W., and Preedy, V.R. (2003) Generation of aldehyde-derived protein modifications in ethanol-exposed heart. *Alcohol Clin Exp Res*, **27**, 1987–1992.
5 Preedy, V.R., Patel, V.B., Why, H.J., Corbett, J.M., Dunn, M.J., and Richardson, P.J. (1996) Alcohol and the heart: biochemical alterations. *Cardiovasc Res*, **31**, 139–147.
6 Lieber, C.S. (2000) Ethnic and gender differences in ethanol metabolism. *Alcohol Clin Exp Res*, **24**, 417–418.
7 Lieber, C.S. (1994) Hepatic and metabolic effects of ethanol: pathogenesis and prevention. *Ann Med*, **26**, 325–330.
8 Riveros-Rosas, H., Julian-Sanchez, A., and Pina, E. (1997) Enzymology of ethanol and acetaldehyde metabolism in mammals. *Arch Med Res*, **28**, 453–471.
9 Yin, S.J., Han, C.L., Lee, A.I., and Wu, C.W. (1999) Human alcohol dehydrogenase family. Functional classification, ethanol/retinol

metabolism, and medical implications. *Adv Exp Med Biol*, **463**, 265–274.

10 Bosron, W.F., Ehrig, T., and Li, T.K. (1993) Genetic factors in alcohol metabolism and alcoholism. *Semin Liver Dis*, **13**, 126–135.

11 Keilin, D. and Hartree, E.F. (1945) Properties of catalase. Catalysis of coupled oxidation of alcohols. *Biochem J*, **39**, 293–301.

12 Williamson, J.R., Scholz, R., Browning, E.T., Thurman, R.G., and Fukami, M.H. (1969) Metabolic effects of ethanol in perfused rat liver. *J Biol Chem*, **244**, 5044–5054.

13 Moser, H.W. and Moser, A.B. (1996) Very long-chain fatty acids in diagnosis, pathogenesis, and therapy of peroxisomal disorders. *Lipids*, **31** (Suppl), S141–S144.

14 Takagi, T., Alderman, J., Gellert, J., and Lieber, C.S. (1986) Assessment of the role of non-ADH ethanol oxidation in vivo and in hepatocytes from deermice. *Biochem Pharmacol*, **35**, 3601–3606.

15 Teschke, R., Hasumura, Y., Joly, J.G., and Lieber, C.S. (1972) Microsomal ethanol-oxidizing system (meos): purification and properties of a rat liver system free of catalase and alcohol dehydrogenase. *Biochem Biophys Res Commun*, **49**, 1187–1193.

16 Koop, D.R. and Casazza, J.P. (1985) Identification of ethanol-inducible p-450 isozyme 3a as the acetone and acetol monooxygenase of rabbit microsomes. *J Biol Chem*, **260**, 13607–13612.

17 Ryan, D.E., Ramanathan, L., Iida, S., Thomas, P.E., Haniu, M., Shively, J.E., Lieber, C.S., and Levin, W. (1985) Characterization of a major form of rat hepatic microsomal cytochrome p-450 induced by isoniazid. *J Biol Chem*, **260**, 6385–6393.

18 Wrighton, S.A., Campanile, C., Thomas, P.E., Maines, S.L., Watkins, P.B., Parker, G., Mendez-Picon, G., Haniu, M., Shively, J.E., Levin, W. *et al.* (1986) Identification of a human liver cytochrome p-450 homologous to the major isosafrole-inducible cytochrome p-450 in the rat. *Mol Pharmacol*, **29**, 405–410.

19 Nelson, D.R., Kamataki, T., Waxman, D.J., Guengerich, F.P., Estabrook, R.W., Feyereisen, R., Gonzalez, F.J., Coon, M.J., Gunsalus, I.C., Gotoh, O. *et al.* (1993) The p450 superfamily: Update on new sequences, gene mapping, accession numbers, early trivial names of enzymes, and nomenclature. *DNA Cell Biol*, **12**, 1–51.

20 Lieber, C.S. (1999) Microsomal ethanol-oxidizing system (meos): the first 30 years (1968-1998)--a review. *Alcohol Clin Exp Res*, **23**, 991–1007.

21 Perozich, J., Nicholas, H., Lindahl, R., and Hempel, J. (1999) The big book of aldehyde dehydrogenase sequences. An overview of the extended family. *Adv Exp Med Biol*, **463**, 1–7.

22 Sophos, N.A., Pappa, A., Ziegler, T.L., and Vasiliou, V. (2001) Aldehyde dehydrogenase gene superfamily: the 2000 update. *Chem Biol Interact*, **130-132**, 323–337.

23 Hoberman, H.D. (1979) Synthesis of 5-deoxy-d-xylulose-1-phosphate by human erythrocytes. *Biochem Biophys Res Commun*, **90**, 757–763.

24 Hoerner, M., Behrens, U.J., Worner, T., and Lieber, C.S. (1986) Humoral immune response to acetaldehyde adducts in alcoholic patients. *Res Commun Chem Pathol Pharmacol*, **54**, 3–12.

25 Niemela, O., Klajner, F., Orrego, H., Vidins, E., Blendis, L., and Israel, Y. (1987) Antibodies against acetaldehyde-modified protein epitopes in human alcoholics. *Hepatology*, **7**, 1210–1214.

26 Thiele, G.M., Tuma, D.J., Willis, M.S., Miller, J.A., McDonald, T.L., Sorrell, M.F., and Klassen, L.W. (1998) Soluble proteins modified with acetaldehyde and malondialdehyde are immunogenic in the absence of adjuvant. *Alcohol Clin Exp Res*, **22**, 1731–1739.

27 Worrall, S., De Jersey, J., Shanley, B.C., and Wilce, P.A. (1989) Ethanol induces the production of antibodies

to acetaldehyde-modified epitopes in rats. *Alcohol Alcohol*, **24**, 217–223.

28 Irving, M.G., Simpson, S.J., Brooks, W.M., Holmes, R.S., and Doddrell, D.M. (1985) Application of the reverse dept polarization-transfer pulse sequence to monitor in vitro and in vivo metabolism of 13c-ethanol by 1h-NMR spectroscopy. *Int J Biochem*, **17**, 471–478.

29 O'Donnell, J.P. (1982) The reaction of amines with carbonyls: its significance in the nonenzymatic metabolism of xenobiotics. *Drug Metab Rev*, **13**, 123–159.

30 Donohue, T.M. Jr, Tuma, D.J., and Sorrell, M.F. (1983) Binding of metabolically derived acetaldehyde to hepatic proteins in vitro. *Lab Invest*, **49**, 226–229.

31 Tuma, D.J., Newman, M.R., Donohue, T.M. Jr, and Sorrell, M.F. (1987) Covalent binding of acetaldehyde to proteins: participation of lysine residues. *Alcohol Clin Exp Res*, **11**, 579–584.

32 Tuma, D.J., Hoffman, T., and Sorrell, M.F. (1991) The chemistry of acetaldehyde-protein adducts. *Alcohol Alcohol Suppl*, **1**, 271–276.

33 Fowles, L.F., Beck, E., Worrall, S., Shanley, B.C., and De Jersey, J. (1996) The formation and stability of imidazolidinone adducts from acetaldehyde and model peptides. A kinetic study with implications for protein modification in alcohol abuse. *Biochem Pharmacol*, **51**, 1259–1267.

34 San George, R.C. and Hoberman, H.D. (1986) Reaction of acetaldehyde with hemoglobin. *J Biol Chem*, **261**, 6811–6821.

35 Klassen, L.W., Tuma, D.J., Sorrell, M.F., McDonald, T.L., DeVasure, J.M., and Thiele, G.M. (1994) Detection of reduced acetaldehyde protein adducts using a unique monoclonal antibody. *Alcohol Clin Exp Res*, **18**, 164–171.

36 Worrall, S., De Jersey, J., and Wilce, P.A. (2000) Comparison of the formation of proteins modified by direct and indirect ethanol metabolites in the liver and blood of rats fed the Lieber-DeCarli liquid diet. *Alcohol Alcohol*, **35**, 164–170.

37 Klassen, L.W., Jones, B.L., Sorrell, M.F., Tuma, D.J., and Thiele, G.M. (1999) Conversion of acetaldehyde-protein adduct epitopes from a nonreduced to a reduced phenotype by antigen processing cells. *Alcohol Clin Exp Res*, **23**, 657–663.

38 Medina, V.A., Donohue, T.M. Jr, Sorrell, M.F., and Tuma, D.J. (1985) Covalent binding of acetaldehyde to hepatic proteins during ethanol oxidation. *J Lab Clin Med*, **105**, 5–10.

39 Behrens, U.J., Hoerner, M., Lasker, J.M., and Lieber, C.S. (1988) Formation of acetaldehyde adducts with ethanol-inducible p450IIE1 in vivo. *Biochem Biophys Res Commun*, **154**, 584–590.

40 Lin, R.C., Smith, R.S., and Lumeng, L. (1988) Detection of a protein-acetaldehyde adduct in the liver of rats fed alcohol chronically. *J Clin Invest*, **81**, 615–619.

41 Zhu, Y., Fillenwarth, M.J., Crabb, D., Lumeng, L., and Lin, R.C. (1996) Identification of the 37-kd rat liver protein that forms an acetaldehyde adduct in vivo as delta 4-3-ketosteroid 5 beta-reductase. *Hepatology*, **23**, 115–122.

42 Worrall, S., De Jersey, J., Shanley, B.C., and Wilce, P.A. (1991) Detection of stable acetaldehyde-modified proteins in the livers of ethanol-fed rats. *Alcohol Alcohol*, **26**, 437–444.

43 Yokoyama, H., Ishii, H., Nagata, S., Kato, S., Kamegaya, K., and Tsuchiya, M. (1993) Experimental hepatitis induced by ethanol after immunization with acetaldehyde adducts. *Hepatology*, **17**, 14–19.

44 Nicholls, R.M., Fowles, L.F., Worrall, S., De Jersey, J., and Wilce, P.A. (1994) Distribution and turnover of acetaldehyde-modified proteins in liver and blood of ethanol-fed rats. *Alcohol Alcohol*, **29**, 149–157.

45 Niemela, O., Juvonen, T., and Parkkila, S. (1991) Immunohistochemical demonstration of

acetaldehyde-modified epitopes in human liver after alcohol consumption. *J Clin Invest*, **87**, 1367–1374.

46 Holstege, A., Bedossa, P., Poynard, T., Kollinger, M., Chaput, J.C., Houglum, K., and Chojkier, M. (1994) Acetaldehyde-modified epitopes in liver biopsy specimens of alcoholic and nonalcoholic patients: localization and association with progression of liver fibrosis. *Hepatology*, **19**, 367–374.

47 Paradis, V., Scoazec, J.Y., Kollinger, M., Holstege, A., Moreau, A., Feldmann, G., and Bedossa, P. (1996) Cellular and subcellular localization of acetaldehyde-protein adducts in liver biopsies from alcoholic patients. *J Histochem Cytochem*, **44**, 1051–1057.

48 Worrall, S., De Jersey, J., Nicholls, R., and Wilce, P. (1993) Acetaldehyde/protein interactions: are they involved in the pathogenesis of alcoholic liver disease? *Dig Dis*, **11**, 265–277.

49 Baraona, E., Di Padova, C., Tabasco, J., and Lieber, C.S. (1987) Red blood cells: a new major modality for acetaldehyde transport from liver to other tissues. *Life Sci*, **40**, 253–258.

50 Baraona, E., DiPadova, C., Tabasco, J., and Lieber, C.S. (1987) Transport of acetaldehyde in red blood cells. *Alcohol Alcohol Suppl*, **1**, 203–206.

51 Lin, R.C., Lumeng, L., Shahidi, S., Kelly, T., and Pound, D.C. (1990) Protein-acetaldehyde adducts in serum of alcoholic patients. *Alcohol Clin Exp Res*, **14**, 438–443.

52 Worrall, S., De Jersey, J., Wilce, P.A., Seppa, K., Hurme, L., and Sillanaukee, P. (1998) Comparison of carbohydrate-deficient transferrin, immunoglobulin a antibodies reactive with acetaldehyde-modified protein and acetaldehyde-modified albumin with conventional markers of alcohol consumption. *Alcohol Clin Exp Res*, **22**, 1921–1926.

53 Stockham, T.L. and Blanke, R.V. (1988) Investigation of an acetaldehyde-hemoglobin adduct in alcoholics. *Alcohol Clin Exp Res*, **12**, 748–754.

54 Niemela, O., Israel, Y., Mizoi, Y., Fukunaga, T., and Eriksson, C.J. (1990) Hemoglobin-acetaldehyde adducts in human volunteers following acute ethanol ingestion. *Alcohol Clin Exp Res*, **14**, 838–841.

55 Niemela, O., Halmesmaki, E., and Ylikorkala, O. (1991) Hemoglobin-acetaldehyde adducts are elevated in women carrying alcohol-damaged fetuses. *Alcohol Clin Exp Res*, **15**, 1007–1010.

56 Niemela, O. and Israel, Y. (1992) Hemoglobin-acetaldehyde adducts in human alcohol abusers. *Lab Invest*, **67**, 246–252.

57 Lin, R.C., Shahidi, S., Kelly, T.J., Lumeng, C., and Lumeng, L. (1993) Measurement of hemoglobin-acetaldehyde adduct in alcoholic patients. *Alcohol Clin Exp Res*, **17**, 669–674.

58 Hazelett, S.E., Liebelt, R.A., and Truitt, E.B. Jr (1993) Improved separation of acetaldehyde-induced hemoglobin. *Alcohol Clin Exp Res*, **17**, 1107–1111.

59 Sillanaukee, P. and Koivula, T. (1990) Detection of a new acetaldehyde-induced hemoglobin fraction Hba1ach by cation exchange liquid chromatography. *Alcohol Clin Exp Res*, **14**, 842–846.

60 Sillanaukee, P., Seppa, K., and Koivula, T. (1991) Association of a haemoglobin-acetaldehyde adduct with questionnaire results on heavy drinkers. *Alcohol Alcohol*, **26**, 519–525.

61 Gross, M.D., Gapstur, S.M., Belcher, J.D., Scanlan, G., and Potter, J.D. (1992) The identification and partial characterization of acetaldehyde adducts of hemoglobin occurring in vivo: a possible marker of alcohol consumption. *Alcohol Clin Exp Res*, **16**, 1093–1103.

62 Gross, M.D., Hays, R., Gapstur, S.M., Chaussee, M., and Potter, J.D. (1994) Evidence for the formation of multiple types of

acetaldehyde-haemoglobin adducts. *Alcohol Alcohol*, **29**, 31–41.

63 Chen, H.M., Scott, B.K., Braun, K.P., and Peterson, C.M. (1995) Validated fluorimetric HPLC analysis of acetaldehyde in hemoglobin fractions separated by cation exchange chromatography: three new peaks associated with acetaldehyde. *Alcohol Clin Exp Res*, **19**, 939–944.

64 Hurme, L., Seppa, K., Rajaniemi, H., and Sillanaukee, P. (1998) Chromatographically identified alcohol-induced haemoglobin adducts as markers of alcohol abuse among women. *Eur J Clin Invest*, **28**, 87–94.

65 Itala, L., Seppa, K., Turpeinen, U., and Sillanaukee, P. (1995) Separation of hemoglobin acetaldehyde adducts by high-performance liquid chromatography-cation-exchange chromatography. *Anal Biochem*, **224**, 323–329.

66 Esterbauer, H., Schaur, R.J., and Zollner, H. (1991) Chemistry and biochemistry of 4-hydroxynonenal, malonaldehyde and related aldehydes. *Free Radic Biol Med*, **11**, 81–128.

67 Crawford, D.L., Yu, T.C., and Sinnhuber, R.O. (1966) Reaction of malondialdehyde with glycine. *J Agric Food Chem*, **14**, 182–184.

68 Nair, V., Cooper, C.S., Vietti, D.E., and Turner, G.A. (1986) The chemistry of lipid peroxidation metabolites: crosslinking reactions of malondialdehyde. *Lipids*, **21**, 6–10.

69 Beppu, M., Murakami, K., and Kikugawa, K. (1986) Fluorescent and cross-linked proteins of human erythrocyte ghosts formed by reaction with hydroperoxylinoleic acid, malondialdehyde and monofunctional aldehydes. *Chem Pharm Bull (Tokyo)*, **34**, 781–788.

70 Buttkus, H. (1968) Reaction of cysteine and methionine with malondialdehyde. *J Am Oil Chem Soc*, **46**, 88–93.

71 Clot, P., Tabone, M., Arico, S., and Albano, E. (1994) Monitoring oxidative damage in patients with liver cirrhosis and different daily alcohol intake. *Gut*, **35**, 1637–1643.

72 Lecomte, E., Herbeth, B., Pirollet, P., Chancerelle, Y., Arnaud, J., Musse, N., Paille, F., Siest, G., and Artur, Y. (1994) Effect of alcohol consumption on blood antioxidant nutrients and oxidative stress indicators. *Am J Clin Nutr*, **60**, 255–261.

73 Houglum, K., Filip, M., Witztum, J.L., and Chojkier, M. (1990) Malondialdehyde and 4-hydroxynonenal protein adducts in plasma and liver of rats with iron overload. *J Clin Invest*, **86**, 1991–1998.

74 Kamimura, S., Gaal, K., Britton, R.S., Bacon, B.R., Triadafilopoulos, G., and Tsukamoto, H. (1992) Increased 4-hydroxynonenal levels in experimental alcoholic liver disease: association of lipid peroxidation with liver fibrogenesis. *Hepatology*, **16**, 448–453.

75 Li, C.J., Nanji, A.A., Siakotos, A.N., and Lin, R.C. (1997) Acetaldehyde-modified and 4-hydroxynonenal-modified proteins in the livers of rats with alcoholic liver disease. *Hepatology*, **26**, 650–657.

76 Niemela, O., Parkkila, S., Yla-Herttuala, S., Villanueva, J., Ruebner, B., and Halsted, C.H. (1995) Sequential acetaldehyde production, lipid peroxidation, and fibrogenesis in micropig model of alcohol-induced liver disease. *Hepatology*, **22**, 1208–1214.

77 Ohhira, M., Ohtake, T., Matsumoto, A., Saito, H., Ikuta, K., Fujimoto, Y., Ono, M., Toyokuni, S., and Kohgo, Y. (1998) Immunohistochemical detection of 4-hydroxy-2-nonenal-modified-protein adducts in human alcoholic liver diseases. *Alcohol Clin Exp Res*, **22**, 145S–149S.

78 French, S.W., Wong, K., Jui, L., Albano, E., Hagbjork, A.L., and Ingelman-Sundberg, M. (1993) Effect of ethanol on cytochrome p450 2e1 (cyp2e1), lipid peroxidation, and serum protein adduct formation in relation to liver pathology

pathogenesis. *Exp Mol Pathol*, **58**, 61–75.

79 Haberland, M.E., Fong, D., and Cheng, L. (1988) Malondialdehyde-altered protein occurs in atheroma of watanabe heritable hyperlipidemic rabbits. *Science*, **241**, 215–218.

80 Yla-Herttuala, S., Palinski, W., Rosenfeld, M.E., Parthasarathy, S., Carew, T.E., Butler, S., Witztum, J.L., and Steinberg, D. (1989) Evidence for the presence of oxidatively modified low density lipoprotein in atherosclerotic lesions of rabbit and man. *J Clin Invest*, **84**, 1086–1095.

81 Palinski, W., Rosenfeld, M.E., Yla-Herttuala, S., Gurtner, G.C., Socher, S.S., Butler, S.W., Parthasarathy, S., Carew, T.E., Steinberg, D., and Witztum, J.L. (1989) Low density lipoprotein undergoes oxidative modification in vivo. *Proc Natl Acad Sci U S A*, **86**, 1372–1376.

82 Niemela, O., Parkkila, S., Yla-Herttuala, S., Halsted, C., Witztum, J.L., Lanca, A., and Israel, Y. (1994) Covalent protein adducts in the liver as a result of ethanol metabolism and lipid peroxidation. *Lab Invest*, **70**, 537–546.

83 Holvoet, P., Perez, G., Zhao, Z., Brouwers, E., Bernar, H., and Collen, D. (1995) Malondialdehyde-modified low density lipoproteins in patients with atherosclerotic disease. *J Clin Invest*, **95**, 2611–2619.

84 Sparrow, C.P., Parthasarathy, S., and Steinberg, D. (1989) A macrophage receptor that recognizes oxidized low density lipoprotein but not acetylated low density lipoprotein. *J Biol Chem*, **264**, 2599–2604.

85 Lung, C.C., Fleisher, J.H., Meinke, G., and Pinnas, J.L. (1990) Immunochemical properties of malondialdehyde-protein adducts. *J Immunol Methods*, **128**, 127–132.

86 Tuma, D.J., Thiele, G.M., Xu, D., Klassen, L.W., and Sorrell, M.F. (1996) Acetaldehyde and malondialdehyde react together to generate distinct protein adducts in the liver during long-term ethanol administration. *Hepatology*, **23**, 872–880.

87 Kearley, M.L., Patel, A., Chien, J., and Tuma, D.J. (1999) Observation of a new nonfluorescent malondialdehyde-acetaldehyde-protein adduct by 13c NMR spectroscopy. *Chem Res Toxicol*, **12**, 100–105.

88 Tuma, D.J., Kearley, M.L., Thiele, G.M., Worrall, S., Haver, A., Klassen, L.W., and Sorrell, M.F. (2001) Elucidation of reaction scheme describing malondialdehyde-acetaldehyde-protein adduct formation. *Chem Res Toxicol*, **14**, 822–832.

89 Xu, D., Thiele, G.M., Beckenhauer, J.L., Klassen, L.W., Sorrell, M.F., and Tuma, D.J. (1998) Detection of circulating antibodies to malondialdehyde-acetaldehyde adducts in ethanol-fed rats. *Gastroenterology*, **115**, 686–692.

90 Jurgens, G., Lang, J., and Esterbauer, H. (1986) Modification of human low-density lipoprotein by the lipid peroxidation product 4-hydroxynonenal. *Biochim Biophys Acta*, **875**, 103–114.

91 Casini, A., Cunningham, M., Rojkind, M., and Lieber, C.S. (1991) Acetaldehyde increases procollagen type i and fibronectin gene transcription in cultured rat fat-storing cells through a protein synthesis-dependent mechanism. *Hepatology*, **13**, 758–765.

92 Pares, A., Potter, J.J., Rennie, L., and Mezey, E. (1994) Acetaldehyde activates the promoter of the mouse alpha 2(i) collagen gene. *Hepatology*, **19**, 498–503.

93 Parola, M., Pinzani, M., Casini, A., Albano, E., Poli, G., Gentilini, A., Gentilini, P., and Dianzani, M.U. (1993) Stimulation of lipid peroxidation or 4-hydroxynonenal treatment increases procollagen alpha 1 (i) gene expression in human liver fat-storing cells. *Biochem Biophys Res Commun*, **194**, 1044–1050.

94 Tsukamoto, H. (1993) Oxidative stress, antioxidants, and alcoholic liver fibrogenesis. *Alcohol*, **10**, 465–467.

95 Kuthan, H. and Ullrich, V. (1982) Oxidase and oxygenase function of the microsomal cytochrome p450 monooxygenase system. *Eur J Biochem*, **126**, 583–588.

96 Fridovich, I. (1989) Oxygen radicals from acetaldehyde. *Free Radic Biol Med*, **7**, 557–558.

97 Nakano, M., Kikuyama, M., Hasegawa, T., Ito, T., Sakurai, K., Hiraishi, K., Hashimura, E., and Adachi, M. (1995) The first observation of o2- generation at real time in vivo from non-kupffer sinusoidal cells in perfused rat liver during acute ethanol intoxication. *FEBS Lett*, **372**, 140–143.

98 Albano, E., Tomasi, A., Goria-Gatti, L., and Dianzani, M.U. (1988) Spin trapping of free radical species produced during the microsomal metabolism of ethanol. *Chem Biol Interact*, **65**, 223–234.

99 Clot, P., Bellomo, G., Tabone, M., Arico, S., and Albano, E. (1995) Detection of antibodies against proteins modified by hydroxyethyl free radicals in patients with alcoholic cirrhosis. *Gastroenterology*, **108**, 201–207.

100 Knecht, K.T., Adachi, Y., Bradford, B.U., Iimuro, Y., Kadiiska, M., Xuang, Q.H., and Thurman, R.G. (1995) Free radical adducts in the bile of rats treated chronically with intragastric alcohol: Inhibition by destruction of kupffer cells. *Mol Pharmacol*, **47**, 1028–1034.

101 Knecht, K.T., Bradford, B.U., Mason, R.P., and Thurman, R.G. (1990) In vivo formation of a free radical metabolite of ethanol. *Mol Pharmacol*, **38**, 26–30.

102 Moncada, C., Torres, V., Varghese, G., Albano, E., and Israel, Y. (1994) Ethanol-derived immunoreactive species formed by free radical mechanisms. *Mol Pharmacol*, **46**, 786–791.

103 Moore, D.R., Reinke, L.A., and McCay, P.B. (1995) Metabolism of ethanol to 1-hydroxyethyl radicals in vivo: detection with intravenous administration of alpha-(4-pyridyl-1-oxide)-n-t-butylnitrone. *Mol Pharmacol*, **47**, 1224–1230.

104 Albano, E., Clot, P., Comoglio, A., Dianzani, M.U., and Tomasi, A. (1994) Free radical activation of acetaldehyde and its role in protein alkylation. *FEBS Lett*, **348**, 65–69.

105 Gonthier, B., Jeunet, A., and Barret, L. (1991) Electron spin resonance study of free radicals produced from ethanol and acetaldehyde after exposure to a fenton system or to brain and liver microsomes. *Alcohol*, **8**, 369–375.

106 Puntarulo, S. and Cederbaum, A.I. (1989) Temperature dependence of the microsomal oxidation of ethanol by cytochrome p450 and hydroxyl radical-dependent reactions. *Arch Biochem Biophys*, **269**, 569–575.

107 Rajasinghe, H., Jayatilleke, E., and Shaw, S. (1990) DNA cleavage during ethanol metabolism: Role of superoxide radicals and catalytic iron. *Life Sci*, **47**, 807–814.

108 Boh, E.E., Baricos, W.H., Bernofsky, C., and Steele, R.H. (1982) Mitochondrial chemiluminescence elicited by acetaldehyde. *J Bioenerg Biomembr*, **14**, 115–133.

109 Nakao, L.S., Kadiiska, M.B., Mason, R.P., Grijalba, M.T., and Augusto, O. (2000) Metabolism of acetaldehyde to methyl and acetyl radicals: In vitro and in vivo electron paramagnetic resonance spin-trapping studies. *Free Radic Biol Med*, **29**, 721–729.

110 Albano, E., Clot, P., Morimoto, M., Tomasi, A., Ingelman-Sundberg, M., and French, S.W. (1996) Role of cytochrome p4502e1-dependent formation of hydroxyethyl free radical in the development of liver damage in rats intragastrically fed with ethanol. *Hepatology*, **23**, 155–163.

111 Dupont, I., Lucas, D., Clot, P., Menez, C., and Albano, E. (1998) Cytochrome p4502e1 inducibility and hydroxyethyl radical formation among alcoholics. *J Hepatol*, **28**, 564–571.

112 Nakao, L.S. and Augusto, O. (1998) Nucleic acid alkylation by free radical metabolites of ethanol. Formation

of 8-(1-hydroxyethyl)guanine and 8-(2-hydroxyethyl)guanine adducts. *Chem Res Toxicol*, **11**, 888–894.

113 Schuessler, H. (1981) Reactions of ethanol and formate radicals with ribonuclease a and bovine serum albumin in radiolysis. *Int J Radiat Biol Relat Stud Phys Chem Med*, **40**, 483–492.

114 Clot, P., Parola, M., Bellomo, G., Dianzani, U., Carini, R., Tabone, M., Arico, S., Ingelman-Sundberg, M., and Albano, E. (1997) Plasma membrane hydroxyethyl radical adducts cause antibody-dependent cytotoxicity in rat hepatocytes exposed to alcohol. *Gastroenterology*, **113**, 265–276.

115 Robin, M.A., Le Roy, M., Descatoire, V., and Pessayre, D. (1997) Plasma membrane cytochromes p450 as neoantigens and autoimmune targets in drug-induced hepatitis. *J Hepatol*, **26** (Suppl 1), 23–30.

116 Meyerhof, O., Lohmann, K., and Schuster, P. (1936) *Biochem Z*, **286**, 319–335.

117 Lumeng, L. and Durant, P.J. (1985) Regulation of the formation of stable adducts between acetaldehyde and blood proteins. *Alcohol*, **2**, 397–400.

118 Al-Abed, Y., Mitsuhashi, T., Li, H., Lawson, J.A., FitzGerald, G.A., Founds, H., Donnelly, T., Cerami, A., Ulrich, P., and Bucala, R. (1999) Inhibition of advanced glycation end-product formation by acetaldehyde: Role in the cardioprotective effect of ethanol. *Proc Natl Acad Sci U S A*, **96**, 2385–2390.

119 Kolber, M.A. and Terabayashi, H. (1991) Cytotoxic t lymphocytes can be generated against acetaldehyde-modified syngeneic cells. *Alcohol Alcohol Suppl*, **1**, 277–280.

120 Terabayashi, H. and Kolber, M.A. (1990) The generation of cytotoxic t lymphocytes against acetaldehyde-modified syngeneic cells. *Alcohol Clin Exp Res*, **14**, 893–899.

121 Israel, Y., Hurwitz, E., Niemela, O., and Arnon, R. (1986) Monoclonal and polyclonal antibodies against acetaldehyde-containing epitopes in acetaldehyde-protein adducts. *Proc Natl Acad Sci U S A*, **83**, 7923–7927.

122 Tsukamoto, H., Rippe, R., Niemela, O., and Lin, M. (1995) Roles of oxidative stress in activation of kupffer and ito cells in liver fibrogenesis. *J Gastroenterol Hepatol*, **10** (Suppl 1), S50–S53.

123 Hoerner, M., Behrens, U.J., Worner, T.M., Blacksberg, I., Braly, L.F., Schaffner, F., and Lieber, C.S. (1988) The role of alcoholism and liver disease in the appearance of serum antibodies against acetaldehyde adducts. *Hepatology*, **8**, 569–574.

124 Worrall, S., De Jersey, J., Shanley, B.C., and Wilce, P.A. (1990) Antibodies against acetaldehyde-modified epitopes: presence in alcoholic, non-alcoholic liver disease and control subjects. *Alcohol Alcohol*, **25**, 509–517.

125 Trudell, J.R., Ardies, C.M., and Anderson, W.R. (1990) Cross-reactivity of antibodies raised against acetaldehyde adducts of protein with acetaldehyde adducts of phosphatidyl-ethanolamine: possible role in alcoholic cirrhosis. *Mol Pharmacol*, **38**, 587–593.

126 Trudell, J.R., Ardies, C.M., Green, C.E., and Allen, K. (1991) Binding of anti-acetaldehyde igg antibodies to hepatocytes with an acetaldehyde-phosphatidylethanolamine adduct on their surface. *Alcohol Clin Exp Res*, **15**, 295–299.

127 Koskinas, J., Kenna, J.G., Bird, G.L., Alexander, G.J., and Williams, R. (1992) Immunoglobulin a antibody to a 200-kilodalton cytosolic acetaldehyde adduct in alcoholic hepatitis. *Gastroenterology*, **103**, 1860–1867.

128 Chancerelle, Y., Mathieu, J., and Kergonou, J.F. (1998) Antibodies against malondialdehyde-modified proteins. Induction and ELISA measurement of specific antibodies. *Methods Mol Biol*, **108**, 111–118.

129 Rolla, R., Vay, D., Mottaran, E., Parodi, M., Traverso, N., Arico, S., Sartori, M., Bellomo, G., Klassen, L.W., Thiele, G.M., Tuma, D.J.,

and Albano, E. (2000) Detection of circulating antibodies against malondialdehyde-acetaldehyde adducts in patients with alcohol-induced liver disease. *Hepatology*, **31**, 878–884.

130 Lecoeur, S., Bonierbale, E., Challine, D., Gautier, J.C., Valadon, P., Dansette, P.M., Catinot, R., Ballet, F., Mansuy, D., and Beaune, P.H. (1994) Specificity of in vitro covalent binding of tienilic acid metabolites to human liver microsomes in relationship to the type of hepatotoxicity: comparison with two directly hepatotoxic drugs. *Chem Res Toxicol*, **7**, 434–442.

131 Boelsterli, U.A. (1993) Specific targets of covalent drug-protein interactions in hepatocytes and their toxicological significance in drug-induced liver injury. *Drug Metab Rev*, **25**, 395–451.

132 Smith, S.L., Jennett, R.B., Sorrell, M.F., and Tuma, D.J. (1992) Substoichiometric inhibition of microtubule formation by acetaldehyde-tubulin adducts. *Biochem Pharmacol*, **44**, 65–72.

133 Nicholls, R., De Jersey, J., Worrall, S., and Wilce, P. (1992) Modification of proteins and other biological molecules by acetaldehyde: adduct structure and functional significance. *Int J Biochem*, **24**, 1899–1906.

134 Anni, H., Pristatsky, P., and Israel, Y. (2003) Binding of acetaldehyde to a glutathione metabolite: mass spectrometric characterization of an acetaldehyde-cysteinylglycine conjugate. *Alcohol Clin Exp Res*, **27**, 1613–1621.

135 Kera, Y., Komura, S., Kiriyama, T., and Inoue, K. (1985) Effects of gamma-glutamyltranspeptidase inhibitor and reduced glutathione on renal acetaldehyde levels in rats. *Biochem Pharmacol*, **34**, 3781–3783.

136 Willis, M.S., Thiele, G.M., Tuma, D.J., and Klassen, L.W. (2003) T cell proliferative responses to malondialdehyde-acetaldehyde haptenated protein are scavenger receptor mediated. *Int Immunopharmacol*, **3**, 1381–1399.

137 Duryee, M.J., Klassen, L.W., Freeman, T.L., Willis, M.S., Tuma, D.J., and Thiele, G.M. (2003) Chronic ethanol consumption impairs receptor-mediated endocytosis of MAA-modified albumin by liver endothelial cells. *Biochem Pharmacol*, **66**, 1045–1054.

138 Duryee, M.J., Klassen, L.W., Freeman, T.L., Willis, M.S., Tuma, D.J., and Thiele, G.M. (2004) Lipopolysaccharide is a cofactor for malondialdehyde-acetaldehyde adduct-mediated cytokine/chemokine release by rat sinusoidal liver endothelial and kupffer cells. *Alcohol Clin Exp Res*, **28**, 1931–1938.

139 Kharbanda, K.K., Todero, S.L., Shubert, K.A., Sorrell, M.F., and Tuma, D.J. (2001) Malondialdehyde-acetaldehyde-protein adducts increase secretion of chemokines by rat hepatic stellate cells. *Alcohol*, **25**, 123–128.

140 Thiele, G.M., Wegter, K.M., Sorrell, M.F., Tuma, D.J., McDonald, T.L., and Klassen, L.W. (1994) Specificity of n-ethyl lysine of a monoclonal antibody to acetaldehyde-modified proteins prepared under reducing conditions. *Biochem Pharmacol*, **48**, 183–189.

5
Iron from Meat Produces Endogenous Procarcinogenic Peroxides

Denis E. Corpet, Françoise Guéraud, and Peter J. O'Brien

Iron is an essential mineral for all living cells and organisms. Iron deficiency remains a large public health threat, particularly in young women and in developing countries. However, iron overload, either due to long-term excess intake of red meat, multiple transfusions, or hereditary hemochromatosis, is associated with toxic effects. Iron overload increases the risk of several chronic diseases that are frequent in affluent countries, including cancer of the colorectum. These toxic effects may be attributed to oxidized biomolecules formed endogenously by the pro-oxidative properties of iron, which catalyze the oxidation of biomolecules. These oxidized biomolecules, including hydroxyl and superoxide radicals, amino acid carbonyl adducts, carboxymethyllysine; and other advanced glycation end products (AGEs, formed via methylglyoxal or glyceraldehyde); and malondialdehyde, 4-hydroxy-2-nonenal, and other advanced lipoxydation end products (ALEs), may thus be considered as endogenous toxins.

5.1
Introduction

Dietary iron is essential to build the body's oxygen carriers (blood hemoglobin and muscle myoglobin) and the vital enzymes in the electron transport chain that make ATP, carriers such as cytochrome *c*, and enzymes such as catalase and xanthine oxidase. Iron deficiency is the most widespread nutritional disorder in the world, notably among children and premenopausal women. It results in anemia. Heme iron from meat is much better absorbed than nonheme iron from plant food (25% vs. 1–10%) [1]. As a result of the limited absorption, the major part of ingested iron is not absorbed and reaches the large intestine, where it may exert unwanted pro-oxidant effects [2].

Red and processed meat consumption increases the risk of colorectal cancer, according to meta-analyses of epidemiological studies [1, 3]. Meat eaters may also be at risk for lung, liver, and pancreas cancer [2]. If meat intake causes cancer, it could be due to carcinogens present in the meat, for example, heterocyclic amines and polycyclic aromatic hydrocarbons in grilled meat and N-nitrosated

Endogenous Toxins. Diet, Genetics, Disease and Treatment.
Edited by Peter J. O'Brien and W. Robert Bruce
Copyright © 2010 WILEY-VCH Verlag GmbH & Co. KGaA, Weinheim
ISBN: 978-3-527-32363-0

compounds in cured meat [4–6]. Alternatively, meat intake could induce the formation of procarcinogenic endogenous toxins in the gut or in the body, such as secondary bile acids, amino acid bacterial metabolites, endogenous N-nitrosated compounds, and heme-induced free radicals and aldehydes. We now briefly review four hypotheses relating meat and iron consumption with carcinogenesis through the latter, endogenous toxins:

1. Fatty meat intake increases the secretion of bile acids in the gut. Gut bacteria deconjugate and dehydroxylate the bile acids, leading to more cytotoxic and genotoxic acids, notably lithocholic acid. However, as fat intake is not associated, *per se*, to colorectal cancer risk in humans or in animal models [7], it is thus unlikely that fat-induced bile acids are the major cause of cancer in the gut [8].
2. High-protein meals lead to excess amino acids entering the large intestine, where bacterial enzymes produce toxic metabolites including ammonia, cresol, indican, and amines. These small MW molecules promoted experimental cancer in rats when they were infused intrarectally. However, they do not appear to play an important role in cancer promotion when brought more physiologically to the colon by a high-protein diet [9].
3. Red meat intake increased fecal excretion of *N*-nitroso compounds (NOCs) in volunteers [10]. Most tested NOCs have been shown to be carcinogenic in rodents, and dietary red meat intake increases DNA adduct, O6-carboxymethyl guanine, formation. However, it is not clear that if meat-induced NOCs are human carcinogens. NOC yield is clearly linked to heme intake, because chicken meat does not induce fecal NOC in contrast to hemoglobin provided as black pudding [11]. A likely source of the increased intestinal NOC concentration is the formation of nitrosyl heme [12].
4. Alternatively, dietary iron may contribute to colon cancer risk through the production of reactive oxygen species (ROS). Indeed, heme iron from red meat induces peroxidation of polyunsaturated fats in the diet and systemically *in vivo*, resulting in mutagenic and toxic aldehydes. Consistent with this notion is the observation that hemochromatosis patients who have very high body iron stores are at a very high risk for liver cancer [13]. It is not known yet whether cancer risk increases with transfusional iron overload and other conditions of increased iron load.

The present chapter deals only with this fourth hypothesis. After reviewing molecular mechanisms of iron toxicity, we examine how heme iron could produce procarcinogenic effects in the gut and inside the body, through endogenous toxins.

5.2
Toxic Effects of Iron: Molecular Mechanisms

Several molecular mechanisms may explain the toxic effects of iron overload due to high red meat diet, multiple transfusions, or hereditary hemochromatosis. Iron and copper can catalyze many chemical reactions through their redox properties. The reactions that take place in the body are relevant not only to cancer but also to human health. The reactions we describe in detail in this chapter include Fenton chemistry, oxidation of proteins, protein carbonylation, advanced glycation end products (AGEs) formation, lipid peroxidation and advanced lipoxidation end products (ALEs) formation, and oxidation of nucleic acids and other macromolecules or biomolecules.

5.2.1
Fe-/Cu-Catalyzed Fenton Chemistry

Iron and copper are essential for life but they also participate in the conversion of hydrogen peroxide to ROS by Fenton chemistry. Fenton in the 1890s and Haber Weiss in the 1930s suggested that ferrous iron was oxidized by hydrogen peroxide to form hydroxyl radicals, ferric iron, and a hydroxyl anion by the following equation [14]:

$$Fe^{2+} + H_2O_2 \longrightarrow Fe^{3+} + {}^{\bullet}OH + OH^-$$

An equation was introduced by Haber and Weiss and later by Beauchamp and Fridovich in which chain reactions involving superoxide radicals reacted with H_2O_2 to form hydroxyl radicals, catalyzed by transition metals as follows:

$$O_2^- + H_2O_2 \longrightarrow O_2 + {}^{\bullet}OH + OH^-$$

However, the Haber Weiss reaction seems to occur only in the gas phase and not in aqueous solutions under physiological conditions [14, 15]. In any case, superoxide radicals formed by ferrous iron autoxidation are rapidly converted to H_2O_2 (catalyzed by superoxide dismutase) and the H_2O_2 is then converted to H_2O and O_2 (catalyzed by catalase). Hydroxyl radical reactions are diffusion controlled. That is, the radicals interact with targets in the vicinity of their generation. Fe(III) or Cu(II), when reduced by ascorbate or NAD(P)H, bind to specific metal binding sites on proteins and react with H_2O_2 to generate ${}^{\bullet}OH$ which then attacks neighboring amino acid residues to form protein-free radicals that can directly oxidize glutathione [16]. The nature of the oxidizing species in aqueous Fenton chemistry is still being debated and likely includes iron in higher oxidation states in addition to hydroxyl radicals. Fenton chemistry forms very reactive oxidizing species that oxidatively destroy organic compounds (e.g., proteins undergo fragmentation, cross-linking, carbonyl formation, peroxide formation) or hydroxylates them (e.g., protein tyrosine hydroxylation or benzene hydroxylation to phenol) [17].

5.2.2
Fe-/Cu-Catalyzed Oxidation of Proteins

The cellular and protein location of available redox active iron determines the specificity of H_2O_2-mediated oxidation in a site-specific manner. Protein histidine, proline, arginine, lysine, methionine, cysteine are oxidized to free radicals by transition metal dependent Fenton radicals. Protein aromatic amino acids, by contrast, are the preferred targets for oxidation by hydroxyl radicals, such as those generated by γ-radiolysis, which do not require catalysis by transition metals. Protein tyrosine, on the one hand, is oxidized by hydroxyl radicals through proton abstraction to form phenoxyl radicals which then undergo radical recombination and isomerization to form dityrosine [17]. Protein tryptophan, on the other hand, is destroyed by hydroxyl radicals to form hydroxytryptophans or by N-formylkynurenine to kynurenine. Similarly, histidine is converted to oxo-histidine and aspartate, and phenylalanine undergoes hydroxylation to o- and m-tyrosines. These oxidative changes are readily followed by fluorescence (dityrosine and tryptophan) or by HPLC, after pronase digestion (tyrosine and bityrosine), and are easily prevented by hydroxyl radical scavengers [18–20]. Dityrosine is useful as a specific marker for selective proteolysis, because after its formation *in vivo*, catalyzed by a Fenton reaction, dityrosine is released by subsequent proteolysis. The major products of protein tyrosine oxidation released by pronase digestion were also due to hydroxylation by hydroxyl radicals to form dopamine, dopamine quinone, and 5,6-dihydroxyindol [19, 21].

5.2.3
Fe-/Cu-Catalyzed Protein Carbonylation

Hydroxyl radicals also oxidize protein amino acids to carbonyls. They can be detected with 2,4-dinitrophenylhydrazine (DNPH), which reacts with carbonyl groups to form dinitrophenylhydrazones with an absorbance maxima at 360–390 nm [18–20]. Protein arginine or proline form glutamic semialdehyde; lysine is converted to aminoadipic semialdehyde. The former product is more common than the latter, but the two semialdehydes are the main carbonyl products of iron- or copper-catalyzed protein oxidation. Tissue protein oxidative carbonylation accumulates with age and also increases in Alzheimer's disease, cataractogenesis, amyotrophic lateral sclerosis, and respiratory distress syndrome [18, 20]. Protein amino acids can also covalently bind carbonyl products formed during sugar dicarbonyl autoxidation and thereby form amino acid : carbonyl adducts [18–21]. In particular, dicarbonyls readily form adducts with cysteine, lysine, and arginine. Lipid peroxide decomposition products also form adducts with cysteine, lysine, or histidine [18, 19, 21]. Cu-catalyzed protein carbonylation requiring ascorbic acid has been largely attributed to the oxidative degradation of ascorbate to glyoxal, although protein oxidation by ROS could also contribute to the protein carbonylation as catalase prevented protein carbonylation [22]. Lens proteins or crystallins incubated with ascorbate caused cataractous crystallin formation through protein glycation and carboxymethyllysine (CML) formation [23].

5.2.4
Fe-/Cu-Catalyzed Advanced Glycation End Products (AGEs) Formation

Thermolysis of sugars with proteins causes a rapid browning reaction, the Maillard reaction. Even at $37\,^\circ$C, incubation of amino acids with glucose for several weeks results in this reaction. The reaction is initiated by reversible Schiff base condensation of the acyclic aldehyde form with free amino groups. The aldimine Schiff base formed then undergoes an irreversible Amadori rearrangement requiring oxygen that is catalyzed by Fe or Cu to form ketoamine Amadori products, collectively called *advanced glycation end products*. According to studies with glutamine synthetase, the site-specific mechanism proposed is the binding of FeII to the amino group of lysine which can then react with H_2O_2 to generate a hydroxyl radical that oxidizes that lysine. It is also proposed that lysines at the iron binding site would be the most vulnerable to lysine AGE formation [18, 20]. Fe chelators also prevented AGE formation [23, 24]. The most abundant AGE is probably CML in which the carboxymethyl radical arises from fragmentation of the attached glucose through an ene-diol intermediate. Rearrangement, dehydration, and condensation reactions may also be involved in forming the irreversibly cross-linked fluorescent AGE derivatives. Protein CML also complexed Cu and oxidized ascorbate [24]. AGEs formed by long-lived proteins (e.g., lens crystallin and collagen), which results in impaired protein functions, are of particular interest to biochemists researching the molecular basis of diabetic complications such as vascular complications (retinopathy). Ferrous iron readily catalyzed nonenzymatic glycosylation of type I collagen (Asn-COOH, Gln-NH2) and D-glucose to irreversibly form AGEs [25]. Alternative pathways include glucose autoxidation to glyoxal and glycolaldehyde, which formed Schiff bases with lysine side chains and then formed CML [26, 27]. Methylglyoxal (pyruvaldehyde) formed from glycolytic triose phosphates also formed Schiff bases and then CML, resulting in mitochondrial damage [28].

Glyceraldehyde is another reactive carbonyl intermediate of glycolysis, which is formed from fructose, catalyzed by fructokinase to form fructose-1P, then converted to glyceraldehyde catalyzed by aldolase B. The AGEs formed by glyceraldehyde are known as *TAGEs* (*toxic advanced glycation end products*), as their toxicity has been demonstrated. They may be responsible for the pathogenesis of Alzheimer's disease, perhaps by killing cortical neuronal cells through AGE formation with the microtubule-associated tau protein of the neurofibrillary tangles [29]. TAGE formation has been postulated to contribute to other neurodegenerative diseases such as Huntington's disease, Freidrich's ataxia disease, and myotonic dystrophy [29]. AGEs have also been linked to heart disease, skin aging, and rheumatoid arthritis [29]. AGEs or TAGEs exert their action through a specific AGE receptor or TAGE receptors that are mostly located on the cell surface. The TAGE receptor activates a cascade of intracellular reactions that produce proinflammatory cytokines and increases oxidative stress thereby causing vasoconstriction, platelet accumulation, and damage to the small blood vessels in the kidney and the retina of the eye. Receptors are also found in endothelial cells, smooth muscle, immune cells, lung, liver, and peripheral blood cells. Most AGEs are probably inert and simply accumulate with age without harmful effects. Indeed, some AGE receptors may activate NADPH oxidase and form low levels of oxygen radicals that activate the antioxidant response element (ARE) of several hundred ARE-regulated antioxidant genes that increase cellular resistance to oxidative stress [30].

5.2.5
Fe-/Cu-Catalyzed Lipid Peroxidation and Advanced Lipoxidation End Products (ALEs)

Polyunsaturated fatty acids readily autoxidize when initiated by Fe^{2+} or Cu^{1+} with H_2O_2, oxygen, or radicals such as phenoxyl radical. The fatty acid hydroperoxides formed then undergo breakdown to malondialdehyde (MDA), acrolein, glyoxal, 4-hydroxy-2-nonenal, and 4-oxo-2-nonenal. They readily form adducts with some protein amino acids and cause intra- and intermolecular protein cross-linking. Advanced lipoxidation products were formed from these lipid peroxidation or unsaturated fatty acid hydroperoxide carbonyl decomposition products [31].

Oxysterol formed from cholesterol is likely responsible for the oxidation of low density lipoprotein (LDL) to form OxLDL, a biomarker for atherosclerosis. OxLDL is also likely responsible for foam cell formation and cytotoxicity that contributes to atherosclerosis. This is probably caused by an ALE formed when 4-hydroxynonenal (HNE) and MDA form Schiff base adducts with the LDL apoB 100 protein lysine. This Schiff base prevents OxLDL binding to the LDL receptor but enables OxLDL to bind to the macrophage scavenger (Gieseg 2009, this book). Macrophage cytotoxicity induced by OxLDL was also attributed to intralysosomal redox active iron released by lysosomal rupture [32]. Recently, diabetic kidneys were shown with antibodies to contain the ALE MDA–lysine located in the macrophage-derived foam cells that had accumulated in diabetic kidneys. Proinflammatory proteins were also significantly induced and monocyte binding to vascular smooth muscle and endothelial cells was increased. This could suggest that diabetic kidney damage involved vascular complications caused by monocyte ALE [33].

HNE cytotoxicity may partly involve inactivation of glyceraldehyde-3-phosphate dehydrogenase by covalent attachment to cysteine and lysine possibly involving cross-linking reactions [34].

5.2.6
Fe-/Cu-Catalyzed Oxidation of Nucleic Acids

Hydroxyl radicals and ferryl iron species can oxidize nucleic acid bases. The most frequently used biomarker of DNA oxidation by hydroxyl radicals is 8-hydroxyguanosine (8-oxodG), which has been found in urine, organs, and cells. 8-oxodG has been shown to be highly mutagenic, creating G→T transversions, a consequence of DNA misreading. Disease states involving an overload of iron frequently develop hepatocellular carcinoma, and a relationship between elevated body iron stores and increased risk of cancer has been established. With idiopathic hemochromatosis the incidence of hepatocarcinoma is 30%. It is preceded by liver fibrosis associated with lipid peroxidation and then cirrhosis.

Treatment of Wistar rats with ferric nitrilotriacetate induces renal carcinoma and causes renal DNA–protein cross-linking [35]. This likely occurs as a result of hydroxyl radical formation from a Fenton reaction involving Fe^{2+} nitrilotriacetate +

H_2O_2. With *in vitro* studies, hepatocytes were loaded with Fe by preincubating hepatocytes with 10 or 100 μM ferric nitrilotriacetate for 4–48 hours. GC–MS analysis of the Fe loaded hepatocytes identified oxidized DNA purine bases, including 8-oxo-guanine, xanthine, fapy-adenine, 2-oxo-adenine, and a lesser number of oxidized DNA pyrimidines, including 5-OHMe-uracil, 5-OH-uracil, 5-OH-cytosine. Under these conditions, the levels of free MDA, used as a measure of lipid peroxidation, increased substantially and levels of 8-oxoDG reached as high as 271 modified bases per 10^6 DNA bases. The formation of both biomarkers was inhibited by α-tocopherol suggesting that 8-oxoDG was caused by ROS formed by lipid peroxidation [36]. The flavonoid, myricetin, prevented biomarker formation by stimulating DNA repair [37]. Whether α-tocopherol or myricetin prevents ferric nitrilotriacetate-induced liver cancer *in vivo* is not known. Efforts to stimulate DNA repair pathways and increased release of oxidized purines into the media by base excision were not sufficient to prevent the intracellular accumulation of oxidized pyrimidines [38]. DNA cleavage occurred when 10 μM FeIII was incubated with CML, an AGE, and was prevented by the ROS scavenger pyridoxal phosphate [39].

In vitro, the lipid peroxyl radicals resulting from Fe-/Cu-catalyzed lipid peroxidation effectively cleave DNA, and aldehydic lipid oxidation products such as MDA, acrolein, HNE, or 4-oxo-nonenal can react with exocyclic amino groups of DNA bases giving mutagenic DNA adducts, which might thus contribute to carcinogenesis [40–43].

5.2.7
Role of Lysosomal Fe/Cu in Oxidative Stress Cytotoxicity

Cu^{1+} is more active than Fe^{2+} in causing oxidative stress–induced cytotoxicity, but the concentration of Fe is much higher than Cu in the cell. To reduce toxicity, free iron or copper in the cell is maintained at a very low concentration by intracellular storage proteins such as transferrin, ferritin, or chaperone proteins. Liver lysosomes are the major store of redox active iron likely because cellular cytochromes are endocytosed and digested by acid-activated cathepsins to release Fe. Lysosomes in the Kupffer cells of the liver are also major stores of copper in metallothionein. Metallothionein contains many cysteine residues that bind Cu and Zn in the cytosol. When liver becomes loaded with Cu as in Wilson's disease or as in the Long-Evans cinnamon rat, the Cu loaded metallothionein is taken up by the lysosomes [44]. The lysosomal membrane contains a H^+-ATPase that maintains the lysosomal pH at around 5.0. Iron chelators that are weak bases accumulate in the lysosomes and can specifically prevent oxidative stress cytotoxicity. The free iron in lysosomes makes lysosomes very susceptible to oxidative stress. To offset this, the lysosomes may autophagocytose iron binding proteins, for example, non-Fe-saturated ferritin or metallothioneins [45]. Cytotoxicity caused by Cu or ROS induced by quinone redox cycling can be prevented by endocytosis inhibitors or lysosomal ionophores [46].

5.2.8
Role of ROS and Iron in Cellular Redox Signaling that Could Cause Cell Transformation

ROS at low concentrations plays a redox signaling role in cell growth and division by activating transcriptional factors, such as AP-1, NF-kB, HIF-1, and p53. Several dozens of transcriptional factors can be activated when cultured cells are exposed to increased H_2O_2 concentrations [47]. These factors control the expression of genes whose protein products could participate in complex signal transduction leading to cell transformation. Tyrosine phosphatases, peroxyredoxins, and IkB kinase contain sensitive cysteines in their active sites that vary in their ease of oxidation by H_2O_2, which could modulate important signaling pathways.

Iron is also involved in cell signaling. For instance, fluctuations in intracellular catalytically active iron levels seem to play crucial roles in important signaling pathways connected with inflammatory processes. Chronic inflammatory diseases such as atherosclerosis, pulmonary fibrosis, Parkinson's disease, and chronic liver disease are often accompanied by iron accumulation. This was originally presumed to form toxic hydroxyl radicals by Fentons reaction that caused tissue damage resulting in inflammation. However, emerging evidence indicates that the effects are more complex than simple oxidant injury and involve mechanisms in which the expression of proinflammatory cytokines is induced by enhanced iron-dependent signaling. Inflammatory cytokines induce hepcidin in hepatocytes, which increases iron storage. This also increases the ability of hepcidin to bind to ferroportin for its internalization and decrease cellular iron export. Furthermore, it has been shown that proinflammatory cytokines are also induced by enhanced iron-dependent signaling. Recently discovered was a transient rise in intracellular iron complexes activated IkB kinase when hepatic macrophages were treated with lipopolysaccharide (LPS) or tumor necrosis factor-α [48].

5.3
Procarcinogenic Effects of Iron: *In vitro* Studies

5.3.1
Iron in Food

Refined vegetable oils are slowly oxidized in air, yielding hydroperoxides (LOOH). Iron decomposes those lipid peroxides to yield alkylperoxyl radicals (LOO$^\bullet$). Heme iron is 10–100 times more effective than inorganic iron to do so [49]. Hemoproteins such as myoglobin and hemoglobin found in meat are thus potent inducers of lipid peroxidation in foods [50, 51]. Secondary products of lipid peroxidation such as aldehydes can in turn covalently bind histidine residues of hemoproteins, inducing modifications of their tertiary structure. This increases exposure of heme to unusual ligands, such as lipids, thereby amplifying the whole process [52, 53].

5.3.2
Iron in Cell Culture

Heme iron induces DNA strand breaks in HT-29 cells and in primary human colonocytes [54]. Hemoglobin accelerates the growth of HT-29 cells through production of reactive oxygen radicals [55]. Heme iron reaching the gut might thus be both an initiator of DNA mutations and a promoter of initiated cells growth: this is the definition of a complete carcinogen, but the *in vivo* demonstration of heme carcinogenicity has not been demonstrated yet. High doses ($>100\,\mu M$) of the lipid peroxidation-induced aldehyde HNE is genotoxic to microadenoma colon cell line LT97 [56], while apoptotic effects and cell growth inhibition on human colon cancer cell lines were observed with lower doses [57, 58]. However, this apoptotic effect can be observed to have different effects in mouse colon cells depending on whether or not they are mutated on the Apc gene. This gives a survival advantage to mutated cells under pro-oxidant conditions. As Apc mutations are an early event in human colon carcinogenesis, secondary lipid peroxidation products could at least partly explain the promoting effects of heme-containing diets observed *in vivo* [59].

5.4
Procarcinogenic Effects of Iron in the Gut

5.4.1
Free Radicals in Gut Lumen

As reported above, Babbs showed that Fe/Cu catalyzed the Fenton reaction, leading to free radical production in human stools, and speculated that this mechanism could be a cause of colon cancer [60].

5.4.2
Hemin Effect in the Colon of Rats

Hemin added to a diet given to rats strongly increases proliferation of colon mucosa and fecal water cytotoxicity. Nonheme iron does not produce the same toxic effects. Calcium or chlorophyll added to hemin diet fully abolishes its toxic effects, likely because they bind to heme and precipitate or inactivate it in the gut lumen [61–63]. However, human beings eat hemoglobin and myoglobin not hemin: hemin is a free ferric porphyrin, with a freely exchangeable axial chloride group. Hemin is thus more reactive in diet, and toxic in rats, than blood hemoglobin or meat myoglobin that contains chlorine-free protein-bound heme [64].

5.4.3
Iron-Induced Lipid Peroxidation Products

Biomarkers of fat peroxidation increase in feces and urine of rodents, and in urine of volunteers, eating red meat, or black pudding [65, 66]. The biomarkers MDA, HNE, and HNE's major urinary metabolite 1,4-dihydroxynonane mercapturic acid (DHN-MA) may be representative of several other undetected aldehydes. MDA and HNE are also formed in foods with heme and PUFA [50]. MDA and HNE are mutagenic and cytotoxic, and their selective toxicity to Apc mutated cells *in vitro* may explain a promoting effect *in vivo* [59]. In addition, chronic exposure to high levels of dietary iron fortification increases lipid peroxidation in the mucosa of the rat large intestine [67]. In volunteers, iron sulfate supplements or a high red meat diet increases the formation of ROS in human feces, which supports Babbs' early hypothesis [68, 69]. In the authors' opinion, heme-induced aldehydes and ROS should be considered not only as biomarkers of meat intake but also as true endogenous toxins.

5.4.4
Effect of Heme Iron on Experimental Carcinogenesis

Carcinogenesis in rats is promoted by hemin, hemoglobin, and red meat intake, provided the diet contains 5% polyunsaturated oil but little calcium [40, 64, 65]. In contrast, diets supplemented with 0.1–0.5% Fe (using three different iron salts) do not promote chemical carcinogenesis in rats [70], although dietary iron at higher, overload concentrations (3.5% Fe-fumarate in diet) increases the number of tumors per carcinogen-induced mouse, and the tumor incidence in mice with sodium dextran sulfate-induced chronic colitis [71, 72]. This heme-induced promotion is associated with increased lipoperoxidation and cytotoxicity of fecal water, and is linked to urinary DHN-MA excretion [64, 65]. It is thus possible, but not formally demonstrated, that heme-induced aldehydes and/or ROS are a cause of colorectal cancer promotion.

5.4.5
Carcinogenesis Prevention

Prevention of heme-induced carcinogenesis is possible with dietary agents. Food components that bind heme iron and those that can block lipoperoxidation can inhibit heme-associated carcinogenicity. The addition of calcium, antioxidant mix, or olive oil to the diet inhibits hemin-induced lipoperoxidation, cytotoxicity, and promotion of carcinogenesis in rats [64]. Similarly, dietary calcium simultaneously normalizes beef meat–induced lipoperoxidation, cytotoxicity, and promotion of carcinogenesis in rats [73]. The risk of eating meat without chewing on the bones might thus be overcome by eating a yogurt after each steak!

5.5
Procarcinogenic Effects of Iron Inside the Body

5.5.1
Iron Concentration is Tightly Limited in Body Fluids

Iron transport and metabolism are strictly regulated to reduce the likelihood of cells being exposed to free iron and so to oxidative damage. When body iron stores are high, the absorption from the gut is much reduced. Most iron in body fluids and living tissues is bound to specialized proteins, such as transferrin and ferritin, which prevents its involvement in free radical generation. Thus, little or no free iron is present in the plasma or in the cytoplasm of cells [1, 60]. However, a hereditary disease that is very frequent (1/150), hemochromatosis, is characterized by excessive absorption of dietary iron, resulting in a pathological increase in total body iron stores. Iron overload can also be a consequence of multiple transfusions.

5.5.2
Iron Overload Causes Severe Diseases

Excess iron accumulates in tissues and organs disrupting their normal function, likely through oxidative stress. The most susceptible organs include the liver, adrenal glands, the heart, and the pancreas: patients can present with cirrhosis, adrenal insufficiency, heart failure, or diabetes. Individuals with no overt hemochromatosis such as heterozygous carriers of the mutation might have excess iron body stores involved in the causation of many of the chronic diseases that are linked to a Western lifestyle. These diseases might thus be gathered under the term *ferrotoxic disease* [74]. Important biological damages are also observed in patients following transfusions of large quantities of free hemoglobin into the circulation [75]. Large excess levels of iron in hemochromatosis often leads to cirrhosis, and put patients at a very high risk for liver cancer, and possibly increased risk of colorectal and hematologic cancers: Nearly one-third of such patients with hemochromatosis and cirrhosis eventually develop liver cancer. Patients with β-thalassemia rely on regular blood transfusions that can result in hepatic iron overload but also infection with hepatitis B or C viruses. Hepatocellular carcinoma occurs in 2% of β-thalassemia patients [76]. Hepatic etheno-adduct levels were significantly elevated in primary hemochromatosis patients and in iron storage disease or Wilson's disease (a copper storage disease) [42]. These promutagenic adducts can be generated by secondary lipid oxidation products such as HNE and could contribute to the increased risk of cancer observed in those patients [42]. Congestive heart failure and heart arrhythmias are also often seen in hemochromatosis patients, but these life-threatening conditions can be reversed with treatments that reduce excess iron stores [13]. Patients are advised not to eat red meat or take iron containing multivitamin pills. Deferoxamine, an iron chelator, is often the drug of choice, associated with periodic phlebotomies for decreasing body iron stores.

5.5.3
Colorectal Cancer and Iron Overload

Colorectal cancer risk is clearly linked to body iron stores in many epidemiological studies [77, 78], but not all, for example, Nurses' Health Study results do not support the association [79]. However, results from a recent randomized, single blinded clinical trial of stored iron reduction through repeated phlebotomies appear compelling: the 636 patients who were randomly assigned to iron reduction, followed for an average of 4.5 years, were found to have a lower rate of visceral cancer occurrence (hazard ratio = 0.65, 95% confidence interval = 0.43–0.97; $P = 0.04$) than the 641 patients in the control arm [74]. In this study, colorectal cancer risk seems to be associated with excess iron concentration in the gut and in the blood, but further clinical studies are needed to confirm this.

5.6
General Conclusion

Iron is an essential mineral to all living cells and organisms. Iron deficiency remains a large public health threat, particularly in young women and in developing countries. However, iron overload, due to long-term excess intake of red meat, multiple transfusions, or hereditary hemochromatosis, is associated with toxic effects. Iron overload increases the risk of several chronic diseases that are frequent in affluent countries, including cancer of the colorectum. These toxic effects may be attributed to oxidized biomolecules formed endogenously by the pro-oxidative properties of iron, which catalyze the oxidation of biomolecules. These oxidized biomolecules may thus be considered as endogenous toxins.

References

1 WCRF (2007) *Food, Nutrition, Physical Activity, and the Prevention of Cancer: A Global Perspective*, WCRF and American Institute for Cancer Research, Washington, DC, pp. 1–537.

2 Cross, A.J., Leitzmann, M.F., Gail, M.H., Hollenbeck, A.R., Schatzkin, A., and Sinha, R. (2007) A prospective study of red and processed meat intake in relation to cancer risk. *PLoS Med*, **4**, e325.

3 Santarelli, R.L., Pierre, F., and Corpet, D.E. (2008) Processed meat and colorectal cancer: a review of epidemiologic and experimental evidence. *Nutr Cancer*, **60**, 131–144.

4 Sugimura, T., Wakabayashi, K., Nakagama, H., and Nagao, M. (2004) Heterocyclic amines: Mutagens/carcinogens produced during cooking of meat and fish. *Cancer Sci*, **95**, 290–299.

5 Sinha, R., Kulldorff, M., Gunter, M.J., Strickland, P., and Rothman, N. (2005) Dietary benzo[a]pyrene intake and risk of colorectal adenoma. *Cancer Epidemiol Biomarkers Prev*, **14**, 2030–2034.

6 Mirvish, S.S., Haorah, J., Zhou, L., Hartman, M., Morris, C.R., and Clapper, M.L. (2003) N-nitroso compounds in the gastrointestinal tract of rats and in the feces of mice with induced colitis or fed hot dogs or

beef. *Carcinogenesis*, **24**, 595–603.

7 Willett, W.C. (2001) Diet and cancer: one view at the start of the millennium. *Cancer Epidemiol Biomarkers Prev*, **10**, 3–8.

8 Bruce, W.R. (1987) Recent hypotheses for the origin of colon cancer. *Cancer Res*, **47**, 4237–4242.

9 Corpet, D.E., Yin, Y., Zhang, X.M., Remesy, C., Stamp, D., Medline, A., Thompson, L., Bruce, W.R., and Archer, M.C. (1995) Colonic protein fermentation and promotion of colon carcinogenesis by thermolyzed casein. *Nutr Cancer Int J*, **23**, 271–281.

10 Bingham, S.A., Pignatelli, B., Pollock, J.R.A., Ellul, A., Malaveille, C., Gross, G., Runswick, S., Cummings, J.H., and Oneill, I.K. (1996) Does increased endogenous formation of N-nitroso compounds in the human colon explain the association between red meat and colon cancer? *Carcinogenesis*, **17**, 515–523.

11 Cross, A.J., Pollock, J.R.A., and Bingham, S.A. (2003) Haem, not protein or inorganic iron, is responsible for endogenous intestinal n-nitrosation arising from red meat. *Cancer Res*, **63**, 2358–2360.

12 Kuhnle, G.G. and Bingham, S.A. (2007) Dietary meat, endogenous nitrosation and colorectal cancer. *Biochem Soc Trans*, **35**, 1355–1357.

13 Ellervik, C., Birgens, H., Tybjaerg-Hansen, A., and Nordestgaard, B.G. (2007) Hemochromatosis genotypes and risk of 31 disease endpoints: meta-analyses including 66,000 cases and 226,000 controls. *Hepatology*, **46**, 1071–1080.

14 Koppenol, W.H. (2001) The Haber-Weiss cycle--70 years later. *Redox Rep*, **6**, 229–234.

15 Blanksby, S.J., Bierbaum, V.M., Ellison, G.B., and Kato, S. (2007) Superoxide does react with peroxides: direct observation of the Haber-Weiss reaction in the gas phase. *Angew Chem Int Ed Engl*, **46**, 4948–4950.

16 Nauser, T., Koppenol, W.H., and Gebicki, J.M. (2005) The kinetics of oxidation of GSH by protein radicals. *Biochem J*, **392**, 693–701.

17 Davies, M.J. and Dean, R.T. (1997) *Radical-Mediated Protein Oxidation: From Chemistry to Medicine*, Oxford University Press.

18 Stadtman, E.R. and Levine, R.L. (2003) Free radical-mediated oxidation of free amino acids and amino acid residues in proteins. *Amino Acids*, **25**, 207–218.

19 Davies, K.J., Delsignore, M.E., and Lin, S.W. (1987) Protein damage and degradation by oxygen radicals. II. Modification of amino acids. *J Biol Chem*, **262**, 9902–9907.

20 Requena, J.R., Chao, C.C., Levine, R.L., and Stadtman, E.R. (2001) Glutamic and aminoadipic semialdehydes are the main carbonyl products of metal-catalyzed oxidation of proteins. *Proc Natl Acad Sci U S A*, **98**, 69–74.

21 Giulivi, C., Traaseth, N.J., and Davies, K.J. (2003) Tyrosine oxidation products: analysis and biological relevance. *Amino Acids*, **25**, 227–232.

22 Shangari, N., Chan, T.S., Chan, K., Huai, W.S., and O'Brien, P.J. (2007) Copper-catalyzed ascorbate oxidation results in glyoxal/AGE formation and cytotoxicity. *Mol Nutr Food Res*, **51**, 445–455.

23 Saxena, P., Saxena, A.K., Cui, X.L., Obrenovich, M., Gudipaty, K., and Monnier, V.M. (2000) Transition metal-catalyzed oxidation of ascorbate in human cataract extracts: possible role of advanced glycation end products. *Invest Ophthalmol Vis Sci*, **41**, 1473–1481.

24 Saxena, A.K., Saxena, P., Wu, X., Obrenovich, M., Weiss, M.F., and Monnier, V.M. (1999) Protein aging by carboxymethylation of lysines generates sites for divalent metal and redox active copper binding: relevance to diseases of glycoxidative stress. *Biochem Biophys Res Commun*, **260**, 332–338.

25 Xiao, H., Cai, G., and Liu, M. (2007) Fe2+-catalyzed non-enzymatic glycosylation alters collagen conformation during AGE-collagen formation in vitro. *Arch Biochem Biophys*, **468**, 183–192.

26 Reddy, S., Bichler, J., Wells-Knecht, K.J., Thorpe, S.R., and Baynes, J.W. (1995) N epsilon-(carboxymethyl)lysine is a dominant advanced glycation end product (AGE) antigen in tissue proteins. *Biochemistry*, **34**, 10872–10878.

27 Glomb, M.A. and Monnier, V.M. (1995) Mechanism of protein modification by glyoxal and glycolaldehyde, reactive intermediates of the Maillard reaction. *J Biol Chem*, **270**, 10017–10026.

28 Brownlee, M. (2001) Biochemistry and molecular cell biology of diabetic complications. *Nature*, **414**, 813–820.

29 Sato, T., Shimogaito, N., Wu, X., Kikuchi, S., Yamagishi, S., and Takeuchi, M. (2006) Toxic advanced glycation end products (TAGE) theory in Alzheimer's disease. *Am J Alzheimers Dis Other Demen*, **21**, 197–208.

30 Lyakhovich, V.V., Vavilin, V.A., Zenkov, N.K., and Menshchikova, E.B. (2006) Active defense under oxidative stress. The antioxidant responsive element. *Biochemistry (Mosc)*, **71**, 962–974.

31 Fu, M.X., Requena, J.R., Jenkins, A.J., Lyons, T.J., Baynes, J.W., and Thorpe, S.R. (1996) The advanced glycation end product, Nepsilon-(carboxymethyl)lysine, is a product of both lipid peroxidation and glycoxidation reactions. *J Biol Chem*, **271**, 9982–9986.

32 Li, W., Yuan, X.M., and Brunk, U.T. (1998) OxLDL-induced macrophage cytotoxicity is mediated by lysosomal rupture and modified by intralysosomal redox-active iron. *Free Radic Res*, **29**, 389–398.

33 Shanmugam, N., Figarola, J.L., Li, Y., Swiderski, P.M., Rahbar, S., and Natarajan, R. (2008) Proinflammatory effects of advanced lipoxidation end products in monocytes. *Diabetes*, **57**, 879–888.

34 Uchida, K. and Stadtman, E.R. (1993) Covalent attachment of 4-hydroxynonenal to glyceraldehyde-3-phosphate dehydrogenase. A possible involvement of intra- and intermolecular cross-linking reaction. *J Biol Chem*, **268**, 6388–6393.

35 Toyokuni, S., Mori, T., Hiai, H., and Dizdaroglu, M. (1995) Treatment of Wistar rats with a renal carcinogen, ferric nitrilotriacetate, causes DNA-protein cross-linking between thymine and tyrosine in their renal chromatin. *Int J Cancer*, **62**, 309–313.

36 Morel, I., Hamon-Bouer, C., Abalea, V., Cillard, P., and Cillard, J. (1997) Comparison of oxidative damage of DNA and lipids in normal and tumor rat hepatocyte cultures treated with ferric nitrilotriacetate. *Cancer Lett*, **119**, 31–36.

37 Abalea, V., Cillard, J., Dubos, M.P., Sergent, O., Cillard, P., and Morel, I. (1999) Repair of iron-induced DNA oxidation by the flavonoid myricetin in primary rat hepatocyte cultures. *Free Radic Biol Med*, **26**, 1457–1466.

38 Abalea, V., Cillard, J., Dubos, M.P., Anger, J.P., Cillard, P., and Morel, I. (1998) Iron-induced oxidative DNA damage and its repair in primary rat hepatocyte culture. *Carcinogenesis*, **19**, 1053–1059.

39 Suji, G. and Sivakami, S. (2007) DNA damage during glycation of lysine by methylglyoxal: assessment of vitamins in preventing damage. *Amino Acids*, **33**, 615–621.

40 Sawa, T., Akaike, T., Kida, K., Fukushima, Y., Takagi, K., and Maeda, H. (1998) Lipid peroxyl radicals from oxidized oils and heme-iron: implication of a high-fat diet in colon carcinogenesis. *Cancer Epidemiol Biomarkers Prev*, **7**, 1007–1012.

41 Marnett, L.J. (2000) Oxyradicals and DNA damage. *Carcinogenesis*, **21**, 361–370.

42 Bartsch, H. and Nair, J. (2004) Oxidative stress and lipid peroxidation-derived DNA-lesions in inflammation driven carcinogenesis. *Cancer Detect Prev*, **28**, 385–391.

43 Blair, I.A. (2008) DNA adducts with lipid peroxidation products. *J Biol Chem*, **283**, 15545–15549.

44 Klein, D., Lichtmannegger, J., Heinzmann, U., Muller-Hocker, J., Michaelsen, S., and Summer, K.H. (1998) Association of copper to metallothionein in hepatic lysosomes of Long-Evans cinnamon (LEC) rats during the development of hepatitis [see comments]. *Eur J Clin Invest*, **28**, 302–310.

45 Kurz, T. and Brunk, U.T. (2009) Autophagy of HSP70 and chelation of lysosomal iron in a non-redox-active form. *Autophagy*, **5**, 93–95.

46 Pourahmad, J., Ross, S., and O'Brien, P.J. (2001) Lysosomal involvement in hepatocyte cytotoxicity induced by Cu(2+) but not Cd(2+). *Free Radic Biol Med*, **30**, 89–97.

47 Galaris, D., Skiada, V., and Barbouti, A. (2008) Redox signaling and cancer: the role of "labile" iron. *Cancer Lett*, **266**, 21–29.

48 Chen, L., Xiong, S., She, H., Lin, S.W., Wang, J., and Tsukamoto, H. (2007) Iron causes interactions of TAK1, p21ras, and phosphatidylinositol 3-kinase in caveolae to activate IkappaB kinase in hepatic macrophages. *J Biol Chem*, **282**, 5582–5588.

49 O'Brien, P.J. (1969) Intracellular mechanisms for the decomposition of a lipid peroxide. I. Decomposition of a lipid peroxide by metal ions, heme compounds, and nucleophiles. *Can J Biochem*, **47**, 485–492.

50 Gasc, N., Tache, S., Rathahao, E., Bertrand-Michel, J., Roques, V., and Gueraud, F. (2007) 4-hydroxynonenal in foodstuffs: heme concentration, fatty acid composition and freeze-drying are determining factors. *Redox Rep*, **12**, 40–44.

51 Kanner, J. (1994) Oxidative processes in meat and meat products: quality implications. *Meat Sci*, **36**, 169–189.

52 Lynch, M.P. and Faustman, C. (2000) Effect of aldehyde lipid oxidation products on myoglobin. *J Agric Food Chem*, **48**, 600–604.

53 Baron, C.P. and Andersen, H.J. (2002) Myoglobin-induced lipid oxidation. a review. *J Agric Food Chem*, **50**, 3887–3897.

54 Glei, M., Klenow, S., Sauer, J., Wegewitz, U., Richter, K., and Pool-Zobel, B.L. (2006) Hemoglobin and hemin induce DNA damage in human colon tumor cells HT29 clone 19A and in primary human colonocytes. *Mutat Res*, **594**, 162–171.

55 Lee, R.A., Kim, H.A., Kang, B.Y., and Kim, K.H. (2006) Hemoglobin induces colon cancer cell proliferation by release of reactive oxygen species. *World J Gastroenterol*, **12**, 5644–5650.

56 Schaeferhenrich, A., Beyerselhmeyer, G., Festag, G., Kuechler, A., Haag, N., Weise, A., Liehr, T., Claussen, U., Marian, B., Sendt, W., Scheele, J., and PoolZobel, B.L. (2003) Human adenoma cells are highly susceptible to the genotoxic action of 4-hydroxy-2-nonenal. *Mutat Res Fundam Mol Mech Mutagen*, **526**, 19–32.

57 West, J.D., Ji, C., Duncan, S.T., Amarnath, V., Schneider, C., Rizzo, C.J., Brash, A.R., and Marnett, L.J. (2004) Induction of apoptosis in colorectal carcinoma cells treated with 4-hydroxy-2-nonenal and structurally related aldehydic products of lipid peroxidation. *Chem Res Toxicol*, **17**, 453–462.

58 Cerbone, A., Toaldo, C., Laurora, S., Briatore, F., Pizzimenti, S., Dianzani, M.U., Ferretti, C., and Barrera, G. (2007) 4-Hydroxynonenal and PPARgamma ligands affect proliferation, differentiation, and apoptosis in colon cancer cells. *Free Radic Biol Med*, **42**, 1661–1670.

59 Pierre, F., Tache, S., Gueraud, F., Rerole, A.L., Jourdan, M.L., and Petit, C. (2007) Apc mutation induces resistance of colonic cells to lipoperoxide-triggered apoptosis induced by faecal water from haem-fed

60 Babbs, C.F. (1990) Free radicals and the etiology of colon cancer. *Free Radic Biol Med*, **8**, 191–200.

61 Sesink, A.L.A., Termont, D.S.M.L., Kleibeuker, J.H., and Vandermeer, R. (1999) Red meat and colon cancer: the cytotoxic and hyperproliferative effects of dietary heme. *Cancer Res*, **59**, 5704–5709.

62 Sesink, A.L.A., Termont, D.S.M.L., Kleibeuker, J.H., and Vandermeer, R. (2001) Red meat and colon cancer: dietary haem-induced colonic cytotoxicity and epithelial hyperproliferation are inhibited by calcium. *Carcinogenesis*, **22**, 1653–1659.

63 de Vogel, J., Jonker-Termont, D.S., van Lieshout, E.M., Katan, M.B., and Van der Meer, R. (2005) Green vegetables, red meat and colon cancer: chlorophyll prevents the cytotoxic and hyperproliferative effects of haem in rat colon. *Carcinogenesis*, **26**, 387–393.

64 Pierre, F., Tache, S., Petit, C.R., Van der Meer, R., and Corpet, D.E. (2003) Meat and cancer: haemoglobin and haemin in a low-calcium diet promote colorectal carcinogenesis at the aberrant crypt stage in rats. *Carcinogenesis*, **24**, 1683–1690.

65 Pierre, F., Freeman, A., Tache, S., Van der Meer, R., and Corpet, D.E. (2004) Beef meat and blood sausage promote the formation of azoxymethane-induced mucin-depleted foci and aberrant crypt foci in rat colons. *J Nutr*, **134**, 2711–2716.

66 Pierre, F., Peiro, G., Tache, S., Cross, A.J., Bingham, S.A., Gasc, N., Gottardi, G., Corpet, D.E., and Gueraud, F. (2006) New marker of colon cancer risk associated with heme intake: 1,4-dihydroxynonane mercapturic Acid. *Cancer Epidemiol Biomarkers Prev*, **15**, 2274–2279.

67 Lund, E.K., Fairweather-Tait, S.J., Wharf, S.G., and Johnson, I.T. (2001) Chronic exposure to high levels of dietary iron fortification increases lipid peroxidation in the mucosa of the rat large intestine. *J Nutr*, **131**, 2928–2931.

68 Erhardt, J.G., Lim, S.S., Bode, J.C., and Bode, C. (1997) A diet rich in fat and poor in dietary fiber increases the in vitro formation of reactive oxygen species in human feces. *J Nutr*, **127**, 706–709.

69 Lund, E.K., Wharf, S.G., Fairweather-Tait, S.J., and Johnson, I.T. (1999) Oral ferrous sulfate supplements increase the free radical-generating capacity of feces from healthy volunteers. *Am J Clin Nutr*, **69**, 250–255.

70 Gershbein, L.L., Rezai, V.K., Amirmokri, E., and Rao, K.C. (1993) Adenocarcinoma production in rats administered 1,2-dimethylhydrazine and fed iron salt and guar gum diets. *Anticancer Res*, **13**, 2027–2030.

71 Seril, D.N., Liao, J., Ho, K.L.K., Warsi, A., Yang, C.S., and Yang, G.Y. (2002) Dietary iron supplementation enhances DSS-induced colitis and associated colorectal carcinoma development in mice. *Dig Dis Sci*, **47**, 1266–1278.

72 Siegers, C.P., Bumann, D., Trepkau, H.D., Schadwinkel, B., and Baretton, G. (1992) Influence of dietary iron overload on cell proliferation and intestinal tumorigenesis in mice. *Cancer Lett*, **65**, 245–249.

73 Pierre, F., Santarelli, R., Tache, S., Gueraud, F., and Corpet, D.E. (2008) Beef meat promotion of dimethylhydrazine-induced colorectal carcinogenesis biomarkers is suppressed by dietary calcium. *Br J Nutr*, **99**, 1000–1006.

74 Edgren, G., Nyren, O., and Melbye, M. (2008) Cancer as a ferrotoxic disease: are we getting hard stainless evidence? *J Natl Cancer Inst*, **100**, 976–977.

75 Everse, J. and Hsia, N. (1997) The toxicities of native and modified hemoglobins. *Free Radic Biol Med*, **22**, 1075–1099.

76 Mancuso, A., Sciarrino, E., Concetta, R., and Maggio, A. (2006) A prospective study of hepatocellular

carcinoma incidence in thalassemia. *Hemoglobin*, **30**, 119–124.

77 Nelson, R.L., Davis, F.G., Sutter, E., Sobin, L.H., Kikendall, J.W., and Bowen, P. (1994) Body iron stores and risk of colonic neoplasia. *J Natl Cancer Inst*, **86**, 455–460.

78 Nelson, R.L. (2001) Iron and colorectal cancer risk: human studies. *Nutr Rev*, **59**, 140–148.

79 Chan, A.T., Ma, J., Tranah, G.J., Giovannucci, E.L., Rifai, N., Hunter, D.J., and Fuchs, C.S. (2005) Hemochromatosis gene mutations, body iron stores, dietary iron, and risk of colorectal adenoma in women. *J Natl Cancer Inst*, **97**, 917–926.

Molecular Toxicology Mechanisms of Dietary Endogenous Toxins

6
Short Chain Sugars as Endogenous Toxins
Ludmil T. Benov

Short chain sugars can be defined as monosaccharides containing fewer than five carbon atoms. Because the carbon skeleton of these sugars is too short to permit cyclization, they predominantly exist in an open chain form, in which the reactive carbonyl group is unprotected. As a result, such sugars are much more reactive. This is the chemical basis for their potential toxicity. One consequence of the inability of the short chain sugars to hide their reactive carbonyl groups via cyclization, is direct reactions with proteins and nucleic acids, damaging those molecules. Another consequence is their propensity to autoxidize in air. Formation of an enediolate intermediate is the rate-limiting step in this process. Once formed, the enediol oxidizes spontaneously to an α-ketoaldehyde via a process involving free radicals. Superoxide, hydrogen peroxide, hydroxyl, and α-hydroxyalkyl radicals are likely intermediates in this process. It can proceed as a chain reaction where the superoxide can be both an initiator and a propagator. The enediol subjected to sequential one-electron oxidations produces a monoradical first, and then a very unstable biradical that rearranges to a final product, $\alpha\beta$-dicarbonyl. Dicarbonyls in turn are able to cause or enhance oxidative stress. Thus, short chain sugars can be a source of free radicals in living organisms, but are also vulnerable targets for direct free direct attack, leading to more reactive and harmful products.

6.1
Definition and Properties

Short chain sugars can be defined as monosaccharides containing fewer than five carbon atoms. The term *"short chain carbohydrates"* used in nutrition applies to short chain oligosaccharides [1], and therefore does not include short chain sugars. The main chemical difference between long chain and short chain sugars depends on their ability to form cyclic hemiacetals. Long chain sugars exist in solution predominantly in cyclic, pyranose, or furanose forms in which the reactive carbonyl group is involved in the formation of the ring, which suppresses its reactivity. For example, straight chain glucose is converted in solution to a pyranose form, in which the carbonyl on C-1 is cyclized with the hydroxyl on C-5 to form a

Endogenous Toxins. Diet, Genetics, Disease and Treatment.
Edited by Peter J. O'Brien and W. Robert Bruce
Copyright © 2010 WILEY-VCH Verlag GmbH & Co. KGaA, Weinheim
ISBN: 978-3-527-32363-0

hemiacetal ring. This prevents the glucose carbonyl from reacting with neighboring molecules. Because the carbon skeleton of the short chain sugars is too short to permit cyclization, they predominantly exist in an open chain form, in which the reactive carbonyl group is unprotected. As a consequence, such sugars are much more reactive. This is the chemical basis for their potential toxicity.

6.2
Short Chain Sugars and Reactive Oxygen Species

One of the consequences of the inability of the short chain sugars to form cyclic hemiacetals is their propensity to autoxidize in air. Formation of an enediolate intermediate is the rate-limiting step in this process. The importance of enolization was first noticed by Robertson et al. [2] while investigating cyanide-catalyzed oxidation of α-hydroxyaldehydes. It has been later confirmed that monosaccharides autoxidize if present in significant proportion as enediols in a solution [3–5]. Because the tautomerization to an enediol is a prerequisite for monosaccharides autoxidation and tautomerization can occur only if sugar is in an open chain form, it becomes obvious why short chain sugars are more prone to oxidation than pentoses and hexoses. Once formed, the enediolate intermediate is spontaneously oxidized to an α-ketoaldehyde via a process involving free radicals [2, 6]. Superoxide, hydrogen peroxide, hydroxyl, and α-hydroxyalkyl radicals are the likely intermediates in this process [3, 4]. Investigations on the mechanism of this oxidation revealed that it can proceed as a chain reaction in which superoxide can serve both as an initiator and as a propagator [6]. The process of short chain sugar autoxidation is schematically presented in Figure 6.1. It starts with the tautomerization of the open chain forms of aldoses or ketoses (I) to the corresponding enediols (II). The enediol is then subjected to sequential one-electron oxidations to first produce a monoradical (III), and then a very unstable biradical (IV). The biradical rearranges to a final product, $\alpha\beta$-dicarbonyl (V) [6–8]. The one-electron oxidation can be caused slowly by dioxygen yielding O_2^- or more rapidly by O_2^- yielding H_2O_2. It thus appears that autoxidation of short chain sugars can be a source of superoxide radical in living organisms and that short chain sugars are also vulnerable targets for direct superoxide attack.

It was first reported in 1982 that trioses are mutagenic on *Salmonella typhimurium* [9]. A year later Yamaguchi and Nakagawa [10] found that reactive oxygen species (ROS) were implicated. Further investigations revealed that erythrose, glyceraldehyde, dihydroxyacetone, and glycolaldehyde (GOA) were more toxic to superoxide dismutase (SOD)-deficient *Escherichia coli* mutants than to the respective parental strain under aerobic conditions [8, 11]. It has also been noticed that glyceraldehyde, dihydroxyacetone, their corresponding phosphates, and GOA greatly enhanced the mutation rates of the SOD-deficient *E. coli* and killed the mutant strain much more than the SOD-replete parent [12, 13]. It has also been established that the SOD-deficient mutants accumulate more $\alpha\beta$-dicarbonyls than the parents [8]. GOA was capable of inducing the superoxide-responding *soxRS* regulon of *E. coli* [14] and

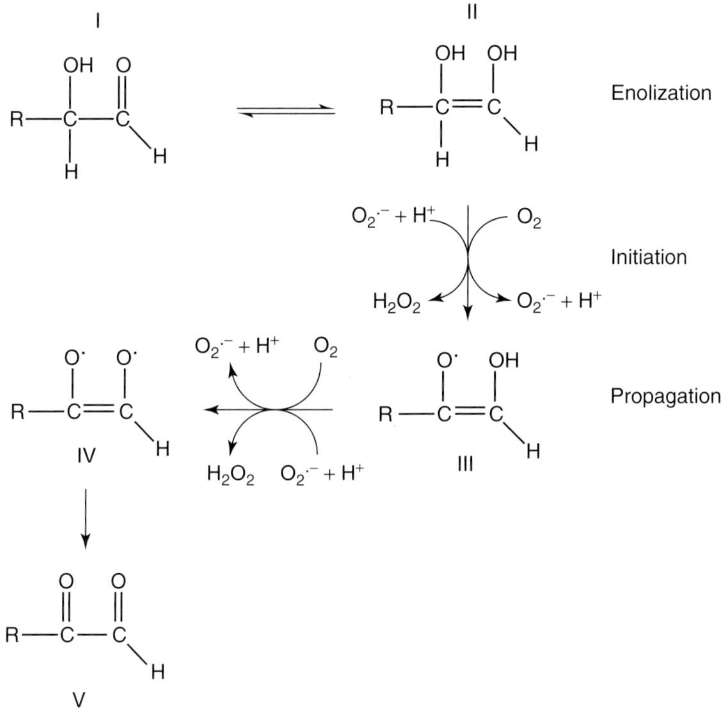

Fig. 6.1 Autoxidation of short chain sugars. Adapted from Mashino, T., Fridovich, I. *Arch Biochem Biophys* 1987, **254**, 547–551; Okado-Matsumoto, A., Fridovich, I. *J Biol Chem* 2000, **275**, 34853–34857.

led to oxidative stress [15] and apoptosis [16] in human breast cancer cells. All these effects were suppressed by a cell-permeable SOD mimic. Increased generation of ROS due to excess short chain sugars has been observed in mammalian cells. Recently, Takahashi *et al.* [17] reported that exposure of pancreatic islets to excess D-glyceraldehyde resulted in increased ROS production in the islets. The authors suggested that enolization of glyceraldehyde and interaction of intermediate enediol radical anion with oxygen generated superoxide radical by a mechanism proposed by Hunt *et al.* [5]. In an earlier work, Krauss *et al.* [18] have shown that high glucose concentration (25 mM) increased mitochondrial superoxide production in pancreatic islets from mice. It therefore appears that sugars can stimulate both mitochondrial and nonmitochondrial pathways of superoxide production, which may play an important role in the damage of beta cells observed in type-two diabetes. All these findings suggest that superoxide and short chain sugars can synergize in causing toxicity. One of the functions of SOD therefore is to protect such metabolites against superoxide attack. On the other hand, accumulation of short chain sugar metabolites would inevitably lead to increased production of ROS and dicarbonyls, both with proven deleterious effects.

6.3
Dicarbonyls

Short chain sugars can be converted to reactive dicarbonyls by both oxidative and nonoxidative mechanisms. It has been demonstrated that degradation of glucose and fructose releases a variety of products including glucosone, 3-deoxyglucosone, 3-deoxyxylosone, tetrosone, triosone, 3-deoxytetrosone, glyoxal, and methylglyoxal (MG), some of which (glyoxal and MG) can also be formed from glyceraldehyde as an intermediate [19]. Among sugar-derived dicarbonyls, MG has been most extensively studied (for reviews see [20, 21]). In mammalian cells, it is generated from triosephosphates by the cleavage of phosphate [22], from amino acids, threonine and glycine, via aminoacetone, and from acetone [20]. An important cellular source of MG is the glycolytic enzyme triosephosphate isomerase [22]. On the basis of the kinetic parameters for the elimination reaction catalyzed by triosephosphate isomerase and cellular concentrations of the enzyme and triosephosphates, Richard [23] calculated velocities of 0.1 and 0.4 mM per day respectively for the spontaneous and triosephosphate isomerase-catalyzed production of MG in rat tissues. Another highly reactive dicarbonyl, glyoxal, can be formed by the autoxidation of GOA [8]. Both MG and glyoxal can be released from glycation products (reviewed in [24]). As mentioned earlier, oxidative stress, that is, increased production of superoxide can lead to increased formation of dicarbonyls [25]. A range of monosaccharides release glyoxal if exposed to hydroxyl radicals produced by the Fenton reaction [26]. On the other hand, dicarbonyls can cause or increase oxidative stress. They can react with the amidine group of arginine residues to form derivatives of imidazole [27]. Such derivatives have been considered to be responsible for various physiological [28, 29] and pathological effects of dicarbonyls, including vascular complications in diabetes [30] and nonenzymatic cross-linking of tissue proteins, leading to oxidative stress [31]. Dicarbonyls can also react nonenzymatically with nucleotides in nucleic acids and can induce mutations [32, 33].

6.4
Direct Reactions with Biomolecules

Another consequence of the inability of the short chain sugars to hide their reactive carbonyl groups via cyclization is direct reactions with other molecules. A variety of stable end products, known as advanced glycation end products (AGEs), are formed through the Maillard reaction [34]. It consists of covalent binding of the carbonyl groups of reducing sugars to amino groups of biomolecules, causing cross-linking, inactivation, and denaturation of these molecules [35, 36]. Major targets are proteins, because of their abundance in the cell [37]. Adducts are readily formed with Lys and Arg side chains and the N-terminal amino groups, acting as nucleophiles. For a typical nonenzymatic glycation reaction to occur, the carbonyl group of a reducing sugar must first react with the free amino groups of a protein to form an amino alcohol (carbinolamine) that dehydrates to a Schiff base salt, which then undergoes an Amadori rearrangement to generate a stable ketoamine product [34]. The process has been shown to occur *in vivo* and has been blamed for some of the complications of diabetes [38–40]. Nonenzymatic glycation has been also implicated in aging, atherosclerosis, renal, eye, and neurological diseases [41–45] as well as in the cross-linking of proteins [35, 36], a phenomenon which has been shown to trigger cellular injury responses [46–48]. Glycation can alter enzymes activities, receptor recognition, and can initiate harmful inflammatory and autoimmune responses [49, 50].

Other amino acid residues can also be modified by sugars. The reversible addition of carbonyl compound to a thiol group to give a thiohemiacetal has long been known [51]. Because the thiol group of cysteine is a powerful nucleophile, it has been proposed that cysteine residues can also be the targets of glycation [52], and that was indeed shown to be the case [53]. Loss of essential thiol groups due to glycation was found to be responsible for the inactivation of glyceraldehyde 3-phosphate dehydrogenase (GAPDH), glutathione reductase (GR), lactate dehydrogenase, and cysteine proteases, by low molecular weight carbonyls, for example, MG, glyoxal, and GOA [54, 55].

Another protein modification resulting from the interaction of amino acid residues with short chain sugars and dicarbonyls is the formation of protein carbonyls [56]. It has been demonstrated that protein carbonyl formation due to aldehyde incorporation is a fast and efficient process that can increase carbonyl levels in proteins as effectively as oxidative mechanisms [56]. In addition, short chain sugars can indirectly promote protein carbonylation by increasing ROS production.

However, proteins are not the only cellular target for short chain sugars and sugar-derived carbonyls. Recent studies have demonstrated that similar to proteins, DNA is susceptible to attack by sugars [57–59], which affects the structure, stability, and conformation of DNA molecules [60–64]. *In vivo*, AGE-linked DNA has been found in the nuclei of epithelial cells, mesangial cells, and endothelial cells of the glomeruli in patients with diabetic nephropathy [65].

6.5
Short Chain Sugars and Their Sources

Most of the biologically important short chain α-hydroxyaldehydes are either intermediates or are derived from intermediates of the normal metabolism. In addition, they can be produced by alternative pathways or as a result of fragmentation of higher sugars or glycation intermediates.

6.5.1
Trioses and Triosephosphates

Major metabolic sources of trioses and triosephosphate include the glycolytic pathway, the polyol pathway, and fructose metabolism [66, 67]. During glycolysis,

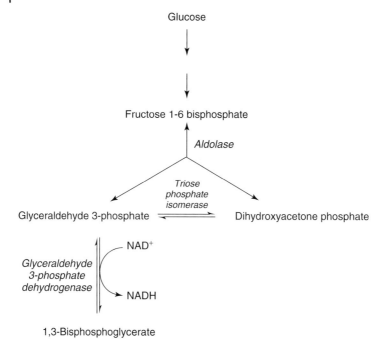

Fig. 6.2 Glyceraldehyde 3-phosphate dehydrogenase and triosephosphate isomerase in glucose metabolism.

glucose is converted by the enzymes of the Embden–Meyerhoff pathway to fructose 1,6-bisphosphate, which in turn is split by the aldolase to the triose phosphates, glyceraldehyde 3-phosphate (GA3P) and dihydroxyacetone phosphate (DHAP). Only GA3P is further metabolized enzymatically to 1,3-bisphosphoglycerate and finally to pyruvate (Figure 6.2). The interconversion of the two triosephosphates is catalyzed by the enzyme triosephosphate isomerase. Normally, the cellular content of triosephosphates is very low because the equilibrium of the aldolase reaction is far toward fructose 1,6-bisphosphate [68]. It has been further argued that the total concentration of GAPDH, triosephosphate isomerase, and aldolase active sites is nearly 10-fold greater than the steady-state concentration of GA3P. If one assumes that the actual binding constants for GA3P for the various enzymes are similar to or smaller than the K_m values, it is highly probable that a very large proportion of the measurable GA3P is bound *in vivo*, and thus protected as the aldehyde. Therefore, at any given moment, the pool of free GA3P would be very small [69].

The steady state concentration of trioses and triosephosphates can rise as a consequence of their increased production or decreased utilization. The increased flux of glucose through the glycolytic and the polyol pathways is assumed to be one of the main reasons for the increased production of these short chain sugars. Decreased use of triosephosphates can be attributed to the low activity or inhibition of GAPDH and triosephosphate isomerase.

6.6
Glyceraldehyde 3-phosphate Dehydrogenase

D-GAPDH (EC 1.2.1.12) is a ubiquitous enzyme displaying a variety of functions [70–74]. It plays an important role in glycolysis and gluconeogenesis by reversibly catalyzing the oxidation and phosphorylation of D-GA3P to 1,3-bisphosphoglycerate in the presence of inorganic phosphate and NAD^+ (Figure 6.2). NAD^+ availability is therefore a limiting factor for the GAPDH activity. Because the enzyme contains a highly reactive thiol at its active site (cysteine 149) [75], it is sensitive to modification by a variety of compounds with subsequent loss of enzyme activity [76, 77]. These include aldehydes, sugars, and dicarbonyls [54, 77–81], and reactive oxygen and nitrogen species [82–84]. GA3P and DHAP can serve as a source of reactive dicarbonyls. MG (pyruvaldehyde) will be produced by elimination of phosphate, and hydroxypyruvaldehyde phosphate, a dicarbonyl similar to MG, by autoxidation [20, 22, 27, 85]. Inhibition of GAPDH has been reported to cause accumulation of dicarbonyls that are derived from triosephosphates [86–89]. Dicarbonyls, at intracellular concentrations close to those found *in vivo*, can in turn modify and inactivate GAPDH [90]. GAPDH can also be glycated and inactivated by its substrate, GA3P, and by DHAP [44].

6.7
Triosephosphate Isomerase

Triosephosphate isomerase (EC 5.3.1.1) catalyzes the interconversion of D-GA3P and DHAP, an essential step in glycolytic and gluconeogenic metabolism (Figure 6.2). The most apparent biochemical alteration reported in triosephosphate isomerase-deficient patients is the increase (~20-fold) in cellular concentration of DHAP [91–93], leading to hemolytic anemia and progressive neuromuscular degeneration [91, 93]. These consequences have been attributed to the spontaneous conversion of DHAP into dicarbonyls, mainly MG, causing DNA and protein damage through the formation of AGEs [93, 94].

6.7.1
Glycolaldehyde

GOA, the simplest sugar, can be produced nonenzymatically by sugar fragmentation [95] and by the oxidation of ascorbic acid in the presence of iron salts [96]. Enzymatically, GOA is produced by glycolate reductase in microorganisms as a precursor of vitamin B6 [97–99]. GOA and other aldehydes are released into the extracellular medium when alveolar macrophages are exposed to nitrogen dioxide [100]. The exact mechanism of this GOA production has not been investigated, but the authors speculate that it is derived from the decomposition of polyunsaturated fatty acid peroxides. Increased concentrations of GOA were found in the basolateral fluid from guinea pig tracheobronchial epithelial cell monolayers exposed to NO_2.

Calculations show that cellular GOA levels could reach up to 3 mM during the 60-minute exposure period [101]. At sites of inflammation, GOA is generated by activated leukocytes [102] in a pathway that begins with myeloperoxidase, secreted by activated neutrophils. Myeloperoxidase catalyzes hypochlorous acid (HOCl) formation from H_2O_2 and chloride. The HOCl oxidizes α-amino acids to aldehydes with L-serine being extensively oxidized to GOA [102]. Such *in vivo* production of GOA and other reactive aldehydes catalyzed by myeloperoxidase may play an important pathogenic role by generating AGE products, which damage tissues at sites of inflammation [103].

GOA is known to react with protein amino acid residues to give an aldimine, which is capable of adding amino or thiol groups leading to covalent cross-linking of proteins and of DNA [104]. Using antibodies specific for GOA-modified proteins, Nagai *et al.* isolated a novel GOA-derived structure [105], identified as a GOA-pyridine. Immunohistochemical studies demonstrated the accumulation of GOA-pyridine in human atherosclerotic lesions [105]. GOA can modify Arg, Lys, and Trp residues of the apoB protein of LDL much faster than glucose [49] and is responsible for the generation of AGEs that are recognized by macrophage scavenger receptors [106]. Modification of bovine serum albumin by GOA was shown to decrease leptin expression in mouse adipocytes and thus might be related to reduction of the insulin sensitivity in metabolic syndrome [107]. Further investigations suggested that the interaction of GOA-modified proteins with adypocyte CD36 membrane receptor induced oxidative stress, which actually led to the inhibition of leptin expression by these cells [108, 109].

In addition to reacting with proteins, GOA is also capable of modifying nucleic acids and can act as a mutagen [12]. *In vitro* incubation of human peripheral mononuclear blood cells with GOA caused a dose-dependent increase in DNA-protein cross-links as well as DNA single-strand breaks [110].

6.7.2
Short Chain Sugars in Diabetes and Aging

6.7.2.1 Diabetes

Increased production of short chain sugars can be a consequence of increased flux of glucose through the glycolytic and the polyol pathways under conditions of hyperglycemia, most often related to diabetes [111–113]. Triosephosphate levels in diabetes are also elevated due to the inhibition of GAPDH [38, 114, 115]. Hyperglycemia-induced overproduction of mitochondrial superoxide, causing inactivation of GAPDH by activating poly(ADP-ribose) polymerase, is considered as the triggering mechanism for the activation of the pathways leading to diabetic complications [116]. Normalizing mitochondrial superoxide production blocks those pathways [117]. At the same time, glycation of mitochondrial proteins leads to the increase in superoxide production [118].

Glycation and formation of AGE is one of the major pathways causing diabetic complications. Although hexoses have been presumed to play a primary role in glycation and cross-linking, recent studies indicate that shorter chain sugars are more potent than glucose in forming AGEs. As mentioned before, glucose possesses a very low reactivity toward protein amino groups due to the fact that sugar is predominantly present in the nonreactive ring form [58] (only in the open chain form is the aldehyde group free to react), whereas short chain glycolytic intermediates, that are denied the possibility of cyclization, are much more reactive [68, 119, 120]. It has been demonstrated that reactivity is proportional to the percentage of sugar present in the open chain form [119–121]. Indeed, the three-carbon sugars GA3P, DHAP, and their dephosphorylated forms, as well as glyceraldehyde and DHA, are potent glycation and cross-linking agents [112, 122–125]. Their cross-linking abilities are reportedly three times higher than those of erythrose and threose, and eight times higher than that of ribose [126]. GA3P has been described as the most reactive glycation intermediate found to be elevated in diabetic cells [39] and as a "potential toxic triose" capable of inhibiting various enzymes [123, 124, 127–129].

Together with triosephosphates, glyceraldehyde is considered as an extremely potent glycating agent. It has been estimated that at concentrations close to physiological, it produces substantial amounts of AGEs [125]. Glyceraldehyde can be transported or can leak passively across the plasma membrane. It can react nonenzymatically with intracellular and extracellular proteins to form AGEs. It has been estimated that the relative browning activity of glyceraldehyde, which reflects an early stage of the Maillard reaction, can be about 2000-fold greater than that of glucose [130]. Recent reports indicate that AGEs derived from glyceraldehyde (Glycer-AGEs) exert strong neurotoxicity [131, 132]. Glycer-AGEs can induce apoptotic cell death in cultured cortical neuronal cells and have been detected in the cytosol of neurons from human Alzheimer's disease brains [133], which suggests that Glycer-AGEs may play a role in the pathogenesis of Alzheimer's disease [134]. In addition, glyceraldehyde shows a far stronger promotion of radical formation than glucose [130, 135]. Usui et al. [135, 136] reported that a glyceraldehyde-derived pyridinium compound (GLAP) depressed the intracellular glutathione level and increased ROS production. GLAP is considered as a specific AGE derived from glyceraldehyde [135, 137], similar to GOA-pyridine mentioned earlier [105]. Interaction of GA3P or glyceraldehyde with lysine in vitro was shown to produce GLAP [138, 139]. GLAP has been identified in the plasma protein and the tail tendon collagen of streptozotocin-induced diabetic rats [138]. It has been suggested that increase in the GLAP level reflects an increase in the glyceraldehyde and GA3P levels. Therefore, GLAP might be considered as a biomarker for reduced activity of the glyceraldehyde-related enzymes in diabetes, consequently leading to diabetic complications [138, 140].

6.7.2.2 Aging, Glycation, and Dietary Restriction

Increased accumulation of AGEs has been observed in aging, and both AGE accumulation and aging are accelerated by diabetes [39, 41, 141]. In contrast, dietary restriction suppresses AGEs accumulation, delays aging, and at least in laboratory animals, extends life span [142–145]. It has recently been proposed that dietary restriction limits the flux of glucose through the glycolytic pathway and thus decreases the production of harmful glycolytic intermediates, which in turn extends the life span [146]. Trioses and triosephosphates are the most reactive and hence, the most deleterious. Furthermore, they are precursors of MG. Their production will eventually depend on the rate and persistence of glycolytic activity [147, 148] and the activity of GAPDH [149]. GAPDH activity in turn depends on the availability of NAD^+ as a cofactor; however, ad libitum feeding conditions decrease NAD^+ availability [150]. This decreases the rate of GA3P metabolism by GAPDH, thus increasing GA3P and DHAP steady state concentration. In contrast, dietary restriction decreases the flow of glycolytic intermediates and makes NAD^+ available, which improves triosephosphate metabolism by GAPDH [150]. It has been further proposed that the conversion of triosephosphates to MG is the actual cause of age-related dysfunction and decreased life span [146]. The increase in free radical-mediated damage, which occurs during ad libitum feeding compared to dietary restriction, has been attributed to the generation of ROS following its reaction with biomolecules, and to the activation of NAD(P)H-oxidase [150]. However, Kalapos [151] pointed out that production of triosephosphates is not solely dependent on the rate of glycolysis. It will be also influenced by the pentosephosphate pathway and the activity of α-glycerophosphate dehydrogenase [151]. Furthermore, methyglyoxal production might be even increased under starvation and dietary restriction [151]. Nevertheless, it is logical to expect that reactive α-hydroxycarbonyls and dicarbonyls that are derived from glucose metabolism have a significant impact on age-related molecular damage, and that dietary restriction affects the production of such metabolites.

Acknowledgment

The author thanks Dr. Irwin Fridovich (Duke University Medical Center) for critically reading the manuscript and his helpful suggestions.

References

1 Cummings, J.H. and Stephen, A.M. (2007) Carbohydrate terminology and classification. *Eur J Clin Nutr*, **61**, S5–S18.

2 Robertson, P. Jr, Fridovich, S.E., Misra, H.P., and Fridovich, I. (1981) Cyanide catalyzes the oxidation of α-hydroxyaldehydes and related compounds: monitored as the reduction of dioxygen, cytochrome c, and nitroblue tetrazolium. *Arch Biochem Biophys*, **207**, 282–289.

3 Thornalley, P., Wolff, S., Crabbe, J., and Stern, A. (1984) The autoxidation of glyceraldehyde and other simple monosaccharides under physiological conditions catalysed by buffer ions. *Biochim Biophys Acta Gen Subj*, **797**, 276–287.

4 Thornalley, P.J. and Stern, A. (1984) The production of free radicals during the autoxidation of monosaccharides by buffer ions. *Carbohydr Res*, **134**, 191–204.

5 Hunt, J.V., Dean, R.T., and Wolff, S.P. (1988) Hydroxyl radical production and autoxidative glycosylation. Glucose autoxidation as the cause of protein damage in the experimental glycation model of diabetes mellitus and ageing. *Biochem J*, **256**, 205–212.

6 Mashino, T. and Fridovich, I. (1987) Mechanism of the cyanide-catalyzed oxidation of α-ketoaldehydes and α-ketoalcohols. *Arch Biochem Biophys*, **252**, 163–170.
7 Mashino, T. and Fridovich, I. (1987) Superoxide radical initiates the autoxidation of dihydroxyacetone. *Arch Biochem Biophys*, **254**, 547–551.
8 Okado-Matsumoto, A. and Fridovich, I. (2000) The role of α,β-dicarbonyl compounds in the toxicity of short chain sugars. *J Biol Chem*, **275**, 34853–34857.
9 Yamaguchi, T. (1982) Mutagenicity of trioses and methyl glyoxal on Salmonella typhimurium. *Agric Biol Chem*, **46**, 849–851.
10 Yamaguchi, T. and Nakagawa, K. (1983) Mutagenicity of and formation of oxygen radicals by trioses and glyoxal derivatives. *Agric Biol Chem*, **47**, 2461–2465.
11 Benov, L. and Fridovich, I. (1998) Superoxide dependence of the toxicity of short chain sugars. *J Biol Chem*, **273**, 25741–25744.
12 Benov, L. and Beema, A.F. (2003) Superoxide-dependence of the short chain sugars-induced mutagenesis. *Free Radic Biol Med*, **34**, 429–433.
13 Benov, L., Beema, A.F., and Sequeira, F. (2003) Triosephosphates are toxic to superoxide dismutase-deficient Escherichia coli. *Biochim Biophys Acta Gen Subj*, **1622**, 128–132.
14 Benov, L. and Fridovich, I. (2002) Induction of the soxRS regulon of Escherichia coli by glycolaldehyde. *Arch Biochem Biophys*, **407**, 45–48.
15 Al-Enezi, K.S., Alkhalaf, M., and Benov, L.T. (2006) Glycolaldehyde induces growth inhibition and oxidative stress in human breast cancer cells. *Free Radic Biol Med*, **40**, 1144–1151.
16 Al-Maghrebi, M.A., Al-Mulla, F., and Benov, L.T. (2003) Glycolaldehyde induces apoptosis in a human breast cancer cell line. *Arch Biochem Biophys*, **417**, 123–127.
17 Takahashi, H., Tran, P.O., LeRoy, E., Harmon, J.S., Tanaka, Y., and Robertson, R.P. (2004) D-Glyceraldehyde causes production of intracellular peroxide in pancreatic islets, oxidative stress, and defective β cell function via non-mitochondrial pathways. *J Biol Chem*, **279**, 37316–37323.
18 Krauss, S., Zhang, C.-Y., Scorrano, L., Dalgaard, L.T., St-Pierre, J., Grey, S.T., and Lowell, B.B. (2003) Superoxide-mediated activation of uncoupling protein 2 causes pancreatic? cell dysfunction. *J Clin Invest*, **112**, 1831–1842.
19 Usui, T., Yanagisawa, S., Ohguchi, M., Yoshino, M., Kawabata, R., Kishimoto, J., Arai, Y., Aida, K., Watanabe, H., and Hayase, F. (2007) Identification and determination of α-dicarbonyl compounds formed in the degradation of sugars. *Biosci Biotechnol Biochem*, **71**, 2465–2472.
20 Kalapos, M.P. (1999) Methylglyoxal in living organisms: chemistry, biochemistry, toxicology and biological implications. *Toxicol Lett*, **110**, 145–175.
21 Kalapos, M.P. (2008) The tandem of free radicals and methylglyoxal. *Chem Biol Interact*, **171**, 251–271.
22 Richard, J.P. (1993) Mechanism for the formation of methylglyoxal from triosephosphates. *Biochem Soc Trans*, **21**, 549–553.
23 Richard, J.P. (1991) Kinetic parameters for the elimination reaction catalyzed by triosephosphate isomerase and an estimation of the reaction's physiological significance. *Biochemistry*, **30**, 4581–4585.
24 Grillo, M.A. and Colombatto, S. (2008) Advanced glycation end-products (AGEs): involvement in aging and in neurodegenerative diseases. *Amino Acids*, **35**, 29–36.
25 Abordo, E.A., Minhas, H.S., and Thornalley, P.J. (1999) Accumulation of α-oxoaldehydes during oxidative stress: a role in cytotoxicity. *Biochem Pharmacol*, **58**, 641–648.
26 Manini, P., La Pietra, P., Panzella, L., Napolitano, A., and d'Ischia, M. (2006) Glyoxal formation by

27 Thornalley, P.J. (1996) Pharmacology of methylglyoxal: formation, modification of proteins and nucleic acids, and enzymatic detoxification – A role in pathogenesis and antiproliferative chemotherapy. *Gen Pharmacol*, **27**, 565–573.

28 Nagaraj, R.H., Biswas, A., Miller, A., Oya-Ito, T., and Bhat, M. (2008) The other side of the Maillard reaction. *Ann N Y Acad Sci*, **1126**, 107–112.

29 Speer, O., Morkunaite-Haimi, S., Liobikas, J., Franck, M., Hensbo, L., Linder, M.D., Kinnunen, P.K.J., Wallimann, T., and Eriksson, O. (2003) Rapid suppression of mitochondrial permeability transition by methylglyoxal: role of reversible arginine modification. *J Biol Chem*, **278**, 34757–34763.

30 Bourajjaj, M., Stehouwer, C.D.A., Van Hinsbergh, V.W.M., and Schalkwijk, C.G. (2003) Role of methylglyoxal adducts in the development of vascular complications in diabetes mellitus. *Biochem Soc Trans*, **31**, 1400–1402.

31 Odani, H., Shinzato, T., Usami, J., Matsumoto, Y., Frye, E.B., Baynes, J.W., and Maeda, K. (1998) Imidazolium crosslinks derived from reaction of lysine with glyoxal and methylglyoxal are increased in serum proteins of uremic patients: Evidence for increased oxidative stress in uremia. *FEBS Lett*, **427**, 381–385.

32 Kalapos, M.P. (1994) Methylglyoxal toxicity in mammals. *Toxicol Lett*, **73**, 3–24.

33 Rahman, A., Shahabuddin, A., and Hadi, S.M. (1990) Formation of strand breaks and interstrand cross-links in DNA by methylglyoxal. *J Biochem Toxicol*, **5**, 161–166.

34 Hodge, J.E. (1955) The Amadori rearrangement. *Adv Carbohydr Chem*, **10**, 169–205.

35 Grandhee, S.K. and Monnier, V.M. (1991) Mechanism of formation of the Maillard protein cross-link pentosidine. Glucose, fructose, and ascorbate as pentosidine precursors. *J Biol Chem*, **266**, 11649–11653.

36 Chellan, P. and Nagaraj, R.H. (2001) Early glycation products produce pentosidine cross-links on native proteins. Novel mechanism of pentosidine formation and propagation of glycation. *J Biol Chem*, **276**, 3895–3903.

37 Davies, M.J. (2005) The oxidative environment and protein damage. *Biochim Biophys Acta Proteins Proteomics*, **1703**, 93–109.

38 Brownlee, M. (2005) The pathobiology of diabetic complications – A unifying mechanism. *Diabetes*, **54**, 1615–1625.

39 Bunn, H.F. and Higgins, P.J. (1981) Reaction of monosaccharides with proteins: possible evolutionary significance. *Science*, **213**, 222–224.

40 Monnier, V.M., Sell, D.R., Dai, Z., Nemet, I., Collard, F., and Zhang, J. (2008) The role of the Amadori product in the complications of diabetes. *Ann N Y Acad Sci*, **1126**, 81–88.

41 Bunn, H.F. (1981) Non-enzymatic glycosylation of protein: a form of molecular aging. *Schweiz Med Wochenschr J Suisse Med*, **111**, 1503–1507.

42 Price, C.L. and Knight, S.C. (2007) Advanced glycation: a novel outlook on atherosclerosis. *Curr Pharm Des*, **13**, 3681–3687.

43 Sell, D.R., Strauch, C.M., Shen, W., and Monnier, V.M. (2008) Aging, diabetes, and renal failure catalyze the oxidation of lysyl residues to 2-aminoadipic acid in human skin collagen: evidence for metal-catalyzed oxidation mediated by β-dicarbonyls. *Ann N Y Acad Sci*, **1126**, 205–209.

44 Vlassara, H. and Palace, M.R. (2002) Diabetes and advanced glycation endproducts. *J Intern Med*, **251**, 87–101.

45 Vlassara, H. and Palace, M.R. (2003) Glycoxidation: the menace of diabetes and aging. *Mt Sinai J Med*, **70**, 232–241.

46 Daniels, B.S. and Hauser, E.B. (1992) Glycation of albumin, not glomerular basement membrane, alters

47 Vlassara, H., Striker, L.J., Teichberg, S., Fuh, H., Li, Y.M., and Steffes, M. (1994) Advanced glycation end products induce glomerular sclerosis and albuminuria in normal rats. *Proc Natl Acad Sci U S A*, **91**, 11704–11708.

48 Brownlee, M. (1994) Glycation and diabetic complications. *Diabetes*, **43**, 836–841.

49 Knott, H.M., Brown, B.E., Davies, M.J., and Dean, R.T. (2003) Glycation and glycoxidation of low-density lipoproteins by glucose and low-molecular mass aldehydes: formation of modified and oxidized particles. *Eur J Biochem*, **270**, 3572–3582.

50 Yamamoto, Y., Yonekura, H., Watanabe, T., Sakurai, S., Li, H., Harashima, A., Myint, K.M., Osawa, M., Takeuchi, A., Takeuchi, M., and Yamamoto, H. (2007) Short-chain aldehyde-derived ligands for RAGE and their actions on endothelial cells. *Diabetes Res Clin Pract*, **77**, S30–S40.

51 Ratner, S. and Clarke, H.T. (1937) The action of formaldehyde upon cysteine. *J Am Chem Soc*, **59**, 200–206.

52 Zeng, J.M. and Davies, M.J. (2006) Protein and low molecular mass thiols as targets and inhibitors of glycation reactions. *Chem Res Toxicol*, **19**, 1668–1676.

53 Zeng, J. and Davies, M.J. (2005) Evidence for the formation of adducts and S-(carboxymethyl)cysteine on reaction of α-dicarbonyl compounds with thiol groups on amino acids, peptides, and proteins. *Chem Res Toxicol*, **18**, 1232–1241.

54 Morgan, P.E., Dean, R.T., and Davies, M.J. (2002) Inactivation of cellular enzymes by carbonyls and protein-bound glycation/glycoxidation products. *Arch Biochem Biophys*, **403**, 259–269.

55 Zeng, J.M., Dunlop, R.A., Rodgers, K.J., and Davies, M.J. (2006) Evidence for inactivation of cysteine proteases by reactive carbonyls via glycation of active site thiols. *Biochem J*, **398**, 197–206.

56 Adams, S., Green, P., Claxton, R., Simcox, S., Williams, M.V., Walsh, K., and Leeuwenburgh, C. (2001) Reactive carbonyl formation by oxidative and non-oxidative pathways. *Front Biosci*, **6**, A17–A24.

57 Li, Y.Y., Dutta, U., Cohenford, M.A., and Dain, J.A. (2007) Nonenzymatic glycation of guanosine 5′-triphosphate by glyceraldehyde: an in vitro study of AGE formation. *Bioorg Chem*, **35**, 417–429.

58 Li, Y.Y., Cohenford, M.A., Dutta, U., and Dain, J.A. (2008) The structural modification of DNA nucleosides by nonenzymatic glycation: an in vitro study based on the reactions of glyoxal and methylglyoxal with 2′-deoxyguanosine. *Anal Bioanal Chem*, **390**, 679–688.

59 Li, Y., Cohenford, M.A., Dutta, U., and Dain, J.A. (2008) In vitro nonenzymatic glycation of guanosine 5′-triphosphate by dihydroxyacetone phosphate. *Anal Bioanal Chem*, **392**, 1–8.

60 Dutta, U., Cohenford, M.A., and Dain, J.A. (2005) Nonenzymatic glycation of DNA nucleosides with reducing sugars. *Anal Biochem*, **345**, 171–180.

61 Mullokandov, E.A., Franklin, W.A., and Brownlee, M. (1994) DNA damage by the glycation products of glyceraldehyde 3-phosphate and lysine. *Diabetologia*, **37**, 145–149.

62 Thornalley, P.J. (2008) Protein and nucleotide damage by glyoxal and methylglyoxal in physiological systems – Role in ageing and disease. *Drug Metabol Drug Interact*, **23**, 125–150.

63 Thornalley, P.J. (2003) Protecting the genome: defence against nucleotide glycation and emerging role of glyoxalase I overexpression in multidrug resistance in cancer chemotherapy. *Biochem Soc Trans*, **31**, 1372–1377.

64 Kang, Y., Edwards, L.G., and Thornalley, P.J. (1996) Effect of

methylglyoxal on human leukaemia 60 cell growth: modification of DNA, G1 growth arrest and induction of apoptosis. *Leuk Res*, **20**, 397–405.

65 Li, H., Nakamura, S., Miyazaki, S., Morita, T., Suzuki, M., Pischetsrieder, M., and Niwa, T. (2006) N2-carboxyethyl-2′-deoxyguanosine, a DNA glycation marker, in kidneys and aortas of diabetic and uremic patients. *Kidney Int*, **69**, 388–392.

66 Sillero, M.A., Sillero, A., and Sols, A. (1969) Enzymes involved in fructose metabolism in lir and the glyceraldehyde metabolic crossroads. *Eur J Biochem*, **10**, 345–350.

67 Sato, T., Iwaki, M., Shimogaito, N., Wu, X.G., Yamagishi, S., and Takeuchi, M. (2006) TAGE (toxic AGEs) theory in diabetic complications. *Curr Mol Med*, **6**, 351–358.

68 Fridovich, I. (2008) Deleterious effects due to glucose or to triose phosphates. *Free Radical Biol Med*, **44**, 1970.

69 Reynolds, S.J., Yates, D.W., and Pogson, C.I. (1971) Dihydroxyacetone phosphate. Its structure and reactivity with -glycerophosphate dehydrogenase, aldolase and triose phosphate isomerase and some possible metabolic implications. *Biochem J*, **122**, 285–297.

70 Sirover, M.A. (1997) Role of the glycolytic protein, glyceraldehyde-3-phosphate dehydrogenase, in normal cell function and in cell pathology. *J Cell Biochem*, **66**, 133–140.

71 Sirover, M.A. (2005) New nuclear functions of the glycolytic protein, glyceraldehyde-3-phosphate dehydrogenase, in mammalian cells. *J Cell Biochem*, **95**, 45–52.

72 Mazzola, J.L. and Sirover, M.A. (2003) Subcellular localization of human glyceraldehyde-3-phosphate dehydrogenase is independent of its glycolytic function. *Biochim Biophys Acta*, **1622**, 50–56.

73 Sirover, M.A. (1999) New insights into an old protein: the functional diversity of mammalian glyceraldehyde-3-phosphate dehydrogenase. *Biochim Biophys Acta*, **1432**, 159–184.

74 Sirover, M.A. (1996) Minireview. Emerging new functions of the glycolytic protein, glyceraldehyde-3-phosphate dehydrogenase, in mammalian cells. *Life Sci*, **58**, 2271–2277.

75 Schuppekoistinen, I., Moldeus, P., Bergman, T., and Cotgreave, I.A. (1994) S-Thiolation of human endothelial-cell glyceraldehyde-3-phosphate dehydrogenase after hydrogen-peroxide treatment. *Eur J Biochem*, **221**, 1033–1037.

76 Dimmeler, S., Lottspeich, F., and Brune, B. (1992) Nitric-oxide causes Adp-ribosylation and inhibition of glyceraldehyde-3-phosphate dehydrogenase. *J Biol Chem*, **267**, 16771–16774.

77 Uchida, K. and Stadtman, E.R. (1993) Covalent attachment of 4-hydroxynonenal to glyceraldehyde-3-phosphate dehydrogenase – a possible involvement of intramolecular and intermolecular cross-linking reaction. *J Biol Chem*, **268**, 6388–6393.

78 Halder, J., Ray, M., and Ray, S. (1993) Inhibition of glycolysis and mitochondrial respiration of Ehrlich ascites-carcinoma cells by methylglyoxal. *Int J Cancer*, **54**, 443–449.

79 Lee, H.J., Howell, S.K., Sanford, R.J., and Beisswenger, P.J. (2005) Methylglyoxal can modify GAPDH activity and structure. *Ann NY Acad Sci*, **1043**, 135–145.

80 Leoncini, G., Maresca, M., and Buzzi, E. (1989) Inhibition of the glycolytic pathway by methylglyoxal in human-platelets. *Cell Biochem Funct*, **7**, 65–70.

81 Novotny, M.V., Yancey, M.F., Stuart, R., Wiesler, D., and Peterson, R.G. (1994) Inhibition of glycolytic-enzymes by endogenous aldehydes – a possible relation to diabetic neuropathies. *Biochim Biophys Acta Mol Basis Dis*, **1226**, 145–150.

82 Hurst, R.D., Azam, S., Hurst, A., and Clark, J.B. (2001) Nitric-oxide-induced inhibition of glyceraldehyde-3-phosphate dehydrogenase may mediate reduced endothelial cell monolayer integrity in an in vitro model blood-brain barrier. *Brain Res*, **894**, 181–188.

83 Morgan, P.E., Dean, R.T., and Davies, M.J. (2002) Inhibition of glyceraldehyde-3-phosphate dehydrogenase by peptide and protein peroxides generated by singlet oxygen attack. *Eur J Biochem*, **269**, 1916–1925.

84 Souza, J.M. and Radi, R. (1998) Glyceraldehyde-3-phosphate dehydrogenase inactivation by peroxynitrite. *Arch Biochem Biophys*, **360**, 187–194.

85 Abordo, E.A., Minhas, H.S., and Thornalley, P.J. (1999) Accumulation of α-oxoaldehydes during oxidative stress: a role in cytotoxicity. *Biochem Pharmacol*, **58**, 641–648.

86 Beisswenger, P.J., Howell, S.K., Smith, K., and Szwergold, B.S. (2003) Glyceraldehyde-3-phosphate dehydrogenase activity as an independent modifier of methylglyoxal levels in diabetes. *Biochim Biophys Acta Mol Basis Dis*, **1637**, 98–106.

87 Beisswenger, P.J., Howell, S.K., Nelson, R.G., Mauer, M., and Szwergold, B.S. (2003) α-oxoaldehyde metabolism and diabetic complications. *Biochem Soc Trans*, **31**, 1358–1363.

88 Cogoli-Greuter, M. and Christen, P. (1981) Formation of hydroxypyruvaldehyde phosphate in human erythrocytes. *J Biol Chem*, **256**, 5708–5711.

89 Best, L. and Thornalley, P.J. (1999) Trioses and related substances: tools for the study of pancreatic β-cell function. *Biochem Pharmacol*, **57**, 583–588.

90 Lee, H.J., Howell, S.K., Sanford, R.J., and Beisswenger, P.J. (2005) Methylglyoxal can modify GAPDH activity and structure. *Ann NY Acad Sci*, **1043**, pp. 135–145.

91 Schneider, A.S. (2000) Triosephosphate isomerase deficiency: historical perspectives and molecular aspects. *Best Pract Res Clin Haematol*, **13**, 119–140.

92 Tanaka, K.R. and Zerez, C.R. (1990) Red-cell enzymopathies of the glycolytic pathway. *Semin Hematol*, **27**, 165–185.

93 Ahmed, N., Battah, S., Karachalias, N., Babaei-Jadidi, R., Horanyi, M., Baroti, K., Hollan, S., and Thornalley, P.J. (2003) Increased formation of methylglyoxal and protein glycation, oxidation and nitrosation in triosephosphate isomerase deficiency. *Biochim Biophys Acta*, **1639**, 121–132.

94 Gnerer, J.P., Kreber, R.A., and Ganetzky, B. (2006) wasted away, a Drosophila mutation in triosephosphate isomerase, causes paralysis, neurodegeneration, and early death. *Proc Natl Acad Sci U S A*, **103**, 14987–14993.

95 Novotny, O., Cejpek, K., and Velisek, J. (2007) Formation of α-hydroxycarbonyl and α-dicarbonyl compounds during degradation of monosaccharides. *Czech J Food Sci*, **25**, 119–130.

96 Mlakar, A., Batna, A., Dudda, A., and Spiteller, G. (1996) Iron (II) ions induced oxidation of ascorbic acid and glucose. *Free Radic Res*, **25**, 525–539.

97 Tani, Y., Nishise, H., Morita, H., and Yamada, H. (1984) Vitamin B6 biosynthesis and glycolate reductase in Flavobacterium sp. 238-7. *J Nutr Sci Vitaminol*, **30**, 415–420.

98 Morita, H., Tani, Y., and Ogata, K. (1978) Control of glycolaldehyde dehydrogenase in vitamin B6 biosynthesis in Escherichia coli B. *Agric Biol Chem*, **42**, 69–73.

99 Sakai, A., Katayama, K., Katsuragi, T., and Tani, Y. (2001) Glycolaldehyde-forming route in Bacillus subtilis in relation to vitamin B6 biosynthesis. *J Biosci Bioeng*, **91**, 147–152.

100 Robinson, T.W., Forman, H.J., and Thomas, M.J. (1995) Release of aldehydes from rat alveolar macrophages exposed in vitro to

low concentrations of nitrogen dioxide. *Biochim Biophys Acta Lipids Lipid Metab*, **1256**, 334–340.
101 Robison, T.W., Zhou, H., and Kim, K.-J. (1996) Generation of glycolaldehyde from guinea pig airway epithelial monolayers exposed to nitrogen dioxide and its effect on sodium pump activity. *Environ Health Perspect*, **104**, 852–856.
102 Anderson, M.M., Hazen, S.L., Hsu, F.F., and Heinecke, J.W. (1997) Human neutrophils employ the myeloperoxidase-hydrogen peroxide-chloride system to convert hydroxy-amino acids into glycolaldehyde, 2-hydroxypropanal, and acrolein. A mechanism for the generation of highly reactive α-hydroxy and α,β-unsaturated aldehydes by phagocytes at sites of inflammation. *J Clin Invest*, **99**, 424–432.
103 Anderson, M.M., Requena, J.R., Crowley, J.R., Thorpe, S.R., and Heinecke, J.W. (1999) The myeloperoxidase system of human phagocytes generates N(epsilon)-(carboxymethyl)lysine on proteins: a mechanism for producing advanced glycation end products at sites of inflammation. *J Clin Invest*, **104**, 103–113.
104 Acharya, A.S. and Manning, J.M. (1983) Reaction of glycolaldehyde with proteins: latent crosslinking potential of α-hydroxyaldehydes. *Proc Natl Acad Sci U S A*, **80**, 3590–3594.
105 Nagai, R., Hayashi, C.M., Xia, L., Takeya, M., and Horiuchi, S. (2002) Identification in human atherosclerotic lesions of GA-pyridine, a novel structure derived from glycolaldehyde-modified proteins. *J Biol Chem*, **277**, 48905–48912.
106 Nagai, R., Matsumoto, K., Ling, X., Suzuki, H., Araki, T., and Horiuchi, S. (2000) Glycolaldehyde, a reactive intermediate for advanced glycation end products, plays an important role in the generation of an active ligand for the macrophage scavenger receptor. *Diabetes*, **49**, 1714–1723.
107 Unno, Y., Sakai, M., Sakamoto, Y.-I., Kuniyasu, A., Nagai, R., Nakayama, H., and Horiuchi, S. (2005) Glycolaldehyde-modified bovine serum albumin downregulates leptin expression in mouse adipocytes via a CD36-mediated pathway. *Ann NY Acad Sci*, **1043**, 696–701.
108 Horiuchi, S., Unno, Y., Usui, H., Shikata, K., Takaki, K., Koito, W., Sakamoto, Y.-I., Nagai, R., Makino, K., Sasao, A., Wada, J., and Makino, H. (2005) Pathological roles of advanced glycation end product receptors SR-A and CD36. *Ann NY Acad Sci*, **1043**, 671–675.
109 Unno, Y., Sakai, M., Sakamoto, Y.-I., Kuniyasu, A., Nakayama, H., Nagai, R., and Horiuchi, S. (2004) Advanced glycation end products-modified proteins and oxidized LDL mediate down-regulation of leptin in mouse adipocytes via CD36. *Biochem Biophys Res Commun*, **325**, 151–156.
110 Hengstler, J.G., Fuchs, J., Gebhard, S., and Oesch, F. (1994) Glycolaldehyde causes DNA-protein crosslinks: a new aspect of ethylene oxide genotoxicity. *Mutat Res Fundam Mol Mech Mutagen*, **304**, 229–234.
111 Thornalley, P.J., Jahan, I., and Ng, R. (2001) Suppression of the accumulation of triosephosphates and increased formation of methylglyoxal in human red blood cells during hyperglycaemia by thiamine in vitro. *J Biochem*, **129**, 543–549.
112 Hamada, Y., Araki, N., Koh, N., Nakamura, J., Horiuchi, S., and Hotta, N. (1996) Rapid formation of advanced glycation end products by intermediate metabolites of glycolytic pathway and polyol pathway. *Biochem Biophys Res Commun*, **228**, 539–543.
113 Travis, S.F., Morrison, A.D., Clements, R.S. Jr, Winegrad, A.I., and Oski, F.A. (1971) Metabolic alterations in the human erythrocyte produced by increases in glucose concentration. The role of the polyol pathway. *J Clin Invest*, **50**, 2104–2112.
114 Knight, R.J., Kofoed, K.F., Schelbert, H.R., and Buxton, D.B. (1996) Inhibition of

glyceraldehyde-3-phosphate dehydrogenase in post-ischaemic myocardium. *Cardiovasc Res*, **32**, 1016–1023.
115 Du, X.L., Edelstein, D., Rossetti, L., Fantus, I.G., Goldberg, H., Ziyadeh, F., Wu, J., and Brownlee, M. (2000) Hyperglycemia-induced mitochondrial superoxide overproduction activates the hexosamine pathway and induces plasminogen activator inhibitor-1 expression by increasing Sp1 glycosylation. *Proc Natl Acad Sci U S A*, **97**, 12222–12226.
116 Du, X.L., Matsumura, T., Edelstein, D., Rossetti, L., Zsengeller, Z., Szabo, C., and Brownlee, M. (2003) Inhibition of GAPDH activity by poly(ADP-ribose) polymerase activates three major pathways of hyperglycemic damage in endothelial cells. *J Clin Invest*, **112**, 1049–1057.
117 Nishikawa, T., Edelstein, D., Du, X.L., Yamagishi, S., Matsumura, T., Kaneda, Y., Yorek, M.A., Beebe, D., Oates, P.J., Hammes, H.P., Giardino, I., and Brownlee, M. (2000) Normalizing mitochondrial superoxide production blocks three pathways of hyperglycaemic damage. *Nature*, **404**, 787–790.
118 Rosca, M.G., Mustata, T.G., Kinter, M.T., Ozdemir, A.M., Kern, T.S., Szweda, L.I., Brownlee, M., Monnier, V.M., and Weiss, M.F. (2005) Glycation of mitochondrial proteins from diabetic rat kidney is associated with excess superoxide formation. *Am J Physiol Renal Physiol*, **289**, F420–F430.
119 Syrovy, I. (1994) Glycation of myofibrillar proteins and ATPase activity after incubation with eleven sugars. *Physiol Res*, **43**, 61–64.
120 Syrovy, I. (1994) Glycation of albumin: reaction with glucose, fructose, galactose, ribose or glyceraldehyde measured using four methods. *J Biochem Biophys Methods*, **28**, 115–121.
121 Tessier, F.J., Monnier, V.M., Sayre, L.M., and Kornfield, J.A. (2003) Triosidines: novel Maillard reaction products and cross-links from the reaction of triose sugars with lysine and arginine residues. *Biochem J*, **369**, 705–719.
122 Acharya, A.S., Cho, Y.J., and Manjula, B.N. (1988) Cross-linking of proteins by aldotriose: reaction of the carbonyl function of the keto amines generated in situ with amino groups. *Biochemistry*, **27**, 4522–4529.
123 Vander Jagt, D.L., Robinson, B., Taylor, K.K., and Hunsaker, L.A. (1992) Reduction of trioses by NADPH-dependent aldo-keto reductases. Aldose reductase, methylglyoxal, and diabetic complications. *J Biol Chem*, **267**, 4364–4369.
124 Fitzgerald, C., Swearengin, T.A., Yeargans, G., McWhorter, D., Cucchetti, B., and Seidler, N.W. (2000) Non-enzymatic glycosylation (or glycation) and inhibition of the pig heart cytosolic aspartate aminotransferase by glyceraldehyde 3-phosphate. *J Enzyme Inhib*, **15**, 79–89.
125 Tessier, F.J., Monnier, V.M., Sayre, L.M., and Kornfield, J.A. (2003) Triosidines: novel Maillard reaction products and cross-links from the reaction of triose sugars with lysine and arginine residues. *Biochem J*, **369**, 705–719.
126 Prabhakaram, M. and Ortwerth, B.J. (1994) Determination of glycation crosslinking by the sugar-dependent incorporation of. *Anal Biochem*, **216**, 305–312.
127 Swearengin, T.A., Fitzgerald, C., and Seidler, N.W. (1999) Carnosine prevents glyceraldehyde 3-phosphate-mediated inhibition of aspartate aminotransferase. *Arch Toxicol*, **73**, 307–309.
128 Murata, T., Miwa, I., Toyoda, Y., and Okuda, J. (1993) Inhibition of glucose-induced insulin secretion through inactivation of glucokinase by glyceraldehyde. *Diabetes*, **42**, 1003–1009.
129 Solomon, L.R. and Crouch, J.Y. (1989) Studies on the mechanism of acetaldehyde-mediated inhibition of aspartate aminotransferase (GOT) in

130 Takeuchi, M., Makita, Z., Bucala, R., Suzuki, T., Koikie, T., and Kameda, Y. (2000) Immunological evidence that non-carboxymethyllysine advanced glycation end-products are produced from short chain sugars and dicarbonyl compounds in vivo. *Mol Med*, **6**, 114–125.

131 Takeuchi, M., Bucala, R., Suzuki, T., Ohkubo, T., Yamazaki, M., Koike, T., Kameda, Y., and Makita, Z. (2000) Neurotoxicity of advanced glycation end-products for cultured cortical neurons. *J Neuropathol Exp Neurol*, **59**, 1094–1105.

132 Choei, H., Sasaki, N., Takeuchi, M., Yoshida, T., Ukai, W., Yamagishi, S., Kikuchi, S., and Saito, T. (2004) Glyceraldehyde-derived advanced glycation end products in Alzheimer's disease. *Acta Neuropathol*, **108**, 189–193.

133 Choei, H., Sasaki, N., Takeuchi, M., Yoshida, T., Ukai, W., Yamagishi, S.-I., Kikuchi, S., and Saito, T. (2004) Glyceraldehyde-derived advanced glycation end products in Alzheimer's disease. *Acta Neuropathol*, **108**, 189–193.

134 Takeuchi, M. and Yamagishi, S.-I. (2008) Possible involvement of advanced glycation end-products (AGEs) in the pathogenesis of Alzheimer's disease. *Curr Pharm Des*, **14**, 973–978.

135 Usui, T., Shimohira, K., Watanabe, H., and Hayase, F. (2004) Detection and determination of glyceraldehyde-derived advanced glycation end product. *Biofactors*, **21**, 391–394.

136 Usui, T., Shizuuchi, S., Watanabe, H., and Hayase, F. (2004) Cytotoxicity and oxidative stress induced by the glyceraldehyde-related Maillard reaction products for HL-60 cells. *Biosci Biotechnol Biochem*, **68**, 333–340.

137 Usui, T. and Hayase, F. (2003) Isolation and identification of the 3-hydroxy-5'hydroxymethylpyridinium compound as a novel advanced glycation end product on glyceraldehyde-related Maillard reaction. *Biosci Biotechnol Biochem*, **67**, 930–932.

138 Usui, T., Shimohira, K., Watanabe, H., and Hayase, F. (2007) Detection and determination of glyceraldehyde-derived pyridinium-type advanced glycation end product in streptozotocin-induced diabetic rats. *Biosci Biotechnol Biochem*, **71**, 442–448.

139 Usui, T., Ohguchi, M., Watanabe, H., and Hayase, F. (2008) The formation of argpyrimidine in glyceraldehyde-related glycation. *Biosci Biotechnol Biochem*, **72**, 568–571.

140 Hayase, F., Usui, T., Ono, Y., Shirahashi, Y., Machida, T., Ito, T., Nishitani, N., Shimohira, K., and Watanabe, H. (2008) Formation mechanisms of melanoidins and fluorescent pyridinium compounds as advanced glycation end products. *Ann N Y Acad Sci*, **1126**, 53–58.

141 Peppa, M., Uribarri, J., and Vlassara, H. (2008) Aging and glycoxidant stress. *Hormones (Athens, Greece)*, **7**, 123–132.

142 Novelli, M., Masiello, P., Bombara, M., and Bergamini, E. (1998) Protein glycation in the aging male Sprague-Dawley rat: effects of antiaging diet restrictions. *J Gerontol A Biol Sci Med Sci*, **53**, B94–B101.

143 Sell, D.R. (1997) Ageing promotes the increase of early glycation Amadori product as assessed by?-N-(2-furoylmethyl)-L-lysine (furosine) levels in rodent skin collagen. The relationship to dietary restriction and glycoxidation. *Mech Ageing Dev*, **95**, 81–99.

144 Suji, G. and Sivakami, S. (2004) Glucose, glycation and aging. *Biogerontology*, **5**, 365–373.

145 Vlassara, H. (2005) Advanced glycation in health and disease: role of the modern environment. *Ann N Y Acad Sci*, **1043**, 452–460.

146 Hipkiss, A.R. (2006) On the mechanisms of ageing suppression by

dietary restriction – is persistent glycolysis the problem? *Mech Ageing Dev*, **127**, 8–15.
147 Beisswenger, P.J., Howell, S.K., O'Dell, R.M., Wood, M.E., Touchette, A.D., and Szwergold, B.S. (2001) α-Dicarbonyls increase in the postprandial period and reflect the degree of hyperglycemia. *Diabetes Care*, **24**, 726–732.
148 Nemet, I., Turk, Z., Duvnjak, L., Car, N., and Varga-Defterdarovic, L. (2005) Humoral methylglyoxal level reflects glycemic fluctuation. *Clin Biochem* **38**, 379–383.
149 Ahmed, N., Babaei-Jadidi, R., Howell, S.K., Thornalley, P.J., and Beisswenger, P.J. (2005) Glycated and oxidized protein degradation products are indicators of fasting and postprandial hyperglycemia in diabetes. *Diabetes Care*, **28**, 2465–2471.
150 Hipkiss, A.R. (2008) Energy metabolism, altered proteins, sirtuins and ageing: converging mechanisms? *Biogerontology*, **9**, 49–55.
151 Kalapos, M.P. (2007) Can ageing be prevented by dietary restriction? *Mech Ageing Dev*, and **128**, 227–228.

7
Fructose-Derived Endogenous Toxins

Peter J. O'Brien, Cynthia Y. Feng, Owen Lee, Q. Dong, Rhea Mehta, Jeff Bruce, and W. Robert Bruce

High sugar/fructose or fat diets are believed to contribute to the tissue damage and origin of many chronic diseases. This is because they result in the formation of endogenous reactive carbonyl species (RCS) or, with the glycation of proteins, advanced glycation end products (AGEs). RCS and AGEs are elevated in diabetes mellitus. They inhibit pancreatic insulin secretion, injure cells, and cause diabetic complications (e.g., vascular, renal failure). For example, methylglyoxal (MG), formed from the glycolytic metabolite, dihydroxyacetone phosphate, is an RCS, which glycates proteins, nucleotides, and phospholipids, thereby damaging the proteome, lipidome, and genome, and contributing to the aging process. Nonalcoholic fatty liver disease (NAFLD, hepatic steatosis) affects 10–24% of the general population and 57–74% of obese individuals, and is a common cause of the abnormal steatohepatitis (NASH), a life-theatening form of NAFLD. A two-hit hypothesis has been proposed for NASH with the first hit being hepatic steatosis, and the second hit being oxidative stress coming from inflammation or exposure to a toxic xenobiotic. High sugar/fructose or fat diets fed to rodents caused steatosis, hepatic AGEs, and hepatocyte cell death. We recently showed that hepatocytes became very susceptible to fructose when exposed to continuous nontoxic H_2O_2 oxidative stress. Furthermore, glyoxal, a major fructose oxidation product, also markedly increased hepatocyte susceptibility to H_2O_2, causing cell death. H_2O_2 also sensitized hepatocytes to cytokines such as those released by inflammatory-immune cells, and caused cytotoxicity. Other endogenous reactive carbonyls associated with oxidative stress induction have been identified on the fructose-glyceraldehyde-oxalate (kidney stones) metabolic pathway for the first time. Glyceraldehyde, dihydroxy-acetone, or glycolaldehyde were rapidly oxidized partly to glyoxal by superoxide radicals or an iron/H_2O_2 Fenton reaction, whereas fructose was oxidized more slowly. Glyoxylate, another carbonyl intermediate, depleted GSH and caused oxidative stress cytotoxicity. The effectiveness of the various detoxifying carbonyl metabolizing enzymes, for preventing hepatocyte cytotoxicity induced by these endogenous reactive carbonyls, was also compared.

Endogenous Toxins. Diet, Genetics, Disease and Treatment.
Edited by Peter J. O'Brien and W. Robert Bruce
Copyright © 2010 WILEY-VCH Verlag GmbH & Co. KGaA, Weinheim
ISBN: 978-3-527-32363-0

7.1
Introduction

Concern over the high chronic consumption of fructose comes from several directions: (i) Sugar and, in particular, fructose consumption has increased markedly in recent human history, so that rapidly metabolized food components now make up a large fraction of food energy in the "developed" world. (ii) There is evidence that sugars could be involved in the etiology of a wide range of human chronic diseases including obesity, nonalcoholic fatty liver disease (NAFLD), nonalcoholic steatohepatitis (NASH), and components of the metabolic syndrome. (iii) Animal experimental studies have demonstrated that dietary sugars can affect animal models for a number of these chronic diseases. In particular, studies of the development of NASH have demonstrated the importance of interactions (two hits) in the development of the toxicity of excess energy substrates with oxidative stress [2]. (iv) Sugars can be oxidatively degraded to form reactive products that can damage cells. In particular, carbonyls formed from fructose can react with long-lived macromolecules resulting in long-term consequences for cellular survival and genetic stability [2].

This chapter briefly reviews the evidence relating to the current state of fructose consumption in the "developed" world, the possible diseases associated with its consumption, then focuses on a model of the interaction of cellular exposure to sugar and oxidative stress. The chemistry involved is discussed, and concluded with a discussion of the possible importance and mitigation of these processes.

7.2
Recent Increased Consumption of Fructose

Diets have changed substantially through human history. Perhaps the most remarkable recent change has been the introduction of the refined carbohydrates, sucrose and fructose. Plant sugars were first discovered in sugarcane, in about 500 BC in Polynesia while sugar beet was first recognized as another source of sugar in 1747 by Marggraf, who succeeded in extracting sugar from beets. Europeans at the time were using relatively small quantities of honey as their main sweetener. Honey contains about 40% fructose, 30% glucose, and a trace of sucrose, whereas cane sugar and beet sugar contain 15–20% sucrose. With the development of refining methods, sucrose availability from sugarcane and beets increased from an average of about 20 g/day per individual in the mid-nineteenth-century Europe to about 90 g/day per individual in 1970. With the development of enzymatic column methods for preparing fructose and glucose from corn syrup by Takasaki and Tanabe, high-fructose corn syrup (HFCS) began to replace sucrose, because of its lower cost. At present, HFCS is consumed at a similar daily rate as sucrose with an average total of about 106 g/day in 2006 (www.ers.usda.gov/data/foodcomposition/foodguideindex.htm). Sucrose intake decreased during this time period; so, while consumption of both fructose and glucose has increased, the ratio of fructose to glucose has not changed much [3]. Sugar availability though, has increased in several waves since then, the most recent being from 1970 to the present, of about 18%, to where added sugars represent an average of about 20% of calories. This is of course, only one of the many dietary changes that have occurred in recent years. Added fats (salad oil, cooking oil, and beef tallow) increased 55% in this period, and now represent 24% of calories. Sugars, however, are the major rapidly metabolized energy substrate in the diet, and individual consumption of sugars can greatly exceed these average numbers.

Table 7.1 shows the food content of fructose, glucose, and sucrose expressed as grams per 100 grams. Foods consumed by humans have likely always contained some fructose from fruits and vegetables. At present, foods with the highest content of fructose, glucose, sucrose, expressed as grams per 100 grams, include honey, syrups, and 90 or 55% fructose HFCSs. Dried fruits that are also high in sugars include raisins, currants, or dates. The lowest sugar content in dried fruits was in dried apricots and peaches. The highest sugar content in fresh fruits was in apples, pears, grapes, with the lowest being limes, apricots, peaches, grapefruits, plums, rhubarbs, and tomatoes. The sugar content of vegetables was lower and mostly due to sucrose, for example, beetroot, carrot, sweet potato, and was lowest in sweet onion. Thus, HFCSs can contain 10 times more fructose than that found in the highest fructose containing fruit (apples, grapes). The order of sweetness

Tab. 7.1 Food content of fructose, glucose, sucrose expressed as grams per 100 grams

	Fructose (g/100 g)	Glucose (g/100 g)	Sucrose (g/100 g)
Honey	42.4	33.8	1.5
90% fructose corn syrup	72.5	7.2	0
55% fructose corn syrup	42.2	33.8	0
Raisins or currants	37.4	31.2	0.8
Dates	31.3	33.4	4
Dried apricots	12.2	20.3	16
Peaches	15.6	15.8	13.2
Apples	7.6	2.3	3.3
Pears	6.4	1.9	3.1
Grapes	7.6	6.5	0.5
Limes	0.2	0.2	0
Apricots	0.7	1.6	5.2
Peaches	1.3	1.2	5.6
Grapefruit	1.2	1.3	3.4
Plums	1.8	2.7	3
Rhubarb or tomatoes	1.1	1.4	>1
Beetroots	0.1	0.1	6.5
Carrots	0.6	0.6	3.6
Sweet potatoes	0.7	1.0	2.5
Sweet onions	2.0	2.3	0.7

of monosaccharides relative to sucrose is fructose (solid 173 or solution 117) > sucrose (100) > glucose (74) > galactose (33) > lactose [4].

Fructose can also exist in the diet as fructans and dietary sugar alcohols. Fructans (fructo-oligosaccharides) are polymers of fructose molecules that are not absorbed by the intestine but are metabolized by anaerobic colonic bacteria. Inulin promotes the growth of benign intestinal bacteria. As such it can be referred to as a *prebiotic*. The three types are inulin (linear linked by $\beta(2\rightarrow1)$ glycosidic bonds, levan (linear linked by $\beta(2\rightarrow6)$ glycosidic bonds, graminan branched fructans linked by both bonds and may contain 20–2000 fructose units. Hydrolysis may yield fructo-oligosaccharides with less than 10 fructose units. Inulin is used by plants to store energy as it contains a third to quarter of the food energy of sugar. Inulin ingestion by humans does not affect blood glucose or insulin levels. It also does not increase plasma triglyceride levels unlike fructose. Fructans are found in the diet mostly as wheat and onions, but also in leeks, green beans, asparagus, artichokes, rye grain, chocolate, and cheese spread [5]. Some traditional diets contain up to 20 g fructans per day. However, dietary restriction of fructans to less than 0.5 g/serving caused symptomatic improvement of patients with irritable bowel improvement [6]. Shorter oligo-fructose dietary fibers are used as sweetener supplements for calorie reduction or preventing yeast infections and can be added to milk products. They are fermented by intestinal bacteria to form hydrogen and carbon dioxide gases.

Dietary sugar alcohols (e.g., sorbitol) are not well absorbed in the intestine, but are rapidly metabolized by hepatic sorbitol dehydrogenase/NAD^+ to form fructose/NADH. However, some tissues use aldose reductase to convert glucose to sorbitol (the polyol pathway) that could contribute to fructose levels *in vivo*. Sorbitol is found in small concentrations in stone fruits (apricot, nectarine, peach, plum) and berries (blueberry).

7.3
Health Concerns Associated with High Chronic Consumption of Fructose

7.3.1
Sugar and Sugar Intolerances

Sucrose on ingestion is hydrolyzed to glucose and fructose catalyzed by sucrose-isomaltase, an $\alpha(1-4)$ glucosidase that is located on the brush border of the small intestine, particularly the jejunum. A congenital sucrose-isomaltase deficiency is much rarer than a lactase deficiency, and treatment requires lifelong adherence to a strict sucrose-free diet. Sugars added to the diet can be decreased, but sucrose is also a constituent of some fruits.

A low fructose diet is particularly important for those patients suffering from hereditary fructose intolerance (HFI, lack of hepatic aldolase B), who are at risk of fructose-mediated liver and kidney damage. Low fructose/fructan diets are also useful for patients with fructose malabsorption (lack the enterocyte fructose transporter) and suffering from fructose-induced abdominal pain, nausea, bloating, and diarrhea. Patients with irritable bowel syndrome (IBS) also obtained relief from dietary fructose/fructan restriction [7].

Fructose is sometimes recommended to diabetic patients as a sweetener substitute for sucrose, because fructose unlike glucose does not require insulin to enter cells, so that moderate fructose intake does not affect glucose levels. Fructose, thus, has a low glycemic index. However, the benefits of fructose for diabetic patients have not been shown. Furthermore, a prospective study of sugar intake and the risk of type 2 diabetes in women, shows that the intake of sugars (sucrose, glucose, fructose, lactose) did not appear to play a role in primary prevention of type 2 diabetes [8].

7.3.2
Fructose and Obesity – Nonalcoholic Liver Disease (NAFLD)

Obesity in the population, over the past 35 years, has risen concurrently with sugars and fats added to the food supply. The increase in obesity coincided with an increase in the consumption of sugars, carbohydrates, and fats. Fructose is mainly ingested as sucrose and as HFCS-55. Sucrose- or fructose-enriched diets caused a sustained elevation of plasma triglycerides in the human population, unlike glucose ingestion, which suggested that chronic overconsumption of fructose could contribute to atherogenesis and cardiovascular disease [9]. Caloric overconsumption is also implicated in NAFLD, which now affects 10–24% of the general population and 57–74% of obese individuals.

Sugar consumption is also a common factor in the development of NASH, a more life-threatening form of NAFLD in which liver inflammation is involved. As most patients with fatty liver do not get hepatitis, a two-hit hypothesis was proposed for NASH, with the first hit being fatty liver (hepatic steatosis) and the second hit being oxidative stress coming from inflammation or exposure to a toxic xenobiotic. The second hit was based on an increase in inflammation biomarkers in those patients with hepatitis [1]. NASH was also associated with an approximately 40% sustained increase in serum advanced glycation end product (AGE)-2 levels [10], formed from glyceraldehyde (GA) (Figure 7.1). Glucose is metabolized by most cells, whereas fructose is mostly metabolized *in vivo* by the liver. Fructose also stimulates liver glucose uptake [11]. A hepatic stress response and systemic inflammation was also elicited. Fructose-induced obesity is also associated with components of the metabolic syndrome including glucose intolerance, insulin resistance, and atherogenic dyslipidemia, as well as with hypertension.

Others, however, have suggested uric acid as the endogenous toxin that causes hypertension by releasing chemotactic and inflammatory substances and decreasing endothelial nitric oxide. They also attribute the increased fructose consumption of African-Americans to explain their predisposition to cardiorenal disease [13]. Fructose, because of high fructokinase activity rapidly uses up ATP to form uric acid by ADP/AMP catabolism; but this does not occur with glucose.

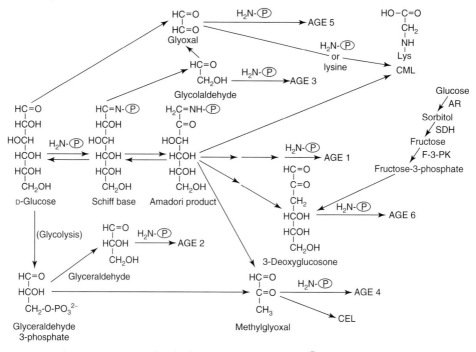

Fig. 7.1 Formation of eight distinct AGEs *in vivo*. $H_2N-Ⓟ$, free amino residue of protein; AR, aldose reductase; SDH, sorbitol dehydrogenase; F-3-PK, fructose-3-phosphokinase CML, carboxy methyllysine; CEL, carboxyethyl lysine. Adapted from [38, 39].

Sucrose- and fructose-enriched diets also caused hepatic insulin resistance in rats independently of obesity that involved c-Jun N-terminal kinase (JNK) activation. In rodents, increased dietary fructose for 4–12 weeks induces many aspects of the metabolic syndrome including hypertension. For example, 4% fructose in drinking water for 11 weeks induced hypertension and increased platelet cytosolic free calcium, as well as kidney aldehyde conjugates in Wistar–Kyoto rats [14]. They attributed the hypertension to protein carbonyl formation caused by increased kidney AGE formation by aldehyde protein conjugates, for example, AGE formation by GA and methylglyoxal (MG).

7.4
Sugars as a Source of Endogenous Reactive Carbonyl Formation and AGEs

7.4.1
Fructose versus Glucose Metabolism by Glycolysis to Carbonyl Metabolites

Fructose hepatic metabolism differs from glucose in being more readily converted to glycerol and triglyceride, thus favoring lipogenesis [15]. Fructose also undergoes glycolysis faster than glucose. As shown in Figure 7.1, glycolysis of glucose involves phosphorylation to glucose 6-phosphate catalyzed by hexokinase or glucokinase. An isomerase then catalyzes fructose 6-phosphate formation and phosphofructokinase, the glycolysis rate limiting step, then catalyzes fructose 1,6-bisphosphate formation. Aldolase then catalyzes the dihydroxyacetone phosphate (DHAP) and glyceraldehyde 3-phosphate (GAP) formation before the latter is oxidized, catalyzed by GA dehydrog enase to form NADH and ATP. Hepatic methylglyoxal synthase (MGS) catalyzes the conversion of DHAP to MG (Km 0.76 mM) and phosphate [16]. The synthase catalyzes the elimination reaction of enediolate form of DHAP (but not GAP) leading to inorganic phosphate and MG enol, which tautomerizes to MG [17]. This elimination reaction is suppressed in triosephosphate isomerase [18], which catalyzes the isomerization of DHAP and GAP bidirectionally without forming MG. The MG is detoxified by glyoxalase I, which protects glycolysis from inactivation by MG. MG may also be formed by spontaneous phosphate elimination from autoxidation of the enediolate form of DHAP or GAP [19, 20].

Glycolysis of fructose, on the other hand, starts with fructokinase in the liver, kidney, and intestine, which catalyzes fructose phosphorylation to fructose 1-phosphate. Rat liver homogenates catalyze fructose phosphorylation 10 times more readily than glucokinase-catalyzed glucose phosphorylation to glucose 6-phosphate [21], largely due to fructokinase having a low Km for D-fructose (0.1–0.29 mM) [22], whereas glucokinase has a high Km for glucose of 5–20 mM, and a high Vmax [23]. Aldolase B then catalyzes DHAP and GA formation, thereby bypassing the slow phosphofructokinase step (Figure 7.2). GA is mostly autoxidized to glyoxal [24] or can be converted to MG by Ga kinase, triose phosphate isomerase followed by MGS. Triglycerides and phospholipids are not only formed from DHAP via glycerol 3-phosphate, but also from GA via glycerol and glycerol 3-phosphate [25].

7.4.2
Fructose Oxidation to Endogenous Genotoxic Carbonyl Products and Protein Advanced Glycation End Products (AGEs)

Glucose (known as *grape sugar*) and galactose are aldose sugars. Ordinarily crystalline glucose is cyclic in the α-form. In solution, hexoses and pentoses exist in the cyclic hemiacetal form and an equilibrium is established among the open-chain form and the two anomers (α,β). They are converted in solution mostly into a pyranose form. This is a six-membered ring consisting of five carbon and one oxygen atom, in which the carbonyl forms the ring, thereby preventing it from reacting with biomolecules. In contrast, fructose (known as *fruit sugar*), sorbose, and tagatose are ketose sugars, and are more readily converted to a furanose form that is a five-membered furan ring with a reactive carbonyl group that can more readily act as a reducing agent. The aldehyde group is free to act as a reducing agent only in the open-chain form, and ketones do not act as reductants. Indeed, in solution, in addition to six-member rings (pyranose), fructose forms five-member rings (furanose) (Figure 7.3). In basic solution, ketoses are also reducing sugars.

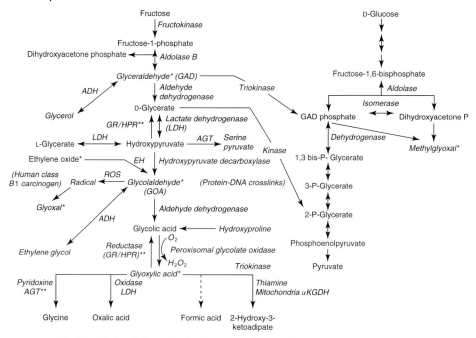

Fig. 7.2 Carbonyl formation by fructose-oxalate path or glucose-pyruvate glycolysis. ADH: alcohol dehydrogenase; AGT: alanine glyoxylate transaminase.

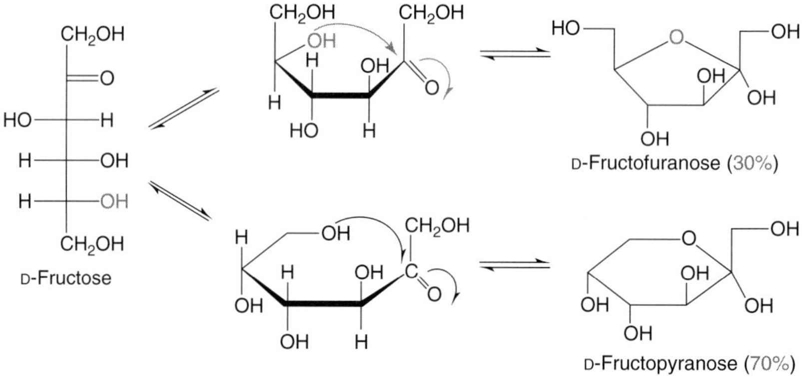

Fig. 7.3 Cyclic hemiketal fructose isomers in aqueous solution.

A ketone that has a H atom on the C adjacent to the carbonyl C, undergoes a rearrangement to an enediol. The enediol rearranges to give an aldose, which is susceptible to oxidation (i.e., can act as a reducing sugar), for example, Maillard reaction. Seventy percentage of fructose exists as D-fructopyranose with 30% existing as D-fructofuranose (Figure 7.3).

The percentage of a given sugar present in the reactive acyclic open-chain form seems to determine the *in vitro* protein glycation rate. The percentage of fructose in

Tab. 7.2 Lens protein AGE formation by 5 mM sugars in phosphate buffer pH 7.4 under N_2 at 37 °C for 10 days [28]

Sugars + lens protein	Protein AGEs	Protein cross-linking
Glucose	4.2	0.2
Fructose	5.2	0.2
Galactose	7.1	0.4
Ribose	15.8	11.3
Erythrose	54.7	63.1
Glyceraldehyde	88.0	75.0
Dihydroxyacetone	25.7	37.1
Glyceraldehyde-3-P	108.5	59.9

the open-chain form is 0.7%, whereas for glucose it is 0.002%. The fructose reactive carbonyl (ketone) present is thus 350-fold more than the glucose reactive carbonyl (aldehyde). Proteins are 10-fold more rapidly fructated than glycated (AGE) [26, 27]. Short chain sugars are however, much more reactive as their carbon skeleton is too short to permit cyclization to a hemiacetal ring; so, they exist mainly in an open-chain form in which the reactive carbonyl group is unprotected. The reactivity of sugars for protein glycation are therefore: GA > ribose > arabinose > fructose > xylose > galactose > mannose > glucose. As shown in Table 7.2, lens protein AGE formation and protein cross-linking by 5 mM sugars, in phosphate buffer pH 7.4 under N_2 at 37 °C for 10 days, follows this order of reactivity [28]. However, the rate of Schiff base formation or browning, often do not follow the same order [29].

Short chain sugars such as glycolaldehyde (GOA) readily autoxidize under physiological conditions as their open-chain form at a slightly alkaline pH tautomerizes to an enediol (see Figure 1 of [30]). Enediol formation was responsible for the reducing properties of sugars that includes reducing oxygen to superoxide radicals and mono and biradical intermediates; hydrogen peroxide, hydroxyl radicals, and α-hydroxyalkyl radicals were also formed, which could initiate and propagate a chain reaction. The more reactive the carbonyl, the more the carbonyl was susceptible to autoxidation. Autoxidation was prevented by iron chelators and autoxidation of some sugars were also readily catalyzed by Fe or Cu. Raman spectroscopy showed that Cu(II) coordinated to the C-1 and C-2 OH groups of fructose, which preferentially complexed the *furanose* form of the fructose, whereas Fe(III) complexes the *pyranose* form. Glucose was fivefold less readily oxidized than fructose, suggesting that Cu(II) poorly complexes the aldohexose form of glucose [31–33].

In phosphate buffer pH 7.0, fructose was 15-fold more readily oxidized than glucose catalysed by Cu(II), which was sevenfold more effective than Fe(III). The products found were HCHO, lactone > glyoxal [34]. Fructose was also much more susceptible to oxidation by Fenton's (H_2O_2/Fe(II) hydroxyl radicals) than glucose. H_2O_2, dihydroxyacetone (DHA), GA, GOA, and glyoxal were formed [24]. A comparison of the susceptibility of simple sugars to oxidation by Fenton's was reported as deoxyribose > sorbitol > fructose, sucrose ≫ glucose > maltose [35]. Fructose was also much more effective than glucose > sucrose for increasing methyl linoleate autoxidation [36].

AGEs of long-lived proteins (serum albumin, lens crystallin, or extracellular matrix collagen) are elevated in people with type 2 diabetes. Even short-lived (120 days) hemoglobin can accumulate AGEs at lower levels that are increased in type 2 diabetes. The atherogenesis rate can be accelerated by low-density lipoprotein (LDL) AGEs that form foam cells, which become trapped in the vascular walls [12]. Persistent hyperglycemia causes formation of irreversible bonds between the aldehyde groups of reducing sugars (glucose, other reducing sugars, and particularly fructose) and protein amino groups to form a reversible or irreversible Schiff base, which slowly rearranges giving Amadori early glycation adducts (Figure 7.1). These adducts then undergo a series of rearrangement, dehydration, condensation, oxidation, and cyclization reactions with protein amino groups leading to the formation of AGEs, known as the *browning reaction*. Only a few endogenous glycated protein repair mechanisms have been identified. However, AGEs are more readily formed in foods, particularly during prolonged cooking at high temperatures, for example, the frying/grilling or cooking of foods containing protein and fructose. AGEs are divided into two categories: those that cross-link with plasma lipids and tissue lipoproteins (e.g., pentosidine and vesperlysines A) and those that do not cross-link (e.g., carboxymethyllysine, imidazolones, and argpyrimidine). Diabetic atheromatoses may develop as a result of AGE formation that cross-link collagen, intracellular proteins, cell membranes, phospholipids, lipoproteins, and DNA. AGEs exert their action through a specific RAGE receptor for AGEs or toxic advanced glycation end product (TAGE) receptor for TAGEs that are located on the cell surface. The TAGE receptor activates a cascade of intracellular reactions that produce proinflammatory cytokines and increases oxidative stress, thereby causing vasoconstriction, platelet accumulation, and damage to the small blood vessels in the kidney and the retina of the eye. AGEs have also been linked to heart disease, skin aging, rheumatoid arthritis, and Alzheimer's disease (AD). Consequently, diabetic patients should be informed about possible harmful effects of AGEs and the advisability of decreasing dietary AGEs [37].

As shown in Figure 7.1, AGE-1 has been attributed to glucose-derived AGEs or CML, while AGE-2 has been attributed to a rearrangement of GA addition products, and has been detected immunohistochemically in both senile plaques and neurofibrillary tangles in AD. Other AGEs include AGE-3 formed from GOA, AGE-4 from MG, AGE-5 from glyoxal, and AGE-6 from deoxyglucosone. AGE-2 and AGE-3 are predominantly associated with toxicity and are called *toxic advanced glycation end products*. TAGEs can glycate amyloid β protein, and H_2O_2 markedly enhances its aggregation *in vitro*. Paired helical filaments are formed by tau glycation and hyperphosphorylation. Interactions between GA TAGEs and their receptors, RAGEs, may also play a role in the pathogenesis of AD and diabetic complications including nephropathy, neuropathy, and retinopathy. TAGE also shows toxicity toward cortical neuronal cells and is implicated in early neurotoxicity associated with AD [38]. AGE-2 and AGE-3 have the strongest binding affinity to RAGE, of the immunoglobulin superfamily. Binding of AGEs to RAGE initiates cellular signals that activate NF-kB, which results in transcription of proinflammatory factors. TAGE also induces oxidative stress associated with Amadori product transition metal catalyzed autoxidation and indirectly by cytokine, for example, TNF, which inhibits mitochondrial electron transport. RAGE also evokes vascular inflammation such as that occurring in accelerated atherosclerosis in diabetes that can be prevented by nateglinide, a meglitinide class of blood glucose-lowering drug [39]. RAGE receptors and scavenger receptors are found in liver sinusoidal cells such as Kupffer and endothelial cells. NASH was also associated with an approximately 40% sustained increase in serum AGE-2 levels [10].

7.4.3
Rodent Models for Fatty Liver (Steatosis) and NASH (Steatohepatitis)

A rodent model for NASH involving abnormally high-fructose or high-fat diets was administered for eight weeks to several months. With a sucrose-rich diet (60% sucrose) rats also became obese and developed steatosis in which fat was accumulated periportally. Hepatic enzymes generating NADPH were induced, for example, malic dehydrogenase, 6-phosphogluconate dehydrogenase. Gamma glutamyl transpeptidase, aspartate transaminase, and lactate dehydrogenase were also similarly induced in the perivenous region. Hepatic triglyceride synthesis was increased and VLDL clearance was impaired. These effects were attributed to fructose as they also occurred with dietary fructose, but not glucose or glycerol.

Ten percentage of fructose in drinking water for 48 hours induced hepatic fatty acid synthase and de novo fatty acid and triglyceride synthesis in rats. Male Wistar rats were more vulnerable than male SD rats, while male Zucker fatty (fa/fa) were particularly sensitive to steatosis induced by feeding sucrose or 1% orotic acid. Myo-inositol, phytic acid, or clofibrate prevented fructose-induced hepatic steatosis [40]. Taurine was also effective [41]. A high-sucrose diet containing 1% orotic acid administered to mice for four weeks induced obesity, steatosis, increased liver triglycerides, hepatocyte apoptosis, and fibrosis. Orotic acid caused the accumulation of triglycerides in hepatocytes partly by inhibiting VLDL synthesis/secretion. This NASH model was prevented by an IKK2 inhibitor. IKK2 is a protein, which is a component of a cytokine-activated intracellular pathway involved in triggering immune response as well as mediating inflammation and insulin resistance [42]. A recent example of a high fructose diet administered for 16 weeks to mice was similar to commonly consumed fast food by using HFCS in their drinking water under conditions that promoted sedentary behavior. Administering trans fats in their chow instead, was also effective. These diets promoted obesity, liver steatosis, insulin resistance, and liver injury including necroinflammatory changes, fibrosis, heptomegaly, and increased hepatic triglycerides. Also increased were plasma transaminase activities and cholesterol levels. Glucose tolerance and impaired fasting glucose developed within two to four weeks [43].

A second rodent model used to define NASH mechanisms included hyperphagic leptin-deficient *ob/ob* mice (with an *ob* gene mutation) also developed steatosis (fatty liver), diabetes, obesity, but did not develop hepatic injury and fibrosis because leptin was required. A third rodent model for defining NASH mechanisms used methionine/choline deficient diets to induce steatosis and hepatic injury that was preceded by decreased GSH and S-adenosylmethionine levels, thereby increasing oxidative stress. This oxidative stress resulted in TNFα induction and increased other proinflammatory cytokines. However, this model was not associated with insulin resistance and the rodents lost weight. A fourth rodent model involved leptin-resistant type 2 diabetic db/db mice that were fed a methionine choline deficient diet for four weeks and developed insulin resistance, steatosis, liver focal inflammation, hepatocyte necrosis, and fibrosis. Steatosis was attributed to increased hepatic fatty acid uptake and decreased VLDL secretion [44].

7.4.4
The Two-Hit Hypothesis for NASH Hepatotoxicity

7.4.4.1 Introduction

Steatohepatitis inducing drugs such as 4,4'-diethylamino ethoxyhexestrol, amiodarone, or perhexiline accumulate in mitochondria, decrease mitochondrial potential and respiration, which inhibits hepatocyte β-oxidation of fatty acids causing steatosis (fat accumulation). This mitochondrial respiratory inhibition resulted in reactive oxygen species (ROS) formation, lipid peroxidation, and ATP depletion [45]. A two-hit hypothesis was proposed for NASH [1] with the first hit being hepatic steatosis that was dependent on insulin resistance, and the second hit being oxidative stress (lipid peroxidation) initiated by ROS. They also suggested that the risk of steatohepatitis would increase with high dietary polyunsaturated fat, iron, or alcohol or a hepatic CYP2E1 induction [1]. Others have suggested other factors that would increase NASH would include excessive availability of free fatty acids, overexpression of proinflammatory cytokines, and adipokines that would lead to inflammation and oxidative stress. Insulin resistance may be a possible source of the second hit as resistance occurs more frequently in those with more severe metabolic disturbances [46]. Polymorphisms of two genes encoding for fatty infiltration of the liver and manganese superoxide dismutase have also provided genetic evidence for the two-hit hypothesis [47].

7.4.4.2 The First Hit: Steatosis Mechanisms and Fructose

The steatosis mechanism for the first hit would likely be more effective with fructose than glucose, as fructose bypasses the glucose regulatory enzyme phosphofructokinase, thereby causing the fructose-derived carbons to more readily undergo lipogenesis than glucose [48]. A high-sucrose diet also increased lipogenesis by repartitioning fatty acids to esterification away from fatty acid β-oxidation possibly by increasing malonyl-CoA, an inhibitor of carnitine palmitoyltransferase-1 [49]. Endogenous glyoxals formed from sucrose metabolism or oxidation could also contribute as they inhibit respiratory complex III that could result in an inhibition of fatty acid β-oxidation [50].

7.4.4.3 The Second Hit: Hepatocyte Inflammation Oxidative Stress Model with Fructose

Rodent studies have shown that drug-induced liver toxicity *in vivo* is markedly enhanced if liver inflammation is induced by endotoxin before drug administration. This inflammation enhanced toxicity has been suggested as a theory to explain why some drugs cause idiosyncratic liver toxicity [51, 52]. This hepatotoxicity can be prevented by agents that inhibit or remove inflammatory cells (Kupffer cells or leukocytes) in the liver [53]. Inflammation has been shown to markedly increase hepatotoxicity induced by allyl alcohol or ethanol, and was attributed to the reactive carbonyl metabolites as a result of oxidation catalyzed by alcohol dehydrogenase (ADH) and CYP2E1 [54]. Inflammatory cells on activation release cytokines and H_2O_2. Furthermore, low levels of H_2O_2 markedly increased allyl alcohol-induced hepatocyte cytotoxicity [55].

In our studies, fructose only became cytotoxic at 1.5 M, whereas glucose was not toxic at this concentration. However, 12 mM fructose caused 50% cell death in 2 hours if the hepatocytes were exposed to a noncytotoxic dose of H_2O_2 continuously generated by glucose and glucose oxidase (Table 7.3). Fructose was more toxic than glucose in the presence of nontoxic concentrations of t-butylhydroperoxide. The cytotoxic mechanism involved oxidative stress as protein carbonylation, ROS, H_2O_2, HCHO formation preceded cytotoxicity, and cytotoxicity was prevented by radical scavengers, lipid antioxidants, and ROS scavengers [56]. It was proposed that the highly potent Fenton-derived ROS, catalyze the oxidation of fructose and particularly its carbonyl metabolites GOA, DHA, and GA, or its oxidation product, glyoxal. The carbon radicals and the glyoxal formed compromise the cell's resistance to H_2O_2. Desferoxamine, a ferric chelator, prevented fructose/H_2O_2

Tab. 7.3 Inflammatory conditions (low noncytotoxic H_2O_2 concentration) causes fructose to become toxic (>66-fold)

Treatment	LD_{50} (mM)	LD_{50} (with *H_2O_2) (mM)	LD_{50} ratio: fold toxicity increase
Fructose	1000 (not glucose)	15	66
Glyceraldehyde	50	8	6
Dihydroxyacetone	500	15	20
Hydroxyacetone	12	1	12
Glycolaldehyde	25	0.7	40
Glyoxal	5	0.025	200
Glycolic acid	40	4	10
Glyoxylic acid	8	4	2
Oxalic acid	6	1.3	1
Methylglyoxal	21	3	7

Glycolaldehyde was likely oxidized to glyoxal by ROS from H_2O_2. Methylglyoxal was detoxified by hepatocytes much more rapidly than glyoxal.

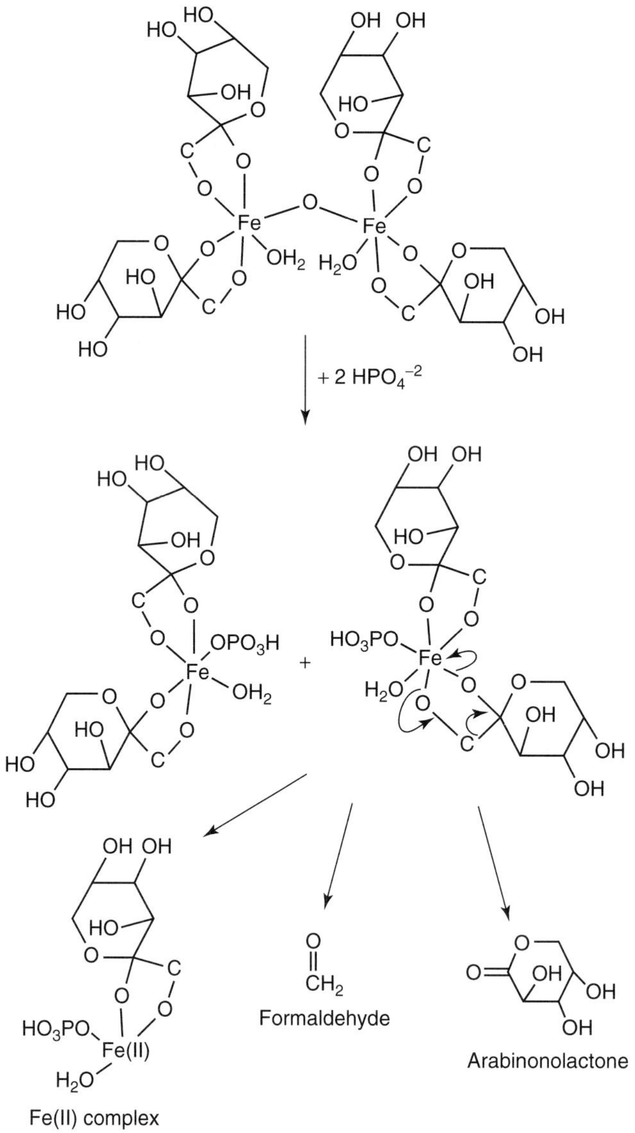

Fig. 7.4 Iron fructose complexes [32].

cytotoxicity, protein carbonylation, and formaldehyde formation, whereas, low Fe concentrations markedly increased cytotoxicity, protein carbonylation, and formaldehyde formation [56]. The mechanism likely involved a fructose ferric complex whose structure has recently been described (Figure 7.4).

These hepatocyte studies showed that fructose was much more effective than other sugars at causing both hits 1 and 2 and suggest that fructose is more likely to cause steatohepatitis than other sugars.

7.4.4.4 The Second Hit: ROS Formation Mechanisms
The oxidative stress mechanism for the second hit could result from ROS which may arise in four different ways.

1. A high-fructose diet induces a hepatic response through the c-Jun N-terminal kinase/activator protein-1 (JNK/AP1) pathways similar to that observed for the inflammatory cytokine TNF-α [57], which is known to upregulate inflammatory genes causing oxidative stress, for example, iNOS and NADPH oxidase [58].
2. Mitochondria are the other major sources of ROS in cells because electrons can leak and reduce oxygen as they pass through the electron transport chain. In particular, glycated mitochondria isolated from the kidney of diabetic rats readily formed ROS and exhibited oxidative damage that was prevented by aminoguanidine, a glyoxal trap. Two-dimensional electrophoresis identified the MG modified protein as respiratory complex III that was carbonylated and caused an inhibition of respiration, and extensive ROS formation. This was prevented by aminoguanidine and suggests that dicarbonyls were responsible. Prolonged ROS formation could be expected to contribute to diabetic nephropathy [50] as well as decreased insulin secretion by pancreatic islet β cells [59]. Glyoxal-induced hepatocyte cytotoxicity was attributed to mitochondrial toxicity that increased several orders of magnitude by low noncytotoxic doses of H_2O_2 [60], so that glyoxals would be the most likely NAFLD endogenous hepatotoxin activated by inflammation. MG formed by glycolysis also glycates mitochondrial protein arginines to form hydroimidazoles, for example, respiratory complex III resulting in ROS formation and mitochondrial toxicity. Other mitochondrial toxins formed from fructose include glyoxylate, the GOA metabolite, which condenses with mitochondrial oxaloacetate to form oxalomalate, a citric acid cycle inhibitor. Oxalate, the end product of fructose metabolism, forms calcium oxalate monohydrate with calcium, which inhibits mitochondrial respiration and induces the mitochondrial permeability transition [61]. Recently, the incidence of kidney stones was found to correlate with dietary fructose intake [62].
3. Activated Kupffer cells and infiltrating neutrophils are major sources of ROS in the liver as a result of activation of phagocytic vacuole or plasma membrane NADPH oxidase.
4. Endotoxins (bacterial lipopolysaccharide from the cell wall of gram negative bacteria in the gut) are another source of oxidative stress as they also activate the NADPH oxidase activity of inflammatory cells (see Section 7.4.4.3). It can result in endotoxemia as a result of tissue inflammation including inflammation of the liver.

7.4.4.5 Fructose-Induced *In Vivo* Liver Toxicity Studies in Rats
Infusion of rats, for an hour with 0.55–2.2 mM fructose caused hepatotoxicity following an accumulation of fructose 1-phosphate, a precipitous fall in hepatic ATP, edema, and degranulation of rough endoplasmic reticulum [63]. Fructose

(250 mg/kg body weight) injected intravenously to human subjects caused a marked decrease in hepatic ATP and phosphate, and increased fructose phosphates for 20 minutes [64]. Fructose taken by human subjects with heredity fructose intolerance (aldolase B deficiency) can progress to liver and kidney toxicity [65]. Consumption of sucrose (equivalent to 25–30% of total calories) by male subjects for 18 days significantly increased serum transaminases, an indicator of liver toxicity [66]. The molecular mechanism for fructose-induced hepatotoxicity has not been investigated. High fructose diets also induced fatty liver in rats with the fat induced being more saturated than unsaturated. However, hepatotoxicity was associated with oxidative stress and/or activation of the inflammatory pathways in the liver.

7.5
Rat Hepatocyte Studies on Endogenous Toxins Formed by Fructose Metabolism and/or Oxidation

7.5.1
Introduction

Another theory that could participate or contribute to fructose/H_2O_2 cytotoxicity would involve fructose enzymic metabolism to other carbonyl metabolites, which in turn could be activated by H_2O_2. The metabolic pathway of fructose to oxalate (the end product) is shown in Figure 7.2. The carbonyl metabolites on this pathway are all potential endogenous cytotoxins particularly when the target cells are exposed to low H_2O_2 concentrations. In the following, the formation of these carbonyl intermediates (Figure 7.2), their detoxication by metabolizing enzymes, their toxicity properties alone versus their toxicity in inflammation models have been described for each carbonyl metabolite as well as oxalic acid.

7.5.2
Glyceraldehyde (GA)

7.5.2.1 GA Enzymic Formation
D-GA is formed from fructose or glucose by glycolysis. GA is also formed from fructose formed by the polyol pathway or the fructokinase/aldolase B pathway. In the latter, fructose unlike glucose forms GA in two steps involving fructose 1-phosphate (formed by fructokinase) and enolase B. Glucose forms GA via GA-phosphate in five steps.

7.5.2.2 GA Detoxication Metabolizing Enzymes
As shown in Figure 7.2, four enzymes metabolize or detoxify GA.

1. Aldehyde reductase (AKR1A1) and aldose reductase (AKR1B1) detoxify GA in cells to form glycerol [67]. In hepatocytes or leukocytes, aldehyde reductase is the major enzyme that detoxifies GA [68].

2. ADH, in rat liver and human liver, is the second most active enzyme metabolizing GA and catalyzing GA reduction to form glycerol [69].
3. GA dehydrogenase, aldehyde dehydrogenase (ALDH)1, in the cytosol, catalyzes GA oxidation to D-glycerate and is the third most active GA metabolizing enzyme [69]. D-Glycerate can then be phosphorylated to form 2P-glycerate, which can be converted to pyruvate and is thus a detoxication. Alternatively, D-glycerate can be reduced to hydroxypyruvate catalyzed by lactate dehydrogenase and hydroxypyruvate, which undergoes a decarboxylase catalyzed decarboxylation to form GOA, a mutagen.
4. GA kinase catalyzes GA phosphorylation by ATP to form GA-3P.

7.5.2.3 GA Toxicity Mechanisms

The carbonyl group of DL-GA glycates the free amino groups on a protein to form a Schiff base imine much more rapidly than does fructose, which in turn, is more reactive than glucose. The protein imine undergoes an Amadori rearrangement to form a stable ketoamine that with time converts to AGEs. Proteins particularly affected are long-lived proteins such as serum albumin, lens crystalline, and extracellular matrix collagen. DL- GA was a much more effective promoter of AGE formation than fructose largely because the fructose carbonyl forms an unreactive hemiketal ring structure. GA AGEs identified in streptozotocin-induced diabetic rats were MG-H1, an MG-derived hydroimidazolone formed with arginine residues and GLAP, GA-derived pyridinium compound formed with lysine residues in plasma protein [70]. GA AGEs, that is, AGE-2s were cytotoxic and inhibited cell proliferation of HL60 cells as well as induced ROS formation and decreased GSH levels. Antioxidants prevented this toxicity [71].

7.5.2.4 GA Oxidative Stress Enhanced Toxicity: Inflammation Model

As shown in Table 7.3, GA-induced hepatocyte cytotoxicity was increased sixfold by continuous low nontoxic concentrations of H_2O_2 generated by glucose/glucose oxidase. GA also caused DNA oxidation and browning to form 8-oxodG, particularly if low levels of H_2O_2 were present and may provide a cause for diabetes-associated carcinogenesis [72]. Deoxyguanosine is readily glycated by GA (unlike glucose) to form nucleoside AGEs [73]. GA can also be cytotoxic at concentrations higher than 2 mM over a 2-hour incubation period inhibited glucose stimulated insulin secretion of rat pancreatic islets. Over a 24-hour period, cellular insulin levels were decreased and intracellular ROS and H_2O_2 were increased. It was suggested that GA autoxidation was the source of ROS rather than mitochondrial ROS [74]. GA at 10 mM depleted hepatocyte ATP presumably as a result of triokinase activity catalyzing GA phosphate formation [75]. In addition, L-GA inhibits glycolysis. Erythrocyte GSH levels were depleted at 10 mM GA presumably due to oxidation or adduct formation, oxyhemoglobin was oxidized, NAD(P)H was oxidized, and the hexose mono phosphate shunt was inhibited [76]. The oxidative stress changes were attributed to ROS formation by transition metal catalyzed GA autoxidation to the ketoaldehyde hydroxypyruvaldehyde, which undergoes transition metal catalyzed autoxidation even more readily than GA [77].

Glyoxal is formed by GA oxidation catalyzed by Fenton's reagent (H_2O_2/Fe) [24]. MG, another more reactive endogenous dicarbonyl toxin formed from D-GAP that can be formed from GA catalyzed by triokinase. Intracellular levels of MG were markedly increased by H_2O_2[78], presumably as a result of inhibition of the enzyme GAP dehydrogenase by H_2O_2 [79].

7.5.3
Dihydroxyacetone (DHA)

7.5.3.1 DHA Enzymic or Nonenzymic Formation
These enzymes are located on the glycolytic metabolic pathway (Figure 7.2). DHAP and GA phosphate are formed from fructose 1,6-bisphosphate catalyzed by the lyase activity of aldolase. Triosephosphate isomerase also catalyzes the interconversion of DHAP and D-GA phosphate. Glycerol is oxidized to DHA and GA catalyzed by Fenton's [80]. P450 also catalyzed glycerol oxidation to formaldehyde [81].

7.5.3.2 DHA Enzymic Metabolism or Oxidation
DHA can be phosphorylated by ATP to DHAP catalyzed by DHA kinase. It is more rapid than the triokinase that catalyzes GA phosphorylation. Steady state cellular DHA concentrations are also higher than GA. However, toxic MG is formed from DHAP catalyzed by MGS. DHA or its phosphate is partly oxidized to glyoxal catalyzed by the Fenton reaction [24].

7.5.3.3 DHA Toxicity Mechanisms
Keratinocytes incubated with 25 mM DHA for 24 hours induced cytoplasmic budding, chromatin condensation, and cell detachment. DNA strand breaks were also induced and prevented by antioxidants [82]. DHA is a browning agent found in sunless tanning lotions and involves a Maillard reaction with protein amino groups. DHA readily undergoes a superoxide radical catalyzed autoxidation to form glyoxal [24], whereas DHAP carries out a β-elimination of phosphate to form MG. However, DHAP is much more toxic to cells lacking SOD, likely because superoxide radicals inactivate glyceraldehyde phosphate dehydrogenase (GAPDH), thereby increasing MG and superoxide to toxic levels [83].

7.5.3.4 DHA Oxidative Stress Enhanced Toxicity: Inflammation Model
As shown in Table 7.3, DHA-induced hepatocyte cytotoxicity was increased 20-fold by continuous low nontoxic concentrations of H_2O_2 generated by glucose/glucose oxidase.

7.5.4
Glycolaldehyde (GOA)

7.5.4.1 GOA Enzymic Formation
GOA is formed from *hydroxypyruvate* (a glycolysis GA metabolite) catalyzed by hepatic hydroxypyruvate decarboxylase by a nonoxidative process that releases

CO_2. The only irreversible step of the fructose/glucose to oxalate pathway is the GOA oxidation to glycolic acid catalyzed by ALDHs. This prevents genotoxic GOA and glyoxal formation. GOA is also formed in the plasma from serine catalyzed by neutrophil myeloperoxidase (MPO)/H_2O_2/chloride [84].

7.5.4.2 GOA Detoxication Metabolizing Enzymes
Two enzyme classes that detoxify GOA:

1. GOA is detoxified by oxidation to glycolic acid by NAD^+ catalyzed by human liver cytosolic aldehyde dehydrogenase ALDH1, an irreversible reaction. This enzyme can be inhibited by disulfiram, chloral hydrate, or cyanamide [85]. This enzyme was also found in human erythrocytes, and has a Km of 0.6 µM and Vm of 0.9 µmol × min^{-1} × mg $protein^{-1}$ for GOA and Km of 15.6 µM and Vm of 0.7 µmol × $min^{×1}$. µmol ×. min^{-1}.mg $protein^{-1}$ for acetaldehyde [86]. GOA oxidation by rat liver is also catalyzed by betaine ALDH in the cytosol E2 (ALDH9) and mitochondria E3 (ALDH9) to form glycolic acid. Betaine aldehyde is the best substrate. Cimetidine is a substrate and inhibitor for E3 and human liver cytosol E2 [87].
2. GOA is also detoxified by reduction by NADH to ethylene glycol that is catalyzed by ADH, as the dehydrogenase inhibitor pyrazole increased GOA cytotoxicity [56]. The reverse reaction, that is, ethylene glycol oxidation to GOA catalyzed by ADH has been demonstrated and ethylene glycol poisoning is treated with the inhibitor methylpyrazole. Ethanol is also an effective antidote since ADH has a greater affinity for ethanol than for ethylene glycol [88].
3. GOA may not be a substrate for aldehyde reductase or aldose reductase as no such activity has been reported.

Three enzymes that detoxify hydroxypyruvate are as follows:

1. Reduction to D-glycerate by NADPH catalyzed by cytosolic hydroxypyruvate reductase. This is the same enzyme as D-glycerate dehydrogenase or glyoxylate reductase, which is mutated in patients with primary hyperoxaluria (PH) type II [89].
2. Reduction to L-glycerate catalyzed by cytosolic L-lactate dehydrogenase/NADH.
3. Transamination by alanine to serine and pyruvate catalyzed by alanine serine:pyruvate transaminase (SPT), a pyridoxal phosphate requiring enzyme. This is the same reversible enzyme as alanine:glyoxylate transaminase (AGT) that is found in both rat liver peroxisomes and mitochondria [90]. Incubation of hepatocytes with >2 mM hydroxypyruvate and NADH generators forms

glucose by gluconeogenesis as a result of 2-phosphoglycerate and pyruvate formation [91]. AGT is mutated in PH type I.

7.5.4.3 GOA Toxicity Mechanisms

GOA glycated serum albumin rapidly to form C-2-imine cross-links and N-(carboxymethyl) lysine by a Maillard reaction [92]. Purified GAPDH (a thiol dependent enzyme) was inhibited by MG more than by glyoxal or GOA, which was concomitant with a loss of thiol groups. However, glutathione reductase and lactate dehydrogenase were inhibited by MG or glyoxal, but not GOA [93]. GOA also inhibited cathepsin B in macrophage lysates [94]. GOA induced apoptosis and inactivated GAPDH, glucose 6-phosphate dehydrogenase, and Cu/Zn SOD in human breast cancer cells [95]. GOA or MG also caused a loss of Arg, Lys, Trp protein amino acids when incubated with LDL apo B [96]. This resulted in cholesterol and cholesterol ester accumulation in macrophage cells, but not smooth muscle or endothelial cells. LDL glycation instead of LDL oxidation is therefore sufficient to cause rapid lipid accumulation by macrophage cells to form foam cells [12]. GOA at 100 µM inhibited the growth of human breast cancer cells, caused apoptosis, induced p53 expression increased ROS, lipid peroxidation, protein carbonylation. Glyoxal induced a similar type of oxidative stress in hepatocytes [97] like GOA (submitted) and suggests that GOA cytoxicity could be mediated by glyoxal. Furthermore, dicarbonyl traps such as aminoguanidine protected cells against GOA.

GOA reacts with proteins to form GOA-pyridine, which has been found to accumulate in tubular epithelial cells and the mesangium of kidneys, and so could play a role in glomerular damage in diabetic and atherosclerotic end-organ damage. GOA-pyridine has also been described in foam cells atherosclerotic plaques [98].

7.5.4.4 GOA Oxidative Stress Enhanced Toxicity: Inflammation Model

As shown in Table 7.3, GOA or hydroxypyruvate-induced hepatocyte cytotoxicity increased 40- or 12-fold respectively by continuous low nontoxic concentrations of H_2O_2 generated by glucose/glucose oxidase. GOA toxicity likely involves oxidation to a reactive species by superoxide radicals or by ROS produced by a H_2O_2/Fe Fenton's reaction to form a GOA radical, which reacts with oxygen to form glyoxal [99] by the following reaction:

$$OHCH_2CHO + OH^\bullet \longrightarrow OHC^6\bullet HCHO \text{(glycolaldehyde radical)}$$
$$OHC^\bullet HCHO + O_2 \longrightarrow glyoxal + HO_2^\bullet \tag{7.1}$$

Evidence for their toxicity role is that GOA inhibition of *Escherichia coli* growth was oxygen dependent and was prevented by MnTM-2-PyP, a superoxide dismutase mimic [100]. If GOA reacted with ROS in the cell to form glyoxal, then GOA would be able to inhibit glutathione reductase and lactate dehydrogenase in the cell as a result of forming the reactive glyoxal intermediate.

7.5.5
Glyoxal (G) and Methylglyoxal (MG)

7.5.5.1 G and MG Enzymic and Nonenzymic Formation

As shown in Figure 7.5, MG is mostly formed endogenously from the glycolytic intermediate DHA-phosphate catalyzed by MGS. GAP also forms MG nonenzymically in a phosphate buffer [20]. Both of these triose phosphates, induced mutations and killed superoxide dismutase-deficient *E. coli* much more than the parental superoxide dismutase-proficient strains. This suggests that the superoxide initiated triose phosphate autoxidation to yield superoxide, hydroxypyruvaldehyde phosphate, and MG [83].

MG was also the product formed when ethanol-inducible CYP2E1 catalyzed the oxidation of the ketone bodies, acetone and acetol, which increase during diabetes [101]. The MPO that is released from immune cells during inflammation, also catalyzed the oxidation by radical mechanisms of the ketone body, acetoacetate or the amino acid serine to form MG [102].

MG and glyoxal formation when hepatocytes were incubated with fructose was more effective than glucose, which was more effective than acetol. However, MG formation when vascular smooth muscle cells (that contain semicarbazide-sensitive Cu-amine oxidase (SSAO)) were incubated for 3 hours with 25 mM metabolites were aminoacetone $\gg$ fructose $>$ D-glucose $>$ acetol $>$ sucrose $>$ L-glucose with mannitol not forming MG. A similar order of metabolite effectiveness was found for MG-induced AGE induction and nitric oxide [103]. SSAO activity is also high in endothelial cells and adipose cells.

Aminoacetone is formed from threonine metabolism, particularly during nutritional deprivation (diabetes) catalyzed by threonine cleavage complex (aminoacetone synthetase), which then undergoes oxidative deamination by oxygen catalyzed by SSAO to form MG, ROS, and NH_4^+. SSAO is an ectoenzyme on the surface of the plasma membrane or plasma SSAO (probably shed from SSAO cells) with SSAO Kms of 19–125 µM aminoacetone. Endothelial SSAO was overexpressed, and plasma SSAO activity was increased causing ROS and MG formation, in patients with diabetic complications, vascular disorders, atherosclerosis, congestive heart failure, and AD. Aminoacetone-induced pancreatic β-cell death was attributed to oxidative stress. Aminoacetone also formed MG by a Fe or Cu catalyzed autoxidation [104]. Another threonine metabolism route is catalyzed by threonine dehydrogenase/NAD^+/CoA to NADH, acetyl CoA, and glycine.

Glyoxal and MG are also formed in the atmospheric oxidation of anthropogenic aromatic hydrocarbons. Glyoxal, more than MG, is also formed by the autoxidation of sugars and unsaturated lipids *in vivo*. Glyoxal is also found in cooked food or coffee with lower levels in ozonated drinking water or cigarette smoke [97, 105].

7.5.5.2 Gloxal and MG Detoxification Metabolizing Enzymes

Both glyoxals are highly water soluble as both aldehyde groups are hydrated (Equation 7.1). Glyoxal in aqueous solution exists mostly as 98% dihydrate, 1–2% dimers, 0.5% monohydrate, 0.005% unhydrated dicarbonyl [106]. Glyoxal has a

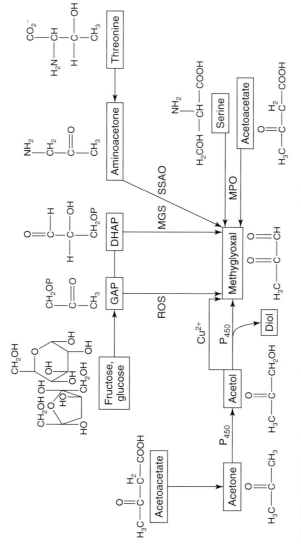

Fig. 7.5 Methylglyoxal sources. ROS, reactive oxygen species; MGS, methylglyoxal synthase; SSAO, semicarbazide-sensitive amine oxidase; MPO, myeloperoxidase; P450, ethanol-inducible cytochrome P450-2E1.

hydration constant of 2.2×10^5, whereas, MG is less hydrated with a hydration constant of 2.7×10^3, and largely exists as the monohydrate that retains a reactive carbonyl moiety, which is much less reactive than glyoxal monohydrate [107, 108]. This could explain the paradox of why MG is less toxic toward hepatocytes than glyoxal in spite of its greater lipophilicity, as this would be offset by MG being the better substrate than glyoxal for detoxifying metabolizing enzymes [109].

MG or G are detoxified by four classes of carbonyl metabolizing enzymes:

1. **Glyoxalase I and II:** MG and GSH form a reversible thiohemiacetal adduct nonenzymically, which rapidly undergoes a disproportionation reaction catalyzed by cytosolic glyoxalase I in cells to form D-lactoylglutathione. The latter undergoes a glyoxalase II (located in the mitochondria) catalyzed hydrolysis forming D-lactate and GSH. Presumably glyoxalase catalyzes the conversion of glyoxal to glycolate. Gloxalase I does not function if cell GSH is oxidized as a result of oxidative stress; so, MG accumulates in the cell [110]. Glyoxalase I and II are ubiquitous Zn containing metalloenzyme enzymes found in most animal tissues, plants, microorganisms, and prokaryotes. Phenylglyoxal is a poorer substrate than MG, but *in vivo* it is excreted in the urine as mandelic acid (Dakin, H.D. and Dudley, H.W. *Journal of Biological Chemistry*, 1913 **14**, 423–431). Ironically, glyoxal has not yet been compared to MG as a glyoxalase substrate, but has been shown to form glycolate catalyzed by GSH and a rat liver extract [111]. Other dicarbonyls, for example, 3-deoxyglucosone, another glucose oxidation product, was not a glyoxalase substrate and was mostly detoxified by reduction by NADPH and aldehyde reductase [78].

2. **Aldehyde/aldose reductases AKR1B1, AKR1A1+NAD(P)H:** Both reductases catalyze the reduction of MG to hydroxyacetone (acetol) and then to lactaldehyde, but are dependent on the presence of glutathione, presumably to overcome the highly hydrated state of MG. These enzymes, unlike glyoxalase, detoxify all of the aldehydes elevated in diabetes, for example, MG, lactaldehyde, and acetol. The liver contains little aldose reductase activity but has high NADPH aldehyde reductase activity, which is likely also the major detoxifying enzyme in rat or human liver for detoxifying deoxyglucosone to form 3-deoxyfructose. Deoxyglucosone is an active intermediate in glucose and fructose-mediated protein cross-linking. The amino acid sequence for purified 3-deoxyglucosone reductase isolated from rat liver was identical to rat and human liver aldehyde reductase [112, 113]. MG and phenylglyoxal were also good substrates.

3. **Betaine aldehyde dehydrogenase ALDH9+NAD$^+$:** Betaine ALDH catalyzes MG oxidation to pyruvate. Betaine aldehyde though is a much better substrate [114].

4. **2-Oxoaldehyde dehydrogenase+NADP$^+$** catalyzes the oxidation of MG to pyruvate, while 2-deoxyglucosone is oxidized to 3-deoxygluconic acid. This dehydrogenase is activated by 2-aminopropanolamine optimally at pH 9. ALDH2 does not oxidize MG, as ALDH only binds unhydrated aldehydes [114].

The rate of metabolism of glyoxals by hepatocyte metabolizing enzymes can be determined from the rate of hepatocyte protein carbonyl disappearance. This was found to be much faster for methylglyoxal than glyoxal. In contrast, GOA addition to hepatocytes resulted in protein carbonylation that increased with time over an hour or two, indicating that protein carbonylation was probably caused by glyoxal formation (results not shown).

The order of hepatocyte cytotoxicity induced by glyoxals were phenylglyoxal (2 mM) > glyoxal (5 mM) ≫ MG (22 mM). The surprising hepatocyte resistance to MG reflects the rapid metabolism of MG catalyzed by glyoxalase and aldehyde reductase [109]. The LD_{50} (2 hours) concentration for MG found toward glyoxalase inhibited hepatocytes (i.e., GSH depleted) was 10 mM, toward glyoxalase inhibited hepatocytes (using 200 µM sorbinil) was 12 mM, and together was 3 mM [109]. On the other hand, the LD_{50} (18 hours) for MG was 0.24 mM toward human leukemia HL60 cells in culture, whereas neutrophils were resistant to MG [115]. However, both neutrophils and HL60 cells contain the sorbitol polyol pathway and higher aldehyde reductase activity than aldose reductase that reduces D-glucuronate and D-GA [68].

7.5.5.3 Glyoxal and MG Toxicity Mechanisms

The order of effectiveness of carbonyls for protein thiol reactivity using papain was MG > Glyoxal > GOA ≫ hydroxyacetone [94]. G incubated with hepatocytes for 60 minutes caused DNA single strand breaks, but not DNA cross-linking. They were also detected at 2 hours in the liver *in vivo* following the oral administration of a single dose of 200 mg/kg, but not in other tissues. The breaks reached a maximum around 9 hours after exposure but were almost fully repaired by 24 hours [116]. Glyoxal is also a direct mutagen.

As shown in Table 7.3, the order of cytotoxicity toward hepatocytes found for fructose and its metabolites were as follows: glyoxal > glyoxalate > hydroxypyruvate ≫ GOA > GA > MG > DHA ≫ fructose > glucose. This order may reflect the electrophilicity of the carbonyls formed and would correlate with lower more negative ELUMO and molecular refractivity value (the less bulky). Previously we reported that glyoxal-induced hepatocyte cytotoxicity could be attributed to mitochondrial oxidative stress toxicity as mitochondrial membrane potential was decreased and cellular oxidative stress (ROS formation, lipid peroxidation, GSH oxidation) leading to cytotoxicity occurred. Cytosolic GSH reductase was inhibited. Cytotoxicity was increased by sorbinil, an aldehyde reductase inhibitor or by depleting hepatocyte GSH. The hepatocytes could be rescued from mitochondrial toxicity and cytotoxicity by trapping the glyoxal with aminoguanidine, penicillamine, cysteine, or by adding NAD(P)H generating glycolytic substrates (xylitol, sorbitol, fructose) or desferoxamine (ferric chelator) or antioxidants (quercetin or butylated hydroxyanisole) [61]. Extensive protein carbonylation occurred on addition of the glyoxal to hepatocytes, but surprisingly the cells and the mitochondria could be rescued by equimolar aminoguanidine, penicillamine, and cysteine. Furthermore, most or all of the protein carbonylation disappeared indicating that the protein carboxylation was mostly due to reversible Schiff base formation with unmetabolized glyoxal.

However, antioxidants or ROS scavengers (hydralazine, TEMPOL, mannitol) added at 1 hour partly rescued the hepatocytes from cytotoxicity and partly decreased ROS formation, but only decreased protein carbonylation a little suggesting that protein oxidation did not contribute much to protein carbonylation [117]. Added thiamine (vitamin B_1) prevented ROS formation, lipid peroxidation, GSH oxidation, loss of NADPH, and decreased mitochondrial potential and cytotoxicity. On the other hand, decreasing hepatocyte thiamine with oxythiamine decreased transketolase activity and increased all of these toxic effects, except mitochondrial potential loss [118].

CHO cells incubated with 1 mM MG for 42 hours accumulated MG and D-lactate (glyoxalase metabolite) by binding to proteins (reversibly and irreversibly). Examples of MG mediated apoptotic cell death mechanisms include L929 cells, which was mediated by TNF-binding to cell surface receptors and mitochondrial ROS, but was caspase independent. Human leukemia U937 cell death involved a mitochondrial cytochrome c-mediated caspase c activation [19].

7.5.5.4 Glyoxal and MG Oxidative Stress Enhanced Cytotoxicity: Inflammation Model

As shown in Table 7.3, glyoxal hepatocyte toxicity was increased several orders of magnitude (200-fold) by low noncytotoxic doses of H_2O_2 (hepatocyte inflammation model) [60]. Fructose and the metabolites GA and DHA cytotoxicity were also markedly increased by H_2O_2 [56] that was partly attributed to glyoxal formation from the oxidation of fructose, GA, and DHA by Fenton ROS radicals formed by Fe/H_2O_2 [24]. However, in the hepatocyte inflammation model, when hepatocytes were exposed to a nontoxic H_2O_2 generating system there was a large increase in the toxicity of fructose and most of the metabolites in the fructose to oxalic acid ($\times 1.3$) pathway. The fold increase in toxicity induced by H_2O_2 is given in brackets: glyoxal ($\times 250$) > fructose ($\times 100$) > GOA ($\times 40$) > DHA ($\times 20$) > hydroxypyruvate ($\times 12$) > glycolate ($\times 10$) > MG ($\times 7$), GA ($\times 6$) > glyoxylate ($\times 2$) > oxalic acid ($\times 1.5$) and likely does not reflect ease of metabolite oxidation by H_2O_2/Fe, but instead reflects a combination of the following:

1. ease of radical and electrophile formation (e.g., GOA readily oxidized by superoxide or H_2O_2/Fe).
2. inactivation of hepatocyte carbonyl detoxification system by carbonyl radicals/electrophiles. For example, lactaldehyde, product of MG reductase and GA are strong noncompetitive inhibitors of glyoxalase I [119]. MG inhibits GOA or lactaldehyde oxidation catalyzed by cytosolic or mitochondrial ALDH [120].
3. inactivation of antioxidant or other cellular enzymes by radicals/electrophiles, thereby increasing hepatocyte susceptible to oxidative stress toxicity, for example, MG inhibits glutathione reductase [93].

7.5.6
Glycolate as a Peroxisomal ROS Source

7.5.6.1 Glycolate (GL) Enzymic Formation
Glycolate is formed from GOA catalyzed by NAD^+/ALDH2. Glycolate is also the major metabolite of glyoxal catalyzed by glyoxalase.

7.5.6.2 GL Detoxification Metabolizing Enzymes
Glycolate is the least toxic intermediate on the GA to oxalic acid metabolic pathway and was toxic at 20 mM. It is metabolized by peroxisomal glycolate oxidase to form glyoxalate and H_2O_2.

7.5.6.3 GL Toxicity
Glycolate cytotoxicity was markedly increased if hepatocyte catalase was inhibited by catalase inhibitors (azide or cyanamide) with a glycolate toxicity now of LD_{50} (2 hours) = 0.5 mM. This 40-fold increase in toxicity can be attributed to the H_2O_2 formed by the oxidase [121]. Glyoxal cytotoxicity was also markedly increased in catalase inhibited hepatocytes presumably caused by glyoxalase catalyzed glyoxal metabolism to glycolate.

7.5.6.4 GL Oxidative Stress Enhanced Toxicity: Inflammation Model
As shown in Table 7.3, glycolate-induced hepatocyte cytotoxicity was markedly increased 10-fold by continuous low nontoxic concentrations of H_2O_2 generated by glucose/glucose oxidase. The cytotoxic mechanism is likely due to the additive effects of the H_2O_2 generated by peroxisomal glycolate oxidase and extracellular glucose oxidase.

7.5.7
Glyoxylate as a Carbonyl Toxin

7.5.7.1 Glyoxylate (GX) Enzymic Formation
Glyoxylate is formed by the oxidation of *glycolate* in peroxisomes catalyzed by glycolate oxidase/oxygen. H_2O_2 is also formed, which is detoxified by catalase in the peroxisomes. Glycolate is formed from GOA catalyzed by NAD^+/ALDH. Glyoxalate and pyruvate are also formed from the mitochondrial degradation or catabolism of *hydroxyproline* estimated at 2–3 g/day. Hydroxyproline is a major component of the protein collagen and is formed posttranslationally following protein synthesis catalyzed by proline hydroxylase and ascorbic acid in the lumen of endoplasmic reticulum. Plasma endogenous hydroxyproline levels are 10–15 µM. The mean daily glycolate intake is 33 mg. Most of the glyoxylate, however, is detoxified by conversion to glycine catalyzed by AGT, thereby, minimizing toxic oxalate formation. Ingestion of collagen containing meats (ground beef contains up to 7% collagen) or gelatin (a collagen containing bone product) also increases plasma hydroxyproline, urinary oxalate, and glycolate [122]. Glyoxalate is also formed from *ethylene glycol*, which is oxidized, catalyzed by ADH to form GOA, which can then

be oxidized to glycolate catalyzed by ALDH. Ethylene glycol nephrotoxicity has been attributed to GOA and glyoxalate by causing ATP depletion, enzyme inactivation, and plasma membrane phospholipid deterioration [123]. However, others believe calcium oxalate also contributes to the nephrotoxicity.

7.5.7.2 GX Detoxication Metabolizing Enzymes
Glyoxalate is detoxified by four enzymes and cysteine:

1. Glyoxalate is detoxified by conversion to glycine catalyzed by liver peroxisomal AGT, which requires pyridoxal (vitamin B_6) as a coenzyme. Mutations on the AGT gene cause PH1 [124].
2. Glyoxalate is reduced to glycolate catalyzed by mitochondrial and cytosolic glyoxylate reductase. The action of NADPH/glyoxalate reductase in the hepatocytes stimulates CO_2, D-glycerate formation, and the pentose phosphate pathway when hepatocytes are incubated with glyoxalate or hydroxypyruvate [125].
3. Glyoxalate is reduced to glycolate catalyzed by NADH and lactate dehydrogenase. Glyoxalate is also oxidized to oxalate catalyzed by peroxisomal glyoxylate oxidase or by NAD^+ and lactate dehydrogenase.
4. Glyoxalate is decarboxylated by 2-oxo-glutarate catalyzed by a carboligase to form 2-hydroxy-3-oxoadipate and CO_2. This reaction occurs in mitochondria and is associated with the E1 subunit of the multienzyme complex, a thiamine-dependent 2-oxoglutarate dehydrogenase [126].
5. Glyoxalate is detoxified by forming nonenzymically, an adduct with cysteine, and is used as therapy to prevent toxic oxalic acid formation in hyperoxaluria.

7.5.7.3 GX Toxicity Mechanisms
Glyoxylate is a two-carbon aldo acid with a free reactive carbonyl electrophile; so, it is much more toxic than D-GA and more readily glycates proteins or glucosamine > lysine. Cysteine forms a cysteine-glyoxylate adduct and can be used to prevent oxalic acid formation from glyoxylate in rats *in vivo* or in hepatocytes [126]. Glyoxalate also forms complexes with urea, oxaloacetate, and certain keto acids [127]. However, unlike glyoxal, glyoxalate does not glycate arginine [128]. Fructose is a much less reactive carbonyl as the fructose carbonyl forms an internal ring structure. Glyoxalate at 2 mM is a mitochondrial toxin that inhibits state 3 respiration (particularly succinate and NADH dehydrogenase substrates) by forming hemimercaptals with protein thiols as dithiothreitol prevented the inhibition [129].

Glyoxalate also interferes with mitochondrial ATP generation as it reacts with the citric acid cycle oxaloacetate, an intermediate of the citric acid cycle, thereby depleting oxaloacetate to form oxalomalate. Oxalomalate is a potent competitive inhibitor of isocitrate dehydrogenase of the citric acid cycle, which would also result in the inhibition of this cycle [130]. Removal of oxaloacetate could also inhibit gluconeogenesis.

A third toxic mechanism is the twofold increase in glyoxylate cytotoxicity, found if catalase inhibited hepatocytes were used instead of normal hepatocytes. This cytotoxicity can be attributed to H_2O_2 formed by peroxisomal glycollate oxidase in response to the glycolate formed by glyoxylate reductase.

7.5.7.4 Glyoxylate Oxidative Stress Enhanced Cytotoxicity: Inflammation Model

As shown in Table 7.3, glyoxylate-induced hepatocyte cytotoxicity was increased twofold by continuous low nontoxic concentrations of H_2O_2 generated by glucose/glucose oxidase. This was likely caused by increased ROS formation (dichlorofluorescein assay) and H_2O_2 formation (Fox assay) as a result of glyoxalate reduction catalyzed by peroxisomal NADH/glyoxylate reductase to glycolate that rapidly formed H_2O_2 catalyzed by glycolate oxidase.

7.5.8
Oxalic Acid

7.5.8.1 Oxalic Acid Enzymic Formation

The order of effectiveness of U-C^{14} sugars at 5–20 mM for inducing oxalate formation by isolated rat hepatocytes was fructose > glycerol > xylitol > sorbitol > glucose [131]. Glyoxylate was also twice as effective as glycolic acid [132]. This explains the striking correlation between the incidence of kidney stones and amount of fructose/sucrose intake in a prospective study involving a combined 48 years of follow up involving nearly 200 000 female and 45 985 male nurses. Table sugar, sugar sweetened soft drinks, orange juice, apple juice, cake were in the higher fructose group [62]. Dietary oxalic acid comes from seeds, spinach, beets, rhubarb; but few edible plants are low in oxalic acid [133].

7.5.8.2 Oxalic Acid Detoxication Metabolizing Enzymes and Toxicogenetic Basis

Primary hyperoxaluria type 1(PH1) occurs in up to 3 million people worldwide. PH1 is a genetic mutation of the *AGT* gene resulting in deficiency of peroxisomal AGT as a result of a trafficking mislocation to the mitochondria where it is metabolically ineffective. This causes increased tissue glyoxylate, glycolate, and oxalic acid levels. Diagnosis involves low AGT activity, high urinary oxalate-creatinine ratio, high plasma oxalate levels. High urinary glycolate occurs in 66% of patients. Oxalic acid forms insoluble calcium salts that accumulate in the kidney and other organs of patients. Deposition of calcium oxalate in the renal pelvis/urinary tract or the renal parenchyma is a risk for recurrent nephrolithiasis, nephrolithiasis, or nephrocalcinosis. Eventually oxalosis occurs, that is, an accumulation of oxalate, particularly, in the bone marrow and kidneys as well as in the retina, myocardium, and peripheral nerves. This is followed by death from end-stage renal disease. PH1 is often associated with metabolic acidosis indicative of calcium oxalate causing mitochondrial toxicity [134]. Some affected infants show an improvement in renal function with high-dose pyridoxine (vitamin B_6) therapy.

PH2 has an incidence of 1 : 120 000, which is 20-fold rarer than PH1 and is caused by a lack of the reductase enzyme normally found in the liver and leukocytes that catalyze the reduction of hydroxypyruvate to D-glycerate, the reduction of glyoxylate to glycolate, and the oxidation of D-glycerate to hydroxypyruvate. PH2 results from a gene mutation for cytosolic glyoxylate reductase/hydroxypyruvate reductase (reverse reaction is catalyzed by D-glycerate dehydrogenase). Normally, this reductase would be expected to detoxify glyoxylate and hydroxypyruvate. The loss of liver reductase activity would increase hydroxypyruvate and L-glycerate levels and would explain the increase in urinary or plasma hydroxypyruvate, glyoxylate, oxalic acid. Glyoxalate is oxidized to oxalic acid catalyzed by cytosolic lactate dehydrogenase. Toxic calcium oxalate complex deposits accumulate in the kidney and a case of restrictive cardiomyopathy has been reported [135].

7.5.8.3 Oxalic Acid Toxicity Mechanisms: Hepatic Endogenous Metabolite and Kidney Toxin

Kidney stone disease affects about 1–3% of the population and is believed to result from oxalic acid, an oxidation product of glyoxylate (a fructose metabolite) that is catalyzed by liver peroxisomal glycolate oxidase or by NAD^+ and liver cytosolic lactate dehydrogenase. Although there are metabolizing enzymes for glyoxalate, there are no enzymes metabolizing oxalic acid. Oxalate is therefore, generally regarded as a useless end product of sugar metabolism as the oxalic acid released from the liver complexes calcium in the blood to form an insoluble complex that readily crystallizes. The calcium oxalate crystals can then form deposits in the kidney that obstruct ducts, cause inflammatory reactions, and damage cells. Calcium oxalate monohydrate crystals were more effective than calcium oxalate at inducing cytotoxicity (>3 mM) to cultured human proximal tubule cells or increasing the permeability transition of isolated rat kidney mitochondria [136]. Calcium oxalate (6–8 mM), added to isolated rat hepatocytes caused 50% cytotoxicity in 2 hours (Table 7.3). However, GOA or glyoxylate were less toxic [137], whereas, the opposite results were obtained by other researchers [123]. Distal tubular epithelial cells treated with crystals caused mitochondrial ROS formation [138]. Oxidative stress and lipid peroxidation occurred in hyperoxaluric rat kidney or calcium oxalate or crystal exposed cultured cells, and cytotoxicity was prevented by antioxidants [139]. Furthermore, antioxidants or hydroxyl radical scavengers prevented urolithiasis in rats *in vivo* [140]. Calcium oxalate accumulation caused urolithiasis and/or nephrocalcinosis, and eventually kidney stones. Following kidney failure, the body no longer excretes oxalic acid and the increasing hyperoxaluria results in systemic oxalosis causing damage to the eyes, blood vessels, bones, muscles, heart, and other major organs.

7.5.8.4 Oxalic Acid Oxidative Stress Enhanced Toxicity: Inflammation Model

As shown in Table 7.3, oxalate-induced hepatocyte cytotoxicity was increased 33% by continuous low nontoxic concentrations of H_2O_2 generated by glucose/glucose oxidase. This increase in H_2O_2 toxicity likely arose because calcium oxalate is a mitochondrial toxin [134], which would compromise hepatocyte susceptibility to H_2O_2.

7.6
Cancer Risk and Genotoxicity of Fructose or Carbonyl Metabolites

As noted in the section above, fructose and its carbonyl metabolites are frequently genotoxic. In addition, plasmid studies show that fructose and its phosphate metabolites can modify plasmid DNA faster than glucose and its phosphate metabolites. Furthermore, 110 mM fructose inhibited colony forming ability, induced mutation in the thymidine kinase gene, caused internucleosomal DNA cleavage, and induced DNA single strand breaks in L5178Y mouse lymphoma cells. The dicarbonyl trap glycation inhibitor aminoguanine also inhibited internucleosomal DNA cleavage. The order of genotoxicity found was fructose 6-phosphate > fructose > glucose 6-phosphate > glucose > glucose 1-phosphate [141].

Exposure of calf thymus DNA to MG at physiological concentrations and temperature, formed the major stable adduct N^2-(1-carboxyethyl)-2'deoxyguanosine in the minor-groove of the DNA without causing steric hindrance. This adduct was also detected in urine samples of healthy subjects and was increased in the kidney and aorta cells of patients with diabetic nephropathy and uremic atherosclerosis. This adduct was weakly mutagenic and polymerase IV was the major DNA polymerase responsible for bypassing this lesion *in vivo*. This polymerase also bypasses the DNA adduct formed with benzopyrene-7,8-diol-9,10-epoxide. Human melanoma cells had an endogenous MG : DNA adduct level of one lesion per 10^7 nucleosides, which was increased on treatment of the cells with glucose or MG [142].

The evident genotoxicity and cytotoxicity of the metabolites and oxidative products of fructose suggest that exposure to sugar might well be associated with risk, perhaps as food-associated metabolites and oxidation products acting as carcinogens. Epidemiology evidence (WCRF/AICR review of epidemiology case–control and cohort studies) does indicate that for "sugary drinks" there was a "probably increased risk" of "weight gain, overweight and obesity," but provides no evidence of association with cancer risk. Furthermore, "foods containing sugar" provided "limited suggestive increased risk" for colorectal cancer risk, but for no other cancers. We suggest that the apparent safety of sugar consumption with relation to cancer is a consequence of the importance of the "second hit," of the importance of the oxidative products. That is, as in NAFLD, sugar can provide calories, but as with NASH, serious disease is only evident with a simultaneous exposure to an oxidative stress such as suggested above (Section 7.4.4.3). Such interactions between fructose exposure and oxidative stress might explain the high-fructose or sucrose intakes with increased pancreatic cancer risk in patients with underlying insulin resistance (overweight or obesity) [139, 140], with colorectal cancer risk [143], and breast cancer risk [141, 142]. Also, GLUT5 fructose transporter expression is much higher in highly proliferated cancer cells than in normal cells [144]. Further studies are required to determine if such interactions play an important role in cancer as they apparently do in liver disease.

7.7
Disease Prevention by Fruits and Vegetables versus Fructose Concern

Epidemiology evidence [145] does show that people whose diets are rich in plant foods such as fruits and vegetables have a lower risk of getting cancers of the mouth, pharynx, larynx, esophagus, stomach, lung, and there is also suggestive evidence for colon, pancreas, ovary, endometrium, and prostate. As noted above, they are also less likely to get diabetes, heart disease, and hypertension. A diet low in energy dense foods and sugary drinks also helps to reduce calorie intake and control weight. To help prevent these cancers and other chronic diseases, experts recommend 4–13 servings of fruits and vegetables daily, depending on energy needs. This includes two to five servings of fruits and two to eight servings of vegetables, with special emphasis on dark-green and orange vegetables and legumes [147]. This chapter suggests that endogenous toxins and even genotoxins may be associated with a diet that is high-fructose or sucrose- or fructose-enriched corn syrup, suggesting that a moderate consumption of fructose would be prudent. However, fruits containing fructose do not appear to confer risk, presumably because they are also rich in antioxidants and ROS scavenging polyphenols. As described above, fructose or fructose metabolite toxicity requires an inflammatory or oxidative stress, a second hit, before cytotoxicity can occur. Indeed, fructose or carbonyl metabolite cytotoxicity was prevented by antioxidants or ROS scavenging polyphenols [56].

7.8
Conclusions

NAFLD affects 10–24% of the general population and 57–74% of obese individuals, and can cause liver toxicity (NASH) in some patients. A two-hit theory has been proposed for those who develop NASH liver toxicity in which hit 1 is steatosis, a fatty liver and hit 2 is oxidative stress. Even though the hepatocytes were not fatty hepatocytes, the oxidative stress was found to markedly increase (100-fold) hepatocyte cytotoxicity induced by high fructose thus providing strong supporting evidence for the second hit theory. The prevention of fructose/H_2O_2 cytotoxicity by iron chelators and the marked increase in cytotoxicity by iron indicates that fructose was oxidized by ROS formed by a Fe/H_2O_2 Fenton reaction. Fructose readily complexes Fe, and the structure of the fructose–ferric complex has been determined. Another part of the marked increase in fructose cytotoxicity by H_2O_2 is that H_2O_2 inhibited GAPDH [79]. This increased GAP levels and cellular endogenous MG levels [78]. MG at physiological levels has also been shown inhibit GAPDH activity by modifying GAPDH lysine larginine. This may therefore occur *in vivo* [148]. Antioxidant enzymes were also inactivated by MG and glyoxal, thereby increasing hepatocyte susceptibility to H_2O_2 [93]. The short-chain sugar metabolites GA, DHA, GOA were formed from fructose catalyzed by enzymes of intermediary metabolism (glycolysis for the first two metabolites) or by a Fe/H_2O_2

Fenton reaction for oxidizing fructose or GA, DHA to GOA, glyoxal, MG, and HCHO [24, 34, 146]. The cytotoxicity and genotoxicity of these compounds may be involved in other chronic diseases associated with energy dense diets containing fructose in the presence of oxidative stress.

References

1 Day, C.P. and James, O.F. (1998) Steatohepatitis: a tale of two "hits"? *Gastroenterology*, **114**, 842–845.
2 Lee, O., Bruce, W.R., Dong, Q., Bruce, J., Mehta, R., and O'Brien, P.J. (2009) Fructose and carbonyl metabolites as endogenous toxins. *Chem Biol Interact*, **178**, 332–339.
3 Anderson, G.H. (2007) Much ado about high-fructose corn syrup in beverages: the meat of the matter. *Am J Clin Nutr*, **86**, 1577–1578.
4 Forshee, R.A., Storey, M.L., Allison, D.B., Glinsmann, W.H., Hein, G.L., Lineback, D.R., Miller, S.A., Nicklas, T.A., Weaver, G.A., and White, J.S. (2007) A critical examination of the evidence relating high fructose corn syrup and weight gain. *Crit Rev Food Sci Nutr*, **47**, 561–582.
5 Prosky, L. and Hoebregs, H. (1999) Methods to determine food inulin and oligofructose. *J Nutr*, **129**, 1418S–1423S.
6 Shepherd, S.J. and Gibson, P.R. (2006) Fructose malabsorption and symptoms of irritable bowel syndrome: guidelines for effective dietary management. *J Am Diet Assoc*, **106**, 1631–1639.
7 Shepherd, S.J., Parker, F.C., Muir, J.G., and Gibson, P.R. (2008) Dietary triggers of abdominal symptoms in patients with irritable bowel syndrome: randomized placebo-controlled evidence. *Clin Gastroenterol Hepatol*, **6**, 765–771.
8 Janket, S.J., Manson, J.E., Sesso, H., Buring, J.E., and Liu, S. (2003) A prospective study of sugar intake and risk of type 2 diabetes in women. *Diabetes Care*, **26**, 1008–1015.
9 Stanhope, K.L., Griffen, S.C., Bair, B.R., Swarbrick, M.M., Keim, N.L., and Havel, P.J. (2008) Twenty-four-hour endocrine and metabolic profiles following consumption of high-fructose corn syrup-, sucrose-, fructose-, and glucose-sweetened beverages with meals. *Am J Clin Nutr*, **87**, 1194–1203.
10 Hyogo, H., Yamagishi, S., Iwamoto, K., Arihiro, K., Takeuchi, M., Sato, T., Ochi, H., Nonaka, M., Nabeshima, Y., Inoue, M., Ishitobi, T., Chayama, K., and Tazuma, S. (2007) Elevated levels of serum advanced glycation end products in patients with non-alcoholic steatohepatitis. *J Gastroenterol Hepatol*, **22**, 1112–1119.
11 Wei, Y. and Pagliassotti, M.J. (2004) Hepatospecific effects of fructose on c-jun NH2-terminal kinase: implications for hepatic insulin resistance. *Am J Physiol Endocrinol Metab*, **287**, E926–E933.
12 Brown, B.E., Dean, R.T., and Davies, M.J. (2005) Glycation of low-density lipoproteins by methylglyoxal and glycolaldehyde gives rise to the in vitro formation of lipid-laden cells. *Diabetologia*, **48**, 361–369.
13 Johnson, R.J., Segal, M.S., Sautin, Y., Nakagawa, T., Feig, D.I., Kang, D.H., Gersch, M.S., Benner, S., and Sanchez-Lozada, L.G. (2007) Potential role of sugar (fructose) in the epidemic of hypertension, obesity and the metabolic syndrome, diabetes, kidney disease, and cardiovascular disease. *Am J Clin Nutr*, **86**, 899–906.
14 Vasdev, S., Gill, V., and Singal, P. (2007) Role of advanced glycation end products in hypertension and atherosclerosis: therapeutic implications. *Cell Biochem Biophys*, **49**, 48–63.

15 Havel, P.J. (2005) Dietary fructose: implications for dysregulation of energy homeostasis and lipid/carbohydrate metabolism. *Nutr Rev*, **63**, 133–157.

16 Ray, S. and Ray, M. (1981) Isolation of methylglyoxal synthase from goat liver. *J Biol Chem*, **256**, 6230–6233.

17 Marks, G.T., Harris, T.K., Massiah, M.A., Mildvan, A.S., and Harrison, D.H. (2001) Mechanistic implications of methylglyoxal synthase complexed with phosphoglycolohydroxamic acid as observed by X-ray crystallography and NMR spectroscopy. *Biochemistry*, **40**, 6805–6818.

18 Richard, J.P. (1991) Kinetic parameters for the elimination reaction catalyzed by triosephosphate isomerase and an estimation of the reaction's physiological significance. *Biochemistry*, **30**, 4581–4585.

19 Kingkeohoi, S. and Chaplen, F.W. (2005) Analysis of methylglyoxal metabolism in CHO cells grown in culture. *Cytotechnology*, **48**, 1–13.

20 Phillips, S.A. and Thornalley, P.J. (1993) The formation of methylglyoxal from triose phosphates. Investigation using a specific assay for methylglyoxal. *Eur J Biochem*, **212**, 101–105.

21 Vestling, C.S., Mylroie, A.K., Irish, U., and Grant, N.H. (1950) Rat liver fructokinase. *J Biol Chem*, **185**, 789–801.

22 Raushel, F.M. and Cleland, W.W. (1977) Bovine liver fructokinase: purification and kinetic properties. *Biochemistry*, **16**, 2169–2175.

23 Storer, A.C. and Cornish-Bowden, A. (1976) Kinetics of rat liver glucokinase. Co-operative interactions with glucose at physiologically significant concentrations. *Biochem J*, **159**, 7–14.

24 Manini, P., La Pietra, P., Panzella, L., Napolitano, A., and d'Ischia, M. (2006) Glyoxal formation by Fenton-induced degradation of carbohydrates and related compounds. *Carbohydr Res*, **341**, 1828–1833.

25 Murray, K., Granner, D.K., Mayes, P.A., and Rodwell, V.W. (2003) *Harper's Illustrated Biochemistry*, Lange Medical Books/McGraw-Hill, New York.

26 Suarez, G., Rajaram, R., Oronsky, A.L., and Gawinowicz, M.A. (1989) Nonenzymatic glycation of bovine serum albumin by fructose (fructation). Comparison with the Maillard reaction initiated by glucose. *J Biol Chem*, **264**, 3674–3679.

27 McPherson, J.D., Shilton, B.H., and Walton, D.J. (1988) Role of fructose in glycation and cross-linking of proteins. *Biochemistry*, **27**, 1901–1907.

28 Swamy, M.S., Tsai, C., Abraham, A., and Abraham, E.C. (1993) Glycation mediated lens crystallin aggregation and cross-linking by various sugars and sugar phosphates in vitro. *Exp Eye Res*, **56**, 177–185.

29 Syrovy, I. (1994) Glycation of albumin: reaction with glucose, fructose, galactose, ribose or glyceraldehyde measured using four methods. *J Biochem Biophys Methods*, **28**, 115–121.

30 Okado-Matsumoto, A. and Fridovich, I. (2000) The role of alpha, beta-dicarbonyl compounds in the toxicity of short chain sugars. *J Biol Chem*, **275**, 34853–34857.

31 Jiang, Z.Y., Woollard, A.C., and Wolff, S.P. (1990) Hydrogen peroxide production during experimental protein glycation. *FEBS Lett*, **268**, 69–71.

32 Chapelle, S. and Verchère, J.-F. (1995) Tungstate complexes of aldoses and ketoses of the lyxo series. Multinuclear NMR evidence for chelation by one or two oxygen atoms borne by the side chain of the furanose ring. *Carbohydr Res*, **277**, 39–50.

33 Charley, P.J., Sarkar, B., Stitt, C.F., and Saltman, P. (1963) Chelation of iron by sugars. *Biochim Biophys Acta*, **69**, 313–321.

34 Lawrence, G.D., Mavi, A., and Meral, K. (2008) Promotion by phosphate of Fe(III)- and Cu(II)-catalyzed autoxidation of fructose. *Carbohydr Res*, **343**, 626–635.

35 Morelli, R., Russo-Volpe, S., Bruno, N., and Lo Scalzo, R. (2003) Fenton-dependent damage to carbohydrates: free radical scavenging activity of some simple sugars. *J Agric Food Chem*, **51**, 7418–7425.

36 Mabrouk, A. and Dugan, L. (1961) Kinetic investigation into glucose-, fructose-, and sucrose-activated autoxidation of methyl linoleate emulsion. *J Am Oil Chem Soc*, **38**, 692–695.

37 Xanthis, A., Hatzitolios, A., Koliakos, G., and Tatola, V. (2007) Advanced glycosylation end products and nutrition – a possible relation with diabetic atherosclerosis and how to prevent it. *J Food Sci*, **72**, R125–R129.

38 Sato, T., Shimogaito, N., Wu, X., Kikuchi, S., Yamagishi, S., and Takeuchi, M. (2006) Toxic advanced glycation end products (TAGE) theory in Alzheimer's disease. *Am J Alzheimers Dis Other Demen*, **21**, 197–208.

39 Kitahara, Y., Takeuchi, M., Miura, K., Mine, T., Matsui, T., and Yamagishi, S. (2008) Glyceraldehyde-derived advanced glycation end products (AGEs). A novel biomarker of postprandial hyperglycaemia in diabetic rats. *Clin Exp Med*, **8**, 175–177.

40 Koteish, A. and Diehl, A.M. (2001) Animal models of steatosis. *Semin Liver Dis*, **21**, 89–104.

41 Chen, S.W., Chen, Y.X., Shi, J., Lin, Y., and Xie, W.F. (2006) The restorative effect of taurine on experimental nonalcoholic steatohepatitis. *Dig Dis Sci*, **51**, 2225–2234.

42 Beraza, N., Malato, Y., Vander, B.S., Liedtke, C., Wasmuth, H.E., Dreano, M., de Vos, R., Roskams, T. and Trautwein, C. (2008) Pharmacological IKK2 inhibition blocks liver steatosis and initiation of non-alcoholic steatohepatitis. *Gut*, **57**, 655–663.

43 Tetri, L.H., Basaranoglu, M., Brunt, E.M., Yerian, L.M. and Neuschwander-Tetri, B.A. (2008) Severe NAFLD with hepatic necroinflammatory changes in mice fed trans fats and a high-fructose corn syrup equivalent. *Am J Physiol Gastrointest Liver Physiol*, **295**, G987–G995.

44 Rinella, M.E., Elias, M.S., Smolak, R.R., Fu, T., Borensztajn, J., and Green, R.M. (2008) Mechanisms of hepatic steatosis in mice fed a lipogenic methionine choline-deficient diet. *J Lipid Res*, **49**, 1068–1076.

45 Berson, A., De Beco, V., Letteron, P., Robin, M.A., Moreau, C., El Kahwaji, J., Verthier, N., Feldmann, G., Fromenty, B., and Pessayre, D. (1998) Steatohepatitis-inducing drugs cause mitochondrial dysfunction and lipid peroxidation in rat hepatocytes. *Gastroenterology*, **114**, 764–774.

46 Marra, F. (2004) NASH: are genes blowing the hits? *J Hepatol*, **40**, 853–856.

47 Namikawa, C., Shu-Ping, Z., Vyselaar, J.R., Nozaki, Y., Nemoto, Y., Ono, M., Akisawa, N., Saibara, T., Hiroi, M., Enzan, H., and Onishi, S. (2004) Polymorphisms of microsomal triglyceride transfer protein gene and manganese superoxide dismutase gene in non-alcoholic steatohepatitis. *J Hepatol*, **40**, 781–786.

48 Rutledge, A.C. and Adeli, K. (2007) Fructose and the metabolic syndrome: pathophysiology and molecular mechanisms. *Nutr Rev*, **65**, S13–S23.

49 Roberts, R., Bickerton, A.S., Fielding, B.A., Blaak, E.E., Wagenmakers, A.J., Chong, M.F., Gilbert, M., Karpe, F., and Frayn, K.N. (2008) Reduced oxidation of dietary fat after a short term high-carbohydrate diet. *Am J Clin Nutr*, **87**, 824–831.

50 Rosca, M.G., Mustata, T.G., Kinter, M.T., Ozdemir, A.M., Kern, T.S., Szweda, L.I., Brownlee, M., Monnier, V.M., and Weiss, M.F. (2005) Glycation of mitochondrial proteins from diabetic rat kidney is associated with excess superoxide formation. *Am J Physiol Renal Physiol*, **289**, F420–F430.

51 Roth, R.A., Luyendyk, J.P., Maddox, J.F., and Ganey, P.E. (2003) Inflammation and drug idiosyncrasy – is

there a connection? *J Pharmacol Exp Ther*, **307**, 1–8.
52 Shaw, P.J., Hopfensperger, M.J., Ganey, P.E., and Roth, R.A. (2007) Lipopolysaccharide and trovafloxacin coexposure in mice causes idiosyncrasy-like liver injury dependent on tumor necrosis factor-alpha. *Toxicol Sci*, **100**, 259–266.
53 Ganey, P.E., Luyendyk, J.P., Maddox, J.F., and Roth, R.A. (2004) Adverse hepatic drug reactions: inflammatory episodes as consequence and contributor. *Chem Biol Interact*, **150**, 35–51.
54 Roth, R.A., Harkema, J.R., Pestka, J.P., and Ganey, P.E. (1997) Is exposure to bacterial endotoxin a determinant of susceptibility to intoxication from xenobiotic agents? *Toxicol Appl Pharmacol*, **147**, 300–311.
55 O'Brien, P.J., Tafazoli, S., Chan, K., Mashregi, M., Mehta, R., and Shangari, N. (2007) A hepatocyte inflammation model for carbonyl induced liver injury: drugs, diabetes, solvents, chlorination. *Enzymol Mol Biol Carbonyl Metab*, **13**, 105–112.
56 Feng, C., and O'Brien, P.J. (2009) Hepatocyte inflammation model for cytotoxicity research: Fructose or glycoaldehyde as a source of endogeneous toxins. *Archives of Physiology and Biochemistry*, **115(2)**, 105–111.
57 Kelley, G.L., Allan, G., and Azhar, S. (2004) High dietary fructose induces a hepatic stress response resulting in cholesterol and lipid dysregulation. *Endocrinology*, **145**, 548–555.
58 Morgan, M.J., Kim, Y.S., and Liu, Z.G. (2008) TNFalpha and reactive oxygen species in necrotic cell death. *Cell Res*, **18**, 343–349.
59 Sakai, K., Matsumoto, K., Nishikawa, T., Suefuji, M., Nakamaru, K., Hirashima, Y., Kawashima, J., Shirotani, T., Ichinose, K., Brownlee, M., and Araki, E. (2003) Mitochondrial reactive oxygen species reduce insulin secretion by pancreatic beta-cells. *Biochem Biophys Res Commun*, **300**, 216–222.

60 Shangari, N., Chan, T.S., Popovic, M., and O'Brien, P.J. (2006) Glyoxal markedly compromises hepatocyte resistance to hydrogen peroxide. *Biochem Pharmacol*, **71**, 1610–1618.
61 Shangari, N. and O'Brien, P.J. (2004) The cytotoxic mechanism of glyoxal involves oxidative stress. *Biochem Pharmacol*, **68**, 1433–1442.
62 Taylor, E.N. and Curhan, G.C. (2008) Fructose consumption and the risk of kidney stones. *Kidney Int*, **73**, 207–212.
63 Yu, D.T., Burch, H.B., and Phillips, M.J. (1974) Pathogenesis of fructose hepatotoxicity. *Lab Invest*, **30**, 85–92.
64 Oberhaensli, R.D., Galloway, G.J., Taylor, D.J., Bore, P.J., and Radda, G.K. (1986) Assessment of human liver metabolism by phosphorus-31 magnetic resonance spectroscopy. *Br J Radiol*, **59**, 695–699.
65 Gaby, A.R. (2005) Adverse effects of dietary fructose. *Altern Med Rev*, **10**, 294–306.
66 Porikos, K.P. and Van Itallie, T.B. (1983) Diet-induced changes in serum transaminase and triglyceride levels in healthy adult men. Role of sucrose and excess calories. *Am J Med*, **75**, 624–630.
67 Vander Jagt, D.L., Hunsaker, L.A., Robinson, B., Stangebye, L.A., and Deck, L.M. (1990) Aldehyde and aldose reductases from human placenta. Heterogeneous expression of multiple enzyme forms. *J Biol Chem*, **265**, 10912–10918.
68 Fukase, S., Sato, S., Mori, K., Secchi, E.F., and Kador, P.F. (1996) Polyol pathway and NADPH-dependent reductases in dog leukocytes. *J Diabetes Complications*, **10**, 304–313.
69 Heinz, F., Lamprecht, W., and Kirsch, J. (1968) Enzymes of fructose metabolism in human liver. *J Clin Invest*, **47**, 1826–1832.
70 Usui, T., Shimohira, K., Watanabe, H., and Hayase, F. (2007) Detection and determination of glyceraldehyde-derived pyridinium-type advanced glycation end product in streptozotocin-induced diabetic

rats. *Biosci Biotechnol Biochem*, **71**, 442–448.

71 Usui, T., Shizuuchi, S., Watanabe, H., and Hayase, F. (2004) Cytotoxicity and oxidative stress induced by the glyceraldehyde-related Maillard reaction products for HL-60 cells. *Biosci Biotechnol Biochem*, **68**, 333–340.

72 Murata, M., Mizutani, M., Oikawa, S., Hiraku, Y., and Kawanishi, S. (2003) Oxidative DNA damage by hyperglycemia-related aldehydes and its marked enhancement by hydrogen peroxide. *FEBS Lett*, **554**, 138–142.

73 Dutta, U., Cohenford, M.A., and Dain, J.A. (2005) Nonenzymatic glycation of DNA nucleosides with reducing sugars. *Anal Biochem*, **345**, 171–180.

74 Takahashi, H., Tran, P.O., LeRoy, E., Harmon, J.S., Tanaka, Y., and Robertson, R.P. (2004) D-Glyceraldehyde causes production of intracellular peroxide in pancreatic islets, oxidative stress, and defective beta cell function via non-mitochondrial pathways. *J Biol Chem*, **279**, 37316–37323.

75 Maswoswe, S.M., Daneshmand, F., and Davies, D.R. (1986) Metabolic effects of D-glyceraldehyde in isolated hepatocytes. *Biochem J*, **240**, 771–776.

76 Thornalley, P.J. and Stern, A. (1984) The effect of glyceraldehyde on red cells. Haemoglobin status, oxidative metabolism and glycolysis. *Biochim Biophys Acta*, **804**, 308–323.

77 Thornalley, P., Wolff, S., Crabbe, J., and Stern, A. (1984) The autoxidation of glyceraldehyde and other simple monosaccharides under physiological conditions catalysed by buffer ions. *Biochim Biophys Acta*, **797**, 276–287.

78 Abordo, E.A., Minhas, H.S., and Thornalley, P.J. (1999) Accumulation of alpha-oxoaldehydes during oxidative stress: a role in cytotoxicity. *Biochem Pharmacol*, **58**, 641–648.

79 Little, C. and O'Brien, P.J. (1969) Mechanism of peroxide-inactivation of the sulphydryl enzyme glyceraldehyde-3-phosphate dehydrogenase. *Eur J Biochem*, **10**, 533–538.

80 Laurie, V.F. and Waterhouse, A.L. (2006) Oxidation of glycerol in the presence of hydrogen peroxide and iron in model solutions and wine. Potential effects on wine color. *J Agric Food Chem*, **54**, 4668–4673.

81 Rashba-Step, J., Step, E., Turro, N.J., and Cederbaum, A.I. (1994) Oxidation of glycerol to formaldehyde by microsomes: are glycerol radicals produced in the reaction pathway? *Biochemistry*, **33**, 9504–9510.

82 Petersen, A.B., Wulf, H.C., Gniadecki, R., and Gajkowska, B. (2004) Dihydroxyacetone, the active browning ingredient in sunless tanning lotions, induces DNA damage, cell-cycle block and apoptosis in cultured HaCaT keratinocytes. *Mutat Res*, **560**, 173–186.

83 Benov, L., Beema, A.F., and Sequeira, F. (2003) Triosephosphates are toxic to superoxide dismutase-deficient Escherichia coli. *Biochim Biophys Acta*, **1622**, 128–132.

84 Anderson, M.M., Hazen, S.L., Hsu, F.F., and Heinecke, J.W. (1997) Human neutrophils employ the myeloperoxidase-hydrogen peroxide-chloride system to convert hydroxy-amino acids into glycolaldehyde, 2-hydroxypropanal, and acrolein. A mechanism for the generation of highly reactive alpha-hydroxy and alpha, beta-unsaturated aldehydes by phagocytes at sites of inflammation. *J Clin Invest*, **99**, 424–432.

85 Kraemer, R.J. and Deitrich, R.A. (1968) Isolation and characterization of human liver aldehyde dehydrogenase. *J Biol Chem*, **243**, 6402–6408.

86 Henehan, G.T. and Tipton, K.F. (1992) Steady-state kinetic analysis of aldehyde dehydrogenase from human erythrocytes. *Biochem J*, **287** (Pt 1), 145–150.

87 Pietruszko, R. and Chern, M. (2001) Betaine aldehyde dehydrogenase

from rat liver mitochondrial matrix. *Chem Biol Interact*, **130-132**, 193–199.

88 Barceloux, D.G., Krenzelok, E.P., Olson, K., and Watson, W. (1999) American academy of clinical toxicology practice guidelines on the treatment of ethylene glycol poisoning. Ad hoc committee. *J Toxicol Clin Toxicol*, **37**, 537–560.

89 Cramer, S.D., Ferree, P.M., Lin, K., Milliner, D.S., and Holmes, R.P. (1999) The gene encoding hydroxypyruvate reductase (GRHPR) is mutated in patients with primary hyperoxaluria type II. *Hum Mol Genet*, **8**, 2063–2069.

90 Ishikawa, K., Kaneko, E., and Ichiyama, A. (1996) Pyridoxal 5′-phosphate binding of a recombinant rat serine: pyruvate/alanine:glyoxylate aminotransferase. *J Biochem*, **119**, 970–978.

91 Williamson, D.H. and Ellington, E.V. (1975) Hydroxypyruvate as a gluconeogenic substrate in rat hepatocytes. *Biochem J*, **146**, 277–279.

92 Glomb, M.A. and Monnier, V.M. (1995) Mechanism of protein modification by glyoxal and glycolaldehyde, reactive intermediates of the Maillard reaction. *J Biol Chem*, **270**, 10017–10026.

93 Morgan, P.E., Dean, R.T., and Davies, M.J. (2002) Inactivation of cellular enzymes by carbonyls and protein-bound glycation/glycoxidation products. *Arch Biochem Biophys*, **403**, 259–269.

94 Zeng, J., Dunlop, R.A., Rodgers, K.J., and Davies, M.J. (2006) Evidence for inactivation of cysteine proteases by reactive carbonyls via glycation of active site thiols. *Biochem J*, **398**, 197–206.

95 Al Maghrebi, M.A., Al Mulla, F., and Benov, L.T. (2003) Glycolaldehyde induces apoptosis in a human breast cancer cell line. *Arch Biochem Biophys*, **417**, 123–127.

96 Knott, H.M., Brown, B.E., Davies, M.J., and Dean, R.T. (2003) Glycation and glycoxidation of low-density lipoproteins by glucose and low-molecular mass aldehydes. Formation of modified and oxidized particles. *Eur J Biochem*, **270**, 3572–3582.

97 O'Brien, P.J., Siraki, A.G., and Shangari, N. (2005) Aldehyde sources, metabolism, molecular toxicity mechanisms, and possible effects on human health. *Crit Rev Toxicol*, **35**, 609–662.

98 Greven, W.L., Waanders, F., Nagai, R., van den Heuvel, M.C., Navis, G. and van Goor, H. (2005) Mesangial accumulation of GA-pyridine, a novel glycolaldehyde-derived AGE, in human renal disease. *Kidney Int*, **68**, 595–602.

99 Butkovskaya, N.I., Pouvesle, N., Kukui, A. and Bras, G.L. (2006) Mechanism of the OH-initiated oxidation of glycolaldehyde over the temperature range 233-296 K. *J Phys Chem A*, **110**, 13492–13499.

100 Benov, L. and Fridovich, I. (1998) Superoxide dependence of the toxicity of short chain sugars. *J Biol Chem*, **273**, 25741–25744.

101 Koop, D.R. and Casazza, J.P. (1985) Identification of ethanol-inducible P-450 isozyme 3a as the acetone and acetol monooxygenase of rabbit microsomes. *J Biol Chem*, **260**, 13607–13612.

102 Harrison, J.E. and Saeed, F.A. (1983) Radical acetoacetate oxidation by myeloperoxidase, lactoperoxidase, prostaglandin synthetase, and prostacyclin synthetase: implications for atherosclerosis. *Biochem Med*, **29**, 149–163.

103 Dhar, A., Desai, K., Kazachmov, M., Yu, P., and Wu, L. (2008) Methylglyoxal production in vascular smooth muscle cells from different metabolic precursors. *Metabolism*, **57**, 1211–1220.

104 Sartori, A., Garay-Malpartida, H.M., Forni, M.F., Schumacher, R.I., Dutra, F., Sogayar, M.C., and Bechara, E.J. (2008) Aminoacetone, a putative endogenous source of methylglyoxal, causes oxidative stress and death to insulin-producing RINm5f cells. *Chem Res Toxicol*, **21**, 1841–1850.

105 Usui, T., Yanagisawa, S., Ohguchi, M., Yoshino, M., Kawabata, R., Kishimoto, J., Arai, Y., Aida, K., Watanabe, H., and Hayase, F. (2007) Identification and determination of alpha-dicarbonyl compounds formed in the degradation of sugars. *Biosci Biotechnol Biochem*, **71**, 2465–2472.

106 Thornalley, P.J., Yurek-George, A., and Argirov, O.K. (2000) Kinetics and mechanism of the reaction of aminoguanidine with the alpha-oxoaldehydes glyoxal, methylglyoxal, and 3-deoxyglucosone under physiological conditions. *Biochem Pharmacol*, **60**, 55–65.

107 Betterton, E.A. and Hoffman, R.M. (1998) Henry's law constant of some environmentally important aldehydes. *Environ Sci Technol*, **32**, 1415–1418.

108 Loeffler, K.W., Koehler, C.A., Paul, N.M., and De Haan, D.O. (2006) Oligomer formation in evaporating aqueous glyoxal and methyl glyoxal solutions. *Environ Sci Technol*, **40**, 6318–6323.

109 Shangari, N., Poon, R.J., and O'Brien, P.J. (2006) Hepatocyte methylglyoxal (MG) resistance is overcome by inhibiting aldo-keto reductases and glyoxalase I catalysed MG metabolism. *Enzymol Mol Biol Carbonyl Metab*, **12**, 266–275.

110 Vander Jagt, D.L., Hassebrook, R.K., Hunsaker, L.A., Brown, W.M., and Royer, R.E. (2001) Metabolism of the 2-oxoaldehyde methylglyoxal by aldose reductase and by glyoxalase-I: roles for glutathione in both enzymes and implications for diabetic complications. *Chem Biol Interact*, **130-132**, 549–562.

111 Kun, E. (1952) A study on the metabolism of glyoxal in vitro. *J Biol Chem*, **194**, 603–611.

112 Kanazu, T., Shinoda, M., Nakayama, T., Deyashiki, Y., Hara, A., and Sawada, H. (1991) Aldehyde reductase is a major protein associated with 3-deoxyglucosone reductase activity in rat, pig and human livers. *Biochem J*, **279** (Pt 3), 903–906.

113 Takahashi, M., Fujii, J., Teshima, T., Suzuki, K., Shiba, T., and Taniguchi, N. (1993) Identity of a major 3-deoxyglucosone-reducing enzyme with aldehyde reductase in rat liver established by amino acid sequencing and cDNA expression. *Gene*, **127**, 249–253.

114 Vander Jagt, D.L. (2008) Methylglyoxal, diabetes mellitus and diabetic complications. *Drug Metabol Drug Interact*, **23**, 93–124.

115 Ayoub, F.M., Allen, R.E., and Thornalley, P.J. (1993) Inhibition of proliferation of human leukaemia 60 cells by methylglyoxal in vitro. *Leuk Res*, **17**, 397–401.

116 Ueno, H., Nakamuro, K., Sayato, Y., and Okada, S. (1991) DNA lesion in rat hepatocytes induced by in vitro and in vivo exposure to glyoxal. *Mutat Res*, **260**, 115–119.

117 Mehta, R., Wong, L., and O'Brien, P.J. (2009) Cytoprotective mechanisms of carbonyl scavenging drugs in isolated rat hepatocytes. *Chem Biol Interact*, **178**, 317–323.

118 Shangari, N., Mehta, R., and O'Brien, P.J. (2007) Hepatocyte susceptibility to glyoxal is dependent on cell thiamin content. *Chem Biol Interact*, **165**, 146–154.

119 Biswas, S., Bhattacharjee, S., Ray, M., and Ray, S. (1996) Interaction of aldehydes with glyoxalase I and the status of several aldehyde metabolizing enzymes of Ehrlich ascites carcinoma cells. *Mol Cell Biochem*, **165**, 9–16.

120 Izaguirre, G., Kikonyogo, A., and Pietruszko, R. (1998) Methylglyoxal as substrate and inhibitor of human aldehyde dehydrogenase: comparison of kinetic properties among the three isozymes. *Comp Biochem Physiol B Biochem Mol Biol*, **119**, 747–754.

121 Siraki, A.G., Pourahmad, J., Chan, T.S., Khan, S., and O'Brien, P.J. (2002) Endogenous and endobiotic induced reactive oxygen species formation by isolated hepatocytes. *Free Radic Biol Med*, **32**, 2–10.

122 Knight, J., Jiang, J., Assimos, D.G., and Holmes, R.P. (2006) Hydroxyproline ingestion and urinary oxalate

and glycolate excretion. *Kidney Int*, **70**, 1929–1934.

123 Poldelski, V., Johnson, A., Wright, S., Rosa, V.D., and Zager, R.A. (2001) Ethylene glycol-mediated tubular injury: identification of critical metabolites and injury pathways. *Am J Kidney Dis*, **38**, 339–348.

124 Danpure, C.J. and Rumsby, G. (2004) Molecular aetiology of primary hyperoxaluria and its implications for clinical management. *Expert Rev Mol Med*, **6**, 1–16.

125 Van Schaftingen, E., Draye, J.P., and Van Hoof, F. (1989) Coenzyme specificity of mammalian liver D-glycerate dehydrogenase. *Eur J Biochem*, **186**, 355–359.

126 Bais, R., Rofe, A.M., and Conyers, R.A. (1991) Investigations into the effect of glyoxylate decarboxylation and transamination on oxalate formation in the rat. *Nephron*, **57**, 460–469.

127 Gupta, S.C. and Dekker, E.E. (1984) Malyl-CoA formation in the NAD-, CoASH-, and alpha-ketoglutarate dehydrogenase-dependent oxidation of 2-keto-4-hydroxyglutarate. Possible coupled role of this reaction with 2-keto-4-hydroxyglutarate aldolase activity in a pyruvate-catalyzed cyclic oxidation of glyoxylate. *J Biol Chem*, **259**, 10012–10019.

128 Dutta, U., Cohenford, M.A., Guha, M., and Dain, J.A. (2007) Non-enzymatic interactions of glyoxylate with lysine, arginine, and glucosamine: a study of advanced non-enzymatic glycation like compounds. *Bioorg Chem*, **35**, 11–24.

129 Lucas, M. and Pons, A.M. (1975) Influence of glyoxylic acid on properties of isolated mitochondria. *Biochimie*, **57**, 637–645.

130 Ingebretsen, O.C. (1976) Mechanism of the inhibitory effect of glyoxylate plus oxaloacetate and oxalomalate on the NADP-specific isocitrate dehydrogenase. *Biochim Biophys Acta*, **452**, 302–309.

131 Rofe, A.M., James, H.M., Bais, R., Edwards, J.B., and Conyers, R.A. (1980) The production of (14C) oxalate during the metabolism of (14C) carbohydrates in isolated rat hepatocytes. *Aust J Exp Biol Med Sci*, **58**, 103–116.

132 Rofe, A.M., Chalmers, A.H., and Edwards, J.B. (1976) [14C]oxalate synthesis from [U-14C]glyoxylate and [1-14C]glycollate in isolated rat hepatocytes. *Biochem Med*, **16**, 277–283.

133 Holmes, R.P., Goodman, H.O., and Assimos, D.G. (2001) Contribution of dietary oxalate to urinary oxalate excretion. *Kidney Int*, **59**, 270–276.

134 McMartin, K.E. and Wallace, K.B. (2005) Calcium oxalate monohydrate, a metabolite of ethylene glycol, is toxic for rat renal mitochondrial function. *Toxicol Sci*, **84**, 195–200.

135 Schulze, M.R., Wachter, R., Schmeisser, A., Fischer, R., and Strasser, R.H. (2006) Restrictive cardiomyopathy in a patient with primary hyperoxaluria type II. *Clin Res Cardiol*, **95**, 235–240.

136 Guo, C. and McMartin, K.E. (2005) The cytotoxicity of oxalate, metabolite of ethylene glycol, is due to calcium oxalate monohydrate formation. *Toxicology*, **208**, 347–355.

137 Guo, C., Cenac, T.A., Li, Y., and McMartin, K.E. (2007) Calcium oxalate, and not other metabolites, is responsible for the renal toxicity of ethylene glycol. *Toxicol Lett*, **173**, 8–16.

138 Khand, F.D., Gordge, M.P., Robertson, W.G., Noronha-Dutra, A.A., and Hothersall, J.S. (2002) Mitochondrial superoxide production during oxalate-mediated oxidative stress in renal epithelial cells. *Free Radic Biol Med*, **32**, 1339–1350.

139 Thamilselvan, S., Khan, S.R., and Menon, M. (2003) Oxalate and calcium oxalate mediated free radical toxicity in renal epithelial cells: effect of antioxidants. *Urol Res*, **31**, 3–9.

140 Selvam, R. (2002) Calcium oxalate stone disease: role of lipid peroxidation and antioxidants. *Urol Res*, **30**, 35–47.

141 Levi, B. and Werman, M.J. (2003) Fructose and related phosphate derivatives impose DNA damage

and apoptosis in L5178Y mouse lymphoma cells. *J Nutr Biochem*, **14**, 49–60.
142 Yuan, B., Cao, H., Jiang, Y., Hong, H., and Wang, Y. (2008) Efficient and accurate bypass of N2-(1-carboxyethyl)-2′-deoxyguanosine by DinB DNA polymerase in vitro and in vivo. *Proc Natl Acad Sci U S A*, **105**, 8679–8684.
143 Higginbotham, S., Zhang, Z.F., Lee, I.M., Cook, N.R., Giovannucci, E., Buring, J.E., and Liu, S. (2004) Dietary glycemic load and risk of colorectal cancer in the Women's Health Study. *J Natl Cancer Inst*, **96**, 229–233.
144 Douard, V. and Ferraris, R.P. (2008) Regulation of the fructose transporter GLUT5 in health and disease. *Am J Physiol Endocrinol Metab*, **295**, E227–E237.
145 World Cancer Research Fund/American Institute for Cancer Research (2007) *Food, Nutrition, Physical Activity, and the Prevention of Cancer: a Global Perspective*, AICR.
146 Novotný, O., Cejpek, K., and Velíšek, J. (2008) Formation of α-hydroxycarbonyl and α-dicarbonyl compounds during degradation of monosaccharides. *Czech J Food Sci*, **25**, 119–130.
147 Krebs-Smith, S.M. and Kantor, L.S. (2001) Choose a variety of fruits and vegetables daily: understanding the complexities. *J Nutr*, **131**, 487S–501S.
148 Lee, H.J., Howell, S.K., Sanford, R.J., and Beisswenger, P.J. (2005) Methyl glyoxal can modify GAPDH activity and structure. *Ann N Y Acad Sci*, **1043**, 135–145.

8
Glyceraldehyde-Related Reaction Products

Teruyuki Usui, Hirohito Watanabe, and Fumitaka Hayase

Proteins are modified by carbonyl compounds in the Maillard reaction, and irreversible products are formed as advanced glycation end products (AGEs). The AGEs may cause a variety of chronic diseases. Toxic Maillard reaction products might be the endogenous toxins. Glyceraldehyde (GLA)-derived AGEs are epitopes of GLA-modified proteins, which dose-dependently inhibit proliferation of HL-60 cells, Caco-2 cells, and PC12 cells. In addition, GLA-modified proteins induce cell death in differentiated PC12 cells. Fluorescent pyridinium-type, non-fluorescent imidazolone-type, and fluorescent pyrimidine-type AGEs exist in the GLA-modified proteins, and a fluorescent pyridinium-type AGE named *GLAP* (glyceraldehyde-derived pyridinium compound) is most toxic. GLAP induces the production of reactive oxygen species (ROS), and is detected in plasma protein and tail tendon collagen in streptozotocin-induced diabetic rats. GLAP might be a biomarker in diabetic complications. GLA-related Maillard reaction products cause cytotoxicity and oxidative stress as endogenous toxins. GLA is increased in chronic diseases, such as diabetic complications and chronic renal disease. The GLA-related metabolism is important in the progression of chronic diseases.

8.1
Maillard Reaction

Similar to post-translational modification, proteins are modified by carbonyl compounds in the Maillard reaction [1]. The Maillard reaction is a nonenzymatic reaction between amino compounds (proteins, peptides, amino acids, phospholipids, etc.) and carbonyl compounds (reducing sugar, degradation products of sugars, secondary products by peroxidation of lipid, ascorbic acid, dehydroascorbic acid, sterol, etc.) [2]. In human body, the major carbonyls are formed enzymically from glucose as reducing sugar, with the short chain carbonyls being formed by the degradation of glucose by autoxidation. As shown in Figure 8.1, carbonyls attack the amino group of protein, lysine or arginine residues. Schiff base (imine) and Amadori rearrangement products (ketoamine) are formed as early stage Maillard products. The early products form other carbonyl intermediates, and the carbonyls

Endogenous Toxins. Diet, Genetics, Disease and Treatment.
Edited by Peter J. O'Brien and W. Robert Bruce
Copyright © 2010 WILEY-VCH Verlag GmbH & Co. KGaA, Weinheim
ISBN: 978-3-527-32363-0

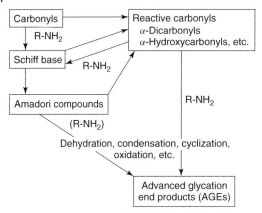

Fig. 8.1 Major pathway of the Maillard reaction.

further react with lysine and arginine residues and the following Maillard processes are accelerated [3]. At the advanced stage of the reaction, irreversible products are formed by dehydration, condensation, cyclization, oxidation, and so on. Modification of proteins by Maillard reaction is also called *glycation*, and the irreversible products are named *advanced glycation end products* (*AGEs*). Figure 8.2 shows the major AGEs [4, 5]. Most AGEs are formed by the modification of lysine and arginine residues in the proteins. Pyrraline, carboxymethyllysine (CML), carboxyethyllysine (CEL), carboxymethylarginine (CMA), 3-deoxyglucosone (3DG)-derived imidazolinones, and MG-H1 are formed as the nonfluorescent adduct-type AGEs and argpyrimidine, glyceraldehyde-derived pyridinium (GLAP), and GA-pyridine are formed as the fluorescent adduct-type AGEs. Furthermore, MOLD, GOLD, and glucosepane are formed as nonfluorescent cross-linked AGEs and pentosidine, crossline, and pyrropyridine are formed as fluorescent cross-linked AGEs.

8.2
Maillard-Related Diseases

The Maillard reaction may cause a variety of chronic diseases. AGEs are detected in diabetes, diabetic complications (such as diabetic nephropathy and retinopathy), atherosclerosis, chronic renal failure, and so on [6, 7]. Additionally, AGEs are also detected in the tissue of the neurodegenerative diseases such as Alzheimer's disease and Parkinson's disease [8, 9]. Maillard reaction has been presumed to be involved in progression of each disease.

Hypotheses for the physiological relevance of Maillard reaction in diseases are as follows. (i) Intracellular enzymes are modified by Maillard reaction, and their activities are inhibited. (ii) Reactive oxygen species (ROS) generated from Maillard intermediates induce oxidative stress injuries. (iii) AGEs also cause other biological effects that are expressed following their binding to an AGE receptor. Binding to the AGE receptor is causally related to the progression of diabetic complications. Toxic Maillard reaction products could then be considered as endogenous toxins.

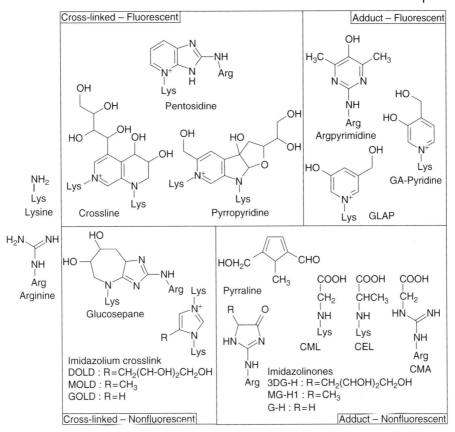

Fig. 8.2 Chemical structure of major AGEs.

8.3
Glyceraldehyde-Modified Protein

Glyceraldehyde is a hydroxy-carbonyl compound, which is produced as glyceraldehyde-3-phosphate in hexose monophosphate pathway. The concentration of free glyceraldehyde increases in the metabolic diseases such as diabetes. Recent studies suggested that the short chain carbonyls, including glyceraldehyde, are cytotoxic. Especially, the modification of proteins by glyceraldehyde is associated with the progression of chronic diseases.

Glyceraldehyde-related metabolism might be important, as neurodegeneration is induced by the inhibition of glyceraldehyde-3-phosphate dehydrogenase [10]. Furthermore, glyceraldehyde-modified bovine serum albumin (GLA-BSA) induced cell death in oligodendrocytes such as glial cells, and the cytotoxicity was not inhibited by the known anti -AGE antibodies [11]. Moreover, it is reported that the cytotoxicity of neurodegenerative AGE fractions, collected in the diabetic nephropathy patients with hemodialysis therapy, is completely inhibited by anti-glyceraldehyde-modified BSA antibodies [11]. Hence, glyceraldehyde-modified proteins might be associated with neurodegenerative diseases, such as Alzheimer's disease and Parkinson's disease. Thus, there are the novel AGEs in GLA-BSA. Moreover, GLA-BSA affects cell functions in Schwann cells, mesangial cells, and retinal pericytes [12–14]. GLA-modified proteins may be formed in the progression of a variety of chronic diseases. Here, we introduce the experimental data for the cytotoxicity induced by GLA-BSA on HL-60 cells, Caco-2 cells, and PC12 cells.

8.3.1
Cytotoxicity for HL-60 Cells

GLA-modified proteins inhibit the proliferation of human promyelocytic HL-60 cells. Figure 8.3 shows the cell numbers when cells were exposed to GLA-BSA- or BSA-containing medium at 37 °C for 96 hours. GLA-BSA

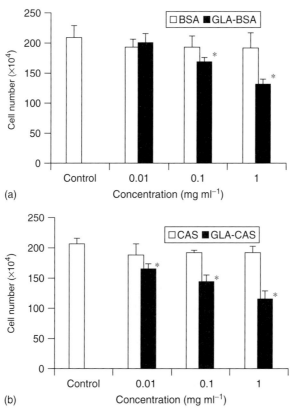

Fig. 8.3 Inhibitory effects of glyceraldehyde-modified proteins on the proliferation of HL-60 cells. HL-60 cells were cultured in glyceraldehyde-modified protein containing medium at 37 °C for 96 hours: (a) glyceraldehyde-modified BSA and (b) glyceraldehyde-modified casein. $*p < 0.05$ (vs. control).

dose-dependently inhibited proliferation of HL-60 cells, whereas no activity was detected with BSA (Figure 8.3a). Similar experiments were conducted with glyceraldehyde-modified casein (GLA-CAS). GLA-CAS inhibited cell proliferation similarly to the effects of GLA-BSA (Figure 8.3b).

8.3.2
Cytotoxicity for Caco-2 Cells

Caco-2 cells are cell line derived from human carcinoma. GLA-modified proteins also inhibited Caco-2 cell proliferation. Caco-2 cells were exposed to GLA-BSA- or BSA-containing medium at 37 °C for 96 hours, and cell numbers were assessed.

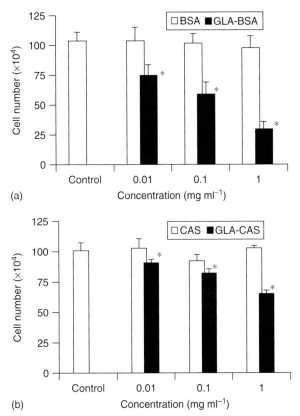

Fig. 8.4 Inhibitory effects of glyceraldehyde-modified proteins on the proliferation of Caco-2 cells. Caco-2 cells were cultured in glyceraldehyde-modified protein containing medium at 37 °C for 96 hours: (a) glyceraldehyde-modified BSA and (b) glyceraldehyde-modified casein. $*p < 0.05$ (vs. control).

GLA-BSA dose-dependently inhibited proliferation of Caco-2 cells, whereas no activity was detected with BSA (Figure 8.4a). GLA-CAS also inhibited cell proliferation similarly to GLA-BSA (Figure 8.4b).

8.3.3
Cytotoxicity for PC12 Cells

PC12 cells are cell line established from rat pheochromocytoma. Cells are differentiated into neuronal cells by nerve growth factor (NGF), which is commonly used as neuron model. Here, PC12 cells and NGF-treated (differentiated) PC12 cells were used in cytotoxicity test. Figure 8.5 shows the cytotoxicity of GLA-modified proteins compared with NGF toward PC12 cells. PC12 cells were exposed to GLA-BSA- or BSA-containing medium at 37 °C for 72 hours, and cell numbers were assessed.

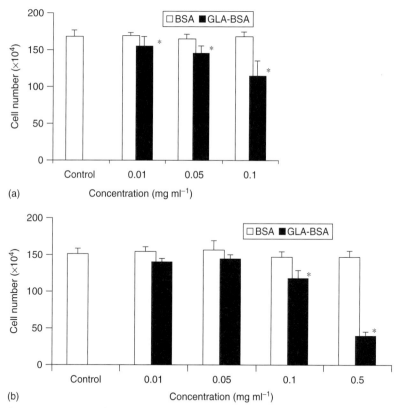

Fig. 8.5 Cytotoxicity of glyceraldehyde-modified BSA on PC12 cells and NGF-treated (differentiated) PC12 cells. (a) PC12 cells cultured in GLA-BSA- or BSA-containing medium at 37 °C for 72 hours. (b) Differentiated PC12 cells cultured in GLA-BSA- or BSA-containing medium at 37 °C for 96 hours. $*p < 0.05$ (vs. control).

GLA-BSA dose-dependently inhibited proliferation of PC12 cells, whereas no activity was detected with BSA (Figure 8.5a). On the other hand, GLA-BSA induced cell death in NGF-treated PC12 cells. As shown in Figure 8.5b, PC12 cells were exposed to GLA-BSA- or BSA-containing medium at 37 °C for 24 hours, and cell numbers were assessed. GLA-BSA dose-dependently induced cell death in NGF-treated PC12 cells, whereas no activity was detected with BSA (Usui and Hayase, unpublished data).

GLA-BSA inhibited cell proliferation of PC12 cells and induced cell death in NGF-treated PC12 neuronal cells. As evident from the cytotoxicity for PC12 cells, the cytotoxicity of GLA-modified protein shows significant differences between differentiated cells and nondifferentiated cells. It is suggested that these differences were due to differences in susceptibility to oxidative stress, for example, neuronal cells are susceptible to oxidative stress. Thus, the cytotoxic-AGE epitope in glyceraldehyde-modified proteins induces oxidative stress. There is considerable interest in the identification of AGEs.

8.4
Glyceraldehyde-Derived AGEs

A glyceraldehyde-derived AGE was formed from glyceraldehyde and lysine residue of proteins as a novel AGE. The compound was originally isolated from glyceraldehyde and N-α-acetyl-lysine, and named *GLAP* compound [15]. Moreover, the compound was detected by LC–MS in plasma protein and tail tendon collagen in streptozotocin-induced diabetic rats [16]. The compound might be a biomarker in diabetic complications. The chemical structure of GLAP is shown in Figure 8.6, and is a fluorescent pyridinium-type AGE. MG-H1 (nonfluorescent imidazolone-type AGE) and argpyrimidine (fluorescent pyrimidine-type AGE) are the other types of glyceraldehyde-derived AGEs [17, 18]. These compounds are known products as methylglyoxal-derived AGEs. It has been reported that MG-H1

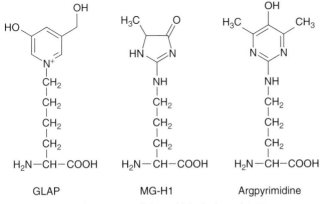

Fig. 8.6 Chemical structure of glyceraldehyde-derived AGEs.

levels increased in the hemoglobin and plasma during triose isomerase deficiency, whereas argpyrimidine increased in the lens protein of cataracts.

8.5
Cytotoxicity and Oxidative Stress

Figure 8.7 shows the effects of GLAP and MG-H1 on the proliferation of HL-60 cells. Cells were exposed to GLAP- or MG-H1-containing medium at 37 °C for 96 hours. GLAP dose-dependently inhibited the proliferation of HL-60 cells (Figure 8.7a), whereas no activity was detected with MG-H1 (Figure 8.7b) [19]. The cytotoxicity of argpyrimidine was tested only in LLC-PK1 cells, and no activity was detected with argpyrimidine (Usui and Hayase, unpublished data).

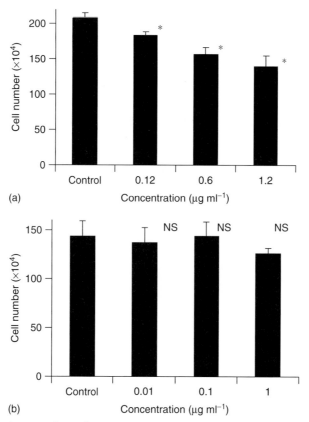

Fig. 8.7 Effects of GLAP and MG-H1 on the proliferation of HL-60 cells. HL-60 cells cultured in (a) GLAP- or (b) MG-H1-containing medium at 37 °C for 96 hours. $*p < 0.05$ (vs. control). NS: not significant.

Glyceraldehyde-modified proteins showed more toxicity than methylglyoxal-modified proteins. GLAP, glyceraldehyde-derived AGE, is more toxic than MG-H1 and argpyrimidine, AGEs formed from both glyceraldehyde and methylglyoxal. It is suggested that GLAP might be a toxic epitope of glyceraldehyde-modified protein.

As shown in Figure 8.8, N-acetylcysteine (NAC) and PDTC, acting as antioxidants, suppressed GLAP-induced cytotoxicity in HL-60 cells. In addition, GLAP decreased intracellular glutathione level as shown in Figure 8.9. Moreover, GLAP induced the production of ROS. Dichlorofluorescein (DCFH)-loaded HL-60 cells were exposed to GLAP at 37 °C for 15 minutes. ROS was detected as dichlorofluorescein (DCF) fluorescence using flow cytometry. Figure 8.10 shows the histogram of the fluorescence peak channel. Each GLAP concentration induced ROS production in HL-60 cells. The ROS production was dose dependent [19]. Hence, it became

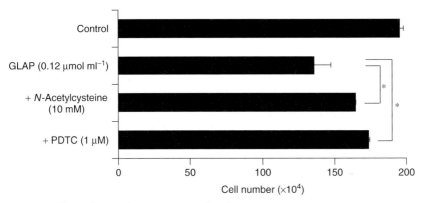

Fig. 8.8 Effects of antioxidants on GLAP-induced cytotoxicity in HL-60 cells. HL-60 cells were cultured in GLAP-containing medium with antioxidant at 37 °C for 96 hours. $*p < 0.05$ (vs. GLAP).

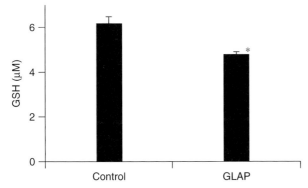

Fig. 8.9 Depression of intracellular GSH level by GLAP in HL-60 cells. HL-60 cells were exposed to GLAP-containing medium at 37 °C for 5 hours. $*p < 0.05$ (vs. control).

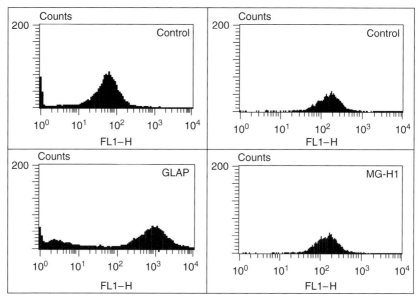

Fig. 8.10 Effects of GLAP and MG-H1 on ROS production in HL-60 cells. H L-60 cells were exposed to GLAP-containing HEPES buffered solution at 37 °C for 15 minutes.

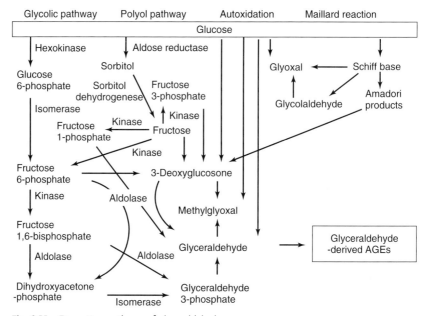

Fig. 8.11 Formation pathway of glyceraldehyde.

clear that GLAP is one of the cytotoxic AGEs of glyceraldehyde-modified proteins, which caused oxidative stress. However, the receptor for GLAP has not been found.

Binding of AGEs to AGE receptor is a biomarker for recognizing endogenous products. AGE receptors that have been reported include RAGE (receptor for advanced glycation end product), OST48/galectin 3/80K-H, SR-A, CD36, SRB, LOX1, and FEEL-1/2 [20–27]. The physiological effects of AGEs are considered to be expressed when AGE binds to an AGE receptor. However, epitope research is slow, so only little is known about the relationship between chemical structure of AGEs and progression of disease.

In this chapter, we introduced the glyceraldehyde-related Maillard reaction products that cause cytotoxicity and/or oxidative stress and act as endogenous toxins. The pathways forming glyceraldehyde and other carbonyls that act as AGE-precursors in the living body are summarized in Figure 8.11. Glyceraldehyde was increased by the activation of the polyol pathway and progression of the Maillard reaction and oxidation in the living body. Thus, glyceraldehyde is increased in chronic diseases, such as chronic complications of diabetes. Glyceraldehyde-related metabolism plays an important role in the progression of chronic diseases.

References

1 Kato H. and Hayase F. (1992) Post-translational modification of proteins by advanced amino-carbonyl reaction, in *The Post-translational Modification of Proteins* (eds S. Tuboi N. Taniguchi, and N. Katunuma, Japan Scientific Societies Press & CRC Press, pp. 237–251.

2 Maillard, L.C. (1912) Action des acides amines sur les sucres; formation des melanoidines par voie methodique. *C R Acad Sci*, **154**, 66–68.

3 Hodge, J.E. (1953) Chemistry of browning reaction in model systems. *J Agric Food Chem*, **1**, 928–943.

4 Schleicher E., Somoza V., and Schieberle P. (eds) (2008) The Maillard reaction. *Ann N Y Acad Sci*, **1126**, 1–340.

5 Baynes J.W., Monnier V.H., Ames J.M., and Thorpe S.R. (eds) (2005) The Maillard reaction. *Ann N Y Acad Sci*, **1043**, 1–954.

6 Ulrich, P. and Cerami, A. (2001) Protein glycation, diabetes, and aging. *Recent Prog Horm Res*, **56**, 1–21.

7 Hammes, H.-P. (2003) Pathysiological mechanisms of diabetic angiopathy. *J Diabetes Complications*, **17**, 16–19.

8 Sasaki, N., Fukatsu, R., Tsuzuki, K., Hayashi, Y., Yoshida, T., Fujii, N., Koike, T., Wakayama, I., Yanagihara, R., Garruto, R., Amano, N., and Makita, Z. (1998) Advanced glycation end products in Alzheimer's disease and other neurodegenerative diseases. *Am J Pathol*, **153**, 1149–1155.

9 Castellani, R., Smith, M.A., Richey, P.J., and Petty, G. (1996) Glycoxidation and oxidative stress in Parkinson's disease and diffuse Lewy body disease. *Brain Res*, **737**, 195–200.

10 Nakazawa, M., Uehara, T., and Nomura, Y. (1997) Koningic acid (a potent glyceraldehyde-3-phosphate dehydrogenase inhibitor)-induced fragmentation and condensation of DNA in NG108-15 cells. *J Neurochem*, **68**, 2493–2499.

11 Takeuchi, M., Bucala, R., Suzuki, T., Ohkubo, T., Yamazaki, M., Koike, T., Kameda, Y., and Makita, Z. (2000) Neurotoxicity of advanced glycation end-products for cultured cortical

neurons. *J Neuropathol Exp Neurol*, **59**, 1094–1105.

12 Yamagishi, S., Inagaki, Y., Okamoto, T., Amano, S., Koga, K., Takeuchi, M., and Makita, Z. (2002) Advanced glycation end products-induced apoptosis and overexpression of vascular endothelial growth factor and monocyte chemoattractant protein-1 in human cultured mesangial cells. *J Biol Chem*, **277**, 20309–20315.

13 Yamagishi, S., Amano, S., Inagaki, Y., Okamoto, T., Koga, K., Sasaki, N., Yamamoto, H., Takeuchi, M., and Makita, Z. (2002) Advanced glycation end products-induced apoptosis and overexpression of vascular endothelial growth factor in bovine retinal pericytes. *Biochem Biophys Res Commun*, **290**, 973–978.

14 Sekido, H., Suzuki, T., Jomori, T., Takeuchi, M., Yabe-Nishimura, C., and Yagihashi, S. (2004) Reduced cell replication and induction of apoptosis by advanced glycation end products in rat Schwann cells. *Biochem Biophys Res Commun*, **320**, 241–248.

15 Usui, T. and Hayase, F. (2003) Isolation and identification of the 3-hydroxy-5-hydroxymethyl-pyridinium compound as a novel advanced glycation end product on glyceraldehyde-related Maillard reaction. *Biosci Biotechnol Biochem*, **67**, 930–932.

16 Usui, T., Shimohira, K., Watanabe, H., and Hayase, F. (2007) Detection and determination of glyceraldehyde-derived pyridinium-type advanced glycation end product in streptozotocin-induced diabetic rats. *Biosci Biotechnol Biochem*, **71**, 442–448.

17 Usui, T., Watanabe, H., and Hayase, F. (2006) Isolation and identification of 5-methyl-imidazolin-4-one derivative as glyceraldehyde-derived advanced glycation end product. *Biosci Biotechnol Biochem*, **70**, 1496–1498.

18 Usui, T., Ohguchi, M., Watanabe, H., and Hayase, F. (2008) Formation of argpyrimidine in glyceraldehyde-related glycation. *Biosci Biotechnol Biochem*, **72**, 568–571.

19 Usui, T., Shizuuchi, S., Watanabe, H., and Hayase, F. (2004) Cytotoxicity and oxidative stress induced by the glyceraldehyde-related Maillard reaction products for HL-60 cells. *Biosci Biotechnol Biochem*, **68**, 333–340.

20 Neeper, M., Schmidt, A.M., Brett, J. et al. (1992) Cloning and expression of a cell surface receptor for advanced glycosylation end products of proteins. *J Biol Chem*, **267**, 14998–15004.

21 Schmidt, A.M., Yan, S.D., and Stern, D.M. (2001) The multiligand receptor RAGE as a progression factor amplifying immune and inflammatory responses. *J Clin Invest*, **108**, 949–955.

22 Yang, S., Makita, Z., Horii, Y. et al. (1991) Two novel rat liver membrane proteins that bind advanced glycosylation endproducts: relationship to macrophage scavenger receptor for glucose modified proteins. *J Exp Med*, **174**, 515–534.

23 Araki, N., Higashi, T., Mori, T. et al. (1995) Macrophage scavenger receptor mediates the endocytic uptake of advanced glycation end-products of the Maillard reaction. *Eur J Biochem*, **230**, 408–415.

24 Ohgami, N., Nagai, R., Ikemoto, M. et al. (2001) CD36, a member of the class B scavenger receptor family, as a receptor for advanced glycation end products. *J Biol Chem*, **276**, 3195–3202.

25 Ohgami, N., Nagai, R., Miyazaki, A. et al. (2001) Scavenger receptor class B type I-mediated reverse cholesterol transport is inhibited by advanced glycation end products. *J Biol Chem*, **276**, 13348–13355.

26 Jono, T., Miyazaki, A., Nagai, R. *et al.* (2002) Lectin-like oxidized low density lipoprotein receptor-1 (LOX-1) serves as an endothelial receptor for advanced glycation end products (AGE). *FEBS Lett*, **511**, 170–174.

27 Tamura, Y., Adachi, H., Osuga, J.I. *et al.* (2003) FEEL-1 and FEEL-2 are endocytic receptors for advanced glycation end products. *J Biol Chem*, **278**, 12613–12617.

9
Estrogens as Endogenous Toxins
Jason Matthews

Estrogens have key roles in the development and maintenance of normal sexual and reproductive function. Although traditionally presumed to be a female hormone, estrogens also exert a vast range of biological effects in the cardiovascular, skeletal, immune, and central nervous systems. There are three natural estrogens: 17β-estradiol (E2), estrone (E1), and estriol (E3); there are also a number of metabolic derivatives. In addition to the physiological role of estrogens, E2 and E1, but not E3, induce tumors in various organs of several laboratory animal species. In humans, a firm link between female reproductive history and increased risk of developing cancer in the breast and endometrium has been established. Women exposed to estrogen, through early menarche, late menopause, and/or prolonged hormone replacement therapy have a higher risk of developing certain types of hormone-dependent cancers. Estrogens have been implicated as complete carcinogens in breast and other tissues. Estrogens are also presumed to play an important role in the etiology of prostate cancer. It is likely that there are multiple overlapping mechanisms of estrogen carcinogenesis. The current dogma states that the mechanism by which estrogens cause an increase in breast cancer is presumed to be through prolonged activation of the estrogen receptor leading to increased cell proliferation. However, tissue-specific synthesis, cellular oxidative metabolism by several specific cytochrome P450 isoforms, peroxidases, and the generation of reactive oxygen species (ROS) are all important mediators of E2-mediated carcinogenesis. This has led to an alternative hypothesis to E2-mediated carcinogenesis, which is based on the findings that estrogens are converted to genotoxic metabolites and inducing ROS resulting in direct and indirect DNA damage. The oxidative stress and genotoxic metabolites generated by estrogen metabolism are presumed to act in concert with ER-mediated signaling pathways to promote DNA damage and altered expression of genes responsible for estrogen-mediated toxicity. Therefore, a balance between activation and estrogen metabolism is important to prevent the endotoxic effects of estrogens, particularly their role in breast and other human cancers. This chapter discusses the role of estrogens and reactive estrogen metabolites as endogenous toxins particularly their role in the etiology of breast cancer and other hormone-dependent diseases.

Endogenous Toxins. Diet, Genetics, Disease and Treatment.
Edited by Peter J. O'Brien and W. Robert Bruce
Copyright © 2010 WILEY-VCH Verlag GmbH & Co. KGaA, Weinheim
ISBN: 978-3-527-32363-0

9.1
Introduction

Epidemiological studies have firmly established a clear link between reproductive history and increased risk of developing cancer in the endometrium and breast [1]. Women exposed to estrogen, through early menarche, late menopause, and/or prolonged hormone replacement therapy (HRT) have a higher risk of developing certain types of hormone-dependent cancers [2]. These adverse health effects elicited by estrogens place it in the growing list of endogenous toxins; understanding its mechanism of toxicity is critical for effectively treating and controlling disease. A number of environmental chemicals are suggested to contribute to breast cancer; however, no single contaminant has been definitively identified to cause breast cancer [3]. Moreover, the multiple benefits associated with estrogen replacement therapy, including decrease in coronary heart disease, osteoporosis, Alzheimer's disease, and relief of postmenopausal symptoms, were proposed to justify their prolonged therapy. The release of Women's Health Initiative Study has resulted in a reevaluation of this treatment regime for postmenopausal women [4, 5]. Data released from the National Cancer Institute in the United States showed that age-related incidence rate in breast cancer fell by 6.7% in 2003 compared to the previous year. The drop was attributed to a drop in the use of HRT. These alarming findings highlight the need to fully understand the deleterious effects of estrogens and their carcinogenic potential.

The mechanisms of estrogen-mediated carcinogenesis remain incompletely understood and most likely result from a combination of a number of different mechanisms [3, 6, 7]. The standard paradigm describing a mechanism for estrogen-dependent cancer is that cell proliferation is provided by 17β-estradiol (E2) or other endogenous estrogens such as estrone (E1). E2 or E1 binding and activation estrogen receptors (ERs) stimulate the transcription of genes involved in cell proliferation. The increased cell replication leads to an increase in the chances of spontaneous mutations via error-prone DNA repair [6]. When one of the mutations occurs in a critical gene required for cell proliferation, DNA repair, apoptosis, or angiogenesis, a neoplastic transformation occurs. Thus, the key contribution of E2 is the stimulation of breast epithelial cell proliferation via a receptor-mediated mechanism of action. However, an important aspect of estrogen toxicology is its tissue-specific synthesis and cellular oxidative metabolism by several specific cytochrome P450 (CYP) isoforms and various peroxidases. Mounting evidence suggests that oxidative estrogen metabolites contribute to estrogen-mediated carcinogenesis [2]. An alternative hypothesis to estrogen-mediated carcinogenesis is based on the findings that estrogens are converted to genotoxic metabolites that directly and indirectly damage DNA. This pathway involves certain CYP enzymes that hydrolyze estrogens to 4-hydroxylated estrogen metabolites. These compounds are further converted either spontaneously or enzymatically to quinones, which bind DNA resulting in depurinating adduct formation. In addition to the direct effects of estrogen metabolites on DNA, estrogen metabolism increases the levels of reactive oxygen species (ROS) that could result in increased DNA, lipid, and additional cellular damage. In this chapter, we consider the role of estrogens as endogenous toxins and identify key metabolites and pathways that are responsible for the estrogen-mediated toxicities.

9.2
Estrogen Synthesis

The biosynthesis of estrogens (E2 and E1) occurs through a sequential series of oxidation reactions catalyzed by various CYP and other enzymes [8] (Figure 9.1). Very few tissues contribute substantially to *de novo* steroidogenesis, which occurs primarily in the adrenal glands, the gonads, and the placenta. E2 is the most potent estrogen and primary endogenous estrogen in mammals and nonmammalian vertebrates. E1 and E3 (estriol) are much weaker estrogens in terms of their binding and activation of the ER, but may have important tissue-specific roles. The ovaries are the predominant estrogen generating tissues in nonpregnant and premenopausal women. Androstenedione and testosterone are key intermediates, which are converted to E1 and E2, respectively, by the enzyme CYP C19 or aromatase. 17β-Hydroxysteroid dehydrogenase catalyzes the interconversion between E1 and E2. Local expression of aromatase can result in high tissue levels of E2. It has been reported that pre- and postmenopausal women have similar E2 levels in breast tissues, despite postmenopausal women having 50–100-fold lower serum E2 levels [9, 10]. This is presumed to be due to the local metabolism of testosterone to E2 by aromatase, increased activity of sulfatases that mobilize E1-sulfate stores to E1, which is then converted to E2. Therefore, alterations in the expression levels of aromatase as well as other enzymes involved in the synthesis of estrogens can significantly impact local E2 concentrations. During pregnancy, the fetal-placental unit produces large amounts of estrogens. Adipose tissue is also a very important source of estrogens in postmenopausal women. There is a strong correlation between obesity and increased risk of estrogen-related diseases such as breast cancer [11]. Although estrogens are historically associated with female reproduction, estrogens have a number of critical nonreproductive roles and are also important signaling molecules in males [12]. Increased levels of testosterone have been reported in women with breast tumors compared to healthy controls [1]. Estrogens are involved in male reproduction and implicated in the etiology of prostate cancer and other human cancers [13].

9.3
Receptor-Mediated Estrogen Signaling

Estrogen signaling is mediated by binding and activation of at least three distinct ligand-activated receptors; ERα, ERβ, and a G-protein-coupled receptor (GPER or GPR30).

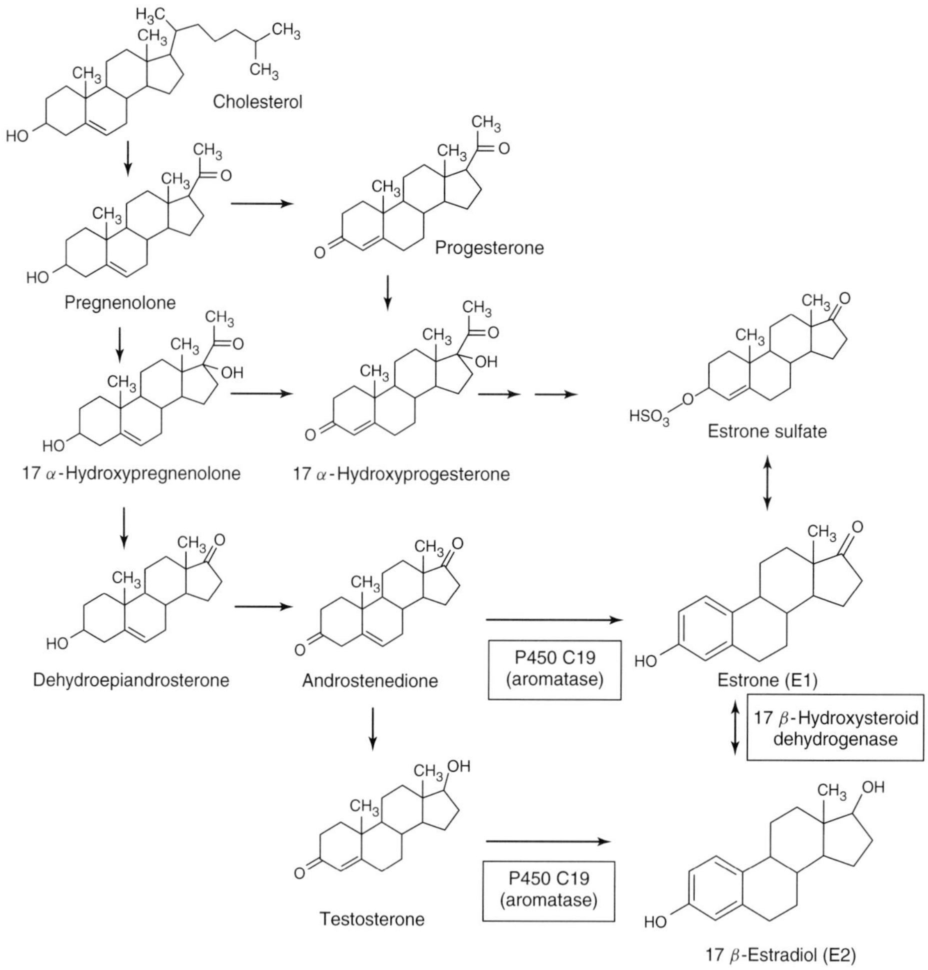

Fig. 9.1 Biosynthetic pathways of natural estrogens, 17β-estradiol (E2) and estrone (E1).

9.3.1
Estrogen Receptor α and β

ERα was first described in the 1950s, and was presumed to be the sole mediator of estrogen signaling. The discovery of a second estrogen-sensitive, ligand-activated transcription factor, ERβ, in the 1997 caused a reevaluation of estrogen-mediated signal transduction [14]. ERα and ERβ are members of the nuclear receptor family of transcription cellular receptors and the products of two distinct genomic loci located on different chromosomes [15]. Similar to other members of this superfamily, ERs contain the evolutionarily conserved structural and functional organization typical of other family members [16]. ERs contain distinct functional domains: the amino-terminal activation function 1 (AF1) domain, the centrally located DNA-binding domain (DBD), and the carboxy-terminal ligand-binding domain (LBD), which also contains the ligand-dependent activation function 2 (AF2) region [16]. ERs are located in the nucleus as monomers, ligand binding causes homodimerization and subsequent DNA binding. ER homodimers bind to specific DNA sequences termed *estrogen responsive elements* (*EREs*; GGTCAnnnTGACC) located in the regulatory region of target genes (Figure 9.2). Alternatively, ERs can interact with other transcription factor through protein–protein interactions to modulate gene expression through the so-called tethering pathway. Once bound to DNA, either directly via EREs or through tethering to other transcription factors, ERs recruit additional accessory proteins known as *coregulators* resulting in changes in gene expression.

The characterization of mice lacking ERα, ERβ, or both has revealed that both subtypes have overlapping but also unique biological roles [17]. Disruption of ERα is not lethal, but both female and male mice are infertile. Disruption of the ERβ is also nonlethal, female mice are fertile, but display reduced fertility and exhibit a number of distinct phenotypes compared to ERα null mice [18]. ERα and ERβ exhibit similar affinity for E2 and DNA binding affinity for ERE sequence; however, their ability to activate target gene expression exhibits context and cell-type specificity [19]. Microarray and genome-wide binding studies have identified a number of subtype-specific target genes, adding another level of complexity to estrogen-mediated signaling pathway. Moreover, for some genes, particularly those involved in cell proliferation, ERα and ERβ can have opposite actions [20, 21]. In general, activation of ERα promotes cell proliferation, whereas ERβ does not [22]. The effects of estrogens on cell proliferation are dependent on the actual ratio of ERα : ERβ expression levels in cells or tissues. The opposing

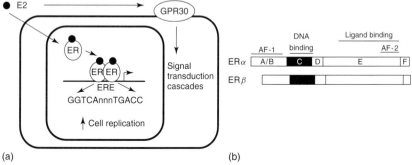

Fig. 9.2 Receptor-mediated estrogen signaling pathway. (a) Estrogen such as 17β-estradiol (E2) diffuse into the cell and into the nucleus where they bind one of two members of the nuclear receptor superfamily of ligand-activated transcription factors, estrogen receptor α (ERα) or ERβ. The receptor–ligand complex binds to specific DNA sequences termed *estrogen responsive elements* (*EREs*) similar to the 5' sequences GGTCAnnnTGACC of the vitellogenin A2 gene. ERs regulate a number of genes involved in cell proliferation. Estrogens also bind to membrane-bound receptors (GPR30) that mediate the reported rapid nongenomic action of these compounds. (b) Modular domain structure of ERα and ERβ, which is characteristic of nuclear receptor family members. AF-1, activation function-1; AF-2 activation function-2.

effect of ERα and ERβ on proliferation has fueled the interest in uncovering the roles of these receptors in diseases such as colon, prostate, breast, and uterus cancers [23].

9.3.2
ER-Mediated Mechanism of Cell Proliferation

The historic view of estrogens and their role in cancer and other diseases has largely focused on the role of estrogen in cell proliferation [6, 24]. The view is that estrogens increase cell proliferation in breast and other tissues, and this proliferation is the main contributing factor to cancer. The contribution of the increase in proliferation to breast cancer or other estrogen-dependent cancer can be manifested by two different mechanisms. In the first mechanism, the increase in cell proliferation induced by estrogens is expected to cause an increase in spontaneous errors associated with DNA replication. In the second mechanism, after mutations have occurred by other processes, estrogens drive and enhance the proliferation of the mutant clones leading to a neoplastic growth.

Estrogens are known to regulate the expression of a number of cell cycle genes, growth factors, and genes regulating cell proliferation. Estrogens regulate the expression of epidermal growth factor family of receptor and ligands, including transforming growth factor-α and epidermal growth factor receptor. Conversely, growth factors are also key regulators of ER function. Estrogens and growth factors exert their main effects on cell cycle through the regulation of cyclin D1, c-Myc, and Bcl-2 family members (Figure 9.2). The increased expression of these key cell cycle regulators enhances cell replication and possibly the incorporation of errors into the newly synthesized DNA.

9.3.3
Rapid Nongenomic Effects of Estrogens

In contrast to the genomic effects described above, estrogens are known to produce effects within the time span of seconds to minutes. This rapid nongenomic action of estrogens occurs via membrane-bound receptors. A lot of controversy concerning the mechanism of action of nongenomic estrogen-dependent signaling exists, and a consensus has not been reached [25, 26]. Two different alternatives have been proposed one involving the classic ERs and the other a distinct membrane-bound receptor. There is considerable evidence that the classic ERs can associate with plasma membranes [26]. ERs can associate with multiprotein complexes at caveolae, which is presumed to be initiated by direct interaction with caveolin-1, and by association with these complexes, E2 appears to induce second messenger-mediated signal transduction [27]. Much of the current focus on membrane-specific receptor has focused on GPER. GPER localizes to the plasma membrane and specifically binds E2 [28]. Binding of E2 activates GPER resulting in calcium mobilization and synthesis of second messengers such as phosphatidylinositol 3,4,5-triphosphate and adenyl cyclase [28]. One of the best characterized nongenomic actions of estrogen is its ability to induce the phosphorylation and activation of endothelial nitric oxide synthase (eNOS) [29, 30]. The activation of eNOS results in the generation of nitric oxide (NO) production enhancing endothethial-dependent vasodilation. Although the role of estrogen in nongenomic responses is well established, the possible role of GPERs and the mechanisms governing this pathway await further investigation. At present, it is unclear how membrane-bound receptors contribute to negative physiological or endotoxic effects of estrogens. As with other receptor systems, any excessive or untimely activation of this pathway is expected to exert a negative influence on cell signaling.

9.4
DNA Damage Induced by Estrogen

Mounting evidence suggests that the oxidative estrogen metabolites are important contributors to estrogen-induced carcinogenesis and estrogen toxicity. A substantial body of evidence exists suggesting that toxic and carcinogenic activities of E2 and E1 are mediated by the highly reactive estrogen quinones that are the metabolites mediating the toxic effects of estrogens [7].

Normal cellular processes result in the formation and degradation of ROS [31]. These chemicals are crucial for life as a number of biochemical reactions require them. All cells are equipped with a multiple level antioxidant control system that protects the cell from damage. However, under conditions of excessive oxidant generation, cellular defenses are overwhelmed and the negative effects of ROS are evident. Oxidants directly induce modification of genetic material, resulting in a number of different types of oxidized DNA bases [31]. Some of these modifications are mutagenic, while others induce DNA hypomethylation resulting in reduced gene expression levels. In the following sections, we consider how oxidative estrogen metabolites and estrogen-mediated generation of ROS contribute to estrogen-mediated toxicity and carcinogenesis.

9.5
Oxidative Metabolism of Estrogen

The oxidative metabolism of estrogen has been extensively studied as part of the regulation and control of estrogen action [32]. The potential role that the E2/E1 metabolites and ROS have in estrogen-dependent cancer has gained significant interest due to studies of estrogen-mediated carcinogenesis in the male Syrian kidney tumor model where modulation of metabolic oxidation increases E2-induced kidney tumor formation through the generation of apurinic DNA lesions [33]. E2/E1-DNA adducts play a major role in tumor initiation in the oncogenic Harvey(ras) mutations in mouse skin papillomas, preneoplastic mouse skin, and a preneoplastic rat mammary tumor models [34, 35]. Cavalieri and colleagues have identified specific oxidative metabolites of estrogen that can interact with DNA [36]. Interestingly, treatment of Syrian hamsters with α-naphthoflavone, an inhibitor of CYP that catalyzes estrogen hydroxylation [37, 38] or the antioxidant ascorbic acid, decreased estrogen-dependent kidney tumor formation [33]. These findings are inconsistent with the previous and widely accepted model of ER-mediated induction of cell proliferation producing an increase in spontaneous errors. According to this view, inhibition of estrogen metabolism should increase tumor formation since E2 metabolites are less hormonally active than the parent compound [6]. Results from these studies have led to a new mechanism of cancer initiation by E1/E2, which is mediated by the oxidative metabolism of estrogens.

9.5.1
Formation of Catechol Estrogens

E2/E1 is oxidized to a number of different hydroxylated metabolites through the catalytic activities of different CYP450 isoforms [32]. Unlike other endogenous steroids, estrogens have an aromatic A-ring, which when oxidized forms catechols and later very reactive quinones. The catechol estrogens (CEs), 2- or 4-hydroxylated E2 (4-OHE2) or E1, are the major oxidized estrogen metabolites in most mammalian species, and are the precursors to more reactive intermediates. The formation of CEs is catalyzed mainly by CYP1A1/1A2, CYP1B1, and CYP3A4 (Figure 9.3) [39]. The tissue-specific expression of these enzymes in many ways dictates the tissues-specific generation of CEs. In humans, 2-hydroxylation of E1/E2 is the major metabolic oxidation of estrogen and is catalyzed by CYP3A4 and CYP1A2 in the liver and by CYP1A1 in extrahepatic tissue. These enzymes convert 80–85% of E2 to 2-hydroxylated E2 (2-OHE2), but due to their low specificity they also generate 15–20% 4-OHE2 [39]. CYP1B1 is the enzyme responsible for the 4-hydroxylation of E1/E2 observed in several human tissues [40].

9.5.2
4-Hydroxylated Oxidation of Estrogens by CYP1B1

The generation and metabolic activation of 4-OHE2/E1 has received considerable scientific interest due to their role as toxic metabolites and as important contributing

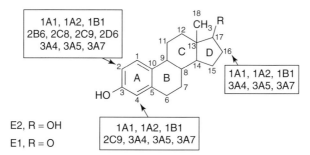

Fig. 9.3 Summary of the hydroxylation of E2 and E1 by human cytochrome P450 isoforms. Isoforms shown in larger and bold font designate the cytochrome P450 isoform with the highest activity for catalyzing the indicated reaction.

factors in the etiology of breast cancer [2, 41]. E2 and E1 are metabolized to 4-OHE2 or 4-OHE1 catalyzed by CYP1B1, a ubiquitously expressed enzyme and the major estrogen hydroxylating enzyme in human breast tissue.

CYP1B1 was isolated for the differential display of dioxin-treated human keratinocyte immortal cell line [42]. Dioxin or 2,3,7,8-tetrachlorodibenzo-*p*-dioxin (TCDD) is a potent environmental contaminant that binds and activates the aryl hydrocarbon receptor (AHR). The AHR is a ligand-activated transcription factor that binds a number of xenobiotic and endogenous compounds; however, a definitive endogenous ligand of AHR remains elusive [43]. After binding a ligand, the AHR heterodimerizes with aryl hydrocarbon receptor nuclear translocator (ARNT) and binds to specific DNA sequences, termed *aryl hydrocarbon receptor responsive elements (AHREs)*, located in the regulatory region of target gene promoters leading to changes in gene expression (Figure 9.4) [44]. Characterization of the CYP1B1 upstream regulatory region identified three AHR response elements (TnGCGTG) that determine the AHR responsiveness of this gene [42]. Treatment of human breast cancer cells with TCDD results in elevated levels of CYP1B1 mRNA and resulted in a greater then 10-fold increase in the rates of hydroxylation at position C2 and C4 [45]. These studies show that activation of the AHR and subsequent induction of CYP1B1 mRNA and protein levels results in increased E2 metabolism and 4-OHE2 generation. More recent studies have also reported that AHR plays an important role in the high basal levels of CYP1B1 in human breast cancer cell lines

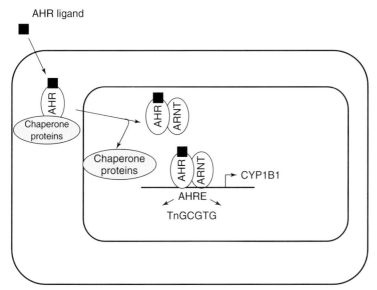

Fig. 9.4 Aryl hydrocarbon receptor (AHR)–mediated mechanism of action. AHR ligand diffuses into the cytosol and binds to the AHR. Ligand binding results in the dissociation of chaperone proteins and nuclear translocation. In the nucleus, AHR binds to its heterodimerization partner aryl hydrocarbon nuclear translocator (ARNT). The heterodimer complex binds AHREs (TnGCGTG) located in the 5' regulatory region of genes, including CYP1B1.

[46]. Thus an important mediator of the endotoxic effects of E2 is the regulation of CYP1B1 expression and its modulation by activators of the AHR.

The relative expression levels of CYP1B1 and CYP1A1 will determine the relative ratio of 4-OHE2 and 2-OHE2, respectively, in tissues. In humans, normal breast tissue, breast tumors, and the prostate CYP1B1 is constitutively expressed, suggesting that a low level 4-OHE2 is present in all tumors originating from these tissues [47]. CYP1A1 mRNA is weakly expressed in most human breast tumor donors and is detectable in the majority of 12 prostate tissues donor set examined [48, 49]. Studies have shown that elevated levels of 4-OHE2 and CYP1B1 mRNA and protein are detected in breast tissue [46]. These results are consistent with the hypothesis that increased expression levels CYP1B1 may contribute to the oxidative metabolism of E2 and other estrogens. The local increased generation of hydroxylated E2 metabolites predisposes the tissue for toxicity and carcinogenesis. Belous and colleagues have demonstrated *in vitro* CYP1B1-mediated, E2-induced DNA adduct formation, further implicating these enzyme in mediating the toxic effects of estrogens [50]. A functional ERE has been identified in the CYP1B1 proximal promoter region [51], and high density microarray studies have reported that both CYP1A1 and CYP1B1 are inducible by E2 [52]. Studies from the author's laboratory and others using chromatin immunoprecipitation assays demonstrate ligand-dependent activation of AHR induces recruitment of ER to the CYP1B1 promoter region further supporting the notion that ER regulates the expression of CYP1B1 [53, 54]. Careful analysis of genetically modified mouse models for ERα will be important for future experiments to clarify the role of ERα in the regulation of these key enzymes involved in the oxidative metabolism of estrogens.

The specific formation of the 4-OHE2 estrogens is of significant interest, since it is carcinogenic, whereas 2-OHE2 is not in a Syrian hamster kidney tumor model [33]. Moreover, specific enzymes that convert E2 to 4-OHE2 are expressed in rodent tissues that develop tumors following chronic exposure to estrogen [47]. In humans, the conversion of E2 to 4-OHE2 is detected in uterus and in benign and malignant mammary tumors, as well as in the prostate [55, 56]. In tissues resistant to estrogen-induced tumor formation, the 2-OHE2 metabolite predominates [40]. One study reported that the ratio of 4-OHE2 to 2-OHE2 was 4 : 1 in human breast cancer extracts and microsomes isolated from breast cancer tissue [57]. These studies led to the conclusion that elevated levels of 4-OHE2 in tissues prone to E2-induced tumor formation (i.e., breast and uterus) may result in increased incidences of tumors and toxic outcomes.

9.5.3
Direct DNA Damage by Estrogen Metabolites

Under normal conditions, CEs are inactivated extrahepatically by O-methylation catalyzed by catechol-O-methyltransferases (COMTs) or in the liver through conjugation reactions, such as glucuronidation and sulfation [7]. COMT is also an important line of defense in preventing the oxidation of CE to quinones by methylating the 2- or 4-hydroxyl group. It has been reported, however, that 4-OHE2 is relatively resistant to methylation, whereas 2-OHE2 undergoes rapid methylation by COMT [58]. Moreover, 2-MeE2 inhibits the ability of COMT to methylate 4-OHE2 [58]. The differential ability of COMT to methylate 2-OHE2 and 4-OHE2 is presumed to contribute to the reduced carcinogenic activity of the former and strong carcinogenic activity of the latter. Some, but not all, studies have reported an increase in the relative risk of breast cancer in women expressing low levels of COMT [59, 60]. The 4-OHE2/E1 metabolites exhibit significantly greater carcinogenic potency compared to the borderline carcinogen, 2-OHE2/E1. Despite the differences in the activities of COMT on CEs, differences in carcinogenic potency between 2-OHE2 and 4-OHE2 are difficult to assign; both compounds have similar redox potentials [61]. Studies of immortalized breast epithelial cells, MCF10F, show that inhibition of COMT blocked methoxylation of CEs and resulted in a three- to four-fold increase in DNA adduct formation [62]. If the cellular protective defense system is overwhelmed by excess formation of hydroxylated E2(E1) metabolites and other conjugation reactions are insufficient, then 2-OH E2(E1) and 4-OH E2(E1) are further oxidized either chemically or enzymatically to reactive E2(E1)-2,3-semiquinone (E2(E1)-2,3-SQ) and E2(E1)-3,4-semiquinone (E2(E1)-3,4-SQ), respectively. In the presence of molecular oxygen, E2(E1)-2,3-SQ and E2(E1)-3,4-SQ are further metabolized to E2(E1)-2,3-quinone (E2(E1)-2,3-Q) and E2-3,4-quinone (E2(E1)-3,4-Q), respectively. CE-quinones can be inactivated through reaction with glutathione (GSH) catalyzed by glutathione-S-transferases. However, if the GSH reaction is incomplete CE-quinones readily bind to DNA to form stable and depurinating DNA adducts (Figure 9.5). NAD(P)H-dependent quinone oxidoreductase (NQO) is also presumed to protect against estrogen carcinogenesis by reducing CE-quinones [63], although other investigators question the protective role of NQO [64]. The E2(E1)-2,3-SQ is an extremely weak carcinogen and mainly forms stable DNA adducts, though formation of depurinating DNA adducts has also been reported, such as 2-hydroxyestradiol-6(α,β)-N3-adenine (2-OHE2-6-N3Ade) [61]. The E2(E1)-3,4-Q readily binds DNA through Michael addition to form depurinating adducts, such as 4-hydroxyestradiol-1(α,β)-N7-guanine (4-OHE1(E2)-1-N7Gua) and 4-hydroxyestradiol-1(α,β)-N3-adenine (4-OHE1(E2)-1-N3Ade) (Figure 9.5) [36]. The depurinating 4-OHE2-N3-Ade adduct was detected at higher levels in women with breast cancer compared to control women [65]. Moreover, the 4-OHE1(E2)-1-N3Ade depurinating adduct is also excreted in the urine of men with prostate cancer, whereas it is virtually undetectable in healthy males [55]. Error-prone repair of apurinic sites in DNA may generate the critical mutations needed to initiate estrogen-dependent cancer formation. The depurinating CE-quinone DNA adducts have been proposed as possible biomarkers for early detection of breast cancer risk and responsiveness to preventative treatment [56].

Therefore, one emerging hypothesis is that the CE-quinone carcinogenic potency is through the formation of depurinating DNA adducts. In rat mammary gland and mouse skin model systems exposed to E2(E1)-3,4-Q, the 4-OHE1(E2)-1-N3Ade and 4-OHE1(E2)-1-N7Gua adducts were formed. The 4-OHE1(E2)-1-N3Ade adduct

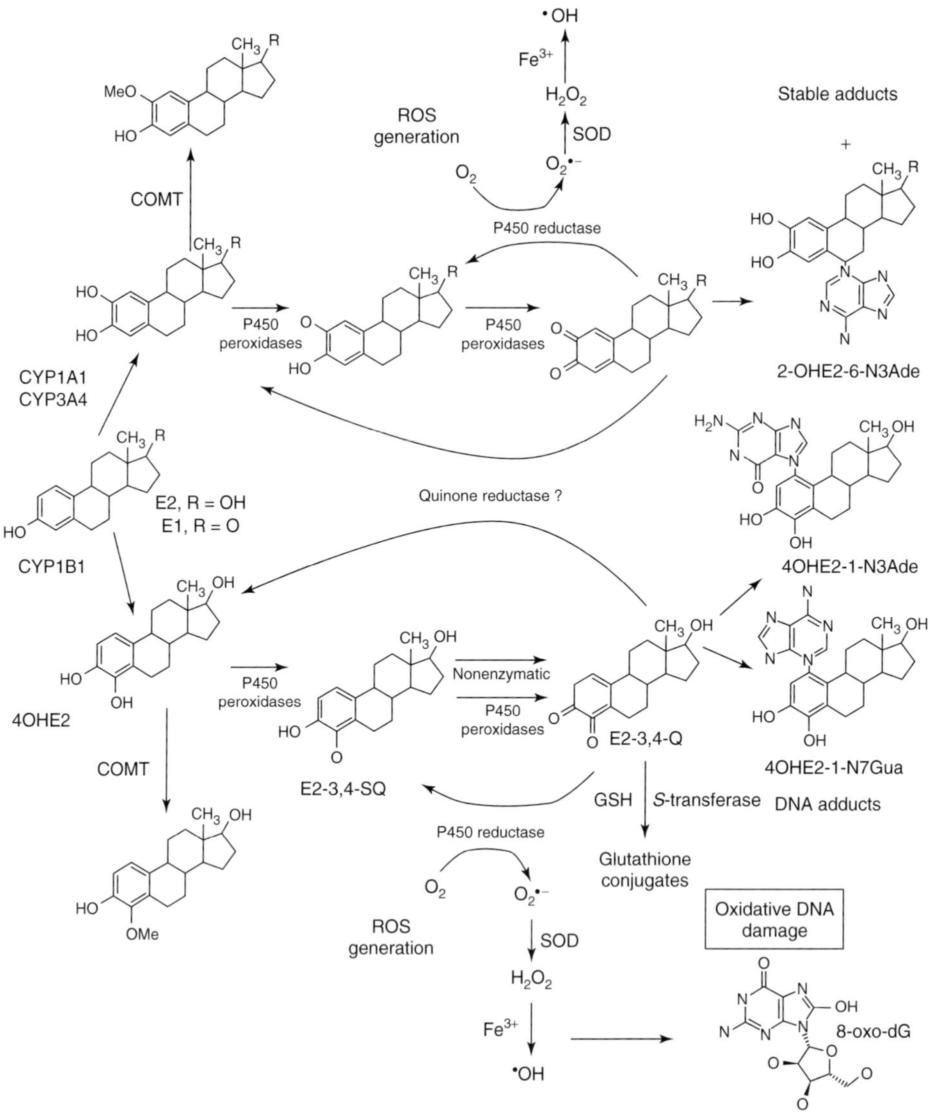

Fig. 9.5 Oxidative metabolism of 17β-estradiol (E2) and estrone (E1). The oxidative metabolism of E2 leads to the production of reactive oxygen species (ROS) leading to oxidative DNA damage. 4OHE2, 4-hydroxylated-E2; CYP1B1, cytochrome P450 1B1; COMT, catechol-O-methyltransferase; GSH, glutathione; SOD, superoxide dismutase; E2-3,4-SQ, E2-3,4-semiquinone, E2-3,4-Q, E2-3,4-quinone; 4OHE2-1-N7Gua, 4-hydroxyestradiol-1(α,β)-N7-guanine; 4OHE2-1-N3Ade, 4-hydroxyestradiol-1(α,β)-N3-adenine; 8-oxo-dG, 8-oxo-7,8-dihydro-2'-deoxyguanosine; 2-OHE2-6-N3Ade, 2-hydroxyestradiol-6(α,β)-N3-adenine. The role of quinone reductase has been questioned; this is denoted by the question mark (see text for details). Modified from [54].

results in rapid depurination, leading to premutagenic apurinic sites, whereas the 4-OHE1(E2)-1-N7Gua adduct is more stable allowing for efficient DNA repair [61]. These data lend support to the hypothesis that DNA adduct formation is the principal characteristic governing the carcinogenic potential of 4OHE2.

9.5.4
Indirect DNA Damage and ROS Generation

One of the main criticisms of studies suggesting that E2(E1)-3,4-Q is the most important modulator of the estrogen-mediated toxicities is that it is generally eliminated and the CEs do not surpass the serum levels of E2, which in non-pregnant females range from 40 to 350 pg ml^{-1} [41]. Oxidative stress has been shown to play an important role in the development of estrogen-induced mammary cancer [7]. Oxidative DNA damage reportedly increased in breast cancer tissues compared to normal tissue [66]. Serum markers for oxidative damage also increased in women with breast cancer [7]. Moreover, E2 has been reported to translocate 8-oxoguanine DNA glycosylase protein (OGG1) within nuclei and to other subcellular compartments in treated rats [67]; the base excision repair marker 8-oxo-7,8-dihydro-2'-deoxyguanosine (8-oxo-dG) also increased in the urine of breast cancer patients [68]. The generation of CE and the subsequent redox cycling resulting from enzymatic reduction of the CE-quinones to CE-semiquinone and back to CE-quinones by autoxidation results in the formation of superoxide anions (O_2^-). Superoxide can then be transformed to H_2O_2 by superoxide dismutase (SOD). In the presence of reduced transition metal ions, such as Fe^{3+}, H_2O_2 is converted to the extremely reactive hydroxyl radical ($^\bullet$OH). H_2O_2 readily crosses cellular and nuclear membranes where it can result in site-specific oxidation of DNA bases [31]. Redox cycling generation of the CE-quinone will occur in the presence of molecular oxygen; thus low levels of CE-E2 may produce significant amounts of ROS, resulting in indirect estrogen-mediated DNA and cell toxicity. Hydroxyl radicals not only oxidize DNA bases but also cause lipid peroxidation. Oxidized lipids then serve as cofactors for estrogen metabolism and further redox cycling between CE-semiquinone and CE-quinone metabolites resulting in further ROS production [7].

Estrogens are well-established regulators of immune response and are known to influence immune-mediated diseases [69]. Physiological doses of E2 induce the expression of interleukin 1α (IL-1α) resulting in the initiation of growth factor, chemotaxis, and cytokines signaling cascades [70]. The increase in chemotaxic factors causes the infiltration of macrophages, which when activated release other cytokines, ROS, and reactive nitrogen species. Macrophages produce NO, which can readily interact with O_2^- to produce peroxynitrite, a very reactive oxidant [7]. Macrophages are present in normal breast tissue, but in breast cancer these cells can represent 50% of breast tumor mass [71]. Macrophages are also significant sources of estrogens since they express CYP19 (aromatase), which converts circulating androgens to estrogens. This creates a cyclic process in which estrogen stimulates macrophages that in turn produce estrogens, and in the presence of CYP1B1 this could potentially lead to an increase in 4-OHE2 and reactive E2-3,4-Q levels. Estrogens also promote the release of hypochlorite/hypochlorous acid from polymorphonuclear leukocytes (PMN); 2OHE1/E2 are inhibitors of PMN activity, suggesting this to be one of the protective properties of 2-hydroxylated estrogen catechols [72].

The oxidative stress generated by estrogens is presumed to act in concert with ER-mediated signaling pathways to promote DNA damage, increase in lipid peroxidation, and alter the expression of genes responsible for controlling cell cycle and proliferation [73]. The upregulation of antioxidant enzymes has been shown to precede tumor development and to occur in a manner dependent on the duration of exposure to E2 [74].

9.6
Role for Estrogen Receptor in ROS Generation

In addition to inducing ROS, E2 has also been reported to regulate the expression of GSH peroxidases, SOD, and glucose-6-phosphate through an ER-dependent manner [41]. There is some controversy as to the role of ER in mediating the oxidative damage resulting from prolonged E2 treatment. Recent studies have reported increased E2-mediated, peroxide-induced DNA damage in ER-positive Michigan Cancer Foundation 7 (MCF-7) cells; however, no adverse affects were observed in ER-negative MDA-MB-231 cells [41]. The data suggest that E2 can induce sensitivity to oxidative DNA damage through an ER-mediated mechanism of action. In contrast, studies in estrogen receptor α knockout mice (ERKO) carrying the wingless-type MMTV integration site family, member 1 (Wnt-1) (ERKO/Wnt-1 model) proto-oncogene provide important evidence for the genotoxic role of estrogen metabolites in cancer initiation [75]. In the ERKO/Wnt-1 model, mice treated with estrogen developed significantly more mammary tumors than control mice [76]. Moreover, mammary tumors also developed in mice treated with both E2 and the pure antiestrogen (ER antagonist) ICI 182,780 [73]. These data suggest that E2-induced tumor formation occurs in the absence of ER; however, it should be noted that the ERKO/Wnt-1 mouse line was generated from an ERKO that expressed a truncated form of ERα and the expression levels of ERβ cannot be ruled out and may simply be limited by current technologies.

Thus, it is possible that ERs may contribute to the observed increases in tumor formation in the ERKO/Wnt-1 model. Increase cell proliferation itself can lead to increase in sensitivity to DNA damage. Furthermore, it is equally possible that ERs are increasing the local concentration of estrogens at a "hot spot" in susceptible genes close to specific EREs. Thus the combination of ER function and generation of toxic DNA adducts as well as ROS induced DNA damaged are all contributing factors to breast cancer and endogenous toxicity induced by estrogens. Aromatase inhibitors should be effective therapeutic against ER-negative breast tumors that are refractory to antiestrogen treatment [77]. Since blocking aromatase activity also prevents the activation of ER receptors, clinical trials and experimental models assessing the effectiveness of preventing relapse of ER-negative breast tumors will be important future studies to address the role of estrogen metabolites in the breast cancer. Although it is attractive to suggest that the generation of CEs and ROS are causal in the breast carcinogenesis, the combination of ER activation and generation of reactive estrogen metabolites most likely work in concert either in an additive or synergistic fashion.

9.6.1
ER in the Mitochondria

In the previous section, we discussed the carcinogenic potential of CE-E2-dependent ROS generation; however, chronic treatment of E2 profoundly increases energy production of mitochondria and suppresses the generation of ROS [78]. The E2-dependent decrease in ROS production and presumably decrease in DNA damage is presumed to be an important contributing factor to increase the longevity of women. Estrogens are important regulators of endothelial function; endothelial dysfunction is an age-related process that is the result of oxidative stress [79]. Recent evidence suggests that estrogen affects mitochondrial function by modifying the expression of mitochondrial and nuclear genes [78]. ERα and ERβ are expressed in mitochondria and E2 treatment leads to increased expression of mitochondrial MnSOD, cytochrome c levels, and nuclear respiratory factor-1; a key regulator of nuclear encoded mitochondrial genes [79]. These findings demonstrate that E2 both positively and negatively affects the generation of ROS; this is an important point to consider when evaluating the overall toxic effects of endogenous estrogens.

9.7
Treatment of Estrogen-Dependent Diseases

Tamoxifen has been the treatment of choice for hormone-responsive and ER-positive breast tissue for the past 30 years. Tamoxifen is a selective estrogen receptor modulator (SERM), since it exhibits tissue-specific actions on ER function. In breast tissue, tamoxifen exhibits antagonistic ER activity, whereas in bone and uterine tissue tamoxifen behaves as an ER agonist [80]. Although, tamoxifen maintains function in tissues such as bone, the increase in incidences of uterine cancer associated with exposure to tamoxifen prompted the search for alternative therapies, such as aromatase inhibitors. Aromatase (CYP19) is the key enzyme responsible for converting circulating testosterone to E2. Regulation of aromatase levels can have dramatic effects on local E2 levels. MCF-7 cells overexpressing aromatase readily and efficiently convert testosterone to estrogen [73]. MCF-7 cells endogenously express high constitutive levels of CYP1B1 and after 24 hour incubation with testosterone the 4-OHE1(E2)-1-N7Gua, 4-OHE1(E2)-1-N3Ade DNA adducts, and E2-3,4-Q were detected [73].

Treatment with the aromatase inhibitor letrozole prevented the formation of E2 and its downstream metabolites [75]. Aromatase inhibitors are proving to be very powerful therapy for the treatment of breast cancer. In a number of clinical trials, aromatase inhibitors are proving to be more effective at delaying the progression of disease than tamoxifen [81, 82]. Like SERMs, aromatase inhibitors have been known for over 30 years, but first generation compounds were associated with severe side effects due to their lack of specificity for aromatase. Third generation aromatase inhibitors are extremely potent inhibitors of aromatase (1000–10 000-fold more potent than the first generation aminoglutethimide) [77]. The potential advantages of aromatase inhibitors are that they block estrogen synthesis, which would reduce ER activity but also prevent the generation of harmful catechol estrogen metabolites. Results from clinical trials comparing the effects of Arimidex (aromatase inhibitor) and tamoxifen, alone or in combination (known as the *Arimidex tamoxifen, alone or in combination (ATAC) trial*), after completion of five years of adjuvant treatment for breast cancer, suggest that Arimidex to be the preferred initial adjuvant therapy for postmenopausal women with hormone-responsive breast cancer [81]. At present, it is unclear whether the aromatase inhibitors are protecting the patient from DNA damaging metabolites or if it is simply reducing the levels of estrogens thereby reducing the activity of ER. Further clinical trials examining the ability of aromatase inhibitors to prevent or delay the recurrence of ER-negative or SERM-resistant breast cancer will be important studies to determine the role of metabolites in breast cancer formation.

9.8
Conclusions

Estrogens have important biological roles in female/male reproduction, immune system, and brain; however, prolonged exposure to estrogens is associated with a number of diseases including breast, uterine, and prostate cancers. Imbalances in estrogen synthesis and metabolism through alterations in the expression or activity of aromatase (CYP19) and CYP1B1 have significant influence on the local concentration of estrogens and the generation of toxic estrogen metabolites and ROS. Current data lend support to the existences of two separate but not necessarily mutually exclusive mechanisms contributing to the endotoxic effects of E1/E2: (i) direct activation of ERs by estrogens and subsequent increase in cell proliferation and (ii) the generation of toxic and reactive estrogen metabolites that (i) directly bind DNA and (ii) indirectly damage DNA through the generation of ROS (Figure 9.6). It will be important to properly assess the role of ER in E2-mediated oxidative DNA damage and if ER directs reactive estrogen metabolites to specific genomic loci causing site-specific DNA damage. Oxidative DNA base modifications and DNA repair have been studied for more than two decades; however, data are reported in fractional values, and no integration of genome-wide information is made. Recently, Akatsuka and colleagues used DNA immunoprecipitation techniques to demonstrate that the localization of oxidative DNA damage is not random [83]. Applying genome-wide profiling for oxidized DNA bases and DNA repair enzymes will allow for the evaluation of genomic region or site-specific modifications and

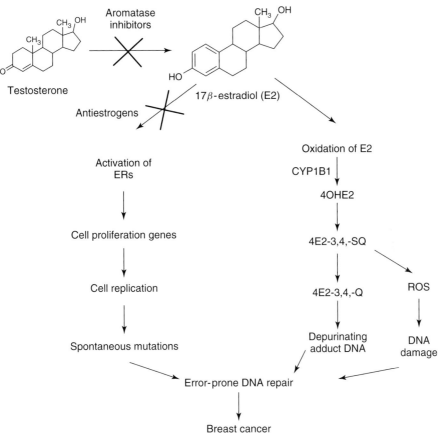

Fig. 9.6 Schematic of the pathways by which estrogens cause breast cancer. The pharmacological antiestrogens, such as Tamoxifen, block receptor-mediated pathways, whereas aromatase inhibitors prevent the formation of oxidative estrogen metabolites, reactive oxygen species generation, and receptor-mediated activity. CYP1B1, cytochrome P450 1B1; E2-3,4-SQ, E2-3,4-semiquinone; E2-3,4-Q, E2-3,4-quinone. Modified from [69].

will be important future experiments to determine the role of oxidative DNA damage in E2-mediated toxicity.

References

1 Yager, J.D. (2000) Endogenous estrogens as carcinogens through metabolic activation. *J Natl Cancer Inst Monogr*, **27**, 67–73.
2 Key, T., Appleby, P., Barnes, I., and Reeves, G. (2002) Endogenous sex hormones and breast cancer in postmenopausal women: reanalysis of nine prospective studies. *J Natl Cancer Inst*, **94**, 606–616.
3 Wolff, M.S., Collman, G.W., Barrett, J.C., and Huff, J. (1996) Breast cancer and environmental risk factors: epidemiological and experimental

findings. *Annu Rev Pharmacol Toxicol*, **36**, 573–596.
4 Anderson, G.L., Limacher, M., Assaf, A.R. et al. (2004) Effects of conjugated equine estrogen in postmenopausal women with hysterectomy: the Women's Health Initiative randomized controlled trial. *J Am Med Assoc*, **291**, 1701–1712.
5 Rossouw, J.E., Anderson, G.L., Prentice, R.L. et al. (2002) Risks and benefits of estrogen plus progestin in healthy postmenopausal women: principal results from the Women's Health Initiative randomized controlled trial. *J Am Med Assoc*, **288**, 321–333.
6 Feigelson, H.S. and Henderson, B.E. (1996) Estrogens and breast cancer. *Carcinogenesis*, **17**, 2279–2284.
7 Cavalieri, E., Frenkel, K., Liehr, J.G., Rogan, E., and Roy, D. (2000) Estrogens as endogenous genotoxic agents – DNA adducts and mutations. *J Natl Cancer Inst Monogr*, **27**, 75–93.
8 Minneman, K. and Wecker, L. (2005) *Brody's Human Pharmacology Molecular to Clinical*, 4th edn, Elsevier Mosby, Philadelphia.
9 Chetrite, G.S., Cortes-Prieto, J., Philippe, J.C., Wright, F., and Pasqualini, J.R. (2000) Comparison of estrogen concentrations, estrone sulfatase and aromatase activities in normal, and in cancerous, human breast tissues. *J Steroid Biochem Mol Biol*, **72**, 23–27.
10 Pasqualini, J.R., Chetrite, G., Blacker, C. et al. (1996) Concentrations of estrone, estradiol, and estrone sulfate and evaluation of sulfatase and aromatase activities in pre- and postmenopausal breast cancer patients. *J Clin Endocrinol Metab*, **81**, 1460–1464.
11 Majed, B., Moreau, T., Senouci, K., Salmon, R.J., Fourquet, A., and Asselain, B. (2008) Is obesity an independent prognosis factor in woman breast cancer? *Breast Cancer Res Treat*, **111**, 329–342.
12 Carreau, S. (2000) Estrogens and male reproduction. *Folia Histochem Cytobiol*, **38**, 47–52.
13 Carruba, G. (2007) Estrogen and prostate cancer: an eclipsed truth in an androgen-dominated scenario. *J Cell Biochem*, **102**, 899–911.
14 Kuiper, G.G., Enmark, E., Pelto-Huikko, M., Nilsson, S., and Gustafsson, J.A. (1996) Cloning of a novel receptor expressed in rat prostate and ovary. *Proc Natl Acad Sci USA*, **93**, 5925–5930.
15 Enmark, E. and Gustafsson, J.A. (1999) Oestrogen receptors – an overview. *J Intern Med*, **246**, 133–138.
16 Nilsson, S., Makela, S., Treuter, E. et al. (2001) Mechanisms of estrogen action. *Physiol Rev*, **81**, 1535–1565.
17 Couse, J.F. and Korach, K.S. (1999) Estrogen receptor null mice: what have we learned and where will they lead us? *Endocr Rev*, **20**, 358–417.
18 Couse, J.F., Curtis Hewitt, S., and Korach, K.S. (2000) Receptor null mice reveal contrasting roles for estrogen receptor α and β in reproductive tissues. *J Steroid Biochem Mol Biol*, **74**, 287–296.
19 Nilsson, S. and Gustafsson, J.A. (2000) Estrogen receptor transcription and transactivation: basic aspects of estrogen action. *Breast Cancer Res*, **2**, 360–366.
20 Matthews, J., Wihlen, B., Tujague, M., Wan, J., Strom, A., and Gustafsson, J.A. (2006) Estrogen receptor (ER) β modulates ERα-mediated transcriptional activation by altering the recruitment of c-Fos and c-Jun to estrogen-responsive promoters. *Mol Endocrinol*, **20**, 534–543.
21 Lindberg, M.K., Moverare, S., Skrtic, S. et al. (2003) Estrogen Receptor (ER)-β Reduces ERα-regulated gene transcription, supporting a "Ying Yang" relationship between ERα and ERβ in mice. *Mol Endocrinol*, **17**, 203–208.
22 Matthews, J. and Gustafsson, J.A. (2003) Estrogen signaling: a subtle balance between ERα and ERβ. *Mol Interv*, **3**, 281–292.
23 Gustafsson, J.A. (2006) ERβ scientific visions translate to

clinical uses. *Climacteric*, **9**, 156–160.
24 Dickson, R.B. and Stancel, G.M. (2000) Estrogen receptor-mediated processes in normal and cancer cells. *J Natl Cancer Inst Monogr*, **27**, 135–145.
25 Hasbi, A., O'Dowd, B.F., and George, S.R. (2005) A G protein-coupled receptor for estrogen: the end of the search? *Mol Interv*, **5**, 158–161.
26 Levin, E.R. (2002) Cellular functions of plasma membrane estrogen receptors. *Steroids*, **67**, 471–475.
27 Chambliss, K.L., Yuhanna, I.S., Anderson, R.G., Mendelsohn, M.E. and Shaul, P.W. (2002) ERβ has nongenomic action in caveolae. *Mol Endocrinol*, **16**, 938–946.
28 Revankar, C.M., Cimino, D.F., Sklar, L.A., Arterburn, J.B., and Prossnitz, E.R. (2005) A transmembrane intracellular estrogen receptor mediates rapid cell signaling. *Science*, **307**, 1625–1630.
29 McNeill, A.M., Kim, N., Duckles, S.P., Krause, D.N., and Kontos, H.A. (1999) Chronic estrogen treatment increases levels of endothelial nitric oxide synthase protein in rat cerebral microvessels. *Stroke*, **30**, 2186–2190.
30 McNeill, A.M., Zhang, C., Stanczyk, F.Z., Duckles, S.P., and Krause, D.N. (2002) Estrogen increases endothelial nitric oxide synthase via estrogen receptors in rat cerebral blood vessels: effect preserved after concurrent treatment with medroxyprogesterone acetate or progesterone. *Stroke*, **33**, 1685–1691.
31 De Bont, R. van and Larebeke, N. (2004) Endogenous DNA damage in humans: a review of quantitative data. *Mutagenesis*, **19**, 169–185.
32 Zhu, B.T. and Conney, A.H. (1998) Functional role of estrogen metabolism in target cells: review and perspectives. *Carcinogenesis*, **19**, 1–27.
33 Li, J.J. and Li, S.A. (1987) Estrogen carcinogenesis in Syrian hamster tissues: role of metabolism. *Fed Proc*, **46**, 1858–1863.
34 Chakravarti, D., Mailander, P.C., Li, K.M. et al. (2001) Evidence that a burst of DNA depurination in SENCAR mouse skin induces error-prone repair and forms mutations in the H-ras gene. *Oncogene*, **20**, 7945–7953.
35 Li, K.M., Todorovic, R., Devanesan, P. et al. (2004) Metabolism and DNA binding studies of 4-hydroxyestradiol and estradiol-3,4-quinone in vitro and in female ACI rat mammary gland in vivo. *Carcinogenesis*, **25**, 289–297.
36 Cavalieri, E.L., Stack, D.E., Devanesan, P.D. et al. (1997) Molecular origin of cancer: catechol estrogen-3,4-quinones as endogenous tumor initiators. *Proc Natl Acad Sci USA*, **94**, 10937–10942.
37 Peter Guengerich, F., Chun, Y.J., Kim, D., Gillam, E.M., and Shimada, T. (2003) Cytochrome P450 1B1: a target for inhibition in anticarcinogenesis strategies. *Mutat Res*, **523-524**, 173–182.
38 Tassaneeyakul, W., Birkett, D.J., Veronese, M.E. et al. (1993) Specificity of substrate and inhibitor probes for human cytochromes P450 1A1 and 1A2. *J Pharmacol Exp Ther*, **265**, 401–407.
39 Lee, A.J., Cai, M.X., Thomas, P.E., Conney, A.H., and Zhu, B.T. (2003) Characterization of the oxidative metabolites of 17β-estradiol and estrone formed by 15 selectively expressed human cytochrome p450 isoforms. *Endocrinology*, **144**, 3382–3398.
40 Jefcoate, C.R., Liehr, J.G., Santen, R.J. et al. (2000) Tissue-specific synthesis and oxidative metabolism of estrogens. *J Natl Cancer Inst Monogr*, **27**, 95–112.
41 Mobley, J.A. and Brueggemeier, R.W. (2004) Estrogen receptor-mediated regulation of oxidative stress and DNA damage in breast cancer. *Carcinogenesis*, **25**, 3–9.
42 Sutter, T.R., Tang, Y.M., Hayes, C.L. et al. (1994) Complete cDNA sequence of a human dioxin-inducible mRNA identifies a new gene subfamily of cytochrome P450 that maps

to chromosome 2. *J Biol Chem*, **269**, 13092–13099.

43 Denison, M.S. and Nagy, S.R. (2003) Activation of the aryl hydrocarbon receptor by structurally diverse exogenous and endogenous chemicals. *Annu Rev Pharmacol Toxicol*, **43**, 309–334.

44 Hankinson, O. (1995) The aryl hydrocarbon receptor complex. *Annu Rev Pharmacol Toxicol*, **35**, 307–340.

45 Lu, F., Zahid, M., Saeed, M., Cavalieri, E.L. and Rogan, E.G. (2007) Estrogen metabolism and formation of estrogen-DNA adducts in estradiol-treated MCF-10F cells. The effects of 2,3,7,8-tetrachlorodibenzo-p-dioxin induction and catechol-O-methyltransferase inhibition. *J Steroid Biochem Mol Biol*, **105**, 150–158.

46 Yang, X., Solomon, S., Fraser, L.R. et al. (2008) Constitutive regulation of CYP1B1 by the aryl hydrocarbon receptor (AhR) in pre-malignant and malignant mammary tissue. *J Cell Biochem*, **104**, 402–417.

47 Murray, G.I., Melvin, W.T., Greenlee, W.F., and Burke, M.D. (2001) Regulation, function, and tissue-specific expression of cytochrome P450 CYP1B1. *Annu Rev Pharmacol Toxicol*, **41**, 297–316.

48 Ragavan, N., Hewitt, R., Cooper, L.J. et al. (2004) CYP1B1 expression in prostate is higher in the peripheral than in the transition zone. *Cancer Lett*, **215**, 69–78.

49 Shimada, T., Hayes, C.L., Yamazaki, H. et al. (1996) Activation of chemically diverse procarcinogens by human cytochrome P-450 1B1. *Cancer Res*, **56**, 2979–2984.

50 Belous, A.R., Hachey, D.L., Dawling, S., Roodi, N. and Parl, F.F. (2007) Cytochrome P450 1B1-mediated estrogen metabolism results in estrogen-deoxyribonucleoside adduct formation. *Cancer Res*, **67**, 812–817.

51 Tsuchiya, Y., Nakajima, M., Kyo, S., Kanaya, T., Inoue, M. and Yokoi, T. (2004) Human CYP1B1 is regulated by estradiol via estrogen receptor. *Cancer Res*, **64**, 3119–3125.

52 Frasor, J., Stossi, F., Danes, J.M., Komm, B., Lyttle, C.R., and Katzenellenbogen, B.S. (2004) Selective estrogen receptor modulators: discrimination of agonistic versus antagonistic activities by gene expression profiling in breast cancer cells. *Cancer Res*, **64**, 1522–1533.

53 Matthews, J. and Gustafsson, J.A. (2006) Estrogen receptor and aryl hydrocarbon receptor signaling pathways. *Nucl Recept Signal*, **4**, e016.

54 Matthews, J., Wihlen, B., Thomsen, J., and Gustafsson, J.A. (2005) Aryl hydrocarbon receptor-mediated transcription: ligand-dependent recruitment of estrogen receptor α to 2,3,7,8-tetrachlorodibenzo-p-dioxin-responsive promoters. *Mol Cell Biol*, **25**, 5317–5328.

55 Markushin, Y., Gaikwad, N., Zhang, H. et al. (2006) Potential biomarker for early risk assessment of prostate cancer. *Prostate*, **66**, 1565–1571.

56 Gaikwad, N.W., Yang, L., Muti, P. et al. (2008) The molecular etiology of breast cancer: evidence from biomarkers of risk. *Int J Cancer*, **122**, 1949–1957.

57 Liehr, J.G. and Ricci, M.J. (1996) 4-Hydroxylation of estrogens as marker of human mammary tumors. *Proc Natl Acad Sci USA*, **93**, 3294–3296.

58 Weisz, J., Clawson, G.A. and Creveling, C.R. (1998) Biogenesis and inactivation of catecholestrogens. *Adv Pharmacol*, **42**, 828–833.

59 Hamajima, N., Matsuo, K., Tajima, K. et al. (2001) Limited association between a catechol-O-methyltransferase (COMT) polymorphism and breast cancer risk in Japan. *Int J Clin Oncol*, **6**, 13–18.

60 Lavigne, J.A., Helzlsouer, K.J., Huang, H.Y. et al. (1997) An association between the allele coding for a low activity variant of catechol-O-methyltransferase and the risk for breast cancer. *Cancer Res*, **57**, 5493–5497.

61 Cavalieri, E., Chakravarti, D., Guttenplan, J. et al. (2006) Catechol estrogen quinones as initiators of breast and other human cancers: implications for biomarkers of susceptibility and cancer prevention. *Biochim Biophys Acta*, **1766**, 63–78.

62 Zahid, M., Saeed, M., Lu, F., Gaikwad, N., Rogan, E., and Cavalieri, E. (2007) Inhibition of catechol-O-methyltransferase increases estrogen-DNA adduct formation. *Free Radic Biol Med*, **43**, 1534–1540.

63 Gaikwad, N.W., Rogan, E.G. and Cavalieri, E.L. (2007) Evidence from ESI-MS for NQO1-catalyzed reduction of estrogen ortho-quinones. *Free Radic Biol Med*, **43**, 1289–1298.

64 Chandrasena, R.E., Edirisinghe, P.D., Bolton, J.L., and Thatcher, G.R. (2008) Problematic detoxification of estrogen quinones by NAD(P)H-dependent quinone oxidoreductase and glutathione-S-transferase. *Chem Res Toxicol*, **21**, 1324–1329.

65 Markushin, Y., Zhong, W., Cavalieri, E.L. et al. (2003) Spectral characterization of catechol estrogen quinone (CEQ)-derived DNA adducts and their identification in human breast tissue extract. *Chem Res Toxicol*, **16**, 1107–1117.

66 Musarrat, J., Arezina-Wilson, J., and Wani, A.A. (1996) Prognostic and aetiological relevance of 8-hydroxyguanosine in human breast carcinogenesis. *Eur J Cancer*, **32A**, 1209–1214.

67 Araneda, S., Pelloux, S., Radicella, J.P. et al. (2005) 8-oxoguanine DNA glycosylase, but not Kin17 protein, is translocated and differentially regulated by estrogens in rat brain cells. *Neuroscience*, **136**, 135–146.

68 Tagesson, C., Kallberg, M., Klintenberg, C. and Starkhammar, H. (1995) Determination of urinary 8-hydroxydeoxyguanosine by automated coupled-column high performance liquid chromatography: a powerful technique for assaying in vivo oxidative DNA damage in cancer patients. *Eur J Cancer*, **31A**, 934–940.

69 Carlsten, H. (2005) Immune responses and bone loss: the estrogen connection. *Immunol Rev*, **208**, 194–206.

70 Basak, S., Dubanchet, S., Zourbas, S., Chaouat, G. and Das, C. (2002) Expression of pro-inflammatory cytokines in mouse blastocysts during implantation: modulation by steroid hormones. *Am J Reprod Immunol*, **47**, 2–11.

71 Lewis, C.E., Leek, R., Harris, A., and McGee, J.O. (1995) Cytokine regulation of angiogenesis in breast cancer: the role of tumor-associated macrophages. *J Leukoc Biol*, **57**, 747–751.

72 Jansson, G. (1991) Oestrogen-induced enhancement of myeloperoxidase activity in human polymorphonuclear leukocytes – a possible cause of oxidative stress in inflammatory cells. *Free Radic Res Commun*, **14**, 195–208.

73 Yue, W., Wang, J.P., Li, Y. et al. (2005) Tamoxifen versus aromatase inhibitors for breast cancer prevention. *Clin Cancer Res*, **11**, 925s–930s.

74 Mense, S.M., Remotti, F., Bhan, A. et al. (2008) Estrogen-induced breast cancer: alterations in breast morphology and oxidative stress as a function of estrogen exposure. *Toxicol Appl Pharmacol*, **232**, 78–85.

75 Yue, W., Santen, R.J., Wang, J.P. et al. (2003) Genotoxic metabolites of estradiol in breast: potential mechanism of estradiol induced carcinogenesis. *J Steroid Biochem Mol Biol*, **86**, 477–486.

76 Bocchinfuso, W.P., Lindzey, J.K., Hewitt, S.C. et al. (2000) Induction of mammary gland development in estrogen receptor-α knockout mice. *Endocrinology*, **141**, 2982–2994.

77 Brodie, A., Sabnis, G., and Jelovac, D. (2006) Aromatase and breast cancer. *J Steroid Biochem Mol Biol*, **102**, 97–102.

78 Duckles, S.P., Krause, D.N., Stirone, C., and Procaccio, V. (2006) Estrogen and mitochondria: a new paradigm

79 Stirone, C., Duckles, S.P., Krause, D.N., and Procaccio, V. (2005) Estrogen increases mitochondrial efficiency and reduces oxidative stress in cerebral blood vessels. *Mol Pharmacol*, **68**, 959–965.

80 Katzenellenbogen, B.S., Choi, I., Delage-Mourroux, R. *et al.* (2000) Molecular mechanisms of estrogen action: selective ligands and receptor pharmacology. *J Steroid Biochem Mol Biol*, **74**, 279–285.

81 Howell, A., Cuzick, J., Baum, M. *et al.* (2005) Results of the ATAC (Arimidex, Tamoxifen, Alone or in Combination) trial after completion of 5 years' adjuvant treatment for breast cancer. *Lancet*, **365**, 60–62.

82 Howell, A. (2006) The 'Arimidex', Tamoxifen, Alone or in Combination (ATAC) Trial: a step forward in the treatment of early breast cancer. *Rev Recent Clin Trials*, **1**, 207–215.

83 Akatsuka, S., Aung, T.T., Dutta, K.K. *et al.* (2006) Contrasting genome-wide distribution of 8-hydroxyguanine and acrolein-modified adenine during oxidative stress-induced renal carcinogenesis. *Am J Pathol*, **169**, 1328–1342.

10
Reactive Oxygen Species, Hypohalites, and Reactive Nitrogen Species in Liver Pathophysiology

Hartmut Jaeschke

Reactive oxygen species, hypochlorous acid/hypochlorite, and peroxynitrite are potent oxidants, which can damage proteins, lipids, and DNA. As unavoidable by-products of aerobic life, these compounds are generally well controlled by multiple levels of antioxidant defense systems. On the other hand, hypochlorite and peroxynitrite are intentionally formed by enzymes such as NADPH oxidases in combination with myeloperoxidase (neutrophils) and nitric oxide synthase (macrophages), respectively. Under these conditions, these aggressive oxidants are a critical part of the host defense system. However, damage to mitochondria or excessive activation of inflammatory cells can lead to enhanced formation of these oxidants. An increased oxidant stress, in particular in the presence of impaired antioxidant defense, can cause or at least significantly contribute to organ injury. This chapter discusses the sources and pathophysiological relevance of reactive oxygen species, hypochlorous acid/hypochlorite, and peroxynitrite in various liver diseases processes, including ischemia-reperfusion injury, endotoxemia, and acetaminophen hepatotoxicity. A better understanding of the nature of the oxidants, the location of the oxidant stress, and the potential targets within the cell will allow to specifically target detrimental effects without affecting the use of these oxidants in host defense mechanisms.

10.1
Introduction

Reactive oxygen species (ROS) and their derivatives are generated intentionally for various cellular functions but are also continuously formed as by-products of aerobic life. However, these ROS are generally well controlled by multiple antioxidant defense systems within cells and the vasculature [1, 2]. In contrast, there is increasing evidence that a whole spectrum of ROS is involved in the pathophysiology of numerous disease processes. In this case, enhanced formation of ROS, impairment of antioxidant defense mechanisms, or a combination of both has been observed. Despite the overwhelming support for a critical role of ROS in the disease pathophysiology, especially in experimental models, antioxidant-based

Endogenous Toxins. Diet, Genetics, Disease and Treatment.
Edited by Peter J. O'Brien and W. Robert Bruce
Copyright © 2010 WILEY-VCH Verlag GmbH & Co. KGaA, Weinheim
ISBN: 978-3-527-32363-0

therapeutic approaches have routinely failed in clinical trials. An important reason for this failure is that in most cases the specific nature of the ROS, the timing of their formation, and the specific location where they are formed is only incompletely understood. It is therefore not surprising that unspecific antioxidants, whose properties are not well defined, have only a limited chance of success. Thus, it is critical to develop a better understanding of the mechanisms of ROS-mediated injury processes. This chapter discusses evidence for the formation and pathophysiological relevance of various ROS, hypohalites, and reactive nitrogen species in liver disease models.

10.2
Reactive Oxygen Species

10.2.1
Nature of Reactive Oxygen Species

The most frequently formed ROS in cells is superoxide, which represents a one-electron reduction product of molecular oxygen (Figure 10.1). In the absence of nitric oxide (NO), most superoxide dismutates either spontaneously or catalyzed by superoxide dismutases (SODs) to hydrogen peroxide and oxygen. Hydrogen peroxide is only a moderate oxidant, but it can be reduced by ferrous iron (Fe^{2+}) to the highly reactive hydroxyl radical (Fenton reaction), which can react with every molecule in its immediate environment to propagate the radical reaction (Figure 10.1). The Fenton reaction generally initiates the peroxidation of unsaturated fatty acids in membranes. Ferrous iron is also needed for the propagation of lipid peroxidation (LPO) by catalyzing the reduction of hydroperoxy fatty acids. Because of the destructive potential of LPO, ferric iron (Fe^{3+}) is kept tightly bound to proteins including the storage protein ferritin [3]. However, superoxide and other reductants can reductively mobilize some of the iron, which can become available to catalyze the Fenton reaction [4]. To avoid this, superoxide is removed by SODs, which include SOD1 (Cu/Zn-SOD) in the cytosol and SOD2 (Mn-SOD) in mitochondria [5]. In addition, hydrogen peroxide is detoxified by glutathione (GSH) peroxidase [6] or the peroxiredoxin/thioredoxin system [7] in the cytosol and mitochondria and potentially by catalase in peroxisomes [8]. Further ways to protect is the binding of all redox-active metals in storage proteins and the presence of chain-breaking antioxidants such as vitamin E in lipid membranes. Together, this multitier defense system is very effective in preventing ROS-mediated cell injury, especially LPO.

10.2.2
Sources of ROS and Their Pathophysiological Relevance

Low levels of superoxide are formed continuously by mitochondria in the respiratory chain. Although there is evidence that the ubiquinone–cytochrome b complex (complex III) can generate superoxide, the NADH dehydrogenase (complex I) is

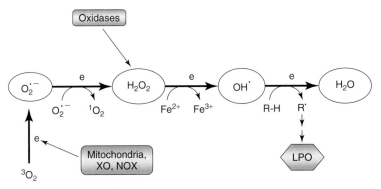

Fig. 10.1 Formation of reactive oxygen species by one-electron reduction steps (see text for details). NOX, NADPH oxidases; XO, xanthine oxidase; LPO, lipid peroxidation.

the most prominent source under normal and pathophysiological conditions [9]. Owing to the large numbers of mitochondria in individual cells and the high capacity, mitochondria are quantitatively the most important source of superoxide within liver cells. The potential danger from the continuous superoxide formation in mitochondria is illustrated by the fact that Mn-SOD-deficient mice are not viable [10], and even heterozygous mice suffer from mitochondrial dysfunction [11] and are highly susceptible to xenobiotics that target mitochondria [12]. An important aspect of mitochondria-induced superoxide formation is the fact that it is dependent on the oxygen concentration, which means that hyperoxic conditions generally lead to higher superoxide formation [13]. Other intracellular sources of superoxide can include the enzyme xanthine oxidase (XO) [14]. However, the capacity of this enzyme to produce superoxide is limited by the substrate availability. Therefore, the pathophysiological relevance of XO has been questioned [15]. The microsomal cytochrome P450 system can also release superoxide during the metabolism of xenobiotics [16]. Cyp2E1 is considered an important source of oxidant stress during the metabolism of ethanol [17]. However, no evidence for reactive oxygen formation during the metabolism of acetaminophen (APAP), another substrate for Cyp2E1, was found [18, 19]. Thus, the potential superoxide formation by Cyps may depend on the isoform of cytochrome P450 involved and on the substrate that is being metabolized – it may require chronic treatment and/or induction of the enzyme and impairment of antioxidant defense systems. Part of the P450 system is cytochrome b_5 and cytochrome P450 reductase, which generally transfer electrons from NAD(P)H to substrates of the P450 cycle. Certain chemicals, for example, paraquat or diquat, can be reduced by these reductases. The resulting unstable radical transfers the electron to molecular oxygen thereby forming superoxide and regenerating the chemical, which can be reduced again. These "redox-cycling agents" can produce large amounts of superoxide within cells causing very selective ROS-mediated liver injury [20]. Another class of compounds that can induce oxidant stress is quinones [21, 22]. In contrast to diquat toxicity, which induces oxidant stress without protein thiol alkylation [23], thiol alkylation is a prominent feature of quinone toxicity [21, 22]. Nevertheless, cell injury induced by certain quinones such as menadione is presumed to be mainly caused by superoxide formation [24].

In addition to the unintentional formation of ROS during aerobic metabolism, cells have enzymes that generate ROS as primary reaction products. A number of oxidases, for example, amino acid oxidases and fatty acid oxidases, are located in peroxisomes [25]. The oxidases generate hydrogen peroxide, which is detoxified by the high levels of catalase in peroxisomes [25]. Thus, in general, ROS generated by these oxidases do not escape the peroxisomal compartment. However, exposure to peroxisome proliferators causes massive induction of fatty acid oxidases, but only a minor increase in catalase activity [8]. Under these conditions, peroxisomes can contribute to the cellular oxidant stress. Liver cancer in rodents is attributed to this peroxisomal oxidant stress [26].

A family of enzymes generating superoxide as the primary reaction product are the NOX family of NADPH oxidases, which are transmembrane proteins that use electrons derived from NADPH on the cytosolic side and transfer them to molecular oxygen either in intracellular vesicles or to the outside of the cell [27]. There are currently seven NOX family members identified; various combinations of these enzymes are expressed in all liver cell types [27]. The best studied isoform is NOX2, which is the prototypic NADPH oxidase found in phagocytic cells (e.g., neutrophils, Kupffer cells, and other tissue macrophages). The main function of NOX2 is to generate ROS for microbial cell killing by phagocytes. Although the function of NOX enzymes beyond their role in host defense is not well understood, all liver cells expressing these enzymes are able to take up material by phagocytosis or endocytosis. The ROS generated in the phagosomes or endosomes may primarily aid in digestion of the foreign material. In addition, it is hypothesized that a low level of oxidant stress generated by these NOX enzymes may be involved in the redox regulation of gene expression in various cell types [27]. However, the NOX-derived ROS formation by Kupffer cells and neutrophils has been shown to contribute to liver injury in various pathophysiologies. This includes acute injury processes during alcohol consumption [28], endotoxemia [29], hemorrhagic shock [30], and ischemia-reperfusion injury [31]. Furthermore, NOX-induced oxidant stress may contribute to chronic liver diseases such as alcoholic and nonalcoholic steatohepatitis [32] and liver fibrosis [33]. If Kupffer cells are activated by bacterial products or activated complement factors, ROS is released into the extracellular space and then mainly hydrogen peroxide diffuses into the target cells, for example, hepatocytes and endothelial cells, and causes an intracellular oxidant stress [34]. This oxidant stress can be scavenged not only by GSH in the vascular space [35, 36] but also by intracellular antioxidant defense systems [34, 37]. Kupffer cells generate predominantly hydrogen peroxide, which is a relatively mild oxidant. As GSH released from hepatocytes can scavenge this oxidant directly, a Kupffer cell–induced oxidant stress causes no relevant tissue damage in a healthy liver [38]. In most cases where a Kupffer cell–mediated oxidant stress contributes to the pathology, additional factors are involved, including impairment of antioxidant defense systems or exposure to additional stress (e.g., ischemia) [31, 36].

10.2.3
Mechanisms of ROS-Mediated Cell Death

In contrast to some of the *in vitro* observations with cultured hepatocytes [24], excessive intracellular ROS formation *in vivo* does not cause apoptosis but triggers necrotic cell death [39]. Although c-jun N-terminal kinase (JNK) is activated by the oxidant stress *in vivo*, JNK inhibitors do not protect under these conditions [39]. In the intact liver, hepatocytes are able to effectively detoxify large amounts of ROS [19, 40]. Massive LPO as a result of excessive ROS formation is a very rare event *in vivo* and generally requires impairment of antioxidant defense systems [41, 42]. As an example, APAP overdose induces a mitochondrial oxidant stress but no LPO in normal animals [43, 44]. However, LPO is the dominant mechanism of cell death in vitamin E–deficient animals [45]. As a result, vitamin E supplementation attenuates APAP-induced liver injury only in the deficient mice but not in normal animals [44, 46]. Moderate oxidant stress more likely results in mitochondrial dysfunction and opening of the mitochondrial membrane permeability transition pore (MPT), which triggers the collapse of the membrane potential and eventually leads to necrotic cell death [47]. The MPT pore opening is preceded by NAD(P)H oxidation, mitochondrial oxidant stress, and iron mobilization and can be blocked with specific interventions at these sites [47]. More recent evidence suggests that oxidant stress causes lysosomal instability with transfer of iron into the mitochondria [48]. Although the oxidant stress can cause the MPT directly, activated JNK can also contribute to this effect, which explains why JNK inhibitors or deletion of JNK can protect against various pathophysiological conditions involving ROS formation [49, 50]. Overall, the current data suggest that massive ROS formation by extra- or intracellular sources and subsequent severe LPO, which directly causes cell death, is an event that rarely occurs during any pathophysiological conditions *in vivo*. In contrast, oxidant stress causes disturbances of cellular homeostasis, lysosomal instability, and eventually mitochondrial dysfunction. Mitochondria can propagate the original insult by generating ROS, formation of MPT pores with collapse of the mitochondrial membrane potential, swelling with release of intermembrane proteins – which can promote caspase activation leading to apoptotic cell death or translocate to the nucleus and promote DNA degradation – which together with loss of ATP formation and ion homeostasis lead to necrotic cell death [51, 52].

10.3
Reactive Nitrogen Species

10.3.1
Sources of Peroxynitrite

Nitric oxide is generated by several forms of nitric oxide synthase (NOS). The constitutively expressed endothelial nitric oxide synthase (eNOS, NOS1) and the inducible nitric oxide synthase (iNOS, NOS2) are most important for liver cells [53].

Basal formation of NO by eNOS functions mainly as a vasodilator in the vasculature [54]. Excess NO formation by iNOS can lead to systemic hypotension and shock during endotoxemia and sepsis [55]. Under some circumstances, for example, hepatic ischemia-reperfusion, when there is increased vasoconstriction due to endothelin formation and upregulation of endothelin receptors, iNOS-derived NO is critical for maintaining liver blood flow [56]. Macrophages can generate superoxide through NOX2 and NO through iNOS at the same time. Both radicals react with diffusion-limited speed to form peroxynitrite (ONOO$^-$), which is a potent oxidant and nitrating agent [57]. Peroxynitrite is used by macrophages during host defense functions because hydrogen peroxide is not very effective for bacterial killing. The problem with generating such a potent oxidant is that host cells can also be injured by peroxynitrite-induced oxidation and nitration of proteins, lipids, and DNA [58, 59]. However, the high reactivity with sulfhydryl groups also means that GSH is a potent scavenger of peroxynitrite [60]. Thus, as long as sufficient GSH is available in the cell, the detrimental impact of peroxynitrite is limited. Peroxynitrite can also be detoxified by peroxiredoxin [61] and glutathione peroxidase 1 (Gpx1) [62], although the relevance of Gpx1 for the metabolism of peroxynitrite *in vivo* has been questioned [60]. Despite evidence for a potential involvement of peroxynitrite in many pathophysiological processes [53, 58, 59, 63], the effective spontaneous reaction with GSH may be the reason why in most liver diseases peroxynitrite appears to have only a limited impact. In addition, the critical role of NO as vasodilator in the liver vasculature, especially in disease states such as ischemia-reperfusion injury, can override the potential detrimental effect of peroxynitrite [64]. Similar to reperfusion injury, alcohol-induced liver injury may involve peroxynitrite to a limited extent [65].

10.3.2
Peroxynitrite and Acetaminophen Hepatotoxicity

An example where peroxynitrite may be the most critical oxidant is APAP hepatotoxicity (Figure 10.2). Here, NO is not needed to counteract vasoconstriction [66] and GSH levels are dramatically depleted before peroxynitrite is formed in those endothelial cells and hepatocytes that eventually undergo necrosis [43, 67]. As indicated by nitrotyrosine protein adduct formation, peroxynitrite is generated inside mitochondria due to the formation of superoxide in the mitochondrial matrix [68]. Peroxynitrite is clearly a critical mediator of the toxicity [60, 69, 70] directly causing mitochondrial DNA damage [68] and opening of MPT pores, which leads to necrotic cell death [71]. Interestingly, despite the substantial oxidant stress and peroxynitrite formation in mitochondria, there is only limited loss of protein sulfhydryl groups and only a very limited impairment of mitochondrial protein function [72]. APAP-induced nuclear DNA fragmentation is caused by lysosomal DNase1 [73] and mitochondria-derived endonuclease G and apoptosis-inducing factor (AIF) [74]. Although the early release of the mitochondrial intermembrane

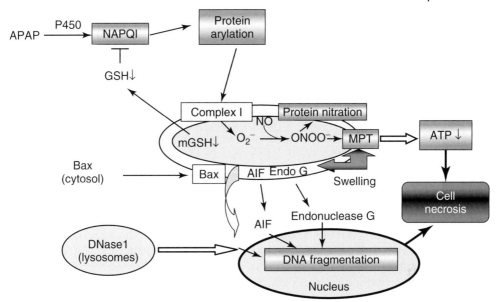

Fig. 10.2 Mechanisms of acetaminophen (APAP) hepatotoxicity. The formation of the reactive metabolite N-acetyl-p-benzoquinone imine (NAPQI) leads to cytosolic and mitochondrial GSH depletion and protein arylation, which triggers superoxide formation by components of the mitochondrial respiratory chain. Superoxide and nitric oxide (NO) form peroxynitrite (ONOO$^-$), which nitrates proteins and causes opening of the mitochondrial membrane permeability transition pore (MPT). Release of endonuclease G and apoptosis-inducing factor (AIF) is caused initially by the formation of Bax pores in the outer membrane and later by MPT-mediated mitochondrial swelling and rupture of the outer membrane. AIF and endonuclease G together with DNase1 from lysosomes are responsible for the extensive DNA fragmentation. Extensive mitochondrial dysfunction and failure and massive nuclear DNA damage are critical for necrotic cell death.

proteins and their nuclear translocation are triggered by Bax pores in the outer mitochondrial membrane, peroxynitrite-induced MPT and mitochondrial swelling are responsible for the later release of endonuclease G and AIF [75]. Despite the clear evidence for a role of peroxynitrite in APAP hepatotoxicity *in vivo*, the source of NO is controversial. Inhibitors of iNOS were shown to be protective [76, 77] or without effect [44, 78]. Similarly, eNOS or iNOS gene knockout mice showed less injury in some studies [79, 80] but not in others [81]. These data support the hypothesis that multiple sources may contribute to NO formation during APAP hepatotoxicity and that the enhanced superoxide formation mainly determines mitochondrial peroxynitrite formation. Supplying precursors for GSH synthesis, which accelerate the recovery of mitochondrial GSH levels, is the most effective way to scavenge peroxynitrite and to protect against APAP-induced liver injury [60, 69, 70].

10.4
Hypohalites

10.4.1
Sources of Hypochlorous Acid

Monocytes and in particular neutrophils contain high levels (up to 5% of total protein) of the enzyme myeloperoxidase (MPO) stored in azurophilic granules [82]. Although MPO is generally not found in tissue macrophages, there is evidence for the presence of MPO in human Kupffer cells [83]. Upon activation of phagocytes, NOX2 is assembled at the membrane of intracellular vesicles, which either fuse with phagosomes/endosomes or with the cell membrane [27]. In addition, the azurophilic granules release their contents including MPO into the same compartment [84]. NOX2-derived superoxide dismutates to hydrogen peroxide, which is used by MPO together with Cl^- to generate hypochlorous acid/hypochlorite ($HOCl/OCl^-$) [82, 85]. HOCl is a potent oxidant that can react with proteins and lipids [86]. Some of the reaction products such as chloramines are by themselves strong oxidants that can cause extensive tissue injury [87]. Chlorine gas (Cl_2) derived from HOCl can act as a chlorinating agent for cholesterol and tyrosine residues of proteins [88]. Chlorotyrosine (CT) can be used as a specific biomarker for MPO activity [89]. HOCl can also oxidize α-amino acids to reactive aldehydes, which can cause protein cross-linking and formation of advanced glycation end products (AGEs) [90, 91].

10.4.2
Hypochlorite and Host Defense Functions

MPO-derived HOCl and its secondary products are the key mediators of the antimicrobial host defense function of phagocytes. Microbial killing is caused by HOCl together with other components of the azurophilic granules such as proteases (elastase and cathepsins) and a number of proteins with direct antimicrobial activity (e.g., defensins) [84]. Although it is difficult to assess the importance of every single component released into the phagosome, the fact that animals deficient in a functional NADPH oxidase or MPO are highly susceptible to infections indicate that the MPO/HOCl system is critical for this host defense function [84]. On the other hand, there is increasing evidence that neutrophil-derived HOCl can contribute to a variety of disease processes in all organs including the liver [82, 92, 93].

10.4.3
Hypochlorite in Liver Pathophysiology

Neutrophils have been shown to contribute to liver injury during ischemia-reperfusion [94], endotoxemia [95], obstructive cholestasis [96], alcoholic hepatitis [97, 98], sepsis, [99], and various chemical toxicities including concanavalin A [100], halothane [101], and α-naphthylisothiocyanate [102, 103].

The role of neutrophils in APAP toxicity is controversial [104]. Neutrophil-induced liver injury starts with the recruitment into the liver vasculature. In contrast to many other organs, neutrophils relevant for the injury accumulate in sinusoids (capillaries) and not in postsinusoidal venules [105]. A large number of inflammatory mediators can trigger neutrophil recruitment into the liver including cytokines, CXC chemokines, complement factors, and platelet activating factor [106]. However, neutrophils in the vasculature are generally primed for ROS formation but not fully activated and therefore do not cause liver injury [107, 108]. For injury to occur, neutrophils need to transmigrate and adhere to hepatocytes [105]. The transmigration process, which depends on adhesion molecules on neutrophils and endothelial cells [109], requires a distress signal from outside the vasculature [93]. The adhesion to the target cell triggers full activation with degranulation and initiation of a long-lasting oxidant stress through NOX2 [92]. Direct evidence for a neutrophil-mediated oxidant stress was obtained during the later phase of ischemia-reperfusion injury [110, 111], endotoxemia [29, 37], and obstructive cholestasis [96, 112]. This oxidant stress depends on the transmigration and the β_2-integrin-mediated adhesion to the target [96, 112, 113]. Interestingly, despite the fact that the ROS are released in the extracellular space, albeit in close proximity to the target, there is clear evidence for an intracellular oxidant stress in the target cells (Figure 10.3). This includes increased GSSG formation as indicator for hydrogen peroxide [37] and immunostaining for CT [29, 96, 112] and hypochlorous acid–modified proteins as indicator for hypochlorite generation *in vivo* [111, 114]. These data suggest that neutrophil-derived oxidants diffused into the target cells and created an intracellular oxidant stress, which was mainly responsible for the injury [93]. Although proteases are important for the injury process *in vivo* [106], their major role may be to facilitate the transmigration process and potentially processing of proinflammatory mediators rather than causing injury directly [93].

The mechanisms of hypochlorous acid–mediated injury have been investigated in various cultured cells but not in hepatocytes. HOCl-mediated effects include dose-dependent oxidation of sulfhydryl groups, which causes inactivation of plasma membrane transporters and enzymes leading to loss of ion homeostasis in P388D1 cells [115]. HOCl can also trigger mitochondrial Bax translocation with release of intermembrane proteins such as AIF and endonuclease G, which cause nuclear DNA damage in chondrocytes [116]. On the other hand, HOCl triggered calpain activation, lysosomal rupture, and cathepsin release in cortical neurons [117]. HOCl appears to effectively inactivate caspases and therefore does not induce apoptosis [116]. Because many of the effects are clearly dependent on the cell type, the dose, and other experimental conditions, these mechanisms cannot be directly extrapolated to hepatocytes. However, similar to other ROS, HOCl is unlikely to kill directly by massive oxidant stress but is more likely disturbing intracellular signaling mechanisms leading to cell death. In this respect, it was demonstrated that activation of JNK [118], lysosomal release of cathepsins [119], activation of calpains [120], and mitochondrial dysfunction [121–123] are critical events in hepatocytes during the neutrophilic inflammatory response after ischemia. As with

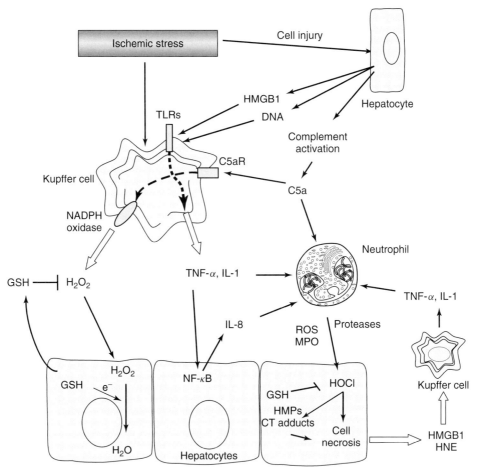

Fig. 10.3 Neutrophil-mediated hypochlorous acid formation during hepatic ischemia-reperfusion injury. The inflammatory response is initiated by early cell injury, which causes release of intracellular proteins, for example, high mobility group box protein 1, and DNA fragments. These cellular components trigger complement activation in plasma and the activation of Kupffer cells. While activation of the complement receptor (C5aR) triggers reactive oxygen formation by NADPH oxidase, stimulation of toll-like receptors leads to cytokine formation, especially tumor necrosis factor-α (TNF-α) and interleukin-1 (IL-1). Cytokines, chemokines (e.g., IL-8), and complement factors (C5a) induce neutrophil activation and recruitment into the liver. After extravasation, the neutrophil generates ROS and releases proteases and myeloperoxidase (MPO), which generates hypochlorous acid (HOCl). HOCl diffuses into the target cells and creates an intracellular stress as indicated by chlorotyrosine (CT) protein adducts and hypochlorous acid–modified proteins (HMPs). The oxidant stress triggers cell necrosis, which is responsible for release of cell contents (e.g., HMGB1) and lipid peroxidation products (HNE, hydroxynonenal). These injury products further activate Kupffer cells and amplify the inflammatory response and neutrophil-mediated cell killing.

all *in vivo* effects, it needs to be kept in mind that these signaling mechanisms can reflect more than just the response to a neutrophil-mediated oxidant stress.

10.5
Summary

Various ROS, hypochlorite, and peroxynitrite are critical mediators of liver injury under a variety of pathophysiological conditions. However, as with all disease processes, a bigger effort needs to be made to better define under what circumstances and at what time during the disease these oxidants are being generated, to identify the specific oxidants involved, to locate the source(s) where they originate from, to identify the cellular compartment(s) that are affected, and to identify the potential targets of these highly reactive molecules. Only if this information is available for the animal model and if it can be assumed with a certain degree of confidence that this also happens in the human disease process, can specific interventions be targeted toward these reactive mediators. However, it needs to be kept in mind that all of these endogenously generated reactive species, hypochlorite, and peroxynitrite are essential for the vital host defense functions of phagocytes and some of them may also be involved in cellular signal transduction mechanisms. Potentially successful intervention strategies need to consider both beneficial and detrimental effects of these reactive intermediates and selectively target only the damaging effects.

References

1. Sies, H. (1993) Strategies of antioxidant defense. *Eur J Biochem*, **215**, 213–219.
2. Jaeschke H. (2007) Antioxidant defense in liver injury: oxidative stress, antioxidant defense and liver injury, in *Drug-induced Liver Disease*, 2nd edn (eds N. Kaplowitz and L.D. DeLeve), Marcel Dekker, Inc., New York, pp. 33–48.
3. Theil, E.C. (2003) Ferritin: at the crossroads of iron and oxygen metabolism. *J Nutr*, **133** (Suppl 1), 1549S–1553S.
4. Reif, D.W. (1992) Ferritin as a source of iron for oxidative damage. *Free Radic Biol Med*, **12**, 417–427.
5. Slot, J.W., Geuze, H.J., Freeman, B.A., and Crapo, J.D. (1986) Intracellular localization of the copper-zinc and manganese superoxide dismutases in rat liver parenchymal cells. *Lab Invest*, **55**, 363–371.
6. Brigelius-Flohé, R. (2006) Glutathione peroxidases and redox-regulated transcription factors. *Biol Chem*, **387**, 1329–1335.
7. Rhee, S.G., Chae, H.Z., and Kim, K. (2005) Peroxiredoxins: a historical overview and speculative preview of novel mechanisms and emerging concepts in cell signaling. *Free Radic Biol Med*, **38**, 1543–1552.
8. Calabrese, E.J. and Canada, A.T. (1989) Catalase: its role in xenobiotic detoxification. *Pharmacol Ther*, **44**, 297–307.
9. Murphy, M.P. (2009) How mitochondria produce reactive oxygen species. *Biochem J*, **417**, 1–13.
10. Li, Y., Huang, T.T., Carlson, E.J., Melov, S., Ursell, P.C., Olson, J.L., Noble, L.J., Yoshimura, M.P., Berger, C., Chan, P.H., Wallace, D.C., and Epstein, C.J. (1995) Dilated cardiomyopathy and neonatal lethality

11 Williams, M.D., Van Remmen, H., Conrad, C.C., Huang, T.T., Epstein, C.J., and Richardson, A. (1998) Increased oxidative damage is correlated to altered mitochondrial function in heterozygous manganese superoxide dismutase knockout mice. *J Biol Chem*, **273**, 28510–28515.

12 Ong, M.M., Wang, A.S., Leow, K.Y., Khoo, Y.M., and Boelsterli, U.A. (2006) Nimesulide-induced hepatic mitochondrial injury in heterozygous Sod2(+/−) mice. *Free Radic Biol Med*, **40**, 420–429.

13 Kussmaul, L. and Hirst, J. (2006) The mechanism of superoxide production by NADH:ubiquinone oxidoreductase (complex I) from bovine heart mitochondria. *Proc Natl Acad Sci U S A*, **103**, 7607–7612.

14 Kooij, A. (1994) A re-evaluation of the tissue distribution and physiology of xanthine oxidoreductase. *Histochem J*, **26**, 889–915.

15 Jaeschke, H. (2002) Xanthine oxidase-induced oxidant stress during hepatic ischemia-reperfusion: are we coming full circle after 20 years? *Hepatology*, **36**, 761–763.

16 Coon, M.J. (2005) Cytochrome P450: nature's most versatile biological catalyst. *Annu Rev Pharmacol Toxicol*, **45**, 1–25.

17 Lu, Y. and Cederbaum, A.I. (2008) CYP2E1 and oxidative liver injury by alcohol. *Free Radic Biol Med*, **44**, 723–738.

18 Lauterburg, B.H., Smith, C.V., Hughes, H., and Mitchell, J.R. (1984) Biliary excretion of glutathione and glutathione disulfide in the rat. Regulation and response to oxidative stress. *J Clin Invest*, **73**, 124–133.

19 Smith, C.V. and Jaeschke, H. (1989) Effect of acetaminophen on hepatic content and biliary efflux of glutathione disulfide in mice. *Chem Biol Interact*, **70**, 241–248.

20 Smith, C.V., Hughes, H., Lauterburg, B.H., and Mitchell, J.R. (1985) in mutant mice lacking manganese superoxide dismutase. *Nat Genet*, **11**, 376–381.

Oxidant stress and hepatic necrosis in rats treated with diquat. *J Pharmacol Exp Ther*, **235**, 172–177.

21 O'Brien, P.J. (1991) Molecular mechanisms of quinone cytotoxicity. *Chem Biol Interact*, **80**, 1–41.

22 Bolton, J.L., Trush, M.A., Penning, T.M., Dryhurst, G., and Monks, T.J. (2000) Role of quinones in toxicology. *Chem Res Toxicol*, **13**, 135–160.

23 Smith, C.V. (1987) Effect of BCNU pretreatment on diquat-induced oxidant stress and hepatotoxicity. *Biochem Biophys Res Commun*, **144**, 415–421.

24 Conde de la Rosa, L., Schoemaker, M.H., Vrenken, T.E., Buist-Homan, M., Havinga, R., Jansen, P.L., and Moshage, H. (2006) Superoxide anions and hydrogen peroxide induce hepatocyte death by different mechanisms: involvement of JNK and ERK MAP kinases. *J Hepatol*, **44**, 918–929.

25 Schrader, M. and Fahimi, H.D. (2006) Peroxisomes and oxidative stress. *Biochim Biophys Acta*, **1763**, 1755–1766.

26 Yu, S., Rao, S., and Reddy, J.K. (2003) Peroxisome proliferator-activated receptors, fatty acid oxidation, steatohepatitis and hepatocarcinogenesis. *Curr Mol Med*, **3**, 561–572.

27 Bedard, K. and Krause, K.H. (2007) The NOX family of ROS-generating NADPH oxidases: physiology and pathophysiology. *Physiol Rev*, **87**, 245–313.

28 Kono, H., Rusyn, I., Yin, M., Gäbele, E., Yamashina, S., Dikalova, A., Kadiiska, M.B., Connor, H.D., Mason, R.P., Segal, B.H., Bradford, B.U., Holland, S.M., and Thurman, R.G. (2000) NADPH oxidase-derived free radicals are key oxidants in alcohol-induced liver disease. *J Clin Invest*, **106**, 867–872.

29 Gujral, J.S., Hinson, J.A., Farhood, A., and Jaeschke, H. (2004) NADPH oxidase-derived oxidant stress is critical for neutrophil cytotoxicity

during endotoxemia. *Am J Physiol Gastrointest Liver Physiol*, **287**, G243–G252.
30. Lehnert, M., Arteel, G.E., Smutney, O.M., Conzelmann, L.O., Zhong, Z., Thurman, R.G., and Lemasters, J.J. (2003) Dependence of liver injury after hemorrhage/resuscitation in mice on NADPH oxidase-derived superoxide. *Shock*, **19**, 345–351.
31. Jaeschke, H. (2003) Molecular mechanisms of hepatic ischemia-reperfusion injury and preconditioning. *Am J Physiol Gastrointest Liver Physiol*, **284**, G15–G26.
32. De Minicis, S. and Brenner, D.A. (2008) Oxidative stress in alcoholic liver disease: role of NADPH oxidase complex. *J Gastroenterol Hepatol*, **23** (Suppl 1), S98–103.
33. De Minicis, S. and Brenner, D.A. (2007) NOX in liver fibrosis. *Arch Biochem Biophys*, **462**, 266–272.
34. Bilzer, M., Jaeschke, H., Vollmar, A.M., Paumgartner, G., and Gerbes, A.L. (1999) Prevention of Kupffer cell-induced oxidant injury in rat liver by atrial natriuretic peptide. *Am J Physiol*, **276**, G1137–G1144.
35. Jaeschke, H. and Farhood, A. (1991) Neutrophil and Kupffer cell-induced oxidant stress and ischemia-reperfusion injury in rat liver. *Am J Physiol*, **260**, G355–G362.
36. Liu, P., McGuire, G.M., Fisher, M.A., Farhood, A., Smith, C.W., and Jaeschke, H. (1995) Activation of Kupffer cells and neutrophils for reactive oxygen formation is responsible for endotoxin-enhanced liver injury after hepatic ischemia. *Shock*, **3**, 56–62.
37. Jaeschke, H., Ho, Y.S., Fisher, M.A., Lawson, J.A., and Farhood, A. (1999) Glutathione peroxidase-deficient mice are more susceptible to neutrophil-mediated hepatic parenchymal cell injury during endotoxemia: importance of an intracellular oxidant stress. *Hepatology*, **29**, 443–450.
38. Jaeschke, H. (1992) Enhanced sinusoidal glutathione efflux during endotoxin-induced oxidant stress in vivo. *Am J Physiol*, **263**, G60–G68.
39. Hong, J.Y., Lebofsky, M., Farhood, A., and Jaeschke, H. (2009) Oxidant stress-induced liver injury in vivo: role of apoptosis, oncotic necrosis and c-jun N-terminal kinase activation. *Am J Physiol Gastrointest Liver Physiol*, **296**, G572–G581.
40. Jaeschke, H., Smith, C.V., and Mitchell, J.R. (1988) Reactive oxygen species during ischemia-reflow injury in isolated perfused rat liver. *J Clin Invest*, **81**, 1240–1246.
41. Jaeschke, H. and Benzick, A.E. (1992) Pathophysiological consequences of enhanced intracellular superoxide formation in isolated perfused rat liver. *Chem Biol Interact*, **84**, 55–68.
42. Mathews, W.R., Guido, D.M., Fisher, M.A., and Jaeschke, H. (1994) Lipid peroxidation as molecular mechanism of liver cell injury during reperfusion after ischemia. *Free Radic Biol Med*, **16**, 763–770.
43. Knight, T.R., Kurtz, A., Bajt, M.L., Hinson, J.A., and Jaeschke, H. (2001) Vascular and hepatocellular peroxynitrite formation during acetaminophen toxicity: role of mitochondrial oxidant stress. *Toxicol Sci*, **62**, 212–220.
44. Knight, T.R., Fariss, M.W., Farhood, A., and Jaeschke, H. (2003) Role of lipid peroxidation as a mechanism of liver injury after acetaminophen overdose in mice. *Toxicol Sci*, **76**, 229–236.
45. Wendel, A. and Feuerstein, S. (1981) Drug-induced lipid peroxidation in mice–I. Modulation by monooxygenase activity, glutathione and selenium status. *Biochem Pharmacol*, **30**, 2513–2520.
46. Jaeschke, H., Knight, T.R., and Bajt, M.L. (2003) The role of oxidant stress and reactive nitrogen species in acetaminophen hepatotoxicity. *Toxicol Lett*, **144**, 279–288.
47. Nieminen, A.L., Byrne, A.M., Herman, B., and Lemasters, J.J. (1997) Mitochondrial permeability transition in hepatocytes induced by

t-BuOOH: NAD(P)H and reactive oxygen species. *Am J Physiol*, **272**, C1286–C1294.

48 Uchiyama, A., Kim, J.S., Kon, K., Jaeschke, H., Ikejima, K., Watanabe, S., and Lemasters, J.J. (2008) Translocation of iron from lysosomes into mitochondria is a key event during oxidative stress-induced hepatocellular injury. *Hepatology*, **48**, 1644–1654.

49 Hanawa, N., Shinohara, M., Saberi, B., Gaarde, W.A., Han, D., and Kaplowitz, N. (2008) Role of JNK translocation to mitochondria leading to inhibition of mitochondria bioenergetics in acetaminophen-induced liver injury. *J Biol Chem*, **283**, 13565–13577.

50 Theruvath, T.P., Snoddy, M.C., Zhong, Z., and Lemasters, J.J. (2008) Mitochondrial permeability transition in liver ischemia and reperfusion: role of c-Jun N-terminal kinase 2. *Transplantation*, **85**, 1500–1504.

51 Jaeschke, H., Gores, G.J., Cederbaum, A.I., Hinson, J.A., Pessayre, D., and Lemasters, J.J. (2002) Mechanisms of hepatotoxicity. *Toxicol Sci*, **65**, 166–176.

52 Lemasters, J.J., Qian, T., He, L., Kim, J.S., Elmore, S.P., Cascio, W.E., and Brenner, D.A. (2002) Role of mitochondrial inner membrane permeabilization in necrotic cell death, apoptosis, and autophagy. *Antioxid Redox Signal*, **4**, 769–781.

53 Pacher, P., Beckman, J.S., and Liaudet, L. (2007) Nitric oxide and peroxynitrite in health and disease. *Physiol Rev*, **87**, 315–424.

54 Nishida, J., McCuskey, R.S., McDonnell, D., and Fox, E.S. (1994) Protective role of NO in hepatic microcirculatory dysfunction during endotoxemia. *Am J Physiol*, **267**, G1135–G1141.

55 Pullamsetti, S.S., Maring, D., Ghofrani, H.A., Mayer, K., Weissmann, N., Rosengarten, B., Lehner, M., Schudt, C., Boer, R., Grimminger, F., Seeger, W., and Schermuly, R.T. (2006) Effect of nitric oxide synthase (NOS) inhibition on macro- and microcirculation in a model of rat endotoxic shock. *Thromb Haemost*, **95**, 720–727.

56 Wang, Y., Lawson, J.A., and Jaeschke, H. (1998) Differential effect of 2-aminoethyl-isothiourea, an inhibitor of the inducible nitric oxide synthase, on microvascular blood flow and organ injury in models of hepatic ischemia-reperfusion and endotoxemia. *Shock*, **10**, 20–25.

57 Squadrito, G.L. and Pryor, W.A. (1998) Oxidative chemistry of nitric oxide: the roles of superoxide, peroxynitrite, and carbon dioxide. *Free Radic Biol Med*, **25**, 392–403.

58 Denicola, A. and Radi, R. (2005) Peroxynitrite and drug-dependent toxicity. *Toxicology*, **208**, 273–288.

59 Radi, R., Cassina, A., Hodara, R., Quijano, C., and Castro, L. (2002) Peroxynitrite reactions and formation in mitochondria. *Free Radic Biol Med*, **33**, 1451–1464.

60 Knight, T.R., Ho, Y.S., Farhood, A., and Jaeschke, H. (2002) Peroxynitrite is a critical mediator of acetaminophen hepatotoxicity in murine livers: protection by glutathione. *J Pharmacol Exp Ther*, **303**, 468–475.

61 Trujillo, M., Ferrer-Sueta, G., and Radi, R. (2008) Kinetic studies on peroxynitrite reduction by peroxiredoxins. *Methods Enzymol*, **441**, 173–196.

62 Sies, H., Sharov, V.S., Klotz, L.O., and Briviba, K. (1997) Glutathione peroxidase protects against peroxynitrite-mediated oxidations. A new function for selenoproteins as peroxynitrite reductase. *J Biol Chem*, **272**, 27812–27817.

63 Szabó, C., Ischiropoulos, H., and Radi, R. (2007) Peroxynitrite: biochemistry, pathophysiology and development of therapeutics. *Nat Rev Drug Discov*, **6**, 662–680.

64 Wang, Y., Mathews, W.R., Guido, D.M., Farhood, A., and Jaeschke, H. (1995) Inhibition of nitric oxide synthesis aggravates reperfusion injury after hepatic ischemia and endotoxemia. *Shock*, **4**, 282–288.

65 Arteel, G.E. (2003) Oxidants and antioxidants in alcohol-induced liver disease. *Gastroenterology*, **124**, 778–790.

66 Ito, Y., Bethea, N.W., Abril, E.R., and McCuskey, R.S. (2003) Early hepatic microvascular injury in response to acetaminophen toxicity. *Microcirculation*, **10**, 391–400.

67 Hinson, J.A., Pike, S.L., Pumford, N.R., and Mayeux, P.R. (1998) Nitrotyrosine-protein adducts in hepatic centrilobular areas following toxic doses of acetaminophen in mice. *Chem Res Toxicol*, **11**, 604–607.

68 Cover, C., Mansouri, A., Knight, T.R., Bajt, M.L., Lemasters, J.J., Pessayre, D., and Jaeschke, H. (2005) Peroxynitrite-induced mitochondrial and endonuclease-mediated nuclear DNA damage in acetaminophen hepatotoxicity. *J Pharmacol Exp Ther*, **315**, 879–887.

69 Bajt, M.L., Knight, T.R., Farhood, A., and Jaeschke, H. (2003) Scavenging peroxynitrite with glutathione promotes regeneration and enhances survival during acetaminophen-induced liver injury in mice. *J Pharmacol Exp Ther*, **307**, 67–73.

70 James, L.P., McCullough, S.S., Lamps, L.W., and Hinson, J.A. (2003) Effect of N-acetylcysteine on acetaminophen toxicity in mice: relationship to reactive nitrogen and cytokine formation. *Toxicol Sci*, **75**, 458–467.

71 Kon, K., Kim, J.S., Jaeschke, H., and Lemasters, J.J. (2004) Mitochondrial permeability transition in acetaminophen-induced necrosis and apoptosis of cultured mouse hepatocytes. *Hepatology*, **40**, 1170–1179.

72 Andringa, K.K., Bajt, M.L., Jaeschke, H., and Bailey, S.M. (2008) Mitochondrial protein thiol modifications in acetaminophen hepatotoxicity: effect on HMG-CoA synthase. *Toxicol Lett*, **177**, 188–197.

73 Napirei, M., Basnakian, A.G., Apostolov, E.O., and Mannherz, H.G. (2006) Deoxyribonuclease 1 aggravates acetaminophen-induced liver necrosis in male CD-1 mice. *Hepatology*, **43**, 297–305.

74 Bajt, M.L., Cover, C., Lemasters, J.J., and Jaeschke, H. (2006) Nuclear translocation of endonuclease G and apoptosis-inducing factor during acetaminophen-induced liver cell injury. *Toxicol Sci*, **94**, 217–225.

75 Bajt, M.L., Farhood, A., Lemasters, J.J., and Jaeschke, H. (2008) Mitochondrial bax translocation accelerates DNA fragmentation and cell necrosis in a murine model of acetaminophen hepatotoxicity. *J Pharmacol Exp Ther*, **324**, 8–14.

76 Gardner, C.R., Heck, D.E., Yang, C.S., Thomas, P.E., Zhang, X.J., DeGeorge, G.L., Laskin, J.D., and Laskin, D.L. (1998) Role of nitric oxide in acetaminophen-induced hepatotoxicity in the rat. *Hepatology*, **27**, 748–754.

77 Kamanaka, Y., Kawabata, A., Matsuya, H., Taga, C., Sekiguchi, F., and Kawao, N. (2003) Effect of a potent iNOS inhibitor (ONO-1714) on acetaminophen-induced hepatotoxicity in the rat. *Life Sci*, **74**, 793–802.

78 Hinson, J.A., Bucci, T.J., Irwin, L.K., Michael, S.L., and Mayeux, P.R. (2002) Effect of inhibitors of nitric oxide synthase on acetaminophen-induced hepatotoxicity in mice. *Nitric Oxide*, **6**, 160–167.

79 Gardner, C.R., Laskin, J.D., Dambach, D.M., Sacco, M., Durham, S.K., Bruno, M.K., Cohen, S.D., Gordon, M.K., Gerecke, D.R., Zhou, P., and Laskin, D.L. (2002) Reduced hepatotoxicity of acetaminophen in mice lacking inducible nitric oxide synthase: potential role of tumor necrosis factor-α and interleukin-10. *Toxicol Appl Pharmacol*, **184**, 27–36.

80 Salhanick, S.D., Orlow, D., Holt, D.E., Pavlides, S., Reenstra, W., and Buras, J.A. (2006) Endothelially derived nitric oxide affects the severity of early acetaminophen-induced

hepatic injury in mice. *Acad Emerg Med*, **13**, 479–485.

81. Michael, S.L., Mayeux, P.R., Bucci, T.J., Warbritton, A.R., Irwin, L.K., Pumford, N.R., and Hinson, J.A. (2001) Acetaminophen-induced hepatotoxicity in mice lacking inducible nitric oxide synthase activity. *Nitric Oxide*, **5**, 432–441.

82. Malle, E., Buch, T., and Grone, H.J. (2003) Myeloperoxidase in kidney disease. *Kidney Int*, **64**, 1956–1967.

83. Brown, K.E., Brunt, E.M., and Heinecke, J.W. (2001) Immunohistochemical detection of myeloperoxidase and its oxidation products in Kupffer cells of human liver. *Am J Pathol*, **159**, 2081–2088.

84. Nauseef, W.M. (2007) How human neutrophils kill and degrade microbes: an integrated view. *Immunol Rev*, **219**, 88–102.

85. Winterbourn, C.C., Vissers, M.C., and Kettle, A.J. (2000) Myeloperoxidase. *Curr Opin Hematol*, **7**, 53–58.

86. Pitt, A.R. and Spickett, C.M. (2008) Mass spectrometric analysis of HOCl- and free-radical-induced damage to lipids and proteins. *Biochem Soc Trans*, **36**, 1077–1082.

87. Bilzer, M. and Lauterburg, B.H. (1991) Effects of hypochlorous acid and chloramines on vascular resistance, cell integrity, and biliary glutathione disulfide in the perfused rat liver: modulation by glutathione. *J Hepatol*, **13**, 84–89.

88. Hazen, S.L., d'Avignon, A., Anderson, M.M., Hsu, F.F., and Heinecke, J.W. (1998) Human neutrophils employ the myeloperoxidase-hydrogen peroxide-chloride system to oxidize alpha-amino acids to a family of reactive aldehydes. Mechanistic studies identifying labile intermediates along the reaction pathway. *J Biol Chem*, **273**, 4997–5005.

89. Winterbourn, C.C. and Kettle, A.J. (2000) Biomarkers of myeloperoxidase-derived hypochlorous acid. *Free Radic Biol Med*, **29**, 403–409.

90. Hazen, S.L., Hsu, F.F., d'Avignon, A., and Heinecke, J.W. (1998) Human neutrophils employ myeloperoxidase to convert alpha-amino acids to a battery of reactive aldehydes: a pathway for aldehyde generation at sites of inflammation. *Biochemistry*, **37**, 6864–6873.

91. Anderson, M.M., Requena, J.R., Crowley, J.R., Thorpe, S.R., and Heinecke, J.W. (1999) The myeloperoxidase system of human phagocytes generates Nepsilon-(carboxymethyl)lysine on proteins: a mechanism for producing advanced glycation end products at sites of inflammation. *J Clin Invest*, **104**, 103–113.

92. Jaeschke, H. and Hasegawa, T. (2006) Role of neutrophils in acute inflammatory liver injury. *Liver Int*, **26**, 912–919.

93. Jaeschke, H. (2006) Mechanisms of Liver Injury. II. Mechanisms of neutrophil-induced liver cell injury during hepatic ischemia-reperfusion and other acute inflammatory conditions. *Am J Physiol Gastrointest Liver Physiol*, **290**, G1083–G1088.

94. Jaeschke, H., Farhood, A., and Smith, C.W. (1990) Neutrophils contribute to ischemia/reperfusion injury in rat liver in vivo. *FASEB J*, **4**, 3355–3359.

95. Jaeschke, H., Farhood, A., and Smith, C.W. (1991) Neutrophil-induced liver cell injury in endotoxin shock is a CD11b/CD18-dependent mechanism. *Am J Physiol*, **261**, G1051–G1056.

96. Gujral, J.S., Farhood, A., Bajt, M.L., and Jaeschke, H. (2003) Neutrophils aggravate acute liver injury during obstructive cholestasis in bile duct-ligated mice. *Hepatology*, **38**, 355–363.

97. Bautista, A.P. (2002) Neutrophilic infiltration in alcoholic hepatitis. *Alcohol*, **27**, 17–21.

98. Jaeschke, H. (2002) Neutrophil-mediated tissue injury in alcoholic hepatitis. *Alcohol*, **27**, 23–27.

99 Molnar, R.G., Wang, P., Ayala, A., Ganey, P.E., Roth, R.A., and Chaudry, I.H. (1997) The role of neutrophils in producing hepatocellular dysfunction during the hyperdynamic stage of sepsis in rats. *J Surg Res*, **73**, 117–122.

100 Bonder, C.S., Ajuebor, M.N., Zbytnuik, L.D., Kubes, P., and Swain, M.G. (2004) Essential role for neutrophil recruitment to the liver in concanavalin A-induced hepatitis. *J Immunol*, **172**, 45–53.

101 You, Q., Cheng, L., Reilly, T.P., Wegmann, D., and Ju, C. (2006) Role of neutrophils in a mouse model of halothane-induced liver injury. *Hepatology*, **44**, 1421–1431.

102 Dahm, L.J., Schultze, A.E., and Roth, R.A. (1991) An antibody to neutrophils attenuates alpha-naphthylisothiocyanate-induced liver injury. *J Pharmacol Exp Ther*, **256**, 412–420.

103 Kodali, P., Wu, P., Lahiji, P.A., Brown, E.J., and Maher, J.J. (2006) ANIT toxicity toward mouse hepatocytes in vivo is mediated primarily by neutrophils via CD18. *Am J Physiol Gastrointest Liver Physiol*, **291**, G355–G363.

104 Jaeschke, H. (2008) Innate immunity and acetaminophen-induced liver injury: why so many controversies? *Hepatology*, **48**, 699–701.

105 Chosay, J.G., Essani, N.A., Dunn, C.J., and Jaeschke, H. (1997) Neutrophil margination and extravasation in sinusoids and venules of liver during endotoxin-induced injury. *Am J Physiol*, **272**, G1195–G1200.

106 Jaeschke, H. and Smith, C.W. (1997) Mechanisms of neutrophil-induced parenchymal cell injury. *J Leukoc Biol*, **61**, 647–653.

107 Jaeschke, H., Bautista, A.P., Spolarics, Z., and Spitzer, J.J. (1991) Superoxide generation by Kupffer cells and priming of neutrophils during reperfusion after hepatic ischemia. *Free Radic Res Commun*, **15**, 277–284.

108 Jaeschke, H., Farhood, A., Bautista, A.P., Spolarics, Z., and Spitzer, J.J. (1993) Complement activates Kupffer cells and neutrophils during reperfusion after hepatic ischemia. *Am J Physiol*, **264**, G801–G809.

109 Jaeschke, H. (1997) Cellular adhesion molecules: regulation and functional significance in the pathogenesis of liver diseases. *Am J Physiol*, **273**, G602–G611.

110 Jaeschke, H., Bautista, A.P., Spolarics, Z., and Spitzer, J.J. (1992) Superoxide generation by neutrophils and Kupffer cells during in vivo reperfusion after hepatic ischemia in rats. *J Leukoc Biol*, **52**, 377–382.

111 Hasegawa, T., Malle, E., Farhood, A., and Jaeschke, H. (2005) Generation of hypochlorite-modified proteins by neutrophils during ischemia-reperfusion injury in rat liver: attenuation by ischemic preconditioning. *Am J Physiol Gastrointest Liver Physiol*, **289**, G760–G767.

112 Gujral, J.S., Liu, J., Farhood, A., Hinson, J.A., and Jaeschke, H. (2004) Functional importance of ICAM-1 in the mechanism of neutrophil-induced liver injury in bile duct-ligated mice. *Am J Physiol Gastrointest Liver Physiol*, **286**, G499–G507.

113 Jaeschke, H., Farhood, A., Bautista, A.P., Spolarics, Z., Spitzer, J.J., and Smith, C.W. (1993) Functional inactivation of neutrophils with a Mac-1 (CD11b/CD18) monoclonal antibody protects against ischemia-reperfusion injury in rat liver. *Hepatology*, **17**, 915–923.

114 Hasegawa, T., Ito, Y., Wijeweera, J., Liu, J., Malle, E., Farhood, A., McCuskey, R.S., and Jaeschke, H. (2007) Reduced inflammatory response and increased microcirculatory disturbances during hepatic ischemia-reperfusion injury in steatotic livers of ob/ob mice. *Am J Physiol Gastrointest Liver Physiol*, **292**, G1385–G1395.

115 Schraufstätter, I.U., Browne, K., Harris, A., Hyslop, P.A., Jackson, J.H., Quehenberger, O., and Cochrane, C.G. (1990) Mechanisms

of hypochlorite injury of target cells. *J Clin Invest*, **85**, 554–562.
116 Whiteman, M., Chu, S.H., Siau, J.L., Rose, P., Sabapathy, K., Schantz, J.T., Cheung, N.S., Spencer, J.P., and Armstrong, J.S. (2007) The pro-inflammatory oxidant hypochlorous acid induces Bax-dependent mitochondrial permeabilisation and cell death through AIF-/EndoG-dependent pathways. *Cell Signal*, **19**, 705–714.
117 Yap, Y.W., Whiteman, M., Bay, B.H., Li, Y., Sheu, F.S., Qi, R.Z., Tan, C.H., and Cheung, N.S. (2006) Hypochlorous acid induces apoptosis of cultured cortical neurons through activation of calpains and rupture of lysosomes. *J Neurochem*, **98**, 1597–1609.
118 Uehara, T., Bennett, B., Sakata, S.T., Satoh, Y., Bilter, G.K., Westwick, J.K., and Brenner, D.A. (2005) JNK mediates hepatic ischemia reperfusion injury. *J Hepatol*, **42**, 850–859.
119 Baskin-Bey, E.S., Canbay, A., Bronk, S.F., Werneburg, N., Guicciardi, M.E., Nyberg, S.L., and Gores, G.J. (2005) Cathepsin B inactivation attenuates hepatocyte apoptosis and liver damage in steatotic livers after cold ischemia-warm reperfusion injury. *Am J Physiol Gastrointest Liver Physiol*, **288**, G396–G402.
120 Kohli, V., Gao, W., Camargo, C.A. Jr, and Clavien, P.A. (1997) Calpain is a mediator of preservation-reperfusion injury in rat liver transplantation. *Proc Natl Acad Sci U S A*, **94**, 9354–9359.
121 Elimadi, A., Settaf, A., Morin, D., Sapena, R., Lamchouri, F., Cherrah, Y., and Tillement, J.P. (1998) Trimetazidine counteracts the hepatic injury associated with ischemia-reperfusion by preserving mitochondrial function. *J Pharmacol Exp Ther*, **286**, 23–28.
122 Theruvath, T.P., Zhong, Z., Pediaditakis, P., Ramshesh, V.K., Currin, R.T., Tikunov, A., Holmuhamedov, E., and Lemasters, J.J. (2008) Minocycline and N-methyl-4-isoleucine cyclosporin (NIM811) mitigate storage/reperfusion injury after rat liver transplantation through suppression of the mitochondrial permeability transition. *Hepatology*, **47**, 236–246.
123 Eismann, T., Huber, N., Shin, T., Kuboki, S., Galloway, E., Wyder, M., Edwards, M.J., Greis, K.D., Shertzer, H.G., Fisher, A.B., and Lentsch, A.B. (2009) Peroxiredoxin-6 protects against mitochondrial dysfunction and liver injury during ischemia-reperfusion in mice. *Am J Physiol Gastrointest Liver Physiol*, **296**, G266–G274.

Part Two
Genetics: Endogenous Toxins Associated with Inborn Errors of Metabolism

11
Oxalate and Primary Hyperoxaluria
Christopher J. Danpure

Oxalate is a potentially life-threatening end product of metabolism. Too much oxalate absorbed from the diet or synthesized within the body can lead to the deposition of insoluble calcium oxalate in the kidney and urinary tract. This can cause kidney dysfunction and, in severe cases, end-stage renal disease. Although idiopathic calcium oxalate kidney stone disease is common, well-characterized monogenic disorders caused by increased oxalate synthesis and excretion are rare. Two of the best studied autosomal recessive calcium oxalate kidney stone diseases are primary hyperoxaluria type 1 (PH1) and type 2 (PH2), which are caused by mutations in the genes encoding the intermediary metabolic enzymes alanine:glyoxylate aminotransferase (AGT) and glyoxylate reductase/hydroxypyruvate reductase (GRHPR), respectively. Elucidation of the molecular etiology of PH1 and PH2 has gone a long way in explaining their pathophysiology. It has also had a major effect on their clinical management. Numerous mutations have been identified in PH1, and many can be related to specific enzyme phenotypes, such as loss of AGT catalytic activity, AGT aggregation, and accelerated degradation, and, most surprisingly of all, a remarkable protein trafficking defect in which AGT is mistargeted from its normal location in the peroxisomes of hepatocytes to the mitochondria.

11.1
Oxalate as an Endogenous Toxin

In humans, oxalate is a metabolic end product of no known use. In fact, when present in excess, oxalate is actually detrimental mainly due to the poor solubility of its calcium salt. The inappropriate crystallization of calcium oxalate (CaOx) causes problems for complex physiological life forms as it can easily obstruct tubular lumens, disrupt intercellular and intracellular interactions and communications, or kill the cells within which or next to which crystallization occurs. Because oxalate can be removed from the body to any significant extent only by urinary excretion, it is the kidney that suffers the brunt of untoward CaOx deposition. Oxalate is a cumulative poison. It is the chronic deposition of insoluble CaOx over weeks, months, or even years that leads eventually to impaired kidney function. If the CaOx accretions, which often take the form of calculi in the kidney and urinary tract (nephro/urolithiasis) or diffuse crystalline deposits in the renal parenchyma (nephrocalcinosis), are not removed in a timely fashion, then irreversible kidney failure can ensue.

Oxalate itself is not particularly toxic; it is just that there is an awful lot of it around. Some plant foods, such as rhubarb and spinach, contain large amounts. Meat contains very little. Luckily, dietary oxalate is usually complexed with calcium and poorly absorbed. On a mixed diet, the average adult human typically consumes in the order of 200 mg of oxalate per day, absorbs 10 mg per day, manufactures 20 mg per day, and excretes 30 mg per day.

Even under normal circumstances, urinary oxalate can crystallize out as CaOx. These crystals are usually very small and are passed in the urine with no adverse consequences. However, the situation can easily change if urinary oxalate concentration increases. This can occur for a variety of reasons, such as increased absorption or increased endogenous synthesis. Increased absorption can occur due to increased ingestion, mucosal disease, or major intestinal surgery. Increased endogenous synthesis can occur due to increased ingestion and absorption of dietary oxalate precursors or due to genetic disorder.

Although CaOx kidney stones are very common, in the great majority of cases their causes are unknown. Family studies suggest that idiopathic CaOx kidney stone disease is multifactorial with both genetic and environmental components. In contrast, a great deal is known about the molecular etiology and pathophysiology of a group of rare monogenic CaOx kidney stone diseases called the *primary hyperoxalurias*.

11.2
Hereditary Oxalate Overproduction – The Primary Hyperoxalurias

11.2.1
Clinical Characteristics

Primary hyperoxaluria is the term given to a group of rare genetic disorders characterized by a marked increase in synthesis and urinary excretion of oxalate, and the deposition of insoluble CaOx in the kidney and urinary tract. Only two of the primary hyperoxalurias have been properly characterized, namely, primary hyperoxaluria type 1 (PH1, MIM259900) and primary hyperoxaluria type 2 (PH2, MIM260000) [1–5]. Increased urinary oxalate excretion is often, but not always, accompanied by increased excretion of glycolate in PH1 and L-glycerate in PH2. The presence or absence of hyperglycolic aciduria or hyper-L-glyceric aciduria appears to be of no pathological consequence, as it is the oxalate, or rather the low solubility of CaOx, which causes all the problems. CaOx deposition can take the form of nephro/urolithiasis or nephrocalcinosis.

At the clinical level, PH1 and PH2 are heterogeneous diseases. For example, the first symptom of PH1, which might be renal colic or hematuria, can occur anywhere between the first few months of life to as late as the fifth or sixth decade. Although there is no clear-cut distinction, the infantile form of PH1, in particular, tends to be characterized by nephrocalcinosis and a rapid rate of progression, whereas the later onset childhood form is more likely to be characterized by nephro/urolithiasis and a slower rate of progression. Untreated patients will succumb eventually to end-stage kidney failure, after which the effects of the increased synthesis of oxalate are exacerbated by the failure to remove it from the body. This results in the deposition of CaOx throughout the body (systemic oxalosis). Although PH2 is generally thought to be a somewhat milder disease than PH1, kidney failure in PH2 can still be the long-term consequence of CaOx deposition.

11.2.2
Enzyme and Metabolic Defects

Despite the similar pathophysiology, PH1 and PH2 have completely different molecular etiologies. PH1 is caused by a deficiency of the liver-specific pyridoxal-phosphate-dependent peroxisomal enzyme alanine:glyoxylate aminotransferase (AGT, EC 2.6.1.44) [6–9]. PH2 is caused by a deficiency of the NAD(P)H-dependent enzyme glyoxylate reductase (GR, EC 1.1.1.26) [10]. GR is more widely expressed than AGT, although most of it is also found in the liver. Although GR is mainly located in the cytosol, a small amount, as yet unquantified, is also found in the mitochondria [10–12].

AGT is a homodimeric enzyme of 86 kDa, which catalyzes the transamination of glyoxylate to glycine, using alanine as the amino donor (Reaction 1 in Table 11.1). The failure of this reaction in PH1 allows glyoxylate to diffuse through the peroxisomal membrane into the cytosol where it is oxidized to oxalate, catalyzed by lactate dehydrogenase (LDH) (Reaction 5), and reduced to glycolate, catalyzed by GR (Reaction 3) (Figure 11.1). The resulting concomitant hyperoxaluria and hyperglycolic aciduria are characteristic of PH1. AGT also catalyzes the transamination of pyruvate to alanine, using serine as the amino donor (Reaction 2). This reaction is of no obvious pathological consequence, but nevertheless results in the enzyme called serine:pyruvate aminotransferase (SPT) in some quarters.

GR is a homodimeric enzyme of 73 kDa, which catalyzes the reduction of glyoxylate to glycolate using either NADPH or NADH (Reaction 3 in Table 11.1). Similar to the situation in PH1, the absence of GR in PH2 allows the glyoxylate to be oxidized to oxalate, again catalyzed by cytosolic LDH (Reaction 5). GR also catalyzes another reaction, namely, the reduction of hydroxypyruvate to D-glycerate (Reaction 4). The absence of this reaction in PH2 allows the hydroxypyruvate to be reduced to L-glycerate catalyzed by LDH (Reaction 6). Because of the two reactions of GR, it is often referred to as glyoxylate reductase/hydroxypyruvate reductase (GRHPR).

Although increased excretion of glycolate and D-glycerate can be useful in the differential diagnosis of PH1 and PH2, it is not completely reliable as a significant number of PH1 and PH2 patients have isolated hyperoxaluria. The most accurate differential diagnosis depends on the assay of AGT and GR catalytic activities in liver biopsies (see below).

Glyoxylate is generally recognized as the most important immediate metabolic precursor of oxalate in humans. However, what is much less certain is which is the most important metabolic precursor of glyoxylate. Dietary carbohydrates and their derivatives, such as glycolate, and amino acids, such as glycine and hydroxyproline, have all been implicated. Clearly, the nature of the diet, especially whether it is mainly vegetarian or contains a high percentage of animal products, is very important. For example, glycolate is a by-product of photorespiration and is found mainly in plants. On the other hand, hydroxyproline is derived mainly from the breakdown of collagen, which is found mainly in animals.

The dietary dependence of the glyoxylate precursors has led to a remarkable finding in mammalian evolution in which the intracellular compartmentalization of AGT has changed on numerous occasions under the influence of dietary selection pressure. Glycolate and glycine can both be metabolized to glyoxylate within hepatocyte peroxisomes, catalyzed by glycolate oxidase (GO) and D-amino acid oxidase (DAO), respectively (Reactions 7 and 8 in Table 11.1) (Figure 11.1).

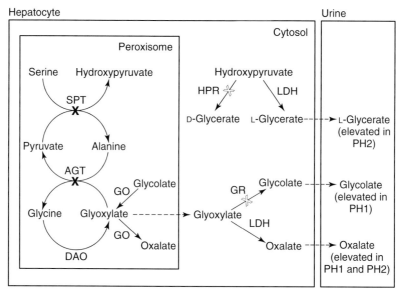

Fig. 11.1 Some of the hepatocyte metabolic pathways important in primary hyperoxaluria. Enzymes: AGT, alanine:glyoxylate aminotransferase; DAO, D-amino acid oxidase; GO, glycolate oxidase; GR, glyoxylate reductase; HPR, hydroxypyruvate reductase; LDH, lactate dehydrogenase; SPT, serine:pyruvate aminotransferase. AGT and SPT are the same enzyme, as are GR and HPR. AGT and SPT are deficient in PH1 and GR and HPR are deficient in PH2. Solid arrows, metabolic conversions; dashed arrows, membrane transport/diffusion. Although only the exodus of glyoxylate from the peroxisome is shown, it is likely that the peroxisomal membrane is permeable to most of the metabolites shown. Solid cross, deficiency in PH1; open cross, deficiency in PH2.

Tab. 11.1 Enzyme and catalytic reactions important in PH1 and PH2

Reactions catalyzed by AGT (SPT) in peroxisomes
1. Alanine + glyoxylate → pyruvate + glycine
2. Serine + pyruvate → hydroxypyruvate + alanine

Reactions catalyzed by the enzyme GRHPR in the cytosol/mitochondria
3. Glyoxylate + NAD(P)H + H$^+$ → glycolate + NAD(P)$^+$
4. Hydroxypyruvate + NAD(P)H + H$^+$ → D-glycerate + NAD(P)$^+$

Reactions catalyzed by the enzyme LDH in the cytosol
5. Glyoxylate + NAD$^+$ → oxalate + NADH + H$^+$
6. Hydroxypyruvate + NADH + H$^+$ → L-glycerate + NAD$^+$

Flavin-dependent oxidations that produce glyoxylate in the peroxisomes
7. Glycolate + O$_2$ → glyoxylate + H$_2$O$_2$ (GO, FMN)
8. Glycine + O$_2$ + H$_2$O → glyoxylate + H$_2$O$_2$ + NH$_3$ (DAO, FAD)

The metabolic route from hydroxyproline to glyoxylate is rather more complicated, but the last reaction, catalyzed by ketohydroxyglutarate aldolase (KHA), takes place in the mitochondria. In order for AGT to be able to minimize the oxidation of glyoxylate to oxalate, by transaminating it to glycine instead, it has to be localized to the same organelle in which the glyoxylate is formed. As a result, herbivores tend to have peroxisomal AGT, whereas carnivores tend to have mitochondrial AGT. Unsurprisingly, omnivores tend to have AGT in both peroxisomes and mitochondria [13–17].

11.2.3
Crystal Structures of AGT and GRHPR

The crystal structures of both AGT and GRHPR have led to invaluable insights into the normal and abnormal workings of both enzymes. AGT has a structure similar to that of other PLP-dependent transferases (PDB: 1H0C) [18]. It crystallizes as an intimate dimer (Figure 11.2), each subunit of which can be divided into two main structural domains: (i) a larger N-terminal domain of 282 amino acids, which contains most of the active site and dimerization interface and (ii) a small C-terminal domain of 110 amino acids, which contains the atypical C-terminal type 1 peroxisomal targeting sequence (PTS1) [19] and an internal ancillary PTS1A [20]. The first 20 or so amino acids of the N-terminal domain make up an N-terminal extension that wraps over the surface of the opposing subunit. The large dimerization interface explains the high stability of dimeric AGT. Although it has been suggested that the unusual N-terminal extension might play a role in the dimerization process, it is unlikely to contribute much to the overall stability of the dimer once formed. Confirming previous biochemical studies [21–23], the crystal structure of AGT shows that each subunit contains one functionally independent PLP-biding site forming a Schiff base on Lys209.

The crystal structure of GRHPR (PDB: 2GCG) [24] has provided significant insights into its rather unusual substrate specificity. For example, it explains why,

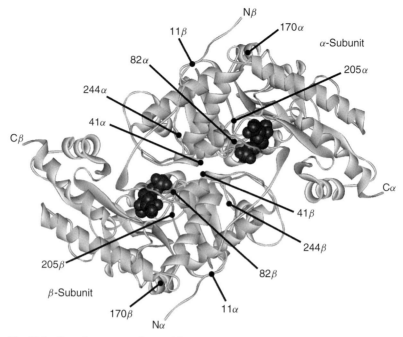

Fig. 11.2 Crystal structure of normal human AGT (PDB 1H0C). Large dark gray balls, pyridoxal phosphate; small black circles, the locations of the residues discussed in the text.

although GRHPR is a typical D-2-hydroxyacid dehydrogenase, it can use glyoxylate and hydroxypyruvate as substrates, but not the closely related pyruvate.

11.2.4
Genes, Mutations, and Polymorphisms

The gene encoding AGT (i.e., *AGXT*) is located on chromosome 2q37.3 and consists of 11 exons spanning about 10 kb [25]. The open reading encodes a polypeptide of 392 amino acids [26]. The gene encoding GRHPR (i.e., *GRHPR*) is located in the pericentromeric region of chromosome 9 and consists of nine exons spanning about 9 kb [27–29]. The open reading frame of *GRHPR* encodes a polypeptide of 328 amino acids.

Two important polymorphic variants of AGT have been identified encoded by the "major" and "minor" alleles. The frequencies of these alleles are population specific. For example, in European and North American populations the minor allele is found with a frequency of 15–20%, whereas in Far Eastern populations it is found at only 2–3% [30, 31]. The polypeptide encoded by the minor allele differs from that encoded by the major allele in two respects, namely, Pro11Leu and Ile340Met amino acid replacements. Unlike most nonsynonymous polymorphisms, the Pro11Leu replacement, in particular, has some significant effects on the properties of AGT (see below).

Over 100 mutations have been found so far in the *AGXT* gene, including a wide variety of missense and nonsense mutations, small and large insertions and deletions, and splice-site mutations (for a review of some of them see [32]). Although most missense mutations in AGT are individually rare, a few are quite common. By far the most common is a Gly170Arg amino acid replacement [31] (Table 11.2). This has an allelic frequency of 30–40% in European and North American PH1 patients. The second most common is an Ile244Thr amino acid replacement with an allelic frequency of about ∼9% [33].

The Pro11Leu polymorphism has a frequency of about 50% in European and North American PH1 patients, which is much higher than that is found in the normal population [34]. This is a consequence of the fact that many of the mutations in AGT in PH1, including some of the most common (e.g., Gly170Arg and Ile244Thr), segregate with the minor allele.

Far fewer mutations have been identified in the *GRHPR* gene, simply due to the smaller number of patients analyzed [3]. Therefore, it is impossible to draw any sensible conclusions about relative frequencies, either globally or in specific populations.

11.2.5
Genotype–Phenotype Correlations

Many of the missense mutations in AGT are associated with very specific enzyme phenotypes (Table 11.2). In addition, many of those that segregate with the minor allele functionally interact with the Pro11Leu polymorphism [22, 35]. Some of the better studied enzymic consequences of amino acid replacements in AGT are described below (refer to Figure 11.2 for structural context).

11.2.5.1 **Pro11Leu**

The normal Pro11Leu replacement characteristic of the minor allele is remarkable because, unlike most other polymorphisms in most other genes which tend to be neutral, it has significant effects on the properties of AGT. Not only does it sensitize AGT to the untoward effects of many mutations, such as Gly170Arg, Ile244Thr, and Gly41Arg (see below), but also in a number of cases it has more of an effect than does the mutation. For example, Pro11Leu on its own interferes with AGT dimerization, particularly at high temperatures. It decreases specific catalytic activity of purified recombinant AGT by two-thirds [22]. And finally, when expressed homozygously, it redirects 5–10% of the enzyme away from the peroxisomes toward the mitochondria in human hepatocytes *in situ* [31]. The Pro11Leu replacement generates an N-terminal mitochondrial targeting sequence (MTS), partly because it replaces a helix-breaking amino acid with a helix-forming one [31, 36]. This makes it more likely that the N-terminus of AGT might fold into an α-helical conformation, compatible with the optimal requirements for an MTS, in the right intracellular

Tab. 11.2 Some of the better studied missense mutations and polymorphisms and their effects on the properties of AGT

Amino acid replacement	Allelic frequency in PH1 patients[a] (%)	Overt enzyme phenotype (normal = +++)		Effect on the properties of AGT	
		Catalytic activity	Immuno-reactivity	On the background of the major allele[b]	On the background of the minor allele[c]
Pro11Leu[d]	50	+++	+++	not applicable	Delays dimerization, decreases catalytic activity, redirects 5–10% to mitochondria, sensitizes AGT to the effects of many PH1-specific mutations
Gly41Arg[e]	<1	−	±	Markedly decrease catalytic activity	Abolishes catalytic activity, aggregates into intraperoxisomal cores
Gly82Glu[f]	<1	−	+++	Interferes with PLP binding and abolishes catalytic activity	Unknown[g]
Gly170Arg[h]	30	++	++	None known	Inhibition of dimerization, redirects 90–95% to mitochondria
Ser205Pro[f]	<1	−	−	Accelerated degradation	Unknown[g]
Ile244Thr[i]	9	±	±	None known	Aggregation and accelerated degradation
Ile340Met[d]	50	+++	+++	not applicable	None

[a] Approximate allelic frequency in European and North American PH1 patients.
[b] Residue 11 = Pro.
[c] Residue 11 = Leu.
[d] Components of the normal minor allele.
[e] Found on both the major and minor alleles.
[f] Found only on the major allele.
[g] Probably the same effect as on the major allele.
[h] Found only on the minor allele.
[i] Found mainly on the minor allele.

environment. Also, the newly generated Leu-X-X-Leu-Leu motif matches more closely with the optimum consensus motif for interaction with the mitochondrial import receptor TOM20 [37, 38]. Although the Pro11Leu-generated MTS is very effective when attached to reporter proteins, such as GFP, it is functionally weak when attached to AGT [39]. This is demonstrated by its ability to target only a small proportion of AGT to mitochondria, even when the PTS1 is removed. Residue 11 sits right in the middle of the N-terminal extension [18] (Figure 11.2). Even though the polymorphic N-terminal 20 amino acids have the potential to form an MTS and target AGT to mitochondria, they are prevented from doing so efficiently probably because the N-terminus is trapped by binding to the surface of the opposing subunit. This would make it unavailable for interaction with TOM20. In addition, even polymorphic AGT rapidly dimerizes into a stable conformation not compatible with mitochondrial import. Rapid, almost irreversible, dimerization does not interfere with the peroxisomal import of AGT because these organelles can take up fully folded, multimerized, cofactor-bound proteins [40, 41]. On the other hand, the mitochondrial import machinery can deal with only unfolded or loosely folded polypeptides [42, 43].

11.2.5.2 Gly170Arg

Because of its high frequency in PH1, it is not surprising that Gly170Arg was the first disease-specific mutation to be identified in PH1 [31]. However, what was surprising was the finding that patients homozygous for Gly170Arg + Pro11Leu were only relatively mildly depleted in AGT catalytic activity and immunoreactivity. Typically, such patients have 10–30% of the normal levels [44], but it can go as high as 72% [35]. This would not in itself be enough to explain the disease, particularly as some asymptomatic carriers can have AGT levels as low as 21% of the mean normal level [44]. Instead, PH1 in these patients is due to an unparalleled protein trafficking defect in which normally peroxisomal AGT is mistargeted to the mitochondria [45]. Mistargeted AGT remains catalytically active in the mitochondria but is metabolically inefficient, presumably because the majority of its substrate (i.e., glyoxylate) is synthesized in the peroxisomes rather than in the mitochondria (see Figure 11.1). Various *in vitro* biochemical experiments indicate that the contribution of the Pro11Leu replacement to the overall AGT trafficking defect is much greater than that of the Gly170Arg replacement; so much so that the latter appears to be innocuous in the absence of the former, at least *in vitro* [22]. The Gly170Arg replacement provides no further mitochondrial targeting information beyond that provided by the Pro11Leu polymorphism. In addition, it does not directly interfere with peroxisomal targeting. The increased functional efficiency of the polymorphic MTS, due to the extra presence of Gly170Arg, seems to be related to their combined effect on AGT dimerization, the rate of which is markedly decreased, even at normal temperatures [46].

The exact effect of the Gly170Arg replacement is unclear. On its own, it has a relatively small effect on the structure of AGT (see PDB 1J04). It does not obviously affect folding or dimerization. However, it does interact synergistically with the Pro11Leu polymorphism to inhibit dimerization. One possible explanation for this is that the two replacements together decrease the affinity with which the N-terminal extension binds to the surface of the opposing subunit, with Pro11Leu altering the conformation of the extension and Gly170Arg disrupting the surface binding sites [18]. This "double whammy" would effectively prevent the N-terminal extension from binding with the dual effect of exposing the MTS and inhibiting dimerization, both of which would encourage mitochondrial import.

11.2.5.3 Ile244Thr

Ile244Thr is the second most common missense mutation in PH1 [32]. Like Gly170Arg, it too segregates with the minor allele and functionally interacts with the Pro11Leu polymorphism. Unlike Gly170Arg, however, it has been found, albeit rarely, on the background of the major allele [35]. When expressed in eukaryotic or prokaryotic expression systems, AGT containing Ile244Thr + Pro11Leu is unstable and readily aggregates [22, 47]. In the livers of PH1 patients, it is associated with a complete loss or low levels of AGT catalytic activity and immunoreactivity.

Analysis of the crystal structure of AGT (Figure 11.2) suggests that the replacement of Ile244 by Thr would only generate a small defect in helix packing. However, it might indirectly affect the inter-subunit binding of the N-terminal clamp as suggested for Gly170Arg. This could destabilize AGT enough to aggregate and be rapidly degraded. Why this does not result in mistargeting to the mitochondria, due to the presence of the Pro11Leu polymorphism, is unclear.

11.2.5.4 Gly41Arg

Gly41Arg is a rare mutation associated with complete loss of AGT catalytic activity and marked reduction of AGT immunoreactivity. Gly41Arg has been found in three, compound heterozygote patients on the background of the minor *AGXT* allele [48]. In addition, there has been at least one case reported in the literature of this mutation being present in the homozygous state on the background of the major *AGXT* allele [49]. At least in the former situation, Gly41Arg results in the aggregation of AGT into core-like structures in the peroxisomal matrix. When expressed in *Escherichia coli*, AGT containing both Gly41Arg and Pro11Leu aggregates into inclusion bodies. However, when expressed with Gly41Arg alone the protein is partially soluble, although its specific catalytic activity is markedly reduced [22]. This partial functional linkage between Gly41Arg and Pro11Leu is compatible with the findings, admittedly in a very small number of cases, that patients with Gly41Arg alone are more mildly affected than those with both Gly41Arg and Pro11Leu.

Analysis of the crystal structure of AGT shows that Gly41 sits right in the middle of the dimerization interface and interacts with its counterpart in the other subunit (Figure 11.2). Replacement of the smallest amino acid (i.e., Gly) with one of the largest (i.e., Arg) would prevent the formation of the interface and prevent dimerization. This would inevitably destabilize AGT leading to its aggregation and accelerated degradation. This again fits exactly with the enzymatic phenotype associated with the Gly41Arg mutation (i.e., loss of catalytic activity, very low immunoreactivity, and intraperoxisomal aggregation).

11.2.5.5 Gly82Glu

Unlike the above mutations, Gly82Glu segregates with the major, rather than minor, allele [50]. It is associated with complete loss of AGT catalytic activity, but normal levels of peroxisomal AGT immunoreactivity [6, 8]. Purified recombinant AGT containing Gly82Glu is perfectly stable and correctly targeted but it fails to bind the cofactor PLP, thereby explaining its loss of activity [22]. However, recent work has shown that, although PLP binding is greatly reduced, the main effect of Gly82Glu is to prevent the efficient transaldimination of the PLP form of the enzyme as well as the efficient conversion of AGT–PMP into AGT–PLP [51].

The crystal structure of AGT shows that Gly82 is situated right in the middle of the catalytic site where it makes hydrogen bond contact with the cofactor. Its replacement by the much larger amino acid Glu would certainly be expected to interfere with PLP and PMP binding, as well as interfering with transaldimination, and so on.

11.2.5.6 Ser205Pro

Ser205Pro is a rare mutation identified in Japanese PH1 patients [52]. It has also been found only on the major allele. AGT containing Ser205Pro is highly unstable and prematurely degraded [53], a finding that is compatible with the absence of both AGT immunoreactivity and catalytic activity in homozygous PH1 patients. The crystal structure (Figure 11.2) shows that the replacement of Ser205 by Pro would necessitate a large conformational change in the backbone of one of the β-strands of AGT that would completely disrupt the main chain hydrogen bonding of the central β-sheet. This would disrupt AGT folding and lead to accelerated degradation.

11.2.5.7 Other Mutations

Most of the other missense mutations found in PH1 have not be studied in as much detail as those described above. Nevertheless, it seems that the highly significant contribution made by the Pro11Leu polymorphism to the disease process is widespread. For example, a recent study analyzed the catalytic activities of 13 different recombinant AGTs, each containing a different newly discovered missense mutation [35]. Out of the eight AGTs containing mutations associated with the minor allele, six had much lower catalytic activities when expressed on the minor, compared to the major, allele. However, in none of the AGTs containing the five missense mutations associated with the major allele did the background allele make any difference.

Although the relationship between genotype and enzyme phenotype in PH1 is in many cases well understood, the same cannot be said of the relationship between genotype and clinical phenotype. Patients with identical mutations can have completely different disease characteristics, as manifested by severity and rate of progression [54]. Notwithstanding this, there is a tendency for patients with the Gly170Arg mutation (and of course the Pro11Leu polymorphism) to have a milder disease on average than other PH1 patients. However, this might simply be due to the fact that these patients are more likely to be pyridoxine responsive (see below). Although PH1 is indisputably a monogenic disease, its severity and rate of progression is clearly influenced by other factors, probably both genetic and environmental.

11.2.6
Diagnosis

Prior to the discovery of the enzyme defects, formal diagnosis of PH1 and PH2 was dependent on the presence of concomitant hyperoxaluria and hyperglycolic aciduria, in the case of PH1, or hyperoxaluria and hyper-L-glyceric aciduria, in the case of PH2 [2]. This caused problems for two main reasons. First, urinary metabolite analysis is rendered meaningless with approaching renal failure. Secondly, not all PH1 and PH2 patients, respectively, have elevated excretion of glycolate [55] and L-glycerate [56]. In fact, not all patients with concomitant hyperoxaluria and hyperglycolic aciduria turn out to have PH1 [57]. Diagnosis based on isolated hyperoxaluria required more detailed clinical analysis of the extent and nature of the CaOx deposition, as well as exclusion of other causes for the elevated oxalate excretion, such as enteric hyperabsorption. Knowing that AGT and GRHPR dysfunction are responsible for PH has enabled much better diagnostic procedures to be introduced. Enzyme analysis of samples obtained by liver biopsy can diagnose all forms of PH1 [58, 59] and PH2 [12, 60] currently recognized, irrespective of the state of kidney function. Recently, it has even been shown that PH2 can be diagnosed by measuring GRHPR activity in peripheral blood leukocytes [61].

In families already known to carry PH1 or PH2, and in whom the disease-causing mutations are known, disease in further family members can more often than not also be diagnosed by mutational analysis of DNA isolated from peripheral blood leukocytes. Recently, it has been estimated that, even without prior family information, complete gene sequencing or limited exon-specific sequencing is a practical proposition for PH1 diagnosis [35, 62, 63].

Prenatal diagnosis of PH1 was first carried out successfully in the late 1980s by AGT assay following fetal liver biopsy in the second trimester [64, 65]. Although this procedure had the potential to identify all forms of PH1, it has now been supplanted by DNA (mutation and/or linkage) analysis in the first trimester [66].

11.2.7
Treatment

The classic treatments for PH1 and PH2 are similar to those of other (idiopathic) CaOx kidney stone diseases, such as good hydration while kidney function is good, and the use of CaOx crystallization inhibitors, such as citrate and magnesium salts [4, 5] (Figure 11.3). Following kidney failure, patients have to be dialyzed and, when available, the kidney has to be transplanted. Neither of these are long-term solutions. Unfortunately, oxalate does not dialyze efficiently enough, so that the problems associated with excessive oxalate synthesis are exacerbated by the failure of its efficient removal [67]. This can lead to the deposition of CaOx throughout the body. Although under the right circumstances the transplanted kidney can last several years, it will inevitably succumb to CaOx deposition as did the original kidney. This is because the basic defect is not in the kidney, but in the liver (particularly in the case of PH1).

It has been known for many years that a proportion of PH1, but not PH2, patients responded well to the administration of pharmacological doses of pyridoxine (vitamin B_6) [68] (Figure 11.3). Pyridoxine is metabolized to PLP, the essential cofactor for AGT. Although it is likely that the latter fact explains the efficacy of pyridoxine in some patients, the exact mechanism is unclear. It has been known since the early 1990s that the success of pyridoxine was related to the presence of residual levels of AGT catalytic activity and immunoreactivity [55]. Significant residual AGT is found mainly in PH1 patients in whom the enzyme is mistargeted from the peroxisomes to the mitochondria. The association between AGT mistargeting and pyridoxine responsiveness has been more formally established recently due to the finding that the latter is related to the presence of the Gly170Arg mutation [69, 70]. Why these patients in particular should be responsive is unclear. However, it may be related to increased AGT stability, correction of AGT compartmentalization, or increased transport of pyridoxine/pyridoxal phosphate into the mitochondria, thereby shifting the equilibrium from the inactive apo-enzyme to the active holo-enzyme. There are also reports that patients expressing other mutations, such as Phe152Ile, are also pyridoxine responsive [71]. Interestingly, like Gly170Arg, Phe152Ile also segregates on the minor allele and has been implicated in partial mistargeting of AGT to the mitochondria [48].

The discovery that PH1 was caused by a deficiency of AGT, and that AGT was expressed mainly in hepatocytes, opened up the possibility that liver transplantation could be used as a rather specialized form of enzyme replacement therapy (ERT) (Figure 11.3). In this context, liver transplantation is an almost perfect form of ERT because it can supply the body's total requirement for AGT, in the correct cell (hepatocyte), the correct subcellular location (peroxisomes), and in the correct relationship with its substrate (glyoxylate) [72]. Several hundred liver transplantations have been carried out for PH1 worldwide [73–76]. Because most PH1 patients receiving liver transplants are in renal failure, more often than not the kidneys are transplanted as well. Although it is clear that liver transplantation can provide a "cure" for PH1, at least in metabolic terms, it can take a long time for this cure to be realized, especially in terms of urinary oxalate excretion. This is thought to be owing to the slow remobilization of CaOx deposited throughout the body, particularly the bones, during periods of compromised renal function and extended periods of dialysis.

Although liver transplantation in PH1 can be more readily understood as a specialized form of ERT, it might better be considered as a rather specialized form of gene therapy. The long-term success of the procedure results not from the AGT enzyme reintroduced with the transplanted liver, which will be turned over with a half life of one to three days, but rather from the introduction of a normal *AGXT* gene which will continue to produce normal AGT hopefully for many years.

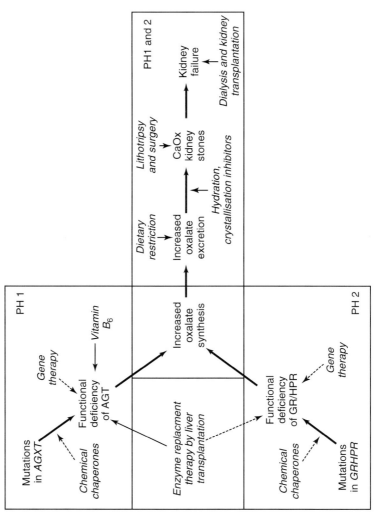

Fig. 11.3 Current treatments for PH and potential treatments for the future. Solid large arrows, causal pathways in the molecular etiology and pathophysiology of PH; solid small arrows, treatments currently used in PH; dashed small arrows, potential treatments of the future.

11.3
Cytotoxicity of Oxalate, Calcium Oxalate, and Related Metabolites

Although chronic cumulative toxicity of oxalate, leading to the progressive deposition of insoluble CaOx in the kidney and urinary tract, is the obvious cause of all the problems in PH1 and PH2, there is some evidence that its acute cytotoxicity might also play a role. Calculations of the fluid dynamics of urine flow suggest that once CaOx microcrystals have formed they are unlikely to be able to grow fast enough to be trapped within the kidney or urinary tract and, therefore, should be swept out and passed in the urine as crystalluria [77]. In fact, CaOx crystalluria is a common finding even in the normal population. Adherence of CaOx microcrystals to the walls of the tubules and collecting ducts would be required to allow the crystals to grow to a size that would be too big to continue on the journey. Such adherence is thought to be mediated by damage to the tubule and collecting duct walls caused either by the CaOx crystals themselves or by the high concentrations of the oxalate anion [78–80]. Although most experiments testing this have really been designed to account for CaOx stone formation in idiopathic CaOx kidney stone disease, the principles involved could also be relevant to PH1 and PH2.

It is clear that renal epithelial cell injury and inflammation, mediated by the production of free radicals, particularly reactive oxygen species, and the development of oxidative stress play a major role in CaOx stone formation [81–84]. However, the specific molecular mechanisms involved are much less clear. This is partly because of the biphasic concentration-dependent effects of oxalate, which at low concentrations is a mitogen [85] and at high concentrations a toxin [82]. It is also due to the plethora of effects on numerous signaling pathways [86] characteristic of both necrosis and apoptosis [87], including altered membrane surface properties and cellular lipids, changes in the expression of a number of genes, formation of reactive oxygen species, activation of phospholipase A2, which in turn increases the levels of arachidonic acid and lysophosphatidylcholine [86]. At least some of the cellular pathology caused by oxalate and CaOx crystals appears to be due to mitochondrial dysfunction, as overtly manifested by decreased respiration, increased swelling, and decreased inner membrane potential difference [88, 89].

Studies into the mechanism of ethylene glycol (EG) nephrotoxicity show that at least some of the metabolic precursors of oxalate, including glycolaldehyde and glyoxylate, but not glycolate, are cytotoxic in addition to the CaOx crystals themselves [90–92]. Glyoxylate, but not glycolate, also appears to be responsible for at least some of the mitochondrial damage found [93].

11.3.1
Indirect Glycolate Toxicity and Screening for Chemotherapeutic Agents

Although the relative importance of the oxalate anion and CaOx crystals as renal cytotoxins is still a matter of debate, one thing that is generally agreed is that some of the metabolic precursors of oxalate, such as glyoxylate and glycolaldehyde, are toxic to cells. Although this may not present any problem, except in the case of

EG poisoning, due to their much lower concentrations than oxalate, the potential cytotoxicity of glyoxylate has been put to good use in the development of a cell-based system that might be of use in the screening of possible chemotherapeutic agents for PH1 and PH2 [94].

Unlike hepatocytes, Chinese hamster ovary (CHO) cells do not express AGT, GRHPR, or GO. They are highly resistant to glycolate which can be added to the culture medium to a level of 2 mM without obvious effect. However, if the cells are stably transformed to express GO, then glycolate shows marked cytotoxicity at 20 µM and below [94]. However, cotransformation with either normal AGT or GRHPR or both AGT and GRHPR prevents, or at least attenuates, this glycolate cytotoxicity. The most likely explanation for this is that glycolate is only toxic to GO-expressing cells because it is converted to glyoxylate. It is the glyoxylate, rather than the glycolate, that is cytotoxic, and its removal by AGT (to glycine) and GRHPR (back to glycolate) allows the cells to survive. Neither mutant AGT nor mutant GRHPR would be expected to be able to protect GO-expressing cells from "indirect" glycolate toxicity. Cell survival could then be used to test the ability of potential therapeutic agents that might either stabilize AGT and GRHPR or inhibit GO. Any such agent could provide the lead for the development of clinically useful therapeutics for PH1 and PH2.

11.4
Conclusions

In humans, oxalate is an unavoidable metabolic end product of no known use. It is the poor solubility of its calcium salt makes it potentially life-threatening, especially when present in elevated concentrations. Oxalate is an endogenous toxin due to elevated synthesis from mainly dietary and other precursors, such as glycolate and hydroxyproline, and an exogenous toxin due to extreme dietary consumption or absorption. The hereditary metabolic disorders PH1 and PH2 amply demonstrate the extreme consequences of too much oxalate.

References

1 Barratt, T.M. and Danpure, C.J. (1999) Hyperoxaluria, in *Pediatric Nephrology* (eds T.M. Barratt, E.D. Avner, and W.E. Harmon), Williams & Wilkins, Baltimore, pp. 609–619.

2 Danpure, C.J. (2001) Primary hyperoxaluria, in *The Metabolic and Molecular Bases of Inherited Disease*, vol. II (eds C.R. Scriver, A.L. Beaudet, W.S. Sly, D. Valle, B. Childs, K.W. Kinzler, and B. Vogelstein), McGraw-Hill, New York, pp. 3323–3367.

3 Danpure, C.J. and Rumsby, G. (2004) Molecular aetiology of primary hyperoxaluria and its implications for clinical management. *Expert Rev Mol Med*, 1–16.

4 Danpure, C.J. and Smith, L.H. (1996) The primary hyperoxalurias, in *Kidney Stones: Medical and Surgical Management* (eds F.L. Coe, M.J. Favus, C.Y. Pak, J.H. Parks, and G.M. Preminger), Lippincott-Raven, Philadelphia, pp. 859–881.

5 Watts, R.W. and Danpure, C.J. (2003) Disorders of oxalate metabolism, in *Oxford Textbook of Medicine* (eds D.A. Warrell, T.M. Cox, and J.D. Firth), OUP, Oxford, pp. 134–138.

6 Cooper, P.J., Danpure, C.J., Wise, P.J., and Guttridge, K.M. (1988) Immunocytochemical localization of human hepatic alanine: glyoxylate aminotransferase in control subjects and patients with primary hyperoxaluria type 1. *J Histochem Cytochem*, **36**, 1285–1294.

7 Danpure, C.J. and Jennings, P.R. (1986) Peroxisomal alanine:glyoxylate aminotransferase deficiency in primary hyperoxaluria type I. *FEBS Lett*, **201**, 20–24.

8 Danpure, C.J. and Jennings, P.R. (1988) Further studies on the activity and subcellular distribution of alanine:glyoxylate aminotransferase in the livers of patients with primary hyperoxaluria type 1. *Clin Sci (Lond)*, **75**, 315–322.

9 Kamoda, N., Minatogawa, Y., Nakamura, M., Nakanishi, J., Okuno, E., and Kido, R. (1980) The organ distribution of human alanine-2-oxoglutarate aminotransferase and alanine-glyoxylate aminotransferase. *Biochem Med*, **23**, 25–34.

10 Mistry, J., Danpure, C.J., and Chalmers, R.A. (1988) Hepatic D-glycerate dehydrogenase and glyoxylate reductase deficiency in primary hyperoxaluria type 2. *Biochem Soc Trans*, **16**, 626–627.

11 Cregeen, D.P., Williams, E.L., Hulton, S., and Rumsby, G. (2003) Molecular analysis of the glyoxylate reductase (GRHPR) gene and description of mutations underlying primary hyperoxaluria type 2. *Hum Mutat*, **22**, 497.

12 Giafi, C.F. and Rumsby, G. (1998) Kinetic analysis and tissue distribution of human D-glycerate dehydrogenase/glyoxylare reductase and its relevance to the diagnosis of primary hyperoxaluria type 2. *Ann Clin Biochem*, **35**, 104–109.

13 Birdsey, G.M., Lewin, J., Cunningham, A.A., Bruford, M.W., and Danpure, C.J. (2004) Differential enzyme targeting as an evolutionary adaptation to herbivory in carnivora. *Mol Biol Evol*, **21**, 632–646.

14 Birdsey, G.M., Lewin, J., Holbrook, J.D., Simpson, V.R., Cunningham, A.A., and Danpure, C.J. (2005) A comparative analysis of the evolutionary relationship between diet and enzyme targeting in bats, marsupials and other mammals. *Proc R Soc B*, **272**, 833–840.

15 Danpure, C.J., Fryer, P., Jennings, P.R., Allsop, J., Griffiths, S., and Cunningham, A. (1994) Evolution of alanine:glyoxylate aminotransferase 1 peroxisomal and mitochondrial targeting. A survey of its subcellular distribution in the livers of various representatives of the classes Mammalia, Aves and Amphibia. *Eur J Cell Biol*, **64**, 295–313.

16 Danpure, C.J., Guttridge, K.M., Fryer, P., Jennings, P.R., Allsop, J., and Purdue, P.E. (1990) Subcellular distribution of hepatic alanine:glyoxylate aminotransferase in various mammalian species. *J Cell Sci*, **97** (Pt 4), 669–678.

17 Holbrook, J.D., Birdsey, G.M., Yang, Z., Bruford, M.W., and Danpure, C.J. (2000) Molecular adaptation of alanine:glyoxylate aminotransferase targeting in primates. *Mol Biol Evol*, **17**, 387–400.

18 Zhang, X., Roe, S.M., Hou, Y., Bartlam, M., Rao, Z., Pearl, L.H., and Danpure, C.J. (2003) Crystal structure of alanine:glyoxylate aminotransferase and the relationship between genotype and enzymatic phenotype in primary hyperoxaluria type 1. *J Mol Biol*, **331**, 643–652.

19 Motley, A., Lumb, M.J., Oatey, P.B., Jennings, P.R., De Zoysa, P.A., Wanders, R.J., Tabak, H.F., and Danpure, C.J. (1995) Mammalian alanine/glyoxylate aminotransferase 1 is imported into peroxisomes via the PTS1 translocation pathway. Increased degeneracy and context specificity of the mammalian

PTS1 motif and implications for the peroxisome-to-mitochondrion mistargeting of AGT in primary hyperoxaluria type 1. *J Cell Biol*, **131**, 95–109.

20 Huber, P.A., Birdsey, G.M., Lumb, M.J., Prowse, D.T., Perkins, T.J., Knight, D.R., and Danpure, C.J. (2005) Peroxisomal import of human alanine:glyoxylate aminotransferase requires ancillary targeting information remote from its C terminus. *J Biol Chem*, **280**, 27111–27120.

21 Ishikawa, K., Kaneko, E., and Ichiyama, A. (1996) Pyridoxal 5′-phosphate binding of a recombinant rat serine: pyruvate/alanine:glyoxylate aminotransferase. *J Biochem (Tokyo)*, **119**, 970–978.

22 Lumb, M.J. and Danpure, C.J. (2000) Functional synergism between the most common polymorphism in human alanine:glyoxylate aminotransferase and four of the most common disease-causing mutations. *J Biol Chem*, **275**, 36415–36422.

23 Oda, T., Miyajima, H., Suzuki, Y., and Ichiyama, A. (1987) Nucleotide sequence of the cDNA encoding the precursor for mitochondrial serine:pyruvate aminotransferase of rat liver. *Eur J Biochem*, **168**, 537–542.

24 Booth, M.P., Conners, R., Rumsby, G., and Brady, R.L. (2006) Structural basis of substrate specificity in human glyoxylate reductase/hydroxypyruvate reductase. *J Mol Biol*, **360**, 178–189.

25 Purdue, P.E., Lumb, M.J., Fox, M., Griffo, G., Hamon-Benais, C., Povey, S., and Danpure, C.J. (1991) Characterization and chromosomal mapping of a genomic clone encoding human alanine:glyoxylate aminotransferase. *Genomics*, **10**, 34–42.

26 Takada, Y., Kaneko, N., Esumi, H., Purdue, P.E., and Danpure, C.J. (1990) Human peroxisomal L-alanine: glyoxylate aminotransferase. Evolutionary loss of a mitochondrial targeting signal by point mutation of the initiation codon. *Biochem J*, **268**, 517–520.

27 Cramer, S.D., Ferree, P.M., Lin, K., Milliner, D.S., and Holmes, R.P. (1999) The gene encoding hydroxypyruvate reductase (GRHPR) is mutated in patients with primary hyperoxaluria type II. *Hum Mol Genet*, **8**, 2063–2069.

28 Rumsby, G. and Cregeen, D.P. (1999) Identification and expression of a cDNA for human hydroxypyruvate/glyoxylate reductase. *Biochim Biophys Acta*, **1446**, 383–388.

29 Webster, K.E., Ferree, P.M., Holmes, R.P., and Cramer, S.D. (2000) Identification of missense, nonsense and deletion mutations in the GRHPR gene in patients with primary hyperoxaluria type II (PH2). *Hum Genet*, **107**, 176–185.

30 Caldwell, E.F., Mayor, L.R., Thomas, M.G., and Danpure, C.J. (2004) Diet and the frequency of the alanine:glyoxylate aminotransferase Pro11Leu polymorphism in different human populations. *Hum Genet*, **115**, 504–509.

31 Purdue, P.E., Takada, Y., and Danpure, C.J. (1990) Identification of mutations associated with peroxisome-to-mitochondrion mistargeting of alanine/glyoxylate aminotransferase in primary hyperoxaluria type 1. *J Cell Biol*, **111**, 2341–2351.

32 Coulter-Mackie, M.B. and Rumsby, G. (2004) Genetic heterogeneity in primary hyperoxaluria type 1: impact on diagnosis. *Mol Genet Metab*, **83**, 38–46.

33 von Schnakenburg, C. and Rumsby, G. (1997) Primary hyperoxaluria type 1: a cluster of new mutations in exon 7 of the AGXT gene. *J Med Genet*, **34**, 489–492.

34 Tarn, A.C., von Schnakenburg, C., and Rumsby, G. (1997) Primary hyperoxaluria type 1: diagnostic relevance of mutations and polymorphisms in the alanine:glyoxylate aminotransferase gene (AGXT). *J Inherit Metab Dis*, **20**, 689–696.

35 Williams, E. and Rumsby, G. (2007) Selected exonic sequencing of the

AGXT gene provides a genetic diagnosis in 50% of patients with primary hyperoxaluria type 1. *Clin Chem*, **53**, 1216–1221.

36 Purdue, P.E., Allsop, J., Isaya, G., Rosenberg, L.E., and Danpure, C.J. (1991) Mistargeting of peroxisomal L-alanine:glyoxylate aminotransferase to mitochondria in primary hyperoxaluria patients depends upon activation of a cryptic mitochondrial targeting sequence by a point mutation. *Proc Natl Acad Sci U S A*, **88**, 10900–10904.

37 Abe, Y., Shodai, T., Muto, T., Mihara, K., Torii, H., Nishikawa, S., Endo, T., and Kohda, D. (2000) Structural basis of presequence recognition by the mitochondrial protein import receptor Tom20. *Cell*, **100**, 551–560.

38 Muto, T., Obita, T., Abe, Y., Shodai, T., Endo, T., and Kohda, D. (2001) NMR identification of the Tom20 binding segment in mitochondrial presequences. *J Mol Biol*, **306**, 137–143.

39 Lumb, M.J., Drake, A.F., and Danpure, C.J. (1999) Effect of N-terminal alpha-helix formation on the dimerization and intracellular targeting of alanine:glyoxylate aminotransferase. *J Biol Chem*, **274**, 20587–20596.

40 Glover, J.R., Andrews, D.W., and Rachubinski, R.A. (1994) Saccharomyces cerevisiae peroxisomal thiolase is imported as a dimer. *Proc Natl Acad Sci U S A*, **91**, 10541–10545.

41 McNew, J.A. and Goodman, J.M. (1994) An oligomeric protein is imported into peroxisomes in vivo. *J Cell Biol*, **127**, 1245–1257.

42 Chen, W.J. and Douglas, M.G. (1987) The role of protein structure in the mitochondrial import pathway. Unfolding of mitochondrially bound precursors is required for membrane translocation. *J Biol Chem*, **262**, 15605–15609.

43 Eilers, M. and Schatz, G. (1988) Protein unfolding and the energetics of protein translocation across biological membranes. *Cell*, **52**, 481–483.

44 Danpure, C.J., Jennings, P.R., Fryer, P., Purdue, P.E., and Allsop, J. (1994) Primary hyperoxaluria type 1: genotypic and phenotypic heterogeneity. *J Inherit Metab Dis*, **17**, 487–499.

45 Danpure, C.J., Cooper, P.J., Wise, P.J., and Jennings, P.R. (1989) An enzyme trafficking defect in two patients with primary hyperoxaluria type 1: peroxisomal alanine/glyoxylate aminotransferase rerouted to mitochondria. *J Cell Biol*, **108**, 1345–1352.

46 Leiper, J.M., Oatey, P.B., and Danpure, C.J. (1996) Inhibition of alanine:glyoxylate aminotransferase 1 dimerization is a prerequisite for its peroxisome-to-mitochondrion mistargeting in primary hyperoxaluria type 1. *J Cell Biol*, **135**, 939–951.

47 Santana, A., Salido, E., Torres, A., and Shapiro, L.J. (2003) Primary hyperoxaluria type 1 in the Canary Islands: a conformational disease due to I244T mutation in the P11L-containing alanine:glyoxylate aminotransferase. *Proc Natl Acad Sci U S A*, **100**, 7277–7282.

48 Danpure, C.J., Purdue, P.E., Fryer, P., Griffiths, S., Allsop, J., Lumb, M.J., Guttridge, K.M., Jennings, P.R., Scheinman, J.I., Mauer, S.M., and Davidson, N.O. (1993) Enzymological and mutational analysis of a complex primary hyperoxaluria type 1 phenotype involving alanine:glyoxylate aminotransferase peroxisome-to-mitochondrion mistargeting and intraperoxisomal aggregation. *Am J Hum Genet*, **53**, 417–432.

49 Pirulli, D., Puzzer, D., Ferri, L., Crovella, S., Amoroso, A., Ferrettini, C., Marangella, M., Mazzola, G., and Florian, F. (1999) Molecular analysis of hyperoxaluria type 1 in Italian patients reveals eight new mutations in the alanine: glyoxylate aminotransferase gene. *Hum Genet*, **104**, 523–525.

50 Purdue, P.E., Lumb, M.J., Allsop, J., Minatogawa, Y., and Danpure, C.J.

(1992) A glycine-to-glutamate substitution abolishes alanine:glyoxylate aminotransferase catalytic activity in a subset of patients with primary hyperoxaluria type 1. *Genomics*, **13**, 215–218.

51 Cellini, B., Bertoldi, M., Montioli, R., Paiardini, A., and Borri, V.C. (2007) Human wild-type alanine:glyoxylate aminotransferase and its naturally occurring G82E variant: functional properties and physiological implications. *Biochem J*, **408**, 39–50.

52 Nishiyama, K., Funai, T., Katafuchi, R., Hattori, F., Onoyama, K., and Ichiyama, A. (1991) Primary hyperoxaluria type I due to a point mutation of T to C in the coding region of the serine:pyruvate aminotransferase gene. *Biochem Biophys Res Commun*, **176**, 1093–1099.

53 Nishiyama, K., Funai, T., Yokota, S., and Ichiyama, A. (1993) ATP-dependent degradation of a mutant serine: pyruvate/alanine:glyoxylate aminotransferase in a primary hyperoxaluria type 1 case. *J Cell Biol*, **123**, 1237–1248.

54 Hoppe, B., Danpure, C.J., Rumsby, G., Fryer, P., Jennings, P.R., Blau, N., Schubiger, G., Neuhaus, T., and Leumann, E. (1997) A vertical (pseudodominant) pattern of inheritance in the autosomal recessive disease primary hyperoxaluria type 1: lack of relationship between genotype, enzymic phenotype, and disease severity. *Am J Kidney Dis*, **29**, 36–44.

55 Danpure, C.J. (1991) Molecular and clinical heterogeneity in primary hyperoxaluria type 1. *Am J Kidney Dis*, **17**, 366–369.

56 Rumsby, G., Sharma, A., Cregeen, D.P., and Solomon, L.R. (2001) Primary hyperoxaluria type 2 without L-glycericaciduria: is the disease under-diagnosed? *Nephrol Dial Transplant*, **16**, 1697–1699.

57 Van Acker, K.J., Eyskens, F.J., Espeel, M.F., Wanders, R.J., Dekker, C., Kerckaert, I.O., and Roels, F. (1996) Hyperoxaluria with hyperglycoluria not due to alanine:glyoxylate aminotransferase defect: a novel type of primary hyperoxaluria. *Kidney Int*, **50**, 1747–1752.

58 Danpure, C.J., Jennings, P.R., and Watts, R.W. (1987) Enzymological diagnosis of primary hyperoxaluria type 1 by measurement of hepatic alanine: glyoxylate aminotransferase activity. *Lancet*, **1**, 289–291.

59 Rumsby, G., Weir, T., and Samuell, C.T. (1997) A semiautomated alanine:glyoxylate aminotransferase assay for the tissue diagnosis of primary hyperoxaluria type 1. *Ann Clin Biochem*, **34** (Pt 4), 400–404.

60 Rumsby, G. (2006) Is liver analysis still required for the diagnosis of primary hyperoxaluria type 2? *Nephrol Dial Transplant*, **21**, 2063–2064.

61 Knight, J., Holmes, R.P., Milliner, D.S., Monico, C.G., and Cramer, S.D. (2006) Glyoxylate reductase activity in blood mononuclear cells and the diagnosis of primary hyperoxaluria type 2. *Nephrol Dial Transplant*, **21**, 2292–2295.

62 Monico, C.G., Rossetti, S., Schwanz, H.A., Olson, J.B., Lundquist, P.A., Dawson, D.B., Harris, P.C., and Milliner, D.S. (2007) Comprehensive mutation screening in 55 probands with type 1 primary hyperoxaluria shows feasibility of a gene-based diagnosis. *J Am Soc Nephrol*, **18**, 1905–1914.

63 Rumsby, G. (2005) An overview of the role of genotyping in the diagnosis of the primary hyperoxalurias. *Urol Res*, **33**, 318–320.

64 Danpure, C.J., Jennings, P.R., Penketh, R.J., Wise, P.J., Cooper, P.J., and Rodeck, C.H. (1989) Fetal liver alanine: glyoxylate aminotransferase and the prenatal diagnosis of primary hyperoxaluria type 1. *Prenat Diagn*, **9**, 271–281.

65 Danpure, C.J., Jennings, P.R., Penketh, R.J., Wise, P.J., and Rodeck, C.H. (1988) Prenatal exclusion of primary hyperoxaluria type 1. *Lancet*, **1**, 367.

66 Danpure, C.J. and Rumsby, G. (1996) Strategies for the prenatal diagnosis

67. Watts, R.W., Veall, N., and Purkiss, P. (1984) Oxalate dynamics and removal rates during haemodialysis and peritoneal dialysis in patients with primary hyperoxaluria and severe renal failure. *Clin Sci*, **66**, 591–597.
68. Gibbs, D.A. and Watts, R.W. (1970) The action of pyridoxine in primary hyperoxaluria. *Clin Sci*, **38**, 277–286.
69. Monico, C.G., Olson, J.B., and Milliner, D.S. (2005) Implications of genotype and enzyme phenotype in pyridoxine response of patients with type I primary hyperoxaluria. *Am J Nephrol*, **25**, 183–188.
70. Monico, C.G., Rossetti, S., Olson, J.B., and Milliner, D.S. (2005) Pyridoxine effect in type I primary hyperoxaluria is associated with the most common mutant allele. *Kidney Int*, **67**, 1704–1709.
71. van Woerden, C.S., Groothoff, J.W., Wijburg, F.A., Annink, C., Wanders, R.J., and Waterham, H.R. (2004) Clinical implications of mutation analysis in primary hyperoxaluria type 1. *Kidney Int*, **66**, 746–752.
72. Danpure, C.J. (1991) Scientific rationale for hepatorenal transplantation in primary hyperoxaluria type 1, in *Transplantation and Clinical Immunology*, vol. 22 (ed. J.L. Touraine), Excerpta Medica, Amsterdam, pp. 91–98.
73. de Pauw, L., Gelin, M., Danpure, C.J., Vereerstraeten, P., Adler, M., Abramowicz, D., and Toussaint, C. (1990) Combined liver-kidney transplantation in primary hyperoxaluria type 1. *Transplantation*, **50**, 886–887.
74. Watts, R.W., Calne, R.Y., Rolles, K., Danpure, C.J., Morgan, S.H., Mansell, M.A., Williams, R., and Purkiss, P. (1987) Successful treatment of primary hyperoxaluria type I by combined hepatic and renal transplantation. *Lancet*, **2**, 474–475.
75. Watts, R.W., Danpure, C.J., de Pauw, L., Toussaint (1991) Combined liver-kidney and isolated liver transplantations for primary hyperoxaluria type 1: the European experience. The European Study Group on Transplantation in Hyperoxaluria Type 1. *Nephrol Dial Transplant*, **6**, 502–511.
76. Watts, R.W., Morgan, S.H., Danpure, C.J., Purkiss, P., Calne, R.Y., Rolles, K., Baker, L.R., Mansell, M.A., Smith, L.H., and Merion, R.M. (1991) Combined hepatic and renal transplantation in primary hyperoxaluria type I: clinical report of nine cases. *Am J Med*, **90**, 179–188.
77. Finlayson, B. and Reid, F. (1978) The expectation of free and fixed particles in urinary stone disease. *Invest Urol*, **15**, 442–448.
78. Hackett, R.L., Shevock, P.N., and Khan, S.R. (1994) Madin-Darby canine kidney cells are injured by exposure to oxalate and to calcium oxalate crystals. *Urol Res*, **22**, 197–203.
79. Hackett, R.L., Shevock, P.N., and Khan, S.R. (1990) Cell injury associated calcium oxalate crystalluria. *J Urol*, **144**, 1535–1538.
80. Thamilselvan, S. and Khan, S.R. (1998) Oxalate and calcium oxalate crystals are injurious to renal epithelial cells: results of in vivo and in vitro studies. *J Nephrol*, **11** (Suppl 1), 66–69.
81. Khan, S.R. (2005) Hyperoxaluria-induced oxidative stress and antioxidants for renal protection. *Urol Res*, **33**, 349–357.
82. Scheid, C., Koul, H., Hill, W.A., Luber Narod, J., Jonassen, J., Honeyman, T., Kennington, L., Kohli, R., Hodapp, J., Ayvazian, P., and Menon, M. (1996) Oxalate toxicity in LLC-PK1 cells, a line of renal epithelial cells. *J Urol*, **155**, 1112–1116.
83. Scheid, C., Koul, H., Hill, W.A., Luber Narod, J., Kennington, L., Honeyman, T., Jonassen, J., and Menon, M. (1996) Oxalate toxicity in LLC-PK1 cells: role of free radicals. *Kidney Int*, **49**, 413–419.
84. Thamilselvan, S., Byer, K.J., Hackett, R.L., and Khan, S.R. (2000) Free radical scavengers, catalase and

superoxide dismutase provide protection from oxalate-associated injury to LLC-PK1 and MDCK cells. *J Urol*, **164**, 224–229.

85 Koul, H., Kennington, L., Nair, G., Honeyman, T., Menon, M., and Scheid, C. (1994) Oxalate-induced initiation of DNA synthesis in LLC-PK1 cells, a line of renal epithelial cells. *Biochem Biophys Res Commun*, **205**, 1632–1637.

86 Jonassen, J.A., Kohjimoto, Y., Scheid, C.R., and Schmidt, M. (2005) Oxalate toxicity in renal cells. *Urol Res*, **33**, 329–339.

87 Miller, C., Kennington, L., Cooney, R., Kohjimoto, Y., Cao, L.C., Honeyman, T., Pullman, J., Jonassen, J., and Scheid, C. (2000) Oxalate toxicity in renal epithelial cells: characteristics of apoptosis and necrosis. *Toxicol Appl Pharmacol*, **162**, 132–141.

88 Cao, L.C., Honeyman, T.W., Cooney, R., Kennington, L., Scheid, C.R., and Jonassen, J.A. (2004) Mitochondrial dysfunction is a primary event in renal cell oxalate toxicity. *Kidney Int*, **66**, 1890–1900.

89 McMartin, K.E. and Wallace, K.B. (2005) Calcium oxalate monohydrate, a metabolite of ethylene glycol, is toxic for rat renal mitochondrial function. *Toxicol Sci*, **84**, 195–200.

90 Poldelski, V., Johnson, A., Wright, S., Rosa, V.D., and Zager, R.A. (2001) Ethylene glycol-mediated tubular injury: identification of critical metabolites and injury pathways. *Am J Kidney Dis*, **38**, 339–348.

91 Guo, C., Cenac, T.A., Li, Y., and McMartin, K.E. (2007) Calcium oxalate, and not other metabolites, is responsible for the renal toxicity of ethylene glycol. *Toxicol Lett*, **173**, 8–16.

92 Guo, C. and McMartin, K.E. (2005) The cytotoxicity of oxalate, metabolite of ethylene glycol, is due to calcium oxalate monohydrate formation. *Toxicology*, **208**, 347–355.

93 Hirose, M., Tozawa, K., Okada, A., Hamamoto, S., Shimizu, H., Kubota, Y., Itoh, Y., Yasui, T., and Kohri, K. (2008) Glyoxylate induces renal tubular cell injury and microstructural changes in experimental mouse. *Urol Res*, **36**, 139–147.

94 Behnam, J.T., Williams, E.L., Brink, S., Rumsby, G., and Danpure, C.J. (2006) Reconstruction of human hepatocyte glyoxylate metabolic pathways in stably transformed Chinese-hamster ovary cells. *Biochem J*, **394**, 409–416.

12
Pathophysiology of Endogenous Toxins and Their Relation to Inborn Errors of Metabolism and Drug-Mediated Toxicities
Vangala Subrahmanyam

Unmanageable toxic events are one of the major reasons for terminating drug development during preclinical and phase 1 clinical trials. In addition, postmarketing idiosyncratic toxicity in humans resulted either in the withdrawal of some drugs from the market or black box warnings limiting their use. In many cases, the mechanisms of preclinical or idiosyncratic toxicity in humans are not well understood and efforts to identify the primary targets of cell injury remains to be elusive. Furthermore, the injured cell typically dies either via necrosis or via apoptosis, irrespective of the cell type, tissues, or organs. Thus, predictive toxicogenomic research primarily lighted up genes involved in drug metabolism and oxidative stress providing little clues for tissue specific mechanisms of toxicity. Moreover, the currently used "clinical chemistry panel" in toxicology studies is not adequate to comprehensively delineate the intricacies behind the toxicities observed in several different organs. There is some evidence indicating that the pathophysiology of drug-mediated toxicity mimics that of inborn errors of metabolism (IEM). IEM's comprise a group of disorders in which a defect in a single gene causes a clinically significant block in a metabolic pathway or metabolite transport resulting in the accumulation of substrates or metabolites leading to serious toxicities. This review is intended to highlight the underlying mechanisms of the endogenous toxicities mediated by accumulated substrates and explore useful links to drug-mediated toxicities wherever possible. Moreover, it is hypothesized that the primary event in drug toxicity is the inhibition of enzymes/transport proteins involved in the intermediary metabolism and transport of carbohydrates, fatty acids, amino acids, proteins, and nucleic acids. Thus, the accumulated substrates could be potential biomarkers for several drug-induced toxicities. Therefore, understanding the mechanisms of IEM-induced toxicities could help in understanding the drug-mediated toxicities also. Metabonomics, a rapidly evolving tool to understand IEMs may be applied as "expanded clinical chemistry panel" during preclinical and clinical safety evaluation. There is substantial evidence that accumulated substrates and/or metabolites due to genetic deficiency of enzymes or transporters can induce oxidative stress, interfere with mitochondrial energy production, and further metabolize to reactive intermediates that covalently interact with cellular macromolecules (e.g.,

Endogenous Toxins. Diet, Genetics, Disease and Treatment.
Edited by Peter J. O'Brien and W. Robert Bruce
Copyright © 2010 WILEY-VCH Verlag GmbH & Co. KGaA, Weinheim
ISBN: 978-3-527-32363-0

lipid peroxidation products: malondialdehyde, 4-hydroxynonenal), thus resulting in tissue injury. Some examples of endogenous metabolites with documented evidence of producing oxidative cell injury include glucose, cis-4-decenoic acid, 5-oxoproline, 3-hydroxy-3-methylglutaric acid. Some of these examples are also reviewed in this chapter.

12.1
Introduction

In recent years, drug discovery and development research to market novel therapeutic agents has become a very "high risk" business with shrinking pipelines from all major pharma companies [1, 2]. One of the reasons for high failure rate is that many NCEs result in unmanageable toxic adverse effects [2–4]. Managing the risk would be possible if appropriate biomarkers of toxicity are available. However, the mechanistic understanding and the clinical significance of preclinical toxicity observed in animal models are severely hampered by the lack of appropriate biomarkers [5, 6]. Unfortunately, the currently used clinical chemistry panel in toxicology studies is not adequate for toxicities observed in several different organs with different underlying mechanisms [5]. Several laboratories in academia and the pharmaceutical industry are actively engaged in biomarker research to find novel useful markers to predict drug toxicity and relate preclinical safety issues to clinical situation. A recent special edition of Toxicology journal has several chapters on the current status of various toxicity biomarkers and their applications [7–10] and interested readers may consult those articles.

I recently hypothesized [5] that many of the drug-mediated toxic adverse effects may mimic the toxicities observed with inborn errors of metabolism (IEM). IEMs are a group of disorders in which a single gene defect in an enzyme or a transport protein can block the metabolic pathway or the transport resulting [5, 11–16] in Figure 12.1:

1. toxic accumulation of substrates and/or metabolites behind the block;
2. toxic accumulation of intermediates from alternative metabolic pathways;
3. defects in mitochondrial energy production or utilization due to deficiency of the products beyond the block;
4. combination of all of the above metabolic deviations.

The accumulated substrates or metabolites appear to produce pathophysiology similar to several drug-induced toxicities [5]. Therefore, the accumulated metabolites are referred to as *endogenous toxins*. In this review, an attempt is made to find some examples from literature where drugs inhibit metabolic enzymes or transporters resulting in toxicities similar to IEM. In addition, this chapter also reviews the literature, wherever available, on the mechanisms of toxicity induced by accumulated metabolites.

The major categories of IEMs include organic acidemias, metabolic acidosis, urea cycle defects, disorders of carbohydrate, amino acid, fatty acid and nucleic

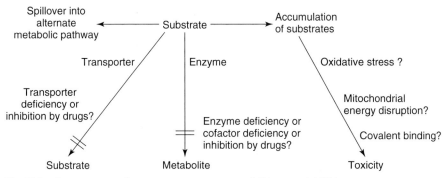

Fig. 12.1 Consequences of an enzyme or transporter deficiency or inhibition.

acid metabolism, lysosomal storage disorders, and peroxisomal disorders ([5] and references within). Additional inherited disorders are linked to defects in cytochrome P450 enzymes catalyzing cholesterol, bile acid, steroid, and vitamin D3 synthesis and metabolism. In recent years, genetic defects in transporters that mediate the transport of endogenous metabolites and drugs have also been described with well-defined clinical pathology [16]. The significance of these enzymes and toxicological outcomes of their genetic deficiencies, a brief link to human genetic defects, and associated diseases were described in a previous review article by the author [5]. This chapter primarily focuses on reviewing the mechanisms of pathophysiology of accumulated substrates/metabolites and attempts to find the similarities and differences in pathophysiological mechanisms for various diseases. Moreover, this chapter cites some relevant examples in the literature for various drugs that inhibit enzymes or transporters involved in intermediary metabolism and associated pathophysiology to determine if the accumulated substrates/metabolites could serve as potential biomarkers of drug-mediated toxic side effects.

12.2 Pathophysiology of Hepatobiliary System

The primary physiological function of bile secretion is to excrete biowaste from the body [16–20]. However, bile also aids in the digestion of food. Recent studies reveal that bile acids are also signaling molecules that activate several nuclear receptors and regulate many physiological pathways and processes to maintain bile acid and cholesterol homeostasis [21]. Bile formation in liver occurs by a concerted action of membrane transporters in hepatocytes and cholangiocytes (Figure 12.2) [22]. The primary site of bile production is tiny biliary canaliculi, embedded between adjacent hepatocytes. Bile components are synthesized within hepatocytes and secreted across their canalicular (apical) cell membranes into the biliary canaliculi. Alternatively, liver cells may extract and take up bile forming compounds from the sinusoidal blood stream via transport across their basolateral (sinusoidal) membranes. After intracellular uptake, sequestration, biotransformation, or conjugation,

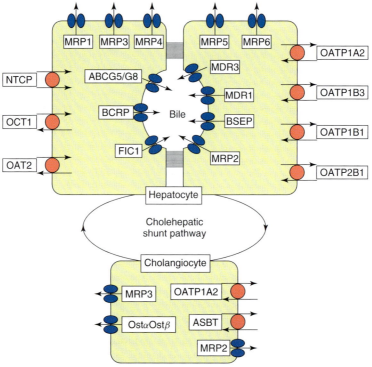

Fig. 12.2 Bile salt transporters in human liver and cholangiocytes. Efflux transporters (blue symbols): BSEP, bile salt export pump; MDR, multidrug resistance protein; MRP, multidrug resistance-associated protein; ABCG5/8; BCRP, breast cancer resistance protein; Ost_/Ost_. Uptake transporters (red symbols): ASBT, apical sodium dependent bile salt transporter; NTCP, sodium taurocholate cotransporting polypeptide; OATP, organic anion transporting polypeptide; OCT, organic cation transporter; OAT, organic anion transporter. (Reproduced from [20].)

they produce metabolites that are amenable for secretion across the canalicular membrane. The secretory products such as bile acids and glutathione (GSH) are excreted at concentrations that are high enough to generate osmotic water flow into the biliary canaliculi. In addition, lipid vesicles are detached from the apical membrane to form biliary micelles composed of phospholipids (PLs), cholesterol, and bile salts. Plasma proteins are also secreted into bile via vesicular transport and by apical exocytosis. Bile is then drained into the branches of intrahepatic bile ductules that converge to the common hepatic bile duct. Inorganic anions such as bicarbonate and chloride are secreted into bile, a function performed by cholangiocytes.

12.2.1
Arginine Vasopressin and Bile Flow

The regulation of bile formation and secretion appears to be more complex than what was described above in the literature. It has been shown that arginine vasopressin (AV) elicits elaborate intercellular Ca^{2+} waves in the liver [23, 24]. These waves propagate across hepatocyte plates following a lobular gradient in V1a vasopressin receptor density. The changes in this receptor distribution have been shown to control Ca^{2+} wave propagation and bile flow. Although basal circulating vasopressin levels do not play a major role in the regulation of V1a receptor expression, increased vasopressin concentrations within physiological limits for 24 hours can abolish the lobular gradient in V1a receptor. In animals in which the V1a receptor gradient was abolished, intercellular Ca^{2+} waves were impaired due to the equalization of Ca^{2+} responses in the various zones of the lobule. In the isolated perfused liver, the early increase in vasopressin-induced bile flow observed in control rats was much smaller if the V1a receptor density gradient was abolished. These findings suggest that V1a vasopressin receptor distribution controls intercellular Ca^{2+} wave propagation and bile flow. Therefore, any drugs/chemicals that interfere with V1a receptors (agonists and antagonists) may have effects on bile secretion and flow, the mechanisms of which are poorly understood.

12.3
Inborn Errors of Bile Acid/Salt Transporter Deficiencies

Many transport systems for secretion of substances (bile aids, GSH, PL, and conjugates) have now been identified in both membrane domains of hepatocytes [16–20] (Figure 12.2 and Table 12.1). The transporters on the basolateral and canalicular side actually transport biowaste into liver and process for excretion into bile. Liver contains two types of transporters, primarily related to their function of either uptake or efflux. Uptake transporters are expressed on the basolateral or sinusoidal side of liver to increase the uptake of drugs into liver for metabolism and excretion. The efflux transporters are expressed on the apical or canalicular side of liver and facilitate the excretion of drugs and their metabolites into bile.

Genetic deficiency of some of the transporters can result in accumulation of bile in canalicular space and related toxicities (Table 12.2). The molecular bases for several forms of cholestatic liver diseases have been defined and are briefly discussed below ([5, 17, 25] and the references within). Since these diseases have been reviewed in detail [17, 25], interested readers can consult these reviews for specific references.

12.3.1
Progressive Familial Cholestasis Type 1 (PFIC1) and Benign Recurrent Intrahepatic Cholestasis (BRIC)

Progressive familial intrahepatic cholestasis (PFIC) represents a clinically important group of cholestatic disorders of infancy that often progresses to cirrhosis. PFIC 1 is a rare autosomal recessive disorder in which cholestasis with impaired excretion of chenodeoxycholic acid into bile was observed. PFIC1 is characterized by jaundice, severe itching, high serum bile salt concentration, low biliary bile

Tab. 12.1 Endogenous and exogenous substrates for ABC transporters

Name	Endogenous substrate	Exogenous substrates	Inhibitors
MDR1	Estrogen glucuronide conjugates (estradiol, estriol), endorphin, glutamate, steroids (cortisol, aldosterone, corticosterone), beta-amyloid, 1-O-alkyl-2-acetyl-sn-glycero-3-phosphocholine (generically platelet-activating factor, PAF)	ActinomycinD, colchicine, daunorubicin, digoxin, docetaxel, doxorubicin, epirubicin, etoposide, fexofenadine, indinavir, mitomycin C, mitoxantrone, paclitaxel, teniposide, vincristine, vinblastine	Ritonavir, cyclosporine, verapamil, erythromycin, ketocoanzole, itraconazole, quinidine, elacridar (GF 120918) LY335979, valspodar (PSC 833)
MDR3	Phosphatidylcholine and other phospholipids	Digoxin, paclitaxel, vinblastine	–
MRP1	Estradiol-17beta(beta-D-glucuronide) glutathione, glutathione S-conjugate leukotriene C4, glucuronosyl bilirubin	Adefovir, indinavir, anthracyclines, colchicine, etoposide, heavy metals (arsenite, arsenate, antimonials), vincristine, vinblastine, paclitaxel	MK571, biricodar (VX710)
MRP2	Estradiol-17beta(beta-D-glucuronide), glutathione, glutathione S-conjugate Leukotriene C4, glucuronosyl bilirubin glucuronides	Indinavir, cisplatin, CPT-11, doxorubicin, etoposide, methotrexate, SN-38, vincristine, vinblastine	Cyclosporine
MRP3	S-(2,4-dinitrophenyl) glutathione	Cisplatin, doxorubicin, etoposide, methotrexate, teniposide, vincristine,	MK571
MRP4	Glucuronide and glutathione conjugates	Methotrexate, nucleotide analogs, PMEA	MK571
MRP5	Glutamate and phosphate conjugates	Doxorubicin, methotrexate, nucleotide analogs, topotecan,	–
MRP6	Cyclic nucleotides (cAMP, cGMP), glutathione conjugate	Doxorubicin, etoposide, teniposide	
MRP7	17beta-estradiol-(17-beta-D-glucuronide)	??	LeukotrieneC4

Tab. 12.1 Continued

Name	Endogenous substrate	Exogenous substrates	Inhibitors
MRP8	17beta-estradiol-(17-beta-D-glucuronide), leukotriene C4, cyclic nucleotides, glutathione, glucuronide, and sulfate-conjugated substrates such as LTC4, estrone-3-sulfate, and bile acids glycocholate, taurocholate	5'-Fluorouracil, 5'-fluoro-2'-deoxyuridine, 5'-fluoro-5'-deoxyuridine, PMEA[a]	–
BCRP	Heme or porphyrin	Bisantrene, camptothecin, daunorubicin, doxorubicin, epirubicin, flavopiridol, mitoxantrone, rosuvastatin, S-38, sulfasalazine, topotecan, rosuvastatin	Elacridar (GF120918), biricodor (VX710)
BSEP	Bile salts, taurocholate, glycocholate, taurochenodeoxycholate, tauroursodeoxycholate	Vinblastine, pravastatin	–
NTCP	Bile salts	Rosuvastatin	
OAT2	–	Zidovudine	
OATP1A2	–	–	
OATP1B1	Thyroxine	Rifampin, rosuvastatin, methotrexate, fluvastatin, pravastatin, simvastatin	
OATP1B3	–	Digoxin, methotrexate, rifampin	
OATP2B1	–	Dehydroepiandrosterone-sulfate (DHEAS) and estrone-3-sulfate (E3S), atorvastatin, fluvastatin	

[a] PMEA, 2',3'-dideoxycytidine 9'-(2'-hosphonylmethoxynyl)adenine.
?? = Not Known

salt concentration, and low serum -glutamyltranspeptidase. Onset in infancy leads to progressive hepatic fibrosis and death. Another inherited liver disease, benign recurrent intrahepatic cholestasis (BRIC), has also been mapped to the same region, suggesting that both PFIC1 and BRIC are caused by mutations in a single gene, which has been termed as *FIC1* (for *familial intrahepatic cholestasis* 1). BRIC, unlike PFIC1, manifests itself during adolescence or early adulthood and is characterized by recurrent episodes of cholestasis.

Tab. 12.2 Genetic defects in transporters and related disorders

Transporter gene mutation	Substrate	Clinical feature/disorder	Biomarker
PFIC[a] type I	?	PFIC type 1 (steatorrhea, diarrhea, jaundice, hepatosplenomegaly, and fatal liver failure within the first 10 years of life)	GGT in serum? Bile salts in serum and bile
BSEP (sPGP)	Bile salts	PFIC type 2 (similar to PFIC type 1, without watery diarrhea)	GGT in serum? Bile salts in serum and bile; phospholipid and cholesterol in bile?
MDR3/MDR2	Phosphatidyl-choline	PFIC type 3 (bile duct proliferation and cirrhosis; portal hypertension; hepatosplenomegaly; pruritus	GGT in serum
MRP2 (cMOAT)	Amphipathic drugs (anionic and neutral)	Dubin-Johnson syndrome (nonhemolytic hyperbilirubinemia; lysosomal accumulation of black pigment)	Bilirubin; urinary coproporphyrin
ABCG5	Phytosterols	Sitosterolemia (tendon xanthomas; accelerated atherosclerosis; hemolytic episodes; arthritis; arthralgia)	Sitosterol in bile and plasma

[a] PFIC indicates progressive familial intrahepatic cholestasis; GGT, gamma-glutamyltranspeptidase; BSEP, bile salt export pump; sPGP, sister P-glycoprotein; MDR, multi drug resistance protein; cMOAT, multi specific organic anion transporter; ABCG, adenosine 5'-triphosphate-binding cassette-Type G protein.
? = Not Known.

12.3.2
Progressive Familial Cholestasis Types 2 and 3 (PFIC2 and PFIC3)

Among the hereditary forms of human cholestasis, two other subtypes of progressive familial intrahepatic cholestasis (PFIC2 and PFIC3) exist. PFIC2 and PFIC3 have been mapped to mutations in genes of bile salt exporter protein (BSEP) and multidrug resistance p-glycoprotein 3 (MDR3), respectively. PFIC2 has been characterized by normal -glutamyltranspeptidase levels and amorphous bile with neonatal hepatitis. PFIC3 is characterized by high serum, glutamyltranspeptidase levels, absence of biliary PLs, normal serum bile salt concentrations and the presence of extensive bile duct proliferation, cirrhosis, and fibrosis.

12.3.3
Dubin-Johnson Syndrome

Dubin-Johnson syndrome is a rare autosomal recessive liver disorder characterized by chronic conjugated hyperbilirubinemia with benign jaundice and a typical dark pigment accumulation in liver parenchymal cells. Patients also have impaired hepatobiliary transport of nonbile salt organic anions and amphiphilic anionic conjugates from hepatocytes into the bile. Selective absence of the canalicular bilirubin transporter multidrug resistance associated protein (MRP2) has been demonstrated in hepatocytes from patients with Dubin-Johnson syndrome.

12.3.4
Sitosterolemia

Sitosterolemia is a rare inherited sterol storage lipid metabolic disease characterized by the accumulation of plant sterols (mainly beta-sitostero and campesterol and stigmasterol) in blood and tissues. Enhanced sterol absorption along with reduced biliary sterol secretion and decreased cholesterol synthesis are the major causes of the ingested sterol accumulation. Sitosterolemia is characterized by tendon and tuberous xanthomas, and strong propensity toward premature coronary atherosclerosis. Sitosterolemia occurs in humans with mutations in transporter gene ABCG5. Mutations in another gene ABCG8 results in the loss of biliary cholesterol secretion, but biliary sitosterol secretion appeared to be preserved.

12.4
Drugs and Cholestatic Liver Disease

Examples of drugs that have been reported to cause cholestasis in humans include amiodarone, amitriptyline, ampicillin, ampicillin/clavulanic acid, anabolic steroids, captopril, carbamazepine, cordarone, chlorpromazine, cyclosporine A, erythromycin, estradiol, glybenclamide, haloperidol, imipramine, naproxen, phenytoin, rifampin, rifamycinSV, sulfamethoxazole, trimethoprim, tetracycline, oral contraceptives, among many others [26–28].

Most patients with drug-induced cholestasis will recover fully within weeks after discontinuing the drug, but jaundice, itching, and abnormal liver tests can last for months in some patients after stopping the drug. Occasionally, patients can develop chronic cholestatic liver disease and liver failure.

Drug-induced cholestasis can manifest either acutely following single doses or chronically following multiple doses of drug treatment ([29] and the references within). In both cases, the cholestasis can be of either intrahepatic or extrahepatic origin. The extrahepatic cholestasis can occur as a secondary event to biliary sludge or calculi (e.g., octreotide, sulindac), bile duct destruction (e.g., ajmaline, chlorpropamide, terbinafine, amoxicillin-clavulanate), pancreatitis (e.g., codeine, isoniazid, riluzole), or pancreatitis and hepatitis (e.g., trimethoprimsulfamethoxazole). The mechanism of drug-induced cholestasis was presumed to be an idiosyncratic reaction of immunologic origin. However, in recent years, evidence has accumulated attributing it to the inhibition (or down-regulation) of drug transporters mimicking the cholestasis induced by genetic deficiency of transporters (Table 12.2) [30–34]. Hepatotoxicity associated with bosentan [35] and troglitazone [36] has been attributed, in part, to the inhibition of bile salt export pump, BSEP. Cyclosporine A [37], on the other hand, was shown to internalize the BSEP in isolated rat hepatocyte couplets. Nevertheless, in rats, single i.v. doses of troglitazone, bosentan, or cyclosporine A resulted in increased plasma/serum bile acid concentrations.

12.4.1
Bosentan Hepatotoxicity – Role of Inhibition of Bile Acid/Salt Transporters

Bosentan, a dual antagonist of endothelin (ET-A&B) receptors, was the first ET receptor antagonist approved for the management of pulmonary arterial hypertension [38]. During clinical trials and postmarketing monitoring, ~11% of patients treated with bosentan developed hepatotoxicity and had elevated serum transaminases less than threefold normal levels (discussed in [35]). As a result, all bosentan-treated patients must undergo frequent biochemical monitoring of liver enzymes. Furthermore, in patients with predisposing factors such as liver fibrosis, potential for hepatotoxicity needs to be taken into consideration.

There has been considerable interest in determining the mechanisms underlying bosentan-induced hepatotoxicity. However, one of the perplexing problems was that bosentan was not hepatotoxic in preclinical animal models and therefore the clinical appearance of hepatotoxicity appears to be idiosyncratic in nature [35]. It has been proposed that bosentan-induced liver injury is due to accumulation of bile acids in the hepatocyte. Bosentan inhibits BSEP/bsep in both rats and humans [35]. Inhibition of BSEP as a mechanism for hepatotoxicity is also consistent with the reported observations in patients, where serum bile acids were increased before the development of hepatotoxicity during clinical trials with bosentan. Moreover, bosentan hepatotoxicity was quite common in patients treated with glyburide, another BSEP inhibitor. In support of this hypothesis, the intravenous administration of bosentan to rats also increased serum bile acids presumably through inhibition of Bsep, but bosentan-treated rats do not develop hepatotoxicity. Another potential explanation was provided by the recent observation that bosentan is a potent inhibitor of Na+-dependent taurocholate uptake in rat hepatocytes (IC50 5.4 μM) but a poor inhibitor of Na+-dependent taurocholate uptake in human hepatocytes (IC50 30 μM) [39]. In addition, bosentan was a more potent inhibitor of taurocholate uptake by rat Na+dependent taurocholate cotransporting polypeptide (Ntcp/Slc10a1) (IC50 0.71 μM) than human sodium taurocholate cotransporting polypeptide (NTCP) (SLC10A1) (IC50 24 μM) expressed in HEK293 cells. Na+-dependent uptake of taurocholate is mediated by uptake transporter Ntcp and NTCP in rat and human hepatocytes. Thus, bosentan is a more potent inhibitor of Ntcp than NTCP, and this should result in less intrahepatocyte accumulation of bile acids in rats during bosentan treatment [39]. The Ntcp/Slc10a1 is the only Na+-dependent bile acid transporter in hepatocytes and is responsible for the majority of bile acid uptake from sinusoidal blood. Bosentan inhibition of Ntcp-mediated bile acid uptake could counterbalance Bsep inhibition, reduce intrahepatocyte bile acid accumulation, and thereby prevent hepatotoxicity in the rat. Preferential inhibition by bosentan of rat Ntcp versus human NTCP would therefore provide a potential explanation for the rat resistance to bosentan-induced hepatotoxicity. Disparities between rat and human Ntcp/NTCP substrate specificities have been observed, including a difference in affinity for taurocholate uptake by Ntcp (Km $25\,\mu M$) and NTCP (Km $6\,\mu M$) [40, 41] indicating that species differences in inhibition of rat Ntcp/human NTCP by bosentan are plausible mechanisms for species differences in hepatotoxicity. Moreover, Ho et al. [42] reported that rosuvastatin is a substrate for NTCP but not for Ntcp.

12.4.2
Troglitazone Hepatotoxicity-Role of Bile Salt Export Pump (BSEP/Bsep)

Troglitazone is an insulin sensitizer of thiazolidinedione class for the treatment of type 2 noninsulin dependent diabetes mellitus [43]. This drug was introduced in 1997 and withdrawn in 2000 due to several cases of fulminant hepatic failure [44]. In some patients treated with troglitazone and who had hepatic failure, cholestasis was described. Several mechanisms have been proposed to explain troglitazone-induced hepatotoxicity [36, 45–49] in humans and one of them is inhibition of BSEP [36]. Troglitazone is extensively metabolized in the liver by sulfation, glucuronidation, and oxidation [47]. In patients with hepatic impairment, troglitazone sulfate was found to accumulate about fourfold in plasma, with a threefold increase in half-life. Studies in rats reproduced the cholestatic effects of troglitazone. Troglitazone sulfate undergoes biliary excretion and accounted for up to 60% of the dose in rats. Consequently, a strong reduction in bile flow was observed in isolated, perfused rat livers treated with troglitazone. The reduction of bile flow has been attributed to the inhibition of BSEP/Bsep. In isolated rat liver plasma membrane preparations, troglitazone, and troglitazone sulfate competitively inhibited the ATP-dependent taurocholate transport mediated by Bsep, with an apparent Ki value of 1.3 and $0.23\,\mu M$, respectively [36]. Troglitazone sulfate is accumulated in rat liver tissue indicating that the hepatobiliary export of this conjugated metabolite is an important rate limiting step in the elimination process. In addition, the inhibition of Bsep by troglitazone sulfate may result in the accumulation of bile acids resulting in cholestasis. Because the cholestatic events are also observed in humans, it has been proposed that cholestasis is perhaps one of the events associated in fulminant hepatitis observed in humans.

12.4.3
Cholestatic Hepatotoxicity Associated with Ethinyl Estradiol: Role of MRP2

Ethinyl estradiol (EE) is a derivative of estradiol. EE is an orally bioactive estrogen used in most oral contraceptives. Estradiol-17beta-D glucuronide (E217G) is a major metabolite that appears to be responsible for cholestasis [50, 51]. Administration of metabolite induces immediate and profound but transient cholestasis in rats when administered as a single bolus dose. Mottino *et al.* [51] found that an initial dose of E217G (15 mmol/kg iv) followed by five subsequent doses of 7.5 mmol/kg from 60 to 240 minutes induced a sustained 40–70% decrease in bile flow. E217G decreases bile flow and bile salt secretion in the rat in a reversible, dose-dependent manner.

The mechanism by which E217G induces cholestasis is multifactorial. E217G induces endocytic internalization of the Bsep (Abcb11) and of Mrp2 (Abcc2), thus abrogating their function. Bsep mediates the concentrative transport of bile salts across the canalicular membrane, thus generating the bile salt dependent component of bile flow, while Mrp2 mediates the transport of GSH and of numerous GSH and glucuronide conjugates into bile and is thus considered to be crucial for the generation of bile salt-independent bile flow. Inhibition of Bsep-mediated transport of bile salts by E217G has also been proposed as a mechanism by which it induces cholestasis.

Kan *et al.* [52] proposed an increase in paracellular permeability as a contributor to cholestasis, in response to E217G. Tight junction disruption is expected to impair formation of bile flow, both by dissipating bile-to-plasma osmotic gradients and by disturbing the apical-basolateral intramembrane diffusion barrier, which is critical for the maintenance of proper localization of apical versus basolateral transporters. Tight junctional proteins zonula occludens-1 (ZO-1) and occluding and the canalicular transporter Mrp2 were severely disrupted under sustained cholestatic conditions. Endocytic internalization of Mrp2 to the peri-canalicular region was apparent in 20 minutes after a single E217G administration; however, Mrp2 was found to be more deeply internalized and partially redistributed to the basolateral membrane under sustained cholestasis. In conclusion, acute E217G-induced cholestasis increased permeability of the tight junction, while sustained cholestasis provoked a significant redistribution of ZO-1, occludin, and Mrp2 in addition to increased permeability of the tight junction. This altered tight junction integrity and likely contributes to impaired bile secretion and may be causally related to changes in Mrp2 and Bsep localization.

12.5
Phospholipases and Phospholipidosis

Phospholipidosis is a lipid storage disorder in which excess PLs are accumulated within cells [53]. Although phospholipidosis is considered as an adaptive response, recent evidence indicates that phospholipidosis occurring in hepatocytes, kidneys, and lungs can result in failure of liver, kidney, and respiratory function, respectively. Phospholipidosis was documented as a side effect with more than 50 drugs on the market [53]. In addition, phospholipase A2 (PLA2) deficiency has been associated with phospholipidosis [54, 55].

Phospholipases are enzymes that hydrolyze PLs. Several types of phospholipases including phospholipase A1, phospholipase B (PLB), phospholipase C (PLC), and phospholipase D (PLD) are present in mammalian tissues with varied functions [53, 56]. These enzymes may be placed into two broad classes, acyl hydrolases and phosphodiesterases.

- Acyl hydrolases:
  - Phospholipase A1 (PLA1), which hydrolyzes the lipid at the sn-1-acyl ester bond;
  - PLA2, which acts at the sn-2 position;
  - PLB which acts at both the sn-1 and sn-2 positions.
- Phosphodiesterases:
  - PLC, which cleaves the glycerol-phosphate bond;
  - PLD, which removes the base from the phospholipid.

The superfamily of PLA2 enzymes currently consists of at least 15 groups classified into many subgroups and includes five distinct types of enzymes:

- secreted phospholipase A2 (sPLA2)
- cytosolic phospholipase A2 (cPLA2)
- Ca^{2+} independent phospholipase A2 (iPLA2)
- platelet-activating factor acetylhydrolases (PAF-AHs)
- lysosomal phospholipase A2 (LPLA2) [54, 55]
- lysosomal phospholipase A2 (aiPLA2) [57–59].

12.5.1
Lysosomal PhospholipaseA2 Deficiency and Phospholipidosis

LPLA2 deficiency has recently been associated with phospholipidosis [54, 55]. A marked accumulation of PLs, in particular of phosphatidylethanolamine and phosphatidylcholine (PC), was found in the alveolar macrophages, the peritoneal macrophages, and the spleens of LPLA2 deficient (Lpla2−/−) mice. Ultrastructural examination of the alveolar and peritoneal macrophages of Lpla2−/− mouse revealed the appearance of foam cells with lamellar inclusion bodies, a hallmark of cellular phospholipidosis. Increased lung surfactant phospholipid and splenomegaly were observed in one-year-old mice. Thus, a deficiency of LPLA2 results in foam cell formation, surfactant lipid accumulation, splenomegaly, and phospholipidosis in mice. In contrast, a deficiency of cPLA2 deficiency has not been associated with phospholipidosis.

Another aiPLA2 was identified [57, 58], which is also involved in lung phospholipid metabolism. This enzyme appears to play an important role in the degradation of internalized dipalmitoylphosphatidylcholine (DPPC) by rat lungs. However, this enzyme appears to be a bifunctional protein with both PLA2 and GSH peroxidase activities and is identical to piridoredoxin. These findings are very interesting and more detailed investigations on structure–activity relationships of mammalian phospholipases are required to understand their physiological functions and pathophysiology associated with their deficiencies. In the knockout mouse with aiPLA2 deficiency, lysosomal PLA2 activity (pH 4; zero Ca^{2+}) was reduced by 97% in lungs and 90% in alveolar macrophages. Content of total PLs, PC, and disaturated phosphatidylcholine (DSPC) in bronchoalveolar lavage fluid, lamellar bodies, and lung homogenate was 25–77% greater in knockout mice compared to wild-type mice at 10 weeks of age and increased progressively with age. Degradation of 3HDPPC was significantly decreased at 2 hour postinfusion into knockout mice. These results confirm that aiPLA2 has a major role in the degradation of lung DPPC. To date, it is not documented whether aiPLA2 deficiency plays any role in the induction of phospholipidosis.

Drug-induced phospholipidosis: Several cationic ampiphilic drugs have been shown to induce phospholipidosis [53]. It has been proposed that inhibition of lysosomal phospholipases is the primary event in inducing phospholipidosis. Examples include aminoglycoside antibiotic-induced nephrotoxicity, amiodarone-induced pulmonary toxicity.

12.5.2
Aminoglycoside Antibiotic-Induced Phospholipidosis and Nephrotoxicity

Nephrotoxicity induced by aminoglycoside antibiotics [60] was attributed to the inhibition of phospholipases A and C in the kidney [61–63]. Accumulation of aminoglycosides and PLs in the lysosomes is a prominent and early feature of aminoglycoside nephrotoxicity and is histologically characterized by the presence of numerous multilamellar bodies in kidney proximal tubule cells. Several aminoglycosides including amikacin, dibekacin, gentamicin, and tobramycin inhibited both phospholipases A1, A2, and PLC activities in soluble lysosomal fraction of kidney cortex. The accumulation of PLs in lysosomes of kidney cortex, an early and pervasive feature of acute aminoglycoside nephrotoxicity, appears to be due to the inhibition of lysosomal phospholipases.

Inhibition of phospholipase activities by drugs could be used as a screening tool in early drug discovery to identify potential phospholipidosis liabilities [61–63]. In addition, measurement of phospholipid accumulation in cell-based assays (e.g., monocytes, lymphocytes, HepG2 cells) using fluorescent probes can also be used as a potential screening tool during early drug discovery phase [64–66]. Metabolic profiling (metabonomics) also offers huge potential to highlight biomarkers and mechanisms in support of phospholipidosis in preclinical drug development [64, 67].

Tab. 12.3 Some examples of accumulated metabolites during in born deficiencies of enzymes/proteins

Accumulated metabolite	Enzyme/protein deficiency
5-Oxoproline	Oxoprolinase; cystathionine β lyase
Cis-4-decenoic acid	Mitochondrial medium chain Acyl CoA dehydrogenase (MCAD)
3-Hydroxy-3-methylglutaric acid	Mitochondrial 3-hydroxy-3-methylglutaryl-CoA lyase
Phenylalanine	Phenylalanine hydroxylase
Glucose	Insulin

12.6
Mechanisms of Toxicity Associated with Accumulated Metabolites during Inborn Errors of Metabolism

There is substantial evidence that substrates and metabolites that are accumulated during the IEM produce oxidative stress, disruption of mitochondrial energy products among many other effects. Examples (Table 12.3) include cis-4-decenoic acid [68, 69], 3-hydroxy-3-methylglutaric acid [70, 71], oxoproline [72], phenylalanine [73, 74], and glucose [75, 76] among many others. In the following two examples, hyperphenylalaninemia and hyperglycemia are reviewed in detail with regard to oxidative stress, disruption of mitochondrial energy production, and their relevance to associated pathophysiology.

12.7
Pathophysiology of Phenylketonuria and Hyperphenylalaninemia

Phenylketonuria (PKU) is caused by a mutated gene for the enzyme phenylalanine hydroxylase (PAH), which converts the amino acid phenylalanine to tyrosine and other essential metabolites in the body [11]. A rarer form of the disease occurs with a defect in the biosynthesis or recycling of the cofactor tetrahydrobiopterin (BH4) by the patient, although PAH activity is normal. This cofactor BH4 is essential for PAH activity. PAH deficiency causes a spectrum of disorders including classic PKU and hyperphenylalaninemia (a less severe accumulation of phenylalanine).

PKU is an autosomal recessive genetic disorder [77, 78]. Each parent must have at least one mutated allele of the gene for PAH, and the child must inherit two mutated alleles, one from each parent. The PAH gene is located on chromosome 12q23.2 [77, 78]. More than 500 disease-causing mutations have been found in the PAH gene [78]. The incidence of PKU in North America is about 1 in 16 000 births [79], but the incidence varies widely in different human populations from 1 in 4500 births among the population of Ireland [80] to fewer than 1 in 33 000 in China [81], 1 in 212 535 in Thailand [82], or 1 in 100 000 in Finland [83]. Untreated children are normal at birth, but fail to attain early developmental milestones, develop microcephaly, and demonstrate progressive impairment of cerebral function. Hyperactivity, EEG abnormalities and seizures, and severe learning disabilities are the major clinical problems that appear later in life. A "musty" odor of skin, hair, sweat, and urine (due to phenyl acetate accumulation) and a tendency to hypopigmentation and eczema are also observed [77, 78].

The enzyme PAH converts the amino acid phenylalanine into the amino acid tyrosine. In the deficiency of PAH, phenylalanine accumulates and tyrosine is absent. Phenylalanine accumulation disrupts brain development, leading to mental retardation. Excessive phenylalanine can be metabolized into phenyl ketones, a transaminase pathway with glutamate. Metabolites include phenyl acetate, phenyl lactate, and phenyl pyruvate [84–86]. Detection of phenyl ketones in the urine is used in the diagnosis of PKU. Phenylalanine is a large, neutral amino acid (LNAA) and competes for transport across the blood-brain barrier (BBB) via the large neutral amino acid transporter (LNAAT) [87, 88]. Excessive phenylalanine in the blood saturates the transporter and significantly decreases the levels of other LNAAs in the brain, thus affecting protein [89] and neurotransmitter synthesis [90]. Supplementing the patients with LLNA decreased the phenylalanine levels in brain in patients, and considered as one of the treatment options [91]. In recent years, it has been shown that phenylalanine accumulation can result in oxidative stress [73, 74, 92–99] and mitochondrial damage [94], a hallmark process of cellular degeneration.

12.7.1
Oxidative Stress, Mitochondrial Damage, and Antioxidant Status in Phenylketonuria

Sirtori *et al.* [73] evaluated various oxidative stress parameters, namely, thiobarbituric acid-reactive species (TBA-RS) and total antioxidant reactivity (TAR) in the plasma of PKU patients. The activities of the antioxidant enzymes catalase (CAT), superoxide dismutase (SOD), and glutathione peroxidase (GSH-Px) were also measured in erythrocytes from these patients. It was observed that phenylketonuric patients present a significant increase of plasma TBA-RS measurement, indicating a stimulation of lipid peroxidation, as well as a decrease of plasma TAR, reflecting a deficient capacity to rapidly handle an increase of reactive species. The results also showed a decrease of erythrocyte GSH-Px activity. Although these authors found no correlation between phenylalanine levels in blood and oxidative stress [74], it is likely that accumulation of secondary metabolites could also contribute to the oxidative stress processes. In addition, Colome *et al.* [96–98] found that lipophilic antioxidant, ubiquinone -10 (Coenzyme Q) levels was lowered in serum and lymphocytes of PKU patients suggesting that increased lipid peroxidation is perhaps a consequence of low Coenzyme Q levels.

PKU-induced oxidative stress has also been reproduced in experimental animals [92–95]. Martinez-cruz *et al.* [87] directly administered l-phenylalanine (300 mg/kg) and 50 mg/kg of p-chlorophenylalanine to pregnant rats and examined the effects of maternal hyperphenylalaninemia on the morphological and biochemical development of pup rat brain and cerebellum. Mitochondrial damage in brain and liver of pups associated with increased formation of malondialdehyde and 4-hydroxynonenal, indicative of oxidative stress, was observed. Further, these authors also demonstrated that coadministration of melatonin, vitamin, and vitamin C prevented the oxidative stress and mitochondrial damage in the brain and liver of these pups. In another study [95], these authors monitored a number of oxidative stress markers such as Ehrlich adducts formation, lipid peroxidation, reduced and oxidized GSH, and the activities of GSH-Px and GSH reductase, heme-oxigenase-1, and mitogen-activated protein kinase 1/2 (MAPK 1/2). Phenylalanine accumulation strongly increased most of these oxidative stress markers and induced significant morphological damage. These authors also found that daily administration of melatonin (20 mg/kg BW), vitamin E (30 mg/kg BW), and vitamin C (30 mg/kg BW) until delivery prevented the oxidative biomolecular damage in the rat brain and cerebellum. Kienzle Hagen et al. [93] also found that the activities of antioxidant enzymes CAT and GSH-Px were reduced in rat brain during experimental hyperphenylalaninemia. Thus, the phenylalanine accumulation in the brain can result in extensive oxidative damage in brain and perhaps promotes the brain damage resulting in mental retardation in PKU patients.

12.7.2
Glucose-Induced Oxidative Stress Mediated Toxicity: Implications in Diabetes

Oxidative stress has been implicated to play an important role in the pathogenesis of diabetic neuropathy [75] and diabetic nephropathy [76]. Recent studies have shown that high glucose levels induce cellular oxidative stress by increasing reactive oxygen species (ROS) generation [75, 76, 100–109]. Overproduction of ROS and/or impaired antioxidant defense in poorly managed diabetes could contribute to endothelial and vascular dysfunction [103, 109]. Studies have also shown that elevated glucose levels, per se, enhance PKC activation, cause oxidative stress and augment membrane lipid peroxidation in glomeruli [104], suggesting that high glucose might play a role in the induction of glomerular lipid peroxidation. Hyperglycemia-induced mitochondrial superoxide formation might be the unifying pathway that activates PKC activity, increases intracellular sorbitol, and advanced glycated endproduct (AGE) formation and subsequently induces endothelial cell dysfunction.

AGEs interact with specific receptors and binding proteins to influence the expression of growth factors and cytokines, including transforming growth factor (TGF) and connective tissue growth factor (CGF), thereby regulating the growth and proliferation of the various renal cell types. The observation that hyperglycemia does indeed cause an upregulation of the expression of fibronectin, TGF, and basic fibroblast growth factor in human peritoneal mesothelial cells further suggests that hyperglycemia may indeed induce these effects via oxidative stress. The inappropriate expression of these growth factors would have deleterious effects on cellular functions and thus may contribute to the pathogenesis of various diabetic complications. Among other changes that occur during chronic hyperglycemia is the activation of NADPH-dependent aldose reductase, which affects the polyol pathway and diminishes the quantity of NADPH that is available for the reduction of oxidized GSH by GSH reductase. The deleterious effects of the increased levels of ROS that are observed in hyperglycemia are very likely to be further amplified by the decreased capacity of the cellular antioxidant defense system under this condition. Alterations in transporter function may occur as a consequence of hyperglycemia, because proteins and lipids are also subject to oxidant injury. Thus, oxidative stress may induce the dysfunction of apical transporters in renal proximal tubule cells.

The metabolism of excessive intracellular glucose may involve a number of processes. One consequence of excessive intracellular glucose levels is an increased rate of oxidative phosphorylation under hyperglycemic conditions. In addition, hyperglycemia may result in the activation of NADPH oxidase, the production of superoxide anion, hydrogen peroxide (H_2O_2), lipid peroxidation, and a decrease in antioxidants CAT and GSH. Another consequence is that the inhibition of sodium-dependent glucose cotransport system (SGLT) with increase in glucose autoxidation, nonenzymatic glycation of proteins, the formation of AGEs, and overproduction of ROS by mitochondria are the potential effects of hyperglycemia-induced oxidative stress. High glucose levels directly stimulate ROS generation, p38 MAPK phosphorylation, and ANG gene expression in immortalized renal proximal tubular cells. The addition of antioxidants, inhibitors of mitochondrial oxidation, and SB 203580 blocked the stimulatory effect of high glucose, implicating mitochondrial ROS and the p38MAPK signal transduction pathway in the upregulation of renal ANG gene expression under hyperglycemic conditions.

12.7.3
Drugs-Induced Hyperglycemia

Rosiglitazone and other antidiabetic drugs decrease the hyperglycemia and the associated oxidative stress, thus preventing diabetic complications. However, many other drugs that are not prescribed for diabetes can increase glucose levels in blood [111–114] include the following: beta blockers, thiazide diuretics, corticosteroids, niacin, pentamidine, protease inhibitors, and some antipsychotic agents.

- atypical antipsychotics, particularly olanzapine and clozapine
- corticosteroids
- diazoxide
- intravenous dextrose
- diuretics
- epinephrine
- estrogens
- isoniazid
- lithium
- phenothiazines
- phenytoin

- salicylates (acute toxicity)
- triamterene
- tricyclic antidepressants.

12.7.4
Antipsychotic-Induced Hyperglycemia and Diabetes

Olanzapine is an atypical antipsychotic, approved by the FDA in 1996, for the treatment of schizophrenia and depressive episodes associated with bipolar disorder. It was also approved as part of the fluoxetine (Symbyax) formulation, in 2003, for the treatment of acute manic episodes and the maintenance treatment for bipolar disorder in 2004.

Of all the atypical antipsychotics, olanzapine and clozapine are most likely to induce weight gain [116]. All atypical antipsychotic manufacturers are required to include a warning about the risk of hyperglycemia and diabetes associated with the use of these drugs. Although the mechanisms of olanzapine-induced hyperglycemia and diabetes are not well understood, recent evidence suggests that olanzapine may directly affect adipocyte function, promoting fat deposition. There are some reports of olanzapine-induced diabetic ketoacidosis [117]. Some data suggest that olanzapine may decrease insulin sensitivity [118].

12.8
Conclusions

As evident from above, several drugs, perhaps, may induce their toxic adverse events by altering the intermediary metabolic processes by inhibiting certain enzymes, receptors, or transport proteins. Therefore, a detailed understanding of IEM and associated pathology can help to link drug-induced toxicities with a specific endogenous metabolic pathway. Thus, more mechanistic studies involving metabonomic profiling should be introduced during drug discovery and development. Recent advances in highly sensitive accurate mass spectrometry can help to elucidate the metabonomic profiles in plasma and urine during preclinical or clinical drug development.

References

1 Yang, S.Q. (2003) Productivity Challenge of the Pharmaceutical Industry and its Implications for its Researchers, http://www.biopharm.us/htm/Quarterly/Dec2003/Productivity.pdf.
2 Kola, I. and Landis, J. (2004) Perspective: can the pharmaceutical industry reduce the attrition rates? *Nat Rev Drug Discov*, **3**, 711–716.
3 Rishton, G.M. (2005) Failure and success in modern drug discovery: guiding principles in the establishment of high probability of success drug discovery organizations. *Med Chem*, **1**, 519–527.

4 Sankar, U. (2005) The delicate toxicity balance in drug discovery. *Scientist*, **19**, 32.
5 Subrahmanyam, V. and Tonelli, A. (2007) Biomarkers, metabonomics, and drug development: can inborn errors of metabolism help in understanding drug toxicity. *AAPS J*, **9**, E284–E297.
6 Robertson, D.G. (2005) Metabonomics in toxicology: a review. *Toxicol Sci*, **85**, 809–822.
7 Collings, F.B. and Vaidya, V.S. (2008) Novel technologies for the discovery and quantitation of biomarkers of toxicity. *Toxicology*, **245**, 167–174.
8 Lock, E.A. and Bonventre, J.V. (2008) Biomarkers in translation: past, present and future. *Toxicology*, **245**, 163–166.
9 Mendrick, D.L. (2008) Genomic and genetic biomarkers of toxicity. *Toxicology*, **245**, 175–181.
10 Goodsaid, F.M., Frueh, F.W., and Mattes, W. (2008) Strategic paths for biomarker qualification. *Toxicology*, **245**, 219–223.
11 Garrod, A.E. (1923) Inborn Errors of Metabolism, http://www.esp.org/books/garrod/inborn-errors/facsimile/. Date accessed, 27 June, 2007.
12 Epstein, C.J. (1996) Genetic disorders and birth defects, in *Rudolph's Pediatrics* (eds A.M. Rudolph, J.I.E. Hoffman, and C.D. Rudolph), McGraw Hill, New York, pp. 265–374.
13 Burton, B.K. (1998) Inborn errors of metabolism in infancy: a guide to diagnosis. *Pediatrics*, **102**, e69.
14 Raghuveer, T.S., Garg, U., and Graf, W.D. (2006) Inborn errors of metabolism in infancy: an update. *Am Fam Physician*, **73**, 1981–1990.
15 Baumgartner, C. and Baumgartner, D. (2006) Biomarker discovery, disease classification, and similarity query processing on high-throughput MS/MS data of inborn errors of metabolism. *J Biomol Screen*, **11**, 90–99.
16 Trauner, M. and Boyer, J.L. (2003) Bile salt transporters: molecular characterization, function and regulation. *Physiol Rev*, **83**, 633–671.
17 Kosters, A. and Karpen, S.J. (2008) Bile acid transporters in health and disease. *Xenobiotics*, **38**, 1043–1071.
18 Chang, J.Y.L. (2002) Bile acid regulation of gene expression: roles of nuclear hormone receptors. *Endocr Rev*, **23**, 443–463.
19 St-Pierre, M.V., Kullack-Ublick, G.A., Hagenbuch, B., and Meier, P. (2001) Transport of bile acids in hepatic and non-hepatic tissues. *J Exp Biol*, **204**, 1673–1686.
20 Pauli-Magnus, C. and Meier, P. (2006) Hepatobiliary transporters and drug-induced cholestasis. *Hepatology*, **44**, 778–787.
21 Chiang, J.Y.L. (2003) Bile acid regulation of hepatic physiology III. Bile acids and nuclear receptors. *Am J Physiol Gastrointest Liver Physiol*, **284**, G349–G356.
22 Zsembery, A., Thalhammer T., and Graf, J. (2000) Bile formation: a concerted action of membrane transporters in hepatocytes and cholangiocytes. *News Physiol Sci*, **15**, 6–11.
23 Tran, D., Stelly, N., Trodjmann, T., Durroux, T., Dufour, M.N., Forchioni, A., Seyer, R., Claret, M., and Guillon, G. (1999) Distribution of signaling molecules involved in vasopressin-induced Ca2+ mobilization in rat hepatocyte multiplets. *J Histochem Cytochem*, **47**, 601–616.
24 Serrière, V., Berthon, B., Boucherie, S., Jacquemin, E., Guillon, G., Claret, M., and Tordjmann, T. (2001) Vasopressin receptor distribution in the liver controls calcium wave propagation and bile flow. *FASEB J*, **15**, 1484–1486.
25 Arrese, M., Ananthanarayanan, M., and Suchy, F.J. (1998) Hepatobiliary transport: molecular mechanisms of development and cholestasis. *Pediatr Res*, **44**, 141–147.
26 Bissell, D.M., Gores, G.J., Laskin, D.L., and Hoofnagle, J.H. (2001)

Drug-induced liver injury: mechanisms and test systems. *Hepatology*, **33**, 1009–1013.

27 Bohan, A. and Boyer, J.L. (2002) Mechanisms of hepatic transport of drugs: implications for cholestatic drug reactions. *Semin Liver Dis*, **22**, 123–136.

28 Chitturi, S. and George, J. (2002) Hepatotoxicity of commonly used drugs: nonsteroidal anti-inflammatory drugs, antihypertensives, antidiabetic agents, anticonvulsants, lipid-lowering agents, psychotropic drugs. *Semin Liver Dis*, **22**, 169–183.

29 Ishak, K.G. (2003) Drug-Induced Cholestasis. United States and Canadian Academy of Pathology, http://www.uscap.org/newindex.htm?92nd/companion08h2.htm.

30 McRae, M.P., Lowe, C.M., Tian, X., Bourdet, D.L., Ho, R.H., Leake, B.F., Kim, R.B., Brouwer, K.L.R., and Kashuba, A.D.M. (2006) Ritonavir, saquinavir and efavirenz, but not nevirapine, inhibits bile acid transport in human and rat hepatocytes. *J Pharmacol Exp Ther*, **318**, 1068–1075.

31 Couture, L., Nash, J.A., and Turgeon, J. (2006) The ATP- binding cassette transporters and their implication in drug disposition: a special look at the heart. *Pharmacol Rev*, **54**, 244–258.

32 Kostrubsky, V.A.E., Strom, S.C., Hanson, J., Urda, E., Rose, K., Burliegh, J., Zocharski, P., Cai, H., Sinclair, J.F., and Sahi, J. (2003) Evaluation of hepatotoxic potential of drugs by inhibition of bile acid transport in cultured primary human hepatocytes and intact rats. *Toxicol Sci*, **76**, 220–228.

33 Mita, S., Suzuki, H., Akita, H., Hayashi, H., Onuki, R., Hoffman, A.F., and Sugiyama, Y. (2006) Inhibition of bile acid transport across Na/Taurocholate cotransporting polypeptide (SLC10A1) and bile salt export pump (ABCB 11)-coexpressing LLC-PK1 cells by cholestasis-inducing drugs. *Drug Metab Dispos*, **34**, 1575–1581.

34 Suchy, F.J. and Ananthanarayanan, M. (2006) Bile salt excretory pump: biology and pathobiology. *J Pediat Gasgtroenterol Nutr*, **43**, S10–S16.

35 Fattinger, K., Funk, C., Pantze, M., Weber, C., Reichen, J., Stieger, B., and Meier, P.J. (2001) The endothelin antagonist bosentan inhibits the canalicular bile salt export pump: a potential mechanism for hepatic adverse reactions. *Clin Pharmacol Ther*, **69**, 223–231.

36 Funk, C., Ponelle, C., Scheurmann, G., and Pantze, M. (2001) Cholestatic potential of troglitazone as a possible factor contributing to troglitazone-induced hepatotoxicity: in vivo and in vitro interaction at the canalicular bile salt export pump. *Mol Pharmacol*, **59**, 627–635.

37 Roman, I.D., Fernandez-Moreno, M.D., Fueyo, J.A., Roma, M.G., and Coleman, R. (2003) Cyclosporin A induced internalization of the bile salt export pump in isolated rat hepatocyte couplets. *Toxicol Sci*, **71**, 276–281.

38 Cohen, H., Chahine, C., Hui, A., and Mukherji, R. (2004) Bosentan therapy for pulmonary arterial hypertension. *Am J Health Syst Pharm*, **61**, 1107–1119.

39 Leslie, E.M., Watkins, P.B., Kim, R.B., and Brouwer, K.L.R. (2007) Differential inhibition of rat and human Na+-dependent taurocholate cotransporting polypeptide (NTCP/SLC10A1) by bosentan: A mechanism for species differences in hepatotoxicity. *J Pharmacol Exp Ther*, **321**, 1170–1178.

40 Hagenbuch, B. and Meier, P.J. (1994) Molecular cloning, chromosomal localization, and functional characterization of a human liver Na_/bile acid cotransporter. *J Clin Invest*, **93**, 1326–1331.

41 Hagenbuch, B., Stieger, B., Foguet, M., Lubbert, H., and Meier, P.J. (1991) Functionalexpression cloning and characterization of the hepatocyte Na+/bile acid cotransportsystem. *Proc Natl Acad Sci U S A*, **88**, 10629–10633.

42 Ho, R.H., Tirona, R.G., Leake, B.F., Glaeser, H., Lee, W., Lemke, C.J., Wang, Y., and Kim, R.B. (2006) Drug and bile acid transporters in rosuvastatin hepatic uptake: function, expression, and pharmacogenetics. *Gastroenterology*, **130**, 1793–1806.

43 Parker, J.C. (2002) Troglitazone: the discovery and development of a novel therapy for the treatment of Type 2 diabetes mellitus. *Adv Drug Deliv Rev*, **54**, 1173–1197.

44 Schwartz, S. and Quick, W. (2000) Rezulin® (troglitazone) was voluntarily withdrawn from the market by the manufacturer on March 21, http://www.diabetesmonitor.com/rezulin.htm.

45 Jaeschke, H. (2007) Troglitazone hepatotoxicity: are we getting closer to understanding idiosyncratic liver injury? *Toxicol Sci*, **97**, 1–3.

46 Nozawa, T., Sugurra, S., Nakajima, M., Goto, A., Yokoi, T., Nezu, J.-I., Tsuji, A., and Tamai, I. (2004) Involvement of organic anion transporting polypeptides in the transport of troglitazone sulfate: implications for understanding troglitazone hepatotoxicity. *Drug Metab Dispos*, **32**, 291–294.

47 Kostrubsky, V.E., Sinclair, J.F., Ramachandran, V., Venkataramanan, R., Wen, Y.H., Kindt, E., Galchev, V., Rose, K., Sinz, M., and Strom, S.C. (2000) The role of conjugation in hepatotoxicity of troglitazone in human and porcine hepatocyte cultures. *Drug Metab Dispos*, **28**, 1192–1197.

48 Smith, M.T. (2003) Mechanisms of troglitazone hepatotoxicity. *Chem Res Toxicol*, **16**, 679–687.

49 Tettey, J.N., Maggs, J.L., Rapeport, W.G., Pirmohamed, M., and Park, B.K. (2001) Enzyme-induction dependent bioactivation of troglitazone and troglitazone quinone in vivo. *Chem Res Toxicol*, **14**, 965–974.

50 Sanchez-Pozzi, E.J., Crocenzi, F.A., Pellegrino, J.M., Catania, V.A., Luquita, M.G., Roma, G., Rodriguez-Garay, E.A., and Mottino, A.D. (2003) Ursodeoxycholate reduces ethinylestradiol glucuronidation in the rat: prevention of estrogen-induced cholestasis. *J Pharmacol Exp Ther*, **306**, 279–286.

51 Mottino, A.D., Hoffman, T., Crocenzi, F.A., Sanchez Pozzi, E.J., Roma, M.G., and Vore, M. (2007) Disruption of function and localization of tight junctional structures and MRP2 in sustained estradiol-17beta-D-glucuronide-induced cholestasis. *Am J Physiol Gastrointest Liver Physiol*, **293**, G391–G402.

52 Kan, K.S., Monte, M., Parslow, R.A., and Coleman, R. (1989) Oestradiol 17 beta-glucuronide increases tight-junctional permeability in rat liver. *Biochem J*, **261**, 297–300.

53 Reasor, M.J. and Kacew, S. (2001) Drug-induced phospholipidosis: are there functional consequences? *Exp Biol Med*, **226**, 825–830.

54 Hiraoka, M., Abe, A., Lu, Y., Yang, K., Han, X., Gross, R.W., and Shayman, J.A. (2006) Lysosomal phospholipase A2 and phospholipidosis. *Mol Cell Biol*, **26**, 6139–6148.

55 Abe, A., Hiraoka, M., and Shayman, J.A. (2007) A role for lysosomal phospholipase A2 in drug induced phospholipidosis. *Drug Metab Lett*, **1**, 49–53.

56 Wilson, P.A., Gardner, S.D., Lambie, N.M., Commans, S.A., and Crowther, D.J. (2006) Characterization of the human patatin-like phospholipase family. *J Lipid Res*, **47**, 1940–1949.

57 Fisher, A.B., Dodia, C., Feinstein, S.I., and Ho, Y.S. (2005) Altered lung phospholipids metabolism in mice with targeted deletion of lysosomal-type phospholipase A2. *J Lipid Res*, **46**, 1248–1256.

58 Wang, X., Phelan, S.A., Forsman-Semb, K., Taylor, E.F., Petros, C., Brown, A., Lerner, C.P., and Paigen, B. (2003) Mice with targeted mutation of peroxiredoxin 6 develop normally but are susceptible to oxidative stress. *J Biol Chem*, **278**, 25179–25190.

59 Fisher, A.B. and Dodia, C. (2001) Lysosomal-type PLA2 and turnover

of alveolar DPPC. *Am J Physiol Lung Cell Mol Physiol*, **280**, L748–L754.

60. Feldman, S., Wang, M.Y., and Kaloyanides, G.J. (1982) Aminoglycosides induce a phospholipidosis in the renal cortex of the rat: an early manifestation of nephrotoxicity. *J Pharmacol Exp Ther*, **220**, 514–520.

61. Hostetler, K.Y. and Hall, L.B. (1982) Aminoglycoside antibiotics inhibit lysosomal phospholipase A and C from rat liver in vitro. *Biochim Biophys Acta*, **710**, 506–509.

62. Hostetler, K.Y. and Hall, L.B. (1972) Inhibition of kidney lysosomal phospholipases A and C by aminoglycoside antibiotics: possible mechanism of aminoglycoside toxicity. *Proc Natl Acad Sci U S A*, **79**, 1663–1667.

63. Carlier, M.B., G Laurent, G., Claes, P.J., Vanderhaeghe, H.J., and Tulkens, P.M. (1983) Inhibition of lysosomal phospholipases by aminoglycoside antibiotics: in vitro comparative studies. *Antimicrob Agents Chemother*, **23**, 440–449.

64. Monteith, D.K., Morgan, R.E., and Halstead, B. (2006) In vitro assays and biomarkers for drug-induced phospholipidosis. *Expert Opin Drug Metab Toxicol*, **2**, 687–696.

65. Natalie, M., Margino, S., Erik, H., Annelieke, P., Geert, V., and Philippe, V. (2009) A 96-well flow cytometric screening assay for detecting in vitro phospholipidosis induction in the drug discovery phase. *Toxicol In Vitro*, **23**, 217–226.

66. Fujimura, H., Dekura, E., Kurabe, M., Shimazu, N., Koitabashi, M., and Toriumi, W. (2007) Cell-based fluorescence assay for evaluation of new-drugs potential for phospholipidosis in an early stage of drug development. *Exp Toxicol Pathol*, **58** (6), 375–382.

67. Clarke, C.J. and Haselden, J.N. (2008) Metabolic profiling as a tool for understanding mechanisms of toxicity. *Toxicol Pathol*, **36**, 140–147.

68. Schuck, P.F., Ceolato, P.C., Ferreira, G.C., Tonin, A., Leipnitz, G., Dutra-Filho, C.S., Latini, A., and Wajner, M. (2007) Oxidative stress induction by cis-4-decenoic acid: relevance for MCAD deficiency. *Free Radic Res*, **41**, 1261–1272.

69. de Assiss, D.R., de Cassia Maria, R., Rosa, R.B., Schuck, P.F., Ribeiro, C.A.J., da Costa Ferreira, G., Dutra-Filho, C.S., de Souza Wyse, A.T., Wannmacher, C.M.D., Perry, M.L.S., and Wajner, M. (2004) Inhibition of energy metabolism in cerebral cortex of young rats by the medium-chain fatty acids accumulating in MCAD deficiency. *Brain Res*, **1030**, 141–151.

70. Leipnitz, G., Seminotti, B., Amaral, A.U., de Bortoli, G., Solano, A., Schuck, P.F., Wyse, A.T., Wannmacher, C.M., Latini, A., and Wajner, M. (2008) Induction of oxidative stress by the metabolites accumulating in 3-methylglutaconic aciduria in cerebral cortex of young rats. *Life Sci*, **82**, 652–662.

71. Leipnitz B., Seminotti B., Haubrich J., Dalcin M.B., Dalcin K.B., Solano A., de Bortoli G., Rosa R.B., Amaral A.U., Dutra-Filho C.S., Latini A., and Wajner M. (2008) Evidence that 3-hydroxy-3-methylglutaric acid promotes lipid and protein oxidative damage and reduces the nonenzymatic antioxidant defenses in rat cerebral cortex. *J Neurosci Res*, **86**, 683–694.

72. Pederzolli, C.D., Sgaravatti, A.M., Braum, C.A., Prestes, C.C., Zorzi, G.K., Sgarbi, M.B., Wyse, A.T., Wannmacher, C.M., Wajner, M. and Dutra-Filho, C.S. (2007) 5-oxoproline reduces non-enzymatic antioxidant defenses in vitro in rat brain. *Metab Brain Dis*, **22**, 51–65.

73. Sirtori, L.R., Dutra-Filho, C.S., Fitarelli, D., Sitta, A., Haeser, A., Barschak, A.G., Wajner, M., Coelho, D.M., Llesuy, S., Belló-Klein, A., Giugliani, R., Deon, M., and Vargas, C.R. (2005) Oxidative stress in patients with phenylketonuria. *Biochim Biophys Acta*, **1740**, 68–73.

74. Sitta, A., Barschak, A.G., Deon, M., Terroso, T., Pires, R., Giugliani, R., Dutra-Filho, C.S., Wajner, M., and

Vargas, C.R. (2006) Investigation of oxidative stress parameters in treated phenylketonuric patients. *Metab Brain Dis*, **21**, 287–296.

75 Russell, J.W., Golovoy, D., Vincent, A.M., Mahendru, P., Olzmann, J.A., Mentzer, A., and Feldman, E.L. (2002) High glucose-induced oxidative stress and mitochondrial dysfunction in neurons. *FASEB J*, **16** (13), 1738–1748.

76 Wolf, G. and Ziyadeh, F.N. (1999) Molecular mechanisms of diabetic renal hypertrophy. *Kidney Int*, **56**, 393–405.

77 Scriver, C.R. (2007) The PAH gene, phenylketonuria, and a paradigm shift. *Hum Mutat*, **28**, 831–845.

78 Williams, R.A., Mamotte, C.D., and Burnett, J.R. (2008) Phenylketonuria: an inborn error of phenylalanine metabolism. *Clin Biochem Rev*, **29**, 31–41.

79 Harding, C. (2008) Progress toward cell-directed therapy for phenylketonuria. *Clin Genet*, **74**, 97–104.

80 O'Neill, C.A., Eisensmith, R.C., Croke, D.T., Naughten, E.R., Cahalane, S.F., and Woo, S.L. (1994) Molecular analysis of PKU in Ireland. *Acta Paediatr Suppl*, **407**, 43–44.

81 Jiang, J., Ma, X., Huang, X., Pei, X., Liu, H., Tan, Z., and Zhu, L. (2003) A survey for the incidence of phenylketonuria in Guangdong, China. *Southeast Asian J Trop Med Public Health*, **34** (Suppl 3), 185.

82 Pangkanon, S., Ratrisawadi, V., Charoensiriwatana, W., Techasena, W., Boonpuan, K., Srisomsap, C., and Svasti, J. (2003) Phenylketonuria detected by the neonatal screening program in Thailand. *Southeast Asian J Trop Med Public Health*, **34** (Suppl 3), 179–181.

83 Guldberg, P., Henriksen, K.F., Sipilä, I., Güttler, F., and de la Chapelle, A. (1995) Phenylketonuria in a low incidence population: molecular characterisation of mutations in Finland. *J Med Genet*, **32**, 976–978.

84 Alvarez Dominguez, L., Campistol Plana, J., Ribes Rubio, A., and Riverola de Vecina, A.T. (1992) Phenylalanine metabolites in hyperphenylalaninemic children. *An Esp Pediatr*, **36**, 371–374.

85 Koepp, P. (1976) Urinary phenylalanine metabolites in hyperphenylalaninemia. *Klin Wochenschr*, **54**, 1047–1053.

86 Kitagawa, T., Smith, B.A., and Brown, E.S. (1975) Gas-liquid chromatography of phenylalanine and its metabolites in serum and urine of various hyperphenylalaninemic subjects, their relatives, and controls. *Clin Chem*, **21**, 735–740.

87 van Spronsen, F.J., Hoeksma, M. and Reijngoud, D.J. (2009) Brain dysfunction in phenylketonuria: is phenylalanine toxicity the only possible cause? *J Inherit Metab Dis*, **32**, 46–51.

88 Pietz, J., Kreis, R., Rupp, A., Mayatepek, E., Rating, D., Boesch, C., and Bremer, H.J. (1999) Large neutral amino acids block phenylalanine transport into brain tissue in patients with phenylketonuria. *J Clin Invest*, **103**, 1169–1178.

89 Hoeksma, M., Reijngoud, D.J., Pruim, J., de Valk, H.W., Paans, A.M., and van Spronsen, F.J. (2009) Phenylketonuria: high plasma phenylalanine decreases cerebral protein synthesis. *Mol Genet Metab*, **96**, 177–182.

90 Hyland K. (2007) Inherited disorders affecting dopamine and serotonin: critical neurotransmitters derived from aromatic amino acids. *J Nutr* **137** (6) (Suppl 1), 1568S–1572S.

91 Moats, R.A., Moseley, K.D., Koch, R., and Nelson, M. Jr (2003) Brain phenylalanine concentrations in phenylketonuria: research and treatment of adults. *Pediatrics*, **112** (6 Pt 2), 1575–1579.

92 Ercal, N., Aykin-Burns, N., Gurer-Orban, H., and McDonald, J.D. (2002) Oxidative stress in a phenylketonuric animal model. *Free Radic Biol Med*, **32**, 906–911.

93 Kienzle Hagen, M.E., Pederzolli, C.D., Sparavatti, A.M., Bridi, R., Wajner, M., Wanmacher, C.M., Wyse, A.T., and Dutra-Filho, C.S.

(2002) Experimental phenylalaninemia provokes oxidative stress in rat brain. *Biochim Biophys Acta*, **1586**, 344–352.

94 Martinez-Cruz, F., Osuna, C., and Guerrero, J.M. (2005) Mitochondrial damage induced by fetal hyperphenylalaninemia in the rat brain and liver: its prevention by melatonin, Vitamin E, and Vitamin C. *Neurosci Lett*, **39**, 1–4.

95 Martinez-Cruz, F., Pozo, D., Osuna, C., Espinar, A., Marchante, C., and Guerrero, J.M. (2002) Oxidative stress induced by phenylketonuria in the rat: Prevention by melatonin, vitamin E, and vitamin C. *J Neurosci Res*, **69**, 550–558.

96 Colomé, C., Artuch, R., Vilaseca, M.A., Sierra, C., Brandi, N., Cambra, F.J., Lambruschini, N., and Campistol, J. (2002) Ubiquinone-10 content in lymphocytes of phenylketonuric patients. *Clin Bichem*, **35**, 81–84.

97 Colome, C., Artuch, R., Vilaseca, M.A., Sierra, C., Brandi, N., Lambruschini, R., Cambra, F.J., and Campistol, J. (2003) Lipophilic antioxidants in patients with phenylketonuria. *Am J Clin Nutr*, **77**, 185–188.

98 Hargreaves, I.P. (2007) Coenzyme Q10 in phenylketonuria and mevalonic aciduria. *Mitochondrion*, **7** (Suppl), S175–S180.

99 Sierra, C., Vilaseca, M.A., Moyano, D., Brandi, N., Campistol, J., Lambruschini, N., Cambra, F.J., Deulofeu, R., and Mira, A. (1998) Antioxidant status in hyperphenylalaninemia. *Clin Chim Acta*, **276**, 1–9.

100 Ho, H.J., Lee, Y.J., Park, S.H., Lee, J.H., and Taub, M. (2005) High glucose-induced oxidative stress inhibits Na+/glucose cotransporter activity in renal proximal tubule cells. *Am J Physiol Renal Physiol*, **288**, F988–F996.

101 Sharpe, P.C., Liu, W.-H., Yue, K.K.M., McMaster, D., Catherwood, M.A., McGinty, A.M., and Trimble, E.R. (1998) Glucose-induced oxidative stress in vascular contractile cells. *Diabetes*, **47**, 801–809.

102 Vincent, A.M., Olzmann, J.A., Brownlee, M., Sivitz, W.I., and Russell, J.W. (2004) Uncoupling proteins prevent glucose-induced neuronal oxidative stress and programmed cell death. *Diabetes*, **53**, 726–734.

103 Li, M., Arsher, P.M., Liang, P., Russell, J.C., Sobel, B.E., and Eukagawa, N.K. (2001) High glucose concentrations induce oxidative damage to mitochondrial DNA in explanted vascular smooth muscle cells. *Exp Biol Med*, **226**, 450–457.

104 Talier, I., Yarkoni, M., Bashan, N., and Eldar-Finkelman, H. (2003) Increased glucose uptake promotes oxidative stress and PKC-d activation in adipocytes of obese, insulin-resistant mice. *Am J Physiol Endocrinol Metab*, **285**, E295–E302.

105 Chung, S.S.M., Ho, E.C.M., Lam, K.S.L., and Chung, S.K. (2003) Contribution of polyol pathway to diabetes-induced oxidative stress. *J Am Soc Nephrol*, **14**, S233–S236.

106 Kaneto, H., Nakatani, Y., Kawamori, D., Miyatsuka, T., and Matsuoka, T.A. (2004) Involvement of oxidative stress and the JNK pathway in glucose toxicity. *Rev Diabet Stud*, **1**, 165–174.

107 Catherwood, M.A., Powell, L.A., Anderson, P., McMaster, D., Sharpe, P.C., and Trimble, E.R. (2002) Glucose-induced oxidative stress in mesangial cells. *Kidney Int*, **61**, 599–608.

108 Tsuneki, H., Sekizaki, N., Suzuki, T., Kobayashi, S., Wada, T., OkAMOTO, t., Kimura, I., and Sasaoka, T. (2007) Coenzyme Q10 prevents high glucose-induced oxidative stress in human umbilical vein endothelial cells. *Eur J Pharmacol*, **566**, 1–10.

109 Ammar, R.F., Gutterman, D.D., Brooks, L.A., and Dellsperger, K.C. Jr (2000) Free radicals mediate endothelial dysfunction of coronary arterioles in diabetes. *Cardiovasc Res*, **47**, 596–601.

110 Newcomer, J.W. (2004) Abnormalities of glucose metabolism associated with atypical antipsychotic drugs. *J Clin Psychiatry*, **65** (Suppl 18), 36–46.

111 Bugajski, J. and Lech, J. (1979) Effects of neuroleptics on blood glucose, free fatty acids and liver glycogen levels in the rat. *Pol J Pharmacol Pharm*, **31**, 45–58.

112 Ferner, R.E. (1992) Drug-induced diabetes. *Baillieres Clin Endocrinol Metab*, **6**, 849–866.

113 Pandit, M.K., Burke, J., Guftason, A.B., Minocha, A., and Peiris, A.N. (1993) Drug induced disorders of glucose tolerance. *Ann Intern Med*, **118**, 529–539.

114 Yasuhara, D., Nakahara, T., Harada, T., and Inui, A. (2007) Olanzapine-induced hyperglycemia in anorexia nervosa. *Am J Psychiatry*, **164**, 528–529.

115 Meatherall, R. and Younes, J. (2002) Fatality from olanzapine induced hyperglycemia. *J Forensic Sci*, **47**, 893–896.

116 Wirshing, D.A., Wirshing, W.C., Kysar, L., and Berisford, M.A. (1999) Novel antipsychotics: comparison of weight gain liabilities. *J Clin Psychol*, **60**, 358–363.

117 Fulbright, A.R. (2006) Complete resolution of olanzapine-induced diabetic ketoacidosis (abstract). *J Pharm Pract*, **19**, 255–258.

118 Sacher, J., Mossaheb, N., Spindelegger, C., Klein, N., Geiss-Granadia, T., Sauermann, R., Lackner, E., Joukhadar, C., Müller, M., and Kasper, S. (2008) Effects of olanzapine and ziprasidone on glucose tolerance in healthy volunteers. *Neuropsychopharmacology*, **33**, 1633–1641.

13
Mechanisms of Toxicity in Fatty Acid Oxidation Disorders
J. Daniel Sharer

The oxidation of fatty acids (FAO) for energy is a fundamental biological process. Organelles within human cells, primarily mitochondria and peroxisomes, contain distinct enzyme systems for metabolizing a variety of fatty acids. Though individually rare, disorders of these pathways together represent a significant health issue. Elucidation of the toxic mechanisms that underlie these conditions is important for the development of effective therapeutic modalities. Mitochondrial fatty acid oxidation disorders (FAODs) are characterized by profound energy depletion, with secondary suppression of the Kreb's cycle and gluconeogenesis, and additional inhibition of FAO due to sequestration and urinary losses of carnitine. Many mitochondrial FAODs also result in accumulation of toxic metabolites that have specific damaging effects on tissues and organs such as the heart, liver, kidney, and the central nervous system. In contrast to the mitochondrial FAODs, the disease phenotypes of single-enzyme peroxisomal FAODs appear to be entirely mediated by metabolite toxicity, without any appreciable energy deprivation. In particular, most peroxisome FAODs are characterized by accumulation of very long chain fatty acids, which damage several cell types within the central nervous system. This review focuses on both major areas of FAO within human cells, with particularly emphasis on the underlying pathophysiological mechanisms that have been delineated for the disorders of each type.

13.1
Introduction

The oxidative degradation of fatty acids is an essential source of energy for heterotrophic organisms, including humans. Because fats are highly reduced and minimally hydrated, they provide the most efficient form of stored energy available to the body, which is of particular importance under conditions of fasting stress [1]. The vast majority of fatty acid oxidation (FAO) that is associated with energy production occurs within mitochondria, which utilizes a relatively efficient β-oxidation process to produce three separate metabolic fuels: acetyl-Coenzyme A (CoA),

Endogenous Toxins. Diet, Genetics, Disease and Treatment.
Edited by Peter J. O'Brien and W. Robert Bruce
Copyright © 2010 WILEY-VCH Verlag GmbH & Co. KGaA, Weinheim
ISBN: 978-3-527-32363-0

ketone bodies (3-OH butyrate and acetoacetate), and reduced nicotinamide adenine dinucleotide (NADH) and flavin adenine dinucleotide ($FADH_2$). Mitochondria can accommodate a significant range of different fatty acids as substrates, including saturated, monounsaturated, and polyunsaturated fatty acids ranging in size from 4 to approximately 20 carbons in length. However, very long chain fatty acids (VLCFAs) (C_{22}–C_{26}) and certain branched-chain acyl compounds are not recognized by the mitochondrial FAO machinery and are instead metabolized in peroxisomes, which contain enzymes that carry out an alternative form of β-oxidation, as well as α-oxidation. In addition, a third, relatively minor ω-FAO system has also been characterized in microsomes derived from the endoplasmic reticulum.

Because FAO is a fundamentally important biological process, it is hardly surprising that mutational impairment of the enzymes that catalyze these pathways can produce severe disease phenotypes. Indeed, nearly all of the genes that comprise the mitochondrial FAO system are associated with an inherited metabolic disease [2, 3]. Similarly, many of the genes that encode peroxisomal FAO enzymes are also known to cause a growing list of often severe conditions that typically involve the brain and central nervous system [4]. While fatty acid oxidation disorders (FAODs) have been the focus of significant research in recent years, the pathophysiological mechanisms underlying these diseases are complex and remain incompletely understood. Mitochondrial β-oxidation disorders certainly result in a profoundly energy deficient state, but this generalization fails to describe the complex metabolic interrelationships that appear to mediate this physiological condition [2]. Furthermore, FAODs are associated with the accumulation of a variety of compounds that can have specific toxic effects on various tissues and processes within the body, and therefore produce or aggravate a number of the clinical effects associated with these diseases [5]. It should be noted that many of these metabolites have become useful as biomarkers for screening and diagnosing a number of FAODs (and other inherited metabolic disorders as well). In the United States and many other countries, population-scale newborn screening programs largely based on high-throughput tandem mass spectrometry are now becoming well established. These programs allow for the presymptomatic detection of 30 or more metabolic disorders, including many mitochondrial FAODs, thus facilitating early diagnosis and treatment of these conditions. As a result of these programs, a significant percentage of new patients are being diagnosed in the absence of clinical symptoms, improving outcomes but also occasionally raising issues concerning differentiation of apparently benign "biochemical phenotypes" and true disease states.

The aim of this review is to describe the FAO systems and associated diseases that have been characterized in humans, with particular emphasis on the mechanisms of toxicity that have been defined for these disorders. Both mitochondrial and peroxisomal oxidation pathways and diseases are presented in separate sections. Because there is considerable pathophysiological redundancy within each category, these mechanisms will not be described for each individual disorder, but rather will be considered together at the end of each section. Because the pathophysiology of these diseases typically involves the combined effects of specific toxic metabolites and "toxic processes," descriptions of both are also included where applicable. It should also be noted that, due to space limitations, many of the relevant genes, gene products, metabolites, and clinical characteristics associated with each process are described only in general terms; for more complete descriptions of these topics, the reader is encouraged to refer the many excellent reviews that are available on these specific subjects.

13.2
Mitochondrial β-Oxidation

Mitochondrial β-oxidation of fatty acids is a fundamental process for the production of energy, particularly when glucose and other dietary fuel sources become limiting, such as during fasting or in response to illness or increased/prolonged physical exertion. It should be noted that utilization of fat for energy is not limited to these conditions; in certain tissues, such as cardiac muscle, catabolism of fat is a significant and constant source of energy at all times [5]. However, during prolonged fasting, the majority (~80%) of metabolic energy requirements for the entire body are transitioned to FAO [6]. This process entails two distinct phases: first, stored fatty acids in adipocytes must be mobilized and ultimately transported (using a conserved carnitine shuttle mechanism) across the inner mitochondrial membrane (IMM) into the mitochondrial matrix of metabolically active cells. Once inside the matrix, fatty acids are then oxidized by a sequential, enzyme-catalyzed process to produce acetyl-CoA, which may be employed to drive ATP production via the TCA cycle or may be used to synthesize ketone bodies; ketogenesis is favored in hepatocytes and these compounds are a particularly important source of energy for certain tissues during fasting. As previously mentioned, disorders involving nearly every gene in this process have been described (Table 13.1) and the combined incidence of mitochondrial β-oxidation disorders has been reported to be on the order of 1 : 8000 [7]. In general, metabolic decompensation leading to disease manifestations in these patients is often precipitated by fasting/intercurrent illness and is typically characterized by evidence of severe energy depletion (hypoketotic hypoglycemia). As is the case for most of the conditions described herein, initial diagnosis may be based on measurement of abnormal levels of free carnitine and/or one or more specific acylcarnitine esters. More complete diagnosis often depends on measurement of enzyme activity in cultured cells and/or direct molecular analysis of the associated gene. Treatment in most of these disorders involves avoidance of fasting to prevent metabolic reliance on FAO as a major source of energy. In addition, many of these conditions cause secondary carnitine deficiency, which may necessitate therapeutic carnitine supplementation.

13.2.1
Mobilization and Activation of Long Chain Fatty Acids

The majority of fatty acids in humans is saturated and unsaturated C16 and C18 species [8] that are stored in adipocytes as triacylglycerol. Under conditions of

Tab. 13.1 Mitochondrial FAO disorders in humans

Protein	Gene Chromosome OMIM	General biochemical characteristics	Proposed mechanisms of toxicity
OCTN2 organic cation transporter (carnitine transporter)	SLC22A5 5q31.1 OMIM #603377	Very low plasma carnitine; Hypoketotic hypoglycemia	Energy depletion
Carnitine palmitoyl transferase 1A (CPT 1; liver isoform)	CPT1A 11q13 OMIM #600528	Reduced long chain acylcarnitines; hypoketotic hypoglycemia	Energy depletion; free fatty acids; acyl-CoAs
Carnitine acylcarnitine translocase (CACT)	SLC25A20 3p21.31 OMIM #212138	Increased long chain plasma acylcarnitines (C16–C18), low plasma carnitine; hypoketotic hypoglycemia	Energy depletion; long chain acylcarnitines; free fatty acids; acyl-CoAs
Carnitine palmitoyl transferase 2 (CPT 2)	CPT2 1p32 OMIM # 600652	Increased long chain plasma acylcarnitines (C16–C18), low plasma carnitine; hypoketotic hypoglycemia	Energy depletion; long chain acylcarnitines; free fatty acids; acyl-CoAs
Very long chain acyl-CoA dehydrogenase (VLCAD)	ACADVL 17p13 OMIM #609575	Hypoketotic hypoglycemia; increased long chain acylcarnitines (C14–C18), low plasma carnitine; increased dicarboxylic acids	Energy depletion; long chain acylcarnitines; free fatty acids; acyl-CoAs
Medium chain acyl-CoA dehydrogenase (MCAD)	ACADM 1p31 OMIM #607008	Hypoketotic hypoglycemia; increased medium chain acylcarnitines and acylglycines; increased dicarboxylic acids	Energy depletion; medium chain acylcarnitines; free fatty acids; acyl-CoAs; cis-decenoate
Short chain acyl-CoA dehydrogenase (SCAD)	ACADS 12q22-qter OMIM #606885	Increased ethylmalonate and butyrylcarnitine, increased ketones	Ethylmalonate and butyrate toxicity
Long chain 3-hydroxyacyl-CoA dehydrogenase (LCHAD)	HADHA 2p23 OMIM #600890	Hypoketotic hypoglycemia; increased 3-hydroxy acylcarnitines; low plasma carnitine; increased dicarboxylic acids; maternal liver disease	Energy depletion; 3-hydroxy acylcarnitines; 3-hydroxy fatty acids and acyl-CoAs
3-hydroxyacyl-CoA dehydrogenase (aka M/SCHAD)	HADH 4q22–q26 OMIM #601609	Hypoglycemia; increased ketones; increased 3-hydroxy butyrylcarnitine	3-hydroxy butyrylcarnitine
3-ketoacyl-CoA thiolase	HADHB 2p23 OMIM #143450	Increased ketones; increased saturated and unsaturated dicarboxylic acids	(?)

Tab. 13.1 Continued

Protein	Gene Chromosome OMIM	General biochemical characteristics	Proposed mechanisms of toxicity
Mitochondrial trifunctional protein (MTP)	HADHA/HADHAB 2p23 OMIM # 609015	Hypoketotic hypoglycemia; increased 3-hydroxy acylcarnitines; low plasma carnitine; increased dicarboxylic acids	Energy depletion; 3-hydroxy acylcarnitines; 3-hydroxy fatty acids and acyl-CoAs
Electron transport flavoprotein/electron transport flavoprotein dehydrogenase (ETF/ETFDH)	ETFA/ETFB/ETFDH 15q25–q26/15q23–q25/4q32-qter OMIM # 608053/130410/231675	Hypoketotic hypoglycemia; increased acylcarnitines, fatty acids, acyl-CoAs; increased organic acid metabolites	Energy depletion; multiple fatty acid intermediate toxicities

fasting or related stress, the body will initially utilize available sources of glucose (stored glycogen) as long as available, but as this material is depleted, falling blood glucose levels will ultimately trigger the secretion of hormones such as glucagon, epinephrine, and adrenocorticotropic hormone [1]. These compounds bind specific receptors on the surface of adipocytes and trigger cAMP-mediated activation of hormone-sensitive lipases, which degrade triacylglycerol into constituent glycerol (which enters the glycolytic pathway) and free fatty acids. The liberated fatty acids are released into the blood stream, where they form soluble complexes with serum albumin, and are then transported across the plasma membrane of hepatocytes or myocytes via a family of tissue-specific fatty acid transport proteins [9]. Once inside the cell, free fatty acids are converted into acyl-CoA esters by chain length–specific acyl-CoA synthetases that are localized to the outer mitochondrial membrane [10]. Formation of acyl-CoA esters serves as an activation step that renders fatty acids accessible to the mitochondrial β-oxidation machinery. It should be noted that eukaryotic cells maintain distinct mitochondrial and cytosolic CoA pools [11]; the former is primarily utilized for oxidation of fatty acids, organic acids, and certain amino acids, while the majority of the cytosolic CoA pool is directed toward fatty acid biosynthesis. This compartmentalization of CoA pools appears to be an important consideration in terms of the toxicity associated with mitochondrial FAODs (see below).

13.2.2
Carnitine-Mediated Transport

The enzymes that comprise the β-oxidation pathway are sequestered in the mitochondrial matrix. Because fatty acids larger than 12 carbons in length cannot freely cross the IMM [2, 6], a specific transport process is required to facilitate metabolism of long chain fatty acids, the most abundant form of these molecules. This is accomplished by the carnitine shuttle system that utilizes the small, hydrophilic compound carnitine as a carrier molecule for transport of long chain fatty acids across the IMM. Derived from both dietary sources (meat or dairy products) and endogenous synthesis, circulatory carnitine is actively transported across the plasma membrane by tissue-specific carnitine transport proteins that maintain a relatively high intracellular concentration to facilitate FAO [12].

Under physiological conditions promoting FAO (i.e., fasting), fatty acids are actively transported into the mitochondrial matrix. Therefore, because CoA esters are inhibited from crossing cellular membranes [11], the first step in carnitine-mediated transport involves replacement of CoA with carnitine, a reaction catalyzed by carnitine palmitoyltransferase 1 (CPT 1) on the outer mitochondrial membrane. This process produces a long chain acylcarnitine ester that is subsequently transported across the IMM and into the matrix by the carnitine acylcarnitine translocase (CACT). Once inside, acylcarnitines are converted back to their corresponding acyl-CoAs by CPT 2, a reaction that draws from the matrix CoA pool and results in the formation of free carnitine, which is subsequently translocated back to the intermembrane space via CACT. This process results in the localization of acyl-CoAs inside the mitochondrial matrix, thus rendering them available as substrates for the enzymes of the β-oxidation spiral.

13.2.2.1 Plasma Membrane Carnitine Transporter

The *SLC22A5* gene at 5q31.1 encodes the OCTN2 organic cation transporter (OMIM 603377), which predominates in kidney, heart, and skeletal muscle [13]. Additional carnitine transport activities have been described [14, 15], including a lower affinity carnitine transport mechanism in hepatocytes [16].

Mutational impairment of OCTN2 causes primary carnitine deficiency, an autosomal recessive FAOD with a frequency of approximately 1 : 40 000 in Japan [17]. In this condition, defective OCTN2 function results in practical ablation of the renal threshold for carnitine, resulting in urinary losses that severely deplete tissue pools to the extent that there is generalized inhibition of FAO [2, 6]. Symptoms in severely affected patients are typically precipitated by an intercurrent illness in the first year or two of life and include lethargy, hypoketotic hypoglycemia, hyperammonemia, and sudden death [2, 6]. Milder forms of the disease are associated with later onset of symptoms that may include exercise intolerance, muscle weakness, and progressive cardiomyopathy. Initial diagnosis is based on laboratory demonstration of very low levels of free plasma carnitine and may be confirmed by enzyme studies in cultured fibroblasts or by molecular analysis of *SLC22A5*. Oral carnitine supplementation is typically an effective therapeutic intervention in these patients.

13.2.2.2 Carnitine Palmitoyl Transferase Type 1 (CPT 1)

Two tissue-specific isoforms of CPT 1 have been described to date (CPT 1A and CPT 1B). These activities are embedded in the mitochondrial outer membrane and catalyze the formation of acylcarnitine esters from acyl-CoAs and free carnitine, the first step in the carnitine-mediated transport process. CPT 1A (11q13; OMIM 600528) is predominantly expressed in the liver and is of primary clinical relevance [18], while CPT 1B (22qter; OMIM 601987) is abundant in heart and skeletal muscle [19]. Recently, a third CPT 1 isoform has also been described (CPT 1C; 19q13.33; [20, 21]).

Regulation of mitochondrial β-oxidation is afforded by malonyl-CoA, an intermediate of cytosolic lipid biosynthesis that directly binds and represses CPT 1 activity [2, 6]. In the postprandial state, when insulin levels are high and malonyl-CoA is abundant, FAO is inhibited due to CPT 1 repression and lipid synthesis is favored. However, under conditions of fasting, the ratios of glucagon: insulin and AMP: ATP increase, stimulating hormone-regulated lipolysis and activation of the AMP-activated kinase (AMPK), respectively. AMPK phosphorylates and inactivates acetyl-CoA carboxylase (AC), the malonyl-CoA biosynthetic enzyme, thus malonyl-CoA levels decline, CPT 1 activity increases, and FAO flux is augmented [22, 23]. Indeed, AMPK functions in a variety of states, including glucose deprivation, exercise, hypoxia, and ischemia, that are associated with depletion of ATP and therefore promote FAO.

CPT 1A deficiency predominantly results in hepatocellular disease, though cases that included some myopathic symptoms have been described [24]. CPT 1 deficiency has also been associated with maternal liver disease [25], a condition that may be caused by other FAODs as well (see below). Metabolic decompensation in CPT 1 deficiency is often precipitated by fasting/intercurrent illness and results in lethargy progressive to coma, vomiting, seizures, and hepatomegaly. Laboratory findings include hypoketotic hypoglycemia and increased levels of total and free carnitine in the presence of reduced long chain acylcarnitines and acylglycines [2, 6]. Diagnosis may be based on measurement of CPT 1 activity in cultured fibroblasts or direct molecular analysis. Therapy focuses on avoidance of fasting.

13.2.2.3 Carnitine Acylcarnitine Translocase (CACT)

The carnitine acylcarnitine translocase (OMIM 212138) is encoded by *SLC25A20* at 3p21.31 and resides in the IMM [26]. In addition to importing acylcarnitines into the mitochondrial matrix for β-oxidation, CACT also translocates free carnitine molecules across the IMM to maintain adequate pools in the respective compartments.

CACT deficiency may present as a severe, early onset form or a milder, later onset condition. Symptoms include chronic hepatocellular disease, hyperammonemia, seizures, coma, cardiomyopathy, and skeletal myopathy [2, 6]. Biochemical findings include reduced levels of free carnitine and elevated concentrations of long chain (C_{16}–C_{18}) acylcarnitines, which accumulate due to blocked transport across the IMM. Confirmatory diagnosis involves measurement of residual enzyme activity in fibroblasts or direct molecular testing. Therapy involves avoidance of fasting and supplementation of carnitine.

13.2.2.4 Carnitine Palmitoyl Transferase Type 2 (CPT 2)

CPT 2 is located at 1p32 (OMIM 600652). The transferase, a homotetramer in active form, localizes to the inner (matrix) face of the IMM and catalyzes the exchange of carnitine for CoA on fatty acids within the mitochondrial matrix, thus completing the carnitine shuttle process in mitochondrial FAO [27]. In contrast to CPT 1, CPT 2 is not regulated by malonyl-CoA and there appears to be only one functional isoform of the protein [28].

Three separate types of clinical manifestations associated with CPT 2 deficiency are a severe neonatal form, an intermediate childhood onset form, and a mild, adult onset form [2, 6]. The neonatal form is typically lethal and may present with congenital anomalies, renal cysts, cardiomyopathy, and hepatomegaly. Manifestations of the later onset form of the disease can include hypotonia, exercise intolerance, muscle pain, and myoglobinuria. Laboratory findings may include increased long chain acylcarnitines and reduced levels of free carnitine; biochemical distinction between CACT deficiency and CPT 2 deficiency can only be made by specific enzyme analysis. A relatively common point mutation (439C > T) has been reported in cases of the late onset myopathic form of the disease [29].

13.2.3
β-Oxidation Enzymes

Following completion of carnitine-mediated transport across the IMM, acyl-CoA molecules become substrates for the sequential enzyme activities that collectively facilitate the β-oxidation process [1, 6]. These activities include a group of chain length–specific acyl-Coenzyme A dehydrogenases (ACDHs; linked to the electron transport chain (ETC) via the electron transport flavoprotein/oxidoreductase system), enoyl hydratases, 3-hydroxyacyl-CoA dehydrogenases, and ketothiolase. The last three enzymatic activities of the process are contained on a multiprotein complex known as the *mitochondrial trifunctional protein* (MTP), which is composed of four α and four β subunits. The first step of mitochondrial β-oxidation involves oxidation of the α,β carbon linkage of an acyl-CoA molecule, producing *trans*-delta2-enoyl CoA. The two electrons captured by this reaction form reduced FADH2, and then are transferred via the electron transfer flavoprotein (ETF)/electron transfer flavoprotein dehydrogenase (ETFDH) to CoQ_{10} for entry into the ETC upstream of complex III. Newly formed *trans*-enoyl-CoA is then hydrated to produce a 3-hydroxyacyl-CoA compound, which is subsequently oxidized again by a 3-hydroxyacyl-CoA dehydrogenase to produce β-ketoacyl-CoA. The electrons obtained by this second dehydrogenation reaction produce reduced NADH and are ultimately transferred to the ETC via reduction of complex I. β-ketoacyl-CoA is then cleaved at the α–β bond by ketothiolase, liberating acetyl-CoA and an acyl-CoA molecule that has been shortened by two carbons. This $n − 2$ acyl-CoA can then return to the first step of β-oxidation; hence, this is a circular process that continues until the original acyl-CoA molecule has been progressively broken down into constituent acetyl-CoAs that may be used to produce ATP via the TCA cycle and oxidative phosphorylation or may be directed toward production of ketone bodies in the liver (there are also several disorders involving the enzymes that catalyze ketogenesis and ketone metabolism; these are beyond the scope of this review). The final product of even chain FAO is acetyl-CoA; odd chain lengthy fatty acids ultimately produce a molecule of propionyl (C_3)-CoA, which is converted to succinyl-CoA and enters the TCA cycle. Monounsaturated and polyunsaturated fatty acids must undergo additional isomerase and reductase reactions to become compatible with the enzymes of the β-oxidation pathway. The individual β-oxidation enzymes and related disorders are described in more detail in the following sections.

13.2.3.1 Very Long Chain Acyl-CoA Dehydrogenase (VLCAD)

Very long chain acyl-Coenzyme A dehydrogenase (VLCAD) is encoded by *ACADVL* on chromosome 17p13 (OMIM 609575) and is distinct from the other mitochondrial ACDHs (see below) in terms of its structure and localization [30, 31]. Uniquely associated with the IMM (as opposed to being a soluble matrix protein), VLCAD is a 154 kDa homodimer, while other ACDHs are homotetramers [2, 6]. The substrate specificity of VLCAD includes acyl-CoA molecules ranging from 12 to 18 carbons in length, which overlaps with both long and medium chain ACDHs (below), but is most active toward palmitoyl-CoA (C_{16}) [2, 6].

VLCAD deficiency can be subdivided into a lethal neonatal form, a childhood form, and an adolescent/adult myopathic form. Biochemically characterized by the presence of elevated long chain acylcarnitines (C_{14}–C_{18}), reduced levels of free carnitine, and increased medium chain dicarboxylic acids, VLCAD deficiency in its most severe form results in cardiomyopathy, hepatocellular dysfunction, lethargy/coma, and/or sudden death. Milder forms of the disease are characterized by exercise intolerance, muscle pain, and myoglobinuria (progressive to renal failure in some cases).

13.2.3.2 Long Chain Acyl-CoA Dehydrogenase (LCAD)

The human *ACADL* gene on 2q34–q35 encodes the long chain acyl-Coenzyme A dehydrogenase (LCAD) monomer [32]; as mentioned above, all ACDHs save VLCAD are matrix-soluble homotetramers. The substrate specificity of LCAD is reported to include acyl-CoAs from 12 to 18 carbons in length, like VLCAD [2]. The functional interrelationship between VLCAD and LCAD remains poorly defined.

Once presumed to be a cause of FAO disease in humans, subsequent analysis has demonstrated that cases of the so-called LCAD deficiency were actually caused by mutations in *ACADVL* (VLCAD); to date, a true case of LCAD deficiency is yet to be described [33]. This presents an interesting contrast to the situation in mice, in which ablation of LCAD activity produces a phenotype that is most comparable to human VLCAD deficiency [34]. Functional overlap between the VLCAD, LCAD, and medium chain acyl-Coenzyme A dehydrogenase (MCAD) in humans may be sufficient to make LCAD deficiency an ambiguous/silent condition. Alternatively, there is evidence to suggest that LCAD may play an important role in early embryonic development, such that dysfunction of the enzyme is incompatible with life [35].

13.2.3.3 Medium Chain Acyl-CoA Dehydrogenase (MCAD)

MCAD deficiency is the most common disorder of FAO in humans, with an estimated incidence rate of ~1 : 12 000–15 000 live births in the United States [36, 37]. This rate is associated with a point mutation (985A > G) that is relatively common (homozygous in 81% of MCAD deficiency cases [36]) amongst individuals of northern European descent.

MCAD is encoded by *ACADM* on 1p31 (OMIM 607008). The active homotetrameric enzyme is soluble and recognizes acyl-CoA molecules of 4–12 carbons in length [2].

As with many disorders of mitochondrial FAO, MCAD deficiency is often precipitated by prolonged fasting/intercurrent illness. Metabolic decompensation in MCAD-deficient individuals is typically associated with lethargy progressive to coma, hypotonia, seizures, vomiting, and liver dysfunction. The first episode may result in sudden death. Hypoketotic hypoglycemia is common, ammonia levels are often moderately elevated, and the patient may be acidotic. Biochemical laboratory findings may include elevations of medium chain length (C_6–C_{10}) acylcarnitines and acylglycines and increased dicarboxylic acids. During periods of good health, acylcarnitine levels in particular have been reported to be normal in some cases. Confirmatory diagnosis is widely available through direct molecular analysis. Treatment, as with most FAODs, involves early detection (i.e., newborn screening) and avoidance of fasting.

13.2.3.4 Short Chain Acyl-CoA Dehydrogenase (SCAD)

The short chain acyl-Coenzyme A dehydrogenase (SCAD) monomer is encoded by *ACADS* on 12q22-qter (OMIM 606885). The mature enzyme is a homotetramer, like other soluble mitochondrial ACDHs, and is highly homologous to MCAD and LCAD (30–40% sequence identity [2]). The substrate specificity for SCAD encompasses acyl-CoAs of four and six carbons in length.

Since it was first described, SCAD deficiency has been the center of some controversy in terms of its clinical importance [38, 39]. The majority of individuals diagnosed with SCAD deficiency is now ascertained by newborn screening with a characteristic biochemical phenotype (elevated butyrylcarnitine and ethylmalonic acid (EMA)) and possess allelic variants with relatively mild functional effects and remain largely asymptomatic [38]. In contrast, rare cases of complete SCAD deficiency have been reported with a disease phenotype that may be severe, and can include certain symptoms (i.e., ketosis) that are quite distinct from most other FAODs. Other manifestations of SCAD deficiency may include seizures, failure to thrive, developmental delay, hypoglycemia, myopathy, and hypotonia [38].

13.2.3.5 3-Hydroxy Acyl CoA Dehydrogenases

Long chain 3-hydroxyacyl-Coenzyme A dehydrogenase (LCHAD) is encoded on the *HADHA* (2p23; OMIM 600890) subunit of the MTP (see below) and appears to recognize 3-hydroxyacyl-CoA molecules between 14 and 18 carbons in length [2]. There is also an additional 3-hydroxyacyl-CoA dehydrogenase activity encoded by *HADH* on 4q22–q26 (OMIM 601609). Commonly known as medium/short chain

3-hydroxyacyl-CoA dehydrogenase (M/SCHAD), the relatively broad chain length specificity of this enzyme extends from C4 to C16 [2].

As its name implies, the condition known as *isolated LCHAD deficiency* [6, 40] may be phenotypically distinct from complete MTP deficiency (see below). Clinical manifestations associated with defective LCHAD activity can include both acute and chronic hepatocellular dysfunction, lethargy/coma, hypoketotic hypoglycemia, hyperammonemia, cardiomyopathy, skeletal myopathy, peripheral neuropathy, seizures, and sudden death. Key biochemical findings include reduced free carnitine levels, increased dicarboxylic acids, and increased concentrations of long chain 3-hydroxyacylcarnitine species. The increased prevalence of LCHAD deficiency relative to MTP deficiency can be explained by the presence of a common mutation (1528G > C) in the LCHAD region of *HADHA* [41]. Additional information pertaining to the MTP is provided below.

Dysfunctional M/SCHAD activity is often a diagnostic challenge [2, 6]. The condition is phenotypically indistinct, with reports of ketotic hypoglycemia, cardiomyopathy, episodic rhabdomyolysis, and liver disease. Laboratory findings may include increased 3-hydroxybutyrylcarnitine and moderate elevations of dicarboxylic acids.

An important clinical aspect of LCHAD deficiency is the role of this disorder in the severe maternal liver conditions – acute fatty liver of pregnancy (AFLP) and hemolysis with elevated liver enzyme activity and low platelets (HELLP) syndrome [6, 42]. A significant number of these conditions have been associated with fetal LCHAD deficiency, in which the obligate heterozygote maternal FAO capacity is unable to cope up with the excessive load of fatty acid intermediates that occurs late in the third trimester of gestation [6]. This scenario may result in the development of potentially fatal maternal steatosis and liver failure. Although defective LCHAD activity is the FAO enzyme most commonly associated with AFLP and HELLP, deficiencies of CPT 1, MCAD, MTP, and SCAD have also been associated with these conditions [25, 43–45].

13.2.3.6 Mitochondrial β-Ketothiolase

Completion of a single cycle of β-oxidation involves thiolytic cleavage of the 3-ketoacyl-CoA $\alpha-\beta$ linkage by 3-ketoacyl-CoA thiolase, which is encoded by the β subunit of the MTP (*HADHB*; 2p23; OMIM 143450). This activity is specific for long chain 3-ketoacyl-CoA species. A distinct medium chain 3-ketoacyl-CoA thiolase activity has been purified from a single patient with a fatal deficiency of this protein [46], but the cognate gene has not yet been reported. In addition, there is a third, short chain–specific thiolase activity that is variously known as *β-ketothiolase, short chain 3-ketoacyl-CoA thiolase*, or *acetyl-CoA acetyltransferase* 1 (*ACAT1;* 11q22.3-q23.1; OMIM 607809), which facilitates the final step in isoleucine catabolism [2, 6].

The single case [46] of medium chain 3-ketoacyl-CoA thiolase deficiency that has been reported involved an infant male presenting with vomiting, acidosis, liver disease, rhabdomyolysis, and myoglobinuria. Laboratory findings included elevations of lactate, 3-OH butyrate, and saturated and unsaturated C_6–C_{16} dicarboxylic acids (particularly C_{12} and C_{16} species). Dysfunction of β-ketothiolase (ACAT1) is associated with ketosis and acidosis and results in accumulation of 2-methyl-3-hydroxybutyric acid, 2-methylacetoacetic acid, and tiglylglycine in the urine [4].

13.2.3.7 Mitochondrial Trifunctional Protein (MTP)

As mentioned previously for LCHAD and 3-ketoacyl-CoA thiolase, above, the MTP is a multiprotein complex composed of four α and four β subunits (encoded by *HADHA* and *HADHB*, respectively). It is associated with the IMM and contains the final three enzyme activities that comprise mitochondrial β-oxidation (hydratase, 3-hydroxy ACDA, and thiolase).

Mutations in either the α or β subunit can cause MTP deficiency [47], which appears to be significantly rarer than the isolated LCHAD deficiency and results in the reduction of all three enzymatic activities. The clinical presentation of the disease primarily involves cardiac and skeletal myopathy, though liver disease and neuropathy have also been described [47].

13.2.3.8 Electron Transfer Flavoprotein (ETF)/Electron Transfer Flavoprotein Dehydrogenase (ETFDH)

ETF and ETFDH represent a fundamental, multiple pathway intersection where electrons gathered by at least nine different FAD-dependent mitochondrial dehydrogenases (including those that function in mitochondrial FAO) are transferred to the ETC to facilitate oxidative phosphorylation [4]. ETF is a heterodimer composed of an α (*ETFA*; 15q25–q26; OMIM 608053) and a β subunit (*ETFB*; 15q23–q25; OMIM 130410). The *ETFDH* gene is located at 4q32-qter (OMIM 231675).

Dysfunction of the ETF/ETFDH system results in a phenotypically diverse condition known as *multiple acyl-CoA dehydrogenase (MAD) deficiency* (also known as *glutaric acidemia type II* [48]). Mutations in *ETFA*, *ETFB*, or *ETFDH* can cause MAD deficiency, which has been classified with two distinct clinical phenotypes: a lethal neonatal/infantile form characterized by congenital anomalies, renal cysts, hepatocellular dysfunction, and/or cardiomyopathy and a milder form (sometimes characterized by responsiveness to riboflavin supplementation) with episodic manifestations that include lethargy, liver disease, myopathy, hypoglycemia, exercise intolerance, and occasionally psychiatric behaviors. Biochemical findings typically and characteristically include elevations of multiple organic acids (especially glutarate and ethylmalonate) and acylcarnitine species.

13.2.4
Mechanisms of Toxicity Associated with Disorders of Mitochondrial β-Oxidation

The pathophysiological mechanisms associated with dysfunction of mitochondrial FAO can be divided into two general categories: (i) depletion of cellular energy and (ii) endogenous toxicity; the latter results either directly from accumulation of specific damaging compounds or indirectly as toxic processes, for example, sequestration of free CoA. For the purpose of completeness, this review considers the physiologic effects of both specific toxic metabolites and toxic processes (Figure 13.1) in the pathogenesis of these disorders. It should be noted that there are, in certain cases, limited experimental data available to support some of the proposed mechanisms that are described here. In these cases, further research will be necessary to more clearly define the pathophysiology underlying these conditions.

13.2.4.1 Energy Depletion

Because the fundamental role of FAO is to generate energy, a key consequence of dysfunction in the pathway is profound energy depletion. While fats are utilized for energy under a variety of physiological conditions, FAO becomes the predominant source of energy for the entire body under conditions of prolonged fasting. Therefore, nearly all FAODs are commonly characterized by fasting intolerance, which results in hypoketotic hypoglycemia as hepatic glycogen stores are depleted and both gluconeogenesis and ketogenesis are impaired (see below). Energy depletion is a consequence of most disorders of mitochondrial FAO, with the apparent exception of defects in short chain fatty acid metabolism.

The effects of FAOD-associated energy depletion on different organs and tissues in the body are considerable and widespread (Figure 13.1). Because of their relatively significant dependence on FAO for fuel, organs such as the heart, skeletal muscle, liver, and kidney are typically and often seriously affected, resulting in cardiomyopathy, myopathy, hepatocellular dysfunction, and renal tubule acidosis [2, 49]. Ketone bodies are vitally important fuels for the central nervous system in particular; hence, FAODs that inhibit ketogenesis can result in significant encephalopathy. Impaired ketone synthesis may also contribute to the pathogenesis of FAODs in other, less obvious ways; for example, ketones appear to inhibit proteolysis of endogenous protein, thus conserving lean body mass during fasting [50]. The deficiency of ketone bodies, which is characteristic of many FAODs, may therefore allow widespread catabolism of lean body mass, thus exacerbating many of the tissue and organ-specific disease processes associated with these conditions [49].

Another important consequence of energy depletion in FAODs is decreased production of acetyl-CoA, a centrally important molecule in metabolic energy interconversion pathways. This concept was eloquently discussed by Roe and Ding [2], who considered the possible mechanism(s) underlying acetyl-CoA depletion and the resulting metabolic consequences (Figure 13.1). In this scenario, the accumulation of acyl-CoAs, which is common to many FAODs, would result in depletion of the intramitochondrial-free CoA pool. This would inhibit other reactions that require free CoA, including conversion of pyruvate to acetyl-CoA, which would inhibit flux through the TCA cycle (the TCA cycle also appears to be negatively affected by other processes as well; see below). Reduced acetyl-CoA synthesis would then further inhibit the TCA cycle, gluconeogenesis, and FAO, largely because of reduced levels of citrate, which provides a link between mitochondrial and cytoplasmic acetyl-CoA. In mitochondria, low acetyl-CoA levels inhibit citric acid

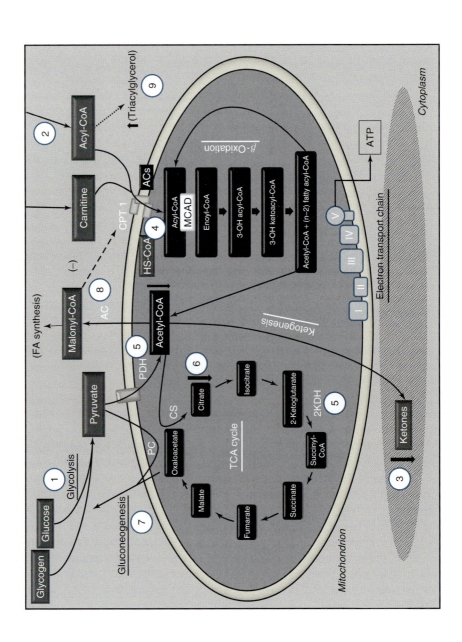

Fig. 13.1 Widespread pathological consequences associated with a single gene disorder of mitochondrial β-oxidation. This hypothetical illustration (adapted from [2].) depicts compromised cellular metabolism in the context of homozygous MCAD deficiency (see text for additional details). Metabolic decompensation is typically precipitated by fasting/intercurrent illness. Under these conditions, energy demands can initially be supported by oxidation of dietary and stored glucose (1). Normally, as these supplies are exhausted, stored fat is mobilized for energy production, resulting in increased levels of circulating fatty acids, which are taken up into cells and converted to acyl-CoAs (2). However, due to impaired β-oxidation, ketone production is inhibited (3), thus glucose levels continue to fall, promoting hypoglycemia. In addition to this primary effect, other metabolic interrelationships associated with dysfunctional FAO produce secondary inhibition of the TCA cycle and gluconeogenesis, and additional FAO defects as well. In this scenario, as decompensation progresses, there is accumulation of various acyl-CoA intermediates, which sequesters free CoA and inhibits other intramitochondrial reactions that require CoA activation (4). In addition, the increased acyl-CoA : CoA ratio represses both pyruvate and 2-ketoglutarate dehydrogenase activities (PDH and 2KDH (5). This results in an overall decrease in flux through the TCA cycle by preventing acetyl-CoA production and also by blocking the transition from 2-ketoglutarate to succinyl-CoA. As a result, synthesis of citrate by citrate synthase (CS) is impeded (6), which has important metabolic effects. By controlling the availability of cytosolic oxaloacetate, reduced concentrations of citrate can directly inhibit gluconeogenesis (7); reduced pyruvate carboxylase (PC) activity due to lack of acetyl-CoA may also contribute to impaired gluconeogenesis. In addition, citrate is the source of the cytosolic acetyl-CoA utilized by acetyl-CoA carboxylase (AC) to produce malonyl-CoA, which controls mitochondrial FAO through CPT 1 repression (8). Citrate is also the primary activator of AC. Therefore, when citrate is limiting, repression of CPT 1 activity is absent, thus allowing for uncontrolled transport of acyl intermediates into mitochondria, further exacerbating the effects of this process. In addition, the inhibition of acyl-CoA oxidation (both by the primary MCAD enzyme defect and also generalized inhibition caused by acyl-CoA intermediate accumulation) ultimately causes fatty acids to be assimilated back into triacylglycerol, which then accrues in the local tissue (9), particularly liver.

synthesis, which both directly impairs the TCA cycle and also limits the availability of acetyl-CoA in the cytosol. In gluconeogenesis, cytosolic acetyl-CoA is required to produce oxaloacetate from pyruvate (via pyruvate carboxylase (PC)) which is then converted to phosphoenolpyruvate, thus low levels of citrate also inhibit this process by limiting cytosolic acetyl-CoA. Moreover, insufficient citrate results in unchecked entry of fatty acids into the mitochondria (even as fatty acids and intermediates accumulate upstream of the FAO block) by inhibiting malonyl-CoA synthesis, which would normally negatively regulate CPT 1 activity. Therefore, dwindling supplies of acetyl-CoA may promote a vicious cycle of worsening hypoglycemia and continued, potentially toxic accumulation of fatty acids, acyl-CoAs, and other FAO intermediates, leading to further depletion of acetyl-CoA, free CoA pools, and other compounds (e.g., carnitine; see below).

13.2.4.2 Accumulation of Toxic Compounds

Although the problem of energy depletion is an almost universal consideration amongst FAODs, these conditions are also complicated by the effects of accumulating toxic acyl compounds upstream of the defective enzyme step. These substances represent true endogenous toxins and include excess free fatty acids, acyl-CoAs, acylcarnitines, and other molecules which may produce or aggravate some of the most significant clinical characteristics associated with these diseases.

Fasting stress in nearly all FAODs is associated with excessive mobilization of fatty acids, which accumulate and subsequently undergo re-esterification into triglycerides due to impaired mitochondrial catabolism. This process of excessive, inappropriate fat deposition may be injurious to tissues and organs such as the liver, heart, and kidneys, resulting in steatosis in hepatocytes [51] and cardiomyopathy. Also of particular interest in terms of site-specific toxicity are the long chain acylcarnitines that accumulate secondary to dysfunction of CACT, CPT 2, or VLCAD. In conjunction with these diseases, these compounds have been found to disturb calcium homeostasis and damage the sarcolemma, causing heartbeat abnormalities and a high incidence of sudden death [5, 52–55].

Metabolite toxicity also appears to underlie some of the classic CNS disturbances associated with FAODs; these effects have primarily been described with regard to MCAD deficiency, but may be true for other FAODs as well. Encephalopathy and coma are important aspects of the disease phenotype in MCAD deficiency, and have been reported even in the absence of obvious hypoglycemia [49, 56]. Experimental evidence suggests that medium and long chain fatty acyl compounds are injurious to the brain, consistent with the observations associated with metabolic decompensation in MCAD deficiency [2, 49]. Encephalopathic effects have been demonstrated in response to acyl compounds of three or more carbons in length. Furthermore, octanoate (C_8) has been shown to damage astrocytes and neuronal mitochondria, and also inhibits transport of acyl-CoAs out of the CNS by the choroid plexus anion transporter [57], thus allowing further accumulation of these potentially harmful molecules. Another potential encephalopathic compound that accumulates in MCAD deficiency is *cis*-decenoate ($C_{10:1}$), which has been shown to specifically inhibit cytochrome *c* oxidase activity [58]; one of several examples of FAOD-associated mitochondrial toxicity.

Although it is likely that the full repertoire of endogenous toxins associated with FAODs remains to be fully appreciated, there are several additional molecules known that appear to act as endogenous toxins in certain FAODs. For example, the disease phenotype in both isolated LCHAD deficiency and generalized MTP deficiency includes peripheral neuropathy and retinopathy, symptoms that are otherwise essentially unknown amongst FAODs [59, 60]. These effects may be mediated by the long chain 3-hydroxyacyl-CoAs that accumulate in these disorders. These molecules have been demonstrated to cause secondary inhibition of the mitochondrial ETC [59, 60] and may also directly disturb the function of retinal epithelial cells or interfere with docosahexaenoic acid (DHA) synthesis, which is abundant in retinal photoreceptors [61].

Specific endogenous toxicity also appears to play a major role in the pathogenesis of SCAD deficiency. Although the majority of cases of this condition appear to be benign, the clinical and biochemical phenotypes associated with rare symptomatic forms of SCAD deficiency are of interest due to the unusual toxic compounds that accumulate [2, 38]. Because of the positioning of this particular enzyme step in terms of overall mitochondrial FAO, the disease phenotype of severe SCAD deficiency is unique amongst FAODs, presenting in many ways like a classic inborn error of organic acid metabolism [2]. Indeed, impaired energy production does not appear to be nearly as significant in short chain disorders, possibly because the location of the SCAD-catalyzed reaction near the end of the β-oxidation spiral does not significantly impair gluconeogenesis and ketogenesis and/or because of compensatory overlap in substrate specificity with MCAD [2, 38]. Instead, it appears likely that the clinical symptoms of severe SCAD deficiency are largely secondary to direct toxicity associated with the metabolites that accumulate in this condition. Severe SCAD deficiency is associated with neuropathy and developmental delay, which are relatively rare findings amongst other FAODs, and it has been theorized that toxic accumulation of EMA and butyric acid may be an important contributor to the SCAD deficiency phenotype [38]. EMA has been shown to specifically inhibit cerebral creatine kinase activity in rats (but not in skeletal or cardiac muscle) and has also been found to inhibit the respiratory chain [62, 63]. Interestingly, the effects of butyric acid may be exerted at the level of gene expression, as this compound seems to promote histone deacetylation [64]. In addition, research involving SCAD deficiency pathogenesis has also implicated the possible role of protein aggregation in the phenotype of this disease. Many point mutations identified in SCAD deficiency appear to promote variable denaturation and accumulation of the mutant enzyme within the mitochondrial matrix, a pathophysiological process that has been implicated in other neurological disorders [38].

Another FAOD with complex and intriguing toxic pathophysiology is MAD deficiency. The fundamentally important mechanism of electron transfer that is rendered dysfunctional in this disorder results in the accumulation of multiple toxic or potentially toxic FAO intermediates, including fatty acids/fatty-CoAs and virtually all acylcarnitine species [4]. Furthermore, inhibition of additional FAD-dependent dehydrogenases that participate in the metabolism of leucine, isoleucine, lysine, and tryptophan can result in accumulation of compounds such as isovaleryl-CoA, glutaryl-CoA, 2-methylbutyryl-CoA, and related metabolites [4]. Although technically beyond the scope of this review, it should be noted that several of these compounds are associated with toxic effects on specific areas of the brain and central nervous system [65].

Many FAODs can also produce mild to moderate hyperammonemia, an effect that is presumed to be primarily due to impaired N-acetylglutamate production caused by depletion of acetyl-CoA [4]. There is also experimental evidence suggesting that transiently reduced urea cycle enzyme expression in early childhood may aggravate this situation [66]. Although ammonia levels do not generally reach the extremes observed in primary urea cycle disorders and certain other conditions, hyperammonemia remains an important pathophysiological consideration given the sensitivity of neurons to this substance [67]. Ammonia freely diffuses across the blood–brain barrier and is rapidly converted into glutamine via glutamine synthetase; the resulting osmotic force imposed by the accumulation of glutamine can cause astrocyte swelling and cytotoxic edema [67]. In addition, hyperammonemia has been reported to adversely affect multiple other aspects of central nervous system physiology, resulting in depressed energy production, inhibited axon development, proteolysis, mitochondrial degradation, and alterations of multiple neurotransmitter systems [67].

Finally, as a general consequence of fatty acid accumulation due to impaired mitochondrial β-oxidation, an increased amount of fatty acids are shunted toward microsomal ω-oxidation, which results in the cytochrome P450-dependent production of dicarboxylic acids [4]. These small molecules have been reported to uncouple mitochondrial oxidative phosphorylation and the respiratory chain [68, 69], thus providing an additional mechanism of toxicity in mitochondrial FAODs.

13.2.4.3 Sequestration of Important Biological Compounds

Disorders of mitochondrial FAO typically result in accumulations of various acyl-CoA esters, as well as other intermediates. Because cells maintain separate free CoA pools in the cytosol and mitochondria, accumulation of unutilized acyl-CoAs appears to reduce the availability of free CoA for other mitochondrial processes [2]. This situation likely further inhibits flux through the TCA cycle and gluconeogenesis by inhibiting enzymes such as pyruvate dehydrogenase and 2-ketoglutarate dehydrogenase, as described above. Furthermore, octanoyl-CoA that accumulates in MCAD deficiency and MAD deficiency has been shown to impair succinyl-CoA ligase activity [70], which would further inhibit flux through the TCA cycle. Similarly, as previously described, energy depletion associated with disorders of mitochondrial FAO results in the reduction of acetyl-CoA, which impairs the TCA cycle as well as gluconeogenesis [2].

In addition to acyl-CoAs, most FAODs (with the exception of CPT 1 deficiency) are also characterized by accumulations of various acylcarnitines, which gather first in the mitochondrial matrix but then undergo retrograde transport into the cytosol and ultimately appear in the circulatory system. As described above, increased concentrations of acylcarnitines have been demonstrated to have specific toxic effects, but the essentially irreversible formation of these carnitine esters also has additional consequences [49]. Accumulation of acylcarnitines reduces the amount of free carnitine available for FAO and, as acylcarnitine levels rise, the excretion of these compounds in the urine also increases, further reducing the free carnitine pool. Furthermore, increased concentrations of acylcarnitines can also inhibit reabsorption of free carnitine from the renal tubules [49]. The net effect is secondary carnitine deficiency, which is a common consequence of many FAODs (and many organic acid disorders as well) that often necessitates carnitine supplementation therapy to avoid further FAO impairment. Primary carnitine deficiency caused by *OCTN2* carnitine transporter mutations also results in significant urinary losses and extremely low levels of blood and tissue carnitine, which can cause generalized FAO inhibition [4].

13.3
Peroxisomal Fatty Acid Oxidation

Peroxisomes are ubiquitous intracellular organelles that play an indispensable role in metabolism in general, and are particularly important in processing of fatty acids (e.g., VLCFAs and branched-chain fatty acids (BCFAs)) that are not accepted by the mitochondrial β-oxidation pathway or other systems [71]. Indeed, more than half of the enzymes that reside in peroxisomes are presumed to function in fatty acid metabolism [72]. Disorders of peroxisome function (Table 13.2), typically involve the brain and CNS, are often severe and progressive, and have an estimated incidence that is likely in excess of 1 : 20 000 [73]. These diseases can be conveniently categorized as (i) peroxisome biogenesis disorders (PBDs), which typically involve *PEX* genes that function for general organelle assembly and function, and (ii) single peroxisomal enzyme disorders [74]. The PBDs, which are exemplified by Zellweger syndrome, result in loss or impairment of multiple peroxisome enzyme functions, including FAO. In an attempt to avoid redundancy, this review focuses on mechanisms of toxicity associated with single enzyme disorders of peroxisomal β- and α-oxidation, though these mechanisms are also present to varying degrees in the biogenesis disorders.

13.3.1
β-Oxidation

Certain fatty acids and fatty acid derivatives cannot be metabolized efficiently by mitochondrial β-oxidation enzymes, but are accepted by the peroxisomal β-oxidation pathway for chain shortening [71]. These substrates include VLCFAs (primarily C_{22}–C_{26} species), long chain dicarboxylic acids, eicosanoids, prostaglandins, and pristanic acid (derived from α-oxidation of phytanic acid, see below) [75]. The overall process of peroxisomal β-oxidation is similar to the mitochondrial pathway, with some key differences [71]. First, peroxisomal oxidation is not linked to ATP synthesis; the initial enzyme activity in the pathway (acyl-CoA oxidase) transfers electrons directly to oxygen, creating H_2O_2, which is in turn converted to water and oxygen by peroxisomal catalase. Enoyl-CoA hydratase and 3-hydroxyacyl-CoA dehydrogenase activities are carried out by the peroxisomal bifunctional protein (BP), followed by thiolytic cleavage to liberate acetyl-CoA. A second difference between the two processes is that entry of fatty acids into peroxisomes is not dependent on a carnitine transport mechanism, but rather appears to involve the function of ATP binding cassette (ABC) transport proteins, including the X-linked adrenoleukodystrophy (*XALD*) gene ALDP. Once an actively processing acyl-CoA molecule has been reduced to approximately eight carbons in length, it ceases to be an effective substrate for peroxisomal enzymes and is converted to a carnitine ester for transport and complete oxidation in a mitochondrion [71]. Several disorders of the peroxisomal β-oxidation pathway have now been described. In general, essentially all peroxisomal disorders are characterized by elevations in VLCFA levels, which

Tab. 13.2 Peroxisomal FAO disorders in humans

Protein	Gene Chromosome OMIM	General biochemical characteristics	Proposed mechanisms of toxicity
β-oxidation			
ACOX1 (acyl-CoA oxidase)	SCOX 17q25 OMIM # 609751	Increased VLCFAs	VLCFA toxicity; impaired DHA synthesis
D-Bifunctional protein (DBP)[a]	HSD17B4 5q2 OMIM # 601860	Increased VLCFAs, pristanate, phytanate, and bile acid intermediates	VLCFA toxicity; pristanate and phytanate toxicity
ALDP (adrenoleukodystrophy protein)	ABCD1 Xq28 OMIM # 300371	Increased VLCFAs	VLCFA toxicity
α-oxidation			
PCH (phytanoyl-CoA hydroxylase)	PYH 10pter-p11.2 OMIM # 602026	Increased phytanate	Phytanate toxicity
AMACR (2-methyl-acyl-CoA racemase)[a]	AMACR 5p13.2-q11.1 OMIM # 604489	Increased pristanate, DHCA, and THCA	Pristanate toxicity; bile acid intermediate toxicity (?)

[a] Overexpression has also been linked to cancer of the breast, colon, and prostate (see text).

make this a convenient (but nonspecific) marker of peroxisome functionality; additional metabolites may be present or absent in specific disorders [71]. Additional steps that may be required to verify a specific diagnosis in the case of suspected peroxisome dysfunction include direct measurement of a particular enzyme activity and/or direct molecular analysis. Treatment options for peroxisome disorders are largely symptomatic in most cases, though more specific alternatives do exist in some cases.

13.3.1.1 Acyl-CoA Oxidase

Two separate peroxisomal acyl-CoA oxidase activities have been described: the *ACOX1* gene product (SCOX; 17q25; OMIM 609751), which is specific for very long straight-chain fatty acids and eicosanoids, and the *ACOX2* gene (BCOX; 3p14.3; OMIM 601641), which is specific for branched-chain and 2-methyl BCFAs (it has been reported that a pristanoyl-CoA oxidase gene, *ACOX3*, is also present in humans, but it appears to be silent after birth [76]). The SCOX enzyme is a homodimer of 72 kDa proteins that are proteolytically cleaved to form mature 51 and 21 subunits [77]. To date, all reports of acyl-CoA oxidase deficiency have involved SCOX.

Acyl-CoA oxidase deficiency, which is also known as *pseudoneonatal adrenoleukodystrophy* (NALD), is characterized biochemically by accumulation of VLCFAs. Clinical findings include failure to thrive, hypotonia, seizures, delayed psychomotor development, visual impairment, hearing loss, neurological regression, and progressive demyelination [78].

13.3.1.2 Bifunctional Protein

The second and third steps (hydration of enoyl-CoA to form L/D-3-hydroxy acyl-CoA, followed by oxidation to form 3-ketoacyl-CoA) of peroxisomal β-oxidation are catalyzed by the D- and L-bifunctional proteins (D/LBP; aka multifunctional proteins). D-bifunctional protein (DBP) is encoded by *HSD17B4* at 5q2 (OMIM 601860), while L-bifunctional protein is the product of *EHHADH* (3q27; OMIM 607037). The DBP recognizes both straight-chain and 2-methyl BCFAs, and also participates in processing cholesterol for the synthesis of bile acids, two activities that the LBP lacks [71].

DBP deficiency (no confirmed cases of LBP deficiency have been reported) is characterized clinically by severe hypotonia and seizures, failure to achieve developmental milestones, dysmorphia, and hepatomegaly [79]. These patients also often show disordered neuronal migration. This severe single peroxisome enzyme disorder is clinically similar to the general phenotype associated with the PBDs; the presence of normal plasmalogen levels in DBP deficiency is a key distinguishing feature [71]. Biochemical findings include elevated VLCFAs, pristanic acid, and phytanic acid, along with increased levels of various bile acid intermediates [71].

13.3.1.3 ABCD1

XALD is a disorder of peroxisomal VLCFA metabolism caused by dysfunction of the ABC transport protein ALDP [80]. This 80 kDa membrane protein forms homo- or heterodimers with one of at least three other similar ABC family proteins and is encoded by the *ABCD1* gene on Xq28 (OMIM 300371). While the precise function of this protein and its pathophysiological role in XALD remains unclear, available evidence continues to suggest it is responsible for energy-dependent import of VLCFAs across the peroxisomal membrane [80].

XALD includes a wide spectrum of clinical phenotypes involving both hemizygous males and carrier females. The classical and most common form of the condition is known as the *childhood cerebral form* (CCER), which is characterized by onset of symptoms at 3–10 years of age, progressive behavioral, cognitive, and neurological decline, and inflammatory brain demyelination, often leading to complete disability within three years of onset [80]. Additional symptoms may include peripheral adrenal insufficiency (Addison's disease), corticospinal tract dysfunction, and cortical blindness. As with many other peroxisomal disorders, the biochemical hallmark of the disease is accumulation of VLCFAs, particularly unbranched, saturated C_{26} species. Treatment with Lorenzo's oil (4 : 1 glyceryl trioleate and glyceryl trierucate) may slow the progress of symptoms if administered early in the course of the disorder [81].

13.3.2
α-Oxidation

The second peroxisomal FAO pathway is based on sequential oxidation of the α-carbon of fatty acids, resulting in release of the terminal carboxyl group as CO_2. This α-oxidation process is important in the metabolism of 2-methyl branched-chain fatty acids (2MBCFAs) that cannot be degraded through β-oxidation [71]. The most important 2MBCFA in this context is phytanic acid, which is derived from dietary sources, particularly dairy products and ruminant fats, following gut bacterial conversion of chlorophyll to phytol [71, 82]. The general α-oxidation process [71] involves activation of a fatty acid to a CoA ester and transport across the peroxisomal membrane. The acyl-CoA is then hydroxylated, forming an unstable 2-hydroperoxy fatty acid that either spontaneously decarboxylates to form a chain-shortened fatty aldehyde and CO_2 or undergoes a peroxidation reaction to form a 2-hydroxy fatty acid. In the specific case of phytanic acid metabolism, 2-hydroxyphytanoyl-CoA undergoes several additional reactions, ultimately forming pristanoyl-CoA, which is then further oxidized via peroxisomal and ultimately mitochondrial β-pathways and the TCA cycle [71].

13.3.2.1 Phytanoyl-CoA Hydroxylase

Phytanoyl-CoA hydroxylase (PCH) is encoded by *PHYH* on 10pter-p11.2 (OMIM 602026). The protein is 41.2 kDa in size and catalyzes the initial step of peroxisomal α-oxidation of phytanoyl-CoA, producing 2-hydroxyphytanoyl-CoA.

Deficiency of PCH activity causes Refsum disease, which is characterized clinically by the presence of peripheral neuropathy, cerebellar ataxia, retinitis pigmentosa, and elevated concentrations of protein in the cerebrospinal fluid [71, 83]. The biochemical hallmark of the disease is a systemic increase in phytanic acid levels, though this characteristic is not unique to Refsum disease; the compound is also present in several of the peroxisomal biogenesis disorders, which also typically exhibit elevations of VLCFAs and other compounds in addition to phytanate [84]. Because this compound is exclusively obtained from dietary sources (ruminant meat, dairy products, and fish), restricted intake of these foods provides an effective therapeutic option for patients, along with plasmapheresis [71].

13.3.2.2 2-Methyl-acyl-CoA Racemase (AMACR)

The product of the 2-methyl-acyl-CoA racemase (*AMACR*) gene (5p13.2–q11.1; OMIM 604489) catalyzes conversion of 2MBCFAs (i.e., pristanoyl-CoA and bile acid intermediates) into their (S)-stereoisomeric conformation following α-oxidation, as (R)-isomers cannot be utilized in the β-oxidation pathway [85].

AMACR deficiency, also known as *congenital bile acid synthesis deficiency*, has been associated with three distinct phenotypes: adult onset sensorimotor neuropathy;

an infantile condition characterized by vitamin deficiency, coagulopathy, and liver disease; and childhood onset bile acid synthesis deficiency [86, 87]. The biochemical phenotype – elevations of pristanic acid and the bile acid intermediates dihydroxycholestanoic acid (DHCA) and trihydroxycholestanoic acid (THCA), but with normal levels of $C_{26:0}$ – is consistent with the fact that AMACR does not function in VLCFA β-oxidation, but instead acts to regulate entry of α-oxidized compounds into the β-pathway [85]. Furthermore, multiple studies have demonstrated a link between BCFA metabolism in general, and AMACR in particular, and certain forms of cancer, particularly that of the prostate [85].

13.3.3
Mechanisms of Toxicity Associated with Disorders of Peroxisomal FAO

Peroxisomal FAO defects are fundamentally distinct from the mitochondrial FAODs in that energy depletion does not appear to be a significant source of pathogenesis in these conditions. Instead, the disease phenotypes of these disorders seem to be entirely due to the toxic effects of one or more compounds that accumulate secondary to the blocked enzyme-catalyzed reaction step. In general, the brain and central nervous system are the tissues most prominently affected by these endogenous toxins, though the liver and other tissues may also be involved [71].

13.3.3.1 Peroxisomal β-Oxidation Defects

Virtually, all single gene β-oxidation defects result in systemic elevations of VLCFAs; this finding is characteristic of the biogenesis disorders as well [71, 84]. The biological effects associated with chronically increased concentrations of VLCFAs have been most extensively characterized in the case of XALD [80], but these findings will likely be broadly applicable to the other peroxisomal disorders with this characteristic.

The predominant long chain fatty acid species that are found in XALD are $C_{24:0}$ and $C_{26:0}$; $C_{22:0}$ is also prominent [80]. When present at high levels, these molecules promote inflammatory demyelination, death of oligodendrocytes and astrocytes, and mitochondrial damage [80, 88]. The cytotoxic effects on neurons appear to be mediated by $C_{26:0}$ in particular, which causes disturbances in Ca^{2+} homeostasis, resulting in increased intracellular Ca^{2+} levels [88]. Mitochondria may be particularly sensitive to $C_{22:0}$, which appears to inhibit the ETC, resulting in increased reactive oxygen species (ROS) production, decreased IMM potential, and sensitization to permeability transition and apoptosis [88].

Elevated VLCFA concentrations are also prominently involved in the clinical picture for acyl-CoA oxidase deficiency and DBP deficiency, but the differing positions of these enzymes in the β-oxidation pathway result in somewhat distinct biochemical and clinical phenotypes [77, 86]. Clinical manifestations are typically seen at a much earlier stage for both these disorders relative to XALD (sometimes involving development *in utero*), consistent with a more acute disease process. It is interesting to note that oxidase deficiency is often associated with pigmentary retinopathy, a characteristic that may be related to impaired synthesis of DHA, an important component of the retina that requires SCOX activity for synthesis [77] (inhibition of DHA synthesis has also been proposed as a mechanism for retinopathy associated with LCHAD deficiency, see above). This is consistent with the fact that retinopathy is not observed in XALD, in which there is marked VLCFA elevation but SCOX-dependent DHA synthesis is not impaired [77]. In cases of DBP deficiency, the disease phenotype may be so severe as to be initially difficult to distinguish from a PBD, with liver and sometimes renal involvement, in addition to severely impaired CNS function and disordered neuronal migration [71, 85, 86]. There are a number of different compounds (pristanic acid, phytanic acid, leukotrienes, and bile acid intermediates) that accumulate in this disorder, in addition to VLCFAs, and the individual and/or synergistic toxicities that may contribute to the phenotype of this condition have yet to be clearly delineated. The physiological effects of phytanate accumulation have been the subject of several studies involving α-oxidation disorders (see below).

13.3.3.2 Peroxisomal α-Oxidation Defects

Disorders of the α-oxidation pathway typically result in accumulation of excess phytanic acid, which is also a characteristic of the PBDs and certain single enzyme defects of the β-oxidation pathway (see above). Phytanic acid toxicity has been demonstrated in ciliary ganglion cells and astrocytes, and it may promote Ca^{2+}-mediated apoptosis in Purkinje cells [82]. In mitochondria, phytanic acid causes depolarization of the inner membrane, inhibits ATP/ADP exchange via the adenine nucleotide translocase (ANT), and sensitizes for permeability transition [89, 90]. The ETC is also affected: phytanate has been shown to uncouple complex I, much like rotenone, resulting in generalized inhibition of the respiratory chain and subsequent increased production of ROS [91]. This is consistent with observations that neuronal and retinal pigment cells that are rich in mitochondria are primarily affected in Refsum disease [92]. In addition, phytanic acid has also been demonstrated to inhibit isoprenoid metabolism and disrupt the fundamental processes of protein prenylation and Ca^{2+} homeostasis [82, 92].

In recent years, dietary BCFAs have been the subject of considerable scientific scrutiny because of the apparent link between these compounds and the development of certain forms of cancer [85, 93]. Overexpression of certain peroxisomal FAO enzymes (particularly AMACR, but also DBP) has been detected in cancer of the breast, colon, and particularly the prostate [85], and an AMACR-based immunoassay is in use for the diagnosis of prostate cancer in biopsy samples [85, 94]. Furthermore, the presence of certain AMACR polymorphisms has been associated with an increased risk for both prostate and colon cancer [95, 96], and there are also functionally important AMACR splice variants that have been identified in prostate cancer [97]. Although these findings suggest that BCFAs are carcinogenic, the pathophysiological mechanism of this effect has not been determined. It has been speculated that metabolism of BCFAs may cause DNA damage via the production of ROS [85]. An alternative mechanism proposes that BCFA metabolites may be ligands for receptors in potential carcinogenic signaling pathways; phytanate and other BCFAs have been shown to bind members of the peroxisome-proliferator activator receptor (PPAR) family [98], which regulates expression of FAO enzymes and may also be capable of regulating proliferation of cancerous cells to some extent [99].

13.4 Conclusions

In summary, metabolism of fatty acids is a fundamental biological process that is essential for normal physiological functions. Human cells have several distinct enzyme systems for metabolizing different forms of fatty acids, which primarily reside in mitochondria and peroxisomes. Though individually rare, disorders of these pathways together represent a significant health issue. As the molecular and biochemical characteristics that underlie these disorders are elucidated, it is anticipated that more effective treatment options will become available for individuals with these conditions. A key consideration in this regard is detailed elucidation of the toxic mechanisms that are associated with the disease phenotypes of different FAODs.

Most disorders of mitochondrial FAO are characterized by profound energy depletion (hypoketotic hypoglycemia in response to fasting), with secondary suppression of the TCA cycle and gluconeogenesis due to depletion of acetyl-CoA and free CoA. In addition to the primary enzyme defect, FAO is further inhibited by sequestration and urinary losses of carnitine. Many mitochondrial FAODs also result in toxic accumulation of various metabolites, including fatty acids and acyl-CoAs, acylcarnitines, 3-OH acyl-CoAs/acylcarnitines, and ammonia. These compounds have been demonstrated to have specific damaging effects on tissues and organs such as the heart, liver, kidney, and the central nervous system.

In contrast to the mitochondrial FAODs, the disease phenotypes of single enzyme peroxisomal FAODs appear to be entirely produced by metabolite toxicity, without any appreciable influence from energy deprivation. Nearly all peroxisome FAODs are characterized by accumulation of VLCFAs, which are cytotoxic to oligodendrocytes and astrocytes and cause progressive CNS demyelination. Specific disorders of β- and α-oxidation result in elevations of other destructive compounds, such as phytanic acid, which has been shown to disrupt mitochondrial electron transport and other essential processes in multiple CNS cell types.

As the pathophysiological mechanisms underlying these diseases are elucidated, the ultimate goal is to apply this knowledge for the identification of new, more effective treatment options. Although the only true cure available for genetic disorders remains gene replacement therapy, effective therapeutic strategies can provide significant improvements in the quality of life in many cases. Avoidance of fasting remains critical for management of most mitochondrial FAODs. In addition, diets enriched in specific fatty acids appear to be beneficial in certain disorders, such as medium chain triglycerides for long chain FAODs, triheptanoin for CPT 2 deficiency [100], DHA for LCHAD deficiency [61] and peroxisome disorders [101], and Lorenzo's oil for XALD. Furthermore, fibrates and other PPAR activators have shown potential efficacy in the treatment of certain mitochondrial FAODs by increasing the expression of FAO enzymes [102]. In the future, additional therapeutic modalities that are currently in use or under investigation for other metabolic disorders (enzyme replacement therapy, substrate reduction therapy, molecular chaperone therapy, suppression of nonsense mutations, etc.) could theoretically be of some benefit with regard to certain FAODs.

Acknowledgments

The author would like to thank Dr. Phillip A. Wood for critical reading of the manuscript.

References

1 Garrett, R. and Grisham, C. (2007) *Biochemistry*, 3rd edn, Thomson Brooks/Cole.
2 Roe, C. and Ding, J. (2001) Mitochondrial FAO disorders, in *The Online Metabolic and Molecular Bases of Inherited Disease* 8th ed. (eds D. Valle, A. Beaudet, B. Vogelstein, K. Kinzler, S. Antonarakis, and A. Ballabio), McGraw-Hill.
3 Wood, P. (1999) Defects in mitochondrial β-oxidation of fatty acids. *Curr Opin Lipidol*, **10**, 107–112.
4 Valle, D., Beaudet, A., Vogelstein, B., Kinzler, K., Antonarakis, S., and Ballabio, A. (2001) 8th ed. (eds) *The Online Metabolic and Molecular Bases of Inherited Disease*, McGraw-Hill.
5 Treem, W. (2000) New developments in the pathophysiology, clinical spectrum, and diagnosis of disorders of fatty acid oxidation. *Curr Opin Pediatr*, **12**, 463–468.
6 Rinaldo, P., Matern, D., and Bennett, M. (2002) Fatty acid oxidation disorders. *Annu Rev Physiol*, **64**, 477–502.
7 Hoffmann, G., von Kries, R., Klose, D., Lindner, M., Schulze, A., Muntau, A., Röschinger, W., Liebl, B., Mayatepek, E., and Roscher, A. (2004) Frequencies of inherited organic acidurias and disorders of mitochondrial fatty acid transport and oxidation in Germany. *Eur J Pediatr*, **163**, 76–80.
8 Malcom, G., Bhattacharyya, A., Velez-Duran, M., Guzman, M., Oalmann, M., and Strong, J. (1989) Fatty acid composition of adipose tissue in humans: differences between subcutaneous sites. *Am J Clin Nutr*, **50**, 288–291.
9 Berk, P. and Stump, D. (1999) Mechanisms of cellular uptake of long chain free fatty acids. *Mol Cell Biochem*, **192**, 17–31.
10 Watkins, P. (2007) Very-long-chain acyl-CoA synthetases. *J Biol Chem*, **25**, 1773–1777.
11 Ramsay, R. and Zammit, V. (2004) Carnitine acyltransferases and their influence on CoA pools in health and disease. *Mol Aspects Med*, **25**, 475–493.
12 Bremer, J. (1990) The role of carnitine in intracellular metabolism. *J Clin Chem Clin Biochem*, **28**, 297–301.
13 Tamai, I., Ohashi, R., Nezu, J., Yabuuchi, H., Oku, A., Shimane, M., Sai, Y., and Tsuji, A. (1998) Molecular and functional identification of sodium ion-dependent, high affinity human carnitine transporter OCTN2. *J Biol Chem*, **273**, 20378–20382.
14 Enomoto, A., Wempe, M., Tsuchida, H., Shin, H., Cha, S., Anzai, N., Goto, A., Sakamoto, A., Niwa, T., Kanai, Y., Anders, M., and Endou, H. (2002) Molecular identification of

a novel carnitine transporter specific to human testis: insights into the mechanism of carnitine recognition. *J Biol Chem*, **277**, 36262–36271.

15 Longo, N., Amat di San Filippo, C., and Pasquali, M. (2006) Disorders of carnitine transport and the carnitine cycle. *Am J Med Genet C Semin Med Genet*, **142C**, 77–85.

16 Scaglia, F., Wang, Y., and Longo, N. (1999) Functional characterization of the carnitine transporter defective in primary carnitine deficiency. *Arch Biochem Biophys*, **364**, 99–106.

17 Koizumi, A., Nozaki, J., Ohura, T., Kayo, T., Wada, Y., Nezu, J., Ohashi, R., Tamai, I., Shoji, Y., Takada, G., Kibira, S., Matsuishi, T., and Tsuji, A. (1999) Genetic epidemiology of the carnitine transporter OCTN2 gene in a Japanese population and phenotypic characterization in Japanese pedigrees with primary systemic carnitine deficiency. *Hum Mol Genet*, **8**, 2247–2254.

18 Bonnefont, J., Demaugre, F., Prip-Buus, C., Saudubray, J., Brivet, M., Abadi, N., and Thuillier, L. (1999) Carnitine palmitoyltransferase deficiencies. *Mol Genet Metab*, **68**, 424–440.

19 Verderio, E., Cavadini, P., Montermini, L., Wang, H., Lamantea, E., Finocchiaro, G., DiDonato, S., Gellera, C., and Taroni, F. (1995) Carnitine palmitoyltransferase II deficiency: structure of the gene and characterization of two novel disease-causing mutations. *Hum Mol Genet*, **4**, 19–29.

20 Zammit, V. (2008) Carnitine palmitoyltransferase 1: central to cell function. *IUBMB Life*, **60**, 347–354.

21 Price, N., van der Leij, F., Jackson, V., Corstorphine, C., Thomson, R., Sorensen, A., and Zammit, V. (2002) A novel brain-expressed protein related to carnitine palmitoyltransferase I. *Genomics*, **80**, 433–442.

22 Dyck, J. and Lopaschuk, G. (2002) Malonyl CoA control of fatty acid oxidation in the ischemic heart. *J Mol Cell Cardiol*, **34**, 1099–1109.

23 Folmes, C. and Lopaschuk, G. (2007) Role of malonyl-CoA in heart disease and the hypothalamic control of obesity. *Cardiovasc Res*, **73**, 278–287.

24 Olpin, S., Allen, J., Bonham, J., Clark, S., Clayton, P., Calvin, J., Downing, M., Ives, K., Jones, S., Manning, N., Pollitt, R., Standing, S., and Tanner, M. (2001) Features of carnitine palmitoyltransferase type I deficiency. *J Inherit Metab Dis*, **24**, 35–42.

25 Innes, A., Seargeant, L., Balachandra, K., Roe, C., Wanders, R., Ruiter, J., Casiro, O., Grewar, D., and Greenberg, C. (2000) Hepatic carnitine palmitoyltransferase I deficiency presenting as maternal illness in pregnancy. *Pediatr Res*, **47**, 43–45.

26 Huizing, M., Wendel, U., Ruitenbeek, W., Iacobazzi, V., IJlst, L., Veenhuizen, P., Savelkoul, P., van den Heuvel, L., Smeitink, J., Wanders, R., Trijbels, J., and Palmieri, F. (1998) Carnitine-acylcarnitine carrier deficiency: identification of the molecular defect in a patient. *J Inherit Metab Dis*, **21**, 262–267.

27 Bonnefont, J., Taroni, F., Cavadini, P., Cepanec, C., Brivet, M., Saudubray, J., Leroux, J., and Demaugre, F. (1996) Molecular analysis of carnitine palmitoyltransferase II deficiency with hepatocardiomuscular expression. *Am J Hum Genet*, **58**, 971–978.

28 Britton, C., Schultz, R., Zhang, B., Esser, V., Foster, D., and McGarry, J. (1995) Human liver mitochondrial carnitine palmitoyltransferase I: characterization of its cDNA and chromosomal localization and partial analysis of the gene. *Proc Natl Acad Sci U S A*, **92**, 1984–1988.

29 Taroni, F., Verderio, E., Dworzak, F., Willems, P., Cavadini, P., and DiDonato, S. (1993) Identification of a common mutation in the carnitine palmitoyltransferase II gene in familial recurrent myoglobinuria patients. *Nat Genet*, **4**, 314–320.

30 Aoyama, T., Souri, M., Ushikubo, S., Kamijo, T., Yamaguchi, S., Kelley, R.I., Rhead, W., Uetake, K., Tanaka, K., and Hashimoto, T. (1995) Purification of human very-long-chain acyl-coenzyme A dehydrogenase and characterization of its deficiency in seven patients. *J Clin Invest*, **95**, 2465–2473.

31 Aoyama, T., Souri, M., Ueno, I., Kamijo, T., Yamaguchi, S., Rhead, W.J., Tanaka, K., and Hashimoto, T. (1995) Cloning of human very-long-chain acyl-coenzyme A dehydrogenase and molecular characterization of its deficiency in two patients. *Am J Hum Genet*, **57**, 273–283.

32 Indo, Y., Yang-Feng, T., Glassberg, R., and Tanaka, K. (1991) Molecular cloning and nucleotide sequence of cDNAs encoding human long-chain acyl-CoA dehydrogenase and assignment of the location of its gene (ACADL) to chromosome 2. *Genomics*, **11**, 609–620.

33 Yamaguchi, S., Indo, Y., Coates, P., Hashimoto, T., and Tanaka, K. (1993) Identification of very-long-chain acyl-CoA dehydrogenase deficiency in three patients previously diagnosed with long-chain acyl-CoA dehydrogenase deficiency. *Pediatr Res*, **34**, 111–113.

34 Kurtz, D., Rinaldo, P., Rhead, W., Tian, L., Millington, D., Vockley, J., Hamm, D., Brix, A., Lindsey, J., Pinkert, C., O'Brien, W., and Wood, P. (1998) Targeted disruption of mouse long-chain acyl-CoA dehydrogenase gene reveals crucial roles for fatty acid oxidation. *Proc Natl Acad Sci U S A*, **95**, 15592–15597.

35 Berger, P. and Wood, P. (2004) Disrupted blastocoele formation reveals a critical developmental role for long-chain acyl-CoA dehydrogenase. *Mol Genet Metab*, **82**, 266–272.

36 Wang, S., Fernhoff, P., Hannon, W., and Khoury, M. (1999) Medium chain acyl-CoA dehydrogenase deficiency: human genome epidemiology review. *Genet Med*, **1**, 332–339.

37 Bodman, M., Smith, D., Nyhan, W., and Naviaux, R. (2001) Medium-chain acyl coenzyme A dehydrogenase deficiency: occurrence in an infant and his father. *Arch Neurol*, **58**, 811–814.

38 Jethva, R., Bennett, M., and Vockley, J. (2008) Short-chain acyl-coenzyme A dehydrogenase deficiency. *Mol Genet Metab*, **95**, 195–200.

39 Waisbren, S., Levy, H., Noble, M., Matern, D., Gregersen, N., Pasley, K., and Marsden, D. (2008) Short-chain acyl-CoA dehydrogenase (SCAD) deficiency: an examination of the medical and neurodevelopmental characteristics of 14 cases identified through newborn screening or clinical symptoms. *Mol Genet Metab*, **95**, 39–45.

40 Wanders, R.J., IJlst, L., van Gennip, A.H., Jakobs, C., de Jager, J.P., Dorland, L., van Sprang, F.J., and Duran, M. (1990) Long-chain 3-hydroxyacyl-CoA dehydrogenase deficiency: identification of a new inborn error of mitochondrial fatty acid β-oxidation. *J Inherit Metab Dis*, **13**, 311–314.

41 IJlst, L., Wanders, R., Ushikubo, S., Kamijo, T., and Hashimoto, T. (1994) Molecular basis of long-chain 3-hydroxyacyl-CoA dehydrogenase deficiency: identification of the major disease-causing mutation in the α-subunit of the mitochondrial trifunctional protein. *Biochim Biophys Acta*, **1215**, 347–350.

42 Browning, M., Levy, H., Wilkins-Haug, L., Larson, C., and Shih, V. (2006) Fetal fatty acid oxidation defects and maternal liver disease in pregnancy. *Obstet Gynecol*, **107**, 115–120.

43 Santos, L., Patterson, A., Moreea, S., Lippiatt, C., Walter, J., and Henderson, M. (2007) Acute liver failure in pregnancy associated with maternal MCAD deficiency. *J Inherit Metab Dis*, **30**, 103.

44 Blish, K. and Ibdah, J. (2005) Maternal heterozygosity for a mitochondrial trifunctional protein mutation as a cause for liver disease

in pregnancy. *Med Hypotheses*, **64**, 96–100.
45 Bok, L., Vreken, P., Wijburg, F., Wanders, R., Gregersen, N., Corydon, M., Waterham, H., and Duran, M. (2003) Short-chain Acyl-CoA dehydrogenase deficiency: studies in a large family adding to the complexity of the disorder. *Pediatrics*, **112**, 1152–1155.
46 Kamijo, T., Indo, Y., Souri, M., Aoyama, T., Hara, T., Yamamoto, S., Ushikubo, S., Rinaldo, P., Matsuda, I., Komiyama, A., and Hashimoto, T. (1997) Medium chain 3-ketoacyl-coenzyme A thiolase deficiency: a new disorder of mitochondrial fatty acid β-oxidation. *Pediatr Res*, **42**, 569–576.
47 den Boer, M., Dionisi-Vici, C., Chakrapani, A., van Thuijl, A., Wanders, R., and Wijburg, F. (2003) Mitochondrial trifunctional protein deficiency: a severe fatty acid oxidation disorder with cardiac and neurologic involvement. *J Dev Behav Pediatr*, **142**, 684–689.
48 Amendt, B. and Rhead, W. (1986) The multiple acyl-coenzyme A dehydrogenation disorders, glutaric aciduria type II and ethylmalonic-adipic aciduria: mitochondrial fatty acid oxidation, acyl-coenzyme A dehydrogenase, and electron transfer flavoprotein activities in fibroblasts. *J Clin Invest*, **78**, 205–213.
49 Olpin, S. (2004) Implications of impaired ketogenesis in fatty acid oxidation disorders. *Prostaglandins Leukot Essent Fatty Acids*, **70**, 293–308.
50 Felig, P., Sherwin, R., and Palaiologos, G. (1978) Ketone utilisation and ketone-amino acid interactions in starvation and diabetes, in *Biochemical and Clinical Aspects of Ketone Body Metabolism* (eds H. Soeling and C. Seufert), Thieme, Stuttgart.
51 Fromenty, B. and Pessayre, D. (1995) Inhibition of mitochondrial β-oxidation as a mechanism of hepatotoxicity. *Pharmacol Ther*, **67**, 101–154.
52 Fischbach, P., Corr, P., and Yamada, K. (1992) Long-chain acylcarnitine increases intracellular Ca++ and induces arrhythmias after depolarization in adult ventricular myocytes. *Circulation*, **96**, 748–749.
53 Corr, P., Creer, M., and Yamada, K. (1989) Prophylaxis of early ventricular fibrillation by inhibition of acylcarnitine accumulation. *J Clin Invest*, **83**, 927–936.
54 Mak, I., Kramer, J., and Weglicki, W. (1986) Potentiation of free radical-induced lipid peroxidative injury to sarcolemmal membranes by lipid amphiphiles. *J Biol Chem*, **26**, 1153–1157.
55 Yamada, K., Kanter, E., and Amit, N.A. (2000) Long-chain acylcarnitine induces Ca2+ efflux from the sarcoplasmic reticulum. *J Cardiovasc Pharmacol*, **36**, 14–21.
56 Wajner, M., Assis, D., Streck, E., Zugno, A., Ribeiro, C., Schuck, P., Leipnitz, G., Palcin, K., Wannmacher, C., Dutra Filho, C., and Wyse, A. (2002) Evidence that octanoic acid compromises brain energy metabolism. *J Inherit Metab Dis*, **25**, 149.
57 Kim, C., O'Tuama, L., Mann, J. and Roe, C. (1983) Effect of increasing carbon chain length on organic acid transport by the choroid plexus: a potential factor in Reye's syndrome. *Brain Res*, **259**, 340–343.
58 Sharpe, M., Clark, J., and Heales, S. (1999) Cytochrome C oxidase inhibition by cis-4-decenoic acid (C10:1). An important mechanism in medium chain acyl-CoA dehydrogenase (MCAD) deficiency? *J Inherit Metab Dis*, **22** (Pt 10), 23.
59 Tyni, T., Paetau, A., Strauss, A., Middleton, B., and Kivela, T. (2004) Mitochondrial fatty acid β-oxidation in the human eye and brain: implications for the retinopathy of long-chain 3-hydroxyacyl-CoA dehydrogenase deficiency. *Pediatr Res*, **56**, 744–750.

60 Tyni, T., Johnson, M., Eaton, S., Pourfarzam, M., Andrews, R., and Turnbull, D. (2002) Mitochondrial fatty acid β-oxidation in the retinal pigment epithelium. *Pediatr Res*, **52**, 595–600.

61 Gillingham, M., Van Calcar, S., Ney, D., Wolff, J., and Harding, C. (1999) Dietary management of long-chain 3-hydroxyacyl-CoA dehydrogenase deficiency (LCHADD). A case report and survey. *J Inherit Metab Dis*, **22**, 123–131.

62 Corydon, M., Gregersen, N., Lehnert, W., Ribes, A., Rinaldo, P., Kmoch, S., Christensen, E., Kristensen, T., Andresen, B., Bross, P., Winter, V., Martinez, G., Neve, S., Jensen, T., Bolund, L., and Kølvraa, S. (1996) Ethylmalonic aciduria is associated with an amino acid variant of short chain acyl-coenzyme A dehydrogenase. *Pediatr Res*, **39**, 1059–1066.

63 Barschak, A., Ferreira, G., Andre, K., Schuck, P., Viegas, C., Tonin, A., Filho, C., Wyse, A., Wannmacher, C., Vargas, C., and Wajner, M. (2006) Inhibition of the electron transport chain and creatine kinase activity by ethylmalonic acid in human skeletal muscle. *Metab Brain Dis*, **21**, 11–19.

64 Chen, J., Faller, D., and Spanjaard, R. (2003) Short-chain fatty acid inhibitors of histone deacetylases: promising anticancer therapeutics? *Curr Cancer Drug Targets*, **3**, 219–236.

65 Kölker, S., Koeller, D., Okun, J., and Hoffmann, G. (2004) Pathomechanisms of neurodegeneration in glutaryl-CoA dehydrogenase deficiency. *Ann Neurol*, **55**, 7–12.

66 Hinsdale, M., Hamm, D., and Wood, P. (1996) Effects of short-chain acyl-CoA dehydrogenase deficiency on developmental expression of metabolic enzyme genes in the mouse. *Biochem Mol Med*, **57**, 106–115.

67 Gropman, A., Summar, M., and Leonard, J. (2007) Neurological implications of urea cycle disorders. *J Inherit Metab Dis*, **30**, 865–869.

68 Tonsgard, J. and Getz, G. (1985) Effect of Reye's syndrome serum on isolated chinchilla liver mitochondria. *J Clin Invest*, **76**, 816–825.

69 Passi, S., Picardo, M., Nazzaro-Porro, M., Breathnach, A., Confaloni, A., and Serlupi-Crescenzi, G. (1984) Antimitochondrial effect of saturated medium chain length (C8C13) dicarboxylic acids. *Biochem Pharmacol*, **33**, 103–108.

70 Parker, W., Haas, R., Stumpf, D., and Eguren, L. (1983) Effects of octanoate on rat brain and liver mitochondria. *Neurology*, **33**, 1374–1377.

71 Wanders, R., Barth, P., and Heymans, H. (2001) Chapter 130: single peroxisomal enzyme deficiencies, in *The Online Metabolic and Molecular Bases of Inherited Disease*, 8th ed. (eds D. Valle, A. Beaudet, B. Vogelstein, K. Kinzler, S. Antonarakis, and A. Ballabio), McGraw-Hill.

72 Baumgart, E., Vanhooren, J., Fransen, M., Marynen, P., Puype, M., Vandekerckhove, J., Leunissen, J., Fahimi, H., Mannaerts, G., and Van Veldhoven, P. (1996) Molecular characterization of the human peroxisomal branched-chain acyl-CoA oxidase: cDNA cloning, chromosomal assignment, tissue distribution, and evidence for the absence of the protein in Zellweger syndrome. *Proc Natl Acad Sci U S A*, **93**, 13748–13753.

73 Chedrawi, A. and Clark, G. (2007) Peroxisomal Disorders. WebMD emedicine.

74 Wanders, R. (2004) Peroxisomes, lipid metabolism, and peroxisomal disorders. *Mol Genet Metab*, **83**, 16–27.

75 Poirier, Y., Antonenkov, V., Glumoff, T., and Hiltunen, J. (2006) Peroxisomal β-oxidation – a metabolic pathway with multiple functions. *Biochim Biophys Acta*, **1763**, 1413–1426.

76 Vanhooren, J., Marynen, P., Mannaerts, G., and Van Veldhoven, P. (1997) Evidence for the existence of a pristanoyl-CoA oxidase gene in man. *Biochem J*, **325**, 593–599.

77 Ferdinandusse, S., Denis, S., Hogenhout, E., Koster, J., van Roermund, C., IJlst, L., Moser, A., Wanders, R., and Waterham, H. (2007) Clinical, biochemical, and mutational spectrum of peroxisomal acyl-coenzyme A oxidase deficiency. *Hum Mutat*, **28**, 904–912.

78 Carrozzo, R., Bellini, C., Lucioli, S., Deodato, F., Cassandrini, D., Cassanello, M., Caruso, U., Rizzo, C., Rizza, T., Napolitano, M.L., Wanders, R., Jakobs, C., Bruno, C., Santorelli, F., Dionisi-Vici, C., and Bonioli, E. (2008) Peroxisomal acyl-CoA-oxidase deficiency: two new cases. *Am J Med Genet*, **146A**, 1676–1681.

79 Huyghe, S., Mannaerts, G., Baes, M., and Van Veldhoven, P. (2006) Peroxisomal multifunctional protein-2: the enzyme, the patients and the knock-out mouse model. *Biochim Biophys Acta*, **1761**, 973–994.

80 Moser, H., Kirby, D., Smith, P., Watkins, J., and Moser, A. (2001) Chapter 131: X-linked adrenoleukodystrophy, in *The Online Metabolic and Molecular Bases of Inherited Disease*. 8th ed. (eds D. Valle, A. Beaudet, B. Vogelstein, K. Kinzler, S. Antonarakis, and A. Ballabio), McGraw-Hill.

81 Moser, H., Moser, A., Hollandsworth, K., Brereton, N., and Raymond, G. (2007) "Lorenzo's oil" therapy for X-linked adrenoleukodystrophy: rationale and current assessment of efficacy. *J Mol Neurosci*, **33**, 105–113.

82 Reiser, G., Schonfeld, P., and Kahlert, S. (2006) Mechanism of toxicity of the branched-chain fatty acid phytanic acid, a marker of Refsum disease, in astrocytes involves mitochondrial impairment. *Int J Dev Neurosci*, **24**, 113–122.

83 Wanders, R., Jakobs, C., and Skjeldal, O. (2001) Chapter 132: refsum disease, in *The Online Metabolic and Molecular Bases of Inherited Disease*, 8th ed. (D. Valle, A. Beaudet, B. Vogelstein, K. Kinzler, S. Antonarakis, and A. Ballabio), McGraw-Hill.

84 Gould, S., Raymond, G., and Valle, D. (2001) Chapter 129: the peroxisome biogenesis disorders, in *The Online Metabolic and Molecular Bases of Inherited Disease* 8th ed. (eds D. Valle, A. Beaudet, B. Vogelstein, K. Kinzler, S. Antonarakis, and A. Ballabio), McGraw-Hill.

85 Lloyd, M., Darley, D., Wierzbicki, A., and Threadgill, M. (2008) Alpha-Methylacyl-CoA racemase – an 'obscure' metabolic enzyme takes centre stage. *FEBS J*, **275**, 1089–1102.

86 Ferdinandusse, S., Denis, S., Clayton, P., Graham, A., Rees, J.E., Allen, J., McLean, B., Brown, A., Vreken, P., Waterham, H., and Wanders, R. (2000) Mutations in the gene encoding peroxisomal α-methylacyl-CoA racemase cause adult-onset sensory motor neuropathy. *Nat Genet*, **24**, 188–191.

87 Setchell, K., Heubi, J., Bove, K., O'Connell, N., Brewsaugh, T., Steinberg, S., Moser, A., and Squires, R. (2003) Liver disease caused by failure to racemize trihydroxycholestanoic acid: gene mutation and effect of bile acid therapy. *Gastroenterology*, **124**, 217–232.

88 Hein, S., Schonfeld, P., Kahlert, S., and Reiser, G. (2008) Toxic effects of X-linked adrenoleukodystrophy associated, very long chain fatty acids on glial cells and neurons from rat hippocampus in culture. *Hum Mol Genet*, **17**, 1750–1761.

89 Komen, J., Distelmaier, F., Koopman, W., Wanders, R., Smeitink, J., and Willems, P. (2007) Phytanic acid impairs mitochondrial respiration through protonophoric action. *Cell Mol Life Sci*, **64**, 3271–3281.

90 Schonfeld, P., Kahlert, S., and Reiser, G. (2004) In brain mitochondria the branched-chain fatty acid phytanic acid impairs energy transduction and sensitizes for permeability transition. *Biochem J*, **383**, 121–128.

91 Schönfeld, P. and Reiser, G. (2006) Rotenone-like action of the

branched-chain phytanic acid induces oxidative stress in mitochondria. *J Biol Chem*, **281**, 7136–7142.

92 Wierzbicki, A. (2007) Peroxisomal disorders affecting phytanic acid α-oxidation: a review. *Biochem Soc Trans*, **35**, 881–886.

93 Thornburg, T., Turner, A., Chen, Y., Vitolins, M., Chang, B., and Xu, J. (2006) Phytanic acid, AMACR and prostate cancer risk. *Future Oncol*, **2**, 213–223.

94 Jiang, Z., Woda, B., Rock, K., Xu, Y., Savas, L., Khan, A., Pihan, G., Cai, F., Babcook, J., Rathanaswami, P., Reed, S., Xu, J., and Fanger, G. (2001) P504S: a new molecular marker for the detection of prostate carcinoma. *Am J Surg Pathol*, **25**, 1397–1404.

95 Levin, A., Zuhlke, K., Ray, A., Cooney, K., and Douglas, J. (2007) Sequence variation in α-methylacyl-CoA racemase and risk of early-onset and familial prostate cancer. *Prostate*, **67**, 1507–1513.

96 Daugherty, S., Platz, E., Shugart, Y., Fallin, M., Isaacs, W., Chatterjee, N., Welch, R., Huang, W., and Hayes, R. (2007) Variants in the α-methylacyl-CoA racemase gene and the association with advanced distal colorectal adenoma. *Cancer Epidemiol Biomarkers Prev*, **16**, 1536–1542.

97 Mubiru, J., Shen-Ong, G., Valente, A., and Troyer, D. (2004) Alternative spliced variants of the α-methylacyl-CoA racemase gene and their expression in prostate cancer. *Gene*, **327**, 89–98.

98 Hostetler, H., Kier, A., and Schroeder, F. (2006) Very long-chain and branched-chain fatty acyl-CoAs are high affinity ligands for the peroxisome proliferator activated receptor alpha (PPAR-a). *Biochemistry*, **45**, 7669–7681.

99 Kopelovich, L., Fay, J., Glazer, R., and Crowell, J. (2002) Peroxisome proliferation-activated receptor modulators as potential chemopreventive agents. *Mol Cancer Ther*, **1**, 357–363.

100 Roe, C., Yang, B., Brunengraber, H., Roe, D., Wallace, M., and Garritson, B. (2008) Carnitine palmitoyltransferase II deficiency: successful anaplerotic diet therapy. *Neurology*, **71**, 260–264.

101 Martinez, M. (2000) The fundamentals and practice of docosahexaenoic acid therapy in peroxisomal disorders. *Curr Opin Clin Nutr Metab Care*, **3**, 101–108.

102 Djouadi, F., Aubey, F., Schlemmer, D., Ruiter, J., Wanders, R., Strauss, A., and Bastin, J. (2005) Bezafibrate increases very-long-chain acyl-CoA dehydrogenase protein and mRNA expression in deficient fibroblasts and is a potential therapy for fatty acid oxidation disorders. *Hum Mol Genet*, **14**, 2695–2703.

14
Homocysteine as an Endogenous Toxin in Cardiovascular Disease

Sana Basseri, Jennifer Caldwell, Shantanu Sengupta, Arun Kumar, and Richard C. Austin

Atherosclerosis is the underlying cause of cardiovascular disease (CVD), with risk factors including hypercholesterolemia, obesity, hypertension, and insulin resistance. Hyperhomocysteinemia (HHcy) has also been recognized as an independent risk factor for CVD. While severe forms of HHcy are caused by rare genetic defects involved in homocysteine (Hcy) metabolism, the most common cause of HHcy is due to folate and/or vitamin B deficiency. Several genetic and dietary animal models have provided insight into the mechanisms through which HHcy contributes to atherogenesis and ultimately CVD. Clinical and population-based studies have evaluated the potential of folate and vitamin B supplementation as a therapeutic approach to reduce plasma total homocysteine (tHcy), atherogenesis, and the resulting cardiovascular outcomes. Therefore, it is well established that elevated levels of tHcy can behave as an endogenous toxin to modulate CVD. Future studies designed to specifically lower tHcy levels will provide important insights into how this thiol-containing amino acid contributes to CVD and its complications.

14.1
Cardiovascular Disease Risk

In 2005, over 80 million Americans were diagnosed with cardiovascular disease (CVD) [1], and in the same year approximately 17 million people worldwide died from CVD [2]. With the rising incidence of CVD in developing countries, as well as the increasing rate of diabetes and obesity in North America and Europe, CVD is expected to become the leading cause of death globally within the next decade [3].

14.2
Overview of Atherosclerosis

Atherosclerosis, the major underlying cause of CVD, is a complex, chronic process that is initiated at sites of endothelial cell (EC) injury and results in the accumulation of lipids within the arterial wall [4, 5]. Lesion development occurs as early as the first decade of life and typically presents as a fatty streak. Although the fatty streak enlarges during adolescence and adulthood growing in size and complexity, most patients remain asymptomatic until the formation of an advanced plaque [6]. Atherosclerotic plaque instability and subsequent thrombosis are the leading cause of morbidity and mortality in acute coronary syndromes including unstable angina, myocardial infarction, and stroke. Although a combination of genetic and environmental factors can predispose an individual to atherosclerosis, the major risk factors contributing to accelerated lesion development include elevated levels of low-density lipoprotein (LDL), decreased levels of high-density lipoproteins (HDLs), obesity, hypertension, insulin resistance, and hyperhomocysteinemia (HHcy) [7, 8]. Interestingly, arterial vessel bifurcations, which are areas of high turbulence, are particularly susceptible to the development of atherosclerosis [9–11].

14.2.1
Lesion Development and Disease Progression

The early atherosclerotic lesion has been correlated with fibrin deposition [12]; however, the initiation of atherosclerosis is presumed to be the response of the ECs lining the arterial wall to injury [13]. The endothelium of the blood vessel is normally a highly selective barrier having antithrombotic properties; however, injury disrupts this protective lining allowing the unregulated entry of LDL into the intima. Following its movement into the intima, LDL is trapped within the subendothelial cell space and undergoes oxidation. This accumulation of oxLDL in turn induces ECs to produce adhesion and chemotactic molecules that attract circulating monocytes [14]. Monocytes then bind to the endothelium and migrate into the intima where they differentiate into macrophages in response to the secretion of monocyte colony-stimulating factor (M-CSF) by ECs [15]. The scavenger receptors SR-A and CD36 expressed on the surface of mature macrophages allow the uptake of oxLDL, thereby inducing foam cell formation [16, 17]. The accumulation of foam cells within the intima leads to the formation of a fatty streak and its progression to an advanced plaque occurs as numerous cytokines and inflammatory mediators are released by cells within the plaque.

Inflammation contributes to all stages of atherosclerotic lesion development. Macrophages release various cytokines that induce smooth muscle cells (SMC) to migrate from the media into the intimal layer [18, 19]. As the plaque progresses, EC dysfunction and apoptosis increase vascular permeability, thereby promoting the migration and proliferation of intimal SMC [20]. The necrotic core is characteristic of advanced lesion development and is the pivotal point leading to plaque instability. Apoptosis of foam cells, macrophages, and SMC results in the accumulation of lipids, cellular debris, and cholesterol crystals within the lesion core [21, 22]. Importantly, this apoptotic process results in the secretion of matrix-degrading metalloproteinases, which directly contributes to plaque instability and rupture [22–24]. Following plaque rupture, activated tissue factor (TF) on the surface of lesion resident cells, including macrophages, foam cells, and SMC, is exposed to circulating coagulation factors that culminate in rapid thrombin generation and clot formation. [25–27]. Formation of the thrombus clot can induce many acute coronary symptoms through the occlusion of local blood vessels or embolism at distal sites [28].

14.2.2
Contribution of Atherosclerosis to Cardiovascular Disease

The advanced atherosclerotic plaque contributes to CVD in many ways. Large advanced plaques can cause stenosis of the blood vessel, thereby impeding or obstructing blood flow to tissues and organs. Stenosis of the coronary arteries can deprive the heart of oxygen and ultimately result in myocardial infarction. Atherosclerotic plaque rupture and subsequent thrombosis can also cause complete vessel stenosis blocking blood flow to vital organs, including the lungs (pulmonary embolism) or brain (stroke) [5]. Currently, there are a number of major risk factors that accelerate lesion development and enhance plaque rupture. These include hypercholesterolemia, hypertension, obesity, and diabetes. Numerous clinical studies have also demonstrated that HHcy is considered to be an independent risk factor for the development of atherosclerosis (discussed in detail below).

14.3
Homocysteine Metabolism

Homocysteine (Hcy) is derived from the demethylation of the essential amino acid methionine [29]. Demethylation of methionine to Hcy occurs through two intermediate compounds: S-adenosylmethionine (S-AM) and S-adenosylhomocysteine (S-AH) (Figure 14.1). Once formed, Hcy is converted into cysteine via the transsulfuration pathway. This pathway consists of the enzymes cystathionine β-synthase (CBS) and cystathionine γ-lyase (CGL), which require vitamin B_6 as a cofactor [29]. Alternatively, by using vitamin B_{12} as a cofactor, Hcy can be remethylated via the methyl donors betaine or 5–10 methylene tetrahydrofolate (remethylation pathways) which results in the formation of methionine [30]. Activation of this pathway allows for methionine conservation during periods of low dietary protein intake [31].

14.4
Hyperhomocysteinemia and CVD

The concept that Hcy is an independent risk factor for CVD was first introduced in 1969 by Kilmer McCully [32]. He observed that children with a genetic defect in vitamin B_{12} metabolism had extensive arterial disease similar to children with CBS deficiency [32]. Because both defects resulted in dramatically elevated total homocysteine (tHcy) levels, McCully hypothesized that HHcy must be the cause of

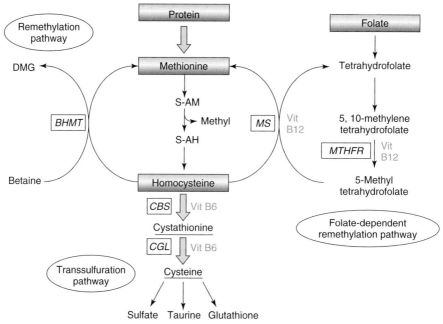

Fig. 14.1 Homocysteine/methionine metabolic pathway. Homocysteine can be metabolized via two alternative pathways. In the liver and kidney, Hcy is irreversibly converted into cysteine through the transsulfuration pathway or it can be remethylated to form methionine. The methyl group is provided by betaine or 5–10 methylene tetrahydrofolate (Vitamin B12-dependent). The transsulfuration pathway requires vitamin B6 for cystathionine β-synthase (CBS) and cystathionine γ-lyase (CGL)-mediated homocysteine catabolism. Abbreviations: 5,10-methylenetetrahydrofolate reductase (MTHFR); betaine homocysteine methyltransferase (BHMT); dimethylglycine (DMG); methionine synthase (MS); S-adenosylmethionine (S-AM); S-adenosylhomocysteine (S-AH).

vascular disease in these children [32]. In 1976, Wilcken and Wilcken provided the first clinical evidence that elevated tHcy levels due to abnormal Hcy metabolism was associated with CVD [33]. Since this seminal finding, numerous epidemiological and animal studies have provided strong evidence for an association between elevated tHcy levels and increased risk of CVD [31, 34–41]. However, it remains controversial whether this relationship is causal [42].

Plasma tHcy refers to a combination of approximately 1% free, 70–80% protein-bound, and 20–30% reduced or oxidized forms of Hcy in the blood [31, 42, 43]. Normal tHcy levels are between 5 and 15 µmol/l [43]. Mild, moderate, and severe HHcy reflect fasting tHcy concentrations of 16–30, 31–100, and >100 µmol/l, respectively [43]. Hcy has been recognized as an independent risk factor for CVD and a number of genetic and environmental factors can contribute to its metabolism and the development of HHcy (Table 14.1).

In addition to CVD, HHcy has been shown to be associated with many other diseases. In patients with renal failure where protein metabolism is affected,

Tab. 14.1 Genetic and environmental factors influencing plasma homocysteine levels

Genetics	Cystathionine β-synthase deficiency
	Inborn errors of vitamin B absorption, transport, and/or metabolism
	Polymorphisms in *MTHFR* and other genes involved in folate and vitamin B metabolism
Drugs	Vitamin B antagonists (i.e., nitrates and estrogenic hormones)
	Folate antagonists (i.e., methotrexate)
	Antacids
	Sex hormones and androgens
Lifestyle/age-related	Menopause
	Smoking
	Lack of exercise
	Aging Male gender Muscle mass
Diseases	Renal failure
	Hypothyroidism
	Cancer
	Diabetes
	Psoriasis
	Inflammatory bowel disease
Dietary	High caffeine intake
	High alcohol consumption
	Low vitamin B/folate intake

HHcy is considered to be an independent risk factor for CVD [44]. In cases of hypothyroidism, decreased folate concentrations were found to be linked to mild HHcy [45]. Certain malignancies [31], end-stage diabetes [42], and severe psoriasis [30] have also been associated with high levels of plasma Hcy. Many studies have shown links between lifestyle factors and HHcy and these include smoking, high caffeine intake, alcohol abuse, physical inactivity, and reduced vitamin intake [30, 31, 42, 46–48]. Even certain medications such as folate antagonists, vitamin B_6 antagonists [31], antacids [42], and sex hormones/androgens [30] have been found to increase plasma tHcy levels. In addition, several physiological factors such as aging, reduced glomerular filtration rate, increased muscle mass, and menopause are associated with HHcy [30].

14.5
Genetic Causes of Hyperhomocysteinemia

The most common form of HHcy results from a genetic mutation in the enzyme cystathionine β synthase (1 in 100 000 live births) [31, 33]. As shown in Figure 14.1, CBS is required to metabolize Hcy to cystathionine. Patients with autosomal recessive CBS deficiency suffer from premature atherosclerosis and thrombotic disease, skeletal and central nervous system abnormalities, and ocular problems

[31]. If left untreated, these individuals have fasting tHcy levels that are 40 times above normal and 50% of them will experience CVD complications by the age of 30 [31, 33]. The heterozygous form of CBS deficiency is more frequently observed (about 1 in 150 people) and these individuals usually have slightly elevated levels of tHcy levels [31].

Mutations in the 5,10-methylenetetrahydrofolate reductase (MTHFR) gene are also known to cause HHcy [29, 35]. Currently, over 33 different MTHFR mutations have been identified that lead to an elevation in tHcy levels [29]. The MTHFR enzyme is important for the remethylation of Hcy through the folate cycle (Figure 14.1). Clinical manifestations of this disorder include neurologic and vascular defects [49]. Kang et al. [50] reported a variant form of MTHFR that is thermolabile under specific heat inactivation conditions that can lead to HHcy. A thermolabile MTHFR enzyme is produced when a point mutation converts the nucleotide cysteine to a thymine at position 677 of the *MTHFR* gene (*MTHFR* 677TT polymorphism) [51]. This results in the conversion of the amino acid alanine to valine [31] and reduces the MTHFR protein activity by half [50]. Individuals with the *MTHFR* 677TT polymorphism have moderate HHcy and are at increased risk of CVD under conditions of low dietary folate intake [35].

Although it is established that *MTHFR* 677TT polymorphism increases plasma tHcy levels [31, 52–54], many studies demonstrated no increased CVD risk associated with this mutation [52–54]. In a meta-analysis by Klerk and others, the results of 23 studies conducted over three years [35] found that individuals with the *MTHFR* 677TT genotype had an increased risk of CVD especially when folate levels were low [35]. Another recent meta-analysis study suggested a causal relationship between high tHcy levels and CVD [40]. This was based on odds ratios provided by 72 studies where there was a mutation in the MTHFR gene as well as 20 prospective studies where Hcy and disease risk were measured [40].

There are some very rare genetic forms of severe HHcy and these include complete loss of MTHFR through autosomal recessive inheritance [55], deficiency or reduced activity of methionine synthase (MS) involved in converting Hcy back to methionine (Figure 14.1), and autosomal recessive forms of vitamin B_{12} metabolic defects [31, 55].

14.6
Recent Studies

The initial findings in the late 1960s and early 1970s led to many studies that examined the role of Hcy on vascular biology and cardiovascular outcomes. To determine the relationship between HHcy and CVD, a combination of cell culture and animal models have been utilized to examine the mechanisms by which HHcy influences the development and progression of vascular disease. In addition, a vast number of clinical and population-based studies have been performed to better determine the role Hcy plays in CVD and its complications.

14.6.1
Mouse Models of Hyperhomocysteinemia and Atherosclerosis

A number of genetic mouse models of HHcy exist and include mice with targeted deletion of genes encoding the enzymes required for Hcy metabolism, namely, CBS, MTHFR, or MS [56]. Mice with targeted deletion of genes for apolipoprotein E (ApoE$^{-/-}$) or the low-density lipoprotein receptor (LDLR$^{-/-}$) are most commonly used in atherosclerosis studies [3–5, 7]. ApoE$^{-/-}$ mice develop spontaneous atherosclerosis and have severe hypercholesterolemia. The use of ApoE$^{-/-}$ mice with dietary and/or genetically induced HHcy has provided insight into the causal role of HHcy in the pathogenesis of atherosclerosis [36, 37].

Several dietary approaches have also been studied and are often combined with genetic mouse models of atherosclerosis and/or HHcy. These diet-induced HHcy models include feeding mice diets (i) high in methionine, (ii) low in folate, vitamin B_{12} and/or vitamin B_6 with or without high methionine content, (iii) low in choline, or (iv) supplementation of the drinking water with Hcy [56]. A diet rich in methionine would directly increase circulating Hcy levels (Figure 14.1). Deficiency in folate and/or vitamin B_{12} prevents the folate-dependent remethylation of Hcy, while decreased vitamin B_6 results in the accumulation of Hcy as a result of defective CBS-mediated catabolism of Hcy into cystathionine. Reduced dietary choline prevents the betaine homocysteine methyltransferase (BHMT)-dependent remethylation of Hcy. Importantly, reduced dietary intake of vitamin B_6, B_{12}, and/or folate are prevalent in patients with HHcy and CVD. Therefore, these dietary interventions in mice can provide data relevant to humans with CVD.

A major limitation to dietary or genetic models of HHcy is that they are often accompanied by changes in other metabolites that may directly or indirectly influence vascular disease independent of Hcy [56]. Tissue-specific knockout or knockdown of *Cbs*, *Mthfr*, or *Bhmt* may serve as useful experimental approaches for future study. In addition, development of mouse models with inducible promoters of gene products involved in Hcy metabolism would enable a better distinction between specific and nonspecific effects of HHcy on vascular pathophysiology and CVD [57].

14.6.2
Animal Study Findings

A study on MTHFR knockout mice was conducted to determine the effect of MTHFR deficiency on CVD [58]. Deficiency in MTHFR resulted in 1.6–10 fold higher Hcy levels as compared to wild-type littermates [58]. These mice also had decreased DNA methylation, increased S-adenosylhomocysteine concentrations, and unusual lipid deposition in the aorta indicating an association between HHcy and vascular disease [58]. Results from a study in rats indicated that four to five weeks of a folate-deplete diet resulted in moderate to severe HHcy and showed decreased relaxation of coronary resistance vessels, increased carotid artery permeability, and stiffening [38]. Furthermore, the arterial stiffness and permeability induced in these rats and mice was the result of superoxide anion production and reduction in nitric oxide (NO) bioavailability [59]. In addition to reducing NO bioavailability, superoxide can react with NO to form peroxynitrite, which in turn leads to the oxidation of lipids and other molecules.

Findings by Dayal and colleagues indicated that $Cbs^{+/-}$ mice fed a high methionine/low folate diet developed HHcy [60]. These heterozygous mice had enhanced thrombosis caused by oxidative stress or defective protein C activation [60]. Previous studies have shown a proatherogenic role of HHcy in $apoE^{-/-}$ mice [36, 37]. However, a recent study by Zhou et al. examined the potential of HHcy as an independent risk factor for atherosclerosis [61]. Wild-type C57BL/6 mice that were fed a high methionine diet developed HHcy, but this was not enough to cause atherosclerosis [61]. When HHcy was accompanied by inflammation and/or high plasma VLDL concentrations, accelerated atherosclerotic lesion development was observed [61]. These findings suggest that HHcy can accelerate lesion development only in the presence of additional risk factors that induce inflammation and/or alter plasma lipid levels.

14.6.3
Clinical Trials Examining the Effects of Folate and B Vitamin Supplementation on Cardiovascular Disease

The leading cause of HHcy is vitamin B deficiency due to inadequate intake, reduced absorption, or drug interactions [62]. The B vitamins, folate, vitamin B_6, vitamin B_{12}, and riboflavin are all intimately involved in the metabolism of Hcy and deficiencies in each of these are responsible for increased plasma Hcy levels. The recommended daily intake of B vitamins is 400 µg; however, the general population ingests less than 300 µg per day [63]. The major dietary sources of B vitamins include green vegetables, cereals, fruits, meat, fish, eggs, and dairy products. Unfortunately, up to 90% of B vitamins are lost during food processing due to the sensitivity of folate to heat, light, and prolonged storage [64]. Age-related decreases in gastric acid secretion and slight increases in gastric pH afflict 30–40% of the elderly population and can result in the inadequate absorption of vitamin B_{12} [65, 66]. Taken as a whole, many studies have examined the potential of vitamin B supplementation to reduce tHcy levels and ultimately decrease the incidence of CVD. Although vitamin B supplementation is an accepted therapeutic strategy to lower tHcy levels [40, 67], the secondary effects of these therapies on lowering the incidence of CVD have not been confirmed [68–70]. It is widely recognized that these recent studies lack sufficient statistical power to detect an association between tHcy levels and CVD [68], and therefore a relationship between vitamin B supplementation and a decreased risk of CVD should not be ruled out.

14.6.3.1 Folate
Folate is a cofactor for MS, an enzyme involved in the remethylation of Hcy to methionine. Folate is the major determinant of tHcy and a meta-analysis has demonstrated that supplementation with 0.5–5 mg/day can lower tHcy by 20–25%

[71]. Of particular interest to food fortification policies in North America is that doses of 0.2–0.4 mg/day can achieve a maximal reduction of tHcy in young healthy adults [72, 73]. However, individuals with established heart disease require 0.8 mg/day of folate to attain the same reduction [74]. Unfortunately, recent studies have uncovered some adverse effects of overexposure to higher doses of folate including reduced natural killer cell cytotoxicity [75] and an increased risk of colorectal adenomas and noncolorectal cancers [76]. Preliminary evidence suggests that individuals with established heart disease may benefit from lower doses of folate when administered over longer treatment durations [77], thereby reducing the risks associated with unmetabolized folate in plasma.

In the HOPE-2 study, a large randomized controlled clinical trial, Lonn and colleagues investigated the effect of a set dosage of folate, vitamin B_6, and B_{12} on patients over 55 years of age who had vascular disease or diabetes [78]. On the basis of the relative risk values for death from cardiovascular complications (RR = 0.96, 0.81–1.13), myocardial infarction (RR = 0.98, 0.85–1.14), and stroke (RR = 0.75, 0.59–0.97), the results indicated that patients with vascular disease did not benefit from B vitamin supplementation when compared to the placebo group [78]. These studies suggest that lowering plasma tHcy levels with folate supplementation does not benefit patients with existing CVD. It also suggests that folate supplementation may somehow reduce the beneficial effects of Hcy lowering by modulating additional cellular pathways that affect vascular disease progression (i.e., enhanced smooth muscle cell proliferation, EC dysfunction, macrophage foam cell formation).

14.6.3.2 Vitamin B_{12}

In addition to folate, vitamin B_{12} is also a cofactor for the enzyme MS. Although vitamin B_{12} is a less useful determinant of tHcy as compared to folate, once folate status is optimized, vitamin B_{12} is the primary nutritional determinant of tHcy [79–81]. Vitamin B_{12} is of particular importance to North Americans because folate fortification of cereal grains results in folate optimization in the majority of the population. Meta-analysis demonstrates that folate alone typically results in a 25% lowering of tHcy; however, with the addition of vitamin B_{12} supplementation, tHcy is lowered by an additional 7% [71–74, 82,]. Furthermore, when patients on regular vitamin B_{12} supplementation or with renal failure were excluded from this study, high doses of vitamin B_{12} conferred a 21% reduction in the risk of CVD [83]. Overall, these studies suggest that therapeutic supplementation of folate and vitamin B_{12} may be more effective than folate alone at lowering the risk of CVD through tHcy reduction.

14.6.3.3 Vitamin B_6

Vitamin B_6 is a cofactor for CBS, the enzyme responsible for the transsulfuration of Hcy to cysteine. Although many studies examining the role of vitamin B_6 supplementation on tHcy levels report contradictory results, there is some evidence that low dose vitamin B_6 can slightly lower tHcy (7.5%) in healthy adults who have been optimized for folate and riboflavin intakes [84]. One 14- year longitudinal study demonstrated that an increased intake of vitamin B_6 (0.3 mg/day) could decrease the risk of developing heart disease [85]. Additionally, retrospective and prospective case-control studies have established that low plasma vitamin B_6 levels confer an increased risk of coronary artery disease, peripheral vascular disease, and stroke [86–89]. However, four randomized, case-control trials did not confirm a reduced risk of recurrent cardiovascular events with vitamin B_6 therapy [90], though it should be noted that this study is acknowledged to be statistically underpowered to detect a significant reduction in CVD risk. Interestingly, vitamin B_6 is also required for the metabolism of n−3 polyunsaturated fatty acids [91], which may suppress other atherogenic mechanisms including platelet aggregation [92] and the proliferation of ECs [93]. Furthermore, low plasma vitamin B_6 levels are related to chronic inflammation [94], an established risk factor for the development of atherosclerosis and CVD.

14.6.3.4 Riboflavin

In addition to folate, the coenzymatic form of riboflavin, FAD, is required for the proper function of the MTHFR enzyme. Although observational studies investigating riboflavin status and tHcy in individuals with various MTHFR genotypes have reported inconsistent results [95–97], there are promising studies examining individuals with the 677TT variant. It is well established that the MTHFR 677TT variant has reduced activity due to the loss of the riboflavin-derived FAD cofactor [98]. Interestingly, a meta-analysis examining individuals with the 677TT variant suggests a 14–21% increased risk of CVD [35, 40, 99]; however, this excess risk is significant in Europe but not in North America. Additionally, among healthy Europeans with the 677TT variant, two studies found riboflavin to be the determinant of tHcy [95, 100], whereas the Framingham cohort found low folate and not riboflavin to be limiting [96]. These discrepancies are likely attributed to folate fortification in North America [35, 99]. Finally, emerging research has confirmed riboflavin supplementation as an independent modifier of tHcy in individuals with the 677TT variant representing a new gene–nutrient interaction. This study demonstrates that riboflavin supplementation can decrease tHcy by 22% in healthy 677TT individuals and by as much as 40% in those 677TT individuals with low riboflavin status [97].

14.6.4
Effect of Lifestyle on tHcy Levels

In addition to dietary folate and B vitamin supplementation, several physiological and lifestyle factors have been associated with HHcy; however, the results remain conflicting. In 2003, a cross-sectional study demonstrated a positive correlation between HHcy and lifestyle factors such as smoking, alcohol consumption, increasing age, male gender, and systolic blood pressure [46]. However, the same survey did not find a correlation between lower levels of tHcy and high vitamin B_{12} and serum folate levels [46]. This population-based study illustrated that supplement use had no significant impact on tHcy concentrations [46]. A US population-based study showed that consumption of milk, yogurt, cold breakfast cereals, peppers, and cruciferous vegetables had beneficial effects on serum tHcy levels [101]. This outcome was suggested to be due to higher intake of folate and riboflavin, though they found no association between vitamin B_{12} and vitamin B_6 supplement use and lower tHcy levels [101]. Similarly, a large prospective study conducted in Norway demonstrated that an individual's lifestyle such as weight change, smoking, and vitamin supplementation over a period of six years had an effect on tHcy levels [102]. However, the changes failed to show a strong association with tHcy as previous cross-sectional studies had shown [47, 48].

14.7
Potential Mechanisms through which Hyperhomocysteinemia Contributes to Atherogenesis

There are a number of hypotheses that may explain the link between HHcy, atherosclerosis and/or CVD. These include N- and S-homocysteinylation of proteins, oxidative stress, endoplasmic reticulum (ER) stress, inflammation, and SMC proliferation. In circulation, Hcy is present in both reduced and oxidized forms. The free, reduced form with a free sulfhydryl (-SH) group represents less than 1% of plasma tHcy [103]. The low-molecular weight oxidized forms including homocystine (homodimer of Hcy) and Hcy-cysteine mixed disulfide accounts for about 10–20% of plasma tHcy. However, greater than 80% of Hcy in the circulation is protein bound. Homocysteinylation of proteins has been proposed to be one of the mechanisms of Hcy toxicity. Hcy forms stable covalent bonds with protein cysteine residues (Protein-S-Homocysteinylation) [104]. Alternatively, Hcy-thiolactone that is formed during the proofreading of methionyl tRNA synthase can bind to ε amine group of protein lysyl residues (Protein-N-homocysteinylation) [105].

14.7.1
Protein-S-Homocysteinylation

Hcy is considered to be a reactive thiol amino acid having the ability to bind free cysteine residues or cleave accessible cysteine disulfide bonds. This reactive characteristic of Hcy can potentially modulate the structure and/or function of other proteins, a phenomenon often referred to as the *"molecular targeting hypothesis"* [106]. Protein cysteine residues play a significant role in protein folding and influence the quaternary structure through the formation of intra/intermolecular disulfide bonds. Several cysteine residues are known to be present in the catalytic domain of many proteins that are involved in cell signaling, transcriptional regulation, and trafficking [107]. Thus, interaction of Hcy with critical protein cysteine residues might disrupt protein function and eventually cellular metabolism. As a result, cysteine-rich proteins with exposed free cysteine residues have a higher propensity to react with circulating thiols similar to Hcy. For example, the mechanism of homocysteinylation of albumin has been studied in detail [104]. Albumin has three domains with 17 intrachain disulfide bonds formed between 34 cysteine residues. The only free cysteine in albumin (Cys^{34}) is situated in a partially protected site that has an abnormally low pKa (pKa ~5) [108]. Thus, under physiological conditions, Cys^{34} exists as a thiolate anion which is highly reactive [109]. This is reflected from the fact that although albumin is secreted into the circulation by the liver in the thiolate anion form [110], one-third of albumin molecules in circulation carry disulfide-bonded thiols, at Cys^{34} [109]. It has been proposed that albumin thiolate anion first attacks cystine (the homodimer of cysteine that is formed due to metal catalyzed autoxidation of cysteine) to form albumin-Cys^{34}-S–S-cysteine. Hcy then binds to albumin in two steps. In the first step, the sulfhydryl group of Hcy preferentially attacks the sulfur of albumin bound cysteine rather than the sulfur of albumin Cys^{34} to predominantly form Hcy-cysteine mixed disulfide and albumin thiolate anion. In the second step, albumin thiolate anion selectively attacks the Hcy sulfur atom in Hcy-cysteine mixed disulfide bond rather than the cysteine sulfur to predominantly form albumin-Cys^{34}-S–S-Hcy [104].

Step 1

$$\text{Hcy-S}^- + \text{Albumin-Cys}^{34}\text{-S-S-Cys}$$
$$\longrightarrow \text{Hcy-S-S-Cys} + \text{Albumin-Cys}^{34}\text{- S}^-$$

Step 2

$$\text{Albumin-Cys}^{34}\text{-S}^- + \text{Hcy-S-S-Cys}$$
$$\longrightarrow \text{Albumin-Cys}^{34}\text{-S-S-Hcy} + \text{Cys-S}^-$$

In addition to albumin, there are currently 19 proteins that have been reported to bind Hcy (Table 14.2) [111]. Homocysteinylation of these proteins in some cases have been reported to modulate the function of the protein. For instance, Hcy binds to cysteine residues of fibronectin localized to the C-terminal region within and adjacent to the fibrin binding domain [112]. The fibrin-binding domain contains numerous intrachain disulfide bonds. Binding of Hcy to fibronectin inhibited its capacity to bind fibrin by 62% [112]. The binding of fibronectin to fibrin plays an important role in thrombosis and wound healing. Similarly, Hcy binds to the cysteine residues in the EGF domain of fibrillin, an extracellular matrix protein and alters the structure of the protein, leading to its increased susceptibility to proteolytic cleavage [113]. Further, it has been proposed that Hcy binds to the apoB component of LDL and increasing concentrations of Hcy in plasma leads to higher S-homocysteinylation of LDL *in vivo* [114]. Apart from the extracellular proteins, Hcy also binds to intracellular proteins and it has recently been shown that Hcy targets the cysteine-rich intracellular protein metallothionein, impairs its zinc binding function, and inhibits the capacity to scavenge superoxide radicals [115]. The zinc released from metallothionein increases the formation of reactive oxygen

Tab. 14.2 Proteins that interact with homocysteine

Serial no.	Protein ID	Protein name	PDB ID
1	APOB_Human	Apolipoprotein B	Not available
2	FA5_Human	Coagulation factor V	1CZS
3	FA12_Human	Coagulation factor XII	Not available
4	TTHY_Human	Transthyretin	1Z7J
5	ALBU_Human	Albumin	2BXQ
6	FBN1_Human	Fibrillin 1	1EMN
7	NOTC1_Human	Notch homolog 1	1TOZ
8	TGFB1_Human	Transforming growth factor, beta 1	1KLA
9	FINC_Human	Fibronectin 1	1E88
10	ANXA2_Human	Annexin A2	1W7B
11	MMP2_Human	Matrix metallopeptidase 2	1CK7
12	MT2_Human	Metallothionein 2A	1MHU
13	TPA_Human	Plasminogen activator	1TPM
14	TRBM_Human	Thrombomodulin	1DX5
15	APOA_Human	Lipoprotein, Lp(a)	1JFN
16	TF_Human	Coagulation factor III	1AHW
17	PROC_Human	Protein C	1AUT
18	FIBB_Human	Fibrinogen beta chain	1FZA
19	FA7_Human	Coagulation factor VII	1WSS

species (ROS). Binding of Hcy to metallothionein also promotes the transient expression of Egr-1, a protein that is involved in atherosclerosis [115].

Recently, an attempt has been made to predict the potential protein targets of Hcy based on structural and physicochemical properties of protein cysteine residues [111]. Parameters such as cysteine content, solvent accessibility of cysteine residues, and dihedral strain energies and pKa of these cysteines were used to predict the proteins that could bind Hcy. Further, the validity of these parameters was successfully tested on proteins that have already been shown to bind Hcy experimentally.

14.7.2
Protein-N-Homocysteinylation

Hcy-thiolactone, which is formed from Hcy by methionyl-tRNA synthetase, can bind to numerous plasma and intracellular proteins. The mechanism of binding of Hcy-thiolactone to proteins (N-homocysteinylation) involves acylation of ε amino group of protein lysine residues by the activated carboxyl group of Hcy-thiolactone [105]. Protein-N-homocysteinylation was first observed *in vitro* [116] and later in humans [117]. Several plasma proteins including human serum albumin, γ-globulin, LDL, HDL, transferrin, antitrypsin, and fibrinogen, have been reported to undergo N-homocysteinylation [118]. The rate of N-homocysteinylation of proteins is proportional to the number of lysine residues in the protein [105]. In contrast to S-homocysteinylation, N-homocysteinylation of proteins leads to an additional free thiol group, which might result in the formation of "new" protein disulfide bonds or protein multimerization. Thus, N-homocysteinylation of proteins also results in the modulation of protein structure, and as a consequence its function. For instance, N-homocysteinylation of fibrinogen forms clots that are more resistant to lysis than the clots formed from native fibrinogen [119], which may contribute to an increased risk of CVD. N-homocysteinylated LDL induces LDL aggregation and a higher uptake by macrophages [120]. Further, homocysteinylation of LDL induces functional alterations and oxidative damage in human ECs [121]. Ferretti and colleagues reported that homocysteinylation of LDL (Hcy-LDL) increases lipid peroxidation and oxidative stress in ECs. This would explain the increased atherogenic effects of Hcy-LDL [121]. In addition, N-homocysteinylation may cause EC dysfunction by causing protein damage, an event that can trigger atherosclerotic plaque development [122, 123]. Likewise, plasma HDL is also sensitive to homocysteinylation and this may result in decreased protective function of HDL against oxidative damage [124]. N-Homocysteinylation also leads to the production of anti-N-Hcy-protein autoantibodies [122], which were found to be increased in the sera of male stroke patients, suggesting a potential role in atherogenesis/CVD [125]. N-Homocysteinylation of plasma proteins may therefore be a key factor in the development and progression of vascular disease [126].

14.7.3
Nitric Oxide Production and Oxidative Stress

Oxidative stress occurs when the antioxidant capacity of the cell is unable to cope with reactive oxygen intermediates such as peroxides and free radicals, thereby resulting in damage to cellular proteins, lipids, and DNA. The thiol group of Hcy readily undergoes auto-oxidation, forming Hcy-thiolactone which leads to N-homocysteinylation of other proteins [116, 117]. N-Homocysteinylation of proteins has been shown to be cytotoxic [121, 122] and leads to the formation of ROS including hydroxyl radicals, hydrogen peroxide, superoxide anions, and peroxynitrite [127]. Superoxide anions and peroxynitrite both lead to reduced nitric oxide (NO) production through a mechanism involving the oxidation of eNOS sulfhydryl groups. Hcy itself is also cytotoxic to cultured ECs and adversely affects endothelial function and viability [128, 129]. In ECs, elevated levels of Hcy decreases the bioavailability of NO, a potent vasodilator [130, 131]. Under normal conditions, NO exerts antiatherogenic effects and decreased bioavailability of NO due to an increased tHcy concentration can lead to vasoconstriction and vascular disease. The bioavailability of NO is reduced in the presence of elevated Hcy levels despite the normal expression of endothelial NO synthase [132]. However, reduced function and expression of the cellular arginine transporter protein decreases NO formation [133]. *In vivo*, HHcy has also been demonstrated to promote EC dysfunction through the reduced bioavailability of NO [134]. Elevated Hcy levels decrease the activity of glutathione peroxidase, an antioxidant enzyme, resulting in increased accumulation of ROS. It has been proposed that NO can bind to molecular oxygen and other oxygen-free radicals, the concentration of which increases in the presence of elevated Hcy levels due to the reduced activity of glutathione peroxidase [135, 136]. Interaction of NO with these oxygen-free radicals results in the formation of peroxynitrite (ONOO$^-$), resulting in decreased NO levels. Additionally, superoxide and peroxynitrite through the modification of tissues can generate lipid peroxides which contribute to plasma membrane instability and cell lysis [137]. *In vitro*, peroxynitrite has been shown to induce EC apoptosis through three major mechanisms: a NO-dependent pathway [138], activation of the mitogen activated protein (MAP) kinase pathways ERK and JNK [139], and an ER stress-mediated pathway [140]. Further, under physiological conditions, Hcy can also react with NO to form S-nitrosohomocysteine which may also contribute to the decrease in NO bioavailability.

14.7.4
Smooth Muscle Cell Proliferation

SMC proliferation within the intima is a hallmark characteristic of an advancing atherosclerotic plaque. To examine the ability of Hcy to induce SMC proliferation, a study performed *in vitro* treated rat aortic SMCs with Hcy and demonstrated an increase in DNA synthesis as well as the cell cycle regulators cyclins D1 and A [141, 142]. Additionally, human SMC stimulated with Hcy demonstrate DNA hypomethylation (an indicator of cellular proliferation), as well as increased collagen generation [143]. *In vivo* studies in animals infused intravenously or supplemented with dietary Hcy further demonstrate the ability of Hcy to induce SMC proliferation by measuring either myointimal cell proliferation [144] or intimal hyperplasia (luminal stenosis) [145]. Interestingly, the latter study demonstrated that a dose-dependent correlation between plasma tHcy levels and percent stenosis exists.

14.7.5
Endothelial Cell Proliferation

In contrast to Hcy-induced SMC proliferation, Hcy inhibits the growth of ECs [141, 146]. The inhibition of vascular EC growth in the presence of Hcy is associated with an increase in the level of S-adenosylhomocysteine. Conversion of Hcy to S-adenosylhomocysteine was associated with decreases in p21ras carboxymethylation and a reduction in MAP kinase activity [146]. Hcy-induced inhibition of EC growth might also be mediated through transcriptional downregulation of fibro-blast growth factor-2 involving G protein-mediated pathway associated with altered promoter DNA methylation [147]. Further, PI3K/Akt/FOXO signaling pathway might play an important role in mediating cell cycle G1 arrest in ECs in the presence of Hcy.

14.7.6
ER Stress

N-Homocysteinylation is a potential mechanism by which excess plasma Hcy and Hcy-thiolactone levels may lead to cellular toxicity [126]. It has been hypothesized that HHcy and *N*-homocysteinylation can cause ER stress due to the accumulation of damaged and misfolded proteins in the ER [126, 148, 149]. ER stress may then activate the unfolded protein response (UPR) pathways that lead to protein degradation and an increase in ER chaperones such as GRP78, GRP94, and calnexin, which assist in the proper folding of newly synthesized proteins in the ER. Hcy treatment induces the expression of the endoplasmic reticulum chaperone GRP78/BiP and two novel ER stress-response genes, RTP and Herp [150, 151]. It also alters the expression of genes responsive to ER stress (i.e., GADD45, GADD153, ATF-4, YY1). Hcy also impairs protein processing and secretion of vWF via the ER [152].

ER stress itself has been shown to be involved in the development of atherosclerosis and the pathogenesis of CVD risk factors such as obesity, insulin resistance, and type 2 diabetes [140, 153–156]. Hcy has also been shown to activate sterol regulatory element-binding proteins (SREBPs), ER membrane bound transcription factors, which are involved in lipid and cholesterol metabolism possibly through ER stress pathways [153, 157]. Normally, the expression and activity of SREBPs are regulated by SREBP cleavage activation protein (SCAP). However, it is believed that Hcy circumvents this mechanism, maintaining the cells in a sterol-starved state although lipids continue to accumulate. Hcy also leads to the increased production of cholesterol and secretion of apoB-100 in HepG2 cells [158]. The induction of cholesterol synthesis is mediated by HMG-CoA reductase, which catalyzes the rate-limiting step in cholesterol biosynthesis. In cultured cells, HMG-CoA reductase activity is suppressed by cholesterol derived from LDL. However, Hcy abolishes the feedback inhibition by LDL. It has been postulated that Hcy causes an upregulation of HMG-CoA reductase synthesis at the transcriptional and/or translational level in HepG2 cells. ER stress-mediated lipid dysregulation may explain the lipid and cholesterol-ester accumulation in the macrophages during the early stages of atherosclerosis, which leads to foam cell formation [153, 157]. In addition, activation of ER stress/UPR pathways has been observed in all stages of atherosclerotic lesion development in ApoE$^{-/-}$ mice [159]. Thus, Hcy-mediated ER stress directly regulates the intracellular concentration of cholesterol and this could be one of the mechanisms for increased hepatic steatosis and atherosclerosis in hyperhomocysteinemic individuals.

14.7.7
Vascular Matrix Studies

The proliferation of SMCs and synthesis of extracellular matrix are important determinants in lesion development and plaque stability. In cultured rabbit and human SMCs, Hcy treatment enhances collagen synthesis [143] whether this impacts lesion stability or thrombogenicity is currently unknown.

14.7.8
Findings from Yeast Studies

The budding yeast *Saccharomyces cerevisiae* is one of the model systems of choice for studies in molecular medicine. The methionine metabolism pathway in *S. cerevisiae* is similar to that in humans with some minor variations. Addition of Hcy to a wild-type yeast strain inhibits their growth in a dose-dependent manner. Transcriptional profiling of yeast treated with Hcy revealed that genes coding for antioxidant enzymes such as glutathione peroxidase, catalase, and superoxide dismutase were down-regulated [160]. Similar downregulation of antioxidant genes in the presence of Hcy were also observed in cell culture studies [161]. Further, it was observed that addition of Hcy resulted in an increase in the expression of *KAR2* (karyogamy 2) gene, a well-known marker of ER stress [160]. It was thus proposed that the inhibition of yeast growth may be due to ER stress. Transcriptional profiling revealed that genes involved in one carbon metabolism, glycolysis, and serine biosynthesis were upregulated in the presence of Hcy suggesting that cells attempt to reduce the intracellular concentration of Hcy by up-regulating the genes involved in its metabolism [160].

It has also been suggested that accumulation of Hcy in *Saccharomyces pombe* causes defects in purine biosynthesis [162]. Disruption of MS (the enzyme that converts Hcy to methionine) results in Hcy accumulation and is closely correlated with defective purine biosynthesis in *S. pombe*. Similarly, depletion of S-adenosylhomocysteine hydrolase, an essential enzyme that hydrolyzes S-adenosylhomocysteine to Hcy, results in profound changes in cellular lipid composition, suggesting a tight interaction between lipid metabolism and S-adenosylhomocysteine function. Recently, it has been shown that Hcy leads to the accumulation of S-adenosylhomocysteine through the S-adenosylhomocysteine hydrolase catalyzed reaction and this leads to decreased *de novo* phosphatidyl choline synthesis. This decrease is accompanied by an increase in triacyl glycerol levels, again indicating a strong relation between cellular lipid homeostasis and S-adenosylhomocysteine [163, 164].

14.7.9
Inflammation

Atherosclerosis is a disease characterized by chronic inflammation, and interestingly Hcy has been demonstrated to induce the production of several proinflammatory cytokines. *In vitro* studies using ECs demonstrated that Hcy treatment can activate the transcription factor NF-κB, which leads to the induction of chemokines, cytokines, hemopoetic growth factors, and leukocyte adhesion molecules, all of which contribute to vascular inflammation and atherogenesis [165]. Specifically, this study found an induction of interleukin-8 (IL-8), and monocyte chemoattractant protein-1 (MCP-1) expression, which are responsible for increased monocyte attachment to the endothelium and migration into the intima, critical steps in the development of atherosclerotic lesions [166]. Later, this result was supported by a prospective human population study showing that plasma IL-8 levels are directly correlated with an increased risk of future CVD in men and women [167]. *In vivo*, the atherosclerotic lesions of hyperhomocysteinemic apoE$^{-/-}$ mice have increased NF-κB activation, proinflammatory mediator expression, and cytokine generation [168]. Additionally, dietary HHcy-induced mice fed an atherogenic diet displayed enlarged atherosclerotic lesions with increased leukocyte infiltration further exemplifying the promotion of atherosclerosis through Hcy-induced inflammation [61]. Furthermore, an observational study in humans over 65 years of age demonstrated that plasma tHcy is correlated with elevated inflammatory cytokines such as IL-1 receptor antagonist and IL-6 [169].

Hcy also induces expression of VEGF, presumably via the activation of NF-κB [170]. VEGF has been found to be expressed in activated macrophages, ECs, and SMCs in human coronary atherosclerotic lesions, but not in normal arteries [171]. Further, adhesion of human monocytes to Hcy-stimulated ECs was significantly upregulated [172]. This was accompanied with an increased expression of vascular endothelial adhesion molecule (VCAM). The interaction of monocytes and ECs is an early event in atherogenesis. These studies suggest that the proinflammatory effects of Hcy may have important implications in atherogenesis.

14.8
Conclusions

Atherosclerosis is the underlying cause of CVD, with risk factors including elevated LDL, decreased HDL, obesity, hypertension, and insulin resistance. HHcy has also been recognized as an independent risk factor for CVD. Although severe forms of HHcy are due to genetic defects involved in Hcy metabolism, the leading cause of HHcy is folate and/or vitamin B deficiency. Several genetic and dietary animal models have provided insight into the mechanism through which HHcy contributes to atherogenesis and ultimately CVD. Clinical- and population-based studies have evaluated the potential of folate and vitamin B supplementation as a therapeutic approach to reduce tHcy, atherogenesis, and the resulting cardiovascular outcomes. Therefore, it is well established that elevated levels of tHcy can behave as an endogenous toxin contributing to CVD. Future studies that are designed to specifically lower plasma tHcy levels will provide important insights into how this thiol-containing amino acid modulates CVD and its complications.

References

1. American Heart Association (2008). Cardiovascular Disease Statistics: 2003, http://www.americanheart.org/presenter.jhtml?identifier=4478 (Last updated November 26).
2. World Health Organization (2007). Cardiovascular Diseases, http://www.who.int/mediacentre/factsheets/fs317/en/print.html (Last updated February 2007).
3. Hansson, G.K. (2005) Inflammation, atherosclerosis, and coronary artery disease. *N Engl J Med*, **352** (16), 1685–1695.
4. Goldschmidt-Clermont, P.J., Creager, M.A., Losordo, D.W., Lam, G.K., Wassef, M., and Dzau, V.J. (2005) Atherosclerosis 2005: recent discoveries and novel hypotheses. *Circulation*, **112** (21), 3348–3353.
5. Lusis, A.J. (2000) Atherosclerosis. *Nature*, **407** (6801), 233–241.
6. Davies, J.R., Rudd, J.H., and Weissberg, P.L. (2004) Molecular and metabolic imaging of atherosclerosis. *J Nucl Med*, **45** (11), 1898–1907.
7. Anand, S.S., Yusuf, S., Vuksan, V., Devanesen, S., Teo, K.K., Montague, P.A., Kelemen, L., Yi, C., Lonn, E., Gerstein, H., Hegele, R.A., and McQueen, M. (2000) Differences in risk factors, atherosclerosis, and cardiovascular disease between ethnic groups in Canada: the Study of Health Assessment and Risk in Ethnic groups (SHARE). *Lancet*, **356** (9226), 279–284.
8. McGill, H.C. Jr, McMahan, C.A., Zieske, A.W., Sloop, G.D., Walcott, J.V., Troxclair, D.A., Malcom, G.T., Tracy, R.E., Oalmann, M.C., and Strong, J.P. The Pathobiological Determinants of Atherosclerosis in Youth (PDAY) Research Group (2000) Associations of coronary heart disease risk factors with the intermediate lesion of atherosclerosis in youth. *Arterioscler Thromb Vasc Biol*, **20** (8), 1998–2004.
9. Aars, H. and Solberg, L.A. (1971) Effect of turbulence on the development of aortic atherosclerosis. *Atherosclerosis*, **13** (2), 283–287.
10. Helderman, F., Segers, D., de Crom, R., Hierck, B.P., Poelmann, R.E., Evans, P.C., and Krams, R. (2007) Effect of shear stress on vascular inflammation and plaque development. *Curr Opin Lipidol*, **18** (5), 527–533.
11. Ku, D.N., Giddens, D.P., Zarins, C.K., and Glagov, S. (1985) Pulsatile flow and atherosclerosis in the human carotid bifurcation. Positive correlation between plaque location and low oscillating shear stress. *Arteriosclerosis*, **5** (3), 293–302.
12. Lou, X.J., Boonmark, N.W., Horrigan, F.T., Degen, J.L., and Lawn, R.M. (1998) Fibrinogen deficiency reduces vascular accumulation of apolipoprotein(a) and development of atherosclerosis in apolipoprotein(a) transgenic mice. *Proc Natl Acad Sci U S A*, **95** (21), 12591–12595.
13. Ross, R. (1993) The pathogenesis of atherosclerosis: a perspective for the 1990s. *Nature*, **362** (6423), 801–809.
14. Takei, A., Huang, Y., and Lopes-Virella, M.F. (2001) Expression of adhesion molecules by human endothelial cells exposed to oxidized low density lipoprotein. Influences of degree of oxidation and location of oxidized LDL. *Atherosclerosis*, **154** (1), 79–86.
15. Rajavashisth, T.B., Andalibi, A., Territo, M.C., Berliner, J.A., Navab, M., Fogelman, A.M., and Lusis, A.J. (1990) Induction of endothelial cell expression of granulocyte and macrophage colony-stimulating factors by modified low-density lipoproteins. *Nature*, **344** (6263), 254–257.
16. Rahaman, S.O., Lennon, D.J., Febbraio, M., Podrez, E.A., Hazen, S.L., and Silverstein, R.L. (2006) A CD36-dependent signaling cascade is necessary for macrophage foam cell formation. *Cell Metab*, **4** (3), 211–221.
17. Zhao, Z., de Beer, M.C., Cai, L., Asmis, R., de Beer, F.C., de Villiers,

W.J., and van der Westhuyzen, D.R. (2005) Low-density lipoprotein from apolipoprotein E-deficient mice induces macrophage lipid accumulation in a CD36 and scavenger receptor class A-dependent manner. *Arterioscler Thromb Vasc Biol*, **25** (1), 168–173.

18 Haque, N.S., Fallon, J.T., Pan, J.J., Taubman, M.B., and Harpel, P.C. (2004) Chemokine receptor-8 (CCR8) mediates human vascular smooth muscle cell chemotaxis and metalloproteinase-2 secretion. *Blood*, **103** (4), 1296–1304.

19 Jovinge, S., Hultgardh-Nilsson, A., Regnstrom, J., and Nilsson, J. (1997) Tumor necrosis factor-α activates smooth muscle cell migration in culture and is expressed in the balloon-injured rat aorta. *Arterioscler Thromb Vasc Biol*, **17** (3), 490–497.

20 Choy, J.C., Granville, D.J., Hunt, D.W., and McManus, B.M. (2001) Endothelial cell apoptosis: biochemical characteristics and potential implications for atherosclerosis. *J Mol Cell Cardiol*, **33** (9), 1673–1690.

21 Bjorkerud, S. and Bjorkerud, B. (1996) Apoptosis is abundant in human atherosclerotic lesions, especially in inflammatory cells (macrophages and T cells), and may contribute to the accumulation of gruel and plaque instability. *Am J Pathol*, **149** (2), 367–380.

22 Clarke, M.C., Figg, N., Maguire, J.J., Davenport, A.P., Goddard, M., Littlewood, T.D., and Bennett, M.R. (2006) Apoptosis of vascular smooth muscle cells induces features of plaque vulnerability in atherosclerosis. *Nat Med*, **12** (9), 1075–1080.

23 Depre, C., Wijns, W., Robert, A.M., Renkin, J.P., and Havaux, X. (1997) Pathology of unstable plaque: correlation with the clinical severity of acute coronary syndromes. *J Am Coll Cardiol*, **30** (3), 694–702.

24 Luan, Z., Chase, A.J., and Newby, A.C. (2003) Statins inhibit secretion of metalloproteinases-1, -2, -3, and -9 from vascular smooth muscle cells and macrophages. *Arterioscler Thromb Vasc Biol*, **23** (5), 769–775.

25 Ardissino, D., Merlini, P.A., Bauer, K.A., Bramucci, E., Ferrario, M., Coppola, R., Fetiveau, R., Lucreziotti, S., Rosenberg, R.D., and Mannucci, P.M. (2001) Thrombogenic potential of human coronary atherosclerotic plaques. *Blood*, **98** (9), 2726–2729.

26 Hatakeyama, K., Asada, Y., Marutsuka, K., Sato, Y., Kamikubo, Y., and Sumiyoshi, A. (1997) Localization and activity of tissue factor in human aortic atherosclerotic lesions. *Atherosclerosis*, **133** (2), 213–219.

27 Toschi, V., Gallo, R., Lettino, M., Fallon, J.T., Gertz, S.D., Fernandez-Ortiz, A., Chesebro, J.H., Badimon, L., Nemerson, Y., Fuster, V., and Badimon, J.J. (1997) Tissue factor modulates the thrombogenicity of human atherosclerotic plaques. *Circulation*, **95** (3), 594–599.

28 Arbustini, E., Grasso, M., Diegoli, M., Pucci, A., Bramerio, M., Ardissino, D., Angoli, L., de Servi, S., Bramucci, E., Mussini, A. et al. (1991) Coronary atherosclerotic plaques with and without thrombus in ischemic heart syndromes: a morphologic, immunohistochemical, and biochemical study. *Am J Cardiol*, **68** (7), 36B–50B.

29 Castro, R., Rivera, I., Blom, H.J., Jakobs, C., and Tavares de Almeida, I. (2006) Homocysteine metabolism, hyperhomocysteinaemia and vascular disease: an overview. *J Inherit Metab Dis*, **29** (1), 3–20.

30 Hankey, G.J., Eikelboom, J.W., Ho, W.K., and van Bockxmeer, F.M. (2004) Clinical usefulness of plasma homocysteine in vascular disease. *Med J Aust*, **181** (6), 314–318.

31 Hankey, G.J. and Eikelboom, J.W. (1999) Homocysteine and vascular disease. *Lancet*, **354** (9176), 407–413.

32 McCully, K.S. (1969) Vascular pathology of homocysteinemia: implications for the pathogenesis of arteriosclerosis. *Am J Pathol*, **56** (1), 111–128.

33 Wilcken, D.E. and Wilcken, B. (1976) The pathogenesis of coronary artery disease. A possible role for methionine metabolism. *J Clin Invest*, **57** (4), 1079–1082.

34 Devlin, A.M., Arning, E., Bottiglieri, T., Faraci, F.M., Rozen, R., and Lentz, S.R. (2004) Effect of Mthfr genotype on diet-induced hyperhomocysteinemia and vascular function in mice. *Blood*, **103** (7), 2624–2629.

35 Klerk, M., Verhoef, P., Clarke, R., Blom, H.J., Kok, F.J., and Schouten, E.G. (2002) MTHFR 677C-->T polymorphism and risk of coronary heart disease: a meta-analysis. *J Am Med Assoc*, **288** (16), 2023–2031.

36 Zhou, J., Møller, J., Danielsen, C.C, Bentzon, J., Ravn, H.B., Austin, R.C., and Falk, E. (2001) Dietary supplementation with methionine and homocysteine promotes early atherosclerosis but not plaque rupture in ApoE-deficient mice. *Arterioscler Thromb Vasc Biol*, **21** (9), 1470–1476.

37 Wang, H., Jiang, X., Yang, F., Gaubatz, J.W., Ma, L., Magera, M.J., Yang, X., Berger, P.B., Durante, W., Pownall, H.J., and Schafer, A.I. (2003) Hyperhomocysteinemia accelerates atherosclerosis in cystathionine beta-synthase and apolipoprotein E double knock-out mice with and without dietary perturbation. *Blood*, **101** (10), 3901–3907.

38 Symons, J.D., Zaid, U.B., Athanassious, C.N., Mullick, A.E., Lentz, S.R., and Rutledge, J.C. (2006) Influence of folate on arterial permeability and stiffness in the absence or presence of hyperhomocysteinemia. *Arterioscler Thromb Vasc Biol*, **26** (4), 814–818.

39 Tayama, J., Munakata, M., Yoshinaga, K., and Toyota, T. (2006) Higher plasma homocysteine concentration is associated with more advanced systemic arterial stiffness and greater blood pressure response to stress in hypertensive patients. *Hypertens Res*, **29** (6), 403–409.

40 Wald, D.S., Law, M., and Morris, J.K. (2002) Homocysteine and cardiovascular disease: evidence on causality from a meta-analysis. *Br Med J*, **325** (7374), 1202.

41 Wang, H., Jiang, X., Yang, F., Gaubatz, J.W., Ma, L., Magera, M.J., Yang, X., Berger, P.B., Durante, W., Pownall, H.J., and Schafer, A.I. (2003) Hyperhomocysteinemia accelerates atherosclerosis in cystathionine beta-synthase and apolipoprotein E double knock-out mice with and without dietary perturbation. *Blood*, **101** (10), 3901–3907.

42 Kaul, S., Zadeh, A.A., and Shah, P.K. (2006) Homocysteine hypothesis for atherothrombotic cardiovascular disease: not validated. *J Am Coll Cardiol*, **48** (5), 914–923.

43 Refsum, H., Smith, A.D., Ueland, P.M., Nexo, E., Clarke, R., McPartlin, J., Johnston, C., Engbaek, F., Schneede, J., McPartlin, C., and Scott, J.M. (2004) Facts and recommendations about total homocysteine determinations: an expert opinion. *Clin Chem*, **50** (1), 3–32.

44 van Guldener, C. (2006) Why is homocysteine elevated in renal failure and what can be expected from homocysteine-lowering? *Nephrol Dial Transplant*, **21** (5), 1161–1166.

45 Ozmen, B., Ozmen, D., Parildar, Z., Mutaf, I., Turgan, N., and Bayindir, O. (2006) Impact of renal function or folate status on altered plasma homocysteine levels in hypothyroidism. *Endocr J*, **53** (1), 119–124.

46 Ganji, V. and Kafai, M.R. (2003) Demographic, health, lifestyle, and blood vitamin determinants of serum total homocysteine concentrations in the third National Health and Nutrition Examination Survey, 1988–1994. *Am J Clin Nutr*, **77** (4), 826–833.

47 Nygard, O., Refsum, H., Ueland, P.M., Stensvold, I., Nordrehaug, J.E., Kvale, G., and Vollset, S.E. (1997) Coffee consumption and plasma total homocysteine: the Hordaland Homocysteine Study. *Am J Clin Nutr*, **65** (1), 136–143.

48 Nygard, O., Refsum, H., Ueland, P.M., and Vollset, S.E. (1998) Major lifestyle determinants

of plasma total homocysteine distribution: the Hordaland Homocysteine Study. *Am J Clin Nutr*, **67** (2), 263–270.

49 Motulsky, A.G. (1996) Nutritional ecogenetics: homocysteine-related arteriosclerotic vascular disease, neural tube defects, and folic acid. *Am J Hum Genet*, **58** (1), 17–20.

50 Kang, S.S., Zhou, J., Wong, P.W., Kowalisyn, J., and Strokosch, G. (1988) Intermediate homocysteinemia: a thermolabile variant of methylenetetrahydrofolate reductase. *Am J Hum Genet*, **43** (4), 414–421.

51 Frosst, P., Blom, H.J., Milos, R., Goyette, P., Sheppard, C.A., Matthews, R.G., Boers, G.J., den Heijer, M., Kluijtmans, L.A., van den Heuvel, L.P. et al. (1995) A candidate genetic risk factor for vascular disease: a common mutation in methylenetetrahydrofolate reductase. *Nat Genet*, **10** (1), 111–113.

52 Brattstrom, L., Wilcken, D.E., Ohrvik, J., and Brudin, L. (1998) Common methylenetetrahydrofolate reductase gene mutation leads to hyperhomocysteinemia but not to vascular disease: the result of a meta-analysis. *Circulation*, **98** (23), 2520–2526.

53 McQuillan, B.M., Beilby, J.P., Nidorf, M., Thompson, P.L., and Hung, J. The Perth Carotid Ultrasound Disease Assessment Study (CUDAS) (1999) Hyperhomocysteinemia but not the C677T mutation of methylenetetrahydrofolate reductase is an independent risk determinant of carotid wall thickening. *Circulation*, **99** (18), 2383–2388.

54 Verhoef, P., Kok, F.J., Kluijtmans, L.A., Blom, H.J., Refsum, H., Ueland, P.M., and Kruyssen, D.A. (1997) The 677C→T mutation in the methylenetetrahydrofolate reductase gene: associations with plasma total homocysteine levels and risk of coronary atherosclerotic disease. *Atherosclerosis*, **132** (1), 105–113.

55 Scheuner, M.T. (2003) Genetic evaluation for coronary artery disease. *Genet Med*, **5** (4), 269–285.

56 Dayal, S. and Lentz, S.R. (2008) Murine models of hyperhomocysteinemia and their vascular phenotypes. *Arterioscler Thromb Vasc Biol*, **28** (9), 1596–1605.

57 Wang, L., Jhee, K.H., Hua, X., DiBello, P.M., Jacobsen, D.W., and Kruger, W.D. (2004) Modulation of cystathionine beta-synthase level regulates total serum homocysteine in mice. *Circ Res*, **94** (10), 1318–1324.

58 Chen, Z., Karaplis, A.C., Ackerman, S.L., Pogribny, I.P., Melnyk, S., Lussier-Cacan, S., Chen, M.F., Pai, A., John, S.W., Smith, R.S., Bottiglieri, T., Bagley, P., Selhub, J., Rudnicki, M.A., James, S.J., and Rozen, R. (2001) Mice deficient in methylenetetrahydrofolate reductase exhibit hyperhomocysteinemia and decreased methylation capacity, with neuropathology and aortic lipid deposition. *Hum Mol Genet*, **10** (5), 433–443.

59 Mullick, A.E., Zaid, U.B., Athanassious, C.N., Lentz, S.R., Rutledge, J.C., and Symons, J.D. (2006) Hyperhomocysteinemia increases arterial permeability and stiffness in mice. *Am J Physiol Regul Integr Comp Physiol*, **291** (5), R1349–R1354.

60 Dayal, S., Wilson, K.M., Leo, L., Arning, E., Bottiglieri, T., and Lentz, S.R. (2006) Enhanced susceptibility to arterial thrombosis in a murine model of hyperhomocysteinemia. *Blood*, **108** (7), 2237–2243.

61 Zhou, J., Werstuck, G.H., Lhotak, S., Shi, Y.Y., Tedesco, V., Trigatti, B., Dickhout, J., Majors, A.K., DiBello, P.M., Jacobsen, D.W., and Austin, R.C. (2008) Hyperhomocysteinemia induced by methionine supplementation does not independently cause atherosclerosis in C57BL/6J mice. *FASEB J*, **22** (7), 2569–2578.

62 Stanger, O., Herrmann, W., Pietrzik, K., Fowler, B., Geisel, J., Dierkes, J., and Weger, M. (2004) Clinical use and rational management of homocysteine, folic acid, and B vitamins in cardiovascular and thrombotic diseases. *Z Kardiol*, **93** (6), 439–453.

63 de Bree, A., van Dusseldorp, M., Brouwer, I.A., van het Hof, K.H., and Steegers-Theunissen, R.P. (1997) Folate intake in Europe: recommended, actual and desired intake. *Eur J Clin Nutr*, **51** (10), 643–660.

64 Schroeder, H.A. (1971) Losses of vitamins and trace minerals resulting from processing and preservation of foods. *Am J Clin Nutr*, **24** (5), 562–573.

65 Malinow, M.R., Bostom, A.G., and Krauss, R.M. (1999) Homocyst(e)ine, diet, and cardiovascular diseases: a statement for healthcare professionals from the Nutrition Committee, American Heart Association. *Circulation*, **99** (1), 178–182.

66 Selhub, J., Jacques, P.F., Wilson, P.W., Rush, D., and Rosenberg, I.H. (1993) Vitamin status and intake as primary determinants of homocysteinemia in an elderly population. *J Am Med Assoc*, **270** (22), 2693–2698.

67 The Homocysteine Studies Collaboration (2002) Homocysteine and risk of ischemic heart disease and stroke: a meta-analysis. *J Am Med Assoc*, **288** (16), 2015–2022.

68 B-Vitamin Treatment Trialists' Collaboration (2006) Homocysteine-lowering trials for prevention of cardiovascular events: a review of the design and power of the large randomized trials. *Am Heart J*, **151** (2), 282–287.

69 Bonaa, K.H., Njolstad, I., Ueland, P.M., Schirmer, H., Tverdal, A., Steigen, T., Wang, H., Nordrehaug, J.E., Arnesen, E., and Rasmussen, K. (2006) Homocysteine lowering and cardiovascular events after acute myocardial infarction. *N Engl J Med*, **354** (15), 1578–1588.

70 Toole, J.F., Malinow, M.R., Chambless, L.E., Spence, J.D., Pettigrew, L.C., Howard, V.J., Sides, E.G., Wang, C.H., and Stampfer, M. (2004) Lowering homocysteine in patients with ischemic stroke to prevent recurrent stroke, myocardial infarction, and death: the Vitamin Intervention for Stroke Prevention (VISP) randomized controlled trial. *J Am Med Assoc*, **291** (5), 565–575.

71 Homocysteine Lowering Trialists' Collaboration (1998) Lowering blood homocysteine with folic acid based supplements: meta-analysis of randomised trials. *Br Med J*, **316** (7135), 894–898.

72 Daly, S., Mills, J.L., Molloy, A.M., Conley, M., McPartlin, J., Lee, Y.J., Young, P.B., Kirke, P.N., Weir, D.G., and Scott, J.M. (2002) Low-dose folic acid lowers plasma homocysteine levels in women of child-bearing age. *Q J Med*, **95** (11), 733–740.

73 Ward, M., McNulty, H., McPartlin, J., Strain, J.J., Weir, D.G., and Scott, J.M. (1997) Plasma homocysteine, a risk factor for cardiovascular disease, is lowered by physiological doses of folic acid. *Q J Med*, **90** (8), 519–524.

74 Wald, D.S., Bishop, L., Wald, N.J., Law, M., Hennessy, E., Weir, D., McPartlin, J., and Scott, J. (2001) Randomized trial of folic acid supplementation and serum homocysteine levels. *Arch Intern Med*, **161** (5), 695–700.

75 Troen, A.M., Mitchell, B., Sorensen, B., Wener, M.H., Johnston, A., Wood, B., Selhub, J., McTiernan, A., Yasui, Y., Oral, E., Potter, J.D., and Ulrich, C.M. (2006) Unmetabolized folic acid in plasma is associated with reduced natural killer cell cytotoxicity among postmenopausal women. *J Nutr*, **136** (1), 189–194.

76 Cole, B.F., Baron, J.A., Sandler, R.S., Haile, R.W., Ahnen, D.J., Bresalier, R.S., McKeown-Eyssen, G., Summers, R.W., Rothstein, R.I., Burke, C.A., Snover, D.C., Church, T.R., Allen, J.I., Robertson, D.J., Beck, G.J., Bond, J.H., Byers, T., Mandel, J.S., Mott, L.A., Pearson, L.H., Barry, E.L., Rees, J.R., Marcon, N., Saibil, F., Ueland, P.M., and Greenberg, E.R. (2007) Folic acid for the prevention of colorectal adenomas: a randomized clinical trial. *J Am Med Assoc*, **297** (21), 2351–2359.

77 Law, M. (2000) Fortifying food with folic acid. *Semin Thromb Hemost*, **26** (3), 349–352.

78 Lonn, E., Yusuf, S., Arnold, M.J., Sheridan, P., Pogue, J., Micks, M., McQueen, M.J., Probstfield, J., Fodor, G., Held, C., and Genest, J. Jr (2006) Homocysteine lowering with folic acid and B vitamins in vascular disease. *N Engl J Med*, **354** (15), 1567–1577.

79 Johnson, M.A., Hawthorne, N.A., Brackett, W.R., Fischer, J.G., Gunter, E.W., Allen, R.H., and Stabler, S.P. (2003) Hyperhomocysteinemia and vitamin B-12 deficiency in elderly using Title IIIc nutrition services. *Am J Clin Nutr*, **77** (1), 211–220.

80 Liaugaudas, G., Jacques, P.F., Selhub, J., Rosenberg, I.H., and Bostom, A.G. (2001) Renal insufficiency, vitamin B(12) status, and population attributable risk for mild hyperhomocysteinemia among coronary artery disease patients in the era of folic acid-fortified cereal grain flour. *Arterioscler Thromb Vasc Biol*, **21** (5), 849–851.

81 Robertson, J., Iemolo, F., Stabler, S.P., Allen, R.H., and Spence, J.D. (2005) Vitamin B12, homocysteine and carotid plaque in the era of folic acid fortification of enriched cereal grain products. *Can Med Assoc J*, **172** (12), 1569–1573.

82 Homocysteine Lowering Trialist's Collaboration (2005) Dose-dependent effects of folic acid on blood concentrations of homocysteine: a meta-analysis of the randomized trials. *Am J Clin Nutr*, **82** (4), 806–812.

83 Spence, J.D., Bang, H., Chambless, L.E., and Stampfer, M.J. (2005) Vitamin intervention for stroke prevention trial: an efficacy analysis. *Stroke*, **36** (11), 2404–2409.

84 McKinley, M.C., McNulty, H., McPartlin, J., Strain, J.J., Pentieva, K., Ward, M., Weir, D.G., and Scott, J.M. (2001) Low-dose vitamin B-6 effectively lowers fasting plasma homocysteine in healthy elderly persons who are folate and riboflavin replete. *Am J Clin Nutr*, **73** (4), 759–764.

85 Rimm, E.B., Willett, W.C., Hu, F.B., Sampson, L., Colditz, G.A., Manson, J.E., Hennekens, C., and Stampfer, M.J. (1998) Folate and vitamin B6 from diet and supplements in relation to risk of coronary heart disease among women. *J Am Med Assoc*, **279** (5), 359–364.

86 Chasan-Taber, L., Selhub, J., Rosenberg, I.H., Malinow, M.R., Terry, P., Tishler, P.V., Willett, W., Hennekens, C.H., and Stampfer, M.J. (1996) A prospective study of folate and vitamin B6 and risk of myocardial infarction in US physicians. *J Am Coll Nutr*, **15** (2), 136–143.

87 Folsom, A.R., Nieto, F.J., McGovern, P.G., Tsai, M.Y., Malinow, M.R., Eckfeldt, J.H., Hess, D.L., and Davis, C.E. (1998) Prospective study of coronary heart disease incidence in relation to fasting total homocysteine, related genetic polymorphisms, and B vitamins: the Atherosclerosis Risk in Communities (ARIC) study. *Circulation*, **98** (3), 204–210.

88 Kelly, P.J., Shih, V.E., Kistler, J.P., Barron, M., Lee, H., Mandell, R., and Furie, K.L. (2003) Low vitamin B6 but not homocyst(e)ine is associated with increased risk of stroke and transient ischemic attack in the era of folic acid grain fortification. *Stroke*, **34** (6), e51–e54.

89 Robinson, K., Arheart, K., Refsum, H., Brattstrom, L., Boers, G., Ueland, P., Rubba, P., Palma-Reis, R., Meleady, R., Daly, L., Witteman, J., and Graham, I. European COMAC Group (1998) Low circulating folate and vitamin B6 concentrations: risk factors for stroke, peripheral vascular disease, and coronary artery disease. *Circulation*, **97** (5), 437–443.

90 Clarke, R., Lewington, S., Sherliker, P., and Armitage, J. (2007) Effects of B-vitamins on plasma homocysteine concentrations and on risk of cardiovascular disease and dementia. *Curr Opin Clin Nutr Metab Care*, **10** (1), 32–39.

91 Tsuge, H., Hotta, N., and Hayakawa, T. (2000) Effects of vitamin B-6 on (n–3) polyunsaturated fatty acid

metabolism. *J Nutr*, **130** (Suppl 2S), 333S–334S.

92. Chang, S.J., Chang, C.N., and Chen, C.W. (2002) Occupancy of glycoprotein IIb/IIIa by B-6 vitamers inhibits human platelet aggregation. *J Nutr*, **132** (12), 3603–3606.

93. Matsubara, K., Matsumoto, H., Mizushina, Y., Lee, J.S., and Kato, N. (2003) Inhibitory effect of pyridoxal 5′-phosphate on endothelial cell proliferation, replicative DNA polymerase and DNA topoisomerase. *Int J Mol Med*, **12** (1), 51–55.

94. Friso, S., Jacques, P.F., Wilson, P.W., Rosenberg, I.H., and Selhub, J. (2001) Low circulating vitamin B(6) is associated with elevation of the inflammation marker C-reactive protein independently of plasma homocysteine levels. *Circulation*, **103** (23), 2788–2791.

95. Hustad, S., Ueland, P.M., Vollset, S.E., Zhang, Y., Bjorke-Monsen, A.L., and Schneede, J. (2000) Riboflavin as a determinant of plasma total homocysteine: effect modification by the methylenetetrahydrofolate reductase C677T polymorphism. *Clin Chem*, **46** (8 Pt 1), 1065–1071.

96. Jacques, P.F., Kalmbach, R., Bagley, P.J., Russo, G.T., Rogers, G., Wilson, P.W., Rosenberg, I.H., and Selhub, J. (2002) The relationship between riboflavin and plasma total homocysteine in the Framingham Offspring cohort is influenced by folate status and the C677T transition in the methylenetetrahydrofolate reductase gene. *J Nutr*, **132** (2), 283–288.

97. McNulty, H., Dowey le, R.C., Strain, J.J., Dunne, A., Ward, M., Molloy, A.M., McAnena, L.B., Hughes, J.P., Hannon-Fletcher, M., and Scott, J.M. (2006) Riboflavin lowers homocysteine in individuals homozygous for the MTHFR 677C->T polymorphism. *Circulation*, **113** (1), 74–80.

98. Yamada, K., Chen, Z., Rozen, R., and Matthews, R.G. (2001) Effects of common polymorphisms on the properties of recombinant human methylenetetrahydrofolate reductase. *Proc Natl Acad Sci U S A*, **98** (26), 14853–14858.

99. Lewis, S.J., Ebrahim, S., and Davey Smith, G. (2005) Meta-analysis of MTHFR 677C->T polymorphism and coronary heart disease: does totality of evidence support causal role for homocysteine and preventive potential of folate? *Br Med J*, **331** (7524), 1053.

100. McNulty, H., McKinley, M.C., Wilson, B., McPartlin, J., Strain, J.J., Weir, D.G., and Scott, J.M. (2002) Impaired functioning of thermolabile methylenetetrahydrofolate reductase is dependent on riboflavin status: implications for riboflavin requirements. *Am J Clin Nutr*, **76** (2), 436–441.

101. Ganji, V. and Kafai, M.R. (2004) Frequent consumption of milk, yogurt, cold breakfast cereals, peppers, and cruciferous vegetables and intakes of dietary folate and riboflavin but not vitamins B-12 and B-6 are inversely associated with serum total homocysteine concentrations in the US population. *Am J Clin Nutr*, **80** (6), 1500–1507.

102. Nurk, E., Tell, G.S., Vollset, S.E., Nygard, O., Refsum, H., Nilsen, R.M., and Ueland, P.M. (2004) Changes in lifestyle and plasma total homocysteine: the Hordaland Homocysteine Study. *Am J Clin Nutr*, **79** (5), 812–819.

103. Mansoor, M.A., Svardal, A.M., and Ueland, P.M. (1992) Determination of the in vivo redox status of cysteine, cysteinylglycine, homocysteine, and glutathione in human plasma. *Anal Biochem*, **200** (2), 218–229.

104. Sengupta, S., Chen, H., Togawa, T., DiBello, P.M., Majors, A.K., Budy, B., Ketterer, M.E., and Jacobsen, D.W. (2001) Albumin thiolate anion is an intermediate in the formation of albumin-S-S-homocysteine. *J Biol Chem*, **276** (32), 30111–30117.

105. Jakubowski, H. (1999) Protein homocysteinylation: possible mechanism underlying pathological consequences of elevated homocysteine levels. *FASEB J*, **13** (15), 2277–2283.

106 Jacobsen, D.W. (2001) Cellular mechanism of homocysteine pathogenesis in atherosclerosis, in *Homocysteine in Health and Disease* (eds R. Carmel and D.W. Jacobsen), Cambridge University Press, Cambridge, p. 436.

107 Barford, D. (2004) The role of cysteine residues as redox-sensitive regulatory switches. *Curr Opin Struct Biol*, **14** (6), 679–686.

108 Narazaki, R., Hamada, M., Harada, K., and Otagiri, M. (1996) Covalent binding between bucillamine derivatives and human serum albumin. *Pharm Res*, **13** (9), 1317–1321.

109 Carter, D.C. and Ho, J.X. (1994) Structure of serum albumin. *Adv Protein Chem*, **45**, 153–203.

110 Peters, T. Jr (1996) *All about Albumin: Biochemistry, Genetics and Medical Applications*, Academia Press, San Diego, pp. 51–54.

111 Sundaramoorthy, E., Maiti, S., Brahmachari, S.K., and Sengupta, S. (2008) Predicting protein homocysteinylation targets based on dihedral strain energy and pKa of cysteines. *Proteins*, **71** (3), 1475–1483.

112 Majors, A.K., Sengupta, S., Willard, B., Kinter, M.T., Pyeritz, R.E., and Jacobsen, D.W. (2002) Homocysteine binds to human plasma fibronectin and inhibits its interaction with fibrin. *Arterioscler Thromb Vasc Biol*, **22** (8), 1354–1359.

113 Hubmacher, D., Tiedemann, K., Bartels, R., Brinckmann, J., Vollbrandt, T., Batge, B., Notbohm, H., and Reinhardt, D.P. (2005) Modification of the structure and function of fibrillin-1 by homocysteine suggests a potential pathogenetic mechanism in homocystinuria. *J Biol Chem*, **280** (41), 34946–34955.

114 Zinellu, A., Zinellu, E., Sotgia, S., Formato, M., Cherchi, G.M., Deiana, L., and Carru, C. (2006) Factors affecting S-homocysteinylation of LDL apoprotein B. *Clin Chem*, **52** (11), 2054–2059.

115 Barbato, J.C., Catanescu, O., Murray, K., DiBello, P.M., and Jacobsen, D.W. (2007) Targeting of metallothionein by L-homocysteine: a novel mechanism for disruption of zinc and redox homeostasis. *Arterioscler Thromb Vasc Biol*, **27** (1), 49–54.

116 Jakubowski, H. (1997) Metabolism of homocysteine thiolactone in human cell cultures. Possible mechanism for pathological consequences of elevated homocysteine levels. *J Biol Chem*, **272** (3), 1935–1942.

117 Jakubowski, H. (2002) Homocysteine is a protein amino acid in humans. Implications for homocysteine-linked disease. *J Biol Chem*, **277** (34), 30425–30428.

118 Perla-Kajan, J., Twardowski, T., and Jakubowski, H. (2007) Mechanisms of homocysteine toxicity in humans. *Amino Acids*, **32** (4), 561–572.

119 Sauls, D.L., Lockhart, E., Warren, M.E., Lenkowski, A., Wilhelm, S.E., and Hoffman, M. (2006) Modification of fibrinogen by homocysteine thiolactone increases resistance to fibrinolysis: a potential mechanism of the thrombotic tendency in hyperhomocysteinemia. *Biochemistry*, **45** (8), 2480–2487.

120 Naruszewicz, M., Mirkiewicz, E., Oleszewski, A.J., and McCully, K.S. (1994) Thiolation of low-density lipoprotein by homocysteine-thiolactone causes increased aggregation and interaction with cultured macrophages. *Nutr Metab Cardiovasc Dis*, **4**, 70–77.

121 Ferretti, G., Bacchetti, T., Moroni, C., Vignini, A., Nanetti, L., and Curatola, G. (2004) Effect of homocysteinylation of low density lipoproteins on lipid peroxidation of human endothelial cells. *J Cell Biochem*, **92** (2), 351–360.

122 Jakubowski, H. (2006) Pathophysiological consequences of homocysteine excess. *J Nutr*, **136** (Suppl 6), 1741S–1749S.

123 Perla-Kajan, J., Stanger, O., Luczak, M., Ziolkowska, A., Malendowicz, L.K., Twardowski, T., Lhotak, S., Austin, R.C., and Jakubowski, H. (2008) Immunohistochemical detection of N-homocysteinylated proteins in humans and mice. *Biomed Pharmacother*, **62** (7), 473–479.

124 Ferretti, G., Bacchetti, T., Marotti, E., and Curatola, G. (2003) Effect of homocysteinylation on human high-density lipoproteins: a correlation with paraoxonase activity. *Metabolism*, **52** (2), 146–151.

125 Undas, A., Perla, J., Lacinski, M., Trzeciak, W., Kazmierski, R., and Jakubowski, H. (2004) Autoantibodies against N-homocysteinylated proteins in humans: implications for atherosclerosis. *Stroke*, **35** (6), 1299–1304.

126 Lawrence de Koning, A.B., Werstuck, G.H., Zhou, J., and Austin, R.C. (2003) Hyperhomocysteinemia and its role in the development of atherosclerosis. *Clin Biochem*, **36** (6), 431–441.

127 Loscalzo, J. (1996) The oxidant stress of hyperhomocyst(e)inemia. *J Clin Invest*, **98** (1), 5–7.

128 de Groot, P.G., Willems, C., Boers, G.H., Gonsalves, M.D., van Aken, W.G., and van Mourik, J.A. (1983) Endothelial cell dysfunction in homocystinuria. *Eur J Clin Invest*, **13** (5), 405–410.

129 Wall, R.T., Harlan, J.M., Harker, L.A., and Striker, G.E. (1980) Homocysteine-induced endothelial cell injury in vitro: a model for the study of vascular injury. *Thromb Res*, **18** (1-2), 113–121.

130 Romerio, S.C., Linder, L., Nyfeler, J., Wenk, M., Litynsky, P., Asmis, R., and Haefeli, W.E. (2004) Acute hyperhomocysteinemia decreases NO bioavailability in healthy adults. *Atherosclerosis*, **176** (2), 337–344.

131 Stamler, J.S., Osborne, J.A., Jaraki, O., Rabbani, L.E., Mullins, M., Singel, D., and Loscalzo, J. (1993) Adverse vascular effects of homocysteine are modulated by endothelium-derived relaxing factor and related oxides of nitrogen. *J Clin Invest*, **91** (1), 308–318.

132 Upchurch, G.R. Jr, Welch, G.N., Fabian, A.J., Freedman, J.E., Johnson, J.L., Keaney, J.F. Jr, and Loscalzo, J. (1997) Homocyst(e)ine decreases bioavailable nitric oxide by a mechanism involving glutathione peroxidase. *J Biol Chem*, **272** (27), 17012–17017.

133 Jin, L., Caldwell, R.B., Li-Masters, T., and Caldwell, R.W. (2007) Homocysteine induces endothelial dysfunction via inhibition of arginine transport. *J Physiol Pharmacol*, **58** (2), 191–206.

134 Lentz, S.R., Sobey, C.G., Piegors, D.J., Bhopatkar, M.Y., Faraci, F.M., Malinow, M.R., and Heistad, D.D. (1996) Vascular dysfunction in monkeys with diet-induced hyperhomocyst(e)inemia. *J Clin Invest*, **98** (1), 24–29.

135 Stamler, J.S., Simon, D.I., Osborne, J.A., Mullins, M.E., Jaraki, O., Michel, T., Singel, D.J., and Loscalzo, J. (1992) S-nitrosylation of proteins with nitric oxide: synthesis and characterization of biologically active compounds. *Proc Natl Acad Sci U S A*, **89** (1), 444–448.

136 Stamler, J.S., Singel, D.J., and Loscalzo, J. (1992) Biochemistry of nitric oxide and its redox-activated forms. *Science*, **258** (5090), 1898–1902.

137 Ferretti, G., Bacchetti, T., Negre-Salvayre, A., Salvayre, R., Dousset, N., and Curatola, G. (2006) Structural modifications of HDL and functional consequences. *Atherosclerosis*, **184** (1), 1–7.

138 Williams, H.M., Lippok, H., and Doherty, G.H. (2008) Nitric oxide and peroxynitrite signalling triggers homocysteine-mediated apoptosis in trigeminal sensory neurons in vitro. *Neurosci Res*, **60** (4), 380–388.

139 Levrand, S., Pacher, P., Pesse, B., Rolli, J., Feihl, F., Waeber, B., and Liaudet, L. (2007) Homocysteine induces cell death in H9C2 cardiomyocytes through the generation of peroxynitrite. *Biochem Biophys Res Commun*, **359** (3), 445–450.

140 Dickhout, J.G., Hossain, G.S., Pozza, L.M., Zhou, J., Lhotak, S., and Austin, R.C. (2005) Peroxynitrite causes endoplasmic reticulum stress and apoptosis in human vascular endothelium: implications in atherogenesis. *Arterioscler Thromb Vasc Biol*, **25** (12), 2623–2629.

141 Tsai, J.C., Perrella, M.A., Yoshizumi, M., Hsieh, C.M., Haber, E., Schlegel, R., and Lee, M.E. (1994) Promotion of vascular smooth muscle cell growth by homocysteine: a link to atherosclerosis. *Proc Natl Acad Sci U S A*, **91** (14), 6369–6373.

142 Tsai, J.C., Wang, H., Perrella, M.A., Yoshizumi, M., Sibinga, N.E., Tan, L.C., Haber, E., Chang, T.H., Schlegel, R., and Lee, M.E. (1996) Induction of cyclin A gene expression by homocysteine in vascular smooth muscle cells. *J Clin Invest*, **97** (1), 146–153.

143 Majors, A., Ehrhart, L.A., and Pezacka, E.H. (1997) Homocysteine as a risk factor for vascular disease. Enhanced collagen production and accumulation by smooth muscle cells. *Arterioscler Thromb Vasc Biol*, **17** (10), 2074–2081.

144 Harker, L.A., Harlan, J.M., and Ross, R. (1983) Effect of sulfinpyrazone on homocysteine-induced endothelial injury and arteriosclerosis in baboons. *Circ Res*, **53** (6), 731–739.

145 Southern, F.N., Cruz, N., Fink, L.M., Cooney, C.A., Barone, G.W., Eidt, J.F., and Moursi, M.M. (1998) Hyperhomocysteinemia increases intimal hyperplasia in a rat carotid endarterectomy model. *J Vasc Surg*, **28** (5), 909–918.

146 Wang, H., Yoshizumi, M., Lai, K., Tsai, J.C., Perrella, M.A., Haber, E., and Lee, M.E. (1997) Inhibition of growth and p21ras methylation in vascular endothelial cells by homocysteine but not cysteine. *J Biol Chem*, **272** (40), 25380–25385.

147 Chang, P.Y., Lu, S.C., Lee, C.M., Chen, Y.J., Dugan, T.A., Huang, W.H., Chang, S.F., Liao, W.S., Chen, C.H., and Lee, Y.T. (2008) Homocysteine inhibits arterial endothelial cell growth through transcriptional downregulation of fibroblast growth factor-2 involving G protein and DNA methylation. *Circ Res*, **102** (8), 933–941.

148 Austin, R.C., Lentz, S.R., and Werstuck, G.H. (2004) Role of hyperhomocysteinemia in endothelial dysfunction and atherothrombotic disease. *Cell Death Differ*, **11** (Suppl 1), S56–S64.

149 Dickhout, J.G., Sood, S.K., and Austin, R.C. (2007) Role of endoplasmic reticulum calcium disequilibria in the mechanism of homocysteine-induced ER stress. *Antioxid Redox Signal*, **9** (11), 1863–1873.

150 Kokame, K., Agarwala, K.L., Kato, H., and Miyata, T. (2000) Herp, a new ubiquitin-like membrane protein induced by endoplasmic reticulum stress. *J Biol Chem*, **275** (42), 32846–32853.

151 Kokame, K., Kato, H., and Miyata, T. (1996) Homocysteine-respondent genes in vascular endothelial cells identified by differential display analysis. GRP78/BiP and novel genes. *J Biol Chem*, **271** (47), 29659–29665.

152 Lentz, S.R. and Sadler, J.E. (1993) Homocysteine inhibits von Willebrand factor processing and secretion by preventing transport from the endoplasmic reticulum. *Blood*, **81** (3), 683–689.

153 Marciniak, S.J. and Ron, D. (2006) Endoplasmic reticulum stress signaling in disease. *Physiol Rev*, **86** (4), 1133–1149.

154 Ozcan, U., Cao, Q., Yilmaz, E., Lee, A.H., Iwakoshi, N.N., Ozdelen, E., Tuncman, G., Gorgun, C., Glimcher, L.H., and Hotamisligil, G.S. (2004) Endoplasmic reticulum stress links obesity, insulin action, and type 2 diabetes. *Science*, **306** (5695), 457–461.

155 Ozcan, U., Yilmaz, E., Ozcan, L., Furuhashi, M., Vaillancourt, E., Smith, R.O., Gorgun, C.Z., and Hotamisligil, G.S. (2006) Chemical chaperones reduce ER stress and restore glucose homeostasis in a mouse model of type 2 diabetes. *Science*, **313** (5790), 1137–1140.

156 Werstuck, G.H., Khan, M.I., Femia, G., Kim, A.J., Tedesco, V., Trigatti, B., and Shi, Y. (2006) Glucosamine-induced endoplasmic reticulum dysfunction is associated with accelerated atherosclerosis

in a hyperglycemic mouse model. *Diabetes*, **55** (1), 93–101.

157 Werstuck, G.H., Lentz, S.R., Dayal, S., Hossain, G.S., Sood, S.K., Shi, Y.Y., Zhou, J., Maeda, N., Krisans, S.K., Malinow, M.R., and Austin, R.C. (2001) Homocysteine-induced endoplasmic reticulum stress causes dysregulation of the cholesterol and triglyceride biosynthetic pathways. *J Clin Invest*, **107** (10), 1263–1273.

158 Karmin, O., Lynn, E.G., Chung, Y.H., Siow, Y.L., Man, R.Y., and Choy, P.C. (1998) Homocysteine stimulates the production and secretion of cholesterol in hepatic cells. *Biochim Biophys Acta*, **1393** (2–3), 317–324.

159 Zhou, J., Lhotak, S., Hilditch, B.A., and Austin, R.C. (2005) Activation of the unfolded protein response occurs at all stages of atherosclerotic lesion development in apolipoprotein E-deficient mice. *Circulation*, **111** (14), 1814–1821.

160 Kumar, A., John, L., Alam, M.M., Gupta, A., Sharma, G., Pillai, B., and Sengupta, S. (2006) Homocysteine- and cysteine-mediated growth defect is not associated with induction of oxidative stress response genes in yeast. *Biochem J*, **396** (1), 61–69.

161 Outinen, P.A., Sood, S.K., Pfeifer, S.I., Pamidi, S., Podor, T.J., Li, J., Weitz, J.I., and Austin, R.C. (1999) Homocysteine-induced endoplasmic reticulum stress and growth arrest leads to specific changes in gene expression in human vascular endothelial cells. *Blood*, **94** (3), 959–967.

162 Fujita, Y., Ukena, E., Iefuji, H., Giga-Hama, Y., and Takegawa, K. (2006) Homocysteine accumulation causes a defect in purine biosynthesis: further characterization of Schizosaccharomyces pombe methionine auxotrophs. *Microbiology*, **152** (Pt 2), 397–404.

163 Malanovic, N., Streith, I., Wolinski, H., Rechberger, G., Kohlwein, S.D., and Tehlivets, O. (2008) S-adenosyl-L-homocysteine hydrolase, key enzyme of methylation metabolism, regulates phosphatidylcholine synthesis and triacylglycerol homeostasis in yeast: implications for homocysteine as a risk factor of atherosclerosis. *J Biol Chem*, **283** (35), 23989–23999.

164 Tehlivets, O., Hasslacher, M., and Kohlwein, S.D. (2004) S-adenosyl-L-homocysteine hydrolase in yeast: key enzyme of methylation metabolism and coordinated regulation with phospholipid synthesis. *FEBS Lett*, **577** (3), 501–506.

165 Zeng, X.K., Guan, Y.F., Remick, D.G., and Wang, X. (2005) Signal pathways underlying homocysteine-induced production of MCP-1 and IL-8 in cultured human whole blood. *Acta Pharmacol Sin*, **26** (1), 85–91.

166 Poddar, R., Sivasubramanian, N., DiBello, P.M., Robinson, K., and Jacobsen, D.W. (2001) Homocysteine induces expression and secretion of monocyte chemoattractant protein-1 and interleukin-8 in human aortic endothelial cells: implications for vascular disease. *Circulation*, **103** (22), 2717–2723.

167 Boekholdt, S.M., Peters, R.J., Hack, C.E., Day, N.E., Luben, R., Bingham, S.A., Wareham, N.J., Reitsma, P.H., and Khaw, K.T. (2004) IL-8 plasma concentrations and the risk of future coronary artery disease in apparently healthy men and women: the EPIC-Norfolk prospective population study. *Arterioscler Thromb Vasc Biol*, **24** (8), 1503–1508.

168 Hofmann, M.A., Lalla, E., Lu, Y., Gleason, M.R., Wolf, B.M., Tanji, N., Ferran, L.J. Jr, Kohl, B., Rao, V., Kisiel, W., Stern, D.M., and Schmidt, A.M. (2001) Hyperhomocysteinemia enhances vascular inflammation and accelerates atherosclerosis in a murine model. *J Clin Invest*, **107** (6), 675–683.

169 Gori, A.M., Corsi, A.M., Fedi, S., Gazzini, A., Sofi, F., Bartali, B., Bandinelli, S., Gensini, G.F., Abbate, R., and Ferrucci, L. (2005) A proinflammatory state is associated with

hyperhomocysteinemia in the elderly. *Am J Clin Nutr*, **82** (2), 335–341.

170 Maeda, M., Yamamoto, I., Fujio, Y., and Azuma, J. (2003) Homocysteine induces vascular endothelial growth factor expression in differentiated THP-1 macrophages. *Biochim Biophys Acta*, **1623** (1), 41–46.

171 Inoue, M., Itoh, H., Ueda, M., Naruko, T., Kojima, A., Komatsu, R., Doi, K., Ogawa, Y., Tamura, N., Takaya, K., Igaki, T., Yamashita, J., Chun, T.H., Masatsugu, K., Becker, A.E., and Nakao, K. (1998) Vascular endothelial growth factor (VEGF) expression in human coronary atherosclerotic lesions: possible pathophysiological significance of VEGF in progression of atherosclerosis. *Circulation*, **98** (20), 2108–2116.

172 Silverman, M.D., Tumuluri, R.J., Davis, M., Lopez, G., Rosenbaum, J.T., and Lelkes, P.I. (2002) Homocysteine upregulates vascular cell adhesion molecule-1 expression in cultured human aortic endothelial cells and enhances monocyte adhesion. *Arterioscler Thromb Vasc Biol*, **22** (4), 587–592.

15
Uric Acid Alterations in Cardiometabolic Disorders and Gout

Renato Ippolito, Ferruccio Galletti, and Pasquale Strazzullo

Uric acid (UA) is the end product of purine catabolism in humans, and its levels are a function of the balance between the breakdown of purines and the rate of UA excretion. UA has a singlet oxygen scavenger activity and antioxidant properties in preventing lipid peroxidation. Nevertheless, UA produces free radicals of its own, either alone or in combination with peroxynitrite, thus increasing low-density lipoprotein (LDL) lipid oxidation and promoting reactive oxygen species (ROS) production in differentiated adipocytes. In vascular smooth muscle cell (VSMC), UA activates proinflammatory and proliferative pathways, while in endothelial cells it decreases NO bioavailability and inhibits cell migration and proliferation. As a consequence, hyperuricemia plays a complex and enigmatic role in the prevention and development of organ damage that could be related to its paradoxical antioxidant/oxidant power. Higher serum uric acid (SUA) levels, thanks to its antioxidant effects, are associated with a lower risk of Parkinson's disease. On the other hand, elevated SUA levels contribute, through promotion of oxidative stress and endothelial dysfunction, to the development of metabolic and insulin resistance syndrome and hypertension, to the new onset of atherosclerotic cardiovascular disease (CVD) and to the increase of long-term CVD mortality. The reduction of CVD risk and coronary event rate, obtained with the decrease of SUA levels in two interventional trials (Losartan Intervention For Endpoint reduction in hypertension (LIFE) and GREek Atorvastatin and Coronary-heart-disease Evaluation (GREACE)) through losartan and statin treatment respectively, confirm the correlation between SUA levels and CVD.

15.1
Introduction

Uric acid (UA) is the final metabolic product of purine catabolism in humans, because a constitutional transcriptional defect of the enzyme uricase does not allow further degradation of UA to allantoin, like in other species [1]. Purinic nucleotides, the principle constituents of cellular energy stores such as ATP, adenylate

Endogenous Toxins. Diet, Genetics, Disease and Treatment.
Edited by Peter J. O'Brien and W. Robert Bruce
Copyright © 2010 WILEY-VCH Verlag GmbH & Co. KGaA, Weinheim
ISBN: 978-3-527-32363-0

monophosphate (AMP), and guanylate monophosphate (GMP) are degraded by a sequence of chemical reactions in which a 5′-nucleotidase deletes a phosphoric group. Other enzymes also play an important role; thus, adenosine deaminase converts adenosine to inosine, which is then hydrolyzed to hypoxanthine and d-ribose, while guanine deaminase catalyzes the conversion of guanine, a product of guanosine hydrolysis, to xanthine. Finally, xanthine oxidoreductase (XO) catalyzes both the conversion of hypoxanthine to xanthine and that of xanthine to UA.

While some purinic metabolites are recycled by a savage pathway and resynthesize nucleotides, others are eliminated daily in the form of UA. This is excreted in the urine by a complex process that involves glomerular filtration, tubular reabsorption in the early proximal convoluted tubule, tubular secretion at postreabsorptive sites, and possibly again, reabsorption in the late proximal tubule [2]. The UA excretion rate in adults is, approximately, 0.6 g/day, an amount derived from both purine dietary intake, and purine nucleotide and nucleic acid intracellular turnover. More than 90% of filtered urate is reabsorbed by the kidney [3], a process that plays a critical role in the maintenance of an adequate urate plasma concentration. In fact, urate levels as low a 0.2–1.8 mg dl^{-1} are detected in case of defects in tubular reabsorption [4]. The UA plasma level in men is normally higher than in women, in whom UA concentration in plasma decreases during the first part of pregnancy [5] and increases after menopause, in relation to the uricosuric properties of estrogens and their changing production throughout life [6].

15.2
Dual Properties of Uric Acid as Antioxidant or Pro-Oxidant Molecule

Simon and Vunakis, described in the 1960s, the capability of UA to act as a singlet oxygen scavenger [7]. Later on, Ames *et al.* [8] and Howell and Wyngaarden [9] reported that UA was oxidized by hydrogen peroxide in the presence of hematin or methemoglobin. In addition, they noted that urate was about as effective as ascorbate in preventing lipid peroxidation, especially in brain tissue, at concentrations considerably below those normally found in plasma, and hypothesized that UA was protecting not only erythrocytes but also T and B lymphocytes. Muraoka and Miura described the scavenger ability of UA to inhibit lipid peroxidation at the lipid/aqueous boundary [10].

These findings provide the basis for the antioxidant role of UA in the protection of cardiac, vascular, and neural cells from oxidation injury reported by several other authors in more recent studies [11–14]. It has been reported that prevention of lipid peroxidation by UA is more effective in hydrophilic than in lipophilic conditions, and that, in different chemical milieus, like in a hydrophobic environment, UA loses its beneficial antioxidant effect [10], to produce free radicals either alone [15] or in combination with peroxynitrite [16].

Many authors have described an association between UA and increased oxidative stress, thus suggesting a mechanistic pathway linking high serum uric acid (SUA) to greater risk of cardiovascular disease (CVD) and kidney disease [17]. Sautin et al. reported that adipocyte differentiation was associated with an increased uptake of UA and reactive oxygen species (ROS) accumulation, and that elevated UA induced a further increase in intracellular ROS production in differentiated adipocytes, mediated by activation of NADPH oxidase (NOX). This was followed by redox-dependent stress signaling, a decrease in nitric oxide (NO) bioavailability, and oxidative modifications of proteins and lipids [18].

Hyperuricemia is common in subjects with obesity and/or metabolic syndrome (MS) [19–22], and it has been proposed to play a role in the development of insulin resistance (IR) and the associated imbalance of vascular homeostasis, mediated by increased oxidative stress [11, 23]. A causal or contributory role for UA has also been proposed in the development of MS induced by a fructose load [23] and in the occurrence of liposome and low-density lipoprotein (LDL) lipid oxidation [24]. As shown by Santos et al. [16], UA can react with peroxynitrous acid and the radicals generated from its decomposition can produce compounds that will further react with the peroxynitrite anion. Peroxynitrite is also able to oxidize other structurally related purines such as xanthine, hypoxanthine [25], and 8-oxodeoxyguanosine [26, 27]. Moreover Bagnati et al. observed that UA may enhance the peroxidation of human LDL, triggered by copper, in the presence of preformed lipid hydroperoxides [28].

15.3
Effects of Uric Acid on the Arterial Wall Properties and Endothelial Function

Several authors have investigated the effects of UA on vascular smooth muscle cell (VSMC) and endothelial cell function. In VSMC, UA activates pro-inflammatory pathways [29, 30] and stimulates cell proliferation [31–33], whereas in endothelial cells it decreases NO bioavailability [33, 34] and inhibits cell migration and proliferation [33]. Elevated plasma UA seems to potentiate the effects of angiotensin II to induce renal vasoconstriction [35], possibly through angiotensin II type 1 receptor up-regulation in VSMCs [36].

Altogether, these results from experimental studies led to a pleiotropic (multi-effect) toxicity hypothesis of increased UA, essentially related to higher oxidative stress and to endothelial dysfunction.

15.4
Causes of Hyperuricemia and Hypouricemia

Hyperuricemia is defined as a SUA concentration greater than $420\,\mu mol\,l^{-1}$ or $7.0\,mg\,dl^{-1}$. It may occur because of decreased excretion, increased production, or a combination thereof. The possible causes of hyper- and hypouricemia are reported in Table 15.1. In particular, mild or moderate hyperuricemias are very often the consequence of impaired renal UA excretion, associated with enhanced sodium reabsorption in highly prevalent disorders such as obesity [37], impaired

Tab. 15.1 Factors responsible for alterations in serum uric acid concentration

Factors that decrease SUA concentration
   Diet: low-fat dairy product intake
   Drugs: hypouricemic agents (allopurinol, febuxostat, sulfinpyrazone, probenecid, rasburicase), coumarin anticoagulants, estrogens
Factors that increase SUA concentration
   Diet: elevated meat, fish, and alcohol (particularly beer and spirits) intake, fructose, and fructose-rich food
   Drugs: diuretics, low-dose salicylates, pyrazinamide, ethambutol, cytotoxic drugs, lead poisoning
Acute or chronic disorders involving
   Increased purine turnover: myeloproliferative and lymphoproliferative disorders, chronic hemolytic anemia, hemoglobinopathies, thalassemia, secondary polycythemia
   Increased purine synthesis: glucose-6-phosphate dehydrogenase deficiency, Lesch–Nyhan syndrome
   Reduced renal excretion: obesity, hypertension, chronic renal disease, hyperparathyroidism, hypothyroidism, sickle cell anemia

---

glucose tolerance and type 2 diabetes [38, 39], and hypertension [40, 41]. Moreover, a high consumption of fructose or fructose-rich food is another significant cause of hyperuricemia [42–44].

15.5
Hyperuricemia, Gout, and Their Treatment

Asymptomatic hyperuricemia is common but, when the local solubility limits of UA are exceeded, monosodium urate crystal deposition occurs in the joints, kidneys, and soft tissues inducing clinical manifestations that include arthritis, soft tissue masses (i.e., tophi), nephrolithiasis, and urate nephropathy. A complete list of consequence of hyperuricemia is provided in Table 15.2.

Gout is defined as a crystal arthropathy, caused by an inflammatory response to the deposition of monosodium urate crystals, whose important features are arthritis or periarthritis attacks and bone erosion following deposition of monosodium urate crystals in soft tissues, joints, bone, and tendons. Although the prevalence of the disease increases with age, it may occasionally occur in patients younger than 30 years [45, 46]. It is more common in men than in women during young and middle age, but at about age 60, its prevalence in women approaches that in men [47, 48]. Identifying urate crystals in the fluid from an affected joint is the definitive diagnostic test for the diagnosis of gout. Since this approach is applicable to only few patients with suggestive symptoms of gout [49], the guidelines of the American College of Rheumatology [50] are often used in practice for the clinical diagnosis of gout without joint aspiration (Table 15.3).

Tab. 15.2 Consequences of hyperuricemia

Gout
Uric acid renal stones (nephrolithiasis)
Uric acid nephropathy (results from deposition of large amounts of UA crystals in the renal collecting ducts, pelvis, and ureters, i.e., in tumor lysis Syndrome)
Urate nephropathy (consequence of monosodium urate crystal deposition in the renal interstitium)

Tab. 15.3 American College of Rheumatology preliminary criteria for the clinical diagnosis of gout

Six or more of these criteria are needed to make a diagnosis

More than one attack of acute arthritis
Maximum inflammation developed within 1 d
Attack of monoarthritis
Redness over joints
Painful or swollen first metatarso-phalangeal joint
Unilateral attack on first metatarso-phalangeal joint
Unilateral attack on tarsal joint
Tophus (proven or suspected)
Hyperuricemia
Asymmetric swelling within a joint at X-ray
Subcortical cysts without erosions at X-ray
Joint fluid culture negative for organisms during attack

Patients affected by gout need symptomatic treatment for acute arthritis attack and chronic UA lowering treatment for preventing recurrent episodes and joints deformation. Dietary recommendations that could reduce SUA levels are reduced intake of purine-rich foods (i.e., all meats, including organ meats and meat extracts, seafood, yeast and yeast extracts, and vegetables such as peas, beans, lentils, asparagus, spinach, mushrooms), alcohol restriction, reduced calorie intake in case of overweight, reduction in saturated fat and relatively higher unsaturated fat intake, moderate carbohydrate restriction, experimental fish oil and plant seed oil dietary supplementation [51].

Pharmacotherapy of acute gout includes various anti-inflammatory measures (Table 15.4). Nonsteroidal anti-inflammatory drugs (NSAIDs) [52, 53], in the absence of contraindications, are first line agents for systemic treatment of acute attacks [54]. Oral colchicine administration is also considered a gout attack first line therapy, according to EULAR (European League Against Rheumatism)'s guidelines [54]; some authors suggest a dose of 0.6 mg every hour for up to 3 hours (maximum three doses) [55, 56], while others believe that a dose of 0.5 mg, with

Tab. 15.4 Pharmacotherapy of acute gout

NSAIDs	Systemic corticosteroids
Indomethacin	Prednisone
Naproxen	Methylprednisolone (intra-articular[a] or intramuscular)
Sulindac	Dexamethasone
Selective COX-2 inhibitors	
Rofecoxib	Colchicine[b]
Celecoxib	Corticotropin

[a] Intra-articular therapy may be the treatment of choice if only one or two accessible joints are involved.
[b] Not to exceed cumulative doses > 4 mg/attack; nausea, vomiting, and diarrhea are the most common adverse reactions, and rhabdomyolysis, agranulocytosis, aplastic anemia, or bone marrow suppression are rare adverse reactions; reduce the dosage in older patients and in patients with lower creatinine clearance; IV is a possible administration route, but needs extreme caution.

the same administration profile is more appropriate and less toxic [57]. After the acute attack, low-dose oral colchicine can be used for prophylaxis against further manifestations of acute gout, particularly before the initiation of antihyperuricemic therapy. Indeed, Borstad et al. observed that treating patients with colchicine during initiation of allopurinol therapy reduced the frequency and severity of acute flares as well as recurrence of acute flares during the course of treatment [58].

Intra-articular aspiration and injection of long-acting corticosteroids [59] is another effective and safe treatment for an acute attack [54] and probably a reasonable choice for patients in whom colchicine or NSAIDs are contraindicated.

The pharmacologic options for urate-lowering therapy in patients with chronic gout are reported in Table 15.5. Urate-lowering therapy is indicated in patients with recurrent acute attacks, arthropathy, tophi, or typical radiographic alterations [54]. The therapeutic goal of urate-lowering therapy is to maintain the SUA concentration below the saturation point for monosodium urate (<360 mmol l^{-1}) in order to promote crystal dissolution and prevent crystal formation. Allopurinol is the most commonly used long-term urate-lowering drug. It reduces the production of UA through inhibition of XO. It should be started at a low dose (100 mg daily) and increased by 100 mg every two to four weeks, as required: this dose must be adjusted in patients with renal impairment [54]. Uricosuric agents lower SUA by increasing renal UA excretion. They are a valid alternative to allopurinol in patients with normal renal function and are relatively contraindicated in patients with urolithiasis. Benzbromarone is a possible therapy in patients with mild to moderate renal insufficiency, but it has a moderate hepatotoxicity [60, 61]. Febuxostat is another orally administered nonpurine analog inhibitor of XO; thanks to its hepatic metabolism, it should be considered as a pharmacological

Tab. 15.5 Pharmacotherapy of chronic gout

Xanthine oxidase inhibitors	Recombinant urate oxidase enzyme
Allopurinol	Rasburicase
Febuxostat	Pegloticase[a]
Uricosuric agents	Carbonic anhydrase inhibitor
Probenecid	Acetazolamide
Sulfinpyrazone	
Benzbromarone	

[a] Presently undergoing trials.

option in patients with chronic renal insufficiency or nonresponsive to allopurinol [56, 62–64]. Finally, rasburicase, a recombinant mammalian uricase, has been successfully used for refractory tophaceous gout [65], and pegloticase, a pegylate recombinant uricase, is presently undergoing trials of efficacy and tolerance [66, 67].

15.6
Uric Acid and Cardiovascular Disease

A debate about the predictive or even causative role of SUA with regard to the development of CVD has been going on for decades [68]. The analysis of this problem is confounded and made difficult by the powerful interrelations between SUA and all the CVD risk factors clustering in the so-called MS, namely, abdominal obesity, high blood pressure, hypertriglyceridemia, low HDL cholesterol, fasting hyperglycemia, IR, type 2 diabetes, and renal dysfunction [69, 70]. Prospective epidemiological studies have provided valuable material for the investigation of this issue in different categories of individuals and a recent overview by Baker *et al.* has offered a systematic evaluation of their results [71]. The overview included 19 cohort studies, evaluating SUA levels and their association with various CVD outcomes, and two intervention trials addressing by *post hoc* analyses of the effect of SUA-lowering therapy on CVD risk. Among these prospective cohort studies, all using multivariate analytic methods, 10 were conducted in healthy subjects, 5 in hypertensive patients, and 4 in subjects with high CVD risk. A moderate and independent direct relationship between SUA and CVD was consistently demonstrated in patients at high risk for CVD, but not in those at lower risk. In two intervention trials, the GREek Atorvastatin and Coronary-heart-disease Evaluation (GREACE) [72] and the Losartan Intervention For Endpoint reduction in hypertension (LIFE) [73], a decrease in SUA levels, prompted by statin and losartan administration, respectively, was associated with a significantly reduced rate of CVD events.

On the basis of these results, Baker *et al.* concluded that in general population samples at relatively low risk of CVD, SUA is at best a very weak predictor of

CVD morbidity and mortality, once the effect of known confounders is accounted for. By contrast, SUA appears to be a significant independent predictor of CVD in hypertensive patients and even more so in certain categories of subjects at high or very high CVD risk, namely, diabetic patients [74], stroke survivors [75, 76], patients with congestive heart failure [77], and those with angiographically proven coronary heart disease [78].

Although a true cause–effect relationship cannot be inferred from these epidemiological observations, the net discrepancy between the findings in relatively healthy persons and in high-risk individuals, raises the possibility that excess SUA levels have different meanings in different circumstances. For instance, the association between SUA and hypertension has ancient roots [79]. Although the hyperuricemia associated with severely elevated blood pressure is a well-recognized consequence of progressive nephrosclerosis [80–83], the possible causal role of hyperuricemia in the development of hypertension is less clear. Many clinical and epidemiological studies, since the early 1990s, suggested an independent predictive role of SUA with respect to incident hypertension [84–91]. Other experimental studies purported a possibly causal role of UA for the development of hypertension [92–95]. It has been proposed that UA may have a role in initiating renal arteriolar lesions in selected categories of hypertensive patients, such as Blacks, subjects with gout or severe hyperuricemia, subjects with chronic lead ingestion, severe obesity, MS, and chronic diuretic use [93, 96, 97]. Zoccali et al. observed that SUA concentration in individuals with essential hypertension is associated with endothelial dysfunction independent of conventional risk factors such as IR and CRP [94].

In a randomized, double-blind, placebo-controlled, crossover trial involving 30 adolescents who had newly diagnosed, never-treated stage 1 essential hypertension and SUA levels of 6 mg dl^{-1} or above, Feig et al. observed that the administration of allopurinol, 200 mg twice daily for four weeks, compared with placebo resulted in a significant reduction of blood pressure [98]. These findings support the link between blood pressure and SUA levels.

Elevated SUA has been found to be closely associated with metabolic and IR syndrome [51, 99–101]. A reduced renal excretion of UA could explain the presence of hyperuricemia in these individuals, but there are cases in which hyperuricemia seems to precede the development of hyperinsulinemia [102, 103], obesity [85], or diabetes [104–106]. In this regard, it has been proposed [23, 95, 96] that endothelial dysfunction, induced by high SUA, could be responsible for reduced insulin-sensitivity, as insulin requires endothelial NO to stimulate skeletal muscle glucose uptake.

The most crucial, still open question is whether elevated SUA levels contribute to atherosclerotic CVD. As previously discussed, the concentration of UA in serum may increase for several reasons: high intake of purine-rich foods and alcohol, elevated rates of cell turnover and/or cell death caused by neoplastic disease and/or cytotoxic drugs, impaired renal function with decreased UA clearance and impaired renal UA excretion associated with obesity [37], IR states [38, 39], hypertension [40, 41], and heavy diuretic therapy [107]. Next to these well-known reasons for an increase in SUA levels stands another poorly recognized factor, that is, local ischemia. In this respect, it is worth noting that XO occurs in two different forms. Xanthine dehydrogenase is the prevalent operative form under physiological conditions, and has greater affinity for oxidized nicotinamide adenine dinucleotide (NAD^+) compared to oxygen, as the electron acceptor. Under ischemic conditions, however, in parallel to the degradation of ATP into adenine and xanthine, an extensive conversion of xanthine dehydrogenase to xanthine oxidase takes place. The latter uses molecular oxygen in place of NAD^+ as electron acceptor, and leads to the formation of superoxide anion and hydrogen peroxide in parallel with SUA, as shown by several experimental studies [11]. Thus, one may wonder to what extent this free radical production is the real culprit in the promotion of the inflammatory reaction and of the arterial wall damage commonly attributed to elevated SUA levels. This uncertainty brings us back to the question of the apparently dual effect of UA *per se* as an oxidant/antioxidant agent. There are indeed convincing demonstrations of a substantial role of UA as an antioxidant molecule acting as a free radical scavenger and a chelator of transitional metal ions, which are converted to poorly reactive forms [11]. Accordingly, several studies have reported on the effects of UA administration on various measures of oxidative stress. Thus, intravenous administration of UA, raised serum free radical scavenging capacity to a greater extent than vitamin C did, in healthy volunteers [108], and lowered the oxidative stress associated with acute aerobic exercise in athletes [109, 110]. These observations led to the speculation that the pro-oxidant [111] and pro-inflammatory [30] actions attributed to UA by previous studies, result from the conversion of xanthine dehydrogenase to xanthine oxidase, with subsequent accumulation of oxygen free radicals in parallel with UA production as an effect of ATP degradation under ischemic conditions. If this is the case, the often mild increase in SUA observed in patients with uncomplicated obesity, IR, and/or hypertension, deriving from an altered renal handling of sodium and UA, might be of no major concern. Actually, at present, there is no recommendation of pharmacological treatment of this asymptomatic mild form of hyperuricemia (up to a SUA level $< 10\,mg\,dl^{-1}$). By contrast, SUA elevation in hypertensive and/or diabetic patients, especially in the presence of overt ischemia needs much attention as a marker of oxidative stress and of underlying inflammatory and degenerative alterations in the CVD system [112]. A controlled trial of the effects of SUA reduction in this category of patients is obviously warranted [71, 107, 113, 114].

References

1 Wu, X.W., Muzny, D.M., Lee, C.C., and Caskey, C.T. (1992) Two independent mutational events in the loss of urate oxidase during hominoid evolution. *J Mol Evol*, **34**, 78–84.

2 Capasso, G., Jaeger, Ph., Robertson, W.C., and Unwin, R.J. (2005) Uric acid and the kidney: urate transport, stone disease and progressive renal failure. *Curr Pharm Des*, **11**, 4153–4159.

3 Wyngaarden, J.B. and Kelley, W.N. (1976) *Gout and Hyperuricemia*, Grune & Stratton, New York.

4 Roch-Ramel, F. and Peters, G. (1978) Uric acid, in *Handbook of Experimental Pharmacology*, vol. **51** (eds W.N. Kelley and I.M. Weiner),

5 Boyle, J.A., Campbell, S., Duncan, A.M., Greig, W.R., and Buchanan, W.W.J. (1966) *Clin Pathol*, **19**, 501–503.
6 Nicholls, A., Snaith, M.L., and Scott, J.T. (1973) Effect of oestrogen therapy on plasma and urinary urate levels. *Br Med J*, **1**, 449–451.
7 Simon, M.I. and Van Vunakis, H. (1964) *Arch Biochem Biophys*, **105**, 197–206.
8 Ames, B.N., Cathcart, R., Schwiers, E. and Hochsteint, P. (1981) Uric acid provides an antioxidant defense in humans against oxidant- and radical-caused aging and cancer: a hypothesis. *Proc Natl Acad Sci*, **78** (11), 6858–6862.
9 Howell, R.R. and Wyngaarden, J.B. (1960) *J Biol Chem*, **235** (16), 3544–3550.
10 Muraoka, S. and Miura, T. (2003) Inhibition by uric acid of free radicals that damage biological molecules. *Pharmacol Toxicol*, **93**, 284–289.
11 Glantzounis, G.K., Tsimoyiannis, E.C., Kappas, A.M., and Galaris, D.A. (2005) Uric acid and oxidative stress. *Curr Pharm Des*, **11**, 4145–4151.
12 Stocker, R. and Keaney, J.F. Jr (2004) Role of oxidative modifications in atherosclerosis. *Physiol Rev*, **84**, 1381–1478.
13 Alonso, A., Rodríguez, L.A., Logroscino, G., and Hernán, M.A. (2007) Gout and risk of Parkinson disease: a prospective study. *Neurology*, **69** (17), 1696–1700.
14 De Vera, M., Rahman, M.M., Rankin, J., Kopec, J., Gao, X., and Choi, H. (2008) Gout and the risk of parkinson's disease: a cohort study. *Arthritis Rheum*, **59** (11), 1549–1554.
15 Maples, K.R. and Mason, R.P. (1988) Free radical metabolite of uric acid. *J Biol Chem*, **263**, 1709–1712.
16 Santos, C.X., Anjos, E.I., and Augusto, O. (1999) Uric acid oxidation by peroxynitrite: multiple reactions, free radical formation, and amplification of lipid oxidation. *Arch Biochem Biophys*, **372**, 285–294.
17 Johnson, R.J., Kang, D.H., Feig, D., Kivlighn, S., Kanellis, J., Watanabe, S., Tuttle, K.R., Rodriguez-Iturbe, B., Herrera-Acosta, J., and Mazzali, M. (2003) Is there a pathogenetic role for uric acid in hypertension and cardiovascular and renal disease? *Hypertension*, **41**, 1183–1190.
18 Sautin, Y.Y., Nakagawa, T., Zharikov, S., and Johnson, R.J. (2007) Adverse effects of the classic antioxidant uric acid in adipocytes: NADPH oxidase-mediated oxidative/nitrosative stress. *Am J Physiol Cell Physiol*, **293**, C584–C596.
19 Bonora, E., Targher, G., Zenere, M.B., Saggiani, F., Cacciatori, V., Tosi, F., Travia, D., Zenti, M.G., Branzi, P., Santi, L., and Muggeo, M. The Verona Young Men Atherosclerosis Risk Factors Study (1996) Relationship of uric acid concentration to cardiovascular risk factors in young men. Role of obesity and central fat distribution. *Int J Obes Relat Metab Disord*, **20**, 975–980.
20 Matsuura, F., Yamashita, S., Nakamura, T., Nishida, M., Nozaki, S., Funahashi, T., and Matsuzawa, Y. (1998) Effect of visceral fat accumulation on uric acid metabolism in male obese subjects: visceral fat obesity is linked more closely to overproduction of uric acid than subcutaneous fat obesity. *Metabolism*, **47**, 929–933.
21 Ogura, T., Matsuura, K., Matsumoto, Y., Mimura, Y., Kishida, M., Otsuka, F., and Tobe, K. (2004) Recent trends of hyperuricemia and obesity in Japanese male adolescents, 1991 through 2002. *Metabolism*, **53**, 448–453.
22 Lin, W.-Y., Liu, C.-S., Li, T.-C., Lin, T., Chen, W., Chen, C.-C., Li, C.-I., and Lin, C.-C. (2008) In addition to insulin resistance and obesity. Hyperuricemia is strongly associated with metabolic syndrome using different definitions in Chinese populations: a population-based study (Taichung Community Health Study). *Ann Rheum Dis*, **67**, 432–433.
23 Nakagawa, T., Hu, H., Zharikov, S., Tuttle, K.R., Short, R.A., Glushakova,

O., Ouyang, X., Feig, D.I., Block, E.R., Herrera-Acosta, J., Patel, J.M., and Johnson, R.J. (2006) A causal role for uric acid in fructose-induced metabolic syndrome. *Am J Physiol Renal Physiol*, **290**, F625–F631.
24 Thomas, S.R., Davies, M.J., and Stocker, R. (1998) *Chem Res Toxicol*, **11**, 484–494.
25 Skinner, K.A., White, C.R., Patel, R., Tan, S., Barnes, S., Kirk, M., Darley-Usmar, V., and Parks, D.A. (1998) *J Biol Chem*, **273**, 24491–24497.
26 Uppu, R.M., Cueto, R., Squadrito, G.L., Salgo, M.G., and Pryor, W.A. (1996) *Free Radic Biol Med*, **21**, 407–411.
27 Tretyakova, N.Y., Niles, J.C., Burney, S., Wishnok, J.S., and Tannenbaum, S.R. (1999) *Chem Res Toxicol*, **12**, 459–466.
28 Bagnati, M., Perugini, C., Cau, C., Bordone, R., Albano, E., and Bellomo, G. (1999) When and why a water-soluble antioxidant becomes pro-oxidant during copper-induced low-density lipoprotein oxidation: a study using uric acid. *Biochem J*, **340**, 143–152.
29 Kanellis, J. and Kang, D.H. (2005) Uric acid as a mediator of endothelial dysfunction, inflammation, and vascular disease. *Semin Nephrol*, **25**, 39–42.
30 Kanellis, J., Watanabe, S., Li, J.H., Kang, D.H., Li, P., Nakagawa, T., Wamsley, A., Sheikh-Hamad, D., Lan, H.Y., Feng, L., and Johnson, R.J. (2003) Uric acid stimulates monocyte chemoattractant protein-1 production in vascular smooth muscle cells via mitogen-activated protein kinase and cyclooxygenase- 2. *Hypertension*, **41**, 1287–1293.
31 Kang, D.H., Han, L., Ouyang, X., Kahn, A.M., Kanellis, J., Li, P., Feng, L., Nakagawa, T., Watanabe, S., Hosoyamada, M., Endou, H., Lipkowitz, M., Abramson, R., Mu, W., and Johnson, R.J. (2005) Uric acid causes vascular smooth muscle cell proliferation by entering cells via a functional urate transporter. *Am J Nephrol*, **25**, 425–433.
32 Rao, G.N., Corson, M.A., and Berk, B.C. (1991) Uric acid stimulates vascular smooth muscle cell proliferation by increasing platelet-derived growth factor A-chain expression. *J Biol Chem*, **266**, 8604–8608.
33 Kang, D.H., Park, S.K., Lee, I.K., and Johnson, R.J. (2005) Uric acid-induced c-reactive protein expression: implication on cell proliferation and nitric oxide production of human vascular cells. *J Am Soc Nephrol*, **16**, 3553–3562.
34 Khosla, U.M., Zharikov, S., Finch, J.L., Nakagawa, T., Roncal, C., Mu, W., Krotova, K., Block, E.R., Prabhakar, S., and Johnson, R.J. (2005) Hyperuricemia induces endothelial dysfunction. *Kidney Int*, **67**, 1739–1742.
35 Perlstein, T.S., Gumieniak, O., Hopkins, P.N., Murphey, L.J., Brown, N.J., Williams, G.H., Hollenberg, N.K., and Fisher, N.D. (2004) Uric acid and the state of the intrarenal renin-angiotensin system in humans. *Kidney Int*, **66**, 1465–1470.
36 Kang, D.H., Yu, E.S., Park, J.E., Yoon, K.I., Kim, M.G., Kim, S.J., and Johnson, R.J. (2003) Uric acid induced C-reactive protein (CRP) expression via upregulation of angiotensin type 1 receptors (AT1) in vascular endothelial cells and smooth muscle cells. *J Am Soc Nephrol*, **14**, 136A.
37 Strazzullo, P., Barba, G., Cappuccio, F.P., Siani, A., Trevisan, M., Farinaro, E. *et al.* (2001) Altered renal sodium handling in men with abdominal adiposity and insulin resistance: a link to hypertension. *J Hypertens*, **19**, 2157–2164.
38 Quiñones Galvan, A., Natali, A., Baldi, S., Frascerra, S., Sanna, G., Ciociaro, D. *et al.* (1995) Effect of insulin on uric acid secretion in humans. *Am J Physiol*, **268**, E1e5.
39 Puig, J.G. and Ruilope, L.M. (1999) Uric acid as a cardiovascular risk factor in arterial hypertension. *J Hypertens*, **17**, 869–872.

40 Cappuccio, F.P., Strazzullo, P., Farinaro, E., and Trevisan, M. (1993) Uric acid metabolism and tubular sodium handling: results from a population-based study. *J Am Med Assoc*, **270**, 354–359.

41 Strazzullo, P., Barbato, A., Galletti, F., Barba, G., Siani, A., Iacone, R. *et al.* Results of the Olivetti Heart Study (2006) Abnormalities of renal sodium handling in the metabolic syndrome. *J Hypertens*, **24**, 1633–1639.

42 Gao, X., Qi, L., Qiao, N., Choi, H.K., Curhan, G., Tucker, K.L., and Ascherio, A. (2007) Intake of added sugar and sugar-sweetened drink was associated with serum uric acid concentration. *Hypertension*, **50** (2), 306–312.

43 Choi, J.J.W., Ford, E.S., Gao, X., and Choi, H.K. (2008) Sugar-sweetened soft drinks, diet soft drinks and serum uric acid level – The Third National Health and Nutrition Examination Survey. *Arthritis Rheum*, **59** (1), 109–116.

44 Choi, H.K. and Curhan, G. (2008) Soft drinks, fructose consumption, and the risk of gout in men - A prospective cohort study. *Br Med J*, **336**, 309–312.

45 Resnick, D. and Niwayama, G. (1988) *Diagnosis of Bone and Joint Disorders*, 2nd edn, W.B. Saunders, Philadelphia, pp. 1618–1671.

46 Wright, I.T. (1966) Unusual manifestations of gout. *Australas Radiol*, **70**, 365.

47 Fam, A.G. (1998) Gout in the elderly: clinical presentation and treatment. *Drugs Aging*, **13** (3), 229–243.

48 Agudelo, C.A. and Wise, C.M. (1998) Crystal-associated arthritis. *Clin Geriatr Med*, **14** (3), 495–513.

49 Choi, H.K., Atkinson, K., Karlson, E., Willett, V., and Curhan, G. (2004) Purine-rich foods, dairy intake, and protein intake, and risk of gout in men. *N Engl J Med*, **345**, 981–986.

50 Wallace, S.L., Robinson, H., Masi, A.T., Decker, J.L., McCarty, D.J., and Yu, T.F. (1977) Preliminary criteria for the classification of the acute arthritis of primary gout. *Arthritis Rheum*, **20**, 895–900.

51 Fam, A.G. (2002) Gout, diet, and the insulin resistance syndrome. *J Rheumatol*, **29** (7), 1350–1355.

52 Altman, R.D., Honig, S., Levin, J.M., and Lightfoot, R.W. (1988) Ketoprofen versus indomethacin in patients with acute gouty arthritis: a multicenter, double blind comparative study. *J Rheumatol*, **15**, 1422–1426.

53 Shrestha, M., Morgan, D.L., Moreden, J.M., Singh, R., Nelson, M., and Hayes, J.E. (1995) Randomized double-blind comparison of the analgesic efficacy of intramuscular ketorolac and oral indomethacin in the treatment of acute gouty arthritis. *Ann Emerg Med*, **26**, 682–686.

54 Zhang, W., Doherty, M., Pascual, E., Bardin, T., Barskova, V., Conaghan, P. *et al.* (2006) EULAR evidence based recommendations for gout. Part I: diagnosis. Report of a task force of the Standing Committee for International Clinical Studies Including Therapeutics (ESCISIT). *Ann Rheum Dis*, **65**, 1301–1311.

55 Robert, A. and Terkeltaub, M.D. (2003) Gout. *N Engl J Med*, **349**, 1647–1655.

56 Eggebeen, A.T. (2007) Gout: an update. *Am Fam Physician*, **76**, 801–808, 811–812.

57 Morris, I., Varughese, G., and Mattingly, P. (2003) Colchicine in acute gout. *Br Med J*, **327**, 1275–1276.

58 Borstad, G.C., Bryant, L.R., Abel, M.P., Scroggie, D.A., Harris, M.D., and Alloway, J.A. (2004) Colchicine for prophylaxis of acute flares when initiating allopurinol for chronic gouty arthritis. *J Rheumatol*, **31**, 2429–2432.

59 Groff, G.D., Franck, W.A., and Raddatz, D.A. (1990) Systemic steroid therapy for acute gout: a clinical trial and review of the literature. *Semin Arthritis Rheum*, **19**, 329–336.

60 Arai, M., Yokosuka, O., Fujiwara, K., Kojima, H., Kanda, T., Hirasawa, H. *et al.* (2002) Fulminant hepatic

failure associated with benzbromarone treatment: a case report. *J Gastroenterol Hepatol*, **17**, 625–626.
61 Van Der Klauw, M.M., Houtman, P.M., Stricker, B.H.C., and Spoelstra, P. (1994) Hepatic injury caused by benzbromarone. *J Hepatol*, **20**, 376–379.
62 Becker, M.A., Schumacher, H.R. Jr, Wortmann, R.L. et al. (2005) Febuxostat, a novel nonpurine selective inhibitor of xanthine oxidase: a twenty-eight-day, multicenter, phase II, randomized, doubleblind, placebo-controlled, dose-response clinical trial examining safety and efficacy in patients with gout. *Arthritis Rheum* **52**, 916–923.
63 Becker, M.A., Schumacher, H.R. Jr, Wortmann, R.L., MacDonald, P.A., Eustace, D., Palo, W.A., Streit, J., and Joseph-Ridge, N. (2005) Febuxostat compared with allopurinol in patients with hyperuricemia and gout. *N Engl J Med*, **353**, 23.
64 Moreland, L.W. (2005) Febuxostat – treatment for hyperuricemia and gout? *N Engl J Med*, **353**, 23.
65 Vogt, B. (2005) Urate oxidase (rasburicase) for treatment of severe tophaceous gout. *Nephrol Dial Transplant*, **20**, 431–433.
66 Ganson, N.J., Kelly, S.J., Scarlett, E., Sundy, J.S., and Hershfield, M.S. (2006) Control of hyperuricemia in subjects with refractory gout, and induction of antibody against poly(ethylene glycol) (PEG), in a phase I trial of subcutaneous PEGylated urate oxidase. *Arthritis Res Ther*, **8**, R12.
67 Sundy, J.S., Becker, M.A., Baraf, H.S.B., Barkhuizen, A., Moreland, L.W., Huang, W., Waltrip, R.W. II, Maroli, A.N., and Horowitz, Z. for the Pegloticase Phase 2 Study Investigators (2008) Reduction of plasma urate levels following treatment with multiple doses of pegloticase (polyethylene glycol–conjugated uricase) in patients with treatment-failure gout. *Arthritis Rheum*, **58** (9), 2882–2891.
68 Gertler, M.M., Garn, S.M., and Levine, S.A. (1951) Serum uric acid in relation to age and physique in health and in coronary heart disease. *Ann Intern Med*, **34**, 1421–1431.
69 Menotti, A., Lanti, M., Agabiti-Rosei, E., Caratelli, L., Cavera, G., Dormi, A. et al. (2005) Riskard 2005. New tools for prediction of cardiovascular disease risk derived from Italian population studies. *Nutr Metab Cardiovasc Dis*, **15**, 426–440.
70 Giampaoli, S., Palmieri, L., Mattiello, A., and Panico, S. (2005) Definition of high risk individuals to optimise strategies for primary prevention of cardiovascular diseases. *Nutr Metab Cardiovasc Dis*, **15**, 79–85.
71 Baker, J.F., Krishnan, E., Chen, L., and Schumacher, R. (2005) Serum uric acid and cardiovascular disease: recent developments, and where do they leave us? *Am J Med*, **118**, 816–826.
72 Athyros, V.G., Elisaf, M., Papageorgiou, A.A., Symeonidis, A.N., Pehlivanidis, A.N., Bouloukos, V.I. et al. (2004) Effect of statins versus untreated dyslipidemia on serum uric acid levels in patients with coronary heart disease: a subgroup analysis of the GREek Atorvastatin and Coronary-heart-disease Evaluation (GREACE) Study. *Am J Kidney Dis*, **43** (4), 589–599.
73 Hoieggen, A., Alderman, M.H., Kjeldsen, S.E., Julius, S., Devereux, R.B., De Faire, U. et al. (2004) The impact of serum uric acid on cardiovascular outcomes in the LIFE study. *Kidney Int*, **65**, 1041–1049.
74 Lehto, S., Niskanen, L., Runnema, T., and Laakso, M. (1998) Serum uric acid is a strong predictor of stroke in patients with non-insulin dependent diabetes mellitus. *Stroke*, **29**, 635–639.
75 Wong, K.Y.K., MacWalter, R.S., Fraser, H.W., Crombie, I., Ogsten, S.A., and Struthers, A.D. (2002) Urate predicts subsequent cardiac death in stroke survivors. *Eur Heart J*, **23**, 788–793.

76 Weir, C.J., Muir, S.W., Walters, M.R., and Lees, K.R. (2003) Serum urate as an independent predictor of poor outcome and future vascular events after acute stroke. *Stroke*, **34**, 1951–1957.

77 Anker, S.D., Doehener, W., Rauchhaus, M., Sharma, R., Francis, D., Knosalla, C. et al. (2003) Uric acid and survival in chronic heart failure: validation and application in metabolic, functional and haemodynamic staging. *Circulation*, **107**, 1991–1997.

78 Bickel, C., Ruppreeht, H.J., Blankerberg, S., Rippin, G., Hafner, G., Daunhauer, A. et al. (2002) Serum uric acid as an independent predictor of mortality in patients with angiographically proven coronary artery disease. *Am J Cardiol*, **89**, 1.

79 Mohamed, F.A. (1879) On chronic Bright's disease, and its essential symptoms. *Lancet*, **1**, 399–401.

80 Messerli, F.H., Frohlich, E.D., Dreslinski, G.R., Suarez, D.H., and Aristimuno, G.G. (1980) Serum uric acid in essential hypertension: an indicator of renal vascular involvement. *Ann Intern Med*, **93** (6), 817–821.

81 Messerli, F.H., Frohlich, E.D., Dreslinski, G.R., Suarez, D.H., and Aristimuno, G.G. (1980) Asymptomatic mild hyperuricaemia: an indicator of nephrosclerosis in essential hypertension. *Clin Sci (Lond)*, **59** (6), 409s–410s.

82 Cannon, P.J., Stason, W.B., Demartini, F.E., Sommers, S.C., and Laragh, J.H. (1966) Hyperuricemia in primary and renal hypertension. *N Engl J Med*, **275**, 457–464.

83 Kinsey, D., Walther, R., Sise, H.S., Whitelaw, G., and Smithwick, R. (1961) Incidence of hyperuricemia in 400 hypertensive subjects. *Circulation*, **24**, 972–973.

84 Selby, J.V., Friedman, G.D., and Quesenberry, C.P. Jr (1990) Precursors of essential hypertension: pulmonary function, heart rate, uric acid, serum cholesterol, and other serum chemistries. *Am J Epidemiol*, **131**, 1017–1027.

85 Hunt, S.C., Stephenson, S.H., Hopkins, P.N., and Williams, R.R. (1991) Predictors of an increased risk of future hypertension in Utah. A screening analysis. *Hypertension*, **17**, 969–976.

86 Jossa, F., Farinaro, E., Panico, S., Krogh, V., Celentano, E., Galasso, R., Mancini, M., and Trevisan, M. (1994) Serum uric acid and hypertension: the Olivetti heart study. *J Hum Hypertens*, **8**, 677–681.

87 Taniguchi, Y., Hayashi, T., Tsumura, K., Endo, G., Fujii, S., and Okada, K. (2001) Serum uric acid and the risk for hypertension and type 2 diabetes in Japanese men. The Osaka Health Survey. *J Hypertens*, **19**, 1209–1215.

88 Masuo, K., Kawaguchi, H., Mikami, H., Ogihara, T., and Tuck, M.L. (2003) Serum uric acid and plasma norepinephrine concentrations predict subsequent weight gain and blood pressure elevation. *Hypertension*, **42**, 474–480.

89 Nakanishi, N., Okamoto, M., Yoshida, H., Matsuo, Y., Suzuki, K., and Tatara, K. (2003) Serum uric acid and risk for development of hypertension and impaired fasting glucose or type II diabetes in Japanese male office workers. *Eur J Epidemiol*, **18**, 523–530.

90 Alper, A.B., Chen, W., Yau, L., Srinivasan, S., Hamm, L.L., and Berenson, G. (2005) Childhood uric acid predicts adult blood pressure: the Bogalusa Heart Study. *Hypertension*, **45**, 34–38.

91 Sundström, J., Sullivan, L., D'Agostino, R.B., Levy, D., Kannel, W.B., and Vasan, R.S. (2005) Relations of serum uric aid to longitudinal blood pressure tracking and hypertension incidence in the Framingham Heart Study. *Hypertension*, **45**, 28–33.

92 Mazzali, M., Hughes, J., Kim, Y.G., Jefferson, J.A., Kang, D.H., Gordon, K.L., Lan, H.Y., Kivlighn, S., and Johnson, R.J. (2001) Elevated uric acid increases blood pressure in the rat by a novel crystal-independent

mechanism. *Hypertension*, **38**, 1101–1106.

93 Sànchez-Lozada, L.G., Tapia, E., Santamaria, J., Avila-Casado, C., Soto, V., Nepomuceno, T., Rodriguez-Iturbe, B., Johnson, R.J., and Herrera-Acosta, J. (2005) Mild hyperuricemia induces vasoconstriction and maintains glomerular hypertension in normal and remnant kidney rats. *Kidney Int*, **67**, 237–247.

94 Watanabe, S., Kang, D.H., Feng, L., Nakagawa, T., Kanellis, J., Lan, H., Mazzali, M., and Johnson, R.J. (2002) Uric acid, hominoid evolution and the pathogenesis of salt-sensitivity. *Hypertension*, **40**, 355–360.

95 Zoccali, C., Maio, R., Mallamaci, F., Sesti, G., and Perticone, F. (2006) Uric acid and endothelial dysfunction in essential hypertension. *J Am Soc Nephrol*, **17**, 1466–1471.

96 Sànchez-Lozada, L.G., Tapia, E., Avila-Casado, C., Soto, V., Franco, M., Santamarìa, J., Nakagawa, T., Rodrìguez-Iturbe, B., Johnson, R.J., and Herrera-Acosta, J. (2002) Mild hyperuricemia induces glomerular hypertension in normal rats. *Am J Physiol Renal Physiol*, **283**, F1105–F1110.

97 Johnson, R.J., Segal, M.S., Srinivas, T., Ejaz, A., Mu, W., Roncal, C., Sànchez-Lozada, L.G., Gersch, M., Rodriguez-Iturbe, B., Kang, D.H., and Acosta, J.H. (2005) Essential hypertension, progressive renal disease, and uric acid: a pathogenetic link? *J Am Soc Nephrol*, **16**, 1909–1919.

98 Feig, D.I., Soletsky, B., and Johnson, R.J. (2008) Effect of allopurinol on blood pressure of adolescents with newly diagnosed essential hypertension: a randomized trial. *J Am Med Assoc*, **300**, 924–932.

99 Sowers, J. (2001) Update on the cardiometabolic syndrome. *Clin Cornerstone*, **4** (2), 17–23.

100 Kahn, H.S. and Valdez, R. (2003) Metabolic risks identified by the combination of enlarged waist and elevated triacylglycerol concentration. *Am J Clin Nutr*, **78** (5), 928–934.

101 Meigs, J. (2003) The metabolic syndrome. *Br Med J*, **327**, 61.

102 Nakagawa, T., Tuttle, K.R., Short, R.A., and Johnson, R.J. (2005) Fructose-induced hyperuricemia as a casual mechanism for the epidemic of the metabolic syndrome. *Nat Clin Pract Nephrol*, **1**, 80–86.

103 Carneton, M.R., Fortmann, S.P., Palaniappan, L., Duncan, B.B., Schmidt, M.I., and Chambless, L.E. (2003) Risk factors for progression to incident hyperinsulinemia: the Atherosclerosis Risk in Communities Study, 1987-1998. *Am J Epidemiol*, **158**, 1058–1067.

104 Nakanishi, N., Okamoto, M., Yoshida, H., Matsuo, Y., Suzuki, K., and Tatara, K. (2003) Serum uric acid and risk for development of hypertension and impaired fasting glucose or Typer II diabetes in Japanese male office workers. *Eur J Epidemiol*, **18**, 523–530.

105 Dehghan, A., van Hoek, M., Sijbrands, E.J., Hofman, A. and Witteman, J.C. (2008) High serum uric acid as a novel risk factor for type 2 diabetes. *Diabetes Care*, **31**, 361–362.

106 Chien, K.L., Chen, M.F., Hsu, H.C. *et al.* (2008) Plasma uric acid and the risk of type 2 diabetes in a Chinese community. *Clin Chem*, **54**, 310–316.

107 Reyes, A.J. and Leary, W.P. (2003) The increase in serum uric acid induced by diuretics could be beneficial to cardiovascular prognosis in hypertension: a hypothesis. *J Hypertens*, **21**, 1775–1777.

108 Waring, W.S., Webb, D.J., and Maxwell, S.R. (2001) Systemic uric acid administration increases serum antioxidant capacity in healthy volunteers. *J Cardiovasc Pharmacol*, **38**, 365–371.

109 Hellsten, Y., Svensson, M., Sodin, B., Smith, S., Christensen, A., Richter, E.A. *et al.* (2001) Allantoin formation and urate and glutathione exchange in human muscle during submaximal exercise. *Free Radic Biol Med*, **31**, 1313–1322.

110 Chevion, S., Moran, D.S., Heled, Y., Shani, Y., Regev, G., Abbou, B. et al. (2003) Plasma antioxidant status and cell injury after physical exercise. *Proc Natl Acad Sci U S A*, **100**, 5119–5123.

111 Ward, H.J. (1998) Uric acid as an independent risk factor in the treatment of hypertension. *Lancet*, **352**, 670–671.

112 Ferroni, P., Basili, S., Paoletti, V., and Davì, G. (2006) Endothelial dysfunction and oxidative stress in arterial hypertension. *Nutr Metab Cardiovasc Dis*, **16**, 222–233.

113 Schachter, M. (2005) Uric acid and hypertension. *Curr Pharm Des*, **11**, 4139–4143.

114 Feig, D.I., Kang, D.H., and Johnson, R.J. (2008) Uric acid and cardiovascular risk. *N Engl J Med*, **359**, 1811–1821.

16
Genetic Defects in Iron and Copper Trafficking
Douglas M. Templeton

Defects in iron and copper metabolism can lead to toxic accumulation of these elements in various organs and tissues. Stringent mechanisms to transport them throughout the body have evolved to avoid the adverse consequences of their ability to redox cycle in biological systems. Major disorders discussed for iron are the hereditary hemochromatoses and other sideroses, Friedreich ataxia, sideroblastic anemias, microcytic anemia, hyperferritinemia cataract syndrome, and hypotransferrinemia. Major copper disorders are represented by defects in the *ATP7A/B* genes, leading to Menkes and Wilson's diseases. Other copper toxicoses and other copper-related genes such as those implicated in amyotrophic lateral sclerosis and the Cu,Zn-superoxide dismutase are also discussed. Aceruloplasminemia serves as an interesting bridge between iron and copper, being a defect in a copper-containing enzyme that results in neurological symptoms from excessive iron deposition. A compilation of genes involved in iron and copper pathobiology, cross-referenced to the Online Mendelian Inheritance in Man (OMIM) catalog, is provided.

16.1
Introduction and Terminology

This chapter deals with toxic effects of iron (Fe) and copper (Cu), when they accumulate in toxic quantities due to genetic defects in their handling, generally meaning their transport into or out of the body or the cell. In view of the title of this book, it should be noted that the term *endogenous toxins* is not appropriate for these metallic elements. First, they are not of endogenous origin, not being transmutable substances produced or destroyed within the body; rather, they are acquired from dietary sources or other exposures, and then redistributed within the body. Secondly, they are both essential and toxic elements; but, they are not toxins. A suitable definition of *toxin* has been given as "a poisonous substance produced by a biological organism such as a microbe, animal, plant, or fungus" [1], such as botulinum toxin, tetrodotoxin, pyrrolizidine alkaloids, or amanitin. The term *toxic metal* is itself a deliberately imprecise term, allowing for adverse effects of some essential metallic ions, even at concentrations below those of some nonessential metals. More precisely, this chapter deals with the toxic effects on the body that result when iron or copper are mishandled as a result of genetic defects.

Discussion of individual genes and associated disorders in this chapter can be cross-referenced to the OMIM catalog (*http://www.ncbi.nlm.nih.gov/omim/*). Entry numbers for the genes and diseases considered here are listed in Table 16.1. For orientation, overviews of iron absorption and distribution, and overview of intracellular copper trafficking, are shown schematically in Figures 16.1–16.3.

16.2
Iron

16.2.1
Physiology

Mechanisms both dependent upon and independent of reactive oxygen species (ROS) appear to play a role in iron toxicity. Fe displays a redox potential for the Fe^{3+}/Fe^{2+} redox couple that varies from $+770$ mV for the "bare" aqueous ion, to $+260$ mV for cytochrome c, and even down to -526 mV for transferrin [2]. Thus, iron actively redox cycles under some biological conditions, and generates ROS through Fenton-like reactions [3], and a major feature of iron toxicity is the generation of toxic ROS that react directly with multiple biomolecular targets. Uptake of nontransferrin-bound iron by transporters such as divalent metal transporter (DMT)1 [4, 5], deliver iron initially to unknown ligands. This has given rise to the concept of a transit pool of iron, referred to as the labile iron pool and considered to be the main pool of iron available for Fenton activity [6, 7]. Though only a small portion of total cellular iron, this pool is catalytically active and expands with iron overload [8, 9], giving it principal importance in iron toxicity. However, it is clear that Fe has farther reaching effects on cells, and there is sufficient evidence that it has effects on gene regulation that are mediated by both ROS-dependent and -independent mechanisms [10].

Humans do not have an effective mechanism for iron excretion [11], and daily loss of 1–2 mg of iron to balance intestinal uptake is primarily from sloughing of cells or blood loss. However, iron absorption is tightly regulated. In the duodenal enterocyte, uptake of dietary Fe^{3+} is initiated by action of a duodenal cytochrome b-like ferrireductase, Dcytb, which generates Fe^{2+} for uptake by the divalent metal transporter, DMT1 (also known as Nramp2, encoded by the *SLC11A2* gene) at the basolateral cell surface [11, 12]. The newly acquired iron may become assimilated into the mineral core of ferritin for storage, enter the mitochondrion for heme or iron–sulfur cluster synthesis, or exit from the apical membrane through the Fe^{2+} transporter, ferroportin [13] (Figure 16.1). To deliver Fe^{3+} to transferrin for transport in the blood, the Fe^{2+} must be reoxidized. This is achieved by a ferroxidase named hephaestin. Here the product of the hemochromatosis (*HFE*) gene also becomes involved. HFE [14] is a major histocompatability class I protein. As such, it binds β_2-microglobulin, which directs its expression to the cell surface, where

Tab. 16.1 Genes/proteins discussed in the text are listed alphabetically, separately for iron and copper

Genes	Genetic diseases
Iron	
ALAS2 (301300)	X-linked sideroblastic anemia (300751)
ABCB7 (300135)	X-linked sideroblastic anemia, with ataxia (301310)
DCYTB (605745)	
DMT1/SLC11A2 (600523)	Hypochromic microcytic anemia with iron overload (206100)
Ferritin L chain (134790)	Hereditary hyperferritinemia cataract syndrome (600886)
Ferroportin/SLC40A1 (604653)	Hemochromatosis type IV (HFE4) (606069)
	(?) African siderosis (601195)
Frataxin (FXN) (606829)	Friedreich ataxia (229300)
Hemojuvelin (HJV) (608374)	Hemochromatosis type IIA (HFE2A) (602390)
Hepcidin/HAMP (606464)	Hemochromatosis type IIB (HFE2B) (602390)
Hephaestin (300167)	
HFE (235200)	Hemochromatosis (HFE1) (235200)
Transferrin (190000)	Atransferrinemia/Hypotransferrinemia (209300)
TfR1 (190010)	
TfR2 (604720)	Hemochromatosis type III (HFE3) (604250)
Copper	
ATP7A (300011)	Menkes disease (309400)
	Occipital horn syndrome (304150)
ATP7B (606882)	Wilson disease (277900)
Ceruloplasmin (117700)	Aceruloplasminemia (604290)
Chaperones	
ATOX1 (602270)	
CCS (603864)	
COX17 (604813)	
COPT1 (603085)	
COPT2 (603088)	
CUTC (610101)	
MURR1/COMMD1 (607238)	
SCO1 (603644)	Early onset hepatic failure (603644)
SCO2 (604272)	Cardioencephalomyopathy (604377)
SOD1 (147450)	Familial ALS (105400)
Genes unknown	Endemic Tyrolean infantile cirrhosis (215600)
	Indian childhood cirrhosis (215600)

Associated syndromes and diseases are given in the right column. Numbers in parentheses are the identifiers in the OMIM database.

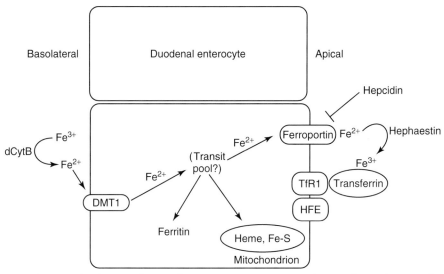

Fig. 16.1 A schematic showing the participants in iron uptake in the gut. Ferric ion in the gut lumen is reduced by Dcytb and taken up at the basolateral side as Fe^{2+} through the Fe^{3+} transporter, DMT1. This iron may reoxidize and be stored as ferritin, or be imported into the mitochondrion for heme and Fe-S cluster synthesis. It may also be delivered to the apical membrane where it is exported through the Fe^{2+} transporter, ferroportin. The nature of the iron in transit is not known. The newly exported Fe^{2+} is reoxidized to Fe^{3+} by hephaestin at the extracellular membrane surface, for incorporation into diferric transferrin. For this purpose, transferrin is recruited to the membrane by its receptor, TfR1, which interacts with the regulatory protein, HFE. Ceruloplasmin replaces hephaestin as the ferroxidase in most other cells. Hepcidin down-regulates ferroportin by direct interaction leading to internalization and degradation.

it forms a complex with the transferrin receptor (TfR). This presumably regulates the positioning of transferrin to acquire the nascent Fe^{3+} resulting from the action of hephaestin. When HFE is deficient, loss of regulation of iron uptake results in iron overload (see type I hemochromatosis, below).

In general, mammalian cells take up iron through receptor mediated endocytosis of transferrin, involving the transferrin receptor, TfR1 [11]. Hepatocytes and erythroid cells express a second transferrin receptor, TfR2. The TfR2 transcript is highly expressed in hepatocytes, and is not regulated by tissue iron status [15]. DMT1 is internalized along with the Fe_2Tf–TfR complex into the endosome, although the precise mechanism that coordinates this is not understood. In the endosome, DMT1 functions to transport the iron that is released from transferrin in the endosome, into the cytosol. Release of iron from nonduodenal cells also involves ferroportin and the functionally coupled ferroxidase, hephaestin, for delivery of Fe^{3+} to transferrin. Systemically, ceruloplasmin serves as the ferroxidase in place of hephaestin.

The erythropoietic system turns over about 25 mg of iron per day in red cell synthesis and degradation (Figure 16.2). The liver peptide, hepcidin, is a key regulator of body iron homeostasis, in that it suppresses iron release from

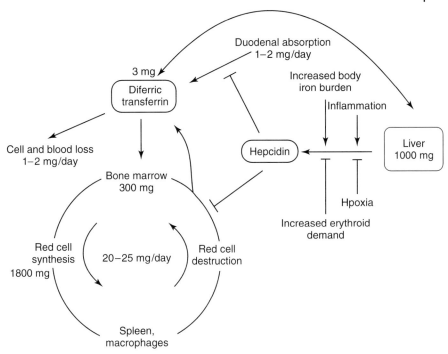

Fig. 16.2 Daily iron traffic showing the central role of diferric transferrin in shuttling iron between the gut, liver, and erythron. Also indicated is the central regulatory role of hepcidin. Numbers in italics are the approximate amount of iron involved in the process or pool in a healthy human adult. Adapted from [11].

reticuloendothelial macrophages [11, 16]. Hepcidin also suppresses duodenal iron uptake by binding ferroportin, resulting in its internalization and degradation [17]. Hepcidin is under control of lipopolysaccharide (i.e., inflammation) and elevated iron stores, both of which increase its secretion from liver. On the other hand, hypoxia and increased erythroid demand suppress hepcidin.

When iron intake is excessive, iron accumulates in the body. In conditions of body iron overload, serum transferrin – normally only about 35% saturated – can become fully saturated. Iron deposits in the soft tissues, notably, the liver and heart, causing severe organ damage and death from cirrhotic liver failure and cardiac dysfunction. Other notable consequences are diabetes from pancreatic iron deposition (e.g., the so-called bronze diabetes of HFE results from skin discoloration due to iron deposition accompanying pancreatic damage), and loss of anterior pituitary function leading to failure to proceed through puberty, or sexual dysfunction, depending on the age at occurrence. A major cause of iron overload from excessive intake is repeated blood transfusions, for example, as occurring in β-thalassemia major, where a failure of β-globin chain synthesis necessitates frequent (often monthly) blood transfusions. After about a year of

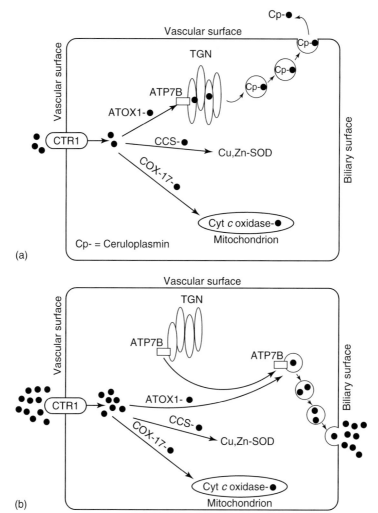

Fig. 16.3 A simplified view of copper trafficking in the hepatocyte. Copper (black dots) enters through the CTR1 transporter and is picked up by specific chaperones discussed in the text. In particular, ATOX1 delivers it to the ATP7B transporter. Under basal copper concentrations (a), copper is assembled into ceruloplasmin and released by vesicles fusing with the membrane at the vascular surface of the cell. When copper levels increase (b), ATP7B translocates to vesicles that fuse with the membrane of the biliary canaliculus to release copper into the bile. In many other cells, ATOX1 delivers copper to ATP7A resident in the TGN under basal conditions, for synthesis of copper-dependent proteins, and shifting to vesicles at the apical cell surface for efflux as copper levels rise. The oxidation states of Cu(I/II) at each step are not indicated. TGN, trans-Golgi network.

such transfusions, iron begins to deposit in parenchymal tissues and initiates toxic changes [18]. Without chelation therapy, symptomatic cardiac disease can occur within 10 years, and the risk of hepatic fibrosis increases above an iron content of 7 mg/g dry weight of liver [18]. In a patient who underwent double organ transplant because of organ failure of iron overload, the heart iron content was 5.8 mg/g dry weight and that of liver, 28.1 mg/g dry weight [19].

16.2.2
Hereditary Hemochromatoses

Iron overload as a result of failure to adequately limit iron absorption in response to body iron stores is especially prevalent in Caucasians, and the disease of hereditary hemochromatosis may affect 1 in 200 individuals of this racial origin. Penetrance is variable, and many cases are undiagnosed. Treatment by periodic phlebotomy is effective; but, if organ damage occurs due to iron overload in untreated individuals, the leading cause of death is liver failure, with a high incidence of hepatocellular carcinoma. In 1996, Feder *et al.* identified a responsible gene as *HFE*, located at chromosome 6p21.3 [20]. We now know of other genetic causes of the disease and classify hereditary hemochromatosis into five variants, denoted as types I, IIA, IIB, III, and IV, with type I being the classical HFE-dependent form.

Most cases of hereditary hemochromatosis are caused by mutations in HFE, and the disease is referred to as hemochromatosis type I (classical HFE). Somewhat redundantly, the gene, its protein product, and the resultant disease are all called HFE. The most common mutation in the *HFE* gene is the homozygous substitution C282Y, but a second missense mutation, H63D, was also found to be enriched in patients heterozygous for the C282Y mutation [20]. The C282Y mutation abolishes its binding to β_2-microglobulin, and thus, its expression on the cell surface, whereas the H63D mutation interrupts neither binding nor cell-surface expression [21]. The inability of HFE protein to negatively regulate the interaction of TfR1 with transferrin at the duodenal apical membrane accounts for increased iron absorption.

Hemochromatosis type II (HFE2) is a juvenile-onset hemochromatosis. In contrast to type I, which has an onset in adulthood, iron accumulation in the juvenile variant begins early in life and becomes symptomatic before 30 years of age. It represents a more severe form of the disease, and cardiac arrhythmias and failure are more prominent causes of death. Hypogonadism is also a common feature. Juvenile hemochromatosis actually results from two distinct mutations. One form, designated type IIA (HFE2A), is due to mutation in a gene called hemojuvelin (*HJV*) located on chromosome 1q21 [22, 23]. HJV protein probably regulates hepcidin expression by binding bone morphogenic protein and suppressing its signaling; mutant HJV decreases hepcidin expression [24]. A number of mutations have been reported in this gene [25], and account for the majority of juvenile hemochromatosis.

A second type of juvenile hemochromatosis designated type IIB (HFE2B) is caused by the gene *HAMP*, encoding hepcidin itself [12, 16], located on chromosome

19q13. Mutations in HAMP cause disruption of the regulation of iron homeostasis by hepcidin and lead directly to iron overload [26, 27], probably through failure to down-regulate ferroportin.

Hemochromatosis type III (HFE3) is caused by mutation of the gene encoding the hepatocyte- and erythroid-specific transferrin receptor, TfR2, which is located on chromosome 7q22 [28, 29]. A number of mutations that result in iron overload have been reported to occur in hemochromatosis patients. Mattman *et al.* [30] studied a group of non-HFE(C282Y) hemochromatosis patients, and identified several sequence variants in the *TFR2* gene, including a homozygous missense Q690P mutation in exon 17.

Hemochromatosis type IV (HFE4) is caused by mutation of the ferroportin/*SLC40A1* gene, located at chromosome 2q32 [31]. The 5′ untranslated region of the ferroportin mRNA contains a functional iron-responsive element (IRE). Wallace *et al.* [32] found heterozygosity for a three base pair (TTG) deletion in exon 5 of the ferroportin gene in an Australian family with autosomal dominant hemochromatosis. The mutation results in deletion of valine at position 162, which creates a loss-of-function mutation and impairs iron homeostasis.

16.2.3
African Siderosis

Iron overload in sub-Saharan Africa, formerly termed Bantu siderosis, occurs in individuals drinking a beer brewed locally in iron vessels. However, not all beer drinkers develop the disorder, and similar iron overload occurs in some individuals who do not drink beer, indicating a genetic component [33], and a possible link to non-HFE-dependent iron overload seen in African-Americans. Suggestively, a Q248H mutation in ferroportin has now been described in Africans and African-Americans with iron overload [34].

16.2.4
Friedreich Ataxia

Friedreich ataxia is an autosomal recessive neurodegenerative disorder involving both the central and peripheral nervous systems. It is characterized by a progressive ataxia, sensory loss, and cardiomyopathy [35, 36]. It is caused by mutation of the gene frataxin (*FXN*) located at chromosome 9q13 [37]. FXN encodes the 210-amino acid mitochondrial protein FXN [38], which is involved in mitochondrial iron metabolism, and the disease is a consequence of mitochondrial iron overload [36, 39, 40]. The major mutation in FXN is the expansion of a GAA trinucleotide repeat occurring in the first intron of the gene, although other mutations have also been reported. Thus, most patients are homozygous for the expansion of a GAA triplet repeat within FXN, but a few show compound heterozygosity for a point mutation in addition to the GAA-repeat expansion [35, 38, 39, 41]. The number of GAA repeats correlates with the age of onset of progressive disease, and clinical variability in Friedreich ataxia is related to the size of the expanded repeat. Up to

1700 repeats have been observed but milder, late-onset disease is associated with shorter expansions [42].

The expanded GAA sequence produces an abnormal and highly thermostable DNA that impairs FXN transcription and results in low-level FXN expression [43, 44]. Gacy et al. [44] suggested that the GAA instability in Friedreich ataxia is a DNA-directed mutation caused by improper DNA structures at the repeat region. The yeast FXN homolog gene, YFH1, has been instrumental in the identification of the human gene and understanding the protein changes involved. Mutations in the FXN gene cause impaired protein translation in both yeast and humans. Wild-type human FXN cDNA can complement the YFH1 protein-deficient yeast with prevention of the mitochondrial iron accumulation and oxidative damage associated with loss of YFH1 [45]. On the other hand, two clinically relevant mutants, G130V and W173G are unable to rescue YFH1 ablations. Both mutants affected protein stability and result in low levels of FXN expression.

In addition to causing mitochondrial oxidative damage, FXN deficiency results in decreased activity of mitochondrial iron–sulfur proteins. It also plays a role in Fe-S cluster assembly and transport and possibly in regulation of energy metabolism and oxidative phosphorylation [40, 46]. This underutilization of iron further contributes to mitochondrial accumulation. FXN has an iron-binding site and has been suggested to act as a mitochondrial iron chaperone protein. Bulteau et al. [47] found that FXN interacts with aconitase in a citrate-dependent fashion, converting the inactive $[3Fe-4S]^{1+}$ enzyme to the active $[4Fe-4S]^{2+}$ form. They suggested that FXN is an iron chaperone protein that protects the aconitase $[4Fe-4S]^{2+}$ cluster from disassembly and promotes enzyme reactivation.

16.2.5
X-Linked Sideroblastic Anemias

X-linked sideroblastic anemia is a hypochromic microcytic anemia characterized by ringed sideroblasts in the marrow, particularly affecting late erythroid precursors. It results in systemic iron overload as a consequence of chronic ineffective erythropoiesis. It is caused by mutations in the δ-aminolevulinic acid synthase 2 (ALAS2) gene, located on the X chromosome. This variant of ALAS is expressed exclusively in the erythroid cells. More than 20 different mutations of this gene have been described [48, 49]. ALAS is the first enzyme in the heme synthetic pathway, and it catalyzes the condensation of glycine and succinyl-CoA to form aminolevulinic acid (ALA). This reaction occurs in the mitochondrion [50]. A defect in the ALAS2 gene inhibits heme synthesis, and consequently, iron is accumulated in the mitochondrion.

Another form of X-linked sideroblastic anemia accompanied by ataxia has been described. It is characterized by an infantile to early childhood onset of nonprogressive cerebellar ataxia and a milder hypochromic, microcytic anemia. It is a recessive disorder caused by gene mutations in the ABCB7 [51, 52]. ABCB7 is a member of the large family of ATP-binding cassette (ABC) transporters. It shares highest sequence similarity with the yeast ATM1 gene, which encodes an

ABC half-transporter located in the mitochondrial inner membrane. ABCB7 was suggested to be a transporter for heme from the mitochondrion to the cytosol [53]. It now appears more likely that it is involved in iron–sulfur cluster export from the mitochondrion, and indeed it may have a more general role in mitochondrial iron homeostasis [54]. In fact, Rouault and Tong [54] have recently reviewed a number of genes related to iron–sulfur cluster synthesis and export (including ABCB7) that are linked to biological defects, mainly in yeast. This is an aspect of iron transport genetics that is surely to have a more recognized role in human disease in the future.

16.2.6
Hypotransferrinemia

Hypotransferrinemia [55], and more rarely atransferrinemia, arise from mutations in the transferrin gene and result in microcytic anemia and iron loading. Impairment of the normal delivery of transferrin-bound iron to tissues for hematopoeisis and iron storage leading to ineffective hematopoeisis is paradoxically accompanied by iron overload, as increased circulating nontransferrin-bound iron finds its way into soft tissues. When iron is not delivered to tissues in a regulated manner, it is unavailable for appropriate hematopoeisis. Mice with congenital hypotransferrinemia given soluble iron salts accumulate radiolabeled iron in the liver and pancreas, in contrast to deposition in the bone marrow and spleen in control animals [56]. This pattern of deposition is similar to that occurring in hemochromatosis [56]. As occurs when transferrin is saturated in states of iron overload, a lack of sufficient transferrin may engage a protective mechanism that clears nontransferrin-bound iron from circulation into the soft tissues [57]. Treatment of hypo- or atransferrinemia is with plasma transfusion.

16.2.7
Hereditary Hyperferritinemia Cataract Syndrome

Some aspects of iron metabolism are regulated at a posttranscriptional level by iron regulatory proteins, IRP1 and IRP2, which bind to IREs in mRNAs [58–60]. With lower iron levels in the cell, binding is favored. In the TfR transcript, multiple copies of the IRE are found in the 3'-untranslated region, where their occupancy by IRPs stabilizes the mRNA, and results in an increase in receptor protein expression. In ferritin mRNA, occupancy of the single IRE in the 5'-untranslated region blocks initiation of translation and protein levels are decreased. Conversely, abundant cellular iron, up-regulates ferritin expression and lowers TfR protein. Families have been identified in Italy, France, Australia, and the United Kingdom, who have a familial inheritance of cataracts with high circulating levels of ferritin. This disorder, termed hereditary hyperferritinemia cataract syndrome, has been found to result from mutation of the 5' IRE of the ferritin L chain [61–63]. Loss of the IRE prevents down-regulation of ferritin L chain synthesis even in low iron conditions, and the excessive production leads to elevated blood levels. This is generally without consequence, except for the formation of cataracts with deposition of crystalline L chain in the lens [64].

16.2.8
Hypochromic Microcytic Anemia with Iron Overload

Two animal models of microcytic anemia with systemic iron overload have been provided by the Belgrade rat, and a microcytic anemic (*mk/mk*) mouse. The same gene is mutated in both models, and it is DMT1/*SLC11A2* [65], which codes the DMT1 transporter described above as involved in the uptake of iron in the gut and release from the endosome. Interestingly, the Belgrade rat shows high serum iron levels with decreased uptake by reticulocytes. When iron chelators are used to bypass the transferrin cycle, hemoglobin synthesis is restored [65]. A human equivalent to these animal models is now known. Patients from Italy and France, with mutations in DMT1/*SLC11A2*, presenting with hypochromic, microcytic anemia, and moderate to severe hepatic iron overload [66, 67] have been reported.

16.2.9
Aceruloplasminemia

An interesting link between iron and copper is provided by the involvement of the copper protein, ceruloplasmin, in iron transport. Human ceruloplasmin is a 132-kDa glycoprotein that was first isolated in 1948 [68], and was believed for many years to function in copper transport and storage. The protein is of the type blue multicopper oxidase that requires at least four copper atoms to function in the four-electron reduction of O_2 to $2H_2O$, via sequential one-electron oxidations of reduced substrates. Ceruloplasmin has two additional copper atoms, for a total copper content of six atoms per protein molecule [69]. In the past, the ability of ceruloplasmin to oxidize Fe^{2+} was not generally considered significant, as it was believed that spontaneous oxidation of Fe^{2+} and subsequent association of Fe^{3+} with transferrin could explain iron transport and delivery. Furthermore, iron metabolism is normal in most patients with Wilson disease, despite decreased ceruloplasmin levels. However, molecular dissection of the role of the Fet3 protein, a homolog of mammalian ceruloplasmin, in iron acquisition by *Saccharomyces cerevisiae*, suggested that parallel pathways could exist in higher animals [70]. In fact, ceruloplasmin functions as a systemic hephaestin [71, 72], that is, as a ferroxidase to generate ferric ion for transport. Revelation of the true function of ceruloplasmin as a ferroxidase enzyme essential for proper iron metabolism explained a long-known connection between copper status and iron utilization [73]. The consequences of decreased ceruloplasmin in Wilson's disease are minimal, but rare occurrences of a complete absence of ceruloplasmin are not.

The identification of several families in Japan and the United States with a gene defect in ceruloplasmin that leads to aceruloplasminemia and severe defects in iron homeostasis has further clarified the true function of this protein [74–76].

Aceruloplasminemic patients present in midlife with neurological symptoms including ataxia, Parkinson-like movement disorder, dystonia, and progressive dementia. Neuronal loss accompanies iron deposition in both neurons and microglia [77], so this is really a disorder characterized as a hemochromatosis, and indeed hepatic iron overload, too, is a feature [78]. The basis of the iron overload is not completely clear. Decreased ferroxidase activity may increase deposition because of increased plasma ferrous iron, and impaired tissue iron efflux is also presumably involved [78].

16.3
Copper

16.3.1
Physiology

Copper is an essential element that, like iron, is toxic if not properly handled. It cycles between Cu(I) and Cu(II) oxidation states in biological systems (the redox potential for $Cu(II/I)_{aq}$ is 130 mV [79]), and so can act as a Fenton catalyst [3]. Nevertheless, it is required by a number of important enzymes, including amine oxidase, ceruloplasmin, hephaestin, Cu,Zn-superoxide dismutase, cytochrome c oxidase, lysyl oxidase, and tyrosinase. Thus, like iron, copper is carefully guided by transporters, binding proteins, and chaperones, from the time of its entry into the gut to its utilization and ultimate excretion into the bile.

O'Halloran and coworkers have estimated that less than one free copper ion per cell exists in yeast, implying an extraordinary overcapacity for chelation of cellular copper [80, 81]. Cellular copper is effectively sequestered by metallothioneins such as the yeast *CUP1* gene product. Thus, the cell is not dependent on diffusion for copper to encounter its enzyme or transporter targets, but rather, exploits a system of chaperones for targeted delivery. These include CCS, which delivers copper to Cu,Zn-superoxide dismutase, ATOX1, which carries copper to the ATP7A/B transporters in the trans-Golgi network, and COX17, which targets copper for insertion into cytochrome c oxidase [82–84].

Although tissue copper storage is dominated by metallothionein [85, 86], this protein is not secreted from cells, and other forms of copper are found in the circulation, predominant among them being ceruloplasmin. About 65% of circulating copper is essentially irreversibly bound to ceruloplasmin [87], while a lesser amount (about 15%) is bound to the N-terminal aspartic acid-alanine-histidine tripeptide of albumin [88]. For many years, ceruloplasmin was viewed as the important transporter of copper in the circulation, and a major player in both transport and storage of copper, while albumin-bound copper was a good candidate for the exchangeable form of Cu^{2+} delivered to tissues. Both roles must now be reformulated.

Normal delivery of injected ^{67}Cu to the livers of analbuminemic rats [89] also casts doubts on an essential role of albumin in delivery of Cu^{2+} to tissues, and suggests rather, a role in buffering plasma Cu^{2+}. Although ceruloplasmin and albumin together account for about 80% of the copper in normal human plasma, the remaining fraction is less well defined. Another protein fraction has been called transcuprein [87], but has not been characterized in any detail. About 5% of plasma Cu^{2+} is bound in low molecular weight complexes. Modeling based on stability constants predicts that histidine and histidine-containing bis-peptide complexes will be important in this fraction [90].

Body copper levels are normally tightly regulated, with a serum concentration of about 10–20 µM in the healthy adult. The mechanism regulating copper uptake in the gut is not well understood, but absorption of available dietary copper can vary from more than 50% in low-copper diets to less than 20% with increased intake [83, 91]. On the other hand, copper excretion is effectively controlled in the liver by the Wilson disease protein, ATP7B. Normally located in the trans-Golgi network where it functions to sequester copper for incorporation into ceruloplasmin, excess cellular copper causes ATP7B to translocate to a cytoplasmic vesicular compartment (Figure 16.3) localized near the biliary canalicular membrane [92]. Elimination of copper into the bile then follows fusion of the vesicles with the cell membrane.

16.3.2
ATP7A/B

Wilson disease is the prototypical copper storage and overload disorder. It is an autosomal recessive disorder with an incidence of 1/50 000–1/100 000 in different populations. Onset in childhood is rare. Also known as hepatolenticular degeneration, it may present with either liver disease or neurological symptoms of ataxia and progressive dementia. Accumulation of copper in the liver due to impaired biliary excretion can also lead to copper deposition in other sites, including the brain. Deposition in the cornea produces a benign condition with the appearance of Kayser–Fleischer rings. If untreated, death ensues from end stage liver disease. Probably the most significant contribution to cell damage is the generation of ROS through Cu(I)/Cu(II) redox cycling when the sequestering capacity, most notably of metallothionein, is exceeded [82]. The main treatment is chelation therapy to lower excessive copper stores.

In contrast to the copper overload of Wilson disease, Menkes disease is a fatal form of copper deficiency. It is a rare X-linked recessive disorder with an estimated incidence of 1/100 000–1/250 000 live births. It is characterized by growth retardation, focal cerebral and cerebellar degeneration with rapidly progressive neurological impairment beginning in the first few weeks of life, connective tissue abnormalities, hypothermia, and unusual "kinky hair" (pili torti). Attempts have been made to prolong life into childhood with parenteral administration of Cu(II)–histidine complex. Milder variants are known, and of particular interest is occipital horn syndrome, also caused by mutation of ATP7A [84]. Although some symptoms overlap with Menkes disease, connective tissue disorders predominate (the disease is also known as X-linked cutis laxa or as Ehlers–Danlos syndrome type IX), and neurological symptoms are milder or even absent.

The genes for both Wilson disease and Menkes disease were cloned in 1993 [93–96]. They encode homologous P-type ATPases that hydrolyze ATP to transport copper ions across membranes. The transporters are denoted ATP7B and ATP7A, respectively. As noted above, ATP7B localizes to the trans-Golgi network in hepatocytes and provides copper for ceruloplasmin synthesis, but in the presence of elevated copper it translocates to a vesicular or canalicular membrane of the liver, to bring about copper excretion. In the same way, ATP7A serves both homeostatic and biosynthetic roles [97], not only providing copper to the trans-Golgi network for synthesis of a number of copper-dependent enzymes in most cells, but also translocating to the plasma membrane to facilitate copper efflux. Therefore, a deficiency in ATP7A is manifest in the enterocyte where dietary copper is sequestered and not released to the circulation, resulting in systemic copper deficiency. Copper deficiency, together with impaired synthesis of copper enzymes requiring ATP7A for copper delivery, such as cytochrome c oxidase, dopamine-β-hydroxylase, lysyl oxidase, peptidylglycine α-amidating monooxygenase, superoxide dismutase, and tyrosinase, account for the pathobiology of Menkes disease [84]. Although ATP7A is expressed in most human tissues, its expression is very low or even undetectable in the liver, where the prominent biosynthetic and homeostatic roles are provided by ATP7B. Thus, mutation of the *ATP7B* gene leads to ceruloplasmin deficiency and loading of the liver with copper.

ATP7A and ATP7B although homologues, appear to have different functions and mechanisms of regulation [98]. Expression of ATP7A in the mouse embryo is ubiquitous, whereas ATP7B is more tissue-specific, found in the liver, intestinal epithelium, bronchial epithelium, and to a lesser extent in the heart and thymus [99]. This suggests that ATP7A serves more of a housekeeping function of copper homeostasis in all tissues, whereas ATP7B has a more specialized function for specific enzyme synthesis (e.g., ceruloplasmin in liver) [98, 99]. This analysis is borne out by kinetic analysis. Both enzymes are P-type ATPases, and their catalytic cycle involves ATP-dependent phosphorylation of an Asp residue, with subsequent hydrolysis of the phosphate bond. Both steps are faster in ATP7A than ATP7B [100]. Differential regulation is also indicated. In cells where the proteins are coexpressed, they are localized to the trans-Golgi network under basal copper conditions, as discussed above, and both translocate with increased copper. In polarized epithelia, ATP7A moves to the basolateral membrane [98]. In contrast, in polarized hepatocytes, ATP7B moves toward the apical membrane and accumulates in subapical vesicles. It has thus been suggested that in cells where ATP7A and ATP7B are coexpressed, their function may be to transport copper across the basolateral and apical membranes, respectively [98]. Both return to the trans-Golgi network when copper levels return to basal. The mechanisms underlying translocation and return are not known.

16.3.3
Other Copper Toxicoses

Several other copper toxicoses are described in the literature, which involve defective copper excretion from the liver. These have been described in reviews by Mercer and coauthors, and Komp and coauthors, among others [82, 84]. Indian childhood cirrhosis results in hepatic copper levels up to 40 times normal. Endemic Tyrolean infantile cirrhosis has been described with a high incidence and probable founder effect in the Tyrol region. It is generally indistinguishable from Indian childhood cirrhosis. They both involve death at an early age from cirrhosis of the liver due to copper overload, without neurological deficit. Each appears to be associated with both excessive copper intake, in addition to impaired excretion. Both disorders have been associated with the practice of boiling milk in brass vessels, and are disappearing as this practice is discouraged. In each case, a mutation in ATP7B has been ruled out, but a genetic component is indicated. The Tyrolean variant has been found to be inherited as an autosomal recessive disorder, and a high rate of consanguinity has been noted in the Indian variety. Homozygosity of a common gene has been proposed. More generally, an idiopathic copper toxicosis has been described that may associate with increased copper in drinking water and appears to result from a genetic predisposition [101]. It is likely closely related, if not identical to, the Indian and Tyrolean variants, and presumably involves a defect in an unrecognized copper transport mechanism.

16.3.4
Amyotrophic Lateral Sclerosis (ALS)

Special mention should be made of the role of copper in amyotrophic lateral sclerosis (ALS). Patients with ALS experience a progressive loss of motor neurons, leading to muscle weakness and death. About 5–15% of cases are familial, and of these, up to 40% are associated with autosomal dominant mutations in the *SOD1* gene encoding Cu,Zn-superoxide dismutase. These cases present an interesting mechanism by which endogenous copper exerts a pathological effect. A gain-of-function mutation leads to an increased production of ROS [102]. Thus, it is suggested that alterations in protein structure, lead in turn to altered copper binding and enhanced Fenton activity at the copper binding site [82, 103, 104]. However, more than 90 mutations in SOD1 have been linked to familial ALS [105], and other mechanisms are also probably involved [106, 107]. These include aberrant folding and protein aggregation, loss of protection of motor neurons from oxidative stress, and formation of aggregates with other proteins (e.g., dynein [105]).

16.3.5
Other Copper-Related Genes

Effects have been documented for genetic variants in a number of other copper transporters and carriers, some associated with a human phenotype and some not. Cytochrome c oxidase is the terminal copper-containing component in the mitochondrial respiratory pathway. Although COX17 is a chaperone delivering copper to the site of synthesis of the enzyme, two other proteins, products of the *SCO1* and *SCO2* genes, are required for assembly of a functional multisubunit complex. Defects in SCO1 cause a rare disorder of early-onset hepatic failure and neurological disease with cytochrome c oxidase deficiency due to impaired copper delivery [108]. SCO2 mutations were found in three patients with cytochrome c oxidase deficiency who presented with a fatal cardioencephalomyopathy in infancy [109]. Cells derived from patients with SCO2 deficiency are rescued from cytochrome c oxidase deficiency by overexpression of the COX17 chaperone, but cells with SCO1 mutations are not [108].

Homologs of the yeast transporters CTR1 and CTR2 have been identified in humans and denoted as COPT1 and COPT2. COPT1 probably encodes a high-affinity copper uptake protein and COPT2 a low-affinity counterpart [110]. Cells overexpressing COPT1 show a dramatic accumulation of copper; those overexpressing COPT2 do not [111]. CUTC, a homolog of mouse and *Escherichia coli* CutC copper transporters, has been cloned from a human fetal brain cDNA library and found to have a widespread expression in human tissues [112]. It is suggested to act as a shuttle protein in copper homeostasis, associated with copper uptake, storage, delivery, and efflux. MURR1/COMMD1 encodes a multifunctional protein that appears to be involved in copper homeostasis and is defective in canine copper toxicosis [113]. No genetic variants of COPT1, COPT2, CUTC, or MURR1/COMMD1 have been identified in humans, to date.

The copper chaperone ATOX1 has been genetically ablated in mice. The mice show symptoms of Menkes disease, with failure to thrive, skin laxity, hypopigmentation, and seizures [114]. Another copper chaperone, CCS, delivers copper to the *SOD1* gene product, Cu,Zn-superoxide dismutase. CCS-null mice show a similar phenotype to SOD1 knockouts, with reduced female fertility and increased sensitivity to paraquat [115]. COX17-null mice develop normally until embryonic day 6.5, but subsequently die between days 8.5 and 10 [116]. No human allelic variants have been described for any of the chaperones ATOX1, CCS, or COX17.

16.4
Summary

Iron and copper are both essential elements that we acquire from the diet. Both redox cycle under physiological conditions, and thus can catalyze the generation of harmful ROS. To avoid this potential toxicity, we have evolved elaborate mechanisms to traffic and sequester these elements, and to chaperone them to sites where they are required. These mechanisms require multiple gene products that can produce a spectrum of overload and deficiency conditions when mutations occur. Body iron status is regulated primarily at the level of intestinal absorption. In contrast, although some control is exerted on copper absorption, the main homeostatic control is at the level of biliary excretion. Paradoxically, some anemias resulting from ineffective absorption or delivery of iron to sites of erythropoiesis, nevertheless are accompanied by systemic iron overload as a consequence of disturbed iron traffic. The homologous copper transporters ATP7A and ATP7B can produce copper deficiency or overload, as a consequence of tissue-specific expression. An important link between iron and copper is provided by the requirement of the ATP7B-dependent copper protein, ceruloplasmin, to act as an essential ferroxidase in the delivery of iron to its transporter, transferrin. The details of the complexity of the handling of both elements that have emerged over the last decade have transformed our understanding of their human biology, and the interdependence they display.

References

1 Duffus, J.H., Nordberg, M., and Templeton, D.M. (2007) Glossary of terms used in toxicology, 2nd edition. *Pure Appl Chem*, **79**, 1153–1341.
2 Harris, W.R. (2002) in *Molecular and Cellular Iron Transport* (ed. D.M. Templeton), Marcel Dekker, New York, pp. 1–40.
3 Halliwell, B. and Gutteridge, J.M.C. (1990) Role of free radicals and catalytic metal ions in human disease: an overview. *Methods Enzymol*, **186**, 1–88.
4 Gunshin, H., Allerson, C.R., Polycarpou-Schwarz, M., Rofts, A., Rogers, J.T., Kishi, F., Hentze, W., Rouault, T.A., Andrews, N.C., and Hediger, M.A. (2001) Iron-dependent regulation of the divalent metal ion transporter. *FEBS Lett*, **509**, 309–316.
5 Gunshin, H. and Hediger, M.A. (2002) in *Molecular and Cellular Iron Transport* (ed. D.M. Templeton), Marcel Dekker, New York, pp. 155–173.
6 Kakhlon, O. and Cabantchik, Z.I. (2002) The labile iron pool: characterization, measurement, and participation in cellular processes. *Free Radic Biol Med*, **33**, 1037–1046.
7 Petrat, F., de Groot, H., Sustmann, R., and Rauen, U. (2002) The chelatable iron pool in living cells: a methodologically defined quantity. *Biol Chem*, **383**, 489–502.
8 Kakhlon, O., Gruenbaum, Y., and Cabantchik, Z.I. (2002) Ferritin expression modulates cell cycle dynamics and cell responsiveness to H-ras-induced growth via expansion of the labile iron pool. *Biochem J*, **363**, 431–436.
9 Zanninelli, G., Loréal, O., Brissot, P., Konijn, A.M., Slotki, I.N., Hider, R.C., and Cabantchik, Z.I. (2002) The labile iron pool of hepatocytes in chronic and acute iron overload and chelator-induced iron deprivation. *J Hepatol*, **36**, 39–46.
10 Templeton, D.M. and Liu, Y. (2003) Genetic regulation of cell function in response to iron overload or chelation. *Biochim Biophys Acta*, **1619**, 113–124.
11 Hentze, M.W., Muckenthaler, M.U., and Andrews, N.C. (2004) Balancing acts: molecular control of mammalian iron metabolism. *Cell*, **117**, 285–297.
12 Anderson, G.J. (2007) Mechanisms of iron loading and toxicity. *Am J Hematol*, **82**, 1128–1131.
13 McKie, A.T., Marciani, P., Rolfs, A., Brennan, K., Wehr, K., Barrow, D., Miret, S., Bomford, A., Peters, T.J., Farzaneh, F., Hediger, M.A., Hentze, M.W., and Simpson, R.J. (2000) A novel duodenal iron-regulated transporter, IREG1, implicated in the basolateral transfer of iron to the circulation. *Mol Cell*, **5**, 299–309.
14 Fleming, R.E., Britton, R.S., Waheed, A., Sly, W.S., and Bacon, B.R. (2002) in *Molecular and Cellular Iron Transport* (ed. D.M. Templeton), Marcel Dekker, New York, pp. 189–205.
15 Fleming, R.E., Migas, M.C., Holden, C.C., Waheed, A., Britton, R.S., Tomatsu, S., Bacon, B.R., and Sly, W.S. (2000) Transferrin receptor 2:

continued expression in mouse liver in the face of iron overload and in hereditary hemochromatosis. *Proc Natl Acad Sci U S A*, **97**, 2214–2219.

16 Ganz, T. and Nemeth, E. (2006) Iron imports. IV. Hepcidin and regulation of body iron metabolism. *Am J Physiol*, **290**, G199–G203.

17 Nemeth, E., Tuttle, M.S., Powelson, J., Vaughn, M.B., Donovan, A., Ward, D.M., Ganz, T., and Kaplan, J. (2004) Hepcidin regulates cellular iron efflux by binding to ferroportin and inducing its internalization. *Science*, **306**, 2090–2093.

18 Olivieri, N.F. (1999) The beta-thalassemias. *N Engl J Med*, **341**, 99–109.

19 Parkes, J.G., Hussain, R.A., Olivieri, N.F., and Templeton, D.M. (1993) Effects of iron loading on uptake, speciation and chelation of iron in cultured myocardial cells. *J Lab Clin Med*, **122**, 36–47.

20 Feder, J.N., Gnirke, A., Thomas, W., Tsuchihashi, Z., Ruddy, D.A., Basava, A., Dormishian, F., Domingo, R. Jr, Ellis, M.C., Fullan, A., Hinton, L.M., Jones, N.L., Kimmel, B.E., Kronmal, G.S., Lauer, P., Lee, V.K., Loeb, D.B., Mapa, F.A., McClelland, E., Meyer, N.C., Mintier, G.A., Moeller, N., Moore, T., and Morikang, E. (1996) A novel MHC class I-like gene is mutated in patients with hereditary haemochromatosis. *Nat Genet*, **13**, 399–408.

21 Feder, J.N., Tsuchihashi, Z., Irrinki, A., Lee, V.K., Mapa, F.A., Morikang, E., Prass, C.E., Starnes, S.M., Wolff, R.K., Parkkila, S., Sly, W.S., and Schatzman, R.C. (1997) The hemochromatosis founder mutation in HLA-H disrupts β_2-microglobulin interaction and cell surface expression. *J Biol Chem*, **272**, 14025–14028.

22 Pissia, M., Polonifi, K., Politou, M., Lilakos, K., Sakellaropoulos, N., and Papanikolaou, G. (2004) Prevalence of the G320V mutation of the HJV gene, associated with juvenile hemochromatosis, in Greece. *Haematologica*, **89**, 742–743.

23 Papanikolaou, G., Samuels, M.E., Ludwig, E.H., MacDonald, M.L.E., Franchini, P.L., Dube, M.-P., Andres, L., MacFarlane, J., Sakellaropoulos, N., Politou, M., Nemeth, E., Thompson, J., Risler, J.K., Zaborowska, C., Babakaiff, R., Radomski, C.C., Pape, T.D., Davidas, O., Christakis, J., Brissot, P., Lockitch, G., Ganz, T., Hayden, M.R., and Goldberg, Y.P. (2004) Mutations in HFE2 cause iron overload in chromosome 1q-linked juvenile hemochromatosis. *Nat Genet*, **36**, 77–82.

24 Babitt, J.L., Huang, F.W., Wrighting, D.M., Xia, Y., Sidis, Y., Samad, T.A., Campagna, J.A., Chung, R.T., Schneyer, A.L., Woolf, C.J., Andrews, N.C., and Lin, H.Y. (2006) Bone morphogenetic protein signaling by hemojuvelin regulates hepcidin expression. *Nat Genet*, **38**, 531–539.

25 Lanzara, C., Roetto, A., Daraio, F., Rivard, S., Ficarella, R., Simard, H., Cox, T.M., Cazzola, M., Piperno, A., Gimenez-Roqueplo, A.P., Grammatico, P., Volinia, S., Gasparini, P., and Camaschella, C. (2004) Spectrum of hemojuvelin gene mutations in 1q-linked juvenile hemochromatosis. *Blood*, **103**, 4317–4321.

26 Roetto, A., Papanikolaou, G., Politou, M., Alberti, F., Girelli, D., Christakis, J., Loukopoulos, D., and Camaschella, C. (2003) Mutant antimicrobial peptide hepcidin is associated with severe juvenile hemochromatosis. *Nat Genet*, **33**, 21–22.

27 Roetto, A., Daraio, F., Porporato, P., Caruso, R., Cox, T.M., Cazzola, M., Gasparini, P., Piperno, A., and Camaschella, C. (2004) Screening hepcidin for mutations in juvenile hemochromatosis: identification of a new mutation (C70R). *Blood*, **103**, 2407–2409.

28 Kawabata, H., Yang, S., Hirama, T., Vuong, P.T., Kawano, S., Gombart, A.F., and Koeffler, H.P. (1999) Molecular cloning of transferrin receptor 2 – a new member of the

transferrin receptor-like family. *J Biol Chem*, **274**, 20826–20832.

29. Kawabata, H., Germain, R.S., Ikezoe, T., Tong, X.J., Green, E.M., Gombart, A.F., and Koeffler, H.P. (2001) Regulation of expression of murine transferrin receptor 2. *Blood*, **98**, 1949–1954.

30. Mattman, A., Huntsman, D., Lockitch, G., Langlois, S., Buskard, N., Ralston, D., Butterfield, Y., Rodrigues, P., Jones, S., Porto, G., Marra, M., De Sousa, M., and Vatcher, G. (2002) Transferrin receptor 2 (TfR2) and HFE mutational analysis in non-C282Y iron overload: identification of a novel TfR2 mutation. *Blood*, **100**, 1075–1077.

31. Abboud, S. and Haile, D.J. (2000) A novel mammalian iron-regulated protein involved in intracellular iron metabolism. *J Biol Chem*, **275**, 19906–19912.

32. Wallace, D.F., Pedersen, P., Dixon, J.L., Stephenson, P., Searle, J.W., Powell, L.W., and Subramaniam, V.N. (2002) Novel mutation in ferroportin1 is associated with autosomal dominant hemochromatosis. *Blood*, **100**, 692–694.

33. Moyo, V.M., Mandishona, E., Hasstedt, S.J., Gangaidzo, I.T., Gomo, Z.A.R., Khumalo, H., Saungweme, T., Kiire, C.F., Paterson, A.C., Bloom, P., MacPhail, A.P., Rouault, T., and Gordeuk, V.R. (1998) Evidence of genetic transmission in African iron overload. *Blood*, **91**, 1076–1082.

34. Gordeuk, V.R., Caleffi, A., Corradini, E., Ferrara, F., Jones, R.A., Castro, O., Onyekwere, O., Kittles, R., Pignatti, E., Montosi, G., Garuti, C., Gangaidzo, I.T., Gomo, Z.A.R., Moyo, V.M., Rouault, T.A., MacPhail, P., and Pietrangelo, A. (2003) Iron overload in Africans and African-Americans and a common mutation in the SCL40A1 (ferroportin 1) gene. *Blood Cells Mol Dis*, **31**, 299–304.

35. Delatycki, M.B., Camakaris, J., Brooks, H., Evans-Whipp, T., Thorburn, D.R., Williamson, R., and Forrest, S.M. (1999) Direct evidence that mitochondrial iron accumulation occurs in Friedreich ataxia. *Ann Neurol*, **45**, 673–675.

36. Palau, F. (2001) Friedreich's ataxia and frataxin: molecular genetics, evolution and pathogenesis. *Int J Mol Med*, **7**, 581–589.

37. Montermini, L., Rodius, F., Pianese, L., Molto, M.D., Cossee, M., Campuzano, V., Cavalcanti, F., Monticelli, A., Palau, F., Gyapay, G., Wenhert, M., Zara, F., Patel, P.I., Cocozza, S., Koenig, M., and Pandolfo, M. (1995) The Friedreich ataxia critical region spans a 150-kb interval on chromosome 9q13. *Am J Hum Genet*, **57**, 1061–1067.

38. Campuzano, V., Montermini, L., Molto, M.D., Pianese, L., Cossee, M., Cavalcanti, F., Monros, E., Rodius, F., Duclos, F., Monticelli, A., Zara, F., Canizares, J., Koutnikova, H., Bidichandani, S.I., Gellera, C., Brice, A., Trouillas, P., De Michele, G., Filla, A., De Frutos, R., Palau, F., Patel, P.I., Di Donato, S., Mandel, J.L., Cocozza, S., Koenig, M., and Pandolfo, M. (1996) Friedreich's ataxia: autosomal recessive disease caused by an intronic GAA triplet repeat expansion. *Science*, **271**, 1423–1427.

39. Delatycki, M.B., Williamson, R., and Forrest, S.M. (2000) Friedreich ataxia: an overview. *J Med Genet*, **37**, 1–8.

40. Calabrese, V., Lodi, R., Tonon, C., D'Agata, V., Sapienza, M., Scapagnini, G., Mangiameli, A., Pennisi, G., Stella, A.M., and Butterfield, D.A. (2005) Oxidative stress, mitochondrial dysfunction and cellular stress response in Friedreich's ataxia. *J Neurol Sci*, **233**, 145–162.

41. Forrest, S.M., Knight, M., Delatycki, M.B., Paris, D., Williamson, R., King, J., Yeung, L., Nassif, N., and Nicholson, G.A. (1998) The correlation of clinical phenotype in Friedreich ataxia with the site of point mutations in the FRDA gene. *Neurogenetics*, **1**, 253–257.

42 Durr, A., Cossee, M., Agid, Y., Campuzano, V., Mignard, C., Penet, C., Mandel, J.L., Brice, A., and Koenig, M. (1996) Clinical and genetic abnormalities in patients with Friedreich's ataxia. *N Engl J Med*, **335**, 1169–1175.

43 Bidichandani, S.I., Ashizawa, T., and Patel, P.I. (1998) The GAA triplet-repeat expansion in Friedreich ataxia interferes with transcription and may be associated with an unusual DNA structure. *Am J Hum Genet*, **62**, 111–121.

44 Gacy, A.M., Goellner, G.M., Spiro, C., Chen, X., Gupta, G., Bradbury, E.M., Dyer, R.B., Mikesell, M.J., Yao, J.Z., Johnson, A.J., Richter, A., Melancon, S.B., and McMurray, C.T. (1998) GAA instability in Friedreich's Ataxia shares a common, DNA-directed and intraallelic mechanism with other trinucleotide diseases. *Mol Cell*, **1**, 583–593.

45 Cavadini, P., Gellera, C., Patel, P.I., and Isaya, G. (2000) Human frataxin maintains mitochondrial iron homeostasis in Saccharomyces cerevisiae. *Hum Mol Genet*, **9**, 2523–2530.

46 Puccio, H. and Koenig, M. (2000) Recent advances in the molecular pathogenesis of Friedreich ataxia. *Hum Mol Genet*, **9**, 887–892.

47 Bulteau, A.L., O'Neill, H.A., Kennedy, M.C., Ikeda-Saito, M., Isaya, G., and Szweda, L.I. (2004) Frataxin acts as an iron chaperone protein to modulate mitochondrial aconitase activity. *Science*, **305**, 242–245.

48 Aoki, Y., Muranaka, S., Nakabayashi, K., and Ueda, Y. (1979) delta-Aminolevulinic acid synthetase in erythroblasts of patients with pyridoxine-responsive anemia: hypercatabolism caused by the increased susceptibility to the controlling protease. *J Clin Invest*, **64**, 1196–1203.

49 Harigae, H., Furuyama, K., Kimura, A., Neriishi, K., Tahara, N., Kondo, M., Hayashi, N., Yamamoto, M., Sassa, S., and Sasaki, T. (1999) A novel mutation of the erythroid-specific delta-aminolaevulinate synthase gene in a patient with X-linked sideroblastic anaemia. *Br J Haematol*, **106**, 175–177.

50 Ponka, P. (1997) Tissue-specific regulation of iron metabolism and heme synthesis: distinct control mechanisms in erythroid cells. *Blood*, **89**, 1–25.

51 Allikmets, R., Raskind, W.H., Hutchinson, A., Schueck, N.D., Dean, M., and Koeller, D.M. (1999) Mutation of a putative mitochondrial iron transporter gene (ABC7) in X-linked sideroblastic anemia and ataxia (XLSA/A). *Hum Mol Genet*, **8**, 743–749.

52 Bekri, S., Kispal, G., Lange, H., Fitzsimons, E., Tolmie, J., Lill, R., and Bishop, D.F. (2000) Human ABC7 transporter: gene structure and mutation causing X-linked sideroblastic anemia with ataxia with disruption of cytosolic iron-sulfur protein maturation. *Blood*, **96**, 3256–3264.

53 Shimada, Y., Okuno, S., Kawai, A., Shinomiya, H., Saito, A., Suzuki, M., Omori, Y., Nishino, N., Kanemoto, N., Fujiwara, T., Horie, M., and Takahashi, E. (1998) Cloning and chromosomal mapping of a novel ABC transporter gene (hABC7), a candidate for X-linked sideroblastic anemia with spinocerebellar ataxia. *J Hum Genet*, **43**, 115–122.

54 Rouault, T.A. and Tong, W.H. (2008) Iron-sulfur cluster biogenesis and human disease. *Trends Genet*, **24**, 398–407.

55 Hayashi, A., Wada, Y., Suzuki, T., and Shimizu, A. (1993) Studies on familial hypotransferrinemia: unique clinical course and molecular pathology. *Am J Hum Genet*, **53**, 201–213.

56 Craven, C.M., Alexander, J., Eldridge, M., Kushener, J.P., Berstein, S., and Kaplan, J. (1987) Tissue distribution and clearance kinetics of non-transferrin-bound iron in the hypotransferrinemic mouse: a

57 Parkes, J.G. and Templeton, D.M. (2002) in *Molecular and Cellular Iron Transport* (ed. D.M. Templeton), Marcel Dekker, New York, pp. 451–466.

58 Eisenstein, R.S. (2000) Iron regulatory proteins and the molecular control of mammalian iron metabolism. *Annu Rev Nutr*, 20, 627–662.

59 Theil, E.C. (2000) Targeting mRNA to regulate iron and oxygen metabolism. *Biochem Pharmacol*, 59, 87–93.

60 Johansson, H.E. and Theil, E.C. (2002) in *Molecular and Cellular Iron Transport* (ed. D.M. Templeton), Marcel Dekker, New York, pp. 237–253.

61 Beaumont, C., Leneuve, P., Devaux, I., Scoazec, J.Y., Berthier, M., Loiseau, M.N., Grandchamp, B., and Bonneau, D. (1995) Mutation in the iron responsive element of the L ferritin mRNA in a family with dominant hyperferritinaemia and cataract. *Nat Genet*, 11, 444–446.

62 Girelli, D., Olivieri, O., Gasparini, P., and Corrocher, R. (1996) Molecular basis for the hereditary hyperferritinemia-cataract syndrome. *Blood*, 87, 4912–4913.

63 Roetto, A., Bosio, S., Gramaglia, E., Barilaro, M.R., Zecchina, G., and Camaschella, C. (2002) Pathogenesis of hyperferritinemia cataract syndrome. *Blood Cells Mol Dis*, 29, 532–535.

64 Mumford, A.D., Cree, I.A., Arnold, J.D., Hagan, M.C., Rixon, K.C., and Harding, J.J. (2000) The lens in hereditary hyperferritinaemia cataract syndrome contains crystalline deposits of L-ferritin. *Br J Ophthalmol*, 84, 697–700.

65 Andrews, N.C. (2002) in *Molecular and Cellular Iron Transport* (ed. D.M. Templeton), Marcel Dekker, New York, pp. 679–697.

66 Iolascon, A., d'Apolito, M., Servedio, V., Cimmino, F., Piga, A., and Camaschella, C. (2006) Microcytic anemia and hepatic iron overload in a child with compound heterozygous mutations in DMT1 (SLC11A2). *Blood*, 107, 349–354.

67 Beaumont, C., Delaunay, J., Hetet, G., Grandchamp, B., de Montalembert, M., and Tchernia, G. (2006) Two new human DMT1 gene mutations in a patient with microcytic anemia, low ferritinemia, and liver iron overload. *Blood*, 107, 4168–4170.

68 Holmberg, C.G. and Laurell, C.B. (1948) Investigations in serum coper. II. Isolation of the copper containing protein, and a description of some of its properties. *Acta Chem Scand*, 2, 550–556.

69 Zaitseva, I., Zaitsev, V., Card, G., Moshkov, K., Bax, B., Ralph, A., and Lindley, P. (1996) The X-ray structure of human serum ceruloplasmin at 3.1 Å: nature of the copper centers. *J Biol Inorg Chem*, 1, 15–23.

70 Ardon, O., Kaplan, J., and Martin, B.D. (2002) in *Cellular and Molecular Iron Transport* (ed. D.M. Templeton), Marcel Dekker, New York, pp. 375–393.

71 Vulpe, C.D., Kuo, Y.M., Murphy, T.L., Cowley, L., Askwith, C., Libina, N., Gitschier, J., and Anderson, G.J. (1999) Hephaestin, a ceruloplasmin homologue implicated in intestinal iron transport, is defective in the sla mouse. *Nat Genet*, 21, 195–199.

72 Anderson, G.J., Frazer, D.M., McKie, A.T., and Vulpe, C.D. (2002) The ceruloplasmin homolog hephaestin and the control of intestinal iron absorption. *Blood Cells Mol Dis*, 29, 367–375.

73 Harris, E.D. (1999) Ceruloplasmin and iron: vindication after 30 years. *Nutrition*, 15, 72–74.

74 Yoshida, K., Furihata, K., Takeda, S., Nakamura, A., Yamamoto, K., Morita, H., Hiyamuta, S., Ikeda, S., Shimizu, N., and Yanagisawa, N. (1995) A mutation in the ceruloplasmin gene is associated with systemic hemosiderosis in humans. *Nat Genet*, 9, 267–272.

75 Harris, Z.L., Klomp, L.J., and Gitlin, J.D. (1998) Aceruloplasminemia: an

inherited neurodegenerative disease with impairment of iron homeostasis. *Am J Clin Nutr*, **67** (Suppl), 972S–977S.

76 Hellman, N.E. and Harris, Z.L. (2002) in *Cellular and Molecular Iron Transport* (ed. D.M. Templeton), Marcel Dekker, New York, pp. 749–760.

77 Morita, H., Ikeda, S., Yamamoto, K., Morita, S., Yoshida, K., Nomoto, S., Kato, M., and Yanagisawa, N. (1995) Hereditary ceruloplasmin deficiency with hemosiderosis: a clinicopathological study of a Japanese family. *Ann Neurol*, **37**, 646–656.

78 Hellman, N.E., Schaefer, M., Gehrke, S., Stegen, P., Hoffman, W.J., Gitlin, J.D., and Stremmel, W. (2000) Hepatic iron overload in aceruloplasminaemia. *Gut*, **47**, 858–860.

79 Ambundo, E.A., Deydier, M.-V., Grall, A.J., Aguera-Vega, N., Dressel, L.T., Cooper, T.H., Heeg, M.J., Ochrymowycz, L.A., and Rorabacher, D.B. (1999) Influence of coordination geometry upon copper(II/I) redox potentials. Physical parameters for twelve copper tripodal ligand complexes. *Inorg Chem*, **38**, 4233–4242.

80 Pufahl, R.A., Singer, C.P., Peariso, K.L., Lin, S.J., Schmidt, P.J., Fahrni, C.J., Culotta, V.C., Penner-Hahn, J.E., and O'Halloran, T.V. (1997) Metal ion chaperone function of the soluble Cu(I) receptor Atx1. *Science*, **278**, 853–856.

81 Rae, T.D., Schmidt, P.J., Pufahl, R.A., Culotta, V.C., and O'Halloran, T.V. (1999) Undetectable intracellular free copper: the requirement of a copper chaperone for superoxide dismutase. *Science*, **284**, 805–808.

82 Llanos, R.M. and Mercer, J.F.B. (2002) The molecular basis of copper homeostasis and copper-related disorders. *DNA Cell Biol*, **21**, 259–270.

83 Bertinato, J. and L'Abbé, M.R. (2004) Maintaining copper homeostasis: regulation of copper trafficking proteins in response to copper deficiency or overload. *J Nutr Biochem*, **15**, 316–322.

84 de Bie, P., Muller, P., Wijmenga, C., and Komp, L.W.J. (2007) Molecular pathogenesis of Wilson and Menkes disease: correlation of mutations with molecular defects and disease phenotypes. *J Med Genet*, **44**, 673–688.

85 Templeton, D.M. and Cherian, M.G. (1991) Toxicological significance of metallothionein. *Methods Enzymol*, **205**, 11–24.

86 Tapia, L., Gonzalez-Aguero, M., Cisternas, M., Suazo, M.F., Cambiazo, V., Uauy, R., and Gonzalez, M. (2004) Metallothionein is crucial for safe intracellular copper storage and cell survival at normal and supra-physiological exposure levels. *Biochem J*, **378**, 617–624.

87 Wirth, P.L. and Linder, M.C. (1985) Distribution of copper among components of human serum. *J Natl Cancer Inst*, **75**, 277–283.

88 Templeton, D.M. (2005) in *Handbook of Elemental Speciation II. Species in the Environment, Food, Medicine and Occupational Health* (eds R. Cornelis, H. Crews, J. Caruso, and K.G. Heumann), John Wiley & Sons, Ltd, Chichester, pp. 638–649.

89 Vargas, E.J., Shoho, A.R., and Linder, M.C. (1994) Copper transport in the Nagase analbuminemic rat. *Am J Physiol*, **267**, G259–G269.

90 May, P. (1995) in *Handbook of Metal-Ligand Interactions in Biological Fluids: Bioinorganic Chemistry* (ed. G. Berthon), Marcel Dekker, New York, pp. 1184–1194.

91 Turnlund, J.R., Keyes, W.R., Peiffer, G.L., and Scott, K.C. (1998) Copper absorption, excretion, and retention by young men consuming low dietary copper determined by using the stable isotope 65Cu. *Am J Clin Nutr*, **67**, 1219–1225.

92 Schaefer, M., Hopkins, R.G., Failla, M.L., and Gitlin, J.D. (1999) Hepatocyte-specific localization and copper-dependent trafficking of the Wilson's disease protein in the liver. *Am J Physiol*, **276**, G639–G646.

93 Chelly, J., Tümer, Z., Tonnesen, T., Petterson, A., Ishikawa-Brush,

Y., Tommerup, N., Horn, N., and Monaco, A.P. (1993) Isolation of a candidate gene for Menkes disease that encodes a potential heavy metal binding protein. *Nat Genet*, **3**, 14–19.

94 Vulpe, C., Levinson, B., Whitney, S., Packman, S., and Gitschier, J. (1993) Isolation of a candidate gene for Menkes disease and evidence that it encodes a copper-transporting ATPase. *Nat Genet*, **3**, 7–13.

95 Mercer, J.F.B., Livingston, J., Hall, B.K., Paynter, J.A., Begy, C., Chandrasekharappa, S., Lockhart, P., Grimes, A., Bhave, M., Siemenack, D., and Glover, T.W. (1993) Isolation of a partial candidate gene for Menkes disease by positional cloning. *Nat Genet*, **3**, 20–25.

96 Bull, P.C., Thomas, G.R., Rommens, J.M., Forbes, J.R., and Cox, D.W. (1993) The Wilson disease gene is a putative copper transporting P-type ATPase similar to the Menkes gene. *Nat Genet*, **5**, 327–337.

97 Bertini, I. and Rosato, A. (2008) Menkes disease. *Cell Mol Life Sci*, **65**, 89–91.

98 Lintz, R. and Lutsenko, S. (2007) Copper-transporting ATPases ATP7A and ATP7B: cousins, not twins. *J Bioenerg Biomembr*, **39**, 403–407.

99 Kuo, Y.M., Gitschier, J., and Packman, S. (1997) Developmental expression of the mouse mottled and toxic milk genes suggests distinct functions for the Menkes and Wilson disease copper transporters. *Hum Mol Genet*, **6**, 1043–1049.

100 Barnes, N., Tsivkovskii, R., Tsivkovskaia, N., and Lutsenko, S. (2005) The Copper-transporting ATPases, Menkes and Wilson disease proteins, have distinct roles in adult and developing cerebellum. *J Biol Chem*, **280**, 9640–9645.

101 Scheinberg, I.H. and Sternlieb, I. (1996) Wilson disease and idiopathic copper toxicosis. *Am J Clin Nutr*, **63**, 842S–845S.

102 Yim, M.B., Kang, J.H., Yim, H.S., Kwak, H.S., Chock, P.B., and Stadtman, E.R. (1996) A gain-of-function of an amyotrophic lateral sclerosis-associated Cu,Zn-superoxide dismutase mutant: an enhancement of free radical formation due to a decrease in Km for hydrogen peroxide. *Proc Natl Acad Sci U S A*, **93**, 5709–5714.

103 Gurney, M.E., Liu, R., Althaus, J.S., Hall, E.D., and Becker, D.A. (1998) Mutant Cu,Zn superoxide dismutase in motor neuron disease. *Age*, **21**, 85–89.

104 Lyons, T.J., Nersissian, A., Huang, H., Yeom, H., Nishida, C.R., Graden, J.A., Gralla, E.B., and Valentine, J.S. (2000) The metal binding properties of the zinc site of yeast copper-zinc superoxide dismutase: implications for amyotrophic lateral sclerosis. *J Biol Inorg Chem*, **5**, 189–203.

105 Zhang, F., Ström, A.-L., Fukada, K., Lee, S., Hayward, L.J., and Zhu, H. (2007) Interaction between familial amyotrophic lateral sclerosis (ALS)-linked SOD1 mutants and the dynein complex. *J Biol Chem*, **282**, 16691–16699.

106 Bruijn, L.I., Houseweart, M.K., Kato, S., Anderson, K.L., Anderson, S.D., Ohama, E., Reaume, A.G., Scott, R.W., and Cleveland, D.W. (1998) Aggregation and motor neuron toxicity of an ALS-linked SOD1 mutant independent from wild-type SOD1. *Science*, **281**, 1851–1854.

107 Valentine, J.S. and Hart, P.J. (2003) Misfolded CuZnSOD and amyotrophic lateral sclerosi. *Proc Natl Acad Sci U S A*, **100**, 3617–3622.

108 Leary, S.C., Kaufman, B.A., Pellecchia, G., Guercin, G.-H., Mattman, A., Jaksch, M., and Shoubridge, E.A. (2004) Human SCO1 and SCO2 have independent, cooperative functions in copper delivery to cytochrome c oxidase. *Hum Mol Genet*, **13**, 1839–1848.

109 Papadopoulou, L.C., Sue, C.M., Davidson, M.M., Tanji, K., Nishino, I., Sadlock, J.E., Krishna, S., Walker, W., Selby, J., Glerum, D.M., Van Coster, R., Lyon, G., Scalais, E., Lebel, R., Kaplan, P., Shanske, S., De Vivo, D.C., Bonilla, E., Hirano, M.,

DiMauro, S., and Schon, E.A. (1999) Fatal infantile cardioencephalomyopathy with COX deficiency and mutations in SCO2, a COX assembly gene. *Nat Genet*, **23**, 333–337.

110 Zhou, B. and Gitschier, J. (1997) hCTR1: a human gene for copper uptake identified by complementation in yeast. *Proc Natl Acad Sci U S A*, **97**, 7481–7486.

111 Moller, L.B., Petersen, C., Lund, C., and Horn, N. (2000) Characterization of the hCTR1 gene: genomic organization, functional expression, and identification of a highly homologous processed gene. *Gene*, **257**, 13–22.

112 Li, J., Ji, C., Chen, J., Yang, Z., Wang, Y., Fei, X., Zheng, M., Gu, X., Wen, G., Xie, Y., and Mao, Y. (2005) Identification and characterization of a novel Cut family cDNA that encodes human copper transporter protein CutC. *Biochem Biophys Res Commun*, **337**, 179–183.

113 de Bie, P., van de Sluis, B., Klomp, L., and Wijmenga, C. (2005) The many faces of the copper metabolism protein MURR1/COMMD1. *J Hered*, **96**, 803–811.

114 Hamza, I., Faisst, A., Prohaska, J., Chen, J., Gruss, P., and Gitlin, J.D. (2001) The metallochaperone Atox1 plays a critical role in perinatal copper homeostasis. *Proc Natl Acad Sci U S A*, **98**, 6848–6852.

115 Wong, P.C., Waggoner, D., Subramaniam, J.R., Tessarollo, L., Bartnikas, T.B., Culotta, V.C., Price, D.L., Rothstein, J., and Gitlin, J.D. (2000) Copper chaperone for superoxide dismutase is essential to activate mammalian Cu/Zn superoxide dismutase. *Proc Natl Acad Sci U S A*, **97**, 2886–2891.

116 Takahashi, Y., Kako, K., Kashiwabara, S., Takehara, A., Inada, Y., Arai, H., Nakada, K., Kodama, H., Hayashi, J., Baba, T., and Munekata, E. (2002) Mammalian copper chaperone Cox17p has an essential role in activation of cytochrome c oxidase and embryonic development. *Mol Cell Biol*, **22**, 7614–7621.

17
Polyglutamine Neuropathies: Animal Models to Molecular Mechanisms
Kelvin Hui and Jeffrey Henderson

Polyglutamine neuropathies represent a devastating group of neurodegenerative disorders that classically include Huntington's disease, dentatorubral pallidoluysian atrophy (DRPLA), Kennedy's disease, and the spinocerebellar ataxias 1, 2, 3, 6, 7, and 17. Clinical manifestations of these disorders and progress that has been made over the past decade toward the creation of murine, *Drosophila*, and *Caenorhabditis elegans* transgenic models are discussed. The strengths and weaknesses of each model to address specific features of polyglutamine neuropathy are examined with respect to specific forms of cellular dysfunction. Increasingly, such studies suggest that the initiating mechanism of polyglutamine-mediated injury lies upstream of those events involved in inclusion body formation, organelle failure, and oxidative stress. In particular, alterations in neural transcription and RNA processing, similar to that previously described for other trinucleotide repeat disorders such as fragile X tremor/ataxia syndrome, dystrophia myotonica 1 and 2, and inherited motor neuropathies such as amyotrophic lateral sclerosis 4 and 6 likely represent critical formative events in disease pathology. The importance of new technologies such as induced pluriporent stem (iPS) cells in enabling researchers to critically examine these early events of CAG triplet expansion pathologies directly in human neurons, and their implications toward future therapies is also discussed.

17.1
Neurobiology of Polyglutamine Diseases

Polyglutamine (polyQ) diseases represent a major subset of pathologic trinucleotide (CAG) expansions within mammalian DNA. These expansions most frequently result in the translation of an altered protein product that ultimately forms insoluble inclusions within the cell nucleus or cytoplasm. Though the mammalian genes affected by CAG triplet expansion represent a structurally diverse group, the pathologic consequences arising from these mutations are largely limited to syndromes affecting function of the central nervous system (CNS). At present, nine classic CAG repeat expansion disorders have been recognized. These include Huntington's disease, dentatorubral pallidoluysian atrophy (DRPLA), spinal bulbar muscular atrophy (SBMA/Kennedy's disease), and spinocerebellar ataxias (SCAs) 1, 2, 3, 6, 7, and 17. With the exception of the X-linked recessive SBMA (Kennedy's disease), all polyQ diseases are autosomal dominant disorders. Clinically, these neurodegenerative disorders most frequently become manifest during the third or fourth decade of life. Although each exhibits considerable case-by-case heterogeneity, in general there is an inverse relationship between the number of CAG repeats and the overall age of onset in both current and successive generations of affected families. Despite some evidence for transcriptional interference and aberrant RNA processing, the preponderance of mammalian data supports a toxic gain-of-function effect resulting from the synthesis of aberrant polyQ protein, resulting in neuronal dysfunction, and eventual cell death within affected neuronal populations. In general, current hypotheses of the mechanism by which CAG tract expansion induces pathologic effects have focused on the potential role by which such changes may alter protein–protein interactions. Specifically, each of the following have been suggested in recent years as potential explanations of the observed neuronal pathology: (i) alteration in normal protein trafficking through formation of amyloid-type (beta-sheet rich) polyQ intracellular inclusions; (ii) altered intracellular cell-signaling through aberrant interaction with expanded polyQ proteins; (iii) sequestration of histone acetyltransferases by polyQ-containing proteins; (iv) induction of oxidative stress by intracellular and extracellular polyQ aggregates; (v) transcriptional interference by CUG-rich mRNA; (vi) alterations in mRNA processing as a result of pathologic CUG repeat interference; and (vii) impairment of ubiquitin-mediated protein degradation (Section 17.3). With respect to ubiquitin–proteasome degradation, the proteasome activator PA28γ has been shown to be highly expressed in neurons. This could exacerbate polyQ-induced pathology due to its ability to suppress chymotrypsin-like proteasome activity. However, no change in disease progression has been observed in PA28γ null mice [1], and several studies of proteasome activity have demonstrated that this activity is unimpaired in polyQ-expressing cells in several neurologic models. Thus, the relative role that ubiquitination plays in the neuropathology of polyQ diseases is presently unclear.

As indicated above, the polyQ diseases exhibit several genetic and mechanistic commonalities. However, due to differences in the nature of the gene affected, they also frequently exhibit significant differences from one another with respect to neuropathology. These characteristics are summarized briefly below for each of the known polyQ disorders.

17.1.1
Huntington's Disease

First identified in 1872, Huntington's disease is perhaps most closely identified with the abnormal involuntary motor movements (dyskinesis – chorea) induced as a consequence of its neurodegenerative effects upon motor circuits. In addition, Huntington's disease is associated with progressive dementia and emotional disorders [2–4]. The disease constitutes a classic example of an autosomal dominant hyperkinetic neurologic disorder, affecting approximately 1 per 10 000 individuals in the North American population. Both sexes appear equally affected by Huntington's disease. The principal postmortem finding has been a reduction in the mass of the caudate and putamen, together with the degeneration of specific basal ganglia circuits (e.g., dorsal lateral prefrontal and lateral orbitofrontal). Successive magnetic resonance imaging (MRI) scans of patients examining the relative volume of the lateral ventricle have been used as a stand-in measure to estimate the degree of caudate/putamen atrophy [5–8]. The clinical manifestations observed are thought to result primarily from an imbalance in three systems: (i) nigrostriatal dopamine input, (ii) intrastriatal cholinergic input, and (iii) GABAergic regulation (striatum to globus pallidus and substantia nigra). The destruction of neurons within groups (ii) and (iii), particularly medium spiny cholinergic and GABAergic projection neurons of the caudate/putamen, results in relative overactivity of the D1 dopaminergic pathway, with resulting hyperkinetic effects through activation of the substantia nigra reticulate [9, 10]. Postmortem cases frequently show plaques and abnormal cells within both the neocortex and caudate nuclei of patients, which stain positive for ubiquitin inclusions [11, 12]. However, it should be noted that experimental studies in several Huntington's disease models in mice have demonstrated that little correlation exists between the formation of intracellular inclusions and the induction of cell death [13–15].

The native huntingtin (HTT) protein is a large (385 kDa) protein on chromosome 4 (4p16.3), which is highly expressed in neurons and testes in humans. It shows little sequence homology to other known proteins and is known to upregulate the expression of brain-derived neurotrophic factor (BDNF). Like many polyQ disease proteins, the polyQ tract is present at the N-terminus, beginning at amino acid 18. Unaffected individuals have between 9 and 35 glutamine residues while affected individuals (mHTT) show 36 or more glutamine residues at this site. The Htt protein is a known target of the cysteine proteases caspase-3 and caspase-6, which critically regulate the process of programmed cell death (PCD) [16, 17]. Work in animal models suggests that cleavage of mHtt by caspase-6 in particular may promote the creation of a toxic protein isoform that ultimately localizes to the cell nucleus, promoting cell death [16].

17.1.2
Kennedy's Disease (SBMA)

SBMA is an adult-onset neuropathy in which the affected gene is the androgen receptor on the X-chromosome at Xq11-q12. Thus, features of this disease in males are similar to androgen insensitivity syndromes, with ensuing gynecomastia and testicular atrophy. While normal individuals possess 11–36 CAG repeats, affected individuals may have 38–62 repeats. Neurons most prominently affected in Kennedy's disease are anterior horn neurons (alpha motor neurons) of the spinal cord, as well as neurons of the cerebellum, pons, and medulla [18, 19]. As a result of progressive motor denervation, SBMA patients develop muscle cramping, difficulty in speaking and swallowing, and progressive weakness particularly in proximal (close to body trunk) limb muscles. Like Huntington's disease, the expanded polyQ tract is located close to the N-terminus of the protein, and the altered protein frequently forms protein aggregates within the cell nuclei [20].

17.1.3
DRPLA/Smith's Disease

DRPLA exhibits autosomal dominant (12p13.31) features similar to Huntington's disease; however, it exhibits a wider range of onset (4–62 years), though patients are typically identified in their twenties. SCA, dementia, and choreoathetosis are cardinal features regardless of the age of onset, while younger patients frequently exhibit myoclonic epilepsy. While normal individuals exhibit 7–25 CAG repeats, affected individuals may exhibit 49–88 CAG repeats. The affected gene has been termed *ATN1/CTG-B37*, and is predominantly expressed within an array of neuronal populations [21]. In affected individuals, atrophin-1 accumulates both as intranuclear inclusions and as diffuse complexes [22, 23]. There is evidence to suggest that the polyQ tract (which resides toward the middle of the protein) may enhance binding to a nuclear arginine–glutamic acid-rich protein designated *RERE* (atrophin-2), and that these proteins may from a molecular complex [24]. Analysis of DRPLA nuclear inclusions have demonstrated the presence of several transcription factors, including CREB and TATA binding protein (TBP), and Sp1, suggesting that it may act as a transcriptional corepressor [25, 26].

17.1.4
Spinocerebellar Ataxia 1 (SCA1)

SCA1 represents an autosomal dominant, adult-onset ataxia located on human chromosome 6p24-p23, which normally clinically manifest itself between ages 30 and 40. Functional and MRI studies have noted pathologic changes to the cerebellum, pons, and olive nuclei. In addition, degeneration of cranial nerve nuclei (IX, X, and XII) is another distinguishing feature, together with dorsal columns and spinocerebellar tract atrophy [27]. Also, abnormal extensor plantar responses (Babinski sign) and choreiform movements may also occur in this disorder. Though Purkinje pathology is the most frequent observation, ataxin-1 is a ubiquitous protein present in both neuronal and non-neuronal cells, which localizes primarily to the cytoplasm. Analyses of ataxin-1 have demonstrated that it can bind to series of proteins regulating transcription. These include the transcriptional repressor Capicua (Cic), the transcriptional corepressor SMRT, and histone deacetylase-3 (HDAC3) [28, 29]. Such findings suggest that pathologic conversion of ataxin-1 may result in altered gene transcription, leading to neuronal pathology. Alternatively, the findings that ataxin-1 can interact with proteins such as A1Up and HSP70 [30] and that CAG expanded ataxin-1 exhibits a threefold greater resistance against proteolytic degradation have led to the suggestion that a primary role of ataxin-1 is to regulate the ubiquitin-proteasome pathways. Finally, it has been shown that ataxin-1 binds to RNA, and that the strength of this binding diminishes as a function of CAG expansion. This has led to the suggestion that the pathologic consequences of ataxin-1 may be related to a role in the regulation of RNA metabolism. With respect to animal models, the insertion of human CAG expanded SCA1 alleles or CAG expansion of the endogenous murine *Atxn1* gene results in a pattern of progressive Purkinje cell degeneration with ataxia in mice, suggesting a potential model for addressing future potential therapies.

17.1.5
SCA2

SCA2 represents an autosomal dominant, adult-onset ataxia on 12q24.1, in which severe olivopontocerebellar atrophy is typically observed (epically this is more extensive than that seen in SCA1 or SCA3). Sensory disturbances are also commonly seen in SCA2 patients. Oculomotor abnormalities, although frequent, are not seen in all SCA2. In particular, upward gaze limitation (as seen in SCA3) is not typically observed. The CAG repeat expansion has been localized to the 5' end of the coding region of the *ATXN2* gene. Genetic analysis suggests that repeat expansions occur most frequently during paternal transmission. Wild-type ataxin-2 exhibits a rather ubiquitous distribution within the body, and contains a domain of 22 consecutive glutamine residues. Like other CAG diseases, triplet expansion beyond a given limit (32 repeats) has been shown to result in neurodegenerative disease in humans. Similarly, mice expressing mutant ataxin-2 with 58 glutamines show functional deficits initiated by loss of Purkinje dendritic arbors followed by cellular loss [31]. Examination of the distribution of ataxin-2 in cells demonstrates that both normal and pathogenic forms of this protein are localized to the rough endoplasmic reticulum (ER)/Golgi apparatus, and that ataxin-2 directly interacts with polyA-binding protein in polyribosomes under normal conditions [32]. These and other data suggest that ataxin-2 is involved in the processing of mRNA and/or the regulation of translation [33]; despite the fact that animals lacking ataxin-2 demonstrate that this protein is not required for normal development or adult survival.

17.1.6
SCA3/Machado-Joseph Disease

Originally described as a form of autosomal dominant inherited ataxia present in descendants of William Machado, a native of the Azores, this disorder manifests itself principally by ataxia, spasticity, and ocular movement abnormalities after age 40. Patients present with widespread fasciculations of the muscles, loss of reflexes in the lower limbs, nystagmus, mild cerebellar tremors and extensor plantar response dystonia or rigidity, and ophthalmoplegia. The gene for SCA3 (*ATXN3*) has been localized to 14q32. Normal individuals show 13–40 CAG repeats, whereas most diagnosed SCA3 patients frequently show repeat number of 68–79. Though the gene for SCA3 is widely expressed within the body, the most common postmortem findings are olivopontocerebellar atrophy at autopsy. In addition, the substantia nigra, pontine nuclei, together with neurons of the vestibular nuclei, cranial nerves, and anterior horn cells have also been observed with variable frequencies in SCA3 patients. Findings such as segregation of protein components of the 26S proteasome complex into mutant ataxin-3 complexes have implicated dysregulation of the ubiquitin–proteasome pathway in the pathogenesis of SCA3. With respect to murine models, the transgenic insertion of mutant forms of ataxin-3 have been shown to result in Purkinje cell degeneration and cerebellar ataxia [34].

17.1.7
SCA6

Expansion of CAG repeats in the voltage-dependent (P/Q type) calcium channel alpha 1A subunit (CACNL1A4) on chromosome 19p13 results in an autosomal dominant, slowly progressive, pure cerebellar ataxia with an onset later than that of the other ataxia described (mean onset 52 years) [35]. SCA6 results from CAG repeat at the 3' end of the coding gene, while point mutations in CACNL1A4 have been shown to be responsible for two additional allelic disorders (episodic ataxia type 2 and familial hemiplegic migraine) [36]. With respect to animal models, the naturally occurring murine mutants *tottering* and *leaner* show defects in the gene of interest [37, 38]. However, the majority of known *tottering* and *leaner* allelic variants is autosomal recessive, and thus dissimilar to SCA6. However, recently, dominant allelic variants have been described [39]. Though controversy remains whether the neuropathology seen in SCA6 is the result of a change in calcium channel function or gain-of-function activities associated with accumulation of polyQ protein, recent murine gene targeting suggests the latter [40].

17.1.8
SCA7

SCA7 is a progressive autosomal dominant disorder localized to 3p13-p12, in which the primary clinical manifestations are cerebellar ataxia and pigmented macular degeneration [41]. Saccadic slowing occurs in the early stages of the disease and tends to develop into almost complete external ophthalmoparesis [42]. Wild-type SCA7 alleles contain between 4 and 20 CAG repeats, whereas 36–306 repeats represent known pathologic alleles. Instability in the repeat number (~12 per generational transmission) has been presumed to account for the marked genetic anticipation seen on the order of 20 years per generation. However, *de novo* mutations can also arise during paternal transmissions of intermediate-sized (28–35) CAG alleles. This is presumed in part to explain the persistence of the disease in spite of the anticipation events that should result in SCA7 extinction within families. Analysis of affected brains at autopsy demonstrates that inclusions are most frequent in the inferior olivary complex, lateral geniculate body, and the substantia nigra [43]. However, inclusions are also seen in other CNS regions, such as regions of the cerebral cortex, not normally considered to be affected in the disease. Ultrastructural analysis has revealed the presence of abnormal mitochondria with irregular cristae in affected neurons [44]. In addition, yeast two-hybrid studies and coimmunoprecipitation experiments have demonstrated an interaction of ataxin-7 with CRX, a nuclear transcription factor predominantly expressed in retinal photoreceptor cells [45]. This may explain components of the observed macular degeneration given that mutations in the CRX gene cause cone/rod dystrophy in humans.

Transgenic mice overexpressing full-length mutant ataxin-7 (Q90) in rod photoreceptors or Purkinje cells exhibited defects in both motor coordination and vision [46]. Among these models, ubiquitinated N-terminal fragments of ataxin-7 accumulate into nuclear inclusions, recruiting distinct chaperone/proteasome subunits.

17.1.9
SCA17

SCA17, also known as Huntington disease-like 4 (HDL4, 6q27), is an autosomal dominant disorder of widely variable onset with respect to clinical presentation (range: 3–75 years, median onset 23 years). The genetic target is TBP, which acts as a general transcription initiator [47]. Pathologic polyQ expansions appear in the range of 45–55 repeats. Signs of functional cerebellar pathology are seen in essentially all cases, with approximately 80% of the patients exhibiting mild to severe cognitive deficits and greater than 60% of patients showing choreic movements and pyramidal signs, bradykinesia and/or dementia [48]. MRI, where performed, demonstrates cortical and cerebellar atrophy virtually in all patients, and neurophysiological examination suggests signs of peripheral nervous system involvement [49]. Oculographic examinations in more than half of patients examined show a distinct pattern of oculomotor abnormalities demonstrated by impairment of smooth pursuit and defects in the saccade accuracy and normal saccade velocity [50]. Hyperreflexia of vestibulo-ocular reflexes is also impaired, as in nystagmus [48]. Postmortem analysis of SCA17 brain tissue demonstrated moderate loss of small neurons with gliosis predominantly in the caudate nucleus and putamen, with similar but less extensive changes in widespread regions of the thalamus, frontal cortex, and temporal cortex [47]. Moderate Purkinje cell loss is seen in the cerebellum. It should be noted that this disease exhibits substantial similarities to DRPLA more than any SCA.

17.2
Animal Models of Polyglutamine Diseases

To elucidate the underlying mechanisms involved in polyQ diseases, the effects of CAG triplet expansion have been examined in numerous animal models. A range of organisms from the nematode *Caenorhabditis elegans* to nonhuman primates such as rhesus macaque [51–54] have been employed as models. In such models, full-length or specific fragment of the polyQ protein in question is expressed in the model organism. This manipulation can be performed in different ways. The first is to ectopically express the normal or mutated protein using ectopic transgenic expression under the control of a desired, frequently tissue-specific promoter to target precise cellular populations *in vivo* [55]. A second strategy is to express the entire gene locus and flanking control elements of interest ectopically using a bacterial artificial chromosome (BAC) or yeast artificial chromosome (YAC) vector. In theory, a BAC/YAC strategy may result in construct expression in a temporal and tissue-specific pattern similar to that seen for the endogenous locus with relative copy number and chromatin position independence at significantly faster production times than that required for normal homologous recombination (see below) [56]. The final strategy involves approaches that target and modify an endogenous DNA locus of interest using vectors containing homologous gene sequences containing the desired modification within. Such procedures have most frequently been performed using murine embryonic stem (ES) cells. The advantage of such an approach is that it is frequently performed as to precisely maintain both distal and local expression cues, resulting in a modified coding locus (gene knock-in), which most closely duplicates the temporal and cell specificity of the endogenous gene *in vivo* [57].

17.2.1
Worm and Fly

Both *C. elegans* and *Drosophila* have been employed to ectopically express both normal and mutant polyQ proteins (either human or their corresponding model organism orthologs) for analyses. Though obvious concerns exist regarding the mechanistic implications of overexpressing mammalian proteins in organisms such as *C. elegans* and *Drosophila*, analysis of the data obtained from such studies with those obtained in higher organisms has highlighted the potential utility of such systems to address some features of polyQ-mediated cellular toxicity. For the majority of studies performed to date, polyQ proteins of interest are expressed in neuronal or muscle cells of *C. elegans* [52, 58] or in neurons of the developing eye of *Drosophila* [51, 59–61]; both of which are well characterized in terms of both functional characteristics and neuroanatomy. In both organisms, the expression of pathologic polyQ proteins has been shown to result in protein aggregation and cellular toxicity [51, 52, 58, 62, 63]. Furthermore, the expression of polyQ disease proteins in the developing eye of *Drosophila* has led to precise functional and cellular mapping of the degenerative changes observed. This system has thus proved informative in detecting early neurodegenerative changes resulting from expression of polyQ proteins [60, 64, 65]. The principal advantage of the above systems is that in the event that they can accurately reproduce a given cellular phenomenon seen in (human) target tissues; such model organisms are far more amenable as *in vivo* high-throughput methods of both functional (e.g., drug screening) and interaction analyses (modification of signal transduction and analysis of modified protein binding partners) than their mammalian counterparts [65–68]. In addition P-element-mediated mutational analysis and small inhibitory

RNA (siRNA) methods of signal modification have been shown to be very robust in both systems. Such methods further extend the range of rapid genome-wide application tools useful in analyzing modifier genes and potential small molecules that may alter polyQ disease pathology [65, 67, 69, 70].

17.2.2
Murine Models

Although *C. elegans* and *Drosophila* may represent good model organisms for relatively rapid genomic analysis of defined polyQ protein traits, one must determine with certainty that the desired feature of interest is faithfully reproduced in these species as it is produced in mammals. Thus, mammalian model organisms (principally murine systems) play a dominant and critical role in deciphering initial features of polyQ-mediated neurotoxicity and pathogenic mechanisms involved. Hence, a large number of polyQ studies are performed in mice due to their fundamental physiologic similarities to humans. Murine studies also allow the performance of robust germline manipulations of each type described above.

Initial studies of the pathogenic mechanisms of polyQ proteins in mice focused on deletion and overexpression studies of human polyQ expanded proteins to determine whether the observed pathology was the result of altered protein expression or was due to dominant changes in protein association. In addition to providing numerous insights into the normal functions of these proteins, the vast majority of experimental evidence now indicates that protein depletion/loss of function *per se* is not a prime cause of polyQ-induced neuropathy [71–74]. Similarly, virtually in all cases, overexpression of wild-type human polyQ proteins does not produce aberrant phenotypes, indicating that CAG expansion within normal limits does not produce neurotoxicity. In contrast, transgenic and gene targeted models expressing pathogenic expanded polyQ proteins under neuronal-specific promoters relevant to human cellular targets frequently show striking similarities observed in human neuropathology. These effects range from neurobehavioral changes to ultrastructural changes such as the development of inclusion bodies in affected neuronal populations [54, 75–77]. As such, these models have provided compelling new tools to decipher the mechanism by which polyQ diseases manifest their neuropathology, as well as an important means to test and validate potential interventions. Despite this, there are a number of issues to consider when addressing relevance of both the significance and utility of animal model-derived information in terms of polyQ-based human neuropathology.

17.2.3
Critical Issues in Model Dependence

Although each of the above models is capable of recapitulating aspects of human polyQ disease pathology, several features are critical in accurately interpreting the significance of any findings. The clear delineation of such factors must be recognized to avoid potential over/misinterpretation of the data obtained. Frequently, the choice of model (cellular or animal) utilized depends on both the complexity and perceived level of mechanistic understanding regarding the phenomenon under study. Within the experimental envelope, there is often a required trade-off between phenomenologic accuracy (e.g., observed neurotoxicity in genetically modified *Drosophila* versus transgenic macaque) and experimental malleability in terms of the scope of the process (e.g., small molecule or whole genome screening in cell lines versus *in vivo* analysis of murine brain). While such trade-offs comprise an aspect of virtually all experimental studies, and are often thought to be minimized through iterative rounds of increasingly complex investigation (e.g., simple organism/cell-based screening followed by *in vivo* mammalian analyses), they are indicated here to highlight the important but too often overlooked problem of inadequate replication of the desired mechanistic or phenomenologic feature as it truly occurs in the target of interest (in this case, human CNS neurons). Inadequate attention to this challenging problem has been considered to be a prime contributor to the rise in experimental therapies that exhibit promising results within cell-based and/or animal studies, only to fail in higher primate and/or latter human clinical trials. Clearly, inadequate or accidental exclusion of an important mechanistic element in the initial phase of any iterative screens cannot be hoped to recover the desired or expected interactors in its latter more complex (reductive) screening stages. Performing large unbiased screens in experimental systems with high mechanistic certainty has traditionally proved intractable. Rather, *in vivo* studies in mammalian system have traditionally emphasized focal investigation on candidate targets based on solid, if circumstantial, evidence (gene linkage studies). Although much has been learned using such a candidate-based approach, it could be argued that such investigation would fail to capture any candidates, which lie beyond the perceived experimental data and model. This rationale is often cited as the basis for the so-called unbiased screening methods. The difficulty of such approaches (besides intrinsic experimental barriers) is that such screens, even when optimally defined, are by nature typically extremely inefficient, resulting in discovery of relatively few additional candidates. The decision whether to employ such procedures is often a function of the perceived significance, which additional candidates would provide and the overall cost and difficulty of performing the indicated screen. Factors highlighted above (phenomenologic adequacy) represent an additional consideration. Recently, however, significant progress has been made toward addressing at least some of these issues (Section 17.4).

A critical concern with any transgenic model of polyQ neurotoxicity is the relative level of protein expression, as both molecular and biophysical studies have highlighted the concentration-dependent nature of these protein aggregates. As such, there are certain advantages to utilizing a BAC/YAC transgenic or homologous knock-in approaches, as they more effectively maintain endogenous expression levels. Despite this, concerns remain regarding the respective molecular mechanisms seen during the relatively rapid onset observed in animal models versus the decades normally required to develop polyQ diseases in human. An example of such expression differences with respect to the rate of pathogenesis may be seen in gene knock-in versus YAC transgenics of mouse Htt models.

Introduction of the mutant human allele via homologous recombination results in the production of murine lines that do not exhibit neurodegeneration during the animals' normal life span, while higher expressing YAC-mediated transgenics do show pathologic changes [78]. Such findings raise questions as to which more appropriately emulates the mechanisms seen in human disease. In practice, recapitulation of a similar neuroanatomic pattern of neural pathology has been taken as evidence of appropriate disease modeling. However, one can easily imagine scenarios where this would not necessarily be the case, such as cases of multiple cellular interferences. Alternatively, if the same cellular processes were being disrupted by a similar mechanism at both high and low levels of mutant protein production (i.e., simple scalar effect), then one could be justified in producing a rapid (month versus decades) murine model of a given human disease simply by increasing the level of mutant protein produced to develop its cognate cellular pathology. Ultimately, it is the validity of such assumptions that must be rigorously tested to determine the relevance of any *in vivo* "model." It is true that when expressed in high enough levels, many of the expanded polyQ mutant proteins identified through disease linkage studies produce pathology in the murine CNS. To determine the validity of current model systems, it is important to address the fundamental biochemical properties of wild-type and expanded polyQ proteins.

17.3
Mechanisms of Polyglutamine-Induced Neural Injury

Molecular and biophysical approaches have been utilized to understand the fundamental mechanisms of pathogenesis of CAG triplet diseases by which polyQ tract expansion may result in neurodegeneration. The results of these studies have suggested several potential molecular mechanisms to explain the induced patterns of neurodegeneration. As indicated above, these include direct aggregate/fibril toxicity, disruption of normal protein signal transduction, aberrant gain of function, induction of oxidative stress, transcriptional interference through depletion of specific protein complexes, impairment of ubiquitin–proteasomal system, and/or alterations in mRNA processing. With exception of the last putative mechanism, each of these relates to aberrant behavior of the expanded polyQ tract. Thus, understanding the underlying biology of polyQ tracts' properties in normal and CAG expanded proteins has been a focus of intense research over the past two decades.

17.3.1
Conformation of Polyglutamine Sequences

Structural elucidation of polyQ sequences of pathogenic protein variants has been a research priority, as it would potentially provide mechanistic insights into potential protein conformations and interactions, as well as aggregation propensity. It has been suggested that pathogenic expanded polyQ sequences may adopt a novel conformation, providing a toxic gain-of-function property to polyQ proteins once they reach a sufficient size, thereby stabilizing the pathogenic conformation. However, since large polyQ sequences alone are highly insoluble, due to their intrinsic tendency to aggregate over time, detailed X-ray crystallographic analyses are difficult. As a result, much of the structural information has been derived from other biophysical methods such as circular dichroism and nuclear magnetic resonance spectroscopy. Several structures have been proposed, but the most consistent among the applied methods is that seen in X-ray fiber diffraction patterns of the peptide ($D_2Q_{15}K_2$) [79]. The proposed structure is a cylindrical β-strands sheet with 20 residues per helical turn. The authors proposed that this structure could potentially explain the pathologic threshold seen with respect to the number of polyQ residues within a given protein sequence as this structure is significantly stabilized when linear polyQ stretches approach 40 or greater residues (equivalent to two successive helical turns) [79]. In addition, the polar zipper around this β-strand tends to facilitate fiber formation as additional strands of polyQ complexes could be recruited into the nucleation core and held tightly by interchain hydrogen bonding [79]. However, it is important to note that more recent evidence suggests that polyQ sequences natively adopt a random coil configuration intracellularly, adopting a β-strand conformation only on fiber aggregation [80–82]. This distinction is important given that there is increasing evidence to suggest that aggregated form of CAG expanded polyQ proteins may not be the primary neurotoxic species.

17.3.2
Polyglutamine Aggregation

As in other neurodegenerative disorders such as Alzheimer's and Parkinson's diseases, aberrant protein aggregates within affected neuronal populations represent as a common morphologic feature of polyQ diseases. A number of molecular and biochemical studies have examined the nature of polyQ protein aggregation. Initially, pure polyQ sequences compared were either "normal" or "pathogenic" based on the number of glutamine residues within each polypeptide. With the exception of SCA2 and SCA6, pathogenic polyQ disease proteins contain repeat lengths of at least 35–40 glutamine residues. To improve the solubility and reduce aggregation of these sequences, many were produced as glutathione S-transferase (GST) or as similar fusion proteins [80, 83–85]. Using such approaches, it was observed that only pathogenic polyQ sequences could form aggregates *in vitro* [80, 84, 86], though this process was shown to depend on time, temperature, and protein concentration [83, 85, 87–90]. Various studies examining aggregation kinetics showed that seeding reduces the lag time of aggregate formation, thus providing evidence that nucleation played a role in regulating protein aggregation and fibril formation. From these observations, it was thus believed that protein aggregation occurring via a nucleation mechanism is a critical step in the formation of the neurotoxic species [84, 88–90]. However, several recent studies have shown that even polyQ sequences within the wild-type range can aggregate, casting doubt on the belief that the primary difference between normal and pathogenic polyQ sequences is strictly related to aggregation propensity [83, 89, 90]. Rather, these data suggest that wild-type polyQ sequences exhibit a potential for aggregation, but require higher concentrations and longer incubation time for nucleation events to occur [83]. This extended nucleation time seen in wild-type proteins may allow cellular systems aimed at detecting aberrant protein conformations sufficient time to degrade such complexes prior to reaching a critical nucleation limit; beyond which the cell's ability to degrade is compromised.

However, in recent years, there has been mounting evidence to suggest that protein context may play a significant contributing role in the propensity of protein aggregation and fibril formation. Specifically, the AXH domain and Josephin domain of ataxin-1 and 3, respectively, have been shown to form aggregates on their own without polyQ sequences [87, 91, 92]. The presence of such domains therefore may have a tendency to shift the equilibrium toward aggregation. How are polyQ sequences involved in regulating such an attraction? Structural studies suggest that while these alternative aggregation domains have the potential to nucleate in isolation forming fibrils, the normal tertiary structures of these proteins tend to minimize such potentials, preventing aggregation. PolyQ expansion may act to destabilize the normal tertiary structure, indirectly favoring protein aggregation and fibril formation. Following this initial aggregation step, polyQ-dependent aggregation may subsequently stabilize these aggregates through the effective increase in local concentrations. Such models have been supported by recent experimental evidence in which aggregation of expanded polyQ proteins involves at least two steps, one is polyQ independent and the other is polyQ-dependent [88–90, 93, 94]. Studies using ataxin-3 found that the first phase of fibril formation is polyQ-independent, involving the Josephin domain [89, 90]. Interestingly, although it was observed that ataxin-3 with no or normal-length polyQ sequence (15Q) formed aggregates with similar kinetics to that seen in pathogenic ataxin-3 variants, only pathogenic ataxin-3 variants (64Q) ultimately formed SDS-resistant aggregates [89, 90]. Such observations indicate that at least two distinct stages can be seen in aggregate formation, a stage which is common to all three ataxin-3 isoforms and a second stage of conversion resulting in the stabilization of SDS-resistant aggregates only in the pathogenic ataxin-3 (64Q) isoform. In addition, it was shown that the second stage of aggregate formation could be inhibited by the QBP1 (polyQ disrupting) peptide, providing further support that this latter stage is polyQ-dependent [90]. This study also demonstrated that normal and pathogenic ataxin-3 aggregates are morphologically distinct [90]. Such findings support the contention that polyQ sequences undergo distinct conformation transitions during the process of protein aggregation. Similar conclusions have been drawn in a study involving a fusion protein of a globular protein (cellular retinoic acid-binding protein, CRABP) and Htt exon 1 [88]. Interestingly, by performing limited trypsinolysis of the aggregates formed, two different protease-resistant cores were formed over time, highlighting a two-stage process with respect to protein aggregation/fibril formation [88]. Can other proteins containing AXH or Josephin domains form aggregates? Analysis of the transcriptional repressor HBP1 that exhibits significant sequence homology to the AXH domain in ataxin-1 was shown to exhibit a significantly different topology compared to that seen in the ataxin-1 AXH domain [91]. Thus, the AXH domain appears capable of adopting alternative conformations, which may favor particular nonnative conformational states depending on local protein context [91]. Such findings highlight both the importance of studying polyQ sequences in their full protein context and of addressing the secondary influence that expanded polyQ tracts may have on surrounding protein conformation. In this respect, the 17-amino-acid flanking sequence (HTT^{NT}) N-terminal to polyQ in the mutant huntingtin exon 1 fragment was recently revealed to initiate an alternative aggregation mechanism [95]. On its own, the HTT^{NT} peptide does not aggregate, but undergoes a conformational change (unfolding) which is polyQ repeat length-dependent when combined with polyQ sequences. This unfolding allows the HTT^{NT} peptide to initiate aggregation in a nucleation-independent manner followed by aggregation into globular oligomers with a core composed of HTT^{NT} and ployQ.

17.3.3
Polyglutamine-Based Disruption of Protein Function

Another mechanism by which polyQ proteins may mediate their neurotoxic effects is through the induction of a change to local protein conformation, which disrupts the normal context of protein–protein interactions or creates pathogenic alternatives. Although a good molecular understanding of the endogenous functions of some polyQ disease proteins (e.g., Ca$_V$2.1 voltage-gated calcium channel in SCA6 [35], TBP in SCA17 [47], and androgen receptor in SBMA [96]) exists, it is only recently that we have begun to understand the intrinsic function of other polyQ protein members. A number of these have been implicated with functions regulating transcription. With respect to Htt, it has been shown that this protein can bind directly to DNA in the absence of other proteins and that this binding can alter DNA conformation, thereby enhancing access for other transcription factors and altering gene transcription [97]. Interestingly, it has been demonstrated that binding of both wild-type and pathogenic Htt isoforms results in the promotion of distinct patterns of gene expression, and that occupation of some gene promoters by Htt is increased in a polyQ-dependent manner, resulting in significant changes in gene expression [97].

Similarly, ataxin-1, ataxin-7, and atrophin-1 have all been implicated in transcription regulation [97–102]. Several recent studies of ataxin-1 have demonstrated that toxicity due to polyQ expansion may involve both loss-of-function and gain-of-function properties. Ataxin-1 has been shown to interact with a DNA binding protein, Cic, which was previously demonstrated to participate in EGFR signaling, cell fate determination, and cerebellar granule cell development [28, 100, 104–106]. This interaction is reduced on polyQ expansion [100]. In addition, ataxin-1 has also been shown to interact with transcription factor Gfi-1 via its AXH domain [104]. Interestingly, AXH domain itself has been demonstrated to be capable of binding to RNA homopolymers, suggesting a putative function at the level of RNA processing as well [100]. Ataxin-7 has been shown to interact with TFTC/STAGA complexes [107, 108], and to possess deubiquitinase activity involved in chromatin remodeling [101]. In *Drosophila*, it has been shown that polyQ expansion of ataxin-7 alters its ability to recruit TFTC/STAGA, resulting in altered chromatin remodeling and transcriptional regulation [101]. Although the precise mechanism is as yet unclear, atrophin-1 has been shown to be a transcriptional corepressor [105, 106, 109]. Thus, in each case, polyQ expansion of these proteins results in altered protein–DNA/RNA interactions, ultimately disrupting the normal pattern of transcriptional activation.

In addition to its influences on gene expression, polyQ expansion in Htt has also been shown to alter calcium signaling and NMDA receptor–mediated excitotoxicity [110, 111]. The mechanism of this effect is presumed to involve trafficking of AMPA and NMDA receptors via Htt interaction with HIP-1 [112–114]. Interestingly, a recent study suggested that these effects require the full-length mutant Htt protein, since disturbance in calcium signaling was not observed in animals expressing a truncated (shortstop) version of mutant Htt [114]. Such observations once again highlight the importance of protein context when examining polyQ disease pathogenesis both in animal models and in biochemical/biophysical experiments.

A deubiquitination activity has been attributed to the Josephin domain of ataxin-3, which was recently shown to interact with valosin-containing protein (VCP/p97) in regulating retrotranslocation of endoplasmic reticulum–associated degradation (ERAD) substrates to the proteasome [115–118]. However, the precise role of ataxin-3 in ERAD is unclear, since two contradictory studies proposed opposing functions of ataxin-3 in retrotranslocation of ERAD substrates for proteasomal degradation [117, 118]. Irrespective of the functions of ataxin-3 in ERAD, ubiquitin-mediated proteasomal degradation and ER stress certainly have been implicated in polyQ diseases as discussed in further detail below.

17.3.4
Toxic Gain of Function in Pathogenic PolyQ Proteins

The dominant nature of virtually all CAG pathologies has led to a suggestion that dominant-negative gain of function may underlie the pathogenic mechanism of these disorders. Therefore, the formation of new pathologic protein–protein and protein–DNA/RNA interactions in expanded polyQ tracts has been investigated. As indicated above, pathogenic Htt has been shown to be capable of interacting with a distinct set of transcription factors compared to the native protein, sequestering glutamine-rich transcription factors, which results in altered gene expression [97]. Similarly, polyQ expanded but not wild-type ataxin-1 has been shown to interact with RBM17, a splicing factor also referred to as *SPF45*, to recruit it into an alternative protein complex [100]. In support of a dominant-negative mechanism, it has been shown that loss of RBM17 in the *Drosophila* eye could suppress the toxicity induced by polyQ expanded ataxin-1 (82Q) [100]. However, it should be noted that the enhanced recruitment of RBM17 with ataxin-1 appears to prevent its complex formation with Cic [100]. Thus, it remains possible that the true mechanism of this effect may relate to a loss of function of Cic.

In addition, numerous studies have demonstrated that expression of mutant Htt results in a change in the function of histone acetyl transferases (HATs), in particular the CREB-binding protein (CBP). The overall change in HAT activity has been suggested to underlie the transcriptional dysregulation seen in Huntington's disease. This has led to the suggestion of a potential therapeutic role for HDAC inhibitors that are now in clinical trials for several polyQ diseases [119].

There have also been suggestions for both mutant Htt and androgen receptor polyQ proteins to directly inhibit axon transport [59, 120–123]. Even polyQ sequences alone (in the absence of associated Htt exon 1 or ataxin-3) appear to inhibit transport [59]. The authors demonstrated the sequestration of axon transporter proteins into polyQ aggregates [59]. It has been shown through siRNA-mediated knockdown studies that one of the endogenous functions of Htt may be regulation of axonal transport. However, it is important to note that a significant fraction of axonal inhibition may also be due to direct blockade through the creation of protein aggregates by expanded polyQ proteins.

17.3.5
Disruption of Ubiquitin-Mediated Proteasomal Degradation and Induction of ER Stress

For more than a decade, the relationship between protein degradation and pathogenic polyQ proteins has been the focus of intensive investigation in large part because proteinaceous inclusion bodies are often seen in affected neuronal populations. Analysis of these inclusion bodies frequently demonstrates the presence of ubiquitinated proteins, part of the cell's natural process of earmarking proteins for degradation by the proteasome. The data from such investigations initially formed the basis for the hypothesis that such long stretches of polyQ-containing proteins might pose a special problem for proteasomal degradation and that sufficiently large nuclear aggregates might effectively be shielded from degradation, resulting in long-term disruption of normal cellular functions and enhanced cellular stress. The observation of ubiquitinated protein suggests that the cell is acting to degrade these aggregates to prevent further toxicity. However, analysis of the ubiquitin-mediated proteasomal degradation system (UPS) in pathogenic polyQ-containing cells suggests that this process may be interrupted through protein aggregation and fibril formation. A recent study demonstrated that the UPS function is impaired both *in vitro* and *in vivo* in the presence of polyQ expanded proteins [124, 125], by measuring the amount of K48-linked ubiquitin isopeptide by mass spectrometry following purification of polyubiquitinated proteins from cell/tissue lysates using the ubiquitin association domain of human ubiquitin 2. Accumulation of K48-linked ubiquitin isopeptide served as a measure of UPS impairment. The authors observed that UPS functions measured in this manner were impaired by the expression of polyQ expanded Htt exon-1 in HEK293 cells, two independent Huntington's disease mouse models (R6/2 and $Hdh^{Q150/Q150}$), and human Huntington's disease cortices compared to their respective controls.

However, how precisely UPS functions are impaired during polyQ disease pathogenesis and the contribution of UPS impairment to the overall pathogenic process remain to be elucidated.

Previous studies have suggested that the UPS impairment is a result of the inability of the eukaryotic proteasome to efficiently digest polyQ sequences [126]. However, such findings have been called into question from recent studies demonstrating that the eukaryotic proteasome is indeed capable of completely digesting polyQ proteins, and that the previous study only failed to detect such degradation products due to issues with the ionization of polyQ sequences, leading to an underestimation of cleavage products by mass spectrometry [127]. Thus, while the eukaryotic proteasome appears capable of degrading polyQ sequences, the ability to degrade full-length pathogenic polyQ proteins is still the subject of debate given the accumulation of polyubiquitinated polyQ disease proteins seen in inclusion bodies. Over the past decade, the role and nature of these inclusion bodies have themselves been subject to debate. Historically, these protein aggregates have been considered to be the principal toxic species, serving as a source for oxidative stress (see below). However, recent findings for a number of pathogenic polyQ proteins have suggested that such macromolecular aggregates may in fact represent a less toxic form of the mutant conformer, and that inclusion bodies may represent the cell's response to attempt to reduce the overall toxicity of such proteins. The majority of studies now point to the mutant protein or low molecular weight aggregates of such proteins as a principal source of cellular injury [128, 129].

With regard to the ER, several studies have demonstrated that ataxin-3 acts in the retrotranslocation of polyubiquitinated proteins from the ER to proteasome for degradation [115, 117, 118]. In these studies, ataxin-3 was observed to interact with a protein critically involved in retrotranslocation, VCP/p97, and that this interaction was critical to retrotranslocation of ERAD substrates. However, there is uncertainty with regard to the functional consequences of this interaction, as independent groups have proposed contradicting outcomes for this protein–protein interaction. Zhong and Pittman [118] have suggested that this interaction disrupts complex formation of p97 with two cofactors (Ufd1/Npl4), resulting in the disruption of retrotranslocation. In contrast, Boeddrich et al. [115] and Wang et al. [117] provided evidence that normal ataxin-3 function is required for proper retrotranslocation. In support of this, Wang et al. [117] demonstrated that expression of ataxin-3 is critical to regulation of retrotranslocation, and that in both siRNA-mediated knockdown and overexpression of ataxin-3, there was a suppression of ERAD and stabilization of ERAD substrates. Hence, it can be concluded that varied results seen in these three studies result from differential expression levels of ataxin-3. Nonetheless, it has been observed that the interaction between ataxin-3 and p97 is enhanced by polyQ expansion, providing a potential mechanism by which polyQ tracts can disrupt ERAD, triggering ER stress, and ultimately neuronal cell death.

Although other polyQ disease proteins have not been shown to directly regulate ERAD in a manner similar to ataxin-3, ER stress induced by protein misfolding, aggregation, and fibril formation can certainly result in neurotoxicity [116]. In this respect, ER stress and contributions by heat shock proteins and chaperones have been examined extensively in polyQ expansion diseases [116, 130–132]. In keeping with this, pharmacologic modulations of heat shock proteins in chaperones have been demonstrated to alleviate toxicity due to polyQ expanded Htt, androgen receptor, and ataxin-3 both *in vitro* and *in vivo* [133–135]. Currently, it remains to be seen whether these findings can be translated into clinically useful features.

17.3.6
CAG Repeats and RNA-Mediated Toxicity

Although much of the work on the mechanisms of polyQ disease pathogenesis have focused on the potential effects of defective proteins to enhance neurodegeneration, it should be noted that several disorders such as dystrophia myotonica (DM1 and DM2), and fragile X tremor/ataxia syndrome (FXTAS) are caused by trinucleotide expansions as a result of altered mRNA processing. Hence, it has been suggested that polyQ diseases may also exhibit an RNA-mediated toxicity component. This has been recently shown for ataxin-3 in *Drosophila*, where it was discovered that a modifier of CUG repeat RNA toxicity, Mbl, was also involved in toxicity due to polyQ expanded ataxin-3 and Htt [64]. The authors used two different methods to demonstrate that toxicity of mutant ataxin-3 was partly due to trinucleotide repeat expansion within the ataxin-3 mRNA. The role of RNA-mediated toxicity was examined by altering the CUG repeat sequence to CUUCUG. This disruption of the overall CUG repeat architecture resulted in reduced toxicity, although the polyQ expansion was still present in the translated protein. The authors further pursued this idea by inserting an untranslated CUG repeat of pathogenic length into the 3' UTR of a transgene. By simply expressing this pathogenic CUG repeat RNA but not the corresponding polyQ protein, neurodegeneration was observed, providing support that expanded CUG repeats can be pathogenic at the RNA level. While the above findings were observed in *Drosophila*, it will be interesting to determine whether these findings can be extended into mammalian systems and whether they will be replicated for other polyQ disease proteins. Although it has been argued that for the majority of pathogenic polyQ proteins, tract expansion occurs within the protein coding sequence of the gene and that polyQ-containing peptides alone have been shown to directly induce neurotoxicity, neither of these features in themselves necessarily rule out potential contribution by CUG tract expansion at the mRNA level.

17.3.7
Questions Remaining

Many issues remain to be resolved with respect to both the commonalities and distinctness of polyQ protein diseases. Perhaps the most fundamental of these is which species constitutes the principle toxic entity in these disorders? Similar to recent work in Alzheimer's disease, a number of lines of investigation now appear to point to the monomer as a principle toxic entity. Does this mean that alterations in molecular signaling constitute the primary dysfunction in these disorders or mechanisms such as aberrant mRNA processing? With respect to specificity, what are the underlying mechanisms that impart specificity with respect to different target neural populations in each of these disorders (given that many of the cognate genes are expressed over much wider population)? What are the factors that regulate phenotypic variability and temporal onset? With respect to the mechanism of neural destruction, a flurry of recent studies have examined the impact of autophagy as a cellular means to alleviate the toxicity by polyQ expanded- and other aggregation-prone proteins [130, 136–141]. In this respect, the interplay between autophagy and PCD in polyQ diseases will need to be addressed. With the advent of new tools such as induced pluripotent stem cells, we now may soon begin to look deeper, experimentally directly examining the affected human neurons in our attempts to understand how CAG trinucleotide expansion induces neuropathology.

17.4
Future Perspectives

Increasingly, experimental evidence suggests that the earliest cell injury events in polyQ diseases arises through the actions of toxic monomeric or oligomeric intermediates rather than macromolecular aggregates, in a manner similar to that previously shown in both Alzheimer's and Parkinson's disease isoforms [142–144]. These intermediates may represent conformational interactions of either full-length protein or some proteolytic fragment thereof. In this respect, it is interesting to note that modification of caspase-7 activity has been shown to reduce the neurotoxicity of ataxin-7 [145]. Such findings imply that aspects of polyQ neuropathy may be somewhat analogous to that seen for β-amyloid (Aβ) activation from the transmembrane amyloid precursor protein [146]. Current evidence also suggests that in majority of the cases, pathologic polyQ expansions induce cellular injury through a toxic gain-of-function. In part, this is suggested both by the autosomal dominant pattern disease transmission and the relatively modest phenotypes observed following gene deletion of a majority of cognate polyQ proteins. If the above suppositions are correct, there are several likely mechanistic implications. The development of macromolecular inclusions and structural disruption of the cell, including mitochondrial/oxidative injury, represent downstream events in the mechanism of injury. This may explain why clinical trials using therapies such as coenzyme Q10 have showed no significant benefit in patients affected with polyQ neuropathies [147]. If pathogenic polyQ proteins operate through a direct toxic gain-of-function, macromolecular inclusions and their ubiquitination would represent an adaptive secondary response by the cell at a point when cellular activity has already been substantially compromised. If this is true, strategies aimed to identify small molecules capable of reducing aberrant polyQ aggregates (as has been done recently in a number of high-throughput screens [148]) and/or strategies aimed at altering proteasome processing *per se* would be of limited utility (being secondary to the underlying injury), and may actually exacerbate cellular damage by increasing the concentrations of the toxic species.

Then, on which should we concentrate our efforts in an attempt to modify the pathogenic effects of mutant polyQ proteins? Although the induction of direct cellular injury through microaggregate pore formation remains a formal possibility, a likely candidate mechanism is the noted ability of polyQ proteins to alter neuronal transcriptional activity through their sequestration of glutamine-rich transcription factors and histone acetyltransferases. Such actions may also underlie the noted regional and cellular specificity of many pathogenic polyQ proteins.

With respect to transcriptional interference, experimental studies have suggested yet another intriguing possibility that such disorders may reflect defects in handling CUG repeats at the level of RNA processing. Work by Li *et al.* [64] in *Drosophila* has demonstrated this possibility as described above (Section 17.3). In the wider context of polyQ diseases, does such a mechanism seem plausible? At present, there are already several well-characterized examples of neuropathies that arise due to problems in RNA processing originating from pathogenic expansion of triplet repeats. These include but are not limited to FXTAS and DM 1 and 2 [149]. Beyond neuropathies involving expansion of trinucleotide repeats, there are also a number of additional neuropathies involving motor control, which have also been traced back to problems in RNA processing (ALS4) [150]. More recently, mutations in the *FUS/TLS* gene, whose protein product is involved in RNA processing, have been identified in familial ALS (ALS6) [151, 152]. Such a mechanism could clearly result in pathogenic changes in the transcriptional machinery of the cell, potentially tying up subtypes of required transcription factors with ensuing cellular dysfunction. In the event of RNA-based pathology, therapeutic approaches such as the stabilization of native protein conformation through polyQ binding peptide such as QBP1, inhibitors of HSP90 designed to upregulate HSP27/HSP70 activity (novobiocin analog A4 and the like), and even mTOR inhibitors (e.g., rapamycin) to enhance levels of cellular autophagy may meet with limited success. Clearly, the most direct route (besides informed genetic screening) would be attempting to alter transcription of the mutant allele. While this is presently beyond the scope of practice, gains have been made in recent years toward the application of allele-specific siRNA knockdown in several experimental systems [153]. Several recent studies have demonstrated different approaches (siRNAs, peptide nucleic acid-peptide conjugates, and morpholino antisense oligonucleotides) to accomplish this task at the cellular level [154–156]. While the first two approaches were designed to reduce the production of mutant proteins, the latter approach was devised to prevent the sequestration of a splice regulator (MBNL1) by expanded CUG repeats in mutant mRNA. Currently, the most likely route for sustained delivery of such an agent to specific neural loci would be through expression of the corresponding shRNA species via an ectopic neural delivery source such as a genetically modified lentiviral vector. It should be noted, however, that use of such vectors alone in humans presently constitutes a significant clinical barrier.

In reality, the situation is likely to be complex, potentially involving components of several of the above mechanisms. Fortunately, to assist us in deciphering the pathogenic mechanisms involved in the development of "polyQ" neuropathies, powerful new tools have been developed. In particular, work over the past several years has demonstrated that the experimental viability of reprogramming easily obtained human somatic cells (frequently fibroblasts) to form induce pluripotent stem (iPS) cells that have been shown to closely mimic many of the properties of ES cells. The availability of such tools has rapidly allowed the creation of iPS libraries for number of genetic diseases, including polyQ neuropathies [157, 158]. Such tools create an opportunity whereby iPS cells could be induced to form human neural (among other) derivatives either through direct *in vitro* manipulation toward neural stem cells or through the potential creation of human–murine celmeras [159]. Though still in its infancy, such a technology will substantially expand our knowledge over the next few years regarding the molecular details of what occurs within living human neurons affected by polyQ neuropathies. Eventually, such approaches may even allow us to recover (through allogeneic transplantation) a measure that is lost to these devastating diseases.

Acknowledgments

The laboratory gratefully acknowledges funding from the National Institute of Health and the National Alliance for Research on Schizophrenia and Depression (NARSAD). We apologize to those authors whose work could not be cited due to space considerations. The authors declare that they have no competing financial interests in the materials presented.

References

1 Bett, J.S. et al. (2006) Proteasome impairment does not contribute to pathogenesis in R6/2 Huntington's disease mice: exclusion of proteasome activator REGgamma as a therapeutic target. *Hum Mol Genet*, **15** (1), 33–44.

2 Cummings, J.L. (1995) Behavioral and psychiatric symptoms associated with Huntington's disease. *Adv Neurol*, **65**, 179–186.

3 Paulsen, J.S. et al. (2001) Neuropsychiatric aspects of Huntington's disease. *J Neurol Neurosurg Psychiatry*, **71** (3), 310–314.

4 Smith, M.A., Brandt, J., and Shadmehr, R. (2000) Motor disorder in Huntington's disease begins as a dysfunction in error feedback control. *Nature*, **403** (6769), 544–549.

5 Aylward, E.H. et al. (2000) Rate of caudate atrophy in presymptomatic and symptomatic stages of Huntington's disease. *Mov Disord*, **15** (3), 552–560.

6 Culjkovic, B. et al. (1999) Correlation between triplet repeat expansion and computed tomography measures of caudate nuclei atrophy in Huntington's disease. *J Neurol*, **246** (11), 1090–1093.

7 Douaud, G. et al. (2006) Distribution of grey matter atrophy in Huntington's disease patients: a combined ROI-based and voxel-based morphometric study. *Neuroimage*, **32** (4), 1562–1575.

8 Mascalchi, M. et al. (2004) Huntington disease: volumetric,

9. Graveland, G.A., Williams, R.S., and DiFiglia, M. (1985) Evidence for degenerative and regenerative changes in neostriatal spiny neurons in Huntington's disease. *Science*, **227** (4688), 770–773.
10. Tang, T.S. et al. (2007) Dopaminergic signaling and striatal neurodegeneration in Huntington's disease. *J Neurosci*, **27** (30), 7899–7910.
11. Becher, M.W. et al. (1998) Intranuclear neuronal inclusions in Huntington's disease and dentatorubral and pallidoluysian atrophy: correlation between the density of inclusions and IT15 CAG triplet repeat length. *Neurobiol Dis*, **4** (6), 387–397.
12. Maat-Schieman, M.L. et al. (1999) Distribution of inclusions in neuronal nuclei and dystrophic neurites in Huntington disease brain. *J Neuropathol Exp Neurol*, **58** (2), 129–137.
13. Arrasate, M. et al. (2004) Inclusion body formation reduces levels of mutant huntingtin and the risk of neuronal death. *Nature*, **431** (7010), 805–810.
14. Reiner, A. et al. (2007) R6/2 neurons with intranuclear inclusions survive for prolonged periods in the brains of chimeric mice. *J Comp Neurol*, **505** (6), 603–629.
15. Slow, E.J. et al. (2005) Absence of behavioral abnormalities and neurodegeneration in vivo despite widespread neuronal huntingtin inclusions. *Proc Natl Acad Sci U S A*, **102** (32), 11402–11407.
16. Graham, R.K. et al. (2006) Cleavage at the caspase-6 site is required for neuronal dysfunction and degeneration due to mutant huntingtin. *Cell*, **125** (6), 1179–1191.
17. Hermel, E. et al. (2004) Specific caspase interactions and amplification are involved in selective neuronal vulnerability in Huntington's disease. *Cell Death Differ*, **11** (4), 424–438.
18. Kennedy, W.R., Alter, M., and Sung, J.H. (1968) Progressive proximal

diffusion-weighted, and magnetization transfer MR imaging of brain. *Radiology*, **232** (3), 867–873.

spinal and bulbar muscular atrophy of late onset. A sex-linked recessive trait. *Neurology*, **18** (7), 671–680.
19. Trifiro, M.A., Kazemi-Esfarjani, P., and Pinsky, L. (1994) X-linked muscular atrophy and the androgen receptor. *Trends Endocrinol Metab*, **5** (10), 416–421.
20. Adachi, H. et al. (2005) Widespread nuclear and cytoplasmic accumulation of mutant androgen receptor in SBMA patients. *Brain*, **128** (Pt 3), 659–670.
21. Nagafuchi, S. et al. (1994) Structure and expression of the gene responsible for the triplet repeat disorder, dentatorubral and pallidoluysian atrophy (DRPLA). *Nat Genet*, **8** (2), 177–182.
22. Schilling, G. et al. (1999) Nuclear accumulation of truncated atrophin-1 fragments in a transgenic mouse model of DRPLA. *Neuron*, **24** (1), 275–286.
23. Yamada, M. et al. (2001) Widespread occurrence of intranuclear atrophin-1 accumulation in the central nervous system neurons of patients with dentatorubral-pallidoluysian atrophy. *Ann Neurol*, **49** (1), 14–23.
24. Yanagisawa, H. et al. (2000) Protein binding of a DRPLA family through arginine-glutamic acid dipeptide repeats is enhanced by extended polyglutamine. *Hum Mol Genet*, **9** (9), 1433–1442.
25. Nucifora, F.C. Jr et al. (2001) Interference by huntingtin and atrophin-1 with cbp-mediated transcription leading to cellular toxicity. *Science*, **291** (5512), 2423–2428.
26. Yamada, M., Tsuji, S., and Takahashi, H. (2000) Pathology of CAG repeat diseases. *Neuropathology*, **20** (4), 319–325.
27. Gilman, S. et al. (1996) Spinocerebellar ataxia type 1 with multiple system degeneration and glial cytoplasmic inclusions. *Ann Neurol*, **39** (2), 241–255.
28. Lam, Y.C. et al. (2006) ATAXIN-1 interacts with the repressor Capicua in its native complex to cause

SCA1 neuropathology. *Cell*, **127** (7), 1335–1347.

29 Tsai, C.C. et al. (2004) Ataxin 1, a SCA1 neurodegenerative disorder protein, is functionally linked to the silencing mediator of retinoid and thyroid hormone receptors. *Proc Natl Acad Sci U S A*, **101** (12), 4047–4052.

30 Davidson, J.D. et al. (2000) Identification and characterization of an ataxin-1-interacting protein: A1Up, a ubiquitin-like nuclear protein. *Hum Mol Genet*, **9** (15), 2305–2312.

31 Huynh, D.P. et al. (2000) Nuclear localization or inclusion body formation of ataxin-2 are not necessary for SCA2 pathogenesis in mouse or human. *Nat Genet*, **26** (1), 44–50.

32 Huynh, D.P. et al. (1999) Expression of ataxin-2 in brains from normal individuals and patients with Alzheimer's disease and spinocerebellar ataxia 2. *Ann Neurol*, **45** (2), 232–241.

33 van de Loo, S. et al. (2009) Ataxin-2 associates with rough endoplasmic reticulum. *Exp Neurol*, **215** (1), 110–118.

34 Chou, A.H. et al. (2008) Polyglutamine-expanded ataxin-3 causes cerebellar dysfunction of SCA3 transgenic mice by inducing transcriptional dysregulation. *Neurobiol Dis*, **31** (1), 89–101.

35 Zhuchenko, O. et al. (1997) Autosomal dominant cerebellar ataxia (SCA6) associated with small polyglutamine expansions in the alpha 1A-voltage-dependent calcium channel. *Nat Genet*, **15** (1), 62–69.

36 Mantuano, E. et al. (2003) Spinocerebellar ataxia type 6 and episodic ataxia type 2: differences and similarities between two allelic disorders. *Cytogenet Genome Res*, **100** (1-4), 147–153.

37 Doyle, J. et al. (1997) Mutations in the Cacnl1a4 calcium channel gene are associated with seizures, cerebellar degeneration, and ataxia in tottering and leaner mutant mice. *Mamm Genome*, **8** (2), 113–120.

38 Fletcher, C.F. et al. (1996) Absence epilepsy in tottering mutant mice is associated with calcium channel defects. *Cell*, **87** (4), 607–617.

39 Xie, G. et al. (2007) Forward genetic screen of mouse reveals dominant missense mutation in the P/Q-type voltage-dependent calcium channel, CACNA1A. *Genes Brain Behav*, **6** (8), 717–727.

40 Watase, K. et al. (2008) Spinocerebellar ataxia type 6 knockin mice develop a progressive neuronal dysfunction with age-dependent accumulation of mutant CaV2.1 channels. *Proc Natl Acad Sci U S A*, **105** (33), 11987–11992.

41 David, G. et al. (1997) Cloning of the SCA7 gene reveals a highly unstable CAG repeat expansion. *Nat Genet*, **17** (1), 65–70.

42 Oh, A.K. et al. (2001) Slowing of voluntary and involuntary saccades: an early sign in spinocerebellar ataxia type 7. *Ann Neurol*, **49** (6), 801–804.

43 Holmberg, M. et al. (1998) Spinocerebellar ataxia type 7 (SCA7): a neurodegenerative disorder with neuronal intranuclear inclusions. *Hum Mol Genet*, **7** (5), 913–918.

44 Cooles, P., Michaud, R., and Best, P.V. (1988) A dominantly inherited progressive disease in a black family characterised by cerebellar and retinal degeneration, external ophthalmoplegia and abnormal mitochondria. *J Neurol Sci*, **87** (2–3), 275–288.

45 La Spada, A.R. et al. (2001) Polyglutamine-expanded ataxin-7 antagonizes CRX function and induces cone-rod dystrophy in a mouse model of SCA7. *Neuron*, **31** (6), 913–927.

46 Yvert, G. et al. (2000) Expanded polyglutamines induce neurodegeneration and trans-neuronal alterations in cerebellum and retina of SCA7 transgenic mice. *Hum Mol Genet*, **9** (17), 2491–2506.

47 Nakamura, K. et al. (2001) SCA17, a novel autosomal dominant cerebellar ataxia caused by an expanded polyglutamine in TATA-binding

protein. *Hum Mol Genet*, **10** (14), 1441–1448.

48 Mariotti, C. et al. (2007) Spinocerebellar ataxia type 17 (SCA17): oculomotor phenotype and clinical characterization of 15 Italian patients. *J Neurol*, **254** (11), 1538–1546.

49 Filla, A. et al. (2002) Early onset autosomal dominant dementia with ataxia, extrapyramidal features, and epilepsy. *Neurology*, **58** (6), 922–928.

50 Hubner, J. et al. (2007) Eye movement abnormalities in spinocerebellar ataxia type 17 (SCA17). *Neurology*, **69** (11), 1160–1168.

51 Warrick, J.M. et al. (1998) Expanded polyglutamine protein forms nuclear inclusions and causes neural degeneration in Drosophila. *Cell*, **93** (6), 939–949.

52 Morley, J.F. et al. (2002) The threshold for polyglutamine-expansion protein aggregation and cellular toxicity is dynamic and influenced by aging in Caenorhabditis elegans. *Proc Natl Acad Sci U S A*, **99** (16), 10417–10422.

53 Yang, S.H. et al. (2008) Towards a transgenic model of Huntington's disease in a non-human primate. *Nature*, **453** (7197), 921–924.

54 Hodgson, J.G. et al. (1999) A YAC mouse model for Huntington's disease with full-length mutant huntingtin, cytoplasmic toxicity, and selective striatal neurodegeneration. *Neuron*, **23** (1), 181–192.

55 Schilling, G. et al. (1999) Intranuclear inclusions and neuritic aggregates in transgenic mice expressing a mutant N-terminal fragment of huntingtin. *Hum Mol Genet*, **8** (3), 397–407.

56 Hodgson, J.G. et al. (1996) Human huntingtin derived from YAC transgenes compensates for loss of murine huntingtin by rescue of the embryonic lethal phenotype. *Hum Mol Genet*, **5** (12), 1875–1885.

57 Wheeler, V.C. et al. (1999) Length-dependent gametic CAG repeat instability in the Huntington's disease knock-in mouse. *Hum Mol Genet*, **8** (1), 115–122.

58 Satyal, S.H. et al. (2000) Polyglutamine aggregates alter protein folding homeostasis in Caenorhabditis elegans. *Proc Natl Acad Sci U S A*, **97** (11), 5750–5755.

59 Gunawardena, S. et al. (2003) Disruption of axonal transport by loss of huntingtin or expression of pathogenic polyQ proteins in Drosophila. *Neuron*, **40** (1), 25–40.

60 Jackson, G.R. et al. (1998) Polyglutamine-expanded human huntingtin transgenes induce degeneration of Drosophila photoreceptor neurons. *Neuron*, **21** (3), 633–642.

61 Takeyama, K. et al. (2002) Androgen-dependent neurodegeneration by polyglutamine-expanded human androgen receptor in Drosophila. *Neuron*, **35** (5), 855–864.

62 Faber, P.W. et al. (1999) Polyglutamine-mediated dysfunction and apoptotic death of a Caenorhabditis elegans sensory neuron. *Proc Natl Acad Sci U S A*, **96** (1), 179–184.

63 Parker, J.A. et al. (2001) Expanded polyglutamines in Caenorhabditis elegans cause axonal abnormalities and severe dysfunction of PLM mechanosensory neurons without cell death. *Proc Natl Acad Sci U S A*, **98** (23), 13318–13323.

64 Li, L.B. et al. (2008) RNA toxicity is a component of ataxin-3 degeneration in Drosophila. *Nature*, **453** (7198), 1107–1111.

65 Bilen, J. and Bonini, N.M. (2007) Genome-wide screen for modifiers of ataxin-3 neurodegeneration in Drosophila. *PLoS Genet*, **3** (10), 1950–1964.

66 Fernandez-Funez, P. et al. (2000) Identification of genes that modify ataxin-1-induced neurodegeneration. *Nature*, **408** (6808), 101–106.

67 Kazemi-Esfarjani, P. and Benzer, S. (2000) Genetic suppression of polyglutamine toxicity in Drosophila. *Science*, **287** (5459), 1837–1840.

68 Warrick, J.M. et al. (1999) Suppression of polyglutamine-mediated neurodegeneration in Drosophila by

the molecular chaperone HSP70. *Nat Genet*, **23** (4), 425–428.
69. Nollen, E.A. et al. (2004) Genome-wide RNA interference screen identifies previously undescribed regulators of polyglutamine aggregation. *Proc Natl Acad Sci U S A*, **101** (17), 6403–6408.
70. Rodrigues, A.J. et al. (2007) Functional genomics and biochemical characterization of the C. elegans orthologue of the Machado-Joseph disease protein ataxin-3. *FASEB J*, **21** (4), 1126–1136.
71. Jun, K. et al. (1999) Ablation of P/Q-type Ca(2+) channel currents, altered synaptic transmission, and progressive ataxia in mice lacking the alpha(1A)-subunit. *Proc Natl Acad Sci U S A*, **96** (26), 15245–15250.
72. Matilla, A. et al. (1998) Mice lacking ataxin-1 display learning deficits and decreased hippocampal paired-pulse facilitation. *J Neurosci*, **18** (14), 5508–5516.
73. Yoo, S.Y. et al. (2003) SCA7 knockin mice model human SCA7 and reveal gradual accumulation of mutant ataxin-7 in neurons and abnormalities in short-term plasticity. *Neuron*, **37** (3), 383–401.
74. Dragatsis, I., Levine, M.S., and Zeitlin, S. (2000) Inactivation of Hdh in the brain and testis results in progressive neurodegeneration and sterility in mice. *Nat Genet*, **26** (3), 300–306.
75. Mangiarini, L. et al. (1996) Exon 1 of the HD gene with an expanded CAG repeat is sufficient to cause a progressive neurological phenotype in transgenic mice. *Cell*, **87** (3), 493–506.
76. Lorenzetti, D. et al. (2000) Repeat instability and motor incoordination in mice with a targeted expanded CAG repeat in the Sca1 locus. *Hum Mol Genet*, **9** (5), 779–785.
77. Watase, K. et al. (2002) A long CAG repeat in the mouse Sca1 locus replicates SCA1 features and reveals the impact of protein solubility on selective neurodegeneration. *Neuron*, **34** (6), 905–919.
78. Van Raamsdonk, J.M., Warby, S.C., and Hayden, M.R. (2007) Selective degeneration in YAC mouse models of Huntington disease. *Brain Res Bull*, **72** (2–3), 124–131.
79. Perutz, M.F. et al. (2002) Amyloid fibers are water-filled nanotubes. *Proc Natl Acad Sci U S A*, **99** (8), 5591–5595.
80. Masino, L. et al. (2002) Solution structure of polyglutamine tracts in GST-polyglutamine fusion proteins. *FEBS Lett*, **513** (2–3), 267–272.
81. Bennett, M.J. et al. (2002) Inaugural Article: a linear lattice model for polyglutamine in CAG-expansion diseases. *Proc Natl Acad Sci U S A*, **99** (18), 11634–11639.
82. Chen, S. et al. (2001) Polyglutamine aggregation behavior in vitro supports a recruitment mechanism of cytotoxicity. *J Mol Biol*, **311** (1), 173–182.
83. Klein, F.A. et al. (2007) Pathogenic and non-pathogenic polyglutamine tracts have similar structural properties: towards a length-dependent toxicity gradient. *J Mol Biol*, **371** (1), 235–244.
84. Bevivino, A.E. and Loll, P.J. (2001) An expanded glutamine repeat destabilizes native ataxin-3 structure and mediates formation of parallel beta-fibrils. *Proc Natl Acad Sci U S A*, **98** (21), 11955–11960.
85. Busch, A. et al. (2003) Mutant huntingtin promotes the fibrillogenesis of wild-type huntingtin: a potential mechanism for loss of huntingtin function in Huntington's disease. *J Biol Chem*, **278** (42), 41452–41461.
86. Marchal, S. et al. (2003) Structural instability and fibrillar aggregation of non-expanded human ataxin-3 revealed under high pressure and temperature. *J Biol Chem*, **278** (34), 31554–31563.
87. Masino, L. et al. (2004) Characterization of the structure and the amyloidogenic properties of the Josephin domain of the polyglutamine-containing protein ataxin-3. *J Mol Biol*, **344** (4), 1021–1035.

88 Ignatova, Z. et al. (2007) In-cell aggregation of a polyglutamine-containing chimera is a multistep process initiated by the flanking sequence. *J Biol Chem*, **282** (50), 36736–36743.

89 Ellisdon, A.M., Pearce, M.C., and Bottomley, S.P. (2007) Mechanisms of ataxin-3 misfolding and fibril formation: kinetic analysis of a disease-associated polyglutamine protein. *J Mol Biol*, **368** (2), 595–605.

90 Ellisdon, A.M., Thomas, B., and Bottomley, S.P. (2006) The two-stage pathway of ataxin-3 fibrillogenesis involves a polyglutamine-independent step. *J Biol Chem*, **281** (25), 16888–16896.

91 de Chiara, C. et al. (2005) The AXH domain adopts alternative folds the solution structure of HBP1 AXH. *Structure*, **13** (5), 743–753.

92 de Chiara, C. et al. (2005) Polyglutamine is not all: the functional role of the AXH domain in the ataxin-1 protein. *J Mol Biol*, **354** (4), 883–893.

93 Nozaki, K. et al. (2001) Amino acid sequences flanking polyglutamine stretches influence their potential for aggregate formation. *Neuroreport*, **12** (15), 3357–3364.

94 Gales, L. et al. (2005) Towards a structural understanding of the fibrillization pathway in Machado-Joseph's disease: trapping early oligomers of non-expanded ataxin-3. *J Mol Biol*, **353** (3), 642–654.

95 Thakur, A.K. et al. (2009) Polyglutamine disruption of the huntingtin exon 1. N terminus triggers a complex aggregation mechanism. *Nat Struct Mol Biol*, **1614**, 380–389.

96 La Spada, A.R. et al. (1991) Androgen receptor gene mutations in X-linked spinal and bulbar muscular atrophy. *Nature*, **352** (6330), 77–79.

97 Benn, C.L. et al. (2008) Huntingtin modulates transcription, occupies gene promoters in vivo, and binds directly to DNA in a polyglutamine-dependent manner. *J Neurosci*, **28** (42), 10720–10733.

98 Bowman, A.B. et al. (2007) Duplication of Atxn1l suppresses SCA1 neuropathology by decreasing incorporation of polyglutamine-expanded ataxin-1 into native complexes. *Nat Genet*, **39** (3), 373–379.

99 Lee, S., Hong, S., and Kang, S. (2008) The ubiquitin-conjugating enzyme UbcH6 regulates the transcriptional repression activity of the SCA1 gene product ataxin-1. *Biochem Biophys Res Commun*, **372** (4), 735–740.

100 Lim, J. et al. (2008) Opposing effects of polyglutamine expansion on native protein complexes contribute to SCA1. *Nature*, **452** (7188), 713–718.

101 Helmlinger, D. et al. (2006) Glutamine-expanded ataxin-7 alters TFTC/STAGA recruitment and chromatin structure leading to photoreceptor dysfunction. *PLoS Biol*, **4** (3), e67.

102 McMahon, S.J. et al. (2005) Polyglutamine-expanded spinocerebellar ataxia-7 protein disrupts normal SAGA and SLIK histone acetyltransferase activity. *Proc Natl Acad Sci U S A*, **102** (24), 8478–8482.

103 Palhan, V.B. et al. (2005) Polyglutamine-expanded ataxin-7 inhibits STAGA histone acetyltransferase activity to produce retinal degeneration. *Proc Natl Acad Sci U S A*, **102** (24), 8472–8477.

104 Tsuda, H. et al. (2005) The AXH domain of Ataxin-1 mediates neurodegeneration through its interaction with Gfi-1/Senseless proteins. *Cell*, **122** (4), 633–644.

105 Haecker, A. et al. (2007) Drosophila brakeless interacts with atrophin and is required for tailless-mediated transcriptional repression in early embryos. *PLoS Biol*, **5** (6), e145.

106 Wood, J.D. et al. (2000) Atrophin-1, the dentato-rubral and pallido-luysian atrophy gene product, interacts with ETO/MTG8 in the nuclear matrix and represses transcription. *J Cell Biol*, **150** (5), 939–948.

107 Strom, A.L., Forsgren, L., and Holmberg, M. (2005) A role for both

wild-type and expanded ataxin-7 in transcriptional regulation. *Neurobiol Dis*, **20** (3), 646–655.
108 Helmlinger, D. et al. (2004) Ataxin-7 is a subunit of GCN5 histone acetyltransferase-containing complexes. *Hum Mol Genet*, **13** (12), 1257–1265.
109 Charroux, B. et al. (2006) Atrophin contributes to the negative regulation of epidermal growth factor receptor signaling in Drosophila. *Dev Biol*, **291** (2), 278–290.
110 Leavitt, B.R. et al. (2006) Wild-type huntingtin protects neurons from excitotoxicity. *J Neurochem*, **96** (4), 1121–1129.
111 Zeron, M.M. et al. (2001) Mutant huntingtin enhances excitotoxic cell death. *Mol Cell Neurosci*, **17** (1), 41–53.
112 Fernandes, H.B. et al. (2007) Mitochondrial sensitivity and altered calcium handling underlie enhanced NMDA-induced apoptosis in YAC128 model of Huntington's disease. *J Neurosci*, **27** (50), 13614–13623.
113 Parker, J.A. et al. (2007) Huntingtin-interacting protein 1 influences worm and mouse presynaptic function and protects Caenorhabditis elegans neurons against mutant polyglutamine toxicity. *J Neurosci*, **27** (41), 11056–11064.
114 Zhang, H. et al. (2008) Full length mutant huntingtin is required for altered Ca^{2+} signaling and apoptosis of striatal neurons in the YAC mouse model of Huntington's disease. *Neurobiol Dis*, **31** (1), 80–88.
115 Boeddrich, A. et al. (2006) An arginine/lysine-rich motif is crucial for VCP/p97-mediated modulation of ataxin-3 fibrillogenesis. *EMBO J*, **25** (7), 1547–1558.
116 Duennwald, M.L. and Lindquist, S. (2008) Impaired ERAD and ER stress are early and specific events in polyglutamine toxicity. *Genes Dev*, **22** (23), 3308–3319.
117 Wang, Q., Li, L., and Ye, Y. (2006) Regulation of retrotranslocation by p97-associated deubiquitinating enzyme ataxin-3. *J Cell Biol*, **174** (7), 963–971.
118 Zhong, X. and Pittman, R.N. (2006) Ataxin-3 binds VCP/p97 and regulates retrotranslocation of ERAD substrates. *Hum Mol Genet*, **15** (16), 2409–2420.
119 Hahnen, E. et al. (2008) Histone deacetylase inhibitors: possible implications for neurodegenerative disorders. *Expert Opin Investig Drugs*, **17** (2), 169–184.
120 Zala, D. et al. (2008) Phosphorylation of mutant huntingtin at S421 restores anterograde and retrograde transport in neurons. *Hum Mol Genet*, **17** (24), 3837–3846.
121 Lee, W.C., Yoshihara, M., and Littleton, J.T. (2004) Cytoplasmic aggregates trap polyglutamine-containing proteins and block axonal transport in a Drosophila model of Huntington's disease. *Proc Natl Acad Sci U S A*, **101** (9), 3224–3229.
122 Szebenyi, G. et al. (2003) Neuropathogenic forms of huntingtin and androgen receptor inhibit fast axonal transport. *Neuron*, **40** (1), 41–52.
123 Piccioni, F. et al. (2002) Androgen receptor with elongated polyglutamine tract forms aggregates that alter axonal trafficking and mitochondrial distribution in motor neuronal processes. *FASEB J*, **16** (11), 1418–1420.
124 Bennett, E.J. et al. (2007) Global changes to the ubiquitin system in Huntington's disease. *Nature*, **448** (7154), 704–708.
125 Bennett, E.J. et al. (2005) Global impairment of the ubiquitin-proteasome system by nuclear or cytoplasmic protein aggregates precedes inclusion body formation. *Mol Cell*, **17** (3), 351–365.
126 Venkatraman, P. et al. (2004) Eukaryotic proteasomes cannot digest polyglutamine sequences and release them during degradation of polyglutamine-containing proteins. *Mol Cell*, **14** (1), 95–104.

127 Pratt, G. and Rechsteiner, M. (2008) Proteasomes cleave at multiple sites within polyglutamine tracts: activation by PA28γ(K188E). *J Biol Chem*, **283** (19), 12919–12925.

128 Nagai, Y. et al. (2007) A toxic monomeric conformer of the polyglutamine protein. *Nat Struct Mol Biol*, **14** (4), 332–340.

129 Li, M. et al. (2007) Soluble androgen receptor oligomers underlie pathology in a mouse model of spinobulbar muscular atrophy. *J Biol Chem*, **282** (5), 3157–3164.

130 Kouroku, Y. et al. (2007) ER stress (PERK/eIF2α phosphorylation) mediates the polyglutamine-induced LC3 conversion, an essential step for autophagy formation. *Cell Death Differ*, **14** (2), 230–239.

131 Kouroku, Y. et al. (2002) Polyglutamine aggregates stimulate ER stress signals and caspase-12 activation. *Hum Mol Genet*, **11** (13), 1505–1515.

132 Thomas, M. et al. (2005) The unfolded protein response modulates toxicity of the expanded glutamine androgen receptor. *J Biol Chem*, **280** (22), 21264–21271.

133 Katsuno, M. et al. (2005) Pharmacological induction of heat-shock proteins alleviates polyglutamine-mediated motor neuron disease. *Proc Natl Acad Sci U S A*, **102** (46), 16801–16806.

134 Sittler, A. et al. (2001) Geldanamycin activates a heat shock response and inhibits huntingtin aggregation in a cell culture model of Huntington's disease. *Hum Mol Genet*, **10** (12), 1307–1315.

135 Waza, M. et al. (2005) 17-AAG, an Hsp90 inhibitor, ameliorates polyglutamine-mediated motor neuron degeneration. *Nat Med*, **11** (10), 1088–1095.

136 Sarkar, S. et al. (2008) Rapamycin and mTOR-independent autophagy inducers ameliorate toxicity of polyglutamine-expanded huntingtin and related proteinopathies. *Cell Death Differ*. **16** (1), 46–56.

137 Yamamoto, A., Cremona, M.L., and Rothman, J.E. (2006) Autophagy-mediated clearance of huntingtin aggregates triggered by the insulin-signaling pathway. *J Cell Biol*, **172** (5), 719–731.

138 Berger, Z. et al. (2006) Rapamycin alleviates toxicity of different aggregate-prone proteins. *Hum Mol Genet*, **15** (3), 433–442.

139 Iwata, A. et al. (2005) HDAC6 and microtubules are required for autophagic degradation of aggregated huntingtin. *J Biol Chem*, **280** (48), 40282–40292.

140 Ravikumar, B. et al. (2004) Inhibition of mTOR induces autophagy and reduces toxicity of polyglutamine expansions in fly and mouse models of Huntington disease. *Nat Genet*, **36** (6), 585–595.

141 Ravikumar, B., Duden, R., and Rubinsztein, D.C. (2002) Aggregate-prone proteins with polyglutamine and polyalanine expansions are degraded by autophagy. *Hum Mol Genet*, **11** (9), 1107–1117.

142 Haass, C. and Selkoe, D.J. (2007) Soluble protein oligomers in neurodegeneration: lessons from the Alzheimer's amyloid β-peptide. *Nat Rev Mol Cell Biol*, **8** (2), 101–112.

143 Lansbury, P.T. and Lashuel, H.A. (2006) A century-old debate on protein aggregation and neurodegeneration enters the clinic. *Nature*, **443** (7113), 774–779.

144 Ross, C.A. and Poirier, M.A. (2005) Opinion: what is the role of protein aggregation in neurodegeneration? *Nat Rev Mol Cell Biol*, **6** (11), 891–898.

145 Young, J.E. et al. (2007) Proteolytic cleavage of ataxin-7 by caspase-7 modulates cellular toxicity and transcriptional dysregulation. *J Biol Chem*, **282** (41), 30150–30160.

146 Gervais, F.G. et al. (1999) Involvement of caspases in proteolytic cleavage of Alzheimer's amyloid-β precursor protein and amyloidogenic A beta peptide formation. *Cell*, **97** (3), 395–406.

147 Stack, E.C., Matson, W.R., and Ferrante, R.J. (2008) Evidence of oxidant damage in Huntington's disease: translational strategies using antioxidants. *Ann NY Acad Sci*, **1147**, 79–92.

148 Rochet, J.C. (2007) Novel therapeutic strategies for the treatment of protein-misfolding diseases. *Expert Rev Mol Med*, **9** (17), 1–34.

149 Gatchel, J.R. and Zoghbi, H.Y. (2005) Diseases of unstable repeat expansion: mechanisms and common principles. *Nat Rev Genet*, **6** (10), 743–755.

150 Chen, Y.Z. et al. (2004) DNA/RNA helicase gene mutations in a form of juvenile amyotrophic lateral sclerosis (ALS4). *Am J Hum Genet*, **74** (6), 1128–1135.

151 Vance, C. et al. (2009) Mutations in FUS, an RNA processing protein, cause familial amyotrophic lateral sclerosis type 6. *Science*, **323** (5918), 1208–1211.

152 Kwiatkowski, T.J. Jr et al. (2009) Mutations in the FUS/TLS gene on chromosome 16 cause familial amyotrophic lateral sclerosis. *Science*, **323** (5918), 1205–1208.

153 van Bilsen, P.H. et al. (2008) Identification and allele-specific silencing of the mutant huntington allele in Huntington's disease patient-derived fibroblasts. *Hum Gene Ther*, **19** (7), 710–719.

154 Wheeler, T.M. et al. (2009) Reversal of RNA dominance by displacement of protein sequestered on triplet repeat RNA. *Science*, **325** (5938), 336–339.

155 Hu, J., et al. (2009) Allele-specific silencing of mutant huntingtin and ataxin-3 genes by targeting expanded CAG repeats in mRNAs. *Nat Biotechnol*, **27** (5), 478–484.

156 Pfister, E.L. et al. (2009) Five siRNAs targeting three SNPs wag provide therapy for three-quarters of Huntington's disease patients. *Curr Biol*, **19** (9), 774–778.

157 Park, I.H. et al. (2008) Disease-specific induced pluripotent stem cells. *Cell*, **134** (5), 877–886.

158 Soldner, F. et al. (2009) Parkinson's disease patient-derived induced pluripotent stem cells free of viral reprogramming factors. *Cell*, **136** (5), 964–977.

159 Henderson, J.T. (2008) Lazarus's gate: challenges and potential of epigenetic reprogramming of somatic cells. *Clin Pharmacol Ther*, **83** (6), 889–893.

17.1.2	Kennedy's Disease (SBMA)	*421*
17.1.3	DRPLA/Smith's Disease	*422*
17.1.4	Spinocerebellar Ataxia 1 (SCA1)	*422*
17.1.5	SCA2	*423*
17.1.6	SCA3/Machado-Joseph Disease	*423*
17.1.7	SCA6	*424*
17.1.8	SCA7	*424*
17.1.9	SCA17	*425*
17.2	Animal Models of Polyglutamine Diseases	*425*
17.2.1	Worm and Fly	*426*
17.2.2	Murine Models	*427*
17.2.3	Critical Issues in Model Dependence	*427*
17.3	Mechanisms of Polyglutamine-Induced Neural Injury	*429*
17.3.1	Conformation of Polyglutamine Sequences	*429*
17.3.2	Polyglutamine Aggregation	*430*
17.3.3	Polyglutamine-Based Disruption of Protein Function	*432*
17.3.4	Toxic Gain of Function in Pathogenic PolyQ Proteins	*433*
17.3.5	Disruption of Ubiquitin-Mediated Proteasomal Degradation and Induction of ER Stress *434*	
17.3.6	CAG Repeats and RNA-Mediated Toxicity	*436*
17.3.7	Questions Remaining	*436*
17.4	Future Perspectives	*437*
	Acknowledgments	*439*
	References	*439*

VOLUME II

Abbreviations *XXXI*

Part Three Examples of Endogenous Toxins Associated with Acquired Diseases or Animal Disease Models *449*

18	**Alcohol-Induced Hepatic Injury** *451*	
	Emanuele Albano	
18.1	Introduction	*451*
18.2	Ethanol Metabolism and Toxicity	*453*
18.2.1	Ethanol Metabolism in the Liver	*453*
18.2.2	Ethanol-Induced Oxidative Stress	*456*
18.2.3	Alcohol Effects on Methionine Metabolism	*459*
18.3	Mechanisms of Alcohol Hepatotoxicity	*460*
18.3.1	Alcoholic Steatosis	*460*
18.3.2	Alcoholic Steatohepatitis	*464*
18.3.2.1	The Mechanism Promoting Inflammatory Reactions in ALD *464*	

14.7.7	Vascular Matrix Studies	364
14.7.8	Findings from Yeast Studies	365
14.7.9	Inflammation	365
14.8	Conclusions	366
	References	367

15   Uric Acid Alterations in Cardiometabolic Disorders and Gout   379
Renato Ippolito, Ferruccio Galletti, and Pasquale Strazzullo

15.1	Introduction	379
15.2	Dual Properties of Uric Acid as Antioxidant or Pro-Oxidant Molecule	380
15.3	Effects of Uric Acid on the Arterial Wall Properties and Endothelial Function	381
15.4	Causes of Hyperuricemia and Hypouricemia	381
15.5	Hyperuricemia, Gout, and Their Treatment	382
15.6	Uric Acid and Cardiovascular Disease	385
	References	387

16   Genetic Defects in Iron and Copper Trafficking   395
Douglas M. Templeton

16.1	Introduction and Terminology	395
16.2	Iron	396
16.2.1	Physiology	396
16.2.2	Hereditary Hemochromatoses	401
16.2.3	African Siderosis	402
16.2.4	Friedreich Ataxia	402
16.2.5	X-Linked Sideroblastic Anemias	403
16.2.6	Hypotransferrinemia	404
16.2.7	Hereditary Hyperferritinemia Cataract Syndrome	404
16.2.8	Hypochromic Microcytic Anemia with Iron Overload	405
16.2.9	Aceruloplasminemia	405
16.3	Copper	406
16.3.1	Physiology	406
16.3.2	ATP7A/B	407
16.3.3	Other Copper Toxicoses	409
16.3.4	Amyotrophic Lateral Sclerosis (ALS)	409
16.3.5	Other Copper-Related Genes	409
16.4	Summary	410
	References	411

17   Polyglutamine Neuropathies: Animal Models to Molecular Mechanisms   419
Kelvin Hui and Jeffrey Henderson

17.1	Neurobiology of Polyglutamine Diseases	419
17.1.1	Huntington's Disease	420

13.2.4.3	Sequestration of Important Biological Compounds 334
13.3	Peroxisomal Fatty Acid Oxidation 335
13.3.1	β-Oxidation 335
13.3.1.1	Acyl-CoA Oxidase 336
13.3.1.2	Bifunctional Protein 337
13.3.1.3	ABCD1 337
13.3.2	α-Oxidation 338
13.3.2.1	Phytanoyl-CoA Hydroxylase 338
13.3.2.2	2-Methyl-acyl-CoA Racemase (AMACR) 338
13.3.3	Mechanisms of Toxicity Associated with Disorders of Peroxisomal FAO 339
13.3.3.1	Peroxisomal β-Oxidation Defects 339
13.3.3.2	Peroxisomal α-Oxidation Defects 340
13.4	Conclusions 341
	Acknowledgments 342
	References 342
14	**Homocysteine as an Endogenous Toxin in Cardiovascular Disease** 349
	Sana Basseri, Jennifer Caldwell, Shantanu Sengupta, Arun Kumar, and Richard C. Austin
14.1	Cardiovascular Disease Risk 349
14.2	Overview of Atherosclerosis 349
14.2.1	Lesion Development and Disease Progression 350
14.2.2	Contribution of Atherosclerosis to Cardiovascular Disease 351
14.3	Homocysteine Metabolism 351
14.4	Hyperhomocysteinemia and CVD 351
14.5	Genetic Causes of Hyperhomocysteinemia 353
14.6	Recent Studies 354
14.6.1	Mouse Models of Hyperhomocysteinemia and Atherosclerosis 355
14.6.2	Animal Study Findings 355
14.6.3	Clinical Trials Examining the Effects of Folate and B Vitamin Supplementation on Cardiovascular Disease 356
14.6.3.1	Folate 356
14.6.3.2	Vitamin B_{12} 357
14.6.3.3	Vitamin B_6 357
14.6.3.4	Riboflavin 358
14.6.4	Effect of Lifestyle on tHcy Levels 358
14.7	Potential Mechanisms through which Hyperhomocysteinemia Contributes to Atherogenesis 359
14.7.1	Protein-S-Homocysteinylation 359
14.7.2	Protein-N-Homocysteinylation 361
14.7.3	Nitric Oxide Production and Oxidative Stress 362
14.7.4	Smooth Muscle Cell Proliferation 363
14.7.5	Endothelial Cell Proliferation 363
14.7.6	ER Stress 364

12.4.2	Troglitazone Hepatotoxicity-Role of Bile Salt Export Pump (BSEP/Bsep) *301*	
12.4.3	Cholestatic Hepatotoxicity Associated with Ethinyl Estradiol: Role of MRP2 *302*	
12.5	Phospholipases and Phospholipidosis *302*	
12.5.1	Lysosomal PhospholipaseA2 Deficiency and Phospholipidosis *303*	
12.5.2	Aminoglycoside Antibiotic-Induced Phospholipidosis and Nephrotoxicity *304*	
12.6	Mechanisms of Toxicity Associated with Accumulated Metabolites during Inborn Errors of Metabolism *305*	
12.7	Pathophysiology of Phenylketonuria and Hyperphenylalaninemia *305*	
12.7.1	Oxidative Stress, Mitochondrial Damage, and Antioxidant Status in Phenylketonuria *306*	
12.7.2	Glucose-Induced Oxidative Stress Mediated Toxicity: Implications in Diabetes *307*	
12.7.3	Drugs-Induced Hyperglycemia *308*	
12.7.4	Antipsychotic-Induced Hyperglycemia and Diabetes *309*	
12.8	Conclusions *309*	
	References *309*	
13	**Mechanisms of Toxicity in Fatty Acid Oxidation Disorders** *317*	
	J. Daniel Sharer	
13.1	Introduction *317*	
13.2	Mitochondrial β-Oxidation *319*	
13.2.1	Mobilization and Activation of Long Chain Fatty Acids *319*	
13.2.2	Carnitine-Mediated Transport *321*	
13.2.2.1	Plasma Membrane Carnitine Transporter *322*	
13.2.2.2	Carnitine Palmitoyl Transferase Type 1 (CPT 1) *322*	
13.2.2.3	Carnitine Acylcarnitine Translocase (CACT) *323*	
13.2.2.4	Carnitine Palmitoyl Transferase Type 2 (CPT 2) *324*	
13.2.3	β-Oxidation Enzymes *324*	
13.2.3.1	Very Long Chain Acyl-CoA Dehydrogenase (VLCAD) *325*	
13.2.3.2	Long Chain Acyl-CoA Dehydrogenase (LCAD) *325*	
13.2.3.3	Medium Chain Acyl-CoA Dehydrogenase (MCAD) *326*	
13.2.3.4	Short Chain Acyl-CoA Dehydrogenase (SCAD) *326*	
13.2.3.5	3-Hydroxy Acyl CoA Dehydrogenases *326*	
13.2.3.6	Mitochondrial β-Ketothiolase *327*	
13.2.3.7	Mitochondrial Trifunctional Protein (MTP) *328*	
13.2.3.8	Electron Transfer Flavoprotein (ETF)/Electron Transfer Flavoprotein Dehydrogenase (ETFDH) *328*	
13.2.4	Mechanisms of Toxicity Associated with Disorders of Mitochondrial β-Oxidation *328*	
13.2.4.1	Energy Depletion *329*	
13.2.4.2	Accumulation of Toxic Compounds *331*	

Part Two Genetics: Endogenous Toxins Associated with Inborn Errors of Metabolism *267*

11 Oxalate and Primary Hyperoxaluria *269*
Christopher J. Danpure
11.1 Oxalate as an Endogenous Toxin *269*
11.2 Hereditary Oxalate Overproduction – The Primary Hyperoxalurias *270*
11.2.1 Clinical Characteristics *270*
11.2.2 Enzyme and Metabolic Defects *271*
11.2.3 Crystal Structures of AGT and GRHPR *273*
11.2.4 Genes, Mutations, and Polymorphisms *274*
11.2.5 Genotype–Phenotype Correlations *275*
11.2.5.1 Pro11Leu *275*
11.2.5.2 Gly170Arg *277*
11.2.5.3 Ile244Thr *278*
11.2.5.4 Gly41Arg *278*
11.2.5.5 Gly82Glu *279*
11.2.5.6 Ser205Pro *279*
11.2.5.7 Other Mutations *279*
11.2.6 Diagnosis *280*
11.2.7 Treatment *280*
11.3 Cytotoxicity of Oxalate, Calcium Oxalate, and Related Metabolites *283*
11.3.1 Indirect Glycolate Toxicity and Screening for Chemotherapeutic Agents *283*
11.4 Conclusions *284*
References *284*

12 Pathophysiology of Endogenous Toxins and Their Relation to Inborn Errors of Metabolism and Drug-Mediated Toxicities *291*
Vangala Subrahmanyam
12.1 Introduction *292*
12.2 Pathophysiology of Hepatobiliary System *293*
12.2.1 Arginine Vasopressin and Bile Flow *294*
12.3 Inborn Errors of Bile Acid/Salt Transporter Deficiencies *295*
12.3.1 Progressive Familial Cholestasis Type 1 (PFIC1) and Benign Recurrent Intrahepatic Cholestasis (BRIC) *295*
12.3.2 Progressive Familial Cholestasis Types 2 and 3 (PFIC2 and PFIC3) *298*
12.3.3 Dubin-Johnson Syndrome *299*
12.3.4 Sitosterolemia *299*
12.4 Drugs and Cholestatic Liver Disease *299*
12.4.1 Bosentan Hepatotoxicity – Role of Inhibition of Bile Acid/Salt Transporters *300*

8.3.3	Cytotoxicity for PC12 Cells	218
8.4	Glyceraldehyde-Derived AGEs	219
8.5	Cytotoxicity and Oxidative Stress	220
	References	223

9	**Estrogens as Endogenous Toxins**	**227**
	Jason Matthews	
9.1	Introduction	228
9.2	Estrogen Synthesis	229
9.3	Receptor-Mediated Estrogen Signaling	229
9.3.1	Estrogen Receptor α and β	230
9.3.2	ER-Mediated Mechanism of Cell Proliferation	232
9.3.3	Rapid Nongenomic Effects of Estrogens	232
9.4	DNA Damage Induced by Estrogen	233
9.5	Oxidative Metabolism of Estrogen	233
9.5.1	Formation of Catechol Estrogens	234
9.5.2	4-Hydroxylated Oxidation of Estrogens by CYP1B1	234
9.5.3	Direct DNA Damage by Estrogen Metabolites	236
9.5.4	Indirect DNA Damage and ROS Generation	239
9.6	Role for Estrogen Receptor in ROS Generation	240
9.6.1	ER in the Mitochondria	241
9.7	Treatment of Estrogen-Dependent Diseases	241
9.8	Conclusions	242
	References	243

10	**Reactive Oxygen Species, Hypohalites, and Reactive Nitrogen Species in Liver Pathophysiology**	**249**
	Hartmut Jaeschke	
10.1	Introduction	249
10.2	Reactive Oxygen Species	250
10.2.1	Nature of Reactive Oxygen Species	250
10.2.2	Sources of ROS and Their Pathophysiological Relevance	250
10.2.3	Mechanisms of ROS-Mediated Cell Death	253
10.3	Reactive Nitrogen Species	253
10.3.1	Sources of Peroxynitrite	253
10.3.2	Peroxynitrite and Acetaminophen Hepatotoxicity	254
10.4	Hypohalites	256
10.4.1	Sources of Hypochlorous Acid	256
10.4.2	Hypochlorite and Host Defense Functions	256
10.4.3	Hypochlorite in Liver Pathophysiology	256
10.5	Summary	259
	References	259

7.5.3.4	DHA Oxidative Stress Enhanced Toxicity: Inflammation Model *190*
7.5.4	Glycolaldehyde (GOA) *190*
7.5.4.1	GOA Enzymic Formation *190*
7.5.4.2	GOA Detoxication Metabolizing Enzymes *191*
7.5.4.3	GOA Toxicity Mechanisms *192*
7.5.4.4	GOA Oxidative Stress Enhanced Toxicity: Inflammation Model *192*
7.5.5	Glyoxal (G) and Methylglyoxal (MG) *193*
7.5.5.1	G and MG Enzymic and Nonenzymic Formation *193*
7.5.5.2	Gloxal and MG Detoxification Metabolizing Enzymes *193*
7.5.5.3	Glyoxal and MG Toxicity Mechanisms *196*
7.5.5.4	Glyoxal and MG Oxidative Stress Enhanced Cytotoxicity: Inflammation Model *197*
7.5.6	Glycolate as a Peroxisomal ROS Source *198*
7.5.6.1	Glycolate (GL) Enzymic Formation *198*
7.5.6.2	GL Detoxification Metabolizing Enzymes *198*
7.5.6.3	GL Toxicity *198*
7.5.6.4	GL Oxidative Stress Enhanced Toxicity: Inflammation Model *198*
7.5.7	Glyoxylate as a Carbonyl Toxin *198*
7.5.7.1	Glyoxylate (Gx) Enzymic Formation *198*
7.5.7.2	Gx Detoxication Metabolizing Enzymes *199*
7.5.7.3	Gx Toxicity Mechanisms *199*
7.5.7.4	Glyoxylate Oxidative Stress Enhanced Cytotoxicity: Inflammation Model *200*
7.5.8	Oxalic Acid *200*
7.5.8.1	Oxalic Acid Enzymic Formation *200*
7.5.8.2	Oxalic Acid Detoxication Metabolizing Enzymes and Toxicogenetic Basis *200*
7.5.8.3	Oxalic Acid Toxicity Mechanisms: Hepatic Endogenous Metabolite and Kidney Toxin *201*
7.5.8.4	Oxalic Acid Oxidative Stress Enhanced Toxicity: Inflammation Model *201*
7.6	Cancer Risk and Genotoxicity of Fructose or Carbonyl Metabolites *202*
7.7	Disease Prevention by Fruits and Vegetables versus Fructose Concern *203*
7.8	Conclusions *203*
	References *204*
8	**Glyceraldehyde-Related Reaction Products** *213*
	Teruyuki Usui, Hirohito Watanabe, and Fumitaka Hayase
8.1	Maillard Reaction *213*
8.2	Maillard-Related Diseases *214*
8.3	Glyceraldehyde-Modified Protein *215*
8.3.1	Cytotoxicity for HL-60 Cells *216*
8.3.2	Cytotoxicity for Caco-2 Cells *217*

6.6	Glyceraldehyde 3-phosphate Dehydrogenase	*159*
6.7	Triosephosphate Isomerase	*159*
6.7.1	Glycolaldehyde	*159*
6.7.2	Short Chain Sugars in Diabetes and Aging	*160*
6.7.2.1	Diabetes	*160*
6.7.2.2	Aging, Glycation, and Dietary Restriction	*161*
	Acknowledgment	*162*
	References	*162*
7	**Fructose-Derived Endogenous Toxins**	*173*
	Peter J. O'Brien, Cynthia Y. Feng, Owen Lee, Q. Dong, Rhea Mehta, Jeff Bruce, and W. Robert Bruce	
7.1	Introduction	*174*
7.2	Recent Increased Consumption of Fructose	*174*
7.3	Health Concerns Associated with High Chronic Consumption of Fructose	*176*
7.3.1	Sugar and Sugar Intolerances	*176*
7.3.2	Fructose and Obesity – Nonalcoholic Liver Disease (NAFLD)	*177*
7.4	Sugars as a Source of Endogenous Reactive Carbonyl Formation and AGEs	*178*
7.4.1	Fructose versus Glucose Metabolism by Glycolysis to Carbonyl Metabolites	*178*
7.4.2	Fructose Oxidation to Endogenous Genotoxic Carbonyl Products and Protein Advanced Glycation End Products (AGEs)	*179*
7.4.3	Rodent Models for Fatty Liver (Steatosis) and NASH (Steatohepatitis)	*183*
7.4.4	The Two-Hit Hypothesis for NASH Hepatotoxicity	*184*
7.4.4.1	Introduction	*184*
7.4.4.2	The First Hit: Steatosis Mechanisms and Fructose	*184*
7.4.4.3	The Second Hit: Hepatocyte Inflammation Oxidative Stress Model with Fructose	*185*
7.4.4.4	The Second Hit: ROS Formation Mechanisms	*187*
7.4.4.5	Fructose-Induced *In Vivo* Liver Toxicity Studies in Rats	*187*
7.5	Rat Hepatocyte Studies on Endogenous Toxins Formed by Fructose Metabolism and/or Oxidation	*188*
7.5.1	Introduction	*188*
7.5.2	Glyceraldehyde (GA)	*188*
7.5.2.1	GA Enzymic Formation	*188*
7.5.2.2	GA Detoxication Metabolizing Enzymes	*188*
7.5.2.3	GA Toxicity Mechanisms	*189*
7.5.2.4	GA Oxidative Stress Enhanced Toxicity: Inflammation Model	*189*
7.5.3	Dihydroxyacetone (DHA)	*190*
7.5.3.1	DHA Enzymic or Nonenzymic Formation	*190*
7.5.3.2	DHA Enzymic Metabolism or Oxidation	*190*
7.5.3.3	DHA Toxicity Mechanisms	*190*

4.13	Future Directions *123*
	Acknowledgments *124*
	References *124*

5	**Iron from Meat Produces Endogenous Procarcinogenic Peroxides** *133*
	Denis E. Corpet, Françoise Guéraud, and Peter J. O'Brien
5.1	Introduction *133*
5.2	Toxic Effects of Iron: Molecular Mechanisms *134*
5.2.1	Fe-/Cu-Catalyzed Fenton Chemistry *135*
5.2.2	Fe-/Cu-Catalyzed Oxidation of Proteins *135*
5.2.3	Fe-/Cu-Catalyzed Protein Carbonylation *136*
5.2.4	Fe-/Cu-Catalyzed Advanced Glycation End Products (AGEs) Formation *136*
5.2.5	Fe-/Cu-Catalyzed Lipid Peroxidation and Advanced Lipoxidation End Products (ALEs) *138*
5.2.6	Fe-/Cu-Catalyzed Oxidation of Nucleic Acids *138*
5.2.7	Role of Lysosomal Fe/Cu in Oxidative Stress Cytotoxicity *139*
5.2.8	Role of ROS and Iron in Cellular Redox Signaling that Could Cause Cell Transformation *140*
5.3	Procarcinogenic Effects of Iron: *In vitro* Studies *140*
5.3.1	Iron in Food *140*
5.3.2	Iron in Cell Culture *141*
5.4	Procarcinogenic Effects of Iron in the Gut *141*
5.4.1	Free Radicals in Gut Lumen *141*
5.4.2	Hemin Effect in the Colon of Rats *141*
5.4.3	Iron-Induced Lipid Peroxidation Products *142*
5.4.4	Effect of Heme Iron on Experimental Carcinogenesis *142*
5.4.5	Carcinogenesis Prevention *142*
5.5	Procarcinogenic Effects of Iron Inside the Body *143*
5.5.1	Iron Concentration is Tightly Limited in Body Fluids *143*
5.5.2	Iron Overload Causes Severe Diseases *143*
5.5.3	Colorectal Cancer and Iron Overload *144*
5.6	General Conclusion *144*
	References *144*

Sub-Part	**Molecular Toxicology Mechanisms of Dietary Endogenous Toxins** *151*

6	**Short Chain Sugars as Endogenous Toxins** *153*
	Ludmil T. Benov
6.1	Definition and Properties *153*
6.2	Short Chain Sugars and Reactive Oxygen Species *154*
6.3	Dicarbonyls *156*
6.4	Direct Reactions with Biomolecules *156*
6.5	Short Chain Sugars and Their Sources *157*
6.5.1	Trioses and Triosephosphates *157*

3.5	Examples and Potential Application of Immuno-Spin Trapping (IST) for the Detection of Macromolecule Radicals 80	
3.5.1	Superoxide-Induced Macromolecule Radicals 80	
3.5.2	Singlet Oxygen–Induced Macromolecule Radicals 81	
3.5.3	Hydroxyl Radical–Induced Macromolecule Radicals 82	
3.5.4	Peroxynitrite, Nitrogen Dioxide Radicals, and Nitric Oxide 83	
3.5.5	Carbonate Radicals 84	
3.5.6	HOCl-Induced Macromolecule Radicals 85	
3.5.7	Thiyl Radical–Induced Macromolecule Radicals 86	
3.5.8	Myeloperoxidase Protein Radical Detection 87	
3.5.9	Heme Protein Radicals–A Unique Exception of Substrate-Induced Protein Radical Formation 88	
3.5.10	Example of *In vivo* Detection of Macromolecule Radicals Using IST 89	
3.5.11	Future Directions 91	
3.6	Immunological Detection of Oxidized Tryptophan Residues in Proteins 92	
	References 92	

4	**Alcohol-Derived Bioadducts** 103	
	Geoffrey M. Thiele and Lynell W. Klassen	
4.1	Introduction 103	
4.2	The Generation of Reactive Metabolites 104	
4.3	Ethanol Metabolites React with Proteins 106	
4.4	Acetaldehyde Adducts 107	
4.4.1	Formation and Types of Adducts 107	
4.4.2	Biological Detection and/or Significance 109	
4.5	Malondialdehyde 111	
4.5.1	Formation and Types of Adducts 111	
4.5.2	Biological Detection and/or Significance 113	
4.6	Malondialdehyde–Acetaldehyde (MAA) Adducts 114	
4.6.1	Formation and Types of Adducts 114	
4.6.2	Biological Detection and/or Significance 114	
4.7	4-Hydroxyalkenals 116	
4.7.1	Formation and Types of Adducts 116	
4.7.2	Biological Detection and/or Significance 116	
4.8	Free Radical–Derived Adducts 117	
4.8.1	Formation and Detection of Adducts 117	
4.8.2	Biological Detection and/or Significance 117	
4.9	5-Deoxy-D-xylulose-1-phosphate (DXP) 118	
4.10	Acetaldehyde–Glucose Amadori Product Adduct 119	
4.11	Immune Responses to Proteins Modified with Alcohol Metabolites 119	
4.12	The Role of Alcohol-Derived Bioadducts in Alcohol-Related Tissue Injury 121	

1.5	Summary 28
	References 28

2 Modification of Cysteine Residues in Protein by Endogenous Oxidants and Electrophiles 43

Norma Frizzell and John W. Baynes

2.1	Introduction 43
2.2	Autoxidation of Cysteine 44
2.3	Glutathione 44
2.3.1	Glutathionylation 45
2.3.2	Glutathione as an Electrophile Trap 45
2.4	Lipoxidation 46
2.4.1	Glyceraldehyde-3-Phosphate Dehydrogenase (GAPDH) 47
2.4.2	Albumin 48
2.4.3	Actin 48
2.4.4	Quantitative Analysis 49
2.5	Maillard Reactions of Cysteine 50
2.5.1	Transglycosylation 50
2.5.2	Carboxyalkylation 51
2.6	Succination 53
2.6.1	Discovery and Metabolic Origin of S-(2-Succinyl)cysteine (2SC) 53
2.6.2	2SC in Skeletal Muscle in Diabetes 55
2.6.3	2SC in Adipose Tissue in Diabetes 55
2.6.4	Overview on Succination of Protein 57
2.7	Conclusion 57
	Acknowledgment 58
	References 58

3 Endogenous Macromolecule Radicals 65

Arno G. Siraki and Marilyn Ehrenshaft

3.1	Introduction 65
3.2	What is a Free Radical? 66
3.2.1	Superoxide Anion 67
3.2.2	Singlet Oxygen 68
3.2.3	Hydrogen Peroxide 68
3.2.4	Peroxynitrite and Carbonate Radicals 69
3.2.5	Nitrogen-Based Radicals 70
3.2.6	Hypochlorous Acid 71
3.2.7	Thiyl Radicals 72
3.2.8	Lipid Peroxidation 73
3.2.9	Summary 73
3.3	Free Radicals as Initiators of Macromolecule Damage 74
3.4	Macromolecule Radical Formation Induced by Small Molecule Free Radicals 78

Contents

VOLUME I

Preface *XXXI*

List of Contributors *XXXIII*

Abbreviations *XLIII*

Part One   Endogenous Toxins Associated with Excessive Sugar, Fat, Meat, or Alcohol Consumption   *1*

Sub-Part   Chemistry and Biochemistry   *3*

1	**Endogenous DNA Damage** *5*	
	Erin G. Prestwich and Peter C. Dedon	
1.1	Introduction *5*	
1.2	Oxidatively Damaged DNA *8*	
1.2.1	Base Oxidation *9*	
1.2.2	Tandem Lesions: Multiple Products Derived from a Single Oxidation Event *12*	
1.2.3	Base Nitration *13*	
1.2.4	Halogenation *14*	
1.2.5	2-Deoxyribose Oxidation *14*	
1.3	DNA Alkylation by Endogenous Electrophiles *15*	
1.3.1	Lipid Peroxidation-Derived Adducts *17*	
1.3.2	Malondialdehyde-Like Adducts *21*	
1.3.3	DNA Glycation *22*	
1.3.4	Estrogen-Derived DNA Adducts *24*	
1.4	DNA and RNA Deamination *25*	
1.4.1	Hydrolytic Deamination *26*	
1.4.2	Nitrosative Chemistry and Inflammation *26*	
1.4.3	Enzymatic Deamination *27*

Endogenous Toxins. Diet, Genetics, Disease and Treatment.
Edited by Peter J. O'Brien and W. Robert Bruce
Copyright © 2010 WILEY-VCH Verlag GmbH & Co. KGaA, Weinheim
ISBN: 978-3-527-32363-0

The Authors

Albano, Emanuele | Austin, Rick | Baynes, John | Beisswenger, Paul | Benov, Ludmil | Bilodeau, Marc
Bruce, Robert | Cani, Patrice | Chan, Tom S. | Corpet, Denis | Danpure, Christopher | Dedon, Peter
Delzenne, Nathalie | Doorn, Jonathan | Eyssen, Gail | Frizzell, Norma | Giacca, Adria | Gieseg, Steven
Hayase, Fumitaka | Henderson, Jeffrey | Herrmann, Wolfgang | Hui, Kelvin | Hyogo, HideyukuJ | Jaeschke, Hartmut
Kim, Hyun-Jung | Klassen, Lyn | Kong, Tony | Labandeira-García, José | Martin, Lisa | Mathews, Jason
Monestier, Marc | Newmark, Harold | O'Brien, Peter | Obeid, Rima | Pamplona, Reinald | Rao, R. K.
Sharer, Daniel | Shiraldi, Michael | Siraki, Arno | Strazzullo, Pasquale | Surh, Young-Joon | Tazuma, Susuma
Templeton, Douglas | Thiele, Geoff | Thompson, Henry | Vangala, Mani | Wright, Jim | Yamagishi, Sho-ichi

The Editors

Prof. Peter J. O'Brien
University of Toronto
Faculty of Pharmacy, Room 1004
College Street 144
Toronto, ON M5S 3M2
Canada

Prof. W. Robert Bruce
University of Toronto
Fitz Gerald Building, Room 342
College Street 150
Toronto, ON M5S 3E2
Canada

■ All books published by Wiley-VCH are carefully produced. Nevertheless, authors, editors, and publisher do not warrant the information contained in these books, including this book, to be free of errors. Readers are advised to keep in mind that statements, data, illustrations, procedural details or other items may inadvertently be inaccurate.

Library of Congress Card No.:
applied for

British Library Cataloguing-in-Publication Data
A catalogue record for this book is available from the British Library.

Bibliographic information published by the Deutsche Nationalbibliothek
The Deutsche Nationalbibliothek lists this publication in the Deutsche Nationalbibliografie; detailed bibliographic data are available on the Internet at <http://dnb.d-nb.de>.

© 2010 WILEY-VCH Verlag GmbH & Co. KGaA, Weinheim

All rights reserved (including those of translation into other languages). No part of this book may be reproduced in any form – by photoprinting, microfilm, or any other means – nor transmitted or translated into a machine language without written permission from the publishers. Registered names, trademarks, etc. used in this book, even when not specifically marked as such, are not to be considered unprotected by law.

Composition Laserwords Private Limited, Chennai, India
Printing Strauss GmbH, Mörlenbach
Bookbinding Litges & Dopf GmbH, Heppenheim
Cover Design Adam Design, Weinheim

Printed in the Federal Republic of Germany
Printed on acid-free paper

ISBN: 978-3-527-32363-0

Endogenous Toxins

Diet, Genetics, Disease and Treatment

Edited by
Peter J. O'Brien and W. Robert Bruce

VOLUME II

WILEY-VCH Verlag GmbH & Co. KGaA

Further Reading

Geacintov, N. E., Broyde, S. (eds.)

The Chemical Biology of DNA Damage

2010
ISBN: 978-3-527-32295-4

Külpmann, W. R. (ed.)

Clinical Toxicological Analysis

Procedures, Results, Interpretation

2009
ISBN: 978-3-527-31890-2

Knasmüller, S., DeMarini, D. M., Johnson, I., Gerhäuser, C. (eds.)

Chemoprevention of Cancer and DNA Damage by Dietary Factors

2008
ISBN: 978-3-527-32058-5

Meyers, R. A. (ed.)

Cancer

From Mechanisms to Therapeutic Approaches

2007
ISBN: 978-3-527-31768-4

Dübel, S. (ed.)

Handbook of Therapeutic Antibodies

2007
ISBN: 978-3-527-31453-9

Brigelius-Flohé, R., Joost, H.-G. (eds.)

Nutritional Genomics

Impact on Health and Disease

2006
ISBN: 978-3-527-31294-8

Endogenous Toxins

Edited by
Peter J. O'Brien and
W. Robert Bruce

18.3.2.2	Role of Immune Reactions in Alcohol-Induced Hepatic Inflammation *466*	
18.3.2.3	Mechanisms in Alcohol-Mediated Hepatocyte Killing *468*	
18.3.2.4	Alcohol and the Formation of Mallory Bodies *469*	
18.3.3	Alcoholic Cirrhosis *469*	
18.3.4	Alcohol and Liver Cancer *472*	
18.4	Conclusions *474*	
	References *475*	
19	**Ethanol-Induced Endotoxemia and Tissue Injury** *485*	
	Radhakrishna K. Rao	
19.1	Alcoholic Endotoxemia *486*	
19.1.1	Endotoxemia in Patients with ALD *486*	
19.1.2	Endotoxemia in Experimental Models of ALD *488*	
19.1.3	Gender Differences in Endotoxemia *489*	
19.1.4	Bacterial Translocation *489*	
19.2	Causes of Alcoholic Endotoxemia *490*	
19.2.1	Delayed Endotoxin Clearance *490*	
19.2.2	Bacterial Overgrowth *491*	
19.2.3	Increased Gut Permeability to Endotoxins *491*	
19.3	Influence of Alcoholic Endotoxemia on Different Organs *493*	
19.3.1	Liver *493*	
19.3.2	Pancreas *495*	
19.3.3	Lung *495*	
19.4	Mechanism of Tissue Damage by Alcoholic Endotoxemia *496*	
19.4.1	Cellular Targets *496*	
19.4.2	Receptors and Signaling *497*	
19.5	Factors that Ameliorate Alcoholic Endotoxemia and Tissue Damage *498*	
19.5.1	Synthetic Drugs *498*	
19.5.2	Dietary Components *499*	
19.5.3	Plant Extracts *499*	
19.5.4	Probiotics *500*	
19.6	Summary and Perspectives *500*	
	Acknowledgments *501*	
	References *501*	
20	**Gut Microbiota, Diet, Endotoxemia, and Diseases** *511*	
	Patrice D. Cani and Nathalie M. Delzenne	
20.1	Introduction *511*	
20.2	Gut Microbiota and Energy Homeostasis *512*	
20.2.1	Gut Microbiota and Energy Harvest *512*	
20.2.1.1	Gut Microbiota and Adipose Tissue Development *513*	
20.2.1.2	Lipogenesis *513*	
20.2.1.3	Specific SCFA Receptors *514*	

20.3	Energy Harvest, Obesity, and Metabolic Disorders: Paradoxes? *514*	
20.4	Role of the Gut Microbiota in the Inflammatory Associated with Obesity *515*	
20.4.1	Metabolic Endotoxemia and Metabolic Disorders *516*	
20.4.2	Metabolic Endotoxemia and Nutritional Intervention *517*	
20.4.3	Metabolic Endotoxemia, Gut Microbiota, and Fatty Liver Diseases *518*	
20.4.4	Selective Changes in Gut Microbiota and NASH *518*	
20.5	Metabolic Endotoxemia and High-Fat Feeding: Human Evidence *519*	
20.6	Conclusion *520*	
	References *520*	
21	**Nutrient-Derived Endogenous Toxins in the Pathogenesis of Type 2 Diabetes at the β-Cell Level** *525*	
	Christine Tang, Andrei I. Oprescu, and Adria Giacca	
21.1	Introduction *525*	
21.2	Acute Effect of Glucose and FFA on Insulin Secretion *526*	
21.3	Insulin Secretory Abnormalities in Type 2 Diabetic Patients *527*	
21.4	β-Cell Glucotoxicity *528*	
21.4.1	Chronic Effect of Glucose on β-Cell Function and Mass *528*	
21.4.2	Level of Impairment of β-Cell Function by Chronic Glucose Exposure *529*	
21.4.2.1	Insulin Gene Transcription *529*	
21.4.2.2	Insulin Biosynthesis *530*	
21.4.2.3	ATP Production *531*	
21.4.2.4	Late Stages of Insulin Secretion *531*	
21.4.2.5	β-Cell Mass *532*	
21.5	β-Cell Lipotoxicity *532*	
21.5.1	Chronic Effect of Free Fatty Acids on β-Cell Function and Mass *532*	
21.5.2	Level of Impairment of β-Cell Function by Chronic Free Fatty Acids Exposure *533*	
21.5.2.1	Insulin Gene Transcription *533*	
21.5.2.2	Insulin Biosynthesis *533*	
21.5.2.3	ATP Production *534*	
21.5.2.4	Late Stages of Insulin Secretion *534*	
21.5.2.5	β-Cell Mass *535*	
21.6	Glucolipotoxicity *535*	
21.7	Oxidative Stress as an Endogenous Toxin *536*	
21.7.1	Reactive Oxygen Species as a Derived Toxin from Chronic Glucose and Fat Exposure *536*	
21.7.2	Sites of Reactive Oxygen Species Generation *537*	
21.7.3	Oxidative Stress, Type 2 Diabetes, and β-Cell Dysfunction *537*	
21.8	Oxidative Stress, β-Cell Glucotoxicity, and Lipotoxicity *538*	
21.8.1	Evidence for a Role of Oxidative Stress in β-Cell Glucotoxicity and Lipotoxicity *538*	

21.8.2	Sites of Oxidative Stress–Induced Impairment of β-Cell Function by Glucotoxicity and Lipotoxicity *540*	
21.8.2.1	Glucose Oxidation and Uncoupling Protein 2 *540*	
21.8.2.2	Insulin Gene Transcription and Insulin Biosynthesis *540*	
21.8.3	Downstream Signaling Mechanisms of Oxidative Stress–Induced Impairment of β-cell Function by Glucotoxicity and Lipotoxicity *541*	
21.8.3.1	JNK *541*	
21.8.3.2	NFκB *541*	
21.8.3.3	β-Cell Insulin Resistance *542*	
21.8.3.4	Endoplasmic Reticulum (ER) Stress *542*	
21.9	Conclusion *542*	
	References *543*	

22 **Endogenous Toxins and Susceptibility or Resistance to Diabetic Complications** *557*
Paul J. Beisswenger

22.1	Introduction and Background *557*
22.1.1	Elevated Glucose, Pathways of Glycation/Oxidation, and Diabetic Complications *557*
22.2	Synthetic Pathways for Glycation Products *558*
22.2.1	Amadori Products *558*
22.2.2	Methylglyoxal (MG) *558*
22.2.3	3-Deoxyglucosone (3DG) *559*
22.2.4	Glyoxal *559*
22.2.5	Synthetic Pathways for Oxidation Products *559*
22.2.6	Formation of Advanced Glycation End-Products (AGEs) *560*
22.3	Enzymatic Deglycation Pathways *560*
22.3.1	Deglycation Systems for Amadori Products *561*
22.3.2	Methylglyoxal (MG) Detoxification *561*
22.3.3	Detoxification Pathways for 3-Deoxyglucosone (3DG) *562*
22.3.4	Endogenous Binders of α-Dicarbonyls as Protective Mechanisms *562*
22.3.5	The Removal of AGEs *562*
22.3.6	Pathways that Lead to Removal of Oxidative Products *563*
22.4	Susceptibility to Diabetic Complications Varies Widely among Individuals *563*
22.4.1	Slow and Rapid Nonenzymatic Glycation and Propensity to Diabetic Complications *564*
22.4.2	Impaired Deglycation of Amadori Products and Diabetic Complications *564*
22.5	Overproduction of α-Dicarbonyls is Characteristic of Individuals Who are Prone to Diabetic Complications *566*
22.5.1	Impaired Degradation of Methylglyoxal and Diabetic Complications *567*
22.5.2	3DG Detoxification and Diabetic Complications *568*

22.6	Oxidative Stress and Propensity to Diabetic Nephropathy 569
22.6.1	Nitrosative Stress and Diabetic Complications 570
22.6.2	Protective Upregulation of Antioxidant Enzymes and Diabetic Complications 570
22.7	Conclusions 571
	References 571
23	**Serum Advanced Glycation End Products Associated with NASH and Other Liver Diseases** 577
	Hideyuki Hyogo, Sho-ichi Yamagishi, and Susumu Tazuma
23.1	Introduction 577
23.2	Formation Pathways of AGEs 578
23.3	AGEs' Association in the Liver 579
23.4	Circulating AGEs and Liver Disease 581
23.4.1	Circulating AGEs in Liver Cirrhosis 581
23.4.2	Circulating AGEs in NASH 581
23.5	Possible Molecular Mechanisms by which the AGEs–RAGE System is Involved in Liver Disease 584
23.5.1	AGEs in Hepatic Sinusoidal Endothelial Cells 584
23.5.2	AGEs in Hepatic Stellate Cells 584
23.5.3	AGEs in Hepatocytes 585
23.5.4	RAGE Involvement in Liver Diseases 588
23.6	Conclusions 589
	Acknowledgment 589
	References 589
24	**Oxidative Stress in the Pathogenesis of Hepatitis C** 595
	Tom S. Chan and Marc Bilodeau
24.1	Hepatitis C 595
24.1.1	Incidence and Economical Burden of Hepatitis C 596
24.1.2	The Hepatitis C Virus 597
24.1.3	Conventional Treatment of Hepatitis C 598
24.2	The Molecular Basis of Oxidative Stress in Hepatitis C 599
24.2.1	Oxidative Stress in HCV 599
24.2.2	ROS Induced by Nonparenchymal Cells of the Liver (Kupffer Cells, Lymphocytes, Stellate Cells) 601
24.2.2.1	Innate and Adaptive Immunity in HCV: the Snowball Effect 601
24.2.2.2	Phagocyte NADPH Oxidase (NOX2) 602
24.2.2.3	Nonphagocyte NOX 603
24.2.3	Pro-oxidant Components of HCV 604
24.2.3.1	Oxidative Stress Induced by HCV Core Protein 605
24.2.3.2	Oxidative Stress Induced by HCV NS5A Protein 606
24.3	The Emerging Role of Antioxidants in the Treatment of HCV 608
24.4	Summary and Conclusions 610
24.4.1	Summary 610

24.4.2	Conclusions	610
	References	610

25		**Oxidized Low Density Lipoprotein Cytotoxicity and Vascular Disease** 619
		Steven P. Gieseg, Elizabeth Crone, and Zunika Amit
25.1		Introduction to Vascular Disease 620
25.2		oxLDL Formation 622
25.3		Toxicity of oxLDL 625
25.4		Types of oxLDL 626
25.5		Cell-Mediated Oxidation and Atherosclerotic Plaques 628
25.6		Endogenous Antioxidants 630
25.7		Mechanism of oxLDL Cytotoxicity 632
25.8		Oxysterol and the Mitochondria 633
25.9		Oxysterols and Calcium 634
25.10		NADPH Oxidase and Apoptosis versus Necrosis 635
25.11		The Future 636
		References 637

26		**Oxidative Stress in Breast Cancer Carcinogenesis** 647
		Lisa J. Martin and Norman Boyd
26.1		Introduction 647
26.2		Markers of Oxidative Stress 648
26.2.1		MDA 649
26.2.2		Isoprostanes 649
26.2.3		Biological Sampling 650
26.3		Oxidative Stress and Breast Cancer 650
26.4		Oxidative Stress and Breast Cancer Risk Factors 651
26.5		Mammographic Density (MD) 652
26.5.1		Mammographic Density and Breast Cancer Risk 653
26.5.2		Mammographic Density and Breast Tissue Composition 654
26.5.3		Age, Other Risk Factors, and Mammographic Density as a Marker of Susceptibility 654
26.5.4		Genetic Factors 655
26.6		Association of Hormones, Mitogens, and Mutagens with Mammographic Density 655
26.6.1		Levels of Blood Hormones and Growth Factors (Mitogens) 658
26.6.2		Urinary MDA and Mammographic Density 659
26.6.3		Urinary Isoprostane and MD 660
26.6.4		Comparison of MDA and Isoprostanes as Markers of Lipid Peroxidation 661
26.7		Potential Mechanisms for the Association of Mitogens and MDA with Mammographic Density and Breast Cancer Risk 662
26.8		Summary 663
		References 664

27		**Lifestyle, Endogenous Toxins, and Colorectal Cancer Risk** *673*
		Gail McKeown-Eyssen, Jeff Bruce, Owen Lee, Peter J. O'Brien, and W. Robert Bruce
	27.1	Introduction *673*
	27.2	Lifestyle Risk Factors for CRC *674*
	27.3	Oxidative Stress Relates to Lifestyle and CRC Risk *676*
	27.4	Energy Excess Relates Lifestyle and CRC Risk *681*
	27.5	Interaction of Toxicity of Energy Excess and Oxidative Stress *683*
	27.6	Future Research *684*
		References *687*
28		**Dopamine-Derived Neurotoxicity and Parkinson's Disease** *695*
		Jose Luis Labandeira-Garcia
	28.1	The Neurotransmitter Dopamine *695*
	28.2	Dopaminergic Alterations – Parkinson's Disease *696*
	28.3	Levodopa Therapy for PD and Progress of Dopaminergic Neuron Degeneration *697*
	28.4	Dopamine Toxicity and Aging *698*
	28.5	Dopamine Toxicity and Animal Models of Parkinsonism *698*
	28.6	Mechanisms of Dopamine Neurotoxicity – Vesicular and Cytosolic Dopamine *700*
	28.7	Neuromelanin Formation and Dopamine Toxicity *701*
	28.8	Interaction between Dopamine Toxicity and Protein Aggregation *702*
	28.9	Receptor-Dependent Dopamine Toxicity, Dopamine-Mediated Excitotoxicity, and Relevance for Huntington's Disease *703*
	28.10	Dopamine and Neuroprotection *704*
	28.11	Conclusion *705*
		Acknowledgments *706*
		References *706*
29		**Dopamine Catabolism and Parkinson's Disease: Role of a Reactive Aldehyde Intermediate** *715*
		Jonathan A. Doorn
	29.1	Dopamine Biosynthesis and Catabolism *715*
	29.2	DOPAL as a Metabolite of DA: Importance to Pharmacology and Toxicology *718*
	29.3	DOPAL Metabolism *719*
	29.4	Mechanisms for Elevation in Levels of DOPAL *720*
	29.5	DOPAL Toxicity *722*
	29.6	DOPAL Reactivity with Proteins *723*
	29.7	Relevance of DOPAL to PD: Summary and "Big Picture" *725*
		Acknowledgments *726*
		References *726*

30	**Tetrahydropapaveroline, an Endogenous Dicatechol Isoquinoline Neurotoxin** 733
	Young-Joon Surh and Hyun-Jung Kim
30.1	Introduction 733
30.2	Biosynthesis of THP 734
30.2.1	Nonenzymatic versus Enantioselective Formation 734
30.2.2	Effects of Alcohol 734
30.3	Neurotoxic Potential of THP 735
30.3.1	Parkinsonism 736
30.3.1.1	Inhibition of Dopamine Biosynthesis 736
30.3.1.2	Inhibition of Dopamine Uptake through the Dopamine Transporter 736
30.3.1.3	Inhibition of Mitochondrial Respiration 737
30.3.1.4	Inhibition of Serotonin Production 738
30.3.2	Implications for L-DOPA Paradox 738
30.3.3	Alcohol Dependence 739
30.4	Biochemical Mechanisms Underlying THP-Induced Neurotoxicity 740
30.4.1	Redox Cycling 740
30.4.2	Oxidative Cell Death 740
30.4.3	DNA Damage 743
30.5	Cellular Protection against THP-Induced Injuries 743
30.6	Concluding Remarks 744
	References 745
31	**Chemically Induced Autoimmunity** 747
	Michael Schiraldi and Marc Monestier
31.1	Introduction 747
31.1.1	Autoimmunity 748
31.2	Human Disease 749
31.2.1	Drug-Induced Lupus 749
31.2.2	Drug-Induced Immune Hemolytic Anemia 751
31.2.3	Primary Biliary Cirrhosis 752
31.2.4	D-Penicillamine-Induced Myasthenia Gravis 752
31.2.5	Mercury and Systemic Autoimmune Disease 753
31.2.6	Hydrocarbon Exposure and Goodpasture's Syndrome 754
31.2.7	Pesticide Exposure and Autoimmunity 755
31.2.8	Eosinophilic Pneumonia 755
31.3	Animal Models of Chemically Induced Autoimmunity 756
31.3.1	Heavy Metals 756
31.3.2	Pristane 758
31.3.3	Phthalate-Induced Autoimmunity 759
31.3.4	D-Penicillamine-Induced Autoimmunity 760
31.4	Conclusion 760
	References 761

32	**Endogenous Toxins Associated with Life Expectancy and Aging** 769
	Victoria Ayala, Jordi Boada, José Serrano, Manuel Portero-Otín, and Reinald Pamplona
32.1	Introduction 769
32.2	Metabolism(s) 771
32.3	The Rate of Generation of Damage Induced by Mitochondrial Free Radicals and Life Span 772
32.4	Structural Components that are Highly Resistant to Oxidative Stress and Life Span 774
32.4.1	The First Antioxidant Defense Line 774
32.4.2	Resistance to Oxidative Damage and Life Span 775
32.5	Oxidative Stress, Aging, and Dietary Restriction 778
32.5.1	Caloric Restriction 778
32.5.2	Protein Restriction 779
32.5.3	Methionine Restriction 781
32.6	Nutritional Considerations on Methionine: A Public Health Matter 783
	Acknowledgments 783
	References 783

Part Four Therapeutics Proposed for Decreasing Endogenous Toxins 787

33	**Therapeutic Potential for Decreasing the Endogenous Toxin Homocysteine: Clinical Trials** 789
	Wolfgang Herrmann and Rima Obeid
33.1	Homocysteine Metabolism 789
33.2	Physiological Determinants of Plasma Homocysteine Level 791
33.3	Causes of Hyperhomocysteinemia 791
33.4	The Role of Vitamin B_{12} (Cobalamin) 792
33.4.1	Cobalamin Metabolism 792
33.4.2	Sources of Cobalamin 794
33.4.3	Absorption, Excretion, and Homeostasis of Cobalamin 794
33.4.4	Cobalamin Deficiency 795
33.5	Folic Acid, Folate (Vitamin B_9) 795
33.5.1	Folate Metabolism 795
33.5.2	Biological Roles of Folate 796
33.5.3	Consequences of Folate Deficiency 797
33.5.4	Cobalamin Deficiency Causes Folate Trap 798
33.6	Vitamin B_6 798
33.6.1	Function 799
33.6.2	Vitamin B_6 Deficiency 800
33.6.2.1	Vitamin B_6 Deficiency, Hyperhomocysteinemia, and Cardiovascular Diseases 800
33.6.2.2	Vitamin B_6 Deficiency and Cognitive as well as Immune Dysfunction 801

33.6.2.3	Disease Treatment *802*	
33.7	Homocysteine (Hcy) as an Endogenous Toxin – Human and Experimental Studies *802*	
33.7.1	Homocysteine Toxicity in the CNS *802*	
33.7.1.1	Mechanisms of Homocysteine Toxicity in the Central Nervous System *803*	
33.7.1.2	Homocysteine in Dementia and Cognitive Decline *807*	
33.7.1.3	Treatment with B Vitamin Supplements *808*	
33.7.1.4	Homocysteine in Patients with Parkinson Disease *810*	
33.7.1.5	Homocysteine in Cerebrovascular Diseases *810*	
33.7.1.6	Reduction of Stroke Risk – Primary and Secondary Prevention Studies *811*	
33.7.2	Homocysteine in the Endothelial System *812*	
33.7.2.1	Homocysteine-Lowering Treatment in Reduction of Cardiovascular Risk *813*	
33.7.2.2	Can Treating HHcy Provide Protection against Cardiovascular Diseases? *813*	
33.7.3	Homocysteine Toxicity in Alcoholic Liver Disease *815*	
33.7.3.1	Fatty Liver and HHcy are Mutual Findings in Chronic Alcoholism *815*	
33.7.4	Homocysteine and Bone Health *816*	
33.7.4.1	Bone Health and Intervention Studies with B Vitamins *817*	
33.7.5	Homocysteine Toxicity in Patients with Renal Diseases *818*	
33.7.5.1	Role of Hyperhomocysteinemia as Vascular Risk Factor in Patients with ESRD *818*	
33.7.5.2	Renal Function and Hyperhomocysteinemia *818*	
33.7.5.3	Reduced Remethylation Pathway in ESRD *819*	
33.7.5.4	Cellular Uptake of Vitamin B_{12} in Patients with Chronic Renal Failure *820*	
33.7.5.5	Response of Homocysteine to Vitamin Treatment in Dialysis Patients *820*	
33.7.5.6	Effect of Homocysteine-Lowering Treatment on Transmethylation *821*	
	References *821*	
34	**Prevention of Oxidative Stress–Induced Diseases by Natural Dietary Compounds: The Mechanism of Actions** *841*	
	Tin Oo Khor, Ka-Lung Cheung, Avantika Barve, Harold L. Newmark, and Ah-Ng Tony Kong	
34.1	Introduction *841*	
34.2	Cytoprotective Effect of Dietary Compounds: Antioxidant Effects *843*	
34.2.1	Phenolic Dietary Compounds *843*	
34.2.1.1	EGCG *844*	
34.2.1.2	Curcumin *846*	
34.2.1.3	Dibenzoylmethane *847*	

34.2.1.4	Resveratrol	847
34.2.1.5	Genistein	848
34.2.2	Isothiocyanates	848
34.2.3	Garlic Organosulfur Compounds (OSCs)	850
34.2.4	Vitamin E	850
34.3	Conclusion Remarks	851
34.4	Tribute to Professor Harold L. Newmark	851
	Acknowledgments	852
	References	852

35 Genotoxicity of Endogenous Estrogens 859
James S. Wright

35.1	Introduction	859
35.2	Estrogens and Women's Health	862
35.3	Causes of Estrogen Genotoxicity	862
35.3.1	Significance of ERα and ERβ: the Estrogen Receptors	862
35.3.2	Quinone Formation is Tumorigenic	863
35.3.3	Experimental Support for the Quinone Hypothesis	865
35.3.4	Factors Affecting Estrogen Concentrations	869
35.4	Enzymatic Reactions that Affect Estrogen Genotoxicity	870
35.4.1	Reactions that Increase Toxicity	870
35.4.1.1	Catechol Formation from E2	870
35.4.1.2	P450 and the Redox Cycle	871
35.4.1.3	Sulfatase Creates More Free Estrogen	871
35.4.1.4	Aromatase Creates More E2 via E1	871
35.4.2	Reactions that Reduce Toxicity	872
35.4.2.1	COMT Deactivates Catechol	872
35.4.2.2	NQO1 Deactivates the Quinone	873
35.4.2.3	GSH, Ascorbate, and Antioxidant Intervention Experiments	873
35.4.2.4	Sulfotransferase Creates Bound Estrogen	874
35.5	Conclusions	874
	Acknowledgment	876
	References	876

36 Design of Nutritional Interventions for the Control of Cellular Oxidation 881
Elizabeth P. Ryan and Henry J. Thompson

36.1	Overview	881
36.2	Antioxidant–Chronic Disease Conundrum Terms	883
36.3	The Design of an Intervention	884
36.3.1	The Oxidizing Species	884
36.3.2	The Macromolecular Target	885
36.3.3	The Target Tissue	888
36.3.4	The Pathogenesis of the Disease	890
36.3.5	The Antioxidants	891

36.4	Designing the Next Generation of Nutritional Interventions 894	
36.4.1	Define Individuals Who are at Increased Risk for Cancer 894	
36.4.1.1	Risk for Oxidative Stress–Induced Mutations 895	
36.4.1.2	Risk for Altered Signal Transduction 895	
36.4.2	Identify Candidate Mechanisms that Accounts for the Abnormal Levels of Oxidative Products 896	
36.4.3	Define a Nutritional Intervention that is Tailored to Correct the Defect 896	
36.5	Overall Principles 897	
36.6	The Broader Perspective 897	
36.7	Summary 898	
	Acknowledgments 900	
	References 900	

Appendix Questions for Discussion 907

1. What are "endogenous toxins?" 907
2. Why consider "alcohol drinking" in a review of endogenous toxins? 907
3. What endogenous toxins are involved in the development of type 2 diabetes? 908
4. What endogenous sugar/fatty acid toxins are involved in the development of nonalcoholic steatohepatitis (NASH)? 909
5. Inflammation and oxidative stress are involved in many disease processes. Are these chronic diseases a direct result of oxidative stress or are they a result of even more reactive endogenous toxins formed with the interaction of oxidative stress with labile metabolites? 910
6. Endogenous toxins resulting from inborn errors of metabolism appear to be involved in an increasing number of diseases. Can dietary modification (supplements or deletions) reduce the levels of these toxins? 911
7. Do differences in endogenous toxins explain the differences in chronic disease incidence with time and across populations in the world? 912

   References 913

Index 915

Abbreviations

$A\beta$	β-amyloid
AA	aromatic amine
AADC	aromatic acid decarboxylase
AAPH	2,2'-azobis(amidinopropane) dihydrochloride
ABC	ATP-binding cassette (transporter)
ABCA1	ATP-binding cassette transporter (member 1)
ABCB7	ATP-binding cassette, sub-family B, member 7
AC	acetyl-CoA carboxylase
ACAT	acetyl-CoA acetyltransferase
ACC	acetyl-CoA carboxylase
ACDH	acyl-Coenzyme A dehydrogenase
ACF	aberrant crypt foci
ACh	acetylcholine
ACOX2	product of the acyl-CoA oxidase family, branched-chain acyl-CoA oxidase
ACR	acrolein
AD	Alzheimer's disease
ADAR	adenosine deaminase
ADAT	adenosine deaminases that act on tRNA
ADH	alcohol dehydrogenase
AdoCbl	adenosylcobalamin
ADP	adenosine diphosphate
AF1	activation function 1
AF2	activation function 2
AFLP	acute fatty liver of pregnancy
2-AG	2-arachidonylglycerol
AGE	advanced glycated endproduct
AGEs-BSA	advanced glycation end products-modified bovine serum albumin
AGT	alanine:glyoxylate transaminase
AHR	aryl hydrocarbon receptor
AHRE	aryl hydrocarbon receptor responsive element
AICAR	5-aminoimidazole-4-carboxamide ribonucleotide

Endogenous Toxins. Diet, Genetics, Disease and Treatment.
Edited by Peter J. O'Brien and W. Robert Bruce
Copyright © 2010 WILEY-VCH Verlag GmbH & Co. KGaA, Weinheim
ISBN: 978-3-527-32363-0

AID	activation-induced cytidine deaminase
AIF	apoptosis-inducing factor
aiPLA2	lysosomal phospholipase A2
Akt	members of protein kinase family, involved in insulin resistance
ALA	aminolevulinic acid
ALAS2	δ-aminolevulinic acid synthase 2
ALD	alcoholic liver disease
ALDH	aldehyde dehydrogenase
ALDP	product of the gene defective in ALD, a membrane transporter of ABC family
ALE	advanced lipoxidation end product
ALS	amyotrophic lateral sclerosis
ALT	alanine transaminase
AMA	antimitochondrial autoantibodies
AMACR	2-methyl-acyl-CoA racemase
AMP	adenylate monophosphate
AMPA	a compound that is a specific agonist for the AMPA receptor
AMPK	AMP-activated kinase
AMPK	AMP-dependent protein kinase
ANA	antinuclear antibodies
ANG	a gene thats product is a potent mediator of new blood vessel formation
ANT	adenine nucleotide translocase
AOM	azoxymethane
AP	apurinic and apyrimidinic sites on DNA
APAP	acetaminophen
Apc	a gene whose dysfunction results in Adenomatous Polyposis Coli
apoB	apolipoprotein B
APOBEC-1	apolipoprotein B mRNA editing catalytic subunit 1
APP	amyloid precursor protein
AR	aldehyde/aldose reductase
ARE	antioxidant response element
ARNT	aryl hydrocarbon receptor nuclear translocator
ASH	alcoholic steatohepatitis
ASK-1	apoptosis signaling kinase-1
ATAC	Arimidex tamoxifen, alone or in combination
ATF4	activating transcription factor 4
ATM1	a mitochondrial inner membrane ABC transporter
ATOX1	encodes a copper chaperone that binds and transports cytosolic copper
ATP	adenosine triphosphate
AV	arginine vasopressin

AXH	a domain of ataxin-1
BAC	bacterial artificial chromosome
BAD	Bcl-2 associated death promotor (apoptotic)
Bak	(Bcl-2 homologous antagonist/killer) a pro-apoptotic gene of the Bcl-2 family
BALB	inbred strain of mouse
BAP	bone alkaline phosphatase
Bax	a pro-apoptotic protein than inactivates Bcl-2
BBB	blood-brain barrier
BCFA	branched-chain fatty acid
Bcl-2	mitochodrial protooncogene encoding a growth factor
BCOX	peroxisomal branched chain acyl-CoA oxidase
BDNF	brain-derived neurotrophic factor
BH4	tetrahydrobiopterin
BHMT	betaine homocysteine methyltransferase
Bid	a cytosolic pro-apoptotic Bcl-2, translocates to mitochondria, activates Bax
BiP	binding immunoglobulin protein
BMD	bone mineral density
BMI	body mass index
BN	brown Norway
BP	bifunctional protein
BRIC	benign recurrent intrahepatic cholestasis
BSA	bovine serum albumin
BSEP	bile salt exporter protein
C/EBP-α	CCAAT/enhancer binding protein-α
c-JUN	in combination with c-Fos, forms the AP-1 early response transcription factor
C-K	cathepsin K
CACNL1A4	mutations in this Cacnl1a4 calcium channel gene is associated with seizures
CACT	carnitine acylcarnitine translocase
CaOx	calcium oxalate
CAT	catalase
Cbl	cobalamin
CBP	CREB-binding protein
CBS	cystathionine-β-synthase
CCER	childhood cerebral
CCS	copper chaperone for superoxide dismutase
CD14	cluster of differentiation 14 cell surface marker protein involved in LPS binding
CD36	cluster of differentiation 36 cell surface marker protein binding many ligands
CDNB	1-chloro-2,4-dinitrobenzene
CE	catechol estrogen

CEBPβ	CCAAT/enhancer-binding protein beta
CEC	S-(carboxyethyl)-cysteine
CEL	carboxyethyl lysine
CGF	connective tissue growth factor
CGL	cystathionine γ-lyase
CHOP	C/EBP homologous protein
ChREBP	carbohydrate responsive element binding protein
CI	chemical ionization, confidence interval
CIA	chemically induced autoimmunity
Cic	Capicua
β-CIT SPECT	[(123)I]beta-CIT (2beta-carbomethoxy-3beta-(4-iodophenyl)tropane) single photon emission computed tomograph
cJNK	c-Jun-terminal kinase
CL	citrate lyase
CLDN1	claudin 1 human gene
CMA	carboxymethylarginine
CMC	S-(carboxymethyl)-cysteine
CML	carboxymethyl lysine
CNS	central nervous system
CoA	coenzyme A
COMMD1	copper metabolism (Murr1) domain containing 1
COMT	catechol O-methyl transferase
COPT1, 2	copper transporter 1, 2
COX	cyclooxygenase
Cp	ceruloplasmin
cPLA2	cytosolic phospholipase A2
CPT	carnitine palmitoyltransferase
CR	caloric restriction
CRC	colorectal cancer
CREB	cAMP responsive element binding protein
CRP	C-reactive protein
CRX	cone-rod homeobox, a nuclear transcription factor in the retina
CS	citrate synthase
CT	chlorotyrosine
CTGF	connective tissue growth factor
CTLA-4	cytotoxic T lymphocyte associated antigen-4
CTR1, 2	copper transporter analogous to COPT
CUP1	copper ion binding, with overexpression protecting from copper excess
CUTC	cutC copper transporter homolog (E. coli)
CVD	cardiovascular disease

CXC	"X" for two N-terminal cysteines of chemokines separated by one amino acid
CYP	cytochrome P450 isoenzyme
CYP2E1	cytochrome mixed-function oxidase system involved in the metabolism of xenobiotics
CYS	cystathionine
DA	dopamine
DAG	diacylglycerol
DAO	D-amino acid oxidase
DAT	dopamine transporter
DBA	inbred strain of mouse
DBD	DNA-binding domain
DBP	D-bifunctional protein
DC	dendritic cell
DCF	dichlorofluorescin
DCFH	dichlorofluorescein
Dcytb	duodenal cytochrome b
3DF	3-deoxyfructose
3DG	3-deoxyglucosone
3-DGA	3-deoxy 2-keto gluconic acid
DHA	dihydroxyacetone
DHA	docosahexaenoic acid
DHAP	dihydroxyacetone phosphate
DHCA	dihydroxycholestanoic acid
DHEAS	dehydroepiandrosterone-sulfate
DHF	dihydrofolate
DHFR	dihydrofolate reductase
DHN-MA	1,4-dihydroxynonane mercapturic acid
DIA	D-penicillamine-induced autoimmunity
DIIHA	drug-induced immune hemolytic anemia
DIL	drug-induced lupus
DM	dystrophia myotonica
DMEM	Dulbecco's modified Eagle's medium
DMG	dimethylglycine
DMPO	5,5-dimethyl-1-pyrroline N-oxide
DMSO	dimethyl sulfoxide
DMT	divalent metal transporter
DN	diabetic nephropathy
7,8-DNP	7,8-dihydroneopterin
DNPH	dinitrophenylhydrazine
dNTP	deoxynucleoside triphosphates
DODE	9,12-dioxo-10(E)-dodecenoic acid
DOLD	3-deoxyglucosone-derived lysine dimer
DOPAC	3,4-dihydroxyphenylacetic acid
DOPAL	3,4-dihydroxyphenylacetaldehyde

DOPET	3,4-dihydroxyphenylethanol
DPD	deoxypyridinoline
DPPC	dipalmitoylphosphatidylcholine
DR	diet restricted
DRPLA	dentatorubral pallidoluysian atrophy
DSPC	disaturated phosphatidylcholine
DT	diaphorosea flavoprotein that reversibly catalyzes the oxidation of NADH or NADPH
dTMP	thymidylate
dUMP	deoxyuridine monophosphate
DXP	5-deoxy-d-xylulose-1-phosphate
DZA	deazaadenosine
E1	estrone
E2	17β-estradiol
E2(E1)-3,4-Q	E2-3,4-quinone
E2(E1)-3,4-SQ	E2(E1)-3,4-semiquinone
E217G	estradiol-17beta-D glucuronide
E3	estriol
EAA	excitatory amino acid
EAE	experimental autoimmune encephalomyelitis
EBP	element (or enhancer) binding protein
EC	endothelial cell
ECD	electrochemical detection
ECM	extracellular matrix
EDE	4,5-epoxy-2(E)-decenal
EE	ethinyl estradiol
EEG	electroencephalogram
EG	ethylene glycol
EGF	epidermal growth factor
EGFR	epidermal growth factor receptor
EGP	early glycation product
Egr-1	early growth response-1
EH	epoxide hydrolase
EIF4	eukaryotic translation initiation factor
ELUMO	energy of the lowest unoccupied molecular orbital
EMA	ethylmalonic acid
e-NOS	constitutively expressed NOS
EP	eosinophilic pneumonia
EPR	electron paramagnetic resonance
EpRE	electrophile-responsive element
ER	endoplasmic reticulum
ER	estrogen receptor
ERα	estrogen receptor α
ERβ	estrogen receptor β
ERAD	endoplasmic reticulum–associated degradation

ERE	estrogen response element
ERK	extracellular signal-regulated kinase (MAPK)
ERKO	estrogen receptor α knockout
ERO1	endoplasmic reticulum oxidoreduction 1
ERT	enzyme replacement therapy
ES	embryonic stem
ESR	electron spin resonance
ESRD	end stage renal disease
EST	expressed sequence tag
ET	endothelin
ETC	electron transport chain
ETF	electron transfer flavoprotein
ETFDH	electron transfer flavoprotein dehydrogenase
EULAR	European League Against Rheumatism
F3P	fructose-3-phosphate
FABP	fatty acid binding protein
FAD	flavin adenine dinucleotide (oxidized)
$FADH_2$	flavin adenine dinucleotide (reduced)
FAO	fatty acid oxidation
FAOD	fatty acid oxidation disorder
FAPγ-G	2,6-diamino-5-formamidopyrimidine
FAS	fatty acid synthase, fetal alcohol syndrome
FDA	Food and Drug Administration, U.S.A.
FEEL-1, -2	endocytic receptors for advanced glycation end products
FFA	free fatty acid
FGF	fibroblast growth factor
FIAF	fasting-induced adipose factor
FICZ	6-formoindolo[3,2-b]carbazole
FL	fructose lysine
FL3P	fructoselysine-3-phosphate
FMN	flavin mononucleotide
FN3K	3-phosphokinase
FN3K	fructosamine-3-kinase
FN3KRP	fructosamine-3-kinase–related protein
FOXO	forkhead box O transcription factors related to insulin signaling pathway
FRL	free radical leak
FXN	frataxin
FXTAS	fragile X tremor/ataxia syndrome
G	glyoxal
GA	glyceraldehyde
GA3P	glyceraldehyde 3-phosphate
GABA	gamma-aminobutyric acid
GADD	growth arrest and DNA damage induced genes
GAP	glyceraldehyde 3-phosphate

GAPDH	glyceraldehyde phosphate dehydrogenase
GBM	glomerular basement membrane
GC/MS	gas chromatography with mass spectrometry
GE	glucose to ethanolamine
GFP	green flourescent protein as a reporter protein
GFR	glomerular filtration rate
GK	the Goto and Kakisaki strain of rats, a spontaneous model of T2DM
GL	glycolate
GLA-BSA	glyceraldehyde-modified bovine serum albumin
GLA-CAS	glyceraldehyde-modified casein
GLAP	glyceraldehyde-derived pyridinium compound
GLUT	glucose transporter
GM-CSF	granulocyte/macrophage colony stimulating factor
GMP	guanylate monophosphate
GO	glycolate oxidase
GOA	glycolaldehyde
GOLD	a glyoxal-derived AGE
GOT	aspartate aminotransferase (AST)
GPR	G-protein coupled receptor
Gpx	glutathione peroxidase
GR	glutathione reductase
GREACE	study on atorvastatin and coronary-heart-disease evaluation
GRHPR	glyoxylate reductase/hydroxypyruvate reductase
GRP	glucose-regulated protein, endoplasmic reticulum chaperone
GS	glutathionyl radical
GS	Goodpasture's syndrome
GSH	glutathione
GSH-Px	glutathione peroxidase
GSIS	glucose-stimulated insulin secretion
GSK3	glycogen synthase kinase, inactivates the enzyme
GSSG	oxidised glutathione
GST	glutathione S-transferase
GTP	guanosine-5'-triphosphate
GX	glyoxylate
GXA	glyoxylic acid
3HAA	3-hydroxyanthranilic acid
HAMP	hepcidin antimicrobial peptide
HAS	human serum albumin
HAT	histone acetyl transferase
HbA1ach	acetaldehyde-induced hemoglobin fraction
HBP1	HMG-box transcription factor 1 protein
HCB	hexachlorobenzene

HCC	hepatocellular carcinoma
HCV	hepatitis C virus
Hcy	homocysteine
HD	Huntington's disease
HDAC	histone deacetylase
HDL	Huntington disease-like
HDL	high-density lipoprotein
HEL	hen egg lysozyme
HELLP	hemolysis with elevated liver enzyme activity and low platelet
HEP	high-ethanol-preferring
HEPES	buffering agent
HER	hydroxyethyl free radical
Herp	homocysteine-responsive endoplasmic reticulum protein
HFCS	high-fructose corn syrup
HFE	major histocompatibility complex-encoded gene
HFE2	hemochromatosis type II
HFE2B	hemochromatosis designated type IIB
HFE3	hemochromatosis type III
HFE4	hemochromatosis type IV
HFI	hereditary fructose intolerance
HGI	hemoglobin glycosylation index
HgIA	mercury-induced autoimmunity
Hh	Hedgehog
HHcy	hyperhomocysteinemia
HI	hydroimidazolone
HIF	hypoxia inducible factor
HIV	human immunodeficiency virus
HJV	hemojuvelin
HMB-1	high mobility group box 1
HMDM	human monocyte–derived macrophage
HMEC	human mammary epithelial cell
HMG	3-hydroxy-3-methylglutaryl-
HMGB	contains a HMG-box domain
HMP	hypochlorous acid–modified protein
HMW	high molecular weight
4-HNE	4-hydroxynonenal
4HNE	4-hydroxy-2-nonenal
HNE	hydroxynonenal
hnRNP	heterogenous nuclear ribonucleoprotein
HO-1	heme oxygenase-1
holoHC	holohaptocorrin
holoTC	holotranscobalamin
HOMA-IR	homeostasis model of insulin resistance

HOPE	a randomized control study of folate, vitamin B6 and B12
HPETE	hydroperoxy eicosatetraenoic acid
HPNE	4-hydroperoxy-2(E)-nonenal
HPODE	13-hydroperoxy-octadecadienoic acid
HPR	hydroxypyruvate reductase
HRT	hormone replacement therapy
HSC	hepatic stellae cells
17β-HSD-1	17β-hydroxysteroid dehydrogenase type 1
Htt	huntingtin protein
Huh-7	hepatocyte-derived cells
HVA	homovanilic acid
IAPP	islet amyloid polypeptide
IARC	International Agency for Research on Cancer
IBD	inflammatory bowel disease
ICAM	intercellular adhesion molecule
ICTP	telopeptide of human collagen I
IDO	indoleamine 2,3-dioxygenase
IEM	inborn errors of metabolism
IFABP	intestinal fatty acid binding protein
IGF	insulin-like growth factor
IGF-1	insulin-like growth factor 1
IGFPB	insulin-like growth factor binding protein
IHA	immune hemolytic anemia
IkB	inhibitor of KappaB
IKK	IKK, inhibitor of KappaB Kinase
IL	interleukin
IL-1α	interleukin 1α
IMM	inner mitochondrial membrane
INF-γ	interferon-γ
INOS	inducible nitric oxide synthase
iNOS	inducible nitric oxide synthetase
IP	intra-peritoneal
iPLA2	independent phospholipase A2
iPS	induce potential stem
IR	insulin resistance
IRE	iron-responsive element
IRP	iron regulatory protein
IRS	insulin receptor substrate
IST	immuno-spin trapping
JNK	c-jun N-terminal kinase
JNK	N-terminal jun kinase
JNK/AP1	c-Jun N-terminal kinase/activator protein-1
K/O	knockout
KAR2	karyogamy 2
KCI	Krebs cycle intermediate

KEAP	human kelch-like ECH-associated protein
KHA	ketohydroxyglutarate aldolase
LA	lipoic acid
LAD	liver alcohol disease (ALD)
LBD	ligand-binding domain
LBP	lipopolysaccharide-binding protein
LC-CoA	long chain acyl CoA
LC-MS	liquid chromatography/mass spectroscopy
LC-MS/MS	liquid chromatography/tandem mass spectroscopy
LCAD	long chain acyl-Coenzyme A dehydrogenase
LCFA-CoA	long chain fatty-CoA
LCHAD	long chain 3-hydroxyacyl-Coenzyme A dehydrogenase
LDH	lactate dehydrogenase
LDL	low density lipoprotein
LDLR	low-density lipoprotein receptor
L-DOPA	L-3,4-dihydroxyphenylalanine
LEW	Lewis rat strain
LFA	lymphocyte function antigen
LIFE	Losartan Intervention For Endpoint reduction in hypertension
LLNA	large, neutral amino acid
LM3α	rabbit alcohol-induced cytochrome enzyme (CYP 2E1)
LNAA	large, neutral amino acid
LNAAT	large neutral amino acid transporter
LPL	lipoprotein lipase
LPLA2	lysosomal phospholipase A2
LPO	lipid peroxidation
LPS	lipopolysaccharide
MAA	malondialdehyde-acetaldehyde
MAA	malonyldialdehyde-acetaldehyde adduct
MAA-Alb	malondialdehyde-acetaldehyde adducts of albumin
MAD	multiple acyl-CoA dehydrogenase
MAO	monoamine oxidase
MAP	mitogen activated protein
MAPK	mitogen-activated protein kinase
MAT	methionine adenosyltransferase
MBG	mean blood glucose
2MBCFA	2-methyl branched-chain fatty acid
MCAD	medium chain acyl-Coenzyme A dehydrogenase
MCAD	mitochondrial medium chain acyl CoA dehydrogenase
MCP	monocyte chemoattractant protein
M-CSF	monocyte colony-stimulating factor
MD	mammographic density
MDA	malondialdehyde
MDR	multidrug resistance

MEC	Ministry of Education
MeCbl	methylcobalamin
MEOS	microsomal ethanol oxidizing system
METH	methamphetamine
5-methylTHF	5-methyltetrahydrofolate
MetR	methionine restriction
MetSO	methionine sulfoxide
2-MF	2-methylene-3(2H)-furanone
5-MF	5-methylene-5(2H)-furanone
MG	methylglyoxal
MG	myasthenia gravis
MG-H1	a methylglyoxal-derived AGE
MGO	methylglyoxal
MGS	methylglyoxal synthase
MHC	major histocompatibility complex
MIP-1	macrophage inflammatory protein 1
MitROS	mitochondrial reactive oxygen species
mLDL	minimally oxidized low density lipoprotein
MMA	methylmalonic acid
MMP	matrix metalloprotease
MnSOD	manganese superoxide dismutase
MOLD	methylglyoxal-derived lysine dimer
MOPS	buffering agent
MPDP	1-methyl-4-phenyl-2,3-dihydropyridinium
MPO	myeloperoxidase
MPP	1-methyl-4-phenyl-pyridinium ion
MPT	membrane permeability transition
MPTP	1-methyl-4-phenyl-1,2,3,6-tetrahydropyridine
MRI	magnetic resonance imagining
MRP	multidrug resistance associated protein
MS	mass spectrometry
MS	metabolic syndrome
MS	methionine synthase
M/SCHAD	medium/short chain 3-hydroxyacyl-CoA dehydrogenase
MSF-1	macrophage stimulating factor 1
mtDNA	mitochondrial DNA
MTHFR	5,10-methylenetetrahydrofolate reductase
mTOR	mammalian target of rifamycin, a serine/threonine protein kinase
MTP	mitochondrial trifunctional protein
MTS	mitochondrial targeting sequence
MTT	an agent used to evaluate mitochondrial reducing activity
MUFA	monounsaturated fatty acyl

MURR	a multifunctional protein involved in copper homeostasis
MW	molecular weight
NAC	N-acetylcysteine
NAD	nicotinamide adenine dinucleotide
NADH	nicotinamide adenine dinucleotide
NADPH	nicotinamide adenine dinucleotide phosphate
NAFLD	nonalcoholic fatty liver disease
NALD	*neonatal adrenoleukodystrophy*
NAPQI	N-acetyl-p-benzoquinone imine
NARSAD	national Alliance for Research on Schizophrenia and Depression
NASH	nonalcoholic steatohepatitis
NBT	nitroblue tetrazolium
NCE	new chemical entities
nDNA	nuclear DNA
Neh	a domain in Nrf2
NF-κB	nuclear factor κB
NFK	N-formylkynurenine
Nfr2	nuclear factor erythroid-2-related factor
NFT	neurofibrillary tangle
NGF	nerve growth factor
NHANES	National Health and Nutrition Examination Survey
NHE	4-hydroxy-2(E)-nonenal
NIH	National Institutes of Health
NK	natural killer cell
NKT	natural killer T cells
NMDA	a synthetic compound agonist at the NMDA receptor
NO	nitric oxide
NOC	N-nitroso compound
NOS	nitric oxide synthase
NOX	NADPH oxidase
NQO	NAD(P)H-dependent quinone oxireductase
Nrf-2	a redox-sensitive transcription factor
NS	not statistically significant
NSAID	nonsteroidal anti-inflammatory drug
3-NT	3-nitrotyrosine
NTCP	sodium taurocholate cotransporting polypeptide
NZB/W	New Zealand Black X White mouse cross F1
OCTN	an organic cation transporter
OGG1	8-oxoguanine DNA glycosylase protein
OGTT	oral glucose tolerance test
6-OHDA	6-hydroxydopamine
8-OHdG	8-hydroxy-2-deoxyguanosine
OHdG	hydroxydeoxyguanosine

4-OHE1(E2)-1-N7Gua	4-hydroxyestradiol-1(α,β)-N7-guanine
4-OHE2	4-hydroxylated E2
2-OHE2	2-hydroxylated E2
2-OHE2-6-N3Ade	2-hydroxyestradiol-6(α,β)-N3-adenine
OMIM	online mendelian inheritance in man
oMLT	oral methionine-loading test
ONE	4-oxononenal
2-OMe E2	2-methoxy E2
OPTIMA	Oxford Project to Investigate Memory and Ageing
OR	odds ratio
OS	oxidative stress
Ox	oxazolone
8-oxo-dG	8-oxo-7,8-dihydro-2'-deoxyguanosine
8-oxo-G	7,8-dihydro-8-oxoguanine
8-oxodG	8-hydroxy-2-deoxyguanosine
8-oxodG	8-hydroxyguanosine
oxLDL	oxidized low density lipoprotein
OxS	oxidative stress
PAA	3-(phenylamino)-alanine
PAF-AH	platelet-activating factor acetylhydrolase
PAH	phenylalanine hydroxylase
PAP	3-phenylamino-1,2-propanediol
2D-PAGE	two-dimensional polyacrylamide gel electrophoresis
PBC	primary biliary cirrhosis
PBD	peroxisome biogenesis disorder
PC	mouse teratoma-derived cell line
PCD	programmed cell death
PCH	phytanoyl-CoA hydroxylase
PD	Parkinson's disease
PDGF	platelet-derived growth factor
PDH	pyruvate dehydrogenase
PDTC	pyrrolidine dithiocarbamate
PEG	polyethylene glycol
PEMT	phosphatidylcholine methyl transferase
PERK	protein kinase-like endoplasmic reticulum kinase
PEX	a group of genes identified as important for peroxisomal synthesis
PFIC	progressive familial cholestasis
PFK	phosphofructokinase
PH	primary hyperoxaluria
PH1	primary hyperoxaluria type 1
PH2	primary hyperoxaluria type 2
PHD	prolyl-4-hydroxylase
PHF	hyperphosphorylation of tau protein
PHS	prostaglandin H synthase

PI3K	phosphoinositide 3-kinase
PIH	pregnancy-induced hypertension
PKB	protein kinase B
PKC	protein kinase C
PKU	phenylketonuria
PL	phospholipid
PL	pyridoxal
PLA1	phospholipase A1
PLA2	phospholipase A2
PLB	phospholipase B
PLC	phospholipase C
PLD	phospholipase D
PLP	pyridoxal 5′-phosphate
PM	pyridoxamine
PME	protein phosphatase methylesterase
PMN	polymorphonuclear leukocytes
PMP	pyridoxamine 5′-phosphate
PN	pyridoxine
polyQ	polyglutamine
PP2A	protein phosphatase 2A
PPAR	peroxisome-proliferator activator receptor
PPAR-α	peroxisome proliferator-activated factor-α
PPAR-γ	peroxisome proliferator-activated receptor γ
PPMT	protein phosphatase methyl transferase
PR	protein restriction
pRB	retinoblastoma protein
Prdx	peroxiredoxin
PS	photosensitizer
PS	presenilin
PTS	peroxisomal targeting sequence
PUFA	polyunsaturated fatty acids
Px	pancreatectomized
QBP1	polyQ disrupting peptide
RA	rheumatoid arthritis
Rac-1	Ras-related C3 botulinum toxin substrate 1
RAE-1	retinoic acid early inducible gene 1
Raf	a serine/threonine kinase frequently mutated in various cancers
RAGE	receptor for advanced glycation endproduct
RBC	red blood cells
RBM17	a splicing factor (SPF45)
RCO	reactive carbonyl compound
RCS	reactive carbonyl species
RERE	arginine-glutamic acid rich nuclear protein (atrophin-2)
RIG-1	a viral helicase

RNS	reactive nitrogen species
ROS	reactive oxygen species
RSNO	nitrosothiol
RT-PCR	reverse transcription polymerase chain reaction
RTP	endoplasmic reticulum stress-response gene
RXR	retinoid X receptor
S-AH	S-adenosylhomocysteine
S-AM	S-adenosylmethionine
SAH	S-adenosylhomocysteine
SAM	S-adenosylmethionine
SAMe	S-adenosylmethionine
SB-203580	an inhibitor of MAP kinase reactivating kinase
SBMA	spinal bulbar muscular atrophy
2SC	S-(2-Succinyl)cysteine
SCA	spinocerebellar ataxia
SCAD	short chain acyl-Coenzyme A dehydrogenase
SCAP	SREBP cleavage activation protein
SCFA	short chain carboxylic acid
SCO1, 2	genes and proteins required for the assembly of cytochrome oxidase
SCOX	straight-chain acyl-CoA oxidase
SCR	scavenger receptor
SD	standard deviation
SDS-PAGE	sodium dodecyl sulfate polyacrylamide gel electrophoresis
SEC	sinusoidal liver endothelial cell
SER	smooth endoplasmic reticulum
SERM	selective estrogen receptor modulator
SGLT	sodium-dependent glucose cotransport
SH	sulfhydryl
SHBG	sex hormone binding globulin
siRNA	small inhibitory RNA
SIRT1	sirtuin 1
SJL	inbred mouse strain
SKH-1	hairless mouse strain
SLC22A5	a membrane transporter gene associated with primary carnitine deficiency
SLE	systemic lupus erythematosus
SMC	smooth muscle cell
SMRT	transcriptional corepressor
SNc	substantia nigra compacta
SNP	single nucleotide polymorphism
SOC	store-operated Ca^{2+} entry channel
SOD	superoxide dismutase
SPF45	a splicing factor (RBM17)

sPLA2	secreted phospholipase A2
SPT	serine:pyruvate aminotransferase
SPT	serine:pyruvate transaminase
SQ	o-semiquinone
SR	scavenger receptor
SR-A	scavenger receptors on the surface of macrophages
sRAGE	soluble receptor for advanced glycation end product
SRB	steroid receptor binding protein
Srx	sulfiredoxin
SSAO	semicarbazide-sensitive Cu-amine oxidase
STAGA	a complex chromatin-acetylating transcription coactivator
STAT	signal transducer and activator of transcription
SUA	serum uric acid
t-PA	tissue plasminogen activator
T2DM	type 2 diabetes mellitus
TAGE	toxic advanced glycation end product
TAR	total antioxidant reactivity
TBA	thiobarbituric acid
TBARS	thiobarbituric acid reactive substance
TBP	TATA binding protein
TC	transcobalamin
TCA	the tricarboxylic acid cycle
TCDD	2,3,7,8-tetrachlorodibenzo-p-dioxin
TCEP	tris(2-carboxyethyl)phosphine
TCR	T cell receptor
TEMPOL	4-hydroxy-2,2,6,6-tetramethylpiperidinyloxy
TER	transepithelial electrical resistance
TF	tissue factor
TfR	transferrin receptor
TFTC	a complex chromatin-acetylating transcription coactivator
TGF	transforming growth factor
TGF-β	transforming growth factor β
TGF-1	transforming growth factor 1
TGN	trans-Golgi network
TH	tyrosine hydroxylase
THCA	trihydroxycholestanoic acid
tHcy	total homocysteine
THIQ	tetrahydroisoquinoline
THP	tetrahydropapaveroline
TIA	transient ischemic attack
TLR	Toll-like receptor
TMX	tetramethylmurexide, a dye used with Mitotracker
TNF	tumor necrosis factor

TNF-α	tumor necrosis factor α
TOM20	a component of the TOM (translocase of outer membrane) complex
TOS	toxic oil syndrome
TPMET	transplasma membrane electron transport
TRAMP	transgenic adenocarcinoma of mouse prostate
TRAP5b	tartrate resistant acid phosphatase 5b
Treg	regulatory T cell
Treg	regulatory T
Trpc-1	TRPC1 transient receptor potential cation channel, subfamily C, member 1
TUNEL	terminal transferase-mediated dUTP nick end labeling
UA	uric acid
UCP	uncoupling protein
UCP2	uncoupling protein 2
UL	tolerable upper intake level
UPR	unfolded protein response
UPS	ubiquitin-mediated proteasomal degradation system
UVB	ultraviolet radiation 290 to 320 nm
UVRR	ultraviolet resonance Raman
VCAM	vascular endothelial adhesion molecule
VCP	valosin-containing protein
VD	voltage-dependent
VEGF	vascular endothelial growth factors
VLCAD	very long chain acyl-Coenzyme A dehydrogenase
VLCFA	very long chain fatty acid
VLDL	very low density lipoprotein
VMAT	vesicular monoamine transporter
VSMC	vascular smooth muscle cell
VTA	ventral tegmental area
vWF	von Willebrand factor
WCRF/AICR	World Cancer Research Fund/American Institute for Cancer Research
Wnt-1	wingless-type MMTV integration site family, member 1
XALD	X-linked adrenoleukodystrophy
XO	xanthine oxidase
XO	xanthine oxidoreductase
YAC	yeast artificial chromosome
ZDF	Zucker diabetic fatty
ZO	zonula occludens

Part Three
Examples of Endogenous Toxins Associated with Acquired Diseases or Animal Disease Models

18
Alcohol-Induced Hepatic Injury

Emanuele Albano

Alcoholic liver disease (ALD) is the most common medical consequence of excessive alcohol intake and represents an important cause for death worldwide. ALD encompasses a broad spectrum of histopathological lesions ranging from steatosis with minimal parenchymal injury to more advanced liver damage, including steatohepatitis and fibrosis/cirrhosis. In recent years, a number of experimental studies have shown that, besides the formation of acetaldehyde, oxidative stress, alterations in methionine metabolism, and endoplasmic reticulum stress are involved in alcohol-mediated injury. These studies have also demonstrated that the development of steatosis, steatohepatitis, and fibrosis are the result of an interference by ethanol with the molecular mechanisms that regulate lipid homeostasis, hepatocyte survival, and inflammatory responses in the liver. In particular, it is increasingly evident that the factors influencing alcohol-mediated hepatic inflammation play a key role in the disease progression. Although the relevance of these findings to the human situation still await more investigations, these new insights open new exciting possibilities for understanding the complexity of alcohol hepatotoxicity and might lead to the development of new therapies for ALD.

18.1
Introduction

The association between excessive consumption of alcoholic beverages and human diseases has been recognized for 3000 years, being mentioned in the Ayur Veda's, a medical text books of the ancient India. Epidemiological studies have demonstrated that alcohol-related diseases are the important causes of morbidity and mortality in most of the well-developed countries. For instance, alcohol-related mortality in the United States and in Western Europe has been estimated to range from 7.9 to 14.3 per 100 000 [1]. These observations are supported by recent data from the World Health Organization that indicate alcohol-related diseases as the third cause for death and disability in most well-developed countries and a leading cause of disease in the developing countries of Eastern Europe, Central and South America, and East Asia [2].

Endogenous Toxins. Diet, Genetics, Disease and Treatment.
Edited by Peter J. O'Brien and W. Robert Bruce
Copyright © 2010 WILEY-VCH Verlag GmbH & Co. KGaA, Weinheim
ISBN: 978-3-527-32363-0

Although several organs including the gastrointestinal tract, the pancreas, the heart, the skeletal muscles, the hemopoietic bone marrow, the testes, and the nervous system are injured by ethanol, alcoholic liver disease (ALD) is the most common medical consequence of excessive alcohol intake. According to recent surveys, ALD accounts for 8% of newly diagnosed liver diseases and for more than 50% of chronic liver diseases [3, 4]. In addition, heavy alcohol consumption is often a comorbidity factor in hepatic injury due to viral infections or obesity [2, 5].

ALD encompasses a broad spectrum of histological features ranging from the accumulation of lipid droplets within the hepatocytes (fatty liver or steatosis) with minimal parenchymal injury to more advanced liver damage, including steatohepatitis and fibrosis/cirrhosis.

Almost all heavy drinkers develop steatosis, whereas 10–35% show various degrees of alcoholic hepatitis and only 8–20% of alcohol abusers progress to cirrhosis [6]. About 15% of the patients with alcoholic cirrhosis also develop hepatocellular carcinoma (HCC) [7]. Hepatic cirrhosis ranks as the twelfth most frequent cause of death in the general population of the United States, and alcohol accounts for 44% of the 28 000 deaths in a year that was ascribed to cirrhosis [5]. Similar data holds for Europe, where the prevalence of liver cirrhosis in the different countries correlates quite well to the national per capita alcohol consumption [8, 9]. The comparison of the death rates for cirrhosis among the different countries shows a significant decline in the last 20 years in some of them (United States, France, Italy, Spain) that is balanced by an appreciable rise in others (United Kingdom, Finland, Denmark, East Europe) [10]. Such different trends are ascribed to the concomitant decrease or increase in alcohol consumption in the same countries [7, 8]. From the clinical point of view, steatosis, steatohepatitis, and fibrosis/cirrhosis represent the evolution of ethanol-induced hepatic injury; however, it is also increasingly clear that the mechanisms responsible for these pathological entities are quite different. At the individual level, there is dose–effect relationship between the daily alcohol intake and the risk of developing ALD. According to a recent meta-analysis, the risk of alcoholic cirrhosis increases 10-fold by increasing alcohol intake from 25 to 100 g/day [11]. However, such an association is not linear as only 6% of the subjects consuming more than 80–90 g of ethanol a day show clinical signs of cirrhosis [12]. Moreover, prospective studies in heavy drinkers suggest that mortality due to alcoholic cirrhosis has a threshold, but not a dose–response effect [13].

It is estimated that alcohol intake exceeding 40–60 g for males and 20–40 g for females is at high risk for developing ALD, but the duration of alcohol abuse, the drinking patterns, and the type of alcoholic beverages have also influence on the risk of developing ALD independently for the amount of ethanol consumed [5, 11]. Chronic alcohol abuse (>80 g/day for more than 10 years) also increases by about fivefold the risk of HCC [14]. The reasons why ALD progresses to more severe disease (e.g., alcoholic hepatitis and cirrhosis) in some patients and not in others is still an open question.

For a long time, it has been appreciated that women develop ALD even with a lower intake of alcohol and, for the same alcohol consumption, the severity of the disease in females is higher than that in males [1]. Earlier studies have ascribed such a gender difference to the higher blood alcohol concentration per volume of alcohol consumed attained in women in relation to a lower body mass index, a higher proportion of fat, and a less efficient alcohol metabolism by the stomach mucosa [15]. More recent animal data suggest that some of the mechanisms of alcohol hepatotoxicity (see below) are enhanced in female. However, the reasons for the gender difference in alcohol hepatotoxicity still remain elusive. Genetic factors are likely to influence the progression of ALD. Studies in twins show that the concordance rate for the presence of alcoholic cirrhosis in homozygous twins is about fourfold higher than that in heterozygous twins [16]. At present, the genetic background contributing to ALD evolution is assumed to account for about 50% of the interindividual variability [17]. The genetic background might also justify some ethnic differences in the evolution of ALD, as there is a high mortality rate for cirrhosis in the United States among Afro-Americans and Hispanics [1]. Unfortunately, the attempts to elucidate the contribution of specific genes by the association analysis of advanced ALD with single nucleotide polymorphisms have so far given conflicting or inconclusive results [17]. These disappointing results are not surprising considering that most of the studies were inadequate to investigate the polygenic interactions that are likely occurring in ALD [17]. Moreover, confounding factors might arise from the nutrition status or the presence of other concomitant diseases that worsen ALD progression [2, 5].

This chapter discusses the biochemical changes induced by ethanol in the liver as well as the current view on the mechanisms that are involved in the pathogenesis of the different hepatic lesions induced by alcohol abuse.

18.2
Ethanol Metabolism and Toxicity

18.2.1
Ethanol Metabolism in the Liver

Ethanol is readily adsorbed from the gastrointestinal tract and is largely metabolized (90–98%) within the liver. With the exception of the stomach, extrahepatic metabolism of ethanol is small. This relative organ specificity justify why ethanol toxicity mostly involves the liver. In human hepatocytes, ethanol is converted to acetaldehyde by the action of alcohol dehydrogenase (ADH), microsomal cytochrome P450, and catalase [18]. The major pathway for hepatic ethanol oxidation involves ADH, an NAD-dependent zinc metalloenzyme for which five different classes have been characterized in human tissues [18]. These enzymes originate from the association of eight different subunits into active dimeric molecules. The hepatic alcohol metabolism largely relies on the Class I isoenzymes (ADH1A/B/C) that have high affinity for ethanol. Several functional genetic polyphormisms of the ADH1 isoenzymes have been identified in humans with different distributions in the various populations [18]. Although the products of the variant ADH1B*2 and ADH1C*2 genes are more active in oxidizing ethanol to acetaldehyde, their presence is not associated with an increased risk of developing ALD [19]. Microsomal ethanol oxidation mainly relies on the action of cytochrome P4502E1 isoenzyme (CYP2E1) with a minor contribution of other isoforms (CYP1A2 and CYP3A4) [18]. The K_m of CYP2E1 for ethanol is about 1 order of magnitude lower than that of ADH and the microsomal ethanol oxidation accounts for about 10% of the overall alcohol elimination [20]. However, chronic alcohol exposure increases CYP2E1 activity 5- to 20-fold through both enzyme stabilization and increased gene expression [20]. In humans, the evaluation of CYP2E1 activity through the measure of the oxidation of the myorelaxant drug chlorzoxazone reveals that an appreciable CYP2E1 induction already occurs in moderate alcohol consumers [21]. Recent evidence indicates that CYP2E1 induction not only involves the microsomal form of the enzyme, but also the two CYP2E1 variants that are present in the mitochondrial matrix (mtCYP2E1s). The two mtCYP2E1s consist of a highly phosphorylated form and a truncated form lacking of the hydrophobic part at the NH_2-terminus [20]. The mtCYP2E1s are catalytically active and, as other mitochondrial CYPs, use adrenodoxin and adrenodoxin reductase for electron transfer [20]. Overall, mtCYP2E1s account for 30–40% of the microsomal enzyme activity and may have important implications in alcohol hepatotoxicity because of their localization [20]. Finally, a small fraction of CYP2E1 (about 10% or the microsomal content) is also expressed on the outer layer of the hepatocyte plasma membranes [22], where it is transported from the Golgi apparatus via the secretory vesicles [23]. Although plasma membrane CYP2E1 is catalytically active, its importance in alcohol toxicity might rely on being a target for allo- and autoimmune reactions (see below). An additional metabolite generated during ethanol oxidation by CYP2E1 is the hydroxyethyl free radical ($CH_3C^{\bullet}HOH$; HER) [22]. HERs are produced by rat liver microsomes at a rate 10 times lower than the ethanol conversion to acetaldehyde, but have a high reactivity toward several cell constituents and alkylate several hepatic proteins including CYP2E1 [22]. One of the consequences of protein alkylation by HER is the stimulation of an immune response characterized by the generation of antibodies that specifically recognize HER-derived epitopes [22]. These antibodies are detectable in the sera of both chronically ethanol-fed rats and ALD patients [22]. Human anti-HER IgG particularly recognize HER-CYP2E1 adducts [24]. In addition, HER alkylation of CYP2E1 has been shown to break the self-tolerance toward CYP2E1 leading to the development of anti-CYP2E1 autoantibodies that are detectable in about 40% of patients with advanced ALD [25]. Interestingly, such an autoimmune reaction is further promoted in individuals carrying a genetic polymorphism (49 AG base exchange) of the gene encoding for the cytotoxic T lymphocyte associated antigen-4 (CTLA-4), a membrane receptor that down-modulates T cell-mediated immune responses [25]. Thus, the antigenic stimulation by HER-modified CYP2E1 in combination with an impaired control of T cell proliferation due to CTLA-4 mutation leads to the development of anti-CYP2E1 autoantibodies. Both allo- and autoreactivity involving CYP2E1 may contribute to hepatic injury by alcohol, as anti-HER IgG recognize HER-CYP2E1 adducts on the outer layer of the plasma membranes of ethanol-treated hepatocytes, which activates antibody-dependent cell-mediated cytotoxicity [22].

In the liver, catalase is mostly localized in the peroxisomes. Besides detoxifying H_2O_2, catalase can oxidize ethanol to acetaldehyde in a H_2O_2-dependent reaction [18]. Although catalase might have a role in alcohol metabolism in the brain, its contribution to ethanol elimination by human livers appears to be negligible [18].

The acetaldehyde that originates from ethanol oxidation is largely detoxified by the action of NAD-dependent aldehyde dehydrogenases (ALDHs) with the formation of acetate. ALDHs are tetramers or dimers that are formed by the same subunits [26]. Among the nineteen ALDHs characterized in humans, only a few are involved in acetaldehyde oxidation, and of these the mitochondrial ALDH2 and the cytosolic ALDH1A are the most efficient enzymes (Km values of about 1 M) with the capacity to detoxify more than 99% of the acetaldehyde generated within the hepatocytes [26]. However, low amounts of acetaldehyde escape detoxification and interact with proteins and nucleic acids [26]. The binding of acetaldehyde to DNA is regarded to play a major role in the carcinogenic effects of alcohol [7]. Also in the case of ALDH, functional genetic variants have been characterized. In particular, 40–50% of Asians have extremely low ALDH2 activity owing to a single amino acid substitution at position 487 (*ALDH2*2*) [17]. People homozygous for *ALDH2*2* are unable to oxidize acetaldehyde and following the ingestion of a small amount of alcohol undergoes the so-called flush syndrome characterized by nausea, vomiting, and facial flushing, whereas heterozygotes have markedly reduced, but still detectable ALDH2 activity [17]. Although heterozygous individuals for the *ALDH2*2* genetic variant who drink alcohol generate threefold higher concentrations of acetaldehyde than homozygous individuals for the wild-type; allele the actual contribution of this polymorphism to the genetic risks of ALD has not been confirmed [18]. Conversely, *ALDH2*2* polymorphism appears to be an important risk factor for alcohol-induced carcinogenesis in many organs of the gastroenteric tract [7].

During acute ethanol intoxication by binge drinking, an increased NADH/NAD ratio can affect hepatic enzymes depending change on these cofactors thus affecting lipid, carbohydrate, and uric acid metabolism [5]. In particular, the excess of NADH decreases fatty acid oxidation and enhances lipogenesis leading to the accumulation of triglycerides within the hepatocytes (see below), while the block of pyruvate conversion to glucose increases lactic acid production causing alcoholic hypoglycemia and acidosis [5]. However, these changes are rapidly reversible and seem to be attenuated during chronic ethanol ingestion. For a long time, acetaldehyde formation and increased NADH/NAD ratio have been considered to be the main factors responsible for the adverse effects of ethanol in the liver. Now, it is increasingly clear that other factors including oxidative stress, inflammatory responses, and interferences with the transduction of intracellular signals are more relevant in the pathogenesis of ALD.

18.2.2
Ethanol-Induced Oxidative Stress

The involvement of oxidative stress in ethanol toxicity was first proposed in the early 1960s, and in the following two decades, this concept has received the support of a number of experimental studies showing that ethanol increases the hepatic production of oxygen-derived free radicals (ROS), hydroxyethyl free radicals, and NO that, by interacting with cellular constituents, cause oxidative modification of nucleic acids and proteins and stimulate the peroxidation of unsaturated lipids ([24, 27] for review). The clinical relevance of these observations was confirmed by demonstrating an increase in oxidative stress biomarkers in both the liver and the serum of patients with ALD [24]. The cause/effect implication of oxidative stress in alcohol hepatotoxicity has been substantiated by experiments using the enteral feeding model of alcohol administration and different dietary combination of fatty acids to modulate liver susceptibility to lipid peroxidation. In this setting, the alcohol combination with a diet rich in highly-oxidizable, unsaturated fatty acids from corn or fish oils causes extensive lipid peroxidation and liver damage, while minor effects are seen when a similar amount of ethanol is given together with poorly-oxidizable, saturated fat [28]. The change to poorly-oxidizable fat also lowers lipid peroxidation and ameliorates liver damage induced after six weeks of combined feeding with ethanol and fish oil [29]. According to these findings, several other studies have demonstrated that the supplementation with antioxidants and free radical scavengers reduces hepatic injury in alcohol-fed rodents [24, 27].

Alcohol-induced oxidative stress is the consequence of the combined effects of an increased generation of ROS by hepatocytes, Kupffer cells, and liver infiltrating granulocytes and the lowering of liver antioxidant defenses [24, 27]. During alcohol intoxication, the CYP2E1-dependent microsomal monoxygenase system and the mitochondrial respiratory chain are the main sources of ROS within the hepatocytes. CYP2E1 has an especially high rate of NADPH oxidase activity that leads to the production of not only hydroxyethyl free radicals but also a large quantity of O_2^- and H_2O_2. In liver microsomes from both humans and alcohol-fed rodents, CYP2E1 content is positively correlated with NADPH oxidase activity and lipid peroxidation. An increased ROS production is also evident in HepG2 hepatoma cells that are stably transfected with *CYP2E1* gene [20]. Thus, the high efficiency of CYP2E1 in reducing oxygen to O_2^- and H_2O_2 could be regarded as one of the key factors contributing to oxidative stress during chronic exposure to alcohol. Consistently, experiments performed using enteral alcohol-fed rats have demonstrated that the induction of CYP2E1 by ethanol is associated with the stimulation of lipid peroxidation, while compounds interfering with CYP2E1 expression significantly decreases oxidative stress and hepatic damage [24]. Although CYP2E1 knockout mice are not protected from alcohol toxicity [20], protein carbonyls and oxidized DNA products are actually lower in CYP2E1-null mice as compared to CYP2E1 expressing mice [30]. Mitochondria represent an important source of ROS in many diseases. During acute ethanol intoxication, an enhanced electron leakage from the complexes I and III of the mitochondrial respiratory chain along with a stimulation of the NADH shuttling to the mitochondria stimulates O_2^- generation [31]. Conversely, an impaired synthesis of mitochondria-encoded constituents of the respiratory chain consequent to oxidative damage of mitochondrial DNA (mtDNA) is likely responsible for the enhanced mitochondrial ROS production following chronic alcohol administration [31]. Indeed, single or multiple deletions of mtDNA are frequent in the liver of alcohol-treated rats and of alcoholic patients [32]. These mtDNA damages also contribute to the impairment of the hepatic respiratory activity caused by alcohol [32]. Furthermore, recent evidence points to the contribution of mtCYP2E1 as a source for mitochondrial ROS [20]. As is discussed in the next chapter, increasing evidence involve inflammation in the pathogenesis of alcohol hepatotoxicity. Therefore, the activation of Kupffer cells and the hepatic infiltration by granulocytes that characterize alcoholic steatohepatitis can be regarded as an important source of ROS. Accordingly, liver depletion of Kupffer cells with gadolinium chloride lowers the hepatic production of O_2^- following ethanol infusion and decreases both liver injury and oxidative stress markers in chronic enteral alcohol-fed rats [27]. A similar protection is also evident in mice that are deficient for ICAM-1 or the phagocyte NADPH oxidase activity ($p47^{phox}$ knockout mice) [27].

The possible contribution of nitric oxide (NO) to alcohol-induced oxidative stress has also received attention, as NO interacts with O_2^- generating the highly oxidizing peroxynitrite ($ONOO^-$) molecule. Chronic ethanol exposure increases hepatic NO generation threefold by inducible nitric oxide synthetase (iNOS) [33]. Wild-type mice treated with the selective iNOS inhibitor N-(3-aminomethyl)benzyl-acetamindine (1400 W) or iNOS knockout mice are protected against oxidative stress, tyrosine nitration, and hepatic injury induced by chronic enteral alcohol feeding [27], suggesting the possible contribution of NO generated by iNOS to alcohol-related oxidative stress. Nonetheless, the actual role of NO in the pathogenesis of oxidative stress induced by alcohol awaits further investigations.

In the presence of trace amounts of transition metals, most frequently iron, O_2^-, and H_2O_2 generate highly reactive hydroxyl radicals ($OH^\bullet$), which are then responsible for the oxidation of biological constituents. Alcohol abuse in humans is often associated with an impaired utilization and/or an increased deposition of iron in the liver [34]. Moreover, even moderate alcohol consumption increases the risk of hepatic iron overload [35]. Although in experimental systems, the presence of iron, particularly of low molecular weight nonprotein iron complexes, exacerbates oxidative damage by ethanol and worsens liver pathology [24], it is not yet clear how ethanol favors iron accumulation in the liver. Over the past decade, the identification of several new proteins involved in the regulation of iron homeostasis has helped to unravel this issue. In particular, it has been shown that ethanol interferes with hepcidin, a liver-produced peptide that play a key role in down-modulating iron absorption in the gut as well as in the release of the metal by macrophages [36]. Hepcidin regulates body iron in response to a variety of stimuli, including iron store repletion and inflammation [36]. Alcohol-induced oxidative stress suppresses the iron-mediated regulation in hepcidin synthesis by interfering with the DNA binding of the hepcidin transcription factor, CCAAT/enhancer binding protein-α

(C/EBP-α) [37]. Thus, in the presence of alcohol, iron is absorbed in excess by the gut and is stored in the liver even in the presence of body iron repletion.

The action of alcohol on liver antioxidant defenses has been matter of extensive investigations. Early studies have shown that a decrease in the liver content of reduced glutathione (GSH) is a common feature in ethanol-fed animals as well as in alcoholic patients independently from the nutritional status or the degree of liver disease [24]. On the other hand, the stimulation of GSH resynthesis by rat supplementation with the GSH precursors L-2-oxothiazolidine-4-carboxylic acid or N-acetylcysteine prevents oxidative stress in the enteral alcohol feed rats [24]. Alcohol-induced GSH depletion mainly involves the mitochondrial pool (mtGSH) in the centrilobular hepatocytes [38]. The selective depletion of the mtGSH pool is the consequence of an increased mtGSH oxidation due to mitochondrial ROS production as well as defects in the GSH transport from cytosol to the mitochondrial matrix due to an increased cholesterol content of mitochondrial membranes that interferes with the activity of carrier proteins [38]. Such an unbalance in mitochondrial cholesterol is the consequence of an increased hepatic synthesis in response to the up-regulation of cholesterol-regulating transcription factor SREPB-2 by endoplasmic reticulum (ER) stress [39]. The action of ethanol on mtGSH homeostasis favors oxidative mitochondrial damage and impairs hepatocyte tolerance to TNF- [38, 39]. Vitamin E (α-tocopherol) is the main lipid-soluble antioxidant in the liver. In both humans and rodents, chronic alcohol intake decreases the liver and the plasma levels of vitamin E and α-tocopherol depletion inversely correlates the detection of lipid peroxidation markers [24]. Supporting the importance of vitamin E in controlling alcohol-induced oxidative stress is the observation that vitamin-E deficient rats have an increased susceptibility to alcohol toxicity [40]. Moreover, upon discontinuation of alcohol feeding, the administration of vitamin E reduces the severity of hepatic lesions [41]. However, a recent randomized, placebo-controlled clinical trial of patients with mild/moderate alcoholic hepatitis has not confirmed a significant effect of vitamin E supplementation in humans [42].

Several studies have investigated the effects of ethanol on the enzymes involved in the detoxification of reactive oxygen species or lipid peroxidation products showing a significant decline in the liver content and enzymatic activity of (Cu-Zn)-superoxide dismutase (SOD-1), catalase, and GSH peroxidase that inversely correlate with the extend of both lipid peroxidation and hepatic injury [43]. Furthermore, SOD-1 knockout mice show increased lipid peroxidation, extensive centrilobular necrosis, and inflammation upon moderate ethanol consumption [44]. Conversely, rodents overexpressing SOD-1 or the mitochondrial form of Mn- SOD-2 are protected against liver injury that is caused by the enteral administration of large amount of alcohol [27]. In humans, about 25% of the Caucasians carry a genetic polymorphism causing an alanine for ^{16}valine substitution in the leader amino acid sequence that is responsible for the mitochondrial localization of SOD-2 [17]. Although the Ala-SOD-2 variant might translocate less efficiently than the Val-SOD-2 to the mitochondria [45], such SOD-2 polymorphisms does not influence oxidative damage in ALD [46]. The importance of endogenous antioxidant defenses in protecting against alcohol hepatotoxicity is supported by the observation that ethanol increases the liver expression of Nfr2 (nuclear factor erythroid-2-related factor), a transcription factor that is responsible for the expression of detoxifying and antioxidant enzymes [47]. Moreover, Nfr2 knockout mice show impaired acetaldehyde detoxification, extensive liver injury, and increased mortality when they are fed with ethanol at doses well tolerated by wild-type mice [48].

18.2.3
Alcohol Effects on Methionine Metabolism

An important consequence of alcohol interaction with the liver is the alteration of methionine metabolism. L-methionine is largely metabolized by the conversion to S-adenosylmethionine (SAMe) in an ATP-dependent reaction catalyzed by the enzyme methionine adenosyltransferase (MAT) [49]. MAT is encoded by two different genes: *MAT1A* and *MAT2A*. *MAT1A* encodes the isoenzymes MATI and MATIII that are responsible for the synthesis SAMe in adult liver, while *MAT2A* encodes the isoenzyme MATII that is mainly active in fetal and regenerating hepatocytes [49]. SAMe is the principal donor of methyl groups for methylation reactions of DNA, RNA biogenic amine, histones, and phospholipids as well as the precursor of cysteine required for GSH synthesis. Indeed, S-adenosylhomocysteine that is formed following the methyl group transfer is first converted to homocysteine and then condenses with serine to generate in sequence cystathionine and cysteine [49]. Homocysteine is also converted back to methionine by the action of methionine synthetase that utilizes methyl groups supplied by the combined action of methyltetrahydrofolate and vitamin B12 [49]. A lowering of hepatic SAMe content has been documented in either the experimental animals that are chronically treated with alcohol or the patients with alcoholic hepatitis [50]. Alcohol affects SAMe formation by impairing MAT and methionine synthetase activities. The action of ethanol on MAT occurs at both transcriptional and post-transcriptional levels as alcohol lowers *MAT1A* gene expression by 50%, while oxidative stress inactivates a key SH group on the protein [50]. Moreover, ethanol interference with the turnover of folate, a key cofactor in methyl group transfer can further worsen SAMe resynthesis [50]. However, the effects of ethanol on methionine metabolism vary with species. In the rats, unlike mice, ethanol induces betaine homocysteine methyltransferase that partially counteracts SAMe deficiency [50]. The decrease in the hepatic SAMe levels reduces DNA methylation, choline synthesis, and interferes with GSH production, while homocysteine accumulates owing to the block of methionine resynthesis from homocysteine [50]. These effects have been proposed to contribute to various extents to the onset of fatty liver, oxidative stress, and hepatocytes apoptosis induced by alcohol [50]. Consistently, SAMe administration attenuates alcohol-induced steatosis, GSH depletion, and transaminase release in rats and improves ALD in humans [50]. Recent evidence has also implicated hyperhomocysteinemia in the pathogenesis of alcohol hepatotoxicity. In mice, a 5- to 10-fold increase in plasma homocysteine is evident following a few weeks of alcohol feeding, which is associated with hepatic steatosis and hepatocyte injury [51]. In these animals, alcohol-induced hyperhomocysteinemia and liver injury can be prevented by dietary supplementation with betaine, that promotes homocysteine conversion to methionine [51]. In humans hyperhomocysteinemia is rather frequent among heavy drinkers [52] in relation to folate deficiency as well as to mutations in 5,10 methylenetetrahydrofolate reductase that affect the regeneration of methyl-tetrahydrofolate for the activity of methionine synthetase [50]. Recent studies demonstrate that in both mice and minipigs, impaired methionine metabolism and homocysteine accumulation induced by chronic alcohol exposure trigger an alteration in the ER known as *ER stress* [51, 53]. Hepatocyte ER is the site of protein folding, lipid and sterol synthesis, and intracellular calcium storage. Increasing evidence indicates that the perturbation of ER functions, namely, the accumulation of unfolded proteins, affects ER functions and activates several sensor proteins that trigger a network of signals collectively termed the unfolded protein response (UPR) [54]. Such a response counteracts ER alterations by reducing protein synthesis and promoting protein refolding and/or degradation. Furthermore, ER stress stimulates the transcription of cytoprotective genes under the control of Nfr2 transcription factor and activates autophagic activity [54]. Nonetheless, an insufficient adaptation to ER stress can trigger proapoptotic signals mediated by the production of the CHOP protein [54]. At present, it is not clear how the impairment of methionine metabolism triggers ethanol-induced ER stress. It has been proposed that the generation of homocysteine-thiolactone might cause protein homocysteinylation leading to the unfolding of the intracellular proteins [51]. The sulfhydryl redox balance in the ER is also critical for the correct protein folding [55]. Thus, GSH depletion and oxidative stress are likely to be involved in causing UPR response. In addition, protein modifications by acetaldehyde and/or aldehydic end products of lipid peroxidation might also contribute to the onset of alcohol-mediated hepatic ER stress [51, 55]. The relevance of ER stress responses to the pathogenesis of alcohol liver injury has emerged from a number of observations showing the implication of ER regulated proteins in the onset of steatosis inflammation and hepatocyte apoptosis (see below).

18.3
Mechanisms of Alcohol Hepatotoxicity

18.3.1
Alcoholic Steatosis

The increase in the hepatocyte content of triglyceride is the most common histological feature of excessive alcohol intake. Alcoholic steatosis mainly includes macrovesicular steatosis, which is characterized by the presence of a medium-sized/large fat droplet in the cytoplasm of the hepatocytes with lateral displacement of the nucleus. Fat accumulation mostly involves the whole hepatic acinus, but may be prominent in the centrilobular areas [6]. Mitochondrial alterations include the presence of enlarged mitochondria (megamitochondria) that are often seen in the hepatocytes of the central zone. Microvesicular steatosis, consisting in the filling of liver cell by small fat droplets, is relatively rare in ALD (0.8–2.3%), but it is often associated with a more severe evolution of the hepatic injury [6]. Alcoholic steatosis is mostly asymptomatic and can be evidenced by hepatomegaly or by echographical signs of "brilliant liver." Although reversible, steatosis is now considered as an important risk factor for the progression of many liver diseases to fibrosis [56]. The effects of alcohol on hepatic lipid metabolism include the stimulation of triglyceride synthesis, the lowering of fatty acids oxidation, and interference with lipoprotein secretion. Early studies suggested that alcohol-induced steatosis is the consequence of a decreased fatty acid oxidation and an enhanced lipogenesis resulting from the increase in NADH/NAD ratio secondary to ethanol and acetaldehyde oxidation [57]. These alterations might possibly contribute to transient triglyceride accumulation after acute alcohol intake by binge drinking, but it is increasingly evident that persistent alcoholic steatosis results from a complex interference by ethanol with the mechanisms controlling lipid metabolism in both the liver and the adipose tissue.

The regulation of hepatic triglyceride synthesis depends upon the increased expression of the enzymes involved in fatty acid synthesis, namely, fatty acid synthetase, acyl-CoA carboxylase, and ATP citrate lyase that are regulated by the transcription factor SREBP-1 (sterol regulatory element binding protein-1) located in the ER. In response to a variety of stimuli, including the abundance of glucose, fat, and sterols, SREBP-1 is proteolytically cleaved to the active form that translocates to the nucleus, inducing the expression of the genes coding for lipogenic enzymes [57, 58]. An increased SREBP-1 activation is evident in mice receiving an ethanol-containing diet, while SREBP-1 knockout mice are protected against ethanol-induced steatosis [57]. Experiments "*in vitro*" have shown that acetaldehyde formation might account for the effect of alcohol on SREBP-1 [57, 58]. ER stress is a known stimulus for SREBP-1 response and dietary betaine supplementation to alcohol-fed mice prevents both SREBP-1 activation and fatty liver [51]. In a similar way, steatosis is associated with ER stress in minipigs receiving alcohol together with a folate-deficient diet [53]. These observations suggest that alcohol-induced ER stress is a trigger for SREBP-1 activity that, in turn, stimulates lipogenesis by inducing the expression of key enzymes of this pathway. The regulation of SREBP-1 stability and activity is also dependent upon its reversible acetylation [57]. Recently, You and coworkers have demonstrated that ethanol feeding of mice down-modulates liver sirtuin 1 (SIRT1), a nuclear protein deacetylase involved in SREBP-1 de-acetylation [59]. SIRT1 inhibition by alcohol increases SREBP-1 acetylation and its transcriptional activity that are reverted by overexpressing SIRT1 [59]. This indicates that ethanol action on SIRT1 can contribute to SREBP-1 stimulation. However, SREBP-1 modulation is not the only mechanism leading to alcoholic steatosis as, differently to mice, rats chronically administered with high doses of ethanol develop fatty liver despite a suppressed SREBP-1 activity [60]. Among the factors that regulate hepatic lipid metabolism a key role is played by AMP-dependent protein kinase (AMPK) that controls lipogenesis by down-modulating SREBP-1 and acyl-CoA carboxylase [57, 58]. The administration of ethanol to mice greatly reduces AMPK activity, while

5-aminoimidazole-4-carboxamide ribonucleotide (AICAR), a known activator of AMPK, prevents alcoholic steatosis and abolishes SREBP-1 activation by ethanol [57, 58].

Among the signals that regulate hepatic AMPK, the adipose tissue–derived cytokine adiponectin has received increasing attention. Adiponectin is a 30 kDa peptide produced by adipocytes, which circulates in three major oligomeric forms: a low molecular weight trimer, a medium size hexamer, and a high molecular weight multimer [61]. Adiponectin interacts with two types of structurally related receptors designated as Adipo R1 and Adipo R2 of which Adipo R2 is predominantly expressed in the liver [61]. Recent observations indicate that alcohol intake lowers circulating adiponectin levels in rodents and interferes with Adipo R2 expression in the hepatocytes [62]. Conversely, the administration of full-length adiponectin ameliorates alcohol-induced steatosis in mice [62]. The decrease in adiponectin secretion observed in ethanol-treated mice is the consequence of an ER stress of the adipocytes triggered by hyperhomocysteinemia [63]. A decrease in the fat adiponectin mRNA expression has also been observed in alcohol-treated rats in relation to enhanced production of fat-derived proinflammatory cytokines by macrophages infiltrating the adipose tissue [64]. Thus, ER stress and inflammation are likely to contribute in decreasing adiponectin secretion during alcohol exposure and the low adiponectinemia may further worsen hepatic steatosis by affecting the AMPK-dependent control of fatty acid metabolism.

Alcohol affects liver fatty acid β-oxidation by interfering with the enzymatic activities of the mitochondria and the peroxisomes. Although the excess of NADH may impair β-oxidation, additional mechanisms are likely to be more relevant. In the liver and skeletal muscles, which greatly depend upon fatty acid oxidation for their energy source, the lipid receptor peroxisome proliferator-activated factor-α (PPAR-α) controls lipid oxidation. PPAR-α interacts with the retinoid X receptor (RXR) and upon binding to fatty acids. The RXR-PPAR-α regulates the expression of the genes that are responsible for fatty acid transport to the mitochondria as well as for both mitochondrial and peroxisomal fatty acid oxidation [65]. Ethanol interferes with the capacity of RXR-PPAR-α to interact with DNA suggesting that the lowering of PPAR-α activity as a cause for the decrease in fatty acids oxidation is associated with alcoholic steatosis [58]. Consistently, PPAR-α agonists ameliorate ethanol-induced fatty liver [58]. The mechanisms by which ethanol impairs PPAR-α are still poorly characterized and might involve acetaldehyde binding [58] as well as oxidative stress mediated by CYP2E1 induction [66]. Adiponectin also regulates fatty acid oxidation by activating AMPK and PPAR-α. Thus, the decrease of adiponectin by alcohol might represent an additional mechanism in depressing PPAR-α activity [62]. Furthermore, by inhibiting acyl-CoA carboxylase, AMPK prevents the formation of malonyl CoA that is a potent inhibitor of lipid oxidation. It is noteworthy that the dietary polyphenol resveratrol ameliorates alcohol-induced steatosis in mice by increasing circulating adiponectin and by stimulating hepatic SIRT-1 and AMPK activities that in turn downregulate SREBP-1 and increase PPAR-α [67]. Appreciable amounts of resveratrol are present in red wines and this might possibly explain the lower risk of ALD observed in wine drinkers as compared to drinkers of beer and spirits [12].

Finally, during chronic alcohol intake, the impairment of mitochondrial functions due to mtDNA mutations and oxidation of mitochondrial proteins further contributes in decreasing fatty acids oxidation [32, 57]. The prevalence of mtDNA deletions is very high (about 85% of the cases) in alcoholics with hepatic microvesicular steatosis, indicating that the loss of the mitochondrial respiratory capacity may represent the main cause for this form of fat accumulation [68]. Although the combination of impaired fatty acid β-oxidation and enhanced hepatic triglyceride synthesis can be seen as the main contributors in the onset of alcoholic steatosis, the effects of alcohol on the hepatic export of lipids through the secretion of very low density proteins (VLDLs) can not be disregarded. The transfer of methyl groups from SAMe to phosphatidylethanolamine catalyzed by phosphatidylethanolamine N-methyltransferase is the key step in the synthesis of phosphatidylcholine that is necessary for VLDL assembly [33]. Accordingly, by maintaining adequate liver SAMe levels, betaine supplementation restores phosphatidylcholine synthesis and reduces alcoholic steatosis [69]. Microsomal triglyceride transfer proteins (MPTs) that catalyze the lipid assembly in VLDL is decreased in the liver of ethanol-fed animals and PPAR-α agonists revert this effect [58]. Moreover, oxidative stress affects VLDL secretion by interfering with lipoprotein glycosylation in the Golgi apparatus [24] as well as by enhancing the intrahepatic degradation of newly synthesized ApoB100 [70].

At present, the relevance of these observations to the pathogenesis of human alcoholic steatosis still await more investigations. In particular, a key issue is whether some of the mechanisms that contribute to fat accumulation in chronic hepatitis C and nonalcoholic fatty liver disease (NAFLD) have any relevance in alcoholic steatosis. Insulin is an important regulator of lipid homeostasis. It stimulates fatty acid deposition in the adipocytes and triglyceride synthesis in the hepatocytes. It is increasingly evident that insulin resistance is important in the pathogenesis of fat accumulation associated with hepatitis C and NAFLD [71]. Recent experimental studies have shown that alcohol interferes with insulin signals both in the adipose tissue and in the liver. In the fat tissues, chronic alcohol intake impairs insulin-mediated suppression of lipolysis, increasing the delivery of free fatty acids to the liver [72]. This effect is important for the onset of steatosis because it stimulates hepatic triglyceride synthesis even in the absence of SREBP-1 activation. In the liver, insulin response varies with the amount of alcohol. The administration of low alcohol doses (4 g/kg daily) to rats enhances hepatic insulin signaling by increasing the association of phosphoinositide 3-kinase (PI3K) with the insulin receptor substrate protein-1 (IRS-1) [73]. This might explain why moderate alcohol consumption ameliorates insulin response in humans possibly accounting for the beneficial effect of sensible drinking on cardiovascular mortality [74]. Conversely, both acute alcohol administration or chronic high doses of ethanol (13 g/kg daily) cause hepatic insulin resistance [57, 73]. Under the latter condition, ER stress induces the formation of TRB3, an inhibitory protein that disrupts insulin signaling downstream to PI3K [73]. Although such an effect down-modulates SREBP-1 [73], it is possible that other metabolic disturbances associated with insulin resistance might still cause SREBP-1-independent hepatic lipid accumulation. Inflammation and, particularly, an increased TNF- production is considered as an important cofactor in causing insulin resistance [71]. Inflammatory signals also decrease adipokine production further contributing to reduce hepatic insulin response [71]. As outlined below, alcohol contributes in different ways to cause liver inflammation and the reduction of TNF- production ameliorates alcohol-promoted steatosis [57]. Finally, oxidative stress might be an additional cause of hepatic insulin resistance during alcohol consumption. Indeed, the activation of stress-responsive kinases by ROS causes serine/threonine phosphorylation of IRS-1 that impairs IRS-1 functions in cells overexpressing CYP2E1 [57]. It is noteworthy that insulin resistance, low adiponectin production, and mild inflammation are common features in obese subjects with NAFLD [71]. These analogies give a possible explanation for the worsening action of obesity in the progression of ALD [75].

18.3.2
Alcoholic Steatohepatitis

The transition from steatosis to steatohepatitis is characterized by the appearance of mixed lobular inflammation with Kupffer cell activation and scattered infiltration of polymorphonuclear leukocytes and mononucleated cells [6]. Hepatocyte ballooning, Mallory body formation, and parenchimal cell death by either necrosis or apoptosis are also evident [6]. These lesions are predominant in centrilobular areas where CYP2E1 induction by alcohol mainly occurs. From the clinical point of view, alcoholic steatohepatitis encompasses a broad spectrum of symptoms ranging from an asyntomatic mild elevation of serum aminotransferase to jaundice, fever, abdominal distress, leukocytosis, and various degrees of hepatic failure. In the severe forms, the parenchimal insufficiency also leads to ascites and coma [76]. The persistence of mild to moderate necroinflammation is now recognized to be the main factor in the progression of ALD to cirrhosis. Indeed, perisinusoidal fibrosis in centrilobular areas is frequently detected in association with necroinflammatory lesions and progressively evolves to the extension of fibrous septa and the development of "chicken wire" fibrosis in perivenular areas [6]. Although the mechanisms responsible for the development of necroinflammation in ALD are still poorly understood, recent findings have opened new areas of investigation.

18.3.2.1 The Mechanism Promoting Inflammatory Reactions in ALD
An important breakthrough in understanding the pathogenesis of alcoholic steatohepatitis has come from the demonstration that chronic alcohol intake promotes hepatic inflammation by increasing the translocation of gut-derived endotoxins to the portal circulation where they stimulate intrahepatic Kupffer cells [77, 78]. Endotoxins are lipopolysaccarides derived from the wall of gram negative bacteria. Physiologically, small amounts of endotoxins are adsorbed from the gut and are transported to the liver where are they are cleared by Kupffer cells. This process is associated with a background secretion of TNF-α, IL-10, and prostanoids

that is responsible for the induction of systemic tolerance to antigens from the portal blood [79]. In chronic alcohol-fed rats as well as in ALD patients, plasma endotoxin content increases several fold over physiological levels, likely reflecting an alcohol-mediated alteration in the permeability to macromolecules of the intestinal epithelium [79]. The interaction of endotoxin with CD14 and the Toll-like receptor 4 (TLR-4) on Kupffer cells stimulates them to produce and release proinflammatory cytokines/chemokines (TNF-, IL-1, IL-6, IL-8/CINC, and macrophage chemotactic protein-1; MCP-1), eicosanoids, ROS, and NO [77, 78]. The CXC chemokines (IL-8/CINC and MCP-1) induce the expression of the adhesion molecules ICAM-1 and VCAM-1 on endothelial cells promoting focal leukocyte infiltration by recruiting circulating neutrophils and monocytes [80]. In patients with ALD, plasma endotoxin content correlates well with the circulating TNF-α levels and the severity of alcoholic hepatitis [78]. Accordingly, the eradication of gut gram negative bacteria with antibiotics prevents endotoxin increase and liver injury in intragastric alcohol-fed rats [78]. Kupffer cell-depleted rats as well as CD14 knockout or TLR-4-deficient mice produces less TNF-α and are resistant to alcohol toxicity. A similar protection against alcohol liver injury is also evident in the mice that are deficient for ICAM-1 or the TNF-α receptor-1 [77, 78]. The interaction between alcohol and the proinflammatory signals triggered by endotoxins is complex. Chronic alcohol intake upregulates the expression of many hepatic TLRs (TLR-2, TLR-4, TLR-6, TLR-6, TLR-8, TLR-9) [81] and sensitizes Kupffer cells to TNF-α production in response to endotoxins lowering the intracellular cAMP content [82]. Ethanol also increases the activity of Erg-1 (early growth response-1), a transcription factor that contributes to the regulation of TNF-α expression [83]. Kupffer cell activation and liver infiltration by neutrophils are important sources of ROS during chronic alcohol intake [80]. In turn, ROS amplifies TNF-α synthesis in Kupffer cells by stimulating the nuclear transcription factor κB (NF-κB), as well as cell signaling through ERK1/2 and p38 MAPK kinases [84, 85]. Hepatocytes also contribute to the release of proinflammatory cytokines/chemokines such as IL-8, MCP-1, and macrophage inflammatory protein 1 (MIP-1). According to Hiriguchi and coworkers, such a response involves IL-6 mediated activation of STAT3 (signal transducer and activator of transcription 3), as alcohol-fed mice deficient in hepatocyte STAT3 produce less MCP-1 and MIP-1 and have milder hepatic inflammation [86]. Moreover, alcohol consumption increases the activity of natural killer T-cells present in the liver that contribute to the proinflammatory stimulation [87]. All these proinflammatory response are likely to be enhanced by the decrease in adiponectin secretion induced by alcohol [62–64]. Indeed, adiponectin has anti-inflammatory actions [88] and its addition to macrophages blunts TNF-α production and enhances the production of anti-inflammatory IL-10 [89]. Osteopontin is another multifunctional cytokine that has been associated with the onset of inflammation in experimental model of both alcoholic and nonalcoholic steatohepatitis [90]. In alcohol-treated rodents, elevated osteopontin levels correlate with hepatic neutrophil infiltration that is prevented by neutralizing, antiosteopontin antibodies [91]. An increase in the hepatic expression of osteopontin mRNA has also been recently reported in patients with alcoholic hepatitis [92]. Because of its capacity to modulate Th-1 responses and lymphocyte survival, osteopontin can be regarded as a possible additional factor in the onset of inflammation in ALD. Interestingly, the gender difference in alcohol hepatotoxicity might rely on the capacity of ethanol to promote inflammation. Using the enteral alcohol feeding model, female rats show hepatic inflammatory infiltrates, ICAM-1 expression by sinusoidal endothelial cells, TNF-α production, and plasma endotoxin levels twofold higher than males [93]. Such a difference is ascribed to an estrogen-dependent increase in gut permeability to endotoxins, as ovariectomy reduces blood endotoxins, TNF-α levels, and hepatic necro-inflammation, whereas this protection is absent in ovariectomized female receiving estrogen replacement [94]. More recently, gender differences in osteopontin production have been implicated in enhancing inflammatory responses in alcohol-treated female rats [92]. Unfortunately, no clinical data is so far available to support the actual contribution of these mechanisms in female susceptibility to ALD.

18.3.2.2 Role of Immune Reactions in Alcohol-Induced Hepatic Inflammation

A key aspect in the evolution of alcoholic steatohepatitis is represented by the factors that contribute to the persistence of proinflammatory responses in the liver. Although heavy drinkers have both an increased susceptibility to infections and depressed innate and acquired immunity [6], early studies have shown that ALD is often associated with immune reactions involving lymphocyte-mediated reactions to autologous hepatocytes and the presence of circulating antibodies targeting hepatocytes obtained from ethanol-treated animals [95]. Moreover, histologically, liver infiltrates containing both $CD8^+$ and $CD4^+$ T-lymphocytes are detectable in about 40% of ALD patients and correlate with the extension of intralobular inflammation, peace-meal necrosis, and septal fibrosis [96]. Liver infiltrating T-lymphocytes in both ALD patients and alcohol-consuming rodents express markers associated with the activation/memory phenotypes and have an increased capacity to secrete proinflammatory cytokines [97], suggesting their possible contribution to orchestrate inflammation during the evolution of ALD. Supporting this view, peripheral blood T-cells from active drinkers with or without ALD show the predominance of a Th-1 pattern (high TNF-, INF-γ (interferon-γ), IL-1), in cytokine production [98]. The cause of these immunological reactions were first ascribed to the formation of new antigens as a result of acetaldehyde binding to proteins, as either experimental animals exposed to alcohol or alcoholic patients display high titers of immunoglobulins reacting with acetaldehyde-protein adducts [99]. However, such an antibody response can not completely explain the immuno-allergic reactions associated with ALD, because antiacetaldehyde antibodies can also be found in patients with liver diseases unrelated to alcohol [100]. More recently, the detection of antibodies-recognizing epitopes derived from protein alkylation by hydroxyethyl radicals has implicated oxidative stress in ALD-induced immunity [22]. Oxidative stress-induced immune responses are not limited to HER adducts. Elevated titers of circulating IgG toward epitopes derived from the modification of proteins by end products of lipid peroxidation such as malonyldialdehyde, 4-hydroxynonenal (4-HNE), and lipid hydroperoxides are also prevalent in patients with advanced ALD

(55–70%) [101]. A further important antigenic stimulus in ALD patients is represented by the formation of condensation products between lysine residues and both malonyldialdehyde and acetaldehyde, known as malonyldialdehyde-acetaldehyde adducts (MAAs) [102]. In about 35% of the patients with advanced ALD, the presence of antibodies against lipid peroxidation-derived antigens is also associated with the detection of CD4$^+$ T-lymphocyte recognizing malonyldialdehyde-derived antigens, indicating that oxidative stress promotes both humoral and cellular immune responses [103]. Finally, patients with alcoholic hepatitis or cirrhosis often have high titers of antiphospholipid antibodies that target oxidized phospholipids, namely, oxidized cardiolipin and phosphatidylserine [101]. The mechanisms responsible for oxidative stress-driven immunological responses in ALD are still unclear. Intraportal lymphoid follicles have the capacity to act as active germinal centers where antigens are presented by professional antigen presenting cells to CD4$^+$ T-lymphocytes leading to B-cell activation, proliferation, and differentiation [101]. It is therefore possible that during the evolution of ALD proteins adducted by HER or lipid peroxidation products are released from damaged hepatocytes and processed in the intraportal lymphoid follicles with the production of antibodies. Hepatic stellate cells (HSCs) might represent an alternative pathway for the presentation of oxidative stress-derived antigens to CD4$^+$ T cell, as HSC are efficient antigen presenting cells [104]. Furthermore, the phagocytosis by dendritic cells of hepatocytes dying because of oxidative injury might lead to the cross-priming of naïve CD8$^+$ T-cells with oxidatively-modified peptides, thus promoting cell-mediated responses. These processes are likely to be enhanced by the Kupffer cell secretion of IL-12 and IL-18 in response to endotoxins and by the impairment of immuno-regulatory mechanisms. Indeed, hepatic steatosis and oxidative stress have been shown lower regulatory T (Tregs) lymphocytes and stimulate Th-1 cytokine production by NKT cells [101].

The involvement of oxidative stress-driven immunity in promoting alcohol-induced hepatic inflammation is supported by experiments performed in enteral alcohol-fed rats with low plasma endotoxins showing that the production of lipid peroxidation-derived antibodies is associated with a sustained increase of TNF-α and IL-12 and histological evidence of necroinflammation [105]. In these animals, the supplementation with the antioxidant N-acetylcysteine reduces oxidative stress, hepatic inflammation, and the immune response triggered by lipid peroxidation [106]. Consistently, heavy drinkers with elevated titers of lipid peroxidation-induced IgG have a fivefold higher prevalence of elevated plasma TNF-α levels than the subjects with these antibodies within the control range [107]. The risk of advanced ALD increases by 11-fold in the heavy drinkers with the combination of high TNF-α and lipid peroxidation-induced antibodies as compared to the subjects with high TNF-α, but no immune responses [107]. Such a role of acquired immunity in fueling hepatic inflammation in ALD does not exclude the importance the innate immunity. Indeed, the powerful activating effects exerted by endotoxins on the innate immune system can be an important stimulus for lymphocyte activation toward oxidative stress-derived antigens. Thus, the combination of an increased response of Kupffer cells to endotoxins, immunological reactions, and adipokine unbalances can be seen as possible cofactors in maintaining chronic hepatic inflammation in ALD.

18.3.2.3 Mechanisms in Alcohol-Mediated Hepatocyte Killing

Parenchymal cell death by either necrosis or apoptosis characterizes the histology of steatohepatitis. In particular, an increase in hepatocyte apoptosis is associated with the severity of liver injury in patients with alcoholic steatohepatitis [108]. Several mechanisms have been proposed to account for the proapoptotic action of ethanol. Ethanol-induced oxidative stress promotes apoptosis by inducing mitochondrial permeability transition (MPT) [109]. In HepG2 cells overexpressing human CYP2E1 gene as well as in cultured rat hepatocytes ethanol-induced apoptosis is prevented by antioxidants as well as by the MTP inhibitor cyclosporine or the antiapoptotic protein Bcl-2 [20, 109]. On the other hand, ethanol-fed mice deficient in the antioxidant enzyme SOD 1 show extensive hepatocellular damage, mitochondrial depolarization, and increased mitochondrial translocation of the MTP-inducing proteins Bax and Bad [110]. Alcohol-induced ER stress might also trigger apoptosis by inducing the production of the proapoptotic CHOP protein [51]. The contribution of inflammation, namely, increased TNF-α production, to alcohol hepatotoxicity has emerged from the observation that rats receiving anti-TNF-antibodies as well as TNF-α receptor 1 (TNF-R1) knockout mice are protected against liver damage induced by enteral alcohol administration [77]. Moreover, in humans, elevated plasma TNF-α levels correlate with the severity and the mortality of alcoholic hepatitis, while the reduction of TNF-α secretion with pentoxifylline has been proposed for the treatment of alcoholic hepatitis [76]. Hepatocytes, as many other cells, are resistant to the proapoptotic action of TNF-α, because TNF-α itself induces prosurvival responses through the activation of antiapoptotic NF-κB-dependent genes and PI3K/PKB/Akt (protein kinase B)-dependent signal cascade [111]. However, growing evidence indicates that ethanol-induced oxidative stress alters such a balance between the pro and antiapoptotic signals. In particular, the selective depletion of mtGSH sensitizes hepatocytes from chronically ethanol-fed rats to TNF-α-induced killing, without interfering with NF-κB response [38]. Indeed, by allowing the oxidation of cardiolipin in the mitochondrial membranes mtGSH depletion promotes Bax-induced MPT and the release of cytochrome c [112]. Additional contributions for enhancing hepatocyte susceptibility to TNF-α-mediated apoptosis might come from an increased intracellular content of S-adenosyl-homocysteine consequent to the impaired methionine metabolism [113] as well as by the activation of the apoptosis signaling kinase-1 (ASK-1) upon the oxidation of its binding proteins thioredoxin [114]. The development of alcohol hepatotoxicity may also be favored by interferences with the network of signals controlling other prosurvival responses in hepatocytes. Indeed, ethanol increases the liver expression of PTEN (phosphatase tensin homologs deleted from chromosome 10) a dual protein/lipid phosphatase that, by degrading inositide-3-phosphate, blocks the transduction of PI3K-dependent survival signals to downstream kinases [115]. Defects in the transduction of IL-6-dependent signals by STAT3 [57] and interference by lipid peroxidation products with hepatocyte ERK1/2, a kinase responsible for transducing antiapoptotic signals [116] have also been implicated in ethanol interference with the hepatocyte survival machinery. Therefore, oxidative mitochondrial damage, ER stress, and abnormal hepatocyte responses such as increased TNF-α production likely contribute to stimulating hepatocyte apoptosis in alcohol-exposed livers. Interestingly, chronic alcohol consumption affects the capacity of hepatocytes to scavenge neighboring apoptotic cells by impairing their recognition through the asialoglycoprotein receptors [117]. Antiphospholipid antibodies associated with ALD have also been shown to specifically bind to oxidized phosphatidylserine on apoptotic cells [118]. It is known that the clearance of apoptotic cells plays an important role in the termination of inflammation as the signals generated by macrophage recognition of apoptotic bodies inhibit the secretion of proinflammatory mediators and stimulate the production of the anti-inflammatory cytokines TGF- and IL-10 [119]. Thus, it is possible that during alcoholic steatohepatitis an impaired disposal of apoptotic hepatocytes might further promote inflammation.

18.3.2.4 Alcohol and the Formation of Mallory Bodies

Mallory bodies are intracellular eosinophilic inclusions consisting of protein aggregates mainly containing cytokeratins 8 and 18 [120]. Although the presence of these intracellular inclusions is not strictly alcohol-specific, their presence is a diagnostic feature of alcoholic and nonalcoholic steatohepatitis [120]. The origin and the significance of Mallory body associated with alcoholic steatohepatitis are still poorly understood. Several studies have reported that alcohol affects protein turnover interfering with their degradation by the proteasome [121]. CYP2E1-mediated ROS production and protein alkylation by lipid peroxidation products are important in this respect [121]. In this context, Bardag-Gorce and coworkers [122] have shown that CYP2E1-expressing Hep2 cells exposed to alcohol form insoluble protein aggregates containing cytokeratins 8 and 18 as a results of oxidative proteasome impairment, suggesting a possible contribution of oxidative mechanisms in the generation of Mallory's bodies.

18.3.3
Alcoholic Cirrhosis

Liver cirrhosis represents the terminal stage of ALD and one of the main causes of death among patients with alcohol abuse. During the progression of ALD, the continuous deposition of collagen rich extracellular matrix (ECM) makes centrilobular "chicken wire" fibrosis to extend to periportal areas with the formation of fibrous septa that encircle islets of hepatic parenchyma eventually leading to micronodular cirrhosis [6]. Hepatic fibrosis is characterized by the presence of myofibroblast-like cells rich in smooth muscle actin, together with mononuclear cell infiltrates and proliferated bile ducts. The disruption of the lobular structure together with neoangiogenesis cause a subversion of the hepatic vascularization with the development of hypertension in the portal tract [6]. The early stage of alcoholic cirrhosis is often clinically asymptomatic, but the persistence of alcohol consumption leads to hepatic insufficiency, jaundice, and encephalopathy. In the recent years, there have been important advances in understanding the mechanisms responsible for the progression of liver fibrosis. Hepatic fibrosis is the consequence of a wound healing response characterized by increased deposition of ECM rich in type 1 collagen to continuous injury and parenchymal cell loss [123]. The transformation of hepatic stellate cells (HSCs), also known as *Ito cells* or *hepatic fat storing cells*, to myofibroblasts-like cells is mainly responsible for the increased ECM production in fibrotic livers [124, 125]. However, it is increasingly evident that myofibroblasts can also originate from portal fibrocytes, bone marrow progenitors, and mesenchymal transitions of epithelial liver cells [125]. In normal livers, HSCs have several functions, such as vitamin A storage, ECM turnover, and sinusoidal blood flow regulation. The transformation of HSCs as well as other cell types to ECM-producing myofibrobasts is the result of a complex network of signals triggered by cell death, inflammation, oxidative stress, and change in ECM composition [104]. Indeed, HSCs can efficiently phagocytose apoptotic bodies and this process is associated with HSC activation [104]. A similar effect is also promoted by the interaction of TLR-4 and TLR-9 with high mobility group box 1 (HMB-1) and nuclear DNA released by necrotic hepatocytes [124], respectively. Kupffer cell activation during chronic inflammation is responsible for the secretion of a variety of cytokines including PDGF (platelet-derived growth factor), TGF-1 (transforming growth factor 1), CTGF (connective tissue growth factor), and MSF-1 (macrophage stimulating factor 1) that are powerful inducers of myofibroblast transition [123–125]. Once activated, HSCs themselves secrete these cytokines along with vascular endothelial growth factors (VEGFs) and fibroblast growth factors (FGFs) to sustain their own proliferation, mobility, and collagen production [104, 124]. Low amounts of ROS and aldehydic end products of lipid peroxidation are well recognized stimuli of hepatic fibrosis, as they can directly trigger intracellular signals leading to myofibroblast transition or amplify the transduction of many profibrogenic extracellular signals [126]. Finally, decreased hepatic matrix degradation due to a reduced production of matrix metalloproteases (MMPs) and/or an increased production of MMP inhibitors also contribute to collagen accumulation [124]. The complexity of all these responses is further enhanced by interactions with lymphocyte-derived mediators, angiotensin II, the nuclear transcription factor, PPAR-γ (peroxisome proliferator-activated receptor γ), and adipokines [104, 123–126]. At present, the processes involved in the development of cirrhosis during chronic alcohol abuse are presumed to be largely similar to those of other chronic liver diseases; however, some specific mechanisms might be more relevant to sustain alcohol-induced fibrogenesis. As previously mentioned, oxidative stress plays an important role in ethanol hepatotoxicity and is implicated in the mechanism of fibrogenesis. Consistently, a number of observations have implicated oxidative damage as an important factor in the development of alcohol-induced hepatic fibrosis. Early studies have shown that liver detection of lipid peroxidation products is associated with TGF-1 production by Kupffer cells and precedes the appearance of the initial signs of fibrosis [24].

Ethanol-stimulated TGF-1 synthesis is particularly evident in the perivenous regions and is abolished by CYP2E1 inhibition with chlormethiazole [127]. Ethanol-induced liver fibrosis is exacerbated by the combined administration of carbonyl iron that greatly enhances oxidative stress and such an effect is closely associated with the promotion of TGF-1 and procollagen-1 mRNA expression in both the whole liver and freshly isolated HSC [127]. These observations are consistent with *in vitro* studies showing that ROS directly triggers collagen synthesis in HSCs overexpressing the CYP2E1 gene [20] as well as with the demonstration that oxidative stress selectively triggers NF-κB-mediated expression of collagen $\alpha 2(1)$ gene in HSCs from ethanol-fed mice by stimulating protein kinase C (PKC), PI3K, and PKB/Akt [128]. Interestingly, MAAs are also capable to induce collagen production by HSC [129].

A further alcohol-specific profibrotic action involves the processes controlling the expansion of HSC populations. Recent evidence indicates that liver NK cells inhibit the development of hepatic fibrosis by selectively killing early activated HSCs. Moreover, by releasing INF-γ, NK cells induce HSC cycle arrest and apoptosis via the action of STAT1 [130, 131]. The selectivity of NK cells for early activated HSCs, but not for quiescent or fully activated HSCs, relies on the fact that the formers express on their surface retinoic acid early inducible gene 1 (RAE-1), a ligand for NK cell receptor NKG2D, and down- modulate class 1 major histocompatibility complex antigens that antagonize NK cell cytotoxicity through both TRAIL engagement and granzyme/perforin secretion [130]. Studies by Gao's group have shown that chronic alcohol intake interferes at different levels with NK cell functions, as ethanol reduces hepatic content of NK cells [130] and impairs their ability to kill HSCs by downregulating the expression of NKG2D, TRAIL, and granzyme/perforin [131]. Furthermore, alcohol blocks the capacity of NK cells to control the number of HSCs through the secretion of INF-γ [131]. Such a latter effect is related to an interference with intracellular signals mediated by STAT1 that control the production of INF-γ [131]. These combined actions of ethanol on NK cells might account for the expansion of matrix-producing HSCs in the centrilobular areas of the liver where alcohol toxicity takes place.

Among the mechanisms possibly responsible for alcoholic fibrosis, it is important to consider the role of ethanol-induced alterations in the adipokines leptin and adiponectin as well as in neurohumoral signals involving endocannabinoids [124, 132]. Activated HSCs selectively express leptin receptors and leptin stimulates HSC survival and has profibrogenic activity stimulating the expression of proinflammatory and angiogenic cytokines [104, 124, 132]. Patients with ALD have increased serum leptin levels that correlate with the severity of fibrosis [133]. Moreover, the combination of high leptin and oxidative stress promote extensive HSC activation and collagen deposition in rats with moderate fatty liver induced by choline deficiency receiving repeated whiskey binges [134]. The profibrogenic action of leptin might be particularly effective in alcohol-damaged livers due to the combined lowering of adiponectin. Indeed, besides its anti-inflammatory action in Kupffer cells [89], adiponectin reduces proliferation and increases apoptosis of cultured HSC [132]. In recent years, the endogenous cannabinoids anandamide and

2-arachidonylglycerol (2-AG) have emerged as important neurohumoral regulators of HSC activities connected with fibrosis [135]. HSCs in injured livers produce large amounts of endocannabinoids and express the two cannabinoid receptors CB1 and CB2 that control HSC functions in opposite ways: CB1 promotes HSC activation and fibrogenesis while CB2 has an antifibrogenic activity [135]. A marked elevation in the circulating levels of the endocannabinoids is evident in cirrhotic patients as well as in experimental model of hepatic fibrosis [135]. Alcohol feeding of mice increases the expression of CB1 receptors and stimulates the synthesis of 2AG by HSCs [136]. Such a paracrine activation of CB1 receptors is evident at early stages of alcohol hepatotoxicity and so far has been associated with the development of steatosis [136]. However, CB1 receptors stimulation might have a relevant impact also in the profibrogenic action of ethanol, as their pharmacological block with the CB1 receptor antagonist rimonabant or the genetic inactivation of CB1 receptors decreases experimental hepatic fibrogenesis [137]. Altogether, these data indicate that chronic inflammation, oxidative stress, impaired NK cell functions, and alterations in adipokine and endocannabinoid secretion might specifically contribute to the evolution of alcohol-induced fibrosis. Further studies are needed to confirm the relevance of these observations to the human situation. Nonetheless, more insights in the molecular effects of ethanol on HSC might offer new tools to control the profibrogenic action of alcohol abuse.

18.3.4
Alcohol and Liver Cancer

Ethanol and acetaldehyde have been recently recognized by the International Agency for Research on Cancer (IARC) as carcinogenic agents in several human tissues including the liver [7]. Epidemiological studies conducted in the United States and Europe indicate ethanol as the most common cause of HCC, accounting for 32–45% of all HCC [14]. The development of HCC in alcohol abusers is almost always associated to the presence of cirrhosis [14]. In this context, alcohol consumption has a synergic action with other causes of cirrhosis, particularly hepatitis C infection [14]. To date, the mechanisms by which alcohol causes HCC are still poorly understood mainly because ethanol does not directly cause liver tumors in experimental animals, but only increases the occurrence of cancers induced by other treatments [14]. Although the binding of acetaldehyde to DNA has an important role in alcohol-mediated carcinogenesis in the upper and lower gastrointestinal tract, its effective metabolism by the hepatocytes makes it unlikely that acetaldehyde makes a significant contribution as a liver carcinogen [7]. On the other hand, recent data point up oxidative stress as a factor responsible for the genotoxic action of alcohol in the liver. In heavy drinkers with fatty liver, and even more in those patients with advanced ALD, the mutagenic adducts 1,N6-ethenodeoxyadenosine and 3,N4-ethenodeoxycytidine formed by the interaction of DNA with the lipid peroxidation products 4-HNE [138] are detectable. A correlation between CYP2E1 levels and DNA adduct formation is also evident in HepG2 cells expressing CYP2E1 [20]. It is noteworthy that 4-HNE-derived DNA adducts involve the codon 249 of human p53 gene that is a unique mutational hot spot in HCC [139]. Moreover, HFE (C282Y) hemochromatosis mutation [140] or Ala-SOD-2 variant [141], that both cause iron overload, increase the incidence of HCC in patients with alcoholic cirrhosis. The loss of heterozygosity at the long arm of chromosome 4 (4q34-35) is often detectable in HCC from patients with long history of alcohol use [142]. Unfortunately, the relationship of this genetic defect with oxidative stress has not been yet established.

During the evolution of liver cirrhosis, preneoplastic lesions, such as enzyme-altered foci and dysplastic nodules, usually precede the appearance of HCC [143]. In this setting, alterations in the processes that control the proliferative activity of hepatic progenitor cells may have an important role in promoting malignant growth, as these cells are likely the precursors of cancer stem cells in HCC [144]. In many pathological condition hepatocyte loss by apoptosis and/or necrosis is counteracted by the proliferation of the surviving hepatocytes in response to cytokines (TNF-, IL-6, and TGF-1) and growth factor (Hepatocyte Growth Factor and Epithelial Growth Factor) released by hepatic mesenchymal cells [145]. During chronic alcohol exposure oxidative stress affects the regenerative capacity of mature hepatocytes through the activation of p38 MAPK and p21 (WAF1/CIP1) [146]. Concomitantly, alcohol liver injury in both rodents and humans promotes the proliferation and accumulation of hepatic progenitor cells, also known as *"oval cells"* [147]. Several studies have shown that oval cells along with immature cholangiocytes and liver myofibroblasts produce and respond to Hedgehog (Hh) ligands, a group of protein mediators that orchestrate cell growth and differentiation during embryogenesis [148]. The hepatic production of Hh ligands is higher in mice fed ethanol together with a high fat diet than in the animals receiving the high fat diet alone [148]. Moreover, the number of Hh-responsive cells increases in the liver of both ethanol-treated mice and patients with severe alcoholic hepatitis [149]. Along the same line, the proliferation and the differentiation of epithelial cells rely on the transcription of genes regulated by the interactions of vitamin A-derived retinoic acid with nuclear receptors (retinoic acid receptors RARα, RARβ, and RARγ) [150]. Chronic alcohol consumption decreases the hepatic content of both vitamin A and retinoic acid [150]. Such depletion is the results of an increased catabolism of retinoic acids to polar metabolites mediated by the induction of CYP2E1 [150]. In the rats the lowering of retinoic acid by chronic ethanol administration is associated with down-regulation of retinoic acid receptors and a dramatic overexpression of transcriptional complex the AP1 that results in a stimulation of hepatic cell proliferation and a decrease in apoptosis [150]. All of these changes are almost completely normalized by the animal supplementation with retinoic acid or by the administration of chlormethiazole, a specific CYP2E1 inhibitor [151]. Consistently, a strong inverse relationship between serum concentrations of vitamin A and the risk of developing HCC has been observed in humans [150].

An additional mechanism by which alcohol might contribute to the development of HCC involves the epigenetic regulation of gene expression through DNA methylation. Approximately 1% of all DNA cytosine residues are methylated by the activity of DNA methyltransferases and gene methylation controls gene expression, as hypermethylation has a silencing effect on gene transcription and hypomethylation results in an increased gene expression [152]. The possible relevance of methylation anomalies to hepatocarcinogenesis comes from a recent report showing that the development of HCC in MAT-1a knockout mice that have a marked deficiency of SAMe, the principal methyl group donor in methylation reactions [153]. As previously mentioned, alcohol affects hepatic synthesis of SAMe [50]. Rats chronically fed with alcohol have a global hypomethylation of hepatic DNA, although the gene methylation pattern appears not to be affected [7].

Altogether these data suggest that during alcoholic cirrhosis, the expansion of Hg-responsive immature cells together with the mutagenic effects induced by oxidative stress and the alcohol-associated depletion of retinoic acid and SAMe might contribute to the development of HCC. Additional molecular events in the pathogenesis of HCC involve abnormalities in Wnt/β-catenin, pRB, and mitogen activated protein kinase pathways [154]. All these alterations have been detected to various extents in alcohol-associated HCC [155]. However, the mechanisms by which ethanol interfere with these different signal pathways remain to be clarified.

18.4
Conclusions

Our knowledge on the adverse effect of alcohol on liver has been greatly increased in recent years by a number of experimental studies showing that the formation of acetaldehyde, oxidative stress, alterations in methionine metabolism, and ER stress play an important role in alcohol hepatotoxicity. These studies have also demonstrated that the development of steatosis, steatohepatitis, and fibrosis is the result of a complex interference of alcohol with the molecular mechanisms that regulate lipid homeostasis, hepatocyte survival, and inflammatory responses in the liver. In addition, growing evidence indicate that alcohol-mediated alteration of the adipose tissue might influence the evolution of liver damage by interfering with adipokine functions. Although the relevance of many of these findings to the human situation still awaits more investigations, it is increasingly evident that several of the new mechanisms implicated in pathogenesis of ALD are common with other liver diseases. This might have important implications in the understanding the interaction of alcohol with obesity, diabetes, and C virus infection that represent important comorbidity conditions in alcohol abusers. Furthermore, these new insights in the multifactorial pathogenesis of ALD open new possibilities in dissecting how gender and genetic factors can influence the disease progression. We might also expect that all these new data can lead to the development of new therapies of ALD. These new inputs in clinical research are urgently needed because of increasing alcohol consumption worldwide.

References

1 Mandayam, S., Jamal, M.M., and Morgan, T.R. (2004) Epidemiology of alcoholic liver disease. *Semin Liver Dis*, **24**, 217–232.
2 Tzukamoto, H. (2007) Conceptual importance of identifying alcoholic liver disease as a lifestyle disease. *J Gastroenterol*, **42**, 603–609.
3 Bell, B.P., Manos, M.M., Zaman, A., Terrault, N., Thomas, A., Navarro, V.J., Dhotre, K.B., Murphy, R.C., Van Ness, G.R., Stabach, N., Robert, M.E., Bower, W.A., Bialik, S.R., and Sofair, A.N. (2008) The epidemiology of newly diagnosed chronic liver disease in gastroenterology practices in the United States: results from population-based surveillance. *Am J Gastroenterol*, **103**, 1–10.
4 John, U. and Hanke, M. (2002) Alcohol-attributable mortality in a high per capita consumption country–Germany. *Alcohol Alcohol*, **37**, 581–585.
5 Zakhari, S. and Kai Li, T. (2007) Determinants of alcohol use and abuse: Impact of quantity and frequency patterns on the liver. *Hepatology*, **46**, 2032–2039.
6 Yip, W.W. and Burt, A.D. (2006) Alcoholic liver disease. *Semin Diagn Pathol*, **23**, 149–160.
7 Seitz, H.K. and Stickel, F. (2007) Molecular mechanisms of alcohol-mediated carcinogenesis. *Nat Rev Cancer*, **7**, 599–612.
8 Sheron, N., Oslen, N., and Gilmore, I. (2008) An evidence based alcohol reduction policy. *Gut*, **57**, 1341–1344.
9 Ramstedt, M. (2001) Per capita alcohol consumption and liver cirrhosis mortality in 14 European countries. *Addiction*, **96** (Suppl 1), S19–S33.
10 Bosetti, C., Levi, F., Lucchini, F., Zatonski, W.A., Negri, E., and La Vecchia, C. (2007) Worldwide mortality from cirrhosis: an update to 2002. *J Hepatol*, **46**, 827–839.
11 Corrao, G., Bagnardi, V., Zambon, A., and La Vecchia, C. (2004) A meta-analysis of alcohol consumption and the risk of 15 diseases. *Prev Med*, **38**, 613–619.
12 Bellentani, S., Saccoccio, G., Costa, G., Tiribelli, C., Manenti, F., Sodde, M., Saveria Crocè, L., Sasso, F., Pozzato, G., Cristianini, G., and Brandi, G. (1977) Drinking habitus as co factor of the risk of alcohol-induced liver damage. *Gut*, **41**, 845–850.
13 Kamper-Jorgensen, M., Gronback, M., Tolstrup, J., and Becker, U. (2004) Alcohol and cirrhosis: dose-response or threshold effect? *J Hepatol*, **41**, 25–30.
14 Morgan, T.R., Mandayam, S., and Jamal, M.M. (2004) Alcohol and hepatocellular carcinoma. *Gastroenterology*, **127**, 587–596.
15 Thomasson, H.R. (1995) Gender differences in alcohol metabolism. Physiological responses to ethanol. *Recent Dev Alcohol*, **12**, 163–179.
16 Reed, T., Page, W.F., Viken, R.J., and Christian, J.C. (1996) Genetic disposition to organ-specific endpoints of alcoholism. *Alcohol Clin Exp Res*, **20**, 1528–1533.
17 Stickel, F. and Österreicher, C.H. (2006) The role of genetic polymorphisms in alcoholic liver disease. *Alcohol Alcohol*, **41**, 209–224.
18 Crabb, D.W. and Liangpunsakul, S. (2007) Acetaldehyde generating enzymes: roles of alcohol dehydrogenase, CYP2E1 and catalase, and speculation on the role of the others enzymes and processes, *Acetaldehyde-Related Pathology; Bridging the Transdisciplinary Divide*, Novartis Foundation Symposium, vol. **286**, John Wiley & Sons, Ltd, Chichester, pp. 4–16.
19 Zintzaras, E., Stefanidis, I., Santos, M., and Vidal, F. (2006) Do alcohol-metabolizing enzyme gene polymorphisms increase the risk of alcoholism and alcoholic liver disease? *Hepatology*, **43**, 352–361.
20 Lu, Y. and Cederbaum, A.L. (2008) CYP2E1 and oxidative liver injury

by alcohol. *Free Radic Biol Med*, **44**, 723–738.

21 Liangpunsakul, S., Kolwankar, D., Pinto, A., Gorski, C.J., Hall, S.D., and Chalasani, N. (2005) Activity of CYP2E1 and CYP3A enzymes in adults with moderate alcohol consumption: a comparison with nonalcoholics. *Hepatology*, **41**, 1144–1150.

22 Albano, E., French, S.W., and Ingelman-Sundberg, M. (1999) Hydroxyethyl radicals in ethanol hepatotoxicity. *Front Biosci*, **4**, 533–540.

23 Neve, E.P. and Ingelman-Sundberg, M. (2008) Intracellular transport and localization of microsomal cytochrome P450. *Anal Bioanal Chem*, **392**, 1075–1084.

24 Albano, E. (2006) Alcohol, oxidative stress and free radical damage. *Proc Nutr Soc*, **65**, 278–290.

25 Vidali, M., Stewart, S.F., Rolla, R., Daly, A.K., Chen, Y., Mottaran, E., Jones, D.E.J., Leathart, J.B., Day, C.P., and Albano, E. (2003) Genetic and epigenetic factors in autoimmune reactions toward cytochrome P4502E1 in alcoholic liver disease. *Hepatology*, **37**, 277–285.

26 Deitrich, R.A., Petersen, D., and Vasiluou, V. (2007) Removal of Acetaldehyde from the body, *Acetaldehyde-Related Pathology; Bridging the Transdisciplinary Divide*, Novartis Foundation Symposium, vol. **286**, John Wiley & Sons, Ltd, Chichester, pp. 23–40.

27 Arteel, G.E. (2003) Oxidants and antioxidants in alcohol-induced liver disease. *Gastroenterology*, **124**, 778–790.

28 Nanji, A.A., Zhao, S., Sadrzadeh, S.M.H., Dannenberg, A.J., Tahan, S.R., and Waxman, D.J. (1994) Markedly enhanced cytochrome P4502E1 induction and lipid peroxidation is associated with severe liver injury in fish oil-treated ethanol-fed rats. *Alcohol Clin Exp Res*, **18**, 1280–1285.

29 Nanji, A.A., Sadrzadeh, S.M.H., Yang, E.K., Fogt, F., Maydani, M., and Dannenberg, A.J. (1995) Dietary saturated fatty acids: A novel treatment for alcoholic liver disease. *Gastroenterology*, **109**, 547–554.

30 Bradford, B.U., Kona, H., Isayama, F., Kosyk, O., Wheeler, M.D., Akiyama, T.E., Bleye, L., Krausz, K.W., Gonzalez, F.J., Koop, D.R., and Rusyn, I. (2005) Cytochrome P450 CYP2E1, but not nicotinamide adenine dinucleotide phosphate oxidase is required for ethanol-induced oxidative DNA damage in rodent liver. *Hepatology*, **41**, 336–344.

31 Bailey, S.M. and Cunningham, C.C. (2002) Contribution of mitochondria to oxidative stress associated with alcohol liver disease. *Free Radic Biol Med*, **32**, 11–16.

32 Hoek, J.B., Cahill, A., and Pastorino, J.G. (2002) Alcohol and mitochondria: a dysfunctional relationship. *Gastroenterology*, **122**, 2049–2063.

33 Chamulitrat, W. and Spitzer, J.J. (1996) Nitric oxide and liver injury in alcohol-fed rats after lipopolysaccaride administration. *Alcohol Clin Exp Res*, **20**, 1065–1070.

34 Irving, M.G., Halliday, J.W., and Powell, L.W. (1988) Association between alcoholism and increased hepatic iron store. *Alcohol Clin Exp Res*, **12**, 7–12.

35 Ioannou, G.N., Dominitz, J.A., Weiss, N.S., Haegerty, P.J., and Wowdley, K.V. (2004) The effect of alcohol consumption on the prevalence of iron overload, iron deficiency and ion deficiency anemia. *Gastroenterology*, **126**, 1293–1301.

36 Ganz, T. and Nemeth, E. (2006) Iron Imports. IV. Hepcidin and the regulation of body iron metabolism. *Am J Physiol*, **290**, G199–G203.

37 Harrison-Findik, D.D., Shafer, D., Klein, E., Timchenko, N.A., Kulaksiz, H., Clemens, D., Fein, E., Andriopulos, B., Pantopulos, K., and Gollan, J. (2006) Alcohol metabolism-mediated oxidative stress down-regulates hepcidin transcription and leads to increased duodenal iron transporter expression. *J Biol Chem*, **281**, 22974–22992.

38 Fernandez-Checa, J.C. and Kaplowitz, N. (2005) Hepatic mitochondrial glutathione: transport and role in disease and toxicity. *Toxicol Appl Pharmacol*, **204**, 263–273.

39 Fernandez, A., Colell, A., Garcia-Ruiz, C., and Fernandez-Checa, J.C. (2008) Cholesterol and sphingolipids in alcohol-induced liver injury. *J Gastroenterol Hepatol*, **23** (Suppl 1), S9–S15.

40 Sadrazadeh, S.M.H., Nanji, A.A., and Meydani, M. (1994) Effect of chronic ethanol feeding on plasma and liver α- and γ-tocopherol levels in normal and vitamin E-deficient rats. *Biochem Pharmacol*, **47**, 2005–2010.

41 Nanji, A.A., Yang, E.K., Fogt, F., Sadrzadeh, S.M.H., and Dannenberg, A.J. (1996) Medium chain triglycerides and vitamin E reduce the severity of established experimental alcoholic liver disease. *J Pharmacol Exp Ther*, **277**, 1694–1700.

42 Mezey, E., Potter, J.J., Rennie-Tankersley, l., Caballeire, J., and Pares, A. (2004) A randomized placebo controlled trial of vitamin E for alcoholic hepatitis. *J Hepatol*, **40**, 40–46.

43 Polavarapu, R., Spitz, D.R., Sim, J.E., Follansbee, M.H., Oberley, L.W., Rahemtulla, A., and Nanji, A.A. (1998) Increased lipid peroxidation and impaired antioxidant enzyme function is associated with pathological liver injury in experimental alcoholic liver disease in rats fed diets high in corn oil and fish oil. *Hepatology*, **27**, 1317–1323.

44 Kessova, I.G., Ho, Y.S., Thung, S., and Cederbaum, A.I. (2003) Alcohol-induced liver injury in mice lacking Cu,Zn-superoxide dismutase. *Hepatology*, **38**, 1136–1145.

45 Sutton, A., Imbert, A., Igoudjil, A., Descatoire, V., Cazanave, S., Pessayre, D., and Degoul, F. (2005) The manganese superoxide dismutase Ala16Val dimorphism modulates both mitochondrial import and mRNA stability. *Pharmacogenet Genomics*, **15**, 311–319.

46 Stewart, S.F., Leathart, J.B., Chen, Y., Daly, A.K., Rolla, R., Mottaran, E., Vay, D., Vidali, M., Day, C.P., and Albano, E. (2002) Valine-alanine manganese superoxide dismutase polymorphism is not associated with alcohol-induced oxidative stress or fibrosis. *Hepatology*, **36**, 1355–1360.

47 Gong, P. and Cederbaum, A.I. (2006) Nrf2 is increased by CYP2E1 in rodent liver HepG2 cells and protects against oxidative stress caused by CYP2E1. *Hepatology*, **43**, 144–153.

48 Lamlé, J., Marhenke, S., Borlak, J., von Wasielewski, R., Eriksson, P.C.J., Geffers, R., Manns, M.P., Yamamoto, M., and Vogel, A. (2008) Nuclear factor-eythroid 2-related factor prevents alcohol-induced fulminant liver injury. *Gastroenterology*, **134**, 1159–1168.

49 Lu, S.C., Tzukamoto, H., and Mato, J.M. (2002) Role of abnormal methionine metabolism in alcoholic liver injury. *Alcohol*, **27**, 155–162.

50 Purohit, V., Abdelmalek, M.F., Barve, S., Benevenga, N.J., Halsted, C.H., Kaplowitz, N., and Kharbanda, K.K. (2007) Role of S-adenosylmethionine, folate and betaine in the treatment of alcoholic liver disease: summary of a symposium. *Am J Clin Nutr*, **86**, 14–24.

51 Kaplowitz, N., Than, T.A., Shinohara, M., and Ji, C. (2007) Endoplasmic reticulum stress and liver injury. *Semin Liver Dis*, **27**, 367–377.

52 Sakuta, H. and Suzuki, T. (2005) Alcohol consumption and plasma homocysteine. *Alcohol*, **37**, 73–77.

53 Esfandiari, F., Villanueva, J.A., Wong, D.H., French, S.W., and Halsted, C.H. (2005) Chronic ethanol feeding and folate deficiency activate hepatic endoplasmic reticulum stress pathway in micropigs. *Am J Physiol Gastrointest Liver Physiol*, **289**, G54–G63.

54 Schröder, M. (2008) Endoplasmic reticulum stress responses. *Cell Mol Life Sci*, **65**, 862–894.

55 Malhotra, J.D. and Kaufman, R.J. (2007) Endoplasmic reticulum stress and oxidative stress: a vicious circle

or a double-edged sword? *Antioxid Redox Signal*, **9**, 2277–2293.
56 Powel, E.E., Jonsson, J.R., and Clouston, A.D. (2005) Steatosis: co-factor in other liver diseases. *Hepatology*, **42**, 5–13.
57 Purohit, V., Gao, B., and Song, B.J. (2009) Molecular mechanisms of alcoholic fatty liver. *Alcohol Clin Exp Res*, **33**, 1–15.
58 Sozio, M. and Crabb, D.W. (2008) Alcohol and lipid metabolism. *Am J Physiol Endocrinol Metab*, **295**, E10–E16.
59 You, M., Liang, X., Ajmo, J.M., and Ness, G.C. (2008) Involvement of mammalian sirtuin 1 in the action of ethanol in the liver. *Am J Physiol Gastrointest Liver Physiol*, **294**, G892–G898.
60 He, L., Simmen, F.A., Ronis, M.J., and Badger, T.M. (2004) Post-transcriptional regulation of sterol regulatory element binding protein-1 by ethanol induces class I alcohol dehydrogenase in rat liver. *J Biol Chem*, **279**, 28113–28121.
61 Berg, A.H., Combs, T.P., and Scherer, P.E. (2002) CRP30/adiponectin: an adipokine regulating glucose and lipid metabolism. *Trends Endocrinol Metab*, **13**, 84–89.
62 Rogers, C.Q., Ajmo, J.M., and You, M. (2008) Adiponectin and alcoholic fatty liver disease. *IUBMB Life*, **60**, 790–797.
63 Song, Z., Zhou, Z., Deaciuc, I., Chen, T., and McClain, C.J. (2008) Inhibition of adiponectin production by homocysteine: a potential mechanism for alcoholic liver disease. *Hepatology*, **47**, 867–879.
64 Kang, L., Sebastian, B.M., Pritchard, M.T., Pratt, B.T., Previs, S.F., and Nagy, L.E. (2007) Chronic ethanol-induced insulin resistance is associated with macrophage infiltration into adipose tissue and altered expression of adiponectin. *Alcohol Clin Exp Res*, **31**, 1581–1588.
65 Yu, S., Rao, S., and Reddy, J.K. (2003) Peroxisome proliferators-activated receptors, fatty acid oxidation, steatohepatitis and hepatocarcinogenesis. *Curr Mol Med*, **3**, 561–572.
66 Lu, Y., Zhuge, J., Wang, X., Bai, J., and Cederbaum, A.I. (2008) Cytochrome P450 2E1 contributes to ethanol-induced fatty liver in mice. *Hepatology*, **47**, 1483–1494.
67 Ajmo, J.M., Liang, X., Rogers, C.Q., Pennock, B., and You, M. (2008) Resveratrol alleviates alcoholic fatty liver in mice. *Am J Physiol Gastrointest Liver Physiol*, **295**, G833–G842.
68 Fromenty, B., Grimbert, S., Mansouri, A., Beaugrand, M., Erlinger, S., Röting, A., and Pessayre, D. (1995) Hepatic mitochondrial DNA deletion in alcoholics: association with microvesicular steatosis. *Gastroenterology*, **108**, 193–200.
69 Kharbanda, K.K., Maillard, M.E., Baldwin, C.R., Beckenhauer, H.C., Sorell, M.F., and Tuma, D.J. (2007) Betaine attenuates alcoholic steatosis by restoring phosphatidylcholine generation via phosphatidylethanolamine methyltransferase pathway. *J Hepatol*, **46**, 314–321.
70 Pan, M., Cederbaum, A.I., Zhang, Y.L., Ginsberg, H.N., Williams, K.J., and Fisher, E.A. (2004) Lipid peroxidation and oxidant stress regulate hepatic apolipoprotein B degradation and VLDL production. *J Clin Invest*, **113**, 1277–1287.
71 Shoelson, S.E., Herrero, L., and Naaz, A. (2007) Obesity, inflammation and insulin resistance. *Gastroenterology*, **132**, 2169–2180.
72 Kang, L., Chen, X., Sebastian, B.M., Pratt, B.T., Bederman, I.R., Alexander, J.C., Previs, S.F., and Nagy, L.E. (2007) Chronic ethanol and triglyceride turnover in white adipose tissue in rats: inhibition of the anti-lipolitic action of insulin after chronic ethanol contributes to increased triglyceride degradation. *J Biol Chem*, **282**, 28465–28473.
73 He, L., Marecki, J.C., Serrero, G., Simmen, F.A., Ronis, M.J., and Badger, T.M. (2007) Dose-dependent

effects of alcohol on insulin signalling: partial explanation for bifasic alcohol impact in human health. *Mol Endocrinol*, **21**, 2541–2550.

74 Bau, P.F., Bau, C.H., Rosito, G.A., Manfroi, W.C., and Fuchs, F.D. (2007) Alcohol consumption, cardiovascular health, and endothelial function markers. *Alcohol*, **41**, 479–488.

75 Diehl, A.M. (2004) Obesity and alcoholic liver disease. *Alcohol*, **34**, 81–87.

76 Tilg, H. and Day, C.P. (2007) Management strategies in alcoholic liver disease. *Nat Clin Pract Gastroenterol Hepatol*, **4**, 24–34.

77 Hines, I.N. and Wheeler, M.D. (2004) Recent advances in alcoholic liver disease III. Role of the innate immune response in alcoholic hepatitis. *Am J Physiol Gastrointest Liver Physiol*, **287**, G310–G314.

78 Rao, R.K., Seth, A., and Sheth, P. (2004) Recent advances in alcoholic liver disease. I Role of intestinal permeability and endotoxemia in alcoholic liver disease. *Am J Physiol Gastrointest Liver Physiol*, **286**, G881–G884.

79 Racanelli, V. and Rehermann, B. (2006) The liver as an immunological organ. *Hepatology*, **3**, S54–S62.

80 Bautista, A.P. (2002) Neutrophilic infiltration in alcoholic hepatitis. *Alcohol*, **7**, 17–21.

81 Gustot, T., Lemmers, A., Moreno, C. et al. (2006) Differential liver sensitization to toll-like receptor pathway in mice with alcoholic fatty liver. *Hepatology*, **43**, 989–1000.

82 Gobejishvili, L., Barve, S., Joshi-Barve, S., Uriarte, S., Song, Z., and McClain, C. (2006) Chronic ethanol-mediated decrease in cAMP primes macrophages to enhanced LPS-inducible NF-κB activity and TNF expression: relevance to alcoholic liver disease. *Am J Physiol Gastrointest Liver Physiol*, **291**, G681–G688.

83 McMullen, M.R., Pritchard, M.T., Wang, Q., Millward, C.A., Croniger, C.M., and Nagy, L.E. (2005) Early growth response-1 transcription factor is essential for ethanol-induced fatty liver in mice. *Gastroenterology*, **128**, 2066–2076.

84 Nagy, L.E. (2003) Recent insights into the role of the innate immune system in the development of alcoholic liver disease. *Exp Biol Med*, **228**, 882–890.

85 Thakur, V., Pritchard, M.T., McMullen, M.R., Wang, Q., and Nagy, L.E. (2006) Chronic ethanol feeding increases activation of NADPH oxidase by liposaccharide in rat Kupffer cells: role of increased reactive oxygen in LPS-stimulated ERK1/2 activation and TNF-α production. *J Leukoc Biol*, **79**, 1348–1356.

86 Horiguchi, N., Wang, L., Mukhopadhyay, P., Park, O., Jeong, W.I., Lafdil, F., Osei-Hyiaman, D., Moh, A., Fu, X.Y., Pacher, P., Kunos, G., and Gao, B. (2008) Cell type-dependent pro-and anti-inflammatory role of signal transducer activator of transcription 3 in alcoholic liver injury. *Gastroenterology*, **134**, 1148–1158.

87 Minagawa, M., Deng, Q., Liu, Z.X., Tzukamoto, H., and Dennert, G. (2004) Activated natural killer T cells induce liver injury by Fas and tumor necrosis factor-α during alcohol consumption. *Gastroenterology*, **126**, 1387–1399.

88 Tilg, H. and Moschen, A.R. (2006) Adipokines: mediators linking adipose tissue, inflammation and immunity. *Nat Rev Immunol*, **6**, 4772–4783.

89 Huang, H., Park, P.H., McMullen, M.R., and Nagy, L.E. (2008) Mechanisms for the anti-inflammatory effects of adiponectin in macrophages. *J Gastroenterol Hepatol*, **23** (Suppl 1), S50–S53.

90 Ramaiah, S.H. and Ritting, S. (2007) Role of osteopontin in regulating hepatic inflammatory responses and toxic liver injury. *Expert Opin Drug Metab Toxicol*, **3**, 519–526.

91 Banerjee, A., Apte, U.M., Smith, R., and Ramaiah, S.K. (2006) Higher

neutrophil infiltration mediated by osteopontin is a likely contributing factor to the increased susceptibility of females to alcoholic liver disease. *J Pathol*, **2008**, 473–485.

92 Seth, D., Gorrell, M.D., Cordoba, S., McCaughan, G.W., and Haber, P.S. (2006) Intrahepatic gene expression in human alcoholic hepatitis. *J Hepatol*, **45**, 306–320.

93 Ikejima, K., Enomoto, N., Iimuro, Y., Ikejima, A., Fang, D., Xu, J., Forman, D.T., Brenner, D.A., and Thurman, R.G. (1998) Estrogen increases sensitivity of hepatic Kupffer cells to endotoxin. *Am J Physiol*, **274**, G669–G676.

94 Yin, M., Ikejima, K., Wheeler, M.D., Bradford, B.U., Seabra, V., Forman, D.T., Sato, N., and Thurman, R.G. (2000) Estrogen is involved in early alcohol-induced liver injury in a rat enteral feeding model. *Hepatology*, **31**, 117–123.

95 Paronetto, F. (1993) Immunologic reactions in alcoholic liver disease. *Semin Liver Dis*, **13**, 183–195.

96 Colombat, M., Charlotte, F., Ratziu, V., and Poyard, T. (2002) Portal lymphocytic infiltrate in alcoholic liver disease. *Hum Pathobiology*, **33**, 1170–1174.

97 Batey, R.G., Cao, Q., and Gould, B. (2002) Lymphocyte-mediated liver injury in alcohol-related hepatitis. *Alcohol*, **27**, 37–41.

98 Laso, J.F., Madruga, I.J., and Orfao, A. (2002) Cytokines and alcohol liver disease, in *Ethanol and the Liver* (eds C.D.I.N. Sherman V.R. Preedy, and R.R. Watson), Taylor and Francis, London, pp. 206–219.

99 Klassen, L.W., Tuma, D., and Sorrell, M.F. (1995) Immune mechanisms of alcohol-induced liver disease. *Hepatology*, **22**, 355–357.

100 Worrall, S., De Jersey, J., Shanley, B.C., and Wilce, P.A. (1990) Antibodies against acetaldehyde-modified epitopes: presence in alcoholic, non-alcoholic liver disease and control subjects. *Alcohol Alcohol*, **25**, 509–517.

101 Vidali, M., Stewart, S.F., and Albano, E. (2008) Interplay between oxidative stress and immunity in the progression of alcohol-mediated liver injury. *Trends Mol Med*, **14**, 63–71.

102 Thiele, G.M., Freeman, T.K., and Klassen, L.W. (2004) Immunological mechanisms of alcoholic liver disease. *Semin Liver Dis*, **24**, 273–287.

103 Stewart, S.F., Vidali, M., Day, C.P., Albano, E., and Jones, D.E.J. (2004) Oxidative stress as a trigger for cellular immune response in patients with alcoholic liver disease. *Hepatology*, **39**, 197–203.

104 Winau, F., Quack, C., Darmoise, A., and Kaufmann, S.H. (2008) Starring stellate cells in liver immunology. *Curr Opin Immunol*, **20**, 68–74.

105 Ronis, M.J., Butura, A., Korourian, S., Shankar, K., Simpson, P., Badeaux, J., Albano, E., Ingelman-Sundberg, M., and Badger, T.M. (2008) Cytokine and chemokine expression associated with steatohepatitis and hepatocyte proliferation in rats fed ethanol via total enteral nutrition. *Exp Biol Med*, **233**, 344–355.

106 Ronis, M.J.J., Butura, A., Sampey, B.P., Prior, R.L., Korourian, S., Albano, E., Ingelman-Sundberg, M., Petersen, D.R., and Badger, T.M. (2005) Effects of N-acetyl cysteine on ethanol-induced hepatotoxicity in rats fed via total enteral nutrition. *Free Radic Biol Med*, **39**, 619–630.

107 Vidali, M., Vietala, J., Occhino, G., Ivaldi, A., Sutti, S., Niemelä, O., and Albano, E. (2008) Immune responses against oxidative stress-derived antigens are associated with increased circulating Tumor Necrosis Factor-α and accelerated liver damage in heavy drinkers. *Free Radic Biol Med*, **45**, 306–311.

108 Natori, S., Rust, C., Stadheim, L.M., Srinivasan, A., Burgart, L.J., and Gores, G.J. (2001) Hepatocyte apoptosis is a pathological feature of human alcoholic hepatitis. *J Hepatol*, **34**, 248–253.

109 Adachi, M. and Ishii, H. (2002) Role of mitochondria in alcoholic

liver injury. *Free Radic Biol Med*, **32**, 487–491.
110 Kessova, I.G. and Cederbaum, A.I. (2007) Mitochondrial alterations in livers of SOD1-/- mice fed alcohol. *Free Radic Biol Med*, **42**, 1470–1480.
111 Schwabe, R.F. and Brenner, D.A. (2006) Mechanisms of liver injury. I. TNF-α-induced liver injury: role of IKK, JNK and ROS pathways. *Am J Physiol Gastrointest Liver Physiol*, **290**, G583–G589.
112 Mari, M., Colell, A., Morales, A., Caballero, F., Moles, A., Fernández, A., Terrones, O., Basañez, G., Antonsson, B., García-Ruiz, C., and Fernández-Checa, J.C. (2008) Mechanism of mitochondrial glutathione-dependent hepatocellular susceptibility to TNF despite NF-κB activation. *Gastroenterology*, **134**, 1507–1520.
113 Song, Z., Zhou, Z., Uriarte, S., Wang, L., Kang, Y.J., Chen, T., Barve, S., and McClain, C.J. (2004) S-adenosylhomocysteine sensitizes to TNF-α hepatotoxicity in mice and liver cells: a possible etiological factor in alcoholic liver disease. *Hepatology*, **40**, 989–997.
114 Hoek, J.B. and Pastorino, J.G. (2004) Cellular signalling mechanisms in alcoholic liver damage. *Semin Liver Dis*, **24**, 257–272.
115 Shulga, N., Hoek, J.B., and Pastorino, J.G. (2005) Elevated PTEN levels account for the increased sensitivity of ethanol-exposed cells to tumor necrosis factor-induced cytotoxicity. *J Biol Chem*, **280**, 9416–9424.
116 Sampey, B.P., Stewart, B.J., and Petersen, D.R. (2007) Ethanol-induced modulation of hepatocellular extracellular signal-regulated kinase 1/2 activity via 4-hydroxynonenal. *J Biol Chem*, **282**, 1925–1937.
117 McVicker, B.L., Tuma, D.J., Kubik, J.A., Hindemith, A.M., Baldwin, C.R., and Casey, C.A. (2002) The effect of ethanol on asialoglycoprotein receptor-mediated phagocytosis of apoptotic cells by rat hepatocytes. *Hepatology*, **36**, 1478–1487.
118 Vay, D., Rigamonti, C., Vidali, M., Mottaran, E., Alchera, E., Occhino, G., Sartori, M., and Albano, E. (2006) Anti-phospholipid antibodies associated with alcoholic liver disease target oxidized phosphatidylserine on apoptotic cell plasma membranes. *J Hepatol*, **44**, 183–189.
119 Krusko, D.V., D'Herde, K., and Vandenabeele, P. (2006) Clearance of apoptotic and necrotic cells and its immunological consequences. *Apoptosis*, **11**, 1709–1726.
120 Zatloukal, K., French, S.W., Stumptner, C., Strnad, P., Harada, M., Toivola, D.M., Cadrin, M., and Omary, M.B. (2007) From Mallory to Mallory-Denk bodies: what, how and why? *Exp Cell Res*, **313**, 2033–2049.
121 Donohue, T.M., Cederbaum, A.I., French, S.W., Barve, S., Gao, B., and Osna, N.A. Jr (2007) Role of the proteasome in ethanol-induced liver pathology. *Alcohol Clin Exp Res*, **31**, 1446–1459.
122 Bardag-Gorce, F., French, B.A., Nan, L., Song, H., Nguyen, S.K., Yong, H., Dede, J., and French, S.W. (2006) CYP2E1 induced by ethanol causes oxidative stress, proteasome inhibition and cytokeratin aggresome (Mallory body-like) formation. *Exp Mol Pathol*, **81**, 191–201.
123 Battaler, R. and Brenner, D.A. (2005) Liver fibrosis. *J Clin Invest*, **115**, 209–218.
124 Friedman, S.L. (2008) Mechanism of hepatic fibrogenesis. *Gastroenterology*, **134**, 1655–1669.
125 Parola, M., Marra, F., and Pinzani, M. (2008) Myofibroblast-like cells and liver fibrogenesis: Emerging concepts in a rapidly moving scenario. *Mol Aspects Med*, **29**, 58–66.
126 Novo, E. and Parola, M. (2008) Redox mechanisms in hepatic chronic wound healing and fibrogenesis. *Fibrogenesis Tissue Repair*, **5**. DOI: 10.1186/1755-1536-1-5.
127 Fang, C., Lindros, K.O., Badger, T.M., Ronis, J.J., and

Ingelman-Sundberg, M. (1998) Zonated expression of cytokines in rat liver: effect of chronic ethanol and cytochrome P4502E1 inhibitor, chlormethiazole. *Hepatology*, **27**, 1304–1310.

128 Nieto, N. (2007) Ethanol and fish oil induce NFκB transactivation of the collagen α 2(1) promoter through lipid peroxidation-driven activation of PKC-PI3K-Akt pathway. *Hepatology*, **45**, 1433–1445.

129 Tuma, D.J. (2002) Role of malondialdehyde-acetaldehyde adducts in liver injury. *Free Radic Biol Med*, **32**, 303–308.

130 Gao, B., Radaeva, S., and Jeong, W. (2007) Activation of NK cells inhibits liver fibrosis: A novel strategy to treat fibrosis. *Expert Rev Gastroenterol Hepatol*, **1**, 173–180.

131 Jeong, W., Park, O., and Gao, B. (2008) Abrogation of anti-fibrotic effects of NK/INF-γ contributes to alcohol acceleration of liver fibrosis. *Gastroenterology*, **134**, 248–258.

132 Wang, J., Brymora, J., and George, J. (2008) Role of adipokines in liver injury and fibrosis. *Expert Rev Gastroenterol Hepatol*, **2**, 47–57.

133 Naveau, S., Perlemuter, G., Chaillet, M., Raynard, B., Balian, A., Beuzen, F., Portier, A., Galanaud, P., Emilie, D., and Chaput, J.C. (2006) Serum leptin in patients with alcoholic liver disease. *Alcohol Clin Exp Res*, **30**, 1422–1428.

134 Nieto, N. and Rojkind, M. (2007) Repeated whiskey binges promotes liver injury in rats fed a choline-deficient diet. *J Hepatol*, **46**, 330–339.

135 Mallat, A. and Lotersztajn, S. (2008) Endocannabinoids and liver disease. 1. Endocannabinoids and their receptors in the liver. *Am J Physiol Gastrointest Liver Physiol*, **294**, G9–G12.

136 Jeong, W.I., Osei-Hyiaman, D., Park, O., Liu, J., Bátkai, S., Mukhopadhyay, P., Horiguchi, N., Harvey-White, J., Marsicano, G., Lutz, B., Gao, B., and Kunos, G. (2008) Paracrine activation of hepatic CB1 receptors by stellate cell-derived endocannabinoids mediates alcoholic fatty liver. *Cell Metab*, **7**, 227–235.

137 Teixeira-Clerc, F., Julien, B., Grenard, P., Tran Van Nhieu, J., Deveaux, V., Li, L., Serriere-Lanneau, V., Ledent, C., Mallat, A., and Lotersztajn, S. (2006) CB1 cannabinoids receptor antagonism: a new strategy for the treatment of liver fibrosis. *Nat Med*, **12**, 671–676.

138 Frank, A., Seitz, H.K., Bartsch, H., Frank, N., and Nair, J. (2004) Immunohistochemical detection of 1, N6-ethenodeoxyadenosine in nuclei of human liver affected by diseases predisposing to hepatocarcinogenesis. *Carcinogenesis*, **25**, 1027–1031.

139 Hu, W., Feng, Z., Eveleigh, J., Iyer, G., Pan, J., Amin, S., Chung, F.L., and Tang, M.S. (2002) The major lipid peroxidation product, trans-4-hydroxy-2-nonenal, preferentially forms DNA adducts at codon 249 of human p53 gene, a unique mutational hot spot in hepatocellular carcinoma. *Carcinogenesis*, **23**, 1781–1789.

140 Nahon, P., Sutton, A., Rufat, P., Ziol, M., Thabut, G., Schischmanoff, P.O., Vidaud, D., Charnaux, N., Couvert, P., Ganne-Carrie, N., Trinchet, J.C., Gattegno, L., and Beaugrand, M. (2008) Liver iron, HFE gene mutations, and hepatocellular carcinoma occurrence in patients with cirrhosis. *Gastroenterology*, **134**, 102–110.

141 Sutton, A., Nahon, P., Pessayre, D., Rufat, P., Poiré, A., Ziol, M., Vidaud, D., Barget, N., Ganne-Carrié, N., Charnaux, N., Trinchet, J.C., Gattegno, L., and Beaugrand, M. (2006) Genetic polymorphisms in antioxidant enzymes modulate hepatic iron accumulation and hepatocellular carcinoma development in patients with alcohol-induced cirrhosis. *Cancer Res*, **66**, 2844–2852.

142 Bluteau, O., Beaudoin, J.C., Pasturaud, P., Belghiti, J., Franco, D., Bioulac-Sage, P., Laurent-Puig, P., and Zucman-Rossi, J. (2002) Specific association between alcohol intake, high grade of differentiation and 4q34-q35 deletions in hepatocellular

carcinomas identified by high resolution allelotyping. *Oncogene*, **21**, 1225–1232.
143 Kojiro, M. and Roskams, T. (2005) Early hepatocellular carcinoma and dysplastic nodules. *Semin Liver Dis*, **25**, 133–142.
144 Sell, S. and Leffert, H.L. (2008) Liver cancer stem cells. *J Clin Oncol*, **26**, 2800–2805.
145 Michalopoulos, G.K. (2007) Liver regeneration. *J Cell Comp Physiol*, **213**, 286–300.
146 Diehl, A.M. (2005) Recent events in alcoholic liver disease V. Effect of ethanol on liver regeneration. *Am J Physiol Gastrointest Liver Physiol*, **288**, G1–G6.
147 Roskams, T., Yang, S.Q., Koteish, A., Durnez, A., DeVos, R., Huang, X., Achten, R., Verslype, C., and Diehl, A.M. (2003) Oxidative stress and oval cell accumulation in mice and humans with alcoholic and nonalcoholic fatty liver disease. *Am J Pathol*, **163**, 1301–1311.
148 Omenetti, A. and Diehl, A.M. (2008) The adventures of sonic hedgehog in the development and repair II. Sonic hedgehog and liver development, inflammation and cancer. *Am J Physiol Gastrointest Liver Physiol*, **294**, G595–G598.
149 Jung, Y., Brown, K.D., Witek, R.P., Omenetti, A., Yang, L., Vandongen, M., Milton, R.J., Hines, I.N., Rippe, R.A., Spahr, L., Rubbia-Brandt, L., and Diehl, A.M. (2008) Accumulation of hedgehog-responsive progenitors parallel alcoholic liver disease in mice and humans. *Gastroenterology*, **134**, 1532–1543.
150 Wang, X.D. (2005) Alcohol, vitamin A, and cancer. *Alcohol*, **35**, 251–258.
151 Liu, C., Chung, J., Seitz, H.K., Russell, R.M., and Wang, X.D. (2002) Chlormethiazole treatment prevents reduced hepatic vitamin A levels in ethanol-fed rats. *Alcohol Clin Exp Res*, **26**, 1703–1709.
152 Baylin, S.B. (2005) DNA methylation and gene silencing in cancer. *Nat Clin Pract Oncol*, **2** (Suppl 1), S4–S11.
153 Santamaria, E., Muñoz, J., Fernandez-Irigoyen, J., Sesma, L., Mora, M.I., Berasain, C., Lu, S.C., Mato, J.M., Prieto, J., Avila, M.A., and Corrales, F.J. (2006) Molecular profiling of hepatocellular carcinoma in mice with a chronic deficiency of hepatic s-adenosylmethionine: relevance in human liver diseases. *J Proteome Res*, **5**, 944–953.
154 Aravalli, R.N., Steer, C.J., and Cressman, E.N.K. (2008) Molecular mechanisms of hepatocellular carcinoma. *Hepatology*, **48**, 2047–2063.
155 Edamoto, Y., Hara, A., Biernat, W., Terracciano, L., Cathomas, G., Riehle, H.M., Matsuda, M., Fujii, H., Scoazec, J.Y., and Ohgaki, H. (2003) Alterations of RB1, p53 and Wnt pathways in hepatocellular carcinomas associated with hepatitis C, hepatitis B and alcoholic liver cirrhosis. *Int J Cancer*, **106**, 334–341.

19
Ethanol-Induced Endotoxemia and Tissue Injury
Radhakrishna K. Rao

Evidence generated from numerous clinical and experimental studies clearly demonstrates that alcohol consumption leads to endotoxemia. The main cause of this endotoxemia appears to be the dysfunctional gut barrier and increased absorption of endotoxins. Although reduced phagocytosis of lipopolysaccharides (LPS) by Kupffer cells and intestinal bacterial overgrowth are still viable causes, they appear to be only minor contributing factors. The role of endotoxins in alcoholic liver disease has been investigated extensively. However, it is likely that endotoxemia also mediates alcohol-induced injury to other organs such as pancreas and lung. Although the Kupffer cell is the initial target, LPS affects other cells such as sinusoidal endothelial cells, stellate cells, monocytes, neutrophils, and hepatocytes. LPS activates toll-like receptor 4 (TLR-4) by interacting with lipopolysaccharide-binding protein (LBP) and CD14 resulting in the activation of NFκB and overexpression of proinflammatory cytokines and chemokines. LPS also induces oxidative stress and proteasome dysfunction leading to steatosis, inflammation, and fibrosis in liver. Synthetic drugs, dietary components, plant extracts, and probiotics ameliorate oxidative stress, inflammation, and endotoxemia induced by ethanol. It appears that factors that act by preventing intestinal permeability and endotoxemia are the most effective factors in preventing ethanol-induced tissue damage. Future studies need to be focused on understanding the more detailed mechanism of synergism between ethanol and endotoxins, ethanol-induced intestinal permeability, and prevention of ethanol-induced permeability and endotoxemia.

A striking observation made during the ongoing journey toward understanding the mechanism of alcoholic liver disease (ALD) is that endotoxemia and endotoxin-mediated alteration of cellular functions play a crucial role in the pathogenesis of ALD. Emerging body of evidence points to the direction of a similar mechanism of endotoxin-mediated tissue injury in the pathogenesis of alcoholic pancreatitis and lung diseases. Endotoxemia may therefore represent a common denominator in the mechanistic equation deriving into various types of alcohol-related diseases. The mechanism of ethanol-induced endotoxemia has been a subject of investigation for decades. A significant body of evidence suggests that there are three likely mechanisms involved in alcohol-induced endotoxemia:

Endogenous Toxins. Diet, Genetics, Disease and Treatment.
Edited by Peter J. O'Brien and W. Robert Bruce
Copyright © 2010 WILEY-VCH Verlag GmbH & Co. KGaA, Weinheim
ISBN: 978-3-527-32363-0

(i) dysfunctional Kupffer cells with reduced ability to detoxify endotoxins, (ii) bacterial overgrowth in the gut leading to excessive generation of endotoxins, and (iii) disruption of intestinal epithelial barrier function and increase in paracellular permeability to endotoxins and bacteria. The initial site of action of absorbed endotoxins is a Kupffer cell that releases cytokines and chemokines. Endotoxin and cytokines/chemokines affect other cells in the liver leading to the state of steatohepatitis and fibrosis. Ethanol and endotoxin work synergistically in this alcoholic liver damage, and possibly in pancreas, lung, and other target tissues. The characteristics of ethanol-induced endotoxemia, its role in pathogenesis of liver, pancreas, and lung diseases, the potential mechanisms involved in the cytotoxic effects of endotoxins, and the factors that ameliorate ethanol-induced endotoxemia and tissue damage are discussed in this chapter. To most part, the discussion in this chapter relates to ALD, as this condition has been studied in greater detail, while alcoholic pancreatitis and lung diseases are emerging areas in alcoholism-related pathology. Additionally, as a growing body of evidence indicates a contribution of endotoxemia in other diseases such as nonalcoholic steatohepatitis, inflammatory bowel disease, and HIV infection, the discussion on endotoxemia in this chapter may be relevant far beyond ALD.

19.1
Alcoholic Endotoxemia

Endotoxins are lipopolysaccharides (LPS) derived from the cell wall of gram-negative bacteria and they are required for the virulence of these pathogens [1]. LPS consist of a nonpolar lipid component and a polar polysaccharide moiety. The circulating endotoxins are derived from dead bacteria or the LPS shed from the cell wall of viable organisms. Normally, hepatic Kupffer cells rapidly detoxify endotoxins. However, when the Kupffer cell function is overwhelmed by a high rate of endotoxin entry and/or the magnitude of its level in blood, the endotoxins escape Kupffer cells and spill into the systemic circulation. They are highly immunogenic and induce production of proinflammatory cytokines such as interleukin (IL)-1 and tumor necrosis factor (TNF)-α. This host response to LPS is the cause of cellular injury, rather than LPS themselves.

19.1.1
Endotoxemia in Patients with ALD

Endotoxin in portal blood was detected in patients with inflammatory bowel disease, cirrhosis, and alcoholic complications, at least three decades ago [2, 3]. Endotoxemia appears to play a role in the pathogenesis of viral hepatitis [4] and surgical procedures such as laparotomy [5]. In patients with viral hepatitis, lactitol reduced plasma endotoxin levels by reducing the number of pathogenic bacteria, while increasing the growth of probiotics [4]. Endotoxemia in ALD was first recognized by the detection of *Escherichia coli* antibodies in plasma of patients with ALD [6]. The levels of anti-*E. coli* antibodies in plasma correlated well with alcohol consumption and alcoholic hepatitis. No microbes were cultured from liver biopsies, indicating the lack of bacterial infection of liver under these conditions. The plasma antibody level was higher in patients with cirrhosis with alcoholic hepatitis compared to that in patients with cirrhosis without alcoholic hepatitis [6]. Increased levels of serum IgA and anti-lipid A antibody were found in plasma of patients with ALD [7]. Lipid A is the core component of different types of LPS and it is the biologically active component of endotoxin. Numerous subsequent studies demonstrated that plasma endotoxin levels in patients with alcoholic cirrhosis are several-fold greater than those in patients with nonalcoholic cirrhosis [8] and healthy subjects [8–15].

The endotoxemia appears to occur during the early stages of ALD [10, 15]. Endotoxemia was detected in alcoholics who showed only minimal symptoms of the disease [15]. The plasma endotoxin levels were similar in patients with intermediate pathology of ALD and patients with severe alcoholic cirrhosis [15]. Similarly, the plasma endotoxin levels were not different in patients with alcoholic cirrhosis with ascites and those without ascites [10], suggesting that plasma endotoxin levels may not serve as an indicator of the severity of ALD. However, one study claimed that a correlation of plasma endotoxin level with the severity of ALD could be demonstrated [11]. One patient with severe hepatitis showed markedly high plasma endotoxin level. Endotoxemia also increased with the progress of disease to the terminal stage. On the other hand, in most survivors, plasma endotoxin levels decreased during the recovery phase. Similarly, the levels of LPS-binding proteins were found to be greater in patients with ALD compared to those in normal subjects. Therefore, the correlation of endotoxemia with the severity of disease may be prominent during the late stages of ALD rather than the early stages.

The normal endotoxin levels in human subjects [8–15] are maintained at a very low level due to the intestinal barrier function and Kupffer cell-mediated phagocytosis of endotoxins in the liver. Endotoxemia is defined as a condition with plasma endotoxin level greater than 2.5 EU (endotoxin unit) ml^{-1} [16]. However, the comparison of plasma endotoxin levels has been difficult due to several factors. Although, in general, 1 ng of endotoxin is considered equivalent to 12 EU, different types of LPS show different degrees of toxic activity. The differences in procedures used in sample preparation and different assay methods used to measure endotoxin can affect the plasma endotoxin values. The endotoxin levels in normal subjects are reported to be 0.3–10.4 pg ml^{-1} [8–15]. Endotoxin concentration in plasma of patients with ALD varied from 8.5 [15] to 206 pg ml^{-1} [11] (Table 19.1). The wide variability of these values may have been caused by the differences in the methods used for sample preparation and endotoxin assay, patient conditions, and possibly, the severity of the disease. However, within an individual study, the endotoxin level in patients with ALD was always 5- to 20-fold greater than that in normal subjects.

Tab. 19.1 Alcohol-induced endotoxemia

Source	Species	Ethanol condition	Plasma endotoxin (pg/ml)		Reference
			Normal	Ethanol-treated	
Patients with ALD					
Fukui et al. 1991	Human	Alcoholic cirrhosis	ND	>20 (systemic)	8
Schafer et al. 1997	Human	Alcoholic hepatitis	2.2	18.1 (systemic)	14
Bode et al. 1997	Human	ALD	ND	24.7 (systemic)	10
Hanck et al. 1998	Human	ALD	2.0	10 (systemic)	12
Fujimoto et al. 2000	Human	Hepatitis and cirrhosis	5.5	206 (systemic)	11
Parlesak et al. 2000	Human	ALD	1.7	18.4 (systemic)	13
Schafer et al. 2002	Human	Hepatitis	2.2	8.5 (systemic)	15
Experimental models of ALD					
Rivera et al. 1998	Rat	Intragastric	0.5	8.4 (systemic)	17
Mathurin et al. 2000	Rat	Intragastric	1.5	43 (portal)	18
Mathurin et al. 2000	Rat	Intragastric	0.0	6.2 (systemic)	18
Keshavarzian et al.	Rat	Intragastric	0.5	1.7 (systemic)	19
Jokelainen et al. 2001	Rat	Intragastric	7.0	40 (systemic)	20
Lambert et al. 2003	Mouse	Intragastric	12	75 (systemic)	21
Lambert et al. 2004	Mouse	Intragastric	15	54 (systemic)	22
Ferrier et al. 2006	Rat	Intragastric	0.3	0.8 (EU/ml)[a]	23
Zhao et al. 2008	Mouse	Intragastric	0.5	3.0 (EU/ml)[a]	24
Nanji et al. 2001	Rat	Chronic feeding	10	84 (systemic)	25
Jarvelainen et al. 1999	Rat	Chronic feeding	3.1	9.3 (systemic)	26
Zuo et al. 2001	Rat	Chronic feeding	48	187 (systemic)	27
Uesugi et al. 2002	Mouse	Chronic feeding	40	130 (systemic)	28
Nanji et al. 2002	Rat	Chronic feeding	10	68 (systemic)	29
Horie et al. 2002	Rat	Chronic feeding	13	22 (portal)	30
Horie et al. 2002	Rat	Chronic feeding	9	19 (systemic)	30
McKim et al. 2003	Mouse	Chronic feeding	21	78 (systemic)	31

[a] Different units of endotoxin levels.
ND, not determined.

19.1.2
Endotoxemia in Experimental Models of ALD

Endotoxemia was also detected in experimentally induced alcoholic liver damage in rats and mice. Both acute [17–19, 23–25] and chronic [18, 20, 27–35] ethanol administration showed an elevated level of endotoxins in plasma. Elevation of plasma endotoxin level by acute ethanol administration suggests that the alcoholic endotoxemia occurs even before the initiation of ALD; these studies, however, have used a single high dose of ethanol. The endotoxemia caused by chronic ethanol feeding demonstrates that the endotoxemia is sustained in alcoholics for long period of time at least until the ethanol consumption is continued. The endotoxemia by acute ethanol administration was reversed in 3 hours after administration, in mice

[24]. Similarly, another study showed that enterally administered LPS appeared in portal blood of rats infused with acute ethanol [18] and the plasma LPS level returned to normalcy in 16 hours. Therefore, ethanol consumption induces transient endotoxemia. However, it is not known whether endotoxemia induced by chronic ethanol feeding can also be reversed quickly. It is likely that endotoxemia in chronic ethanol-fed animals sustains for a longer period of time than the quick reversal of endotoxemia induced by acute ethanol administration.

Strikingly enough, there is a good agreement between the experimental models of ALD and the patients with ALD, as far as the plasma endotoxin levels are concerned. The basal endotoxin levels in rats and mice varied from 6.2 to 13 pg ml^{-1} [17, 18, 22, 29–33, 35]. Ethanol administration increased plasma endotoxin to 43–187 pg ml^{-1} (Table 19.1). The endotoxin levels in ethanol-fed animals were consistently 4- to 10-fold greater than those in control animals. There were no species-dependent differences in the plasma endotoxin in ethanol-fed rats and mice. The acute and chronic ethanol-fed animals showed similar levels of endotoxemia. Therefore, as far as the endotoxemia is concerned, there is a good agreement between different experimental models and patients with ALD. Once again, the wide variability of plasma endotoxin levels reported could be a result of different methods of sample preparation, methods of endotoxin assay, and differences in the experimental conditions applied in different laboratories. Additionally, studies by Mathurin et al. [18] demonstrated that endotoxins in portal blood are sevenfold greater than those in plasma. Similarly, Riviera et al. [17] showed that portal blood endotoxin level was threefold greater than plasma endotoxin. Another study by Horie et al. [30] also indicated a similarly higher level of endotoxin in portal blood compared to plasma, although the difference was not as dramatic as it was in other studies [17, 18].

19.1.3
Gender Differences in Endotoxemia

It is well established that the development of alcohol-induced liver injury in women is more rapid than it is in men, and alcoholic hepatitis progresses more rapidly in women [36]. A similar gender difference in sensitivity to alcoholic liver injury has been established in rats, as plasma endotoxin levels in alcohol-fed female rats were greater than those in alcohol-fed male rats [29, 34]. Such gender difference in severity of endotoxemia in patients with LAD has not been evaluated. However, Kupffer cells in women with ALD are more sensitive to endotoxin-induced cytokine release than Kupffer cells from men with ALD [21].

19.1.4
Bacterial Translocation

Although endotoxemia in ALD has been consistently demonstrated by both clinical and experimental studies, bacterial infection was rarely reported in these patients [37, 38]. However, bacterial translocation from the gut and infection of mesenteric lymph nodes and liver were shown in experimental models of a combination of ethanol and burn injuries. Ethanol and burn injuries work synergistically to cause severe liver injury in rats [39]. This combined effect of ethanol and burn injuries resulted in a dramatic increase in bacterial infection of mesenteric lymph nodes [39, 40]. Therefore, the synergistic effect of other pathological conditions on ethanol-induced endotoxemia is an important area that remains to be explored further. In support of this view, a fairly recent study demonstrated that ethanol and ischemia/reperfusion-induced injury also synergistically elevated plasma endotoxin levels [30].

19.2
Causes of Alcoholic Endotoxemia

The source of endotoxin is the bacteria inhabiting the lumen of colon and terminal ileum. Endotoxins normally penetrate the gut epithelium only in trace amounts due to the presence of the mucosal barrier function, which prevents the diffusion of macromolecules across the epithelium. However, the transepithelial diffusion of endotoxins may be elevated under several pathophysiologic conditions. The mechanism of alcoholic endotoxemia may involve multiple factors. The three major factors that facilitate endotoxemia are: (i) delayed endotoxin clearance, (ii) ethanol-induced bacterial overgrowth, and (iii) disruption of the intestinal mucosal barrier function by ethanol consumption.

19.2.1
Delayed Endotoxin Clearance

One possible cause of alcoholic endotoxemia is ethanol-induced delay in the clearance of endotoxins from the circulation. Experimental studies demonstrated that clearance of LPS from the blood was significantly delayed by ethanol administration [17, 26]. Hepatic Kupffer cells play an important role in detoxification of circulating endotoxins [41, 42]. Endotoxins are phagocytosed by the Kupffer cells located in the sinusoids. Few studies indicated that ethanol impairs the phagocytic function of Kupffer cells [21, 42], and attenuates the endotoxin uptake by these cells [43]. Therefore, endotoxins that escape dysfunctional Kupffer cells spill into the systemic circulation. Furthermore, acute alcohol administration lowers the phagocytic activities of recruited PMNs and Kupffer cells more rapidly in female rats than in male rats [44], which correlates well with the higher sensitivity of female rats to develop alcoholic liver damage. Therefore, it is possible that reduced phagocytic activities of Kupffer cells may contribute to the reduced clearance of endotoxins from the circulation causing endotoxemia. However, this hypothesis is complicated to some extent by the fact that endotoxin-mediated Kupffer cell function is essential for the development of alcoholic liver damage. Therefore, it seems paradoxical that Kupffer cells can be responsible for liver damage, while they are simultaneously responsible for detoxifying the injurious endotoxins. Endotoxin-binding proteins synthesized by hepatocytes, play a defensive role by binding LPS, and preventing its access to LPS receptors, CD14, and toll-like receptor (TLR-4) on Kupffer cells. Studies demonstrate that ethanol significantly reduces the production of endotoxin-binding proteins by hepatocytes [8, 43, 45]. Hepatic production of endotoxin-binding proteins was increased when the Kupffer cells were preincubated with ethanol in a dose-dependent manner [46]. However, when Kupffer cells from chronically ethanol-fed rats were used, the ethanol-induced increase in lipopolysaccharide-binding protein (LBP) production was absent. Therefore, multiple factors may contribute to the delayed clearance of endotoxins from the circulation.

19.2.2
Bacterial Overgrowth

Overgrowth of bacteria in the small bowel in alcoholics was first demonstrated by Bode [47]. Culture of jejunal aspirates from normal subjects and patients with ALD demonstrated that the numbers of both aerobic and anaerobic bacteria were high in alcoholics. The incidence of gram-negative anaerobes was also higher in jejunal aspirates from alcoholics. This was further confirmed by a breath test study [48], which indicated that small bowel bacterial growth was three times greater in alcoholics. Subsequent studies further confirmed the effect of alcohol consumption on bacterial overgrowth in the intestine [48–52]. Bacterial overgrowth certainly increases the volume of the endotoxin source and may contribute to the alcoholic endotoxemia. However, no correlation has been demonstrated between the bacterial overgrowth and severity of liver damage [47, 48]. Therefore, the potential contribution of bacterial overgrowth on alcoholic endotoxemia is unclear.

19.2.3
Increased Gut Permeability to Endotoxins

Gastrointestinal mucosa impedes the diffusion of macromolecules from the gut lumen into the tissue and systemic circulation due to the epithelial barrier function. Endotoxins, therefore, cannot penetrate the gut barrier into circulation. Studies using permeability markers such as polyethyleneglycol, chromium-EDTA, mannitol/lactulose, or sucrose showed that the gastrointestinal permeability to macromolecules is significantly greater in alcoholics when compared to that in normal subjects [19, 47, 48, 53]. Alcohol administration increases gastrointestinal permeability in both normal subjects and patients with ALD [53]. In normal subjects, the acute ethanol induces only a transient increase in intestinal permeability. Although in some patients with chronic alcohol abuse, the intestinal permeability is reduced to normal levels by one to two weeks of sobriety; in many patients the abnormality in gastrointestinal permeability persists even after two weeks of sobriety [53].

It is not clear if permeability is caused by an abnormality in the gastroduodenal barrier function or the intestinal barrier function. Keshavarzian and coworkers showed that acute ethanol intake increases gastroduodenal permeability without affecting the intestinal permeability, while chronic alcohol abuse in alcoholics showed increased intestinal permeability without affecting the gastroduodenal permeability [53]. The location of barrier dysfunction is important in understanding the role of barrier dysfunction in ALD, as the large intestine is the main source of bacterial endotoxins. Another study [54] compared the gastrointestinal permeability changes with liver disease in alcoholic subjects. Both gastroduodenal permeability and intestinal permeability were significantly higher in alcoholics with symptoms of liver injury than those in alcoholics without liver damage, normal subjects, or patients with non-ALD. However, in this study, it was not clear if the permeability changes observed were due to the cause of liver damage or if it was the consequence of liver damage. Several other studies showed no such correlation between the intestinal permeability and the severity of liver disease, as increased intestinal permeability was observed at quite early during the disease [9, 15].

A correlation between intestinal permeability and alcoholic liver damage was also demonstrated in experimental models of ALD using rats and mice. A number of studies using animal models [18, 19, 37, 40, 55] demonstrated that chronic or acute administration of ethanol increases gastrointestinal permeability to macromolecular markers such as mannitol, lactulose, PEG, and dextran as well as LPS. At present, there is no animal model that can mimic symptoms of ALD similar to those seen in humans with ALD. However, several studies showed that ethanol-induced gastrointestinal permeability is associated with elevated plasma endotoxin levels and liver injury [18, 19, 40], indicating that a correlation between intestinal permeability, endotoxemia, and liver injury can be demonstrated in animal models similar to those seen in humans with ALD. Therefore, the results of these experimental studies are relevant to our understanding of the pathogenesis of ALD. It is not clear at present if changes in both gastroduodenal and intestinal permeability are important in causing alcoholic liver injury. In an intragastric model of ethanol exposure, the French and Tsukamoto model of ALD, Mathurin et al. [18] showed that liver injury induced by chronic ethanol administration is associated with the elevated intestinal permeability, while having no influence on gastroduodenal permeability.

Most recent studies have addressed the possible effect of ethanol on paracellular permeability and disruption of epithelial tight junctions in a cell culture model of intestinal epithelium, the Caco-2 cell monolayer [38, 56, 57]. Paracellular permeability was evaluated by measuring transepithelial electrical resistance (TER) and unidirectional flux of extracellular markers such as inulin and mannitol. Ethanol at concentration up to 5% produced no effect on barrier function, but 10% ethanol transiently decreased TER. But acetaldehyde (0.1–0.6 mM), the metabolic product of ethanol, potently reduced TER and increased paracellular permeability to inulin in a dose-dependent manner [38, 57]. The weak effect of ethanol may be attributed to the very low level of expression of alcohol dehydrogenase in Caco-2 cells. A study by Ma et al. [38] indicated that ethanol at 5–10% concentration reduced TER of Caco-2 cell monolayers by 30–50%, and increased mannitol permeability twofold by a myosin light-chain kinase-dependent mechanism. Such a mechanism may exist in the gastroduodenal region where alcohol levels can be expected to be high.

However, there is no evidence for such high levels of alcohol being maintained in the small or large intestine. Increased permeability, in the Caco-2 cell monolayer, by 2.5–15% ethanol was shown in another recent study; however, this permeability was caused by the loss of cell viability [58].

Mounting evidence indicates that ethanol is oxidized to acetaldehyde in the gastrointestinal tract, and suggests that acetaldehyde may contribute to the pathogenesis of alcohol-related diseases [59]. In addition to mucosal alcohol dehydrogenases, intestinal bacteria seem to play a significant role in the oxidation of ethanol to acetaldehyde [59]. The capacity of colonic mucosa and microbes to oxidize acetaldehyde to acetate is low compared to that in other tissues, suggesting a greater ability of the colon to accumulate acetaldehyde. High levels of acetaldehyde appear to be accumulated in the colonic mucosa and lumen in alcoholics. Acetaldehyde disrupts tight junction and increases paracellular permeability by inducing redistribution of tight junction proteins from the intercellular junctions. This effect of acetaldehyde on tight junction and paracellular permeability is mediated by a tyrosine kinase-dependent mechanism [56]. Acetaldehyde does not alter the overall tyrosine kinase activity, but, it effectively inhibits protein tyrosine phosphatase activity in Caco-2 cells. Generation and accumulation of acetaldehyde in the intestinal lumen may play a crucial role in the onset of the cascade of cellular responses that ultimately lead to endotoxemia and liver injury.

19.3
Influence of Alcoholic Endotoxemia on Different Organs

Endotoxemia affects multiple tissues and organs. Alcoholic endotoxemia has been extensively investigated in relation to the pathogenesis of liver disease. Emerging evidence indicates that alcoholic endotoxemia may also play a crucial role in the pathogenesis of diseases of pancreas, heart, lung, and other organs.

19.3.1
Liver

Evidence of LPS-induced hepatic injury has been reported in patients with cirrhosis, autoimmune hepatitis, and primary biliary cirrhosis [60]. The effect of LPS on liver injury was tested in numerous experimental studies. At large doses, LPS induce liver injury by themselves [61–63], while at low doses it potentiates liver injury induced by xenobiotics [64–66]. LPS administration at a dose of 1.0 µg/g body weight in rats increases the expression of TNF-α, IL-1, and IL-6 [67] and induced oxidative stress in the liver [68]. A positive correlation of plasma endotoxin levels with the plasma TNF-α levels was established in patients with ALD [12]. Attenuation of alcohol-induced liver injury by antibiotics and a positive correlation between endotoxemia and TNF-α levels indicate that endotoxemia plays a crucial role in the pathogenesis of alcohol-induced steatosis, hepatitis, and cirrhosis.

LPS by themselves did not mimic ethanol-induced steatosis or hepatitis. However, ethanol and LPS together, synergistically induced liver damage. A significant body of evidence indicates that chronic ethanol feeding sensitizes liver for LPS-induced injury in experimental animals [69–74]. Ethanol potentiates LPS-induced liver injury without altering apoptosis of hepatocytes [73] and exacerbates LPS-induced cytokine release in liver [67, 75]. The mechanism by which ethanol sensitizes liver for LPS-induced injury is unclear. However, few recent studies have provided some insights into this process. First, the effect of ethanol on endotoxemia is fairly quick. The acute administration of ethanol increases the plasma levels of endotoxin and TNF-α in rats as early as 1.5 hours [24]. Ethanol administration for 1–21 hours caused threefold increase in LPS-induced alanine transaminase (ALT) [76] release in mice [77]. The ethanol effect on sensitization of liver for LPS-induced transaminase release requires gut-derived endotoxin, as antibiotics attenuated the sensitization effect [77]. Another study suggested that downregulation of IL-10 is involved in the ethanol-mediated sensitization of LPS-induced liver injury [74], as LPS induce hepatic injury in IL-10 knockout mice even in the absence of ethanol. Both ethanol and LPS impair the activity of endothelial nitric oxide synthase in rat liver, and ethanol and LPS together showed a synergistic effect. Another mechanism may involve ethanol-induced oxidative stress. Chronic ethanol feeding increases LPS-stimulated NADPH oxidase-dependent production of reactive oxygen species (ROS) in Kupffer cells and enhances the production of TNF-α in LPS-treated rats [71]. A most recent study implicated adrenergic stimulation as a potential mechanism in the ethanol sensitization of LPS-induced liver injury. Epinephrine mimicked ethanol-induced sensitization of LPS effect on liver injury [72]. Ethanol and epinephrine combined showed a synergistic effect on LPS-induced injury, while propranalol attenuated ethanol-induced sensitization. Therefore, multiple factors may be involved in the sensitizing effect of ethanol on LPS-induced hepatic injury.

It may be difficult to distinguish whether ethanol potentiates LPS-induced cell injury or if LPS sensitize ethanol-induced tissue damage. LPS at low doses is known to exacerbate liver injury caused by a variety of injurious factors [64, 65, 78]. Ethanol-induced liver necrosis, inflammation, and fibrosis were exacerbated by the enteral administration of LPS [18]. LPS pretreatment enhanced collagen secretion in isolated hepatic stellate cells by both ethanol and acetaldehyde, by diminishing the levels of reduced glutathione and increasing the levels of oxidized glutathione [60]. LPS infusion *in vivo* increases ethanol-induced free radical formation in rats [68]; this was mediated by the activation of Kupffer cells. LPS also increased serum ALT levels in ethanol-fed rats [79]. A recent study demonstrated that malondialdehyde-acetaldehyde adducts of albumin (MAA-Alb)-induced release of cytokines and chemokines from the sinusoidal endothelial cells was increased three to fivefold by LPS administration [80]. Therefore, the synergistic effect of ethanol and LPS may involve both potentiation of the LPS effect by ethanol, as well as exacerbation of ethanol effect by LPS, although the sensitivity of different cell types in liver may vary.

19.3.2
Pancreas

Disruption of intestinal barrier function and bacterial translocation has been implicated in the pathogenesis of acute pancreatitis. Compared to mild attacks, severe attacks of acute pancreatitis in human subjects were associated with significantly higher urinary excretion of intestinal fatty acid binding protein (IFABP), a marker of elevated intestinal permeability [81]. Urinary IFABP was very low in normal subjects compared to patients with mild attacks of pancreatitis. The urinary IFABP levels were positively correlated with intestinal permeability to polyethyleneglycol, a marker of extracellular fluid. Increased gut permeability was observed during the early stages of acute pancreatitis [82, 83]. Therefore, increased intestinal permeability appears to be associated with the pathogenesis of acute pancreatitis. This raised the question of whether endotoxemia plays a role in the pathogenesis of pancreatitis, especially in alcoholic pancreatitis. The effect of LPS on apoptosis of pancreatic acinar cells and its sensitization by ethanol feeding was demonstrated several years ago [84]. Recent studies showed that LPS induce pancreatic injury in rats *in vivo* and that chronic ethanol feeding exacerbates these effects of LPS on the pancreas [85, 86]. LPS in ethanol-fed rats caused necrosis of pancreatic acini [85] and induced pancreatic fibrosis [86]. Acinar necrosis and pancreatic fibrosis were not evident in LPS-induced rats in the absence of ethanol feeding. Therefore, there is significant evidence indicating that endotoxemia and alcohol play important roles in the pathogenesis of pancreatitis. However, this area of research remains relatively unexplored.

19.3.3
Lung

Ethanol administration predisposes rats to edematous lung injury elicited by endotoxemia or sepsis [87]. Ethanol administration depleted reduced glutathione and increased alveolar epithelial permeability resulting in reduced alveolar liquid clearance and depletion of surfactant depletion. Subsequent studies have confirmed the ethanol-induced sensitization of LPS injury of rat lung [88, 89]. These studies also confirmed the glutathione depletion and disruption of alveolar epithelial barrier function in rats fed with ethanol and LPS. Granulocyte/macrophage colony stimulating factor (GM-CSF) is known to restore alveolar epithelial barrier function both *in vivo* and *in vitro* even in the face of acute endotoxemia. Studies by Joshi *et al.* [89] indicated that chronic ethanol reduces the expression of GM-CSF receptor in rat airway epithelium. Recent studies by Zhang *et al.* [90, 91] demonstrated that intestinal endotoxemia plays an important role in hepatopulmonary syndrome. Histological deterioration of lung endotoxemia paralleled the damage to the liver in cirrhotic rats. Therefore, the available information points to the direction of lung injury caused by ethanol-induced endotoxemia. Further studies are needed to substantiate this view.

19.4
Mechanism of Tissue Damage by Alcoholic Endotoxemia

The current understanding of the mechanism of alcoholic endotoxemia and endotoxin effects in the liver is illustrated in Figure 19.1.

19.4.1
Cellular Targets

Endotoxin-induced organ injury involves a variety of cells, including macrophages, monocytes, neutrophils, and platelets. LPS also directly affect sinusoidal endothelial cells, hepatocytes, and stellate cells in the liver. Kupffer cell is the first target of

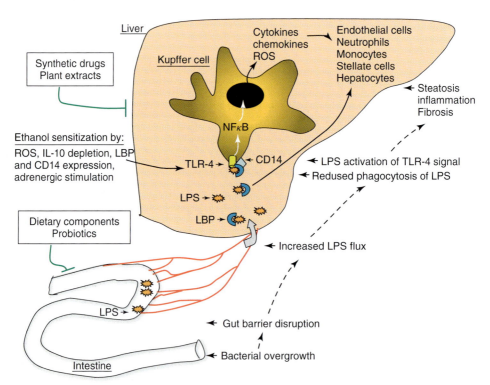

Fig. 19.1 Schematic representation of alcoholic endotoxemia and liver injury. Intestinal microflora is the source of endotoxins. Bacterial overgrowth and gut barrier disruption leads to increased LPS flux into portal vein. The first target of portal LPS is liver. LPS stimulates Kupffer cells by LBP, CD14, and TLR-4-dependent mechanism to induce gene expression and production of cytokines, chemokines, and ROS. The inflammatory mediators affect other cells in the liver. LPS may also directly affect other cell types in the liver, although the sensitivity to LPS may vary. Dietary components such as glutamine, zinc, and oat bran may act in the gut to prevent barrier dysfunction and endotoxemia, while plant extracts and synthetic drugs may act in liver to prevent LPS-induced inflammation and cell injury.

intestinal endotoxins, and play a pivotal role in hepatotoxicity induced by ethanol exposure [92, 93]. Ron Thurman's group first demonstrated that inactivation of Kupffer cells prevents ethanol-induced liver injury in rats [94]. This observation was confirmed by a couple of subsequent studies [95, 96]. Kupffer cell inactivation by gadolinium treatment attenuates ethanol-induced CYP2E1 expression and steatosis [96], and thalidomide prevents ethanol-induced liver injury by preventing the activation of Kupffer cells [95]. LPS rapidly increase intracellular calcium in isolated Kupffer cells [97]. Kupffer cells from women were found to be more sensitive to LPS than those from men [21]. LPS in the presence of MAA-Alb adduct stimulated sinusoidal endothelial cells to release cytokines and chemokines [80]; this effect was potentiated by ethanol administration. LPS pretreatment enhances collagen secretion by hepatic stellate cells exposed to ethanol [60]. Therefore, ethanol-induced endotoxemia involves the activation of numerous cell types leading to a complex process of cellular injury in the liver. A similar mechanism may also be involved in alcoholic pancreatitis, lung disease, and other organ injury.

19.4.2
Receptors and Signaling

Mounting evidence indicates that LPS-induced activation of multiple types of cells leads to the generation of ROS, cytokines, and chemokines. LPS interact with LBP and LBP present LPS to CD14. CD14, a 55-kDa glycoprotein, is bound to the surface of monocytes and most macrophages, including hepatic Kupffer cells [98]. CD14 specifically binds to LPS and interacts with the transmembrane receptor, TLR-4 [99]. Chronic ethanol feeding increases the production of LBP [20, 27] and the interaction between LPS and LBP is essential for ethanol-induced liver damage. Ethanol-induced increase in ALT levels was absent in LBP knockout mice [28]. Therefore, LBP plays an important role in the early stages of alcohol-induced liver injury by enhancing the LPS-induced signal transduction in Kupffer cells. Chronic ethanol has also been shown to dramatically elevate the expression of CD14 [16, 27, 100], and early phase of ethanol-induced liver injury was absent in CD14 knockout mice [16]. Ethanol-induced increase in ALT level, appearance of necrotic foci in liver, and the expression of TNF-α were absent in ethanol-fed CD14 knockout mice. LPS stimulation of human Kupffer cells to produce TNF-α can be prevented by anti-CD14 antibodies [42]. LPS-induced TNF-α production was also absent in Kupffer cells isolated from CD14 knockout mice [42]. Therefore, CD14 plays a pivotal role in ethanol-induced liver injury. The observation made in the experimental studies is supported by a human study demonstrating that CD14 promoter polymorphism correlates with enhanced production of CD14 and the severity of ALD [101].

Evidence indicates that TLR-4 plays an important role in LPS signaling [102] and ethanol-induced liver injury. Ron Thurman's group first demonstrated that ethanol-induced increase in plasma ALT levels and expression of TNF-α were absent in TLR-4 knockout mice [103]. Immunofluorescence localization and RT-PCR of TLR-4 in ethanol-fed rat liver showed that ethanol increases the levels of

TLR-4 protein and mRNA for TLR-4 in Kupffer cells [104]. This observation was supported by a recent study showing that LPS-induced rise in plasma endotoxin levels, steatosis, and triglyceride levels was absent in TLR-4 knockout mice [105]. But, deletion of TLR-2 or MyD88 did not affect ethanol-induced liver injury. In addition to Kupffer cells, LPS affect other types of cells in the liver. LPS induced the activation of NFκB in stellate cells isolated from human liver [106]. LPS also activate N-terminal jun kinase (JNK) and up regulate the release of chemokines and adhesion molecules in stellate cells by a TLR-4-dependent mechanism. Therefore, TLR-4 plays a crucial role in endotoxemia-mediated liver injury in alcoholics.

TLR-4-mediated signal transduction by LPS in different cells such as macrophages (Kupffer cells), monocytes, sinusoidal endothelial cells, and stellate cells leads to synthesis and secretion of a variety of injurious factors, such as cytokines [107–111], chemokines [108, 112], ROS [107, 113], and impaired proteasome [114–118]. One of the major mechanisms associated with TLR-4-mediated dysregulation of cytokines/chemokine production, oxidative stress, and proteasome function is activation of NFκB. Chronic ethanol feeding activates NFκB in rat liver [32, 114–118] and LPS *in vitro* activate NFκB in Kupffer cells. As discussed above, the tissue injury in ALD involves stimulation of a variety of cell types and release of an array of injurious factors. It appears that endotoxemia caused by intestinal barrier dysfunction triggers this complex cellular turbulence and outburst of an array of injurious factors.

19.5
Factors that Ameliorate Alcoholic Endotoxemia and Tissue Damage

A significant body of evidence indicates that alleviation of ethanol-induced endotoxemia by therapeutic intervention can ameliorate liver injury. First, intestinal sterilization by antibiotic treatment in rats attenuated ethanol-induced endotoxemia and liver damage [119]. Several factors were found to be beneficial in diminishing ethanol-induced endotoxemia and liver injury. The factors that were used to control alcohol-induced liver damage include synthetic drugs, dietary components, plant products, and probiotics.

19.5.1
Synthetic Drugs

Peroxisome proliferator-activated receptor-γ (PPARγ) is a ligand-dependent transcription factor. Stimulation of PPARγ has been shown to inhibit the production of inflammatory cytokines in monocytes and macrophages. Pioglitamide, a ligand for PPARγ attenuates ethanol-induced liver injury [120]. Pioglitamide attenuated ethanol-induced increase in ALT and TNF-α and prevented hepatic inflammation and necrosis. However, pioglitamide did not prevent the ethanol-induced generation of lipid peroxides.

Bicyclol is a novel synthetic antihepatitis drug that was shown to be protective against hepatic injury induced by various factors. Bicyclol also attenuates ethanol-induced liver injury [24]. Ethanol-induced increases in plasma ALT, triglyceride, and liver pathology were prevented by bicyclol. One of the mechanisms of bicyclol involved in the prevention of ethanol-induced liver damage is attenuation of ethanol-induced oxidative stress. Bicyclol attenuates ethanol-induced increase in lipid peroxides and restores the antioxidant defense system by preventing the downregulation of glutathione, superoxide dismutase, catalase, glutathione peroxidase, and glutathione reductase. Bicyclol also prevents ethanol-induced endotoxemia and the expression of cytokines and CD14, suggesting that bicyclol may attenuate the ethanol-induced liver injury by preventing ethanol-induced increase in intestinal permeability and endotoxemia.

19.5.2
Dietary Components

Glycine was shown to attenuate hepatic damage induced by various factors that involve endotoxemia. Administration of glycine for two weeks after inducing liver injury by chronic ethanol feeding accelerates the recovery from liver damage. Glycine rapidly reduced liver pathology score and attenuated leukocyte infiltration and TNF-α production by Kupffer cells [121]. Studies have also shown that feeding oat bran or zinc attenuates liver injury induced by acute or chronic ethanol feeding. Zinc supplementation attenuates ethanol-induced increase in plasma ALT and liver pathology [37]. This hepato protective effect of zinc was associated with a reduction of endotoxemia and reduced intestinal permeability, suggesting that zinc affects the intestine and the amelioration of liver injury is the cause of alleviation of endotoxemia. Similarly, oat bran supplementation also prevents liver injury caused by chronic ethanol administration [19]. Once again, oat bran supplementation attenuates ethanol-induced intestinal permeability and endotoxemia.

19.5.3
Plant Extracts

Plant products such as green tea extracts [122] and cocoa extracts [123] have been found to be beneficial in ameliorating ethanol-induced liver injury. Green tea extracts attenuate ethanol-induced increase in ALT and production of ROS and prevent hepatic necrosis. However, green tea extracts do not affect the ethanol-induced steatosis or inflammation. Cocoa extracts, on the other hand, alleviate ethanol-induced increase in hepatic fat accumulation, inflammation, and necrosis [123]. Cocoa extracts also attenuate ethanol-induced production of TNF-α and ROS. Therefore, the protective role of flavanoids present in cocoa extract and green tea extract may involve distinct cellular mechanisms. Baicalein, a traditional Chinese medicine, is a major bioactive flavanoid component of dried roots of huang qin. Bacalein has been shown to attenuate LPS-induced elevation of oxidative stress and neutrophil infiltration into liver and lung [124]. However, its effect on alcohol-induced endotoxemia is unknown.

19.5.4
Probiotics

Probiotics have been considered in the treatment of non-ALDs [125, 126]. A significant body of evidence suggests that probiotics may prove to be beneficial in controlling ALD [127]. A first study in this regard by Nanji et al. [128] demonstrated that feeding *Lactobacillus* reduces the severity of ethanol-induced liver injury in rats. The pathology score dramatically reduced by *Lactobacillus* feeding. Interestingly enough, *Lactobacillus* feeding also prevented ethanol-induced endotoxemia, suggesting that the initial effect of probiotic may be at the level of intestinal mucosa. The hepatoprotective role of probiotics was recently confirmed by other groups [129, 130]. Feeding lactic acid bacteria prevented ethanol-induced increase in the levels of plasma endotoxin and ALT [129, 130]. Another study showed that feeding heat killed *Lactobacillus* significantly inhibited ethanol-induced increase in serum levels of ALT, triglyceride, and total cholesterol [130]. *Lactobacillus* also attenuated ethanol-induced TNF-α, SREBP-1, and SREBP-2 in the liver. *Lactobacillus* treatment effectively works against endotoxin translocation and in the course of acute pancreatitis [131]. A recent clinical study showed that patients with alcoholic liver injury showed an alteration of microfloral populations in the gut [132]; this was reversed by short-term oral supplementation with *Bifidobacterium bifidum* and *Lactobacillus plantarum* 9PA3. The oral supplementation with probiotics also showed an improvement in alcohol-induced liver injury than standard therapy alone [132].

19.6
Summary and Perspectives

Evidence generated from numerous clinical and experimental studies clearly demonstrates that alcohol consumption leads to endotoxemia (Figure 19.1). The main cause of this endotoxemia appears to be the dysfunctional gut barrier and increased absorption of endotoxins. While reduced phagocytosis of LPS by Kupffer cells and intestinal bacterial overgrowth are still viable causes, they appear to be only minor contributing factors. The role of endotoxins in ALD has been investigated extensively. However, it is likely that endotoxemia also mediates alcohol-induced injury to other organs such as the pancreas and lung. Although Kupffer cells are the initial target, LPS affect other cells such as sinusoidal endothelial cells, stellate cells, monocytes, neutrophils, and hepatocytes. LPS activate TLR-4 by interacting with LBP and CD14 resulting in the activation of NFκB and overexpression of proinflammatory cytokines and chemokines. LPS also induce oxidative stress and proteasome dysfunction leading to steatosis, inflammation, and fibrosis in liver. Synthetic drugs, dietary components, plant extracts, and probiotics ameliorate oxidative stress, inflammation, and endotoxemia induced by ethanol. It appears that factors that act by preventing intestinal permeability and endotoxemia are the most effective factors in preventing ethanol-induced tissue damage. Future studies need to be focused on understanding the more detailed mechanism of synergism between ethanol and endotoxins, ethanol-induced intestinal permeability and prevention of ethanol-induced permeability and endotoxemia.

Acknowledgments

Preparation of this chapter was supported by National Institute of Health grants R01-DK55532 and R01-AA12307.

References

1 Fenwick, B.W. (1990) Virulence attributes of the liposaccharides of the HAP group organisms. *Can J Vet Res*, **54** (Suppl), S28–S32.

2 Ansell, J., Widrich, W., Johnson, W., and Fine, J. (1977) Endotoxin and bacteria in portal blood. *Gastroenterology*, **73**, 1190.

3 Liehr, H. (1979) The Limulus assay for endotoxemia, as applied in gastroenterology. *Prog Clin Biol Res*, **29**, 309–320.

4 Chen, C., Li, L., Wu, Z., Chen, H., and Fu, S. (2007) Effects of lactitol on intestinal microflora and plasma endotoxin in patients with chronic viral hepatitis. *J Infect*, **54**, 98–102.

5 Schietroma, M., Carlei, F., Cappelli, S., and Amicucci, G. (2006) Intestinal permeability and systemic endotoxemia after laparotomic or laparoscopic cholecystectomy. *Ann Surg*, **243**, 359–363.

6 Staun-Olsen, P., Bjorneboe, M., Prytz, H., Thomsen, A.C., and Orskov, F. (1983) Escherichia coli antibodies in alcoholic liver disease. Correlation to alcohol consumption, alcoholic hepatitis, and serum IgA. *Scand J Gastroenterol*, **18**, 889–896.

7 Nolan, J.P., DeLissio, M.G., Camara, D.S., Feind, D.M., and Gagliardi, N.C. (1986) IgA antibody to lipid A in alcoholic liver disease. *Lancet*, **1**, 176–179.

8 Fukui, H., Brauner, B., Bode, J.C., and Bode, C. (1991) Plasma endotoxin concentrations in patients with alcoholic and non-alcoholic liver disease: reevaluation with an improved chromogenic assay. *J Hepatol*, **12**, 162–169.

9 Bode, C., Kugler, V., and Bode, J.C. (1987) Endotoxemia in patients with alcoholic and non-alcoholic cirrhosis and in subjects with no evidence of chronic liver disease following acute alcohol excess. *J Hepatol*, **4**, 8–14.

10 Bode, C., Schafer, C., Fukui, H., and Bode, J.C. (1997) Effect of treatment with paromomycin on endotoxemia in patients with alcoholic liver disease–a double-blind, placebo-controlled trial. *Alcohol Clin Exp Res*, **21**, 1367–1373.

11 Fujimoto, M., Uemura, M., Nakatani, Y., Tsujita, S., Hoppo, K., Tamagawa, T., Kitano, H. et al. (2000) Plasma endotoxin and serum cytokine levels in patients with alcoholic hepatitis: relation to severity of liver disturbance. *Alcohol Clin Exp Res*, **24**, 48S–54S.

12 Hanck, C., Rossol, S., Bocker, U., Tokus, M., and Singer, M.V. (1998) Presence of plasma endotoxin is correlated with tumour necrosis factor receptor levels and disease activity in alcoholic cirrhosis. *Alcohol Alcohol*, **33**, 606–608.

13 Parlesak, A., Schafer, C., Schutz, T., Bode, J.C., and Bode, C. (2000) Increased intestinal permeability to macromolecules and endotoxemia in patients with chronic alcohol abuse in different stages of alcohol-induced liver disease. *J Hepatol*, **32**, 742–747.

14 Schafer, C., Greiner, B., Landig, J., Feil, E., Schutz, E.T., Bode, J.C., and Bode, C. (1997) Decreased endotoxin-binding capacity of whole blood in patients with alcoholic liver disease. *J Hepatol*, **26**, 567–573.

15 Schafer, C., Parlesak, A., Schutt, C., Bode, J.C., and Bode, C. (2002) Concentrations of lipopolysaccharide-binding protein, bactericidal/permeability-increasing protein, soluble CD14 and plasma lipids in relation to endotoxaemia in patients with alcoholic liver disease. *Alcohol Alcohol*, **37**, 81–86.

16 O'Malley, C.M., Frumento, R.J., Mets, B., Naka, Y., and Bennett-Guerrero, E. (2004) Endotoxaemia during left ventricular assist device insertion: relationship between risk factors and outcome. *Br J Anaesth*, **92**, 131–133.

17 Rivera, C.A., Bradford, B.U., Seabra, V. and Thurman, R.G. (1998) Role of endotoxin in the hypermetabolic state after acute ethanol exposure. *Am J Physiol*, **275**, G1252–G1258.

18 Mathurin, P., Deng, Q.G., Keshavarzian, A., Choudhary, S., Holmes, E.W., and Tsukamoto, H. (2000) Exacerbation of alcoholic liver injury by enteral endotoxin in rats. *Hepatology*, **32**, 1008–1017.

19 Keshavarzian, A., Choudhary, S., Holmes, E.W., Yong, S., Banan, A., Jakate, S., and Fields, J.Z. (2001) Preventing gut leakiness by oats supplementation ameliorates alcohol-induced liver damage in rats. *J Pharmacol Exp Ther*, **299**, 442–448.

20 Jokelainen, K., Reinke, L.A., and Nanji, A.A. (2001) Nf-κb activation is associated with free radical generation and endotoxemia and precedes pathological liver injury in experimental alcoholic liver disease. *Cytokine*, **16**, 36–39.

21 Muller, C. (2006) Liver, alcohol and gender. *Wien Med Wochenschr*, **156**, 523–526.

22 Lambert, J.C., Zhou, Z., Wang, L., Song, Z., McClain, C.J., and Kang, Y.J. (2004) Preservation of intestinal structural integrity by zinc is independent of metallothionein in alcohol-intoxicated mice. *Am J Pathol*, **164**, 1959–1966.

23 Ferrier, L., Berard, F., Debrauwer, L., Chabo, C., Langella, P., Bueno, L., and Fioramonti, J. (2006) Impairment of the intestinal barrier by ethanol involves enteric microflora and mast cell activation in rodents. *Am J Pathol*, **168**, 1148–1154.

24 Zhao, J., Chen, H., and Li, Y. (2008) Protective effect of bicyclol on acute alcohol-induced liver injury in mice. *Eur J Pharmacol*, **586**, 322–331.

25 Nanji, A.A., Jokelainen, K., Fotouhinia, M., Rahemtulla, A., Thomas, P., Tipoe, G.L., Su, G.L. *et al.* (2001) Increased severity of alcoholic liver injury in female rats: role of oxidative stress, endotoxin, and chemokines. *Am J Physiol Gastrointest Liver Physiol*, **281**, G1348–G1356.

26 Jarvelainen, H.A., Fang, C., Ingelman-Sundberg, M., and Lindros, K.O. (1999) Effect of chronic coadministration of endotoxin and ethanol on rat liver pathology and proinflammatory and anti-inflammatory cytokines. *Hepatology*, **29**, 1503–1510.

27 Zuo, G.Q., Gong, J.P., Liu, C.A., Li, S.W., Wu, X.C., Yang, K., and Li, Y. (2001) Expression of lipopolysaccharide binding protein and its receptor CD14 in experimental alcoholic liver disease. *World J Gastroenterol*, **7**, 836–840.

28 Uesugi, T., Froh, M., Arteel, G.E., Bradford, B.U., Wheeler, M.D., Gabele, E., Isayama, F. *et al.* (2002) Role of lipopolysaccharide-binding protein in early alcohol-induced liver injury in mice. *J Immunol*, **168**, 2963–2969.

29 Nanji, A.A., Su, G.L., Laposata, M., and French, S.W. (2002) Pathogenesis of alcoholic liver disease–recent

advances. *Alcohol Clin Exp Res*, **26**, 731–736.

30. Horie, Y., Yamagishi, Y., Kato, S., Kajihara, M., Tamai, H., Granger, D.N., and Ishii, H. (2002) Role of ICAM-1 in chronic ethanol consumption-enhanced liver injury after gut ischemia-reperfusion in rats. *Am J Physiol Gastrointest Liver Physiol*, **283**, G537–G543.

31. McKim, S.E., Gabele, E., Isayama, F., Lambert, J.C., Tucker, L.M., Wheeler, M.D., Connor, H.D. et al. (2003) Inducible nitric oxide synthase is required in alcohol-induced liver injury: studies with knockout mice. *Gastroenterology*, **125**, 1834–1844.

32. Nanji, A.A., Jokelainen, K., Tipoe, G.L., Rahemtulla, A., and Dannenberg, A.J. (2001) Dietary saturated fatty acids reverse inflammatory and fibrotic changes in rat liver despite continued ethanol administration. *J Pharmacol Exp Ther*, **299**, 638–644.

33. Tipoe, G.L., Liong, E.C., Casey, C.A., Donohue, T.M. Jr, Eagon, P.K., So, H., Leung, T.M., et al. (2008) A voluntary oral ethanol-feeding rat model associated with necroinflammatory liver injury. *Alcohol Clin Exp Res*, **32**, 669–682.

34. Iimuro, Y., Frankenberg, M.V., Arteel, G.E., Bradford, B.U., Wall, C.A., and Thurman, R.G. (1997) Female rats exhibit greater susceptibility to early alcohol-induced liver injury than males. *Am J Physiol*, **272**, G1186–G1194.

35. Zhang, X.G., Yu, C.H., Qing, Y.E., Zhang, Y., Chen, S.H., and Li, Y.M. (2003) Effect of tea polyphenol on liver fibrosis in rats and related mechanism. *Zhongguo Zhong Yao Za Zhi*, **28**, 1070–1072.

36. Schenker, S. (1997) Medical consequences of alcohol abuse: is gender a factor? *Alcohol Clin Exp Res*, **21**, 179–181.

37. Lambert, J.C., Zhou, Z., Wang, L., Song, Z., McClain, C.J., and Kang, Y.J. (2003) Prevention of alterations in intestinal permeability is involved in zinc inhibition of acute ethanol-induced liver damage in mice. *J Pharmacol Exp Ther*, **305**, 880–886.

38. Ma, T.Y., Nguyen, D., Bui, V., Nguyen, H., and Hoa, N. (1999) Ethanol modulation of intestinal epithelial tight junction barrier. *Am J Physiol*, **276**, G965–G974.

39. Kavanaugh, M.J., Clark, C., Goto, M., Kovacs, E.J., Gamelli, R.L., Sayeed, M.M., and Choudhry, M.A. (2005) Effect of acute alcohol ingestion prior to burn injury on intestinal bacterial growth and barrier function. *Burns*, **31**, 290–296.

40. Choudhry, M.A., Fazal, N., Goto, M., Gamelli, R.L., and Sayeed, M.M. (2002) Gut-associated lymphoid T cell suppression enhances bacterial translocation in alcohol and burn injury. *Am J Physiol Gastrointest Liver Physiol*, **282**, G937–G947.

41. Cubero, F.J. and Nieto, N. (2006) Kupffer cells and alcoholic liver disease. *Rev Esp Enferm Dig*, **98**, 460–472.

42. Han, D.W. (2002) Intestinal endotoxemia as a pathogenetic mechanism in liver failure. *World J Gastroenterol*, **8**, 961–965.

43. Fukui, H., Kitano, H., Okamoto, Y., Kikuchi, E., Matsumoto, M., Kikukawa, M., Morimura, M. et al. (1995) Interaction of Kupffer cells to splenic macrophages and hepatocytes in endotoxin clearance: effect of alcohol. *J Gastroenterol Hepatol*, **10** (Suppl 1), S31–S34.

44. Spitzer, J.A. and Zhang, P. (1996) Gender differences in phagocytic responses in the blood and liver, and the generation of cytokine-induced neutrophil chemoattractant in the liver of acutely ethanol-intoxicated rats. *Alcohol Clin Exp Res*, **20**, 914–920.

45. Fukui, H. (2005) Relation of endotoxin, endotoxin binding proteins and macrophages to severe alcoholic liver injury and multiple organ failure. *Alcohol Clin Exp Res*, **29**, 172S–179S.

46. Fukui, H., Kitano, H., Okamoto, Y., Kikuchi, E., Matsumoto, M.,

Kikukawa, M., Morimura, M. et al. (1994) Effect of alcohol on the endotoxin binding protein produced in the liver. *Alcohol Alcohol Suppl*, **29**, 87–91.

47 Bode, J.C., Bode, C., Heidelbach, R., Durr, H.K., and Martini, G.A. (1984) Jejunal microflora in patients with chronic alcohol abuse. *Hepatogastroenterology*, **31**, 30–34.

48 Bode, C., Kolepke, R., Schafer, K., and Bode, J.C. (1993) Breath hydrogen excretion in patients with alcoholic liver disease–evidence of small intestinal bacterial overgrowth. *Z Gastroenterol*, **31**, 3–7.

49 Persson, J. (1991) Alcohol and the small intestine. *Scand J Gastroenterol*, **26**, 3–15.

50 Bode, C. and Bode, J.C. (2003) Effect of alcohol consumption on the gut. *Best Pract Res Clin Gastroenterol*, **17**, 575–592.

51 Bode, J.C. and Bode, C. (2000) Alcohol, the gastrointestinal tract and pancreas. *Ther Umsch*, **57**, 212–219.

52 Morencos, F.C., de las Heras Castano, G., Martin Ramos, L., Lopez Arias, M.J., Ledesma, F., and Pons Romero, F. (1995) Small bowel bacterial overgrowth in patients with alcoholic cirrhosis. *Dig Dis Sci*, **40**, 1252–1256.

53 Keshavarzian, A., Fields, J.Z., Vaeth, J., and Holmes, E.W. (1994) The differing effects of acute and chronic alcohol on gastric and intestinal permeability. *Am J Gastroenterol*, **89**, 2205–2211.

54 Keshavarzian, A., Holmes, E.W., Patel, M., Iber, F., Fields, J.Z., and Pethkar, S. (1999) Leaky gut in alcoholic cirrhosis: a possible mechanism for alcohol-induced liver damage. *Am J Gastroenterol*, **94**, 200–207.

55 Schmidt, K.L., Henagan, J.M., Smith, G.S., and Miller, T.A. (1987) Effects of ethanol and prostaglandin on rat gastric mucosal tight junctions. *J Surg Res*, **43**, 253–263.

56 Atkinson, K.J. and Rao, R.K. (2001) Role of protein tyrosine phosphorylation in acetaldehyde-induced disruption of epithelial tight junctions. *Am J Physiol Gastrointest Liver Physiol*, **280**, G1280–G1288.

57 Rao, R.K. (1998) Acetaldehyde-induced increase in paracellular permeability in Caco-2 cell monolayer. *Alcohol Clin Exp Res*, **22**, 1724–1730.

58 Banan, A., Choudhary, S., Zhang, Y., Fields, J.Z., and Keshavarzian, A. (1999) Ethanol-induced barrier dysfunction and its prevention by growth factors in human intestinal monolayers: evidence for oxidative and cytoskeletal mechanisms. *J Pharmacol Exp Ther*, **291**, 1075–1085.

59 Salaspuro, M. (1996) Bacteriocolonic pathway for ethanol oxidation: characteristics and implications. *Ann Med*, **28**, 195–200.

60 Quiroz, S.C., Bucio, L., Souza, V., Hernandez, E., Gonzalez, E., Gomez-Quiroz, L., Kershenobich, D. et al. (2001) Effect of endotoxin pretreatment on hepatic stellate cell response to ethanol and acetaldehyde. *J Gastroenterol Hepatol*, **16**, 1267–1273.

61 Iimuro, Y., Yamamoto, M., Kohno, H., Itakura, J., Fujii, H., and Matsumoto, Y. (1994) Blockade of liver macrophages by gadolinium chloride reduces lethality in endotoxemic rats–analysis of mechanisms of lethality in endotoxemia. *J Leukoc Biol*, **55**, 723–728.

62 Moulin, F., Pearson, J.M., Schultze, A.E., Scott, M.A., Schwartz, K.A., Davis, J.M., Ganey, P.E. et al. (1996) Thrombin is a distal mediator of lipopolysaccharide-induced liver injury in the rat. *J Surg Res*, **65**, 149–158.

63 Pearson, J.M., Schultze, A.E., Jean, P.A., and Roth, R.A. (1995) Platelet participation in liver injury from gram-negative bacterial lipopolysaccharide in the rat. *Shock*, **4**, 178–186.

64 Barton, C.C., Barton, E.X., Ganey, P.E., Kunkel, S.L., and Roth, R.A. (2001) Bacterial lipopolysaccharide enhances aflatoxin B1 hepatotoxicity in rats by a mechanism that depends

on tumor necrosis factor alpha. *Hepatology*, **33**, 66–73.
65 Sneed, R.A., Grimes, S.D., Schultze, A.E., Brown, A.P., and Ganey, P.E. (1997) Bacterial endotoxin enhances the hepatotoxicity of allyl alcohol. *Toxicol Appl Pharmacol*, **144**, 77–87.
66 Thurman, R.G. II (1998) Alcoholic liver injury involves activation of Kupffer cells by endotoxin. *Am J Physiol*, **275**, G605–G611.
67 Pennington, H.L., Hall, P.M., Wilce, P.A., and Worrall, S. (1997) Ethanol feeding enhances inflammatory cytokine expression in lipopolysaccharide-induced hepatitis. *J Gastroenterol Hepatol*, **12**, 305–313.
68 Chamulitrat, W. and Spitzer, J.J. (1996) Nitric oxide and liver injury in alcohol-fed rats after lipopolysaccharide administration. *Alcohol Clin Exp Res*, **20**, 1065–1070.
69 Karaa, A., Kamoun, W.S., and Clemens, M.G. (2005) Chronic ethanol sensitizes the liver to endotoxin via effects on endothelial nitric oxide synthase regulation. *Shock*, **24**, 447–454.
70 Koteish, A., Yang, S., Lin, H., Huang, X., and Diehl, A.M. (2002) Chronic ethanol exposure potentiates lipopolysaccharide liver injury despite inhibiting Jun N-terminal kinase and caspase 3 activation. *J Biol Chem*, **277**, 13037–13044.
71 Thakur, V., Pritchard, M.T., McMullen, M.R., Wang, Q., and Nagy, L.E. (2006) Chronic ethanol feeding increases activation of NADPH oxidase by lipopolysaccharide in rat Kupffer cells: role of increased reactive oxygen in LPS-stimulated ERK1/2 activation and TNF-α production. *J Leukoc Biol*, **79**, 1348–1356.
72 von Montfort, C., Beier, J.I., Guo, L., Kaiser, J.P., and Arteel, G.E. (2008) Contribution of the sympathetic hormone epinephrine to the sensitizing effect of ethanol on LPS-induced liver damage in mice. *Am J Physiol Gastrointest Liver Physiol*, **294**, G1227–G1234.
73 Deaciuc, I.V., D'Souza, N.B., Burikhanov, R., Lee, E.Y., Tarba, C.N., McClain, C.J., and de Villiers, W.J. (2002) Epidermal growth factor protects the liver against alcohol-induced injury and sensitization to bacterial lipopolysaccharide. *Alcohol Clin Exp Res*, **26**, 864–874.
74 Hill, D.B., Barve, S., Joshi-Barve, S., and McClain, C. (2000) Increased monocyte nuclear factor-κB activation and tumor necrosis factor production in alcoholic hepatitis. *J Lab Clin Med*, **135**, 387–395.
75 Worrall, S. and Wilce, P.A. (1994) The effect of chronic ethanol feeding on cytokines in a rat model of alcoholic liver disease. *Alcohol Alcohol Suppl*, **2**, 447–451.
76 Karaa, A., Thompson, K.J., McKillop, I.H., Clemens, M.G., and Schrum, L.W. (2008) S-adenosyl-L-methionine attenuates oxidative stress and hepatic stellate cell activation in an ethanol-LPS-induced fibrotic rat model. *Shock*, **30**, 197–205.
77 Yamashina, S., Takei, Y., Ikejima, K., Enomoto, N., Kitamura, T., and Sato, N. (2005) Ethanol-induced sensitization to endotoxin in Kupffer cells is dependent upon oxidative stress. *Alcohol Clin Exp Res*, **29**, 246S–250S.
78 Thurman, R.G., Bradford, B.U., Iimuro, Y., Knecht, K.T., Arteel, G.E., Yin, M., Connor, H.D. *et al.* (1998) The role of gut-derived bacterial toxins and free radicals in alcohol-induced liver injury. *J Gastroenterol Hepatol*, **13** (Suppl), S39–S50.
79 Tamai, H., Horie, Y., Kato, S., Yokoyama, H., and Ishii, H. (2002) Long-term ethanol feeding enhances susceptibility of the liver to orally administered lipopolysaccharides in rats. *Alcohol Clin Exp Res*, **26**, 75S–80S.
80 Duryee, M.J., Klassen, L.W., Freeman, T.L., Willis, M.S., Tuma, D.J., and Thiele, G.M. (2004) Lipopolysaccharide is a cofactor for malondialdehyde-acetaldehyde adduct-mediated cytokine/chemokine

release by rat sinusoidal liver endothelial and Kupffer cells. *Alcohol Clin Exp Res*, **28**, 1931–1938.

81 Rahman, S.H., Ammori, B.J., Holmfield, J., Larvin, M., and McMahon, M.J. (2003) Intestinal hypoperfusion contributes to gut barrier failure in severe acute pancreatitis. *J Gastrointest Surg*, **7**, 26–35; discussion 35-26.

82 Ammori, B.J. (2003) Role of the gut in the course of severe acute pancreatitis. *Pancreas*, **26**, 122–129.

83 Liu, H., Li, W., Wang, X., Li, J., and Yu, W. (2008) Early gut mucosal dysfunction in patients with acute pancreatitis. *Pancreas*, **36**, 192–196.

84 Fortunato, F. and Gates, L.K. Jr (2000) Alcohol feeding and lipopolysaccharide injection modulate apoptotic effectors in the rat pancreas in vivo. *Pancreas*, **21**, 174–180.

85 Fortunato, F., Deng, X., Gates, L.K., McClain, C.J., Bimmler, D., Graf, R., and Whitcomb, D.C. (2006) Pancreatic response to endotoxin after chronic alcohol exposure: switch from apoptosis to necrosis? *Am J Physiol Gastrointest Liver Physiol*, **290**, G232–G241.

86 Vonlaufen, A., Xu, Z., Daniel, B., Kumar, R.K., Pirola, R., Wilson, J., and Apte, M.V. (2007) Bacterial endotoxin: a trigger factor for alcoholic pancreatitis? Evidence from a novel, physiologically relevant animal model. *Gastroenterology*, **133**, 1293–1303.

87 Guidot, D.M. and Roman, J. (2002) Chronic ethanol ingestion increases susceptibility to acute lung injury: role of oxidative stress and tissue remodeling. *Chest*, **122**, 309S–314S.

88 Bechara, R.I., Brown, L.A., Eaton, D.C., Roman, J., and Guidot, D.M. (2003) Chronic ethanol ingestion increases expression of the angiotensin II type 2 (AT2) receptor and enhances tumor necrosis factor-α- and angiotensin II-induced cytotoxicity via AT2 signaling in rat alveolar epithelial cells. *Alcohol Clin Exp Res*, **27**, 1006–1014.

89 Joshi, P.C., Applewhite, L., Ritzenthaler, J.D., Roman, J., Fernandez, A.L., Eaton, D.C., Brown, L.A. *et al.* (2005) Chronic ethanol ingestion in rats decreases granulocyte-macrophage colony-stimulating factor receptor expression and downstream signaling in the alveolar macrophage. *J Immunol*, **175**, 6837–6845.

90 Zhang, H.Y., Han, D.W., Su, A.R., Zhang, L.T., Zhao, Z.F., Ji, J.Q., Li, B.H. *et al.* (2007) Intestinal endotoxemia plays a central role in development of hepatopulmonary syndrome in a cirrhotic rat model induced by multiple pathogenic factors. *World J Gastroenterol*, **13**, 6385–6395.

91 Zhang, H.Y., Han, D.W., Zhao, Z.F., Liu, M.S., Wu, Y.J., Chen, X.M., and Ji, C. (2007) Multiple pathogenic factor-induced complications of cirrhosis in rats: a new model of hepatopulmonary syndrome with intestinal endotoxemia. *World J Gastroenterol*, **13**, 3500–3507.

92 Enomoto, N., Ikejima, K., Bradford, B.U., Rivera, C.A., Kono, H., Goto, M., Yamashina, S. *et al.* (2000) Role of Kupffer cells and gut-derived endotoxins in alcoholic liver injury. *J Gastroenterol Hepatol*, **15** (Suppl), D20–D25.

93 Thurman, R.G. (2000) Sex-related liver injury due to alcohol involves activation of Kupffer cells by endotoxin. *Can J Gastroenterol*, **14** (Suppl D), 129D–135D.

94 Adachi, Y., Bradford, B.U., Gao, W., Bojes, H.K., and Thurman, R.G. (1994) Inactivation of Kupffer cells prevents early alcohol-induced liver injury. *Hepatology*, **20**, 453–460.

95 Enomoto, N., Takei, Y., Hirose, M., Ikejima, K., Miwa, H., Kitamura, T., and Sato, N. (2002) Thalidomide prevents alcoholic liver injury in rats through suppression of Kupffer cell sensitization and TNF-α production. *Gastroenterology*, **123**, 291–300.

96 Jarvelainen, H.A., Fang, C., Ingelman-Sundberg, M., Lukkari, T.A., Sippel, H., and Lindros, K.O.

(2000) Kupffer cell inactivation alleviates ethanol-induced steatosis and CYP2E1 induction but not inflammatory responses in rat liver. *J Hepatol*, **32**, 900–910.

97 Enomoto, N., Yamashina, S., Goto, M., Schemmer, P., and Thurman, R.G. (1999) Desensitization to LPS after ethanol involves the effect of endotoxin on voltage-dependent calcium channels. *Am J Physiol*, **277**, G1251–G1258.

98 Haziot, A., Chen, S., Ferrero, E., Low, M.G., Silber, R., and Goyert, S.M. (1988) The monocyte differentiation antigen, CD14, is anchored to the cell membrane by a phosphatidylinositol linkage. *J Immunol*, **141**, 547–552.

99 da Silva Correia, J., Soldau, K., Christen, U., Tobias, P.S., and Ulevitch, R.J. (2001) Lipopolysaccharide is in close proximity to each of the proteins in its membrane receptor complex. transfer from CD14 to TLR4 and MD-2. *J Biol Chem*, **276**, 21129–21135.

100 Yin, M., Bradford, B.U., Wheeler, M.D., Uesugi, T., Froh, M., Goyert, S.M., and Thurman, R.G. (2001) Reduced early alcohol-induced liver injury in CD14-deficient mice. *J Immunol*, **166**, 4737–4742.

101 Jarvelainen, H.A., Orpana, A., Perola, M., Savolainen, V.T., Karhunen, P.J., and Lindros, K.O. (2001) Promoter polymorphism of the CD14 endotoxin receptor gene as a risk factor for alcoholic liver disease. *Hepatology*, **33**, 1148–1153.

102 Poltorak, A., He, X., Smirnova, I., Liu, M.Y., Van Huffel, C., Du, X., Birdwell, D. et al. (1998) Defective LPS signaling in C3H/HeJ and C57BL/10ScCr mice: mutations in Tlr4 gene. *Science*, **282**, 2085–2088.

103 Uesugi, T., Froh, M., Arteel, G.E., Bradford, B.U., and Thurman, R.G. (2001) Toll-like receptor 4 is involved in the mechanism of early alcohol-induced liver injury in mice. *Hepatology*, **34**, 101–108.

104 Zuo, G., Gong, J., Liu, C., Wu, C., Li, S., and Dai, L. (2003) Synthesis of Toll-like receptor 4 in Kupffer cells and its role in alcohol-induced liver disease. *Chin Med J (Engl)*, **116**, 297–300.

105 Hritz, I., Mandrekar, P., Velayudham, A., Catalano, D., Dolganiuc, A., Kodys, K., Kurt-Jones, E. et al. (2008) The critical role of toll-like receptor (TLR) 4 in alcoholic liver disease is independent of the common TLR adapter MyD88. *Hepatology*, **48**, 1224–1231.

106 Paik, Y.H., Schwabe, R.F., Bataller, R., Russo, M.P., Jobin, C., and Brenner, D.A. (2003) Toll-like receptor 4 mediates inflammatory signaling by bacterial lipopolysaccharide in human hepatic stellate cells. *Hepatology*, **37**, 1043–1055.

107 Fernandez, A., Colell, A., Garcia-Ruiz, C., and Fernandez-Checa, J.C. (2008) Cholesterol and sphingolipids in alcohol-induced liver injury. *J Gastroenterol Hepatol*, **23** (Suppl 1), S9–15.

108 Lalor, P.F., Faint, J., Aarbodem, Y., Hubscher, S.G., and Adams, D.H. (2007) The role of cytokines and chemokines in the development of steatohepatitis. *Semin Liver Dis*, **27**, 173–193.

109 McClain, C.J., Hill, D.B., Song, Z., Deaciuc, I., and Barve, S. (2002) Monocyte activation in alcoholic liver disease. *Alcohol*, **27**, 53–61.

110 McClain, C.J., Song, Z., Barve, S.S., Hill, D.B., and Deaciuc, I. (2004) Recent advances in alcoholic liver disease. IV. Dysregulated cytokine metabolism in alcoholic liver disease. *Am J Physiol Gastrointest Liver Physiol*, **287**, G497–G502.

111 Tilg, H., Kaser, A., and Moschen, A.R. (2006) How to modulate inflammatory cytokines in liver diseases. *Liver Int*, **26**, 1029–1039.

112 Bautista, A.P. (2000) Impact of alcohol on the ability of Kupffer cells to produce chemokines and its role in alcoholic liver disease. *J Gastroenterol Hepatol*, **15**, 349–356.

113 Albano, E. (2008) Oxidative mechanisms in the pathogenesis of

alcoholic liver disease. *Mol Aspects Med*, **29**, 9–16.

114 Bardag-Gorce, F., Venkatesh, R., Li, J., French, B.A., and French, S.W. (2004) Hyperphosphorylation of rat liver proteasome subunits: the effects of ethanol and okadaic acid are compared. *Life Sci*, **75**, 585–597.

115 Donohue, T.M. Jr, Cederbaum, A.I., French, S.W., Barve, S., Gao, B., and Osna, N.A. (2007) Role of the proteasome in ethanol-induced liver pathology. *Alcohol Clin Exp Res*, **31**, 1446–1459.

116 Joshi-Barve, S., Barve, S.S., Butt, W., Klein, J., and McClain, C.J. (2003) Inhibition of proteasome function leads to NF-κB-independent IL-8 expression in human hepatocytes. *Hepatology*, **38**, 1178–1187.

117 McClain, C., Barve, S., Joshi-Barve, S., Song, Z., Deaciuc, I., Chen, T. and Hill, D. (2005) Dysregulated cytokine metabolism, altered hepatic methionine metabolism and proteasome dysfunction in alcoholic liver disease. *Alcohol Clin Exp Res*, **29**, 180S–188S.

118 Osna, N.A. and Donohue, T.M. Jr (2007) Implication of altered proteasome function in alcoholic liver injury. *World J Gastroenterol*, **13**, 4931–4937.

119 Adachi, Y., Moore, L.E., Bradford, B.U., Gao, W., and Thurman, R.G. (1995) Antibiotics prevent liver injury in rats following long-term exposure to ethanol. *Gastroenterology*, **108**, 218–224.

120 Ohata, M., Suzuki, H., Sakamoto, K., Hashimoto, K., Nakajima, H., Yamauchi, M., Hokkyo, K. *et al.* (2004) Pioglitazone prevents acute liver injury induced by ethanol and lipopolysaccharide through the suppression of tumor necrosis factor-α. *Alcohol Clin Exp Res*, **28**, 139S–144S.

121 Yin, M., Ikejima, K., Arteel, G.E., Seabra, V., Bradford, B.U., Kono, H., Rusyn, I. *et al.* (1998) Glycine accelerates recovery from alcohol-induced liver injury. *J Pharmacol Exp Ther*, **286**, 1014–1019.

122 Arteel, G.E., Uesugi, T., Bevan, L.N., Gabele, E., Wheeler, M.D., McKim, S.E., and Thurman, R.G. (2002) Green tea extract protects against early alcohol-induced liver injury in rats. *Biol Chem*, **383**, 663–670.

123 McKim, S.E., Konno, A., Gabele, E., Uesugi, T., Froh, M., Sies, H., Thurman, R.G. *et al.* (2002) Cocoa extract protects against early alcohol-induced liver injury in the rat. *Arch Biochem Biophys*, **406**, 40–46.

124 Cheng, P.Y., Lee, Y.M., Wu, Y.S., Chang, T.W., Jin, J.S., and Yen, M.H. (2007) Protective effect of baicalein against endotoxic shock in rats in vivo and in vitro. *Biochem Pharmacol*, **73**, 793–804.

125 Lirussi, F., Mastropasqua, E., Orando, S., and Orlando, R. (2007) Probiotics for non-alcoholic fatty liver disease and/or steatohepatitis. *Cochrane Database Syst Rev*, Jan **24** (1) (Art. No.: CD005165).

126 Medina, J., Fernandez-Salazar, L.I., Garcia-Buey, L., and Moreno-Otero, R. (2004) Approach to the pathogenesis and treatment of nonalcoholic steatohepatitis. *Diabetes Care*, **27**, 2057–2066.

127 Purohit, V., Bode, J.C., Bode, C., Brenner, D.A., Choudhry, M.A., Hamilton, F., Kang, Y.J. *et al.* (2008) Alcohol, intestinal bacterial growth, intestinal permeability to endotoxin, and medical consequences: summary of a symposium. *Alcohol*, **42**, 349–361.

128 Nanji, A.A., Khettry, U., and Sadrzadeh, S.M. (1994) Lactobacillus feeding reduces endotoxemia and severity of experimental alcoholic liver (disease). *Proc Soc Exp Biol Med*, **205**, 243–247.

129 Qing, L. and Wang, T. (2008) Lactic acid bacteria prevent alcohol-induced steatohepatitis in rats by acting on the pathways of alcohol metabolism. *Clin Exp Med*, **8**, 187–191.

130 Segawa, S., Wakita, Y., Hirata, H., and Watari, J. (2008) Oral administration of heat-killed Lactobacillus brevis SBC8803

ameliorates alcoholic liver disease in ethanol-containing diet-fed C57BL/6N mice. *Int J Food Microbiol*, **128**, 371–377.

131 Marotta, F., Barreto, R., Wu, C.C., Naito, Y., Gelosa, F., Lorenzetti, A., and Yoshioka, M. *et al.* (2005) Experimental acute alcohol pancreatitis-related liver damage and endotoxemia: synbiotics but not metronidazole have a protective effect. *Chin J Dig Dis*, **6**, 193–197.

132 Kirpich, I.A., Solovieva, N.V., Leikhter, S.N., Shidakova, N.A., Lebedeva, O.V., Sidorov, P.I., Bazhukova, T.A. *et al.* (2008) Probiotics restore bowel flora and improve liver enzymes in human alcohol-induced liver injury: a pilot study. *Alcohol*, **42**, 675–682.

20
Gut Microbiota, Diet, Endotoxemia, and Diseases
Patrice D. Cani and Nathalie M. Delzenne

The evidence reviewed in this chapter, has largely been made possible by the recent development of powerful tools (high-throughput sequencing methods) that help to delineate the complexity of the gut microbiota. Overall, germ-free mice, probiotic and prebiotic studies suggest that the gut microbiota play a pivotal role in the regulation of energy balance and the development of metabolic disorders. Moreover, the discovery of the role of gut-derived factors, such as lipopolysaccharides (LPSs), in the physiopathology of low-grade inflammation, type 2 diabetes, insulin resistance, and liver diseases is promising. It will be important, nevertheless, to determine whether metabolic endotoxemia induced by high-fat feeding contributes to the development of metabolic disorders associated with obesity. Another interesting point to unravel is whether and how changing gut microbiota impacts on metabolic endotoxemia.

20.1
Introduction

Obesity is a growing epidemic in developed countries and constitutes a major health problem. Obesity is now classically associated with a cluster of metabolic disorders including glucose intolerance, insulin resistance, type 2 diabetes, hypertension, dyslipidemia, fibrinolysis disorders, epithelial dysfunction, atherosclerosis, cardiovascular diseases, nonalcoholic fatty liver diseases (NAFLDs), and nonalcoholic steatohepatitis (NASH) [1, 2]. Most of them are related to glucose homeostasis and to the development of cardiovascular diseases, and probably result from a combination of variable associations of genetic and environmental factors [3–5].

Over the past decade, the physiological processes regulating body weight and energy metabolism, including appetite signals, the central mechanisms of the appetite integration, and gastrointestinal responses to food intake, have received intense investigation [6–9]. Excessive energy intakes and the reduction of physical activity are certainly two environmental factors classically associated with the development of metabolic diseases. The analysis of the nutritional disorders associated with obesity reveals that the adverse health consequences of weight gain and obesity are especially prominent following prolonged periods of positive energy balance and are mostly associated with a high-fat diet ingestion in our Western countries. A fat-enriched diet generates features of metabolic disorders leading to the diseases and is considered the most common triggering event. Finally, combining an increased energy intake with a reduced physical activity surely contributes to the development of obesity and metabolic disorders. However, the existence of complex systems that regulate energy homeostasis requires that this paradigm be considered in a larger context.

Over the last years, unequivocal epidemiological, clinical, and/or experimental evidence have causally linked inflammatory signaling responses to the development of the metabolic disorders associated with obesity. Now it becomes clear that obesity and type 2 diabetes are characterized by a low grade inflammation associated with the development of insulin resistance [10–12]. However, it is more difficult to understand why and how metabolic diseases are so commonly linked to inflammatory processes. *What are the mechanisms by which high-fat diet feeding promotes low grade inflammation? What is the molecular link between high-fat or high-energy feeding and the development of this particular response?* New evidence supports the idea that the increased prevalence of obesity and type 2 diabetes cannot be attributed solely to changes in the human genome, nutritional habits, or the reduction of physical activity in our daily lives [3]. Numerous attempts have been made to determine the triggering factor, which (i) would depend on fat feeding, (ii) trigger a low grade inflammation, and (iii) contribute over a long term period to progressive disease. These questions constitute the core of this chapter.

20.2
Gut Microbiota and Energy Homeostasis

Over the past five years, studies have highlighted some key aspects of the mammalian host-gut microbial relationship. Gut microbiota could now be considered as a "microbial organ" placed within a host organism. In addition to the obvious role of the intestine in the digestion and absorption of nutrients, the human gastrointestinal tract contains a diverse collection of microorganisms, residing mostly in the colon. To date, the human gut microbiota has not been fully described, but it is clear that human gut is home for a complex consortium of 10^{13}–10^{14} bacterial cells and up to 1000 different species. As a whole, the microorganisms that live inside humans are estimated to outnumber human cells by a factor of 10 and represent genomes overall more than 100 times the size of the human genome to become the so-called "metagenome" [13, 14].

20.2.1
Gut Microbiota and Energy Harvest

The gut flora has been recently proposed as an environmental factor involved in the control of body weight and energy homeostasis [15–20]. Thus, the gut microbiota can be considered as an "exteriorized organ" which contributes to our homeostasis with multiple metabolic functions largely diversified [21]. More specifically, the biological functions controlled by the gut microbiota are related to the effectiveness of energy harvest, by the bacteria, of the energy ingested but not digested by the host. This mechanism facilitates the extraction of calories from the ingested dietary substances and helps to store these calories in host adipose tissue for later use. Among the dietary compounds escaping digestion in the upper part of human gastrointestinal tract, polysaccharides constitute the major source of nutrient for the bacteria. Part of these polysaccharides could be transformed into digestible substances such as sugars or short chain carboxylic acids (SCFAs), providing energy substrates. The control of body weight depends on mechanisms subtly controlled over time, and a small daily excess, as low as 1–2% of the daily energy needs, can have important consequences in the long term on body weight and metabolism [22]. Consequently, it has been assumed that all mechanisms modifying the food-derived energy availability should contribute to the balance of the body weight.

20.2.1.1 Gut Microbiota and Adipose Tissue Development

Gordon and colleagues have demonstrated in an elegant series of experiments [13, 15–20, 23] that the mice raised in the absence of microorganisms (germ free or gnotobiotic) had about 40% less total body fat than mice with a normal gut microbiota, even though the latter ate 30% less diet than did the germ-free mice [17, 19]. When the distal gut microbiota from the mice was transplanted into the gnotobiotic mice, this conventionalization produced a 60% increase in body fat content and insulin resistance within two weeks, without any clear changes in food consumption or obvious differences in energy expenditure [19]. These data support the hypothesis that the composition of the gut microbiota affects the amount of energy extracted from the diet.

20.2.1.2 Lipogenesis

To account for the effect of gut microbiota to increase fat mass, the authors proposed two leading theories. The first one supports the role of an increase in the intestinal glucose absorption and energy extraction from nondigestible food and concomitantly higher glycemia and insulinemia, two key metabolic factors regulating lipogenesis. The second theory supports the role of a SCFA receptor (GPR41), which could somehow promote deposition of fat [24–26].

Glucose and insulin are known to promote hepatic *de novo* lipogenesis through the expression of several key enzymes such as acetyl-CoA carboxylase (ACC) and fatty acid synthase (FAS). Backhed *et al.* found that a two weeks conventionalization of germ-free mice is accompanied by a twofold increase in hepatic triglyceride content [19]. Both ACC and FAS are controlled by ChREBP (carbohydrate responsive element binding protein) and SREBP-1 (sterol responsive element binding protein) [27]. Accordingly, the conventionalized mice exhibited an increased hepatic ChREBP and SREBP-1 mRNA levels [19].

Strikingly, the development of the adipose tissue observed in the mice harboring gut microbiota was not explained by the modulation of adipogenesis or lipogenesis. The authors proposed that the adipocyte hypertrophy was merely due to a general increase in the activity of the enzyme lipoprotein lipase (LPL), catalyzing the release of fatty acids from circulating triacylglycerol in lipoproteins, which was taken up by muscle and adipose tissue. The researchers proposed that this phenomenon was the consequence of suppression of the fasting-induced adipose factor (FIAF) in the intestine. FIAF inhibits LPL activity, and a blunted FIAF expression in conventionalized germ-free mice could thus participate to the accumulation of triacylglycerol in the adipose tissue.

20.2.1.3 Specific SCFA Receptors

SCFA have been proposed as signaling molecules; propionate, acetate, and to a lesser extent butyrate and pentanoate have been described as ligands for at least two G protein-coupled receptors GPR41 and GPR43, largely expressed in the distal small intestine, the colon, and adipocytes. Very recently, Samuel et al. demonstrated that both conventionalized raised and germ-free GPR41−/− mice were leaner than their wild-type counterparts and developed less adipose tissue after colonization with a model of fermentative microbial community (composed of *Bacteroides thetaiotaomicron* and *Methanobrevibacter smithii*) [28]. These data support the role of a microbiota-dependent metabolic flux in the regulation of the flow of calories between the diet and the host.

20.3
Energy Harvest, Obesity, and Metabolic Disorders: Paradoxes?

This particular original idea, that the bacteria can contribute to the maintenance of the host body weight, is characterized by several paradoxes: first it is not clear that the small increase in energy extraction can actually contribute to a meaningful body weight gain within a short period, as suggested in the gut microbiota transplantation studies; second we and others have clearly shown that a diet enriched with specific nondigestible fibers decreases body weight, fat mass, and the severity of diabetes [29–33]. These specific nondigestible fibers are known as prebiotics: "a selectively fermented ingredient that allows specific changes in the composition and/or in the activity in the gastrointestinal microflora that confers benefits upon host well-being and health" [34]. Moreover, these prebiotics increase the strains of bacteria able to digest the polysaccharide compounds and provide extra energy for the host as they increase the total mass of bacteria in the colon [35–37].

The results, obtained both in rodents and human, suggested that obesity is associated with an altered composition of gut microbiota, with obese subjects or animals characterized by lower *Bacteroidetes* and more *Firmicutes* than lean [15, 16]. To investigate the relation between gut microbial ecology and body fat mass in humans, Ley et al. studied 12 obese subjects assigned to a low calorie diet (fat or carbohydrate restricted), and found that the ratio of *Bacteroidetes* to *Firmicutes* approached a lean type profile after 52 weeks of diet-induced weight loss. However, this study did not demonstrate that the relative changes in bacterial strain profile lead to the different fates of body weight gain. Very recently, Duncan *et al.* performed a similar study, and found data that do not support the hypothesis that the proportions of *Bacteroidetes* and *Firmicutes* are different between obese and lean subjects [38]. The authors did not detect difference between obese and nonobese individuals in the proportion of *Bacteroidetes* measured in fecal samples, and no significant change in the percentage of *Bacteroidetes* in feces from obese subjects on weight loss diets. These data lend credence to the hypothesis that smaller changes or more specific modulation of the gut microbiota community are involved in the development of obesity. The important consequences of such a study warrant the further investigation of this hypothesis.

Although all the data showing the role of the gut microbiota in the extraction of energy from the diet and the development of obesity and related metabolic disorders are convincing, this theory has never unraveled the interplay between obesity and the obesity-related metabolic disorders and the development of a low grade inflammation. It does not explain the result of experiments in which germ-free or conventionalized mice were maintained on a high-fat/high-refined carbohydrate diet (Western diet). Such a study found that conventionalized mice fed the Western diet not only gained significantly more weight and fat mass but also had higher glycemia and insulinemia than the germ-free mice [17]. Not only were the results opposite to those previously observed with germ-free mice fed a normal chow diet but also the amount of Western diet consumed by germ-free and conventionalized mice was similar and hence had similar fecal energy output. All these data suggest that a bacterially related factor is responsible for the development of diet-induced obesity and diabetes.

We thus turn to the question: *What are the mechanisms by which high-fat diet feeding promotes low grade inflammation? Can we attribute the low grade inflammatory process observed during metabolic diseases to the gut microbiota?* This question will now be considered to understand how the gut microbiota may play an even more important role in the development of metabolic disorders associated with obesity.

20.4
Role of the Gut Microbiota in the Inflammatory Associated with Obesity

On the basis of the recent demonstration that obesity and insulin resistance are associated with a low grade inflammation [10–12], we have postulated another mechanism linking gut microbiota to the development of obesity and metabolic disorders. In the models of high-fat diet–induced obesity, adipose depots express several inflammatory factors such as IL-1, TNF-α, MCP-1, iNOS, and IL-6 [39, 40]. These factors have been causally related with the development of impaired insulin action and insulin resistance. The proinflammatory effect of high-fat diets has mainly been attributed to the inflammatory properties of dietary fatty acids (i.e., palmitic acid). Recently, it has been proposed that such fatty acids trigger inflammatory response by acting via the toll-like receptor-4 (TLR4) signaling in the adipocyte and macrophage, which might contribute to inflammation of adipose tissue in obesity [41–43]. Because high-fat diet–induced type 2 diabetes and obesity are closely associated with a low grade inflammatory state, we have been seeking a bacterially related factor able to trigger the development of high-fat diet–induced obesity, diabetes, and inflammation. The eligible candidate should be an inflammatory compound of bacterial origin, continuously produced within the gut and its absorption/action should be associated with high-fat diet feeding. We hypothesized that the bacterial lipopolysaccharide (LPS) could be the eligible candidate for the following reasons: (i) LPS is a constituent of gram-negative bacteria present in the gut microbiota; (ii) triggers the secretion of proinflammatory cytokines when it binds to the complex of CD14 and TLR4 at the surface of innate immune cells [44]; (iii) continuously produced within the gut by the death of gram-negative bacteria and is physiologically carried into intestinal capillaries through a TLR4-dependent mechanism [45]; (iv) transported from the intestine toward target tissues by a mechanism facilitated by lipoproteins, notably chylomicrons freshly synthesized from epithelial intestinal cells in response to fat feeding [46–50].

20.4.1
Metabolic Endotoxemia and Metabolic Disorders

Therefore, we identified LPS as the triggering factor of the early development of metabolic diseases, the *primum movens* in the cascade of inflammation [51]. Excess dietary fat not only increases systemic exposure to potential proinflammatory free fatty acids and their derivatives, but, as we have recently demonstrated, also facilitates the absorption of highly proinflammatory bacterial LPS from the gut [51, 52]. This new hypothesis provides a new insight into the role played by key gut microbiota-derived products, because the LPS absorbed could affect whole body inflammation and interfere with both metabolism and the function of the immune system. In a series of experiments in mice fed a high-fat diet, we showed that (i) a high-fat diet increases endotoxemia two- to threefold, a level defined as "metabolic endotoxemia"; (ii) fat feeding changes the bacterial populations, which are predominant in the intestinal microbiota, with a marked reduction in *Bifidobacterium* spp. number, and also a reduced *Bacteroides*-related bacteria and *Eubacterium rectale–Clostridium coccoides* group content [51]. We also demonstrated that chronic metabolic endotoxemia, produced by subcutaneously infusing a chronic low dose LPS with an osmotic mimipumps, induces obesity, insulin resistance and diabetes, which was measured by the euglycemic-hyperinsulinemic clamp and oral glucose tolerance test. With LPS receptor knockout mice (CD14−/−) fed a high-fat diet, we showed that metabolic endotoxemia triggers the expression of inflammatory cytokines (TNF-α, IL-1, IL-6, and PAI-1) via a CD14-dependent mechanism. In a study to ascertain the role of the gut microbiota in the development of metabolic endotoxemia following a high-fat diet treatment, we chronically treated high-fat fed mice with antibiotic. The results confirmed the role of LPS in the development of metabolic diseases associated with obesity because antibiotic treatment completely abolished the high-fat diet–induced disorders: metabolic endotoxemia, the development of visceral adipose tissue inflammation, macrophages infiltration, oxidative stress, and metabolic disorders (e.g., glucose intolerance and insulin resistance) [53]. These last experiments clearly demonstrate the contribution of the gut microbiota to the metabolic endotoxemia.

Consistent with our results, recent studies reported that plasma LPS was increased in *ob/ob* and *db/db* mice [54]. Furthermore, polymyxin B treatment, which specifically eliminates gram-negative bacteria and further quenches LPS, diminished hepatic steatosis [55]. However, these studies did not demonstrate that the gut bacteria determine the threshold at which metabolic endotoxemia occurs or that the modulation of gut microbiota in obese and diabetic *ob/ob* mice controls the occurrence of metabolic and inflammatory disorders. To test this hypothesis, we changed the gut microbiota of *ob/ob* mice using antibiotic treatment and found that macrophage infiltration, inflammatory markers, and oxidative stress were reduced in the visceral adipose depots and to a lesser extent in the subcutaneous fat [53]. Together with the earlier results, these findings strongly suggest that the gut microbiota contributes to the metabolic endotoxemia related to high-fat diet feeding.

20.4.2
Metabolic Endotoxemia and Nutritional Intervention

Among the possibilities for selectively modulating gut microflora, prebiotics and probiotics (live bacteria given in oral quantities that allow for colonization of the colon [56]) are the most important. We found that *Bifidobacterium* spp. were markedly reduced during high-fat diet treatment. This is likely important because several studies have shown that these bacteria reduced the intestinal endotoxin levels and improved mucosal barrier function [57–59]. We therefore used prebiotic dietary fibers [60] to specifically increase the gut bifidobacteria content of high-fat diet–treated mice. We found that among the different gut bacteria analyzed, metabolic endotoxemia correlated negatively with the bifidobacteria count [61]. We also found that in the prebiotic-treated mice, *Bifidobacterium* spp. significantly and positively correlated with improved glucose homeostasis and decreased markers of systemic inflammation (decreased metabolic endotoxemia and plasma and adipose tissue proinflammatory cytokines) [61]. All these data reinforce the role of the gut microbiota in the pathophysiological regulation of endotoxemia, and the development of inflammation and related metabolic abnormalities occurring through diabetes/obesity.

A similar recent study reported that changing gut microbiota with the use of probiotics improved the development of high-fat diet–induced obesity and insulin resistance. Unfortunately, the authors did not investigate the relationship between gut microbiota and metabolic endotoxemia in that context [62].

20.4.3
Metabolic Endotoxemia, Gut Microbiota, and Fatty Liver Diseases

The concept that the gut microbiota matters for the pathogenesis of liver disease is not novel. An extensive body of literature, mostly derived from animal studies, highlights the concept that a gut-derived bacterial product can target the liver and cause systemic diseases [63–68]. However, the notion that selective changes in the gut microbiota occurring during obesity might prevent the development of NAFLD and NASH is relatively recent. The specific mechanisms linking gut microbiota, obesity, and liver diseases has thus been more specifically investigated. The gastrointestinal tract appears to play a key role in the pathogenesis of fatty liver diseases and inflammation. LPS constitutes an important component, which is believed to stimulate proinflammatory cytokines. More specifically, it has been suggested that portal endotoxemia is a major risk for inducing hepatic inflammation in alcoholic liver diseases [69] and NAFLD [54]. It has also been suggested that bacteria growth in the small intestine of NASH patients could be promoted by a decreased sIgA (a factor inhibiting adherence of bacteria to the intestinal mucosa). A small bacterial overgrowth can produce endotoxin in the enteric cavity, leading to intestinal mucosal barrier damage and higher endotoxin absorption, finally leading to metabolic endotoxemia [70].

20.4.4
Selective Changes in Gut Microbiota and NASH

We and others have demonstrated that changing the gut microbiota by using prebiotics has a salutary effect on the development of NASH in several animal models (i.e., high-fat diet–induced obesity and genetically obese Zucker *fa/fa* rats) and in human subjects [30, 71–74]. More recently, we have demonstrated that changing the gut microbiota of *ob/ob* mice by using antibiotic significantly reduces hepatic triglyceride accumulation, improves liver function, improves glucose tolerance, and increases adiponectin. Importantly, all these changes were associated with a significantly lower metabolic endotoxemia [75].

Similarly, probiotics protect against high-fat diet–induced NAFLD and hepatic inflammation. It has also been found that changing the gut microbiota by using probiotics significantly suppressed high-fat diet–induced activation of the TNF-α/IKK-β signaling, which is the critical signaling for diet-induced insulin resistance [62]. Along the same line in a model of alcohol-induced metabolic endotoxemia and liver disease, rats fed a probiotic lactobacilli were characterized by a reduced plasma endotoxin levels in addition to a lower liver pathology score [76]. A probiotic mixture (*Bifidobacterium*, *Lactobacillus*, and *Streptococcus thermophilus*) has also been found to decrease liver inflammation in *ob/ob* mice [77]. The administration of TNF-α antibodies has also been found to result in a comparable decrease in inflammation to the prebiotics-induced decrease. While probiotics may prevent further damage to the liver by modulating the gut microbiota, they may also help to stimulate the immune system to produce noninflammatory cytokines and reduce proinflammatory markers. Among the mechanisms, several authors have proposed that changing the gut microbiota can effectively attenuate liver damage and maintain/restore gut barrier function and epithelial function. Other authors have elegantly discussed the role of probiotics in the liver health and overall health [78].

20.5
Metabolic Endotoxemia and High-Fat Feeding: Human Evidence

Even if from a mechanistic point of view the results obtained in rodent models are very encouraging, it remains to be demonstrated that similar mechanisms are observed in humans. Recent data suggest, however, that high-fat feeding induces a metabolic endotoxemia similar to the one observed in rodents. The first study examining the kinetics of baseline endotoxemia in healthy human subjects was recently published by Erridge *et al.* The authors highlighted the putative role of a high-fat meal on the development of metabolic endotoxemia. They found that a high-fat meal induces a metabolic endotoxemia, which fluctuates rapidly in healthy subjects, from a very low concentration at baseline (between 1 and 9 pg ml^{-1}) to concentrations that would be sufficient to induce some degree of cellular activation even in *in vitro* experiments [79]. In addition, they found that the metabolic endotoxemia was sufficient to activate cultured human aortic endothelial cells, and that this endothelial cell activation was likely due to the release of soluble inflammatory mediators, such as TNF-α, from monocytes. Circulating endotoxin levels also increased in patients with type 2 diabetes [80]. The authors found that metabolic endotoxemia was twofold higher in the BMI-, sex-, and age-matched type 2 diabetic patients group than that in the nondiabetic subjects. Furthermore, they found a positive correlation between fasting insulin and metabolic endotoxemia in the whole nondiabetic population, and this correlation persisted after controlling for sex, age, and BMI [80]. Along the same line, we have recently demonstrated that in a large sample of men ($n=211$) from a population-based study, a positive correlation existed between plasma endotoxin levels and energy/or fat intake [81]. Furthermore, a similar metabolic endotoxemia was shown to increase adipose TNF-α and IL-6 concentrations and insulin resistance in healthy volunteers [82]. This study shows for the first time that the confounding factor of the relation between fat intake and metabolic endotoxemia is likely to be energy intake. Taken together, both human studies suggest that diet-induced changes in endotoxemia may bridge the gap between food intake behavior and metabolic diseases in humans. Finally, a pancreatic and gastric lipase inhibitor has been shown to reduce metabolic endotoxemia in individuals with impaired glucose tolerance. These findings reinforce the role of fat feeding (and absorption) in the development of metabolic endotoxemia [83].

20.6
Conclusion

The evidence reviewed in this chapter has largely been made possible by the recent development of powerful tools (high throughput sequencing methods) that help to delineate the complexity of the gut microbiota. Overall, germ-free mice, probiotic, and prebiotic studies suggest that the gut microbiota play a pivotal role in the regulation of energy balance and the development of metabolic disorders. Moreover, the discovery of the role of gut-derived factors, such as LPS, in the physiopathology of low grade inflammation, type 2 diabetes, insulin resistance, and liver diseases is promising. It will be important, nevertheless, to determine whether metabolic endotoxemia induced by high-fat feeding contributes to the development of metabolic disorders associated with obesity. Another interesting point to unravel is whether and how changing gut microbiota impacts metabolic endotoxemia.

References

1 Ogden, C.L., Yanovski, S.Z., Carroll, M.D., and Flegal, K.M. (2007) The epidemiology of obesity. *Gastroenterology*, **132**, 2087–2102.

2 Eckel, R.H., Grundy, S.M., and Zimmet, P.Z. (2005) The metabolic syndrome. *Lancet*, **365**, 1415–1428.

3 Kahn, S.E., Hull, R.L., and Utzschneider, K.M. (2006) Mechanisms linking obesity to insulin resistance and type 2 diabetes. *Nature*, **444**, 840–846.

4 Alberti, K.G., Zimmet, P., and Shaw, J. (2005) The metabolic syndrome – a new worldwide definition. *Lancet*, **366**, 1059–1062.

5 Matarese, G., Mantzoros, C., and La, C.A. (2007) Leptin and adipocytokines: bridging the gap between immunity and atherosclerosis. *Curr Pharm Des*, **13**, 3676–3680.

6 Chaudhri, O.B., Salem, V., Murphy, K.G., and Bloom, S.R. (2008) Gastrointestinal satiety signals. *Annu Rev Physiol*, **70**, 239–255.

7 Levin, B.E. (2006) Central regulation of energy homeostasis intelligent design: how to build the perfect survivor. *Obesity*, **14** (Suppl 5), 192S–196S.

8 Wynne, K., Stanley, S., McGowan, B., and Bloom, S. (2005) Appetite control. *J Clin Endocrinol Metab*, **184**, 291–318.

9 Small, C.J. and Bloom, S.R. (2004) Gut hormones and the control of appetite. *Trends Endocrinol Metab*, **15**, 259–263.

10 Heilbronn, L.K. and Campbell, L.V. (2008) Adipose tissue macrophages, low grade inflammation and insulin resistance in human obesity. *Curr Pharm Des*, **14**, 1225–1230.

11 Hotamisligil, G.S. (2006) Inflammation and metabolic disorders. *Nature*, **444**, 860–867.

12 Wellen, K.E. and Hotamisligil, G.S. (2005) Inflammation, stress, and diabetes. *J Clin Invest*, **115**, 1111–1119.

13 Xu, J., Mahowald, M.A., Ley, R.E., Lozupone, C.A., Hamady, M., Martens, E.C. et al. (2007) Evolution of symbiotic bacteria in the distal human intestine. *PLoS Biol*, **5**, e156.

14 Xu, J. and Gordon, J.I. (2003) Inaugural Article: Honor thy symbionts. *Proc Natl Acad Sci U S A*, **100**, 10452–10459.

15 Ley, R.E., Turnbaugh, P.J., Klein, S., and Gordon, J.I. (2006) Microbial ecology: human gut microbes associated with obesity. *Nature*, **444**, 1022–1023.

16 Turnbaugh, P.J., Ley, R.E., Mahowald, M.A., Magrini, V., Mardis, E.R., and Gordon, J.I. (2006) An obesity-associated gut

microbiome with increased capacity for energy harvest. *Nature*, **444**, 1027–1031.
17 Backhed, F., Manchester, J.K., Semenkovich, C.F., and Gordon, J.I. (2007) Mechanisms underlying the resistance to diet-induced obesity in germ-free mice. *Proc Natl Acad Sci U S A*, **104**, 979–984.
18 Backhed, F., Ley, R.E., Sonnenburg, J.L., Peterson, D.A., and Gordon, J.I. (2005) Host-bacterial mutualism in the human intestine. *Science*, **307**, 1915–1920.
19 Backhed, F., Ding, H., Wang, T., Hooper, L.V., Koh, G.Y., Nagy, A. *et al.* (2004) The gut microbiota as an environmental factor that regulates fat storage. *Proc Natl Acad Sci U S A*, **101**, 15718–15723.
20 Ley, R.E., Backhed, F., Turnbaugh, P., Lozupone, C.A., Knight, R.D., and Gordon, J.I. (2005) Obesity alters gut microbial ecology. *Proc Natl Acad Sci U S A*, **102**, 11070–11075.
21 Jia, W., Li, H., Zhao, L., and Nicholson, J.K. (2008) Gut microbiota: a potential new territory for drug targeting. *Nat Rev Drug Discov*, **7**, 123–129.
22 Hill, J.O. (2006) Understanding and addressing the epidemic of obesity: an energy balance perspective. *Endocr Rev*, **27**, 750–761.
23 Turnbaugh, P.J., Backhed, F., Fulton, L., and Gordon, J.I. (2008) Diet-induced obesity is linked to marked but reversible alterations in the mouse distal gut microbiome. *Cell Host Microbe*, **3**, 213–223.
24 Xiong, Y., Miyamoto, N., Shibata, K., Valasek, M.A., Motoike, T., Kedzierski, R.M. *et al.* (2004) Short-chain fatty acids stimulate leptin production in adipocytes through the G protein-coupled receptor GPR41. *Proc Natl Acad Sci U S A*, **101**, 1045–1050.
25 Brown, A.J., Goldsworthy, S.M., Barnes, A.A., Eilert, M.M., Tcheang, L., Daniels, D. *et al.* (2003) The Orphan G protein-coupled receptors GPR41 and GPR43 are activated by propionate and other short chain carboxylic acids. *J Biochem Mol Biol Biophys*, **278**, 11312–11319.
26 Le, P.E., Loison, C., Struyf, S., Springael, J.Y., Lannoy, V., Decobecq, M.E. *et al.* (2003) Functional characterization of human receptors for short chain fatty acids and their role in polymorphonuclear cell activation. *J Biochem Mol Biol Biophys*, **278**, 25481–25489.
27 Denechaud, P.D., Dentin, R., Girard, J., and Postic, C. (2008) Role of ChREBP in hepatic steatosis and insulin resistance. *FEBS Lett*, **582**, 68–73.
28 Samuel, B.S., Shaito, A., Motoike, T., Rey, F.E., Backhed, F., Manchester, J.K. *et al.* (2008) Effects of the gut microbiota on host adiposity are modulated by the short-chain fatty-acid binding G protein-coupled receptor, Gpr41. *Proc Natl Acad Sci U S A*, **105**, 16767–16772.
29 Cani, P.D., Dewever, C., and Delzenne, N.M. (2004) Inulin-type fructans modulate gastrointestinal peptides involved in appetite regulation (glucagon-like peptide-1 and ghrelin) in rats. *Br J Nutr*, **92**, 521–526.
30 Cani, P.D., Neyrinck, A.M., Maton, N., and Delzenne, N.M. (2005) Oligofructose promotes satiety in rats fed a high-fat diet: involvement of glucagon-like Peptide-1. *Obes Res*, **13**, 1000–1007.
31 Cani, P.D., Daubioul, C.A., Reusens, B., Remacle, C., Catillon, G., and Delzenne, N.M. (2005) Involvement of endogenous glucagon-like peptide-1(7-36) amide on glycaemia-lowering effect of oligofructose in streptozotocin-treated rats. *J Clin Endocrinol Metab*, **185**, 457–465.
32 Cani, P.D., Knauf, C., Iglesias, M.A., Drucker, D.J., Delzenne, N.M., and Burcelin, R. (2006) Improvement of glucose tolerance and hepatic insulin sensitivity by oligofructose requires a functional glucagon-like peptide 1 receptor. *Diabetes*, **55**, 1484–1490.
33 Cani, P.D., Joly, E., Horsmans, Y., and Delzenne, N.M. (2006)

Oligofructose promotes satiety in healthy human: a pilot study. *Eur J Clin Nutr*, **60**, 567–572.

34 Gibson, G.R. and Roberfroid, M.B. (1995) Dietary modulation of the human colonic microbiota: introducing the concept of prebiotics. *J Anim Physiol Anim Nutr (Berl)*, **125**, 1401–1412.

35 Kleessen, B., Hartmann, L., and Blaut, M. (2001) Oligofructose and long-chain inulin: influence on the gut microbial ecology of rats associated with a human faecal flora. *Br J Nutr*, **86**, 291–300.

36 Kolida, S., Meyer, D., and Gibson, G.R. (2007) A double-blind placebo-controlled study to establish the bifidogenic dose of inulin in healthy humans. *Eur J Clin Nutr*, **61**, 1189–1195.

37 Kolida, S., Saulnier, D.M., and Gibson, G.R. (2006) Gastrointestinal microflora: probiotics. *Adv Appl Microbiol*, **59**, 187–219.

38 Duncan, S.H., Lobley, G.E., Holtrop, G., Ince, J., Johnstone, A.M., Louis, P. *et al.* (2008) Human colonic microbiota associated with diet, obesity and weight loss. *Int J Obes*, **32**, 1720–1724.

39 Hotamisligil, G.S., Shargill, N.S., and Spiegelman, B.M. (1993) Adipose expression of tumor necrosis factor-α: direct role in obesity-linked insulin resistance. *Science*, **259**, 87–91.

40 Weisberg, S.P., McCann, D., Desai, M., Rosenbaum, M., Leibel, R.L., and Ferrante, A.W. Jr (2003) Obesity is associated with macrophage accumulation in adipose tissue. *J Clin Invest*, **112**, 1796–1808.

41 Shi, H., Kokoeva, M.V., Inouye, K., Tzameli, I., Yin, H., and Flier, J.S. (2006) TLR4 links innate immunity and fatty acid-induced insulin resistance. *J Clin Invest*, **116**, 3015–3025.

42 Suganami, T., Mieda, T., Itoh, M., Shimoda, Y., Kamei, Y., and Ogawa, Y. (2007) Attenuation of obesity-induced adipose tissue inflammation in C3H/HeJ mice carrying a Toll-like receptor 4 mutation. *Biochem Biophys Res Commun*, **354**, 45–49.

43 Suganami, T., Tanimoto-Koyama, K., Nishida, J., Itoh, M., Yuan, X., Mizuarai, S. *et al.* (2007) Role of the Toll-like receptor 4/NF-kappaB pathway in saturated fatty acid-induced inflammatory changes in the interaction between adipocytes and macrophages. *Arterioscler Thromb Vasc Biol*, **27**, 84–91.

44 Wright, S.D., Ramos, R.A., Tobias, P.S., Ulevitch, R.J., and Mathison, J.C. (1990) CD14, a receptor for complexes of lipopolysaccharide (LPS) and LPS binding protein. *Science*, **249**, 1431–1433.

45 Neal, M.D., Leaphart, C., Levy, R., Prince, J., Billiar, T.R., Watkins, S. *et al.* (2006) Enterocyte TLR4 mediates phagocytosis and translocation of bacteria across the intestinal barrier. *J Clin Lab Immunol*, **176**, 3070–3079.

46 Tomita, M., Ohkubo, R., and Hayashi, M. (2004) Lipopolysaccharide transport system across colonic epithelial cells in normal and infective rat. *Drug Metab Pharmacokinet*, **19**, 33–40.

47 Moore, F.A., Moore, E.E., Poggetti, R., McAnena, O.J., Peterson, V.M., Abernathy, C.M. *et al.* (1991) Gut bacterial translocation via the portal vein: a clinical perspective with major torso trauma. *J Neurotrauma*, **31**, 629–636.

48 Vreugdenhil, A.C., Rousseau, C.H., Hartung, T., Greve, J.W., van't Veer, V., and Buurman, W.A. (2003) Lipopolysaccharide (LPS)-binding protein mediates LPS detoxification by chylomicrons. *J Clin Lab Immunol*, **170**, 1399–1405.

49 Black, D.D., Tso, P., Weidman, S., and Sabesin, S.M. (1983) Intestinal lipoproteins in the rat with D-(+)-galactosamine hepatitis. *J Lipid Res*, **24**, 977–992.

50 Ghoshal, S., Witta, J., Zhong, J., de Villiers, W., and Eckhardt, E. Chylomicrons promote intestinal

absorption of lipopolysaccharides. (2008) *J Lipid Res*, **50**, 1–2.

51 Cani, P.D., Amar, J., Iglesias, M.A., Poggi, M., Knauf, C., Bastelica, D. *et al.* (2007) Metabolic endotoxemia initiates obesity and insulin resistance. *Diabetes*, **56**, 1761–1772.

52 Cani, P.D. and Delzenne, N.M. (2007) Gut microflora as a target for energy and metabolic homeostasis. *Curr Opin Clin Nutr Metab Care*, **10**, 729–734.

53 Cani, P.D., Bibiloni, R., Knauf, C., Waget, A., Neyrinck, A.M., Delzenne, N.M. *et al.* (2008) Changes in gut microbiota control metabolic endotoxemia-induced inflammation in high-fat diet-induced obesity and diabetes in mice. *Diabetes*, **57**, 1470–1481.

54 Brun, P., Castagliuolo, I., Leo, V.D., Buda, A., Pinzani, M., Palu, G. *et al.* (2007) Increased intestinal permeability in obese mice: new evidence in the pathogenesis of nonalcoholic steatohepatitis. *Am J Physiol Gastrointest Liver Physiol*, **292**, G518–G525.

55 Pappo, I., Becovier, H., Berry, E.M. and Freund, H.R. (1991) Polymyxin B reduces cecal flora, TNF production and hepatic steatosis during total parenteral nutrition in the rat. *J Invest Surg*, **51**, 106–112.

56 Hord, N.G. (2008) Eukaryotic-microbiota crosstalk: potential mechanisms for health benefits of prebiotics and probiotics. *Annu Rev Nutr*, **28**, 215–231.

57 Griffiths, E.A., Duffy, L.C., Schanbacher, F.L., Qiao, H., Dryja, D., Leavens, A. *et al.* (2004) In vivo effects of bifidobacteria and lactoferrin on gut endotoxin concentration and mucosal immunity in Balb/c mice. *Dig Dis Sci*, **49**, 579–589.

58 Wang, Z., Xiao, G., Yao, Y., Guo, S., Lu, K., and Sheng, Z. (2006) The role of bifidobacteria in gut barrier function after thermal injury in rats. *J Neurotrauma*, **61**, 650–657.

59 Wang, Z.T., Yao, Y.M., Xiao, G.X., and Sheng, Z.Y. (2004) Risk factors of development of gut-derived bacterial translocation in thermally injured rats. *World J Gastroenterol*, **10**, 1619–1624.

60 Tuohy, K.M., Rouzaud, G.C., Bruck, W.M., and Gibson, G.R. (2005) Modulation of the human gut microflora towards improved health using prebiotics – assessment of efficacy. *Curr Pharm Des*, **11**, 75–90.

61 Cani, P.D., Neyrinck, A.M., Fava, F., Knauf, K., Burcelin, R.G., Tuohy, K.M. *et al.* (2007) Selective increases of bifidobacteria in gut microflora improves high-fat diet-induced diabetes in mice through a mechanism associated with endotoxemia *Diabetologia*, **50** (11), 2374–2383.

62 Ma, X., Hua, J., and Li, Z. (2008) Probiotics improve high fat diet-induced hepatic steatosis and insulin resistance by increasing hepatic NKT cells. *J Hepatol*, **49**, 821–830.

63 Nolan, J.P. (1975) The role of endotoxin in liver injury. *Gastroenterology*, **69**, 1346–1356.

64 Nolan, J.P. and Ali, M.V. (1975) Letter: Kupffer cells and cirrhosis. *Lancet*, **1**, 449.

65 Nolan, J.P., Hare, D.K., McDevitt, J.J., and Ali, M.V. (1977) In vitro studies of intestinal endotoxin absorption. I. Kinetics of absorption in the isolated everted gut sac. *Gastroenterology*, **72**, 434–439.

66 Nolan, J.P. and Leibowitz, A.I. (1978) Endotoxin and the liver. III. Modification of acute carbon tetrachloride injury by polymyxin b – an antiendotoxin. *Gastroenterology*, **75**, 445–449.

67 Nolan, J.P. and Leibowitz, A.I. (1978) Endotoxins in liver disease. *Gastroenterology*, **75**, 765–766.

68 Nolan, J.P. (1979) The contribution of gut-derived endotoxins to liver injury. *Yale J Biol Med*, **52**, 127–133.

69 Adachi, Y., Moore, L.E., Bradford, B.U., Gao, W., and Thurman, R.G. (1995) Antibiotics prevent liver injury in rats following long-term exposure to ethanol. *Gastroenterology*, **108**, 218–224.

70 Li, S., Wu, W.C., He, C.Y., Han, Z., Jin, D.Y., and Wang, L. (2008)

Change of intestinal mucosa barrier function in the progress of non-alcoholic steatohepatitis in rats. *World J Gastroenterol*, **14**, 3254–3258.

71 Daubioul, C., Rousseau, N., Demeure, R., Gallez, B., Taper, H., Declerck, B. *et al.* (2002) Dietary fructans, but not cellulose, decrease triglyceride accumulation in the liver of obese Zucker fa/fa rats. *J Anim Physiol Anim Nutr (Berl)*, **132**, 967–973.

72 Daubioul, C.A., Taper, H.S., De Wispelaere, L.D., and Delzenne, N.M. (2000) Dietary oligofructose lessens hepatic steatosis, but does not prevent hypertriglyceridemia in obese zucker rats. *J Anim Physiol Anim Nutr (Berl)*, **130**, 1314–1319.

73 Daubioul, C.A., Horsmans, Y., Lambert, P., Danse, E., and Delzenne, N.M. (2005) Effects of oligofructose on glucose and lipid metabolism in patients with nonalcoholic steatohepatitis: results of a pilot study. *Eur J Clin Nutr*, **59**, 723–726.

74 Delzenne, N.M., Daubioul, C., Neyrinck, A., Lasa, M., and Taper, H.S. (2002) Inulin and oligofructose modulate lipid metabolism in animals: review of biochemical events and future prospects. *Br J Nutr*, **87**, S255–S259.

75 Membrez, M., Blancher, F., Jaquet, M., Bibiloni, R., Cani, P.D., Burcelin, R.G. *et al.* (2008) Gut microbiota modulation with norfloxacin and ampicillin enhances glucose tolerance in mice *FASEB J*, **22** (7), 2416–2426.

76 Nanji, A.A., Khettry, U., and Sadrzadeh, S.M. (1994) Lactobacillus feeding reduces endotoxemia and severity of experimental alcoholic liver (disease). *Proc Soc Exp Biol Med*, **205**, 243–247.

77 Li, Z., Yang, S., Lin, H., Huang, J., Watkins, P.A., Moser, A.B. *et al.* (2003) Probiotics and antibodies to TNF inhibit inflammatory activity and improve nonalcoholic fatty liver disease. *Hepatology*, **37**, 343–350.

78 O'Sullivan, D.J. (2008) Genomics can advance the potential for probiotic cultures to improve liver and overall health. *Curr Pharm Des*, **14**, 1376–1381.

79 Erridge, C., Attina, T., Spickett, C.M., and Webb, D.J. (2007) A high-fat meal induces low-grade endotoxemia: evidence of a novel mechanism of postprandial inflammation. *Am J Clin Nutr*, **86**, 1286–1292.

80 Creely, S.J., McTernan, P.G., Kusminski, C.M., Fisher, M., da Silva, N.F., Khanolkar, M. *et al.* (2007) Lipopolysaccharide activates an innate immune system response in human adipose tissue in obesity and type 2 diabetes. *Am J Physiol Endocrinol Metab*, **292**, E740–E747.

81 Amar, J., Burcelin, R., Ruidavets, J.B., Cani, P.D., Fauvel, J., Alessi, M.C. *et al.* (2008) Energy intake is associated with endotoxemia in apparently healthy men. *Am J Clin Nutr*, **87**, 1219–1223.

82 Anderson, P.D., Mehta, N.N., Wolfe, M.L., Hinkle, C.C., Pruscino, L., Comiskey, L.L. *et al.* (2007) Innate immunity modulates adipokines in humans. *J Clin Endocrinol Metab*, **92**, 2272–2279.

83 Dixon, A.N., Valsamakis, G., Hanif, M.W., Field, A., Boutsiadis, A., Harte, A. *et al.* (2008) Effect of the orlistat on serum endotoxin lipopolysaccharide and adipocytokines in South Asian individuals with impaired glucose tolerance. *Int J Clin Pract*, **62**, 1124–1129.

21
Nutrient-Derived Endogenous Toxins in the Pathogenesis of Type 2 Diabetes at the β-Cell Level

Christine Tang, Andrei I. Oprescu, and Adria Giacca

Chronically elevated plasma glucose and free fatty acids, as in type 2 diabetes, can behave as endogenous toxins and directly exert toxic effects on β-cells, or indirectly via the generation of reactive oxygen species. Both glucose and free fatty acid (FFA) increase reactive oxygen species production. Oxidative stress can act directly on various stages of the insulin production/release pathway, or can activate downstream signaling pathways to decrease β-cell function and mass. In obese people, at least in predisposed individuals, an elevation in FFA levels can contribute to the development of type 2 diabetes. Once type 2 diabetes is established, glucotoxicity and lipotoxicity and their derived toxins such as oxidative stress can further decrease both β-cell function and mass.

21.1
Introduction

Type 2 diabetes is a major cause of morbidity and mortality worldwide. It is estimated that approximately 150 million people are affected by type 2 diabetes in the year 2000, and that this number could double by the year 2025 [1]. It has long been recognized that insulin resistance precedes the development of hyperglycemia and type 2 diabetes. However, it is increasingly evident that a concurrent decrease in β-cell function is necessary for the development of type 2 diabetes.

Insulin is an anabolic hormone secreted by the β-cells in the pancreas. An acute increase in glucose or free fatty acids (FFAs) stimulates insulin secretion. However, numerous studies have now demonstrated that chronic elevation of glucose and FFA levels can paradoxically act as toxins to decrease both β-cell function and mass. The term β-cell glucotoxicity and β-cell lipotoxicity have been coined to describe the adverse or damaging effect of excessive glucose and FFA on the β-cells, respectively. The combined toxic effect of glucose and FFA on β-cell function has been termed *glucolipotoxicity*.

Glucotoxicity, lipotoxicity, and glucolipotoxicity have been proposed to play a pathogenic role in type 2 diabetes. The underlying concept is that once type 2 diabetes ensues, the increase in glucose and FFA levels, which are characteristics of diabetes, exerts additional toxic effects on the already dysfunctional β-cells, thus further aggravating the diabetic state [2, 3]. This leads to a vicious positive feedback cycle, deteriorating the β-cells. Lipotoxicity has also been proposed to play a role in the onset of type 2 diabetes, at least in predisposed individuals [4–8]. Obese individuals have elevated levels of FFA due to their expanded [9] and more lipolytically active [10] adipose tissue mass. Obesity-associated insulin resistance in these individuals further contributes to elevated FFA levels, due to the lack of antilipolytic action of insulin. Thus, lipotoxicity may impair β-cell function in these individuals to induce the development of diabetes.

In this chapter, evidence for glucotoxicity, lipotoxicity, and glucolipotoxicity in β-cell dysfunction is reviewed. Possible mechanisms, in particular, a central role of oxidative stress, are also discussed.

21.2
Acute Effect of Glucose and FFA on Insulin Secretion

Normally, insulin is released by pancreatic β-cells in response to a variety of secretagogues. Glucose is a major regulator of insulin secretion (Figure 21.1). Glucose enters the β-cells through the GLUT2 transporters that are present in abundant numbers on the plasma membrane. Once inside the β-cells, glucose is phosphorylated by the rate-limiting enzyme glucokinase to glucose-6-phosphate. Glucose-6-phosphate then enters glycolysis and the Krebs (tricarboxylic acid, TCA) cycle to generate ATP. An increase in ATP to ADP ratio stimulates the closure

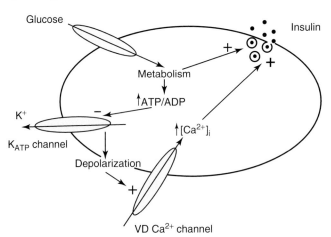

Fig. 21.1 Stimulation of insulin release by glucose. Glucose enters the β-cell and is metabolized to produce ATP. The increased ATP/ADP ratio closes the K_{ATP} channels, which causes membrane depolarization and activation of the VD (voltage-dependent) Ca^{2+} channels. The increase in intracellular Ca^{2+} concentration $[Ca^{2+}]_i$ triggers exocytosis. Exocytosis is also stimulated by glucose metabolism independent of K_{ATP} channels and, in part, also independent of ATP.

of ATP-sensitive potassium (K_{ATP}) channels. This triggers the opening of the voltage-dependent (VD) calcium channels and increased intracellular cytoplasmic calcium concentration, which subsequently causes exocytosis of insulin-containing granules from multiple intracellular stores. Glucose stimulates insulin exocytosis by inducing fusion of a readily releasable pool of insulin granules localized in the plasma membrane, and through the mobilization and trafficking of insulin granules from intracellular storage pools to the membrane [11].

In addition to K_{ATP}-dependent pathway, glucose can stimulate insulin secretion through a K_{ATP}-independent pathway. This ATP independent pathway does not require a change in the concentration of intracellular calcium and is likely accounted for by direct effects on exocytosis of the increase in ATP/ADP ratio and/or putative "coupling factors" derived from glucose or increased by glucose. One of these coupling factors is long chain acyl CoA (LC-CoA) [12]. Glucose is metabolized through glycolysis to generate pyruvate, which is then converted to citrate in the mitochondria. Citrate can be oxidized by the Krebs cycle in the mitochondria or it can be exported to the cytoplasm. In the cytoplasm, citrate can be converted to malonyl-CoA via the sequential action of the enzymes ATP-citrate lyase (CL) and acetyl-CoA carboxylase (ACC). Malonyl-CoA is a potent allosteric inhibitor of the mitochondrial membrane enzyme carnitine palmitoyltransferase-1 (CPT-1), which controls transport of LC-CoA into the mitochondria for oxidation. This inhibition leads to an increase in cytosolic LC-CoA, which can directly stimulate insulin secretion by various mechanisms, including protein acylation [13] and stimulatory effect of protein kinase C (PKC; the majority of PKC isoforms has stimulatory effect on insulin secretion) [14].

In addition to having a direct effect on insulin secretion, glucose can also increase insulin content (which is an important determinant of insulin secretion) by increasing transcription of the proinsulin gene, translation of proinsulin, and conversion of proinsulin to insulin [11].

FFAs, like glucose, are insulin secretagogues. FFA can be oxidized to generate ATP to stimulate insulin secretion. However, increasing evidence shows that most FFA-induced insulin secretion is mediated through LC-CoA (as discussed above). FFA can also stimulate insulin secretion by increasing the concentration of calcium via the G-protein-coupled GPR40 receptor [15].

21.3
Insulin Secretory Abnormalities in Type 2 Diabetic Patients

Normally, a rapid rise in blood glucose level elicits a biphasic insulin secretory response. The first phase of insulin secretion occurs for 5–10 minutes, followed by the second phase. In patients with type 2 diabetes, this first phase insulin secretion to glucose is no longer present. However, insulin response to secretagogues other than glucose is relatively unimpaired. Multiple insulin secretory defects are present, including absence of pulsatility, excess in prohormone secretion, and progressive decrease in insulin secretory capacity with time. These are due to a reduction in both β-cell function and mass. β-cell mass is reduced because of β-cell death induced by the toxic effects of glucose and fat and by amyloid and other endogenous toxins (see [16] for review). Amyloid fibers, which are composed mainly of islet amyloid polypeptide (IAPP; a protein cosecreted with insulin), have been reported in up to 90% of type 2 diabetic subjects compared to 10–13% in nondiabetic subjects [17]. The mechanism of formation of these precipitates is still not entirely clear. It not only results from prolonged β-cell overstimulation by hyperglycemia, but is also favored by inflammation and glucolipotoxicity. Recently, it has been suggested that the toxic form of amyloidogenic proteins is not the amyloid fibers, but rather the small IAPP oligomers [18–20] that can induce β-cell apoptosis. The mechanism of how toxic IAPP oligomers induce β-cell death is unclear, but may be linked to alterations in endoplasmic reticulum (ER) membrane by IAPP oligomers, leading to calcium leakage [21, 22], and subsequent β-cell apoptosis (see [23] for review).

21.4
β-Cell Glucotoxicity

21.4.1
Chronic Effect of Glucose on β-Cell Function and Mass

Hyperglycemia, a hallmark of diabetes, results from defective glucose homeostasis. Plasma glucose levels are regulated by (i) acute and sustained insulin secretion, (ii) stimulatory effect of insulin on glucose uptake by muscle and fat (peripheral insulin sensitivity), and (iii) inhibitory effect of insulin on hepatic glucose production and stimulatory effect of insulin on hepatic glucose utilization (hepatic insulin sensitivity). In subjects with normal β-cell function, a decrease in hepatic/peripheral sensitivity is compensated by an increase in insulin secretion and plasma glucose levels remain normal. However, in individuals with glucose intolerance or type 2 diabetes, the decrease in insulin sensitivity is uncompensated by a corresponding increase in insulin secretion. This leads to progressive elevation in plasma glucose levels.

As discussed above, acute glucose elevation is a major stimulator of insulin secretion. However, chronic glucose exposure can impair β-cell function. The idea that chronic glucose can act as a toxin to cause β-cell dysfunction is not a new one. As early as the 1940s, it was demonstrated that chronic glucose injections could cause diabetes in the partially pancreatectomized (Px) cat [24]. This finding was reproduced in the chronically glucose infused Px dogs [25]. Since then numerous studies have demonstrated that prolonged glucose elevation can impair β-cell function and mass. This has been demonstrated *in vitro* [26–31], *ex vivo* in the perfused pancreas [32, 33], or in freshly isolated islets of glucose infused rats [34, 35]. It is important to note that the toxic effect of prolonged glucose elevation also occurs in humans. For example, Boden *et al.* demonstrated, in humans, that 68

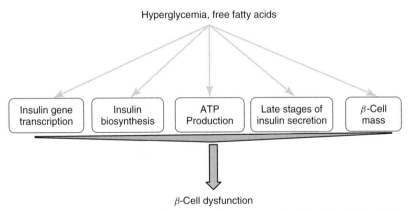

Fig. 21.2 Levels of impairment of β-cell function by chronic glucose and FFA exposure. Chronically elevated levels of glucose and FFA can act at multiple cellular sites to impair β-cell function. This can affect β-cell function at multiple levels, which include impairment in insulin gene transcription, insulin biosynthesis, ATP production, late stages of insulin secretion, that is, exocytosis, and a decrease in β-cell mass.

hours of glucose elevation can impair β-cell function and induce insulin resistance [36].

21.4.2 Level of Impairment of β-cell Function by Chronic Glucose Exposure

Chronic glucose exposure can impair β-cell function at multiple levels, as depicted Figure 21.2.

21.4.2.1 Insulin Gene Transcription

One of the most well-studied mechanisms of β-cell glucotoxicity is the impairing effect of chronic high glucose on insulin gene promoter activity. The transcriptional activity of the insulin promoter is mainly regulated by three enhancer elements: A3, E1, and C1. Mutation of these sites leads to markedly impaired insulin gene transcription [37]. A3, E1, and C1 are bound by the transcription factors Pdx-1, BETA2/NeuroD, and RIPE3b1/MafA, respectively. Early studies demonstrating the effect of chronic glucose on insulin gene transcription were performed in β-cell lines. Prolonged culture of HIT-T15 cells (a β-cell line) in high glucose led to a significant decrease in insulin gene transcription [38], and this was associated with decreased activity of the insulin promoter, and decreased binding of Pdx-1 and MafA to their respective elements [39–41]. In another β-cell line, BetaTC-6 cells, prolonged exposure high glucose led to decreased insulin gene transcription, but this was associated with a decrease in only MafA binding activity [31]. *In vivo* evidence that glucose impairs insulin gene transcription can be observed in animal models of type 2 diabetes. In the 90% partial Px rat model which is hyperglycemic, insulin mRNA, insulin content, glucose-stimulated insulin secretion (GSIS), and

Pdx-1 expression are markedly reduced four weeks post Px [41]. In the Zucker diabetic fatty rat (ZDF rat), a genetic model of type 2 diabetes, onset of diabetes is associated with decreased insulin gene transcription and insulin content [42]. Harmon et al. demonstrated that treatment with troglitazone in ZDF rat to prevent hyperglycemia improved Pdx-1 binding activity, insulin gene transcription, and insulin content [43]. The mechanisms involved are unclear, but may be linked to oxidative stress, and subsequent activation of the JNK pathway (see Section 21.8.3.1 below).

Studies have shown that these same transcription factors that are important for insulin gene transcription are also important in the regulation of other genes. For example, Pdx-1 has been shown to regulate GLUT2 and glucokinase [44]. Both GLUT2 and glucokinase play an important role in the β-cell glucose sensing mechanism, and thus alterations in the expression of these genes could predictably alter GSIS. Indeed, a marked reduction in GLUT2 protein levels has been observed in animal models of type 2 diabetes, such as the Px [41] and ZDF rat [45].

In addition to the effect of glucose on Pdx-1 and MafA, chronic glucose is associated with increased expression of the insulin gene transcriptional repressor CCAAT/enhancer-binding protein beta (CEBPβ) [46, 47] and of the proto-oncogene c-myc [48]. Increase in both transcription factors can further impair insulin gene transcription. c-myc suppresses insulin gene expression by inhibiting NeuroD-mediated transcriptional activation [49], and its increase is postulated to result in a loss of differentiation of β-cells exposed to elevated glucose, which could explain, in part, defective insulin secretion following chronic glucose elevation.

21.4.2.2 Insulin Biosynthesis

Insulin biosynthesis and insulin content are reduced following chronic exposure to high glucose. Although this may be secondary to a decrease in insulin gene transcription (as discussed above), there is evidence that chronic glucose elevation can specifically decrease insulin biosynthesis independent of a decrease in insulin gene transcription. One mechanism proposed for this is the role of ER stress. The ER is involved in folding, processing, and exporting secretory and membrane proteins to the Golgi apparatus. When the demand for proteins exceeds the folding capacity of the ER, ER stress ensues leading to the activation of unfolded protein response, a mechanism that inhibits global protein synthesis, but increases ER chaperones and degradation of misfolded proteins to counteract ER stress. The same mechanism, when chronically activated, also initiates apoptosis of cells unable to cope up with ER stress [50]. That ER stress plays a role in diabetes at the β-cell level is suggested by the early destruction of β-cells in Walcott–Ralliston syndrome, which is due to a mutation in PERK, a major factor in the unfolded protein response [51]. It has been reported that chronic exposure to glucose elevates ER stress markers in vitro [52–55], suggesting that glucose-induced ER stress can decrease global protein translation, and hence insulin biosynthesis. It has been proposed that ER stress in β-cell glucotoxicity is due to increased folding load due to oversecretion of insulin and/or oxidative stress (Section 21.8.3.4).

Some argue that the effect of chronic glucose on β-cell function is not toxic, but rather a reflection of a depletion in insulin stores due to excessive insulin secretion. It is believed that β-cell function can be restored following "β-cell rest." The term *β-cell exhaustion* has been used to describe this phenomenon. Grill *et al.* demonstrated that coinfusion of diazoxide, an inhibitor of K_{ATP} channels, and hence insulin secretion (β-cell rest), with glucose for 48 hours in normal rats prevented glucose-induced β-cell dysfunction [56]. However, β-cell exhaustion as a cause for glucotoxicity has been challenged by Moran *et al.* [57]. In this study, HIT-T15 cells were cultured chronically with high glucose and somatostatin. Somatostatin effectively inhibited insulin secretion. However, β-cells exposed to high glucose and somatostatin still had impaired insulin gene transcription, content, and GSIS, suggesting that β-cell exhaustion is not the primary cause for β-cell glucotoxicity.

21.4.2.3 ATP Production

It is well established that ATP production is essential for GSIS, thus any decrease in ATP production can impair insulin release. ATP production can be decreased by a reduction in glucose oxidation. However, this is unlikely the main site of impairment as chronic glucose exposure is associated with only mild reduction in glucose oxidation [27, 28]. Another more possible mechanism for the decrease in ATP production is the upregulation or activation of uncoupling protein 2 (UCP2) by chronic glucose exposure. Uncoupling proteins (UCPs) are located in the inner mitochondrial membrane, and can uncouple the electrochemical gradient produced by the electron transport chain from ATP synthesis. β-cells express UCP2, the only member of the UCP family located in the pancreatic islet. Overexpression of UCP2 in islets decreases insulin secretion [58]. In contrast, knockout of UCP2 in mice leads to hyperinsulinemia and decreased plasma glucose [59]. These findings suggest that UCP2 is a negative regulator of insulin secretion. Recently, a role of UCP2 in β-cell glucotoxicity has been proposed. However, studies to date investigating UCP2 expression in β-cell glucotoxicity have yielded conflicting results. *In vitro* exposure of isolated islets or INS1 cells to high glucose for at least 48 hours increased [28, 60], decreased [61], or had no effect [62] on UCP2 mRNA and/or protein expression. *In vivo*, partial Px rats [41] and glucose infused rats [63] had increased UCP2 mRNA expression, whereas the hyperglycemic ZDF rats have been reported to have low expression of islet UCP2 mRNA [64]. Interestingly, one study by Krauss *et al.* showed that UCP2 mRNA was unaltered in mouse islets exposed to chronic high glucose. However, UCP2 activity was increased [29] due to activation of UCP2 by superoxide (Section 21.8.2.1). Therefore, although UCP2 appears to play a role in glucotoxicity, this may not result from changes in UCP2 expression.

21.4.2.4 Late Stages of Insulin Secretion

The majority of studies demonstrates that the impairing effect of chronic glucose on insulin secretion is specific for glucose but not other secretagogues, suggesting interference with glucose metabolism or any other downstream mechanisms activated by glucose. The late stage of insulin secretion, that is, exocytosis, is a common event in insulin secretion by all secretagogues. This suggests that chronic glucose is unlikely to impair insulin secretion by affecting the exocytosis of insulin granules. Recently, however, it has been suggested that chronic glucose elevation can affect calcium handling and the late stages of insulin secretion [65–67]. In INS-1 cells, a β-cell line, prolonged exposure to high glucose resulted in higher basal calcium levels. However, insulin secretion was reduced when permealized cells were directly stimulated by calcium [65]. Furthermore, these cells were found to express lower levels of proteins required for calcium-induced exocytosis of insulin, such as SNARE proteins, VAMP-2, and syntaxin 1. VAMP-2 was also reduced in human islets cultured in high glucose for 72 hours. In another study in isolated islets, it was found that chronic glucose reduces insulin secretion by interfering with the exit of insulin via the fusion pore [66].

21.4.2.5 β-Cell Mass

It is now well accepted that β-cell mass is constantly changing in response to changes in metabolic demands. In the presence of insulin resistance, for example, β-cell mass increases to compensate for decreased insulin action. β-cell mass is determined by the balance between β-cell regeneration and β-cell apoptosis. It has been reported that the frequency of β-cell replication is very low in both diabetic and nondiabetic subjects, with no difference between the two. However, frequency of β-cell apoptosis is reported to be significantly higher in pancreatic tissue obtained from patients with type 2 diabetes compared to normal subjects [17]. This imbalance between proliferation and apoptosis in type 2 diabetes leads to a reduction in β-cell mass. Chronic glucose elevation has been reported to induce β-cell death, and has been proposed to act as a secondary force to decrease β-cell mass following the onset of diabetes.

That chronic glucose exposure can increase β-cell apoptosis is supported by *in vitro* and *in vivo* studies. In cultured human islets, graded increases in glucose from 5.5 to 11 mmol l^{-1} glucose and above induced apoptosis in a concentration-dependent fashion [68]. In the desert gerbil, *Psammomys obesus*, a shift from low to high calorie intake increased plasma glucose levels, and this was associated with increased β-cell apoptosis. This toxic effect of glucose was reproduced in primary cultured islets from the *P. obesus* exposed to elevated glucose concentrations [69]. Many of the mechanisms involved in glucose-induced β-cell apoptosis are similar to those involved in β-cell dysfunction. These include oxidative stress, ER stress, and inflammation, which are discussed in later sections.

21.5
β-Cell Lipotoxicity

21.5.1
Chronic Effect of Free Fatty Acids on β-Cell Function and Mass

There is a well-known relationship between obesity and diabetes mellitus. More than 85% of individuals diagnosed with type 2 diabetes are obese [70]. Obese individuals have elevated FFA levels due to their more expanded [9] and lipolytically active [10] adipose tissue mass. The obesity-associated insulin resistance further increases FFA levels due to the lack of the antipolytic action of insulin. Acutely, FFA stimulates insulin secretion (as discussed above). In contrast, chronic elevation of FFA (>24 hours) impairs GSIS. This has been demonstrated *in vitro* in islets [28, 71–75] or β-cell lines [76–79] and by studies *in situ* in the perfused pancreas. *In vivo*, the effect of prolonged FFA elevation on β-cell function has been more controversial. Absolute GSIS evaluated *in vivo* (uncorrected for insulin resistance) was found to be increased [80–82], unchanged [83–85], or decreased [86, 87]. However, in most of these *in vivo* studies, GSIS was not corrected for insulin resistance and failure to do so likely explains the controversial results. It is now generally accepted that "β-cell lipotoxicity" occurs *in vivo*, when GSIS is corrected for insulin resistance (as chronic FFA elevation also induces insulin resistance); however, it is still unclear whether this requires genetic predisposition [6, 7].

21.5.2
Level of Impairment of β-Cell Function by Chronic Free Fatty Acids Exposure

Chronic free fatty acids exposure can impair β-cell function at multiple levels as depicted in Figure 21.2.

21.5.2.1 Insulin Gene Transcription
Similar to glucose, prolonged FFA elevation can impair insulin gene transcription. This effect is specific for palmitate (saturated FFA) but not for oleate, in the presence of high glucose [44, 88–90]. The effect has been attributed to ceramides that decrease MafA expression and Pdx-1 nuclear localization [91, 92]. The exact mechanisms are unclear. In insulin-sensitive tissues, ceramide formation from palmitate inhibits Akt [93]. Akt prevents the translocation of the transcription factor Fox-O1 to the nucleus [94]. In β-cells, Pdx-1 and Fox-O1 exhibit a mutually exclusive pattern of nuclear localization [95]. Thus, inhibition of Akt by ceramide can lead to nuclear translocation of Fox-O1 and nuclear exclusion of Pdx-1, which results in a decrease in insulin gene transcription. Furthermore, ceramide is an activator of JNK [96], and JNK activation has been linked to decreased insulin gene transcription by decreasing binding and nuclear localization of Pdx-1 (Section 21.8.3.1). Ceramides are also known inducers of oxidative stress, and oxidative stress has been linked to β-cell dysfunction (Sections 21.7 and 21.8). However, the role of oxidative stress in this effect has not been investigated.

As discussed in Section 21.4.2.1, Pdx-1 also regulates the gene transcription of GLUT2 and glucokinase. Thus, prolonged exposure to fat, like glucose, can affect GSIS by impairing glucose sensing.

21.5.2.2 Insulin Biosynthesis
In addition to having an effect on insulin gene transcription, fatty acids can impair insulin production by inhibiting proinsulin biosynthesis [75, 97]. This can

occur independent of a decrease in insulin gene transcription. One mechanism that has been proposed for this reduction in insulin biosynthesis is ER stress (Sections 21.4.2.2 and 21.8.3.4). It has been demonstrated that chronic exposure to palmitate, but perhaps not oleate, increases ER stress markers [98–100]. This can possibly be due to the alteration in ER membrane composition by palmitate, as well as by oxidative stress (Sections 21.7 and 21.8). Studies demonstrate that prolonged FFA exposure can also impair the function of the prohormone convertase enzymes (PC2 and PC3) [101], and decrease the expression of carboxypeptidase E [102] to decrease insulin biosynthesis. All three enzymes catalyze the conversion of proinsulin into insulin.

21.5.2.3 ATP Production

One suggested mechanism of β-cell lipotoxicity is a decrease in glucose metabolism due to enhanced rate of fat oxidation (Randle's cycle). This inverse relationship between FFA and glucose metabolism is known to occur in the heart [103]. According to Randle's cycle, an increase in fat oxidation enhances NADH and acetyl-CoA production, leading to the inhibition of pyruvate dehydrogenase (PDH), and thereby glucose oxidation. Increase in FFA oxidation also increases citrate, a metabolite of the Krebs cycle, which can then negatively regulate the rate-limiting glycolytic enzyme phosphofructokinase (PFK), and thereby decreases glucose utilization. That Randle's cycle plays a role in β-cell lipotoxicity is supported by *in vitro* observations that show that FFA inhibits PDH activity [104] and that inhibition of fat oxidation prevents fat-induced β-cell dysfunction [75]. In other studies, however, it was found that fatty acids did not affect β-cell glucose utilization rate [78], or glucose-6-phosphate and citrate levels [105], suggesting that a reduction in glucose metabolism, as predicted by the glucose–fatty acid cycle, is not the main cause of decreased ATP production and blunted GSIS.

Similar to glucose, prolonged exposure to FFA has been shown to increase the expression [106], and possibly the activity of UCP2 [107]. FFA acts as cofactor for the proton transport function of UCP2 [107]. As discussed above in Section 21.4.2.3, UCP2 uncouples the mitochondrial electrochemical gradient, and thus decreases ATP production and insulin secretion. The data on the expression of UCP2 in lipotoxicity are more convincing than the data in glucotoxicity. *In vitro*, exposure of isolated islets or β-cell lines to palmitate or oleate increases UCP2 expression [28, 62, 106, 108]. UCP2-null mice are resistant to the impairment in GSIS induced by palmitate *in vitro* [109] and by high-fat diet [110] *in vivo*. The upregulation of UCP2 by FFA has been linked to the transcription factors PPARγ [28] and sterol regulatory element binding protein (SREBP-1) [108].

21.5.2.4 Late Stages of Insulin Secretion

The majority of experimental data shows that the impairing effect of prolonged FFA elevation on insulin secretion is specific for glucose, and not other secretagogues, suggesting interference with glucose metabolism or any other downstream mechanisms activated by glucose [97, 111]. However, one recent study showed that long-term exposure to lipids can inhibit GSIS downstream of granule fusion with plasma membrane [66].

21.5.2.5 β-Cell Mass

In Caucasians, relative β-cell mass is reduced by 40% in obese subjects with impaired fasting glucose, and by 63% in obese subjects with type 2 diabetes compared to obese normoglycemic subjects [17]. The reduction in β-cell mass has been linked to an increase in apoptosis by prolonged exposure to FFA [17]. In vitro, saturated FFAs induce apoptosis [112–114], whereas unsaturated FFAs are usually protective [112, 114, 115]. Several mechanisms have been proposed to explain FFA-induced β-cell apoptosis. Studies have found that FFA-induced β-cell apoptosis depends in part on the generation of ceramides. These lipid second messengers are involved in the apoptosis response induced by a variety of different triggers, including cytokines, ionizing radiation, and heat shock [116]. Studies have found that ceramide levels are elevated in fat-rich islets from rodents [117, 118], presumably due to the abundance of substrate for *de novo* ceramide synthesis. Treatment with ceramide synthase inhibitors has been shown to prevent FFA-induced β-cell apoptosis in both rodent [118] and human [119] islets. Ceramides can enhance the production of reactive oxygen species (ROS) [120], and free radicals have been implicated as important regulators of the apoptotic pathway [121]. Other mechanisms that have been suggested to be important for β-cell apoptosis induced by prolonged fat exposure include oxidative stress (also non-ceramide mediated) (Sections 21.7 and 21.8) and ER stress (Sections 21.4.2.2 and 21.8.3.4).

21.6
Glucolipotoxicity

Elevations of both glucose and FFA concentrations have been reported to act synergistically to impair β-cell function and mass. This is known as *glucolipotoxicity*. The exact mechanisms of glucolipotoxicity are unclear. One proposed mechanism is the increased generation of intermediate metabolites via elevated FFA esterification. It has been proposed that under circumstances when both glucose and FFA are elevated, cytosolic citrate accumulates due to increased production of acetyl-CoA from glucose and FFA (and also increased production of oxaloacetate from glucose), ATP inhibition of further krebs cycle activity, and citrate exit into the cytosol. Accumulation of cytosolic citrate leads to increased generation of malonyl-CoA (citrate is the precursor of malonyl-CoA). Malonyl-CoA inhibits CPT-1, the enzyme responsible for transport of FFA into the mitochondria. Sustained inhibition of CPT-1 leads to accumulation of long chain fatty-CoA in the cytosol, which is proposed to exert deleterious effects, either directly or via generation of lipid-derived signals such as ceramides or diacylglycerol (DAG), on β-cell function (Figure 21.3). DAG has been shown to activate PKC, in which the isoforms δ [122] and ε [123] have been implicated in β-cell lipotoxicity (the mechanisms involved are unclear, but may include changes to lipid metabolism, insulin content, and insulin gene

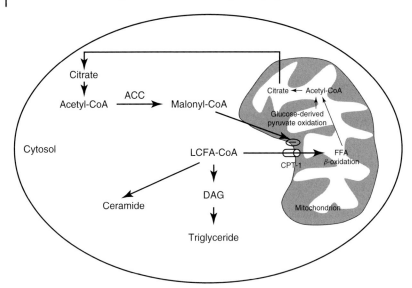

Fig. 21.3 Accumulation of fatty acid esterification products. Under circumstances when both glucose and FFA are elevated, cytosolic citrate accumulates which leads to increased generation of malonyl-CoA (citrate is the precursor of malonyl-CoA) by acetyl-CoA carboxylase (ACC). Malonyl-CoA inhibits carnitine palmitoyltranferase-1 (CPT-1), the enzyme responsible for transport of FFA into the mitochondria. Sustained inhibition of CPT-1 leads to accumulation of long chain fatty-CoA (LCFA-CoA) in the cytosol, which is proposed to exert deleterious effects, either directly or via the generation of lipid-derived signals such as ceramides or diacylglycerol (DAG), on β-cell function.

transcription by PKC activation [123]). This hypothesis was first proposed by Prentki and Corkey, and is known as the *malonyl-CoA/LC-CoA hypothesis* [124].

In addition to switching FFA oxidation to esterification, prolonged glucose can activate the expression of genes involved in lipogenesis [125]. One transcription factor that has been shown to regulate lipogenic gene expression is SREBP-1c [126]. SREBP-1c expression is elevated in islets of animal models of type 2 diabetes [127], and β-cell-specific SREBP-1c transgenic mice have elevated lipogenesis and thus exaggerated lipotoxicity. Studies show that chronic glucose elevation can activate SREBP-1 [128], and dominant negative suppression of SREBP-1c activity protects against β-cell glucolipotoxicity [52].

21.7
Oxidative Stress as an Endogenous Toxin

21.7.1
Reactive Oxygen Species as a Derived Toxin from Chronic Glucose and Fat Exposure

As discussed above (Sections 21.4 and 21.5), various mechanisms have been proposed for the deleterious effect of glucose and fat elevation on β-cell function and mass. These include glucose and FFA effects on β-cell enzymes and transcription factors. Recently, it has been found that many of these mechanisms of glucose and FFA are linked to oxidative stress. Thus, it has been proposed that oxidative stress is the central mechanism of β-cell dysfunction associated with chronic glucose and FFA elevation. In the following sections of this chapter, the role and mechanism of oxidative stress in β-cell gluco- and lipotoxicity are reviewed.

Long-term exposure to glucose and/or FFA can lead to the generation of ROS. Endogenous ROS in physiological concentrations are important for many physiological processes [129] such as gene transcription and leukocyte function. Recent studies also demonstrate that physiological levels of ROS are important for insulin signaling [130] and normal β-cell function [131]. However, when concentrations of ROS reach excessive levels for prolonged periods of time, oxidative stress ensues. Oxidative stress can directly cause functional damage to proteins, enzymes, DNA, and lipids or can indirectly activate various stress-sensitive pathways to alter cellular function.

21.7.2
Sites of Reactive Oxygen Species Generation

The main site of ATP production is the mitochondrial electron transport chain. This process of oxidative phosphorylation is accompanied by ROS production in minute amounts, which is usually buffered by native antioxidant defense mechanisms such as superoxide dismutase (SOD), catalase, and glutathione. However, in states of excess energy (increased glucose and FFA), this balance is tipped toward increased mitochondrial production of ROS. FFA may also amplify mitochondrial ROS production by changes in mitochondrial membrane composition. Other intracellular sources of ROS induced by high glucose exposure include glycosylation, the hexosamine pathway, PKC activation, glucose autoxidation, and increased endoplasmic reticulum oxidoreduction 1 (ERO1) activity due to increased folding in the ER. For FFA exposure, potential pathways include NADPH oxidase, which is activated by PKC [132], ceramides, increased ERO1 activity, peroxisomal oxidases, endoplasmic reticular oxidases, cycloxygenases, and lipoxygenases or hexosamines (reviewed in [133]).

21.7.3
Oxidative Stress, Type 2 Diabetes, and β-Cell Dysfunction

Studies suggest that type 2 diabetic patients are subjected to chronic oxidative stress [3]. Markers of oxidative stress damage are elevated in patients with type 2 diabetes [134–136]. Levels of glutathione, the primary intracellular antioxidant [137], were also found to be decreased in erythrocytes of diabetic patients, a defect ameliorated by improved metabolic control [138]. Oxidative stress has been shown to play a key role in diabetic complications and insulin resistance induced by fat and glucose. More recently, oxidative stress has been suggested to play a role in β-cell dysfunction. Pancreatic β-cells are reported to have intrinsically low levels of antioxidant enzyme expression and activity. SOD-1 and 2, catalase and glutathione peroxidases are expressed in low levels in the β-cells [139, 140] compared to other tissues such as the liver, thus β-cells are particularly vulnerable to oxidative stress. Chronic glucose [42] and fat [141] can increase ROS generation in β-cells, and oxidative stress can result in marked impairment in GSIS [142–145]. Furthermore, many proposed pathways of β-cell glucotoxicity and lipotoxicity are upstream and downstream or both upstream and downstream of oxidative stress. PKC, hexosamine, and ceramide pathways can induce [120, 132, 146] and are activated by oxidative stress [147–149], and the JNK/p38 and IKK/NFkB can be activated by oxidative stress [144, 150]. Together, these findings suggest that glucose and FFA may increase oxidative stress to impair β-cell function.

In animal models of type 2 diabetes, antioxidant treatment has been demonstrated to improve β-cell function. In the ZDF rat, treatment with antioxidants, either N-acetylcysteine (NAC) or aminoguanidine, increased Pdx-1 binding, insulin mRNA levels, and insulin content [43]. In the db/db mice, NAC was effective in preserving insulin content, insulin mRNA, and Pdx-1 protein levels [151]. Moreover, treatment with the antioxidant vitamin E had beneficial effects on glycemic control in GK rats, which was accompanied by improvement in insulin secretion and lower levels of HBA1c [152]. In humans, daily supplementation with 1.5 g of antioxidant taurine for eight weeks had no effect on insulin secretion in patients at high risk for type 2 diabetes [153]; however, intravenous infusion of the antioxidant glutathione in type 2 diabetic patients improved insulin secretion and glucose tolerance [154]. More recently, it has been reported that islets isolated from pancreata of type 2 diabetic cadaveric organ donors have elevated oxidative stress marker levels as well as low levels of glucose-induced insulin secretion *in vitro*. Treatment with the antioxidants significantly improved β-cell function [155].

21.8
Oxidative Stress, β-Cell Glucotoxicity, and Lipotoxicity

21.8.1
Evidence for a Role of Oxidative Stress in β-Cell Glucotoxicity and Lipotoxicity

Solid *in vitro* data have linked oxidative stress to β-cell glucotoxicity. Prolonged exposure to high glucose has been shown to increase peroxide levels in HIT-T15 cells, rodent and human islets *in vitro* [156]. In HIT-T15 cells, a β-cell line, prolong culture in high glucose concentration impaired β-cell function and induced apoptosis, an effect that was prevented by either the antioxidant NAC or aminoguanidine [42]. Treatment with either of these antioxidants in HIT-T15 cells preserved insulin promoter activity, Pdx-1 binding, and levels of proinsulin mRNA [42]. In cultured islets, adenovirally overexpressing the antioxidant enzyme glutathione peroxidase (which catabolizes both hydrogen peroxide and lipid peroxides to hydrogen and water) protected against ribose-induced β-cell toxicity [156]. Ribose is a more reducing sugar than glucose, and can cause glucotoxicity in a shorter period. In another study in cultured islets, overexpression of the catalytic subunit of glutamylcysteine ligase, which regulates the use of cysteine as the rate-limiting substrate to form glutathione (an abundant endogenous antioxidant), increased glutathione levels and protected the islets from oxidative stress [157]. Although the majority of studies indicates that oxidative stress is an important mechanism of β-cell glucotoxicity, there are a few studies that argue against this. A study by Martens et al. found that oxidative stress was suppressed, rather than induced, following exposure of primary β-cells to high glucose [158].

In vivo evidence that chronic glucose impairs β-cell function is mainly derived from studies in animal models of type 2 diabetes, where antioxidants were shown to improve β-cell function [42, 151]. Recently, our laboratory has demonstrated in a more selective in vivo model of β-cell glucotoxicity (i.e., in the absence of hormonal and metabolic changes found in animal models of type 2 diabetes) that prolonged glucose infusion in rats (48 or 96 hours) impairs β-cell function, and that coinfusion of the superoxide dismutase mimetic tempol decreases superoxide levels and prevents glucose-induced β-cell dysfunction [159]. Interestingly, we also tested the antioxidants taurine and NAC. Both these antioxidants decreased islet oxygen species levels, but they did not prevent glucose-induced β-cell dysfunction [159]. The mechanisms involved are unclear. One possibility is that different antioxidants decrease different types of ROS, suggesting that the type and/or site of ROS generation are important. This is consistent with another study by Krauss et al., which demonstrated that overexpression of MnSOD (mitochondrial isoform of superoxide dismutase) but not glutathione peroxidase 1 (cytosolic and mitochondrial antioxidant enzyme) prevented glucotoxicity in cultured islets [29]. Some in vitro studies, however, show that NAC is effective in preventing β-cell glucotoxicity [160, 161]. The reason why NAC is effective in some but not all models of glucotoxicity is unclear, but may be linked to the duration of glucose exposure. It is possible that initially, glucose increases ROS, mainly superoxide in the mitochondria. However, with more chronic exposure, cytosolic ROS such as hydrogen peroxide become important. Future studies in more chronic models of glucotoxicity, that is, in the GK rats (hyperglycemic but not hyperlipidemic model of type 2 diabetes) will help address this question.

Evidence for a causal role of oxidative stress in the β-cell dysfunction induced by lipotoxicity had been negative in a study in cultured rat islets [73], whereas a study by our group in MIN6 cells indicated restoration of insulin content but not insulin secretion by the antioxidant NAC [79]. Furthermore, our laboratory has demonstrated that 48 hours oleate infusion in rat impairs β-cell function, and that coinfusion of the antioxidants NAC, taurine, or tempol reverse oleate induced impairment in β-cell function [162]. More recently, we have shown that oral treatment with the antioxidant taurine in humans prevented lipotoxicity induced by i.v. infusion of intralipid + heparin (fat infusate) [163].

In summary, there is accumulating evidence that oxidative stress is involved in glucose- and fat-induced β-cell dysfunction. However, the mechanisms of oxidative stress–induced β-cell dysfunction are still unclear. In the next two sections, possible sites of impairment of β-cell function and downstream signaling pathways initiated by oxidative stress are reviewed. It is interesting to note that many of the mechanisms that are associated with gluco- and lipotoxicity described above have been shown to involve oxidative stress. Thus, it has been suggested that oxidative stress is the central mechanism of both gluco- and lipotoxicity.

21.8.2
Sites of Oxidative Stress–Induced Impairment of β-Cell Function by Glucotoxicity and Lipotoxicity

21.8.2.1 Glucose Oxidation and Uncoupling Protein 2

In vitro studies show that ROS can decrease glucose oxidation due to inhibition of mitochondrial [142] and glycolytic enzymes [145], and induce uncoupling by activating [29, 107] and/or upregulating [164, 165] UCP2 in β-cells. This suggests that the effect of glucose [29] or FFA [166] on UCP2 may be mediated, at least in part, via the generation of ROS.

21.8.2.2 Insulin Gene Transcription and Insulin Biosynthesis

Other possible sites of action of oxidative stress are at the level of insulin gene transcription and (pro)insulin biosynthesis. As discussed above in Sections 21.4.2.1–21.4.2.2 and 21.5.2.1–21.5.2.2, both glucotoxicity and lipotoxicity are associated with impaired insulin gene transcription and insulin biosynthesis. Studies *in vitro* show that the effect of high glucose to impair insulin gene transcription is due to the induction of oxidative stress, which decreases binding of the transcription factors Pdx-1 [39] and MafA [31, 160] to the insulin promoter. Impairment of insulin gene transcription also occurs after prolonged islet exposure to palmitate, but not oleate, in the presence of high glucose [44, 88–90], and this effect has been attributed to ceramides, which decrease MafA expression and Pdx-1 nuclear localization [91, 92]. However, whether this effect is dependent on oxidative stress is unknown.

Prolonged exposure of human islets to high glucose can decrease (pro)insulin biosynthesis [27]; however, it is unclear whether it is secondary to decreased insulin gene transcription and/or whether it is due to an independent effect of oxidative stress. The hexosamine pathway, which induces oxidative stress [146], impairs (pro)insulin biosynthesis [167] via impairment of the PI3-kinase/Akt/mTOR pathway. Activation of mTOR phosphorylates 4EBP-1, which allows eIF4E to interact with eIF4G [168]. This step is critical to initiate translation. Thus, inhibition of mTOR activation by oxidative stress can impair (pro)insulin biosynthesis. In *in vitro* models of lipotoxicity, (pro)insulin biosynthesis is reduced [75, 97, 101] in the absence of changes in insulin gene expression [97, 101]. Our group has also shown no decrease in insulin gene expression by oleate in MIN6 cells but a decrease in insulin content, which is partially reversed by NAC [79]. These results raise the possibility that fat-induced oxidative stress impairs β-cell function directly at the site of (pro)insulin biosynthesis.

21.8.3
Downstream Signaling Mechanisms of Oxidative Stress–Induced Impairment of β-cell Function by Glucotoxicity and Lipotoxicity

21.8.3.1 JNK

Oxidative stress can have direct effects on β-cell transcription factors [31, 160, 169], such as Fox-O1. Oxidative stress increases Fox-O1 nuclear retention [170, 171], which results in nuclear exclusion of Pdx-1 [95] and consequent inhibition of insulin gene transcription. At least part of the effect of oxidative stress on Fox-O1 is currently presumed to be indirect and mediated by JNK [172, 173]. Suppression of JNK prevents the ROS-induced impairment in insulin gene expression [144], and improves β-cell function in *db/db* mice [174]. Also, islets treated with dominant negative JNK adenovirus are more effective than control islets in ameliorating hyperglycemia after transplantation [144]. It is believed that activation of JNK by glucose induces serine phosphorylation of insulin receptor substrates (IRSs) [167], which impairs insulin/IGF signaling and results in the inhibition of Akt. Inhibition of Akt leads to increased Fox-O1 nuclear localization, decreased Pdx-1 nuclear localization, and thus impaired insulin gene transcription. Furthermore, inhibition of IGF/insulin signaling cascade by JNK can lead to inhibition of the mTOR pathway, which can decrease (pro)insulin biosynthesis [167] (Section 21.8.3.3).

21.8.3.2 NFκB

Oxidative stress is a known activator of IKKβ, which, by phosphorylating the inhibitor IkBα, activates NFκB. The role of NFκB on β-cell function is more controversial than that of JNK, because there are no reports that NFκB inhibits Fox-O1 and insulin gene transcription and NFκB may actually be beneficial for GSIS [175, 176]. However, inhibition of NFκB did prevent the β-cell dysfunction induced by cytokines [177]. It should be noted that models of IKKβ inhibition that, in addition to inhibition of NFκB, also involves inhibition of IRS serine phosphorylation show improvement of β-cell function [178, 179]. Similar controversy relates as to whether NFκB is activated by glucose. It has been reported that glucose activates NFκB by inducing oxidative stress in rat islets [180] and that the IKKβ inhibitor salicylate protects human islets against glucotoxicity [179] and partially restores β-cell function in type 2 diabetes [181]. However, in other studies in rat islets neither glucose nor oxidative stress activated NFκB [182]. The role of IKKβ/NFκB in β-cell lipotoxicity is also controversial. Although fatty acids did not activate NFκB in either INS-1 or primary rat β-cells [98], lipotoxicity was associated with NFκB activation and inhibited by an IKKβ inhibitor in INS-1 β-cells in another study [183]. A recent study further demonstrated that salicylate, an IKKβ inhibitor, improves insulin secretion in obese subjects [184]. Thus, although there is some evidence to suggest that glucose- and fat-induced oxidative stress activates IKKβ–NFκB pathway to impair β-cell function, this needs to be further investigated.

21.8.3.3 β-Cell Insulin Resistance

One target of both JNK and IKKβ/NFκB is the serine/threonine phosphorylation of IRS, which impairs IGF-I/insulin signaling. Insulin and IGF-I receptors are expressed in β-cells and affect β-cell function and mass [185]. Short-term effects of insulin on GSIS can be either inhibitory or stimulatory [186–189], whereas effects on insulin gene transcription and (pro)insulin biosynthesis are stimulatory [190–192] and may serve to prevent depletion of insulin in the actively secreting β-cell. Although there is still debate about the importance of insulin for β-cell function, there is consensus that insulin signaling, in particular the PI3K–Akt cascade, however activated (i.e., by insulin released upon glucose stimulation, directly by glucose or by other hormones [193]) plays an important and mainly stimulatory role in β-cell function and growth. This raises the possibility that known inducers of insulin resistance at peripheral sites, such as oxidative stress, may induce insulin resistance also at the β-cell level. However, there are only a few reports on the effect of prolonged exposure to glucose and FFA on β-cell insulin signaling and the role of oxidative stress in these effects have not been investigated. Regarding glucose, studies in RIN cells show serine 307 phosphorylation of IRS [167] by prolonged glucose exposure and reduced Akt/mTOR/4eIFBP phosphorylation in late-passage glucose-cultured cells [194], effects associated with reduced (pro)insulin biosynthesis. Regarding FFA, oleate inhibited IGF-I-induced Akt activation in INS-1 cells [195], and palmitate led to serine phosphorylation of IRS-1 and IRS-2 in mouse islets [196]. Prolonged exposure to FFA also impaired insulin signaling and increased Fox-O1 nuclear retention, leading to increased β-cell apoptosis [197]. The role of oxidative stress in the induction of "β-cell insulin resistance" to decrease β-cell function is currently unclear.

21.8.3.4 Endoplasmic Reticulum (ER) Stress

As described in Sections 21.4.2.2 and 21.5.2.2, both chronic glucose and FFA elevation are associated with increased ER stress levels. Oxidative stress can induce ER stress via depletion of ER calcium, and altering redox status in the ER. ER stress in turn can induce oxidative stress through increased ERO1p oxidase activity, an enzyme that transfers electrons generated by disulfide formation in the ER to molecular oxygen, and/or by depleting reduced glutathione [198]. Whether chronic glucose and/or FFA exposure generates oxidative stress to induce ER stress or vice versa is unclear.

21.9
Conclusion

Chronically elevated plasma glucose and FFAs, as in type 2 diabetes, can behave as endogenous toxins and directly exert toxic effects on β-cells or indirectly via the generation of ROS (Figure 21.4). In this model, both glucose and FFA increase ROS production. Oxidative stress can act directly on various stages of the insulin

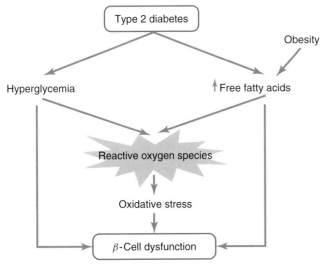

Fig. 21.4 Proposed mechanism for β-cell gluco- and lipotoxicity. Type 2 diabetes is associated with chronically elevated levels of both glucose and free fatty acids (FFA). Chronically elevated glucose and FFA levels can directly exert toxic effect on β-cells or indirectly via the generation of reactive oxygen species. Oxidative stress can act on various stages of insulin production/release pathway or can activate downstream signaling pathways, such as JNK, NFκB, and endoplasmic reticulum stress pathways, to decrease β-cell function. In obesity, at least in predisposed individuals, an elevation in FFA can impair β-cell function, and contribute to the development of type 2 diabetes.

production/release pathway or can activate downstream signaling pathways to decrease β-cell function and mass.

In obese people, at least in predisposed individuals, an elevation in FFA levels can contribute to the development of type 2 diabetes. Once type 2 diabetes is established, glucotoxicity and lipotoxicity and their derived toxins such as oxidative stress can further decrease both β-cell function and mass. It is important to note that many of the proposed mechanisms described above are based on studies *in vitro* and in rodents. The precise role of chronic glucose, FFA, and oxidative stress on β-cell function in humans need to be more extensively investigated.

References

1 Zimmet, P., Alberti, K.G., and Shaw, J. (2001) Global and societal implications of the diabetes epidemic. *Nature*, **414**, 782–787.

2 Poitout, V. and Robertson, R.P. (2008) Glucolipotoxicity: fuel excess and β-cell dysfunction. *Endocr Rev*, **29**, 351–366.

3 Robertson, R.P., Harmon, J., Tran, P.O., and Poitout, V. (2004) β-cell glucose toxicity, lipotoxicity, and chronic oxidative stress in type 2 diabetes. *Diabetes*, **53** (Suppl 1), S119–S124.

4 Lewis, G.F., Carpentier, A., Adeli, K., and Giacca, A. (2002) Disordered fat storage and mobilization in the pathogenesis of insulin resistance and type 2 diabetes. *EndocrRev*, **23**, 201–229.

5 Paolisso, G., Tagliamonte, M.R., Rizzo, M.R. *et al.* (1998) Lowering

fatty acids potentiates acute insulin response in first degree relatives of people with type II diabetes. *Diabetologia*, **41**, 1127–1132.

6 Cusi, K., Kashyap, S., Gastaldelli, A., Bajaj, M., and Cersosimo, E. (2007) Effects on insulin secretion and action of a 48-hour reduction of plasma FFA with acipimox in non-diabetic subjects genetically predisposed to type 2 diabetes. *Am J Physiol Endocrinol Metab*, **292**, E1775–E1781.

7 Kashyap, S., Belfort, R., Gastaldelli, A. et al. (2003) A sustained increase in plasma free fatty acids impairs insulin secretion in nondiabetic subjects genetically predisposed to develop type 2 diabetes. *Diabetes*, **52**, 2461–2474.

8 Boden, G. (2005) Free fatty acids and insulin secretion in humans. *Curr Diab Rep*, **5**, 167–170.

9 Bjorntorp, P., Bergman, H., Varnauskas, E., and Lindholm, B. (1969) Lipid mobilization in relation to body composition in man. *Metabolism*, **18**, 840–851.

10 Rebuffe-Scrive, M., Anderson, B., Olbe, L., and Bjorntorp, P. (1990) Metabolism of adipose tissue in intraabdominal depots in severely obese men and women. *Metabolism*, **39**, 1021–1025.

11 Bratanova-Tochkova, T.K., Cheng, H., Daniel, S. et al. (2002) Triggering and augmentation mechanisms, granule pools, and biphasic insulin secretion. *Diabetes*, **51** (Suppl 1), S83–S90.

12 Prentki, M., Joly, E., El Assaad, W., and Roduit, R. (2002) Malonyl-CoA signaling, lipid partitioning, and glucolipotoxicity: role in β-cell adaptation and failure in the etiology of diabetes. *Diabetes*, **51** (Suppl 3), S405–S413.

13 Haber, E.P., Ximenes, H.M., Procopio, J., Carvalho, C.R., Curi, R., and Carpinelli, A.R. (2003) Pleiotropic effects of fatty acids on pancreatic β-cells. *J Cell Physiol*, **194**, 1–12.

14 Deeney, J.T., Gromada, J., Hoy, M. et al. (2000) Acute stimulation with long chain acyl-CoA enhances exocytosis in insulin-secreting cells (HIT T-15 and NMRI β-cells). *J Biol Chem*, **275**, 9363–9368.

15 Fujiwara, K., Maekawa, F., and Yada, T. (2005) Oleic acid interacts with GPR40 to induce Ca2+ signaling in rat islet β-cells: mediation by PLC and L-type Ca2+ channel and link to insulin release. *Am J Physiol Endocrinol Metab*, **289**, E670–E677.

16 Wajchenberg, B.L. (2007) β-cell failure in diabetes and preservation by clinical treatment. *Endocr Rev*, **28**, 187–218.

17 Butler, A.E., Janson, J., Bonner-Weir, S., Ritzel, R., Rizza, R.A., and Butler, P.C. (2003) β-cell deficit and increased β-cell apoptosis in humans with type 2 diabetes. *Diabetes*, **52**, 102–110.

18 Janson, J., Ashley, R.H., Harrison, D., McIntyre, S., and Butler, P.C. (1999) The mechanism of islet amyloid polypeptide toxicity is membrane disruption by intermediate-sized toxic amyloid particles. *Diabetes*, **48**, 491–498.

19 Mirzabekov, T.A., Lin, M.C., and Kagan, B.L. (1996) Pore formation by the cytotoxic islet amyloid peptide amylin. *J Biol Chem*, **271**, 1988–1992.

20 Jayasinghe, S.A. and Langen, R. (2007) Membrane interaction of islet amyloid polypeptide. *Biochim Biophys Acta*, **1768**, 2002–2009.

21 Orrenius, S., Zhivotovsky, B., and Nicotera, P. (2003) Regulation of cell death: the calcium-apoptosis link. *Nat Rev Mol Cell Biol*, **4**, 552–565.

22 Rizzuto, R., Brini, M., Murgia, M., and Pozzan, T. (1993) Microdomains with high Ca2+ close to IP3-sensitive channels that are sensed by neighboring mitochondria. *Science*, **262**, 744–747.

23 Haataja, L., Gurlo, T., Huang, C.J., and Butler, P.C. (2008) Islet amyloid in type 2 diabetes, and the toxic oligomer hypothesis. *Endocr Rev*, **29**, 303–316.

24 Dohan, F.C. and Lukens, F.D.W. (1948) Experimental diabetes produced by the administration of glucose. *Endocrinology*, **42**, 244–262.

25 Imamura, T., Koffler, M., Helderman, J.H. *et al.* (1988) Severe diabetes induced in subtotally depancreatized dogs by sustained hyperglycemia. *Diabetes*, **37**, 600–609.

26 Davalli, A.M., Ricordi, C., Socci, C. *et al.* (1991) Abnormal sensitivity to glucose of human islets cultured in a high glucose medium: partial reversibility after an additional culture in a normal glucose medium. *J Clin Endocrinol Metab*, **72**, 202–208.

27 Eizirik, D.L., Korbutt, G.S., and Hellerstrom, C. (1992) Prolonged exposure of human pancreatic islets to high glucose concentrations in vitro impairs the β-cell function. *J Clin Invest*, **90**, 1263–1268.

28 Patane, G., Anello, M., Piro, S., Vigneri, R., Purrello, F., and Rabuazzo, A.M. (2002) Role of ATP production and uncoupling protein-2 in the insulin secretory defect induced by chronic exposure to high glucose or free fatty acids and effects of peroxisome proliferator-activated receptor-gamma inhibition. *Diabetes*, **51**, 2749–2756.

29 Krauss, S., Zhang, C.Y., Scorrano, L. *et al.* (2003) Superoxide-mediated activation of uncoupling protein 2 causes pancreatic beta cell dysfunction. *J Clin Invest*, **112**, 1831–1842.

30 Gleason, C.E., Gonzalez, M., Harmon, J.S., and Robertson, R.P. (2000) Determinants of glucose toxicity and its reversibility in the pancreatic islet β-cell line, HIT-T15. *Am J Physiol Endocrinol Metab*, **279**, E997–1002.

31 Poitout, V., Olson, L.K., and Robertson, R.P. (1996) Chronic exposure of betaTC-6 cells to supraphysiologic concentrations of glucose decreases binding of the RIPE3b1 insulin gene transcription activator. *J Clin Invest*, **97**, 1041–1046.

32 Leahy, J.L., Cooper, H.E., Deal, D.A., and Weir, G.C. (1986) Chronic hyperglycemia is associated with impaired glucose influence on insulin secretion. A study in normal rats using chronic in vivo glucose infusions. *J Clin Invest*, **77**, 908–915.

33 Leahy, J.L. and Weir, G.C. (1988) Evolution of abnormal insulin secretory responses during 48-h in vivo hyperglycemia. *Diabetes*, **37**, 217–222.

34 Tang, C., Han, P., Oprescu, A.I. *et al.* (2007) Evidence for a role of superoxide generation in glucose induced B-cell dysfunction in vivo. *Diabetes*, **56**, 2722–2731.

35 Bedoya, F.J. and Jeanrenaud, B. (1991) Insulin secretory response to secretagogues by perifused islets from chronically glucose-infused rats. *Diabetes*, **40**, 15–19.

36 Boden, G., Ruiz, J., Kim, C.J., and Chen, X. (1996) Effects of prolonged glucose infusion on insulin secretion, clearance, and action in normal subjects. *Am J Physiol*, **270**, E251–E258.

37 Weir, G.C., Sharma, A., Zangen, D.H., and Bonner-Weir, S. (1997) Transcription factor abnormalities as a cause of beta cell dysfunction in diabetes: a hypothesis. *Acta Diabetol*, **34**, 177–184.

38 Robertson, R.P., Zhang, H.J., Pyzdrowski, K.L., and Walseth, T.F. (1992) Preservation of insulin mRNA levels and insulin secretion in HIT cells by avoidance of chronic exposure to high glucose concentrations. *J Clin Invest*, **90**, 320–325.

39 Olson, L.K., Redmon, J.B., Towle, H.C., and Robertson, R.P. (1993) Chronic exposure of HIT cells to high glucose concentrations paradoxically decreases insulin gene transcription and alters binding of insulin gene regulatory protein. *J Clin Invest*, **92**, 514–519.

40 Sharma, A., Olson, L.K., Robertson, R.P., and Stein, R. (1995) The reduction of insulin gene transcription in HIT-T15 beta cells chronically exposed to high glucose concentration is associated with the loss of RIPE3b1 and STF-1 transcription

factor expression. *Mol Endocrinol*, **9**, 1127–1134.

41 Laybutt, D.R., Sharma, A., Sgroi, D.C., Gaudet, J., Bonner-Weir, S., and Weir, G.C. (2002) Genetic regulation of metabolic pathways in β-cells disrupted by hyperglycemia. *J Biol Chem*, **277**, 10912–10921.

42 Tanaka, Y., Gleason, C.E., Tran, P.O., Harmon, J.S., and Robertson, R.P. (1999) Prevention of glucose toxicity in HIT-T15 cells and Zucker diabetic fatty rats by antioxidants. *Proc Natl Acad Sci U S A*, **96**, 10857–10862.

43 Harmon, J.S., Gleason, C.E., Tanaka, Y., Oseid, E.A., Hunter-Berger, K.K., and Robertson, R.P. (1999) In vivo prevention of hyperglycemia also prevents glucotoxic effects on Pdx-1 and insulin gene expression. *Diabetes*, **48**, 1995–2000.

44 Gremlich, S., Bonny, C., Waeber, G., and Thorens, B. (1997) Fatty acids decrease IDX-1 expression in rat pancreatic islets and reduce GLUT2, glucokinase, insulin, and somatostatin levels. *J Biol Chem*, **272**, 30261–30269.

45 Orci, L., Ravazzola, M., Baetens, D. et al. (1990) Evidence that down-regulation of β-cell glucose transporters in non-insulin-dependent diabetes may be the cause of diabetic hyperglycemia. *Proc Natl Acad Sci U S A*, **87**, 9953–9957.

46 Lu, M., Seufert, J., and Habener, J.F. (1997) Pancreatic β-cell-specific repression of insulin gene transcription by CCAAT/enhancer-binding protein beta. Inhibitory interactions with basic helix-loop-helix transcription factor E47. *J Biol Chem*, **272**, 28349–28359.

47 Seufert, J., Weir, G.C., and Habener, J.F. (1998) Differential expression of the insulin gene transcriptional repressor CCAAT/enhancer-binding protein beta and transactivator islet duodenum homeobox-1 in rat pancreatic beta cells during the development of diabetes mellitus. *J Clin Invest*, **101**, 2528–2539.

48 Jonas, J.C., Sharma, A., Hasenkamp, W. et al. (1999) Chronic hyperglycemia triggers loss of pancreatic beta cell differentiation in an animal model of diabetes. *J Biol Chem*, **274**, 14112–14121.

49 Kaneto, H., Sharma, A., Suzuma, K. et al. (2002) Induction of c-Myc expression suppresses insulin gene transcription by inhibiting NeuroD/BETA2-mediated transcriptional activation. *J Biol Chem*, **277**, 12998–13006.

50 Zhang, K. and Kaufman, R.J. (2004) Signaling the unfolded protein response from the endoplasmic reticulum. *J Biol Chem*, **279**, 25935–25938.

51 Harding, H.P. and Ron, D. (2002) Endoplasmic reticulum stress and the development of diabetes: a review. *Diabetes*, **51** (Suppl 3), S455–S461.

52 Wang, H., Kouri, G., and Wollheim, C.B. (2005) ER stress and SREBP-1 activation are implicated in β-cell glucolipotoxicity. *J Cell Sci*, **118**, 3905–3915.

53 Elouil, H., Bensellam, M., Guiot, Y. et al. (2007) Acute nutrient regulation of the unfolded protein response and integrated stress response in cultured rat pancreatic islets. *Diabetologia*, **50**, 1442–1452.

54 Lipson, K.L., Fonseca, S.G., Ishigaki, S. et al. (2006) Regulation of insulin biosynthesis in pancreatic beta cells by an endoplasmic reticulum-resident protein kinase IRE1. *Cell Metab*, **4**, 245–254.

55 Hou, Z.Q., Li, H.L., Gao, L., Pan, L., Zhao, J.J., and Li, G.W. (2008) Involvement of chronic stresses in rat islet and INS-1 cell glucotoxicity induced by intermittent high glucose. *Mol Cell Endocrinol*, **291**, 71–78.

56 Sako, Y., Eizirik, D., and Grill, V. (1992) Impact of uncoupling glucose stimulus from secretion on B-cell release and biosynthesis. *Am J Physiol*, **262**, E150–E154.

57 Moran, A., Zhang, H.J., Olson, L.K., Harmon, J.S., Poitout, V., and

Robertson, R.P. (1997) Differentiation of glucose toxicity from beta cell exhaustion during the evolution of defective insulin gene expression in the pancreatic islet cell line, HIT-T15. *J Clin Invest*, **99**, 534–539.

58 Chan, C.B., MacDonald, P.E., Saleh, M.C., Johns, D.C., Marban, E., and Wheeler, M.B. (1999) Overexpression of uncoupling protein 2 inhibits glucose-stimulated insulin secretion from rat islets. *Diabetes*, **48**, 1482–1486.

59 Zhang, C.Y., Baffy, G., Perret, P. et al. (2001) Uncoupling protein-2 negatively regulates insulin secretion and is a major link between obesity, beta cell dysfunction, and type 2 diabetes. *Cell*, **105**, 745–755.

60 Brown, J.E., Thomas, S., Digby, J.E., and Dunmore, S.J. (2002) Glucose induces and leptin decreases expression of uncoupling protein-2 mRNA in human islets. *FEBS Lett*, **513**, 189–192.

61 Roduit, R., Morin, J., Masse, F. et al. (2000) Glucose down-regulates the expression of the peroxisome proliferator-activated receptor-α gene in the pancreatic beta -cell. *J Biol Chem*, **275**, 35799–35806.

62 Li, L.X., Skorpen, F., Egeberg, K., Jorgensen, I.H., and Grill, V. (2002) Induction of uncoupling protein 2 mRNA in β-cells is stimulated by oxidation of fatty acids but not by nutrient oversupply. *Endocrinology*, **143**, 1371–1377.

63 Kassis, N., Bernard, C., Pusterla, A. et al. (2000) Correlation between pancreatic islet uncoupling protein-2 (UCP2) mRNA concentration and insulin status in rats. *Int J Exp Diabetes Res*, **1**, 185–193.

64 Zhou, Y.T., Shimabukuro, M., Koyama, K. et al. (1997) Induction by leptin of uncoupling protein-2 and enzymes of fatty acid oxidation. *Proc Natl Acad Sci U S A*, **94**, 6386–6390.

65 Dubois, M., Vacher, P., Roger, B. et al. (2007) Glucotoxicity inhibits late steps of insulin exocytosis. *Endocrinology*, **148**, 1605–1614.

66 Olofsson, C.S., Collins, S., Bengtsson, M. et al. (2007) Long-term exposure to glucose and lipids inhibits glucose-induced insulin secretion downstream of granule fusion with plasma membrane. *Diabetes*, **56**, 1888–1897.

67 Tsuboi, T., Ravier, M.A., Parton, L.E., and Rutter, G.A. (2006) Sustained exposure to high glucose concentrations modifies glucose signaling and the mechanics of secretory vesicle fusion in primary rat pancreatic β-cells. *Diabetes*, **55**, 1057–1065.

68 Maedler, K., Sergeev, P., Ris, F. et al. (2002) Glucose-induced beta cell production of IL-1beta contributes to glucotoxicity in human pancreatic islets. *J Clin Invest*, **110**, 851–860.

69 Leibowitz, G., Yuli, M., Donath, M.Y. et al. (2001) β-cell glucotoxicity in the Psammomys obesus model of type 2 diabetes. *Diabetes*, **50** (Suppl 1), S113–S117.

70 Frayling, T.M. (2007) A new era in finding Type 2 diabetes genes-the unusual suspects. *Diabet Med*, **24**, 696–701.

71 Patane, G., Piro, S., Rabuazzo, A.M., Anello, M., Vigneri, R., and Purrello, F. (2000) Metformin restores insulin secretion altered by chronic exposure to free fatty acids or high glucose: a direct metformin effect on pancreatic β-cells. *Diabetes*, **49**, 735–740.

72 Elks, M.L. (1993) Chronic perifusion of rat islets with palmitate suppresses glucose-stimulated insulin release. *Endocrinology*, **133**, 208–214.

73 Moore, P.C., Ugas, M.A., Hagman, D.K., Parazzoli, S.D., and Poitout, V. (2004) Evidence against the involvement of oxidative stress in fatty acid inhibition of insulin secretion. *Diabetes*, **53**, 2610–2616.

74 Zhou, Y.P. and Grill, V. (1995) Long term exposure to fatty acids and ketones inhibits B-cell functions in human pancreatic islets of Langerhans. *J Clin Endocrinol Metab*, **80**, 1584–1590.

75 Zhou, Y.P. and Grill, V.E. (1994) Long-term exposure of rat pancreatic islets to fatty acids inhibits

76 Assimacopoulos-Jeannet, F., Thumelin, S., Roche, E., Esser, V., McGarry, J.D., and Prentki, M. (1997) Fatty acids rapidly induce the carnitine palmitoyltransferase I gene in the pancreatic β-cell line INS-1. *J Biol Chem*, **272**, 1659–1664.

77 Brun, T., Assimacopoulos-Jeannet, F., Corkey, B.E., and Prentki, M. (1997) Long-chain fatty acids inhibit acetyl-CoA carboxylase gene expression in the pancreatic β-cell line INS-1. *Diabetes*, **46**, 393–400.

78 Liang, Y., Buettger, C., Berner, D.K., and Matschinsky, F.M. (1997) Chronic effect of fatty acids on insulin release is not through the alteration of glucose metabolism in a pancreatic β-cell line (beta HC9). *Diabetologia*, **40**, 1018–1027.

79 Wang, X., Li, H., De Leo, D. et al. (2004) Gene and protein kinase expression profiling of reactive oxygen species-associated lipotoxicity in the pancreatic β-cell line MIN6. *Diabetes*, **53**, 129–140.

80 Magnan, C., Collins, S., Berthault, M.F. et al. (1999) Lipid infusion lowers sympathetic nervous activity and leads to increased β-cell responsiveness to glucose. *J Clin Invest*, **103**, 413–419.

81 Magnan, C., Cruciani, C., Clement, L. et al. (2001) Glucose-induced insulin hypersecretion in lipid-infused healthy subjects is associated with a decrease in plasma norepinephrine concentration and urinary excretion. *J Clin Endocrinol Metab*, **86**, 4901–4907.

82 Boden, G., Chen, X., Rosner, J., and Barton, M. (1995) Effects of a 48-h fat infusion on insulin secretion and glucose utilization. *Diabetes*, **44**, 1239–1242.

83 Storgaard, H., Jensen, C.B., Vaag, A.A., Volund, A., and Madsbad, S. (2003) Insulin secretion after short- and long-term low-grade free fatty acid infusion in men with increased risk of developing type 2 diabetes. *Metabolism*, **52**, 885–894.

84 Stefan, N., Wahl, H.G., Fritsche, A., Haring, H., and Stumvoll, M. (2001) Effect of the pattern of elevated free fatty acids on insulin sensitivity and insulin secretion in healthy humans. *Horm Metab Res*, **33**, 432–438.

85 Carpentier, A., Mittelman, S.D., Lamarche, B., Bergman, R.N., Giacca, A., and Lewis, G.F. (1999) Acute enhancement of insulin secretion by FFA in humans is lost with prolonged FFA elevation. *Am J Physiol*, **276**, E1055–E1066.

86 Mason, T.M., Goh, T., Tchipashvili, V. et al. (1999) Prolonged elevation of plasma free fatty acids desensitizes the insulin secretory response to glucose in vivo in rats. *Diabetes*, **48**, 524–530.

87 Paolisso, G., Gambardella, A., Amato, L. et al. (1995) Opposite effects of short-and long-term fatty acid infusion on insulin secretion in healthy subjects. *Diabetologia*, **38**, 1295–1299.

88 Briaud, I., Harmon, J.S., Kelpe, C.L., Segu, V.B., and Poitout, V. (2001) Lipotoxicity of the pancreatic β-cell is associated with glucose-dependent esterification of fatty acids into neutral lipids. *Diabetes*, **50**, 315–321.

89 Jacqueminet, S., Briaud, I., Rouault, C., Reach, G., and Poitout, V. (2000) Inhibition of insulin gene expression by long-term exposure of pancreatic beta cells to palmitate is dependent on the presence of a stimulatory glucose concentration. *Metabolism*, **49**, 532–536.

90 Ritz-Laser, B., Meda, P., Constant, I. et al. (1999) Glucose-induced preproinsulin gene expression is inhibited by the free fatty acid palmitate. *Endocrinology*, **140**, 4005–4014.

91 Hagman, D.K., Hays, L.B., Parazzoli, S.D., and Poitout, V. (2005) Palmitate inhibits insulin gene expression by altering PDX-1 nuclear localization and reducing MafA expression in isolated rat islets of Langerhans. *J Biol Chem*, **280**, 32413–32418.

92 Kelpe, C.L., Moore, P.C., Parazzoli, S.D., Wicksteed, B., Rhodes, C.J., and Poitout, V. (2003) Palmitate inhibition of insulin gene expression is mediated at the transcriptional level via ceramide synthesis. *J Biol Chem*, **278**, 30015–30021.

93 Summers, S.A., Garza, L.A., Zhou, H., and Birnbaum, M.J. (1998) Regulation of insulin-stimulated glucose transporter GLUT4 translocation and Akt kinase activity by ceramide. *Mol Cell Biol*, **18**, 5457–5464.

94 Whiteman, E.L., Cho, H., and Birnbaum, M.J. (2002) Role of Akt/protein kinase B in metabolism. *Trends Endocrinol Metab*, **13**, 444–451.

95 Kitamura, T., Nakae, J., Kitamura, Y. et al. (2002) The forkhead transcription factor Foxo1 links insulin signaling to Pdx1 regulation of pancreatic beta cell growth. *J Clin Invest*, **110**, 1839–1847.

96 Mathias, S., Pena, L.A., and Kolesnick, R.N. (1998) Signal transduction of stress via ceramide. *Biochem J*, **335** (Pt 3), 465–480.

97 Bollheimer, L.C., Skelly, R.H., Chester, M.W., McGarry, J.D., and Rhodes, C.J. (1998) Chronic exposure to free fatty acid reduces pancreatic beta cell insulin content by increasing basal insulin secretion that is not compensated for by a corresponding increase in proinsulin biosynthesis translation. *J Clin Invest*, **101**, 1094–1101.

98 Kharroubi, I., Ladriere, L., Cardozo, A.K., Dogusan, Z., Cnop, M., and Eizirik, D.L. (2004) Free fatty acids and cytokines induce pancreatic β-cell apoptosis by different mechanisms: role of nuclear factor-κB and endoplasmic reticulum stress. *Endocrinology*, **145**, 5087–5096.

99 Karaskov, E., Scott, C., Zhang, L., Teodoro, T., Ravazzola, M., and Volchuk, A. (2006) Chronic palmitate but not oleate exposure induces endoplasmic reticulum stress, which may contribute to INS-1 pancreatic β-cell apoptosis. *Endocrinology*, **147**, 3398–3407.

100 Laybutt, D.R., Preston, A.M., Akerfeldt, M.C. et al. (2007) Endoplasmic reticulum stress contributes to beta cell apoptosis in type 2 diabetes. *Diabetologia*, **50**, 752–763.

101 Furukawa, H., Carroll, R.J., Swift, H.H., and Steiner, D.F. (1999) Long-term elevation of free fatty acids leads to delayed processing of proinsulin and prohormone convertases 2 and 3 in the pancreatic β-cell line MIN6. *Diabetes*, **48**, 1395–1401.

102 Jeffrey, K.D., Alejandro, E.U., Luciani, D.S. et al. (2008) Carboxypeptidase E mediates palmitate-induced β-cell ER stress and apoptosis. *Proc Natl Acad Sci U S A*, **105**, 8452–8457.

103 Randle, P.J., Kerbey, A.L., and Espinal, J. (1988) Mechanisms decreasing glucose oxidation in diabetes and starvation: role of lipid fuels and hormones. *Diabetes Metab Rev*, **4**, 623–638.

104 Zhou, Y.P. and Grill, V.E. (1995) Palmitate-induced β-cell insensitivity to glucose is coupled to decreased pyruvate dehydrogenase activity and enhanced kinase activity in rat pancreatic islets. *Diabetes*, **44**, 394–399.

105 Segall, L., Lameloise, N., Assimacopoulos-Jeannet, F. et al. (1999) Lipid rather than glucose metabolism is implicated in altered insulin secretion caused by oleate in INS-1 cells. *Am J Physiol*, **277**, E521–E528.

106 Lameloise, N., Muzzin, P., Prentki, M., and Assimacopoulos-Jeannet, F. (2001) Uncoupling protein 2: a possible link between fatty acid excess and impaired glucose-induced insulin secretion? *Diabetes*, **50**, 803–809.

107 Koshkin, V., Wang, X., Scherer, P.E., Chan, C.B., and Wheeler, M.B. (2003) Mitochondrial functional state in clonal pancreatic β-cells exposed to free fatty acids. *J Biol Chem*, **278**, 19709–19715.

108 Medvedev, A.V., Robidoux, J., Bai, X. et al. (2002) Regulation of the uncoupling protein-2 gene in INS-1

β-cells by oleic acid. *J Biol Chem*, **277**, 42639–42644.

109 Joseph, J.W., Koshkin, V., Saleh, M.C. et al. (2004) Free fatty acid-induced β-cell defects are dependent on uncoupling protein 2 expression. *J Biol Chem*, **279**, 51049–51056.

110 Joseph, J.W., Koshkin, V., Zhang, C.Y. et al. (2002) Uncoupling protein 2 knockout mice have enhanced insulin secretory capacity after a high-fat diet. *Diabetes*, **51**, 3211–3219.

111 Kawai, T., Hirose, H., Seto, Y., Fujita, H., and Saruta, T. (2001) Chronic effects of different fatty acids and leptin in INS-1 cells. *Diabetes Res Clin Pract*, **51**, 1–8.

112 Maedler, K., Spinas, G.A., Dyntar, D., Moritz, W., Kaiser, N., and Donath, M.Y. (2001) Distinct effects of saturated and monounsaturated fatty acids on β-cell turnover and function. *Diabetes*, **50**, 69–76.

113 Maedler, K., Oberholzer, J., Bucher, P., Spinas, G.A., and Donath, M.Y. (2003) Monounsaturated fatty acids prevent the deleterious effects of palmitate and high glucose on human pancreatic β-cell turnover and function. *Diabetes*, **52**, 726–733.

114 El-Assaad, W., Buteau, J., Peyot, M.L. et al. (2003) Saturated fatty acids synergize with elevated glucose to cause pancreatic β-cell death. *Endocrinology*, **144**, 4154–4163.

115 Cnop, M., Hannaert, J.C., Hoorens, A., Eizirik, D.L., and Pipeleers, D.G. (2001) Inverse relationship between cytotoxicity of free fatty acids in pancreatic islet cells and cellular triglyceride accumulation. *Diabetes*, **50**, 1771–1777.

116 Kolesnick, R.N. and Kronke, M. (1998) Regulation of ceramide production and apoptosis. *Annu Rev Physiol*, **60**, 643–665.

117 Shimabukuro, M., Higa, M., Zhou, Y.T., Wang, M.Y., Newgard, C.B., and Unger, R.H. (1998) Lipoapoptosis in β-cells of obese prediabetic fa/fa rats. Role of serine palmitoyl-transferase overexpression. *J Biol Chem*, **273**, 32487–32490.

118 Shimabukuro, M., Zhou, Y.T., Levi, M., and Unger, R.H. (1998) Fatty acid-induced β cell apoptosis: a link between obesity and diabetes. *Proc Natl Acad Sci U S A*, **95**, 2498–2502.

119 Lupi, R., Dotta, F., Marselli, L. et al. (2002) Prolonged exposure to free fatty acids has cytostatic and pro-apoptotic effects on human pancreatic islets: evidence that β-cell death is caspase mediated, partially dependent on ceramide pathway, and Bcl-2 regulated. *Diabetes*, **51**, 1437–1442.

120 Garcia-Ruiz, C., Colell, A., Mari, M., Morales, A., and Fernandez-Checa, J.C. (1997) Direct effect of ceramide on the mitochondrial electron transport chain leads to generation of reactive oxygen species. Role of mitochondrial glutathione. *J Biol Chem*, **272**, 11369–11377.

121 Gottlieb, E., Vander Heiden, M.G., and Thompson, C.B. (2000) Bcl-x(L) prevents the initial decrease in mitochondrial membrane potential and subsequent reactive oxygen species production during tumor necrosis factor α-induced apoptosis. *Mol Cell Biol*, **20**, 5680–5689.

122 Eitel, K., Staiger, H., Rieger, J. et al. (2003) Protein kinase C delta activation and translocation to the nucleus are required for fatty acid-induced apoptosis of insulin-secreting cells. *Diabetes*, **52**, 991–997.

123 Schmitz-Peiffer, C., Laybutt, D.R., Burchfield, J.G. et al. (2007) Inhibition of PKC epsilon improves glucose-stimulated insulin secretion and reduces insulin clearance. *Cell Metab*, **6**, 320–328.

124 Prentki M., Corkey B.E. (1996) Are the β-cell signaling molecules malonyl-CoA and cystolic long-chain acyl-CoA implicated in multiple tissue defects of obesity and NIDDM? *Diabetes* **45**, 273–283.

125 Roche, E., Farfari, S., Witters, L.A. et al. (1998) Long-term exposure

of β-INS cells to high glucose concentrations increases anaplerosis, lipogenesis, and lipogenic gene expression. *Diabetes*, **47**, 1086–1094.

126 Foufelle, F. and Ferre, P. (2002) New perspectives in the regulation of hepatic glycolytic and lipogenic genes by insulin and glucose: a role for the transcription factor sterol regulatory element binding protein-1c. *Biochem J*, **366**, 377–391.

127 Shimomura, I., Bashmakov, Y. and Horton, J.D. (1999) Increased levels of nuclear SREBP-1c associated with fatty livers in two mouse models of diabetes mellitus. *J Biol Chem*, **274**, 30028–30032.

128 Sandberg, M.B., Fridriksson, J., Madsen, L. et al. (2005) Glucose-induced lipogenesis in pancreatic β-cells is dependent on SREBP-1. *Mol Cell Endocrinol*, **240**, 94–106.

129 Rhee, S.G. (2006) Cell signaling. H2O2, a necessary evil for cell signaling. *Science*, **312**, 1882–1883.

130 Goldstein, B.J., Mahadev, K., Wu, X., Zhu, L., and Motoshima, H. (2005) Role of insulin-induced reactive oxygen species in the insulin signaling pathway. *Antioxid Redox Signal*, **7**, 1021–1031.

131 Pi, J., Bai, Y., Zhang, Q. et al. (2007) Reactive oxygen species as a signal in glucose-stimulated insulin secretion. *Diabetes*, **56**, 1783–1791.

132 Inoguchi, T., Li, P., and Umeda, F. et al. (2000) High glucose level and free fatty acid stimulate reactive oxygen species production through protein kinase C–dependent activation of NAD(P)H oxidase in cultured vascular cells. *Diabetes*, **49**, 1939–1945.

133 Evans, J.L., Goldfine, I.D., Maddux, B.A., and Grodsky, G.M. (2002) Oxidative stress and stress-activated signaling pathways: a unifying hypothesis of type 2 diabetes. *Endocr Rev*, **23**, 599–622.

134 Gopaul, N.K., Anggard, E.E., Mallet, A.I., Betteridge, D.J., Wolff, S.P., and Nourooz-Zadeh, J. (1995) Plasma 8-epi-PGF2 alpha levels are elevated in individuals with non-insulin dependent diabetes mellitus. *FEBS Lett*, **368**, 225–229.

135 Nourooz-Zadeh, J., Tajaddini-Sarmadi, J., McCarthy, S., Betteridge, D.J., and Wolff, S.P. (1995) Elevated levels of authentic plasma hydroperoxides in NIDDM. *Diabetes*, **44**, 1054–1058.

136 Shin, C.S., Moon, B.S., Park, K.S. et al. (2001) Serum 8-hydroxyguanine levels are increased in diabetic patients. *Diabetes Care*, **24**, 733–737.

137 Murakami, K., Kondo, T., Ohtsuka, Y., Fujiwara, Y., Shimada, M., and Kawakami, Y. (1989) Impairment of glutathione metabolism in erythrocytes from patients with diabetes mellitus. *Metabolism*, **38**, 753–758.

138 Grinberg, L., Fibach, E., Amer, J., and Atlas, D. (2005) N-acetylcysteine amide, a novel cell-permeating thiol, restores cellular glutathione and protects human red blood cells from oxidative stress. *Free Radic Biol Med*, **38**, 136–145.

139 Lenzen, S., Drinkgern, J., and Tiedge, M. (1996) Low antioxidant enzyme gene expression in pancreatic islets compared with various other mouse tissues. *Free Radic Biol Med*, **20**, 463–466.

140 Tiedge, M., Lortz, S., Drinkgern, J., and Lenzen, S. (1997) Relation between antioxidant enzyme gene expression and antioxidative defense status of insulin-producing cells. *Diabetes*, **46**, 1733–1742.

141 Carlsson, C., Borg, L.A., and Welsh, N. (1999) Sodium palmitate induces partial mitochondrial uncoupling and reactive oxygen species in rat pancreatic islets in vitro. *Endocrinology*, **140**, 3422–3428.

142 Miwa, I., Ichimura, N., Sugiura, M., Hamada, Y., and Taniguchi, S. (2000) Inhibition of glucose-induced insulin secretion by 4-hydroxy-2-nonenal and other lipid peroxidation products. *Endocrinology*, **141**, 2767–2772.

143 Maechler, P., Jornot, L., and Wollheim, C.B. (1999) Hydrogen

peroxide alters mitochondrial activation and insulin secretion in pancreatic beta cells. *J Biol Chem*, **274**, 27905–27913.

144 Kaneto, H., Xu, G., Fujii, N., Kim, S., Bonner-Weir, S., and Weir, G.C. (2002) Involvement of c-Jun N-terminal kinase in oxidative stress-mediated suppression of insulin gene expression. *J Biol Chem*, **277**, 30010–30018.

145 Sakai, K., Matsumoto, K., Nishikawa, T. et al. (2003) Mitochondrial reactive oxygen species reduce insulin secretion by pancreatic β-cells. *Biochem Biophys Res Commun*, **300**, 216–222.

146 Kaneto, H., Xu, G., Song, K.H. et al. (2001) Activation of the hexosamine pathway leads to deterioration of pancreatic β-cell function through the induction of oxidative stress. *J Biol Chem*, **276**, 31099–31104.

147 Du, X.L., Edelstein, D., Rossetti, L. et al. (2000) Hyperglycemia-induced mitochondrial superoxide overproduction activates the hexosamine pathway and induces plasminogen activator inhibitor-1 expression by increasing Sp1 glycosylation. *Proc Natl Acad Sci U S A*, **97**, 12222–12226.

148 Kunisaki, M., Bursell, S.E., Umeda, F., Nawata, H., and King, G.L. (1994) Normalization of diacylglycerol-protein kinase C activation by vitamin E in aorta of diabetic rats and cultured rat smooth muscle cells exposed to elevated glucose levels. *Diabetes*, **43**, 1372–1377.

149 Liu, B. and Hannun, Y.A. (1997) Inhibition of the neutral magnesium-dependent sphingomyelinase by glutathione. *J Biol Chem*, **272**, 16281–16287.

150 Azevedo-Martins, A.K., Lortz, S., Lenzen, S., Curi, R., Eizirik, D.L., and Tiedge, M. (2003) Improvement of the mitochondrial antioxidant defense status prevents cytokine-induced nuclear factor-κB activation in insulin-producing cells. *Diabetes*, **52**, 93–101.

151 Kaneto, H., Kajimoto, Y., Miyagawa, J. et al. (1999) Beneficial effects of antioxidants in diabetes: possible protection of pancreatic β-cells against glucose toxicity. *Diabetes*, **48**, 2398–2406.

152 Ihara, Y., Yamada, Y., Toyokuni, S. et al. (2000) Antioxidant alpha-tocopherol ameliorates glycemic control of GK rats, a model of type 2 diabetes. *FEBS Lett*, **473**, 24–26.

153 Brons, C., Spohr, C., Storgaard, H., Dyerberg, J., and Vaag, A. (2004) Effect of taurine treatment on insulin secretion and action, and on serum lipid levels in overweight men with a genetic predisposition for type II diabetes mellitus. *Eur J Clin Nutr*, **58**, 1239–1247.

154 De Mattia, G., Bravi, M.C., Laurenti, O. et al. (1998) Influence of reduced glutathione infusion on glucose metabolism in patients with non-insulin-dependent diabetes mellitus. *Metabolism*, **47**, 993–997.

155 Del Guerra, S., Lupi, R., Marselli, L. et al. (2005) Functional and molecular defects of pancreatic islets in human type 2 diabetes. *Diabetes*, **54**, 727–735.

156 Tanaka, Y., Tran, P.O., Harmon, J., and Robertson, R.P. (2002) A role for glutathione peroxidase in protecting pancreatic beta cells against oxidative stress in a model of glucose toxicity. *Proc Natl Acad Sci U S A*, **99**, 12363–12368.

157 Tran, P.O., Parker, S.M., LeRoy, E. et al. (2004) Adenoviral overexpression of the glutamylcysteine ligase catalytic subunit protects pancreatic islets against oxidative stress. *J Biol Chem*, **279**, 53988–53993.

158 Martens, G.A., Cai, Y., Hinke, S., Stange, G., Van de Casteele, M., and Pipeleers, D. (2005) Glucose suppresses superoxide generation in metabolically responsive pancreatic beta cells. *J Biol Chem*, **280**, 20389–20396.

159 Tang, C., Han, P., Oprescu, A.I. et al. (2007) Evidence for a role of superoxide generation in glucose-induced β-cell dysfunction in vivo. *Diabetes*, **56**, 2722–2731.

160 Harmon, J.S., Stein, R., and Robertson, R.P. (2005) Oxidative stress-mediated, post-translational loss of MafA protein as a contributing mechanism to loss of insulin gene expression in glucotoxic beta cells. *J Biol Chem*, **280**, 11107–11113.

161 Takahashi, H., Tran, P.O., LeRoy, E., Harmon, J.S., Tanaka, Y., and Robertson, R.P. (2004) D-Glyceraldehyde causes production of intracellular peroxide in pancreatic islets, oxidative stress, and defective beta cell function via non-mitochondrial pathways. *J Biol Chem*, **279**, 37316–37323.

162 Oprescu, A.I., Bikopoulos, G., Naassan, A. *et al.* (2007) Free fatty acid-induced reduction in glucose stimulated insulin secretion evidence for a role of oxidative stress in vitro and in vivo. *Diabetes*, **56**, 2027–2937.

163 Xiao, C., Giacca, A., and Lewis, G.F. (2007) Oral taurine but not N-acetylcysteine ameliorates non-esterified fatty acid-induced impairment in insulin sensitivity and beta cell function in obese and overweight, non-diabetic men, *Diabetologia*. **51**, 139–146.

164 Li, L.X., Skorpen, F., Egeberg, K., Jorgensen, I.H., and Grill, V. (2001) Uncoupling protein-2 participates in cellular defense against oxidative stress in clonal β-cells. *Biochem Biophys Res Commun*, **282**, 273–277.

165 Pecqueur, C., Alves-Guerra, M.C., Gelly, C. *et al.* (2001) Uncoupling protein 2, in vivo distribution, induction upon oxidative stress, and evidence for translational regulation. *J Biol Chem*, **276**, 8705–8712.

166 Echtay, K.S., Esteves, T.C., Pakay, J.L. *et al.* (2003) A signalling role for 4-hydroxy-2-nonenal in regulation of mitochondrial uncoupling. *EMBO J*, **22**, 4103–4110.

167 Andreozzi, F., D'Alessandris, C., Federici, M. *et al.* (2004) Activation of the hexosamine pathway leads to phosphorylation of insulin receptor substrate-1 on Ser307 and Ser612 and impairs the phosphatidylinositol 3-kinase/Akt/mammalian target of rapamycin insulin biosynthetic pathway in RIN pancreatic β-cells. *Endocrinology*, **145**, 2845–2857.

168 Shi, Y., Taylor, S.I., Tan, S.L., and Sonenberg, N. (2003) When translation meets metabolism: multiple links to diabetes. *Endocr Rev*, **24**, 91–101.

169 Boucher, M.J., Selander, L., Carlsson, L., and Edlund, H. (2006) Phosphorylation marks IPF1/Pdx1 protein for degradation by glycogen synthase kinase 3-dependent mechanisms. *J Biol Chem*, **281**, 6395–6403.

170 Frescas, D., Valenti, L., and Accili, D. (2005) Nuclear trapping of the forkhead transcription factor FoxO1 via Sirt-dependent deacetylation promotes expression of glucogenetic genes. *J Biol Chem*, **280**, 20589–20595.

171 Essers, M.A., Vries-Smits, L.M., Barker, N., Polderman, P.E., Burgering, B.M., and Korswagen, H.C. (2005) Functional interaction between β-catenin and FOXO in oxidative stress signaling. *Science*, **308**, 1181–1184.

172 Kawamori, D., Kaneto, H., Nakatani, Y. *et al.* (2006) The forkhead transcription factor Foxo1 bridges the JNK pathway and the transcription factor PDX-1 through its intracellular translocation. *J Biol Chem*, **281**, 1091–1098.

173 Essers, M.A., Weijzen, S., Vries-Smits, A.M. *et al.* (2004) FOXO transcription factor activation by oxidative stress mediated by the small GTPase Ral and JNK. *EMBO J*, **23**, 4802–4812.

174 Kaneto, H., Nakatani, Y., Miyatsuka, T. *et al.* (2004) Possible novel therapy for diabetes with cell-permeable JNK-inhibitory peptide. *Nat Med*, **10**, 1128–1132.

175 Hammar, E.B., Irminger, J.C., Rickenbach, K. *et al.* (2005) Activation of NF-kappaB by extracellular matrix is involved in spreading and glucose-stimulated insulin secretion

of pancreatic beta cells. *J Biol Chem*, **280**, 30630–30637.

176 Norlin, S., Ahlgren, U., and Edlund, H. (2005) Nuclear factor-κB activity in {beta}-cells is required for glucose-stimulated insulin secretion. *Diabetes*, **54**, 125–132.

177 Giannoukakis, N., Rudert, W.A., Trucco, M., and Robbins, P.D. (2000) Protection of human islets from the effects of interleukin-1beta by adenoviral gene transfer of an Ikappa B repressor. *J Biol Chem*, **275**, 36509–36513.

178 Rehman, K.K., Bertera, S., Bottino, R. et al. (2003) Protection of islets by in situ peptide-mediated transduction of the Ikappa B kinase inhibitor Nemo-binding domain peptide. *J Biol Chem*, **278**, 9862–9868.

179 Zeender, E., Maedler, K., Bosco, D., Berney, T., Donath, M.Y., and Halban, P.A. (2004) Pioglitazone and sodium salicylate protect human β-cells against apoptosis and impaired function induced by glucose and interleukin-1beta. *J Clin Endocrinol Metab*, **89**, 5059–5066.

180 Laybutt, D.R., Kaneto, H., Hasenkamp, W. et al. (2002) Increased expression of antioxidant and antiapoptotic genes in islets that may contribute to β-cell survival during chronic hyperglycemia. *Diabetes*, **51**, 413–423.

181 Chen, M. and Robertson, R.P. (1978) Restoration of the acute insulin response by sodium salicylate. A glucose dose-related phenomenon. *Diabetes*, **27**, 750–756.

182 Elouil, H., Cardozo, A.K., Eizirik, D.L., Henquin, J.C., and Jonas, J.C. (2005) High glucose and hydrogen peroxide increase c-Myc and haeme-oxygenase 1 mRNA levels in rat pancreatic islets without activating NFkappaB. *Diabetologia*, **48**, 496–505.

183 Rakatzi, I., Mueller, H., Ritzeler, O., Tennagels, N., and Eckel, J. (2004) Adiponectin counteracts cytokine-and fatty acid-induced apoptosis in the pancreatic β-cell line INS-1. *Diabetologia*, **47**, 249–258.

184 Fernandez-Real, J.M., Lopez-Bermejo, A., Ropero, A.B. et al. (2008) Salicylates increase insulin secretion in healthy obese subjects. *J Clin Endocrinol Metab*, **93**, 2523–2530.

185 Kulkarni, R.N. (2002) Receptors for insulin and insulin-like growth factor-1 and insulin receptor substrate-1 mediate pathways that regulate islet function. *Biochem Soc Trans*, **30**, 317–322.

186 Aspinwall, C.A., Lakey, J.R., and Kennedy, R.T. (1999) Insulin-stimulated insulin secretion in single pancreatic beta cells. *J Biol Chem*, **274**, 6360–6365.

187 Borge, P.D., Moibi, J., and Greene, S.R. et al. (2002) Insulin receptor signaling and sarco/endoplasmic reticulum calcium ATPase in β-cells. *Diabetes*, **51** (Suppl 3), S427–S433.

188 Persaud, S.J., Asare-Anane, H., and Jones, P.M. (2002) Insulin receptor activation inhibits insulin secretion from human islets of Langerhans. *FEBS Lett*, **510**, 225–228.

189 Khan, F.A., Goforth, P.B., Zhang, M., and Satin, L.S. (2001) Insulin activates ATP-sensitive K(+) channels in pancreatic β-cells through a phosphatidylinositol 3-kinase-dependent pathway. *Diabetes*, **50**, 2192–2198.

190 Kulkarni, R.N., Bruning, J.C., Winnay, J.N., Postic, C., Magnuson, M.A., and Kahn, C.R. (1999) Tissue-specific knockout of the insulin receptor in pancreatic beta cells creates an insulin secretory defect similar to that in type 2 diabetes. *Cell*, **96**, 329–339.

191 Leibiger, B., Leibiger, I.B., Moede, T. et al. (2001) Selective insulin signaling through A and B insulin receptors regulates transcription of insulin and glucokinase genes in pancreatic beta cells. *Mol Cell*, **7**, 559–570.

192 Xu, G.G. and Rothenberg, P.L. (1998) Insulin receptor signaling in the β-cell influences insulin gene expression and insulin content: evidence for autocrine β-cell regulation. *Diabetes*, **47**, 1243–1252.

193 Dickson, L.M., Lingohr, M.K., McCuaig, J. et al. (2001) Differential activation of protein kinase B and p70(S6)K by glucose and insulin-like growth factor 1 in pancreatic β-cells (INS-1). *J Biol Chem*, **276**, 21110–21120.

194 Hribal, M.L., Perego, L., Lovari, S. et al. (2003) Chronic hyperglycemia impairs insulin secretion by affecting insulin receptor expression, splicing, and signaling in RIN beta cell line and human islets of Langerhans. *FASEB J*, **17**, 1340–1342.

195 Wrede, C.E., Dickson, L.M., Lingohr, M.K., Briaud, I., and Rhodes, C.J. (2003) Fatty acid and phorbol ester-mediated interference of mitogenic signaling via novel protein kinase C isoforms in pancreatic β-cells (INS-1). *J Mol Endocrinol*, **30**, 271–286.

196 Solinas, G., Naugler, W., Galimi, F., Lee, M.S., and Karin, M. (2006) Saturated fatty acids inhibit induction of insulin gene transcription by JNK-mediated phosphorylation of insulin-receptor substrates. *Proc Natl Acad Sci U S A*, **103**, 16454–16459.

197 Martinez, S.C., Tanabe, K., Cras-Meneur, C., Abumrad, N.A., Bernal-Mizrachi, E., and Permutt, M.A. (2008) Inhibition of Foxo1 protects pancreatic islet β-cells against fatty acid and endoplasmic reticulum stress-induced apoptosis. *Diabetes*, **57**, 846–859.

198 Zhang, K. and Kaufman, R.J. (2008) From endoplasmic-reticulum stress to the inflammatory response. *Nature*, **454**, 455–462.

22
Endogenous Toxins and Susceptibility or Resistance to Diabetic Complications

Paul J. Beisswenger

Substantial evidence is accumulating that susceptibility to diabetic complications relates to the capacity of biological protective systems to protect against the cellular toxicity of glycation and oxidation products. The currently available data on this important area of medical investigation has been reviewed to provide the reader with the background to understand new data, as it accrues, in this rapidly changing field.

22.1
Introduction and Background

One hundred and eighty million people worldwide now suffer from the devastating effects of diabetes, including 22 million people in the United States and 48 million in Europe, and the number is rising rapidly. The annual cost of diabetes in this country, mostly related to treating its complications, was $172 billion in 2007. Since the discovery of insulin, people rarely die of diabetes. They now die of the complications of the disease, with diabetes being the leading cause of kidney failure, blindness, and amputations in the United States. Diabetes is also a major cause of heart attack and stroke, which cause the greatest number of diabetes related deaths.

22.1.1
Elevated Glucose, Pathways of Glycation/Oxidation, and Diabetic Complications

High glucose levels can lead to cellular dysfunction by increasing the levels of glycation and oxidative damage to cellular and plasma proteins. One of the most important processes associated with hyperglycemia is nonenzymatic glycation, which has gained increasing acceptance as a significant mechanism contributing to the tissue damage in diabetes [1–4]. Nonenzymatic glycation involves a complex series of chemical reactions that lead to the formation of early glycation products (EGPs), including Amadori products such as fructose lysine (FL), N-terminal fructosyl compounds, and their reactive Schiff's bases. Other EGPs

Endogenous Toxins. Diet, Genetics, Disease and Treatment.
Edited by Peter J. O'Brien and W. Robert Bruce
Copyright © 2010 WILEY-VCH Verlag GmbH & Co. KGaA, Weinheim
ISBN: 978-3-527-32363-0

include highly reactive α-dicarbonyl compounds (glyoxal, methylglyoxal (MG), and 3-deoxyglucosone (3-DG)), which can be formed by glycolytic intermediates, lipid peroxidation, and the degradation of glycated proteins.

Oxidative stress (OS) is also initiated in diabetes and can produce direct tissue damage or lead to activation of the major pathways that produce diabetic complications [5–8]. Glucose-induced OS can lead to increased production of superoxide and free hydroxyl radicals, as well as peroxides. These reactive products can lead to direct tissue damage or to activation of the major pathways that can produce diabetic complications through oxidation of the enzyme, glyceraldehyde 3-phosphate dehydrogenase (GAPDH) in the glycolytic pathway, among others [5–8].

22.2
Synthetic Pathways for Glycation Products

Although nonenzymatic processes are responsible for the chemical reactions associated with immediate glycation reactions, enzymatic factors are known to control the production of the major precursors responsible for the formation of EGPs and advanced glycation end-products (AGEs). Thus, processes are determined by genetic factors that potentially vary among individuals, and can regulate the levels of Amadori and dicarbonyl products, while ultimately determining the levels of specific AGEs related to these precursors.

22.2.1
Amadori Products

The formation of Amadori products initially involves spontaneous reactions between reducing sugars, such as glucose, and amines on arginine or lysine. The initial chemical reaction is fairly rapid (minutes) and involves the formation of an unstable Schiff's base, which then undergoes a slower (hours) Amadori rearrangement to a stable ketoamine. The extent of this reaction depends not only on the nature and concentration of the reagents (sugars and amines), but also on temperature and time.

22.2.2
Methylglyoxal (MG)

MG is formed mainly by the degradation of triose phosphates, and also by the metabolism of ketone bodies, threonine degradation, and the fragmentation of glycated proteins. MG is directly toxic to tissues, is a precursor of AGEs [9, 10], and activates multiple biochemical pathways [8, 11] that further increase glycation and activate pathways that can lead to additional cellular dysfunction [12–16].

22.2.3
3-Deoxyglucosone (3DG)

3DG was initially shown to be produced from the Amadori product, FL [17], or from auto-oxidation of glucose [6]. A major source of this potent glycating agent is fructose-3-phosphate (F3P), which is produced from FL by a specific fructosamine-3-kinase (FN3K) [18], that is, discussed further below.

22.2.4
Glyoxal

Glyoxal, another two-carbon α-dicarbonyl, has received somewhat less attention in clinical studies than MG or 3DG, although it is a precursor of carboxymethyl lysine (CML), one of the earliest AGEs to be identified and quantified. Glyoxal can be produced from oxidation of glucose in the presence of trace metals, and *in vitro* experiments have implicated glyoxal as an intermediate in the browning and cross-linking of proteins by glucose under oxidative conditions [19].

22.2.5
Synthetic Pathways for Oxidation Products

Hyperglycemia can increase oxidative stress through both nonenzymatic and enzymatic processes [20]. Glucose can undergo auto-oxidation to free oxygen radicals in the presence of heavy metals [21], and glycated proteins (Amadori products or Schiff's bases [22]) can further undergo glycoxidation resulting in increased generation of reactive oxygen species. Mitochondrial generation of reactive oxygen species [5] is also a major source of superoxide and other free oxygen radicals. Hyperglycemia increases the production of reactive oxygen species by generating a proton gradient and a high electrochemical potential difference across the inner mitochondrial membrane. This results in overproduction of electron donors by the TCA cycle, which in turn, causes a marked increase in the production of superoxide by endothelial cells [23]. Excessive glucose metabolism can also produce reducing equivalents that fuel oxidative phosphorylation in mitochondria, the byproducts of which include free radicals such as superoxide anion [20].

Brownlee has proposed that this overproduction of superoxide by the mitochondrial electron-transport chain can activate four of the pathways proposed to lead to diabetic microvascular complications, and has written an excellent review that describes the data supporting this hypothesis [11]. Furthermore, increased flux through the sorbitol pathway is associated with above-normal superoxide formation [24] resulting from the generation of reducing equivalents in the form of unbound cytosolic NADH, which provides the source of electrons needed by several enzyme systems to generate superoxide.

In diabetes, free radical formation, regardless of the cause, results in damaged protein and mitochondrial DNA, which can have deleterious effects on the microvasculature. Substantial data is also available on the generation of nitrosative stress by hyperglycemia. Elevated glucose can lead to the simultaneous overgeneration of NO and superoxide, and lead to the overproduction the peroxynitrite anion, a potentially toxic reaction product in its own right [25]. Peroxynitrite anion produces cytotoxicity by oxidizing sulfhydryl groups in proteins, and initiating lipid peroxidation, as well as oxidation of amino acids such as tyrosine, affecting many signal transduction pathways [26]. The production of peroxynitrite can be indirectly inferred by the presence of the more stable end product, nitrotyrosine, which can serve as a marker of this reaction in the clinical setting as well.

These programmed synthetic mechanisms could regulate production of EGPs, oxidation products, and ultimately advanced glycation and oxidation products in a given individual.

22.2.6
Formation of Advanced Glycation End-Products (AGEs)

Glycation of proteins occurs in all tissues and is a complex series of parallel and sequential reactions. Early stage reactions lead to the formation of EGP, FL, and other fructosamines, and later stage reactions form AGEs [9]. FL degrades slowly to form AGEs, while glyoxal, MG, and 3-DG are also potent glycating agents that react with lysyl and arginyl groups on proteins to form AGEs directly.

Methods that measure specific AGEs and stable oxidation products have now been developed and in recent collaborative studies we have shown the highest *in vivo* AGEs, quantitatively, are hydroimidazolones (HIs), derived from arginine residues modified by glyoxal, MG, and 3DG and include – G-H1, MG-H1, and 3DG-H, respectively [27–29]. The concentrations of these products are remarkably high *in vivo*, with up to 26% of cellular proteins having such a modification. Other quantifiable AGEs, most of which are also elevated in diabetes, include lysine-derived CML, carboxyethyl lysine (CEL), and specific lysine-derived cross-links GOLD (glyoxal-derived lysine dimer), MOLD (methylglyoxal-derived lysine dimer), and DOLD, (3-deoxyglucosone-derived lysine dimer) [10]. Quantitative markers of oxidation and nitration can also be measured, and include methionine sulfoxide (MetSO) and *N*-formylkynurenine (NFK), formed by the oxidation of methionine and tryptophan respectively [30–32], while 3-nitrotyrosine (3-NT) and dityrosine [33] represent markers of nitration damage to proteins.

22.3
Enzymatic Deglycation Pathways

Pathways exist that are capable of detoxifying most glycation and oxidation products, and variable activity could potentially lead to protection or susceptibility to complications by leading to increased or decreased accumulation of their precursors. More specifically, pathways have been identified that can remove FL (fructosamine-3-kinase or (FN3K)), α-dicarbonyls (glyoxalase, MG, and 3DG reductases), and oxidation products (catalase, superoxide dismutases) [34, 35].

22.3.1
Deglycation Systems for Amadori Products

Deglycation systems for Amadori products are known, based on the discovery of a unique enzyme, FN3K, to deglycate protein-bound FL [18]. The seminal work on this enzyme was done in our laboratory by Dr. Szwergold, who has characterized and purified this enzyme to homogeneity [36]. It effectively removes Amadori adducts from glycated proteins and regenerates free lysine groups. FN3K can play an active role in preventing glycation in virtually every mammalian tissue that has been examined to date [37], and variable FN3K activity could potentially play an important role in determining susceptibility to diabetic complications.

FN3K removes fructose from lysine residues by phosphorylating FL residues on position 3 of the fructose moiety to fructoselysine-3-phosphate (FL3P) in the presence of ATP. This destabilizes the fructose-amine linkage, leading to spontaneous decomposition of FL3P to lysine, 3-DG, and inorganic phosphate [3, 5]. Because 3DG is a potent glycating sugar in its own right, the potential benefits of this system are dependent on the presence of processes that can subsequently remove this α-dicarbonyl. Enzymatic mechanisms that can achieve this are discussed below under "Detoxification Pathways for 3-Deoxyglucosone (3DG)."

Recently, an FN3K isoform, fructosamine-3-kinase–related protein (FN3KRP), has also been identified at a position 8 kb upstream of the *FN3K* gene on chromosome 17q25.3. FN3KRP has a 65% amino acid identity with FN3K and an identical genomic organization, suggesting that it is also involved in enzymatic deglycation. Unlike FN3K, however, FN3KRP does not phosphorylate FLs, but can phosphorylate other unusual ketosamines such as psicosamine and ribulosamine [38]. Because neither psicosamine nor ribulosamine are found in nature, the actual physiological substrate of FN3KRP remains unknown, but a reasonable assumption can be made that FN3KRP plays a role in deglycation of an as yet unidentified adduct of an aldose with a primary or secondary amine (probably a ketosamine). Finally, analysis of available DNA sequence data in both the genomic and expressed sequence tag (EST) databases revealed the presence of at least one gene highly homologous to FN3K in all vertebrates, as well as genes with lesser degrees of homology in many simpler organisms, such as *Arabidopsis thaliana* and *Caenorhabditis elegans*. Interestingly, no FN3K-homologous sequences have been detected in insects or yeasts.

22.3.2
Methylglyoxal (MG) Detoxification

MG is primarily catalyzed by the glyoxalase system, which leads to the generation of inert D-lactate as an end product [39, 40]. The glyoxalase system is present in the cytosol of all cells and catalyzes the conversion of reactive, acyclic α-oxoaldehydes into the corresponding α-hydroxyacids. It is composed of two enzymes, glyoxalase

I and glyoxalase II, and requires a catalytic amount of reduced glutathionine (GSH). Glyoxalase I catalyzes the isomerization of the hemithioacetal, formed spontaneously from α-oxoaldehyde (RCOCHO) and GSH, into S-2 hydroxyacylglutathione derivatives, while glyoxalase II catalyzes the conversion of S-2-hydroxyacylglutathione derivatives into α-hydroxyacids, and re-forms GSH consumed in the glyoxalase I-catalyzed reaction step. The major physiological substrate for glyoxalase I is MG, which can accumulate markedly when glyoxalase I is inhibited *in situ* by cell-permeable glyoxalase I inhibitors and by depletion of GSH [41, 42]. Other substrates are glyoxal (formed by lipid peroxidation and the fragmentation of glycated proteins), hydroypyruvaldehyde ($HOCH_2COCHO$), and 4,5-doxovalerate. MG can also be detoxified by an aldehyde reductase, leading to the production of acetol [43]; but this pathway appears to be minor relative to glyoxalase.

22.3.3
Detoxification Pathways for 3-Deoxyglucosone (3DG)

3DG detoxification to more innocuous products is also important to protect [44] against cellular damage by excessive glycation. Because the production of 3DG and its related AGEs is inevitable as long as glucose is utilized as a primary energy source, both oxidative and reductive pathways exist for its breakdown. These pathways can catabolize 3DG to harmless substances such as 3-deoxy 2-keto gluconic acid (3-DGA) or 3-deoxyfructose (3DF). The reductive pathway, mediated by aldose reductase, results in the production of 3DF, while a less well-characterized oxidative pathway catalyzed by oxoaldehyde dehydrogenase, leads to the production of 3-DGA. The major detoxification route for 3DG appears to occur by reduction to 3DF [45–47], and the potential importance of this pathway has been shown by the elevation of 3DF in diabetes [48], presumably secondary to the increased 3DG levels found in this condition [49–51].

22.3.4
Endogenous Binders of α-Dicarbonyls as Protective Mechanisms

Protection from the damaging effects of MG, 3DG, and glyoxal can be attenuated by endogenous and exogenous nucleophilic compounds that can react with these toxins and prevent glycation of endogenous cellular proteins [52, 53].

22.3.5
The Removal of AGEs

AGE-modified proteins are chemically damaged proteins, and as with other damaged molecules, such as oxidized DNA or lipoproteins, biological mechanisms have evolved for their recognition, and degradation. A number of cell surface AGE receptors have now been identified and appear to play an important role in the uptake and catabolism of AGE-proteins in plasma, multiple cell types, and the extracellular matrix. The best characterized among these are RAGE (receptor for advanced glycation end-product) [54, 55] and the macrophage scavenger receptor [56]. RAGE is widely distributed among cell types, including endothelial and smooth muscle cells and macrophages. RAGE amplifies the adverse effect of AGEs on cell viability, because binding of AGE-proteins to RAGE on cell surfaces induces an intracellular oxidative stress response *in vitro*, characterized by increased nuclear factor kappa-B (NFκB), and TNF-α levels [31, 32].

AGEs can also be removed and catabolized by circulating macrophages utilizing AGE receptor mediated processes [57]. This family of AGE receptors can bind to specific AGEs, and subsequently internalize and degrade to innocuous products. Thus, as opposed to RAGE, they can serve a useful function by removing proteins modified by glycation or oxidation. It has been documented that proteins modified by glycation, oxidation, and nitration are degraded by cellular proteolysis, releasing free protein glycation, oxidation, and nitration glycation adducts to maintain the quality and functional integrity of proteins. This cellular proteolysis liberates the glycated, oxidized, and nitrated amino acids as free adducts [27, 58] into blood plasma and excretes the products in urine. The changes in plasma concentrations and urinary excretion of glycation, oxidation, and nitration-free adducts may reflect tissue damage in diabetes, and can provide new markers indicative of the damaging effects of hyperglycemia.

22.3.6
Pathways that Lead to Removal of Oxidative Products

Multiple enzymatic pathways exist to protect against oxidative stress. Catalase, superoxide dismutases, and glutathione-peroxidase are some of the key enzymes that protect against the excessive hyperglycemia-associated oxidative stress associated with diabetes [6, 34, 59]. It has also been recognized that upregulation of these antioxidant enzymes may protect against oxidative stress associated with diabetes [34]. It has also been postulated that endogenous and exogenous antioxidants, such as reduced glutathione, carnosine, and, other nucleophilic receptor compounds can play an important role in protection against oxidative stress [60–62].

22.4
Susceptibility to Diabetic Complications Varies Widely among Individuals

It is well recognized that individuals with diabetes vary considerably in their propensity to develop diabetic nephropathy [63–65]. Thus, some individuals can develop diabetic nephropathy despite good glycemic control, while only a substantial subset (20–25%) of those with T1DM develop diabetic nephropathy despite the fact that the majority are exposed to at least some degree of long-term hyperglycemia. Family studies show strong concordance for the risk of diabetic nephropathy [63, 65–69], as well as concordance for the severity and patterns of diabetic glomerular structural lesions among type 1 diabetic sibling pairs [64], consistent with very important genetic contributions to risk. Further support for a variable response of the kidney to hyperglycemia comes from work showing that other less well-defined factors are of equal or greater importance to glycemia in influencing nephropathy risk [70].

22.4.1
Slow and Rapid Nonenzymatic Glycation and Propensity to Diabetic Complications

There is increasing support for the hypothesis that the degree of individual glycation/oxidation varies for a given blood glucose concentration, and that those with higher levels have a greater propensity to diabetic complications. A good example of this is that some individuals at the same mean blood glucose levels (MBG) have consistently higher A_{1c} levels (rapid glycators) and others consistently lower A_{1c} levels (slow glycators) [71–74]. A glycosylation index (HGI) (defined as the ratio of A_{1c} to blood glucose) has been used to document this significant between-individual variation in hemoglobin glycation among T1DM patients [75]. This index varies significantly between individual patients and is consistent in individuals, providing evidence that biological variation in A_{1c} is distinct from that attributable to MBG. This biological variation in A_{1c} has also been linked to both macro- and microvascular diabetes related pathology, with convincing evidence coming from the observation that variation in A_{1c}, which is distinct from that attributable to MBG, is a strong predictor of risk for diabetes complications in T1DM in the DCCT [76]. Therefore, elevated A_{1c} in diabetes could reflect two major components, one being glucose levels (MBG), and the other, a collection of as yet undetermined factors responsible for biological variation in A_{1c}.

22.4.2
Impaired Deglycation of Amadori Products and Diabetic Complications

It is apparent that less deglycation of Amadori products could lead to increased tissue glycation and in turn increase propensity to diabetic complications. Enzymes such as FN3K and FN3KRP, which appear to be involved in cellular repair, maintenance, and defense against toxins, have diverse transcriptional regulation, but are generally considered to be housekeeping genes. Because all cells are exposed to glucose and thus, can potentially be affected by the deleterious effects of the Maillard reaction, FN3K and FN3KRP could act as important and ubiquitous cellular defense mechanisms and could possibly be inducible by diabetes-associated conditions. Many housekeeping genes are constitutively active, while others can be upregulated on exposure to their toxic substrates. We have characterized some aspects of the function of the *FN3K* and *FN3KRP* genes, by analyzing their promoters in silico, and examining their transcriptional profile in representative human tissues. Our results indicate that the highest levels of FN3K expression are found in tissues known to be susceptible to glycation and diabetic complications, such as kidney and neurons. In addition to analyzing steady state levels of the two mRNAs, we have examined expression of these two genes in cultured fibroblasts subjected to conditions that mimic the cellular stresses encountered in diabetes such as hyperglycemia, hyperinsulinemia, increased interleukin 1 (IL-1), or NFκB activation. These experiments were performed because it is known that even some genes that are expressed constitutively may be induced further in response to external stimuli [14, 15]. Our results indicate, however, that neither FN3K nor FN3KRP responded to conditions associated with diabetes in this fibroblast system, and can therefore, appear to be purely housekeeping genes.

Our studies have shown that FN3K is essential for cell viability [77], suggesting that decreased intrinsic activity in diabetes could increase the risk of diabetic complications. We have shown that 72 hours of exposure of skin fibroblasts to siRNA, with an optimal oligonucleotide chosen as GGGAGAAGUUGAAGGAG-GATT, spanning the junction of exon 3 and 4 of the *FN3K* gene, decreased the amount of FN3K mRNA in the exposed cells by 75% relative to cells treated with scrambled siRNA [77, 78]. Concomitantly, we found that cell growth was similarly inhibited by 70% over three to four days, suggesting that the full protective effect of FN3K was necessary for cell viability.

Subsequent work with FN3K nondiabetic knockout (K/O) mice has provided further information on the protective role of FN3K against tissue glycation [79]. The K/O mouse showed 2.5-fold increase in hemoglobin glycation and increased levels of FL in all tissues that were examined. These animals, however, did not demonstrate any physical or histological tissue changes over their 16-month life span, relative to wild-type animals. Producing the diabetic state in these K/O animals will be necessary to determine if the development of diabetic complications are accelerated in the absence of FN3K activity.

Recent studies in human diabetic populations have also provided further information on the possible role of FN3K in the propensity to diabetic nephropathy. These studies have shown that cultured fibroblasts from subjects with biopsy documented progression of nephropathy have significant decreases in FN3K mRNA relative to nephropathy progressors and nondiabetic controls, and that exposure to elevated glucose concentrations *in vitro* significantly decreases FN3K mRNA message in all groups [80].

Although demonstrating decreased FN3K activity in human populations would confirm the significance of these observations, accurate direct quantification of enzymatic activity has been difficult to achieve. Our studies of the relationship between Amadori products and products reflecting FN3K activity *in vivo* however, have been informative. As noted above, we have postulated that variable activity of FN3K could play a role in susceptibility to diabetic nephropathy, and that one would expect to see evidence of reduced activity of this enzyme in those with susceptibility to diabetic complications. To address this question, we have used an indirect measure of FN3K activity by determining the production of 3DG, a major by-product of FL breakdown. HbA_{1c} quantitatively reflects levels of both FL and fructose-valine residues *in vivo*, so we examined the relationship between HbA_{1c} and 3DG in clinical samples to estimate the FN3K activity *in vivo*. In such an analysis, a high $HbA_{1c}/3DG$ ratio indicates a relatively low FN3K activity, while a low $HbA_{1c}/3DG$ ratio indicates higher FN3K activity.

When we examined the mean HbA_{1c}/3DG ratio in a population of 110 type 1 diabetic subjects and compared it to nondiabetic controls, we found that this ratio was in fact significantly higher in diabetic subjects. In addition, we have also observed that the slope of the regression analysis of HbA_{1c} versus 3DG was steeper for the control subjects relative to those with type 1 diabetes (0.56 µmol vs 0.010/%HbA_{1c}), indicating higher 3DG levels relative to their Amadori precursor in normoglycemic subjects than in the diabetic group, which could be explained by decreased FN3K activity in diabetes.

Furthermore, when we analyzed the relationship between HbA_{1c}/3DG ratios and progression of DN (fractional mesangial volume) for the "Natural History" population, we also observed that those with greater progression over five years had a significantly higher ratio, suggesting that lower FN3K activity was associated with greater progression of nephropathy in this population.

22.5
Overproduction of α-Dicarbonyls is Characteristic of Individuals Who Are Prone to Diabetic Complications

As previously discussed, we and others have postulated that the appearance of the toxic AGE precursors, MG and 3DG, is ultimately controlled by enzymatic mechanisms [39, 50, 81]. Both genetic and environmental factors could determine the production and detoxification of glycation products and potentially account for variable complication rates observed when individuals are exposed to variable degrees of hyperglycemia.

Our recent data has indicated that activation of pathways that determine the production of α-dicarbonyl compounds occurs in patients who are susceptible to diabetic nephropathy [35]. Since variable production of α-dicarbonyl compounds can result in increased levels of specific AGEs related to them, overproduction of these products is likely to influence propensity to diabetic complications. Our hypothesis is supported by studies showing that variable production of MG and 3DG occurs in individuals with type 1 diabetes, and that a direct relationship exists between levels of α-dicarbonyls and diabetic nephropathy in several well-characterized study populations [35].

More specifically, the role of α-dicarbonyl metabolism in the pathogenesis of diabetic nephropathy has been examined by investigating dicarbonyl production from *in vivo* samples obtained from the NHS "Natural History of diabetic Nephropathy Study," a group of 240 relatively healthy subjects with type 1 diabetes, where the degree of nephropathy was documented by electron microscopy of renal biopsy specimens spaced five years apart [82, 83]. In these studies we found that the rate of increase of glomerular basement membrane (GBM) thickness, the primary early marker of diabetic nephropathy, was significantly associated with blood and urinary levels of MG ($p = 0.03$), its major metabolite, D-lactate ($p = 0.03$), and 3DG ($p = 0.004$).

We have subsequently confirmed these observations in a second study population consisting of 45 Pima Indians with type 2 diabetes, who underwent careful clinical evaluation and renal biopsies to quantify their degree of diabetic nephropathy. Our measurements of plasma levels of MG in this population showed a strong relationship with GBM width ($R=0.37$, $p=0.016$) and reduction in epithelial podocyte number ($p=0.02$). Both of these findings remained significant when adjusted for HbA_{1c} or GFR by multiple regression analysis. On the basis of these findings we have concluded that diabetic kidney disease, specifically GBM thickening, and loss of epithelial podocytes, is significantly related to elevated MG and may play a role in the high risk of diabetic kidney disease experienced by Pima Indians.

To further investigate the role of cellular MG production in nephropathy progression, we examined cryopreserved erythrocytes from subjects in the NHS with nephropathy progression or nonprogression (upper and lower quintiles of GBM thickening). Cells were incubated for 48 hours with 30 mM glucose, and intracellular MG and D-lactate production was measured. These incubations clearly showed that subjects in the upper quintile of GBM thickening produced significantly higher levels of MG than those in the lower quintile ($p=0.04$). These findings are of particular significance as they are observed at a time when there is virtually no clinical evidence of nephropathy, and when nephropathy was unlikely to have an impact on cellular metabolism.

From these results, we conclude that progression of diabetic nephropathy is significantly related to elevated dicarbonyl stress. This data also suggests that dysregulation of dicarbonyl homeostasis can contribute to the pathogenesis of diabetic nephropathy, and supports our hypothesis that this variable capacity for MG production contributes to individual susceptibility to diabetic nephropathy.

22.5.1
Impaired Degradation of Methylglyoxal and Diabetic Complications

The glyoxalase system is an efficient enzymatic detoxification system preventing the accumulation of MG and MG-derived AGEs. MG can also be detoxified by an aldehyde reductase, leading to the production of acetol [43]; but, this pathway appears to be minor relative to glyoxalase. Recent quantitative analysis of MG-arginine derived advanced glycation adducts of cellular and extracellular proteins indicates that these imidazolones are quantitatively the most prevalent adducts formed in diabetes, and can account for modification of up to 25% cellular arginine residues [27, 28]. Convincing experimental evidence that glyoxalase I can suppress the formation of AGEs *in vitro* came from studies of endothelial cells in normoglycemic and hyperglycemic culture conditions [84]. In these studies, hyperglycemia induced increases in the concentrations of MG, D-lactate, and cellular protein AGEs. Overexpression of glyoxalase I in these cells totally prevented the increase in MG and cellular protein AGEs, and increased the concentration of D-lactate, indicating that glyoxalase I could potentially play a critical role in suppressing the formation of MG and the resulting protein AGEs.

Studies in diabetic and control populations have shown a high degree of variability in the activity of MG detoxification pathways, and *in vivo* studies suggest that the diabetic state may be associated with impairment of glyoxalase activity [85]. It has also been observed that both D-lactate and acetol levels are elevated in diabetes indicating that hyperglycemia significantly increases flux through these pathways [81, 86], although it is not yet clear if decreased flux is associated with propensity to diabetic complications. As a matter of fact, studies of glyoxalase 1 and 2 in RBCs of subjects with diabetes showed increased activity of both in those with diabetes, and no difference in those with or without diabetic complications [87]. Other studies have confirmed these observations and observed no differences in glyoxalase activity in diabetic subjects with accelerated diabetic nephropathy relative to those who were protected [35].

In diabetes, depletion of GSH is common, providing another potential mechanism for accumulation of MG and its related AGEs. Glyoxalase I is a GSH-dependent enzyme, making its activity proportional to the cellular concentration of GSH. Experimental depletion of GSH, by oxidative or nonoxidative mechanisms (exposure to cytotoxic levels of hydrogen peroxide or the glutathione transferase substrate CDNB (1-chloro-2,4-dinitrobenzene)), can induce marked accumulation of MG, and induce cytotoxicity [87]. GSH is also decreased in people with diabetic nephropathy [35] suggesting that *in vivo* perturbations could lead to decreased glyoxalase activity, even if the enzymes are intrinsically functional, and that impaired detoxification of MG could be at least partially responsible for the increased MG levels observed in individuals with diabetic complications.

22.5.2
3DG Detoxification and Diabetic Complications

3-DG can also have deleterious effects on cell function, therefore its effective metabolism to more innocuous products can protect against excessive intra and extracellular glycation. The major detoxification route for 3DG is by reduction to 3DF [46, 47], and the activity of this pathway is shown by the elevation of 3DF in diabetes [48]. Under-activity of these pathways could lead to accumulation of 3DG and its associated AGEs, and lead to the development of diabetic complications observed in individuals who are prone to renal and retinal complications [85, 86]. This hypothesis is supported by observations suggesting that metabolism of 3DG to 3DF is variably impaired in type 1 diabetes [48]. It has also been shown that a high degree of individual variation in rates of 3DG detoxification occurs among diabetic subjects, which was initially suggested by the high intersubject coefficient of variation of ±96% for tissue levels of the major 3DG detoxification product 3DF, particularly in subjects with type 1 diabetes, indicating quite variable production and/or detoxification of 3DG. Studies of the activity of 3DG degradative pathways have also shown highly predictable production of 3DF from 3DG in nondiabetic subjects, while studies comparing urinary levels of 3DG and 3DF in diabetic populations have shown a different picture. Variable and decreased rates of 3DG detoxification are apparent in diabetic subjects with both type 1 and type 2 diabetes, and could enhance 3DG accumulation and subsequent toxicity in some individuals.

22.6
Oxidative Stress and Propensity to Diabetic Nephropathy

Several translational studies have shown associations between oxidative stress and diabetic complications [31, 35, 88] and are consistent with previous basic studies that have proposed such associations [6, 89]. The potential mechanisms for this increase in oxidative stress could include factors such as increased mitochondrial generation of reactive oxygen species [5], decreased activity of protective mechanisms [34], or increased generation of reactive oxygen species from Amadori products or Schiff's bases [22].

In studies performed on a population with type 1 diabetes and biopsy documented progression or nonprogression of diabetic nephropathy, we also found that the stable lipid oxidation product, 8-isoprostane, was significantly associated with progression, as defined by rates of GBM thickening, while levels of a surrogate marker for oxidative stress (reduced GSH) were inversely related to this parameter [35]. These associations remained significant when adjusted for other important variables such as glycemic control, age, and duration of diabetes. The concurrent increases in oxidative stress seen in our translational studies were not reflected by glycemic control as reflected by A_{1c}, and it is possible that other factors could be responsible for oxidative stress in diabetes and result in progression of diabetic nephropathy.

In subsequent studies, we have investigated the degree of oxidative stress in cultured skin fibroblasts from another unique population with type 1 diabetes and equal numbers of subjects with clinical and biopsy proven diabetic nephropathy, or protection from significant progression as well as nondiabetic controls [80]. Perhaps the most important observation of these studies was that cells from nephropathy progressors showed greater degrees of oxidative stress than controls under both 5 and 25 mM glucose incubation conditions. This was not true for nephropathy nonprogressors, whose levels of cellular oxidative stress did not significantly differ from controls. We also observed that GAPDH activity was significantly lower in cells from nephropathy progressors than in controls, in both 5 and 25 mM glucose, but no differences were seen between nephropathy nonprogressors and nondiabetic controls. In contrast, there were no significant differences between diabetic nephropathy groups and controls in GAPDH mRNA, even when the cells were incubated in 25 mM glucose. These findings support the Brownlee hypothesis in human populations with type 1 diabetes, and suggest that increased oxidative stress may play a role in the development of diabetic nephropathy in susceptible individuals. The observation that the greatest differences in oxidative stress were seen between nephropathy progressors and controls, even under normal glucose concentrations (5 mM), also suggests that factors leading to increased OS represents an intrinsic cellular characteristic in those with diabetes, who are susceptible to development of diabetic nephropathy. It is not clear if these findings can be explained by genetic characteristics of nephropathy prone or resistant individuals, or if it represents a memory effect from prior *in vivo* glucose exposure.

22.6.1
Nitrosative Stress and Diabetic Complications

Several pieces of evidence support a direct role of hyperglycemia in the overgeneration of nitrotyrosine in diabetes, and it is possible that this by-product of the potent oxidative product, peroxynitrite, could also play a role in the development of diabetic complications. Nitrotyrosine is readily detected in the artery wall of monkeys during hyperglycemia [90] as well as in the plasma of healthy subjects during hyperglycemic clamp or OGTT [91–93]. Hyperglycemia is also accompanied by nitrotyrosine deposition in a perfused working heart from rats, and it is related to unbalanced production of NO and superoxide, through iNOS overexpression [94]. Nitrotyrosine formation can lead to the development of endothelial dysfunction in both healthy subjects and in coronaries of perfused hearts [95, 96], and has also been shown to be directly toxic to endothelial cells [97]. Further studies are needed to confirm that long-term elevation of nitrotyrosine is associated with the development of diabetic complications in the clinical setting.

22.6.2
Protective Upregulation of Antioxidant Enzymes and Diabetic Complications

Other important mechanisms can protect against the excessive hyperglycemia-associated oxidative stress, including upregulation of antioxidant enzymes such as catalase, superoxide dismutases, and glutathione-peroxidase. These compensatory mechanisms can protect against oxidative stress associated with diabetes, while tissue damage could be amplified by failure of this upregulation to occur. Support for failure of this protective mechanism in subjects susceptible to diabetic complications comes from studies showing that cultured skin fibroblasts from subjects with diabetic nephropathy fail to up-regulate enzyme activity and mRNA message of the antioxidant enzymes, catalase, and glutathione-peroxidase relative to subjects with T1DM without diabetic nephropathy [34]. Moreover, it has also been shown that antioxidant gene expression in skin fibroblasts is concordant in TIDM sibling pairs, who demonstrate remarkable similar renal histological changes [64] and is, at least in part, under genetic control [59]. However, confirmatory studies that this mechanism plays a role in the development of DN in larger populations are needed.

22.7 Conclusions

Substantial evidence is accumulating that susceptibility to diabetic complications relates to the capacity of biological protective systems to protect against the cellular toxicity of glycation and oxidation products. I have attempted to review the currently available data on this important area of medical investigation to provide the reader with the background to understand new data as it accrues in this rapidly changing field.

References

1. Bucala, R. and Cerami, A. (1992) Advanced glycosylation: chemistry, biology, and implications for diabetes and aging. *Adv Pharmacol*, **23**, 1–34.
2. Brownlee, M. (1992) Glycation products and the pathogenesis of diabetic complications. *Diabetes Care*, **15** (12), 1835–1843.
3. Monnier, V.M. (1989) The Maillard reaction in aging, diabetes and nutrition. *Prog Clin Biol Res*, **304**, 1–22.
4. Bunn, H.F. and Higgins, P.J. (1981) Reaction of monosaccharides with proteins. Possible evolutionary significance. *Science*, **213**, 222–224.
5. Brownlee, M. (2005) The pathobiology of diabetic complications: a unifying mechanism. *Diabetes*, **54** (6), 1615–1625.
6. Baynes, J. and Thorpe, S. (1999) Role of oxidative stress in diabetic complications: a new perspective on an old paradigm. *Diabetes*, **48**, 1–9.
7. Hink, U., Li, H., Mollnau, H. et al. (2001) Mechanisms underlying endothelial dysfunction in diabetes mellitus. *Circ Res*, **88** (2), 2.
8. Nishikawa, T., Edelstein, D., Du, X. et al. (2000) Normalizing mitochondrial superoxide production blocks three pathways of hyperglycemic damage. *Nature*, **404**, 787–790.
9. Thornalley, P.J. (1999) Clinical significance of glycation. *Clin Lab*, **45**, 263–273.
10. Thornalley, P.J., Battah, S., Ahmed, N. et al. (2003) Quantitative screening of advanced glycation endproducts in cellular and extracellular proteins by tandem mass spectrometry. *Biochem J*, **375** (Pt 3), 581–592.
11. Brownlee, M. (2001) Biochemistry and molecular cell biology of diabetic complications. *Nature*, **414** (6865), 813–820.
12. Du, X., Matsumura, T., Edelstein, D. et al. (2003) Inhibition of GAPDH activity by poly(ADP-ribose) polymerase activates three major pathways of hyperglycemic damage in endothelial cells [see comment]. *J Clin Invest*, **112** (7), 1049–1057.
13. Soriano, F.G., Virag, L., and Szabo, C. (2001) Diabetic endothelial dysfunction: role of reactive oxygen and nitrogen species production and poly(ADP-ribose) polymerase activation [Review] [92 Refs]. *J Mol Med*, **79** (8), 437–448.
14. Hammes, H.P., Lin, J., Renner, O. et al. (2002) Pericytes and the pathogenesis of diabetic retinopathy. *Diabetes*, **51** (10), 3107–3112.
15. Huang, C., Kim, Y., Caramori, M.L. et al. (2002) Cellular basis of diabetic nephropathy: II. The transforming growth factor-β system and diabetic nephropathy lesions in type 1 diabetes. *Diabetes*, **51** (12), 3577–3581.
16. Hirata, C., Nakano, K., Nakamura, N. et al. (1997) Advanced glycation end products induce expression of vascular endothelial growth factor by retinal muller cells. *Biochem Biophys Res Commun*, **236**, 712–715.
17. Thornalley, P., Langborg, A., and Minhas, H. (1999) Formation of glyoxal, methylglyoxal and 3-deoxyglucosone in the glycation of

proteins by glucose. *Biochem J*, **344**, 109–116.

18 Delpierre, G., Rider, M., Collard, F. et al. (2000) Identification, cloning, and heterologous expression of a mammalian fructosamine-3-kinase. *Diabetes*, **49**, 1627–1634.

19 Wells-Knecht, K.J., Zyzak, D.V., Litchfield, J.E., Thorpe, S.R., and Baynes, J.W. (1995) Mechanism of autoxidative glycosylation: identification of glyoxal and arabinose as intermediates in the autoxidative modification of proteins by glucose. *Biochemistry*, **34** (11), 3702–3709.

20 Tilton, R.G. (2002) Diabetic vascular dysfunction: links to glucose-induced reductive stress and vegf [Review] [175 Refs]. *Microsc Res Tech*, **57** (5), 390–407.

21 Wolff, S.P., Basca, Z.A., and Hunt, J.V. (1989) Autoxidative glycosylation: free radicals and glycation theory, *The Maillard Reaction in Aging, Diabetes, and Nutrition*, Liss, New York, pp. 259–275.

22 Baynes, J.W. and Thorpe, S.R. (2000) Glycoxidation and lipoxidation in atherogenesis [Review] [73 Refs]. *Free Radic Biol Med*, **28** (12), 1708–1716.

23 Rosen, P., Du, X., and Sui, G.Z. (2001) Molecular mechanisms of endothelial dysfunction in the diabetic heart [Review] [47 Refs]. *Adv Exp Med Biol*, **498**, 75–86.

24 Bravi, M.C., Pietrangeli, P., Laurenti, O. et al. (1997) Polyol pathway activation and glutathione redox status in non-insulin-dependent diabetic patients. *Metab Clin Exp*, **46** (10), 1194–1198.

25 Beckman, J.S. and Koppenol, W.H. (1996) Nitric oxide, superoxide, and peroxynitrite: the good, the bad, and ugly [Review] [109 Refs]. *Am J Physiol*, **271** (5 Pt 1), C1424–C1437.

26 Estevez, A.G., Spear, N., Pelluffo, H. et al. (1999) Examining apoptosis in cultured cells after exposure to nitric oxide and peroxynitrite [Review] [43 Refs]. *Methods Enzymol*, **301**, 393–402.

27 Ahmed, N., Babaei-Jadidi, R., Thornalley, P.J., Howell, S.K., and Beisswenger, P.J. (2005) Increased plasma and urinary methylglyoxal-derived hydroimidazolone in type 1 diabetic patients. *Ann N Y Acad Sci*, **1043**, 928.

28 Ahmed, N., Babaei-Jadidi, R., Howell, S., Thornalley, P., and Beisswenger, P. (2005) Glycated and oxidised protein degradation products are indicators of fasting and postprandial hyperglycemia in diabetes. *Diabetes Care*, **28**, 2465–2471.

29 Ahmed, N., Babaei-Jadidi, R., Howell, S., Beisswenger, P., and Thornalley, P. (2005) Degradation products of proteins damaged by glycation, oxidation and nitration in clinical type 1 diabetes. *Diabetologia*, **48**, 1590–1603.

30 Geibauf, A., Van Wickern, B., Simat, T., Steinhart, H., and Esterbauer, H. (1996) Formation of N-formylkynurenine suggests the involvement of apolipoprotein B-100 centered tryptophan radicals in the initiation of LDL lipid peroxidation. *FEBS Lett*, **389**, 136–140.

31 Wells-Knecht, M.C., Lyons, T.J., McCance, D.R., Thorpe, S.R., and Baynes, J.W. (1997) Age-dependent increase in ortho-tyrosine and methionine sulfoxide in human skin collagen is not accelerated in diabetes. Evidence against a generalized increase in oxidative stress in diabetes. *J Clin Invest*, **100** (4), 839–846.

32 Geibauf, A., Van Wickern, B., Simat, T., Steinhart, H., and Esterbauer, H. (1996) Formation of N-formylkynurenine suggests the involvement of apolipoprotein B-100 centered tryptophan radicals in the initiation of LDL lipid peroxidation. *FEBS Lett*, **389**, 136–140.

33 Gaut, J.P., Byun, J., Tran, H.D., Heinecke, J.W. (2002) Artifact-free quantification of free 3-chlorotyrosine, 3-bromotyrosine, and 3-nitrotyrosine in human plasma by electron capture-negative chemical ionization gas chromatography mass spectrometry and liquid

chromatography-electrospray ionization tandem Mass Spectrometry. *Anal Biochem*, **300** (2), 252–259. [Erratum Appears in Anal Biochem 2002 May 15;304(2):275].

34. Ceriello, A., Morocutti, A., Mercuri, F. et al. (2000) Defective intracellular antioxidant enzyme production in type 1 diabetic patients with nephropathy. *Diabetes*, **49** (12), 2170–2177.

35. Beisswenger, P., Drummond, K., Nelson, R. et al. (2005) Susceptibility to diabetic nephropathy is related to dicarbonyl and oxidative stress. *Diabetes*, **54**, 3274–3281.

36. Szwergold, B.S., Howell, S., and Beisswenger, P.J. (2001) Human fructosamine-3-kinase: purification, sequencing, substrate specificity, and evidence of activity in vivo. *Diabetes*, **50** (9), 2139–2147.

37. Delpierre, G., Collard, F., Fortpied, J., and Van Schaftingen, E. (2002) Fructosamine 3-kinase is involved in an intracellular deglycation pathway in human erythrocytes. *Biochem J*, **365** (Pt 3), 801–808.

38. Collard, F., Wiame, E., Bergans, N. et al. (2004) Fructosamine 3-kinase-related protein and deglycation in human erythrocytes. *Biochem J*, **382** (Pt 1), 137–143.

39. Thornalley, P. (1990) The glyoxalase system: new developments towards functional characterization of a metabolic pathway fundamental to biological life. *Biochem J*, **269** (1), 1–11.

40. Thornalley, P. (2003) Glyoxalase I–structure, function and a critical role in the enzymatic defence against glycation [Review] [23 Refs]. *Biochem Soc Trans*, **31**, 1343–1348.

41. Thornalley, P. (1993) Modification of the glyoxalase system in disease processes and prospects for therapeutic strategies. *Biochem Soc Trans*, **21** (2), 531–534.

42. Abordo, E.A., Minhas, H.S., and Thornalley, P.J. (1999) Accumulation of alpha-oxoaldehydes during oxidative stress: a role in cytotoxicity. *Biochem Pharmacol*, **58** (4), 641–648.

43. Vander Jagt, D.L., Robinson, B., Taylor, K.K., and Hunsaker, L.A. (1992) Reduction of trioses by NADPH-dependent aldo-keto reductases. Aldose reductase, methylglyoxal, and diabetic complications. *J Biol Chem*, **267** (7), 4364–4369.

44. Hipkiss, A. and Chana, H. (1998) Carnosine protects proteins against methylglyoxal-mediated modifications. *Biochem Biophys Res Commun*, **248**, 28–32.

45. Wells-Knecht, K., Lyons, T., McCance, D. et al. (1994) 3-deoxyfructose concentrations are increased in human plasma and urine in diabetes. *Diabetes*, **43** (9), 1152–1156.

46. Fujii, E., Iwase, H., Ishii-Karakasa, I., Yajima, Y., and Hotta, K. (1995) The presence of 2-keto-3-deoxygluconic acid and oxoaldehyde dehydrogenase activity in human erythrocytes. *Biochem Biophys Res Commmun*, **210** (3), 852–857.

47. Oimomi, M., Hata, F., Igaki, N. et al. (1989) Purification of alpha-ketoaldehyde dehydrogenase from the human liver and its possible significance in the control of glycation. *Experientia*, **45** (5), 463–466.

48. Lal, S., Szwergold, B., Walker, M. et al. (1998) Production and metabolism of 3-deoxyglucosone in humans, in *The Maillard Reaction in Foods and Medicine* (eds J. O'Brien et al.), Royal Society of Chemistry, Cambridge, pp. 291–297.

49. Yamada, H., Miyata, S., Igaki, N. et al. (1994) Increase in 3-deoxyglucosone levels in diabetic rat plasma. *J Biol Chem*, **269**, 20275–20280.

50. Beisswenger, P., Lal, S., Howell, S. et al. (1998) The role of 3-deoxyglucosone and the activity of its degradative pathways in the etiology of diabetic microvascular disease, in *The Maillard Reaction in Foods and Medicine* (eds J. O'Brien et al.), Royal Society of Chemistry, Cambridge, pp. 298–303.

51 Schindhelm, R.K., Alssema, M., Scheffer, P.G. et al. (2007) Fasting and postprandial glycoxidative and lipoxidative stress are increased in women with type 2 diabetes. *Diabetes Care*, **30** (7), 1789–1794.

52 Szwergold, B., Howell, S., and Beisswenger, P. (2005) Transglycation-a potential new mechanism for deglycation of schiff's bases. *Ann N Y Acad Sci*, **1043**, 845–864.

53 Wondrak, G.T., Cervantes-Laurean, D., Roberts, M.J. et al. (2002) Identification of alpha-dicarbonyl scavengers for cellular protection against carbonyl stress. *Biochem Pharmacol*, **63** (3), 361–373.

54 Ramasamy, R., Yan, S.F., Herold, K., Clynes, R., and Schmidt, A.M. (2008) Receptor for advanced glycation end products: fundamental roles in the inflammatory response: winding the way to the pathogenesis of endothelial dysfunction and atherosclerosis. *Ann N Y Acad Sci*, **1126**, 7–13.

55 Yan, S.F., Ramasamy, R., and Schmidt, A.M. (2008) Mechanisms of disease: advanced glycation end-products and their receptor in inflammation and diabetes complications [Review] [52 Refs]. *Nat Clin Pract Endocrinol Metab*, **4** (5), 285–293.

56 Araki, N., Higashi, T., Mori, T. et al. (1995) Macrophage scavenger receptor mediates the endocytic uptake and degradation of advanced glycation end products of the Maillard reaction. *Eur J Biochem*, **230**, 408–415.

57 Vlassara, H., He, C., Koschinsky, T., and Buenting, C. (2001) The age-receptor in the pathogenesis of diabetic complications [Review] [60 Refs]. *Diabetes Metab Res Rev*, **17** (6), 436–443.

58 Goldberg, A.L. (2003) Protein degradation and protection against misfolded or damaged proteins [Review] [45 Refs]. *Nature*, **426** (6968), 895–899.

59 Huang, C., Kim, Y., Caramori, M.L. et al. (2006) Diabetic nephropathy is associated with gene expression levels of oxidative phosphorylation and related pathways. *Diabetes*, **55** (6), 1826–1831.

60 Babizhayev, M., Seguin, M.-C., Gueynej, J. et al. (1994) L-carnosine (-alanyl-L-histidine) and carcinine F-alanylhistamine) act as natural antioxidants with hydroxyl-radical-scavenging and lipid-peroxidase activities. *Biochem J*, **304**, 509–516.

61 Da Ros, R., Assaloni, R., and Ceriello, A. (2004) Antioxidant therapy in diabetic complications: what is new? [Review] [107 Refs]. *Curr Vasc Pharmacol*, **2** (4), 335–341.

62 Dumaswala, U.J., Zhuo, L., Mahajan, S. et al. (2001) Glutathione protects chemokine-scavenging and antioxidative defense functions in human RBCs. *Am J Physiol Cell Physiol*, **280** (4), C867–C873.

63 DCCT Research Group (1997) Clustering of long-term complications in families with diabetes in the diabetes control and complications trial. *Diabetes*, **46**, 1829–1839.

64 Fioretto, P., Steffes, M.W., Barbosa, J. et al. (1999) Is diabetic nephropathy inherited? Studies of glomerular structure in type 1 diabetic sibling pairs. *Diabetes*, **48** (4), 865–869.

65 Seaquist, E.R., Goetz, F.C., Rich, S., and Barbosa, J. (1989) Familial clustering of diabetic kidney disease: evidence for genetic susceptibility to diabetic nephropathy. *N Engl J Med*, **320**, 1161–1165.

66 Prager, T.C., Wilson, D.J., and Avery, G.D. (1981) Vitreous fluorophotometry: identification of sources of variability. *Invest Ophthalmol Vis Sci*, **21**, 854–864.

67 Pettitt, D., Saad, M., Bennett, P., Nelson, R., and Knowler, W. (1990) Familial predisposition to renal disease in two generations of Pima Indians with type 2 (non-insulin-dependent) diabetes mellitus. *Diabetologia*, **33** (7), 438–443.

68 Borch-Johnsen, K., Norgaard, K., Hommel, E. et al. (1992) Is diabetic

nephropathy an inherited complication? *Kidney Int*, **41** (4), 719–722.
69 Krolewski, A.J., Warram, J.H., Kahn, R., Kahn, L.I., and Kahn, C.R. (1987) Epidemiologic approach to the etiology of type I diabetes mellitus and its complications. *N Engl J Med*, **18**, 267–273.
70 Caramori, M.L., Kim, Y., Huang, C. et al. (2002) Cellular basis of diabetic nephropathy: 1. Study design and renal structural-functional relationships in patients with long-standing type 1 diabetes. *Diabetes*, **51** (2), 506–513. [Erratum Appears in Diabetes 2002 Apr;51(4):1294]
71 Rohlfing, C., Wiedmeyer, H.M., Little, R. et al. (2002) Biological variation of glycohemoglobin. *Clin Chem*, **48** (7), 1116–1118.
72 Kilpatrick, E.S., Maylor, P.W., and Keevil, B.G. (1998) Biological variation of glycated hemoglobin. Implications for diabetes screening and monitoring [see comments]. *Diabetes Care*, **21** (2), 261–264.
73 Gould, B.J., Davie, S.J., and Yudkin, J.S. (1997) Investigation of the mechanism underlying the variability of glycated haemoglobin in non-diabetic subjects not related to glycaemia. *Clin Chim Acta*, **260** (1), 49–64.
74 Yudkin, J., Forrest, R., Jackson, C. et al. (1990) Unexplained variability of glycated haemoglobin in non-diabetic subjects not related to glycaemia. *Diabetologia*, **33**, 208–215.
75 Hudson, P.R., Child, D.F., Jones, H., and Williams, C.P. (1999) Differences in rates of glycation (glycation index) may significantly affect individual HbA1c results in type 1 diabetes [see comment]. *Ann Clin Biochem*, **36** (Pt 4), 451–459.
76 McCarter, R.J., Hempe, J.M., Gomez, R., and Chalew, S.A. (2004) Biological variation in HbA1c predicts risk of retinopathy and nephropathy in type 1 diabetes [see comment]. *Diabetes Care*, **27** (6), 1259–1264.
77 Conner, J., Beisswenger, P., and Szwergold, B. (2005) Some clues as to the regulation, expression, function, and distribution of fructosamine-3-kinase and fructosamine-3-kinase-related protein. *Ann N Y Acad Sci*, **1043**, 824–836.
78 Conner, J., Beisswenger, P., and Szwergold, B. (2004) The expression of the genes for fructosamine-3-kinase and fructosamine-3-kinase-related-protein appears to be constitutive and unaffected by environmental signals. *Biochem Biophys Res Commun*, **323**, 932–936.
79 Veiga-Da-Cunha, M., Jacquemin, P., Delpierre, G. et al. (2006) Increased protein glycation in fructosamine 3-kinase-deficient mice. *Biochem J*, **399**, 257–264.
80 Beisswenger, P., Howell, S., Szwergold, B. et al. (2008) Progression of diabetic nephropathy is predicted by increased oxidative stress and decreased deglycation. *Diabetologia*, **51** (Suppl 1), S24.
81 Beisswenger, P., Howell, S., Smith, K., and Szwergold, B. (2003) Glyceraldehyde-3-phosphate dehydrogenase activity as an independent modifier of methylglyoxal levels in diabetes. *Biochem Biophys Acta*, **1637**, 98–106.
82 Drummond, K., Mauer, M. G. International Diabetic Nephropathy Study (2002) The early natural history of nephropathy in type 1 diabetes: II. Early renal structural changes in type 1 diabetes. *Diabetes*, **51** (5), 1580–1587.
83 Drummond, K.N., Kramer, M.S., Suissa, S. et al. (2003) Effects of duration and age at onset of type 1 diabetes on preclinical manifestations of nephropathy. *Diabetes*, **52** (7), 1818–1824.
84 Shinohara, M., Thornalley, P.J., Giardino, I. et al. (1998) Overexpression of glyoxalase-I in bovine endothelial cells inhibits intracellular advanced glycation endproduct formation and prevents hyperglycemia-induced increases in macromolecular endocytosis. *J Clin Invest*, **101** (5), 1142–1147.
85 Beisswenger, P., Touchette, A., and Howell, S. (1997) The effect

of hyperglycemia and diabetes on methylglyoxal and its detoxification pathways (Abstract). *Diabetologia*, **40** (Suppl 1), I-V, 588.

86 Beisswenger, P., Jean, S., Brinck-Johnsen, T., Siegel, A., and Cavender, J. (1995) Acetol production is increased in early diabetic nephropathy (Abstract). *Diabetes*, **44** (Suppl 1), 23A.

87 McLellan, A.C., Thornalley, P.J., Benn, J., and Sonksen, P.H. (1994) Glyoxalase system in clinical diabetes mellitus and correlation with diabetic complications. *Clin Sci*, **87** (1), 21–29.

88 Yu, Y., Thorpe, S.R., and Jenkins, A.J. et al. (2006) Advanced glycation end-products and methionine sulphoxide in skin collagen of patients with type 1 diabetes. *Diabetologia*, **49** (10), 2488–2498.

89 Rosen, P., Nawroth, P., King, G. et al. (2001) The role of oxidative stress in the onset and progression of diabetes and its complications: a summary of a congress series sponsored by UNESCO-MCBN, the American Diabetes Association and the German Diabetes Society. *Diabetes Metab Res Rev*, **17**, 189–212.

90 Pennathur, S., Wagner, J., Leeuwenburgh, C., Litwak, K., and Heinecke, J. (2001) A hydroxyl radical-like species oxidizes cynomolgus monkey artery wall proteins in early diabetic vascular disease. *J Clin Invest*, **107**, 853–860.

91 Ceriello, A., Brown, W., Le, N. et al. (2007) Post-prandial increases in inflammatory markers and oxidative stress are significantly related to increases in post-prandial hyperglycemia, but not to increases in post-prandial insulin, in patients with type 2 diabetes. *Diabetologia*, **50** (Suppl 1), S500.

92 Ceriello, A., Quagliaro, L., Catone, B. et al. (2002) Role of hyperglycemia in nitrotyrosine postprandial generation. *Diabetes Care*, **25** (8), 1439–1443.

93 Ceriello, A., Quagliaro, L., Piconi, L. et al. (2004) Effect of postprandial hypertriglyceridemia and hyperglycemia on circulating adhesion molecules and oxidative stress generation and the possible role of simvastatin treatment. *Diabetes*, **53** (3), 701–710.

94 Ceriello, A. (2005) Postprandial hyperglycemia and diabetes complications: is it time to treat? [Review] [97 Refs]. *Diabetes*, **54** (1), 1–7.

95 Ceriello, A., Taboga, C., Tonutti, L. et al. (2002) Evidence for an independent and cumulative effect of postprandial hypertriglyceridemia and hyperglycemia on endothelial dysfunction and oxidative stress generation: effects of short- and long-term simvastatin treatment. *Circulation*, **106**, 1211–1218.

96 Marfella, R., Quagliaro, L., Nappo, F., Ceriello, A., and Giugliano, D. (2001) Acute hyperglycemia induces an oxidative stress in healthy subjects (letter). *J Clin Invest*, **108**, 635–636.

97 Mihm, M., Jing, L., and Bauer, J. (2000) Nitrotyrosine causes selective vascular endothelial dysfunction and DNA damage. *J Cardiovasc Pharmacol*, **36**, 182–187.

23
Serum Advanced Glycation End Products Associated with NASH and Other Liver Diseases

Hideyuki Hyogo, Sho-ichi Yamagishi, and Susumu Tazuma

Proteins are modified by carbonyl compounds in the Maillard reaction, and irreversible products are formed as advanced glycation end products (AGEs). The AGEs may cause a variety of chronic diseases. Toxic Maillard reaction products might be the endogenous toxins. Glyceraldehyde (GLA)-derived AGEs are epitopes of GLA-modified proteins which dose-dependently inhibit proliferation of HL-60 cells, Caco-2 cells, and PC12 cells. In addition, GLA-modified proteins induce cell death in differentiated PC12 cells. Fluorescent pyridinium-type, non-fluorescent imidazolone-type, and fluorescent pyrimidine-type AGEs exist in the GLA-modified proteins, and a fluorescent pyridinium-type AGE named *GLAP* (glyceraldehyde-derived pyridinium compound) is most toxic. GLAP induces the production of reactive oxygen species (ROS), and is detected in plasma protein and tail tendon collagen in streptozotocin-induced diabetic rats. GLAP might be a biomarker in diabetic complications. GLA-related Maillard reaction products cause cytotoxicity and oxidative stress as endogenous toxins. GLA is increased in chronic diseases, such as diabetic complications and chronic renal disease. GLA-related metabolism is important in the progression of chronic diseases.

23.1
Introduction

Nonalcoholic fatty liver disease (NAFLD) encompasses a broad spectrum of conditions, ranging from simple steatosis to nonalcoholic steatohepatitis (NASH). While simple steatosis seems to be a benign and nonprogressive condition, NASH is recognized as a potentially progressive disease that can lead to cirrhosis, liver failure, and hepatocellular carcinoma (HCC) [1–4]. In Western countries, the prevalence of NASH in the general population ranges from 1 to 5%, and that of NAFLD ranges from 15 to 39% [5, 6]. In Japan, a quarter of Japanese adults have become overweight, approximately 20% of Japanese adults have NAFLD, and about 1% of those are estimated to have NASH as well [7, 8]. Thus, the prevalence of NAFLD and NASH is increasing and becoming a major target disease not only in Western countries but also in Japan.

Endogenous Toxins. Diet, Genetics, Disease and Treatment.
Edited by Peter J. O'Brien and W. Robert Bruce
Copyright © 2010 WILEY-VCH Verlag GmbH & Co. KGaA, Weinheim
ISBN: 978-3-527-32363-0

NASH is considered the hepatic manifestation of the metabolic syndrome and is particularly associated with insulin resistance (IR), obesity, hypertension, and abnormalities in glucose and lipid metabolism [9–12]. These underlying metabolic conditions account for the risk of liver fibrosis and advanced liver disease, in addition to the cardiovascular risk. The "two-hit theory" simply and best describes the pathological mechanisms of NASH: first, the accumulation of excessive fat in the liver and secondly, the development of oxidative stress [4, 13]. Although oxidative stress is supposed to play an important role in the transition from steatosis to steatohepatitis and is caused by various conditions, the exact molecular mechanism underlying NASH has not been clarified. It is therefore important to discern what factors modulate the development of NASH.

Reactive derivatives from nonenzymatic glucose–protein condensation reactions, as well as lipids and nucleic acids exposed to reducing sugars, such as glucose or glyceraldehyde, form a heterogeneous group of irreversible adducts called *advanced glycation end products (AGEs)*. AGEs were originally characterized by a yellow-brown fluorescent color and an ability to form cross-links with and between amino groups, but the term is now used for a broad range of advanced products of the glycation process (also called the *Maillard reaction*). The formation and accumulation of AGEs in various tissues are known to progress during normal aging and at an extremely accelerated rate in diabetes mellitus [14]. Recent understandings of this process have confirmed that the interaction of AGEs with their cell surface receptor (RAGE, receptor for advanced glycation end product) may play a role in the pathogenesis of various devastating disorders, including diabetic vascular complications, neurodegenerative disorders, and IR [15–18]. We have recently shown that the AGEs–RAGE-mediated reactive oxygen species (ROS) generation elicits vascular inflammation as well as angiogenesis and alters atherosclerosis-related gene expression, thus participating in the development and progression of diabetic vascular complications [19]. In addition, several reports have suggested the involvement of RAGE in various types of liver diseases [20, 21]. Furthermore, Sebekova *et al.* have recently shown the elevation of circulating AGE levels (fluorescent AGE and *N*-(carboxymethyl)lysine (CML)) in patients with liver cirrhosis, which was ameliorated by liver transplantation [22]. From these observations and because abnormalities in glucose metabolism are highly detectable when the glucose tolerance test are done [23], we speculated that AGEs could be a biomarker of NASH and play a role in the pathogenesis of NASH. In this chapter, we introduce the study about the relevancy of glyceraldehyde-derived AGE, one of the bioactive AGEs, with NASH and the role of AGEs in liver diseases as well.

23.2
Formation Pathways of AGEs

Reducing sugars, including glucose, fructose, and glyceraldehyde, are known to react nonenzymatically with the amino groups of proteins to form reversible Schiff bases and then Amadori products [24, 25]. These early glycation products undergo further complex reactions such as rearrangement, dehydration, and condensation to become irreversibly cross-linked, heterogeneous fluorescent derivatives, called "*AGEs*" [14, 26]. AGEs were originally characterized by a yellow-brown fluorescent color and an ability to form cross-links with and between amino groups [24, 26], but the term is now used for a broad range of advanced products of the glycation process (also called "*Maillard reaction*"), including CML and pyrraline, which show neither color nor fluorescence and are not cross-linked proteins [25]. CML can be formed from both glyoxal and glycolaldehyde by an intramolecular Cannizzaro reaction, a process that is largely independent on glucose autoxidation [25]. The concept of CML as an oxidation marker rather than a glycation product has recently attracted support.

The formation of AGEs *in vitro* and *in vivo* is dependent on the turnover rate of the chemically modified target, the time available, and the sugar concentration. The structures of the various cross-linked AGEs that are generated *in vivo* have not yet been completely determined. Because of their heterogeneity and the complexity of the chemical reactions involved, only some AGEs have been structurally characterized *in vivo*. The structural identity of AGEs with cytotoxic properties remains unknown.

AGEs are formed by a nonenzymatic reaction between ketone group of the glucose molecule or aldehydes and the amino groups of proteins. Recent studies have suggested that AGEs can arise not only from reducing sugars but also from carbonyl compounds derived from the autoxidation of sugars and other metabolic pathways [27–29]. Indeed, we have recently found that glucose, α-hydroxyaldehydes (glyceraldehyde and glycolaldehyde) and dicarbonyl compounds (methylglyoxal, glyoxal, and 3-deoxyglucosone) are actively involved in the process of AGEs formation *in vivo* (Figure 23.1) [25, 30–32].

23.3
AGEs' Association in the Liver

Recent understandings have confirmed that the interaction of AGEs with RAGE may play a role in the pathogenesis of various devastating disorders, including diabetic vascular complications, neurodegenerative disorders, IR, and cancers [15, 19, 33–52]. Further, there is accumulating evidence that AGEs and RAGE interaction elicits oxidative stress and subsequently alters gene expression in various types of cells, including hepatocytes and hepatic stellate cells (HSCs) [20, 53–55].

Liver is not only a target organ but also an important site for clearance and catabolism of circulating AGEs. Indeed, the liver can sequester a number of circulating senescent macromolecules such as AGEs [22, 56, 57]. Clearance of macromolecules is performed by several types of cells, such as resident liver macrophages, Kupffer cells, and liver sinusoidal endothelial cells, that constitute sinusoidal walls in the liver [22, 56, 57]. In the case of AGEs, it is reported

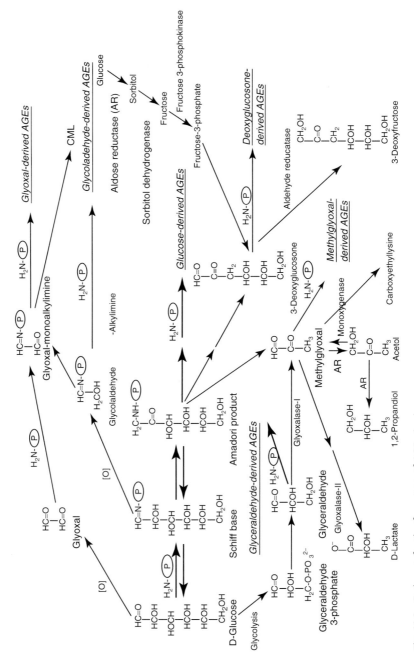

Fig. 23.1 Pathways for the formation of AGEs *in vivo*.

that in rats more than 90% of the intravenously administered *in vitro*-prepared advanced glycation end products-modified bovine serum albumin (AGEs-BSA) is eliminated by sinusoidal liver cells and Kupffer cells [57]. The contribution of hepatic sinusoidal endothelial cells and Kupffer cells to AGEs clearance is about 60 and 25%, respectively, whereas that of hepatocytes is relatively low (less than 15%) [57]. The reports about the role of AGEs–RAGE system in various types of liver diseases such as NASH, liver cirrhosis, and cancers [22, 23, 52] are still limited and further researches are anticipated.

23.4
Circulating AGEs and Liver Disease

23.4.1
Circulating AGEs in Liver Cirrhosis

The concept that catabolism and clearance of circulating AGEs could be impaired by various liver diseases was first introduced by Sebekova *et al.* in 2002 [22]. They measured plasma CML levels in 51 patients with liver cirrhosis (5 of them were followed 36 months after liver transplantation) and 19 healthy controls. The major findings of their study were as follows: (i) plasma CML levels markedly elevated in patients with liver cirrhosis and positively correlated with the severity of the disease; (ii) CML levels were inversely associated with residual liver function of the patients estimated by serum albumin and plasma bilirubin levels; and (iii) plasma CML levels markedly decreased (to about 50% of those before treatment) within three months of liver transplantation [22]. These observations suggest that the liver plays an important role in the removal of circulating AGEs and that hepatic clearance of circulating AGEs is impaired in liver cirrhosis. Similar findings have been shown by Yagmur *et al.* in 2006 [58]. They found that serum CML concentration was significantly higher in patients with liver cirrhosis compared to patients without cirrhosis, and its levels were positively associated with the severity of cirrhosis defined by a Child–Pugh score [58]. These findings suggest that circulating AGEs level may be one of the useful biomarkers for evaluating residual liver function.

23.4.2
Circulating AGEs in NASH

To investigate whether measurement of circulating AGEs level is also a useful tool for discriminating NASH from simple steatosis, we examined serum levels of AGEs (glucose-derived AGE, glyceraldehyde-derived AGE, and CML) in 66 patients with histologically defined NASH, but without liver cirrhosis; 10 with simple steatosis; and in 30 control subjects [23]. We found that serum glyceraldehyde-derived AGE level may play a role in the pathogenesis of NASH and be a biomarker for discriminating NASH from simple steatosis on the basis of the following evidence:

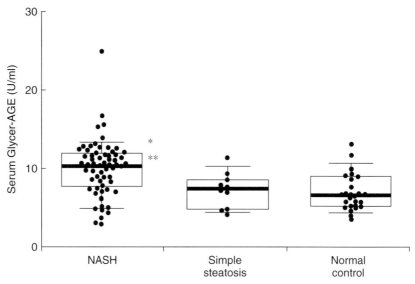

Fig. 23.2 Serum glyceraldehyde-derived advanced glycation end products (AGEs) levels in nonalcoholic steatohepatitis (NASH; $n = 66$), simple steatosis ($n = 10$), and healthy controls ($n = 30$). Boxes contain the values between the 25th and 75th percentiles; bold horizontal line is the median; error bars stretch from the 10th to the 90th percentiles; and individual data are represented by closed circles. Overall significance of the differences between the three groups according to a nonparametric Kruskal–Wallis analysis of variance was $P < 0.0001$. Therefore, the significance of the differences between the groups was determined using Scheff's method. *$P < 0.01$ and **$P < 0.001$, compared to the values for simple steatosis and the control group, respectively.

(i) Serum glyceraldehyde-derived AGE level significantly elevated in NASH patients compared with simple steatosis or healthy controls (Figure 23.2). Receiver operating characteristic curves for circulating glyceraldehyde-derived AGE level revealed that the threshold value for the prediction of NASH was 8.53 U ml^{-1}. At this threshold, the sensitivity was 66.7% and the specificity was 88.9% (Figure 23.3). (ii) Serum glyceraldehyde-derived AGE level was positively correlated with homeostatic model assessment of IR and inversely associated with adiponectin levels (Figure 23.4). (iii) Serum levels of glyceraldehyde-derived AGE were not correlated with the severity of hepatic steatosis or fibrosis, and also, the values were not affected by the status of glucose tolerance pattern detected by 75 g oral glucose tolerance test (Figure 23.5). There was no difference in glyceraldehyde-derived AGE level between normal and impaired glucose tolerant patients. (iv) Glyceraldehyde-derived AGE was detected in hepatocytes of the patients with NASH, but scarce in simple steatosis. (v) There was no significant difference in CML- or glucose-derived AGE levels among the groups [23]. In addition to these findings, we have recently found that the serum levels of glyceraldehyde-derived AGE are reduced by the treatment of NASH without changing the glucose tolerance pattern (unpublished data). Taken together, circulating glyceraldehyde-derived AGE might become a biomarker for discriminating NASH, an indicator for the effectiveness of treatment, and oxidative

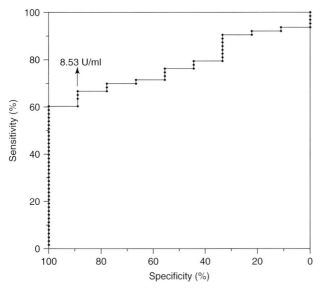

Fig. 23.3 Receiver operating characteristic curves (ROCs) for serum glyceraldehyde-derived advanced glycation end products used to discriminate nonalcoholic steatohepatitis from simple steatosis. Area under the ROC curve was 0.78.

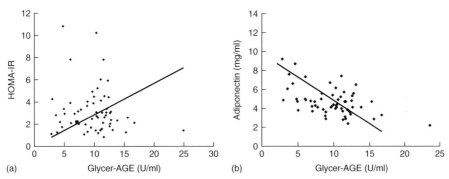

Fig. 23.4 Regression analysis between serum glyceraldehyde-derived nonalcoholic steatohepatitis (NASH) and homeostasis model of insulin resistance (HOMA-IR) (a) and adiponectin (b) in nonalcoholic steatohepatitis patients. Significant correlation were found between serum glyceraldehyde-derived AGE and HOMA-IR ($r = 0.70$, $P < 0.02$) and adiponectin ($r = 0.92$, $P = 0.001$).

stress. Further studies are needed to clarify the underlying mechanism(s) whereby AGEs play a pathophysiological role in NASH. In addition, the possible role of AGEs in cardiovascular complications in NASH is also of interest for future investigations.

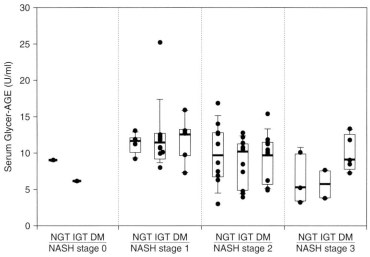

Fig. 23.5 Comparison between serum glyceraldehyde-derived advanced glycation end products (AGEs) levels and glucose tolerance pattern in individual fibrosis stage in nonalcoholic steatohepatitis ($n = 66$). Boxes contain the values between the 25th and 75th percentiles; bold horizontal line is the median; the error bars stretch from the 10th to the 90th percentiles; and individual data are represented by closed circles. DM, diabetes mellitus; IGT, impaired glucose tolerance; NGT, normal glucose tolerance.

23.5
Possible Molecular Mechanisms by Which the AGEs–RAGE System Is Involved in Liver Disease

23.5.1
AGEs in Hepatic Sinusoidal Endothelial Cells

There is a growing body of evidence that the engagement of RAGE with AGEs is shown to elicit oxidative stress and subsequently evoke inflammatory responses in various types of endothelial cells [59–63]. However, the influence of AGEs on liver sinusoidal endothelial cells has not been fully explored. Hansen et al. have shown that AGEs impair the scavenger function of rat hepatic sinusoidal endothelial cells [64], thus disturbing the intracellular transport system of senescent macroprotein derivatives, including AGEs themselves. Therefore, the impairment of scavenger function in hepatic sinusoidal endothelial cells may lead to further elevation of circulating AGEs and oxidized low density lipoprotein levels, which could augment the AGEs–RAGE axis in the body.

23.5.2
AGEs in Hepatic Stellate Cells

HSCs are the main extracellular matrix–producing cells in the liver, and thus play a pivotal role in liver fibrogenesis [20]. The role of RAGE in the spreading and

23.5 Possible Molecular Mechanisms by Which the AGEs–RAGE System Is Involved in Liver Disease

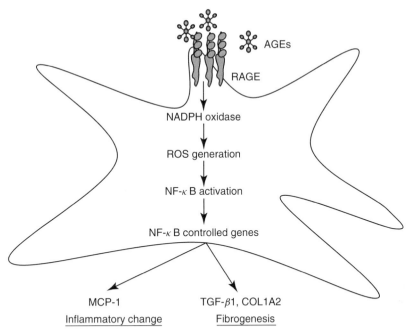

Fig. 23.6 Schematic showing the AGEs–RAGE interaction in hepatic stellate cells.

migration of activated HSCs has been shown by Fehrenbach et al. in 2001 [20]. They showed that HSCs and myofibroblasts expressed RAGE, and its expression increased during activation of HSCs and the process of differentiation to myofibroblasts, and was modulated by transforming growth factor-β1 (TGF-β1). Ligand activation of RAGE led to the formation of ROS and induction of mitogen-activated protein kinase and nuclear factor κB (NF-κB) signaling pathways in HSCs [20].

Further, based upon our findings that serum glyceraldehyde-derived AGEs, one of the circulating AGEs are increased in patients with NASH [23], we explored the effects of AGEs on HSCs. We have found that glyceraldehyde-derived AGEs induce fibrogenesis- and inflammation-related gene and protein expression such as TGF-β1, collagen type I alpha2, and monocyte chemoattractant protein-1 in cultured HSCs via NADPH oxidase-derived ROS generation (Figure 23.6) [55]. These observations could provide us a clue for understanding how glyceraldehyde-derived AGEs could be involved in the pathogenesis of NASH.

23.5.3
AGEs in Hepatocytes

Regarding the effect of AGEs on hepatocytes, we have recently found that AGEs–RAGE interaction stimulates hepatic C-reactive protein (CRP) in human hepatoma cells – Hep3B cells via activation of Rac-1 [65]. Our data suggest that there exist at least two distinct signaling pathways to *CRP* gene induction in the AGEs-exposed Hep3B cells: (i) Rac-1-involved NADPH oxidase–mediated

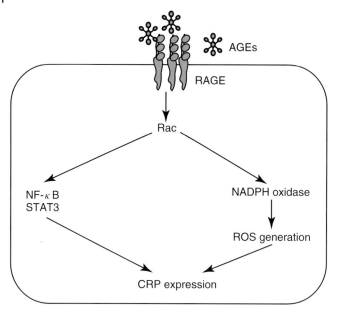

Fig. 23.7 Schematic showing the AGEs-induced CRP expression in hepatocytes.

ROS-dependent pathway and (ii) Rac-1-induced signal transducer and activator of transcription 3- and NF-κB-dependent pathway, which is not directly mediated by ROS [65]. It is conceivable that the early stage of CRP induction by AGEs is ROS independent, whereas the later stage of CRP induction relies on an ROS-mediated pathway (Figure 23.7).

AGEs were also found to increase phosphorylation of insulin receptor substrate-1 (IRS-1) at serine-307 residues, Jun N-terminal kinase (JNK), c-JUN, and IκB kinase in association with decreased IκB levels in Hep3B cells [66]. Overexpression of dominant negative Rac-1 blocked these effects of AGEs on Hep3B cells. Further, an inhibitor of JNK, or curcumin, an inhibitor of NF-κB, also decreased the IRS-1 serine phosphorylation levels in Hep3B cells. In addition, AGEs decreased tyrosine phosphorylation of IRS-1, and subsequently reduced the association of p85 subunit of phosphatidylinositol 3-kinase with IRS-1 and glycogen synthesis in insulin-exposed Hep3B cells, all of which were prevented by an inhibitor of JNK or NF-κB [66]. These findings suggest that AGEs were involved in hepatic IR through the induction of JNK- and IκB kinase-dependent serine phosphorylation of IRS-1 via activation of Rac-1 (Figure 23.8).

We have found that telmisartan, an angiotensin II type 1 receptor blocker with a unique peroxisome proliferators activated receptor-γ (PPAR-γ) modulating ability, but not candesartan, decreased the AGEs-induced RAGE expression, ROS generation, and subsequent CRP production in Hep3B cells (Figure 23.9) [54]. In addition, GW9662, an inhibitor of PPAR-γ, blocked the inhibitory effects of telmisartan on RAGE expression and its downstream signaling in Hep3B cells, and troglitazone and ciglitazone, full agonists of PPAR-γ, mimicked the effects of

23.5 Possible Molecular Mechanisms by Which the AGEs–RAGE System Is Involved in Liver Disease

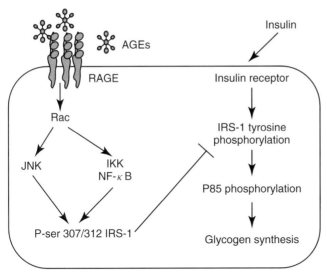

Fig. 23.8 Schematic showing the molecular mechanisms of AGEs-elicited insulin resistance in hepatocytes.

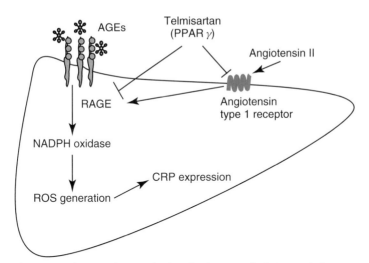

Fig. 23.9 Schematic showing the beneficial aspect of telmisartan in hepatocytes.

telmisartan [54]. These observations suggest that telmisartan could block the AGEs signaling to CRP expression in Hep3B cells through its antioxidative properties via PPAR-γ-mediated RAGE downregulation. Our study indicates a unique beneficial aspect of telmisartan: it may work as an anti-inflammatory agent against AGEs by suppressing RAGE expression via PPAR-γ activation in the liver. Telmisartan, but not candesartan, also blocked the AGEs signaling to hepatic IR, as described above; telmisartan could improve the AGEs-elicited IR in Hep3B cells by inhibiting the serine phosphorylation of IRS-1, at least in part, via the activation of PPAR-γ [67].

23.5.4
RAGE Involvement in Liver Diseases

There are a couple of papers to show the involvement of RAGE in other liver diseases [21, 52, 53, 68]. Zeng et al. have demonstrated that animals treated with soluble receptor for advanced glycation end products (sRAGEs), the extracellular ligand-binding domain of RAGE, displayed increased survival after total hepatic ischemia/reperfusion compared with vehicle treatment, and that blockade of RAGE was highly protective against hepatocellular death and necrosis; in parallel, proliferating cell nuclear antigen was enhanced in livers of mice treated with sRAGE [21]. They concluded that activation of RAGE contributed to the induction of proinflammatory and tissue-destructive processes on hepatic ischemia/reperfusion in mice, and that blockade of RAGE limited immediate deleterious inflammatory mechanisms, and thereby facilitated regenerative potential in the injured liver. In addition, the active role of RAGE was shown in toxin-induced liver injury and hepatectomy models. Ekong et al. have recently found that the blockade of RAGE attenuates acetaminophen-induced hepatotoxicity in mice [53]. The major findings of their studies were as follows: (i) in acetaminophen-administered mice, treatment with sRAGE displayed increased survival compared to vehicle treatment, and markedly decreased hepatic necrosis; (ii) a significant reduction of nitrotyrosine protein adducts was observed in hepatic tissue, in parallel with significantly increased levels of glutathione; and (iii) proregenerative cytokines tumor necrosis factor-α and interleukin-6 were increased. Cataldegirmen et al. have also suggested that RAGE blockade is a novel strategy to promote regeneration in the massively injured liver on the basis of the following evidence [68]: (i) after massive liver resection, the blockade of RAGE significantly increased survival; (ii) remnants retrieved from RAGE-blocked mice displayed increased activated NF-κB, principally in hepatocytes, and enhanced expression of regeneration-promoting cytokines, tumor necrosis factor-α and interleukin-6, and the anti-inflammatory cytokine, interleukin-10; and (iii) hepatocyte proliferation was increased by RAGE blockade, in parallel with significantly reduced apoptosis [68]. Further, Hiwatashi et al. have shown that the expression of RAGE mRNA is lower in normal liver than in hepatitis and highest in HCC [52]. They showed that in HCC, RAGE expression was high in well- and moderately differentiated tumors but declined as tumors dedifferentiated to poorly differentiated HCC. Further, HCC lines resistant to hypoxia were found to have higher levels of RAGE expression, and RAGE transfectant also showed significantly prolonged survival under hypoxia [52]. These findings suggest that HCC during the early stage of tumorigenesis with less blood supply may acquire resistance to stringent hypoxic milieu by hypoxia-induced RAGE expression.

23.6 Conclusions

There is accumulating data to suggest the active role of the AGEs–RAGE axis in liver diseases. Further clinical and experimental studies will be needed to clarify the underlying mechanisms by which the AGEs/RAGE is involved in the development and progression of various liver diseases.

Acknowledgment

This work was supported in part by Grants of Collaboration with Venture Companies Project from the Ministry of Education, Culture, Sports, Science and Technology, Japan (S. Yamagishi) and in part by a grant from the Japanese Ministry of Education, Culture, Sports, Science and Technology and by Grants-in-Aid for Hepatolithiasis from the Ministry of Health, Labour and Welfare of Japan (S. Tazuma).

References

1 Teli, M.R., James, O.F., Burt, A.D. et al. (1995) The natural history of nonalcoholic fatty liver: a follow-up study. *Hepatology*, **22**, 1714–1719.

2 Dam-Larsen, S., Franzmann, M., Andersen, I.B. et al. (2004) Long term prognosis of fatty liver: risk of chronic liver disease and death. *Gut*, **53**, 750–755.

3 Matteoni, C.A., Younossi, Z.M., Gramlich, T. et al. (1999) Nonalcoholic fatty liver disease: a spectrum of clinical and pathological severity. *Gastroenterology*, **116**, 1413–1419.

4 James, O. and Day, C. (1999) Non-alcoholic steatohepatitis: another disease of affluence. *Lancet*, **353**, 1634–1636.

5 Powell, E.E., Cooksley, W.G., Hanson, R. et al. (1990) The natural history of nonalcoholic steatohepatitis: a follow-up study of forty-two patients for up to 21 years. *Hepatology*, **11**, 74–80.

6 Bellentani, S., Saccoccio, G., Masutti, F. et al. (2000) Prevalence of and risk factors for hepatic steatosis in Northern Italy. *Ann Intern Med*, **132**, 112–117.

7 Saibara, T. (2005) Nonalcoholic steatohepatitis in Asia-Oceania. *Hepatol Res*, **33**, 64–67.

8 Yoshiike, N. and Lwin, H. (2005) Epidemiological aspects of obesity and NASH/NAFLD in Japan. *Hepatol Res*, **33**, 77–82.

9 Chitturi, S., Abeygunasekera, S., Farrell, G.C. et al. (2002) NASH and insulin resistance: insulin hyper-secretion and specific association with the insulin resistance syndrome. *Hepatology*, **35**, 373–379.

10 Marchesini, G., Brizi, M., Bianchi, G. et al. (2001) Nonalcoholic fatty liver disease: a feature of the metabolic syndrome. *Diabetes*, **50**, 1844–1850.

11 Hui, J.M. and Farrell, G.C. (2003) Clear messages from sonographic shadows? Links between metabolic disorders and liver disease, and what to do about them. *J Gastroenterol Hepatol*, **18**, 1115–1117.

12 Marchesini, G., Bugianesi, E., Forlani, G. et al. (2003) Nonalcoholic fatty liver, steatohepatitis, and the metabolic syndrome. *Hepatology*, **37**, 917–923.

13 Day, C.P. and James, O.F. (1998) Steatohepatitis: a tale of two

'hits'? *Gastroenterology*, **114**, 842–845.

14 Brownlee, M., Cerami, A., and Vlassara, H. (1988) Advanced glycosylation end products in tissue and the biochemical basis of diabetic complications. *N Engl J Med*, **19**, 1315–1321.

15 Yamagishi, S., Takeuchi, M., Inagaki, Y. et al. (2003) Role of advanced glycation end products (AGEs) and their receptor (RAGE) in the pathogenesis of diabetic microangiopathy. *Int J Clin Pharmacol Res*, **23**, 129–134.

16 Inagaki, Y., Yamagishi, S., Okamoto, T. et al. (2003) Pigment epithelium-derived factor prevents advanced glycation end products-induced monocyte chemoattractant protein-1 production in microvascular endothelial cells by suppressing intracellular reactive oxygen species generation. *Diabetologia*, **46**, 284–287.

17 Yamagishi, S., Inagaki, Y., Amano, S. et al. (2002) Pigment epithelium-derived factor protects cultured retinal pericytes from advanced glycation end product-induced injury through its antioxidative properties. *Biochem Biophys Res Commun*, **296**, 877–882.

18 Miele, C., Riboulet, A., Maitan, M.A. et al. (2003) Human glycated albumin affects glucose metabolism in L6 skeletal muscle cells by impairing insulin-induced insulin receptor substrate (IRS) signaling through a protein kinase C alpha-mediated mechanism. *J Biol Chem*, **278**, 47376–47387.

19 Yamagishi, S. and Imaizumi, T. (2005) Diabetic vascular complications: pathophysiology, biochemical basis and potential therapeutic strategy. *Curr Pharm Des*, **11**, 2279–2299.

20 Fehrenbach, H., Weiskirchen, R., Kasper, M., and Gressner, A.M. (2001) Up-regulated expression of the receptor for advanced glycation end products in cultured rat hepatic stellate cells during transdifferentiation to myofibroblasts. *Hepatology*, **34**, 943–952.

21 Zeng, S., Feirt, N., Goldstein, M. et al. (2004) Blockade of receptor for advanced glycation end product (RAGE) attenuates ischemia and reperfusion injury to the liver in mice. *Hepatology*, **39**, 422–432.

22 Sebeková, K., Kupcová, V., Schinzel, R., and Heidland, A. (2002) Markedly elevated levels of plasma advanced glycation end products in patients with liver cirrhosis – amelioration by liver transplantation. *J Hepatol*, **36**, 66–71.

23 Hyogo, H., Yamagishi, S., Iwamoto, K. et al. (2007) Elevated levels of serum advanced glycation end products in patients with non-alcoholic steatohepatitis. *J Gastroenterol Hepatol*, **22**, 1112–1119.

24 Bucala, R. and Cerami, A. (1992) Advanced glycosylation: chemistry, biology, and implications for diabetes and aging. *Adv Pharmacol*, **23**, 1–34.

25 Takeuchi, M. and Makita, Z. (2001) Alternative routes for the formation of immunochemically distinct advanced glycation end-products in vivo. *Curr Mol Med*, **1**, 305–315.

26 Monnier, V.M. and Cerami, A. (1981) Non-enzymatic browning in vivo: possible process for aging of long-lived proteins. *Science*, **211**, 491–493.

27 Wells-Knecht, K.J., Zyzak, D.V., Litchfield, J.E. et al. (1995) Mechanism of autoxidative glycosylation: identification of glyoxal and arabinose as intermediates in the autoxidative modification of proteins by glucose. *Biochemistry*, **34**, 3702–3709.

28 Thornalley, P.J. (1996) Pharmacology of methylglyoxal: formation, modification of proteins and nucleic acids, and enzymatic detoxification – a role in pathogenesis and antiproliferative chemotherapy. *Gen Pharmacol*, **27**, 565–573.

29 Thornalley, P.J., Langborg, A., and Minhas, H.S. (1999) Formation of glyoxal, methylglyoxal and

3-deoxyglucosone in the glycation of proteins by glucose. *Biochem J*, **344**, 109–116.

30 Takeuchi, M., Makita, Z., Yanagisawa, K. *et al.* (1999) Detection of non-carboxymethyllysine and carboxymethyllysine advanced glycation end products (AGE) in serum of diabetic patients. *Mol Med*, **5**, 393–405.

31 Takeuchi, M., Makita, Z., Bucala, R. *et al.* (2000) Immunological evidence that non-carboxymethyllysine advanced glycation end-products are produced from short chain sugars and dicarbonyl compounds in vivo. *Mol Med*, **6**, 114–125.

32 Takeuchi, M., Yanase, Y., Matsuura, N. *et al.* (2001) Immunological detection of a novel advanced glycation end-product. *Mol Med*, **7**, 783–791.

33 Takeuchi, M. and Yamagishi, S. (2004) TAGE (toxic AGEs) hypothesis in various chronic diseases. *Med Hypotheses*, **63**, 449–452.

34 Takeuchi, M. and Yamagishi, S. (2004) Alternative routes for the formation of glyceraldehydes-derived AGEs (TAGE) in vivo. *Med Hypotheses*, **63**, 453–455.

35 Brownlee, M. (2001) Biochemistry and molecular cell biology of diabetic complications. *Nature*, **414**, 813–820.

36 Smith, M.A., Taneda, S., Richey, P.L., Miyata, S., Yan, S.-D., Stern, D. *et al.* (1994) Advance Maillard reaction end products are associated with Alzheimer disease pathology. *Proc Natl Acad Sci U S A*, **91**, 5710–5714.

37 Vitek, M.P., Bhattacharya, K., Glendening, J.M., Stopa, E., Vlassara, H., Bucala, R. *et al.* (1994) Advanced glycation end products contribute to amyloidosis in Alzheimer disease. *Proc Natl Acad Sci U S A*, **91**, 4766–4770.

38 Sasaki, N., Toki, S., Chowei, H., Saito, T., Nakano, N., Hayashi, Y. *et al.* (2001) Immunohistochemical distribution of the receptor for advanced glycation end products in neurons and astrocytes in Alzheimer's disease. *Brain Res*, **888**, 256–262.

39 Takeuchi, M., Bucala, R., Suzuki, T., Ohkubo, T., Yamazaki, M., Koike, T. *et al.* (2000) Neurotoxicity of advanced glycation end-products for cultured cortical neurons. *J Neuropathol Exp Neurol*, **59**, 1094–1105.

40 Takeuchi, M., Kikuchi, S., Sasaki, N., Suzuki, T., Watai, T., Iwaki, M. *et al.* (2004) Involvement of advanced glycation end-products (AGEs) in Alzheimer's disease. *Curr Alzheimer Res*, **1**, 39–46.

41 Hofmann, S.M., Dong, H.J., Li, Z., Cai, W., Altomonte, J., Thung, S.N. *et al.* (2002) Improved insulin sensitivity is associated with restricted intake of dietary glycoxidation products in the db/db mouse. *Diabetes*, **51**, 2082–2089.

42 Sullivan, C.M., Futers, T.S., Barrett, J.H., Hudson, B.I., Freeman, M.S. and Grant, P.J. (2005) RAGE polymorphisms and the heritability of insulin resistance: the Leeds family study. *Diabetes Vasc Dis Res*, **2**, 42–44.

43 Unoki, H., Bujo, H., Yamagishi, S., Takeuchi, M., Imaizumi, T., and Saito, Y. (2007) Advanced glycation end products attenuate cellular insulin sensitivity by increasing the generation of intracellular reactive oxygen species in adipocytes. *Diabetes Res Clin Pract*, **76**, 236–244.

44 Kuniyasu, H., Chihara, Y., and Kondo, H. (2003) Differential effects between amphoterin and advanced glycation end products on colon cancer cells. *Int J Cancer*, **104**, 722–727.

45 Abe, R., Shimizu, T., Sugawara, H., Watanabe, H., Nakamura, H., Choei, H. *et al.* (2004) Regulation of human malignant melanoma growth and metastasis by AGE-AGE receptor interactions. *J Invest Dermatol*, **122**, 461–467.

46 Ishiguro, H., Nakaigawa, N., Miyoshi, Y., Fujinami, K., Kubota, Y., and Uemura, H. (2005) Receptor for advanced glycation end products (RAGE) and its ligand, amphoterin are overexpressed and associated with prostate cancer development. *Prostate*, **64**, 92–100.

47 Bartling, B., Demling, N., Silber, R.E., and Simm, A. (2006) Proliferative stimulus of lung fibroblasts on lung cancer cells is impaired by the receptor for advanced glycation end-products. *Am J Respir Cell Mol Biol*, **34**, 83–91.

48 Riuzzi, F., Sorci, G., and Donato, R. (2007) RAGE expression in rhabdomyosarcoma cells results in myogenic differentiation and reduced proliferation, migration, invasiveness, and tumor growth. *Am J Pathol*, **171**, 947–961.

49 Lee, S.J. and Lee, K.W. (2007) Protective effect of (-)-epigallocatechin gallate against advanced glycation endproducts-induced injury in neuronal cells. *Biol Pharm Bull*, **30**, 1369–1373.

50 Bakshi, N., Kunju, L.P., Giordano, T., and Shah, R.B. (2007) Expression of renal cell carcinoma antigen (RCC) in renal epithelial and nonrenal tumors: diagnostic Implications. *Appl Immunohistochem Mol Morphol*, **15**, 310–315.

51 Tesarová, P., Kalousová, M., Jáchymová, M., Mestek, O., Petruzelka, L., and Zima, T. (2007) Receptor for advanced glycation end products (RAGE) – soluble form (sRAGE) and gene polymorphisms in patients with breast cancer. *Cancer Invest*, **25**, 720–725.

52 Hiwatashi, K., Ueno, S., Abeyama, K., Kubo, F., Sakoda, M., Maruyama, I. et al. (2008) A novel function of the receptor for advanced glycation end-products (RAGE) in association with tumorigenesis and tumor differentiation of HCC. *Ann Surg Oncol*, **15**, 923–933.

53 Ekong, U., Zeng, S., Dun, H., Feirt, N., Guo, J., Ippagunta, N. et al. (2006) Blockade of the receptor for advanced glycation end products attenuates acetaminophen-induced hepatotoxicity in mice. *J Gastroenterol Hepatol*, **21**, 682–688.

54 Yoshida, T., Yamagishi, S., Nakamura, K., Matsui, T., Imaizumi, T., Takeuchi, M. et al. (2006) Telmisartan inhibits AGE-induced C-reactive protein production through downregulation of the receptor for AGE via peroxisome proliferator-activated receptor-gamma activation. *Diabetologia*, **49**, 3094–3099.

55 Iwamoto, K., Kanno, K., Hyogo, H., Yamagishi, S., Takeuchi, M., Tazuma, S. et al. (2008) Advanced glycation end products enhance the proliferation and activation of hepatic stellate cells. *J Gastroenterol*, **43**, 298–304.

56 Takata, K., Horiuchi, S., Araki, N., Shiga, M., Saitoh, M., and Morino, Y. (1988) Endocytic uptake of nonenzymatically glycosylated proteins is mediated by a scavenger receptor for aldehyde-modified proteins. *J Biol Chem*, **263**, 14819–14825.

57 Smedsrod, B., Melkko, J., Araki, N., Sano, H., and Horiuchi, S. (1997) Advanced glycation end products are eliminated by scavenger-receptor mediated endocytosis in hepatic sinusoidal Kupffer and endothelial cells. *Biochem J*, **322**, 567–573.

58 Yagmur, E., Tacke, F., Weiss, C., Lahme, B., Manns, M.P., Kiefer, P. et al. (2006) Elevation of Nepsilon-(carboxymethyl)lysine-modified advanced glycation end products in chronic liver disease is an indicator of liver cirrhosis. *Clin Biochem*, **39**, 39–45.

59 Sengoelge, G., Födinger, M., Skoupy, S., Ferrara, I., Zangerle, C., Rogy, M. et al. (1998) Endothelial cell adhesion molecule and PMNL response to inflammatory stimuli and AGE-modified fibronectin. *Kidney Int*, **54**, 1637–1651.

60 Yamagishi, S., and Takeuchi, M. (2004) Nifedipine inhibits gene expression of receptor for advanced glycation end products (RAGE) in endothelial cells by suppressing reactive oxygen species generation. *Drugs Exp Clin Res*, **30**, 169–175.

61 Yamagishi, S., Matsui, T., Nakamura, K., and Takeuchi, M. (2005) Minodronate, a nitrogen-containing bisphosphonate, inhibits advanced glycation end product-induced

vascular cell adhesion molecule-1 expression in endothelial cells by suppressing reactive oxygen species generation. *Int J Tissue React*, **27**, 189–195.

62 Zhong, Y., Li, S.H., Liu, S.M., Szmitko, P.E., He, X.Q., Fedak, P.W. et al. (2006) C-Reactive protein upregulates receptor for advanced glycation end products expression in human endothelial cells. *Hypertension*, **48**, 504–511.

63 Yamagishi, S.I., Matsui, T., Nakamura, K., Inoue, H., Takeuchi, M., Ueda, S. et al. (2007) Olmesartan blocks inflammatory reactions in endothelial cells evoked by advanced glycation end products by suppressing generation of reactive oxygen species. *Ophthalmic Res*, **40**, 10–15.

64 Hansen, B., Svistounov, D., Olsen, R., Nagai, R., Horiuchi, S., and Smedsrød, B. (2002) Advanced glycation end products impair the scavenger function of rat hepatic sinusoidal endothelial cells. *Diabetologia*, **45**, 1379–1388.

65 Yoshida, T., Yamagishi, S., Nakamura, K., Matsui, T., Imaizumi, T., Takeuchi, M. et al. (2006) Pigment epithelium-derived factor (PEDF) inhibits advanced glycation end product (AGE)-induced C-reactive protein expression in hepatoma cells by suppressing Rac-1 activation. *FEBS Lett*, **580**, 2788–2796.

66 Yoshida, T., Yamagishi, S., Nakamura, K., Matsui, T., Imaizumi, T., Takeuchi, M. et al. (2008) Pigment epithelium-derived factor ameliorates AGE-induced hepatic insulin resistance in vitro by suppressing Rac-1 activation. *Horm Metab Res*, **40**, 620–625.

67 Yoshida, T., Yamagishi, S., Matsui, T., Nakamura, K., Takeuchi, M., Ueno, T. et al. (2008) Telmisartan, an angiotensin II type 1 receptor blocker, inhibits advanced glycation end product (AGE)-elicited hepatic insulin resistance via peroxisome proliferator-activated receptor-γ activation. *J Int Med Res*, **36**, 237–243.

68 Cataldegirmen, G., Zeng, S., Feirt, N., Ippagunta, N., Dun, H., Qu, W. et al. (2005) RAGE limits regeneration after massive liver injury by coordinated suppression of TNF-alpha and NF-kappaB. *J Exp Med*, **201**, 473–484.

24
Oxidative Stress in the Pathogenesis of Hepatitis C
Tom S. Chan and Marc Bilodeau

Hepatitis C virus (HCV) is a hepatotropic agent affecting as many as 170 million people worldwide. Chronic infection with HCV is characterized by persistent inflammation and oxidative stress in the liver that can ultimately lead to liver cirrhosis, and hence, an increased risk of developing liver cancer. A substantial collection of evidence has suggested that oxidative stress plays an especially important role during the pathogenesis of HCV infection. Both the immune system and the virus may concertedly act to increase the production of reactive oxygen and reactive nitrogen species in the liver. In addition, oxidative stress is also emerging as an integral signaling pathway promoting the progression to cirrhosis. This chapter presents an overview of the recent literature concerning the mechanisms of HCV-associated oxidative stress in the liver contributing to the pathology of this infection.

24.1
Hepatitis C

A virus would not normally be considered an endogenous toxin in the same way as heme, hypochlorite, bile acids, ammonia, or reactive oxygen species (ROS) that are outlined in this book. However, the progression of chronic viral hepatitis shares many attributes of ailments such as (i) autoimmune liver disease, in that the immune system plays a significant role in damaging the liver and (ii) genetic disorders, in that the causative insult to the organ is persistent. Indeed, it is the lengthy course of the disease, which sets chronic viral hepatitis apart from most other viral infections. Immune reactions caused by the presence of the virus leads to the generation of other endogenous toxins in the form of ROS and reactive nitrogen species (collectively referred to in this article with ROS) that are well known to be associated with chronic inflammation.

Endogenous Toxins. Diet, Genetics, Disease and Treatment.
Edited by Peter J. O'Brien and W. Robert Bruce
Copyright © 2010 WILEY-VCH Verlag GmbH & Co. KGaA, Weinheim
ISBN: 978-3-527-32363-0

24.1.1
Incidence and Economical Burden of Hepatitis C

Viral hepatitis accounts for the majority of liver injuries in the world. There are several subtypes of the hepatitis virus designated A, B, C, D, and E. Hepatitis A and E are acute and self-limiting infections, while hepatitis D can only exist as a coinfection with hepatitis B. Similar to the hepatitis C virus (HCV), hepatitis B can take a chronic course potentially leading to many of the same complications during the lifetime of the individual. However, unlike hepatitis B, an effective and readily available vaccine that can stem the spread of the disease does not exist for HCV. Indeed, this virus is characterized by its vast genomic diversity (often referred as genotypes and quasispecies) as well as by its high rate of de novo mutations (caused by the error-prone, virus-associated, RNA polymerase). These characteristics, plus some ill-defined defects in the immune response to the virus have allowed HCV to thwart both the immune system and our best efforts to find a vaccine. As a consequence, HCV is now acknowledged to be a pandemic that is affecting as many as 3.2 million Americans, according to the Center for Disease Control, and 170 million people worldwide [1]. HCV infection is presumed to account for 1 million deaths per year worldwide. The annual economical burden of HCV in the United States has been estimated to be US$1 billion [2].

Transmission of HCV occurs predominantly through blood transfer (the majority of new cases in developed countries are a consequence of injection drug use) [3]. Because of the associated risk factors for new viral infections, HCV is often diagnosed as a coinfection with hepatitis B and/or the human immunodeficiency virus (HIV) [4]. In the latter case, as many as 10–25% of HIV-infected individuals have HCV [5], such that currently, liver failure and its complications account for most cases of mortality in the HIV-infected population [6].

Although 15–40% of individuals who are infected spontaneously clear the virus (most within three months), 60–85% of HCV-infected patients become chronically infected for the rest of their lives if untreated. These individuals are at high risk of gradually developing cirrhosis of the liver. Current antiviral drug therapy is based on a combination of interferon and ribavirin [7]. This form of treatment is costly and can be associated with a number of side effects including hemolytic anemia, fatigue, flulike symptoms, and depression [8, 9]. Furthermore, only around half of those being treated fully respond (clearance of viral RNA from the blood) to treatment. The hepatic sequelae of the infection (especially when cirrhosis has developed) can, however, remain despite adequate antiviral response. Individuals can therefore remain at risk for the development of hepatocellular carcinoma [10]. Individuals who contract the virus and go on to be chronically infected have been shown to have a significantly reduced quality of life in addition to a decreased life span.

Tab. 24.1 Protein components of the hepatitis C virus

Protein	Function	Weight (kDa)	References
C	Viral nucleocapsid protein	21	[14]
E1	Envelope glycoproteins	31	[14]
E2	Envelope glycoproteins	70	[14]
P7	Calcium ion channel	7	[15]
NS2	Serine protease clearing NS2–NS3 junction	23	[14, 16]
NS3	Protease	70	[14]
	RNA helicase		
	Nucleotide phosphatase		
NS4A	Cofactor for NS3	8	[14, 17]
NS4B	Membrane anchor for the replication complex	27	[14, 18]
NS5A	Component of the replication complex	56–58	[14, 19]
NS5B	Viral RNA polymerase	68	[14, 20]

24.1.2
The Hepatitis C Virus

HCV was identified in 1989 by Choo *et al.* as the causative agent in the majority of non-A non-B forms of hepatitis [11, 12]. HCV is a positive, single strand RNA virus belonging to the genus *Hepacivirus* in the Flaviviridae family. The complete virus has been extremely difficult to visualize in humans [13]. Its genome consists of 9600 nucleotides encoding for a precursor polyprotein of approximately 3000 amino acids. The polyprotein is posttranslationally cleaved by both viral and host proteases to form structural proteins (C, E1, E2), and nonstructural proteins (NS2, NS3, NS4A, NS4B, NS5A, and NS5B) and two, small ill-defined proteins (p7 and F). What is currently known about the functions of each of the proteins is listed in Table 24.1.

The life cycle of the HCV virus in hepatocytes is depicted in Figure 24.1. It is unclear how the virus enters hepatocytes, but, recent studies have indicated that the multifunctional membrane proteins CD81 and Claudin-1 can associate with the E2 envelope glycoprotein [21–23]. In addition, viral particles may also directly associate or "piggy back" on apo-B-associated VLDL via the scavenger receptor (SR-BI) into endosomes, whereby internalization is associated with a decrease in endosomal pH [24, 25]. Within the coding regions for glycoproteins E1 and E2, hypervariable sequences 1 and 2 (HVR 1, 2), respectively, can be found [26]. These regions are particularly poorly conserved among virus isolates and have been hypothesized to be a factor in the successful evasion of host immune surveillance during chronic infections [23, 27, 28]. In addition to the genetic diversity of HCV within each genotype, binding of E2 to CD81 in B lymphocytes alters lymphocyte somatic hypermutation necessary for the synthesis of high-affinity antibodies, ultimately decreasing host adaptive immunity against the virus [29, 30].

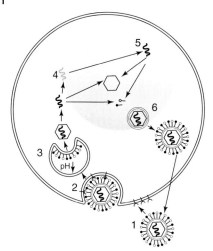

Fig. 24.1 Life cycle of HCV in hepatocytes. E1 and E2 glycoproteins are bound to CD81 and CLDN1 and internalized into endosomes within the hepatocytes. A decrease in endosomal pH leads to the fusion of the viral envelope with the endosomal membrane. Viral capsid releases the RNA into the cytosol, whereby transcription of the HCV polyprotein can take place. Further copies of the positive-strand RNA are produced along with the viral components, which are assembled into the complete viral particle.

The immune response to HCV infection involves many different nonparenchymal cells. HCV infection triggers the innate immune response, a largely nonspecific process directed toward stopping the spread of a pathogen [31]. Recognition of HCV as a pathogen is accomplished through the binding of HCV-derived proteins to membrane-bound pattern recognition receptors known as the toll-like receptors (TLRs), or through the detection of viral RNA via RIG-1 helicase [32]. The resultant production and release of cytokines leads to the activation of natural killer cells that indiscriminantly kill cells in the vicinity of the cytokines. More HCV-specific defense is established through the adaptive immune response, as dendritic cells present viral components to different subsets of lymphocytes, which in turn produce antibodies against viral-derived proteins. Later on, an HCV-specific cytotoxic response involving CD8+, killer T cells will also develop in order to eradicate only the infected cells. Some studies have suggested that chronicity of the infection is caused by a deficiency in the innate and or the adaptive immune responses [33]. Unlike other forms of hepatitis, effective viral clearance by the immune system, when it occurs, does not confer immunity to subsequent infections.

24.1.3
Conventional Treatment of Hepatitis C

Efforts to stem the spread of the disease have been hampered by major difficulties in developing an effective and reliable vaccine because of the virus' diverse genetic variability (six genotypes, each containing many subtypes and quasispecies) [34, 35]. Current HCV therapy is based on a combination of pegylated-interferon and ribavirin. Sustained virological response, as defined by the absence of measurable HCV RNA, six months following the end of therapy, can be achieved in approximately 50% of individuals [36]. Treatment response varies according to HCV genotype (rates being higher with genotypes 2 and 3 versus genotype 1), viral load, severity of liver damage, and other comorbid factors [37–39]. It is also known that higher response rates can be achieved if therapy is administered early after infection (during the acute hepatitis setting). On the basis of the known effects of interferon-α on hepatic immunity against pathogens, it is currently presumed that exogenously administered interferon works through the innate response. The therapeutic mechanism of ribavirin is less clear, as little direct antiviral effects have been demonstrated in chronically infected HCV patients when the drug was administered alone [40, 41].

24.2
The Molecular Basis of Oxidative Stress in Hepatitis C

24.2.1
Oxidative Stress in HCV

Oxidative stress is defined as a state where the levels of ROS surmount the normal antioxidant capacity in a cell resulting in the oxidation of important macromolecules. Numerous studies have described the presence of markers of oxidative stress within the sequelae of the HCV-infected liver. Hepatic levels of the antioxidant, glutathione were previously shown to be significantly decreased in conjunction with marked alterations in the hepatocytes' mitochondrial ultrastructure [42], and increased hepatic markers of oxidized DNA (8-hydroxy deoxyguanosine) and lipid peroxidation (4-hydroxynonenal) were also detected in HCV-infected livers [43].

As the primary organ involved in the detoxification of chemicals, the liver undergoes some of the harshest conditions in the body. Its defense mechanisms and unique ability to regenerate make it a robust organ. However, extensive chemical or immunological stress leads to the production of ROS and oxidative stress in the liver.

Molecular oxygen is inherently a reactive molecule that contains two unpaired electrons. Oxidative stress commences with the energization of molecular oxygen to form the singlet oxygen species or the addition of one or two electrons to form the superoxide anion or hydrogen peroxide, respectively. Both superoxide and hydrogen peroxide can participate in metal-catalyzed formation of the highly reactive hydroxyl radical, which can covalently bind to cellular macromolecules, thereby altering their structure and/or decreasing their function. The role of metal-catalyzed oxidative stress is potentially important in HCV-induced chronic hepatitis, as elevated copper and iron levels have been reported in the fibrotic livers of HCV patients [44, 45].

Under physiological conditions, the majority of ROS are derived from the mitochondrial electron transport chain (ETC). Single electrons that escape from the tightly regulated complexes of the mitochondrial ETC can be transferred to molecular oxygen to form superoxide. This mode of electron transfer normally accounts for approximately 2–3% of the total amount of O_2 consumed by the chain [46]. These minute levels of superoxide can be readily detoxified by cellular levels of antioxidant enzymes such as superoxide dismutase, catalase, glutathione peroxidase, and glutathione-S-transferase and/or directly scavenged by antioxidant chemicals such as vitamin E, vitamin C, and glutathione [47, 48]. Blockages in the flow of electrons using inhibitors against specific complexes (rotenone (blockage at complex I), malonate (blockage at complex II), myxathiazole (blockage at complex III), or antimycin C (blockage at complex III)) through the chain have been shown to stimulate mitochondria ROS formation and promotion of cellular damage and death [49]. Changes in the ultrastructure of mitochondria in HCV-infected livers have been previously described and specific mitochondrial proteins have been found to inhibit cellular respiration and promote mitochondrial-derived ROS [50–52].

Another well-known source of ROS is the endoplasmic reticulum (ER) that accommodates cytochrome P450 mono-oxygenase enzymes that are implicated in ROS generation as well as in drug bioactivation. Among these, cytochrome P450 2E1 (CYP2E1) has been particularly implicated as a source of ROS in the liver [53]. Ethanol-induced upregulation of CYP2E1 is known to generate a significant amount of ROS leading to some of the complications observed in the context of alcoholic liver cirrhosis [54]. Ethanol exposure has also been associated with a more rapid progression to cirrhosis in HCV-infected patients [55]. Furthermore, human hepatoma-derived HepG2 cell lines were found to generate increased levels of ROS in the presence of HCV core protein and CYP2E1 [56]. Indeed, it has been suggested that alcohol-induced oxidative stress, which is mediated largely through the metabolism of ethanol by CYP2E1 or through the toxic effects of acetaldehyde, might further activate the immune system [57].

ROS generated from activated macrophages is also a major concern during inflammatory processes. The majority of the ROS from these cell types is derived from a membrane-bound endosomal complex called nicotinamide adenine dinucleotide phosphate oxidase (NOX). In the presence of a pathogen, macrophages are able to activate NOX that will assemble at the membrane of the pathogen-containing endosome leading to the GTP-assisted transfer of electrons from nicotinamide adenine dinucleotide phosphate (NADPH) to oxygen and the subsequent generation of superoxide. Hydrogen peroxide generated directly from superoxide may further utilize myeloperoxidase to generate hypochlorous acid, a powerful oxidizing agent capable of chlorinating macromolecules [58].

Lastly, overproduction of nitric oxide, a potent vasodilatory agent can lead to protein nitration through the production of peroxynitrite. Peroxynitrite is generated when nitric oxide combines with superoxide. Peroxynitrite has been shown to lead to protein nitration in the liver during inflammation [59]. Inducible nitric oxide synthase (iNOS) has been found to be upregulated during hepatocellular stress, including HCV and hepatitis B-induced infections [50, 60].

24.2.2
ROS Induced by Nonparenchymal Cells of the Liver (Kupffer Cells, Lymphocytes, Stellate Cells)

Oxidative stress is present in all types of chronic liver diseases, whether it is viral or chemical-induced [48]. The characteristic deposition of excessive extracellular matrix, mainly in the form of type 1 collagen, is carried out predominantly by activated hepatic stellate cells [61]. In viral-induced liver diseases, activation of resident Kupffer cells, stellate cells, and fibroblasts by the invading virus leads to a cascade of molecular events presumed to be directed toward controlling the spread of the virus, preventing further liver damage, and/or initiating the tissue reparation processes. These events become detrimental during chronic inflammation, when the production of oxidative stress is augmented by an immune system incapable of clearing the pathogen.

The interaction between the virus and the immune system plays an important role in the generation of ROS in the liver. The innate immune response is characterized partly by the production of inflammatory cytokines necessary to perpetuate and magnify the production of ROS by resident leukocytes.

24.2.2.1 Innate and Adaptive Immunity in HCV: the Snowball Effect

Immune system-mediated oxidative stress in the liver has been shown to play a significant role in the progression to cirrhosis. The inflammatory response against infection acts to magnify the number of ROS-producing leukocytes infiltrating the liver. Recognition of the infection commences with the detection of the HCV virus by cells expressing toll-like receptors (TLRs). TLRs recognize pathogen-associated molecular patterns such as lipopolysaccharide (LPS), flagellin, and foreign RNA. TLR 1 and 6 have been shown to be activated by the viral core and NS3 proteins, respectively, while other evidence suggests that TLR 3 may be activated by the presence of double strand HCV RNA [62–64]. TLR activation promotes a cascade of signaling events resulting in the increased expression of proinflammatory cytokines [65, 66]. Perhaps the most critical role that TLR activation plays is triggering the production of interferon α and β, which are important mediators of both the innate and the adaptive immune responses. Cytokine production is also particularly important toward the recruitment of circulatory leukocytes into the liver, leading to more macrophage-induced ROS formation. In this regard, the serum level of IP-10, a potent chemokine, was shown to be positively correlated with liver cirrhosis and damage [67]. Also, NS5A was found to cause an increase in the expression of IL-8, a potent neutrophil chemoattractant [68].

In addition to increased leukocyte recruitment, oxidative stress can be exacerbated by viral mechanisms that subvert the effectiveness of the innate immune system. For example, TLR 4 was also found to be significantly upregulated in a lymphoblastoid cell line expressing NS5A. As TLR 4 is responsive to LPS-derived from microorganisms, Machida et al. hypothesized that lymphocytes containing NS5A could be more sensitive to physiological levels of LPS [69]. Furthermore, TLR 4 activation by heat-inactivated *Escherichia coli* was shown to cause increased iNOS expression and nitric oxide synthesis, which was responsible for mitochondrial DNA depletion [70].

In addition to the innate immune response, the activation of the adaptive immune response by an antigen-presenting cell and propagation by CD4+ T-helper lymphocytes leads to the production of HCV-specific antibodies by B lymphocytes. Cytotoxic CD8+ lymphocytes can then eradicate the infected cells by direct cytolysis (perforin) or through activating the apoptotic pathway in infected hepatocytes (granzyme B).

24.2.2.2 Phagocyte NADPH Oxidase (NOX2)

In the presence of ongoing liver inflammation, the majority of ROS generated in the liver is derived from activated Kupffer cells and incoming leukocytes. This is supported by an overwhelming amount of evidence showing that liver oxidative stress levels can be significantly ameliorated by selectively inactivating Kupffer cells using gadolinium chloride. Some of the conditions that have been studied include LPS [71] and ethanol-induced inflammation [72]. Consistent with these studies are the findings that 8-OH guanosine and 4-hydroxynonenal protein adducts (two lipid peroxidation products) were positively correlated with CD68+ cells (a specific macrophage/Kupffer cell marker) in a set of 30 HCV-infected patients [43] (Figure 24.2).

Leukocytes produce ROS mainly through the phagocytic isoform of NOX (NOX2, gp91phox) (nonphagocytic isoforms of NOX have been reported in hepatocytes and hepatic stellate cells) [58]. Both nonphagocytic and phagocytic forms of NOX are membrane-bound enzymes whose activities require the association of a series of proteins in order to function. Membrane-bound NOX2 and p22phox are inactive until they are associated with the other cytosolic components of the complex (p47phox, p67phox, p47phox, and Rac) through a signal involving the phosphorylation of p47phox (Figure 24.3) [73, 74]. Activated NOX generates superoxide anions by donating electrons from NADPH to molecular oxygen within the phagosome. Although it is well known that NOX2-generated superoxide plays an integral role in liver disease as an armament against pathogens, it can also signal the expression of Kupffer cell major histocompatibility complex II and FAS ligand [75, 76].

Leukocyte NOX2 has been implicated in the "collateral damage" associated with chronic infections including HCV-mediated liver injury. The persistent production of inflammatory cytokines can therefore magnify leukocyte-induced ROS formation by increasing NOX2 activity [77]. This is supported by evidence that showed that the presence of serum host-derived factors such as interferon-γ, endotoxin, and serum viral core protein were positively correlated with reduced monocyte tolerance during chronic HCV infection. Elevations in TLR-specific signaling complexes in monocytes were associated with this observation, further suggesting that this could contribute to the persistent inflammatory state [78].

Although HCV may stimulate the NOX2-derived ROS production by indirectly promoting the expression of chemotactic cytokines in Kupffer cells, a

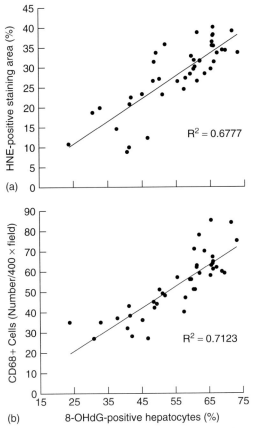

Fig. 24.2 Positive correlation between lipid peroxidation (a) and number of CD68+-expressing leukocytes (b) with oxidative DNA damage (8-OH-deoxyguanosine) in HCV-infected liver. Reprinted with permission [43].

calcium-dependent oxidative burst has also been observed in isolated, healthy, human monocytes when exposed to recombinant NS3 protein. NS3-stimulated NOX2 activity was associated with increased phosphorylation of p47phox [79].

24.2.2.3 Nonphagocyte NOX

Nonphagocytic NOX has been shown to be potentially involved in hepatic stellate cell activation. Nonphagocytic NOX has been found to be constitutively active at a low level in comparison to activated phagocytic NOX2; however, certain isoforms may be chemically stimulated by calcium or angiotensin II [80, 81]. Elevated activation of NOX in response to signaling with angiotensin II [82], platelet derived growth factor [83], or the engulfment of apoptotic bodies [73] has been demonstrated. Angiotensin II-induced stimulation of nonphagocytic NOX led to the redox-associated, profibrogenic increases in the expression of TGF-β1, and procollagen α1 [80]. Although no HSC-derived ROS has been implicated in

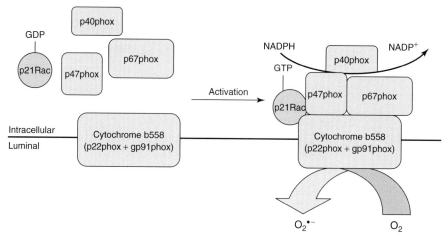

Fig. 24.3 Activation of NADPH oxidase within the endosome or at the plasma membrane leads to the recruitment of p40phox, p47phox, p67phox, and p21Rac to membrane-bound cytochrome b558 complex (composed of p22phox and gp91phox). Electrons from NADPH are channeled through the multimeric complex to molecular oxygen resulting in the formation of superoxide.

cellular damage, HSC activation leads to extracellular matrix deposition, and hence cirrhosis. Very little, however, is known about the role of nonphagocytic NOX in normal hepatocytes, although some evidence suggests that NOX may be an important costimulator of apoptosis under some circumstances [83–86].

24.2.3
Pro-oxidant Components of HCV

HCV replication in hepatocytes leads to the production of 10 individual proteins cleaved from the HCV polyprotein. Data available about the individual functions of each protein is listed in Table 24.1. The core, NS3 and NS5A proteins have been associated with increases in cellular oxidative stress. All of the above, except NS3 (see Section 24.2.4) were found to stimulate the production of oxidative stress within hepatocytes.

Chronically infected individuals with liver cirrhosis have a 1–3% annual risk of developing hepatocellular carcinoma [87]; a statistic that is significantly higher than the nominal rate of under 0.0024%. Moreover, oxidative stress and carcinogenesis have been described in some animal models of HCV infection in the absence of inflammation [88]. This suggests that oxidative stress caused by viral proteins alone may be sufficient to alter the function of critical proto-oncogenes or tumor suppressors, thus, leading to hepatocellular carcinoma. Furthermore, recent evidence shows that HCV proteins including core and NS5A can inhibit TNF-α-induced apoptosis, while causing chromosomal instability [89, 90].

24.2.3.1 Oxidative Stress Induced by HCV Core Protein

HCV core is a 21-kDa protein that is the building block of the HCV capsid containing the viral RNA. Monomers of core proteins have been shown to associate directly with mitochondrial and microsomal fractions in liver [91]. A stretch of 10 amino acids on the hydrophobic C terminus of the HCV core allows the protein to associate with the outer mitochondrial membrane [92]. In both transgenic mice and transfected cells expressing core proteins, elevations in mitochondrial ROS were observed [52, 93]. This data was consistent with a selective decrease in mitochondrial reduced glutathione pool. Core proteins were subsequently found to block electron transport at complex I [93]. Furthermore, core-induced oxidative stress increased the susceptibility to cell death and to glutathione depletion caused by L-buthionine sulfoximine [56]. Despite the experimental evidence that core proteins can contribute to mitochondrial impairment, little clinical evidence exists to corroborate these findings. By isolating hepatocytes from HCV-infected livers, we found a heterogeneity of infection in the hepatocyte population, which indicated that some infected hepatocytes contained more core proteins than others. We observed a corresponding decrease in mitochondrial membrane potential in hepatocytes with highest levels of core proteins (Figure 24.4). Consistent with these findings, Ando *et al.* found that by reducing viral replication in the nonneoplastic PH5CH8 hepatocyte cell line containing the full length HCV replicon, a reduction in mitochondrial redox status and decrease in core-induced oxidative stress occurred [94]. Although it is difficult to envision how core proteins might directly affect mitochondrial electron transport given that the components of the ETC are located on the inner mitochondrial membrane, whereas core proteins are found to be associated with the outer membrane, contact sites between the outer and inner mitochondrial membrane might explain this phenomenon [92]. Alternatively, it is also possible that core proteins may associate with complex I subunits synthesized on the ER prior to being imported to the mitochondria. However, evidence suggests that calcium may play an integral role in inducing HCV infection-associated mitochondrial ROS production. Elevations in ROS production by HCV core proteins were also found to be dependent on mitochondrial calcium influx, as Ru360, a specific inhibitor of the mitochondrial calcium uniporter, abolished mitochondrial ROS production in Huh7 cells expressing core proteins [51].

Core proteins may also stimulate nitric oxide formation through the upregulation of iNOS. This activation of iNOS was found to be NFκB-dependent and may contribute to oxidative stress in cells, as has been shown in the Chang human hepatocyte-derived cell line expressing core protein [50, 95] (Figure 24.5a).

The pathological effects of core protein expression in conjunction with ethanol exposure have also been considered given that ethanol, a more well-known mitochondrial toxin, has been shown to significantly accelerate HCV-induced liver injury [96, 97]. Ethanol-induced mitochondrial ROS production exacerbated core-induced mitochondrial ROS generation leading to increased hepatocellular death. Coexpression of the core protein and CYP2E1, both located on the mitochondria, has been shown to synergistically increase mitochondrial-derived ROS, oxidant-induced cell death, and sensitivity to receptor-mediated apoptosis [98]. Transgenic mice

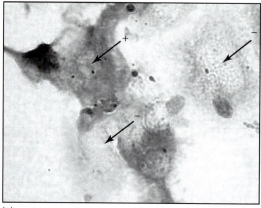

(a)

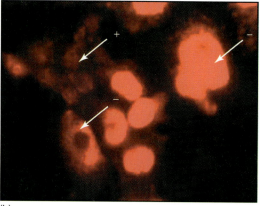

(b)

Fig. 24.4 Expression of HCV core protein is inversely correlated with mitochondrial membrane potential in cultured hepatocytes from HCV-infected humans. (a) HCV core protein was detected with a commercially available antibody coupled with immunohistochemical staining using Dako cytomation kit™. (b) Mitochondrial membrane potential was analyzed by incubating cultured hepatocytes from an explant liver from a HCV-infected donor. Mitotracker red TMX-ROS was incubated with the hepatocytes for 30 minutes prior to washing three times in warm, mitotracker-free DMEM, and fixed with formalin 5%. Analysis of mitochondrial membrane potential was carried out visually using a Zeiss Axioplan II microscope set at 400× magnification. (+) arrows indicate cells staining positive for core protein and (−) arrows indicate cells staining less or negative for core protein.

expressing the core protein have also been shown to produce significantly more ROS in conjunction with ethanol administration [99].

24.2.3.2 Oxidative Stress Induced by HCV NS5A Protein

The mature NS5A protein has a molecular weight of 56 kDa. The exact function of NS5A is currently unknown. However, it is generally acknowledged that NS5A is an important factor for efficient viral replication. It has also been proposed to

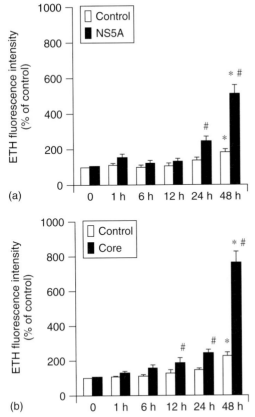

Fig. 24.5 Core and NS5A-induced reactive oxygen species in control cells, CHL-NS5A, or CHL core-expressing cells. (a) CHL-NS5A cells generate significantly more ROS as early as 6 hours following cytokine activation in comparison to control cells. (b) CHL core cells generate significantly more ROS, 48 hours following cytokine activation. *Significantly higher than all other time points ($P < 0.05$). #Significantly different than the control cells at the same time point. Reprinted with permission [50].

serve as a transcriptional activator with oncogenic potential [100]. NS5A aggregates primarily in the perinuclear regions of the hepatocytes on the ER [101].

The association of NS5A with the ER, likely causes ER stress in some form either through its mere accumulation at the ER or through its interactions with host proteins at the ER. NS5A is also associated with mitochondrial ROS production that was found to be accompanied by a rapid increase in calcium efflux from the ER [102]. Calcium transfer from the ER to the mitochondria has been shown to occur in an NS5A-overexpressing cell line. Furthermore, the redistribution of ER calcium to the mitochondria has been associated with mitochondrial oxidative stress and apoptosis [103–105].

Similar to the core protein, mitochondrial oxidative stress induced by NS5A may also increase the production of nitric oxide by upregulating iNOS through the

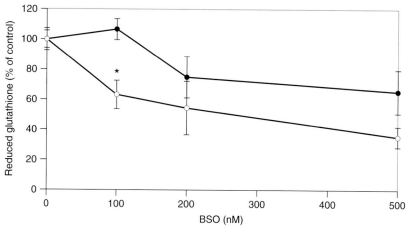

Fig. 24.6 Effect of NS5A on L-buthionine sulfoximine-induced glutathione depletion. Percentage glutathione depletion in Huh-7 cells (●) in comparison to Huh-7-derived 9-13 cells expressing NS5A protein (○). Cells were allowed to grow to confluency in 12-well polystyrene petri dishes. Following one wash with serum-free DMEM, cells were incubated for 24 hours in serum-free DMEM containing various concentrations of L-buthionine sulfoximine. *Significantly different than the control at the same time point as assessed by Student's T test ($n = 3$). Glutathione was analyzed as previously described [107]. 9-13 cells were acquired as a generous gift from Dr Ralf Bartenschlager at the University of Heidelberg, Heidelberg, Germany.

activation of NFκB [106]. A comparison of the ROS-producing potential of core and NS5A in a human-derived hepatocyte cell line showed that NS5A was more effective than core at generating ROS and potentiating cell death (Figure 24.5b). Furthermore, these effects were proportional to the abilities of each protein to stimulate iNOS expression [50]. Our preliminary studies showing that NS5A sensitizes hepatocyte-derived Huh-7 cells against L-buthionine sulfoximine-induced glutathione depletion is consistent with the pro-oxidant effects of this protein (Figure 24.6).

24.3
The Emerging Role of Antioxidants in the Treatment of HCV

In addition to conventional therapeutics that slow the growth and spread of the virus, pilot research projects on preventing oxidative stress-induced liver damage by using antioxidants have been ongoing. In a set of 23 HCV-infected patients, viral load was negatively correlated with the content of plasma antioxidants including β-carotene, retinol, vitamin C, and vitamin E suggesting that antioxidant therapy may ameliorate HCV-induced oxidative stress *in vivo* [108]. The use of antioxidants to reduce liver damage has shown promise as a possible adjuvant to antiviral therapy in the context of HCV infection. A recent randomized, double-blind study of interferon treatment nonresponders receiving intravenous antioxidants including

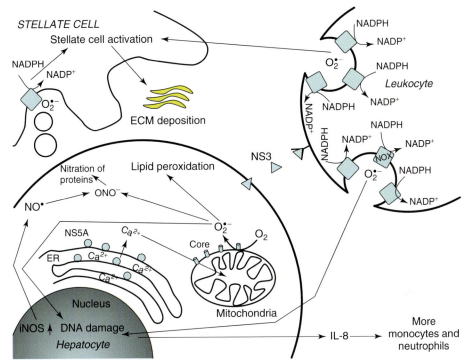

Fig. 24.7 Oxidative stress caused by leukocytes (monocytes and neutrophils) and viral proteins lead to the activation of hepatic stellate cells, macromolecular, and DNA damage. Viral core and NS5A lead to ER stress and the release of calcium to the cytosol. Sequestration of too much calcium by the mitochondria leads to increased ROS production, while core protein itself exacerbates this process by impairing electron transport. NFκβ signaling from the increased ROS results in the upregulation of iNOS and production of nitric oxide (NO*). NO* and superoxide (O_2*^-) can combine to form peroxynitrite (ONO^-), which can damage proteins by nitration. NS3 protein that is not encapsulated by the virus can activate leukocyte oxidative burst directly. The combined production of ROS by leukocytes, hepatocytes, and phagocytosis of apoptotic bodies by stellate cells causes stellate cell activation leading to extracellular matrix deposition and cytokine release (not shown).

10-g ascorbate and 750-mg glutathione showed reduced indices of liver damage and decreased serum viral load in comparison to the placebo group [109]. Despite these promising results, more clinical data is required before recommending the use of antioxidants in HCV-infected individuals.

As research on HCV continues, it is becoming clear that oxidative stress plays a significant role in the progression of the disease. Figure 24.7 illustrates the concerted role of the immune system and the virus itself in contributing to hepatocellular damage leading to cirrhosis and HCC. The available evidence reviewed here suggests that ROS production is part of the pathological process in this type of chronic infection.

24.4
Summary and Conclusions

24.4.1
Summary

In summary, we have reviewed some of the current evidence concerning both the viral and the immunological causes of oxidative stress during chronic HCV infection. These include the induction of pro-oxidant enzymes such as NO synthase and NOX as well as the direct effects of viral-derived proteins on hepatocellular oxidative stress and their synergistic effects on leukocyte-produced ROS. Together, these evidences highly suggest that the production of ROS by both the host immune response and the HCV-infected hepatocytes are true endogenous toxins that cause liver injury by promoting liver cirrhosis and initiating the neoplastic transformation of hepatocytes toward the development of HCC. Figure 24.7 illustrates the concerted role of the immune system and the virus itself in contributing to hepatocellular damage leading to cirrhosis and HCC.

24.4.2
Conclusions

As research on chronic HCV infection continues, it is becoming clear that oxidative stress plays a significant role in the progression toward end-stage liver cirrhosis, not only as an instigator of hepatocellular damage, but also as a profibrotic mediator in both parenchymal and nonparenchymal cells in the liver. The highly mutative characteristics of the virus, also contributes to oxidative liver damage by maintaining a pathologically persistent state of inflammation.

References

1 Brown, R.S. Jr and Gaglio, P.J. (2003) Scope of worldwide hepatitis C problem. *Liver Transpl*, **9**, S10–S13.

2 Kim, W.R. (2002) The burden of hepatitis C in the United States. *Hepatology*, **36**, S30–S34.

3 Aceijas, C. and Rhodes, T. (2007) Global estimates of prevalence of HCV infection among injecting drug users. *Int J Drug Policy*, **18**, 352–358.

4 Balasubramanian, A., Groopman, J.E., and Ganju, R.K. (2008) Underlying pathophysiology of HCV infection in HIV-positive drug users. *J Addict Dis*, **27**, 75–82.

5 Danta, M. and Dusheiko, G.M. (2008) Acute HCV in HIV-positive individuals – a review. *Curr Pharm Des*, **14**, 1690–1697.

6 Weber, R., Sabin, C.A., Friis-Moller, N., Reiss, P., El-Sadr, W.M., Kirk, O., Dabis, F., Law, M.G., Pradier, C., De Wit, S., Akerlund, B., Calvo, G., Monforte, A., Rickenbach, M., Ledergerber, B., Phillips, A.N., and Lundgren, J.D. (2006) Liver-related deaths in persons infected with the human immunodeficiency virus: the D:A:D study. *Arch Intern Med*, **166**, 1632–1641.

7 Kim, W.R., Poterucha, J.J., Hermans, J.E., Therneau, T.M., Dickson, E.R., Evans, R.W., and Gross, J.B. Jr (1997) Cost-effectiveness of 6 and

12 months of interferon-alpha therapy for chronic hepatitis C. *Ann Intern Med*, **127**, 866–874.

8. Dusheiko, G., Nelson, D., and Reddy, K.R. (2008) Ribavirin considerations in treatment optimization. *Antivir Ther*, **13** (Suppl 1), 23–30.

9. Saunders, J.C. (2008) Neuropsychiatric symptoms of hepatitis C. *Issues Ment Health Nurs*, **29**, 209–220.

10. Di Bisceglie, A.M. (1995) Hepatitis C and hepatocellular carcinoma. *Semin Liver Dis*, **15**, 64–69.

11. Choo, Q.L., Kuo, G., Weiner, A.J., Overby, L.R., Bradley, D.W., and Houghton, M. (1989) Isolation of a cDNA clone derived from a blood-borne non-A, non-B viral hepatitis genome. *Science*, **244**, 359–362.

12. Choo, Q.L., Weiner, A.J., Overby, L.R., Kuo, G., Houghton, M., and Bradley, D.W. (1990) Hepatitis C virus: the major causative agent of viral non-A, non-B hepatitis. *Br Med Bull*, **46**, 423–441.

13. De Vos, R., Verslype, C., Depla, E., Fevery, J., Van Damme, B., Desmet, V., and Roskams, T. (2002) Ultrastructural visualization of hepatitis C virus components in human and primate liver biopsies. *J Hepatol*, **37**, 370–379.

14. Grakoui, A., Wychowski, C., Lin, C., Feinstone, S.M., and Rice, C.M. (1993) Expression and identification of hepatitis C virus polyprotein cleavage products. *J Virol*, **67**, 1385–1395.

15. Lin, C., Lindenbach, B.D., Pragai, B.M., McCourt, D.W., and Rice, C.M. (1994) Processing in the hepatitis C virus E2-NS2 region: identification of p7 and two distinct E2-specific products with different C termini. *J Virol*, **68**, 5063–5073.

16. Grakoui, A., McCourt, D.W., Wychowski, C., Feinstone, S.M., and Rice, C.M. (1993) Characterization of the hepatitis C virus-encoded serine proteinase: determination of proteinase-dependent polyprotein cleavage sites. *J Virol*, **67**, 2832–2843.

17. Bartenschlager, R., Lohmann, V., Wilkinson, T., and Koch, J.O. (1995) Complex formation between the NS3 serine-type proteinase of the hepatitis C virus and NS4A and its importance for polyprotein maturation. *J Virol*, **69**, 7519–7528.

18. Gosert, R., Egger, D., Lohmann, V., Bartenschlager, R., Blum, H.E., Bienz, K., and Moradpour, D. (2003) Identification of the hepatitis C virus RNA replication complex in Huh-7 cells harboring subgenomic replicons. *J Virol*, **77**, 5487–5492.

19. Macdonald, A. and Harris, M. (2004) Hepatitis C virus NS5A: tales of a promiscuous protein. *J Gen Virol*, **85**, 2485–2502.

20. Behrens, S.E., Tomei, L., and De Francesco, R. (1996) Identification and properties of the RNA-dependent RNA polymerase of hepatitis C virus. *EMBO J*, **15**, 12–22.

21. Cocquerel, L., Kuo, C.C., Dubuisson, J., and Levy, S. (2003) CD81-dependent binding of hepatitis C virus E1E2 heterodimers. *J Virol*, **77**, 10677–10683.

22. Evans, M.J., von Hahn, T., Tscherne, D.M., Syder, A.J., Panis, M., Wolk, B., Hatziioannou, T., McKeating, J.A., Bieniasz, P.D., and Rice, C.M. (2007) Claudin-1 is a hepatitis C virus co-receptor required for a late step in entry. *Nature*, **446**, 801–805.

23. Kato, N., Sekiya, H., Ootsuyama, Y., Nakazawa, T., Hijikata, M., Ohkoshi, S., and Shimotohno, K. (1993) Humoral immune response to hypervariable region 1 of the putative envelope glycoprotein (gp70) of hepatitis C virus. *J Virol*, **67**, 3923–3930.

24. Maillard, P., Huby, T., Andreo, U., Moreau, M., Chapman, J., and Budkowska, A. (2006) The interaction of natural hepatitis C virus with human scavenger receptor SR-BI/Cla1 is mediated by ApoB-containing lipoproteins. *FASEB J*, **20**, 735–737.

25. Lavillette, D., Bartosch, B., Nourrisson, D., Verney, G., Cosset, F.L., Penin, F., and Pecheur, E.I. (2006) Hepatitis C virus glycoproteins mediate low pH-dependent

26 Hijikata, M., Kato, N., Ootsuyama, Y., Nakagawa, M., Ohkoshi, S., and Shimotohno, K. (1991) Hypervariable regions in the putative glycoprotein of hepatitis C virus. *Biochem Biophys Res Commun*, **175**, 220–228.

27 Kato, N., Ootsuyama, Y., Sekiya, H., Ohkoshi, S., Nakazawa, T., Hijikata, M., and Shimotohno, K. (1994) Genetic drift in hypervariable region 1 of the viral genome in persistent hepatitis C virus infection. *J Virol*, **68**, 4776–4784.

28 Guillou-Guillemette, H., Vallet, S., Gaudy-Graffin, C., Payan, C., Pivert, A., Goudeau, A., and Lunel-Fabiani, F. (2007) Genetic diversity of the hepatitis C virus: impact and issues in the antiviral therapy. *World J Gastroenterol*, **13**, 2416–2426.

29 Machida, K., Kondo, Y., Huang, J.Y., Chen, Y.C., Cheng, K.T., Keck, Z., Foung, S., Dubuisson, J., Sung, V.M., and Lai, M.M. (2008) Hepatitis C virus (HCV)-induced immunoglobulin hypermutation reduces the affinity and neutralizing activities of antibodies against HCV envelope protein. *J Virol*, **82**, 6711–6720.

30 von Hahn, T., Yoon, J.C., Alter, H., Rice, C.M., Rehermann, B., Balfe, P., and McKeating, J.A. (2007) Hepatitis C virus continuously escapes from neutralizing antibody and T-cell responses during chronic infection in vivo. *Gastroenterology*, **132**, 667–678.

31 Bowen, D.G. and Walker, C.M. (2005) Adaptive immune responses in acute and chronic hepatitis C virus infection. *Nature*, **436**, 946–952.

32 Saito, T., Owen, D.M., Jiang, F., Marcotrigiano, J., and Gale, M. Jr (2008) Innate immunity induced by composition-dependent RIG-I recognition of hepatitis C virus RNA. *Nature*, **454**, 523–527.

33 Thimme, R., Lohmann, V., and Weber, F. (2006) A target on the move: innate and adaptive immune escape strategies of hepatitis C virus. *Antiviral Res*, **69**, 129–141.

34 Simmonds, P. (2004) Genetic diversity and evolution of hepatitis C virus–15 years on. *J Gen Virol*, **85**, 3173–3188.

35 Timm, J. and Roggendorf, M. (2007) Sequence diversity of hepatitis C virus: implications for immune control and therapy. *World J Gastroenterol*, **13**, 4808–4817.

36 Bellecave, P. and Moradpour, D. (2008) A fresh look at interferon-alpha signaling and treatment outcomes in chronic hepatitis C. *Hepatology*, **48**, 1330–1333.

37 Kau, A., Vermehren, J., and Sarrazin, C. (2008) Treatment predictors of a sustained virologic response in hepatitis B and C. *J Hepatol*, **49**, 634–651.

38 Moreau, I., Levis, J., Crosbie, O., Kenny-Walsh, E., and Fanning, L.J. (2008) Correlation between pre-treatment quasispecies complexity and treatment outcome in chronic HCV genotype 3a. *Virol J*, **5**, 78.

39 Pawlotsky, J.M., Roudot-Thoraval, F., Bastie, A., Darthuy, F., Remire, J., Metreau, J.M., Zafrani, E.S., Duval, J., and Dhumeaux, D. (1996) Factors affecting treatment responses to interferon-alpha in chronic hepatitis C. *J Infect Dis*, **174**, 1–7.

40 Schinkel, J., de Jong, M.D., Bruning, B., van Hoek, B., Spaan, W.J., and Kroes, A.C. (2003) The potentiating effect of ribavirin on interferon in the treatment of hepatitis C: lack of evidence for ribavirin-induced viral mutagenesis. *Antivir Ther*, **8**, 535–540.

41 Zeuzem, S., Schmidt, J.M., Lee, J.H., von Wagner, M., Teuber, G., and Roth, W.K. (1998) Hepatitis C virus dynamics in vivo: effect of ribavirin and interferon alfa on viral turnover. *Hepatology*, **28**, 245–252.

42 Barbaro, G., Di Lorenzo, G., Asti, A., Ribersani, M., Belloni, G., Grisorio, B., Filice, G., and Barbarini, G. (1999) Hepatocellular mitochondrial alterations in patients with chronic hepatitis C: ultrastructural and biochemical findings. *Am J Gastroenterol*, **94**, 2198–2205.

43 Maki, A., Kono, H., Gupta, M., Asakawa, M., Suzuki, T., Matsuda, M., Fujii, H., and Rusyn, I. (2007) Predictive power of biomarkers of oxidative stress and inflammation in patients with hepatitis C virus-associated hepatocellular carcinoma. *Ann Surg Oncol*, **14**, 1182–1190.

44 Hatano, R., Ebara, M., Fukuda, H., Yoshikawa, M., Sugiura, N., Kondo, F., Yukawa, M., and Saisho, H. (2000) Accumulation of copper in the liver and hepatic injury in chronic hepatitis C. *J Gastroenterol Hepatol*, **15**, 786–791.

45 Shedlofsky, S.I. (1998) Role of iron in the natural history and clinical course of hepatitis C disease. *Hepatogastroenterology*, **45**, 349–355.

46 Boveris, A., Oshino, N., and Chance, B. (1972) The cellular production of hydrogen peroxide. *Biochem J*, **128**, 617–630.

47 Gebhardt, R. (2002) Oxidative stress, plant-derived antioxidants and liver fibrosis. *Planta Med*, **68**, 289–296.

48 Parola, M. and Robino, G. (2001) Oxidative stress-related molecules and liver fibrosis. *J Hepatol*, **35**, 297–306.

49 Adam-Vizi, V. (2005) Production of reactive oxygen species in brain mitochondria: contribution by electron transport chain and non-electron transport chain sources. *Antioxid Redox Signal*, **7**, 1140–1149.

50 Garcia-Mediavilla, M.V., Sanchez-Campos, S., Gonzalez-Perez, P., Gomez-Gonzalo, M., Majano, P.L., Lopez-Cabrera, M., Clemente, G., Garcia-Monzon, C., and Gonzalez-Gallego, J. (2005) Differential contribution of hepatitis C virus NS5A and core proteins to the induction of oxidative and nitrosative stress in human hepatocyte-derived cells. *J Hepatol*, **43**, 606–613.

51 Li, Y., Boehning, D.F., Qian, T., Popov, V.L., and Weinman, S.A. (2007) Hepatitis C virus core protein increases mitochondrial ROS production by stimulation of Ca2+ uniporter activity. *FASEB J*, **21**, 2474–2485.

52 Okuda, M., Li, K., Beard, M.R., Showalter, L.A., Scholle, F., Lemon, S.M., and Weinman, S.A. (2002) Mitochondrial injury, oxidative stress, and antioxidant gene expression are induced by hepatitis C virus core protein. *Gastroenterology*, **122**, 366–375.

53 Caro, A.A. and Cederbaum, A.I. (2004) Oxidative stress, toxicology, and pharmacology of CYP2E1. *Annu Rev Pharmacol Toxicol*, **44**, 27–42.

54 Jimenez-Lopez, J.M. and Cederbaum, A.I. (2005) CYP2E1-dependent oxidative stress and toxicity: role in ethanol-induced liver injury. *Expert Opin Drug Metab Toxicol*, **1**, 671–685.

55 Hezode, C., Lonjon, I., Roudot-Thoraval, F., Pawlotsky, J.M., Zafrani, E.S., and Dhumeaux, D. (2003) Impact of moderate alcohol consumption on histological activity and fibrosis in patients with chronic hepatitis C, and specific influence of steatosis: a prospective study. *Aliment Pharmacol Ther*, **17**, 1031–1037.

56 Wen, F., Abdalla, M.Y., Aloman, C., Xiang, J., Ahmad, I.M., Walewski, J., McCormick, M.L., Brown, K.E., Branch, A.D., Spitz, D.R., Britigan, B.E., and Schmidt, W.N. (2004) Increased prooxidant production and enhanced susceptibility to glutathione depletion in HepG2 cells co-expressing HCV core protein and CYP2E1. *J Med Virol*, **72**, 230–240.

57 Albano, E. (2002) Free radical mechanisms in immune reactions associated with alcoholic liver disease. *Free Radic Biol Med*, **32**, 110–114.

58 De Minicis, S. and Brenner, D.A. (2007) NOX in liver fibrosis. *Arch Biochem Biophys*, **462**, 266–272.

59 Dedon, P.C. and Tannenbaum, S.R. (2004) Reactive nitrogen species in the chemical biology of inflammation. *Arch Biochem Biophys*, **423**, 12–22.

60 Atik, E., Onlen, Y., Savas, L., and Doran, F. (2008) Inducible nitric oxide synthase and histopathological

correlation in chronic viral hepatitis. *Int J Infect Dis*, **12**, 12–15.
61 Bataller, R. and Brenner, D.A. (2001) Hepatic stellate cells as a target for the treatment of liver fibrosis. *Semin Liver Dis*, **21**, 437–451.
62 Chang, S., Dolganiuc, A., and Szabo, G. (2007) Toll-like receptors 1 and 6 are involved in TLR2-mediated macrophage activation by hepatitis C virus core and NS3 proteins. *J Leukoc Biol*, **82**, 479–487.
63 Alexopoulou, L., Holt, A.C., Medzhitov, R., and Flavell, R.A. (2001) Recognition of double-stranded RNA and activation of NF-kappaB by Toll-like receptor 3. *Nature*, **413**, 732–738.
64 Seya, T., Shingai, M., and Matsumoto, M. (2004) Toll-like receptors that sense viral infection. *Uirusu*, **54**, 1–8.
65 Adachi, O., Kawai, T., Takeda, K., Matsumoto, M., Tsutsui, H., Sakagami, M., Nakanishi, K., and Akira, S. (1998) Targeted disruption of the MyD88 gene results in loss of IL-1- and IL-18-mediated function. *Immunity*, **9**, 143–150.
66 Medzhitov, R., Preston-Hurlburt, P., Kopp, E., Stadlen, A., Chen, C., Ghosh, S., and Janeway, C.A. Jr (1998) MyD88 is an adaptor protein in the hToll/IL-1 receptor family signaling pathways. *Mol Cell*, **2**, 253–258.
67 Roe, B., Coughlan, S., Hassan, J., Grogan, A., Farrell, G., Norris, S., Bergin, C., and Hall, W.W. (2007) Elevated serum levels of interferon-gamma -inducible protein-10 in patients coinfected with hepatitis C virus and HIV. *J Infect Dis*, **196**, 1053–1057.
68 Girard, S., Shalhoub, P., Lescure, P., Sabile, A., Misek, D.E., Hanash, S., Brechot, C., and Beretta, L. (2002) An altered cellular response to interferon and up-regulation of interleukin-8 induced by the hepatitis C viral protein NS5A uncovered by microarray analysis. *Virology*, **295**, 272–283.
69 Machida, K., Cheng, K.T., Sung, V.M., Levine, A.M., Foung, S., and Lai, M.M. (2006) Hepatitis C virus induces toll-like receptor 4 expression, leading to enhanced production of beta interferon and interleukin-6. *J Virol*, **80**, 866–874.
70 Suliman, H.B., Welty-Wolf, K.E., Carraway, M.S., Schwartz, D.A., Hollingsworth, J.W., and Piantadosi, C.A. (2005) Toll-like receptor 4 mediates mitochondrial DNA damage and biogenic responses after heat-inactivated E. coli. *FASEB J*, **19**, 1531–1533.
71 Fukuda, M., Yokoyama, H., Mizukami, T., Ohgo, H., Okamura, Y., Kamegaya, Y., Horie, Y., Kato, S., and Ishii, H. (2004) Kupffer cell depletion attenuates superoxide anion release into the hepatic sinusoids after lipopolysaccharide treatment. *J Gastroenterol Hepatol*, **19**, 1155–1162.
72 Koop, D.R., Klopfenstein, B., Iimuro, Y., and Thurman, R.G. (1997) Gadolinium chloride blocks alcohol-dependent liver toxicity in rats treated chronically with intragastric alcohol despite the induction of CYP2E1. *Mol Pharmacol*, **51**, 944–950.
73 Zhan, S.S., Jiang, J.X., Wu, J., Halsted, C., Friedman, S.L., Zern, M.A., and Torok, N.J. (2006) Phagocytosis of apoptotic bodies by hepatic stellate cells induces NADPH oxidase and is associated with liver fibrosis in vivo. *Hepatology*, **43**, 435–443.
74 Ago, T., Nunoi, H., Ito, T., and Sumimoto, H. (1999) Mechanism for phosphorylation-induced activation of the phagocyte NADPH oxidase protein p47(phox). Triple replacement of serines 303, 304, and 328 with aspartates disrupts the SH3 domain-mediated intramolecular interaction in p47(phox), thereby activating the oxidase. *J Biol Chem*, **274**, 33644–33653.
75 Maemura, K., Zheng, Q., Wada, T., Ozaki, M., Takao, S., Aikou, T., Bulkley, G.B., Klein, A.S., and Sun, Z. (2005) Reactive oxygen species are

76 Uchikura, K., Wada, T., Hoshino, S., Nagakawa, Y., Aiko, T., Bulkley, G.B., Klein, A.S., and Sun, Z. (2004) Lipopolysaccharides induced increases in Fas ligand expression by Kupffer cells via mechanisms dependent on reactive oxygen species. *Am J Physiol Gastrointest Liver Physiol*, **287**, G620–G626.

77 Decoursey, T.E. and Ligeti, E. (2005) Regulation and termination of NADPH oxidase activity. *Cell Mol Life Sci*, **62**, 2173–2193.

78 Dolganiuc, A., Norkina, O., Kodys, K., Catalano, D., Bakis, G., Marshall, C., Mandrekar, P., and Szabo, G. (2007) Viral and host factors induce macrophage activation and loss of toll-like receptor tolerance in chronic HCV infection. *Gastroenterology*, **133**, 1627–1636.

79 Bureau, C., Bernad, J., Chaouche, N., Orfila, C., Beraud, M., Gonindard, C., Alric, L., Vinel, J.P., and Pipy, B. (2001) Nonstructural 3 protein of hepatitis C virus triggers an oxidative burst in human monocytes via activation of NADPH oxidase. *J Biol Chem*, **276**, 23077–23083.

80 Bataller, R., Schwabe, R.F., Choi, Y.H., Yang, L., Paik, Y.H., Lindquist, J., Qian, T., Schoonhoven, R., Hagedorn, C.H., Lemasters, J.J., and Brenner, D.A. (2003) NADPH oxidase signal transduces angiotensin II in hepatic stellate cells and is critical in hepatic fibrosis. *J Clin Invest*, **112**, 1383–1394.

81 Jagnandan, D., Church, J.E., Banfi, B., Stuehr, D.J., Marrero, M.B., and Fulton, D.J. (2007) Novel mechanism of activation of NADPH oxidase 5: calcium sensitization via phosphorylation. *J Biol Chem*, **282**, 6494–6507.

82 Griendling, K.K. and Ushio-Fukai, M. (2000) Reactive oxygen species as mediators of angiotensin II signaling. *Regul Pept*, **91**, 21–27.

83 Adachi, T., Togashi, H., Suzuki, A., Kasai, S., Ito, J., Sugahara, K., and Kawata, S. (2005) NAD(P)H oxidase plays a crucial role in PDGF-induced proliferation of hepatic stellate cells. *Hepatology*, **41**, 1272–1281.

84 Reinehr, R., Becker, S., Braun, J., Eberle, A., Grether-Beck, S., and Haussinger, D. (2006) Endosomal acidification and activation of NADPH oxidase isoforms are upstream events in hyperosmolarity-induced hepatocyte apoptosis. *J Biol Chem*, **281**, 23150–23166.

85 Reinehr, R., Becker, S., Eberle, A., Grether-Beck, S., and Haussinger, D. (2005) Involvement of NADPH oxidase isoforms and Src family kinases in CD95-dependent hepatocyte apoptosis. *J Biol Chem*, **280**, 27179–27194.

86 Carmona-Cuenca, I., Roncero, C., Sancho, P., Caja, L., Fausto, N., Fernandez, M., and Fabregat, I. (2008) Upregulation of the NADPH oxidase NOX4 by TGF-beta in hepatocytes is required for its pro-apoptotic activity. *J Hepatol*. **49**(6), 965–976.

87 Koike, K., Tsutsumi, T., Fujie, H., Shintani, Y., and Kyoji, M. (2002) Molecular mechanism of viral hepatocarcinogenesis. *Oncology*, **62** (Suppl 1), 29–37.

88 Moriya, K., Nakagawa, K., Santa, T., Shintani, Y., Fujie, H., Miyoshi, H., Tsutsumi, T., Miyazawa, T., Ishibashi, K., Horie, T., Imai, K., Todoroki, T., Kimura, S., and Koike, K. (2001) Oxidative stress in the absence of inflammation in a mouse model for hepatitis C virus-associated hepatocarcinogenesis. *Cancer Res*, **61**, 4365–4370.

89 Ghosh, A.K., Majumder, M., Steele, R., Meyer, K., Ray, R., and Ray, R.B. (2000) Hepatitis C virus NS5A protein protects against TNF-alpha mediated apoptotic cell death. *Virus Res*, **67**, 173–178.

90 Baek, K.H., Park, H.Y., Kang, C.M., Kim, S.J., Jeong, S.J., Hong, E.K., Park, J.W., Sung, Y.C., Suzuki, T., Kim, C.M., and Lee, C.W. (2006) Overexpression of hepatitis C virus

NS5A protein induces chromosome instability via mitotic cell cycle dysregulation. *J Mol Biol*, **359**, 22–34.

91 Suzuki, R., Sakamoto, S., Tsutsumi, T., Rikimaru, A., Tanaka, K., Shimoike, T., Moriishi, K., Iwasaki, T., Mizumoto, K., Matsuura, Y., Miyamura, T., and Suzuki, T. (2005) Molecular determinants for subcellular localization of hepatitis C virus core protein. *J Virol*, **79**, 1271–1281.

92 Schwer, B., Ren, S., Pietschmann, T., Kartenbeck, J., Kaehlcke, K., Bartenschlager, R., Yen, T.S., and Ott, M. (2004) Targeting of hepatitis C virus core protein to mitochondria through a novel C-terminal localization motif. *J Virol*, **78**, 7958–7968.

93 Korenaga, M., Wang, T., Li, Y., Showalter, L.A., Chan, T., Sun, J., and Weinman, S.A. (2005) Hepatitis C virus core protein inhibits mitochondrial electron transport and increases reactive oxygen species (ROS) production. *J Biol Chem*, **280**, 37481–37488.

94 Ando, M., Korenaga, M., Hino, K., Ikeda, M., Kato, N., Nishina, S., Hidaka, I., and Sakaida, I. (2008) Mitochondrial electron transport inhibition in full genomic hepatitis C virus replicon cells is restored by reducing viral replication. *Liver Int*, **28**, 1158–1166.

95 de Lucas, S., Bartolome, J., Amaro, M.J., and Carreno, V. (2003) Hepatitis C virus core protein transactivates the inducible nitric oxide synthase promoter via NF-kappaB activation. *Antiviral Res*, **60**, 117–124.

96 McCartney, E.M., Semendric, L., Helbig, K.J., Hinze, S., Jones, B., Weinman, S.A., and Beard, M.R. (2008) Alcohol metabolism increases the replication of Hepatitis C virus and attenuates the antiviral action of interferon. *J Infect Dis*. **198**(12), 1766–1775.

97 Szabo, G. (2003) Pathogenic interactions between alcohol and hepatitis C. *Curr Gastroenterol Rep*, **5**, 86–92.

98 Otani, K., Korenaga, M., Beard, M.R., Li, K., Qian, T., Showalter, L.A., Singh, A.K., Wang, T., and Weinman, S.A. (2005) Hepatitis C virus core protein, cytochrome P450 2E1, and alcohol produce combined mitochondrial injury and cytotoxicity in hepatoma cells. *Gastroenterology*, **128**, 96–107.

99 Koike, K., Tsutsumi, T., Miyoshi, H., Shinzawa, S., Shintani, Y., Fujie, H., Yotsuyanagi, H., and Moriya, K. (2008) Molecular basis for the synergy between alcohol and hepatitis C virus in hepatocarcinogenesis. *J Gastroenterol Hepatol*, **23** (Suppl 1), S87–S91.

100 Kato, N., Lan, K.H., Ono-Nita, S.K., Shiratori, Y., and Omata, M. (1997) Hepatitis C virus nonstructural region 5A protein is a potent transcriptional activator. *J Virol*, **71**, 8856–8859.

101 Ide, Y., Zhang, L., Chen, M., Inchauspe, G., Bahl, C., Sasaguri, Y., and Padmanabhan, R. (1996) Characterization of the nuclear localization signal and subcellular distribution of hepatitis C virus nonstructural protein NS5A. *Gene*, **182**, 203–211.

102 Gong, G., Waris, G., Tanveer, R., and Siddiqui, A. (2001) Human hepatitis C virus NS5A protein alters intracellular calcium levels, induces oxidative stress, and activates STAT-3 and NF-kappa B. *Proc Natl Acad Sci U S A*, **98**, 9599–9604.

103 Liu, H., Bowes, R.C. III, van de Water, B., Sillence, C., Nagelkerke, J.F., and Stevens, J.L. (1997) Endoplasmic reticulum chaperones GRP78 and calreticulin prevent oxidative stress, Ca2+ disturbances, and cell death in renal epithelial cells. *J Biol Chem*, **272**, 21751–21759.

104 Pahl, H.L. (1999) Signal transduction from the endoplasmic reticulum to the cell nucleus. *Physiol Rev*, **79**, 683–701.

105 Deniaud, A., Sharaf el dein, O., Maillier, E., Poncet, D., Kroemer, G., Lemaire, C., and Brenner, C. (2008) Endoplasmic reticulum stress induces calcium-dependent permeability transition, mitochondrial outer

membrane permeabilization and apoptosis. *Oncogene*, **27**, 285–299.

106 Waris, G., Livolsi, A., Imbert, V., Peyron, J.F., and Siddiqui, A. (2003) Hepatitis C virus NS5A and subgenomic replicon activate NF-kappaB via tyrosine phosphorylation of IkappaBalpha and its degradation by calpain protease. *J Biol Chem*, **278**, 40778–40787.

107 Chan, T.S., Wilson, J.X., and O'Brien, P.J. (2004) Glycogenolysis is directed towards ascorbate synthesis by glutathione conjugation. *Biochem Biophys Res Commun*, **317**, 149–156.

108 Gronbaek, K., Sonne, J., Ring-Larsen, H., Poulsen, H.E., Friis, H., and Bygum, K.H. (2005) Viral load is a negative predictor of antioxidant levels in hepatitis C patients. *Scand J Infect Dis*, **37**, 686–689.

109 Gabbay, E., Zigmond, E., Pappo, O., Hemed, N., Rowe, M., Zabrecky, G., Cohen, R., and Ilan, Y. (2007) Antioxidant therapy for chronic hepatitis C after failure of interferon: results of phase II randomized, double-blind placebo controlled clinical trial. *World J Gastroenterol*, **13**, 5317–5323.

25
Oxidized Low Density Lipoprotein Cytotoxicity and Vascular Disease

Steven P. Gieseg, Elizabeth Crone, and Zunika Amit

The formation of oxidized low density lipoprotein (oxLDL) within atherosclerotic plaques is a significant event, which appears to drive the transition from fatty streaks to advanced complex plaque by initiating cell death. OxLDL in tissue culture is a potent cytotoxic agent that triggers a number of competing cell death mechanisms. Oxysterols within oxLDL can alter membrane lipid rafts resulting in calcium influx with calpain activation and cytochrome c release. Apoptosis is also activated by oxysterol-induced degradation of the prosurvival protein kinase, AKT. In contrast, CD36 binding of oxLDL causes necrosis through the generation of an intracellular oxidant flux, while excessive uptake of oxLDL triggers lysosome destabilization. The nature and significance of these mechanisms depends on the type of cell under investigation and how the oxLDL is prepared in the laboratory. Similarly, the endogenous protection mechanisms are also dependent on cell type and the oxLDL preparation. γ-Interferon stimulation of macrophages generates the pterin, 7,8-dihydroneopterin, which inhibits oxLDL toxicity in monocyte-like U937 and human monocyte-derived macrophages, but not the monocyte-like THP-1 cells. Cytotoxic and cellular protection of oxLDL, as well oxLDL formation mechanisms both *in vivo* and *in vitro* will be discussed.

Cholesterol is usually described as the major risk factor for the development of blood clots, high blood pressure, heart disease, and stroke. Yet, it is the oxidation of cholesterol carrying particle in the artery wall which appears to drive much of the pathology behind these diseases of the vasculature. Oxidized cholesterol and lipids have a range of properties which activate the body's inflammatory and blood clotting machinery. Most significantly, oxidized cholesterol and the oxidized form of the cholesterol carrying particle, called *oxidized low density lipoprotein (oxLDL)*, are cytotoxic to a range of cells. oxLDL-induced death of immune cells within the artery wall contributes to the growth of deposits, which narrow and block arteries, causing strokes and heart attacks. In this chapter, we examine how oxidized cholesterol particles form *in vivo* and contribute to the development of vascular disease through cytotoxic activity. We also examine how oxLDL is made in the laboratory and explore the various cytotoxic

Endogenous Toxins. Diet, Genetics, Disease and Treatment.
Edited by Peter J. O'Brien and W. Robert Bruce
Copyright © 2010 WILEY-VCH Verlag GmbH & Co. KGaA, Weinheim
ISBN: 978-3-527-32363-0

mechanisms triggered by oxidized cholesterol, endogenous oxidized cholesterol particles.

25.1
Introduction to Vascular Disease

Triacylglycerides, phospholipids, and cholesterol esters are transported within the circulation as water-soluble lipoprotein particles. The oxidation of low density lipoprotein (LDL), the cholesterol carrying lipoprotein particle, in the walls of arteries is a significant step in the development of heart disease, stroke, and often high blood pressure. Collectively these conditions are all termed *vascular disease* as they are caused by the thickening and growth of the artery wall at bends and bifurcations. These thickened areas develop over time into atherosclerotic plaques that reduce the lumen of the blood vessel, so less blood is supplied to organs downstream. The rupture of these plaques results in the formation of blood clots that may block the downstream blood supply altogether. In the case of the coronary arteries supplying the heart, this leads to a heart attack and within the arteries supplying the brain, a stroke. These plaques have a rubbery consistency, and some have areas that are very hard due to sheets of calcium deposits (Figure 25.1). These hard plaques are the basis for the name atherosclerosis, which means "hardening of the arteries." The particular plaque shown in Figure 25.1 was excised from the carotid artery supplying the brain of an 84-year-old patient. What is shown is not the artery but the atherosclerotic plaque itself that has been excised from the lining of the artery. The procedure involves clamping off the bifurcation of the carotid artery, cutting the artery open, and then scraping out the plaque leaving the artery wall intact and clear of obstruction.

Atherosclerotic plaques are often described as cholesterol deposits but are actually collections of cholesterol filled macrophage cells recruited from the blood. The cells enter the artery wall as monocytes and collect in the intimal layer that is between the endothelial cells forming the inner wall of the artery and the underlying basement membrane. The monocytes differentiate into larger macrophage cells within the intimal layer. Macrophages readily accumulate cholesterol esters from oxLDL to form lipid loaded "foam cells," so called because of their foamy appearance under the microscope. Growth of plaques appears to be due to further recruitment into the intimal layer of monocyte cells from the blood and possibly fibroblasts and smooth muscles from the artery wall as well. The death of the cells in the center of the plaque mass leads to the formation of an advanced plaque, with a characteristic necrotic core of cell debris and lipid deposits rich in cholesterol and cholesterol esters [1, 2].

What actually drives the formation and growth of atherosclerotic plaque within artery walls has been a subject of controversy for many decades. Originally viewed as a cholesterol deposition problem due to high plasma cholesterol levels, it is now realized that atherosclerosis is actually a chronic inflammatory process in the arterial intima, which is why phagocytic antigen presenting cells such as

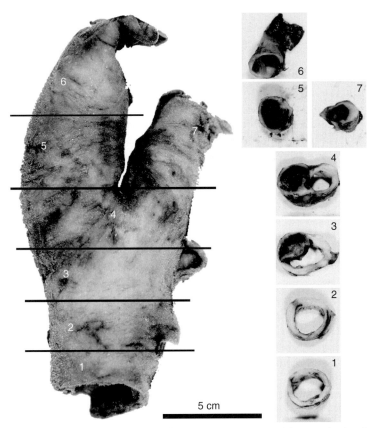

Fig. 25.1 Atherosclerotic plaque. The large structure on the left-hand side is an atherosclerotic plaque, it is not the whole artery. The plaque was surgically removed by cutting open the artery and peeling out the plaque from the inner layer of artery wall. In healthy tissue, this material would normally be only a few cells thick. Note that the plaque goes right around the artery wall and the blood flowed through the center. Blood flowed from the bottom section (in the common carotid artery) which then bifurcated into the external carotid artery (on the right) and the internal carotid artery (on the left) which supplies the brain. The plaque was removed from the left common carotid artery of an 84-year-old nonsmoking patient who presented with transient ischemic attacks. Ultrasound analysis showed 85% stenosis (narrowing) of the vessel. The plaque was sectioned across the lines shown in the photograph. Each section has been placed on its side to give a view through the plaque showing the level of stenosis of the artery lumen in each segment. The number of each segment corresponds to the section shown on the intact plaque on the left-hand side. Note that the blood flow through the left-hand side in section 3 was almost completely blocked by the plaque. This blockage continued through sections 4 and 5 into the internal carotid artery. This is the classic type of advanced lipid-filled plaque with a large necrotic core region and significant levels of calcification. The frosted appearance of the photographs is because the plaque was photographed and sectioned while still frozen.

macrophages are present [3, 4]. Though high blood pressure and plasma cholesterol are known risk factors for atherosclerosis, how they interact with the inflammatory process within the artery wall has been difficult to demonstrate. What does link the various observations is the formation of oxidized lipids, mainly oxLDL within the artery wall.

In the laboratory, oxLDL is chemotactic to monocytes/macrophages, induces macrophage uptake of oxLDL causing the formation of lipid loaded foam cells [5], and is cytotoxic to a range of cells including macrophages [6], smooth muscle cells [7], and endothelial cells [8]. oxLDL also readily binds to extracellular matrix suggesting this enhances its retention within the atherosclerotic plaque [9]. The experimental evidence therefore does support oxLDL to be a key driver of plaque formation.

oxLDL particles along with oxidatively modified lipids and proteins have been measured in atherosclerotic plaques [10–12], as has unoxidized LDL and significant amounts of tocopherol, the main lipophilic antioxidant [13]. The failure of human antioxidant supplementation trials to reduce death as a result of vascular disease has further called into question the role of oxLDL in atherosclerosis [14]. What is clear is oxLDL's toxicity to cells and its presence within plaques. Though oxLDL may not be the initiator of atherosclerosis, it is a key component in the development of complex plaques and subsequent plaque rupture where cell death is a prominent feature. Cell death appears to be the major driving factor in the formation of a necrotic core in plaques, especially plaques with a high lipid content [15]. Cell death has also been implicated in plaque instability leading to plaque rupture and thrombus formation (blood clots) causing strokes and heart attacks [16].

oxLDL is not found in circulation, though antibodies to oxLDL and low levels of lipid peroxides are found in plasma. Plasma, except under very high radical flux, does not support oxLDL formation due to the high concentrations of protein, ascorbate (vitamin C), and urate, which effectively scavenge the oxidative radicals. The site of oxLDL formation therefore appears to be within inflammatory sites such as atherosclerotic plaques where levels of plasma antioxidants are reduced or depleted. The problem has therefore become identifying the conditions under which oxLDL forms and what type of oxLDL is interacting with cells.

25.2
oxLDL Formation

LDL has a high content of polyunsaturated fatty acids (PUFAs), mainly cholesterol esters of linoleic acid and arachidonic acid. These fatty acids react readily and rapidly with oxygen to form lipid peroxides via a peroxyl radical chain reaction (Figure 25.2). With purified LDL, the reaction can be initiated by reduced metals such as iron and copper. LDL has a number of redox active transition metal ion binding sites which appear to be the source of the initiating radical generation

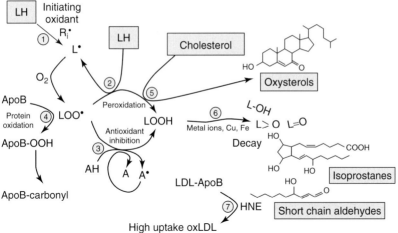

Fig. 25.2 Formation of oxLDL by peroxidation. Polyunsaturated fatty acids (LH) esterified to cholesterol or phospholipids (1) lose a hydrogen to an attacking initiating radical ($R_i^\bullet$). The resulting carbon-centered radical rapidly reacts with oxygen to give a lipid peroxyl radical ($LOO^\bullet$) which can (2) react with more PUFA (LH) to give a lipid hydroperoxide (LOOH) and a new carbon-centered radical ($L^\bullet$), which can initiate a new round of oxidation. If an antioxidant (AH) is present (3) such as α-tocopherol, the lipid peroxyl radical ($LOO^\bullet$) is neutralized to give a lipid hydroperoxide without further lipid radical generation. The resulting tocopherol radical can further react with and neutralize a second lipid peroxyl radical. The lipid peroxyl radicals ($LOO^\bullet$) can also attack the R groups of the amino acid residues in the ApoB100 protein (4) to form protein hydroperoxides (ApoB-OOH) which decay over hours to protein carbonyls. The peroxyl radical ($LOO^\bullet$) and hydroperoxide decay products (6) such as the alkoxyl radicals ($LO^\bullet$) cause (5) secondary oxidative damage to cholesterol generating a range of oxysterols such as 7-ketocholesterol (shown). Much of these oxysterols remain esterified to fatty acids, which may also be oxidized. Breakdown of lipid hydroperoxides to peroxyl ($LOO^\bullet$) and alkoxyl ($LO^\bullet$) radicals generates isoprostanes and short chain aldehydes (6). These lipid-derived aldehydes such as 4-hydroxynonenal (HNE) can react (7) with lysines within the apoB100 protein moiety of LDL, so destroying the LDL receptor binding site and changing the resulting oxLDL into the form recognized by receptors of the macrophage scavenger receptor family.

[17]. The lipid peroxidation reaction driven by the primary radicals ($R_i^\bullet$) can be monitored by measuring conjugated diene formation. This type of analysis shows there is an initial "lag phase" where the lipophilic antioxidants, mainly α-tocopherol (vitamin E), inhibit the peroxidation chain reaction by scavenging the lipid peroxyl radicals [18, 19]. Addition of water-soluble antioxidants such as ascorbate or 7,8-dihydroneopterin (7,8-DNP) (Section 25.6) can extend this lag phase by scavenging (removing and neutralizing) the lipid peroxyl radical (Figure 25.2, reaction 3) [20–22].

The exact mechanism of the initiating radical generation remains controversial both with cell- and transition ion–mediated LDL peroxidation. With copper ions, the initiating radical formation appears to involve the metal ions binding to amino acids on the apolipoprotein B (apoB) protein moiety of the LDL, most likely histidine residues. The bound ions are then reduced by vitamin E, reductive amino acid residues, or preformed hydroperoxides. The reduced metal ions can further react directly with oxygen or peroxides via Fenton reactions [23]. The processes generating the preformed lipid peroxides in the LDL is relatively vague. Studies in Esterbauer's laboratory in the early 1990s failed to find any difference in oxidation rates of LDL prepared from freshly taken blood or LDL left to "age" in the fridge for 24 hours during which lipid peroxide levels climb to one peroxide per particle.

At extremely low radical fluxes (usually created using low copper or iron levels), there are insufficient lipid peroxyl radicals to further react and neutralize the tocopherol radical [24, 25]. In this type of oxidation, the tocopherol radical will then react with the PUFA to generate a new peroxyl radical and regenerate the tocopherol. This type of tocopherol-mediated oxidation is probably responsible for the slow loss of tocopherol from LDL and generation of preformed lipid hydroperoxides on LDL stored in the laboratory fridge. Though the occurrence of this tocopherol-mediated oxidation has been argued as evidence that tocopherol does not inhibit LDL oxidation as it forms part of the propagation radical chain, the removal of antioxidant from LDL results in a rapid uninhibited reaction. So, though tocopherol radicals cause lipid oxidation at low radical fluxes, the reaction is much, much slower than the uninhibited reaction.

Once the antioxidants have been consumed, there is a rapid rise in lipid peroxide formation as the chain reaction proceeds uninhibited. The lipid peroxyl radicals react with the PUFAs to generate more lipid peroxyl radicals. In a competing reaction, the lipid peroxyl radicals also attack the apoB100 protein component of the LDL, forming reactive protein hydroperoxides [26]. The peroxide forms on the R groups of the majority of amino acids, though proline, glutamine, lysine, valine, leucine, and isoleucine are the most susceptible to peroxidation. The decay of the protein peroxides generates well-characterized protein carbonyls, which are often used as a crude marker of protein oxidation. As this chain reaction slows, the decay of the lipid peroxides becomes the dominant reaction, forming oxysterols such as 7-ketocholesterol, isoprostanes, and reactive short chain aldehydes. These late oxidation products appear to give oxLDL its cytotoxic and atherogenic properties. A number of isoprostanes have notable chemotactic activity. The short chain aldehydes, particularly 4-hydroxynonenal (HNE) and malondialdehyde (MDA), react with lysine residues within the apoB100 protein destroying the binding site to the LDL receptor. These derivatizations also make oxLDL a ligand for scavenger receptors (SCRs), which facilitates the rapid and apparently unregulated uptake of oxLDL into the macrophage to form a foam cell [27, 28]. HNE is also very cytotoxic [29] and though significant amounts are generated during oxLDL formation, it is unlikely that any is free due to its reactivity with proteins. Some but not all of the oxysterols are strongly cytotoxic when given to cells in culture [30–32]. Consequently, it is currently hypothesized that oxysterols are the main cytotoxic agent within oxLDL and the possible cause of cellular death within the plaque core.

25.3
Toxicity of oxLDL

It has been known for a long time that oxLDL triggers cell death in a wide number of different cell types [33, 34]. The cytotoxicity of oxLDL appears experimentally as the loss of the various cell viability markers, such as MTT reduction, trypan blue exclusion, and lactate dehydrogenase release, and usually ends with cell lysis. The classic apoptosis marker of DNA fragmentation has been reported in a number of studies along with caspase activation, annexin V staining, and cytochrome c release [35, 36]. These same marks have been observed within complex plaques [2, 15, 37].

The cytotoxicity of oxLDL in *in vitro* experiments generally occurs between 100 and 400 µg ml^{-1} of apoB protein (0.25–2 mg ml^{-1} total mass of LDL) depending on the LDL preparation and type of cells. Though never mentioned in research publications, the LD$_{50}$ for oxLDL can vary from one batch to the next with no apparent reason. Also the presence or absence of serum greatly affects toxicity, possibly due to the various antioxidants within serum and the antiatherogenic effects of HDL [38]. As is discussed later, ascorbate and tocopherol reportedly have protective effects against oxLDL cytotoxicity.

Whether the cell death induced by oxLDL is necrotic or apoptotic is very much dependent on the cell type used and possibly incubation conditions. This difference is highlighted by the response of two similar monocyte-like human cell lines to cytotoxic levels of oxLDL. The human-derived monocyte-like THP-1 cells undergo classic apoptosis with the appearance of annexin V staining morphology due to the flipping of phosphatidylserine to the outer surface of the membrane, caspase-3 activation, and cell shrinkage [39, 40]. In complete contrast, U937 cells rapidly lose their intracellular glutathione resulting in caspase-3 inactivation due to oxidation of the essential free thiol groups in the active site. The failure of caspase-3 appears to lead to default necrosis with cell swelling and lysis with no phosphatidylserine exposure on the cell membrane [39, 41]. Between these two extremes, a wide range of responses have been reported. With monocytes prepared from human blood and macrophages generated via differentiation from these cells, both caspase-dependent [37] and caspase-independent cell death has been reported [42]. In our own laboratory, we have found that oxLDL causes human monocyte–derived macrophages (HMDMs) to rapidly undergo necrosis without caspase-3 activation and a rapid loss of glutathione similar to that reported with U937 cells [43].

In addition to these classic markers of cellular death, oxLDL appears to induce oxidant production in or around the cell. In macrophages, this is clearly seen by the loss of glutathione and the effect of enhancing glutathione levels on cell survival [44, 45]. Superoxide production from both the plasma membrane and cytoplasm has been reported in macrophage, endothelial, and smooth muscle cells [8,46–48]. Some of this oxidant production does appear to be due to the release of cytochrome c from the mitochondria causing mitochondrial dysfunction and oxidant production [47]. The oxidant production, cytochrome c release, and oxLDL-induced cell death are inhibited in endothelial cells with cyclosporine-A [49]. The importance of this oxidant release to cytotoxicity was demonstrated in monomac cells where increased expression of MnSOD decreased the oxLDL cytotoxicity [50].

How a highly oxidized, relatively stable particle such as oxLDL triggers these events is initially difficult to understand. Initial investigation focused on activity being due to further lipid oxidation reactions, but it is now clear that oxLDL-induced oxidant release and cell death is through a more complex mechanism. To fully appreciate these processes, it is necessary to first consider the different types of oxLDL and how they are prepared.

25.4
Types of oxLDL

The balance of the oxidized compounds formed within the oxLDL, and therefore the oxLDL bioactivity/cytotoxicity, is determined by the rate of radical generation (radical flux), reaction period, temperature, and ionic conditions in which the LDL was oxidized. Therefore, the type of oxLDL made in the laboratory is also chosen according to the type of cellular or physiochemical phenomena the researcher is exploring. By limiting or halting LDL oxidation at different stages, it is possible to generate a range of LDL oxidation types. The types of oxLDL range from minimally modified oxLDL with very low levels of peroxide and a significant amount of α-tocopherol present through to heavily oxidized oxLDL (Table 25.1).

The last type of oxidation described above generates "heavily oxidized oxLDL" which is commonly referred to as *oxLDL*. We routinely generate this heavily oxidized oxLDL in the laboratory by oxidizing LDL purified from human plasma with copper chloride in phosphate buffer at pH 7.2 at 37 °C. It is a generally reliable method as the LDL is completely oxidized, so the resulting oxLDL is very stable. It is this heavily oxidized LDL which has the highest oxysterol levels and is the most cytotoxic. Considering the brute force of this method, it is surprising to find considerable variation in oxLDL toxicity from one oxLDL preparation to the next. In our laboratory, we have LD_{50} ranging from 0.5 to 2 mg ml^{-1} LDL total mass (0.1–0.4 mg ml^{-1} LDL protein) from one batch of plasma to the next. Yet we observe no difference in the nature of the cytotoxicity with either monocyte-like U937 or THP-1 cell lines, or monocyte-derived macrophages; just the amount of oxLDL needed to trigger cell death. The cause of the LD_{50} variation is likely to be a result of differences in the lipid composition of donor LDL.

At the other end of the oxidation spectrum is aggregated LDL. Though technically not oxidized, aggregated LDL is rapidly taken up by macrophages. Aggregated LDL has lower solubility and reduced diffusion potential within the plaque making it a primary candidate for the initiation of vascular disease and a source of lipoprotein for oxLDL formation.

Minimally oxidized low density lipoprotein (mLDL) contains peroxidized lipids, usually lipid hydroperoxides, and significant levels of α-tocopherol. It is generated by incubation with lipoxygenase and phospholipase A2 [53, 54] or gentle oxidation methods such as Fe^{2+}/ADP for 3 hours [55]. Though not cytotoxic, mLDL (and

Tab. 25.1 Different types of oxLDL

oxLDL type	Common abbreviation	Preparation	Characteristic	Cytotoxic	Cite
Aggregated	aggLDL	Vortex LDL preparation in glass test tube under argon for 60 s	Strong light absorbing at 680 nm due to the formation aggregates. No or little lipid peroxidation or loss of tocopherol or CoQ	No	[51, 52]
Minimal	mLDL	Enzymatic oxidation with soybean lipoxygenase and phospholipase A2 or 3 h incubation in $FeCl_2$/ADP	Low level of thiobarbituric acid-reactive substances, some cleavage of ApoB100, α-tocopherol present but reduced by 30–50% of unoxidized LDL	No	[53–55]
Moderately	Mod-LDL	Incubate LDL in dialysis bag in 10 mM MOPS buffer with 10 mM copper chloride at 4 °C for 24 h or copper-mediated LDL until 234 nm diene absorbance peaks. Stop reaction with EDTA.	High lipid peroxide level but low oxysterol (7-ketocholesterol). No α-tocopherol present	Yes	[56, 57]
Heavily	oxLDL	Incubate with 200-fold excess of copper ions over LDL for 24 h at 37 °C in phosphate buffer saline pH 7.4.	Colorless, all tocopherol and β-carotene oxidized. Often aggregated during preparation. High in oxysterols, protein carbonyls, isoprostanes content. ApoB100 heavily fragmented on SDS-PAGE analysis	Yes	[6, 39, 56, 58, 59]
Chlorinated	HOCl-oxLDL	Treat LDL with 500-fold excess of HOCl at 37 °C for 30 min	Heavy oxidation of ApoB100 residues of Cys, Met, Tyr, Trp, Lys. lipid oxidation and organic chloramine formation	Yes	[60, 61]

oxLDL) inhibits fibroblast cells [55, 62] and human coronary artery smooth muscle cell proliferation [63]. mLDL has also been found to increase intracellular calcium levels of vascular endothelial cells, though the effect is diminished over time with chronic exposure [55]. There is a possible protective effect against oxLDL when cells are first exposed to mLDL [54].

In between these states of mLDL and heavily oxidized LDL, there is a mass of literature describing a range of LDL oxidation levels. These are described in vague terms such as moderate or mildly oxidized LDL, though some publications consider these two different types as well. Moderately oxidized LDL has been described as the product of the incubation with Cu^{2+} in phosphate buffered saline for 15 hours at 37 °C. This oxLDL is high in oxysterols and is cytotoxic [64]. Changing the redox ion to Fe^{2+} in the same media generates what was termed *mildly oxLDL*, which is less cytotoxic with only half the oxysterol content. Lowering the pH to 5.5 using Fe^{+2} generates very-mildly oxidized LDL, which is not cytotoxic and has a very low sterol content [64]. This appears to be the same as the minimally modified LDL described above.

Some attempt has been made to sort out the ambiguity by creating a more defined form of oxLDL, which is between the minimally and heavily oxidized LDL. This form of oxLDL is made by oxidizing LDL at 4 °C so that the hydroperoxides that form are stable and do not decay further to drive oxysterol formation [56]. The success of the method is partially due to confining the LDL to dialysis bags placed in an excess volume of copper ions. Thus binding of copper ions does not alter the copper concentration within the solution. The resulting hydroperoxide-rich LDL is less cytotoxic as it has very little oxysterols present [57]. This moderately oxidized oxLDL is unstable at 37 °C due to decay of the hydroperoxides, and the resulting radical flux causes further oxysterol formation.

Hypochlorous acid is a major oxidant generated by activated neutrophils and macrophages through the release of the enzyme myeloperoxidase [65–67]. Chlorinated amino acid residues have been detected within atherosclerotic plaques suggesting that HOCl-oxidized LDL may be a significant product of the inflammatory events occurring within the artery wall [68]. In the laboratory, treatment of LDL with hypochlorite causes rapid oxidation of a number of key amino acid residues, PUFA oxidation with a large rise in TBARS, and the loss of α-tocopherol [69]. The resulting HOCl is cytotoxic to cells including macrophages [40, 60].

25.5
Cell-Mediated Oxidation and Atherosclerotic Plaques

Macrophages are considered to be one of the main sources of the oxidative stress triggering oxLDL formation within the plaques. The techniques described in the previous section aim to produce oxLDL similar to that generated *in vivo* by macrophages. There is some scientific basis to this view as two decades of research and argument have shown that a number of cell types, including macrophages, will oxidize LDL to high uptake oxLDL via a transition metal ion–dependent mechanism

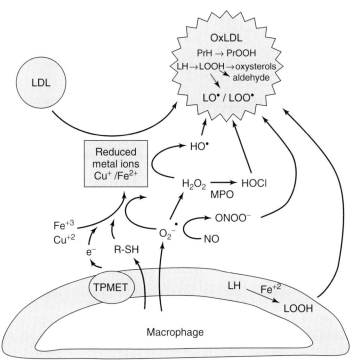

Fig. 25.3 Cell-mediated LDL mechanisms. Peroxidation of LDL to oxLDL results in apoB100 proteins (PrH) and LDL lipids (LH) being oxidized to protein hydroperoxides (PrOOH) and lipid peroxides (LOOH), respectively. Lipid peroxyl (LOO$^\bullet$) and lipid alkoxyl radicals (LO$^\bullet$) can initiate further lipid and protein oxidation. Macrophages release superoxide ($O_2^{\bullet-}$), which can dismutate to hydrogen peroxide. The Fenton reaction between hydrogen peroxide (H_2O_2) and reduced metal ions generates hydroxyl radicals (HO$^\bullet$) which can directly oxidize LDL. The hydrogen peroxide can also be further oxidized to HOCl by macrophage myeloperoxidase (MPO). HOCl reacts with both lipid and protein components of the LDL to generate oxLDL. Macrophages can increase the pool of reduced copper and iron ions through the release of reduced thiols (R-SH) or direct electron transfer via the plasma membrane electron transport complex (TPMET). The binding of the reduced metal ions initiates lipid peroxidation reaction within the LDL. This reaction generally involves copper ions and is enhanced by reduced iron ions. Oxidation of plasma membrane lipids by iron ions forms plasma lipid hydroperoxides which can transfer to the LDL and so increase the level of preformed hydroperoxides on the LDL.

[70, 71]. Traditionally, this is carried out in serum-free Hams-F10 media over a period of days, though changes in the media formulation by manufacturers means the media must be treated to prevent oxidation in the cell-free controls [72]. Macrophage-mediated LDL oxidation appears to involve a number of different cell specific mechanisms, all of which are transition metal ion dependent (Figure 25.3). The majority of these oxidative mechanisms essentially involve the cells initiating or enhancing oxidation by increasing the pool of reduced copper or iron ions through the release of either reducing thiols [73, 74] or direct reduction of the metal

ions via a transplasma membrane election transport mechanism driven by NADPH [75]. Seeding of LDL with preformed hydroperoxides from the macrophage plasma membrane has also been demonstrated to promote LDL oxidation [76]. The role of macrophage superoxide release has been controversial but appears to be involved in some cell types [77]. Peroxynitrite, the product of superoxide and nitric oxide, also reacts with LDL to generate cytotoxic oxLDL [78–80].

The main problem with all these mechanisms is that the presence of serum inhibits oxLDL formation by binding the available redox active metal ions. HDL also has some antioxidant properties that affect the cell's ability to form oxLDL. Atherosclerotic plaques also contain ascorbate and tocopherol, which inhibits or slows oxLDL formation via these mechanisms. However, redox active copper and iron ions have been identified within excised atherosclerotic plaques [81, 82] along with actual oxLDL particles, suggesting there are oxidative permissive regions within plaques. Alternatively, it has been suggested that macrophage secreted lipoxygenases to generate oxLDL as these enzymatic reactions are not inhibited by serum proteins or antioxidants [53]. However, lipid oxidation isomers specific to lipoxygenase activity have been found only in young human lesions while older lesions show only isomers characteristic of metal ion–mediated LDL oxidation [83].

There is one site that has had very little attention and that is the interior of the macrophage cells. J774 mouse macrophages readily take up aggregated nonoxidized LDL that collects in the lysosomes where it is oxidized via an iron ion–dependent mechanism to generate oxysterols [84]. This reaction is also inhibited by water-soluble antioxidants suggesting that a significant amount of the oxLDL detected in plaques originates from within macrophages. This oxLDL is released if the cells die via necrosis, the very cell death process that appears to generate the plaque necrotic core regions. Iron-dependent oxLDL formation with lysosomes has also been shown to cause macrophage cell death through partial rupture of the lysosome [85, 86]. Within the plaque, this released oxLDL along with the iron could drive further extracellular LDL oxidation leading to the formation and growth of advanced plaques.

It should be noted that oxLDL is generated within other sites of inflammation. oxLDL has been isolated from knee-joint synovial fluid of patients with rheumatoid arthritis [87–89]. oxLDL-containing foam cells have also been identified in the joints of arthritis patients suggesting a mechanistic link between arthritis and atherosclerosis [89]. As in vascular disease, early rheumatoid arthritis patients have elevated autoantibody titers against mildly oxLDL [90].

25.6
Endogenous Antioxidants

Macrophages do not just release oxidants. Activated human macrophages, particularly γ-interferon stimulated macrophages and monocytes, release the antioxidants 7,8-DNP and 3-hydroxyanthranilic acid (3HAA) [91]. 7,8-DNP is generated through the enzymatic breakdown of guanosine-5'-triphosphate (GTP)

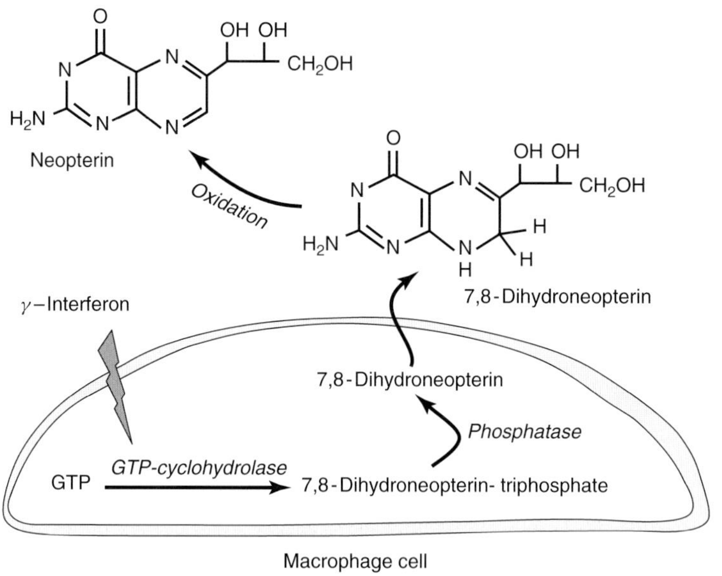

Fig. 25.4 7,8-Dihydroneopterin and neopterin formation. γ-Interferon stimulation of macrophage cells upregulates the intracellular levels of the GTP-cyclohydrolase enzyme, causing increased breakdown of GTP to 7,8-dihydroneopterin-triphosphate. The action of nonspecific phosphatase releases the neutrally charged 7,8-dihydroneopterin which diffuses from the cells where it is oxidized to the high fluorescent neopterin, possibly by HOCl.

by GTP-cyclohydrolase (Figure 25.4). The resulting 7,8-DNP triphosphate in nonprimate macrophages is further metabolized to inducible nitric oxide synthase (iNOS) cofactor 5,6,7,8-tetrahydrobiopterin. Primate macrophages (including human) do not express the required enzymes for 5,6,7,8-tetrahydrobiopterin synthesis. Instead, nonspecific phosphatases remove the charged triphosphate and the resulting 7,8-DNP is released from human macrophages instead of nitric oxide [92, 93]. How 7,8-DNP passes through the plasma membrane is unclear but in our laboratory we have observed it to be a very rapid process. In vivo, 7,8-DNP is oxidized to the highly fluorescent compound neopterin, which has been used clinically as a marker of immune cell activation for a number of years [94–96]. Not surprisingly, plasma neopterin levels markedly elevated in patients suffering from vascular disease [97–99] and neopterin has been measured in micromolar concentrations in atherosclerotic plaques [91, 100]. 7,8-DNP is a potent inhibitor of copper, peroxyl radical, and cell-mediated LDL oxidation through its ability to effectively compete for and neutralize the lipid peroxyl radical [22, 26, 101, 102].

3HAA is generated through degradation of the amino acid tryptophan by the enzyme indoleamine 2,3-dioxygenase (IDO), which is also upregulated by γ-interferon [103]. Like 7,8-DNP, 3HAA inhibits copper and 2,2'-azobis(amidinopropane) dihydrochloride (AAPH)-peroxyl radical-mediated LDL oxidation at micromolar concentrations. At high tryptophan concentrations, interferon-stimulated macrophages generated enough 3HAA to inhibit macrophage-mediated LDL oxidation [104, 105]. Whether 3HAA occurs in plaque is uncertain at present.

How these cellular antioxidants, as well as the thiols, tocopherol, and ascorbic acid, influence the radical flux and therefore the rate and level of LDL oxidation is uncertain, but manipulation of the underlying inflammatory process will require more knowledge of plaque biochemistry [91].

25.7
Mechanism of oxLDL Cytotoxicity

OxLDL is not handled by cells in the same way as unoxidized LDL. oxLDL is not a ligand for the LDL receptor, but instead binds to a class of receptors collectively known as the *scavenger receptors* [106]. Of these scavenger receptors, scavenger receptor-A (SCR-A) and CD36 have received the most attention in the study of both foam cell formation and oxLDL-mediated cell death. The receptor-mediated uptake of oxLDL via the SCR at sublethal concentrations leads to cholesterol ester accumulation. Cholesterol is normally exported from the cells to nascent high density lipoprotein (HDL) apoA1 proteins via ABCA1 transport. Oxysterols within oxLDL impair the ABCA1 transporter causing accumulation of cholesterol within the cells. The oxidized lipids also inhibit normal processing of lipids through the endosomes and lysosomes leading to the accumulation of cholesterol ester filled vesicles within the cell cytoplasm [107, 108]. The oxysterols and oxidized proteins also appear to resist processing by the lysosomal esterases and proteases.

Cell death is triggered in a growing percentage of cells as oxLDL concentration increases. The intensity and speed of the cell death process also intensifies as death mechanisms receive stronger stimulation by components of the oxLDL. This is clearly seen with the human monocyte-like THP-1 cell line where the level of caspase-3 activity and the speed at which caspase-3 is activated increase with increasing oxLDL concentration [39]. Though oxLDL uptake appears to be important in the oxLDL-induced death of some cells, it is not the case in others. OxLDL-driven necrosis through lysosome destabilization will require the uptake of oxLDL, yet this process may be overtaken by more rapid mitochondrial-driven apoptosis mechanisms as incubation proceeds. A close examination of the literature shows a number of competing and overlapping mechanisms which either accelerate or inhibit each other depending on the type of death mechanism stimulated by oxLDL. The scavenger receptors involved also appears to alter cellular response to oxLDL. Antibodies to CD36 do not inhibit oxLDL uptake, but do significantly reduce the level of reactive oxygen species (ROS) production, caspase activation, and cell death [48, 109]. oxLDL generates more ROS in monocyte-like U937 cells than that is generated in THP-1 cells that also have the lowest level of CD36 expression of the two cell lines [110]. Conversely, increased expression and activity of macrophage SRC-A receptor causes decreased oxLDL toxicity in THP-1 cells. In these experiments, SRC-A appears not to be responsible for toxicity and gives some protection by taking up the oxLDL [111]. Interestingly, aggregated oxLDL has been reported to be less cytotoxic as it is not taken up by the CD36 receptor but through the SCR-A receptor which results in oxLDL metabolism [51]. Surprisingly, inhibition of the SCR-A uptake increased the cytotoxicity of the aggregated oxLDL suggesting that with macrophages the cytotoxicity was due to changes in the plasma membrane, possibly mediated through CD36 as is discussed latter.

25.8
Oxysterol and the Mitochondria

The key feature of oxLDL cytotoxicity with all cells is the generation of ROS within the cell. With some cell types, ROS generation is central to the cell death mechanism, as antioxidant treatments inhibit oxLDL cytotoxicity [8, 112], but this is not the case with all cell types [113]. ROS production was initially a surprise as oxLDL is so oxidized that it is unable to support further peroxidation events. Treatment of oxLDL with various antioxidants before addition to cells does not reduce its toxicity or ROS production showing it is cell's own oxidant generating system that is being activated or stimulated by the oxLDL. There are also reports of less oxidized LDL undergoing further oxidation within the cell. Lysosome iron within cells causes hydroperoxide decomposition generating further lipid radicals that destabilize the lysosome so triggering cell death [85]. This may explain some reports of lipoxygenase being involved in the cell death mechanism.

Current evidence suggests that chemical species within oxLDL, which trigger ROS production and cell death, are the oxysterols. Heavily oxidized LDL is rich in oxysterols, which (when given to cells dissolved in DMSO or ethanol) trigger the same types of cell death as those observed with oxLDL. In general, the order of toxicity is 7-ketocholesterol being the most toxic with 7α-hydroxycholesterol, 7β-hydroxycholesterol, 25-hydroxycholesterol, and 26-hydroxycholesterol less toxic to HMDMs [30]. Using oxysterol mixtures similar in concentration to that found in advanced atherosclerotic plaques, 7β-hydroxycholesterol and 7-ketocholesterol have been found to induce caspase activation, ROS production, cellular thiol depletion, permeabilization of lysosomal and mitochondrial membranes, and cell death in U937 cells. 25-Hydroxycholesterol and 26-hydroxycholesterol at plaque concentrations do not appear to cause any of the above cellular events [32].

Oxysterols are very stable and do not generally react to form oxidants and free radicals. Yet, their uptake into a cell whether it is free or within oxLDL triggers a series of cellular events triggering cell death. The central event in all the described death processes is depolarization of the mitochondria due to pore formation. Depolarized mitochondria generate superoxide directly from complex I and possibly through the release of cytochrome c into the cytoplasm [47, 114]. This mitochondrial ROS release can be seen with a number of fluorescent dyes under fluorescence microscopy. The addition of cyclosporine-A blocks oxLDL-induced mitochondrial pore formation, depolarization, and, most importantly, the loss of cytochrome c to the cytoplasm [49, 115]. The chain of events set in process are well documented and easily observed with oxLDL and oxysterol-treated cells. Cytochrome c binds to

Apo-1 and procaspase-9 along with dATP to form the apoptosome. The apoptosome releases active caspase-9 that cleaves procaspase-3 to caspase-3 that goes on to cleave a range of proteins, thereby initiating apoptosis [116, 117].

There is some disagreement on whether the ROS generated is hydrogen peroxide, a lipid peroxide, or superoxide. The fluorescent dyes used are often described as being specific to one of these oxidants, but in reality there is considerable cross-reactivity. The cell's own antioxidant machinary will also dismutase superoxide to hydroperoxide, and hydrogen peroxide can react with low molecular weight iron to generate oxidants further confusing the matter. The exact nature of the oxidant generated is somewhat immaterial as all these oxidants will react and consume essential cellular antioxidants, especially ascorbate and glutathione.

It does not appear that oxysterols or oxLDL directly interact with the mitochondria to trigger the intrinsic apoptosis pathway. Though lipid peroxidation products and oxidation of the lipid plasma membrane of the lysosomes have been implicated in oxLDL-induced cell death, this process does not appear to occur with mitochondrial membranes. A possible explanation is oxysterols may not specifically accumulate within the mitochondrial membranes.

How oxLDL or specific oxysterols initiate mitochondrial dysfunction was until recently unclear, but it has now been shown that the mitochondrial pore formation, depolarization, and cytochrome c release are activated through the Bcl-2 system [116] that is under the control of the mitogen-activated protein kinases (MAPKs) system. In the murine macrophage-like cell line P388D1 and PC12 cells, 25-hydroxycholesterol and 7-ketocholesterol induced degradation of the prosurvival protein kinase AKT (protein kinase B) [118]. How exactly these oxysterols cause this is unclear, but hydrogen peroxide treatment also causes AKT degradation [119]. 25-Hydroxycholesterol also appears to alter the function of other phosphatases/kinases as it causes the dephosphorylation and inhibition of AKT [120, 121]. Without the activity of AKT, glycogen synthase kinase-3β is activated leading to increased cJNK activity with a shift in the balance of the Bcl family of protein from the antiapoptotic Bcl-2 to the proapoptotic activity of Bax and Bak with mitochondrial pore formation and cytochrome c release, caspase activation, and apoptosis. So, in this mechanism, oxysterol action on AKT causes mitochondrial-associated ROS release and apoptosis. Pretreatment with mLDL has been reported to protect macrophages from oxLDL via an upregulation of AKT activity suggesting that inhibition of this apoptosis mechanism is important in foam cell survival within plaques [54]. Blocking this mechanism through the overexpression of Bcl-2 does not remove oxLDL cytotoxicity, but rather changes the mode of cell death from apoptosis to a calcium-dependent necrosis [122].

25.9
Oxysterols and Calcium

OxLDL-induced rise in cytosolic calcium is a common event and apparent trigger for necrosis and apoptosis in a range of cells [122]. Inhibitors of the plasma membrane calcium channels or the calcium-dependent calpain family of proteases block the cytotoxicity of some oxysterols, but not all, suggesting multiple mechanisms occurring with different oxysterols [123]. Activation of calpains within the cytoplasm is considered key event in necrosis and apoptosis. In endothelial HMEC-1 cells, oxLDL causes calpain-dependent proteolysis of cytoskeletal a-fodrin and cleavage of the Bcl-2 protein Bid. Interestingly, caspase-3 is not activated but is instead polyubiquinated, so a loss of 32 kDa procaspase-3 is observed [124]. In support of this, cyclosporine-A blocks or reduces oxLDL-induced caspase-3 activation in endothelial cells (HMEC-1) and THP-1 cells, further demonstrating calpains do not activate caspase-3 but require the activation of caspase-9 via cytochrome c [125, 126]. oxLDL-induced caspase-independent cell death has also been reported in HMDMs [42, 43]. Whether calpains are also responsible for some of the AKT proteolysis remains to be determined.

The rise in calcium due to 7-ketocholesterol treatment of THP-1 cells also induces dephosphorylation of the proapoptotic protein BAD by calcium-dependent phosphatase calcineurin [125, 126]. BAD heterodimerization with antiapoptotic proteins Bcl-2 and Bcl-X causes cytochrome c release through pore formation.

With THP-1 cells, the source of the calcium causing all these effects is the store-operated Ca^{2+} entry channel (SOC). 7-Ketocholesterol collects within the lipid rafts of the plasma membrane causing translocation of the Trpc-1 protein to the lipid raft to form the SOC leading to calcium influx in the cytosol [125]. It is possible that much of the initial effects observed with oxysterols are mediated via lipid raft–associated receptors and channels within the plasma membrane.

25.10
NADPH Oxidase and Apoptosis versus Necrosis

In all the above mechanisms, the ROS production from the mitochondria is relatively modest, causing only a small yet significant reduction in cellular glutathione [114]. Loss of cellular GSH through various inhibitors or derivatizing agents increases cells' sensitivity to oxLDL cytotoxicity while elevating glutathione decreases toxicity [44, 127]. Increasing levels of cellular GSH by over expressing glutathione reductase in mouse macrophages decrease the size of the plaques in mice on a high cholesterol diet, which shows the importance of ROS production and cell death in lesion development [45].

The effect of oxLDL is very different in monocyte-like human-derived U937 cells. OxLDL triggers an intense ROS production causing rapid loss of all cellular glutathione with subsequent oxidation of free thiol groups on many proteins [39]. The oxidation of free thiol within the active site of the various caspases renders them inactive [128] generating a caspase-independent cell death, which looks very much like apoptosis. Though THP-1 cells readily undergo caspase-dependent apoptosis, this mechanism is converted to necrosis when the cells are exposed to high levels of peroxyl radicals. The cells rapidly lose glutathione and the ability to activate caspase-3 [41]. Scavenging of oxLDL-induced ROS production by the water-soluble antioxidant 7,8-DNP reduces the level of cell death in U937 cells [129]. In smooth muscle cells, ROS production is high enough to cause loss of the glycolytic enzyme glyceraldehyde-3-phosphate dehydrogenase (GAPDH), which naturally leads to necrosis as the cells are unable to generate sufficient ATP [48]. The loss of GAPDH is prevented by various antioxidants and NADPH oxidase or lipoxygenase inhibitors. Anti-CD36 antibodies also blocked ROS production and GAPDH loss suggesting activation of NADPH oxidase or lipoxygenase is triggered by oxLDL binding to CD36. As discussed earlier, oxLDL binding to CD36 triggers ROS production in a number of cell types, especially U937 cells that have a high level of CD36 expression [48, 110, 114]. NADPH oxidase is rapidly activated by 7-ketocholesterol at plaque concentrations in the mouse macrophage J774A.1. The effect of 7-ketocholesterol does not appear to be as intense with U937 cells, but is greatly elevated when it is present within oxLDL and is presumably presented to the cell via CD36 [32, 39, 110]. These observations suggest that in cells that express high levels of CD36 and NADPH oxidase, such as monocyte-derived macrophages or U937 cells, the resulting ROS production overtakes the slower mitochondrial-/caspase-dependent cell death or the calcium-dependent calpain mechanisms. If ROS production is high enough, through either reduced intracellular antioxidant levels (e.g., glutathione or ascorbate) or high levels of NADPH oxidase and CD36, the cell cysteine protease will be oxidized meaning the cell will slide into necrosis and cell lysis instead of a controlled apoptosis. How CD36 binding of oxLDL triggers NADPH oxidase activation remains to be determined. What these findings suggest is that anti-inflammatory agents, which specifically target plaques, could reduce the rate of formation of advanced necrotic core regions by inhibiting NADPH oxidase–induced cell death, via scavenging ROS, inhibiting NADPH oxidase, or by reducing the level of CD36 expression in the foam cells.

25.11
The Future

OxLDL formation and cytotoxicity are a clear consequence of inflammatory and apoptosis pathways. It is therefore possible that manipulation of these processes within a plaque would slow or inhibit necrotic core formation. Plaque regression via cellular efflux to plasma Apo-A1 requires the toxic effects of oxLDL to be mitigated within the plaque or inhibition of oxLDL formation. oxLDL clearly has an adverse effect on efflux. Anti-inflammatory drugs have been shown to reduce clinical consequences of vascular disease, but whether this is due to a slowing oxLDL formation or reducing the cytotoxic effects of oxLDL is unknown [130–132].

The most pressing need is for better diagnostic tools to detect what is actually happening inside the plaques well before they become clinically significant. For most patients, the first indication of a vascular problem is usually a critical event triggered by the advanced state of the disease process. We need to develop plasma markers of the biochemical and immunological events associated with complex plaque and necrotic core formation at an early stage. Compounds such as neopterin and 7,8-DNP, which escape from the plaques, already allow the monitoring of immune activity within the total body plaque burden. The addition of plasma markers of LDL oxidation and cellular apoptosis would provide the basis for a biochemical differential diagnosis of growing or unstable plaques.

Though oxLDL formation is a prominent feature of vascular disease, it should not be forgotten that oxidant production is a common event during inflammation so oxLDL generation is likely to be a component of other clinical pathologies. OxLDL has been already found in the cerebrospinal fluid of patients with Alzheimer's disease and synovial fluid from rheumatoid arthritis patients [87]. Whether OxLDL is a significant cytotoxic agent in these other chronic inflammatory diseases remains to be seen.

References

1 Harada, K., Chen, Z., Ishibashi, S., Osuga, J., Yagyu, H., Ohashi, K., Yahagi, N., Shionoiri, F., Sun, L.M., Yazaki, Y., and Yamada, N. (1997) Apoptotic cell death in atherosclerotic plaques of hyperlipidemic knockout mice. *Atherosclerosis*, **135**, 235–239.

2 Hegyi, L., Skepper, J.N., Cary, N.R.B., and Mitchinson, M.J. (1996) Foam cell apoptosis and the development of the lipid core of human atherosclerosis. *J Pathol*, **180**, 423–429.

3 Steinberg, D., Parthasarathy, S., Carew, T.E., Khoo, J.C., and Witztum, J.L. (1989) Beyond cholesterol: modifications of low-density lipoprotein that increase its atherogenicity. *N Engl J Med*, **320**, 915–924.

4 Libby, P., Ridker, P.M., and Maseri, A. (2002) Inflammation and atherosclerosis. *Circulation*, **105**, 1135–1143.

5 Steinbrecher, U.P. (1987) Oxidation of human low density lipoprotein results in derivatization of lysine residues of apolipoprotein B by lipid peroxide decomposition products. *J Biol Chem*, **262**, 3603–3608.

6 Reid, V.C. and Mitchinson, M.J. (1993) Toxicity of oxidised low density lipoprotein towards mouse peritoneal macrophages in vitro. *Atherosclerosis*, **98**, 17–24.

7 Bjorkerud, B. and Bjorkerud, S. (1996) Contrary effects of light and strongly oxidised LDL with promotion of growth versus apoptosis on artery smooth muscle cells, macrophages and fibroblasts. *Arterioscler Thromb Vasc Biol*, **16**, 416–424.

8 Harada-Shiba, M., Kinoshita, M., Kamido, H., and Shimokado, K. (1998) Oxidized low density lipoprotein induces apoptosis in cultured human umbilical vein endothelial cells by common and unique mechanisms. *J Biol Chem*, **273**, 9681–9687.

9 Wang, X.S., Greilberger, J., Ratschek, M., and Jurgens, G. (2001) Oxidative modifications of LDL increase it's binding to extracellular matrix from human aortic intima: influence of lesion development, lipoprotein lipase and calcium. *J Pathol*, **195**, 244–250.

10 Ylaherttuala, S., Palinski, W., Rosenfeld, M.E., Parthasarathy, S., Carew, T.E., Butler, S., Witztum, J.L., and Steinberg, D. (1989) Evidence for the presence of oxidatively modified low-density lipoprotein in atherosclerotic lesions of rabbit and man. *J Clin Invest*, **84**, 1086–1095.

11 Brown, A.J., Leong, S., Dean, R.T., and Jessup, W. (1997) 7-Hydroperoxycholesterol and its products in oxidised low density lipoprotein and human atherosclerotic plaque. *J Lipid Res*, **38**, 1730–1745.

12 Fu, S., Davies, M., Stocker, R., and Dean, R.T. (1998) Evidence for role

of radicals in protein oxidation in advanced human atherosclerotic plaque. *Biochem J*, **333**, 519–525.

13 Suarna, C., Dean, R.T., May, J., and Stocker, R. (1995) Human atherosclerotic plaque contains both oxidised lipids and relatively large amounts of α-tocopherol and ascorbate. *Arterioscler Thromb Vasc Biol*, **15**, 1616–1624.

14 Stocker, R. and Keaney, J.F. (2004) Role of oxidative modifications in atherosclerosis. *Physiol Rev*, **84**, 1381–1478.

15 Hutter, R., Valdiviezo, C., Sauter, B.V., Savontaus, M., Chereshnev, I., Carrick, F.E., Bauriedel, G., Luderitz, B., Fallon, J.T., Fuster, V., and Badimon, J.J. (2004) Caspase-3 and tissue factor expression in lipid-rich plaque macrophages: evidence for apoptosis as link between inflammation and atherothrombosis. *Circulation*, **109**, 2001–2008.

16 Van Der Wal, A.C. and Becker, A.E. (1999) Atherosclerotic plaque rupture: pathologic basis of plaque stability and instability. *Cardiovasc Res*, **41**, 334–344.

17 Gieseg, S.P. and Esterbauer, H. (1994) Low density lipoprotein is saturable by pro-oxidant copper. *FEBS Lett*, **343**, 188–194.

18 Esterbauer, H. and Jurgens, G. (1993) Mechanistic and genetic aspects of susceptibility of LDL to oxidation. *Curr Opin Lipidol*, **4**, 114–124.

19 Puhl, H., Waeg, G., and Esterbauer, H. (1994) Methods to determine oxidation of low density lipoproteins. *Methods Enzymol*, **233**, 425–441.

20 Esterbauer, H., Dieber-Rotheneder, M., Striegl, G., and Waeg, G. (1991) Role of vitamin E in preventing the oxidation of low-density-lipoprotein. *Am J Clin Nutr*, **53**, 314–321.

21 Esterbauer, H., Gieseg, S.P., Giessauf, A. Ziouzenkova, O., and Ramos, P. (1995) Free radicals and oxidative modification of LDL: role of natural antioxidants, in *Atherosclerosis X* (eds F.P. Woodford, J. Davignon,

and A. Sniderman.), Elsevier, Netherlands, 203–208.

22 Gieseg, S.P., Reibnegger, G., Wachter, H., and Esterbauer, H. (1995) 7,8-Dihydroneopterin inhibits low density lipoprotein oxidation in vitro. Evidence that this macrophage secreted pteridine is an antioxidant. *Free Radic Res*, **23**, 123–136.

23 Ziouzenkova, O., Sevanian, A., Abuja, P.M., Ramos, P., and Esterbauer, H. (1998) Copper can promote oxidation of LDL by markedly different mechanisms. *Free Radical Biol Med*, **24**, 607–623.

24 Bowry, V.W., Ingold, K.U., and Stocker, S. (1992) Vitamin E in human low-density lipoprotein, when and how this antioxidant becomes a pro-oxidant. *Biochem J*, **288**, 341–344.

25 Bowry, V.W. and Stocker, R. (1993) Tocopherol-mediated peroxidation. The prooxidant effect of vitamin E on radical-initiated oxidation of human low-density lipoprotein. *J Am Chem Soc*, **115**, 6029–6044.

26 Gieseg, S.P., Pearson, J., and Firth, C.A. (2003) Protein hydroperoxides are a major product of low density lipoprotein oxidation during copper, peroxyl radical and macrophage-mediated oxidation. *Free Radic Res*, **37**, 983–991.

27 Haberland, M.E., Fless, G.M., Scanu, A.M., and Fogelman, A.M. (1992) Malondialdehyde modification of lipoprotein(a) produces avid uptake by human monocyte-macrophages. *J Biol Chem*, **267**, 4143–4151.

28 Hoff, F., O'Neil, J., Chisolm, G.M., Cole, T.B., Quehenberger, O., Esterbauer, H., and Jurgens, G. (1989) Modification of low density lipoproteins with 4-hydroxynonenal induces uptake by macrophages. *Atherosclerosis*, **9**, 538–549.

29 Esterbauer, H. (1993) Cytotoxicity and genotoxicity of lipid-oxidation products. *Am J Clin Nutr*, **57** (Suppl), 779S–786S.

30 Clare, K., Hardwick, S.J., Carpenter, K.L., Weeratunge, N., and

Mitchinson, M.J. (1995) Toxicity of oxysterols to human monocyte-macrophages. *Atherosclerosis*, **118**, 67–75.

31 Colles, S.M., Maxson, J.M., Carlson, S.G., and Chisolm, G.M. (2001) Oxidized LDL-induced injury and apoptosis in atherosclerosis – Potential roles for oxysterols. *Trends Cardiovasc Med*, **11**, 131–138.

32 Larsson, D.A., Baird, S., Nyhalah, J.D., Yuan, X.M., and Li, W. (2006) Oxysterol mixtures, in atheroma-relevant proportions, display synergistic and proapoptotic effects. *Free Radical Biol Med*, **41**, 902–910.

33 Henriksen, T., Evensen, S.A., and Carlander, B. (1979) Injury to human-endothelial cells in culture induced by low-density lipoproteins. *Scand J Clin Lab Invest*, **39**, 361–368.

34 Hessler, J.R., Robertson, A.L., and Chisolm, G.M. (1979) LDL-induced cytotoxicity and its inhibition by HDL in human vascular smooth-muscle and endothelial-cells in culture. *Atherosclerosis*, **32**, 213–229.

35 Lordan, S., O'Callaghan, Y.C., and O'Brien, N.M. (2007) Death-signaling pathways in human myeloid cells by OxLDL and its cytotoxic components 7 Beta-hydroxycholesterol and cholesterol-5 beta,6 beta-epoxide. *J Biochem Mol Toxicol*, **21**, 362–372.

36 Reid, V.C., Hardwick, S.J., and Mitchinson, M.J. (1993) Fragmentation of DNA in P388D1 macrophages exposed to oxidising low density lipoprotein. *FEBS Lett*, **332**, 218–220.

37 Nhan, T.Q., Liles, W.C., Chait, A., Fallon, J.T., and Schwartz, S.M. (2003) The P17 cleaved form of caspase-3 is present within viable macrophages in vitro and in therosclerotic plaque. *Arterioscler Thromb Vasc Biol*, **23**, 1276–1282.

38 Li, W., Yuan, X.M., Olsson, A.G., and Brunk, U.T. (1998) Uptake of oxidised LDL by macrophages results in partial lysosomal enzyme inactivation and relocation. *Arterioscler Thromb Vasc Biol*, **18**, 177–184.

39 Baird, S.K., Hampton, M., and Gieseg, S.P. (2004) Oxidised LDL triggers phosphatidylserine exposure in human monocyte cell lines by both caspase-dependent and independent mechanisms. *FEBS Lett*, **578**, 169–174.

40 Vicca, S., Hennequin, C., Nguyen-Khoa, T., Massy, Z.A., Descamps-Latscha, B., Drueke, T.B., and Lacour, B. (2000) Caspase-dependent apoptosis in THP-1 cells exposed to oxidized low-density lipoproteins. *Biochem Biophys Res Commun*, **273**, 948–954.

41 Kappler, M., Gerry, A.J., Brown, E., Reid, L., Leake, D.S., and Gieseg, S.P. (2007) Aqueous peroxyl radical exposure to THP-1 cells causes glutathione loss followed by protein oxidation and cell death without increased caspase-3 activity. *Biochim Biophys Acta*, **1773**, 945–953.

42 Asmis, R. and Begley, J.G. (2003) Oxidized LDL promotes peroxide-mediated mitochondrial dysfunction and cell death in human macrophages: a caspase-3-independent pathway. *Circ Res*, **92**, E20–E29.

43 Gieseg, S.P., Leake, D.S., Flavall, E.M., Amit, Z., Reid, L., and Yang, Y. (2009) Macrophage antioxidant protection within atherosclerotic plaques. *Front Biosci*, **14**, 1230–1246.

44 Gotoh, N., Graham, A., Niki, E., and Darley-Usmar, V.M. (1993) Inhibition of glutathione synthesis increases the toxicity of oxidized low-density lipoprotein to human monocytes and macrophages. *Biochem J*, **296**, 151–154.

45 Qiao, M., Kisgati, M., Cholewa, J.M., Zhu, W.F., Smart, E.J., Sulistio, M.S., and Asmis, R. (2007) Increased expression of glutathione reductase in macrophages decreases atherosclerotic lesion formation in low-density lipoprotein receptor-deficient mice. *Arterioscler Thromb Vasc Biol*, **27**, 1375–1382.

46 Kinscherf, R., Claus, R., Wagner, M., Gehrke, C., Kamencic, H., Hou, D., Nauen, O., Schmiedt, W., Kovacs, G.,

Pill, J., Metz, J., and Deigner, H.P. (1998) Apoptosis caused by oxidized LDL is manganese superoxide dismutase and p53 dependent. *FASEB J*, **12**, 461–467.

47 Zmijewski, J.W., Moellering, D.R., Le Goffe, C., Landar, A., Ramachandran, A., and Darley-Usmar, V.M. (2005) Oxidized LDL induces mitochondrially associated reactive oxygen/nitrogen species formation in endothelial cells. *Am J Physiol Heart Circ Physiol*, **289**, H852–H861.

48 Sukhanov, S., Higashi, Y., Shai, S.Y., Itabe, H., Ono, K., Parthasarathy, S., and Delafontaine, P. (2006) Novel effect of oxidized low-density lipoprotein: cellular ATP depletion via downregulation of glyceraldehyde-3-phosphate dehydrogenase. *Circ Res*, **99**, 191–200.

49 Walter, D.H., Haendeler, J., Galle, J., Zeiher, A.M., and Dimmeler, S. (1998) Cyclosporin a inhibits apoptosis of human endothelial cells by preventing release of cytochrome c from mitochondria. *Circulation*, **98**, 1153–1157.

50 Shatrov, V.A. and Brune, B. (2003) Induced expression of manganese superoxide dismutase by non-toxic concentrations of oxidized low-density lipoprotein (oxLDL) protects against oxLDL-mediated cytotoxicity. *Biochem J*, **374**, 505–511.

51 Asmis, R., Begley, J.G., Jelk, J., and Everson, W.V. (2005) Lipoprotein aggregation protects human monocyte-derived macrophages from OxLDL-induced cytotoxicity. *J Lipid Res*, **46**, 1124–1132.

52 Khoo, J.C., Miller, E., Mcloughlin, P., and Steinberg, D. (1988) Enhanced macrophage uptake of low-density lipoprotein after self-aggregation. *Arteriosclerosis*, **8**, 348–358.

53 Sparrow, C.P., Parthasarathy, S., and Steinberg, D. (1988) Enzymatic modification of low-density lipoprotein by purified lipoxygenase plus phospholipase-A2 mimics cell-mediated oxidative modification. *J Lipid Res*, **29**, 745–753.

54 Boullier, A., Li, Y.K., Quehenberger, O., Palinski, W., Tabas, I., Witztum, J.L., and Miller, Y.I. (2006) Minimally oxidized LDL offsets the apoptotic effects of extensively oxidized LDL and free cholesterol in macrophages. *Arterioscler Thromb Vasc Biol*, **26**, 1169–1176.

55 Massaeli, H., Austria, J.A., and Pierce, G.N. (1999) Chronic exposure of smooth muscle cells to minimally oxidized LDL results in depressed inositol 1,4,5-trisphosphate receptor density and Ca^{2+} transients. *Circ Res*, **85**, 515–523.

56 Gerry, A.B., Satchella, L., and Leake, D.S. (2007) A novel method for production of lipid hydroperoxide- or oxysterol-rich low-density lipoprotein. *Atherosclerosis*, **197**, 579–587.

57 Gerry, A.B., and Leake, D.S. (2008) A moderate reduction in extracellular pH protects macrophages against apoptosis induced by oxidized low density lipoprotein. *J Lipid Res*, **49**, 782–789.

58 Esterbauer, H., Striegl, G., Puhl, H., and Rotheneder, M. (1989) Continuous monitoring of in vitro oxidation of human low density lipoprotein. *Free Radic Res Commun*, **6**, 67–75.

59 Lenz, M.L., Hughes, H., Mitchell, J.R., Via, D.P., Guyton, J.R., Taylor, A.A., Gotto, A.M., and Smith, C. (1990) Lipid hydroperoxy and hydroxy derivatives in copper-catalyzed oxidation of low density lipoprotein. *J Lipid Res*, **31**, 1043.

60 Vicca, S., Massy, Z.A., Hennequin, C., Rihane, D., Drueke, T.B., and Lacour, B. (2003) Apoptotic pathways involved in U937 cells exposed to LDL oxidized by hypochlorous acid. *Free Radical Biol Med*, **35**, 603–615.

61 Hazell, L.J., van den Berg, J.M., and Stocker, R. (1994) Oxidation of low density lipoprotein by hypochlorite causes aggregation that is mediated by modification of lysine residues rather than lipid oxidation. *Biochem J*, **302**, 297–304.

62 Zettler, M.E., Prociuk, M.A., Austria, J.A., Zhong, G.M., and Pierce, G.N. (2004) Oxidized low-density

lipoprotein retards the growth of proliferating cells by inhibiting nuclear translocation of cell cycle proteins. *Arterioscler Thromb Vasc Biol*, **24**, 727–732.

63 Tanigawa, H., Miura, S., Zhang, B., Uehara, Y., Matsuo, Y., Fujino, M., Sawamura, T., and Saku, K. (2006) Low-density lipoprotein oxidized to various degrees activates ERK 1/2 through Lox-1. *Atherosclerosis*, **188**, 245–250.

64 Carpenter, K.L.H., Challis, I.R., and Arends, M.J. (2003) Mildly oxidised LDL induces more macrophage death than moderately oxidised LDL: roles of peroxidation, lipoprotein-associated phospholipase a(2) and PPAR-Gamma. *FEBS Lett*, **553**, 145–150.

65 Hazen, S.L., Hsu, F.F., Mueller, D.M., Crowley, J.R., and Heinecke, J.W. (1996) Human neutrophils employ chlorine gas as an oxidant during phagocytosis. *J Clin Invest*, **98**, 1283–1289.

66 Kettle, A.J. and Winterbourn, C.C. (1997) Myeloperoxidase: a key regulator of neutrophil oxidant production. *Redox Rep*, **3**, 3–15.

67 Woods, A.A., Linton, S.M., and Davies, M.J. (2003) Detection of HOCl-mediated protein oxidation products in the extracellular matrix of human atherosclerotic plaques. *Biochem J*, **370**, 729–735.

68 Hazen, S.L. and Heinecke, J.W. (1997) 3-Chlorotyrosine, a specific marker of myeloperoxidase-catalyzed oxidation, is markedly elevated in low density lipoprotein isolated from human atherosclerotic intima. *J Clin Invest*, **99**, 2075–2081.

69 Heinecke, J.W. (1999) Mass spectrometric quantification of amino acid oxidation products in proteins: insights into pathways that promote LDL oxidation in the human artery wall. *FASEB J*, **13**, 1113–1120.

70 Steinbrecher, U.P., Parthasarathy, S., Leake, D.S., Witztum, J.L., and Steinberg, D. (1984) Modification of low density lipoprotein by endothelial cells involves lipid peroxidation and degradation of low density lipoprotein phospholipids. *Proc Natl Acad Sci U S A*, **81**, 3883–3887.

71 Jessup, W., Rankin, S.M., De Whalley, C.V., Hoult, R.S., and Scott, J. (1990) Alpha-tocopherol consumption during low density lipoprotein oxidation. *Biochem J*, **265**, 399–405.

72 Firth, C.A. and Gieseg, S.P. (2007) Redistribution of metal ions to control low density lipoprotein oxidation in Ham's F10 medium. *Free Radic Res*, **41** (10), 1109–1115.

73 Sparrow, C.P. and Olszewski, J. (1993) Cellular oxidation of low density lipoprotein is caused by thiol production in media containing transition metal ions. *J Lipid Res*, **34**, 1219–1228.

74 Wood, J.L. and Graham, A. (1995) Structural requirements for oxidisation of low density lipoprotein by thiols. *FEBS Lett*, **366**, 72–74.

75 Baoutina, A., Dean, R.T., and Jessup, W. (2001) Trans-plasma membrane electron transport induces macrophage mediated low density lipoprotein oxidation. *FASEB J*, **15**, 1580–1582.

76 Fuhrman, B., Oikinine, J., and Aviram, M. (1994) Iron induces lipid peroxidation in cultured macrophages, increases their ability to oxidatively modify LDL, and affects their secretory properties. *Atherosclerosis*, **111**, 65–78.

77 Jessup, W., Simpson, J.A., and Dean, R.T. (1993) Does superoxide radical have a role in macrophage-mediated oxidative modification of LDL? *Atherosclerosis*, **99**, 107–120.

78 Darley-Usmar, V.M., Hogg, N., O'Leary, V.J.O., Wilson, M.T., and Moncada, S. (1992) The simultaneous generation of superoxide and nitric oxide can initiate lipid peroxidation in human low density lipoprotein. *Free Radic Res Commun*, **17**, 9–20.

79 Leeuweenburgh, C., Hardy, M.M., Hazen, S.L., Wagner, P., Oh-ishi, S., Steinbrecher, U.P., and Heinecke, J.W. (1997) Reactive nitrogen intermediates promote low density

lipoprotein oxidation in human atherosclerosis in intima. *J Biol Chem*, **272**, 1433–1436.

80 Hazen, S.L., Zhang, R., Shen, Z., Wu, W., Podrez, E.A., MacPherson, J.C., Schmitt, D., Mitra, S.N., Mukhopadhyay, C., Chen, Y., Cohen, P.A., Hoff, H.F., and Abu-Soud, H.M. (1999) Formation of nitric oxide-derived oxidants by myeloperoxidase in monocytes: pathways for monocyte-mediated protein nitration and lipid peroxidation in vivo. *Circ Res*, **85**, 950–958.

81 Smith, C., Mitchinson, M.J., Aruoma, O.I., and Halliwell, B. (1992) Stimulation of lipid peroxidation and hydroxyl-radical generation by the contents of human atherosclerotic lesions. *Biochem J*, **286**, 901–905.

82 Stadler, N., Lindner, R.A., and Davies, M.J. (2004) Direct detection and quantification of transition metal ions in human atherosclerotic plaques: evidence for the presence of elevated levels of iron and copper. *Arterioscler Thromb Vasc Biol*, **24**, 949–954.

83 Kuhn, H., Heydeck, D., Hugou, I., and Gniwotta, C. (1997) In vivo action of 15-lipoxygenase in early stages of human atherogenesis. *J Clin Invest*, **99**, 888–893.

84 Wen, Y.C. and Leake, D.S. (2007) Low density lipoprotein undergoes oxidation within lysosomes in cells. *Circ Res*, **100**, 1337–1343.

85 Li, W., Yuan, X.M., and Brunk, U.T. (1998) OxLDL-induced macrophage cytotoxicity is mediated by lysosomal rupture and modified by intralysosomal redox-active iron. *Free Radic Res*, **29**, 389.

86 Yuan, X.M., Li, W., Brunk, U.T., Dalen, H., Chang, Y.H., and Sevanian, A. (2000) Lysosomal destabilization during macrophage damage induced by cholesterol oxidation products. *Free Radic Biol Med*, **28**, 208–218.

87 Dai, L., Lamb, D.J., Leake, D.S., Kus, M.L., Jones, H.W., Morris, C.J., and Winyard, P.G. (2000) Evidence for oxidised low density lipoprotein in synovial fluid from rheumatoid arthritis patients. *Free Radic Res*, **32**, 479–486.

88 Fairburn, K., Grootveld, M., Ward, R.J., Abiuka, C., Kus, M., Williams, R.B., Winyard, P.G., and Blake, D.R. (1992) Alpha-Tocopherol, lipids and lipoproteins in knee-joint synovial-fluid and serum from patients with inflammatory joint disease. *Clin Sci*, **83**, 657–664.

89 Winyard, P.G., Tatzber, F., Esterbauer, H., Kus, M.L., Blake, D.R., and Morris, C.J. (1993) Presence of foam cells containing oxidized low-density-lipoprotein in the synovial-membrane from patients with rheumatoid-arthritis. *Ann Rheum Dis*, **52**, 677–680.

90 Lourida, E.S., Georgiadis, A.N., Papavasiliou, E.C., Papathanasiou, A.I., Drosos, A.A., and Tselepis, A.D. (2007) Patients with early rheumatoid arthritis exhibit elevated autoantibody titers against mildly oxidized low-density lipoprotein and exhibit decreased activity of the lipoprotein-associated phospholipase A2. *Arthritis Res Ther*, **9**, 9.

91 Gieseg, S.P., Crone, E.M., Flavall, E.A., and Amit, Z. (2008) Potential to inhibit growth of atherosclerotic plaque development through modulation of macrophage neopterin/7,8-dihydroneopterin synthesis. *Br J Pharmacol*, **153**, 627–635.

92 Ziegler, I. (1985) Synthesis and interferon-gamma controlled release of pteridines during activation of human peripheral blood mononuclear cells. *Biochem Biophys Res Commun*, **132**, 404–411.

93 Wachter, H., Fuchs, D., Hausen, A., Reibnegger, G., and Werner, E.R. (1989) Neopterin as marker for activation of cellular immunity: immunologic basis and clinical application. *Adv Clin Chem*, **27**, 81–141.

94 Schroecksnadel, K., Murr, C., Winkler, C., Wirleitner, B., Fuith, L.C., and Fuchs, D. (2004) Neopterin

to monitor clinical pathologies involving interferon-gamma production. *Pteridines*, **15**, 75–90.

95 Flavall, E.A., Crone, E.M., Moore, G.A., and Gieseg, S.P. (2008) Dissociation of neopterin and 7,8-dihydroneopterin from plasma components before HPLC analysis. *J Chromatogr B*, **863**, 167–171.

96 Firth, C.A., Laing, A.D., Baird, S.K., Pearson, J., and Gieseg, S.P. (2008) Inflammatory sites as a source of plasma neopterin: measurement of high levels of neopterin and markers of oxidative stress in pus drained from human abscesses. *Clin Biochem*, **41**, 1078–1083.

97 Schumacher, M., Eder, B., Tatzber, F., Kaufmann, P., Esterbauer, H., and Klein, W. (1992) Neopterin levels in patients with coronary artery disease. *Atherosclerosis*, **94**, 87–88.

98 Adachi, T., Naruko, T., Itoh, A., Komatsu, R., Abe, Y., Shirai, N., Yamashita, H., Ehara, S., Nakagawa, M., Kitabayashi, C., Ikura, Y., Ohsawa, M., Yoshiyama, M., Haze, K., and Ueda, M. (2007) Neopterin is associated with plaque inflammation and destabilisation in human coronary atherosclerotic lesions. *Heart*, **93**, 1537–1541.

99 Ray, K.K., Morrow, D.A., Sabatine, M.S., Shui, A., Rifai, N., Cannon, C.P., and Braunwald, E. (2007) Long-term prognostic value of neopterin a novel marker of monocyte activation in patients with acute coronary syndrome. *Circulation*, **115**, 3071–3078.

100 Firth, C.A., Crone, E.M., Flavall, E.A., Roake, J., and Gieseg, S.P. (2008) Macrophage mediated protein hydroperoxide formation and lipid oxidation in low density lipoprotein is inhibited by the inflammation marker 7,8 dihydroneopterin. *Biochim Biophys Acta Mol Cell Res*, **1783**, 1095–1101.

101 Gieseg, S.P. and Cato, S. (2003) Inhibition of THP-1 cell-mediated low-density lipoprotein oxidation by the macrophage-synthesised pterin, 7,8-dihydroneopterin. *Redox Rep*, **8**, 113–119.

102 Firth, C.A., Yang, Y., and Gieseg, S.P. (2007) Lipid oxidation predominates over protein hydroperoxide formation in human monocyte-derived macrophages exposed to aqueous peroxyl radicals. *Free Radic Res*, **41**, 839–848.

103 Werner-Felmayer, G., Werner, E.R., Fuchs, D., Hausen, A., Reibnegger, G., and Wachter, H. (1990) Neopterin formation and tryptophan degradation by a human myelomonocytic cell line (THP-1) upon cytokine treatment. *Cancer Res*, **50**, 2863–2867.

104 Christen, S., Thomas, S.R., Garner, B., and Stocker, R. (1994) Inhibition by interferon-gamma of human mononuclear cell-mediated low density lipoprotein oxidation: Participation of tryptophan metabolism along the kynurenine pathway. *J Clin Invest*, **93**, 2149–2158.

105 Thomas, S.R., Witting, P.K., and Stocker, R. (1996) 3-Hydroxyanthranilic acid is an efficient, cell-derived co-antioxidant for alpha-tocopherol, inhibiting human low density lipoprotein and plasma lipid peroxidation. *J Biol Chem*, **271**, 32714–32721.

106 Plüddemanna, A., Neyena, C., and Gordon, S. (2007) Macrophage scavenger receptors and host-derived ligands. *Methods*, **43**, 207–217.

107 Maor, I. and Aviram, M. (1994) Oxidised low density lipoprotein leads to macrophage accumulation of unesterified cholesterol as a result of lyosomal trapping of the lipoprotein hydrolysed cholesterol ester. *J Lipid Res*, **35**, 803–819.

108 Van Reyk, D.M. and Jessup, W. (1999) The macrophage in atherosclerosis: modulation of cell function by sterols. *J Leukoc Biol*, **66**, 557–561.

109 Wintergerst, E.S., Jelk, J., Rahner, C., and Asmis, R. (2000) Apoptosis induced by oxidized low density lipoprotein in human monocyte-derived macrophages involves CD36 and activation of

caspase-3. *Eur J Biochem*, **267**, 6050–6058.

110 Nguyen-Khoa, T., Massy, Z.A., Witko-Sarsat, V., Canteloup, S., Kebede, M., Lacour, B., Drueke, T., and Descamps-Latscha, B. (1999) Oxidized low-density lipoprotein induces macrophage respiratory burst via its protein moiety: a novel pathway in atherogenesis? *Biochem Biophys Res Commun*, **263**, 804–809.

111 Liao, H.S., Kodama, T., and Geng, Y.J. (2000) Expression of class A scavenger receptor inhibits apoptosis of macrophages triggered by oxidized low density lipoprotein and oxysterol. *Arterioscler Thromb Vasc Biol*, **8**, 1968–1975.

112 Marchant, C.E., Law, N.S., Vanderveen, C., Hardwick, S.J., Carpenter, K.L.H., and Mitchinson, M.J. (1995) Oxidized low-density-lipoprotein is cytotoxic to human monocyte-macrophages: protection with lipophilic antioxidants. *FEBS Lett*, **358**, 175–178.

113 Harris, L.K., Mann, G.E., Ruiz, E., Mushtaq, S., and Leake, D.S. (2006) Ascorbate does not protect macrophages against apoptosis induced by oxidised low density lipoprotein. *Arch Biochem Biophys*, **455**, 68–76.

114 Ryan, L., O'Callaghan, Y.C., and O'Bien, N.M. (2004) Generation of an oxidative stress precedes caspase activation during 7 beta-hydroxycholesterol-induced apoptosis in U937 cells. *J Biochem Mol Toxicol*, **18**, 50–59.

115 Packer, M.A. and Murphy, M.P. (1994) Peroxynitrite causes calcium efflux from mitochondria which is prevented by cyclosporin A. *FEBS Lett*, **345**, 237–240.

116 Budihardjo, I., Oliver, H., Lutter, M., Luo, X., and Wang, X.D. (1999) Biochemical pathways of caspase activation during apoptosis. *Annu Rev Cell Dev Biol*, **15**, 269–290.

117 Yang, L. and Sinensky, M.S. (2000) 25-Hydroxycholesterol activates a cytochrome c release-mediated caspase cascade. *Biochem Biophys Res Commun*, **278**, 557–563.

118 Rusinol, A.E., Thewke, D., Liu, J., Freeman, N., Panini, S.R., and Sinensky, M.S. (2004) AKT/Protein kinase B regulation of BCL family members during oxysterol-induced apoptosis. *J Biol Chem*, **279**, 1392–1399.

119 Martin, D., Salinas, M., Fujita, N., Tsuruo, T., and Cuadrado, A. (2002) Ceramide and reactive oxygen species generated by H2O2 induce caspase-3-independent degradation of AKT/protein kinase B. *J Biol Chem*, **277**, 42943–42952.

120 Cross, D.A.E., Alessi, D.R., Cohen, P., Andjelkovich, M., and Hemmings, B.A. (1995) Inhibition of glycogen-synthase kinase-3 by insulin-mediated by protein-kinase-B. *Nature*, **378**, 785–789.

121 Choi, Y.K., Kim, Y.S., Choi, I.Y., Kim, S.W., and Kim, W.K. (2008) 25-Hydroxycholesterol induces mitochondria-dependent apoptosis via activation of glycogen synthase kinase-3 beta in PC12 Cells. *Free Radic Res*, **42**, 544–553.

122 Meilhac, O., Escargueil-Blanc, I., Thiers, J.C., Salvayre, R., and Negre-Salvayre, A. (1999) BCL-2 alters the balance between apoptosis and necrosis, but does not prevent cell death induced by oxidized low density lipoproteins. *FASEB J*, **13**, 485–494.

123 Ryan, L., O'callaghan, Y.C., and O'brien, N.M. (2006) Involvement of calcium in 7 beta-hydroxycholesterol and cholesterol-5 beta,6 beta-epoxide-induced apoptosis. *Int J Toxicol*, **25**, 35–39.

124 Porn-Ares, M.I., Saido, T.C., Andersson, T., and Ares, M.P.S. (2003) Oxidized low-density lipoprotein induces calpain-dependent cell death and ubiquitination of caspase 3 in HMEC-1 endothelial cells. *Biochem J*, **374**, 403–411.

125 Berthier, A., Lemaire-Ewing, S., Prunet, C., Monier, S., Athias, A., Bessede, G., De Barros, J.P.P., Laubriet, A., Gambert, P., Lizard, G.,

and Neel, D. (2004) Involvement of a calcium-dependent dephosphorylation of BAD associated with the localization of TRPC-1 within lipid rafts in 7-ketocholesterol-induced THP-1 cell apoptosis. *Cell Death Differ*, **11**, 897–905.

126 Vindis, C., Elbaz, M., Escargueil-Blanc, I., Auge, N., Heniquez, A., Thiers, J.C., Negre-Salvayre, A., and Salvayre, R. (2005) Two distinct calcium-dependent mitochondrial pathways are involved in oxidized LDL-induced apoptosis. *Arterioscler Thromb Vasc Biol*, **25**, 639–645.

127 Lizard, G., Miguet, C., Bessede, G., Monier, S., Gueldry, S., Neel, D., and Gambert, P. (2000) Impairment with various antioxidants of the loss of mitochondrial transmembrane potential and of the cytosolic release of cytochrome c occurring during 7-ketocholesterol-induced apoptosis. *Free Radic Biol Med*, **28**, 743–753.

128 Hampton, M.B., Stamenkovic, I., and Winterbourn, C.C. (2002) Interaction with substrate sensitises caspase-3 to inactivation by hydrogen peroxide. *FEBS Lett*, **517**, 229–232.

129 Baird, S.K., Reid, L., Hampton, M., and Gieseg, S.P. (2005) OxLDL induced cell death is inhibited by the macrophage synthesised pterin, 7,8-dihydroneopterin, in U937 cells but not THP-1 cells. *Biochim Biophys Acta*, **1745**, 361–369.

130 Collaborative GRP Primary Prevention. (2001) Low-dose aspirin and Vitamin E in people at cardiovascular risk: a randomised trial in general practice. *Lancet*, **357**, 89–95.

131 Hanefeld, M., Marx, N., Pfutzner, A., Baurecht, W., Lubben, G., Karagiannis, E., and Forst, T. (2007) Anti-inflammatory effects of pioglitazone and/or sinivastatin in high cardiovascular risk patients with elevated high sensitivity C-reactive protein: the Piostat Study. *J Am Coll Cardiol*, **49**, 290–297.

132 Jasiska, M., Owczarek, J., and Orszulak-Michalak, D. (2007) Statins: a new insight into their mechanisms of action and consequent pleiotropic effects. *Pharmacol Rep*, **59**, 483–499.

26
Oxidative Stress in Breast Cancer Carcinogenesis
Lisa J. Martin and Norman Boyd

Increased oxidative stress is thought to play a role in the development of several chronic diseases including breast cancer, but direct evidence for the effect of oxidative stress on breast cancer development in humans is limited. In this chapter, we discuss some of the issues related to measurement of oxidative stress in human studies and review the evidence for a role of oxidative stress in breast cancer risk. Markers of lipid peroxidation, such as urinary malondialdehyde (MDA) and isoprostanes, appear to be valid markers of oxidative stress if they are measured using appropriate techniques. However, more studies directly comparing different markers and their association with risk factors and disease are needed, because markers may vary in their sensitivity to specify pathways or contexts of oxidative stress. An increased understanding of these relationships will facilitate the interpretation of studies and increase our understanding of the role of oxidative stress in disease. We also describe the association of MDA, a marker of lipid peroxidation, with mammographic density, a strong risk factor for breast cancer, and discuss potential mechanisms for the increased risk of breast cancer associated with mammographic density.

26.1
Introduction

Reactive oxygen species (ROS) are generated endogenously during normal cell metabolism and even low levels of ROS play important roles in many biochemical processes. Oxidative stress occurs when an excess of ROS is produced in relation to antioxidant defenses. Antioxidant defenses include both enzymatic (including superoxide dismutase, glutathione peroxidase, and catalase) and non-enzymatic defenses (including antioxidant vitamins, such as carotenoids and α-tocopherol) [1].

Oxidative stress can result in oxidative damage to DNA, protein, and lipid molecules. DNA damage can result in mutagenesis and increased risk of cancer [2]. Inflammation is also associated with increased ROS and may be an additional

Endogenous Toxins. Diet, Genetics, Disease and Treatment.
Edited by Peter J. O'Brien and W. Robert Bruce
Copyright © 2010 WILEY-VCH Verlag GmbH & Co. KGaA, Weinheim
ISBN: 978-3-527-32363-0

pathway that relates oxidative stress to cancer risk [3]. Lipid peroxidation (oxidative damage to lipids) is particularly important because it leads to alterations in biological properties of cell membranes and to the propagation of free radical reactions that further increases oxidative damage to DNA [4]. ROS and products of lipid peroxidation can increase cell division (mitogenesis) [5], which is also related to increased risk of cancer [6], and itself can lead to an increase in ROS production.

Although it is generally thought that oxidative stress influences the risk of cancer, other chronic diseases, and aging, most evidence of the role of oxidative stress is indirect. Intervention studies of antioxidant vitamin supplementation have provided conflicting results [7].

Research, in humans, on the role of oxidative stress in disease is limited by issues related to the measurement and interpretation of markers of oxidative stress *in vivo* [8]. In this chapter, we discuss some of the issues related to the measurement of oxidative stress in human studies and review the evidence of a role of oxidative stress in breast cancer risk. We also describe the association of malondialdehyde (MDA), a marker of lipid peroxidation, with mammographic density (MD), a strong risk factor for breast cancer, and discuss potential mechanisms for the increased risk of breast cancer associated with MD.

26.2
Markers of Oxidative Stress

Oxidative stress can be measured directly using electron spin resonance to detect oxygen free radicals [9]. However, this method is not applicable to epidemiological studies as free radicals are extremely unstable and the measurement technique is complex. The focus in human studies has been on the detection of products formed from the oxidation of macromolecules, which are more stable than free radicals [9] or on components of the antioxidant defense system. There have been few studies that compare markers of oxidative stress directly in the same samples and it is not known which methods provide the most reliable or accurate information or whether different markers are specific for different types of oxidative insults.

A multilaboratory validation study compared several of these biomarkers in a rat model of oxidative stress induced by carbon tetrachloride poisoning. The time- and dose-dependent responses of several plasma and urinary markers to carbon tetrachloride poisoning were examined. The study concluded that plasma and urinary MDA and isoprostanes and urinary 8-hydroxy-2-deoxyguanosine (8-OHdG) are the most promising markers of *in vivo* oxidative stress [10]. All other markers tested, including plasma markers of oxidative damage to proteins, leukocyte DNA adducts, and DNA strand breaks, did not show a clear relationship with oxidative stress in this model.

8-OHdG is a marker of oxidative DNA damage that is highly mutagenic [11]. After cleavage from DNA as a result of DNA repair, 8-OHdG is excreted in urine. Therefore, the amount excreted in urine is a result of both production and repair and does not reflect oxidative stress alone. Determination of the source of urinary 8-OHdG and standardization of methods for its measurement are currently underway [12].

MDA and isoprostanes are both products of lipid peroxidation produced from the free radical–mediated oxidation of polyunsaturated fatty acids in membranes [4]. These compounds can promote cell proliferation through cell signaling [13] and MDA is a known mutagen [14, 15].

26.2.1
MDA

MDA is a major product of lipid peroxidation and has been the most commonly used marker of this process. MDA is usually not measured directly but is estimated by the measurement of thiobarbituric acid reactive substance (TBARS). The TBARS method involves heating the biological material with thiobarbituric acid (TBA), which produces an adduct that can be measured as the intensity of a pink chromogen by UV or fluorescence. The TBARS method of measuring MDA is easy and inexpensive to perform; however, it has several important limitations for the assessment of *in vivo* lipid peroxidation. A major limitation of this method is that TBARS are not specific for MDA as many other compounds (such as other aldehydes, carbohydrates, ascorbate, and amino acids) can form the same chromogens [4]. Even urine, which is relatively devoid of lipids or lipid peroxides, contains several compounds that react with TBA [8]. The specificity of the method is improved substantially by the measurement of TBA–MDA adducts following HPLC to isolate the TBA–MDA chromogens [16]. However, many of the interfering compounds also form MDA–TBA adducts that are indistinguishable from MDA formed during lipid peroxidation. Another limitation is that a large amount of MDA may be formed artifactually during the incubation stage of the assay, presumably from decomposition of lipid peroxides with the generation of further free radicals [4]. However, this is less of problem in urine samples as urine contains very little lipid material.

A further limitation of MDA as a marker of endogenous lipid production is that diet is a potential source of MDA excreted in urine [17]. MDA can be produced *in vivo* or during storage and processing of food. Finally, MDA is formed as a by-product of thromboxane synthesis in platelets via the cyclooxygenase cascade [18]. For each molecule of thromboxane produced by thromboxane synthase, a molecule of MDA is also produced. Therefore, not all MDA produced in the body reflects lipid peroxidation specifically.

26.2.2
Isoprostanes

Isoprostanes are prostaglandin-like compounds formed from the peroxidation of arachidonic acid [19]. Unlike prostaglandins, which are formed via the action of cyclooxygenase enzymes, isoprostanes are formed nonenzymatically as a result of the free radical–mediated peroxidation of arachidonic acid in cell membranes.

Isoprostanes are cleaved and then circulate in blood and are excreted in urine. They are specific products of lipid peroxidation [20] that can be measured accurately with high sensitivity using mass spectrometry (MS) methods as described by Morrow et al. [19]. Other less labor-intensive methods have been developed such as liquid chromatography MS [21] and immunologic approaches [22, 23], but less is known about their sensitivity and reliability in complex tissues.

Although isoprostanes can be detected in food, dietary isoprostanes do not appear to affect plasma or urinary levels of isoprostane [24], and isoprostane levels are not influenced acutely by the lipid content of the diet [25].

Increased levels of isoprostanes have been observed in many conditions associated with oxidative stress, including cardiovascular, neurological, lung, renal, and liver diseases (see [20, 26] for reviews). In addition, increased isoprostane levels are associated with several risk factors for cardiovascular disease, which are thought to be associated with oxidative stress (e.g., smoking, high cholesterol, diabetes, and obesity) [27, 28].

26.2.3
Biological Sampling

MDA and isoprostanes are frequently measured in serum or plasma. Care must be taken during sample collection to minimize artifactual production, including avoiding light and flash freezing. Measurement of these markers in urine (spot, overnight or 24 hour collections) minimizes their artifactual production during storage/processing because there is virtually no lipid in urine (in contrast to blood or tissue). Twenty-four hours urine collections have the advantage of providing an integrated measure over 24 hours.

26.3
Oxidative Stress and Breast Cancer

The protective effects of higher fruit and vegetable intake and serum antioxidant levels on breast cancer risk seen in some studies [29, 30], and studies showing that genetic polymorphisms in some antioxidant enzymes are associated with breast cancer risk [29], provide indirect evidence of a role of oxidative stress in the development of breast cancer. The association of hormone replacement therapy with breast cancer risk may be modified by catalase genotype suggesting a role for oxidative stress [31].

Direct evidence of an association of oxidative stress with breast cancer risk arises from case–control studies of patients with and without breast cancer. Several small case–control studies have shown that the concentration of plasma MDA or TBARS was elevated in breast cancer patients compared to healthy controls [32–35]. The concentration of DNA adduct, 8-OHdG, was significantly greater in breast tissue from subjects with malignant breast tumors than from those with benign tumors or in tissue from reduction mammoplasty [36, 37]. Levels of MDA–DNA adducts were significantly higher in breast tissue of cancer patients than in the breast tissue of controls without cancer, independent of age, smoking status, and body mass index (BMI) [38].

An early report from the Long Island Breast Cancer Study Project showed a significant trend of increasing breast cancer risk with increasing urinary excretion of isoprostanes measured by immunoassay kit [39], but not with urinary 8-OHdG excretion [39]. However a recent extended report from this study, which included over 1000 cases and controls, showed no association between urinary isoprostane or urinary 8-OHdG and breast cancer risk [40]. In a separate report from the same study, plasma carbonyl levels, a marker of protein oxidation, were positively associated with breast cancer risk, independent of urinary isoprostane excretion [41].

A limitation of all case–control studies cited above is that the markers of oxidative stress were measured in biological samples collected after breast cancer diagnosis, and therefore the higher levels of oxidative stress in cases could be due to the presence of cancer or its treatment, rather than an increased risk for developing breast cancer. In addition, if cases and controls are selected from different sources, any differences between them in the marker of interest could be due to factors other than the risk of disease.

Prospective cohort studies, or case–control studies that are nested within them, can eliminate the effects of disease or treatment on the markers of interest if samples are collected prior to diagnosis. In addition, they ensure that cases and controls are selected from the same source population. Prospective studies of oxidative stress are challenging as blood samples require special handling and storage and urine is not commonly collected in large cohorts. The first prospective study of oxidative stress and breast cancer risk has now been reported in abstract form [42]. Urine samples were collected prior to diagnosis in 436 cases and 852 controls selected from the Shanghai Women's Health Study (nested case–control study). Although there was no overall association between urine isoprostane level and breast cancer risk, in women with high BMI (>29), those with high isoprostane levels had a 10-fold higher risk of developing breast cancer, while in women with BMI less than 23, high isoprostane levels were associated with a reduced risk of developing breast cancer.

26.4
Oxidative Stress and Breast Cancer Risk Factors

Oxidative stress is influenced by many of the environmental, lifestyle, and nutritional variables that are known, or suspected, of influencing risk of breast cancer, such as ethnicity [43], body weight [28, 44], physical activity [45, 46], and energy restriction [47]. Chinese women living in San Francisco ($n = 33$) had 42% higher levels of urinary MDA excretion than Chinese women living in China ($n = 50$) [43], who have a lower risk of breast cancer. Chinese–American women have a lower BMI, and lower excretion of urinary isoprostane than Caucasians [48]. In contrast, higher plasma levels of the less specific marker, TBARS, were observed in Japanese living in Japan compared to Japanese living in the United States or Caucasians [49].

Higher body weight is associated with increased breast cancer risk, at least in postmenopausal women [50], and is consistently associated with lower serum antioxidant levels [51, 52]. Higher body weight is associated with higher levels of isoprostane [28, 44], but is not consistently related to MDA levels [44, 53]. Energy restriction and weight loss may reduce breast cancer risk and oxidative stress [47]. Vigorous physical activity increases oxidative stress acutely [54], whereas chronic moderate levels of activity increase antioxidant activity [45, 46], reduce oxidative stress [55], and are associated with reduced breast cancer risk [56].

Higher intake of vitamin E has been found to be associated with lower urinary excretion of isoprostane [48]. Dietary antioxidant and fruit and vegetable interventions reduce isoprostane and MDA levels and oxidative DNA damage in some [57–60], but not all, studies [61–63]. Higher serum antioxidant levels are associated with lower breast cancer risk [29], and with lower levels of markers of lipid peroxidation and inflammation [44, 64]. Alcohol intake increases both breast cancer risk [50] and plasma isoprostane levels [65]. The effect of dietary fat on breast cancer risk is not clear [66, 67], but decreased fat intake may reduce markers of DNA oxidative damage [68], and markers of lipid peroxidation have been shown to be associated with dietary fat intake in some [48], but not all, studies [25, 44].

In terms of hormonal risk factors known to be associated with breast cancer, markers of oxidative stress are higher in postmenopausal compared to premenopausal women [53, 69] and may be reduced by tamoxifen [54]. Urinary isoprostane excretion was not associated with blood estrogen levels [60], but estrogen and its metabolites have both antioxidant and pro-oxidant effects [54].

26.5
Mammographic Density (MD)

The examination of women who differ substantially in their risk of developing breast cancer, but in whom the disease has not yet developed, is likely to provide useful information about factors associated with breast cancer risk. Extensive MD is a strong risk factor for breast cancer that is common in the population, and therefore a substantial number of cancers occur in women with this risk factor [70]. MD refers to the appearance of the breast on mammography. Stroma and epithelium attenuate X-rays more than fat and appear light, an appearance we refer to as *mammographic density*, while fat appears dark. The extent of MD varies among women (Figure 26.1), reflecting variations in breast tissue composition. In the following sections, we describe the association of MD with breast cancer risk and the evidence that it is associated with mitogenesis and mutagenesis.

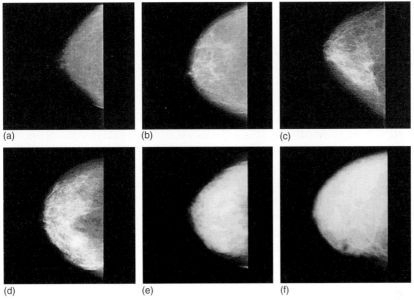

Fig. 26.1 Variations in mammographic density. Six categories of mammographic density based on percentage of the breast occupied by density are shown. (a) 0%; (b) >0 to <10%; (c) 10 to <25%; (d) 25 to <50%; (e) 50 to <75%; (f) ≥75%. (Reprinted from Boyd et al. 2005 [87] with permission.)

26.5.1
Mammographic Density and Breast Cancer Risk

Wolfe first described a qualitative method of classifying variations in the appearance of the mammogram [71, 72], and subsequently quantitative methods of measuring MD were developed, including estimation by radiologists and a computer-assisted method [73]. The percentage of the breast occupied by MD is now recognized as one of the strongest known risk factors for breast cancer [74, 75].

A systematic meta-analysis of 42 studies including data for >14 000 cases and 226 000 non-cases showed that MD was consistently associated with risk of breast cancer [74]. The associations were stronger in studies in the general population, rather than symptomatic women; in studies of incident, rather than prevalent cancer; and in studies using quantitative measures of percent MD, rather than for qualitative measures. The breast cancer risk associated with density did not differ by age, menopausal status, or ethnicity and cannot be explained by the "masking" of cancers by dense tissue [76]. Extensive MD is associated with an increased risk of breast cancer at screening and between screening examinations, and the increase in risk persists for at least 8–10 years from the date of the mammogram used to classify MD [76, 77].

26.5.2
Mammographic Density and Breast Tissue Composition

Studies of the relationship between breast tissue histology and the radiological appearance of the breast (described in detail in [78]) have been done using surgical biopsies or mastectomy specimens. All of these studies have found greater amounts of epithelium and/or stroma to be associated with MD.

Li *et al.* [79] used quantitative microscopy to examine histological features of randomly selected tissue blocks from breast tissue obtained at forensic autopsy, and determined the proportions of the biopsy occupied by cells (estimated by nuclear areas), glandular structures, and collagen. Greater percent MD was associated with a significantly greater total nuclear area, a greater nuclear area of both epithelial and nonepithelial cells, a greater proportion of collagen, and a greater area of glandular structures. Of the tissue components measured, collagen was present in the greatest quantity, was most strongly associated with percent density, and explained 29% of the variance in percent density. Nuclear area and glandular area accounted for between 4 and 7% of the variance in percent density.

26.5.3
Age, Other Risk Factors, and Mammographic Density as a Marker of Susceptibility

The average level of MD declines with increasing age [80, 81], while breast cancer incidence increases with age (Figure 26.2a). This seeming paradox may, however, be resolved by reference to a model of breast cancer incidence proposed by Pike [82], that is based on the concept that the rate of breast tissue "aging" or "exposure," rather than chronological age, is the relevant measure for describing the age-specific incidence of breast cancer. Breast tissue "aging" is thought to be closely related to the mitotic activity of breast epithelial or stem cells and their susceptibility to genetic damage. According to the model shown in Figure 26.2b, the rate of breast tissue "aging" is most rapid at the time of menarche, slows with pregnancy, slows further in the perimenopausal period, and is least after the menopause. After fitting numerical values for these parameters, Pike showed that cumulative exposure to breast tissue "aging," given by the area under the curve in Figure 26.2b, described the age–incidence curve for breast cancer in the United States, shown in Figure 26.2a. Thus, cumulative exposure to breast tissue "aging" and the age-specific breast incidence increase with age, but the rate of increase slows with age, particularly after menopause.

MD shares many of the features of "breast tissue age" and is influenced by similar factors. Detailed descriptions of the associations of risk factors with MD can be found elsewhere [78, 83, 84]. Body size in particular is strongly and inversely associated with MD, and is a risk factor for breast cancer independent of MD [85]. We focus here on the associations of MD with age, parity, menopause, and variables in the Pike model, which are also associated with variations in one or more of the histological features of the breast [79].

In addition to the effects of age referred to above, MD is less extensive in women who are parous, and in those with a larger number of live births. Postmenopausal women have consistently been found to have less extensive MD than premenopausal women, and a longitudinal study of the effects of the menopause on MD showed that percent MD was reduced by about 8% on average over the menopause [86].

All risk factors for breast cancer must ultimately exert their influence by an effect on the breast, and these findings suggest that, for at least some risk factors, this influence includes an effect on the number of cells and the quantity of collagen in the breast that is reflected in differences in MD. The concept of breast tissue age in the Pike model is related to the effects of hormones on the kinetics of breast cells, and the accumulation of genetic damage. MD may reflect cumulative exposure to stimuli to division of breast cells (discussed below), which predisposes them to genetic damage by mutagens.

26.5.4
Genetic Factors

The factors described above account for only 20–30% of the variation in MD observed in the population [84]. We have carried out two independent twin studies in Australia and North America [87]. After adjustment for age, BMI, and reproductive risk factors, the proportion of the residual variation accounted for by additive genetic factors (heritability) was estimated to be 63% (95% CI: 59–67%) in the combined studies [87]. MD thus has the characteristics of a quantitative trait. Several large-scale genome-wide linkage and association studies to determine the genetic determinants of MD are in progress.

26.6
Association of Hormones, Mitogens, and Mutagens with Mammographic Density

The effects of age, risk factors, and genes on breast tissue composition described above are likely to be mediated, at least in part, by one or more of the several endocrine, paracrine, and autocrine mechanisms that regulate the growth and development of breast stroma and epithelium. Variations in exposure, or response, to one or more of these mechanisms may explain the effects that genetic and environmental factors have on differences in breast tissue composition. As we describe in more detail elsewhere [88], the risk of breast cancer associated with MD may arise from the combined effects of cell proliferation, in response to mitogens and the resulting increase in the number of susceptible cells, and genetic damage to cells by mutagens. Figure 26.3a and b, respectively, provide a schematic overview and a more detailed description of aspects of this hypothesis. The evidence of the role of mitogens (blood hormones and growth factors) is described briefly below and the evidence of lipid peroxidation and MDA is discussed in more detail in the following sections.

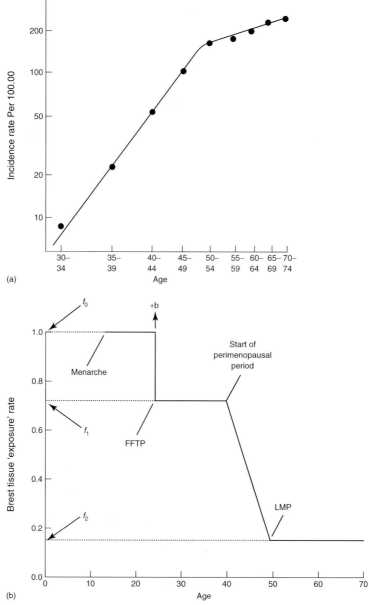

Fig. 26.2 Age, mammographic density, and the incidence of breast cancer. (a) Log–log plot of the age-specific incidence of breast cancer [82]. (b) The Pike model of breast tissue aging. FFTP, first full-term pregnancy and LMP, last menstrual period. "b" is a parameter used in calculating age at menarche (see [82]).

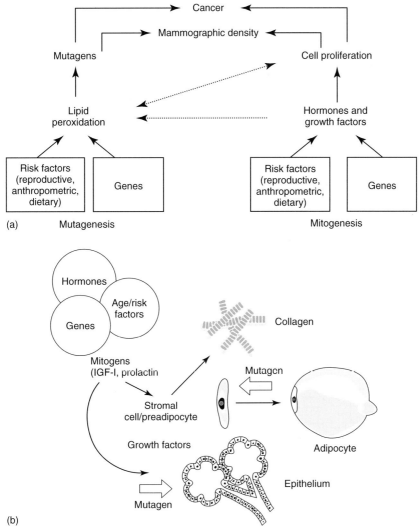

Fig. 26.3 Hypotheses. (a) Schematic summary. We postulate that the combined effects of cell proliferation (mitogenesis) and genetic damage to proliferating cells caused by mutagens (mutagenesis) may underlie the increased risk of breast cancer associated with extensive mammographic density. Mitogenesis and mutagenesis are related processes. Increased cell proliferation increases susceptibility to mutations, and cell death resulting from genetic damage can increase cell proliferation (see text). (b) Biological hypotheses. The tissue components (epithelial cells, stromal cells, collagen, and fat) that are responsible for variations in mammographic density are related to each other in several ways. Stromal fibroblasts produce collagen, and some are preadiopocytes that differentiate into adipocytes. Stromal and epithelial cells influence each other through paracrine growth factors, and both cell types are influenced by endocrine stimuli to cell proliferation (mitogenesis). Genetic damage to either stromal or epithelial cells caused by mutagens (mutagenesis) could initiate carcinogenesis (see text). (Reprinted from Martin and Boyd [88] with permission.)

26.6.1
Levels of Blood Hormones and Growth Factors (Mitogens)

Studies of blood estrogen levels and percent MD have found either no association or an inverse association, with estrogen levels (five of seven studies) or total or free estradiol (seven of eight studies) in premenopausal or postmenopausal women [88]. Progesterone, testosterone, and androstenedione levels have not been shown to be associated with MD. Sex hormone binding globulin (SHBG) has been found to have a significant positive association with MD in two studies after adjustment for other variables, and in four other studies before adjustment [88].

Blood estradiol and testosterone levels have been shown to be related to risk of breast cancer in premenopausal and postmenopausal women [89, 90], but have not consistently been shown to be associated with MD, suggesting that they may influence risk through pathways that are unrelated to density. In support of this idea, Tamimi et al. reported that circulating sex steroid levels and MD are independently associated with breast cancer risk in postmenopausal women [91]. However, it remains possible that other forms of estrogen not measured in these studies, including estrogen metabolites [92], may influence MD and the associated risk of breast cancer.

Combined estrogen–progesterone menopausal hormone therapy, but not estrogen therapy alone, is associated with a small increase in risk of breast cancer [93], and increases MD [94–96]. Percent density is reduced by *tamoxifen* [97] and by a gonadotrophin releasing hormone agonist [98] that reduces exposure to estrogen and progesterone in premenopausal women.

Prolactin levels have been found to be positively associated with MD in postmenopausal women in two studies, and in a further study statistical significance was lost after adjustment for other variables [88]. Prolactin increases cell proliferation and decreases apoptosis in the breast and higher blood prolactin levels have been found to be associated with an increased risk of breast cancer in both pre- and postmenopausal women [99].

Blood growth hormone levels have been found to be positively associated with MD in premenopausal women, but this association became nonsignificant after adjustment for body size [100]. Because growth hormone is one of the factors that influence body size, this may be overadjustment. MD has been found to be positively associated with serum IGF-I levels in premenopausal women, in three of five studies, and one study found an association in postmenopausal women [88]. The extent of immunostaining for IGF-I in breast biopsies is also positively related to MD in the breast from which the biopsy was taken [101]. Blood IGF-I levels have been found to be associated with risk of breast cancer in premenopausal women [102]. IGF-I is a known mitogen for breast epithelium that is produced in the breast stroma, as well as by the liver in response to growth hormone [103], and the administration of growth hormone to aging primates has been shown to induce epithelial proliferation [104]. Genetic factors that influence MD might do so in part through effects on breast mitogens [105]. Twin studies have shown that about 60% of the variance in levels of IGF-I in circulation is explained by heritable factors [106] and that growth hormone secretion is markedly dependent on genetic factors [107].

26.6.2
Urinary MDA and Mammographic Density

We have seen evidence for a positive association between the amount of MD and lipid peroxidation, measured by urinary MDA excretion, in three independent studies [69, 108, 109]. MDA was measured in 24 hours urine collections by HPLC determination of TBARS [110] and MD was measured using a computer-assisted method [73]. In the first study, urinary MDA in premenopausal women with extensive MD (>75%; $n = 30$) was significantly greater than that of women with little MD (>25%; $n = 16$) before ($p = 0.02$) and after adjustment for body weight ($p = 0.0006$). We confirmed this observation in a larger independent sample of premenopausal women ($n = 273$) recruited to represent a wide range of MD [109]. In this study, urinary MDA excretion was 28% higher in the highest quintile of MD compared to the lowest after adjustment for age and BMI ($p = 0.04$).

In the third study, we extended our observations to include pre- and postmenopausal women and additional markers of lipid peroxidation (serum MDA and MDA–DNA adducts), as well as serum antioxidants (α- and β-carotene, α-tocopherol, cryptoanthine, lutein, lycopene, and retinol). There was a trend of increasing urinary MDA excretion across increasing quintiles of MD for all women ($p = 0.01$) and for both premenopausal ($p = 0.02$) and postmenopausal women ($p = 0.13$) after adjustment for age and waist circumference (Table 26.1). Urinary MDA excretion was about 31 and 22% higher in the highest quintile of MD compared to the lowest in pre- and postmenopausal women, respectively. Serum MDA, DNA–MDA adducts and serum antioxidants were not significantly associated with MD after adjustment for weight. Serum and urinary MDA levels positively

Tab. 26.1 Urinary excretion of MDA (mmol d^{-1}) by quintile of percent mammographic density

	Quintiles of percent density					
	1	2	3	4	5	p value for trend
Premenopausal ($n = 160$)	2.76[a]	2.29	2.86	2.75	3.62	0.02
Postmenopausal ($n = 175$)	3.19	3.23	3.67	3.46	3.88	0.13
All women ($n = 335$)	3.02	2.76	3.10	3.38	3.68	0.01

[a] Least square mean of MDA adjusted for age and waist circumference; for All Women, values are also adjusted for menopausal status.
  Data are from Hong et al. [68].
  Reprinted from Martin and Boyd [88] with permission.

Tab. 26.2 Urinary excretion of MDA (mmol d^{-1}) by quintile of percent mammographic density – stratified by median waist circumference

	Quintiles of percent density					
	1	2	3	4	5	p value for trend
Below median waist ($n = 167$)	1.74[a]	2.32	2.52	3.17	3.44	0.0003
Above median waist ($n = 168$)	3.26	3.04	3.33	3.21	3.63	0.68

[a] Least square mean of MDA adjusted for age and waist circumference.
Data are from Hong et al. [68].

correlated ($r = 0.19$; $p = 0.0006$), but the DNA–MDA adduct did not correlate with either serum or urinary MDA levels.

In the study described above, women were selected to reflect a wide range of percent MD. Because weight and body fat measures strongly negatively correlated with percent MD, this process resulted in very wide range of adiposity. For example, BMI ranged from 18 to 54 (interquartile range 6.4) and weight ranged from 40 to 150 kg (interquartile range of 19 kg). To further examine the influence of body size on the relationship between MDA and MD, we stratified the analysis by the median waist circumference. There was a very strong positive association of urinary MDA with percent MD in women with lower waist circumference and no significant relationship in women with a waist circumference above the median (Table 26.2). Similar results were seen with stratification by BMI (data not shown).

26.6.3
Urinary Isoprostane and MD

As isoprostane has been proposed as a sensitive and specific marker of endogenous lipid peroxidation [20], we carried out a pilot study to examine the association of urinary isoprostane levels and MD. We measured urinary isoprostane levels using the MS method [19] in a subset of 77 women (oversampled in the extremes of MD) from our study of urinary MDA (described above). Urinary isoprostane excretion strongly negatively correlated with percent MD ($r = -0.48$; $p < 0.0001$). This negative correlation probably reflects the fact that body fat measures are strongly positively correlated with urinary isoprostane level, but strongly negatively correlated with percent MD. After adjustment for age and waist circumference, the negative association between urinary isoprostane and MD reduced substantially and was not statistically significant ($r = -0.16$; $p = 0.15$). In contrast, in this same subset of 77 women, urinary MDA positively associated with percent MD after adjustment for age and waist circumference ($r = 0.21$; $p = 0.07$). These preliminary results suggest that the positive association with MD may be specific to urinary MDA.

26.6.4
Comparison of MDA and Isoprostanes as Markers of Lipid Peroxidation

As discussed above, MDA and isoprostanes are both considered to be useful markers of lipid peroxidation; however, very few studies have examined the relationship between different markers of oxidative stress. In our data, there was no correlation between urinary isoprostane measured by MS and urinary MDA measured by HPLC ($r = 0.04$; $p = 0.70$; $n = 77$). Urinary isoprostane was negatively associated with age and strongly positively associated with weight, waist circumference, BMI, and serum insulin levels (Table 26.3). In this small subset, urinary MDA did not significantly associate with any of these variables. In the full sample of 333 subjects, urinary MDA showed a positive correlation with these variables, but the associations were much weaker than those seen for isoprostanes.

A large study of 298 healthy adults ranging in age from 19 to 78 years measured both plasma MDA and isoprostanes [44] and reported that these markers correlated only at $r = 0.13$ ($p = 0.05$). In multiple regression models, common associations included higher levels of both markers in women compared to men and an inverse correlation with plasma vitamin C levels. Age did not significantly associate with either marker. The study was designed to look at smoking, a major source of oxidative stress [27, 111], and 46% of the subjects were current smokers. Plasma MDA was higher in smokers compared to nonsmokers and higher in heavier compared to light smokers, but the association of smoking status with plasma isoprostane level was inconsistent. In agreement with our preliminary data above using urinary markers, BMI strongly positive with isoprostane levels but was not associated with MDA.

The strong correlations of urinary isoprostane with several factors associated with lipid peroxidation (Table 26.3) suggest that it might be better marker of lipid peroxidation than MDA. However, based on our data, urinary MDA, but not isoprostanes, is positively associated with MD. Differences in the association of some risk factors with MDA and isoprostanes may reflect different sensitivities of the markers to specific pathways of oxidation. For example, several studies have reported a strong positive correlation of isoprostane levels with measures of obesity

Tab. 26.3 Correlation of urinary isoprostane and MDA with selected factors

Factor	Urinary isoprostane ($n = 77$)	Urinary MDA ($n = 77$)	Urinary MDA ($n = 333$)
Age (yr)	−0.17 (0.13)[a]	0.04 (0.76)	0.12 (0.04)
Body mass index	0.49 (<0.0001)	−0.04 (0.70)	0.09 (0.08)
Weight	0.52 (<0.0001)	0.03 (0.78)	0.14 (0.01)
Waist circumference	0.54 (<0.0001)	0.00 (0.98)	0.14 (0.01)
Insulin	0.33 (0.003)	−0.07 (0.53)	0.08 (0.12)

[a] Spearman correlation coefficient (p value).

such as BMI, which appears to be stronger than that seen with MDA [28, 44] (and our preliminary data) suggesting that isoprostanes may be a better marker than MDA for oxidative stress associated with adiposity. The association of MDA with MD is not explained by adiposity, and the association may be stronger in leaner women.

26.7
Potential Mechanisms for the Association of Mitogens and MDA with Mammographic Density and Breast Cancer Risk

The risk of breast cancer associated with MD may be due to the combined effects cell proliferation (mitogenesis) and damage to DNA of dividing cells (mutagenesis) (Figure 26.2) [88]. MD likely reflects the proliferation of breast epithelium and stroma in response to hormones and growth factors. Breast cancer arises from epithelial cells, and the number and proliferative state of these cells may influence both the radiological density of the breast and the probability of genetic damage that may give rise to cancer. Prolactin and IGF-I have been the breast mitogens in blood found most consistently to be positively associated with MD, and both have been found to be associated with breast cancer risk [112, 113]. Little work has yet been reported on the association of paracrine and autocrine factors that influence breast tissue growth with MD, but one study has shown an association of density with IGF-I in breast tissue [101]. IGF-I is produced in the breast by fibroblasts and preadipocytes [114] and has paracrine activity on breast epithelial cells [115].

The mechanisms responsible for the association of MDA, a mutagenic product of lipid peroxidation, with MD remain to be determined. The processes of mitogenesis and mutagenesis are not independent. Increased cell proliferation may cause an increase in the production of ROS and therefore the increased oxidative stress associated with MD may reflect increased cell proliferation. Alternatively, MDA may promote cell proliferation via cell signaling [5], and therefore lipid peroxidation maybe a cause of increased MD.

Some factors that regulate cell division also influence lipid peroxidation and/or are involved in processes that generate mutagens or that modify their effect. As noted above, the available evidence suggests that the growth hormone–IGF-I axis and prolactin are associated with MD, and that serum estradiol levels are not. However, estrogen is one of the factors that regulates the production of growth hormone and prolactin [116, 117], and other forms of estrogen not measured in the studies conducted to date might be associated with MD. Estrogens can induce lipid peroxidation [118] and may decrease catalase activity [119]. Catechol estrogens, metabolites of estrone and estradiol, can generate ROS through redox recycling [118, 120] and can react with DNA to form adducts [121].

Previous work has shown an association between functional polymorphisms in the *catechol-o-methyl transferase (COMT)* gene and MD and serum levels of IGF-I in premenopausal but not postmenopausal women [122]. The enzyme encoded by *COMT* is involved in the metabolism of catechol estrogens that have both pro-oxidant and antioxidant activity [122], as well as in the metabolism of the catecholamines epinephrine, norepinephrine, and dopamine.

The estrogen metabolizing gene *CYP1A2* may also be associated with both MD and lipid peroxidation [69]. *CYP1A2* activity, assessed by the caffeine metabolic ratio, has been found to be associated with MD in postmenopausal but not premenopausal women, and with serum and urinary MDA levels in premenopausal women. The enzyme encoded by the *CYP1A2* gene metabolizes estrone to 2-OH (which is nonestrogenic) and 4-OH estrone (which is estrogenic and can form ROS). These associations with *COMT* and *CYP1A2* are, however, both based on relatively small numbers of subjects and require confirmation.

Higher MD reflects an increase in the number of epithelial and nonepithelial breast cells (such as fibroblasts), but the largest contributor to MD is collagen [79]. Fibroblasts, in addition to secreting growth factors that activate epithelial receptors, are the primary producers of extracellular matrix components such as collagen, fibronectin, and proteoglycans. In response to inflammation or injury, fibroblasts proliferate, become activated, and produce increased amounts of extracellular matrix resulting in the accumulation of collagen and other matrix components (a process called *fibrosis*) [123]. It is unknown whether this process is related to MD and increased risk of breast cancer. However, chronic inflammation and/or the wound healing response, which involves an increase in extracellular matrix and collagen production, may be involved in the initiation or promotion of cancer [124, 125]. In addition, the presence of breast cancer is associated with reactive stroma, a process that resembles fibrosis [126], which is assumed to promote tumor progression and invasion.

Some of the factors associated with MD are associated with increased collagen production. The products of lipid peroxidation (e.g., MDA, other aldehydes, and isoprostanes) and IGF-1 may be mediators of the increased collagen production seen in hepatic fibrosis [13, 127, 128]. MDA is a marker of lipid peroxidation, but it is also produced during the synthesis of thromboxane [18]. Thromboxane, a prostaglandin and an inflammatory mediator, also plays a role in tissue repair and fibrotic processes [129]. The major cytokine associated with fibrosis is transforming growth factor beta (TGF-β); however, to our knowledge the relationship of TGF-β with MD has not yet been examined.

26.8 Summary

Increased oxidative stress is thought to play a role in the development of several chronic diseases including breast cancer, but direct evidence for effect of oxidative stress on breast cancer development in humans is limited. Some case–control studies provide support for a positive association, but all have measured oxidative stress markers after diagnosis of disease and usually after treatment.

The lack of validated, inexpensive markers for the measurement of *in vivo* oxidative stress is a limitation for epidemiological research in this area. Markers of lipid peroxidation, such as urinary MDA and isoprostanes, appear to be valid markers of oxidative stress if they are measured using appropriate techniques. However, more studies directly comparing different markers and their association with risk factors and disease are needed as markers may vary in their sensitivity to specify pathways or contexts of oxidative stress (such as type of oxidative insult or degree of adiposity). An increased understanding of these relationships will facilitate the interpretation of studies and increase our understanding of the role of oxidative stress in disease.

MD is a strong risk factor for breast cancer, but the biological mechanism of the increased risk is not known. We propose here that the risk of breast cancer associated with MD may be explained by the combined effects of mitogens that influence cell proliferation and the size of the cell population in the breast, and mutagens that influence the probability of genetic damage to those cells. Higher MDA excretion associated with higher MD may be a cause or an effect of increased cell proliferation and collagen production, and the risk of breast cancer may be increased by these processes as well as by mutagenesis.

Ultimately, the risk of breast cancer associated with MD will be explained by an improved understanding of the biology of the breast. However, just as epidemiological methods have identified MD as an important risk factor for breast cancer, whose biology is likely to play an important role in the etiology of the disease, epidemiological approaches may be able to suggest potential pathways and mechanisms that are responsible for risk.

References

1 Mates, J.M., Perez-Gomez, C., and De Castro, I.N. (1999) Antioxidant enzymes and human diseases. *Clin Biochem*, **32**, 595–603.
2 Valko, M., Izakovic, M., Mazur, M., Rhodes, C.J., and Talser, J. (2004) Role of oxygen radicals in DNA damage and cancer incidence. *Mol Cell Biochem*, **266**, 37–56.
3 Pathak, S.K., Sharma, R.A., Steward, W.P., Mellon, J.K., Griffiths, T.R., and Gescher, A.J. (2005) Oxidative stress and cyclooxygenase activity in prostate carcinogenesis: targets for chemopreventive strategies. *Eur J Cancer*, **41**, 61–70.
4 Moore, K. and Roberts, L.J. II (1998) Measurement of lipid peroxidation. *Free Radic Res*, **28**, 659–671.
5 Davies, K.J.A. (1999) The broad spectrum of responses to oxidants in proliferating cells: a new paradigm for oxidative stress. *IUBMB Life*, **48**, 41–47.
6 Ames, B.N., Gold, L.S., and Willett, W.C. (1995) The causes and prevention of cancer. *Proc Natl Acad Sci U S A*, **92**, 5258–5265.
7 Huang, H.Y., Caballero, B., Chang, S., Alberg, A., Semba, R., Schneyer, C., Wilson, R.F., Cheng, T.Y., Prokopowicz, G., Barnes, G.J., Vassy, J., and Bass, E.B. (2006) Multivitamin/mineral supplements and prevention of chronic disease. *Evid Rep Technol Assess*, **139**, 1–117.
8 Halliwell, B. and Whiteman, M. (2004) Measuring reactive species and oxidative damage in vivo and in cell culture: how should you do it and what do the results mean? *Br J Pharmacol*, **142**, 231–255.
9 Kohen, R. and Nyska, A. (2002) Oxidation of biological systems: oxidative stress phenomena, antioxidants, redox reactions, and methods for their quantification. *Toxicol Pathol*, **30**, 620–650.

10 Kadiiska, M.B., Gladen, B.C., Baird, D.D., Germolec, D., Graham, L.B., Parker, C.E., Nyska, A., Wachsman, J.T., Ames, B.N., Basu, S., Brot, N., Fitzgerald, G.A., Floyd, R.A., George, M., Heinecke, J.W., Hatch, G.A., Hensley, K., Lawson, J.A., Marnett, L.J., Morrow, J.D., Murray, D.M., Plastaras, J., Roberts, L.J. II, Rokach, J., Shigenaga, M.K., Sohal, R.S., Sun, J., Tice, R.R., Van Thiel, D.H., Wellner, D., Walter, P.B., Tomer, K.B., Mason, R.P., and Barrett, J.H. (2005) Biomarkers of oxidative stress study II: are oxidation products of lipids, proteins, and DNA markers of CCl4 poisoning. *Free Radic Biol Med*, **38**, 698–710.

11 Dalle-Donne, I., Rossi, R., Colombo, R., Giustarini, D., and Milzani, A. (2006) Biomarkers of oxidative damage in human disease. *Clin Chem*, **52**, 601–623.

12 Cooke, M.S., Olinski, R., and Loft, S. (2008) Measurement and meaning of oxidatively modified DNA lesions in urine. *Cancer Epidemiol Biomarkers Prev*, **17**, 3–14.

13 Comporti, M., Arezzini, B., Signorini, C., Sgherri, C., Monaco, B., and Gardi, C. (2005) F2-isoprostanes stimulate collagen synthesis in activated hepatic stellate cells: a link with liver fibrosis? *Lab Invest*, **85**, 1381–1391.

14 Mukai, F.H. and Goldstein, B.D. (1976) Mutagenicity of malondialdehyde, a decomposition product of peroxidized polyunsaturated fatty acids. *Science*, **191**, 868–869.

15 Basu, A.K. and Marnett, L.J. (1983) Unequivocal demonstration that malondialdehyde is a mutagen. *Carcinogenesis*, **4**, 331–333.

16 Draper, H.H., Squires, E.J., Mahmoodi, H., Wu, J., Agarwal, S., and Hadley, M. (1993) A comparative evaluation of thiobarbituric acid methods for the determination of malondialdehyde in biological materials. *Free Radic Biol Med*, **15**, 353–363.

17 Draper, H.H., Polensek, L., Hadley, M., and McGirr, L.G. (1984) Urinary malondialdehyde as an indicator of lipid peroxidation in the diet and in the tissues. *Lipids*, **19**, 836–843.

18 Wang, L.H., Tsai, A.L., and Hsu, P.Y. (2001) Substrate binding is the rate-limiting step in thromboxane synthase catalysis. *J Biol Chem*, **276**, 14737–14743.

19 Morrow, J.D. and Roberts, L.J. II (1999) Mass spectrometric quantification of F2-isoprostanes in biological fluids and tissues as measure of oxidant stress. *Methods Enzymol*, **300**, 3–12.

20 Milne, G.L., Musiek, E.S., and Morrow, J.D. (2005) F2-isoprostanes as markers of oxidative stress in vivo: an overview. *Biomarkers*, **10**, S10–S23.

21 Liang, Y., Wei, P., Duke, R.W., Reaven, P.D., Harman, S.M., Cutler, R.G., and Heward, C.B. (2003) Quantification of 8-iso-prostaglandin-F(2α) and 2,3-dinor-8-iso-prostaglandin-F(2α) in human urine using liquid chromatography-tandem mass spectrometry. *Free Radic Biol Med*, **34**, 409–418.

22 Proudfoot, J., Barden, A., Mori, T.A., Burke, V., Croft, K.D., Beilin, L.J., and Puddey, I.B. (1999) Measurement of urinary F(2)-isoprostanes as markers of in vivo lipid peroxidation-A comparison of enzyme immunoassay with gas chromatography/mass spectrometry. *Anal Biochem*, **272**, 209–215.

23 Bessard, J., Cracowski, J.L., Stanke-Labesque, F., and Bessard, G. (2001) Determination of isoprostaglandin F2α type III in human urine by gas chromatography-electronic impact mass spectrometry. Comparison with enzyme immunoassay. *J Chromatogr B Biomed Sci Appl*, **754**, 333–343.

24 Gopaul, N.K., Zacharowski, K., Halliwell, B., and Anggard, E.E. (2000) Evaluation of the postprandial effects of a fast-food meal on human plasma F(2)-isoprostane levels. *Free Radic Biol Med*, **28**, 806–814.

25. Richelle, M., Turini, M.E., Guidoux, R., Tavazzi, I., Metairon, S., and Fay, L.B. (1999) Urinary isoprostane excretion is not confounded by the lipid content of the diet. *FEBS Lett*, **459**, 259–262.

26. Montuschi, P., Barnes, P.J., and Roberts, L.J. II (2004) Isoprostanes: markers and mediators of oxidative stress. *FASEB J*, **18**, 1791–1800.

27. Davi, G., Falco, A., and Patrono, C. (2004) Determinants of F2-isoprostane biosynthesis and inhibition in man. *Chem Phys Lipids*, **128**, 149–163.

28. Keaney, J.F. Jr, Larson, M.G., Vasan, R.S., Wilson, P.W., Lipinska, I., Corey, D., Massaro, J.M., Sutherland, P., Vita, J.A., and Benjamin, E.J. (2003) Obesity and systemic oxidative stress: clinical correlates of oxidative stress in the Framingham Study. *Arterioscler Thromb Vasc Biol*, **23**, 434–439.

29. Ambrosone, C.B. (2000) Oxidants and antioxidants in breast cancer. *Antioxid Redox Signal*, **2**, 903–917.

30. Toniolo, P., Van Kappel, A.L., Akhmedkhanov, A., Ferrari, P., Kato, I., Shore, R.E., and Riboli, E. (2001) Serum carotenoids and breast cancer. *Am J Epidemiol*, **153**, 1142–1147.

31. Quick, S.K., Shields, P.G., Nie, J., Platek, M.E., McCann, S.E., Hutson, A.D., Trevisan, M., Vito, D., Modali, R., Lehman, T.A., Seddon, M., Edge, S.B., Marian, C., Muti, P., and Freudenheim, J.L. (2008) Effect modification by catalase genotype suggests a role for oxidative stress in the association of hormone replacement therapy with postmenopausal breast cancer risk. *Cancer Epidemiol Biomarkers Prev*, **17**, 1082–1087.

32. Gonenc, A., Ozkan, Y., Torun, M., and Simsek, B. (2001) Plasma malondialdehyde (MDA) levels in breast and lung cancer patients. *J Clin Pharm Ther*, **26**, 141–144.

33. Ray, G., Batra, S., Shukla, N.K., Deo, S., Raina, V., Ashok, S., and Husain, S.A. (2000) Lipid peroxidation, free radical production and antioxidant status in breast cancer. *Breast Cancer Res Treat*, **59**, 163–170.

34. Huang, Y.L., Sheu, J.Y., and Lin, T.H. (1999) Association between oxidative stress and changes of trace elements in patients with breast cancer. *Clin Biochem*, **32**, 131–136.

35. Suzana, S., Normah, H., Fatimah, A., Nor Fadilah, R., Rohi, G.A., Amin, I., Cham, B.G., Rizal, R.M., and Fairulnizal, M.N. (2008) Antioxidants intake and status, and oxidative stress in relation to breast cancer risk: a case-control study. *Asian Pac J Cancer Prev*, **9**, 343–350.

36. Musarrat, J., Arezina-Wilson, J., and Wani, A.A. (1996) Prognostic and aetiological relevance of 8-hydroxyguanosine in human breast carcinogenesis. *Eur J Cancer*, **32A**, 1209–1214.

37. Li, D., Zhang, W., Zhu, J., Chang, P., Sahin, A., Singletary, E., Bondy, M., Hazra, T., Mitra, S., Lau, S.S., Shen, J., and DiGiovanni, J. (2001) Oxidative DNA damage and 8-hydroxy-2-deoxyguanosine DNA glycosylase/apurinic lyase in human breast cancer. *Mol Carcinog*, **31**, 214–223.

38. Wang, M., Dhingra, K., Hittelman, W.N., Liehr, J.G., de Andrade, M., and Li, D. (1996) Lipid peroxidation-induced putative malondialdehyde-DNA adducts in human breast tissues. *Cancer Epidemiol Biomarkers Prev*, **5**, 705–710.

39. Rossner, P. Jr, Gammon, M.D., Terry, M.B., Agarwal, M., Zhang, F.F., Teitelbaum, S.L., Eng, S.M., Gaudet, M.M., Neugut, A.I., and Santella, R.M. (2006) Relationship between urinary 15-F2t-isoprostane and 8-oxodeoxyguanosine levels and breast cancer risk. *Cancer Epidemiol Biomarkers Prev*, **15**, 639–644.

40. Shen, J., Gammon, M.D., Terry, M.B., Wang, K., Bradshaw, P., Teitelbaum, S.L., Neugut, A.I., and Santella, R.M. (2009) Telomere length, oxidative damage, antioxidants and breast cancer risk. *Int J Cancer*, **124**, 1637–1643.

41 Rossner, P. Jr, Terry, M.B., Gammon, M.D., Agrawal, M., Zhang, F.F., Ferris, J.S., Teitelbaum, S.L., Eng, S.M., Gaudet, M.M., Neugut, A.I., and Santella, R.M. (2007) Plasma protein carbonyl levels and breast cancer risk. *J Cell Mol Med*, **11**, 1138–1148.

42 Dai, Q., Gao, Y.T., Shu, X.O., Yang, G., Milne, G., Cai, Q., Wen, W., Rothman, N., Cai, H., Li, H., Xiang, Y., Chow, W.H., and Zheng, W. (2009) Oxidative stress, obesity, and breast cancer risk: Results from the Shanghai Women's Health Study. *J Clin Oncol*, **27**, 2482–2488.

43 Yeung, K.S., McKeown-Eyssen, G.E., Li, G.F., Glazer, E., Hay, K., Child, P., Gurgin, V., Zhu, S.L., Baptista, J., Aloe, M., Mee, D., Jazmaji, V., Austin, D.F., Li, C.C., and Bruce, W.R. (1991) Comparison of diet and biochemical characteristics of stool and urine between Chinese populations with low and high colorectal cancer rates. *J Natl Cancer Inst*, **83**, 46–50.

44 Block, G., Dietrich, M., Norkus, E.P., Morrow, J.D., Hudes, M., Caan, B., and Packer, L. (2002) Factors associated with oxidative stress in human populations. *Am J Epidemiol*, **156**, 274–285.

45 Karolkiewicz, J., Szczesniak, L., Deskur-Smielecka, E., Nowak, A., Stemplewski, R., and Szeklicki, R. (2003) Oxidative stress and antioxidant defense system in healthy, elderly men: relationship to physical activity. *Aging Male*, **6**, 100–105.

46 Covas, M.I., Elosua, R., Fito, M., Alcantara, M., Coca, L., and Marrugat, J. (2002) Relationship between physical activity and oxidative stress biomarkers in women. *Med Sci Sports Med*, **34**, 814–819.

47 Vincent, H.K. and Taylor, A.G. (2006) Biomarkers and potential mechanisms of obesity-induced oxidant stress in humans. *Int J Obes*, **30**, 400–418.

48 Tomey, K.M., Sowers, M.R., Li, X., McConnell, D.S., Crawford, S., Gold, E.B., Lasley, B. and Randolph, J.F. Jr (2007) Dietary fat subgroups, zinc, and vegetable components are related to urine F2a-Isoprostane concentration, a measure of oxidative stress, in midlife women. *J Nutr*, **137**, 2419.

49 Ito, Y., Shimizu, H., and Yoshimura, T. (1999) Serum concentrations of carotenoids, α-tocopherol, fatty acids, and lipid peroxides among Japanese in Japan, and Japanese and Caucasians in the US. *Int J Vitam Nutr Res*, **69**, 385–395.

50 Key, T.J., Verkasalo, P.K., and Banks, E. (2001) Epidemiology of breast cancer. *Lancet Oncol*, **2**, 133–140.

51 Casso, D., White, E., Patterson, R.E., Agurs-Collins, T., Kooperberg, C., and Haines, P.S. (2000) Correlates of serum lycopene in older women. *Nutr Cancer*, **36**, 163–169.

52 White, E., Kristal, A.R., Shikany, J.M., Wilson, A.C., Chen, C., Mares-Perlman, J.A., Masaki, K.H., and Caan, B.J. (2001) Correlates of serum α- and γ-tocopherol in the Women's Health Initiative. *Ann Epidemiol*, **11**, 136–144.

53 Trevisan, M., Browne, R., Ram, M., Muti, P., Freudenheim, J., Carosello, A.M., and Armstrong, D.G. (2001) Correlates of oxidative stress in the general population. *Am J Epidemiol*, **154**, 348–356.

54 Gago-Dominguez, M., Castelao, J.E., Pike, M.C., Sevanian, A., and Haile, R.W. (2005) Role of lipid peroxidation in the epidemiology and prevention of breast cancer. *Cancer Epidemiol Biomarkers Prev*, **14**, 2829–2839.

55 Schmitz, K.H., Warren, M., Rundle, A.G., Williams, N.I., Gross, M.D., and Kurzer, M.S. (2008) Exercise effect on oxidative stress is independent of change in estrogen metabolism. *Cancer Epidemiol Biomarkers Prev*, **17**, 220–223.

56 Friedenreich, C.M. (2001) Physical activity and cancer prevention: from observational to intervention research. *Cancer Epidemiol Biomarkers Prev*, **10**, 287–301.

57 Dietrich, M., Block, G., Benowitz, N.L., Morrow, J.D., Hudes, M., Jacob,

P. III, Norkus, E.P., and Packer, L. (2003) Vitamin C supplementation decreases oxidative stress biomarker f2-isoprostanes in plasma of non-smokers exposed to environmental tobacco smoke. *Nutr Cancer*, **45**, 176–184.

58 Thompson, H.J., Heimendinger, J., Haegele, A., Sedlacek, S.M., Gillette, C., O'Neill, C., Wolfe, P., and Conry, C. (1999) Effect of increased vegetable and fruit consumption on markers of oxidative cellular damage. *Carcinogenesis*, **20**, 2261–2266.

59 Thompson, H.J., Heimendinger, J., Sedlacek, S., Haegele, A., Diker, A., O'Neill, C., Meinecke, B., Wolfe, P., Zhu, Z., and Jiang, W. (2005) 8-Isoprostane F2α excretion is reduced in women by increased vegetable and fruit intake. *Am J Clin Nutr*, **82**, 768–776.

60 Ide, T., Tsutsui, H., Ohashi, N., Hayashidani, S., Suematsu, N., Tsuchihashi, M., Tamai, H., and Takeshita, A. (2002) Greater oxidative stress in healthy young men compared with premenopausal women. *Arterioscler Thromb Vasc Biol*, **22**, 438–442.

61 van den Bergh, R., van Vliet, T., Broekmans, W.M., Cnubben, N.H., Vaes, W.H., Roza, L., Haenen, G.R., Bast, A., and van den Berg, H. (2001) A vegetable/fruit concentrate with high antioxidant capacity has no effect on biomarkers of antioxidant status in male smokers. *J Nutr*, **131**, 1714–1722.

62 Stewart, R.J., Askew, E.W., McDonald, C.M., Metos, J., Jackson, W.D., Balon, T.W., and Prior, R.L. (2002) Antioxidant status of young children: response to an antioxidant supplement. *J Am Diet Assoc*, **102**, 1652–1657.

63 Block, G., Jensen, C.D., Morrow, J.D., Holland, N., Norkus, E.P., Milne, G.L., Hudes, M., Dalvi, T.B., Crawford, P.B., Fung, E.B., Schumaker, L., and Harmatz, P. (2008) The effect of vitamins C and E on biomarkers of oxidative stress depends on baseline level. *Free Radic Biol Med*, **45**, 377–384.

64 Walston, J., Xue, Q., Semba, R.D., Ferrucci, L., Cappola, A.R., Ricks, M., Guralnik, J., and Fried, L.P. (2006) Serum antioxidants, inflammation, and total mortality in older women. *Am J Epidemiol*, **163**, 18–26.

65 Hartman, T.J., Baer, D.J., Graham, L.B., Stone, W.L., Gunter, E.W., Parker, C.E., Albert, P.S., Dorgan, J.F., Clevidence, B.A., Campbell, W.S., Tomer, K.B., Judd, J.T., and Taylor, P.R. (2005) Moderate alcohol consumption and levels of antioxidant vitamins and isoprostanes in postmenopausal women. *Eur J Clin Nutr*, **59**, 161–168.

66 Prentice, R.L., Cann, B., Chlebowski, R.T., Patterson, R., Kuller, L.H., Ockene, J.K., Margolis, K.L., Limacher, M.C., Manson, J.E., Parker, L.M., Paskett, E., Phillips, L., Robbins, J., Rossouw, J.E., Sarto, G.E., Shikany, J.M., Stefanick, M.L., Thomson, C.A., Van Horn, L., Vitolins, M.Z., Wactawski-Wende, J., Wallace, R.B., Wassertheil-Smoller, S., Whitlock, E., Yano, K., Adams-Campbell, L., Anderson, G.L., Assaf, A.R., Beresford, S.A., Black, H.R., Brunner, R.L., Brzyski, R.G., Ford, L., Gass, M., Hays, J., Heber, D., Heiss, G., Hendrix, S.L., Hsia, J., Hubbell, F.A., Jackson, R.D., Johnson, K.C., Kotchen, J.M., LaCroix, A.Z., Lane, D.S., Langer, R.D., Lasser, N.L., and Henderson, M.M. (2006) Low-fat dietary pattern and risk of invasive breast cancer: the women's health initiative randomized controlled dietary modification trial. *J Am Med Assoc*, **295**, 629–642.

67 Bingham, S.A., Luben, R., Welch, A., Wareham, N., Khaw, K.-T., and Day, N. (2003) Are imprecise methods obscuring a relation between fat and breast cancer? *Lancet*, **362**, 212–214.

68 Djuric, Z., Heilbrun, L.K., Reading, B.A., Boomer, A., Valeriote, F.A., and Martino, S. (1991) Effects of a low-fat diet on levels of oxidative damage to DNA in human peripheral nucleated

blood cells. *J Natl Cancer Inst*, **83**, 766–769.

69 Hong, C.C., Tang, B.K., Rao, V., Agarwal, S., Martin, L., Tritchler, D., Yaffe, M., and Boyd, N.F. (2004) Cytochrome P450 1A2 (CYP1A2) activity, mammographic density, and oxidative stress: a cross-sectional study. *Breast Cancer Res*, **6**, R338–R351.

70 Boyd, N.F., Rommens, J.M., Vogt, K., Hopper, J.L., Lee, V., Yaffe, M.J., and Paterson, M. (2005) Mammographic breast density as an intermediate phenotype for breast cancer. *Lancet*, **6**, 798–808.

71 Wolfe, J.N. (1976) Breast patterns as an index of risk for developing breast cancer. *AJR Am J Roentgenol*, **126**, 1130–1137.

72 Wolfe, J.N. (1976) Risk for breast cancer development determined by mammographic parenchymal pattern. *Cancer*, **37**, 2486–2492.

73 Byng, J.W., Boyd, N.F., Fishell, E., Jong, R.A., and Yaffe, M.J. (1994) The quantitative analysis of mammographic densities. *Phys Med Biol*, **39**, 1629–1638.

74 McCormack, V.A. and dos Santos Silva, I. (2006) Breast density and parenchymal patterns as markers of breast cancer risk: A meta-analysis. *Cancer Epidemiol Biomarkers Prev*, **15**, 1159–1169.

75 Kerlikowske, K. (2007) The mammogram that cried Wolfe. *N Engl J Med*, **356**, 297–300.

76 Boyd, N.F., Guo, H., Martin, L.J., Sun, L., Stone, J., Fishell, E., Jong, R.A., Hislop, G., Chiarelli, A., Minkin, S., and Yaffe, M. (2007) Mammographic density and the risk and detection of breast cancer.\break *N Engl J Med*, **356**, 227–236.

77 Byrne, C., Schairer, C., Wolfe, J., Parekh, N., Salane, M., Brinton, L.A., Hoover, R., and Haile, R. (1995) Mammographic features and breast cancer risk: Effects with time, age, and menopause status. *J Natl Cancer Inst*, **87**, 1622–1629.

78 Boyd, N.F., Lockwood, G.A., Byng, J., Tritchler, D.L., and Yaffe, M. (1998) Mammographic densities and breast cancer risk. *Cancer Epidemiol Biomarkers Prev*, **7**, 1133–1144.

79 Li, T., Sun, L., Miller, N., Nicklee, T., Woo, J., Hulse-Smith, L., Tsao, M., Khokha, R., Martin, L., and Boyd, N.F. (2005) The association of measured breast tissue characteristics with mammographic density and other risk factors for breast cancer. *Cancer Epidemiol Biomarkers Prev*, **14**, 343–349.

80 Maskarinec, G., Pagano, I., Lurie, G., and Kolonel, L.N. (2006) A longitudinal investigation of mammographic density: the multiethnic cohort. *Cancer Epidemiol Biomarkers Prev*, **15**, 732–739.

81 Vachon, C.M., Pankratz, V.S., Scott, C.G., Maloney, S.D., Ghosh, K., Brandt, K.R., Milanese, T., Carston, M.J., and Sellers, T.A. (2007) Longitudinal trends in mammographic percent density and breast cancer risk. *Cancer Epidemiol Biomarkers Prev*, **16**, 921–928.

82 Pike, M.C., Krailo, M.D., Henderson, B.E., Casagrande, J.T., and Hoel, D.G. (1983) "Hormonal" risk factors, "breast tissue age" and the age-incidence of breast cancer. *Nature*, **303**, 767–770.

83 Grove, J.S., Goodman, M.J., Gilbert, F., and Mi, M.P. (1985) Factors associated with mammographic pattern. *Br J Radiol*, **58**, 21–25.

84 Vachon, C.M., Kuni, C.C., and Anderson, K. (2000) Association of mammographically defined percent breast density with epidemiologic risk factors for breast cancer (United States). *Cancer Causes Control*, **11**, 653–662.

85 Boyd, N.F., Martin, L.J., Sun, L., Guo, H., Chiarelli, A., Hislop, G., Yaffe, M., and Minkin, S. (2006) Body size, mammographic density and breast cancer risk. *Cancer Epidemiol Biomarkers Prev*, **15**, 2086–2092.

86 Boyd, N., Martin, L., Stone, J., Little, L., Minkin, S., and Yaffe, M. (2002) A longitudinal study of the effects of menopause on mammographic

features. *Cancer Epidemiol Biomarkers Prev*, **11**, 1048–1053.
87 Boyd, N.F., Dite, G.S., Stone, J., Gunasekara, A., English, D.R., McCredie, M.R.E., Giles, G.G., Tritchler, D., Chiarelli, A., Yaffe, M.J., and Hopper, J.L. (2002) Heritability of mammographic density, a risk factor for breast cancer. *N Engl J Med*, **347**, 886–894.
88 Martin, L.J. and Boyd, N. (2008) Potential mechanisms of breast cancer risk associated with mammographic density: hypotheses based on epidemiological evidence. *Breast Cancer Res*, **10**, 1–14.
89 Key, T., Appleby, P., Barnes, I., and Reeves, G. Endogenous Hormones and Breast Cancer Collaborative Group (2002) Endogenous sex hormones and breast cancer in postmenopausal women: reanalysis of nine prospective studies. *J Natl Cancer Inst*, **94**, 606–616.
90 Eliassen, A.H., Missmer, S.A., Tworoger, S.S., Spiegelman, D., Barbieri, R.L., Dowsett, M., and Hankinson, S.E. (2006) Endogenous steroid hormone concentrations and risk of breast cancer among premenopausal women. *J Natl Cancer Inst*, **98**, 1406–1415.
91 Tamimi, R.M., Byrne, C., Colditz, G.A., and Hankinson, S.E. (2007) Endogenous hormone levels, mammographic density, and subsequent risk of breast cancer in postmenopausal women. *J Natl Cancer Inst*, **99**, 1178–1187.
92 Riza, E., De Stavola, B., Bradlow, H.L., Sepkovic, D.W., Linos, D., and Linos, A. (2001) Urinary estrogen metabolites and mammographic parenchymal patterns in postmenopausal women. *Cancer Epidemiol Biomarkers Prev*, **10**, 627–634.
93 Chlebowski, R.T., Hendrix, S.L., Langer, R., Stefanick, M.L., Gass, M., Lane, D., Rodabough, R.J., Gilligan, M.A., Cyr, M.G., Thomson, C.A., Khandekar, J., Petrovitch, H., and McTiernan, A. (2003) Influence of estrogen plus progestin on breast cancer and mammography in healthy postmenopausal women. The women's health initiative randomized trial. *J Am Med Assoc*, **289**, 3243–3253.
94 Greendale, G.A., Reboussin, B.A., Slone, S., Wasilauskas, C., Pike, M.C., and Ursin, G. (2003) Postmenopausal hormone therapy and change in mammographic density. *J Natl Cancer Inst*, **95**, 30–37.
95 Lundstrom, E., Wilczek, B., von Palffy, Z., Soderqvist, G., and von Schoultz, B. (1999) Mammographic breast density during hormone replacement therapy: differences according to treatment. *Am J Obstet Gynecol*, **181**, 348–352.
96 Rutter, C.M., Mandelson, M.T., Laya, M.B., Seger, D.J., and Taplin, S. (2001) Changes in breast density associated with initiation, discontinuation, and continuing use of hormone replacement therapy. *J Am Med Assoc*, **285**, 171–176.
97 Cuzick, J., Warwick, J., Pinney, E., Warren, R.M.L., and Duffy, S.W. (2004) Tamoxifen and breast density in women at increased risk of breast cancer. *J Natl Cancer Inst*, **96**, 621–628.
98 Spicer, D.V., Ursin, G., Parisky, Y.R., Pearce, J.G., Shoupe, D., Pike, A., and Pike, M.C. (1994) Changes in mammographic densities induced by a hormonal contraceptive designed to reduce breast cancer risk. *J Natl Cancer Inst*, **86**, 431–436.
99 Tworoger, S.S., Eliassen, A.H., Sluss, P., and Hankinson, S.E. (2007) A prospective study of plasma prolactin concentrations and risk of premenopausal and postmenopausal breast cancer. *J Clin Oncol*, **25**, 1482–1488.
100 Boyd, N.F., Stone, J., Martin, L.J., Jong, R., Fishell, E., Yaffe, M., Hammond, G., and Minkin, S. (2002) The association of breast mitogens with mammographic densities. *Br J Cancer*, **87**, 876–882.
101 Guo, Y.P., Martin, L.J., Hanna, W., Banerjee, D., Miller, N., Fishell, E.,

Khokha, R., and Boyd, N.F. (2001) Growth factors and stromal matrix proteins associated with mammographic densities. *Cancer Epidemiol Biomarkers Prev*, **10**, 243–248.

102 Renehan, A.G., Zwahlen, M., Minder, C., O'Dwyer, S.T., Shalet, S.M., and Egger, M. (2004) Insulin-like growth factor (IGF)-I, IGF binding protein-3, and cancer risk: systematic review and meta-regression analysis. *Lancet*, **363**, 1346–1353.

103 Pollak, M.N., Schernhammer, E.S., and Hankinson, S.E. (2004) Insulin-like growth factors and neoplasia. *Nat Rev*, **4**, 505–518.

104 Ng, S.T., Zhou, J., Adesanya, O.O., Wang, J., LeRoith, D., and Bondy, C.A. (1997) Growth hormone treatment induces mammary gland hyperplasia in aging primates. *Nat Med*, **3**, 1141–1144.

105 Dunning, A.M., Healey, C.S., Pharoah, P.D.P., Teare, M.D., Ponder, B.A.J., and Easton, D.F. (1999) A systematic review of genetic polymorphism and breast cancer risk. *Cancer Epidemiol Biomarkers Prev*, **8**, 843–854.

106 Hong, Y., Pedersen, N.L., Brismar, K., Hall, K., and de Faire, U. (1996) Quantitative genetic analyses of insulin-like growth factor I (IGF-I) IGF-binding protein-1 and insulin levels in middle-aged and elderly twins. *J Clin Endocrinol Metab*, **81**, 1791–1797.

107 Mendlewicz, J., Linkowski, P., Kerkhofs, M., Leproult, R., Copinschi, G., and Van Couter, E. (1999) Genetic control of 24-hour growth hormone secretion in man: a twin study. *J Clin Endocrinol Metab*, **84**, 856–862.

108 Boyd, N.F. and McGuire, V. (1990) Evidence of lipid peroxidation in premenopausal women with mammographic dysplasia. *Cancer Lett*, **50**, 31–37.

109 Boyd, N.F., Connelly, P., Byng, J., Yaffe, M., Draper, H., Little, L., Jones, D., Martin, L.J., Lockwood, G., and Tritchler, D. (1995) Plasma lipids, lipoproteins, and mammographic densities. *Cancer Epidemiol Biomarkers Prev*, **4**, 727–733.

110 Bird, R.P., Hung, S.S.O., Hadley, M., and Draper, H.H. (1983) Determination of malondialdehyde in biological materials by high pressure liquid chromatography. *Anal Biochem*, **128**, 240–244.

111 Lykkesfeldt, J. (2007) Malondialdehyde as a biomarker of oxidative damage to lipid caused by smoking. *Clin Chim Acta*, **380**, 50–58.

112 Hankinson, S., Willet, W.C., Colditz, G., Hunter, D., Michaud, D.S., Deroo, B., Rosner, B., Speizer, F., and Pollak, M. (1998) Circulating concentrations of insulin-like growth factor I and risk of breast cancer. *Lancet Oncol*, **351**, 1393–1396.

113 Hankinson, S.E., Willett, W.C., Michaud, D.S., Manson, J.E., Colditz, G.A., Longcope, C., Rosner, B., and Speizer, F.E. (1999) Plasma prolactin levels and subsequent risk of breast cancer in postmenopausal women. *J Natl Cancer Inst*, **91**, 629–634.

114 Bluher, S., Kratzsch, J., and Kiess, W. (2005) Insulin-like growth factor 1, growth hormone and insulin in white adipose tissue. *Best Pract Res Clin Endocrinol Metab*, **19**, 577–587.

115 Giovannucci, E. (2003) Nutrition, insulin, insulin-like growth factors and cancer. *Horm Metab Res*, **35**, 694–704.

116 Veldhuis, J.D., Evans, W.S., Bowers, C.Y., and Anderson, S. (2001) Interactive regulation of postmenopausal growth hormone insulin-like growth factor axis by estrogen and growth hormone-releasing peptide-2. *Endocrine*, **14**, 45–62.

117 Gudelsky, G.A., Nansel, D.D., and Porter, J.C. (1981) Role of estrogen in the dopaminergic control of prolactin secretion. *Endocrinology*, **108**, 440–444.

118 Liehr, J.G. and Roy, D. (1990) Free radical generation by redox cycling of estrogens. *Free Radic Biol Med*, **8**, 415–423.

119 Mobley, J.A. and Brueggemeier, R.W. (2004) Estrogen receptor-mediated

regulation of oxidative stress and DNA damage in breast cancer. *Carcinogenesis*, **25**, 3–9.

120 Yager, J.D. (2000) Endogenous estrogens as carcinogens through metabolic activation. *J Natl Cancer Inst Monogr*, **27**, 67–73.

121 Cavalieri, E., Frenkel, K., Liehr, J.G., Rogan, E., and Roy, D. (2000) Estrogens as endogenous genotoxic agents - DNA adducts and mutations. *J Natl Cancer Inst Monogr*, **27**, 75–93.

122 Hong, C.-C., Thompson, H.J., Jiang, C., Hammond, G.L., Tritchler, D., Yaffe, M., and Boyd, N.F. (2003) Val158Met Polymorphism in catechol-O-methyltransferase (COMT) gene associated with risk factors for breast cancer. *Cancer Epidemiol Biomarkers Prev*, **12**, 838–847.

123 Hinz, B. (2007) Formation and function of the myofibroblast during tissue repair. *J Invest Dermatol*, **127**, 526–537.

124 Marks, F., Furstenberger, G., and Muller-Decker, K. (2007) Tumor promotion as a target of cancer prevention. *Recent Results Cancer Res*, **174**, 537–547.

125 Bhowmick, N.A., Neilson, E.G., and Moses, H.L. (2004) Stromal fibroblasts in cancer initiation and progression. *Nature*, **432**, 332–337.

126 Kalluri, R. and Zeisberg, M. (2006) Fibroblasts in cancer. *Nat Rev Cancer*, **6**, 392–401.

127 Poli, G. (2000) Pathogenesis of liver fibrosis: role of oxidative stress. *Mol Aspects Med*, **21**, 49–98.

128 Gressner, O.A., Weiskirchen, R., and Gressner, A.M. (2007) Biomarkers of hepatic fibrosis, fibrogenesis and genetic pre-disposition pending between fiction and reality. *J Cell Mol Med*, **11**, 1031–1051.

129 Kohyama, T., Liu, X., Wen, F.Q., Kim, H.J., Takizawa, H., and Rennard, S.I. (2002) Potentiation of human lung fibroblast chemotaxis by the thromboxane A(2) analog U-46619. *J Lab Clin Med*, **139**, 43–49.

27
Lifestyle, Endogenous Toxins, and Colorectal Cancer Risk

Gail McKeown-Eyssen, Jeff Bruce, Owen Lee, Peter J. O'Brien, and W. Robert Bruce

The clearest lifestyle risk factors for colorectal cancer are body fatness, abdominal fatness, adult attained height, red meat, processed meat, and alcoholic drinks, each associated with increased risk, and physical activity; foods containing dietary fiber, milk, calcium, and garlic, associated with reduced risk. We review the evidence that these factors increase oxidative stress and energy excess, and that these two factors interact, possibly through the formation of cytotoxic and genotoxic products with the oxidation of early glycolysis metabolites. Biomarkers of the processes involved are identified. These might be used to determine the validity of the general model of carcinogenesis associated with the western diet.

27.1
Introduction

More than one million people around the world develop colorectal cancer (CRC) every year [1]. Most of these cancers occur in the "developed" world. Perhaps not surprisingly, the major risk factors for this cancer are characteristics of the "Western lifestyle," with its industrialized refined oils, fats, and grains, energy-dense red meat diets, as well as sedentary lifestyles [2]. These risk factors have led to two very different ways in which the relationship between lifestyle and the origin of CRC has been understood. In one way, as we describe below, foods such as red and processed meat and/or alcoholic drinks may result in oxidative stress, either directly or by producing subclinical colonic inflammation involving activation of inflammatory cells. The oxidative stress may then increase DNA damage and initiate CRC. In the other way, energy-dense foods may result in excess nutrition that leads to increased energy excess and body fatness, increased plasma levels of insulin and related hormones, and promotion of CRC. Our purpose here is to elaborate more fully the evidence for the "oxidative stress" and the "energy excess" hypotheses, and then to consider them with more recent findings, on how they might suggest a fuller understanding of the carcinogenesis process. Thus, this chapter will first review the lifestyle risk factors for CRC that have been identified from epidemiology studies. It will determine whether these factors could be expected to result in oxidative

Endogenous Toxins. Diet, Genetics, Disease and Treatment.
Edited by Peter J. O'Brien and W. Robert Bruce
Copyright © 2010 WILEY-VCH Verlag GmbH & Co. KGaA, Weinheim
ISBN: 978-3-527-32363-0

stress with or without colonic inflammation, and/or excess energy, and hence result in CRC. Recent studies suggesting that interactions between oxidative stress and energy excess result in the formation of novel cytotoxins and genotoxins will also be reviewed. Finally, the implications for future research of the hypothesized mechanisms relating to lifestyle, endogenous toxins, and CRC will be considered.

27.2
Lifestyle Risk Factors for CRC

The incidence and mortality of CRC is similar in males and females, but they both differ markedly, over 20-fold, from country to country [1, 3]. CRC is highest in North America, Australia/New Zealand, Western Europe, and Japan and lowest in parts of Africa and Asia. "Lifestyle factors" such as type of diet and physical inactivity, rather than "genetic factors" appear to be the major determinants for CRC worldwide. This is evident from migrant studies, which have consistently shown that CRC risk increases for individuals in low-risk populations, when they migrate to "developed" countries and adopt their pattern of living [4]. It is also evident from the time trends of CRC incidence in low-risk populations, which also increased as they adopted the living patterns of the "developed" countries. This pattern has been documented clearly for Japan [5] where increased consumption of fat and red meat was associated with incidence of colon cancer 20 years later.

The nature of the lifestyle factors that could be responsible for CRC has been the center of intense study both in ecologic and analytic case–control and cohort studies. The results have been reviewed and summarized by an expert committee in a recent report of the American Institute for Cancer Research (see Table 27.1) [2, p. 281]. The clearest direct associations with risk, judged as "convincing" and "probable," were for *body fatness, abdominal fatness,* and *adult attained height,* as well as for *red meat, processed meat,* and *alcoholic drinks,* and the clearest inverse associations were for *physical activity, foods containing dietary fiber, milk, calcium,* and *garlic.*

Although the lifestyle risk factors for CRC may act through a variety of mechanisms, it is also possible that two basic mechanisms, oxidative stress and energy excess, acting separately or in combination, could explain associations with risk for all the established lifestyle risk factors. Figure 27.1 shows the relationships of such a general scheme. To establish the validity of such a model, it is first necessary to show that all the established lifestyle risk factors can affect either oxidative stress or energy excess, or both, and then establish that oxidative stress and excess energy could affect CRC risk.

We first argue that oxidative stress is associated directly with markers of energy excess, *body fatness, abdominal fatness, adult attained height,* as well as *red meat, processed meat,* and *alcoholic drinks* ("⇑" in Table 27.1), while *foods containing dietary fiber, garlic, milk,* and *calcium* are associated inversely with oxidative stress ("⇓" in Table 27.1). Similarly, we will argue that energy excess is associated directly with *body fatness, abdominal fatness, adult attained height,* all of which reflect energy

Tab. 27.1 Convincing and probable lifestyle risk factors for CRC as associated with oxidative stress and energy excess

Lifestyle factor *with*	Oxidative stress	Energy excess
Positive association		
Body fatness	⇑	⇑
Abdominal fatness	⇑	⇑
Adult attained height		⇑
Red meat	⇑	⇑
Processed meat	⇑	⇑
Alcoholic drinks	⇑	⇑
Negative association		
Physical activity		⇓
Foods containing dietary fiber	⇓	⇓
Milk	⇓	⇓?
Calcium	⇓	⇓?
Garlic	⇓	

⇑, direct relationship; ⇓; indirect relationship; ⇓(?): less clear indirect relationship, see text.

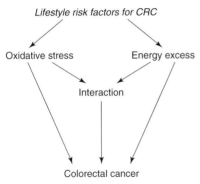

Fig. 27.1 The general scheme relating lifestyle risk factors for CRC with disease through oxidative stress and energy excess. Endogenous biomarkers may provide measures of each of the risks.

availability, as well as with foods such as *red meat* and *processed meat*, which are energy dense, and with *alcoholic drinks*, which provide energy without nutrients. *Physical activity* helps to maintain energy balance, and thus is associated inversely with energy excess. *Foods containing dietary fiber* are low in energy density, and therefore are also associated inversely with excess energy. *Milk* and *calcium* may be associated inversely with energy excess, because they decrease fat absorption. As considered below, most of the lifestyle factors thus affect both oxidative stress and energy excess. The pathways from lifestyle risk factors to oxidative stress and to energy excess, and on to CRC, together with a pathway from the interaction of oxidative stress and energy excess to CRC (Figures 27.2–27.4) will now be considered in more detail.

27.3
Oxidative Stress Relates to Lifestyle and CRC Risk

Figure 27.2 summarizes the evidence that many of the CRC risk factors may affect risk through increased colonic inflammation and oxidative stress. We will first consider the relationship between lifestyle risk factors and oxidative stress, and will then discuss how inflammation and oxidative stress increase the risk of CRC.

There are two potential pathways from lifestyle risk factors to oxidative stress (Figure 27.2). In Pathway 1 (solid lines), *body fatness*, and *abdominal fatness*, taken as reflecting chronic energy excess, are associated with increased inflammation and oxidative stress [6]. The body mass index (BMI), for instance, is associated with C-reactive protein, a marker of generalized inflammation [7], perhaps a consequence of the populations of macrophages and neutrophils associated with fat depots in high-fat diets [8]. The mechanisms underlying the accumulation and activation of these innate immune cells are not known [9]. However, it is known that these immune cells can release reactive oxygen species (ROS) as well as inflammatory cytokines such as tumor necrosis factor-α (TNF-α) and proteolytic enzymes. The major ROS is formed as a burst of readily diffusible H_2O_2 from the superoxide radical anion ($O_2^{\bullet}$), formed through the action of the NADPH oxidase (NOX) enzyme [10].

Dietary factors, *red meat*, and *processed meat*, may also be associated with oxidative stress because they contain heme. Through its content of iron, heme can increase the formation of ROS, including $O_2^{\bullet}$, H_2O_2, and the highly reactive hydroxyl radical (HO$^{\bullet}$). These in turn can form reactive nitrogen species (RNS, e.g., peroxynitrite), and lipids to form potentially toxic lipid oxidation products (e.g., malondialdehyde) [11–14]. It is possible that *calcium* and *milk* (containing calcium) may lower cancer risk through inhibition of the effect of heme by a process not yet fully understood [15].

Foods containing fiber, (i.e., cereals, roots, tubers, and plantains as well as vegetables, fruits, and pulses [2]) may contain micronutrients and other bioactive compounds that could inhibit colon carcinogenesis by decreasing oxidative stress. Evidence of such a role of micronutrients comes primarily from a large number of animal studies that have shown an inhibition of colon carcinogenesis with micronutrients contained in fiber-containing foods, which act as antioxidants. These include vitamin E (β- and γ-tocopherol, which block the propagation of membrane lipid peroxidation [10]), selenium (a component in selenoproteins such as glutathione peroxidase that are a major component of the body's antioxidant system [10]), flavonoids such as quercetin (which are free radical scavengers, e.g., [2, 16, 17]), and other micronutrients [18]. Foods containing fiber are also rich in thiamin, a vitamin required for the formation of NADPH and the reduction of glutathione, a major antioxidant formed in the body [19].

27.3 Oxidative Stress Relates to Lifestyle and CRC Risk | 677

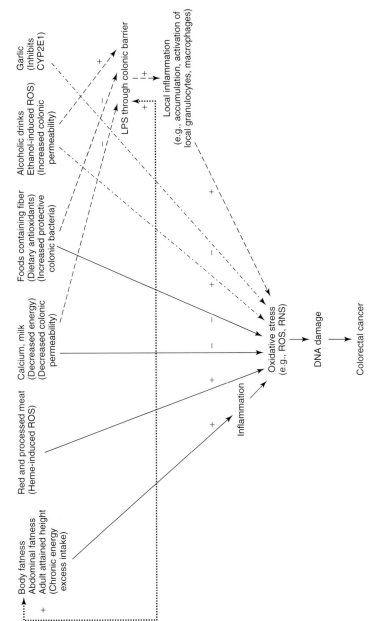

Fig. 27.2 Summary of evidence relating lifestyle factors in two pathways to oxidative stress, DNA damage, and colorectal cancer: in Pathway 1 (solid lines), through chronic energy excess and heme with calcium, milk, and dietary antioxidants protective. In Pathway 2 (dashed lines), indirectly through increased penetration of LPS through the colonic barrier with alcoholic drinks and energy excess, reduced by calcium and milk (containing calcium and vitamin D), resulting in local inflammation, oxidative stress, DNA damage, and CRC. Evidence relating LPS through the colonic barrier and body size measures is shown as a dotted line. The association of alcohol to oxidative stress via the 1-hydroxyethyl radical and CYP2E1, possibly ameliorated by garlic, (lines with alternating dots and dashes) may involve the LPS pathway, but its place in this is unknown.

These observations indicate that increased oxidative stress in the colonic epithelium is associated with (Pathway 1, solid lines in Figure 27.2) *body fatness*, *abdominal fatness*, and *adult attained height* and heme-containing *red* and *processed meat*, an effect possibly ameliorated by *calcium* and *milk* (containing calcium), and *foods containing fiber*, which decrease risk through the action of micronutrient antioxidants that they contain.

Pathway 2 (dashed lines in Figure 27.2) illustrates another, more recently described, possible path from lifestyle risk factors to oxidative stress and CRC involving lipopolysaccharides (LPS, also called endotoxin), which are heat-stable lipid polysaccharide polypeptide complexes formed from the cell walls of dying gram-negative bacteria. Under this mechanism, dietary/lifestyle factors affect the rate of formation of LPS within the intestinal lumen as well as the rate at which LPS pass through or between epithelial cells, through the colonic basement membrane into the mesenchymal tissue. There, LPS could initiate an inflammatory response by activating granulocytes and macrophages, which would ultimately lead to oxidative stress through the processes described for Pathway 1.

The first evidence to implicate such pathways to inflammatory responses came from studies of ethanol toxicity to the liver and the development of alcoholic steatohepatitis (ASH) (reviewed in [20]). Unexpectedly, toxicity in the liver was found to be partially a consequence of the effects of alcohol on the intestinal wall. Ethanol, or its oxidation product, acetaldehyde, decreased the integrity of the intestinal epithelial membrane and increased circulating levels of LPS, associated with the soluble LPS-binding protein [21, 22]. LPS that reached the portal blood stream, induced an innate inflammatory response with Kupffer cells, the resident macrophage-like cells in the liver. It interacted with a receptor complex composed of proteins CD-14 and TLR4, to initiate signaling that resulted in the production of cytokines and H_2O_2 formed by the membrane-bound heme protein NOX [23–25]. This hepatic inflammation induced by alcohol (ASH) in rats could be reduced in several ways: by sterilizing the intestine with antibiotics [26]; by feeding large numbers of benign bacteria (i.e., *Lactobacillus* spp.) and flooding out the deleterious colonic bacterial species (e.g., *Enterococci* spp.) [27]; or by destroying the Kupffer cells or inhibiting the activity of their NOX enzyme or CD-14 protein [25, 28]. Thus, deleterious colonic bacteria, combined with an agent that reduced the integrity of the colonic epithelium and resulted in a "leaky gut," increased the passage of LPS through the colon wall, increased LPS levels in the portal vein, and increased the exposure of the parietal cells in the liver to an inflammatory response and oxidative stress, through the mechanisms described above.

Another piece of the puzzle is not fully understood, although the relationship between alcohol and oxidative stress has been extensively investigated. Numerous experimental studies have shown that the hydroxy radical ($HO^\bullet$) can extract a proton from ethanol through a one-electron oxidation to form the reactive 1-hydroxyethyl radical ($CH_3C^\bullet HOH$). This process may involve the CYP2E1 enzyme [29, 30] and, interestingly, *garlic*, a dietary factor associated with reduced CRC risk, contains diallylsulfide, known to be a specific inhibitor of CYP2E1 [31]. Where this fits in

Pathway 2 involving colonic permeability, LPS, and local inflammation is presently unclear.

In Figure 27.2, we suggest that a mechanism similar to that relating alcohol exposure to ASH could also explain the positive association of *alcoholic drinks* with CRC. Parallel to the findings for liver damage, *alcoholic drinks* may increase colonic permeability, and if deleterious bacteria populated the colon, would then facilitate the passage of LPS through the basement membrane. Here, before reaching the blood stream, it would induce an innate inflammatory reaction in macrophage, granulocytes, and enteroendocrine cells in the local colon mucosa, and would expose the neighboring colonocytes to oxidative stress similar to that observed in the more distal Kupffer cells in the liver. The whole colonic mucosa would also be exposed to LPS that were transported through the portal vein to the liver and escaped to the general circulation.

Energy-dense diets resulting in *body fatness* and *abdominal fatness*, may also be associated with CRC risk through effects on reduced colonic integrity and increased circulating LPS. A high-fat diet, compared with a low-fat diet based on unrefined grains, was found, in an animal model, not only to increase weight but also to increase circulating LPS [32, 33], presumably released from the colon by reduced colonic integrity. When levels of LPS were increased by implanting small LPS pumps so as to produce a "metabolic endotoxemia," weight also increased [32, 33]. Thus, there appears to be a positive association between weight gain and LPS, though the mechanisms involved remain to be understood. That this association involves gut bacteria is suggested by the fact that it has been clearly demonstrated that germ-free animals are resistant to obesity typically induced by energy-dense diets with saturated fat and sucrose [34]. The mechanism responsible for this resistance has been the focus of much interest [35]. Furthermore, levels of LPS were reduced with diets based on oligofructose, a complex fiber that favored the growth of benign bacterial species [32, 33].

There is also some human evidence that diet can influence LPS levels. Fat and energy intakes were directly associated with circulating LPS concentrations among healthy men [36], and men fed a high-fat meal with 50 g of butter had a significant increase in plasma LPS [37]. Furthermore, diets containing refined carbohydrates (sucrose, fructose, and glucose, features of the western diet) have also been associated with elevated plasma LPS concentrations [38]. Dietary fiber as oligofructose-enriched inulin, together with *Lactobacillus* spp. and *Bifidobacterium* spp., provided as a "synbiotic" to patients with previously resected polyps (and cancers) for a period of 12 weeks reduced evidence of CRC risk, reflected by reduced genotoxicity in the epithelial cells and cytotoxicity of fecal water [39]. These results may thus explain the protective effect of these agents, pre- and probiotics, for colon carcinogenesis observed in many previous experimental animal studies [40–42].

Other dietary factors could also affect the integrity of the colonic epithelium. *Calcium* and vitamin D (both found in *milk*) are known to be involved in tight junctions and cell membrane integrity [43, 44]. They may have their protective effects in colon carcinogenesis through their effect on the permeability of the colon to LPS and the resulting reduction of colonic inflammation. Other bacterial components in addition to LPS may pass through a "leaky colon" with a reduced barrier function. Proinflammatory effects have been also observed between fragments of bacterial flagella (flagellin) and enteroendocrine cells [45]; and intact bacteria in the colonic mucosa have been found associated with CRC [46].

Further evidence pointing to the importance of membrane integrity comes from studies of putative colonic precursor lesions, aberrant crypt foci (ACF), and polyps. For instance, increased tight junction permeability of the colonic epithelium with decreased epithelial barrier function was found to precede the development of colonic tumors, suggesting that the defect, presumably with its resulting inflammation and consequent increase in oxidative stress, was involved in the carcinogenesis process [47, 48].

These observations suggested that *alcoholic drinks*, energy-dense foods, *red meat* and *processed meat*, and possibly diets containing refined carbohydrates, may increase colonic permeability and/or bacterial overgrowth and that *foods containing fiber, calcium*, and *milk* may have protective effects by reducing exposure of the colonic mucosa to LPS, reducing subclinical colonic inflammation, and reducing the exposure of colonic epithelial cells to oxidative stress. Both Pathway 1 and Pathway 2 related lifestyle risk factors and oxidative stress. We now discuss how inflammation and oxidative stress could increase the risk of CRC.

CRC has been associated with colonic inflammation and oxidative stress in many studies. Cancer is a very common consequence of inflammatory bowel disease (IBD), with its associated evidence of increased oxidative stress. Individuals with active IBD were found to have reduced plasma antioxidant concentrations [49] and reduced mucosal function of an enzyme sensitive to oxidative stress (glyceraldehyde-3-phosphate dehydrogenase, GAPDH, [50]). CRC not associated with IBD may also be associated with subclinical colonic inflammation. Fecal calprotectin, the fecal concentration of a robust granulocyte marker protein associated with colonic inflammation, has been associated with, and used as a screen for, CRC [51, 52]. Fecal calprotectin, in individuals without known disease, has also been associated with some lifestyle risk factors for CRC: positively with age and obesity, and negatively with fiber consumption and physical activity [53].

The mechanism through which inflammation gives rise to CRC may be extremely complex. As noted above, the inflammation process can give rise not only to oxidative processes such as ROS, RNS, reactive lipid oxidation products, and carbonyls, but also to cytokines, proteases, and other products (reviewed in [54]). Of these processes, the oxidative processes have been the most thoroughly studied, perhaps because they have provided such a large repertoire of toxic DNA damage that could explain the genotoxicity of such exposures. Thus, lipid peroxidation products including 4-hydroxynonenal, malondialdehyde, and acrolein can react directly with DNA bases to form DNA adducts. The frequency of such adducts in organs affected by cancer-prone inflammatory disease can be 1 or 2 orders of magnitude higher than in normal organs [55]. In particular, the DNA adduct of malondialdehyde, M_1-dG, in colonic mucosa was associated with colonic adenoma, precursor of CRC [56].

The DNA of mucosal cells also had increased DNA damage, presumed to have originated from increased oxidative stress (8-hydroxydeoxyguanosine (8-OHdG) [57] and etheno-modified DNA bases, and lipid or carbohydrate-derived adducts [58]).

27.4
Energy Excess Relates Lifestyle and CRC Risk

Figure 27.3 summarizes the evidence that many of the CRC risk factors may affect risk through increased energy excess, resultant obesity and components of the associated metabolic syndrome. We will again first consider the traditional explanation for such an association and then consider a mechanism suggested by the recent studies with LPS.

In Pathway 1 (solid lines), *body fatness* (high BMI) and *abdominal fatness* are taken to reflect excess intake of calories and *adult attained height* to reflect energy supply through the periods of maximum growth through puberty. Utilization of energy substrates for *physical activity* reduces endogenous energy availability so that *physical activity* supports energy balance. Energy density of foods is associated with increased food consumption and the possibility of excess energy consumption [2, p. 324]. *Red meat* and *processed meat* are typically high in energy density, partly because of their high fat content. *Alcoholic drinks* also provide energy, but with little satiety and no reduction in the rate of digestion [59]. In contrast, *foods containing dietary fiber* are typically low in energy density, provide satiety, and have a reduced rate of digestion and of the postprandial glucose and insulin response, thus reducing exposure to insulin's growth stimulating effects [60]. *Calcium* and *milk* (containing calcium) may reduce net energy excess through the binding of intestinal fatty acids as insoluble calcium salts and the resulting reduced absorption of fat, especially saturated fat [61, 62]. The dietary risk factors considered thus influence net energy balance, the long-term effects of which are increased accumulation of fat and adiposity, and the short-term effects are production of increased intermediate metabolites. Both long and short-term effects may be related to the risk of CRC.

In Pathway 2 (dotted lines in Figure 27.3), we note the recent evidence cited earlier, suggesting that body size measures are associated with circulating LPS [33, 36, 63]. Thus, dietary factors that influence LPS may influence body size both through direct effects of energy intake and through effects on LPS.

In both pathways, chronic, long-term energy excess, reflected in adiposity, has been related to components of the metabolic syndrome, which has been associated with increased risk of CRC [64, 65]. The metabolic syndrome is a condition in which the cells of the major organs involved in the processing of metabolites (liver, muscle, and fat cells), become less sensitive to plasma insulin [66]. It is associated with an increase in levels of both insulin and glucose, as well as with increased serum triglycerides and fatty acids. In experimental studies with rodents, energy-dense, high-fat diets increased weight gain, insulin resistance, epithelial cell proliferation, and the development of CRC [67–70]. Epidemiology studies have consistently shown associations of CRC with insulin and C-peptide, and to a less

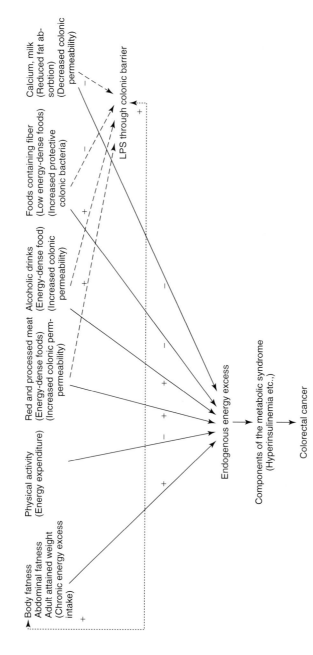

Fig. 27.3 Summary of evidence relating lifestyle factors to energy excess, metabolic syndrome, and colorectal cancer (solid lines). Evidence relating LPS and body size measures as noted in text, is shown as a dotted line.

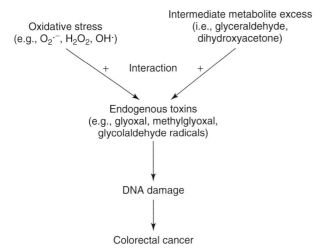

Fig. 27.4 Possible interaction of oxidative stress with intermediate metabolites to form endogenous toxins that result in DNA damage and colorectal cancer that inhibits CYP2E1.

marked degree with glucose, as well as with insulin-like growth factors (IGF-I) and reduced insulin-like growth factor binding proteins (IGFPBs), though the association with triglycerides is unclear [71–78].

In addition to its chronic effects, acute, short-term energy excess might be expected to increase energy substrates and intermediate metabolites, in particular, glyceraldehyde and dihydroxyacetone. As considered below (Figure 27.4), they may act synergistically with oxidative stress, an interaction that may substantially increase the risk of CRC.

27.5
Interaction of Toxicity of Energy Excess and Oxidative Stress

Figures 27.2 and 27.3 illustrate that the pathways we have shown could, independently, relate the lifestyle risk factors for CRC, through oxidative stress and DNA damage or energy excess and features of the metabolic syndrome. However, Lee et al. [79] asked if the toxic consequences of energy excess interacted with those of oxidative stress so that their toxicities were synergistic in a study directed at understanding the "two-hit" etiology of nonalcoholic steatohepatitis (NASH) [80].

To test for such a synergistic toxicity they used the well-characterized in vitro hepatocyte inflammatory model [81–83]. Excess energy was simulated by exposure to fructose, which rapidly floods the metabolic pathways. Oxidative stress was simulated by H_2O_2, generated with a glucose oxidase system, so as to simulate H_2O_2 generated by activated immune cells [84, 85]. They observed a striking, 100-fold increased toxicity of the combined factors [79]. The exact mechanism responsible for this dramatic increase in toxicity is not known. Studies with chemical inhibitors, however, showed that fructose and H_2O_2 could result in cytotoxicity through several possible mechanisms, and that these mechanisms unexpectedly involved early steps in carbohydrate metabolism ([79], Chapter 7).

The carbohydrate intermediate metabolites, glyceraldehyde, dihydroxyacetone, glycolaldehyde, and methylglyoxal, as well as their oxidation product, glyoxal, are known to be genotoxic [86–88] and their mutagenicity is markedly increased in the presence of H_2O_2 [89–95]. For instance, physical chemical studies have also shown that glycolaldehyde can be oxidized by ROS ($OH^{\bullet}$) to form a highly reactive free radical [96, 97]. This could occur through a process similar to the oxidation of ethanol for which it has been surmised for some time that oxidative injury and formation of radicals were involved in ethanol toxicity [29, 30].

Glycolaldehyde is the putative reactive derivative of the known human carcinogens, ethylene oxide, chorethylene oxide, and vinyl carbamate [98–100]. Ethylene oxide exposure has been associated with an excess of lymphomas and leukemias in men and breast cancers in women [101]. It also induced chromosome breaks (micronuclei) in hematopoietic cells [102] and leukemia, breast, brain, and lung tumors in rodents [101].

Taken together, this *in vitro* model makes it reasonable to suggest that oxidative stress and energy excess may act synergistically to increase exposure to reactive genotoxic and cytotoxic compounds, increasing the risk of CRC, as summarized in Figure 27.4.

27.6
Future Research

A model describing the relationship between lifestyle and CRC (Figure 27.5) needs to consider that lifestyle factors for CRC could:

1. increase oxidative stress (e.g., $O_2^{\bullet-}$, H_2O_2, $OH^{\bullet}$), DNA damage, and CRC;
2. increase inflammation, which would then increase oxidative stress;
3. increase the permeability of the colon, local and circulating LPS, and inflammation;
4. increase energy excess, components of the metabolic syndrome, and CRC;
5. through the combined effect of energy excess and oxidative stress, result in the exposure of epithelial cells to cytotoxicity and genotoxicity more marked than that from either stress alone.

Study of the existence and relative importance of these pathways, therefore, provides an exciting opportunity for future research. Each of the paths describing these associations can be assessed with biomarkers, as suggested in Table 27.2.

For use in epidemiology studies, it is necessary to assume that serum and/or red blood cell levels of biomarkers are relevant measures of biologic action; therefore animal model studies supporting this assumption are desirable. Circulatory levels can be used in human studies to examine each of the paths in Figure 27.5. Such

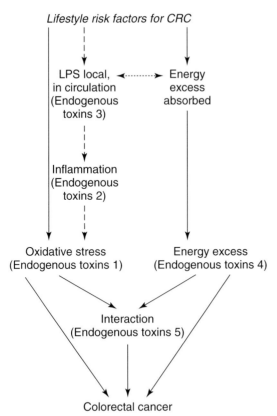

Fig. 27.5 The general scheme (Figure 27.1) modified to include the possible LPS/colonic inflammation/oxidative stress path (dashed line), a possible important relationship between circulating LPS on energy excess absorbed (dotted line with double arrows, because the direction of the association is unclear), and of energy excess on oxidative stress (dotted line). Note that there are five possible groups of endogenous biomarkers on pathways between the lifestyle risk factors and colorectal cancer (Table 27.2).

studies will be facilitated by use of stored blood, as available in many large-scale epidemiological investigations [103–106].

In animal experimental studies, it is first necessary to develop laboratory conditions that encompass the range of identified CRC risk factors, extending the range of dietary variables adopted in previous studies [107]. Once conditions are established in which the risk factors result in relevant pathological changes in the colon, circulating levels can be used to examine each of the paths in Figure 27.5. Appropriate dietary changes could then be tested to determine the most efficacious changes to reduce CRC risk.

In conclusion, we note that the accumulation of extensive effort in defining the mechanism relating lifestyle factors to the development of CRC has revealed important clues to progress.

Tab. 27.2 Possible measures of endogenous toxins

Endogenous toxins or markers	Possible biomarkers
1. Oxidative stress	Lipid oxidation products (e.g., 4-OH-nonenal)
	DNA damage (e.g., 8-OHdG, DNA adducts)
	Protein carbonylation
2. Colonic inflammation	Fecal calprotectin
	Colonic epithelial granulocyte/macrophage quantitation
3. LPS in circulation	Plasma LPS
4. Energy excess	Body fatness
	Abdominal fatness
	Adult attained height
	Plasma insulin
	Plasma glucose/fructose
	Plasma triglycerides
	Plasma glyceraldehyde
	Free fatty acids
5. Interaction products	Plasma glycolaldehyde, glyoxal, AGEs
	Protein/DNA adducts of glycolaldehyde, glyoxal, and acetyl radical

1. As illustrated in the figures here, it is possible that the multiple lifestyle factors involved in CRC do not act independently. Interactions between the risk factors may be important and the interactions may be quite complex. The biomarkers identified in Table 27.2 may, by virtue of their closer proximity to the colon carcinogenesis process, provide a clearer signal of associations with disease than the lifestyle risk factors themselves.
2. A major factor in risk may be total energy consumption leading to endogenous energy excess. In many epidemiology evaluations, dietary data are presented in a form adjusted for total dietary energy, such as food servings or grams of nutrients per 1000 kcal. This may have obscured the effects of a possibly important role of total energy. An evaluation of the relationship between CRC risk and total energy intake from the past, as well as in future studies, especially with the consideration of the potential for mediation by adiposity, may provide insight into this important pathway.
3. It is also possible that an important diet-related endogenous factor has not been identified or has been neglected. One such factor might be LPS (endotoxin), recently identified as a factor in the origin of ASH. Its possible importance in CRC offers exciting research opportunities. If risk for CRC were found to be associated with a circulating endotoxin, as it is with ASH, and if endotoxins were associated with relevant food and/or nutrients, a rapid endpoint for intervention studies would become evident.

References

1 Parkin, D.M., Bray, F., Ferlay, J., and Pisani, P. (2005) Global cancer statistics, 2002. *CA Cancer J Clin*, **55**, 74–108.
2 World Cancer Research Fund/American Institute for Cancer Research (2007) Food, nutrition, physical activity, and the prevention of cancer: a global perspective. *Am Inst Cancer Res*.
3 Segi, M., Kirihara, M., and Sukahara, Y. (1966) Cancer Mortality for Selected Sites in 24 Countries, *Department of Public Health, Tohoku University School of Medicine*.
4 Haenszel, W. and Kurihara, M. (1968) Studies of Japanese migrants. I. Mortality from cancer and other diseases among Japanese in the United States. *J Natl Cancer Inst*, **40**, 43–68.
5 Kono, S. (2004) Secular trend of colon cancer incidence and mortality in relation to fat and meat intake in Japan. *Eur J Cancer Prev*, **13**, 127–132.
6 Wellen, K.E. and Hotamisligil, G.S. (2005) Inflammation, stress, and diabetes. *J Clin Invest*, **115**, 1111–1119.
7 Raitakari, M., Mansikkaniemi, K., Marniemi, J., Viikari, J.S., and Raitakari, O.T. (2005) Distribution and determinants of serum high-sensitive C-reactive protein in a population of young adults: The Cardiovascular Risk in Young Finns Study. *J Intern Med*, **258**, 428–434.
8 Elgazar-Carmon, V., Rudich, A., Hadad, N., and Levy, R. (2008) Neutrophils transiently infiltrate intra-abdominal fat early in the course of high-fat feeding. *J Lipid Res*, **49**, 1894–1903.
9 Park, E., Wong, V., Guan, X., Oprescu, A.I., and Giacca, A. (2007) Salicylate prevents hepatic insulin resistance caused by short-term elevation of free fatty acids in vivo. *J Endocrinol*, **195**, 323–331.
10 Boelsterli, U.A. (2007) *Mechanistic Toxicology*, 2nd edn, CRC Press, Boca Raton.
11 Sesink, A.L., Termont, D.S., Kleibeuker, J.H., and Van der Meer, R. (1999) Red meat and colon cancer: the cytotoxic and hyperproliferative effects of dietary heme. *Cancer Res*, **59**, 5704–5709.
12 Pierre, F., Peiro, G., Tache, S., Cross, A.J., Bingham, S.A., Gasc, N., Gottardi, G., Corpet, D.E., and Gueraud, F. (2006) New marker of colon cancer risk associated with heme intake: 1,4-dihydroxynonane mercapturic acid. *Cancer Epidemiol Biomarkers Prev*, **15**, 2274–2279.
13 Hughes, R., Cross, A.J., Pollock, J.R., and Bingham, S. (2001) Dose-dependent effect of dietary meat on endogenous colonic N-nitrosation. *Carcinogenesis*, **22**, 199–202.
14 Lunn, J.C., Kuhnle, G., Mai, V., Frankenfeld, C., Shuker, D.E., Glen, R.C., Goodman, J.M., Pollock, J.R., and Bingham, S.A. (2007) The effect of haem in red and processed meat on the endogenous formation of N-nitroso compounds in the upper gastrointestinal tract. *Carcinogenesis*, **28**, 685–690.
15 Pierre, F., Santarelli, R., Tache, S., Gueraud, F., and Corpet, D.E. (2008) Beef meat promotion of dimethylhydrazine-induced colorectal carcinogenesis biomarkers is suppressed by dietary calcium. *Br J Nutr*, **99**, 1000–1006.
16 Yang, C.S., Landau, J.M., Huang, M.T., and Newmark, H.L. (2001) Inhibition of carcinogenesis by dietary polyphenolic compounds. *Annu Rev Nutr*, **21**, 381–406.
17 Newmark, H.L., Huang, M.T., and Reddy, B.S. (2006) Mixed tocopherols inhibit azoxymethane-induced aberrant crypt foci in rats. *Nutr Cancer*, **56**, 82–85.
18 Corpet, D.E. and Tache, S. (2002) Most effective colon cancer chemopreventive agents in rats: a systematic review of aberrant crypt foci and tumor data, ranked by potency. *Nutr Cancer*, **43**, 1–21.

19 Bakker, S.J., Hoogeveen, E.K., Nijpels, G., Kostense, P.J., Dekker, J.M., Gans, R.O., and Heine, R.J. (1998) The association of dietary fibres with glucose tolerance is partly explained by concomitant intake of thiamine: the Hoorn Study. *Diabetologia*, **41**, 1168–1175.

20 Purohit, V., Bode, J.C., Bode, C., Brenner, D.A., Choudhry, M.A., Hamilton, F., Kang, Y.J., Keshavarzian, A., Rao, R., Sartor, R.B., Swanson, C., and Turner, J.R. (2008) Alcohol, intestinal bacterial growth, intestinal permeability to endotoxin, and medical consequences: summary of a symposium. *Alcohol*, **42**, 349–361.

21 Jokelainen, K., Matysiak-Budnik, T., Makisalo, H., Hockerstedt, K., and Salaspuro, M. (1996) High intracolonic acetaldehyde values produced by a bacteriocolonic pathway for ethanol oxidation in piglets. *Gut*, **39**, 100–104.

22 Rao, R.K., Seth, A., and Sheth, P. (2004) Recent Advances in Alcoholic Liver Disease I. Role of intestinal permeability and endotoxemia in alcoholic liver disease. *Am J Physiol Gastrointest Liver Physiol*, **286**, G881–G884.

23 Takeuchi, O., Hoshino, K., Kawai, T., Sanjo, H., Takada, H., Ogawa, T., Takeda, K., and Akira, S. (1999) Differential roles of TLR2 and TLR4 in recognition of gram-negative and gram-positive bacterial cell wall components. *Immunity*, **11**, 443–451.

24 Krause, K.H. (2004) Tissue distribution and putative physiological function of NOX family NADPH oxidases. *Jpn J Infect Dis*, **57**, S28–S29.

25 Wheeler, M.D., Kono, H., Yin, M., Nakagami, M., Uesugi, T., Arteel, G.E., Gabele, E., Rusyn, I., Yamashina, S., Froh, M., Adachi, Y., Iimuro, Y., Bradford, B.U., Smutney, O.M., Connor, H.D., Mason, R.P., Goyert, S.M., Peters, J.M., Gonzalez, F.J., Samulski, R.J., and Thurman, R.G. (2001) The role of Kupffer cell oxidant production in early ethanol-induced liver disease. *Free Radic Biol Med*, **31**, 1544–1549.

26 Adachi, Y., Moore, L.E., Bradford, B.U., Gao, W., and Thurman, R.G. (1995) Antibiotics prevent liver injury in rats following long-term exposure to ethanol. *Gastroenterology*, **108**, 218–224.

27 Nanji, A.A., Khettry, U., and Sadrzadeh, S.M. (1994) Lactobacillus feeding reduces endotoxemia and severity of experimental alcoholic liver (disease). *Proc Soc Exp Biol Med*, **205**, 243–247.

28 Adachi, Y., Bradford, B.U., Gao, W., Bojes, H.K., and Thurman, R.G. (1994) Inactivation of Kupffer cells prevents early alcohol-induced liver injury. *Hepatology*, **20**, 453–460.

29 Albano, E. (2008) Oxidative mechanisms in the pathogenesis of alcoholic liver disease. *Mol Aspects Med*, **29**, 9–16.

30 Lu, Y. and Cederbaum, A.I. (2008) CYP2E1 and oxidative liver injury by alcohol. *Free Radic Biol Med*, **44**, 723–738.

31 Wargovich, M.J. (2006) Diallylsulfide and allylmethylsulfide are uniquely effective among organosulfur compounds in inhibiting CYP2E1 protein in animal models. *J Nutr*, **136**, 832S–834S.

32 Cani, P.D., Neyrinck, A.M., Fava, F., Knauf, C., Burcelin, R.G., Tuohy, K.M., Gibson, G.R., and Delzenne, N.M. (2007) Selective increases of bifidobacteria in gut microflora improve high-fat-diet-induced diabetes in mice through a mechanism associated with endotoxaemia. *Diabetologia*, **50**, 2374–2383.

33 Cani, P.D., Bibiloni, R., Knauf, C., Waget, A., Neyrinck, A.M., Delzenne, N.M., and Burcelin, R. (2008) Changes in gut microbiota control metabolic endotoxemia-induced inflammation in high-fat diet-induced obesity and diabetes in mice. *Diabetes*, **57**, 1470–1481.

34 Backhed, F., Manchester, J.K., Semenkovich, C.F., and Gordon, J.I. (2007) Mechanisms underlying the resistance to diet-induced obesity

in germ-free mice. *Proc Natl Acad Sci U S A*, **104**, 979–984.
35. DiBaise, J.K., Zhang, H., Crowell, M.D., Krajmalnik-Brown, R., Decker, G.A., and Rittmann, B.E. (2008) Gut microbiota and its possible relationship with obesity. *Mayo Clin Proc*, **83**, 460–469.
36. Amar, J., Burcelin, R., Ruidavets, J.B., Cani, P.D., Fauvel, J., Alessi, M.C., Chamontin, B., and Ferrieres, J. (2008) Energy intake is associated with endotoxemia in apparently healthy men. *Am J Clin Nutr*, **87**, 1219–1223.
37. Erridge, C., Attina, T., Spickett, C.M., and Webb, D.J. (2007) A high-fat meal induces low-grade endotoxemia: evidence of a novel mechanism of postprandial inflammation. *Am J Clin Nutr*, **86**, 1286–1292.
38. Thuy, S., Ladurner, R., Volynets, V., Wagner, S., Strahl, S., Konigsrainer, A., Maier, K.P., Bischoff, S.C., and Bergheim, I. (2008) Nonalcoholic fatty liver disease in humans is associated with increased plasma endotoxin and plasminogen activator inhibitor 1 concentrations and with fructose intake. *J Nutr*, **138**, 1452–1455.
39. Rafter, J., Bennett, M., Caderni, G., Clune, Y., Hughes, R., Karlsson, P.C., Klinder, A., O'Riordan, M., O'Sullivan, G.C., Pool-Zobel, B., Rechkemmer, G., Roller, M., Rowland, I., Salvadori, M., Thijs, H., Van Loo, J., Watzl, B., and Collins, J.K. (2007) Dietary synbiotics reduce cancer risk factors in polypectomized and colon cancer patients. *Am J Clin Nutr*, **85**, 488–496.
40. Goldin, B.R., Gualtieri, L.J., and Moore, R.P. (1996) The effect of Lactobacillus GG on the initiation and promotion of DMH-induced intestinal tumors in the rat. *Nutr Cancer*, **25**, 197–204.
41. Reddy, B.S., Hamid, R., and Rao, C.V. (1997) Effect of dietary oligofructose and inulin on colonic preoplastic aberrant crypt foci inhibition. *Carcinogenesis*, **18**, 1371–1374.
42. Rowland, I.R., Rumney, C.J., Coutts, J.T., and Lievense, L.C. (1998) Effect of Bifidobacterium longum and inulin on gut bacterial metabolism and carcinogen-induced aberrant crypt foci in rats. *Carcinogenesis*, **19**, 281–285.
43. Kong, J., Zhang, Z., Musch, M.W., Ning, G., Sun, J., Hart, J., Bissonnette, M., and Li, Y.C. (2008) Novel role of the vitamin D receptor in maintaining the integrity of the intestinal mucosal barrier. *Am J Physiol Gastrointest Liver Physiol*, **294**, G208–G216.
44. Froicu, M. and Cantorna, M.T. (2007) Vitamin D and the vitamin D receptor are critical for control of the innate immune response to colonic injury. *BMC Immunol*, **8**, 5.
45. Neish, A.S. (2007) TLRS in the gut. II. Flagellin-induced inflammation and antiapoptosis. *Am J Physiol Gastrointest Liver Physiol*, **292**, G462–G466.
46. Selleri, S., Palazzo, M., Deola, S., Wang, E., Balsari, A., Marincola, F.M., and Rumio, C. (2008) Induction of pro-inflammatory programs in enteroendocrine cells by the Toll-like receptor agonists flagellin and bacterial LPS. *Int Immunol*, **20**, 961–970.
47. Soler, A.P., Miller, R.D., Laughlin, K.V., Carp, N.Z., Klurfeld, D.M., and Mullin, J.M. (1999) Increased tight junctional permeability is associated with the development of colon cancer. *Carcinogenesis*, **20**, 1425–1431.
48. Mullin, J.M., Valenzano, M.C., Trembeth, S., Allegretti, P.D., Verrecchio, J.J., Schmidt, J.D., Jain, V., Meddings, J.B., Mercogliano, G., and Thornton, J.J. (2006) Transepithelial leak in Barrett's esophagus. *Dig Dis Sci*, **51**, 2326–2336.
49. D'Odorico, A., Bortolan, S., Cardin, R., D'Inca', R., Martines, D., Ferronato, A., and Sturniolo, G.C. (2001) Reduced plasma antioxidant concentrations and increased oxidative DNA damage in inflammatory

bowel disease. *Scand J Gastroenterol*, **36**, 1289–1294.

50 McKenzie, S.J., Baker, M.S., Buffinton, G.D., and Doe, W.F. (1996) Evidence of oxidant-induced injury to epithelial cells during inflammatory bowel disease. *J Clin Invest*, **98**, 136–141.

51 Aadland, E. and Fagerhol, M.K. (2002) Faecal calprotectin: a marker of inflammation throughout the intestinal tract. *Eur J Gastroenterol Hepatol*, **14**, 823–825.

52 Tibble, J., Sigthorsson, G., Foster, R., Sherwood, R., Fagerhol, M., and Bjarnason, I. (2001) Faecal calprotectin and faecal occult blood tests in the diagnosis of colorectal carcinoma and adenoma. *Gut*, **49**, 402–408.

53 Poullis, A., Foster, R., Shetty, A., Fagerhol, M.K., and Mendall, M.A. (2004) Bowel inflammation as measured by fecal calprotectin: a link between lifestyle factors and colorectal cancer risk. *Cancer Epidemiol Biomarkers Prev*, **13**, 279–284.

54 Hussain, S.P. and Harris, C.C. (2007) Inflammation and cancer: an ancient link with novel potentials. *Int J Cancer*, **121**, 2373–2380.

55 Nair, U., Bartsch, H., and Nair, J. (2007) Lipid peroxidation-induced DNA damage in cancer-prone inflammatory diseases: a review of published adduct types and levels in humans. *Free Radic Biol Med*, **43**, 1109–1120.

56 Leuratti, C., Watson, M.A., Deag, E.J., Welch, A., Singh, R., Gottschalg, E., Marnett, L.J., Atkin, W., Day, N.E., Shuker, D.E., and Bingham, S.A. (2002) Detection of malondialdehyde DNA adducts in human colorectal mucosa: relationship with diet and the presence of adenomas. *Cancer Epidemiol Biomarkers Prev*, **11**, 267–273.

57 D'Inca, R., Cardin, R., Benazzato, L., Angriman, I., Martines, D., and Sturniolo, G.C. (2004) Oxidative DNA damage in the mucosa of ulcerative colitis increases with disease duration and dysplasia. *Inflamm Bowel Dis*, **10**, 23–27.

58 Nair, J., Gansauge, F., Beger, H., Dolara, P., Winde, G., and Bartsch, H. (2006) Increased etheno-DNA adducts in affected tissues of patients suffering from Crohn's disease, ulcerative colitis, and chronic pancreatitis. *Antioxid Redox Signal*, **8**, 1003–1010.

59 Yeomans, M.R. and Phillips, M.F. (2002) Failure to reduce short-term appetite following alcohol is independent of beliefs about the presence of alcohol. *Nutr Neurosci*, **5**, 131–139.

60 Slavin, J.L. (2005) Dietary fiber and body weight. *Nutrition*, **21**, 411–418.

61 Papakonstantinou, E., Flatt, W.P., Huth, P.J., and Harris, R.B. (2003) High dietary calcium reduces body fat content, digestibility of fat, and serum vitamin D in rats. *Obes Res*, **11**, 387–394.

62 Bendsen, N.T., Hother, A.L., Jensen, S.K., Lorenzen, J.K., and Astrup, A. (2008) Effect of dairy calcium on fecal fat excretion: a randomized crossover trial. *Int J Obes (London)*, **32**, 1816–1824.

63 Cani, P.D., Delzenne, N.M., Amar, J., and Burcelin, R. (2008) Role of gut microflora in the development of obesity and insulin resistance following high-fat diet feeding. *Pathol Biol (Paris)*, **56**, 305–309.

64 McKeown-Eyssen, G. (1994) Epidemiology of colorectal cancer revisited: are serum triglycerides and/or plasma glucose associated with risk? *Cancer Epidemiol Biomarkers Prev*, **3**, 687–695.

65 Giovannucci, E. (1995) Insulin and colon cancer. *Cancer Causes Control*, **6**, 164–179.

66 Reaven, G.M. (1988) Banting lecture 1988. Role of insulin resistance in human disease. *Diabetes*, **37**, 1595–1607.

67 Koohestani, N., Chia, M.C., Pham, N.A., Tran, T.T., Minkin, S., Wolever, T.M., and Bruce, W.R. (1998) Aberrant crypt focus promotion and glucose intolerance: correlation in the rat across diets differing in fat, n-3 fatty acids

and energy. *Carcinogenesis*, **19**, 1679–1684.

68 Tran, T.T., Gupta, N., Goh, T., Naigamwalla, D., Chia, M.C., Koohestani, N., Mehrotra, S., McKeown-Eyssen, G., Giacca, A., and Bruce, W.R. (2003) Direct measure of insulin sensitivity with the hyperinsulinemic-euglycemic clamp and surrogate measures of insulin sensitivity with the oral glucose tolerance test: correlations with aberrant crypt foci promotion in rats. *Cancer Epidemiol Biomarkers Prev*, **12**, 47–56.

69 Tran, T.T., Medline, A. and Bruce, W.R. (1996) Insulin promotion of colon tumors in rats. *Cancer Epidemiol Biomarkers Prev*, **5**, 1013–1015.

70 Corpet, D.E., Jacquinet, C., Peiffer, G., and Tache, S. (1997) Insulin injections promote the growth of aberrant crypt foci in the colon of rats. *Nutr Cancer*, **27**, 316–320.

71 Giovannucci, E. (2001) Insulin, insulin-like growth factors and colon cancer: a review of the evidence. *J Nutr*, **131**, 3109S–3120S.

72 Sandhu, M.S., Dunger, D.B., and Giovannucci, E.L. (2002) Insulin, insulin-like growth factor-I (IGF-I), IGF binding proteins, their biologic interactions, and colorectal cancer. *J Natl Cancer Inst*, **94**, 972–980.

73 Giovannucci, E. (2007) Metabolic syndrome, hyperinsulinemia, and colon cancer: a review. *Am J Clin Nutr*, **86**, s836–s842.

74 Jenab, M., Riboli, E., Cleveland, R.J., Norat, T., Rinaldi, S., Nieters, A., Biessy, C., Tjonneland, A., Olsen, A., Overvad, K., Gronbaek, H., Clavel-Chapelon, F., Boutron-Ruault, M.C., Linseisen, J., Boeing, H., Pischon, T., Trichopoulos, D., Oikonomou, E., Trichopoulou, A., Panico, S., Vineis, P., Berrino, F., Tumino, R., Masala, G., Peters, P.H., van Gils, C.H., Bueno-de-Mesquita, H.B., Ocke, M.C., Lund, E., Mendez, M.A., Tormo, M.J., Barricarte, A., Martinez-Garcia, C., Dorronsoro, M., Quiros, J.R., Hallmans, G., Palmqvist, R., Berglund, G., Manjer, J., Key, T., Allen, N.E., Bingham, S., Khaw, K.T., Cust, A., and Kaaks, R. (2007) Serum C-peptide, IGFBP-1 and IGFBP-2 and risk of colon and rectal cancers in the European Prospective Investigation into Cancer and Nutrition. *Int J Cancer*, **121**, 368–376.

75 Kaaks, R. (2004) Nutrition, insulin, IGF-1 metabolism and cancer risk: a summary of epidemiological evidence. *Novartis Found Symp*, **262**, 247–260.

76 Trevisan, M., Liu, J., Muti, P., Misciagna, G., Menotti, A., and Fucci, F. (2001) Markers of insulin resistance and colorectal cancer mortality. *Cancer Epidemiol Biomarkers Prev*, **10**, 937–941.

77 Nilsen, T.I. and Vatten, L.J. (2001) Prospective study of colorectal cancer risk and physical activity, diabetes, blood glucose and BMI: exploring the hyperinsulinaemia hypothesis. *Br J Cancer*, **84**, 417–422.

78 Schoen, R.E., Tangen, C.M., Kuller, L.H., Burke, G.L., Cushman, M., Tracy, R.P., Dobs, A., and Savage, P.J. (1999) Increased blood glucose and insulin, body size, and incident colorectal cancer. *J Natl Cancer Inst*, **91**, 1147–1154.

79 Lee, O., Bruce, W.R., Dong, Q., Bruce, J., Mehta, R., and O'Brien, P.J. (2009) Fructose and carbonyl metabolites as endogenous toxins. *Chem Biol Interact*, **178**, 332–339.

80 Day, C.P. and James, O.F. (1998) Steatohepatitis: a tale of two "hits"? *Gastroenterology*, **114**, 842–845.

81 Chan, K., Jensen, N.S., Silber, P.M., and O'Brien, P.J. (2007) Structure-activity relationships for halobenzene induced cytotoxicity in rat and human hepatocytes. *Chem Biol Interact*, **165**, 165–174.

82 O'Brien, P.J. and Siraki, A.G. (2005) Accelerated cytotoxicity mechanism screening using drug metabolising enzyme modulators. *Curr Drug Metab*, **6**, 101–109.

83 O'Brien, P.J., Chan, K., and Silber, P.M. (2004) Human and animal hepatocytes in vitro with extrapolation in vivo. *Chem Biol Interact*, **150**, 97–114.

84 Tafazoli, S., Mashregi, M., and O'Brien, P.J. (2008) Role of hydrazine in isoniazid-induced hepatotoxicity in a hepatocyte inflammation model. *Toxicol Appl Pharmacol*, **229**, 94–101.

85 Tafazoli, S. and O'Brien, P.J. (2008) Accelerated cytotoxic mechanism screening of hydralazine using an in vitro hepatocyte inflammatory cell peroxidase model. *Chem Res Toxicol*, **21**, 904–910.

86 Nagao, M., Fujita, Y., Wakabayashi, K., Nukaya, H., Kosuge, T., and Sugimura, T. (1986) Mutagens in coffee and other beverages. *Environ Health Perspect*, **67**, 89–91.

87 Petersen, A.B., Wulf, H.C., Gniadecki, R., and Gajkowska, B. (2004) Dihydroxyacetone, the active browning ingredient in sunless tanning lotions, induces DNA damage, cell-cycle block and apoptosis in cultured HaCaT keratinocytes. *Mutat Res*, **560**, 173–186.

88 Yamaguchi, T. (1982) Mutagenicity of trioses and methyl glyoxal on Salmonella typhimurium. *Agric Biol Chem*, **46**, 849–851.

89 Benov, L. and Beema, A.F. (2003) Superoxide-dependence of the short chain sugars-induced mutagenesis. *Free Radic Biol Med*, **34**, 429–433.

90 Yamaguchi, T. and Nakagawa, K. (1983) Mutagenicity of and formation of oxygen radicals by trioses and glyoxal derivatives. *Agric Biol Chem*, **47**, 2461–2465.

91 Fujita, Y., Wakabayashi, K., Nagao, M., and Sugimura, T. (1985) Implication of hydrogen peroxide in the mutagenicity of coffee. *Mutat Res*, **144**, 227–230.

92 Nukaya, H., Inaoka, Y., Ishida, H., Tsuji, K., Suwa, Y., Wakabayashi, K., and Kosuge, T. (1993) Modification of the amino group of guanosine by methylglyoxal and other alpha-ketoaldehydes in the presence of hydrogen peroxide. *Chem Pharm Bull (Tokyo)*, **41**, 649–653.

93 Tada, A., Wakabayashi, K., Totsuka, Y., Sugimura, T., Tsuji, K., and Nukaya, H. (1996) 32P-Postlabeling analysis of a DNA adduct, an N2-acetyl derivative of guanine, formed in vitro by methylglyoxal and hydrogen peroxide in combination. *Mutat Res*, **351**, 173–180.

94 Al Maghrebi, M.A., Al Mulla, F., and Benov, L.T. (2003) Glycolaldehyde induces apoptosis in a human breast cancer cell line. *Arch Biochem Biophys*, **417**, 123–127.

95 Murata, M., Mizutani, M., Oikawa, S., Hiraku, Y., and Kawanishi, S. (2003) Oxidative DNA damage by hyperglycemia-related aldehydes and its marked enhancement by hydrogen peroxide. *FEBS Lett*, **554**, 138–142.

96 Karunanandan, R., Holscher, D., Dillon, T.J., Horowitz, A., Crowley, J.N., Vereecken, L., and Peeters, J. (2007) Reaction of HO with glycolaldehyde, HOCH2CHO: rate coefficients (240–362 K) and mechanism. *J Phys Chem A*, **111**, 897–908.

97 Steenken, S. and Shulte-Frohlinde, D. (1973) Fragmentation of radicals derived from glycolaldehyde and glyceraldehyde in aqueous solution. An EPR study. *Tetrahedron Lett*, **9**, 653–654.

98 Hengstler, J.G., Fuchs, J., Gebhard, S., and Oesch, F. (1994) Glycolaldehyde causes DNA-protein crosslinks: a new aspect of ethylene oxide genotoxicity. *Mutat Res*, **304**, 229–234.

99 Barbin, A., Bereziat, J.C., Croisy, A., O'Neill, I.K., and Bartsch, H. (1990) Nucleophilic selectivity and reaction kinetics of chloroethylene oxide assessed by the 4-(p-nitrobenzyl)pyridine assay and proton nuclear magnetic resonance spectroscopy. *Chem Biol Interact*, **73**, 261–277.

100 Ballering, L.A., Nivard, M.J., and Vogel, E.W. (1996) Characterization by two-endpoint comparisons of the genetic toxicity profiles of vinyl

chloride and related etheno-adduct forming carcinogens in Drosophila. *Carcinogenesis*, **17**, 1083–1092.

101 Grosse, Y., Baan, R., Straif, K., Secretan, B., El Ghissassi, F., Bouvard, V., Altieri, A., and Cogliano, V. (2007) Carcinogenicity of 1,3-butadiene, ethylene oxide, vinyl chloride, vinyl fluoride, and vinyl bromide. *Lancet Oncol*, **8**, 679–680.

102 Lorenti, G.C., Darroudi, F., Tates, A.D., and Natarajan, A.T. (2001) Induction and persistence of micronuclei, sister-chromatid exchanges and chromosomal aberrations in splenocytes and bone-marrow cells of rats exposed to ethylene oxide. *Mutat Res*, **492**, 59–67.

103 Michels, K.B., Solomon, C.G., Hu, F.B., Rosner, B.A., Hankinson, S.E., Colditz, G.A., and Manson, J.E. (2003) Type 2 diabetes and subsequent incidence of breast cancer in the Nurses' Health Study. *Diabetes Care*, **26**, 1752–1758.

104 Newcomb, P.A., Baron, J., Cotterchio, M., Gallinger, S., Grove, J., Haile, R., Hall, D., Hopper, J.L., Jass, J., Le Marchand, L., Limburg, P., Lindor, N., Potter, J.D., Templeton, A.S., Thibodeau, S., and Seminara, D. (2007) Colon Cancer Family Registry: an international resource for studies of the genetic epidemiology of colon cancer. *Cancer Epidemiol Biomarkers Prev*, **16**, 2331–2343.

105 Riboli, E., Hunt, K.J., Slimani, N., Ferrari, P., Norat, T., Fahey, M., Charrondiere, U.R., Hemon, B., Casagrande, C., Vignat, J., Overvad, K., Tjonneland, A., Clavel-Chapelon, F., Thiebaut, A., Wahrendorf, J., Boeing, H., Trichopoulos, D., Trichopoulou, A., Vineis, P., Palli, D., Bueno-de-Mesquita, H.B., Peeters, P.H., Lund, E., Engeset, D., Gonzalez, C.A., Barricarte, A., Berglund, G., Hallmans, G., Day, N.E., Key, T.J., Kaaks, R., and Saracci, R. (2002) European Prospective Investigation into Cancer and Nutrition (EPIC): study populations and data collection. *Public Health Nutr*, **5**, 1113–1124.

106 Zhou, B.F., Stamler, J., Dennis, B., Moag-Stahlberg, A., Okuda, N., Robertson, C., Zhao, L., Chan, Q., and Elliott, P. (2003) Nutrient intakes of middle-aged men and women in China, Japan, United Kingdom, and United States in the late 1990s: the INTERMAP study. *J Hum Hypertens*, **17**, 623–630.

107 Newmark, H.L., Yang, K., Lipkin, M., Kopelovich, L., Liu, Y., Fan, K., and Shinozaki, H. (2001) A Western-style diet induces benign and malignant neoplasms in the colon of normal C57Bl/6 mice. *Carcinogenesis*, **22**, 1871–1875.

28
Dopamine-Derived Neurotoxicity and Parkinson's Disease
Jose Luis Labandeira-Garcia

A number of studies in our laboratory and many other laboratories have shown that adequate dopamine levels and adequate interaction of dopamine with other major neurotransmitters, particularly glutamate, are crucial for normal functioning of the basal ganglia and other brain areas, and that dysfunctions in this crosstalk may lead to a number of diseases such as schizophrenia, depression, epilepsy, obesity, and others. In addition, several studies have shown that dopamine and levodopa (3,4-dihydroxyphenylalanine) can induce neurotoxicity and neuron death, which may be counteracted by intrinsic neuroprotective molecules or drugs. Dopamine, particularly when acting via dopamine type 2 receptors, also appears necessary for neuroprotective functions and for counteracting neurotoxic effects of other neurotransmitters such as glutamate. It appears that precise regulation of dopamine levels is necessary for neuron survival and function. Disruption of this equilibrium by any of a number of factors or the introduction of additional sources of oxidative stress or neuronal dysfunction may set off a cascade of events that lead to neuron degeneration, and neurodegenerative diseases such as Parkinson's disease.

28.1
The Neurotransmitter Dopamine

The neurotransmitter dopamine (DA) is synthesized by mesencephalic neurons of the substantia nigra compacta (SNc) and ventral tegmental area (VTA), and by some other groups of neurons such as hypothalamic neurons of the arcuate and periventricular nuclei [1]. Neurons located in these areas project dopaminergic axons via four major pathways in which DA acts as a neuromodulator that controls important physiological functions such as voluntary movements, motivated behavior, learning, and hormone production. SNc neurons innervate the striatum through the nigrostriatal pathway and VTA neurons project to medial frontal, cingulated, and entorhinal cortices through the mesocortical pathway and to the nucleus accumbens and other limbic areas through the mesolimbic pathway. Finally, the arcuate and periventricular hypothalamic nuclei project to the hypophysis through the tuberoinfundibular pathway. DA synthesis primarily occurs at the neuronal terminal via the hydroxylation of tyrosine catalyzed by tyrosine hydroxylase (TH) to form L-3,4-dihydroxyphenylalanine (L-DOPA), which is decarboxylated to DA catalyzed by aromatic acid decarboxylase (AADC). DA is then sequestered from the cytosol and packaged into synaptic vesicles by the vesicular dopamine transporter 2 (VMAT2) [2]. Finally, following stimulation, vesicular DA is released to the synaptic terminals by exocytosis. Extracellular DA is recycled back into the terminal by the dopamine transporter (DAT), where it is either repackaged into vesicles by VMAT2 or is metabolized via several pathways that are potential sources of oxidative stress (OS), as described below.

28.2
Dopaminergic Alterations – Parkinson's Disease

Alterations in dopaminergic innervation are known to be involved in a number of diseases, including depression, attention deficit disorders, schizophrenia, epilepsy, pituitary tumors, Huntington's disease (HD), and, particularly, Parkinson's disease (PD). PD is a progressive neurodegenerative disease characterized by progressive degeneration of DA-containing neurons in the SNc and by the presence of intraneuronal proteinaceous cytoplasmic inclusions termed *Lewy bodies* [3]. This leads to a marked deficiency in striatal DA, which causes the major clinical symptoms of PD: akinesia, muscular rigidity, and resting tremor [4, 5]. In PD, clinical signs are usually detected when approximately 50% of nigral neurons and 80% of striatal DA are lost.

Although it has been shown that several genes are mutated or deleted in familial PD, the etiology of sporadic PD, which accounts for most cases of PD, is still unclear. A number of mechanisms have been involved in DA neuron degeneration in PD, including mitochondrial dysfunction, oxidative stress, inflammation, and impairment of the ubiquitin–proteosome system [6]. These pathogenic factors are not mutually exclusive, and one of the key aims of current PD research is to discover the mechanisms involved in possible interactions between these pathways that result in DA neuron degeneration. Several studies have provided evidence that OS plays a major role in all forms of PD [7, 8], and there has been some discussion whether OS is a primary event or a consequence of other pathogenic factors. However, DA degeneration is unquestionably mediated by an overproduction of reactive oxygen species (ROS) and reactive nitrogen species (RNS). ROS are generated as a result of normal metabolism. However, OS occurs when ROS or RNS are excessively produced or insufficiently degraded and overwhelm the protective defense mechanisms of a cell leading to functional impairment and finally cell death [7]. The brain is particularly liable to OS because it consumes about 20% of the total body O_2, and DA nigrostriatal neurons appear particularly vulnerable to OS-derived cell death [9, 10]. A number of potential factors have been involved, including increased iron content, reduced antioxidant capacity, and, particularly, those factors derived from the DA synthesized, released, and metabolized in these neurons. Potential DA neurotoxicity is a major point to understand pathogenesis and progression of PD [11, 12]. DA neurotoxicity is likely to be self-perpetuating, because synaptic DA depletion caused by a decrease in the number of DA neurons leads to a compensatory increase in DA turnover and further neurotoxicity. This chapter focuses on the neurotoxicity of DA and DA-related compounds on the SNc neurons and the potential role of these compounds in the etiology of PD.

It is important to note that a number of studies have suggested that DA metabolism (i.e., the "dopamine oxidative stress hypothesis") does not play a central causative role in PD, and that oxyradical products are not only unique to PD but are also found in other neurodegenerative diseases [13, 14]. The work by Braak et al. [15, 16] is considered to support this view, as a number of nondopaminergic neuronal populations have been observed to be affected by protein aggregates in a uniform sequence (i.e., the Braak PD staging scheme) in which the dorsal vagal nucleus, locus coeruleus, SNc, mesocortex, and neocortex acquire protein aggregates in this temporal order. However, these histological changes are poorly correlated with both neuronal cell loss, which is mostly concentrated in the SNc, and also the major clinical symptoms of PD. It is possible that a number of deleterious factors affect different neuronal populations. These factors may be counteracted by protective defense mechanisms in most of these neurons. However, the protective defense mechanisms may be overwhelmed by additional deleterious factors in neurons already subjected to DA-derived toxicity and lead to DA neuron death (i.e., a "synergistic effect hypothesis").

28.3
Levodopa Therapy for PD and Progress of Dopaminergic Neuron Degeneration

In addition to the possible role of DA toxicity in the pathogenesis and/or progression of PD, it is particularly important to determine if the main therapeutic tool for PD (i.e., levodopa therapy) is toxic to DA neurons and may actually accelerate progression of the disease. L-DOPA treatment to supplement the decreased DA remains the most effective clinical pharmacotherapy for PD. However, as well as being associated with adverse reactions such as wearing-off phenomenon, on–off phenomenon, and dyskinesia, the long-term L-DOPA therapy may also enhance neuronal damage progression [10, 17]. L-DOPA induced increased DA turnover [18], inhibition of complex IV in the mitochondrial electron transport chain [19], increased levels of lipid peroxidation [20], and other neurotoxic effects of L-DOPA have also been observed. This is an important question for clinicians, and potentially affects their decisions regarding treatment of PD patients with L-DOPA [21]. However, several studies have reported an absence of clinical or neuropathological toxicity after chronic oral administration of L-DOPA in animal models of PD [22, 23]. There is, therefore, some controversy regarding the potential neurotoxicity of L-DOPA treatment in PD patients. The ELLDOPA 40-week clinical trial addressed this question with levodopa-treated PD patients being compared to placebo-treated patients after drug washout [24]. Interestingly, clinical progression of levodopa-treated PD patients was not accelerated, and levodopa-treated patients had better clinical scores than placebo-treated patients a month after all drugs were discontinued. However, L-DOPA-treated patients showed an increased rate of decline in striatal DA uptake in the β-CIT SPECT imaging arm of the trial, which demonstrated that placebo treatment was associated with significantly lower loss of striatal DA signal over the 40-week trial [24]. Therefore, it was concluded that "these contradictory findings warrant further investigation into the effect of levodopa on PD" and that "the potential long-term effects of levodopa on PD remain uncertain."

28.4
Dopamine Toxicity and Aging

DA neurotoxicity has been also associated with progressive loss of neurons in the SNc of human brains with advancing age, and losses of more than one-third may occur between 20 and 90 years of age, at a rate of 5–10% loss per decade [25, 26]. PD was once considered to be a form of accelerated aging. However, a number of recent studies have shown that changes in the DA system in PD differ from those induced by aging, although aging-induced changes may increase the vulnerability of DA neurons to toxic damage and increase the risk of developing PD [27, 28]. Thus, cell loss is concentrated in ventrolateral and caudal portions of the SNc in PD, whereas during normal aging, the dorsomedial aspect of the SNc is affected [26]. Furthermore, factors other than DA are presumed to be involved in the loss of DA neurons during aging. We have recently observed age-related decrease in microvascularization and angiogenic factors such as vascular endothelial growth factor (VEGF) in the nigra [29]. As suggested above for PD, neurotoxicity derived from DA metabolism may act synergistically with other factors to induce cell death with advancing age.

28.5
Dopamine Toxicity and Animal Models of Parkinsonism

Some possible neurotoxic effects of DA metabolism have been demonstrated by the use of a number of pharmacological and genetic animal models. One of the first animal models of PD was obtained by administration of methamphetamine. Methamphetamine is a weak base that diffuses through the vesicular membrane and collapses the electrochemical gradient that is required for DA sequestration into vesicles. This leads to accumulation of DA in the cytosol, with a resulting degeneration of DA terminals in the nigrostriatal system [30]. The neurotoxin 6-hydroxydopamine (6-OHDA) is the first DA neurotoxin discovered and is the most commonly used neurotoxin to produce *in vivo* and *in vitro* animal models of PD (Figure 28.1). Formation of ROS by autoxidation of

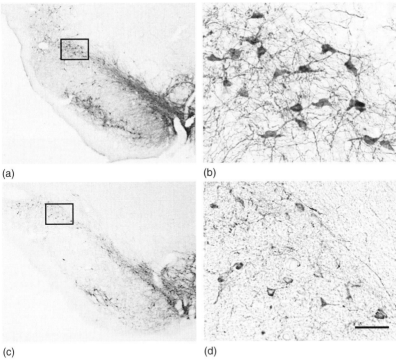

Fig. 28.1 Photomicrographs of dopaminergic neurons (black) in the substantia nigra of a normal rat (a and b), and a degenerated substantia nigra (c and d) after intraventricular injection of the neurotoxin 6-hydroxydopamine (6-OHDA) as rat model of parkinsonism. Higher magnification of the boxed areas in (a) and (c) are shown in (b) and (d), respectively. The dopaminergic neurons were immunostained with antibodies against tyrosine hydroxylase. Scale bar = 300 μm (a and c) and 60 μm (b and d).

6-OHDA in the cytosol is considered to be the main molecular mechanism underlying the neurotoxicity of 6-OHDA [31, 32], although other mechanisms may also be involved [33, 34]. Interestingly, it has been suggested that endogenous 6-OHDA is possibly formed naturally in the brain by nonenzymatic hydroxylation DA in the presence of Fe^{2+} and H_2O_2 [35, 36]. This endogenous 6-OHDA has been proposed as a possible neurotoxic factor in the pathogenesis of PD [37, 38] on the basis of the reported formation of 6-OHDA in the rat brain [39] and also the 6-OHDA accumulation found in PD patients [40]. Introduction of the selective DA neurotoxin 1-methyl-4-phenyl-1,2,3,6-tetrahydropyridine (MPTP) as a PD model further strengthened the DA hypothesis of PD, particularly the free radical hypothesis [41]. In the brain, MPTP is quickly metabolized to 1-methyl-4-phenyl-2,3-dihydropyridinium ($MPDP^+$), which undergoes spontaneous oxidation to its active metabolite 1-methyl-4-phenylpyridinium (MPP^+) ion by astrocytes, and MPP^+ is selectively taken up into DA neurons by the DAT. MPP^+ mediates its toxicity through a number of mechanisms, including inhibition of mitochondrial complex I and the subsequent reduction in ATP and ROS generation

[42, 43]. It has also been suggested that MPP$^+$ induces an increase in extraneuronal DA levels and a subsequent lesion of the DA terminals by DA neurotoxicity [44]. A direct demonstration of the neurotoxic effect of DA was achieved by intrastriatal injection of DA, which induced both pre- and postsynaptic damage of the nigrostriatal pathway [45, 46]. Several *in vitro* studies have shown that DA is toxic to cultured cells [47] via a number of mechanisms, including DNA damage, increased levels of p53 transcription factor, activation of caspase 3 and NF-kB, and downregulation of mitochondrial channels [48–50]. However, this was not confirmed in other studies [51, 52], and it has been suggested that the *in vitro* results may be an artifact of the culture medium [53].

28.6
Mechanisms of Dopamine Neurotoxicity – Vesicular and Cytosolic Dopamine

The results of most of the above-mentioned studies suggest that DA can induce cell toxicity through a number of mechanisms that include both receptor-independent and receptor-dependent mechanisms. DA is a highly reactive molecule that has the capacity to generate ROS such as hydroxyl radical (OH$^\bullet$), superoxide (O$_2^{\bullet-}$), hydrogen peroxide (H$_2$O$_2$), nitric oxide (NO$^\bullet$), and peroxynitrite (HOONO$^-$), as well as neurotoxic quinones through its metabolism in the cytosol [54, 55]. However, DA is taken up directly into synaptic vesicles after synthesis. DA is then protected from oxidation by low pH in the vesicles, which stabilizes the catechol structure and confers a milieu to prevent deprotonation [56, 57].

Sequestration of cytosolic DA into vesicles by VMAT2, feedback inhibition of TH, and conjugation of DA quinones by glutathione (GSH) are considered as major neuroprotective mechanisms developed by DA neurons to attenuate the damage caused by DA-derived OS. More than 90% of intracellular DA is sequestered into vesicles, which constitutes a crucial process to avoid DA neurotoxicity. This is supported by the results of recent studies, which implicate deficits in DA storage, via disruption of VMAT2, with DA neuron degeneration in PD [58, 59]. A genetically or environmentally induced decrease in the ability of vesicles to sequester DA properly may lead to increased cytosolic DA and increased neuron vulnerability [59, 60]. The use of genetically engineered animal models that express reduced levels of VMAT2 has clearly demonstrated the importance of vesicles in the protection of neurons from DA-derived toxicity [61, 62]. Furthermore, studies with VMAT2-deficient mice provide evidence for an interaction between cytoplasmic DA and α-synuclein *in vivo* [58] (see below).

There are two major pathways for metabolizing cytosolic DA and inducing DA-derived oxidative stress. First, cytosolic DA is metabolized via monoamine oxidase (MAO) [55, 63]. Oxidation of DA by MAO forms hydrogen peroxide (H$_2$O$_2$) and 3,4-dihydroxyphenylaldehyde, which is oxidized by aldehyde dehydrogenase to 3,4-dihydroxyphenylacetic acid (DOPAC). Part of the DOPAC is directly eliminated and part is converted to homovanilic acid (HVA) by catechol-*O*-methyltransferase (COMT) (see Chapter 29). Secondly, nonenzymatic and spontaneous autoxidation of

DA and L-DOPA produces superoxide and reactive quinones such as DA quinones or DOPA quinones [54,64–67]. Superoxide is either metabolized into H_2O_2 by superoxide dismutase (SOD) or reacts with NO to generate strongly reactive peroxynitrite. In the DA neurons, where transition metals are abundant, H_2O_2 can react with metals, especially iron, to form the most cytotoxic radical, hydroxyl radical ($OH^\bullet$). In a second step, quinones rearrange or cyclize and oxidize to aminochromes. These may then be polymerized, leading to the formation of neuromelanin or may be conjugated with GSH and reduced by one- or two-electron transfer catalyzed by quinone reductases. The conjugated aminochrome leukoaminochrome–GSH is very stable in contrast to the unconjugated aminochrome reduced forms [7, 68, 69]. Other pathways of DA oxidation have also been reported, and in these pathways, DA quinones are also generated during the enzymatic oxidation of DA by prostaglandin H synthase (COX), lipoxygenase, tyrosinase, and xanthine oxidase [70–73].

In addition to the general oxidative stress induced by ROS and RNS, quinone-derived toxicity has received particular attention as a more specific oxidative stress in DA neurons [65–68, 74,]. It has been suggested that quinones exert cytotoxicity by interacting with a number of bioactive molecules, particularly those proteins that possess cysteine residues. It has been reported that quinones may affect DAT [75] and also proteins involved in causing PD such as parkin or α-synuclein [76–78]. Furthermore, it has been reported that quinones may inactivate TH (the rate-limiting enzyme in catecholamine biosynthesis) [79], induce activation of transcription factors AP-1 and NF-kB or activation of neurodestructive proteases, increase the vulnerability of neurons to glutamate toxicity, and may be involved in other mechanisms that enhance DA neuron degeneration (for review see [65, 74]). It has also been suggested that some of the enzymes that detoxify environmental toxins possibly related to the pathogenesis of PD (such as glutathione S-transferase) also detoxify quinones [74], and that an overload of environmental toxins may interfere with detoxification of quinones and vice versa. Furthermore, DA quinone– or DOPA quinone–induced cytotoxicity may be involved not only in the pathogenesis of PD but also in the neurotoxicity induced by long-term L-DOPA therapy in PD patients.

Therefore, both DA-derived generation of ROS/RNS and quinone toxicity may contribute to neuronal death. It has been reported, however, that quinone toxicity may play a major role, because antioxidants such as α-tocopherol have lesser or no effects against DA-induced cell death, whereas cell death is markedly inhibited by the thiol-containing compounds such as GSH, N-acetylcysteine, and dithiothreitol, which prevent DA quinone from binding to the sulfhydryl groups of cysteine on functional proteins by their conjugation with DA quinone [80, 81].

28.7
Neuromelanin Formation and Dopamine Toxicity

Neuromelanin is an insoluble pigment that accumulates in catecholaminergic neurons of the SNc and locus coeruleus (which are the most highly pigmented neurons in the human brain) during normal aging [82]. The neuromelanin molecule is a complex polymeric molecule composed of melanic, aliphatic, and peptide residues [82–85]. The melanic component includes benzothiazine and indole moieties. The aliphatic component is a distinctive characteristic of neuromelanin not found in other melanins, but the structure of this component is as yet unclear. Interestingly, it has also been reported that the major iron component in catecholaminergic neurons of the SNc and locus ceruleans is the iron–neuromelanin complex [84].

It is usually considered that in PD there is a preferential loss of those neurons that contain neuromelanin [86]. However, this has also been questioned because of a number of conflicting observations: not all nigral DA neurons contain neuromelanin, long-term L-DOPA therapy does not appear to enhance neuromelanin levels in the surviving neurons on postmortem examination, and it has also been reported that those neurons with most neuromelanin are spared and vice versa [87–89]. A number of possible mechanisms for explaining these conflicting observations have been suggested [74].

As described above, neuromelanin formation in DA neurons is generally regarded as the result of enzymatic oxidation or autoxidation of DA [83, 90]. The role of tyrosinase, which catalyzes the conversion of tyrosine to L-DOPA- and to DOPA quinone in melanin synthesis is controversial, as is the role of TH, prostaglandin H synthase, peroxidase, macrophage migration-inhibitory factor, and other enzymes [83, 84, 90]. Neuromelanin could, alternatively, be derived from autoxidation of DA to quinones as described above. Thus, neuromelanin synthesis appears to be induced by an excess of cytosolic DA that is not accumulated in vesicles. Neuromelanin synthesis may lead to removal of DA-derived quinones, thereby preventing DA-derived damage and resulting in a protective process. In addition, it has been observed that neuromelanin binds a number of metals, in particular iron, and its ability to chelate redox-active metals may also prevent neuronal damage [84]. However, it has also been reported that neuromelanin may contribute to DA neuron degeneration by itself or via interaction with iron, catechols, or neurotoxic metabolites [91]. Interestingly, it has been observed that in SNc of PD patients, the extraneuronal granules of neuromelanin released by degenerating neurons contribute to microglial activation [92], and it is known that microglial activation leads to enhanced DA neuron death and progression of PD. However, many other factors are involved in microglial activation in PD [34, 93]. In conclusion, the role of neuromelanin has been debated for a long time, and it appears that neuromelanin biochemistry is much more complex than that associated with autoxidized DA [87, 94], and that it may play a protective or toxic role depending on the cellular context.

28.8
Interaction between Dopamine Toxicity and Protein Aggregation

Accumulation of the protein α-synuclein in the Lewy bodies is a pathological hallmark of PD and diffuse Lewy-body disease [95], and a number of studies have suggested that DA neurons are preferentially susceptible to α-synuclein [96, 97]. Although the association between α-synuclein and neurodegeneration has been extensively demonstrated, the exact pathway by which α-synuclein or its mutants cause degeneration remains unclear. It has been suggested that α-synuclein may play a dual role in DA degeneration as α-synuclein may contribute to enhance the effects of OS through its interaction with the toxic potential of endogenous DA, and may also enhance OS by disrupting DA handling [98–100]. First, DA-derived oxidative damage to α-synuclein can enhance its ability to misfold and aggregate [101, 102]. It has been suggested that α-synuclein is one of the target proteins for DA quinone and L-DOPA quinone [77, 101]. These quinones may react with α-synuclein to form a quinoprotein DA quinone–α-synuclein adduct by coupling to tyrosine residues of α-synuclein and/or by nucleophilic attack on lysine residues, consequently inhibiting the conversion of toxic protofibrils to fibrils [77]; however, alpha-synuclein does not contain any free thiols, therefore reaction with DA quinone might be of question given the rapid rearrangement of the DA quinone [64]. Secondly, α-synuclein may modulate levels of cytoplasmic DA via an interaction with synaptic vesicles. It has been reported that α-synuclein can bind to and permeabilize vesicles, which may result in leakage of DA into the cytosol [103]. Furthermore, α-synuclein may act by decreasing VMAT2 protein or interact with VMAT2 protein causing an increase in cytosolic DA [104, 105].

Interactions between DA and parkin protein have also been suggested. It is well known that in a familial type of PD, inherited mutations of PARK2, the gene encoding parkin, cause selective cell death in DA neurons of the SNc and early PD. However, the role of parkin in the more frequent, sporadic PD has not been totally clarified. Interestingly, it has been shown that DA covalently modifies parkin in living DA cells to increase its insolubility and to inactivate its E3 ubiquitin ligase function [76]. This was confirmed in brains from PD patients, which showed decreased parkin solubility, and in normal brains, which showed catechol modified parkin in the SNc but not in other areas [76]. It has also been suggested that parkin is a DA quinone–quenching molecule that prevents DA quinone–induced pathogenicity [106]. Therefore, cytosolic DA may induce progressive loss of parkin function in DA neurons during aging and PD, and contribute to DA cell death.

28.9
Receptor-Dependent Dopamine Toxicity, Dopamine-Mediated Excitotoxicity, and Relevance for Huntington's Disease

The effects of DA are mediated through its interaction with G-protein-coupled membrane receptors. The DA receptor family can be divided in two subfamilies according to structural and pharmacological properties: the D1-like family (D1 and D5 receptors) and the D2-like family (D2, D3, and D4 receptors). As described above, DA can exert toxic effects through the formation of oxidative metabolites. However, some data suggest that DA toxicity may also occur via DA-receptor-dependent mechanisms. It has been reported that DA-induced cell death is inhibited by D1 receptor antagonists in a human neuroblastoma cell line, whereas D1 agonists induce cytotoxicity [107], and DA receptor stimulation has been shown to induce cytotoxicity in striatal cells [108, 109]. In fact, a number of recent studies have suggested that elevated levels of DA in the striatum may be relevant in the pathogenesis of HD [110].

HD is an inherited neurodegenerative disorder characterized by progressive neuronal death that predominantly affects basal ganglia structures, most notably the striatum [111]. The genetic mutation responsible for the disease produces an expansion of CAG repeats in the gene encoding the cytoplasmic protein huntingtin [112]. However, it is not clear what causes the neuronal death as excitotoxicity, mitochondrial dysfunction, or oxidative stress has been suggested to be involved. The striatum is densely innervated by DA fibers originated in the substantia nigra and contains DA at high concentrations, which may contribute to the particular vulnerability of the striatal neurons in HD [113]. Furthermore, the loss of inhibitory striatal GABAergic neurons may further increase DA levels in the striatum [114]. The results of a number of studies suggest an alteration in DA metabolism in this disease. It has been observed that when treated with L-DOPA, about a third of asymptomatic relatives of patients with HD developed involuntary movements similar to those observed in HD [115], and that the involuntary movements were increased by DA agonists and inhibited by DA antagonists [116].

Furthermore, DA toxicity may be mediated by excitatory amino acids (EAAs) such as glutamate. EAAs bind stereospecifically to a number of receptor subtypes, usually classified as NMDA and non-NMDA receptors, activation of which induces cell death (i.e., excitotoxicity). It has been observed that some striatal lesions mediated by activation of EAA receptors can be attenuated by DA denervation [117, 118]. Elevated DA may induce an increase in synaptic glutamate levels by inhibition of glutamate uptake [119], while D1 receptor activation may increase NMDA-receptor-induced toxicity by activation of Ca^{2+} conductances or phosphorylation of NMDA receptors [120]. However, interactions between DA and glutamate in the striatum appear complex and are still unclear [113,121–123].

28.10
Dopamine and Neuroprotection

Several studies have shown neurotoxic effects of DA or L-DOPA in cell cultures and also in animal models. However, there are also studies showing that DA exerts neuroprotective effects in some circumstances, for example, DA can attenuate striatal neuron damage caused by the non-NMDA-receptor agonist kainate [124]. DA agonists also reduce both NMDA- and kainate-induced neuronal lesions [120, 124]. The neuroprotective effect of DA appears to be exerted through receptor-dependent mechanisms, and particularly through D2-receptor-mediated signaling [125]. D2 agonists protected mesencephalic neurons from glutamate

[126] and L-DOPA-induced cell death [127]. The role of D2 receptors in neuroprotection has also been supported by a number of *in vivo* studies showing that blockage of D2 receptor signaling with D2 antagonists induces apoptosis in striatum and substantia nigra, whereas D2 agonists induce neuroprotection [125, 128]. Similarly, mutant mice lacking D2 receptors are highly susceptible to excitotoxicity-induced neurodegeneration [129]. Furthermore, clinical trials using D2-selective agonists (ropirinole and pramipexole) as an alternative to L-DOPA therapy also support a neuroprotective effect of D2 receptors, as neuroimaging data suggested a decrease in the loss of striatal DA terminals in these patients [130, 131]. However, as in the case of the above-mentioned ELLDOPA study carried out to evaluate the long-term effects of levodopa treatment [24], the neuroimaging results using the DA precursor fluoro-DOPA or the DAT ligand β-CIT have been questioned as these molecules may interfere with DA metabolism and uptake.

The mechanisms involved in the neuroprotective effects of D2 receptor stimulation have not been totally clarified. However, as D2 receptors are expressed presynaptically in DA neurons and postsynaptically in areas receiving DA innervation, D2 receptors may act via two main pathways. First, it is well known that D2 presynaptic autoreceptors regulate DA synthesis and release [132]. Therefore, D2 stimulation may exert neuroprotective effects by decreasing DA levels, which decreases subsequent DA-induced cytotoxicity. Secondly, a number of recent studies suggest that D2 receptor stimulation may induce intracellular signaling cascades leading to neuroprotection [125]. In fact, two different D2 receptor isoforms have been reported: D2S (short) and D2L (long), which may be particularly associated with the autoregulation of DA synthesis/release and postsynaptic neuroprotective effects, respectively [133, 134].

28.11
Conclusion

In conclusion, a number of studies in our and many other laboratories have shown that adequate DA levels and adequate interaction of DA with other major neurotransmitters, particularly glutamate, are crucial for normal functioning of the basal ganglia and other brain areas, and that dysfunctions in this cross talk may lead to a number of diseases such as schizophrenia, depression, epilepsy, obesity, and others. In addition, several studies have shown that DA and L-DOPA can induce neurotoxicity and neuron death, which may be counteracted by intrinsic neuroprotective molecules or drugs. DA, particularly when acting via D2 receptors, also appears necessary for neuroprotective functions and for counteracting neurotoxic effects of other neurotransmitters such as glutamate. It appears that precise regulation of DA levels is necessary for neuron survival and function. Disruption of this equilibrium by any of a number of factors or the introduction of additional sources of oxidative stress or neuronal dysfunction may set off a cascade of events that lead to neuron degeneration and neurodegenerative diseases such as PD.

Acknowledgments

Grant sponsors: Spanish Ministry of Education (MEC), Spanish Ministry of Health (RD06/0010/0013 and Cibernet) and Galician Government (XUGA).

References

1 Carlsson, A., Falck, B., and Hillarp, N.A. (1962) Cellular localization of brain monoamines. *Acta Physiol Scand Suppl*, **56**, 1–28.
2 Eisenhofer, G., Kopin, I.J., and Goldstein, D.S. (2004) Catecholamine metabolism: a contemporary view with implications for physiology and medicine. *Pharmacol Rev*, **56**, 331–349.
3 Carlsson, A. (1959) The occurrence, distribution and physiological role of catecholamines in the nervous system. *Pharmacol Rev*, **11**, 490–493.
4 Hughes, A.J., Daniel, S.E., Kilford, L., and Lees, A.J. (1992) Accuracy of clinical diagnosis of idiopathic Parkinson's disease: a clinico-pathological study of 100 cases. *J Neurol Neurosurg Psychiatry*, **55**, 181–184.
5 Bergman, H. and Deuschl, G. (2002) Pathophysiology of Parkinson's disease: from clinical neurology to basic neuroscience and back. *Mov Disord*, **17** (Suppl 3), 28–40.
6 Olanow, C.W. (2007) The pathogenesis of cell death in Parkinson's disease. *Mov Disord*, **22** (Suppl 17), S335–S342.
7 Berg, D., Youdim, M.B., and Riederer, P. (2004) Redox imbalance. *Cell Tissue Res*, **318**, 201–213.
8 Andersen, J.K. (2004) Oxidative stress in neurodegeneration: cause or consequence. *Nat Med*, **10**, S18–S25.
9 Olanow, C.W. (1990) Oxidation reactions in Parkinson's disease. *Neurology*, **40** (Suppl 3), 32–39.
10 Fahn, S. and Cohen, G. (1992) The oxidant stress hypothesis in Parkinson's disease: evidence supporting it. *Ann Neurol*, **32**, 804–812.
11 Offen, D., Hochman, A., Gorodin, S., Ziv, I., Shirvan, A., Barzilai, A., and Melamed, E. (1999) Oxidative stress and neuroprotection in Parkinson's disease: implications from studies on dopamine-induced apoptosis. *Adv Neurol*, **80**, 265–269.
12 Barzilai, A., Melamed, E., and Shirvan, A. (2001) Is there a rationale for neuroprotection against dopamine toxicity in Parkinson's disease? *Cell Mol Neurobiol*, **21**, 215–235.
13 Ahlskog, J.E. (2007) Beating a dead horse: dopamine and Parkinson disease. *Neurology*, **69**, 1701–1711.
14 Lang, A.E. and Obeso, J.A. (2004) Time to move beyond nigrostriatal dopamine deficiency in Parkinson's disease. *Ann Neurol*, **55**, 761–765.
15 Braak, H., Del Tredici, K., Rüb, U., de Vos, R.A., Jansen Steur, E.N., and Braak, E. (2003) Staging of brain pathology related to sporadic Parkinson's disease. *Neurobiol Aging*, **24**, 197–211.
16 Braak, H., Bohl, J.R., Müller, C.M., Rüb, U., de Vos, R.A., and Del Tredici, K. (2006) Stanley Fahn Lecture 2005: the staging procedure for the inclusion body pathology associated with sporadic Parkinson's disease reconsidered. *Mov Disord*, **21**, 2042–2051.
17 Olanow, C.W. and Jankovic, J. (2005) Neuroprotective therapy in Parkinson's disease and motor complications: a search for a pathogenesis-targeted, disease-modifying strategy. *Mov Disord*, **20** (Suppl 11), S3–S10.
18 Ogawa, N., Tanaka, K., and Asanuma, M. (2000) Bromocriptine markedly suppresses levodopa-induced abnormal increase of dopamine turnover in the parkinsonian striatum. *Neurochem Res*, **25**, 755–800.

19. Pardo, B., Mena, M.A., Casarejos, M.J., Paíno, C.L., and De Yébenes, J.G. (1995) Toxic effects of L-DOPA on mesencephalic cell cultures: protection with antioxidants. *Brain Res*, **5** (682), 133–430.
20. Ogawa, N., Asanuma, M., Kondo, Y., Kawada, Y., Yamamoto, M., and Mori, A. (1994) Differential effects of chronic L-dopa treatment on lipid peroxidation in the mouse brain with or without pretreatment with 6-hydroxydopamine. *Neurosci Lett*, **25** (171), 55–80.
21. Müller, T., Hefter, H., Hueber, R., Jost, W.H., Leenders, K.L., Odin, P., and Schwarz, J. (2004) Is levodopa toxic? *J Neurol*, **251** (Suppl 6), VI/44–VI/46.
22. Datla, K.P., Blunt, S.B., and Dexter, D.T. (2001) Chronic L-DOPA administration is not toxic to the remaining dopaminergic nigrostriatal neurons, but instead may promote their functional recovery, in rats with partial 6-OHDA or FeCl(3) nigrostriatal lesions. *Mov Disord*, **16**, 424–440.
23. Mytilineou, C., Walker, R.H., JnoBaptiste, R., and Olanow, C.W. (2003) Levodopa is toxic to dopamine neurons in an in vitro but not an in vivo model of oxidative stress. *J Pharmacol Exp Ther*, **304**, 792–800.
24. Fahn, S., Oakes, D., Shoulson, I., Kieburtz, K., Rudolph, A., Lang, A., Olanow, C.W., Tanner, C., Marek, K. Parkinson Study Group (2004) Levodopa and the progression of Parkinson's disease. *N Engl J Med*, **9** (351), 2498–2508.
25. McGeer, P.L., Itagaki, S., Akiyama, H., and McGeer, E.G. (1988) Rate of cell death in parkinsonism indicates active neuropathological process. *Ann Neurol*, **24**, 574–576.
26. Fearnley, J.M. and Lees, A.J. (1991) Ageing and Parkinson's disease: substantia nigra regional selectivity. *Brain*, **114**, 2283–2301.
27. Kubis, N., Faucheux, B.A., Ransmayr, G., Damier, P., Duyckaerts, C., Henin, D., Forette, B., Le Charpentier, Y., Hauw, J.J., Agid, Y., and Hirsch, E.C. (2000) Preservation of midbrain catecholaminergic neurons in very old human subjects. *Brain*, **123**, 366–373.
28. Collier, T.J., Lipton, J., Daley, B.F., Palfi, S., Chu, Y., Sortwell, C., Bakay, R.A., Sladek, J.R. Jr and Kordower, J.H. (2007) Aging-related changes in the nigrostriatal dopamine system and the response to MPTP in nonhuman primates: diminished compensatory mechanisms as a prelude to parkinsonism. *Neurobiol Dis*, **26**, 56–65.
29. Villar-Cheda, B., Sousa-Ribeiro, D., Rodriguez-Pallares, J., Rodriguez-Perez, A.I., Guerra, M.J., and Labandeira-Garcia, J.L. (2009) Aging and sedentarism decrease vascularization and VEGF levels in the rat substantia nigra. Implications for Parkinson's disease. *J Cereb Blood Flow Metab*, **29** (2), 230–234.
30. Sulzer, D. and Rayport, S. (1990) Amphetamine and other psychostimulants reduce pH gradients in midbrain dopaminergic neurons and chromaffin granules: a mechanism of action. *Neuron*, **5**, 797–808.
31. Kumar, R., Agarwal, A.K., and Seth, P.K. (1995) Free radical-generated neurotoxicity of 6-hydroxydopamine. *J Neurochem*, **64**, 1703–1707.
32. Soto-Otero, R., Méndez-álvarez, E., Hermida-Ameijeiras, A., Muñoz-Patiño, A., and Labandeira-Garcia, J.L. (2000) Autoxidation and neurotoxicity of 6-hydroxydopamine in the presence of some antioxidants: potential implication in relation to the pathogenesis of Parkinson's disease. *J Neurochem*, **74**, 1605–1612.
33. Hanrott, K., Gudmunsen, L., O'Neill, M.J., and Wonnacott, S. (2006) 6-hydroxydopamine-induced apoptosis is mediated via extracellular auto-oxidation and caspase 3-dependent activation of protein kinase Cdelta. *J Biol Chem*, **281**, 5373–5382.
34. Rodriguez-Pallares, J., Parga, J.A., Muñoz, A., Rey, P., Guerra, M.J., and Labandeira-Garcia, J.L. (2007)

Mechanism of 6-hydroxydopamine neurotoxicity: the role of NADPH oxidase and microglial activation in 6-hydroxydopamine-induced degeneration of dopaminergic neurons. *J Neurochem*, **103**, 145–156.

35 Linert, W., Herlinger, E., Jameson, R.F., Kienzl, E., Jellinger, K., and Youdim, M.B.H. (1996) Dopamine, 6-hydroxydopamine, iron, dioxygen – their mutual interactions and possible implication in the development of Parkinson's disease. *Biochim Biophys Acta*, **1316**, 160–168.

36 Glinka, Y., Gassen, M., Youdim, M.B. (1997) Mechanism of 6-hydroxydopamine neurotoxicity. *J Neural Transm* (Suppl 50), **50**, 55–66.

37 Jellinger, K., Linert, L., Kienzl, E., Herlinger, E., and Youdim, M.B.H. (1995) Chemical evidence for 6-hydroxydopamine to be an endogenous toxic factor in the pathogenesis of Parkinson's disease. *J Neural Transm*, **46**, 297–314.

38 Irwin, I. and Langston, J.W. (1995) Endogenous toxins as potential etiologic agents in Parkinson's disease, in *Etiology of Parkinson's Disease* (eds J.H. Ellenberg, W.C. Koller, and J.W. Langston), Marcel Dekker Inc., New York, pp. 153–201.

39 Senoh, S., Creveling, C.R., Udenfriend, S., and Witkop, B. (1959) Chemical, enzymatic and metabolic studies on the mechanism of oxidation of dopamine. *J Am Chem Soc*, **81**, 6236–6240.

40 Andrew, R., Watson, D.G., Best, S.A., Midgley, J.M., Wenlong, H., and Petty, R.K.H. (1993) The determination of hydroxydopamines and other trace amines in the urine of parkinsonian patients and normal controls. *Neurochem Res*, **18**, 1175–1177.

41 Lotharius, J. and O'Malley, K.L. (2000) The parkinsonism-inducing drug 1-methyl-4-phenylpyridinium triggers intracellular dopamine oxidation. A novel mechanism of toxicity. *J Biol Chem*, **275**, 38581–38588.

42 Javitch, J.A., D'Amato, R.J., Strittmatter, S.M., and Snyder, S.H. (1985) Parkinsonism-inducing neurotoxin, N-methyl-4-phenyl-1,2,3,6-tetrahydropyridine: uptake of the metabolite N-methyl-4-phenylpyridine by dopamine neurons explains selective toxicity. *Proc Natl Acad Sci U S A*, **82**, 2173–2177.

43 Przedborski, S., Jackson-Lewis, V., Djaldetti, R., Liberatore, G., Vila, M., Slovodanka, V., and Almer, G. (2000) The parkinsonian toxin MPTP: action and mechanism. *Restor Neurol Neurosci*, **16**, 135–142.

44 Obata, T. (2002) Dopamine efflux by MPTP and hydroxyl radical generation. *J Neural Transm*, **109**, 1159–1180.

45 Filloux, F. and Townsend, J.J. (1993) Pre- and postsynaptic neurotoxic effects of dopamine demonstrated by intrastriatal injection. *Exp Neurol*, **119**, 79–88.

46 Rabinovic, A.D., Lewis, D.A., and Hastings, T.G. (2000) Role of oxidative changes in the degeneration of dopamine terminals after injection of neurotoxic levels of dopamine. *Neuroscience*, **101**, 67–76.

47 Michel, P.P. and Hefti, F. (1990) Toxicity of 6-hydroxydopamine and dopamine for dopaminergic neurons in culture. *J Neurosci Res*, **26**, 428–435.

48 Daily, D., Barzilai, A., Offen, D., Kamsler, A., Melamed, E., and Ziv, I. (1999) The involvement of p53 in dopamine-induced apoptosis of cerebellar granule neurons and leukemic cells overexpressing p53. *Cell Mol Neurobiol*, **19**, 261–276.

49 Hou, S.T., Cowan, E., Dostanic, S., Rasquinha, I., Comas, T., Morley, P., and MacManus, J.P. (2001) Increased expression of the transcription factor E2F1 during dopamine-evoked, caspase-3-mediated apoptosis in rat cortical neurons. *Neurosci Lett*, **306**, 153–156.

50 Premkumar, A. and Simantov, R. (2002) Mitochondrial voltage-dependent anion channel is involved in dopamine-induced apoptosis. *J Neurochem*, **82**, 345–352.

51. Han, S.K., Mytilineou, C., and Cohen, G. (1996) L-DOPA up-regulates glutathione and protects mesencephalic cultures against oxidative stress. *J Neurochem*, **66**, 501–510.
52. Mena, M.A., Davila, V., and Sulzer, D. (1997) Neurotrophic effects of L-DOPA in postnatal midbrain dopamine neuron/cortical astrocyte cocultures. *J Neurochem*, **69**, 1398–1408.
53. Clement, M.V., Long, L.H., Ramalingam, J., and Halliwell, B. (2002) The cytotoxicity of dopamine may be an artefact of cell culture. *J Neurochem*, **81**, 414–421.
54. Graham, D.G. (1978) Oxidative pathways for catecholamines in the genesis of neuromelanin and cytotoxic quinones. *Mol Pharmacol*, **14**, 633–643.
55. Chen, L., Ding, Y., Cagniard, B., Van Laar, A.D., Mortimer, A., Chi, W., Hastings, T.G., Kang, U.J., and Zhuang, X. (2008) Unregulated cytosolic dopamine causes neurodegeneration associated with oxidative stress in mice. *J Neurosci*, **28**, 425–433.
56. Erickson, J.D., Eiden, L.E., and Hoffman, B.J. (1992) Expression cloning of a reserpine-sensitive vesicular monoamine transporter. *Proc Natl Acad Sci U S A*, **89**, 10993–10997.
57. Miller, G.W., Erickson, J.D., Perez, J.T., Penland, S.N., Mash, D.C., Rye, D.B., and Levey, A.I. (1999) Immunochemical analysis of vesicular monoamine transporter (VMAT2) protein in Parkinson's disease. *Exp Neurol*, **156**, 138–148.
58. Caudle, W.M., Richardson, J.R., Wang, M.Z., Taylor, T.N., Guillot, T.S., McCormack, A.L., Colebrooke, R.E., Di Monte, D.A., Emson, P.C., and Miller, G.W. (2007) Reduced vesicular storage of dopamine causes progressive nigrostriatal neurodegeneration. *J Neurosci*, **27**, 8138–8148.
59. Caudle, W.M., Colebrooke, R.E., Emson, P.C., and Miller, G.W. (2008) Altered vesicular dopamine storage in Parkinson's disease: a premature demise. *Trends Neurosci*, **31**, 303–308.
60. Glatt, C.E., Wahner, A.D., White, D.J., Ruiz-Linares, A., and Ritz, B. (2006) Gain-of-function haplotypes in the vesicular monoamine transporter promoter are protective for Parkinson disease in women. *Hum Mol Genet*, **15**, 299–305.
61. Wang, Y.M., Gainetdinov, R.R., Fumagalli, F., Xu, F., Jones, S.R., Bock, C.B., Miller, G.W., Wightman, R.M., and Caron, M.G. (1997) Knockout of the vesicular monoamine transporter 2 gene results in neonatal death and supersensitivity to cocaine and amphetamine. *Neuron*, **19**, 1285–1296.
62. Mooslehner, K.A., Chan, P.M., Xu, W., Liu, L., Smadja, C., Humby, T., Allen, N.D., Wilkinson, L.S., and Emson, P.C. (2001) Mice with very low expression of the vesicular monoamine transporter 2 gene survive into adulthood: potential mouse model for parkinsonism. *Mol Cell Biol*, **21**, 5321–5331.
63. Hastings, T.G., Lewis, D.A., and Zigmond, M.J. (1996) Role of oxidation in the neurotoxic effects of intrastriatal dopamine injections. *Proc Natl Acad Sci U S A*, **93**, 1956–1961.
64. Tse, D.C., McCreery, R.L., and Adams, R.N. (1976) Potential oxidative pathways of brain catecholamines. *J Med Chem*, **19**, 37–40.
65. Miyazaki, I. and Asanuma, M. (2008) Dopaminergic neuron-specific oxidative stress caused by dopamine itself. *Acta Med Okayama*, **62**, 141–150.
66. Asanuma, M., Miyazaki, I., and Ogawa, N. (2003) Dopamine- or L-DOPA-induced neurotoxicity: the role of dopamine quinone formation and tyrosinase in a model of Parkinson's disease. *Neurotox Res*, **5**, 165–176.
67. Asanuma, M., Miyazaki, I., Diaz-Corrales, F.J., and Ogawa, N. (2004) Quinone formation as dopaminergic neuron-specific oxidative stress in the pathogenesis

of sporadic Parkinson's disease and neurotoxin-induced parkinsonism. *Acta Med Okayama*, **58**, 221–233.

68 Graumann, R., Paris, I., Martinez-Alvarado, P., Rumanque, P., Perez-Pastene, C., Cardenas, S.P., Marin, P., Diaz-Grez, F., Caviedes, R., Caviedes, P., and Segura-Aguilar, J. (2002) Oxidation of dopamine to aminochrome as a mechanism for neurodegeneration of dopaminergic systems in Parkinson's disease. Possible neuroprotective role of DT-diaphorase. *Pol J Pharmacol*, **54**, 573–579.

69 Choi, H.J., Kim, S.W., Lee, S.Y., and Hwang, O. (2003) Dopamine-dependent cytotoxicity of tetrahydrobiopterin: a possible mechanism for selective neurodegeneration in Parkinson's disease. *J Neurochem*, **86**, 143–152.

70 Korytowski, W., Sarna, T., Kalyanaraman, B., and Sealy, R.C. (1987) Tyrosinase-catalyzed oxidation of dopa and related catechol(amine)s: a kinetic electron spin resonance investigation using spin-stabilization and spin label oximetry. *Biochim Biophys Acta*, **924**, 383–392.

71 Rosei, M.A., Blarzino, C., Foppoli, C., Mosca, L., and Coccia, R. (1994) Lipoxygenase-catalyzed oxidation of catecholamines. *Biochem Biophys Res Commun*, **200**, 344–350.

72 Hastings, T.G. (1995) Enzymatic oxidation of dopamine: the role of prostaglandin H synthase. *J Neurochem*, **6**, 919–924.

73 Foppoli, C., Coccia, R., Cini, C., and Rosei, M.A. (1997) Catecholamines oxidation by xanthine oxidase. *Biochim Biophys Acta*, **1334**, 200–206.

74 Smythies, J., De Iuliis, A., Zanatta, L., and Galzigna, L. (2002) The biochemical basis of Parkinson's disease: the role of catecholamine o-quinones: a review-discussion. *Neurotox Res*, **4**, 77–81.

75 Whitehead, R.E., Ferrer, J.V., Javitch, J.A., and Justice, J.B. (2001) Reaction of oxidized dopamine with endogenous cysteine residues in the human dopamine transporter. *J Neurochem*, **76**, 1242–1251.

76 LaVoie, M.J., Ostaszewski, B.L., Weihofen, A., Schlossmacher, M.G., and Selkoe, D.J. (2005) Dopamine covalently modifies and functionally inactivates parkin. *Nat Med*, **11**, 1214–1221.

77 Conway, K.A., Rochet, J.C., Bieganski, R.M., and Lansbury, P.T. Jr (2001) Kinetic stabilization of the alpha-synuclein protofibril by a dopamine-alpha-synuclein adduct. *Science*, **294**, 1346–1349.

78 Volles, M.J., Lee, S.J., Rochet, J.C., Shtilerman, M.D., Ding, T.T., Kessler, J.C., and Lansbury, P.T. Jr (2001) Vesicle permeabilization by protofibrillar alpha-synuclein: implications for the pathogenesis and treatment of Parkinson's disease. *Biochemistry*, **40**, 7812–7819.

79 Kuhn, D.M., Arthur, R.E. Jr, Thomas, D.M., and Elferink, L.A. (1999) Tyrosine hydroxylase is inactivated by catechol-quinones and converted to a redox-cycling quinoprotein: possible relevance to Parkinson's disease. *J Neurochem*, **73**, 1309–1317.

80 Offen, D., Ziv, I., Sternin, H., Melamed, E., and Hochman, A. (1996) Prevention of dopamine-induced cell death by thiol antioxidants: possible implications for treatment of Parkinson's disease. *Exp Neurol*, **141**, 32–39.

81 Muñoz, A.M., Rey, P., Soto-Otero, R., Guerra, M.J. and Labandeira-Garcia, J.L. (2004) Systemic administration of N-acetylcysteine protects dopaminergic neurons against 6-hydroxydopamine-induced degeneration. *J Neurosci Res*, **76**, 551–562.

82 Zucca, F.A., Giaveri, G., Gallorini, M., Albertini, A., Toscani, M., Pezzoli, G., Lucius, R., Wilms, H., Sulzer, D., Ito, S., Wakamatsu, K., and Zecca, L. (2004) The neuromelanin of human substantia nigra: physiological and pathogenic aspects. *Pigment Cell Res*, **17**, 610–6177.

83 Zecca, L., Mecacci, C., Seraglia, R., and Parati, E. (1992) The chemical

characterization of melanin contained in substantia nigra of human brain. *Biochim Biophys Acta*, **1138**, 6–10.
84 Zecca, L., Casella, L., Albertini, A., Bellei, C., Zucca, F.A., Engelen, M., Zadlo, A., Szewczyk, G., Zareba, M., and Sarna, T. (2008) Neuromelanin can protect against iron-mediated oxidative damage in system modeling iron overload of brain aging and Parkinson's disease. *J Neurochem*, **106**, 1866–1875.
85 Wakamatsu, K., Fujikawa, K., Zucca, F.A., Zecca, L., and Ito, S. (2003) The structure of neuromelanin as studied by chemical degradative methods. *J Neurochem*, **86**, 1015–1023.
86 Kastner, A., Hirsch, E.C., Lejeune, O., Javoy-Agid, F., Rascol, O., and Agid, Y. (1992) Is the vulnerability of neurons in the substantia nigra of patients with Parkinson's disease related to their neuromelanin content? *J Neurochem*, **59**, 1080–1089.
87 Halliday, G.M., Ophof, A., Broe, M., Jensen, P.H., Kettle, E., Fedorow, H., Cartwright, M.I., Griffiths, F.M., Shepherd, C.E., and Double, K.L. (2005) Alpha-synuclein redistributes to neuromelanin lipid in the substantia nigra early in Parkinson's disease. *Brain*, **128**, 2654–2664.
88 Hirsch, E., Graybiel, A.M., and Agid, Y.A. (1988) Melanized dopaminergic neurons are differentially susceptible to degeneration in Parkinson's disease. *Nature*, **334**, 345–348.
89 Wakabayashi, K., Tanji, K., Mori, F., and Takahashi, H. (2007) The Lewy body in Parkinson's disease: molecules implicated in the formation and degradation of alpha-synuclein aggregates. *Neuropathology*, **27**, 494–506.
90 Zecca, L., Zucca, F.A., Wilms, H., and Sulzer, D. (2003) Neuromelanin of the substantia nigra: a neuronal black hole with protective and toxic characteristics. *Trends Neurosci*, **26**, 578–580.
91 Zareba, M., Bober, A., Korytowski, W., Zecca, L., and Sarna, T. (1995) The effect of a synthetic neuromelanin on yield of free hydroxyl radicals generated in model systems. *Biochim Biophys Acta*, **1271**, 343–348.
92 Langston, J.W., Forno, L.S., Tetrud, J., Reeves, A.G., Kaplan, J.A., and Karluk, D. (1999) Evidence of active nerve cell degeneration in the substantia nigra of humans years after 1-methyl-4-phenyl-1,2,3,6-tetrahydropyridine exposure. *Ann Neurol*, **46**, 598–605.
93 Block, M.L. and Hong, J.S. (2005) Microglia and inflammation-mediated neurodegeneration: multiple triggers with a common mechanism. *Prog Neurobiol*, **76**, 77–98.
94 Fedorow, H., Pickford, R., Kettle, E., Cartwright, M., Halliday, G.M., Gerlach, M., Riederer, P., Garner, B., and Double, K.L. (2006) Investigation of the lipid component of neuromelanin. *J Neural Transm*, **113**, 735–739.
95 Spillantini, M.G., Schmidt, M.L., Lee, V.M., Trojanowski, J.Q., Jakes, R., and Goedert, M. (1997) Alpha-synuclein in Lewy bodies. *Nature*, **388**, 839–840.
96 Kirik, D., Rosenblad, C., Burger, C., Lundberg, C., Johansen, T.E., Muzyczka, N., Mandel, R.J., and Björklund, A. (2002) Parkinson-like neurodegeneration induced by targeted overexpression of alpha-synuclein in the nigrostriatal system. *J Neurosci*, **22**, 2780–2791.
97 Kirik, D., Annett, L.E., Burger, C., Muzyczka, N., Mandel, R.J., and Björklund, A. (2003) Nigrostriatal alpha-synucleinopathy induced by viral vector-mediated overexpression of human alpha-synuclein: a new primate model of Parkinson's disease. *Proc Natl Acad Sci U S A*, **100**, 2884–2889.
98 Xu, J., Kao, S.Y., Lee, F.J., Song, W., Jin, L.W., and Yankner, B.A. (2002) Dopamine-dependent neurotoxicity of alpha-synuclein: a mechanism for selective neurodegeneration

in Parkinson disease. *Nat Med*, **8**, 600–606.

99 Mazzulli, J.R., Mishizen, A.J., Giasson, B.I., Lynch, D.R., Thomas, S.A., Nakashima, A., Nagatsu, T., Ota, A., and Ischiropoulos, H. (2006) Cytosolic catechols inhibit alpha-synuclein aggregation and facilitate the formation of intracellular soluble oligomeric intermediates. *J Neurosci*, **26**, 10068–10078.

100 Follmer, C., Romão, L., Einsiedler, C.M., Porto, T.C., Lara, F.A., Moncores, M., Weissmüller, G., Lashuel, H.A., Lansbury, P., Neto, V.M., Silva, J.L., and Foguel, D. (2007) Dopamine affects the stability, hydration, and packing of protofibrils and fibrils of the wild type and variants of alpha-synuclein. *Biochemistry*, **46**, 472–482.

101 Norris, E.H., Giasson, B.I., Hodara, R., Xu, S., Trojanowski, J.Q., Ischiropoulos, H., and Lee, V.M. (2005) Reversible inhibition of alpha-synuclein fibrillization by dopaminochrome-mediated conformational alterations. *J Biol Chem*, **280**, 21212–21219.

102 Giasson, B.I., Duda, J.E., Murray, I.V., Chen, Q., Souza, J.M., Hurtig, H.I., Ischiropoulos, H., Trojanowski, J.Q., and Lee, V.M. (2000) Oxidative damage linked to neurodegeneration by selective alpha-synuclein nitration in synucleinopathy lesions. *Science*, **290**, 985–989.

103 Lotharius, J. and Brundin, P. (2002) Impaired dopamine storage resulting from alpha-synuclein mutations may contribute to the pathogenesis of Parkinson's disease. *Hum Mol Genet*, **11**, 2395–2407.

104 Mosharov, E.V., Staal, R.G., Bové, J., Prou, D., Hananiya, A., Markov, D., Poulsen, N., Larsen, K.E., Moore, C.M., Troyer, M.D., Edwards, R.H., Przedborski, S., and Sulzer, D. (2006) Alpha-synuclein overexpression increases cytosolic catecholamine concentration. *J Neurosci*, **26**, 9304–9311.

105 Guo, J.T., Chen, A.Q., Kong, Q., Zhu, H., Ma, C.M., and Qin, C. (2008) Inhibition of vesicular monoamine transporter-2 activity in alpha-synuclein stably transfected SH-SY5Y cells. *Cell Mol Neurobiol*, **28**, 35–47.

106 Machida, Y., Chiba, T., Takayanagi, A., Tanaka, Y., Asanuma, M., Ogawa, N., Koyama, A., Iwatsubo, T., Ito, S., Jansen, P.H., Shimizu, N., Tanaka, K., Mizuno, Y., and Hattori, N. (2005) Common anti-apoptotic roles of parkin and alpha-synuclein in human dopaminergic cells. *Biochem Biophys Res Commun*, **332**, 233–240.

107 Chen, J., Wersinger, C., and Sidhu, A. (2003) Chronic stimulation of D1 dopamine receptors in human SK-N-MC neuroblastoma cells induces nitric-oxide synthase activation and cytotoxicity. *J Biol Chem*, **278**, 28089–28100.

108 Davis, S., Brotchie, J., and Davies, I. (2002) Protection of striatal neurons by joint blockade of D1 and D2 receptor subtypes in an in vitro model of cerebral hypoxia. *Exp Neurol*, **176**, 229–236.

109 Charvin, D., Vanhoutte, P., Pagès, C., Borrelli, E., and Caboche, J. (2005) Unraveling a role for dopamine in Huntington's disease: the dual role of reactive oxygen species and D2 receptor stimulation. *Proc Natl Acad Sci U S A*, **102**, 12218–12223.

110 Cyr, M., Beaulieu, J.M., Laakso, A., Sotnikova, T.D., Yao, W.D., Bohn, L.M., Gainetdinov, R.R., and Caron, M.G. (2003) Sustained elevation of extracellular dopamine causes motor dysfunction and selective degeneration of striatal GABAergic neurons. *Proc Natl Acad Sci U S A*, **100**, 11035–11040.

111 Vonsattel, J.P., Myers, R.H., Stevens, T.J., Ferrante, R.J., Bird, E.D., and Richardson, E.P. Jr (1985) Neuropathological classification of Huntington's disease. *J Neuropathol Exp Neurol*, **44**, 559–577.

112 The Huntington's Disease Collaborative Research Group (1993) A novel gene containing a trinucleotide repeat that is expanded

and unstable on Huntington's disease chromosomes. *Cell*, **72**, 971–983.
113 Jakel, R.J. and Maragos, W.F. (2000) Neuronal cell death in Huntington's disease: a potential role for dopamine. *Trends Neurosci*, **23**, 239–245.
114 Ellison, D.W., Beal, M.F., Mazurek, M.F., Malloy, J.R., Bird, E.D., and Martin, J.B. (1987) Amino acid neurotransmitter abnormalities in Huntington's disease and the quinolinic acid animal model of Huntington's disease. *Brain*, **110**, 1657–1673.
115 Klawans, H.L., Goetz, C.G., Paulson, G.W., and Barbeau, A. (1980) Levodopa and presymptomatic detection of Huntington's disease-eight-year follow-up. *N Engl J Med*, **302**, 1090.
116 Cass, W.A. (1997) Decreases in evoked overflow of dopamine in rat striatum after neurotoxic doses of methamphetamine. *J Pharmacol Exp Ther*, **280**, 105–113.
117 Filloux, F. and Wamsley, J.K. (1991) Dopaminergic modulation of excitotoxicity in rat striatum: evidence from nigrostriatal lesions. *Synapse*, **8**, 281–288.
118 Buisson, A., Pateau, V., Plotkine, M., and Boulu, R.G. (1991) Nigrostriatal pathway modulates striatum vulnerability to quinolinic acid. *Neurosci Lett*, **131**, 257–259.
119 Kerkerian, L., Dusticier, N., and Nieoullon, A. (1987) Modulatory effect of dopamine on high-affinity glutamate uptake in the rat striatum. *J Neurochem*, **48**, 1301–1306.
120 Cepeda, C., Colwell, C.S., Itri, J.N., Chandler, S.H., and Levine, M.S. (1998) Dopaminergic modulation of NMDA-induced whole cell currents in neostriatal neurons in slices: contribution of calcium conductances. *J Neurophysiol*, **79**, 82–94.
121 Guerra, M.J., Liste, I., and Labandeira-Garcia, J.L. (1998) Interaction between the serotonergic, dopaminergic, and glutamatergic systems in fenfluramine-induced Fos expression in striatal neurons. *Synapse*, **28**, 71–82.
122 Liste, I., Muñoz, A., Guerra, M.J., and Labandeira-Garcia, J.L. (2000) Fenfluramine-induced increase in preproenkephalin mRNA levels in the striatum: interaction between the serotonergic, glutamatergic, and dopaminergic systems. *Synapse*, **35**, 182–191.
123 Morari, M., Marti, M., Sbrenna, S., Fuxe, K., Bianchi, C., and Beani, L. (1998) Reciprocal dopamine-glutamate modulation of release in the basal ganglia. *Neurochem Int*, **33**, 383–397.
124 Amano, T., Ujihara, H., Matsubayashi, H., Sasa, M., Yokota, T., Tamura, Y., and Akaike, A. (1994) Dopamine-induced protection of striatal neurons against kainate receptor-mediated glutamate cytotoxicity in vitro. *Brain Res*, **655**, 61–69.
125 Bozzi, Y. and Borrelli, E. (2006) Dopamine in neurotoxicity and neuroprotection: what do D2 receptors have to do with it? *Trends Neurosci*, **29**, 167–174.
126 Sawada, H., Ibi, M., Kihara, T., Urushitani, M., Akaike, A., Kimura, J., and Shimohama, S. (1998) Dopamine D2-type agonists protect mesencephalic neurons from glutamate neurotoxicity: mechanisms of neuroprotective treatment against oxidative stress. *Ann Neurol*, **44**, 110–119.
127 Takashima, H., Tsujihata, M., Kishikawa, M., and Freed, W.J. (1999) Bromocriptine protects dopaminergic neurons from levodopa-induced toxicity by stimulating D(2)receptors. *Exp Neurol*, **159**, 98–104.
128 Mitchell, I.J., Cooper, A.C., Griffiths, M.R., and Cooper, A.J. (2002) Acute administration of haloperidol induces apoptosis of neurones in the striatum and substantia nigra in the rat. *Neuroscience*, **109**, 89–99.
129 Schauwecker, P.E. and Steward, O. (1997) Genetic determinants of susceptibility to excitotoxic cell death:

130 implications for gene targeting approaches. *Proc Natl Acad Sci U S A*, **94**, 4103–4108.
130 Parkinson Study Group (2002) Dopamine transporter brain imaging to assess the effects of pramipexole vs levodopa on Parkinson disease progression. *J Am Med Assoc*, **287**, 1653–1661.
131 Whone, A.L., Watts, R.L., Stoessl, A.J., Davis, M., Reske, S., Nahmias, C., Lang, A.E., Rascol, O., Ribeiro, M.J., Remy, P., Poewe, W.H., Hauser, R.A., and Brooks, D.J. REAL-PET Study Group (2003) Slower progression of Parkinson's disease with ropinirole versus levodopa: the REAL-PET study. *Ann Neurol*, **54**, 93–101.
132 Rouge-Pont, F., Usiello, A., Benoit-Marand, M., Gonon, F., Piazza, P.V., and Borrelli, E. (2002) Changes in extracellular dopamine induced by morphine and cocaine: crucial control by D2 receptors. *J Neurosci*, **22**, 3293–3301.
133 Usiello, A., Baik, J.H., Rougé-Pont, F., Picetti, R., Dierich, A., LeMeur, M., Piazza, P.V., and Borrelli, E. (2000) Distinct functions of the two isoforms of dopamine D2 receptors. *Nature*, **408**, 199–203.
134 Lindgren, N., Usiello, A., Goiny, M., Haycock, J., Erbs, E., Greengard, P., Hokfelt, T., Borrelli, E., and Fisone, G. (2003) Distinct roles of dopamine D2L and D2S receptor isoforms in the regulation of protein phosphorylation at presynaptic and postsynaptic sites. *Proc Natl Acad Sci U S A*, **100**, 4305–4309.

29
Dopamine Catabolism and Parkinson's Disease: Role of a Reactive Aldehyde Intermediate
Jonathan A. Doorn

Numerous studies implicate oxidative stress in the etiology of Parkinson's disease (PD), and oxidative stress is hypothesized to originate from environmental insult(s), such as exposure to pesticides. Individuals with PD were found to have increased oxidative burden and higher levels of lipid peroxidation products such as 4-hydroxy-2-nonenal (4HNE), as compared to controls. Although it is clear that oxidative stress is a factor in PD pathogenesis, the mechanistic link between oxidative stress and dopamine (DA) neuron degeneration is unknown. Furthermore, the cause of the initial selective neurodegeneration in PD, that is, death of DA neurons, is not clear; however, some have suggested that a neurotoxin endogenous to DA neurons, such as DA, is responsible. It is proposed that a reactive intermediate of DA catabolism, that is, 3,4-dihydroxyphenylacetaldehyde (DOPAL), may represent that mechanistic link as the metabolism for this toxic intermediate is significantly inhibited by products of oxidative stress. In addition, DOPAL was demonstrated to be orders of magnitude more toxic to DA cells than DA. Several mechanisms for DOPAL-mediated cytotoxicity have been proposed over the years. Given the reactive groups on DOPAL, it is conceivable that the DA-derived neurotoxin modifies protein amines via the aldehyde or thiols subsequent to catechol oxidation. Therefore, DOPAL could serve as a "chemical trigger" for some stage of PD pathogenesis and be an attractive therapeutic target. In addition, the DOPAL–protein adduct could serve as a biomarker for initiation of events related to PD and provide a mechanism for earlier disease diagnosis. The role of DOPAL in PD warrants further study and may yield novel and innovative means to diagnose and treat/halt PD.

29.1
Dopamine Biosynthesis and Catabolism

Once presumed to be merely an intermediate in the synthesis of norepinephrine, dopamine (DA) is now recognized an important neurotransmitter involved in several vital physiological functions, including coordination of voluntary movement, behavior, reward, and hormonal production [1].

Endogenous Toxins. Diet, Genetics, Disease and Treatment.
Edited by Peter J. O'Brien and W. Robert Bruce
Copyright © 2010 WILEY-VCH Verlag GmbH & Co. KGaA, Weinheim
ISBN: 978-3-527-32363-0

DA synthesis and catabolism occurs primarily at the nerve terminal [2, 3] and begins with the hydroxylation of the amino acid L-tyrosine yielding L-3,4-dihydroxyphenylalanine (L-DOPA), and this reaction is catalyzed by the enzyme tyrosine hydroxylase (TH). This hydroxylation reaction, which requires molecular oxygen and tetrahydrobiopterin, is the rate-limiting step in DA biosynthesis and can be modulated via several factors including the action of autoreceptors and concentration of catecholamines. The enzyme L-aromatic amino acid decarboxylase with the cofactor pyridoxal phosphate catalyzes decarboxylation of L-DOPA to DA. Following biosynthesis of DA, the neurotransmitter is rapidly taken up into vesicles via the action of the vesicular monoamine transporter (VMAT2) for the purpose of synaptic signaling and, importantly, for cellular protection against reactive intermediates that can be generated via free, cytosolic DA.

Catabolism of DA generally takes place in the cells in which the neurotransmitter is produced and occurs for free, nonvesicular DA (i.e., cytosolic) [2–4]. It is presumed that DA catabolism takes place mainly when catecholamines leak, whether passive or induced, from vesicles; however, in addition, reuptake of DA from the synapse may also contribute to this degradation pathway [2, 4].

As shown in Figure 29.1, cytosolic DA undergoes oxidative deamination, a reaction catalyzed by the flavin-containing enzyme monoamine oxidase (MAO), which requires molecular oxygen for activity and resides on the outer mitochondrial membrane in the cytosol [2, 4]. It is known that there are two forms of MAO (i.e., A and B), and the contribution of each toward neurotransmitter catabolism has been extensively studied in various mammalian species [5–7]. Both MAO A and B exist in the mammalian brain in distinct locations, and it is only A that is found in dopaminergic terminals [8–10]. In rodent models, oxidative deamination of DA is primarily due to MAO A under both normal conditions and increased DA levels due to L-DOPA treatment [5]. Despite the lower ratio of MAO A/MAO B in humans and nonhuman primates [8], DA metabolism appears to preferentially involve MAO A in such species [6]. However, the contribution of MAO B toward DA deamination should not be underestimated or minimized as the activity of this isoform is important for catecholamine metabolism, especially after L-DOPA

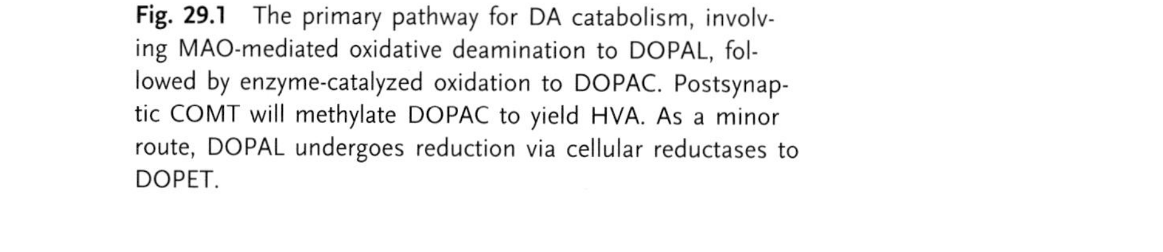

Fig. 29.1 The primary pathway for DA catabolism, involving MAO-mediated oxidative deamination to DOPAL, followed by enzyme-catalyzed oxidation to DOPAC. Postsynaptic COMT will methylate DOPAC to yield HVA. As a minor route, DOPAL undergoes reduction via cellular reductases to DOPET.

treatment [6]. In addition, the effectiveness of MAO B inhibitors (e.g., deprenyl) in treatment of Parkinson's disease (PD) has been demonstrated by numerous clinical studies [11].

Three products result from the MAO reaction involving reduction of molecular oxygen and amine-oxidation of DA: ammonia, hydrogen peroxide, and 3,4-dihydroxyphenylacetaldehyde (DOPAL). It should be noted that oxidative deamination is a two-step process that entails oxidation of the amine to an imine, followed by hydrolysis and rearrangement to an aldehyde, and therefore the oxygen on DOPAL originates from water and not molecular oxygen [12].

Ammonia formed via this enzymatic reaction can be removed via the urea cycle in the mitochondria. Hydrogen peroxide is a reactive oxygen species (ROS) that can cause deleterious consequences if not dealt with via the activity of catalase, catalyzing conversion of the hydrogen peroxide to water and molecular oxygen.

The other product, DOPAL, contains an aldehyde group that can undergo (primarily) oxidation to an acid (3,4-dihydroxyphenylacetic acid (DOPAC)) or reduction to an alcohol (3,4-dihydroxyphenylethanol (DOPET)) catalyzed by aldehyde dehydrogenase (ALDH) or aldehyde/aldose reductase (AR), respectively [2, 4]. DOPAC subsequently undergoes methylation at the 3-OH catalyzed by catechol-O-methyl transferase (COMT), yielding homovanillic acid (HVA). The enzyme COMT requires the cosubstrate S-adenosyl methionine as the methyl donor and is located postsynaptically [13].

Other pathways do exist for DA catabolism: methylation of DA, yielding 3-methoxy-tyramine, and sulfation of DA or DOPAC catalyzed by sulfotransferases [2, 4]. Interestingly, O'Leary and Baughn demonstrated that as a minor reaction, DOPA decarboxylase can catalyze decarboxylation-dependent transamination of L-DOPA directly to DOPAL [14].

Several transporters are involved in the DA catabolic pathway in addition to VMAT2 [2, 4]. The dopamine transporter (DAT) is responsible for the uptake of synaptic DA following the release of vesicular DA. It was found that DOPAL inhibited the uptake of DA but not the neurotransmitter γ-aminobutyric acid into neostriatal synaptosomes, indicating selective interaction of the DA-derived aldehyde with DAT [15]. Subsequent experiments demonstrated that mazindol, a DAT inhibitor, prevented the uptake of tritiated DOPAL into rat neostriatal synaptosomes, which suggests selective uptake of the aldehyde into dopaminergic cells [15]. However, it should be noted that micromolar levels of mazindol are required to inhibit DOPAL uptake (i.e., 5 µM mazindol needed to block ~50% DOPAL uptake).

In addition to the DA transporter, cells contain a sulfonyl-urea-sensitive DOPAC transporter responsible for the rapid efflux of DOPAC [16, 17]. However, it is not known at this point whether or not the transporter is inhibited by DOPAL or participates in efflux of the DA-derived aldehyde.

29.2
DOPAL as a Metabolite of DA: Importance to Pharmacology and Toxicology

As noted above, the oxidative deamination of "biogenic amines" (e.g., the catecholamine DA) catalyzed by MAO yields "biogenic aldehydes." Oxidative deamination was first described in the late 1920s through the 1930s, and by the mid-1930s, it was known that an aldehyde is the product [18–21]. In the following years, minimal work was performed to characterize the biochemistry of DOPAL, in part because the parent compound (i.e., DA) was viewed as merely an intermediate in the biosynthesis of norepinephrine and epinephrine [1]. Therefore, it was not until much time later that the pharmacological and toxicological properties of DOPAL were realized.

A review published in the 1950s proposed DOPAL to be toxic to cells in which it is produced [12]; however, a subsequent report some years later minimized the relevance of these "biogenic aldehydes" to toxicity and disease [22]. Another study demonstrated the production of DOPAL following treatment of epinephrine with acid and speculated that such a rearrangement was a biochemical pathway for catecholamines, but noted that the epinephrine-derived aldehyde was much less pharmacologically active than epinephrine [23].

During the 1970s and 1980s there was a resurgence in DOPAL interest with the evolving field of ethanol addiction as part of what was termed the *biogenic aldehyde hypothesis*. [24] According to the "biogenic aldehyde hypothesis," an increase in steady-state level of a biogenic aldehyde (e.g., DOPAL) yields central nervous system effects similar to those observed for ethanol. It was proposed that acetaldehyde, a product of ethanol metabolism, inhibited ALDH, yielding an increase in DOPAL concentration. Elevated levels of DOPAL were presumed to mediate effects via several mechanisms, including protein modification and/or enzyme inhibition [24, 25], condensation with amines to form pharmacologically active substances (e.g., DA with DOPAL to form tetrahydropapaveroline) [26], and direct action of DOPAL on cell receptors [27].

Subsequent work demonstrated the surprising resistance of cellular aldehyde metabolism to acetaldehyde, and, in particular, that DOPAL oxidation was not inhibited by physiological concentrations of acetaldehyde [28, 29]. In addition, these studies showed the presence of multiple forms of ALDH and AR enzymes that may compensate for DOPAL metabolism should one enzyme or form of the enzyme be inhibited [30, 31].

During the last 10–15 years, several key studies have been published demonstrating DOPAL to be toxic to dopaminergic cells, both *in vitro* and *in vivo* [32, 33]. The data generated implicated the DA-derived aldehyde in neurodegenerative disease, such as PD, to explain the selective degeneration of DA neurons. These studies found that DOPAL was more toxic to DA cells than DA, considered to be an endogenous neurotoxin, and several DA metabolites (e.g., DOPAC and DOPET).

In addition, the concentration of DOPAL was measured in postmortem tissue (i.e., substantia nigra) and found to be $\sim$2 µM [32, 34], which is of concern, given that observed cytotoxicity for the aldehyde has been measured to be as low as

6.6 μM [35]. Should the *in vivo* steady-state level of DOPAL be correct, it may indicate that DA neurons function in a "near-compromised" state and be highly dependent on efficient DA metabolism and trafficking. Furthermore, dopaminergic cells may be highly vulnerable to environmental or pharmacological agents that target DA metabolism or trafficking, yielding elevated levels of DOPAL, and such a mechanism may have implications for PD pathogenesis.

29.3
DOPAL Metabolism

As shown in Figure 29.1, the primary route for metabolism/detoxication of DOPAL is aldehyde oxidation catalyzed by cellular ALDH using NAD+ [2, 4]. Numerous reports indicate that the mitochondrial form(s) of the enzyme may play a predominant role [36–38], suggesting that detoxication of the aldehyde may be dependent on complex I for production of NAD+. Therefore, a decrease in or inhibition of complex I activity has the potential to adversely affect ALDH activity and DOPAL metabolism. Indeed, the pesticide rotenone, a complex I inhibitor, was demonstrated to inhibit DOPAL oxidation in mitochondria [29] and cause a significant elevation in DOPAL upon treatment of PC12 cells [39, 40]. K_m values have been measured for mitochondrial ALDHs and found to be in the sub to low micromolar range, in the range of normal physiological levels of DOPAL [37].

In addition to mitochondrial ALDH participating in DOPAL metabolism, there are multiple other forms of the enzyme that can catalyze oxidation of the aldehyde [37]. Numerous studies have demonstrated the presence of DOPAL-metabolizing ALDHs in both brain mitochondria and cytosol [30, 31]. Such multiplicity may serve as a protecting or compensating mechanism should one form of the enzyme be inhibited (e.g., acetaldehyde) or inactive (e.g., ALDH2 polymorphism in Asian population). K_m values have been measured for cytosolic ALDH and found to be in the sub to low micromolar range, and therefore the cytosolic enzyme may also play a significant role in DOPAL oxidation [37]. Interestingly, a decrease in *ALDH1A1* (cytosolic) gene expression has been observed in PD patients, which raises the question of DOPAL metabolism in disease-compromised individuals [41]. For more information on the topic of ALDH enzymes involved in DOPAL metabolism, the reader is referred to the excellent review of Marchitti *et al.* [37].

In addition to enzyme-catalyzed oxidation, the brain contains AR, which catalyzes the reduction of aldehydes such as DOPAL using the cofactor NADPH [42, 43]. This reaction occurs in the cytosol and is presumed to constitute a minor pathway for metabolism of the aldehyde. AR "prefers" the MAO metabolite of norepinephrine and epinephrine (i.e., 3,4-dihydroxyphenylglycolaldehyde) with the hydroxyl group β-position to the aldehyde [2, 4, 43]. However, the reductases may serve a compensatory role in DOPAL metabolism should the ALDH enzymes be inhibited [44].

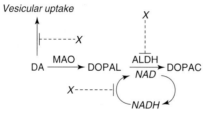

Fig. 29.2 Mechanisms for generation of DOPAL at aberrant levels. The "X" denotes targets in the DA catabolic pathway known to be affected by toxicants and/or oxidative stress.

29.4
Mechanisms for Elevation in Levels of DOPAL

The generation and metabolism of DOPAL is a normal part of the DA pathway, therefore it is of question how or why DOPAL could be generated at levels harmful to dopaminergic cells. As noted above, the measured steady-state concentration of DOPAL (~2 μM) is not much lower than the level at which cytotoxicity has been observed in DA cell models (6.6 μM) [32, 34, 35]. Therefore, a significant insult may not be needed to perturb DOPAL metabolism resulting in an elevated concentration of the aldehyde to the point of being harmful. Several mechanisms are predicted to be responsible for elevated DOPAL levels, as shown in Figure 29.2.

A potentially sensitive target that is predicted to impair DA catabolism is ALDH. As noted above, any disturbances in complex I activity will decrease availability of the cofactor NAD+ and therefore, ALDH-mediated oxidation of DOPAL [39, 40]. In addition, the ALDH enzymes are potently inhibited by the by-products of oxidative stress, such as the lipid peroxidation products 4-hydroxy-2-nonenal (4HNE) and malondialdehyde (MDA), and therefore these agents may impair DOPAL oxidation as shown in Figure 29.2 [45, 46].

Both 4HNE and MDA are major products of lipid peroxidation and often considered to be "gold standards" of oxidative stress [47]. Numerous studies implicate oxidative stress in PD pathogenesis, and elevated levels of 4HNE have been observed in brains of PD patients [48, 49]. The oxidative stress in PD may be the result of exposure to environmental agents such as pesticides [50, 51].

An earlier report demonstrated 4HNE to be a potent, mixed-type inhibitor of rat brain mitochondrial ALDH ($K_i = 0.5$ μM) [46]. Although the active site of ALDH contains a Cys nucleophile, this residue is not covalently modified by 4HNE, an α,β-unsaturated aldehyde, unless the 4HNE concentration is >50 μM [52]. In addition, MDA was found to irreversibly inactivate ALDH (in low micromolar range); however, the mechanism was not reported [45]. Both 4HNE and MDA are lipid and water soluble and have the ability to readily cross biological membranes [47]. Therefore, it is predicted that even low concentrations of 4HNE and/or MDA will inhibit ALDH activity toward DOPAL (Figure 29.2). However, as noted above, there are multiple forms of ALDH in cytosol and mitochondria, and it is of question whether or not all or some of these are adversely affected by lipid peroxidation products [37].

Recent work has confirmed this hypothesis (i.e., lipid aldehyde inhibit ALDH) using isolated rat brain mitochondria and rat striatal synaptosomes [53, 54]. Rat brain mitochondria were supplemented with DA to generate DOPAL *in situ* and treated with physiologic concentrations of 4HNE [53]. It was found that 4HNE inhibited ALDH activity (i.e., DOPAL oxidation) in a dose-dependent manner; however, the lipid peroxidation product had no effect on MAO, even at concentrations up to 150 µM. The latter observation is important as it demonstrates that 4HNE targets the second step in DA catabolism (i.e., ALDH-mediated oxidation) while not affecting the first step (i.e., MAO-catalyzed oxidative deamination).

A subsequent paper further demonstrated 4HNE and MDA-mediated impairment of DOPAL metabolism using rat striatal synaptosomes supplemented with DA to generate DOPAL *in situ* [54]. It was found that both 4HNE and MDA, at physiologic concentrations (i.e., 5–50 µM), inhibited DOPAL oxidation in a dose-dependent manner, yielding a significant elevation in the level of DOPAL over the course of 1 hour. Interestingly, the rise in DOPAL was found to correlate with increased catechol adducts on proteins, with protein modification being dependent on MAO activity. Such a finding indicates that the adducts were derived from the reaction of DOPAL with proteins.

DOPAL may be a self-propagating species as the aldehyde was found to inhibit human recombinant mitochondrial ALDH at higher concentrations (>20 µM) [53]. The mechanism of inhibition was not reported; however, earlier work demonstrated that DOPAL covalently modifies ALDH peptides, suggesting that the DA-derived aldehyde may be an irreversible inactivator at high enough levels [55]. In addition, DOPAL was found to induce dose-dependent release of DA from PC12 cells, which might yield increased turnover of DA and higher levels of the DA-derived aldehyde [56].

In addition, other mechanisms for elevation of DOPAL may exist, including toxicant-mediated release of DA from vesicles or inhibition of VMAT2. Methamphetamine (METH) or pesticides, such as the organochlorine dieldrin, that are known to interfere with DA trafficking may be capable of increasing levels of the DA-derived aldehyde. METH is known to inhibit VMAT2-mediated sequestering of DA into vesicles as well as diffuse the proton gradient necessary for uptake of the catecholamine given its potential to act as a weak base [57–59].

Recent work demonstrated that treatment of dopaminergic cells with dieldrin, a pesticide associated with PD incidence that was shown to selectively damage DA neurons *in vivo*, yields increased DA release and higher extracellular DOPAC, which suggests elevated DOPAL levels [60, 61]. In support of this contention, it has been observed that incubation of dopaminergic PC6-3 cells with low micromolar concentrations of the organochlorine (not causing significant cytotoxicity) results in an elevated extracellular concentration of DOPAL [62].

It should be noted that xenobiotics may cause an elevation in DOPAL via several means, as shown in Figure 29.2. For example, it has been reported that dieldrin causes oxidative stress in dopaminergic cells and has the potential to inhibit the electron transport system [50, 60, 61]. This accumulating evidence suggests that DOPAL may represent a mechanistic link between environmental insult and damage to DA neurons, with relevance to PD.

29.5
DOPAL Toxicity

Numerous reports have demonstrated the toxicity of DOPAL toward dopaminergic cells, with adverse effects being observed at concentrations as low as 6.6 µM [32, 33]. The toxicity of DOPAL has been demonstrated both *in vitro* and *in vivo*, and found to be much greater than that measured for DA. Several mechanisms for DOPAL toxicity have been proposed, which include generation of toxic radicals (i.e., hydroxyl radical), mitochondrial pore transition, and protein modification [32, 33]. Other means for DOPAL to cause cellular dysfunction are possible, such as DNA modification, which has been previously reported [63]. A paper by Li *et al.* demonstrated that in the presence of hydrogen peroxide, a by-product of DA catabolism, DOPAL produced the extremely reactive hydroxyl radical, as determined using salicylic acid as a radical trap [64]. Interestingly, this effect or activity was specific for DOPAL, as neither DA nor DA metabolites (e.g., DOPAC and DOPET) were capable of generating hydroxyl radical.

Kristal *et al.* found that DOPAL acts at the mitochondrial level by inducing permeability transition at physiologic levels [35]. This is in contrast to DA, which had no consistent effects except at very high concentrations (i.e., >500 µM). To further support such a mechanism, it was demonstrated that mitochondrial permeability transition inhibitors (e.g., cyclosporin A) could attenuate or block DOPAL toxicity.

Protein modification by DOPAL has been reported throughout the years and is predicted to cause cellular dysfunction via one or more mechanisms [54, 62, 63, 65–67]. For example, modification of critical residues may impair enzyme activity or result in protein denaturation followed by rapid degradation. It is conceivable that the opposite could occur, thus causing the protein to be resistant to degradation, yielding protein aggregation. In support of such a contention, a recent report demonstrated that DOPAL modifies the neuronal protein α-synuclein, a known component of the Lewy body protein aggregates observed in PD [68]. DOPAL adducts on α-synuclein cause the protein to aggregate and form high molecular weight oligomers.

An earlier study demonstrated that catechol adducts, resulting from modification of proteins by DA, are redox active and capable of generating ROS, especially in the presence of transition metals [69]. The net effect observed was oxidative degradation of proteins with DA adducts. It is conceivable that DOPAL modifications may exert the same deleterious effects on proteins.

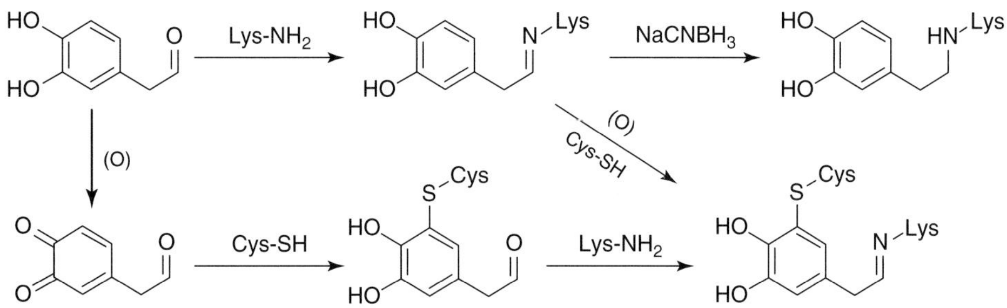

Fig. 29.3 The reaction of DOPAL with protein may involve Lys or Cys residues. Reaction with Lys occurs via the aldehyde to yield a Schiff base that can be stabilized via reduction to an amine. Cys reactivity requires catechol oxidation to a quinone. DOPAL may form intra- and/or intermolecular cross-links via reaction with both Lys and Cys.

29.6
DOPAL Reactivity with Proteins

As shown in Figure 29.3, there are two probable mechanisms for protein adduct formation, involving the aldehyde or catechol of DOPAL. It is known that amines (e.g., Lys) react with aldehydes to form a Schiff base (imine) or possibly an enamine product, with the former being more likely for a primary amine. Of note, Schiff base formation occurs more rapidly in acidic environments, such as pH 5, and the resulting imine is reversible, often requiring reduction to an amine for stability. Therefore, the stability and physiologic relevance of a Schiff base adduct are of question.

Recent work showed that DOPAL reacts with a model peptide (i.e., RKRSRAE) at pH 7.4 to form what appears to be Schiff base adducts based on mass spectrometry data [54]. Inclusion of the reducing agent sodium cyanoborohydride (Figure 29.3) resulted in a shift in adduct mass by 2 Da, which is consistent with a Schiff base. Interestingly, it was reported that the Schiff base adducts were readily observed for experiments performed in pH 7.4 tricine buffer and did not require reduction for stability. Such a finding is consistent with that reported by Helander and Tottmar for the reaction of DOPAL with hemoglobin [67].

Using SDS-PAGE and redox-active nitroblue tetrazolium (NBT) to detect catechol adducts on proteins[70], it was found that DOPAL covalently modifies bovine serum albumin (BSA) with the order of reactivity being DOPAL > DOPAC >> DA = L-DOPA [54]. Such a discovery is significant for several reasons. First, it demonstrates that DOPAL is more reactive than DA, which is often considered to be a highly reactive intermediate and endogenous neurotoxin. Such a finding is consistent with an earlier study showing that DOPAL is much more reactive toward hemoglobin than DA. Secondly, this evidence demonstrates the stability of the DOPAL adduct as it is resistant to the SDS-PAGE process, with or without sodium cyanoborohydride reduction. Thirdly, BSA contains only one free thiol but numerous Lys residues and such evidence suggests that Lys is a significant target of the DA-derived aldehyde. A recent paper reported that DOPAL modifies α-synuclein, yielding protein cross-linking, and such a finding supports amine adduction by DOPAL given that this protein does not contain any free thiols. In addition, it was observed that treatment of BSA with the Lys modifier citraconic anhydride, but not the Cys modifier iodoacetic acid, significantly prevented protein adduction by DOPAL.

As noted previously, an earlier report showed that inhibiting DOPAL oxidation in striatal synaptosomes yielded an increase in catechol adducts on proteins. SDS-PAGE and the redox-sensitive dye (NBT) were used to demonstrate increased protein modification. Interestingly, the catechol adducts were found to be dependent on MAO activity as inclusion of pargyline, an MAO inhibitor, nearly reduced all NBT staining. Such a discovery agrees with an earlier study showing that inclusion of pargyline with experiments involving radiolabeled DA and rat brain homogenates greatly decreased radioactivity associated with tissue. Given these findings and that aldehyde metabolite of DA is far more reactive toward proteins than DA, L-DOPA, or DOPAC, it is evident that the aldehyde group on DOPAL significantly contributes to protein adduction.

Much data have accumulated demonstrating the reactivity of the aldehyde moiety of DOPAL; however, the catechol also has the potential to yield protein modification. Catechols are labile functional groups as they can readily undergo oxidation to a highly reactive ortho-quinone. Autoxidation of DA, which refers to molecular oxygen catalyzed oxidation, is a well-studied example of catechol activation [71–74]. The autoxidation reaction involves two rounds of one-electron oxidation of DA via molecular oxygen, yielding two superoxide anions and one DA quinone. In addition to autoxidation, the enzyme prostaglandin H synthase (PHS), which is found in neuronal tissue, catalyzes oxidation of DA to a quinone [75]. Several members of the cytochrome P450 family can also mediate DA oxidation, for example, P450 1A1, 1A2, and 2D18 [76, 77]. Finally, transition metals such as Mn and Cu have been shown to be redox active toward DA, yielding ROS and the DA quinone [78].

The DA quinone, a soft electrophile, is highly reactive toward soft nucleophiles such as thiols, and numerous studies have demonstrated its ability to modify protein Cys residues [73, 79]. However, this reactive intermediate readily undergoes rearrangement via intramolecular nucleophilic addition of the amine to leukoaminochrome, followed by further autoxidation to aminochrome [80]. The order of reactivity for DA quinone was found to be the following: reaction with thiols > rearrangement via intramolecular amine addition to quinone > reaction with amines [80]. Such a finding indicates the transient and unstable nature of the DA quinone and suggests that if a Cys residue is not in proximity to the activated DA following autoxidation, rearrangement will occur yielding a less protein-reactive form (i.e., aminochrome).

Therefore, it is conceivable that DOPAL undergoes autoxidation to a protein-reactive quinone in a similar manner as DA (Figure 29.3). In support of such a hypothesis, an earlier study demonstrated that inclusion of thiols (reactive toward quinones), ascorbate (quinone reducing agent), or pargyline (MAO inhibitor) decreased the binding of DOPAL to tissue [66]. It is possible that enzymes such as PHS, cytochrome P450, or other factors (e.g., transition metals) catalyze oxidation of DOPAL to the quinone form. However, at this point, the stability and reactivity of the DOPAL quinone and its potential to rearrange are unknown, and work is in progress to address these issues. It was observed that tyrosinase or sodium metaperiodate oxidize DOPAL to a reactive form, with significant absorbance at ~500 nm, which rapidly changed to a more stable product with a lambda max around 400 nm (Anderson, D.G., et al.). Work is in progress to determine the chemical identity of such species, which might include a quinone and/or quinone methide, and characterize protein reactivity of oxidized DOPAL.

As shown in Figure 29.3, the quinone of DOPAL is predicted to react with thiols (e.g., Cys), yielding protein adducts via a thioether linkage. Such an adduct contains an aldehyde, which could subsequently react with Lys to yield an intra- or intermolecular cross-link. The Schiff base product formed via reaction of Lys with DOPAL might undergo autoxidation to the quinone, followed by addition of Cys to form an intra- or intermolecular cross-link. As noted above, the aldehyde and catechol/quinone of DOPAL are reactive, and both may contribute to protein modification as shown in Figure 29.3.

Currently, the *in vivo* protein targets of DOPAL are not known, and of question is whether or not there is specificity in regard to protein adduction. In addition, the subcellular location of DOPAL-modified proteins is not known; however, the MAO-mediated production of the DA-derived aldehyde occurs on the outer mitochondrial membrane [2, 4], and therefore proteins in the mitochondria and cytosol may be targeted. Given that DA and DOPAL are structurally analogous, does DOPAL selectively modify proteins with DA binding sites, for example, TH or VMAT2? Or, does DOPAL randomly react with any nucleophile in proximity?

In addition, dopaminergic cells generating a significant steady state of DOPAL (e.g., measured ~2 μM) may experience background levels of protein adduction [54], and therefore of question is how does the cell deal with DOPAL-modified proteins? As mentioned earlier, does the DOPAL adduct predispose a protein to degradation or make it more resistant?

29.7
Relevance of DOPAL to PD: Summary and "Big Picture"

Numerous studies implicate oxidative stress in the etiology of PD, and the oxidative stress is hypothesized to originate from environmental insult(s), such as exposure to pesticides [50, 51, 60, 81]. Individuals with the disease were found to have increased oxidative burden and higher levels of lipid peroxidation products such as 4HNE, as compared to controls [48, 49]. Although it is clear that oxidative stress is a factor in PD pathogenesis, the mechanistic link between oxidative stress and DA neuron degeneration is unknown.

DOPAL may represent a component of that mechanistic link as the metabolism of this toxic aldehyde is significantly inhibited by-products of oxidative stress and elevation in the concentration of this reactive intermediate is predicted to be damaging to cells. Therefore, DOPAL could serve as a "chemical trigger" for some stage of PD pathogenesis and be an attractive therapeutic target. It is possible that DOPAL acts early on in the disease process with elevated levels of the DA-derived aldehyde serving to initiate pathogenesis. A possible timeline and scenario might include the following: environmental insult → oxidative stress → elevated levels of DOPAL → dysfunction of DA neurons (e.g., protein modification, aggregation, hydroxyl radical, proteasome failure, etc.) → onset or progression of PD. Carbonyl scavengers have found utility in preventing aldehyde-induced toxicity [82–84] and could protect against DOPAL-mediated cell degeneration. Perhaps such agents might prevent PD pathogenesis. In addition, the DOPAL–protein adduct could serve as a biomarker for initiation of events related to PD and provide a mechanism for earlier disease diagnosis. The role of DOPAL in PD warrants further study and may yield novel and innovative means to diagnose and treat/halt PD.

Acknowledgments

This work was supported by a grant from the National Institutes of Health (NIH R01 ES15507).

References

1 Marsden, C.A. (2006) Dopamine: the rewarding years. *Br J Pharmacol*, **147** (Suppl 1), S136–S144.
2 Eisenhofer, G., Kopin, I.J., and Goldstein, D.S. (2004) Catecholamine metabolism: a contemporary view with implications for physiology and medicine. *Pharmacol Rev*, **56**, 331–349.
3 Kopin, I.J. (1964) Storage and metabolism of catecholamines: the role of monoamine oxidase. *Pharmacol Rev*, **16**, 179–191.
4 Elsworth, J.D. and Roth, R.H. (1997) Dopamine synthesis, uptake, metabolism, and receptors: relevance to gene therapy of Parkinson's disease. *Exp Neurol*, **144**, 4–9.
5 Paterson, I.A., Juorio, A.V., Berry, M.D., and Zhu, M.Y. (1991) Inhibition of monoamine oxidase-B by (-)-deprenyl potentiates neuronal responses to dopamine agonists but does not inhibit dopamine catabolism in the rat striatum. *J Pharmacol Exp Ther*, **258**, 1019–1026.
6 Di Monte, D.A., DeLanney, L.E., Irwin, I., Royland, J.E., Chan, P., Jakowec, M.W., and Langston, J.W. (1996) Monoamine oxidase-dependent metabolism of dopamine in the striatum and substantia nigra of L-DOPA-treated monkeys. *Brain Res*, **738**, 53–59.
7 Glover, V., Sandler, M., Owen, F., and Riley, G.J. (1977) Dopamine is a monoamine oxidase B substrate in man. *Nature*, **265**, 80–81.
8 Fowler, C.J., Wiberg, A., Oreland, L., Marcusson, J., and Winblad, B. (1980) The effect of age on the activity and molecular properties of human brain monoamine oxidase. *J Neural Transm*, **49**, 1–20.
9 Westlund, K.N., Denney, R.M., Kochersperger, L.M., Rose, R.M., and Abell, C.W. (1985) Distinct monoamine oxidase A and B populations in primate brain. *Science*, **230**, 181–183.

10. Thorpe, L.W., Westlund, K.N., Kochersperger, L.M., Abell, C.W., and Denney, R.M. (1987) Immunocytochemical localization of monoamine oxidases A and B in human peripheral tissues and brain. *J Histochem Cytochem*, **35**, 23–32.
11. Tetrud, J.W. and Langston, J.W. (1989) The effect of deprenyl (selegiline) on the natural history of Parkinson's disease. *Science*, **245**, 519–522.
12. Blaschko, H. (1952) Amine oxidase and amine metabolism. *Pharmacol Rev*, **4**, 415–458.
13. Mannisto, P.T. and Kaakkola, S. (1999) Catechol-O-methyltransferase (COMT): biochemistry, molecular biology, pharmacology, and clinical efficacy of the new selective COMT inhibitors. *Pharmacol Rev*, **51**, 593–628.
14. O'Leary, M.H. and Baughn, R.L. (1977) Decarboxylation-dependent transamination catalyzed by mammalian 3,4-dihydroxyphenylalanine decarboxylase. *J Biol Chem*, **252**, 7168–7173.
15. Mattammal, M.B., Haring, J.H., Chung, H.D., Raghu, G., and Strong, R. (1995) An endogenous dopaminergic neurotoxin: implication for Parkinson's disease. *Neurodegeneration*, **4**, 271–281.
16. Lamensdorf, I., He, L., Nechushtan, A., Harvey-White, J., Eisenhofer, G., Milan, R., Rojas, E., and Kopin, I.J. (2000) Effect of glipizide on dopamine synthesis, release and metabolism in PC12 cells. *Eur J Pharmacol*, **388**, 147–154.
17. Lamensdorf, I., Hrycyna, C., He, L.P., Nechushtan, A., Tjurmina, O., Harvey-White, J., Eisenhofer, G., Rojas, E., and Kopin, I.J. (2000) Acidic dopamine metabolites are actively extruded from PC12 cells by a novel sulfonylurea-sensitive transporter. *Naunyn Schmiedebergs Arch Pharmacol*, **361**, 654–664.
18. Hare, M.L. (1928) Tyramine oxidase: a new enzyme system in liver. *Biochem J*, **22**, 968–979.
19. Bernheim, M.L. (1931) Tyramine oxidase II: the course of the oxidation. *J Biol Chem*, **93**, 299–309.
20. Kohn, H.I. (1937) Tyramine oxidase. *Biochem J*, **31**, 1693–1704.
21. Richter, D. (1937) Adrenaline and amine oxidase. *Biochem J*, **31**, 2022–2028.
22. Renson, J., Weissbach, H., and Udenfriend, S. (1964) Studies on the biological activities of the aldehydes derived from norepinephrine, serotonin, tryptamine and histamine. *J Pharmacol Exp Ther*, **143**, 326–331.
23. Fellman, J.H. (1958) The rearrangement of epinephrine. *Nature*, **182**, 311–312.
24. Deitrich, R.A. and Erwin, V.G. (1975) Involvement of biogenic amine metabolism in ethanol addiction. *Fed Proc*, **34**, 1962–1968.
25. Tabakoff, B. (1974) Inhibition of sodium, potassium, and magnesium activated ATPases by acetaldehyde and "biogenic" aldehydes. *Res Commun Chem Pathol Pharmacol*, **7**, 621–624.
26. Weiner, H. (1978) Relationship between 3,4-dihydroxyphenylacetaldehyde levels and tetrahydropapaveroline formation. *Alcohol Clin Exp Res*, **2**, 127–131.
27. Sabelli, H.C. and Giardina, W.J. (1970) CNS effects of the aldehyde products of brain monoamines. *Biol Psychiatry*, **2**, 119–139.
28. Tank, A.W. and Weiner, H. (1979) Ethanol-induced alteration of dopamine metabolism in rat liver. *Biochem Pharmacol*, **28**, 3139–3147.
29. Tank, A.W., Weiner, H., and Thurman, J.A. (1981) Enzymology and subcellular localization of aldehyde oxidation in rat liver. Oxidation of 3,4-dihydroxyphenylacetaldehyde derived from dopamine to 3,4-dihydroxyphenylacetic acid. *Biochem Pharmacol*, **30**, 3265–3275.
30. Ryzlak, M.T. and Pietruszko, R. (1987) Purification and characterization of aldehyde dehydrogenase from human brain. *Arch Biochem Biophys*, **255**, 409–418.

31 Ryzlak, M.T. and Pietruszko, R. (1989) Human brain glyceraldehyde-3-phosphate dehydrogenase, succinic semialdehyde dehydrogenase and aldehyde dehydrogenase isozymes: substrate specificity and sensitivity to disulfiram. *Alcohol Clin Exp Res*, **13**, 755–761.

32 Burke, W.J., Li, S.W., Chung, H.D., Ruggiero, D.A., Kristal, B.S., Johnson, E.M., Lampe, P., Kumar, V.B., Franko, M., Williams, E.A., and Zahm, D.S. (2004) Neurotoxicity of MAO metabolites of catecholamine neurotransmitters: role in neurodegenerative diseases. *Neurotoxicology*, **25**, 101–115.

33 Burke, W.J. (2003) 3,4-dihydroxyphenylacetaldehyde: a potential target for neuroprotective therapy in Parkinson's disease. *Curr Drug Targets CNS Neurol Disord*, **2**, 143–148.

34 Burke, W.J., Chung, H.D., and Li, S.W. (1999) Quantitation of 3,4-dihydroxyphenylacetaldehyde and 3,4-dihydroxyphenylglycolaldehyde, the monoamine oxidase metabolites of dopamine and noradrenaline, in human tissues by microcolumn high-performance liquid chromatography. *Anal Biochem*, **273**, 111–116.

35 Kristal, B.S., Conway, A.D., Brown, A.M., Jain, J.C., Ulluci, P.A., Li, S.W., and Burke, W.J. (2001) Selective dopaminergic vulnerability: 3,4-dihydroxyphenylacetaldehyde targets mitochondria. *Free Radic Biol Med*, **30**, 924–931.

36 Hafer, G., Agarwal, D.P., and Goedde, H.W. (1987) Human brain aldehyde dehydrogenase: activity with DOPAL and isozyme distribution. *Alcohol*, **4**, 413–418.

37 Marchitti, S.A., Deitrich, R.A., and Vasiliou, V. (2007) Neurotoxicity and metabolism of the catecholamine-derived 3,4-dihydroxyphenylacetaldehyde and 3,4-dihydroxyphenylglycolaldehyde: the role of aldehyde dehydrogenase. *Pharmacol Rev*, **59**, 125–150.

38 Erwin, V.G. and Deitrich, R.A. (1966) Brain aldehyde dehydrogenase. Localization, purification, and properties. *J Biol Chem*, **241**, 3533–3539.

39 Lamensdorf, I., Eisenhofer, G., Harvey-White, J., Hayakawa, Y., Kirk, K., and Kopin, I.J. (2000) Metabolic stress in PC12 cells induces the formation of the endogenous dopaminergic neurotoxin, 3,4-dihydroxyphenylacetaldehyde. *J Neurosci Res*, **60**, 552–558.

40 Lamensdorf, I., Eisenhofer, G., Harvey-White, J., Nechustan, A., Kirk, K., and Kopin, I.J. (2000) 3,4-Dihydroxyphenylacetaldehyde potentiates the toxic effects of metabolic stress in PC12 cells. *Brain Res*, **868**, 191–201.

41 Mandel, S., Grunblatt, E., Riederer, P., Amariglio, N., Jacob-Hirsch, J., Rechavi, G., and Youdim, M.B. (2005) Gene expression profiling of sporadic Parkinson's disease substantia nigra pars compacta reveals impairment of ubiquitin-proteasome subunits, SKP1A, aldehyde dehydrogenase, and chaperone HSC-70. *Ann N Y Acad Sci*, **1053**, 356–375.

42 Tabakoff, B. and Erwin, V.G. (1970) Purification and characterization of a reduced nicotinamide adenine dinucleotide phosphate-linked aldehyde reductase from brain. *J Biol Chem*, **245**, 3263–3268.

43 Tabakoff, B., Anderson, R., and Alivisatos, S.G. (1973) Enzymatic reduction of "biogenic" aldehydes in brain. *Mol Pharmacol*, **9**, 428–437.

44 Turner, A.J., Illingworth, J.A., and Tipton, K.F. (1974) Simulation of biogenic amine metabolism in the brain. *Biochem J*, **144**, 353–360.

45 Hjelle, J.J., Grubbs, J.H., and Petersen, D.R. (1982) Inhibition of mitochondrial aldehyde dehydrogenase by malondialdehyde. *Toxicol Lett*, **14**, 35–43.

46 Mitchell, D.Y. and Petersen, D.R. (1991) Inhibition of rat hepatic mitochondrial aldehyde dehydrogenase-mediated acetaldehyde oxidation by

trans-4-hydroxy-2-nonenal. *Hepatology*, **13**, 728–734.
47 Esterbauer, H., Schaur, R.J., and Zollner, H. (1991) Chemistry and biochemistry of 4-hydroxynonenal, malonaldehyde and related aldehydes. *Free Radic Biol Med*, **11**, 81–128.
48 Yoritaka, A., Hattori, N., Uchida, K., Tanaka, M., Stadtman, E.R., and Mizuno, Y. (1996) Immunohistochemical detection of 4-hydroxynonenal protein adducts in Parkinson disease. *Proc Natl Acad Sci U S A*, **93**, 2696–2701.
49 Dexter, D.T., Carter, C.J., Wells, F.R., Javoy-Agid, F., Agid, Y., Lees, A., Jenner, P., and Marsden, C.D. (1989) Basal lipid peroxidation in substantia nigra is increased in Parkinson's disease. *J Neurochem*, **52**, 381–389.
50 Drechsel, D.A. and Patel, M. (2008) Role of reactive oxygen species in the neurotoxicity of environmental agents implicated in Parkinson's disease. *Free Radic Biol Med*, **44**, 1873–1886.
51 Jenner, P. (2003) Oxidative stress in Parkinson's disease. *Ann Neurol*, **53** (Suppl 3), S26–S36; discussion S36-28.
52 Doorn, J.A., Hurley, T.D., and Petersen, D.R. (2006) Inhibition of human mitochondrial aldehyde dehydrogenase by 4-hydroxynon-2-enal and 4-oxonon-2-enal. *Chem Res Toxicol*, **19**, 102–110.
53 Florang, V.R., Rees, J.N., Brogden, N.K., Anderson, D.G., Hurley, T.D., and Doorn, J.A. (2007) Inhibition of the oxidative metabolism of 3,4-dihydroxyphenylacetaldehyde, a reactive intermediate of dopamine metabolism, by 4-hydroxy-2-nonenal. *Neurotoxicology*, **28**, 76–82.
54 Rees, J.N., Florang, V.R., Anderson, D.G., and Doorn, J.A. (2007) Lipid peroxidation products inhibit dopamine catabolism yielding aberrant levels of a reactive intermediate. *Chem Res Toxicol*, **20**, 1536–1542.
55 MacKerell, A.D.Jr and Pietruszko, R. (1987) Chemical modification of human aldehyde dehydrogenase by physiological substrate. *Biochim Biophys Acta*, **911**, 306–317.
56 Hashimoto, T. and Yabe-Nishimura, C. (2002) Oxidative metabolite of dopamine, 3,4-dihydroxyphenylacetaldehyde, induces dopamine release from PC12 cells by a Ca2+-independent mechanism. *Brain Res*, **931**, 96–99.
57 Sulzer, D. and Rayport, S. (1990) Amphetamine and other psychostimulants reduce pH gradients in midbrain dopaminergic neurons and chromaffin granules: a mechanism of action. *Neuron*, **5**, 797–808.
58 Riddle, E.L., Fleckenstein, A.E., and Hanson, G.R. (2006) Mechanisms of methamphetamine-induced dopaminergic neurotoxicity. *AAPS J*, **8**, E413–E418.
59 Quinton, M.S. and Yamamoto, B.K. (2006) Causes and consequences of methamphetamine and MDMA toxicity. *AAPS J*, **8**, E337–E347.
60 Kanthasamy, A.G., Kitazawa, M., Kanthasamy, A., and Anantharam, V. (2005) Dieldrin-induced neurotoxicity: relevance to Parkinson's disease pathogenesis. *Neurotoxicology*, **26**, 701–719.
61 Kitazawa, M., Anantharam, V., and Kanthasamy, A.G. (2001) Dieldrin-induced oxidative stress and neurochemical changes contribute to apoptopic cell death in dopaminergic cells. *Free Radic Biol Med*, **31**, 1473–1485.
62 Jinsmaa, Y., Florang, V.R., Rees, J.N., Anderson, D.G., Strack, S., and Doorn, J.A. (2009) Products of Oxidative Stress Inhibit Aldehyde Oxidation and Reduction Pathways in Dopamine Catabolism Yielding Elevated Levels of a Reactive Intermediate. *Chem Res Toxicol*, **22**, 835–841.
63 Mattammal, M.B., Chung, H.D., Strong, R., and Hsu, F.F. (1993) Confirmation of a dopamine metabolite in parkinsonian brain tissue by gas chromatography-mass spectrometry. *J Chromatogr*, **614**, 205–212.

64 Li, S.W., Lin, T.S., Minteer, S., and Burke, W.J. (2001) 3,4-Dihydroxyphenylacetaldehyde and hydrogen peroxide generate a hydroxyl radical: possible role in Parkinson's disease pathogenesis. *Brain Res Mol Brain Res*, **93**, 1–7.

65 Nilsson, G.E. and Tottmar, O. (1987) Biogenic aldehydes in brain: on their preparation and reactions with rat brain tissue. *J Neurochem*, **48**, 1566–1572.

66 Ungar, F., Tabakoff, B., and Alivisatos, S.G. (1973) Inhibition of binding of aldehydes of biogenic amines in tissues. *Biochem Pharmacol*, **22**, 1905–1913.

67 Helander, A. and Tottmar, O. (1989) Reactions of biogenic aldehydes with hemoglobin. *Alcohol*, **6**, 71–75.

68 Burke, W.J., Kumar, V.B., Pandey, N., Panneton, W.M., Gan, Q., Franko, M.W., O'Dell, M., Li, S.W., Pan, Y., Chung, H.D., and Galvin, J.E. (2008) Aggregation of alpha-synuclein by DOPAL, the monoamine oxidase metabolite of dopamine. *Acta Neuropathol*, **115**, 193–203.

69 Akagawa, M., Ishii, Y., Ishii, T., Shibata, T., Yotsu-Yamashita, M., Suyama, K., and Uchida, K. (2006) Metal-catalyzed oxidation of protein-bound dopamine. *Biochemistry*, **45**, 15120–15128.

70 Paz, M.A., Fluckiger, R., Boak, A., Kagan, H.M., and Gallop, P.M. (1991) Specific detection of quinoproteins by redox-cycling staining. *J Biol Chem*, **266**, 689–692.

71 Graham, D.G. (1978) Oxidative pathways for catecholamines in the genesis of neuromelanin and cytotoxic quinones. *Mol Pharmacol*, **14**, 633–643.

72 Graham, D.G., Tiffany, S.M., Bell, W.R.Jr, and Gutknecht, W.F. (1978) Autoxidation versus covalent binding of quinones as the mechanism of toxicity of dopamine, 6-hydroxydopamine, and related compounds toward C1300 neuroblastoma cells in vitro. *Mol Pharmacol*, **14**, 644–653.

73 Hastings, T.G. and Zigmond, M.J. (1994) Identification of catechol-protein conjugates in neostriatal slices incubated with [3H] dopamine: impact of ascorbic acid and glutathione. *J Neurochem*, **63**, 1126–1132.

74 Bisaglia, M., Mammi, S., and Bubacco, L. (2007) Kinetic and structural analysis of the early oxidation products of dopamine: analysis of the interactions with alpha-synuclein. *J Biol Chem*, **282**, 15597–15605.

75 Mattammal, M.B., Strong, R., Lakshmi, V.M., Chung, H.D., and Stephenson, A.H. (1995) Prostaglandin H synthetase-mediated metabolism of dopamine: implication for Parkinson's disease. *J Neurochem*, **64**, 1645–1654.

76 Thompson, C.M., Capdevila, J.H., and Strobel, H.W. (2000) Recombinant cytochrome P450 2D18 metabolism of dopamine and arachidonic acid. *J Pharmacol Exp Ther*, **294**, 1120–1130.

77 Segura-Aguilar, J. (1996) Peroxidase activity of liver microsomal vitamin D 25-hydroxylase and cytochrome P450 1A2 catalyzes 25-hydroxylation of vitamin D3 and oxidation of dopamine to aminochrome. *Biochem Mol Med*, **58**, 122–129.

78 Bindoli, A., Rigobello, M.P., and Deeble, D.J. (1992) Biochemical and toxicological properties of the oxidation products of catecholamines. *Free Radic Biol Med*, **13**, 391–405.

79 LaVoie, M.J., Ostaszewski, B.L., Weihofen, A., Schlossmacher, M.G., and Selkoe, D.J. (2005) Dopamine covalently modifies and functionally inactivates parkin. *Nat Med*, **11**, 1214–1221.

80 Tse, D.C., McCreery, R.L., and Adams, R.N. (1976) Potential oxidative pathways of brain catecholamines. *J Med Chem*, **19**, 37–40.

81 Landrigan, P.J., Sonawane, B., Butler, R.N., Trasande, L., Callan, R., and Droller, D. (2005) Early environmental origins of neurodegenerative disease in later life. *Environ Health Perspect*, **113**, 1230–1233.

82 Burcham, P.C. and Pyke, S.M. (2006) Hydralazine inhibits rapid acrolein-induced protein oligomerization: role of aldehyde scavenging and adduct trapping in cross-link blocking and cytoprotection. *Mol Pharmacol*, **69**, 1056–1065.

83 Burcham, P.C., Fontaine, F.R., Kaminskas, L.M., Petersen, D.R., and Pyke, S.M. (2004) Protein adduct-trapping by hydrazinophthalazine drugs: mechanisms of cytoprotection against acrolein-mediated toxicity. *Mol Pharmacol*, **65**, 655–664.

84 Kaminskas, L.M., Pyke, S.M., and Burcham, P.C. (2004) Reactivity of hydrazinophthalazine drugs with the lipid peroxidation products acrolein and crotonaldehyde. *Org Biomol Chem*, **2**, 2578–2584.

30
Tetrahydropapaveroline, an Endogenous Dicatechol Isoquinoline Neurotoxin

Young-Joon Surh and Hyun-Jung Kim

Tetrahydropapaveroline (THP), a neurotoxic tetrahydroisoquinoline alkaloid formed by condensation between dopamine and dopaldehyde, has been speculated to contribute to the etiology of Parkinson's disease and also to the neurobehavioral abnormalities associated with alcoholism. However, to date, there is a paucity of direct evidence to support such roles. As THP bears two catechol moieties, it may readily undergo oxidation to form an *o*-quinone intermediate with concomitant production of reactive oxygen species, which can cause neuronal cell death and DNA damage. This chapter deals with the current knowledge of neurotoxic effects of this endogenous alkaloid and underlying biochemical mechanisms.

30.1
Introduction

A group of cyclized condensation adducts of biogenic amines with aldehydes, referred to as *mammalian alkaloids*, has been the subject of neurochemical pharmacology over the past few decades [1]. Among these, of particular interest are tetrahydroisoquinolines (THIQs), such as salsolinol (1-methyl-6,7-dihydroxy-1,2,3,4-tetrahydroisoquinoline; SAL) and tetrahydropapaveroline (6,7-dihydroxy-1-(3',4'-dihydroxybenzyl)-1,2,3,4-tetrahydroisoquinoline; THP) derived from dopamine through condensation with acetaldehyde and dopaldehyde (3,4-dihydroxyphenylacetaldehyde), respectively [2].

THP is a potential dopaminergic neurotoxin that has been considered to contribute to the etiology of Parkinson's disease [3–5]. Though present in normal human brain, THP has been detected in a larger quantity in the brain [6] and urine [2, 7] of parkinsonian patients undergoing 3,4-dihydroxyphenylalanine (L-DOPA) therapy [14]. THP has also been detected in the brain of rats given ethanol [8] and/or L-DOPA [9, 10], but not in that of normal rats [8, 9]. This endogenous alkaloid has also been considered to account for the neurobehavioral abnormalities associated with alcoholism and to act as a false neurotransmitter [11].

Endogenous Toxins. Diet, Genetics, Disease and Treatment.
Edited by Peter J. O'Brien and W. Robert Bruce
Copyright © 2010 WILEY-VCH Verlag GmbH & Co. KGaA, Weinheim
ISBN: 978-3-527-32363-0

As THP bears two catechol moieties, the compound may readily undergo redox cycling to produce reactive oxygen species (ROS) as well as toxic quinoids [12]. The present chapter elaborates over 30 years of investigation into neurochemical/ neuropharmacological properties of THP and related THIQs with regard to their possible implications in the etiology of some neuronal disorders with focus on parkinsonism and alcohol dependence.

30.2
Biosynthesis of THP

30.2.1
Nonenzymatic versus Enantioselective Formation

Since its first identification in the urine of parkinsonian patients around 1970 [2], THP has long been considered to be formed spontaneously by nonenzymatic Pictet–Spengler condensation of dopamine with its aldehyde metabolite (3,4-dihydroxyphenylacetaldehyde) produced by monoamine oxidase (MAO) as illustrated in Figure 30.1. *In vitro* studies using tissue homogenates confirmed the formation of THP from dopamine [13, 14]. However, later studies by Sango *et al.* have revealed the stereochemical characteristic of THP identified in human brains [11]. Thus, in all four control human brains examined, only the (S) enantiomer of THP was detected, and the concentrations ranged from 0.12 to 0.22 pmol g^{-1} wet weight of brain tissue. These findings suggest that this endogenous dicatechol isoquinoline is presumably synthesized in the brain by an enzyme-catalyzed reaction. However, the exogenous or peripheral origin of THP found in brain cannot be excluded. Some edible plants adopt the distinct enzymatic pathways for the synthesis of (S)-THP [15], and exogenous THP consumed by man is anticipated to cross the blood–brain barrier [16]. In this context, it is interesting to note that dietary sources of SAL contribute to its detection in biological samples [17], and this may also be the case for THP.

Although minute amounts of THP are likely to be present in normal brain, easy detection of the compound in parkinsonian patients receiving L-DOPA medication and experimental animals treated with L-DOPA seems to be a consequence of the high levels of dopamine derived from its prodrug [4]. Under such circumstances, excess dopamine may inhibit aldehyde dehydrogenase [18], thereby blocking the conversion of 3,4-dihydroxyphenylacetaldehyde into 3,4-dihydroxyphenylacetic acid. The resulting accumulation of a relatively high concentration of 3,4-dihydroxyphenylacetaldehyde, together with the high levels of dopamine derived from L-DOPA administration, may favor the formation of THP [4].

30.2.2
Effects of Alcohol

It was postulated that excess acetaldehyde, secondary to ethanol metabolism, would competitively inhibit the breakdown of dopaldehyde by aldehyde dehydrogenase

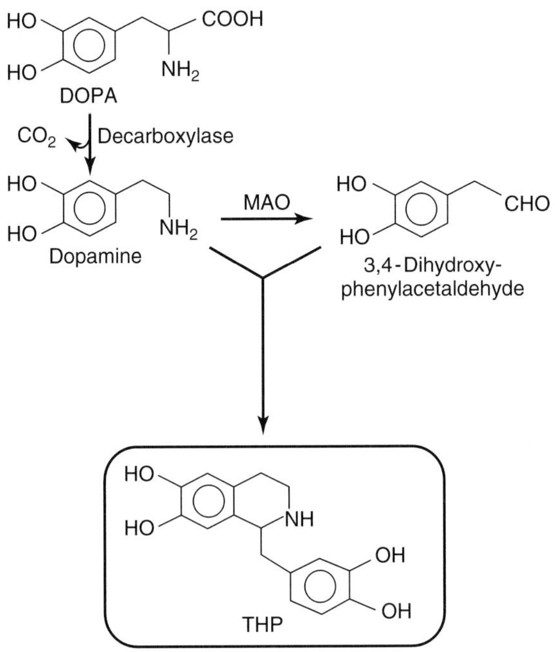

Fig. 30.1 Biosynthesis of THP. Although the Pictet–Spengler condensation reaction of dopamine with its aldehyde metabolite (3,4-dihydroxyphenylacetaldehyde) produced by monoamine oxidase (MAO) yields racemic THP, (S)-enantiomer is predominantly found in the brain of man, suggesting the occurrence of enzymatic synthesis.

[19]. The resultant accumulation of dopaldehyde would lead to an additional condensation reaction with dopamine to augment THP formation [19, 20]. THP was detected in specific brain regions of the rat after acute ethanol administration (3 g kg^{-1} body weight), but not in the same regions of untreated animals. Most brain regions had detectable levels of THP 100 minutes after the animals received ethanol, and the striatum contained the highest concentration of the alkaloid [8]. THP can be detected in the brain of rats not only after acute ethanol administration [8] but also when animals were subjected to alcohol consumption *ad libitum* for a longer period [21]. In the latter study, alcohol administration for 18 months induced formation of the (S) enantiomer of THP only in the striatum of the rat brain.

30.3
Neurotoxic Potential of THP

THP has been speculated to be implicated in the etiology of several human neurological, behavioral, and psychiatric disorders, such as parkinsonism and alcohol addiction [1, 3]. However, to date there is little convincing evidence to support such roles.

30.3.1
Parkinsonism

1-Methyl-4-phenyl-1,2,3,6-tetrahydropyridine (MPTP) is an exogenous dopaminergic neurotoxin considered to produce parkinsonism in humans, monkeys, and various animals. MPTP is converted by MAO to the 1-methyl-4-phenyl-pyridinium ion (MPP$^+$), which is capable of selectively killing the nigrostriatal dopaminergic neurons [6]. Endogenous isoquinoline derivatives, structurally related to MPTP and its active metabolite MPP$^+$, have been found in the brain of patients with Parkinson's disease, and they are suspected to contribute, in part, to dopaminergic neurodegeneration in Parkinson's disease [1–6, 22]. Among them, some THIQs and their N-methylated metabolites were identified in the brain of postmortem specimens of Parkinson's patients, and their levels were about 10 times higher than those in the normal brain [23]. It has also been demonstrated that THIQ administration provokes parkinsonism-like symptoms in monkeys [24]. Although levels of THP and other THIQs have been elevated in the brain and urine of parkinsonian patients, it remains unclear whether this is causally linked to the pathogenesis of the disease or simply a result of the disease progression. The considerable part of THP found in the brain of parkinsonian patients may be related to L-DOPA they intake as an effective symptomatic therapeutic agent, which releases dopamine in the brain, resulting in increased synthesis of THP. However, the possibility of impaired metabolism of THP in patients with Parkinson's disease cannot be excluded [5]. Despite the lack of direct evidence supporting the role of THP in neuronal death in Parkinson's disease, accumulating data from cell culture and animal studies suggest that this endogenous THIQ may act as a dopaminergic neurotoxin [3, 5, 6].

30.3.1.1 Inhibition of Dopamine Biosynthesis
Treatment of PC12 cells with THP significantly decreased the intracellular dopamine content in a concentration-dependent manner [25]. The activity of tyrosine hydroxylase, the rate-determining enzyme for the production of DOPA, was also inhibited by the treatment with THP. In addition, THP had an inhibitory effect on bovine adrenal tyrosine hydroxylase. However, the reduction of dopamine content by THP in PC12 cells was inversed by the antioxidant N-acetyl-L-cysteine (NAC). These results indicate that THP decreases the basal dopamine content and also reduces the increased dopamine content derived from L-DOPA in PC12 cells through inhibition of tyrosine hydroxylase activity, and possibly by evoking oxidative stress [25].

30.3.1.2 Inhibition of Dopamine Uptake through the Dopamine Transporter
THP and some of its derivatives inhibited dopamine uptake through the dopamine transporter (DAT) in human embryonic kidney HEK293 cells [26]. As THP is

considered to be synthesized from dopamine, it may affect dopaminergic neurons through the reuptake system, that is, DAT. To determine whether THP has affinity for DAT, Okada *et al.* examined whether THP and its derivatives could inhibit [^{3}H]dopamine uptake in a cell line that stably expresses DAT [26]. The K_i values of THP and three synthetic derivatives for inhibition of dopamine uptake were almost similar to that of MPP$^+$, an isoquinoline dopaminergic neurotoxin implicated in parkinsonism. These results suggest that THP and its derivatives might be taken through DAT and be involved in Parkinson's disease [26].

The importance of the DAT for selective dopaminergic toxicity was addressed by testing the differential cytotoxicity of THIQ derivatives including SAL and THP as well as MPP$^+$ in non-neuronal and neuronal heterologous expression systems of the *DAT* gene (HEK-293 and mouse neuroblastoma Neuro-2A cells, respectively). Both SAL and THP were among those that have the most toxic potency [22]. Neurotoxic potential of THP is summarized in Table 30.1.

30.3.1.3 Inhibition of Mitochondrial Respiration

Since the discovery of MPTP-induced parkinsonism, MPTP-like toxins, such as THIQs, have been speculated to induce Parkinson's disease. As the neuronal degeneration in MPTP-induced parkinsonism is presumed to be caused by the inhibition of mitochondrial respiration by MPP$^+$, the effects of THIQ-like alkaloids including THP and tetrahydropapaverine on the mitochondrial respiration were investigated using mouse brains [6, 27, 28]. Both compounds significantly inhibited state 3 and state 4 respiration and reduced the respiratory control ratio. Toxic properties of these compounds on mitochondrial respiration were quite similar to that of MPP$^+$. These results support the hypothesis that MPTP- or MPP$^+$-like

Tab. 30.1 Neurotoxic potential of THP

Parkinsonism
Inhibition of dopamine biosynthesis
    Inhibition of tyrosine hydroxylase activity
    Production of oxidative stress
Inhibition of dopamine uptake through dopamine transporter
    Affinity for DAT similar to MPP$^+$
Inhibition of mitochondrial function
    Inhibition of state 3 and state 4 respiration
    Inhibition of the respiratory control ratio
    Inhibition of α-ketoglutarate dehydrogenase
L-DOPA paradox
    Elevated levels of THP found in the urine and brain of Parkinsonian patients on L-DOPA medication
    Aggravation of ROS generation and cytotoxicity of L-DOPA.
Alcohol dependence
    Increased preference for alcohol consumption in animals by direct THP infusion
    THP as an intermediate in morphine and codeine synthesis

substances may be responsible for the nigral degeneration implicated in the etiology of Parkinson's disease [3, 5, 6]

30.3.1.4 Inhibition of Serotonin Production

The inhibitory effects of THP on serotonin biosynthesis in murine mastocytoma P815 cells were investigated. THP decreased serotonin content in a concentration-dependent manner in serotonin-producing mastocytoma cells [29]. Further, the activity of tryptophan hydroxylase was inhibited after treatment with THP in the same cell line. These data imply that THP treatment leads to a decrease in serotonin content by inhibiting tryptophan hydroxylase activity in murine mastocytoma cells [29].

30.3.2
Implications for L-DOPA Paradox

L-DOPA, the natural precursor of dopamine, is the most effective and frequently prescribed drug for controlling symptoms of Parkinson's disease [30]. Nonetheless, the L-DOPA therapy is not curative and the dopaminergic cells continue to die in patients with Parkinson's disease during the L-DOPA therapy [31]. L-DOPA treatment may even hasten the underlying neurodegenerative process [32] and also provoke some side effects [33], a phenomenon dubbed as L-*DOPA paradox*.

The dopaminergic neurotoxicity and other adverse effects of L-DOPA have been suggested to potentially involve THP [3, 5]. Elevated levels of THP have been found in the urine and brain of parkinsonian patients on L-DOPA medication [2, 7, 34] with and without ethanol consumption [2]. Three parkinsonian patients given 250, 750, or 1000 mg of L-DOPA (as Sinemet) daily exhibited 24-hours urinary THP excretion levels of 989, 1017, and 1600 pmol, respectively [7]. THP has also been detected in the brain of rats given L-DOPA with [9] or without [10] ethanol cotreatment. Again, THP was not found in the brain of untreated rats, but levels of 0.42 pmol THP per gram brain were observed in animals given L-DOPA (200 mg kg^{-1}) intraperitoneally 90 minutes prior to decapitation [9]. Intraperitoneal administration of ethanol to L-DOPA-treated rats resulted in significant increases in the THP and dopamine levels in the brain as compared to animals treated with L-DOPA alone. Likewise, the levels of THP as well as SAL in the brain of parkinsonian patients on L-DOPA treatment were apparently elevated after voluntary ingestion of absolute ethanol [2]. Notably, treatment of PC12 cells with L-DOPA induced apoptosis by generating ROS, which was aggravated by THP cotreatment [35].

Therefore, parkinsonian patients treated with L-DOPA for long term need to be monitored for the relationship between plasma concentration of THP and any plausible symptoms of neurotoxicity [25]. Moreover, it would also be worthwhile determining any differences in the levels of THP among untreated parkinsonian subjects, those on L-DOPA medication, and normal individuals.

30.3.3
Alcohol Dependence

THIQs have been postulated to play a role in the pathogenesis of chronic alcoholism [1, 20]. Thus, Collins proposed that possible oxidative metabolites of endogenous THIQs that elevate in the brain of chronic alcoholics are responsible for neuronal damage [1]. The identification of SAL formed in mammalian tissue following exposure to ethanol and its principal metabolite acetaldehyde [1, 36] has stimulated a search for the etiological involvement of this THIQ in addictive states related to chronic alcohol consumption [20].

Possible implications of THIQs in alcohol dependence were inferred from the observation that rats that normally rejected alcohol, following direct delivery of THP into the cerebral ventricle every 15 minutes for up to 12 days, drank ethyl alcohol in increasingly excessive amounts [37]. A subsequent behavioral study with rats injected with THP into brain regions revealed the preference for alcohol up to 10 months after THP infusion [38]. When injected into several cerebral regions of rats including substantia nigra, THP induced or sustained significant increases in alcohol intake [39]. Likewise, microinjection of THP into the ventral tegmental area of anesthetized high-ethanol-preferring (HEP) rats directly altered the function of the pathway of mesolimbic neurons generally and the dopaminergic system specifically. Such a perturbation could account for the induction of alcohol preference [40].

However, the significance of THP formation to addictive states in alcoholic patients is uncertain [20]. As (S)-THP can serve as an intermediate in *de novo* synthesis of more complex alkaloids including morphine and codeine in mammals [11, 21], their endogenous formation as a consequence of overproduction of THP during alcohol consumption may be associated with addiction liability of alcohol [20, 39]. THP was found as a requisite intermediate in the biosynthesis of morphine in the opium poppy *Papaver somniferum* [42]. Notably, morphine and codeine have been identified even in animal tissues [43, 44] and human cerebrospinal fluid [45]. However, the physiologic roles of endogenous morphine and codeine are unknown. Morphine and codeine produce physical dependence as well as analgesia, and these addictive alkaloids, endogenously formed from THP as an intermediate, might subserve alcohol dependence [3, 41]. The complete biosynthetic pathways leading to the formation of these narcotic substances in mammalian species are still unclear, but THP has been suggested as a possible precursor as in the case of plants [34, 41]. If THP is a requisite intermediate for the synthesis of morphine and codeine in mammals and plants, the levels of all these endogenous substances in L-DOPA-treated parkinsonian patients should be higher than those in normal control subjects. Thus, Matsubara and colleagues found significantly elevated urinary morphine and codeine as well as THP levels in parkinsonian patients under L-DOPA therapy compared to those in healthy nondrinker controls [34]. According to their study, there were significant correlations among these three alkaloid levels in the urine. It is also noteworthy that there were very low levels of THP as well as morphine and codeine in the urine of abstinent alcoholics.

30.4
Biochemical Mechanisms Underlying THP-Induced Neurotoxicity

Catechol compounds readily undergo auto- or enzymatic oxidation to form reactive quinoids. o-Quinones can directly modify crucial thiol protein residues of many functionally important proteins, and this has been suggested as a principal toxic mechanism involved in the harmful effects exerted by catechol compounds [46]. The catechol–quinone redox cycling, that is, the one-electron reduction of quinone to semiquinone and the autoxidation of semiquinone to the quinone, is able to release large quantities of superoxide anion that, in turn, can – spontaneously or by superoxide dismutase (SOD) action – be transformed into hydrogen peroxide (H_2O_2). H_2O_2, via the Fenton reaction, is readily decomposed to yield hydroxyl radical, an extremely reactive species with devastating action on practically every cell component and organelle [46].

Likewise, catechol-bearing THIQs can undergo autoxidation or enzymatic oxidation [47] and subsequent generation of reactive quinones via semiquinones may cause the degeneration of dopaminergic neurons and other injuries (Figure 30.2). Although quinoidal forms of THIQs are anticipated to be reactive *per se*, they can be converted back to the parent catechol molecules, with concomitant production of ROS capable of damaging critical cellular molecules, such as DNA, RNA, protein, and membrane lipid. According to a recent study in our laboratory, redox cycling of SAL [48, 49] and THP [50, 51], facilitated in the presence of certain transition metal ions such as Cu^{2+}, lead to ROS overproduction and subsequently oxidative cell death and DNA damage. Therefore, ROS overproduction is likely to contribute to the mechanisms underlying deleterious effects of THP and other THIQs [4, 5].

30.4.1
Redox Cycling

The oxidation chemistry of THIQ analogs was extensively studied by electrochemical approaches. Some reaction products were isolated and identified structurally by a series of spectrometric analyses, and an oxidation mechanism was proposed [47]. As described earlier, THP undergoes oxidation to form an o-quinone intermediate with concomitant production of ROS (Figure 30.2). The presence of ascorbate enhances this process by establishing a redox cycle, which regenerates THP from its quinoidal form. Ascorbate-promoted THP autoxidation caused an increase in the protein carbonyl content, in the presence of iron [5].

30.4.2
Oxidative Cell Death

Dopamine-derived catechol-bearing THIQs are neurotoxic to some extent and might be implicated in the pathogenesis of Parkinson's disease. In human dopaminergic neuroblastoma SH-SY5Y cells, THP caused predominantly necrotic

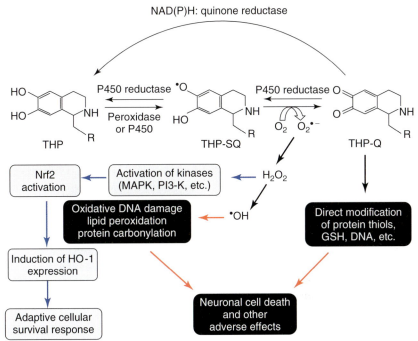

Fig. 30.2 A proposed role for oxidative metabolism of THP in its neurotoxic effects. Cytochrome P450 or peroxidase catalyzes one-electron oxidation of THP to produce o-semiquinone radical (THP–SQ), which in turn oxidized enzymatically or spontaneously to o-quinone (THP–Q). THP–Q can be reduced to THP–SQ by one-electron reductase or back to THP by NAD(P)H:quinone oxidoreductase (DT-diaphorase). Though not illustrated in this figure, THP–Q may tautomerize to yield reactive quinone methide. Although overproduction of ROS can cause neuronal cell death, initial mild oxidative stress induces activation of the redox-sensitive transcription factor Nrf2 and subsequently expression of antioxidant/cytoprotective enzymes, such as HO-1, via the upstream kinase signaling, thereby provoking adaptive survival response (refer to Section 30.5 for further details).

death, whereas papaveroline and N-methyl-papaveroline induced apoptosis as evidenced by typical features of condensed and fragmented nuclei [52]. THP markedly reduced adenosine triphosphate (ATP) level, whereas papaverolines did not, suggesting that the type of cell death (necrosis and apoptosis), induced by these isoquinolines, depends on the concentration of ATP in the cells [52].

The effect of SAL and THP on the viability of melanoma cell lines was investigated. THP appeared to be more cytotoxic than SAL and DOPA [46]. To evaluate the mechanisms of cell loss, cultures were assayed for the formation of DNA ladder and cytosolic nucleosomes under THIQ treatment condition. No elevation of both indicators was observed, suggesting that apoptosis is not likely to be associated with the observed cytotoxicity [46]. THP-induced cytotoxicity in melanoma cells was partially protected by exogenous catalase and SOD, but completely abolished by reduced glutathione (GSH) and NAC, suggesting that THP toxicity was likely due to increased oxidative stress provoked by the generation of ROS and oxidative metabolites [46]. THP exerts toxicity toward the mitochondrial respiratory chain [27, 28]. Mitochondrial metabolism in terms of α-ketoglutarate dehydrogenase activity was inhibited in murine and human melanoma cells after exposure to 100 μM THP [46]. A similar association between the inhibition of α-ketoglutarate dehydrogenase and the impairment of cell viability was observed in studies with untransformed human epithelial keratinocyte cell lines [53].

The effects of THP and SAL on human primary melanocytes were reevaluated by treating the cells with variable concentrations of each THIQ [54]. Both TIQs were well tolerated up to roughly 30 μM, but at higher concentrations, THP became overtly toxic while SAL showed no cytotoxic effect up to 100 μM. A decrease in α-ketoglutarate dehydrogenase activity was also evident following THIQs treatment; a very strong diminution was found in THP-treated cells, whose viability was highly decreased [54].

Rat C6 glioma cells treated with THP underwent both necrotic and apoptotic cell death as assessed by microscopic observation of cells after differential staining with propidium iodide and Hoechst 33342, respectively. THP-induced cell death was accompanied by increased membrane lipid peroxidation and attenuated in the presence of NAC [50]. Furthermore, pretreatment with inhibitors of c-Jun N-terminal kinase and p38 mitogen-activated protein kinase (MAPK) rescued the glioma cells from THP-induced cytotoxicity, suggestive of the involvement of these kinases in THP-induced glioma cell death [50]. Likewise, THP caused cytotoxicity in PC12 cells, which was completely blocked by GSH and NAC. THP-treated PC12 cells exhibited increased intracellular accumulation of ROS and underwent apoptosis as determined by poly(ADP-ribose)polymerase cleavage, an increased ratio of Bax to Bclx$_L$, positive terminal transferase-mediated dUTP nick end labeling (TUNEL), and nuclear fragmentation/condensation [51]. Consistent with these results, exposure of PC12 cells to THP, at concentrations higher than 15 μM, resulted in apoptosis, which appeared to be mediated by ROS [35]. Furthermore, exposure of PC12 cells to THP combined with L-DOPA elicited synergistic effects, further increasing the proportion of TUNEL-positive apoptotic cells.

THP induced apoptosis in the human leukemia (HL-60) cell line, but did not in its hydrogen peroxide (H_2O_2)-resistant clone HP100. THP-induced DNA ladder formation in HL-60 cells was inhibited by a metal chelator. This study demonstrated that THP induced HL-60 cell apoptosis by generating ROS, presumably hydroxyl radical, through a reaction involving H_2O_2 and transition metals [55]. In untransformed human epithelial keratinocytes cells, THP at concentrations higher than 50 μM induced a mild impairment of cell growth, while L-DOPA, even at a 100 μM concentration, was well tolerated by these cells. In contrast, a pronounced toxic effect was exerted by THP on transformed keratinocytes [53].

Taking all the aforementioned findings together, it is evident that THP provokes cytotoxicity in many different cell lines through induction of oxidative stress (Figure 30.2). However, the molecular milieu mediating THP-induced oxidative cell death needs to be uncovered.

30.4.3
DNA Damage

Incubation of THP with phiX174 supercoiled DNA or calf thymus DNA in the presence of cupric ion caused substantial DNA damage as determined by strand scission or formation of 8-oxo-7,8-dihydro-2′-deoxyguanosine, respectively. THP plus copper-induced DNA damage was ameliorated by some ROS scavengers/antioxidants and catalase [50].

THP induced damage to ^{32}P-labeled DNA fragments in the presence of metals. In the presence of Fe(III)EDTA, THP caused DNA damage at every nucleotide [55]. The DNA damage was inhibited by free hydroxyl radical scavengers and catalase. In the presence of Cu(II), THP caused DNA damage mainly at T and G of 5′-TG-3′ sequence. The Cu(II)-mediated DNA damage was attenuated by catalase and the copper chelator, bathocuproine [55]. Table 30.2 highlights the biochemical basis of THP-induced neurotoxicity.

30.5
Cellular Protection against THP-Induced Injuries

The body's antioxidant defense system serves to protect the cells from excess ROS production and comprises both endogenous and exogenous entities [56]. The endogenous antioxidant defense system include not only radical scavengers (e.g., GSH, thioredoxin, bilirubin, and uric acid) but also antioxidant enzymes such as SOD, catalase, glutathione peroxidase, glutathione S-transferase, glutamate-cysteine ligase, and NAD(P)H:quinone oxidoreductase 1. The exogenous antioxidants (e.g., carotenoids, tocopherols, ascorbate, and bioflavonoids) are consumed through diet, especially fruits and vegetables.

As THP bears two catechol moieties, it undergoes autoxidation or enzymatic oxidation to produce ROS, which may contribute to the THP-induced cell death.

Tab. 30.2 Biochemical mechanisms underlying THP-induced neurotoxicity

Redox cycling
    Undergo oxidation to form reactive semiquinone
        and o-quinone intermediates
    ROS production
Oxidative cell death
    Necrosis or apoptosis
    Reduction of ATP levels
DNA damage
    DNA strand scission or base modification (accelerated
        in the presence of transition metal ions)
    Amelioration of DNA damage by ROS scavengers/antioxidants

Although overproduction of ROS is cytotoxic, the initial accumulation of moderate amounts of ROS may provoke a survival response.

Keratinocytes of the human epidermis, a tissue particularly exposed to oxidant stimuli, possess a wide range of antioxidant and detoxifying mechanisms aimed to avoid oxidative damage of the tissue. Foppoli *et al.* demonstrated that THP and L-DOPA upregulated expression of intracellular antioxidant enzymes to a different extent in normal keratinocytes of human epidermal origin with respect to transformed ones [53]. Normal diploid keratinocytes adequately scavenge toxic substances through the coordinated activation of several concurrent pathways. Conversely, in transformed cells, the whole oxidative burden must be neutralized by the limited set of conserved pathways that, accordingly, have to be highly activated [53].

THP treatment caused activation of the redox-sensitive transcription factor nuclear factor κB (NF-κB). Pretreatment of PC12 cells with NF-κB inhibitors, such as L-1-tosylamido-2-phenylethyl chloromethyl ketone and parthenolide, aggravated THP-induced cell death [51]. THP treatment resulted in differential activation of MAPK as well as Akt/protein kinase B, thereby transmitting cell survival or death signals. The redox-sensitive transcription factor Nrf2 is known to regulate the expression of various cellular genes that encode detoxifying and antioxidant enzymes and other defensive proteins against oxidative stress and other noxious conditions [57]. Treatment of PC12 cells with THP increased expression of heme oxygenase-1 (HO-1), a key enzyme that plays an essential role in cellular adaptive survival response to a wide variety of stress [58]. THP-induced cytotoxicity was attenuated by the HO-1 inducer ($SnCl_2$) and exacerbated by pharmacologic inhibition of HO-1 activity [59]. In addition, THP treatment elevated nuclear translocation of Nrf2 and subsequent binding to antioxidant response element (ARE). PC12 cells transfected with dominant-negative Nrf exhibited increased cytotoxicity and decreased HO-1 expression after THP treatment. Together, these findings suggest that Nrf2 activation and subsequent induction of antioxidant/cytoprotective enzymes, particularly HO-1, confer a cellular adaptive response against THP-mediated cell death (Figure 30.2).

30.6
Concluding Remarks

Since the 1970s, THP and other structurally related THIQ alkaloids have been the frequent subjects for investigations in the area of neuropharmacology and neurotoxicology. However, there has been a paucity of evidence supporting the neurotoxic potential of THP in humans. Over the past three decades, there has been enormous progress in understanding the neurochemistry and neuropharmacology of THP, and the implications of this endogenous toxin in the etiology of parkinsonism and alcoholism are now revisited. Additional studies will be necessary to uncover the molecular targets of this neurotoxin, especially those involved in intracellular signaling cascades. In addition, the adverse effects of THP on cellular functions other than neurotoxic activity merit further investigation.

References

1. Collins, M.A. (1982) *Trends Pharmacol Sci*, **3**, 373–375.
2. Sandler, M., Carter, S.B., Hunter, K.R., and Stern, G.M. (1973) *Nature*, **241**, 439–443.
3. Collins, M.A. (2004) *Neurotoxicology*, **25**, 117–120.
4. Soto-Otero, R., Sanmartin-Suarez, C., Sanchez-Iglesias, S., Hermida-Ameijeiras, A., Sanchez-Sellero, I., and Mendez-Alvarez, E. (2006) *J Biochem Mol Toxicol*, **20**, 209–220.
5. McNaught, K.S., Carrupt, P.A., Altomare, C., Cellamare, S., Carotti, A., Testa, B., Jenner, P., and Marsden, C.D. (1998) *Biochem Pharmacol*, **56**, 921–933.
6. Nagatsu, T. (1997) *Neurosci Res*, **29**, 99–111.
7. Cashaw, J.L. (1993) *J Chromatogr*, **613**, 267–273.
8. Cashaw, J.L. (1993) *Alcohol*, **10**, 133–138.
9. Cashaw, J.L. (1987) *Anal Biochem*, **162**, 274–282.
10. Turner, A.J., Baker, K.M., Algeri, S., Erigerio, A., and Garattini, S. (1974) *Life Sci*, **14**, 2247–2257.
11. Sango, K., Maruyama, W., Matsubara, K., Dostert, P., Minami, C., Kawai, M., and Naoi, M. (2000) *Neurosci Lett*, **283**, 224–226.
12. Surh, Y.-J. (1999) *Eur J Clin Invest*, **29**, 650–651.
13. Walsh, M.J., Davis, V.E., and Yamanaka, Y. (1970) *J Pharmacol Exp Ther*, **174**, 388–400.
14. Hotz, P., Stock, K., and Westermann, E. (1964) *Nature*, **203**, 656–658.
15. Rueffer, M., El-Shagi, H., Nagakura, N., and Zenk, M.H. (1981) *FEBS Lett*, **129**, 5–9.
16. Cashaw, J.L. and Geraghty, C.A. (1991) *Alcohol*, **8**, 317–319.
17. Smythe, G.A. and Duncan, M.W. (1985) *Prog Clin Biol Res*, **183**, 77–84.
18. Turan, S.C., Shah, P., and Pietruszko, R. (1989) *Alcohol*, **6**, 455–460.
19. Davis, V.E., Walsh, M.J., and Yamanaka, Y. (1970) *J Pharmacol Exp Ther*, **174**, 401–412.
20. Nace, E.P. (1986) *Alcohol*, **3**, 83–87.
21. Haber, H., Roske, I., Rottmann, M., Georgi, M., and Melzig, M.F. (1997) *Life Sci*, **60**, 79–89.
22. Storch, A., Ott, S., Hwang, Y.I., Ortmann, R., Hein, A., Frensel, S., Matsubara, K., Ohta, S., Wolf, H.U., and Schwarz, J. (2002) *Biochem Pharmacol*, **63**, 909–920.
23. Niwa, T., Takeda, N., Kaneda, N., Hashizume, Y., and Nagatsu, T. (1987) *Biochem Biophys Res Commun*, **144**, 1084–1089.
24. Nagatsu, T. and Yoshida, M. (1988) *Neurosci Lett*, **87**, 178–182.
25. Kim, Y.M., Reed, W., Wu, W., Bromberg, P.A., Graves, L.M., and Samet, J.M. (2006) *Am J Physiol Lung Cell Mol Physiol*, **290**, L1028–L1035.
26. Okada, S., Shimada, S., Sato, K., Kotake, Y., Kawai, H., Ohta, S., Tohyama, H., and Nishimura, T. (1998) *Neurosci Res*, **30**, 87–90.
27. Suzuki, K., Mizuno, Y., and Yoshida, M. (1990) *Neurochem Res*, **15**, 705–710.
28. Morikawa, N., Nakagawa-Hattori, Y., and Mizuno, Y. (1996) *J Neurochem* **66**, 1174–1181.
29. Kim, E.I., Yin, S., Kang, M.H., Hong, J.T., Oh, K.W., and Lee, M.K. (2003) *Neurosci Lett*, **339**, 131–134.
30. Marsden, C.D. (1994) *J Neurol Neurosurg Psychiatry*, **57**, 672–681.
31. Ogawa, N. (1994) *Eur J Neurol*, **34**, 20–28.
32. Fahn, S. (1992) *Ann Neuropsichiatr Psicoanal*, **32**, 804–812.
33. Morris, J.G. (1978) *Clin Exp Neurol*, **15**, 24–50.
34. Matsubara, K., Fukushima, S., Akane, A., Kobayashi, S., and

Shiono, H. (1992) *J Pharmacol Exp Ther*, **260**, 974–978.

35 Lee, J.J., Kim, Y.M., Yin, S.Y., Park, H.D., Kang, M.H., Hong, J.T., and Lee, M.K. (2003) *Biochem Pharmacol*, **66**, 1878–1795.

36 Cohen, G. and Collins, M. (1970) *Science*, **167**, 1749–1751.

37 Myers, R.D. and Melchior, C.L. (1977) *Science*, **196**, 554–556.

38 Dunkan, C. and Deitrich, R.A. (1980) *Pharmacol Biochem Behav*, **13**, 265–281.

39 Myers, R.D. and Privette, T.H. (1989) *Brain Res Bull*, **22**, 899–911.

40 Myers, R.D. and Robinson, D.E. (1999) *Alcohol*, **18**, 83–90.

41 Davis, V.E. and Walsh, M.J. (1970) *Science*, **167**, 1005–1007.

42 Kirby, G.W. (1967) *Science*, **155**, 170–173.

43 Donnerer, J., Cardinale, G., Coffey, J., Lisek, C.A., Jardine, I., and Spector, S. (1987) *J Pharmacol Exp Ther*, **242**, 583–587.

44 Weitz, C.J., Lowney, L.I., Faull, K.F., Feistner, G., and Goldstein, A. (1986) *Proc Natl Acad Sci U S A*, **83**, 9784–9788.

45 Cardinale, G.J., Donnerer, J., Finck, A.D., Kantrowitz, J.D., Oka, K., and Spector, S. (1987) *Life Sci*, **40**, 301–306.

46 De Marco, F., Perluigi, M., Marcante, M.L., Coccia, R., Foppoli, C., Blarzino, C., and Rosei, M.A. (2002) *Biochem Pharmacol*, **64**, 1503–1512.

47 Zhang, F. and Dryhurst, G. (2001) *J Pharm Biomed Anal*, **25**, 181–189.

48 Kim, H.J., Soh, Y., Jang, J.H., Lee, J.S., Oh, Y.J., and Surh, Y.-J. (2001) *Mol Pharmacol*, **60**, 440–449.

49 Jung, Y.-J. and Surh, Y.-J. (2001) *Free Radic Biol Med*, **30**, 1407–1417.

50 Soh, Y., Shin, M.H., Lee, J.S., Jang, J.H., Kim, O.H., Kang, H., and Surh, Y.J. (2003) *Mutat Res*, **544**, 129–142.

51 Shin, M.-H., Jang, J.-H., and Surh, Y.-J. (2004) *Free Radic Biol Med*, **36**, 1185–1194.

52 Maruyama, W., Sango, K., Iwasa, K., Minami, C., Dostert, P., Kawai, M., Moriyasu, M., and Naoi, M. (2000) *Neurosci Lett*, **291**, 89–92.

53 Foppoli, C., De Marco, F., Blarzino, C., Perluigi, M., Cini, C., and Coccia, R. (2005) *Int J Biochem Cell Biol*, **37**, 852–863.

54 Perluigi, M., De Marco, F., Foppoli, C., Coccia, R., Blarzino, C., Marcante, M.L., and Cini, C. (2003) *Biochem Biophys Res Commun*, **305**, 250–256.

55 Kobayashi, H., Oikawa, S., and Kawanishi, S. (2006) *Neurochem Res*, **31**, 523–532.

56 Fisher-Wellman, K. and Bloomer, R.J. (2009) *Dyn Med*, **8**, [1].

57 Motohashi, H. and Yamamoto, M. (2004) *Trends Mol Med*, **10**, 549–557.

58 Otterbein, L.E. and Choi, A.M. (2000) *Am J Physiol Lung Cell Mol Physiol*, **279**, L1029–L1037.

59 Park, S.-H., Jang, J.-H., Li, M.-H., Na, H.-K., Cha, Y.-N., and Surh, Y.-J. (2007) *Antioxid Redox Signal*, **9**, 2075–2086.

31
Chemically Induced Autoimmunity

Michael Schiraldi and Marc Monestier

Since the first report of drug-induced lupus in 1945, knowledge of the autoimmunity-inducing and modulating potential of chemical agents has been fueled by epidemiologic, clinical, and animal studies. Well over one hundred different pharmacotherapeutics ranging the entire spectrum of drug classes, as well as many environmental pollutants and marketed products, have been implicated in the onset or exacerbation of autoimmune manifestations resembling lupus, myasthenia gravis, glomerulonephritis, pemphigus, biliary cirrhosis, Goodpasture's disease, and eosinophilic pneumonia. Several animal models of chemically induced autoimmunity have also been developed, creating a resource for understanding the interactions between chemicals and cells of the immune system. Chemical agents can directly promote lymphocyte and monocyte function, deregulate mechanisms of peripheral tolerance, and alter self molecules to generate immunogenic epitopes. Future research in the field of chemically induced autoimmunity will help physicians and scientists develop alternatives to autoimmune-inducing drugs, shape guidelines for product usage and waste management, and provide insights into naturally occurring autoimmune processes that will fuel the development of novel therapies.

31.1
Introduction

Since Hoffman reported the first case of drug-induced systemic lupus erythematosus (SLE) in 1945 [1], many small molecules, including pharmacologic agents and environmental toxins, have been implicated in the induction of autoimmune responses (immune reactions directed against self without associated symptoms) as well as clinical autoimmune disease. Over 80 therapeutic agents have been reported to cause drug-induced lupus (DIL) [2]. Drug-induced immune hemolytic anemia (DIIHA) has been linked to use of nonsteroidal anti-inflammatory drugs (NSAIDs), antimalarial agents (quinine and quinidine), cephalosporins, penicillin and its derivatives, as well as several other agents [3]. The chelating agent penicillamine can cause a disease clinically identical to myasthenia gravis (MG) in up to 0.5% of patients treated; this same agent results in pemphigus in another subset of patients. Similarly, environmental agents such as mercury, cigarette smoke, and pesticides have been linked to the induction of autoimmune processes ranging from subclinical autoantibody production to severe disease [4].

Research in the area of chemically induced autoimmunity (CIA) is important for several obvious reasons. Therapeutic agents with untoward effects present challenges to physicians, and the identification of agents without such properties will benefit patients. Understanding interactions of environmental toxins with the immune system will aid in establishing guidelines for the use of chemical agents and management of waste products. Finally, studies of CIA in both patient populations and animal models provide insights into the pathogenesis of naturally occurring autoimmune disease. This chapter will address clinically relevant autoimmune manifestations induced de novo by chemicals (as distinct from drug hypersensitivity reactions, which are clinically and mechanistically unrelated), animal models of CIA, and hypotheses of the mechanisms involved in such diseases. Biologic agents intended to modulate immune system function such as cytokine therapies, although reported to induce autoimmunity (presumably by disrupting immune homeostasis), will not be covered.

31.1.1
Autoimmunity

The immune system is generally regarded as a protection from disease-causing pathogens. It is the system of organs, cells, and cellular products responsible for distinguishing nonself molecules from self molecules, and those belonging to commensal microbes and ridding the organism of nonself. Nonself molecules include those belonging to bacteria, viruses as well as neoantigens on the surface of genetically aberrant cancer cells. Autoimmunity results when tolerance for self is broken and a reaction directed against self ensues. For example, in patients with SLE, antibodies directed against nuclear antigens are produced (antinuclear antibodies (ANA)). Autoimmune disease occurs when such reactions to self result in clinically relevant symptoms.

Autoimmunity is a heterogenous group of manifestations including systemic responses (SLE) and tissue or organ-specific reactions (multiple sclerosis). These processes can be mediated by antibodies directed against specific self molecules (autoantigens), immune complexes, as well as cell-mediated cytotoxicity. CIA includes both system and tissue-specific responses and can range from subclinical autoantibody formation to severe disease with manifestations similar to naturally occurring autoimmunity. Classically, remission of autoimmune manifestations upon withdrawal of the inciting agent is a requirement for the diagnosis of CIA, but persistent manifestations, in a subset of patients, do occur.

31.2
Human Disease

31.2.1
Drug-Induced Lupus

SLE is a relatively common multisystem immune complex disease marked by the presence of ANA specific for various antigens, including native DNA and extensive inflammatory lesions. Acute cases, although rare, may precipitate death within weeks to months, but more commonly the disease is characterized by flare-ups and remissions over years or decades with appropriate disease management. A role for genetic predisposition is supported by many studies, but incomplete concordance amongst identical twins (as is the case with autoimmune disease in general) points to the multifactorial nature of this disease.

Cigarette smoke is reported as a risk factor for SLE, but in both humans and rodents the agent decreases the manifestations of disease. This apparent discrepancy is reconciled by a report that the immunosuppressive effects of cigarette smoke are abated with smoking cessation and such patients or animals reach autoantibody levels higher than still-smoking and nonsmoking controls [5]. That smoking contributes to autoimmunity is supported by evidence that antielastin autoimmune responses occur in patients with smoking-induced emphysema [6].

DIL has been correlated with the use of over 80 pharmacologic agents spanning at least 10 major categories of drugs; the prototypic SLE-inducing drugs are hydralazine and procainamide. While these two drugs each induce lupus in 7–13% and 15–20% of patients taking them respectively, the many others reported to cause lupus do so at rates less than 1%. Still, between 15 000 and 30 000 cases of DIL are estimated to occur each year in the United States. Lupus-like symptoms such as fever, weight loss, fatigue, and musculoskeletal complaints are present in about half of affected patients. Hydralazine in particular, is associated with skin rashes, but not the malar rashes typical of SLE. Also in contrast to naturally occurring SLE, DIL rarely includes renal involvement. The syndrome is generally milder than SLE possibly because while the autoantibodies of naturally occurring SLE fix complement, those produced in DIL do not. DIL resolves within days to weeks of discontinuing the drug. Serologic testing reveals ANA, which target the histone-(H2A-H2B)-DNA subnucleosome particle, but not native DNA. In naturally occurring disease, native DNA is targeted in addition to the subnucleosome [2, 7, 8]. Table 31.1 compares and contrasts the characteristics of SLE and DIL.

Therapeutics known to cause DIL are diverse in molecular structure and pharmacodynamics, yet each results in the same clinical and laboratory features. A summary of proposed mechanisms leading to DIL is summarized in Table 31.2. In most cases, months or years of continuous treatment with the inciting agent precede DIL; this point suggests a role for drug metabolism in these untoward effects. Activated neutrophils oxidize all classes of lupus-inducing drugs by enzymatic action of myeloperoxidase (MPO). Within each class there are chemical analogs of lupus-inducing drugs, which are not substrates for MPO [9]. Patients who are

Tab. 31.1 Naturally occurring versus drug-induced lupus

	SLE	DIL
Disease course	Chronic disease marked by flare-ups and remissions	Insidious onset of generally mild symptoms after weeks or months of exposure to inciting drug; remission upon withdrawal of drug
Nuclear targets	Subnucleosome particles and native DNA and	Subnucleosome particles
Complement fixation	Yes	No
Renal involvement	Severe	Rare
Skin involvement	Characteristic malar rash	Rash without malar distribution

Tab. 31.2 Proposed mechanisms of drug-induced lupus

Drugs act as haptens
Noncovalent TCR interactions leading to activation
Drug-induced cytotoxicity releases sequestered antigen
Inhibition of DNA methylation at promoter sequences
    Increased cell adhesion molecule expression (i.e., LFA-1)
    Increased costimulatory molecule expression (i.e., CD70)
Disruption of B-cell receptor editing

phenotypically rapid acetylators of hydralazine and procainamide are less likely to develop DIL because N-acetylation of these drugs competes with N-oxidation [10].

The formation of stable drug- or metabolite-self molecule complexes is one possible mechanism for breakage of tolerance to self. Drugs may serve as haptens for drug-specific T cells [11], or noncovalent interactions with the T-cell receptor can promote activation [12]. Activated T cells can augment autoimmune responses that may have remained subclinical prior to pharmacotherapy or those arising de novo because of the drug. The cytotoxicity of certain metabolites may also contribute to DIL; the metabolism by neutrophils of amodiaquine, carbamazepine, chlorpromazine, clozapine, hydralazine, isoniazid, procainamide, propylthiouracil, quinidine, and sulfonamides yields cytotoxic agents [7]. Experimental evidence is lacking in regards to connecting drug cytotoxicity with autoimmunity, but possible mechanisms include the production of cryptic T-cell autoepitopes or release of nuclear antigens from dying cells capable of eliciting autoimmune responses.

Both procainamide and hydralazine can nonspecifically activate splenocytes [13]. Inhibition of DNA methyltransferase by these agents results in hypomethylation of promoter sequences in T cells, contributing to their activation; increased expression of lymphocyte function antigen-1 (LFA-1), a cell adhesion molecule, is one example of this phenomenon. Overexpression of LFA-1 will promote contact between T cells and antigen presenting cells (APCs), and may lead to activation by normally low-affinity interactions with TCR-self-antigen complexes [14]. Lupus-inducing drugs also induce hypomethylation of the CD70 promoter and subsequent overexpression. Binding of CD70 to CD28 activates the immune system via NFκB and MAPK8/JNK. The same region is hypomethylated in patients with SLE [15]. In both DIL and SLE, hypomethylation of the CD70 promoter is correlated with impaired T-cell protein kinase C delta activation [16].

Intrathymic injection of procainamide results in antichromatin autoantibodies under experimental conditions, supporting a role for perturbations of central immune tolerance in the onset of DIL [17]. B-cell receptor editing is essential for preventing maturing cells whose surface immunoglobulin targets the autoantigen from entering the periphery. A potential mechanism by which hydralazine induces lupus is disruption of this process, as suggested by Mazari et al. [18].

Research in the area of DIL is important for two clear reasons: the development of drugs without DIL-inducing effects would be beneficial; and the recognition of this clinical entity has fueled decades of studies aimed at understanding not only immune system–environment interactions, but the mechanisms behind naturally occurring autoimmune diseases.

31.2.2
Drug-Induced Immune Hemolytic Anemia

Immune hemolytic anemias (IHAs) are a group of disorders clinically suggested by recent onset hemolytic anemia The onset of IHA is a relatively common form of CIA. Fulminant hemolysis with resultant death is rare, but many patients develop subclinical manifestations made evident only by a positive direct Coombs' test, which detects binding of antibody to the red cell surface. When a patient is found to develop a positive Coombs' test, it must be determined whether overt hemolysis is also present. In such cases, the hemolysis is rapid and death may ensue [19].

The first confirmed report of DIIHA described a patient treated for schistosomiasis with stibophen in 1954 [20]. By 1975, the list of agents implicated in DIIHA included 22 pharmacologics as well as insecticides (chlorinated hydrocarbons) [21]. The number of agents today, known to cause DIIHA is over 50 and at least three separate mechanisms can work to yield autoimmunity depending on the agent [19]. A recent report stressed the importance of recognizing NSAIDs as a class of drugs capable of inducing drug-dependent antibodies. Indeed, a single-center study has demonstrated that seven different NSAIDs can yield DIIHA [3].

Several mechanisms may be involved in the onset of DIIHA (Table 31.3). Quinidine and the analgesic phenacetin bind to cell membranes and act as haptens to which antibodies are formed. These antibodies interact with free drugs and form immune complexes that interact with the surface of red blood cells (RBCs), where complement activation leads to hemolysis. In this way, an adaptive immune response to a foreign molecule leads to an inadvertently autoimmune response by components of the innate system. Penicillins as well as cephalosporins interact

Tab. 31.3 Proposed mechanisms of drug-induced immune hemolytic anemia

Drugs act as haptens
Antibody–drug complexes bind RBC membranes and activate complement
Direct chemical modification of RBC membranes create neoantigens

---

directly with red cell membranes and promote the binding of antidrug antibodies to RBCs [22]. Cephalosporins can also chemically modify the membranes of RBCs, which leads to nonspecific protein adsorption by these cells; the binding of complement proteins promotes hemolysis, while antibody binding provides a positive direct Coomb's test. Other drugs, such as α-methyldopa, induce anti-RBC antibodies by still unknown mechanisms [23, 24].

DIIHA presents a challenge because manifestations are rarely clinically evident. However, understanding of this adverse drug reaction is important, and distinguishing a subclinical positive Coomb's test from fulminant hemolysis is imperative when treating patients. The number of drugs capable of inducing immune mediated hemolysis should be appreciated by prescribing physicians and efforts to identify DIIHA as a possible side effect of new drugs is warranted.

31.2.3
Primary Biliary Cirrhosis

Primary biliary cirrhosis (PBC) is a chronic cholestatic liver disease mediated by autoreactive T cells and antimitochondrial autoantibodies (AMA). A role for chemicals in the onset of PBC has been proposed and is supported by epidemiologic and laboratory data [25]. AMA from PBC patients bind autoepitopes modified by various organic chemicals with greater affinity than the native structures [26]. In rabbits and guinea pigs, AMA as well as anti-lipoic acid (LA) antibodies have been induced by treatment with 6-bromohexanoate [27, 28].

The targeting of LA suggests that chemicals similar in structure serve to trigger AMA in PBC. The binding of AMA from PBC patients to various structures mimicking LA have been quantified, and results suggest 2-octynoic acid and 2-nonynoic acid contribute to disease onset. Both of these chemicals are additives in cosmetic products [29, 30]. In C57Bl/6 mice, treatment with 2-octynoic acid led to the development of AMA and biliary ductular disease [31].

31.2.4
D-Penicillamine-Induced Myasthenia Gravis

D-Penicillamine is a chelating agent used to treat patients with Wilson's disease (a genetic disorder resulting in the accumulation of toxic levels of copper), cysteinuria, and rheumatoid arthritis (RA) – patients who have failed conventional therapy. Long-term use of this agent can result in a syndrome clinically identical to MG, an autoimmune disease targeting the neuromuscular junction [32]. Autoantibodies to the acetylcholine (ACh) receptor are detected in 0.5% of individuals treated with D-penicillamine, but titers decrease once the drug is withdrawn. Rare cases are not reversed by cessation of the therapy, but in such cases it is possible that MG would have developed in these patients without D-penicillamine treatment [33, 34]. The onset of D-penicillamine-induced MG has a strong correlation with the DR1 major histocompatibility complex (MHC)-II haplotype [35].

T-cell clones from DR1+ patients with D-penicillamine-induced MG responded to drug-treated mononuclear cells, but not to drug-treated autologous B cells. It has been postulated that T cells recognize DR1 : peptide complexes that are modified by D-penicillamine and that such T cells drive the disease[35].

31.2.5
Mercury and Systemic Autoimmune Disease

Mercury release from natural degassing of the earth's crust, its release as an environmental pollutant, and its presence in man made products, make human exposure unavoidable [36]. The toxic effects of mercury (Hg) on the kidney and central nervous system have been extensively studied [37–39], while the ability of Hg to induce or exacerbate human autoimmune disease is less understood. Table 31.4 summarizes evidence supporting a role for Hg in human autoimmunity. Mercury-containing skin-lightening cream has been a known cause of nephrotic syndrome since 1972 [40]. The sale of such creams is restricted in many countries, but the use of this product remains a reported cause of membranous nephropathy and minimal change disease (nephrotic syndrome due to podocyte effacement visible only with electron microscopy) in some parts of the world. In such cases, immunoglobulin and complement deposits are detected in the glomerulus [41, 42].

Occupational exposure to Hg vapors has been linked to T-cell lymphoproliferation [43] and the production of antilaminin autoantibodies [44]. A recent retrospective study revealed a positive association between exposure to mercury and the development of Wegener's granulomatosis, an autoimmune necrotizing vasculitis associated with the presence of cytoplasm targeting antineutrophil antibodies [45].

Tab. 31.4 Evidence that Hg promotes autoimmunity in humans

Source	Findings
Hg-containing skin creams	Case reports of membranous nephropathy and minimal change disease
Occupational and environmental exposures	T-cell lymphoproliferation, antilaminin, and antinucleolar autoantibodies
Dental amalgam	Controversial; a few animal and clinical studies support an association with autoimmunity
Other	Associated with Wegener's granulomatosis; Hg levels in urine associated with disease severity in a subset of scleroderma patients

Dental amalgam contains elemental mercury and continuously releases small amounts of mercury vapor. Amalgam implantation in SJL mice induces a state of chronic immunostimulation including increased T and B-cell proliferation, and autoantibodies to the nucleolar protein fibrillarin [46]. In humans, the controversy about the safety of dental amalgam is a lengthy and contentious issue. The vast majority of individuals with mercury-containing dental amalgams do not develop any complications, but certain studies have suggested that mercury vapor can be neurotoxic and could contribute to the pathogenesis of multiple sclerosis [47]. A study reported that the removal of dental amalgam from patients with SLE, autoimmune thyroiditis, or multiple sclerosis led to clinical improvements [48]. It would however, be premature to advocate this approach in the treatment of autoimmune diseases.

31.2.6
Hydrocarbon Exposure and Goodpasture's Syndrome

Goodpasture's syndrome (GS) is a tissue-specific autoimmune disease wherein autoantibody targeting of the noncollagenous domain of the α-3 chain of type IV collagen results in inflammatory damage of kidney glomeruli and lung alveoli. Unlike the majority of autoimmune diseases, men are affected more frequently than women; most cases occur in the second or third decade of life. Presenting symptoms usually include hemoptysis and evidence of focal pulmonary consolidations, but this can soon be followed by rapidly progressive glomerulonephritis and subsequent death.

A long-proposed mechanism for the development of GS is exposure to hydrocarbon solvents. Indeed, a case report from 1979 states factually the association between volatile hydrocarbons and this disease, while reporting on a 22-year-old girl recreational drug user who developed GS following prolonged solvent abuse [49]. Yet, even a decade later, there remained no evidence to support the hypothesis of hydrocarbon involvement in the induction of GS [4]. Proponents of the hydrocarbon exposure theory cite consistency among case reports implicating the chemicals [50]. On the other hand, a large-scale review of 700 GS cases over a 31-year period, revealed only a 6% incidence of confirmed associations with exposure [51]. There is no laboratory evidence that hydrocarbons can induce Goodpasture's de novo, but a role for disease propagation by chemical exposure has been demonstrated in rabbits. In naïve rabbits treated with antibasement membrane antibodies, the immunoglobulins did not bind to the alveolar basement membrane. The same antibodies did bind the alveolar basement membranes of rabbits subjected to intrathecal administration of gasoline, secondary to damage to the pulmonary barrier normally preventing such interactions. This study supports the hypothesis that hydrocarbon exposure can precipitate pulmonary hemorrhage in Goodpasture's patients by intermittently allowing autoantibody binding with the alveolar basement membrane, but the incidence of confirmed hydrocarbon exposure in GS is only 6% [52].

Renewed interest in the relationship between hydrocarbons and autoimmunity has been fueled by the discovery that the aryl hydrocarbon receptor (AhR) modulates the differentiation of proinflammatory interleukin-17 (IL-17) producing T helper (Th17) cells, and regulatory T cells (Tregs). This receptor is expressed in most cells and binds promiscuously to both naturally occurring and synthetic hydrocarbon ligands [53]. In CD4+ T cells from mice, receptor expression is limited to Th17 cells, where it is also expressed in humans. Activation of AhR by 6-formoindolo[3,2-b]carbazole (FICZ), a naturally occurring ligand, during Th17 cell development increased cell numbers and promoted cytokine development. By binding to AhR, hydrocarbons may serve as cofactors during the development of autoimmune disease [54]. In the mouse model of MS, experimental autoimmune encephalomyelitis (EAE), AhR ligation by 2,3,7,8-tetrachlorodibenzo-p-dioxin (TCDD), a synthetic dioxin pollutant, induced Treg development and decreased severity of disease. In contrast, ligation of AhR with FICZ prevented Treg development and promoted pathogenesis by increasing Th17 cell numbers [55].

31.2.7
Pesticide Exposure and Autoimmunity

Between 1955 and 1959, over 3000 people were exposed to the fungicide hexachlorobenzene (HCB) via ingestion of contaminated seed grain. Individuals developed hepatic porphyria (porphyria turcica) with cutaneous manifestations including bullous lesions that healed with severe scarring. Other symptoms included splenomegaly, enlarged lymph nodes, and painless arthritis [56]. The hypothesis that the disease resulting from HCB exposure was of autoimmune etiology is supported by a report of elevated serum IgM and IgG levels in individuals occupationally exposed to the pesticide [57]. The mechanism by which HCB induces autoimmunity remains unknown, but recent work implicates tetrachlorobenzoquinone, a reactive metabolite of HCB in disease onset [58].

Epidemiologic evidence suggests that farmers are at an increased risk of developing autoimmunity, and exposure to pesticides plays a likely role. In 2007, a study of a rural population concluded that ANA levels were higher in farmers than control subjects living in urban settings. The prevalence of ANA was also higher during the spring (herbicide application season) than during winter months [59].

Despite epidemiologic and animal model evidence that HCB and other pesticides can cause a syndrome of suspected autoimmune etiology, the body of literature addressing the role of pesticides in autoimmune disease has remained small. Emphasis has instead been placed on the immunosuppressive effects of pesticides, as well as their ability to evoke hypersensitivity and allergic reactions [60].

31.2.8
Eosinophilic Pneumonia

At least 49 pharmacologic agents, as well as pollutants such as cigarette smoke, nickel dust, and World Trade Center dust have been implicated in the onset of eosinophilic pneumonia (EP). The diagnosis of EP is based on an elevation of peripheral eosinophils in the setting of pulmonary consolidations [61]. Notable outbreaks of EP have occurred on two occasions.

In 1981, in Spain, food-grade rapeseed oil that had been denatured with aniline was sold by street vendors leading to an epidemic of what is now known as *toxic oil syndrome* (TOS) [62]. Over 20 000 individuals in Madrid and northwest Spain who ingested the oil developed an acute illness including myalgias, peripheral eosinophilia, and pulmonary infiltrates. More than 300 deaths resulted during the first 20 months of the epidemic, most often from pulmonary hypertension and vascular thrombosis. Severe cases included proliferation of vessel intima with fibrosis and thrombosis [63, 64]. Eight years after the epidemic, it became evident that 20% of affected individuals developed chronic lung disease [65]. Patients in the chronic phase of TOS developed autoantibodies to acute phase reactants including C-reactive protein, α_1-antitrypsin, fibrinogen, and ceruloplasmin [66].

A model of TOS has been induced in B10.S mice by intraperitoneal injection of oleic acid. Treatment leads to IL-1, and IL-6 increases with associated polyclonal activation of B cells. Elevations of serum IgM, IgE, and IgG1 occur, and these antibodies target histones, denatured DNA, and rheumatoid factor [67].

The second notable epidemic was in 1989, when more than 1400 cases of eosinophilia-myalgia were recognized in New Mexico (up to 60 000 total cases in the United States). The causal agent of this outbreak was use of the amino acid L-tryptophan that had been manufactured in Japan as a food supplement [68]. The clinical manifestations of this syndrome are strikingly similar to TOS, thus a shared etiology has been postulated [69, 70]. Samples of L-tryptophan responsible for eosinophilia-myalgia contained 3-(phenylamino)-alanine (PAA), while the so-called "toxic oil" contained 3-phenylamino-1,2-propanediol (PAP). Research aimed at determining if these two compounds are converted to one another have not been fruitful, but a plausible theory is that both compounds belong to a group of chemically related structures capable of inducing disease [71]. In 2007, Martinez-Cabot and Messeguer reported on the generation of quinoneimine intermediates by liver metabolism of both PAA and PAP, suggesting that such metabolites are capable of inducing disease [72].

31.3
Animal Models of Chemically Induced Autoimmunity

31.3.1
Heavy Metals

The heavy metals, mercury, gold, silver, cadmium, and platinum can all induce autoimmunity in susceptible animals. Cadmium administration results in antinuclear antibody production in outbred ICR mice [73] and platinum induces antibodies to several nucleoplasmic antigens [74]. Gold, silver, and mercury induce autoantibodies specific to the nucleolar protein fibrillarin in a genetically restricted manner

[75–77]. Mercury-induced autoimmunity (HgIA) is the most widely studied model of heavy metal-induced autoimmunity and will be the focus of this section.

The first reports of HgIA showed that repeated subtoxic doses of $HgCl_2$ resulted in membranous glomerulonephritis in Wistar rats. Susceptibility to HgIA in rats maps to the RT-1 locus of the MHC class II genes. Inbred rats with the RT-1^n haplotype are highly susceptible; RT-$1^{a,c,b,f,k}$ haplotypes confer intermediate susceptibility, and the RT-1^l haplotype confers resistance [78, 79]. In male brown Norway (BN) rats, $HgCl_2$ leads to polyclonal B and T-cell activation, increased serum immunoglobulin levels, autoantibody production, and glomerulonephritis with immune complex deposition [80, 81]. The autoantibodies target a variety of antigens including DNA, phospholipids, glomerular basement membrane proteins, laminin-1, and thyroglobulin [82, 83]. Disease manifestations are self-limiting, and resolve in four to five weeks, even if $HgCl_2$ injections are continued. After disease resolution, animals are resistant to further $HgCl_2$ challenge, a phenomenon mediated by CD8+ T cells [84–87].

In mice, HgIA is also a genetically restricted disease with susceptibility mapping to the I-A region of the MHC class II locus. The H-2^s haplotype confers the greatest susceptibility, H-2^q and H-2^f mice are less susceptible while H-2^a, H-2^b, and H-2^d mice are resistant [88–91]. The syndrome induced by mercury in mice includes a Th2 biased polyclonal expansion of T and B lymphocytes, increased serum IgG1 and IgE, mild glomerulonephritis, and the production of highly specific autoantibodies targeting fibrillarin [92–95]. As in rats, the polyclonal activation and serum antibody levels decrease within four weeks, but antifibrillarin autoantibodies persist for months after cessation of treatment. In contrast to the rat model of HgIA, mice do not become resistant to subsequent mercury challenge [96]. A comparison of the rat and mice models of HgIA is found in Table 31.5.

Several mechanisms cause or promote HgIA (Table 31.6). Mercury ions (Hg^{2+}) bind thiol, amine, phosphoryl, carboxyl, and hydroxyl groups [97, 98]. These affinities allow mercury to interact with numerous biological molecules and affect their function. An example is the ability of mercury to bind cell surface molecules

Tab. 31.5 Hg-induced autoimmunity in rodents

	Mice	Rats
Genetic susceptibility	I-A locus of MHC class II	RT-1 locus of MHC class II
Targets	Nucleolus (fibrillarin)	DNA, phospholipids, GBM proteins, laminin-1, thyroglobulin
Course	Polyclonal activation is self-limited and resolves in four to five weeks; antifibrillarin response persists for months	Self-limited disease resolving in four to five weeks
Postdisease resistance	No	Yes

Tab. 31.6 Putative mechanisms of HgIA

Mechanism	Example
Promotes receptor aggregation	CD3, CD4, CD45, and Thy-I on T cells leading to deregulated signal transduction
Disruption of self-tolerance	Attenuation of CD95-induced apoptosis
Disruption of cytokine production by GSH depletion	Decreased interferon gamma and increased Il-4
Activation of T helper cells	Disruption of costimulatory interactions (i.e., CD40–CD40L or B7–CD28) prevents Hg-induced autoantibody production
Promotion of autoantigen processing and presentation	In vitro, Hg causes fibrillarin to colocalize with proteasomes
Local cytotoxicity	Hg-induced cell death results in a unique cleavage fragment of fibrillarin not produced by other causes of cell death

and cause receptor aggregation. On T cells, mercury-induced aggregation of CD3, CD4, CD45, and Thy-I results in deregulation of signal transductions pathways; such an effect has also been observed with a B-cell lymphoma line [99, 100].

Mercury may also promote autoimmunity by disrupting mechanisms of tolerance to self. After an immune reaction, the molecule CD95 plays a key role in apoptosis induction for the purpose of lymphocyte depletion. Low concentrations of mercury (5–10 μM) attenuate CD95-induced apoptosis providing a possible escape mechanism for autoreactive lymphocytes [101, 102]. Similar doses of mercury also have proliferative effects on splenocytes from susceptible strains of mice, but no such effect is observed in resistant mice [103]. The lymphoproliferation-inducing effects of mercury have also been observed in humans [104].

The mechanism by which mercury induces polyclonal activation is even less understood than the disruption of tolerance to self. As stated, mercury has a strong affinity for thiol-containing molecules and likely alters the availability of such species in immune cells [105]. Patterns of cytokine expression are partially dependent on glutathione (GSH), the largest source of intracellular thiols. Interferon-γ induction by concavalin A *in vivo* requires GSH, and mercury results in suppression of this cytokine in susceptible rats [106]. When GSH is depleted in mice, T cells stimulated *in vitro* produce decreased interferon-γ and increased Il-4 [107]. The cytokine profile of HgIA is Th2 biased and the depletion of GSH likely contributes to this phenotype.

31.3.2
Pristane

Pristane (2,6,10,14-tetramethylpentadecane) is a hydrocarbon component of mineral oil that has been used to induce models with both arthritis and SLE [108].

In 1994, it was serendipitously discovered that a single intraperitoneal injection of pristane alone resulted in the production of IgG autoantibodies targeting SLE-associated nuclear antigens in BALB/c mice. The high IgG titer results in immune complex glomerulosclerosis, similar to that seen in SLE patients. The mouse strain SJL/J is also susceptible to pristane-induced lupus, but the cytokine profile differs in these animals [109–111].

Pristane treatment results in local production of the proinflammatory cytokine IL-12 by APCs. The production of autoantibodies in these mice correlates with the ability to produce IL-12. Tumor necrosis factor alpha (TNF-α) and IL-6 are increased to a lesser extent [112]. There is expansion of marginal zone B cells, but a decrease in the number of CD1d-expressing natural killer T (NKT) cells and dendritic cells (DCs). CD1d is a MHC-I-like molecule; its deficiency exacerbates pristane-induced lupus in BALB/c mice. The role of CD1d in this animal model suggests that it may play a regulatory role in SLE [113]. Interestingly, NKT cell stimulation with the ligand α-galactosylceramide prevents disease in BALB/c mice, but exacerbates kidney manifestations in SJL/J mice [114].

That IL-12 is normally secreted in response to antigen presentation, and bacterial stimulation acts synergistically with pristane, indicates that the mechanism of disease onset may be increased exposure to microbial products by the oil. In 2005, Mizutani and colleagues explored this hypothesis by injecting germ-free mice with pristane. Autoantibodies developed in over 40% of treated mice, which indicates that a pathogen-independent mechanism can result in pristane-induced lupus [115]. This mechanism most likely involves the induction of apoptosis within the peritoneal cavity and subsequent release of (possibly chemically altered) nuclear antigens [116]. Components of this nuclear material likely interact with the nucleic acid receptor toll-like receptor-7 (TLR-7), which is required for pristane-induced autoantibody production and glomerular disease [117].

One-third of BALB/c mice treated with a single injection of pristane develop joint disease similar to RA. Pristane also induces arthritis in the DA and LEW rat strains [118]. In mice there is a reaction to heat-shock proteins, which also occurs in humans with RA [119–121]. Patients with RA also often produce antibodies to the heterogenous nuclear ribonucleoprotein (hnRNP)-A2, also known as the *RA33 antigen*. Autoantibodies in rats with pristane-induced arthritis also target hnRNP-A2, an antigen considered a primary inducer of the disease [122].

31.3.3
Phthalate-Induced Autoimmunity

Phthalates (o-benzene dicarboxylates) are a group of chemicals widely used in plastic-containing products including medical devices, children's toys, and cosmetics. Study of an antiphthalate antibody (2C3-Ig) revealed significant homology with the anti-DNA autoantibody BV04-01 from the autoimmune prone NZB/W F1 mice. The homology is 98% for the κ-light chain and 70% for the γ-heavy chain. In rodents, repeated exposure of phthalate-conjugated keyhole limpet hemocyanin induces activation of autoreactive B cells and the production of antibodies capable of binding both DNA and phthalate. Although such immune responses occur in BALB/c, NZB, and NZB/W F1 mice, they lead to increased mortality only in the latter [123]. Because NZB/W F1 mice are naturally prone to lupus, these results suggest that phthalates act by worsening a preexisting autoimmune condition.

In a subsequent study, the same authors found that BALB/c and DBA/2 mice overcame phthalate-induced autoimmune reactions by the induction of CD8+ T suppressor cells. By depleting CD8+ cells with antibody treatment, these strains of mice are rendered susceptible to the autoimmune manifestations induced by phthalate. Further attention to this model is warranted, as it is valuable for the study of both mechanisms of molecular mimicry as well as strain-specific differences in susceptibility to CIA [124].

31.3.4
D-Penicillamine-Induced Autoimmunity

As previously noted, D-penicillamine can produce autoimmune reactions in humans. Strain-specific susceptibility to D-penicillamine-induced autoimmunity (DIA), a model of systemic autoimmune disease, has been observed in mice including A/WySn and A.SW strains [125]. Upon intraperitoneal injection, D-penicillamine reacts directly with cell surface molecules to generate antigenic determinants [126]. In BN rats, D-penicillamine administration results in dermatitis, ANA, circulating immune complexes, and linear IgG deposits at the glomerular basement membrane [127].

The immunoglobulin heavy chain variable regions of antibodies created by A.SW mice during DIA are structurally similar to those from mice with HgIA [128]. Additionally, disease can be prevented by prechallenge injection with a low dose of the inciting chemical, also a feature of HgIA [129]. Transfer of this tolerance is mediated by both T and non-T cells [130].

The relevance of animal models to the study of human CIA is exemplified by recent work demonstrating that NSAIDs can modulate DIA in a drug-dependent fashion. Nonselective cyclooxygenase inhibitors increased the manifestations of DIA, but the selective COX-2 inhibitor rofecoxib decreased the incidence of disease [131]. As noted, NSAIDs have been added to the list of drugs capable of inducing DIIHA in humans. This animal model may prove useful in dissecting the mechanisms by which a class of drugs considered anti-inflammatory causes or promotes autoimmune reactions.

31.4
Conclusion

Chemical exposure has been linked to the onset of autoimmune reactions to a growing list of autoantigens in both humans and animals. The list of environmental agents, pharmacologics, and product additives capable of inducing autoimmunity has also expanded greatly since 1945. The study of CIA provides insights into mechanistic and genetic factors involved in disease onset and will continue to be invaluable in shaping guidelines for waste exposure, product usage, and pharmacotherapy. Comprehensive epidemiologic studies are needed to achieve these goals along with translational research aimed at elucidating the mechanisms of disease onset and course. Continued supplementation of human studies with animal models will be especially important in discovering mechanisms behind pathogenesis and possible pharmacologic targets for naturally occurring autoimmune diseases.

References

1 Hoffman, B.J. (1945) Sensitivity towards sulfadiazine resembling acute disseminated lupus erythematosus. *Arch Derm Syphilol*, **51**, 190–192.
2 Vasoo, S. (2006) Drug-induced lupus: an update. *Lupus*, **15**, 757–761.
3 Johnson, S.T., Fueger, J.T., and Gottschall, J.L. (2007) One center's experience: the serology and drugs associated with drug-induced immune hemolytic anemia: a new paradigm. *Transfusion*, **47**, 697–702.
4 Bigazzi, P.E. (1988) Autoimmunity induced by chemicals. *Clin Toxicol*, **26**, 125–156.
5 Rubin, R.L., Hermanson, T.M., Bedrick, E.J., McDonald, J.D., Burchiel, S.W., Reed, M.D., and Sibbitt, W.L. Jr (2005) Effect of cigarette smoke on autoimmunity in murine and human systemic lupus erythematosus. *Toxicol Sci*, **87**, 86–96.
6 Lee, S.H., Goswami, S., Grudo, A., Song, L.Z., Bandi, V., Goodnight-White, S., Green, L., Hacken-Bitar, J., Huh, J., Bakaeen, F., Coxson, H.O., Cogswell, S., Storness-Bliss, C., Corry, D.B., and Kheradmand, F. (2007) Antielastin autoimmunity in tobacco smoking-induced emphysema. *Nat Med*, **13**, 567–569.
7 Rubin, R.L. (2005) Drug-induced lupus. *Toxicology*, **209**, 135–147.
8 Borchers, A.T., Keen, C.L., and Ric, G.E.R.S. (2007) Drug-induced lupus. *Ann NY Acad Sci*, **1108**, 166–182.
9 Rubin, R.L. and Curnutte, J.T. (1989) Metabolism of procainamide to the cytotoxic hydroxylamine by neutrophils activated in vitro. *J Clin Invest*, **83**, 1336–1343.
10 Rubin, R.L. (1994) Role of xenobiotic oxidative metabolism. *Lupus*, **3**, 479–482.
11 Goebel, C., Vogel, C., Wulferink, M., Mittmann, S., Sachs, B., Schraa, S., Abel, J., Degen, G., Uetrecht, J., and Gleichmann, E. (1999) Procainamide, a drug causing lupus, induces prostaglandin H synthase-2 and formation of T cell-sensitizing drug metabolites in mouse macrophages. *Chem Res Toxicol*, **12**, 488–500.
12 Engler, O.B., Strasser, I., Naisbitt, D.J., Cerny, A., and Pichler, W.J. (2004) A chemically inert drug can stimulate T cells in vitro by their T cell receptor in non-sensitised individuals. *Toxicology*, **197**, 47–56.
13 Zhou, Y. and Lu, Q. (2008) DNA methylation in T cells from idiopathic lupus and drug-induced lupus patients. *Autoimmun Rev*, **7**, 376–383.
14 Cornacchia, E., Golbus, J., Maybaum, J., Strahler, J., Hanash, S., and Richardson, B. (1988) Hydralazine and procainamide inhibit T cell DNA methylation and induce autoreactivity. *J Immunol*, **140**, 2197–2200.
15 Lu, Q., Wu, A., and Richardson, B.C. (2005) Demethylation of the same promoter sequence increases CD70 expression in lupus T cells and T cells treated with lupus-Inducing drugs. *J Immunol*, **174**, 6212–6219.
16 Gorelik, G., Fang, J.Y., Wu, A., Sawalha, A.H., and Richardson, B. (2007) Impaired T cell protein kinase Cδ activation decreases ERK

pathway signaling in idiopathic and hydralazine-induced lupus. *J Immunol*, **179**, 5553–5563.

17 Kretz-Rommel, A., Duncan, S.R., and Rubin, R.L. (1999) Autoimmunity caused by disruption of central T cell tolerance: a murine model of drug-induced lupus. *J Clin Invest*, 1888–1896.

18 Mazari, L., Ouarzane, M., and Zouali, M. (2007) Subversion of B lymphocyte tolerance by hydralazine, a potential mechanism for drug-induced lupus. *Proc Natl Acad Sci*, **104**, 6317–6322.

19 DeLoughery, T. (1998) Drug-induced immune hematologic disease. *Immunol Allergy Clin North Am*, **18**, 829–841.

20 Harris, J.W. (1956) Studies on the mechanism of a drug-induced hemolytic anemia. *J Lab Clin Med*, **47**, 760–775.

21 Garratty, G. and Petz, L.D. (1975) Drug-induced immune hemolytic anemia. *Am J Med*, **58**, 398–407.

22 Garratty, G. (1972) Drug-related problems. A seminar on problems encountered in pre-transfusion tests. *Am Assoc Blood Banks*, 33.

23 Gralnick, H.R., Wright, L.D., and McGinniss, M.H. (1967) Coombs' positive reactions associated with sodium cephalothin therapy. *J Am Med Assoc*, **199**, 725–726.

24 Molthan, L., Reidenberg, M.M., and Eichman, M.F. (1967) Positive direct Coombs tests due to cephalothin. *New Engl J Med*, **277**, 123–125.

25 Aftab, A., Stanca, C.M., Ghanim, M.B., Ahmado, I., Branch, A.D., Schiano, T.D., Odin, J.A., and Bach, N. (2006) Increased prevalence of primary biliary cirrhosis near superfund toxic waste sites. *Hepatology*, **43**, 525–531.

26 Long, S.A., Quan, C., Van de Water, J., Nantz, M.H., Kurth, M.J., Barsky, D., Colvin, M.E., Lam, K.S., Coppel, R.L., Ansari, A., and Gershwin, M.E. (2001) Immunoreactivity of organic mimeotopes of the E2 component of pyruvate dehydrogenase: connecting xenobiotics with primary biliary cirrhosis. *J Immunol*, **167**, 2956–2963.

27 Amano, K., Leung, P.S.C., Xu, Q., Marik, J., Quan, C., Kurth, M.J., Nantz, M.H., Ansari, A., Lam, K.S., Zeniya, M., Coppel, R.L., and Gershwin, M.E. (2004) Xenobiotic-induced loss of tolerance in rabbits to the mitochondrial autoantigen of primary biliary cirrhosis is reversible. *J Immunol*, **172**, 6444–6452.

28 Leung, P.S.C., Park, O., Tsuneyama, K., Kurth, M.J., Lam, K.S., Ansari, A.A., Coppel, R.L., and Gershwin, M.E. (2007) Induction of primary biliary cirrhosis in guinea pigs following chemical xenobiotic immunization. *J Immunol*, **179**, 2651–2657.

29 Amano, K., Leung, P.S.C., Rieger, R., Quan, C., Wang, X.M., Jan, S., Yat, F., Kurth, M.J., Nantz, M.H., Ansari, A.A., Lam, K.S., Zeniya, M., Matsuura, E., Coppel, R.L., and Gershwin, M.E. (2005) Chemical xenobiotics and mitochondrial autoantigens in primary biliary cirrhosis: identification of antibodies against a common environmental, cosmetic, and food additive, 2-octynoic acid. *J Immunol*, **174**, 5874–5883.

30 Rieger, R., Leung, P.S.C., Jeddeloh, M.R., Kurth, M.J., Nantz, M.H., Lam, K.S., Barsky, D., Ansari, A.A., Coppel, R.L., Mackay, I.R., and Gershwin, M.E. (2006) Identification of 2-nonynoic acid, a cosmetic component, as a potential trigger of primary biliary cirrhosis. *J Autoimmun*, **27**, 7–16.

31 Wakabayashi, K., Lian, Z.X., Leung, P.S.C., Moritoki, Y., Tsuneyama, K., Kurth, M.J., Lam, K.S., Yoshida, K., Yang, G.X., Hibi, T., Ansari, A.A., Ridgway, W.M., Coppel, R.L., Mackay, I.A., and Gershwin, E. (2008) Loss of tolerance in C57BL/6 mice to the autoantigen E2 subunit of pyruvate dehydrogenase by a xenobiotic with ensuing biliary ductular disease. *Hepatology*, **48**, 531–540.

32. d'Angeljean, J., Morel, E., Feuillet-Fieux, M.N., Raimond, F., Vernet Der Garabedian, B., Jacob, L., and Bach, J.F. (1985) Myasthenie induite par la D-penicillamine. Etude des correlations immuno-cliniques dans 23 cases. *Presse Med*, **14**, 2336–2340.
33. Smith, C.I.E. and Hammarstrom, L. (1985) Immunologic abnormalities induced by D-penicillamine. *Pseudo-Allergic React*, **4**, 138–180.
34. Howard-Lock, H.E., Lock, C.J.L., Mewa, A., and Kean, W.F. (1986) D-penicillamine: chemistry and clinical use in rheumatic disease. *Semin Arthritis Rheum*, **15**, 261–281.
35. Hill, M., Moss, P., Wordsworth, P., Newsom-Davis, J., and Willcox, N. (1999) T cell responses to D-penicillamine in drug-induced myasthenia gravis: recognition of modified DR1:peptide complexes. *J Neuroimmunol*, **97**, 146–153.
36. Clarkson, T.W., Vyas, J.B., and Ballatori, N. (2007) Mechanisms of mercury deposition in the body. *Am J Ind Med*, **50**, 757–764.
37. Sager, P.R., Aschner, M., and Rodier, P.M. (1984) Persistent, differential alterations in developing cerebellar cortex of male and female mice after methylmercury exposure. *Brain Res*, **314**, 1–11.
38. Hammond, A.L. (1971) Mercury in the environment: natural and human factors. *Science*, **171**, 788–789.
39. Daston, G.P., Gray, J.A., Carver, B., and Kavlock, R.J. (1984) Toxicity of mercuric chloride to the developing rat kidney. II. Effect of increased dosages on renal function in suckling pups. *Toxicol Appl Pharmacol*, **74**, 35–45.
40. Barr, R.D., Rees, P.H., Cordy, P.E., Kungu, A., Woodger, B.A., and Cameron, H.M. (1972) Nephrotic syndrome in adult Africans in Nairobi. *Br Med J*, **2**, 131–134.
41. Soo, Y.O.Y., Chow, K.M., Lam, C.W.K., Lai, F.M.M., Szeto, C.C., Chan, M.H.M., and Li, P.K.T. (2003) A whitened face woman with nephrotic sundrome. *Am J Kidney Dis*, **41**, 250–253.
42. Tang, H.L., Chu, K.H., Mak, Y.F., Lee, W., Cheuk, A., Yim, K.F., Fung, K.S., Chan, H.W., and Tong, K.L. (2006) Minimal change disease following exposure to mercury-containing skin lightening cream. *Hong Kong Med J*, **12**, 316–318.
43. Moszczynski, P., Slowinski, S., Ruthowski, J., Bem, S., and Jakus-Stoga, D. (1995) Lymphocytes, T and NK cells, in men occupationally exposed to mercury vapours. *Int J Occup Med Environ Health*, **8**, 49–56.
44. Lauwerys, R., Bernard, A., Roels, H., Buchet, J.P., Gennart, J.P., Mahieu, P., and Foidart, J.M. (1983) Anti-laminin antibodies in workers exposed to mercury vapour. *Toxicol Lett*, **17**, 113–116.
45. Daniel, A., Clarkin, C., Komoroski, J., Brensinger, C.M., and Berlin, J.A. (2004) Wegener's granulomatosis: possible role of environmental agents in its pathogenesis. *Arthritis Care Res*, **51**, 656–664.
46. Hultman, P., Johansson, U., Turley, S.J., Lindh, U., Enestrom, S., and Pollard, K.M. (1994) Adverse immunological effects and autoimmunity induced by dental amalgam and alloy in mice. *FASEB J*, **8**, 1183–1190.
47. Mutter, J., Naumann, J., Walach, H., and Daschner, F. (2005) Amalgam: eine risikobewertung unter bencksichtigung der neuen literatur bis 2005. *Gesundheitswesen*, 204–216.
48. Prochazkova, J., Sterzl, I., Kucerova, H., Bartova, J., and Stejskal, V.D.M. (2004) The beneficial effect of amalgam replacement on health in patients with autoimmunity. *Neuroendocrinol Lett*, **25**, 211–218.
49. Nathan, A.W. and Toseland, P.A. (1979) Goodpasture's syndrome and trichloroethane intoxication. *Br J Clin Pharmacol*, **8**, 284–286.
50. Bombassei, G.J. and Kaplan, A.A. (1992) The association between hydrocarbon exposure and

anti-glomerular basement membrane antibody-mediated disease (Goodpasture's syndrome). *Am J Ind Med*, **21**, 141–153.

51 Shah, M.K. (2002) Outcomes in patients with Goodpasture's syndrome and hydrocarbon exposure. *Renal Failure*, **24**, 545.

52 Yamamoto, T. and Wilson, C.B. (1987) Binding of anti-basement membrane antibody to alveolar basement membrane after intratracheal gasoline instillation in rabbits. *Am J Pathol*, **126**, 497–505.

53 Ho, P.P. and Steinman, L. (2008) The aryl hydrocarbon receptor: a regulator of Th17 and Treg cell development in disease. *Cell Res*, **18**, 605–608.

54 Veldhoen, M., Hirota, K., Westendorf, A.M., Buer, J., Dumoutier, L., Renauld, J.C., and Stockinger, B. (2008) The aryl hydrocarbon receptor links TH17-cell-mediated autoimmunity to environmental toxins. *Nature*, **453**, 106–109.

55 Quintana, F.J., Basso, A.S., Iglesias, A.H., Korn, T., Farez, M.F., Bettelli, E., Caccamo, M., Oukka, M., and Weiner, H.L. (2008) Control of Treg and TH17 cell differentiation by the aryl hydrocarbon receptor. *Nature*, **453**, 65–71.

56 Cam, C. (1958) Cases of skin porphyria related to hexachlorobenzene intoxication. *Saglik Dergisi*, **32**, 215–216.

57 Queiroz, M.L.S., Bincoletto, C., Perlingeiro, R.C.R., Quadros, M.R., and Souza, C.A. (1998) Immunoglobulin levels in workers exposed to hexachlorobenzene. *Human Exp Toxicol*, **17**, 172–175.

58 Ezendam, J., Vissers, I., Bleumink, R., Vos, J.G., and Pieters, R. (2003) Immunomodulatory effects of tetrachlorobenzoquinone, a reactive metabolite of hexachlorobenzene. *Chem Res Toxicol*, **16**, 688–694.

59 Semchuk, K.M., Rosenberg, A.M., McDuffie, H.H., Cessna, A.J., Pahwa, P., and Irvine, D.G. (2007) Antinuclear antibodies and bromoxynil exposure in a rural sample. *J Toxicol Environ Health A*, **70**, 638–657.

60 Holsapple, M.P. (2002) Autoimmunity by pesticides: a critical review of the state of the science. *Toxicol Lett*, **127**, 101–109.

61 Solomon, J. and Schwarz, M. (2006) Drug-, toxin-, and radiation therapy-induced eosinophilic pneumonia. *Semin Respir Crit Care Med*, 192–198.

62 Kilbourne, E.M., Rigau-Perez, J.G., Heath, C.W. Jr, Zach, M.M., Falk, H., Martin-Marcos, M., and de Carlos, A. (1983) Clinical epidemiology of toxic-oil syndrome. Manifestations of a new illness. *N Engl J Med*, **309**, 1408–1414.

63 Borda, I.A., Kilbourne, E.M., de la Paz, M.P., Ruiz-Navarro, M.D., Sanchez, R.G., and Falk, H.E.N.R. (1993) Mortality among people affected by toxic oil syndrome. *Int J Epidemiol*, **22**, 1077–1084.

64 Borda, I.A., Philen, R.M., de la Paz, M.P., de la Camara, A.G., Ruiz-Navarro, M.D., Ribota, O.G., Soldevilla, J.A., Terracini, B., Peta, S.S., Leal, C.F., and Kilbourne, E.M. (1998) Toxic oil syndrome mortality: the first 13 years. *Int J Epidemiol*, **27**, 1057–1063.

65 Alonso-Ruiz, A., Calabozo, M., Perez-Ruin, F., and Mancebo, L. (1993) Toxic oil syndrome. A long-term follow-up of a cohort of 332 patients. *Medicine (Baltimore)*, **72**, 285–295.

66 Bell, S.A., Du Clos, T.W., Khursigara, G., Picazo, J.J., and Rubin, R.L. (1995) Autoantibodies to cryptic epitopes of c-reactive protein and other acute phase proteins in the toxic oil syndrome. *J Autoimmun*, **8**, 293–303.

67 Bell, S.A., Hobbs, M.V., and Rubin, R.L. (1992) Isotype-restricted hyperimmunity in a murine model of the toxic oil syndrome. *J Immunol*, **148**, 3369–3376.

68 Silver, R.M., Heyes, M.P., Maize, J.C., Quearry, B., Vionnet-Fuasset, M., and Sterberg, E.M. (1990) Scleroderma, fasciitis, and eosinophilia

associated with the ingestion of tryptophan. *New Engl J Med*, **322**, 874–881.

69 Mayeno, A.N., Benson, L.M., Naylor, S., Colberg-Beers, M., Puchalski, J.T., and Gleich, G.J. (1995) Biotransformation of 3-(Phenylamino)-1,2-propanediol to 3-(Phenylamino)alanine: a chemical link between toxic oil syndrome and eosinophilia-myalgia syndrome. *Chem Res Toxicol*, **8**, 911–916.

70 Philen, R.M. and Hill, R.H. Jr (1993) 3-(Phenylamino)alanine–a link between eosinophilia-myalgia syndrome and toxic oil syndrome? *Mayo Clin Proc*, **68**, 197–200.

71 de la Paz, M.P., Philen, R.M., and Borda, I.A. (2001) Toxic oil syndrome: the perspective after 20 Years. *Epidemiol Rev*, **23**, 231–247.

72 Martínez-Cabot, A. and Messeguer, A. (2007) Generation of quinoneimine intermediates in the bioactivation of 3-(N-Phenylamino)alanine (PAA) by human liver microsomes: a potential link between eosinophilia-myalgia syndrome and toxic oil syndrome. *Chem Res Toxicol*, **20**, 1556–1562.

73 Ohsawa, M., Takahashi, K., and Otsuka, F. (1988) Induction of anti-nuclear antibodies in mice orally exposed to cadmium at low concentrations. *Clin Exp Immunol*, **73**, 98–102.

74 Chen, M., Hemmerich, P., and von Mikecz, A. (2002) Platinum-induced autoantibodies target nucleoplasmic antigens related to active transcription. *Immunobiology*, **206**, 474–483.

75 Hultman, P., Enestrom, S., Turley, S.J., and Pollard, K.M. (1994) Selective induction of anti-fibrillarin autoantibodies by silver nitrate in mice. *Clin Exp Immunol*, **96**, 285–291.

76 Hultman, P., Ganowiak, K., Turley, S.J., and Pollard, K.M. (1995) Genetic susceptibility to silver-induced anti-fibrillarin autoantibodies in mice. *Clin Immunol Immunopathol*, **77**, 291–297.

77 Johansson, U., Hansson-Georgiadis, H., and Hultman, P. (1992) Murine silver-induced autoimmunity: silver shares induction of antinucleolar antibodies with mercury, but causes less activation of the immune system. *Int Arch Allergy Immunol*, **113**, 432–443.

78 Goldman, M., Druet, P., and Gleichmann, E. (1991) TH2 cells in systemic autoimmunity: insights from allogeneic diseases and chemically-induced autoimmunity. *Immunol Today*, **12**, 223–227.

79 Druet, E., Sapin, C., Günther, E., Feingold, N., and Druet, P. (1977) Mercuric chloride-induced anti-glomerular basement membrane antibodies in the rat Genetic control. *Eur J Immunol*, **7**, 348–351.

80 Druet, P., Druet, E., Potdevin, F., and Sapin, C. (1978) Immune type glomerulonephritis induced by HgCl2 in the Brown Norway rat. *Ann Inst Pasteur Immunol*, **129C**, 777–792.

81 Druet, P., Pelletier, L., Hirsch, F., Rossert, J., Pasquier, R., Druet, E., and Sapin, C. (1988) Mercury-induced autoimmune glomerulonephritis in animals. *Contrib Nephrol*, **61**, 120–130.

82 Pusey, C.D., Bowman, C., Morgan, A., Weetman, A.P., Hartley, B., and Lockwood, C.M. (1990) Kinetics and pathogenicity of autoantibodies induced by mercuric chloride in the brown Norway rat. *Clin Exp Immunol*, **81**, 76–82.

83 Marriott, J.B., Qasim, F., and Oliveira, D.B.G. (1994) Anti-phospholipid antibodies in the mercuric chloride treated Brown Norway rat. *J Autoimmun*, **7**, 457–467.

84 Mathieson, P.W., Tapleton, K.J., Oliveira, D.B.G., and Lockwood, M. (1991) Immunoregulation of mercuric chloride-induced autoimmunity in Brown Norway rats: a role for CD8+ T cells revealed by in vivo depletion studies. *Eur J Immunol*, **21**, 2105–2109.

85 Bowman, C., Mason, D.W., Pusey, C.D., and Lockwood, C.M. (1984)

Autoregulation of autoantibody synthesis in mercuric chloride nephritis in the Brown Norway rat I. A role for T suppressor cells. *Eur J Immunol*, **14**, 464–470.

86 Castedo, M., Pelletier, L., Rossert, J., Pasquier, R., Villarroya, H., and Druet, P. (1993) Mercury-induced autoreactive anti-class II T cell line protects from experimental autoimmune encephalomyelitis by the bias of CD8+ antiergotypic cells in Lewis rats. *J Exp Med*, **177**, 881–889.

87 Pelletier, L., Rossert, J., Pasquier, R., Vial, M.C., and Druet, P. (1990) Role of CD8+ T cells in mercury-induced autoimmunity or immunosuppression in the rat. *Scand J Immunol*, **31**, 65–74.

88 Hultman, P., Bell, L.J., Enestrom, S., and Pollard, K.M. (1992) Murine susceptibility to mercury: I. Autoantibody profiles and systemic immune deposits in inbred, congenic, and intra-H-2 recombinant strains. *Clin Immunol Immunopathol*, **65**, 98–109.

89 Mirtcheva, J., Pfeiffer, C., De Bruijn, J.A., Jacquesmart, F., and Gleichmann, E. (1989) Immunological alterations inducible by mercury compounds. III. H-2A acts as an immune response and H-2E as an immune suppression locus for HgCl2 induced antinucleolar autoantibodies. *Eur J Immunol*, **19**, 2257–2261.

90 Robinson, C.J., Balazs, T., and Egorov, I.K. (1986) Mercuric chloride-, gold sodium thiomalate-, and D-penicillamine-induced antinuclear antibodies in mice. *Toxicol Appl Pharmacol*, **86**, 159–169.

91 Hansson, M. and Abedi, V.A.L.U. (2003) Xenobiotic metal-induced autoimmunity: mercury and silver differentially induce antinucleolar autoantibody production in susceptible H-2s, H-2q and H-2f. *Clin Exp Immunol*, **131**, 405–414.

92 Bagentose, L.M., Salgame, P., and Monestier, M. (1999) Murine mercury-induced autoimmunity: a model of chemically related autoimmunity in humans. *Immunol Res*, **20**, 67–78.

93 Hultman, P., Bell, L.J., Enestrom, S., and Pollard, K.M. (1993) Murine susceptibility to mercury: II. Autoantibody profiles and renal immune deposits in hybrid, backcross, and H-2d congenic mice. *Clin Immunol Immunopathol*, **68**, 9–20.

94 Pollard, K.M., Hultman, P., and Kono, D.H. (2005) Immunology and genetics of induced systemic autoimmunity. *Autoimmun Rev*, **4**, 282–288.

95 Monestier, M., Losman, M.J., Novick, K.E., and Aris, J.P. (1994) Molecular analysis of mercury-induced antinucleolar antibodies in H-2S mice. *J Immunol*, **152**, 667–675.

96 Hultman, P., Turley, S.J., Enestrom, S., Lindh, U., and Pollard, M.K. (1996) Murine genotype influences the specificity, magnitude and persistence of murine mercury-induced autoimmunity. *J Autoimmun*, **9**, 139–149.

97 Oram, P.D., Fang, X., Fernando, Q., Letkeman, P., and Letkeman, D. (1996) The formation constants of mercury(II)-glutathione complexes. *Chem Res Toxicol*, **9**, 709–712.

98 Passow, H., Rothstein, A., and Clarkson, T.W. (1961) The general pharmacology of the heavy metals. *Pharmacol Rev*, **13**, 185–224.

99 Nakashima, I., Pu, M.Y., Nishizaki, A., Rosila, I., Ma, L., Katano, Y., Ohkusu, K., Rahman, S.M., Isobe, K., and Hamaguchi, M. (1994) Redox mechanism as alternative to ligand binding for receptor activation delivering disregulated cellular signals. *J Immunol*, **152**, 1064–1071.

100 McCabe, M.J., Santini, R.P., and Rosenspire, A.J. (1999) Low and non-toxic levels of ionic mercury interfere with the regulation of cell growth in the WEHI-231 B-cell lymphoma. *Scand J Immunol*, **50**, 233–241.

101 Whitekus, M.J., Santini, R.P., Rosenspire, A.J., and McCabe, M.J. Jr. (1999) Protection against

CD95-mediated apoptosis by inorganic mercury in Jurkat T cells. *J Immunol*, **162**, 7162–7170.
102 Ziemba, S.E., McCabe, M.J. Jr, and Rosenspire, A.J. (2005) Inorganic mercury dissociates preassembled Fas/CD95 receptor oligomers in T lymphocytes. *Toxicol Appl Pharmacol*, **206**, 334–342.
103 Jiang, Y. and Moller, G. (1995) In vitro effects of HgCl2 on murine lymphocytes. I. Preferable activation of CD4+ T cells in a responder strain. *J Immunol*, **154**, 3138–3146. [published erratum appears in J Immunol 1996 Apr 15;156(8): following 3088].
104 Caron, G.A., Poutala, S., and Provost, T.T. (1970) Lymphocyte transformation induced by inorganic and organic mercury. *Intl Arch Allergy Appl Immunol*, **37**, 76–87.
105 Bagentose, L.M., Salgame, P., and Monestier, M. (1999) Cytokine regulation of a rodent model of mercuric chloride-induced autoimmunity. *Environ Health Perspect*, **107**, 807–810.
106 van der Meide, P.H., de Labie, M.C., Botman, C.A.D., van Bennekom, W.P., Olsson, T., Aten, J., and Weening, J.J. (1993) Mercuric chloride down-regulates T cell interferon-gamma production in Brown Norway but not in Lewis rats: role of glutathione. *Eur J Immunol*, **23**, 675–681.
107 Peterson, J.D., Herzenberg, L.A., Vasquez, K., and Waltenbaugh, C. (1998) Glutathione levels in antigen-presenting cells modulate Th1 versus Th2 response patterns. *Proc Natl Acad Sci*, **95**, 3071–3076.
108 Anderson, P.N. and Potter, M.I.C.H. (1969) Induction of plasma cell tumours in BALB/c mice with 2,6,10,14-tetramethylpentadecane (pristane). *Nature*, **222**, 994–995.
109 Satoh, M. and Reeves, W.H. (1994) Induction of lupus-associated autoantibodies in BALB/c mice by intraperitoneal injection of pristane. *J Exp Med*, **180**, 2341–2346.
110 Satoh, M., Kumar, A., Kanwar, Y.S., and Reeves, W.H. (1995) Anti-nuclear antibody production and immune-complex glomerulonephritis in BALB/c mice treated with pristane. *Proc Natl Acad Sci*, **92**, 10934–10938.
111 Satoh, M., Treadwell, E.L., and Reeves, W.H. (1995) Pristane induces high titers of anti-Su and anti-nRNP/Sm autoantibodies in BALB/c mice Quantitation by antigen capture ELISAs based on monospecific human autoimmune sera. *J Immunol Methods*, **182**, 51–62.
112 Satoh, M., Kuroda, Y., Yoshida, H., Behney, K.M., Mizutani, A., Akaogi, J., Nacionales, D.C., Lorenson, T.D., Rosenbauer, R.J., and Reeves, W.H. (2003) Induction of lupus autoantibodies by adjuvants. *J Autoimmun*, **21**, 1–9.
113 Yang, J.Q., Singh, A.K., Wilson, M.T., Satoh, M., Stanic, A.K., Park, J.J., Hong, S., Gadola, S.D., Mizutani, A., Kakumanu, S.R., Reeves, W.H., Cerundolo, V., Joyce, S., Van Kaer, L., and Singh, R.R. (2003) Immunoregulatory role of CD1d in the hydrocarbon oil-induced model of lupus nephritis. *J Immunol*, **171**, 2142–2153.
114 Avneesh, K.S., Jun, Q.Y., Vrajesh, V.P., Jie, W., Chyung, R.W., Sebastian, J., Ram, R.S., and Luc, V.K. (2005) The natural killer T cell ligand alpha-galactosylceramide prevents or promotes pristane-induced lupus in mice. *Eur J Immunol*, **35**, 1143–1154.
115 Mizutani, A., Shaheen, V.M., Yoshida, H., Akaogi, J., Kuroda, Y., Nacionales, D.C., Yamasaki, Y., Hirakata, M., Ono, N., Reeves, W.H., and Satoh, M. (2005) Pristane-induced autoimmunity in germ-free mice. *Clin Immunol*, **114**, 110–118.
116 Calvani, N., Caricchio, R., Tucci, M., Sobel, E.S., Silvestris, F., Tartaglia, P., and Richards, H.B. (2005) Induction of apoptosis by the hydrocarbon oil pristane: implications for pristane-induced lupus. *J Immunol*, **175**, 4777–4782.

117 Savarese, E., Steinberg, C., Pawar, R.D., Reindl, W., Akira, S., Anders, J., and rug, A. (2008) Requirement of toll-like receptor 7 for pristane-induced production of autoantibodies and development of murine lupus nephritis. *Arthritis Rheum*, **58**, 1107–1115.

118 Joe, B. and Wilder, R.L. (1999) Animal models of rheumatoid arthritis. *Mol Med Today*, **5**, 369.

119 Thompson, S.J., Rook, G.A.W., Brealey, R.J., Van Der-Zee, R., and Elson, C.J. (1990) Autoimmune reactions to heat-shock proteins in pristane-induced arthritis. *Eur J Immunol*, **20**, 2479–2484.

120 Sharif, M., Worrall, J.G., Singh, B., Gupta, R.S., Lydyard, P.M., Lambert, C., McCulloch, J., and Rook, G.A. (1992) The development of monoclonal antibodies to the human mitochondrial 60-kd heat-shock protein, and their use in studying the expression of the protein in rheumatoid arthritis. *Arthritis Rheum*, **35**, 1427–1433.

121 van Eden, W. (2008) Immunity to heat shock proteins and arthritic disorders. *Infect Dis Obstet Gynecol*, **7**, 49–54.

122 Hoffmann, M.H., Tuncel, J., Skriner, K., Tohidast-Akrad, M., Turk, B., Pinol-Roma, S., Serre, G., Schett, G., Smolen, J.S., Holmdahl, R., and Steiner, G. (2007) The rheumatoid arthritis-associated autoantigen hnRNP-A2 (RA33) is a major stimulator of autoimmunity in rats with pristane-induced arthritis. *J Immunol*, **179**, 7568–7576.

123 Lim, S.Y. and Ghosh, S.K. (2003) Autoreactive responses to an environmental factor: 1. phthalate induces antibodies exhibiting anti-DNA specificity. *Immunology*, **110**, 482–492.

124 Lim, S.Y. and Ghosh, S.K. (2005) Autoreactive responses to environmental factors: 3. Mouse strain-specific differences in induction and regulation of anti-DNA antibody responses due to phthalate-isomers. *J Autoimmun*, **25**, 33–45.

125 Robinson, C.J.G., Balazs, T., and Egorov, I.K. (1986) Mercuric chloride-, gold sodium thiomalate-, and D-penicillamine-induced antinuclear antibodies in mice. *Toxicol Appl Pharmacol*, **86**, 159.

126 O'Donnell, C.A., Foster, A.L., and Coleman, J.W. (1991) Penicillamine and penicillin can generate antigenic determinants on rat peritoneal cells in vitro. *Immunology*, **72**, 571–576.

127 Tournade, H., Pelletier, L., Pasquier, R., Vial, M.C., Mandet, C., and Druet, P. (1990) D-penicillamine-induced autoimmunity in Brown-Norway rats. Similarities with HgCl2-induced autoimmunity. *J Immunol*, **144**, 2985–2991.

128 Monestier, M., Novick, K.E., and Losman, M.J. (1994) D-penicillamine- and quinidine-induced antinuclear antibodies in A.SW (H-2s) mice: similarities with autoantibodies in spontaneous and heavy metal-induced autoimmunity. *Eur J Immunol*, **24**, 723–730.

129 Masson, M.J. and Uetrecht, J.P. (2004) Tolerance induced by low dose D-penicillamine in the Brown Norway rat model of drug-induced autoimmunity is immune-mediated. *Chem Res Toxicol*, **17**, 82–94.

130 Seguin, B., Masson, M.J., and Uetrecht, J. (2004) D-penicillamine-induced autoimmunity in the Brown Norway rat: role for both T and non-T splenocytes in adoptive transfer of tolerance. *Chem Res Toxicol*, **17**, 1299–1302.

131 Seguin, B., Teranishi, M., and Uetrecht, J.P. (2003) Modulation of-penicillamine-induced autoimmunity in the Brown Norway rat using pharmacological agents that interfere with arachidonic acid metabolism or synthesis of inducible nitric oxide synthase. *Toxicology*, **190**, 267–278.

32
Endogenous Toxins Associated with Life Expectancy and Aging

Victoria Ayala, Jordi Boada, José Serrano, Manuel Portero-Otín, and Reinald Pamplona

The basic chemical process underlying the aging process was first put forward as the free radical theory of aging in 1956: the reaction of active free radicals, physiologically produced within an organism itself by cellular constituents, initiates the changes associated with aging. The involvement of free radicals in aging is related to their fundamental role in the origin and evolution of life. The specific composition of cellular macromolecules (carbohydrates, lipids, DNA, and proteins) in long-lived animal species gives them an intrinsically high resistance to oxidative injury that contributes to the superior life span of these species. Long-lived species also show low rates of reactive oxygen species (ROS) generation and oxidative molecular damage. Caloric restriction, a nutritional intervention that extends life span, further decreases mitochondrial ROS production and oxidative damage due to the decreased intake of dietary proteins. These effects of protein restriction are due to the lowered methionine intake of protein and caloric restricted animals.

32.1
Introduction

When Jeanne Calment died in a nursing home in southern France in 1997, she was 122 years old; the longest-living human ever documented. But Calment's status will change in forthcoming decades if the predictions of biogerontologists and demographers are borne out. Life span extension in species from yeast to mice and extrapolation from life expectancy trends in humans have convinced scientists that humans will routinely live beyond 100 or 110 years. Today, 1 in 10 000 people in industrialized countries hold centenarian status. By 2025, the United Nations anticipates that there will be 822 million people in the world aged 65 and over [1, 2]. The elderly population will have grown by a factor of 2.5 between 1990 and 2025. This is faster than the total population growth, resulting in the world's elderly population increasing from 6.2 to 9.7%.

The World Health Organization classifies persons who are 60–75 years of age as elderly, 76–90 as old, and 90 years and older as very old. Individuals aged 80–85

Endogenous Toxins. Diet, Genetics, Disease and Treatment.
Edited by Peter J. O'Brien and W. Robert Bruce
Copyright © 2010 WILEY-VCH Verlag GmbH & Co. KGaA, Weinheim
ISBN: 978-3-527-32363-0

Tab. 32.1 Brief glossary of aging-related terms (modified from [3])

Term	Definition
Aging	Latin "aetas," age or lifetime – the condition of becoming old
Average life span	The average of individual life spans for members of a group (cohort) of the same birth date
Biomarkers	Biologic indicators (morphologic, functional, and behavioral) specific to old age
Geriatrics	Greek "geron," old man, and "iatros," healer – a medical speciality dealing with the problems and diseases of the elderly
Gerontology	Greek "geron," old man, and "logos," knowledge – the study of aging as a physiological process and the problems of old age
Life expectancy	The average amount of time of life remaining for a population whose members all have the same birth date
Life span	The duration of the life of an individual/organism in a particular environment and/or under specific circumstances
Longevity	Long duration of an individual's life; the condition of being "long lived" is also often used as a synonym for life span
Maximum life span	The length of life of the longest-lived individual member of a species
Senescence	Latin "senex," old man – the condition of being old, used interchangeably with aging

and older are also called *"old-old."* Centenarians are persons 100 years old and older. In Table 32.1 are listed a brief glossary of aging-related terms.

The rise in life expectancy related to improvements in health is among the most remarkable demographic changes of the past century [1, 2, 4–6]. For the world as a whole, life expectancy more than doubled from around 30 years in 1900 to 65 years by 2000, and it is projected to rise to 81 years by the end of this century. Most of the historical rise reflects declines in infant and child mortality due to public health interventions related to drinking water and sanitation and medical interventions such as vaccination and the use of antibiotics. By contrast, the life expectancy gains observed over the past few decades (especially in high-income countries) and which are projected into the future, are predominantly associated with reductions in age-specific death rates at the middle and older ages. These reductions are typically associated with improvements in medical technology, lifestyle changes, and income growth.

What structural components and physiological mechanisms determine the aging process? For instance, why can human beings reach 122 years, whereas rats only live at most up to four years? "How much can human life span be extended?" was one of the questions featured by the journal *Science* recently on the occasion of its 125th anniversary. It addresses what the journal regards as one of the frontiers of science for the next 25 years. An answer to the question will probably emerge from many different investigative avenues, with possibly the most important being an examination of how nature determines the diverse and distinctive maximum life spans throughout the animal kingdom.

The postreproductive phase of life of virtually all cellular species is characterized by the progressive decline in the efficiency of maximum physiological functions [7]. Consequently, the ability to maintain homeostasis is correspondingly attenuated – leading eventually to an increased risk of developing cancer, and to neurodegenerative and cardiovascular diseases – increasing the chances of death. Any theory that explains aging must fit in with four main characteristics of this natural process: aging is progressive, endogenous, irreversible, and deleterious (in the sense that it damages the soma) for the individual [7]. First, the progressive character of aging means that the cause(s) of aging must be present during the whole life span: in both young and old. Secondly, as aging is an endogenous process, exogenous factors (e.g., UV rays and dietary antioxidants) are not causes of aging though they may interact with endogenous causes enhancing or mitigating their effects. The endogenous character of aging means that the rate of aging of different animal species, and therefore their maximum life span potentials, is genotypically determined, not dependent upon the environment. Hence, different animal species age at widely different rates in similar environments. In contrast, the mean life span (frequently and wrongly termed *longevity*) or life expectancy, which is calculated from the amount of time that each individual lives is mainly determined by the environment and to a lesser extent by the genotype (genetic determination of mean life span in humans is commonly agreed to be around 30%). This is the reason why many environmental factors such as smoking, the amount of saturated fat consumed, an unbalanced diet, a sedentary life, and possibly the action of antioxidants are so important in determining the age of death. Conversely, no matter what an elephant eats or does, it will never age in 2 years like a healthy rat does, and no diet will make a mouse survive for 85 years. Thus, mortality should not be confused with aging, even though advancing age increases the probability of death. In relation to this, the interindividual variation in the time lived for a given species (mainly environmentally determined) should not be confused with interspecies variation in life span (which is genetically determined). Although aging seems to be a multicausal process, it is perhaps mainly due to a small number of principal causes with major effects.

In this chapter, we update the available evidence concerning the mitochondrial oxidative stress theory of aging [8–10] and focus on comparative and dietary restriction (DR) models and the underlying mechanisms involved. In addition, we highlight two main characteristics that link slow animal aging to oxidative stress: (i) low generation of endogenous damage and (ii) a macromolecular composition that is highly resistant to oxidative modification.

32.2
Metabolism(s)

Oxygen appeared in significant amounts (about 1%) in the earth's atmosphere some 2700 million years ago, and geological evidence suggests that this was due to the photosynthetic activity of the blue-green algae. As this photosynthetic activity splits water to obtain their essential requirement for hydrogen atoms, microorganisms released tonnes of oxygen into the atmosphere. First eukaryotes with mitochondria appeared between 2300 and 2000 million years ago when rise in oxygen levels to about 5–18% took place. The slow and steady rise in atmospheric oxygen concentrations was accompanied by the formation of the ozone layer in the stratosphere. Both oxygen and the ozone layer acted as critical filters against the intense solar ultraviolet light striking the surface of the Earth. When the earth's atmosphere changed to the oxygen-rich state that we know today, the anaerobic life forms existing at that time adapted, died, or retreated to places where little or no oxygen was present. Organisms that evolved the capacity to cope with oxygen did best, because they could simultaneously evolve to use oxygen for efficient energy production and for other oxidation reactions [11]. Aerobic life demanded metabolic innovations that, in turn, allowed the emergence of complex multicellular organisms.

In this scenario of evolutionary change, living cells emerged as open systems, exchanging matter and energy with their surrounding and extracting and channeling energy to maintain themselves in a dynamic steady state distant from equilibrium. Properties that distinguish living organisms, among others, are a high degree of chemical complexity and molecular organization, defined functions for each of their components, and regulated interactions among them. The overall network of interconnected reaction sequences that interconvert cellular metabolites constitutes cellular metabolism [12]. Metabolism is adapted to achieve balance and economy. So, as a rule, chemical reactions in living cells are under strict enzyme control and conform to a tightly regulated metabolic program. So, one of the attractors involved in biomolecular evolution is the minimizing of unwanted side reactions. However, uncontrolled and potentially deleterious endogenous reactions occur, even under physiological conditions. Golubev [13] has termed these reactions *the other side of metabolism*. Aging, in this chemical context, could be viewed as an entropic process – the result of chemical side reactions that chronically and cumulatively degrades the function of biological systems. Death, the endpoint in the kinetic process of aging, represents the threshold of damage sufficient to compromise the normal function of an essential physiological subsystem or its ability to survive a challenge to that function [14].

32.3
The Rate of Generation of Damage Induced by Mitochondrial Free Radicals and Life Span

About 85–90% of oxygen is used by the mitochondria; these organelles are the major source of energy (as adenosine triphosphate (ATP) molecule) in aerobic organisms. Electrons from reduced substrates move from complexes I and II of the electron transport chain through complexes III and IV to oxygen, forming water and causing protons to be pumped across the mitochondrial inner membrane. The proton motive force set up by proton pumping drives protons back through the

ATP synthase in the inner membrane, forming ATP from their precursors ADP (adenosine diphosphate) and phosphate [12]. There are two major physiological side reactions that are significant here: (i) electrons leak from the respiratory chain and react with oxygen to form free radicals and (ii) pumped protons leak back across the inner membrane, diverting the conserved energy away from ATP biosynthesis and into heat production. A free radical is any molecule capable of independent existence that contains one or more unpaired electrons [15]. They are extremely reactive and have damaging effects. The fact that mitochondria are responsible for free radical generation suggests that they can play a causal role in the progressive process that is aging.

In this context, aerobic life demands antioxidant defenses. An antioxidant is "any substance that when present at low concentrations compared to those of an oxidizable substrate significantly delays or prevents oxidation of that substrate." An antioxidant either reacts with an oxidant and neutralizes it or regenerates other molecules capable of reacting with the oxidant. *Oxidative damage* is a broad term used to cover the attack upon biological molecules by free radicals. Cellular protection against oxidative damage includes both the elimination of reactive oxygen species ROS and repairing damage, with antioxidants constituting a fundamental line of this defense [15]. Although antioxidants may protect against various age-related diseases, they do not seem, however, to control the rate of aging based on the following evidence (reviewed in [10]): (i) long-lived species, including both invertebrates and vertebrates, constitutively have lower (not higher) tissue levels of antioxidant enzymes and of low molecular weight endogenous antioxidants than short-lived ones; (ii) experimentally increasing tissue antioxidants through dietary supplementation, pharmacological induction, or transgenic techniques sometimes moderately increases mean life span but does not change maximum life span; and (iii) animals in which genes codifying for particular antioxidant enzymes are knocked out may show different pathologies, but their rates of aging do not seem to be affected.

The strong negative correlation between endogenous tissue antioxidants and maximum life span suggests that the rate of endogenous free radical production *in vivo* must be much lower in long-lived than in short-lived animals. If long-lived animals had high rates of ROS production together with their very low levels of endogenous antioxidants, their cells would not be able to maintain their oxidative stress balance. Decreasing mitochondrial reactive oxygen species (MitROS) production instead of increasing antioxidants or repair systems makes sense when considered from the viewpoint of the evolution of life span among species. It would be very inefficient to generate large amounts of ROS and, afterwards, try to intercept them before they reach biomolecules, or even worse, try to repair biomolecules after they have been heavily damaged. This makes even more sense if we take into account (i) the high energetic cost of continuously maintaining high levels of antioxidant and repair molecules in tissues and (ii) the capacity of all kinds of animals (short- and long lived) to temporarily induce these protective molecules when needed in larger amounts. All the research published in this field has found that the rate of MitROS production is lower in the tissues of long-lived than in those of short-lived animal species [10, 16]. Recent studies have also found much lower rates of ROS generation in human than in rat brain mitochondria [17]. This characteristic thus explains why endogenous cellular antioxidants correlate negatively with maximum life span across species: long-lived animals have constitutively lower levels of antioxidants because they produce ROS at a low rate. Therefore, antioxidant levels reflect, under physiological conditions, an adaptation to the rate of cellular free radical production.

This fact is possible because the percentage of total electron flow in the respiratory chain directed to MitROS production (the percentage of free radical leak, %FRL) is lower in long-lived animals. This means that their respiratory chains transport electrons more efficiently avoiding univalent electron leaks to oxygen upstream of mitochondrial complex IV. Interestingly, total rates of reactive oxygen species ROS generation vary between species in a way not linked to their differences in oxygen consumption, indicating that MitROS production is not a simple by-product of mitochondrial respiration. Instead, it is regulated independent of O_2 consumption in many different physiologic situations, tissues, and animal species [18, 19].

32.4
Structural Components that are Highly Resistant to Oxidative Stress and Life Span

Molecular damage caused by oxidation is one of the natural consequences of aerobic life. Classically, cellular protection against oxidative damage includes free radical elimination and repair/turnover systems. These are considered as the first and the second lines of defense, respectively. Recent studies, however, support the notion of another line of defense based on the inherent susceptibility of macromolecules to oxidative damage [19]. This susceptibility (defined as the ease with which macromolecules suffer an oxidative injury) is intrinsically associated with the specific chemical composition of carbohydrates, lipids, DNA, and proteins.

32.4.1
The First Antioxidant Defense Line

In the context of oxidative stress, the following structural components are significantly highly susceptible to oxidative damage: (i) Carbohydrates reactivity, referring very particularly to monosaccharides of biological interest, is dependent on the extent to which it exists in the open (carbonyl) configuration (glycolytic intermediates the most unstable and reactive monosaccharides) rather than in the ring (hemiacetal or hemiketal) structure [20]. (ii) Highly unsaturated fatty acids in cell membranes are the macromolecules most susceptible to oxidative damage in cells, and this sensitivity increases as a function of the number of double bonds they contain. This means that saturated fatty acyl (SFA) and monounsaturated fatty acyl (MUFA) chains are essentially resistant to peroxidation, whereas polyunsaturated fatty acids (PUFAs) are easily damaged [21, 22]. (iii) Of the four nucleobases, guanine has the lowest oxidation potential and is thus generally most easily oxidized

[23]. (iv) Finally, methionine residues from proteins are, among the amino acids, most susceptible to oxidation by free radicals [24].

On the basis of these premises, the following available evidence is shown in living systems (i) Glucose has emerged as the most important carrier of energy from cell to cell in animal species, precisely because it is the slowest reacting carbohydrate. From experimental data obtained by Bunn and Higgins [20], it is evident that other sugars (e.g., ribose) are of the order of 100 times more reactive in the browning reaction (also named *Maillard reaction*; [25, 26]) than glucose. Thus, the lower the plasma and cellular concentration of these highly reactive sugars, the less the biological stress [27]. For instance, a look at the plasma concentrations of reducing sugars and related compounds offers an insight into the type of reactants that are expected to modify plasma protein *in vivo*. As expected, the most abundant sugar is glucose (concentration in the range of millimolar). Methylglyoxal, free pentoses, and other sugars are present in much lower concentrations (µM). In a similar way, concentrations of intracellular glycolytic intermediates are also in the micromolar range, and are overall tightly controlled. (ii) Highly unsaturated fatty acids (more than two double bonds) are on average the least abundant fatty acids in cell membranes [21, 22]. (iii) Guanine is the least abundant nucleotide in mitochondrial DNA (mtDNA) [28]. (iv) Finally, (iv) methionine is the amino acid that on average has the smallest percentage presence in cellular proteins [29–31].

In summary, we can infer from the available evidence that aerobic life evolved by reducing the relative abundance of those structural components that are highly susceptible to oxidative damage thus conferring to the macromolecules a higher structural stability and lower susceptibility to oxidative stress (Table 32.2).

32.4.2
Resistance to Oxidative Damage and Life Span

A low intracellular level of glycolytic intermediates would protect other molecules (e.g., aminophospholipids, DNA, and proteins) against glycoxidation-derived damage. The nonenzymatic chemical reaction that occurs between reactive sugars and amino and sulfhydryl groups leads to the formation of Maillard reaction products

Tab. 32.2 Structural components that are highly resistant to oxidative stress and its relationship to life span

Cellular component	Structural component highly susceptible to oxidative damage	Content (%) from compositional analysis	Content in long-lived versus short-lived species
Carbohydrates	Glycolytic intermediates	↓	↓
Lipids	Polyunsaturated fatty acids	↓	↓
Nucleic acids	Guanosine	↓	↓
Proteins	Methionine residues	↓	↓

Note: References are given in Sections 32.4.1 and 32.4.2.

in vivo (the Maillard reaction is also commonly known as the *glycation* reaction). Oxidative reactions, and by inference, oxidative stress and ROS, catalyze the chemical modification of amino acid residues in proteins (cysteine, histidine, and lysine as the main targets) through the Maillard reaction [14, 26]. Reactive carbonyl species such as glyoxal and methylglyoxal, formed endogenously during oxidation of carbohydrates, have been identified as intermediates in the formation of advanced glycoxidation end products (AGEs). During the Maillard reaction, these carbonyl or aldehydic compounds react nonenzymatically with extra- and intracellular proteins to form protein adducts and cross-links such as vesperlysine, crossline, pentosidine, N-ε-(carboxymethyl)lysine (CML), and N-ε-(carboxyethyl)lysine (CEL), among others, which contribute to protein structure and function deterioration due to the formation of such irreversible molecules, a hallmark of the aging process [25, 26, 32]. In a similar way, carbonyl compounds can modify the DNA generating specific products [33], guanine being the base exhibiting the highest modification rate. However, there is a lack of studies evaluating the relevance of these compounds in aging and life span.

Highly unsaturated fatty acids (more than two double bonds) are on average the least abundant fatty acids in cell membranes [21, 22]. A low degree of unsaturation (measured as the double bond index and the peroxidizability index) of cell membranes is a general characteristic of long-lived species [22]. Animal species (including humans and species with an exceptional longevity such as the naked mole-rat) with a high maximum life span have a low degree of membrane fatty acid unsaturation based on the redistribution between types of PUFA without any alteration in the total (percentage) PUFA content and the average chain length [22]. This protects macromolecules (aminophospholipids, DNA, and proteins) against lipoxidation-derived damage. In agreement with this, it has been found that long-lived animals (birds and mammals, including humans) have a lower degree of *in vitro* and *in vivo* lipid peroxidation and lipoxidation-derived protein damage than short-lived ones [22, 34, 35].

Free radical attack of PUFAs generates hydroperoxides and endoperoxides that can undergo fragmentation to produce a broad range of reactive carbonyl compounds [36]. These endogenous carbonyl compounds, analogous to those derived from oxidation of carbohydrates, have unique properties. For instance, compared with ROS, reactive carbonyl compounds have a much longer half-life (minutes instead of microseconds or nanoseconds). Further, the noncharged structure of aldehydes allows them to migrate with relative ease through hydrophobic membranes and hydrophilic cytosolic media, thereby extending damage far from the ROS production site. On the basis of these features, the carbonyl compounds can be more destructive than free radicals and may have far-reaching damaging effects both within and outside membranes. These carbonyl compounds react with nucleophilic groups in proteins, DNA, and aminophospholipids [22, 25] resulting in their nonenzymatic chemical modification (Figure 32.1). Thus, lipid peroxidation should not be perceived solely in a "damage to lipids" scenario. It should also be considered as a significant endogenous source of damage to other cellular macromolecules. These aldehydes were initially believed to produce only cytotoxic

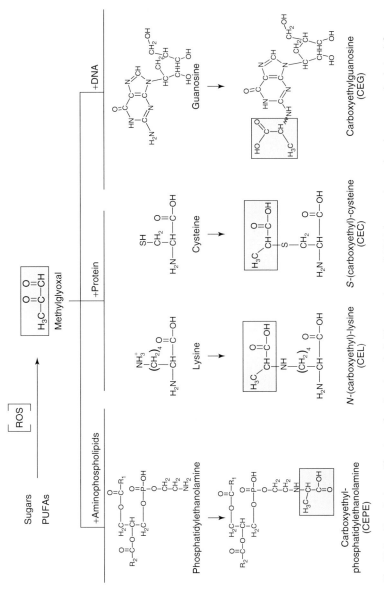

Fig. 32.1 Chemical modification of aminophospholipids, proteins, and DNA by reactive carbonyl compounds derived from lipid peroxidation and reducing sugars. Shown are examples of molecular adducts (advanced glycoxidation end products, AGEs) generated by the reactive carbonyl compound methylglyoxal.

effects associated with oxidative stress. But, there is increasing evidence that these compounds can also have specific signaling roles during normal function [22]. For instance, the superoxide-initiated activation of proton conductance by mitochondrial uncoupling proteins has been shown to be mediated by the lipid peroxidation product 4-hydroxy-*trans*-2-nonenal and its homologs [37, 38].

Guanine is the least abundant nucleotide in mtDNA [28]. In addition, the longer the maximum life span of a species (including invertebrates, birds, and mammals), the lower the mtDNA free energy (a physical property of the double-stranded DNA molecule, which measures the binding energy between the two DNA strands, and can be interpreted as a measure of the susceptibility of mtDNA to mutation) [28]. Consequently, the mtDNA sequence composition expresses a highly resistant life span adaptation. In agreement with this, long-lived mammals and birds have lower steady-state levels of 8-oxodG (8-oxo-7,8-dihydro-2′deoxyguanosine, a specific marker of DNA oxidation) in their mtDNA [10, 39] and lower rates of urinary excretion of 8-oxodG than short-lived ones [40]. It has also been observed that the rate of accumulation of mtDNA oxidation–derived mutations with age is much slower in humans than in mice [41].

Finally, methionine is the amino acid that on average has the smallest percentage presence in cellular proteins [19, 29–31]. In addition, the longer the life span of a species, the lower its tissue protein methionine content and the less the protein damage derived from glycoxidation and lipoxidation, because the longer the longevity of a species, the higher the content of proteins resistant to oxidative damage [19, 29–31].

In summary, we can infer from the available evidence that long-lived species evolved by reducing the relative abundance of those structural components that are highly susceptible to oxidative damage thus conferring to the macromolecules a lower susceptibility to oxidative damage (Table 32.2).

32.5
Oxidative Stress, Aging, and Dietary Restriction

32.5.1
Caloric Restriction

The comparative studies described above strongly suggest a causal relationship between MitROS production, oxidative molecular damage, and the rate of aging. However, correlation does not necessarily indicate a cause–effect relationship; experimental studies are needed for confirmation. Caloric restriction (CR) is the experimental intervention described in most detail that slows down aging and increases maximum longevity [42]. It is therefore very interesting to study whether CR can change the rate of MitROS generation. Much research has shown that changes in antioxidant levels during CR are inconsistent [10, 43] and cannot explain the increase in life span due to this dietary manipulation. The effect of CR on MitROS production, however, has been studied intensively in rodents, especially in rats [10, 39]. These studies consistently demonstrate that CR (usually applying

Tab. 32.3 Summary of changes in mitochondrial ROS generation, oxidative molecular damage, and maximum life span in caloric, protein, and methionine restriction in animal models

	Dietary restriction	Protein restriction	Methionine restriction
Mitochondrial free radical production	↓	↓	↓
Carbohydrates concentration	↓	↓	↓
Membrane unsaturation[a]	↓	↓	↓
Oxidation-derived mtDNA damage[b]	↓	↓	↓
Glyco- and lipoxidation-derived mtDNA damage	nd	nd	nd
Oxidation-derived protein damage[c]	↓	↓	↓
Glyco- and lipoxidation-derived protein damage[d,e]	↓	↓	↓
Maximum life span	↑	↑	↑

↑, increase; ↓, decrease; nd: not determined.
[a] Measured as double bond index and peroxidizability index.
[b] Measured as 8-oxo-7,8-dihydro-2′deoxyguanosine.
[c] Measured as the specific protein carbonyls glutamic semialdehyde and aminoadipic semialdehyde.
[d] Measured as the advanced glycoxidation end product N-ε-(carboxyethyl)lysine.
[e] Measured as the advanced lipoxidation end products N-ε-(carboxymethyl)lysine and N-ε-(malondialdehyde)lysine.
Note: References are given in Section 32.5.

40% CR) significantly decreases the rate of MitROS generation in rat tissues including heart, brain, skeletal muscle, liver, and kidney [10, 19]. In addition, the decrease in MitROS generation observed in DR rats was accompanied by significant decreases in 8-oxodG levels either in mtDNA alone or in mtDNA and nuclear DNA (nDNA) depending on the tissue studied. It was also accompanied by decreases in membrane unsaturation and damage to tissue and mitochondrial proteins resulting from oxidation, glycoxidation, and lipoxidation [10, 21, 22, 25, 35, 39].

In summary, lowering the rate of MitROS production seems to be a very conservative mechanism, which has evolved both within (CR) and between species. It allows both long-lived species and CR animals to decrease steady-state oxidative damage to macromolecules, as well as the rate of aging (Table 32.3).

32.5.2
Protein Restriction

As described in the previous section, much research has consistently found that CR decreases MitROS production and oxidative damage to macromolecules. However, the specific dietary factor that causes these beneficial changes remained unknown. A systematic series of studies was recently performed on experimental animals to answer this question. It is commonly believed that all the antiaging effect of CR is due to the decreased caloric intake and not due to decreases in specific dietary components. The few studies that we are aware of on the subject do not support the possibility of life-long carbohydrate or lipid restriction increasing rodent longevity. According to the two published studies in which dietary lipid restriction was applied to Fischer 344 rats, their life span remained unchanged [44, 45]. The two available studies of carbohydrate restriction led to minor and contradictory changes in rat life span [46, 47]. In contrast, reconsideration of classic studies of protein restriction (PR) performed on rats and mice shows that PR (the mean degree of PR applied was 66.7%) increases maximum life span [48]. The magnitude of the increase in life span (around 20%) was, however, lower than that typically found in 40% CR (about a 40% increase). Thus, assuming proportionality between the life extension and the degree of restriction (as is known to occur in CR), the increase in maximum life span expected at 40% PR would be 11.5%. The decrease in protein intake therefore seems to be responsible for around one-third of the total life span extension effect of CR in rodents. A significant PR life span extension effect, which is lower than the total CR effect, agrees with the widely accepted notion that aging has more than one single main cause.

It seems that, even though CR is one important mechanism underlying the effects of different DR regimes on life span and disease susceptibility, at least some beneficial effects of DR regimens may result from a mechanism other than an overall reduction in caloric intake. One such possible mechanism is the stimulation of cellular stress resistance pathways, which are strongly induced by intermittent (alternate-day) fasting – DR regimes that maintain overall food intake and body weight [49]. A recent study in Drosophila also suggests that the life span extension effect of CR is not simply due to the number of calories [50]. The study did not use chemically defined diets, but the results were compatible with a specific role of the decreased protein intake in life span extension. This was further supported by the observation that decreasing the amount of dietary casein from 4 to 2% and from 2 to 1% or 0.5% increases the maximum life span of Drosophila. Furthermore, CR and PR share many other common effects in mammals. These include transitory decreases in metabolic rate; boosting of cell-mediated immunity; decreased IGF-1; increased hepatic protection against xenobiotics; and decreased incidence of tumors, glomerulosclerosis, or chronic nephropathy, and cardiomyopathy [48].

But what is responsible for the decrease in endogenous MitROS generation in CR? Is the decrease in MitROS production due to PR, or is it due to restriction of some other dietary components, or simply the overall reduction in calories? Recent research suggests answers to these questions. The effect of PR on MitROS generation and oxidative damage was studied in rats, without changing the amount of other dietary components eaten per day. It was found that 40% PR decreases MitROS production, specifically at complex I (the main mitochondrial generator of free radicals), lowers the %FRL (see Section 32.3), and decreases the concentration of 8-oxodG in mtDNA [51]. It also decreases specific markers of protein oxidative modification, membrane unsaturation, and complex I content in rat liver mitochondria and tissue [52] (Table 32.3).

Interestingly, the magnitude, direction of change, mechanisms, and site of action of many of these changes are very similar to those found in CR. In the 40% PR studies, an 8.5% CR in the PR group was unavoidable because that is the caloric content of the proteins themselves that were withheld from the experimental animals. However, we recently found that a total food restriction that decreased the caloric intake by precisely 8.5% does not decrease MitROS generation, %FRL, or 8-oxodG in mtDNA in rat liver [53]. There is, however, an improvement in membrane unsaturation and oxidation-derived protein damage by decreasing them [53]. Thus, the decrease in MitROS generation and oxidative mtDNA damage observed during 40% PR was specifically due to the lower protein consumption of the PR group, not due to its 8.5% lower caloric intake.

The possible effects of lipid or carbohydrate restriction without changing the intake of other dietary components were also studied. It was found that neither lipid nor carbohydrate restriction changes MitROS production or oxidative damage to mtDNA in rat liver [48]. In agreement with a role for MitROS generation in aging, neither lipid nor carbohydrate restriction seems to increase maximum life span in rats. All these, together with the decrease in MitROS generation and increase in life span caused by PR, demonstrate that proteins are the dietary component responsible for the decreases in MitROS generation and oxidative damage that take place in CR and possibly for part (about 30%) of the increase in maximum life span during CR. The rest of the effect of CR on life span (the other two-thirds) must be due to other mechanisms, which may or may not depend on oxidative stress.

32.5.3
Methionine Restriction

As dietary protein is responsible for the decrease in MitROS generation and oxidative damage during CR, the next step was to establish if some specific protein component(s) are responsible. As it was known that L-methionine restriction (MetR), like PR, increases maximum life span [54–56], it was logical to suspect that dietary methionine could be involved. It is known that MetR increases maximum life span in rats and mice independent of energy restriction, although until recently there have been no clues as to what molecular mechanisms mediate this effect. Decreased methionine ingestion could be responsible for the PR-induced increase in longevity and for part of the life span extension effect of CR. In the mouse study, 65% MetR led to significant increases in maximum life span (at least 10%) when the unfinished life span experiment was published [54]. This would agree in general terms with the mean increase in life span at 40% PR calculated above from the different PR studies (an 11.5% increase) [48]. The rat MetR studies found a 44% increase [55] and an 11% increase [56] in maximum life span at 80% MetR, which would be equivalent to a mean 14% increase at 40% MetR. This value is also within the range reported for MetR mice and for PR in rats and mice. The MetR studies available suggest that the reduced dietary intake of this amino acid could be responsible for all the life span extension effect of PR. The decrease in methionine intake seems to be responsible for around one-third (11–14% increase in life span) of the life extension effect of CR in rodents (approximately a 40% increase in longevity).

Regarding oxidative stress, only changes in liver and blood glutathione (GSH), the major cellular thiol or redox buffer, have been studied in MetR models [55]. Thus, we decided to study the effect of MetR on MitROS generation and oxidative molecular damage. We found that the protocol of 80% MetR (without CR), as well as that of 40% MetR also decrease MitROS generation (mainly at complex I), FRL, 8-oxodG in mtDNA, complex I content, specific markers of protein oxidative modification, and membrane unsaturation in rat heart and liver mitochondria [57, 58]. These decreases were dose dependent and stronger at 80% than at 40% MetR [58]. Strikingly, the pattern of many of these changes was very similar to those previously found in CR and PR. Both the occurrence and the intensity of the changes at 40% MetR (without changing other dietary components) strongly suggest that MetR may be responsible for 100% of the decrease in MitROS generation and oxidative stress that occur at 40% PR and 40% CR, and possibly for all (PR) or part (CR) of the life span extension effect (Table 32.3).

Any treatment that increases maximum life span must also decrease the incidence of age-related degenerative diseases [48]. It is known that excessive methionine dietary supplementation damages many vital organ systems and increases tissue oxidative stress. Thus, methionine supplementation increases plasma hydroperoxides and LDL-cholesterol, raises liver iron, lipid peroxidation, conjugated dienes, is hepatotoxic, alters liver antioxidant enzymes and GSH, and decreases vitamin E levels in liver and heart. It also raises plasma, heart, and aortic homocysteine levels leading to angiotoxicity, mitochondrial degeneration in arterial smooth muscle cells and accelerated aging of the rat vascular system, induces hypertension and coronary disease, and seems to accelerate brain aging [48]. Interestingly, the negative effects observed in rats fed high protein (50%) or high methionine (2%) diets for two years are similar. Thus, the high methionine and protein content of the western diet could predispose people in the West to cardiovascular and other degenerative diseases. Most interestingly, the influence of dietary methionine on age-related changes is not limited to excess methionine dietary supplementation. Recent studies have found that the same MetR protocol that increased rodent life span slows cataract development, minimizes age-related changes in T cells, increases macrophage migration inhibition factor, and lowers serum glucose, IGF-I, and insulin levels in mice [54]. It also decreases visceral fat mass (by around 70%), plasma insulin, IGF-1, and insulin response to glucose, and avoids the age-related increases in blood cholesterol and triglycerides in rats [59]. The marked decrease in visceral fat mass described for MetR is striking and suggests that such a change, typical in CR animals, is not necessarily linked to a decreased caloric dietary intake, contrary to common belief.

In summary, the association between methionine dietary intake and deleterious changes appears through a wide range of dietary concentrations covering both MetR and supplementation below and above optimum dietary levels.

32.6
Nutritional Considerations on Methionine: A Public Health Matter

The capacity of MetR and PR to decrease endogenous MitROS generation, oxidative molecular damage and many diseases, and to increase life span is most interesting becasue these interventions are more easily practicable for humans than CR is. CR is a difficult manipulation for human populations due to (i) the marked difficulty in modifying acquired nutritional habits in human adults, (ii) the high risk of malnutrition, and (iii) possible decreases in acute resistance to the normally stressful human living conditions. The last two reasons present even bigger problems in the case of children and the elderly. At present, western human populations consume levels of dietary protein that are three to fourfold higher than the recommended values (0.5–0.75 g kg^{-1} body weight per day). Therefore, there is ample room for safely decreasing the amount of protein ingested. A lack of negative effects of diets containing as little as 0.5 g of protein per kilogram per day for periods of up to one year in human adult males was described in 1909 [60]. Decreasing only the ingestion of protein or even of a single molecule (methionine) through emphasizing the intake of foods that are particularly low in methionine (e.g., fruit, legumes, vegetables), without having to decrease the total ingestion of calories and the overall food intake, is much easier than CR for most people. In any case, more studies are needed to confirm the human health beneficial effects of an MetR diet.

Acknowledgments

Research performed by the authors of this chapter has been supported by grants awarded to R. Pamplona from the Spanish Ministry of Science and Education, the Spanish Ministry of Health, and the regional government of Catalonia.

References

1. United Nations (2002) *World Population Ageing, 1950–2050*, United Nations, New York.
2. United Nations (2002) *Report of the Second World Assembly on Aging*, United Nations, Madrid.
3. Carey, J.R. (2003) *Longevity: The Biology and Demography of Life Span*, Princeton University Press, Princeton.
4. Martin, L.G. and Preston, S.H. (1994) *Demography of Aging*, National Academic Press, Washington, DC.
5. Eberstadt, N. and Groth, H. (2007) *Europe's Coming Demographic Challenge. Unlocking the Value of Health*, The American Enterprise Institute Press, Washington, DC.
6. Magnus, G. (2008) *The Age of Aging: How Demographics are Changing the Global Economy and our World*, John Wiley & Sons, Inc., New York.
7. Strehler, B.L. (1962) *Time, Cells and Aging*, Academic Press, New York.
8. Harman, D. (1972) The biological clock: the mitochondria? *J Am Geriatr Soc*, **20**, 145–147.
9. Miquel, J., Economos, A.C., Fleming, J., and Johnson, J.E. Jr (1980) Mitochondrial role in cell aging. *Exp Gerontol*, **15**, 575–591.

10 Sanz, A., Pamplona, R., and Barja, G. (2006) Is the mitochondrial free radical theory of aging intact? *Antioxid Redox Signal*, **8**, 582–599.

11 Lane, N. (2002) *Oxygen. The Molecule that Made the World*, Oxford University Press, Oxford.

12 Nelson, D.L. and Cox, M.M. (2005) *Lehninger Principles of Biochemistry*, WH Freeman and Company, New York.

13 Golubev, A.G. (1996) The other side of metabolism. *Biokhimiia*, **61**, 2018–2039.

14 Baynes, J.W. (2000) From life to death – the struggle between chemistry and biology during aging: the Maillard reaction as an amplifier of genomic damage. *Biogerontology*, **1**, 235–246.

15 Halliwell, B. and Gutteridge, J.M.C. (2004) *Free Radicals in Biology and Medicine*, Oxford University Press, New York.

16 Lambert, A.J., Boysen, H.M., Buckingham, J.A., Yang, T., Podlutsky, A., Austad, S.N., Kunz, T.H., Buffenstein, R., and Brand, M.D. (2007) Low rates of hydrogen peroxide production by isolated heart mitochondria associate with long maximum lifespan in vertebrate homeotherms. *Aging Cell*, **6**, 607–618.

17 Kudin, A.P., Bimpong-Buta, N.Y., Vielhaber, S., Elger, C.E., and Kunz, W.S. (2004) Characterization of superoxide producing sites in isolated brain mitochondria. *J Biol Chem*, **279**, 4127–4135.

18 Barja, G. (2007) Mitochondrial oxygen consumption and reactive oxygen species production are independently modulated: implications for aging studies. *Rejuvenation Res*, **10**, 215–224.

19 Pamplona, R. and Barja, G. (2007) Highly resistant macromolecular components and low rate of generation of endogenous damage: two key traits of longevity. *Ageing Res Rev*, **6**, 189–210.

20 Bunn, H.F. and Higgins, P.J. (1981) Reaction of monosaccharides with proteins: possible evolutionary significance. *Science*, **213**, 222–224.

21 Hulbert, A.J., Pamplona, R., Buffenstein, R., and Buttemer, W. (2007) Life and death: metabolic rate, membrane composition and life span. *Physiol Rev*, **87**, 1175–1213.

22 Pamplona, R. (2008) Membrane phospholipids, lipoxidative damage and molecular integrity: a causal role in aging and longevity. *Biochim Biophys Acta*, **1777**, 1249–1262.

23 Bjelland, S. and Seeberg, E. (2003) Mutagenicity, toxicity and repair of DNA base damage induced by oxidation. *Mutat Res*, **531**, 37–80.

24 Stadtman, E.R., Moskovitz, J., and Levine, R.L. (2003) Oxidation of methionine residues of proteins: biological consequences. *Antioxid Redox Signal*, **5**, 577–582.

25 Portero-Otín, M. and Pamplona R. (2006) Is endogenous oxidative protein damage envolved in the aging process? in *Protein Oxidation and Disease* (ed. J. Pietzsch), Research Signpost, Trivandrum, Kerala, pp. 91–142.

26 Thorpe, S.R. and Baynes, J.W. (2003) Maillard reaction products in tissue proteins: new products and new perspectives. *Amino Acids*, **25**, 275–281.

27 Monnier, V.M., Sell, D.R., Nagaraj, R.H., and Miyata, S. (1991) Mechanisms of protection against damage mediated by the Maillard reaction in aging. *Gerontology*, **37**, 152–165.

28 Samuels, D.C. (2005) Life span is related to the free energy of mitochondrial DNA. *Mech Ageing Dev*, **126**, 1123–1129.

29 Portero-Otín, M., Requena, J.R., Bellmunt, M.J., Ayala, V., and Pamplona, R. (2004) Protein nonenzymatic modifications and proteasome activity in skeletal muscle from the short-lived rat and long-lived pigeon. *Exp Gerontol*, **39**, 1527–1535.

30 Ruiz, M.C., Ayala, V., Portero-Otín, M., Requena, J.R., Barja, G., and Pamplona, R. (2005) Protein methionine content and MDA-lysine

adducts are inversely related to maximum life span in the heart of mammals. *Mech Ageing Dev*, **126**, 1106–1114.
31 Pamplona, R., Portero-Otín, M., Sanz, A., Ayala, V., Vasileva, E., and Barja, G. (2005) Protein and lipid oxidative damage and complex I content are lower in the brain of budgerigards and canaries than in mice. Relation to aging rate. *AGE J*, **27**, 267–280.
32 Sell, D.R., Lane, M.A., Johnson, W.A., Masoro, E.J., Mock, O.B., Reiser, K.M., Fogarty, J.F., Cutler, R.G., Ingram, D.K., Roth, G.S., and Monnier, V.M. (1996) Longevity and the genetic determination of collagen glycoxidation kinetics in mammalian senescence. *Proc Natl Acad Sci U S A*, **93**, 485–490.
33 Dutta, U., Cohenford, M.A., and Dain, J.A. (2005) Nonenzymatic glycation of DNA nucleosides with reducing sugars. *Anal Biochem*, **345**, 171–180.
34 Pamplona R. and Barja G. (2003) Aging rate, free radical production, ans constitutive sensitivity to lipid peroxidation: insights from comparative studies, in *Biology of Aging and its Modulation Series*, Aging at the Molecular Level, vol. 1 (ed. T. Van Zglinicki), Kluwer Academic Publisher, New York, pp. 47–64.
35 Pamplona, R., Barja, G., and Portero-Otín, M. (2002) Membrane fatty acid unsaturation, protection against oxidative stress, and maximum life span: a homeoviscous-longevity adaptation? *Ann N Y Acad Sci*, **959**, 475–490.
36 Esterbauer, H., Schaur, R.J., and Zollner, H. (1991) Chemistry and biochemistry of 4-hydroxynonenal, malonaldehyde and related aldehydes. *Free Radic Biol Med*, **11**, 81–128.
37 Echtay, K.S., Esteves, T.C., Pakay, J.L., Jekabsons, M.B., Lambert, A.J., Portero-Otín, M., Pamplona, R., Vidal-Puig, A.J., Wang, S., Roebuck, S.J., and Brand, M.D. (2003) A signalling role for 4-hydroxy-2-nonenal in regulation of mitochondrial uncoupling. *EMBO J*, **22**, 4103–4110.
38 Brand, M.D., Affourtit, C., Esteves, T.C., Green, K., Lambert, A.J., Miwa, S., Pakay, J.L., and Parker, N. (2004) Mitochondrial superoxide: production, biological effects, and activation of uncoupling proteins. *Free Radic Biol Med*, **37**, 755–767.
39 Barja, G. (2004) Aging in vertebrates and the effect of caloric restriction: a mitochondrial free radical production-DNA damage mechanism? *Biol Rev*, **79**, 235–251.
40 Foksinski, M., Rozalski, R., Guz, J., Ruszowska, B., Sztukowska, P., Ptwowarski, M., Klungland, A., and Olinski, R. (2004) Urinary excretion of DNA repair products correlates with metabolic rates as well as with maximum life pans of different mammalian species. *Free Radic Biol Med*, **37**, 1449–1454.
41 Wang, E., Wonq, A., and Cortopassi, G. (1997) The rate of mitochondrial mutagenesis is faster in mice than in humans. *Mutat Res*, **377**, 157–166.
42 Masoro, E.J. (2005) Overview of caloric restriction and ageing. *Mech Ageing Dev*, **126**, 913–922.
43 Sohal, R.S., Ku, H.H., Agarwal, S., Forster, M.J., and Lal, H. (1994) Oxidative damage, mitochondrial oxidant generation and antioxidant defenses during aging and in response to food restriction in the mouse. *Mech Ageing Dev*, **74**, 121–133.
44 Iwasaki, K., Gleiser, C.A., Masoro, E.J., McMahan, C.A., Seo, E.J., and Yu, B.P. (1988) Influence of the restriction of individual dietary components on longevity and age-related disease of Fisher rats: the fat component and the mineral component. *J Gerontol*, **43**, B13–B21.
45 Shimokawa, I., Higami, Y., Yu, B.P., Masoro, E.J., and Tikeda, T. (1996) Influence of dietary components on occurrence of and mortality due to neoplasms in male F344 rats. *Aging Clin Exp Res*, **8**, 254–262.
46 Ross M.H. (1976) Nutrition and longevity in experimental animals, in

47 Khorakova, M., Deil, Z., Khausman, D., and Matsek, K. (1990) Effect of carbohydrate-enriched diet and subsequent food restriction on life prolongation in Fisher 344 male rats. *Fiziol Zh*, **36**, 16–21.

48 Pamplona, R. and Barja, G. (2006) Mitochondrial oxidative stress, aging and caloric restriction: the protein and methionine connection. *Biochim Biophys Acta*, **1757**, 496–508.

49 Anson, R.M., Guo, Z., de Cabo, R., Iyun, T., Rios, M., Hagepanos, A., Ingram, D.K., Lane, M.A., and Mattson, M.P. (2003) Intermittent fasting dissociates beneficial effects of dietary restriction on glucose metabolism and neuronal resistance to injury from calorie intake. *Proc Natl Acad Sci U S A*, **100**, 6216–6220.

50 Mair, W., Piper, M.D.W., and Partridge, L. (2005) Calories do not explain extension of life span by dietary restriction in Drosophila. *PLOS Biol*, **3**, 1305–1311.

51 Sanz, A., Caro, P., and Barja, G. (2004) Protein restriction without strong caloric restriction decreases mitochondrial oxygen radical production and oxidative DNA damage in rat liver. *J Bioenerg Biomembr*, **36**, 545–552.

52 Ayala, V., Naudí, A., Sanz, A., Caro, P., Portero-Otin, M., Barja, G., and Pamplona, R. (2006) Dietary protein restriction decreases oxidative protein damage, peroxidizability index, and mitochondrial complex I content in rat liver. *J Gerontol Biol Sci Med Sci*, **62A**, 352–360.

53 Gómez, J., Caro, P., Naudí, A., Portero-Otin, M., Pamplona, R., and Barja, G. (2007) Effect of 8.5% and 25% caloric restriction on mitochondrial free radical production and oxidative stress in rat liver. *Biogerontology*, **8**, 555–566.

54 Miller, R.A., Buehner, G., Chang, Y., Harper, J.M., Sigler, R., and Smith-Wheelock, M. (2005) Methionine-deficient diet extends mouse lifespan, slows immune and lens aging, alters glucose, T4, IGF-I and insulin levels, and increases hepatocyte MIF levels and stress resistance. *Aging Cell*, **4**, 119–125.

55 Richie, J.P. Jr, Leutzinger, Y., Parthasarathy, S., Malloy, V., Orentreich, N., and Zimmerman, J.A. (1994) Methionine restriction increases blood glutathione and longevity in F344 rats. *FASEB J*, **8**, 1302–1307.

56 Orentreich, N., Matias, J.R., DeFelice, A., and Zimmerman, J.A. (1993) Low methionine ingestion by rats extends life span. *J Nutr*, **123**, 269–274.

57 Sanz, A., Caro, P., Ayala, V., Portero-Otin, M., Pamplona, R., and Barja, G. (2006) Methionine restriction decreases mitochondrial oxygen radical generation and leak as well as oxidative damage to mitochondrial DNA and proteins. *FASEB J*, **20**, 1064–1073.

58 Caro, P., Gómez, J., López-Torres, M., Sánchez, I., Naudí, A., Jove, M., Pamplona, R., and Barja, G. (2008) Forty percent and eighty percent methionine restriction decrease mitochondrial ROS generation and oxidative stress in rat liver. *Biogerontology*, **9**, 183–196.

59 Malloy, V.L., Krajcik, R.A., Bailey, S.J., Hristopoulos, G., Plummer, J.D., and Orentreich, N. (2006) Methionine restriction decreases visceral fat mass and preserves insulin action in aging male Fischer 344 rats independent of energy restriction. *Aging Cell*, **5**, 305–314.

60 Chittenden, R.H. (1909) *The Nutrition of Man*, Heinemann, London.

Part Four
Therapeutics Proposed for Decreasing Endogenous Toxins

33
Therapeutic Potential for Decreasing the Endogenous Toxin Homocysteine: Clinical Trials

Wolfgang Herrmann and Rima Obeid

Homocysteine (Hcy) is a nonprotein forming, sulphur-containing amino acid of methionine metabolism. First reports about Hcy as a pathogenic agent in patients with homocystinuria, a congenital disorder of Hcy metabolism, were made in the early 1960s. This metabolic defect results in a marked increase of plasma Hcy and significant homocystine excretion in the urine [1]. Patients with severe hyperhomocysteinemia (HHcy) are usually mentally retarded and develop severe atherosclerotic vascular occlusions and skeletal deformaties at an early age [1, 2]. Moderate HHcy is thought to be an independent risk factor for cardiovascular disease, responsible for about 10% of the total risk [3, 4]. In recent treatment trials with B-vitamins a significant reduction in stroke risk, up to 30%, has been reported. HHcy is also thought to be a risk factor for neurodegenerative diseases (e.g. vascular dementia, Alzheimer disease and cognitive disorders), osteoporotic fractures and complications of pregnancy. Older people frequently have HHcy as a result of vitamin deficiencies.

33.1
Homocysteine Metabolism

Hcy is a product in intermediary metabolism and is located at the intersection of two metabolic pathways: the remethylation of Hcy to methionine and the transsulfuration of Hcy to cystathionine (CYS) (Figure 33.1). In the first path, the remethylation rate is dependant on the dietary methionine supply. During remethylation Hcy gains a methyl group from 5-methyltetrahydrofolate (5-methylTHF) and forms methionine. The reaction is catalyzed by the enzyme methionine synthase, which is dependent on vitamin B_{12}. The methyl group can also be provided by betaine (originates from dietary choline), though this reaction is principally limited to the liver. Methionine is transformed into *S*-adenosylmethionine (SAM), which acts as a universal methyl group donor [5–7]. During the demethylation of SAM *S*-adenosylhomocysteine (SAH) is formed, which releases Hcy, catalyzed by SAH hydrolases. SAH is a potent competitive inhibitor of many transmethylases and

Endogenous Toxins. Diet, Genetics, Disease and Treatment.
Edited by Peter J. O'Brien and W. Robert Bruce
Copyright © 2010 WILEY-VCH Verlag GmbH & Co. KGaA, Weinheim
ISBN: 978-3-527-32363-0

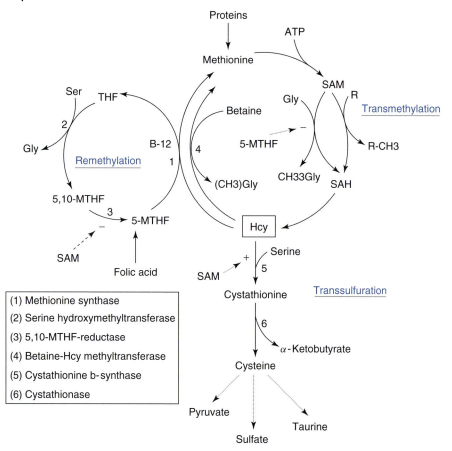

Fig. 33.1 Homocysteine metabolic pathways. Hcy, homocysteine; SAH, S-adenosylhomocysteine; SAM, S-adenosylmethionine.

in this way can inhibit numerous transmethylation reactions. Methionine is an essential amino acid. If it is limiting, choline can become a nutritional requirement. In the second path, the catabolic transsulfuration pathway, the vitamin B_6-dependent cystathionine-β-synthase (CBS) condenses Hcy with serine to form CYS. CYS is then hydrolyzed to cysteine and α-ketobutyrate by cystathionase, again which is dependent on vitamin B_6. SAM plays an important role in the regulation between the Hcy remethylation and transsulfuration [8]. If the kidney function is normal, moderate HHcy, mostly results due to enzyme deficiency or folate, vitamin B_{12}, or vitamin B_6 deficiency.

The most important carrier of methyl groups is SAM. Some of the important reactions in which SAM is involved are as follows:

- Methylation of DNA and RNA. DNA- and RNA. Methylases use SAM as a source of methyl groups. A major target of methylases is the 5 position of

cytosine of DNA. The degree of methylation is correlated with the transcriptional activity. Globin genes, for example, are highly methylated in nonerythroid cells but not in erythroid cells.
- The conversion of epinephrine to norepinephrine is also catalyzed by an *N*-methyl transferase that uses SAM.

33.2
Physiological Determinants of Plasma Homocysteine Level

Age and gender are two main determinants of Hcy concentrations in humans. Hcy concentrations are higher in infants (mean (95%CI) 5.1; 4.6–5.6 µmol l^{-1}) than in children older than one year (mean (95%CI) 4.6; 4.2–5.1 µmol l^{-1}) [9, 10]. Vitamin B$_{12}$ is the primary modulator of Hcy concentration in infants less than one year old. In children older than one year, folate becomes more important as a modulator of the Hcy level. Age is the most important influencing factor of the Hcy level in children older than one year as well as in adults. Children older than one year have lower Hcy concentrations than adolescents (5–10 µmol l^{-1}) and from about 10 years old onward boys have a higher plasma Hcy concentration than girls. Elevated Hcy levels in children can be hereditary or acquired [9, 10].

In adults, plasma concentration of Hcy generally increases with age [11]. Young men (30–40 years) normally have higher Hcy level than women (difference about 2 µmol l^{-1}). The difference between the sexes can be explained by the effect of estrogen in women. This also explains the increase in plasma Hcy during the menopause when estrogen levels decline. The age-dependent elevation of the Hcy level is mainly due to the physiological decrease in kidney function. Hcy levels increase linearly up to age 60 or 65, but at older ages it increases on average by about 10%, or 1 µmol l^{-1}, per decade. Age-dependent reference ranges are not recommended because the levels in older people can be lowered by vitamin supplementation [12, 13]. Increased Hcy concentrations with age could be related to renal insufficiency or low intake and low concentration of folate, vitamin B$_6$, or vitamin B$_{12}$.

Plasma concentrations of Hcy decrease during pregnancy. The lowest concentrations are detected during the second trimester. In the third trimester, Hcy levels increase though they remain lower than prepregnancy levels. Women receiving folate supplementation during pregnancy have lower concentrations of Hcy compared to unsupplemented women.

33.3
Causes of Hyperhomocysteinemia

The intracellular Hcy concentration is kept low by catabolism and by a cellular Hcy export into the plasma. Plasma Hcy is primarily metabolized (remethylized) by the kidney (about 70%) and only a small portion is excreted with urine. Moderate HHcy (fasting Hcy between >12 and 30 µmol l^{-1}) can have different causes: enzyme deficiency (lack of methylenetetrahydrofolate reductase (MTHFR)), B vitamin deficiency (folate, B$_{12}$, or B$_6$), or renal insufficiency [14, 15]. Higher HHcy (between 30 and 100 µmol l^{-1}) is primarily due to enzyme defects (e.g., heterozygous deficiency of CBS) or chronic renal failure. Severe HHcy (>100 µmol l^{-1}) is usually related to homozygous enzyme defects with severe loss of enzyme activity. The severe form is very rare compared to moderate HHcy.

Hcy measurement after a standardized oral methionine-loading test (oMLT) can diagnose subjects with disorders in the transsulfuration pathway (e.g., with vitamin B$_6$ or heterozygous CBS deficiency), even when subjects have normal fasting levels of Hcy.

Vitamin deficiencies are by far the most common cause of HHcy. Deficiency can be developed by lack of supply, decreased absorption in the gastrointestinal tract, increased requirements (pregnancy and AIDS), and by interaction with some medications. People who ingest a vegetarian diet, older people, pregnant women, patients with kidney diseases, malabsorption (inflammatory intestinal diseases), and patients with tumors are among the risk groups for vitamin deficiencies of clinical relevance. Furthermore, alcohol abuse and certain medications are additional factors for developing vitamin deficiencies [16, 17].

Folate deficiency is the most frequent vitamin deficiency in central Europe and is caused by lack of absorption, low intake, or by destruction of food folate content through storage or cooking. The recommended dietary allowance of folate is 400 µg day^{-1}, which increases to 500 µg day^{-1} in pregnant and lactating women.

In contrast to folate, vitamin B$_{12}$ is stable and usually absorbed effectively, but vitamin B$_{12}$ deficiency remains a major health problem in some risk groups. Deficiency is very frequent in older people, mostly a result of insufficient intestinal absorption. Vitamin B$_{12}$ malabsorption can affect 30–40% of older people due to age-related disorders such as decreased stomach acid secretion, lack of intrinsic factor, or pernicious anemia. Animal products, especially red meat, are major sources of vitamin B$_{12}$ in the human diet. For this reason, vegetarians frequently have vitamin B$_{12}$ deficiency and develop HHcy [18, 19].

33.4
The Role of Vitamin B$_{12}$ (Cobalamin)

33.4.1
Cobalamin Metabolism

Prior to the discovery of cobalamin (Cbl; vitamin B$_{12}$) pernicious anemia was a fatal disease. In the early 1920s Minot and Murphy demonstrated that they were able to cure pernicious anemia by whole liver extract. Later, it was shown that

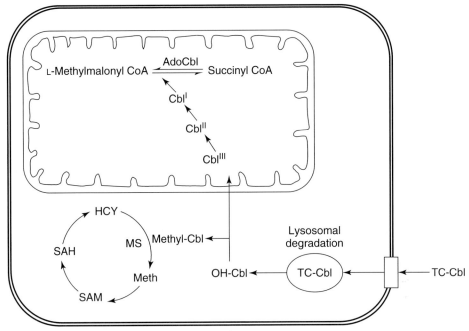

Fig. 33.2 The metabolic pathways enhanced by cobalamin and inherited disorders of cobalamin. Hcy, homocysteine; Met, methionine; SAH, S-adenosylhomocysteine; SAM, S-adenosylmethionine; methyl-Cbl, methyl cobalamin; AdoCbl, adenosyl cobalamin.

liver was an important source of Cbl, and Cbl was isolated, crystallized and its structure was identified. Cbl belongs to a group of compounds of similar chemical structure, but completely different biological functions. Cbl consists of a corrinoid molecule with cobalt in the center of this molecule. The synthetic forms of Cbl are cyanocobalamin and hydroxycobalamin. There are only two forms of Cbl that have biological activity as cofactors in enzyme reactions (Figure 33.2). They are adenosylcobalamin (AdoCbl) and methylcobalamin (MeCbl) [20]. In mammalian metabolism, Cbl is required for only two key enzymatic reactions. The first reaction occurs in the cytosol and involves synthesis of methionine from Hcy. This reaction is an important mechanism in detoxification of Hcy and it provides the precursor molecule for methyl group donation. Methionine synthase and its cofactor MeCbl catalyze this reaction. The second pathway takes place in the mitochondria and involves isomerization of methylmalonyl-CoA to succinyl-CoA. This reaction is catalyzed by methylmalonyl-CoA mutase with AdoCbl as a cofactor [21]. The latter reaction is part of the catabolism of odd-chained fatty acids, cholesterol, and several amino acids. The excess of methylmalonyl-CoA is converted into methylmalonic acid (MMA). Cbl deficiency thus leads to MMA elevation, a sensitive marker for this deficiency. A reduced flux through the methylmalonyl-CoA mutase reaction is presumed to contribute to neurological tissue damage.

33.4.2
Sources of Cobalamin

Cbl synthesis is very complex and restricted to certain strains of bacteria. Neither humans nor animals are able to synthesize this complex molecule. Neither plants not fungi are presumed to synthesize or use the vitamin as well. Animals can get Cbl by consuming foods contaminated with Cbl synthesizing bacteria, thereby incorporating the vitamin into their body organs. Foods of animal source are the only natural source of Cbl in human diet. The main sources of Cbl in human diet are meat (2–5 µg/100 g), fish (2–8 µg/100 g), milk (1.5 µg/100 ml), cheese (1–2 µg/100 g), and egg (2 µg/100 g) [22]. Cbl is essential for normal maturation and development of all DNA synthesizing cells including blood cells and cells of the central nervous system. The ultimate source of Cbl in human diet is from foods contaminated with B_{12}-synthesizing bacteria. Not all forms of Cbl formed by microbes are metabolically active for mammalian cells. Some naturally occurring forms of corrinoids have similar structure but no biological roles in humans. These are produced by some algae (spirulina) and are termed *analogs*. Cbl analogs may even block the normal metabolism of the vitamin [23].

33.4.3
Absorption, Excretion, and Homeostasis of Cobalamin

Cbl is bound to food proteins and must be released before the vitamin can be absorbed. This is achieved by the action of gastric acid and proteolytic enzymes in the stomach. In the stomach Cbl is captured by haptocorrin, an R-binder protein made in the saliva and stomach. In the upper small intestine, pancreatic enzymes and an alkaline pH degrade the haptocorrin–Cbl complex. The free vitamin is then captured by intrinsic factor, another B_{12}-binding protein. The intrinsic factor–Cbl complex is transported to the terminal ileum where the complex is recognized and internalized by specific membrane receptors of the enterocytes (intrinsic factor receptor). The receptor-mediated absorption of Cbl is a saturable process and a maximal amount of 3 µg of the vitamin per meal can be internalized via this pathway. After absorption of Cbl–intrinsic factor complex into the enterocytes, the complex is degraded and Cbl is transferred to a third binding protein, transcobalamin (TC).

TC is synthesized within the enterocytes and is the only binder that can deliver Cbl to cells via the TC receptor. The TC–Cbl complex is released into the portal circulation and is subsequently recognized by TC receptors that are expressed by all cell types. The part of Cbl that is bound to TC is named *holotranscobalamin* (holoTC) [24]. Only 6–20% of total plasma Cbl is present as holoTC [25]. The remaining part of Cbl is bound to haptocorrin and is called *holohaptocorrin* (holoHC) [26]. Although haptocorrin binds almost 80% of total plasma Cbl, the functions of this protein have not been well investigated.

A considerable amount of Cbl is secreted into the bile. Two-thirds of the secreted Cbl in the bile is reabsorbed in the ileum. The liver contains most of the body's Cbl. It is calculated that 2–3 mg of Cbl is stored in the liver [27]. The kidney and the brain are also two important organs in which Cbl is stored. The kidney can release Cbl in the case of short-term depletion of the vitamin. Cbl excreted into the urine and this can be reabsorbed in the proximal tubules via a specific receptor (megalin) [28]. The major route by which Cbl is lost from the body is through the feces.

33.4.4
Cobalamin Deficiency

Frank cobalamin deficiency is common worldwide [29, 30]. In developed countries, Cbl deficiency is found in patients with malabsorption, intestinal resection, and in those who do not ingest a sufficient amount of the vitamin [18, 31]. Cbl deficiency is common in elderly people and in those who ingest a strict vegetarian diet. Cbl deficiency has been reported in children, middle age, and elderly people. Preclinical Cbl deficiency is the state in which metabolic evidence of insufficiency exists without symptoms of anemia or neurological complications. Measurement of serum concentrations of metabolic markers allows the detection of a large number of asymptomatic subjects who have this vitamin deficiency.

33.5
Folic Acid, Folate (Vitamin B$_9$)

33.5.1
Folate Metabolism

Vitamin B$_9$, also called *pteroylglutamic acid* (Figure 33.3), folate, or folacin, is essential for the metabolic processes in all living cells. Folic acid is necessary for the synthesis of nucleic acids and the formation of heme, the pigmented, iron-carrying component of the hemoglobin, in red blood cells. Folate deficiency can impair the maturation of red blood cells, resulting in macrocytic anemia. Folic acid is known chemically as *pteroylmonoglutamic acid*. A part of its molecular structure represents pteroinic acid, which consists of pteridin and *p*-aminobenzoic acid. Natural sources of folate contain this vitamin in the polyglutamates form.

Folate is a carrier of one-carbon fragments, which it transfers to various biochemical targets. MethylTHF and methylene-THF are the two forms of folate of great biological significance.

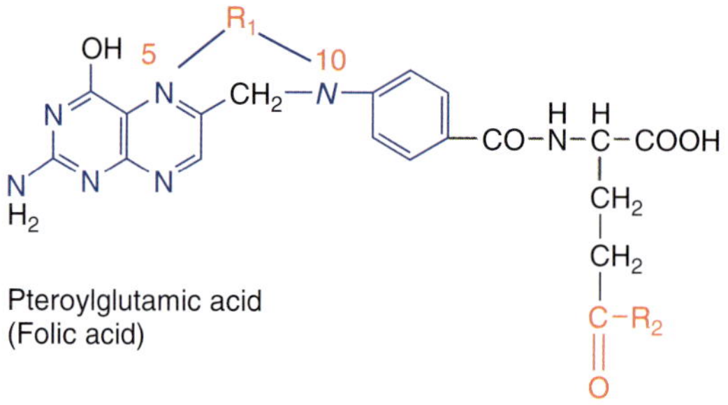

Fig. 33.3 Folic acid and its derivatives.

R1	R2
5-Methyl	–OH
5-Formyl	Glutaminic acid
5-Formimino	(–glutamyl)$_n$
10-Formyl	
5,10-Methenyl	
5,10-Methylene	

33.5.2
Biological Roles of Folate

The fundamental biochemical disturbance in folate metabolism is seen in the deficiency of the methylene form (N^5,N^{10}-methylene tetrahydrofolate). The key reaction is the methylation of deoxyuridine monophosphate (dUMP) to generate thymidylate (dTMP), which is needed for DNA synthesis. One of the obvious consequences of this deficiency is megaloblastic anemia, that is, a deficit in the generation of mature red blood cells. The transfer includes the CH_2 group plus hydrogen from the cofactor itself, that is, the cofactor is effectively oxidized in the process.

1. Thymidylate synthase: dUMP + methylene-THF → dTMP + dihydrofolate (DHF). This means DHF must be rereduced to THF.
2. Dihydrofolate reductase (DHFR): DHF + 2NADPH → THF + 2NADP.

On the one hand, folate deficiency causes decreased thymidylate synthase reaction, thus causing arrest of cellular growth mostly seen in rapidly growing cells like red blood cells. On the other hand, this biological function of folate is used in anticancer therapy, where folate antagonist (methotrexate) is used to inhibit DHFR. THF gets its methylene group from serine to form methylene-THF. An

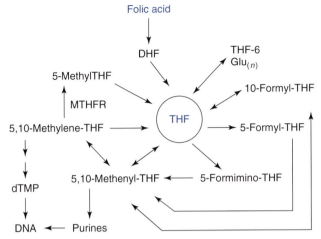

Fig. 33.4 Folate metabolism and the role of folate in one-carbon metabolism.

important biological role of folate is in the production of methyl groups needed for many methylation reactions in the cell. Folate metabolism is shown in Figure 33.4.

33.5.3
Consequences of Folate Deficiency

Folic acid is essential during early pregnancy for the development of neural tube that forms the brain and spinal cord. Folate deficiency is a risk factor for poor pregnancy outcome such as neural tube defects, prematurity, and low birth weight. Folate supplementation before or at early pregnancy caused a marked decrease in neural tube defects. There are some indications that folic acid use may also reduce the risk for other birth defects, such as cleft lip and palate and certain congenital heart defects. Low folate intake in Europe [32] implies that a large part of the population does not ingest the required daily amount of folate.

Folic acid may also play a role in protecting against some forms of cancer and stroke. Older people (above 60) and those who consume large amounts of alcohol need to take even more folic acid supplements. Decreased blood folic acid levels are associated with an increased risk of colon cancer in women. Long-term supplementation with folic acid from a multivitamin has been found, in one large population study, to be associated with a reduced risk of colon cancer [33]. However, 15 years of supplementation was necessary before a significant reduction in colon cancer risk became apparent [34]. In that study, folate from dietary sources alone was associated with a modest reduction in the risk of colon cancer.

Folate and vitamin B_{12} metabolisms are interrelated. Therefore, folic acid may mask vitamin B_{12} deficiency (i.e., pernicious anemia) causing recovery of the hematological symptoms, thereby allowing the deficiency to proceed and cause neurological manifestations. Physicians should be aware of this problem, especially in population from countries where food is fortified with folate.

33.5.4
Cobalamin Deficiency Causes Folate Trap

Because of the role of Cbl in folate metabolism, Cbl deficiency can cause a secondary folate deficiency. Cbl deficiency inhibits the activity of methionine synthase, and causes the retention of N5-methylTHF (5-MTHF). 5-MTHF is trapped because the transfer of the methyl group is inhibited (Figure 33.1). The loss of THF forms due to the inability to use 5-MTHF is referred to as the *methyl trap (folate trap)*. In this case, 5-MTHF cannot be recycled, which causes a secondary deficiency of folate especially the forms needed in DNA synthesis. The level of folate in serum or plasma of Cbl-deficient subjects may be normal to highly normal; however, it is mostly 5-MTHF. Folate and Cbl deficiency have similar clinical (megaloblastic red cells) and metabolic (elevation of tHcy) signs. This explains why Cbl-deficient subjects were frequently treated with folate. This treatment may even relieve the hematological symptoms. Nevertheless, neurological signs of Cbl deficiency may be worsen. Figure 33.5 shows this phenomenon in strict vegetarians compared to omnivorous subjects.

33.6
Vitamin B$_6$

Vitamin B$_6$ is a water-soluble vitamin that was first isolated in the 1930s. There are three traditionally considered forms of vitamin B$_6$: pyridoxal (PL), pyridoxine (PN),

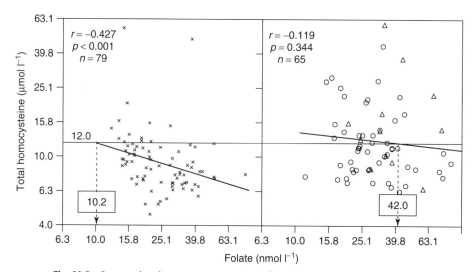

Fig. 33.5 Scatter plot showing concentrations of tHcy in relation to serum folate in omnivorous (left) and vegetarians (right). Numbers on the axis are anti-Log.

and pyridoxamine (PM). The phosphate ester derivative pyridoxal 5′-phosphate (PLP) is the principal coenzyme form and has importance in human metabolism.

33.6.1
Function

Vitamin B_6 must be delivered through the diet because humans cannot synthesize it. PLP plays a vital role in the function of approximately 100 enzymes that catalyze essential chemical reactions in the human body [35, 36]. For example, PLP functions as a coenzyme for glycogen phosphorylase, an enzyme that catalyzes the release of glucose from stored glycogen. Much of the PLP in the human body is found in muscle bound to glycogen phosphorylase. PLP is also a coenzyme for reactions generating glucose from amino acids in gluconeogenesis. In the brain, the synthesis of the neurotransmitter serotonin from the amino acid tryptophan is catalyzed by a PLP-dependent enzyme. Other neurotransmitters, such as dopamine, norepinephrine and gamma-aminobutyric acid (GABA), are also synthesized using PLP-dependent enzymes.

- **Red blood cell formation and function**: PLP functions as a coenzyme in the synthesis of heme, an iron-containing component of hemoglobin. Hemoglobin is found in red blood cells and is critical to the blood's ability to transport oxygen throughout the body. Both PL and PLP are able to bind to the hemoglobin molecule and affect its ability to pick up and release oxygen. However, the impact of this on normal oxygen delivery to tissues is not known.
- **Niacin formation**: PLP is a coenzyme for a critical reaction in the synthesis of niacin from tryptophan; thus, adequate vitamin B_6 decreases the requirement for dietary niacin. The human requirement for niacin (another B vitamin) can be met in part by the conversion of the essential amino acid tryptophan to niacin, as well as through dietary intake.
- **Hormone function**: Steroid hormones, such as estrogen and testosterone, exert their effects in the body by binding to steroid hormone receptors in the nucleus of the cell and altering gene transcription. PLP binds to steroid receptors in a manner that inhibits the binding of steroid hormones, thereby decreasing their effects. The binding of PLP to steroid receptors for estrogen, progesterone, testosterone, and other steroid hormones suggests that the vitamin B_6 status of an individual may have implications for diseases affected by steroid hormones, including breast cancer and prostate cancers.
- **Nucleic acid synthesis**: PLP serves as a coenzyme for a key enzyme involved in the mobilization of single-carbon functional groups (one-carbon metabolism). Such reactions are involved in the synthesis of nucleic acids. The effect of vitamin B_6 deficiency on the function of the immune system may be partly related to the role of PLP in one-carbon metabolism.

33.6.2
Vitamin B$_6$ Deficiency

Severe deficiency of vitamin B$_6$ is uncommon. Alcoholics are presumed to be at risk of vitamin B$_6$ deficiency due to low dietary intakes and impaired metabolism of the vitamin. Abnormal electroencephalogram (EEG) patterns have been noted in some studies of vitamin B$_6$ deficiency. Other neurologic symptoms noted in severe vitamin B$_6$ deficiency include irritability, depression, and confusion; additional symptoms include inflammation of the tongue, sores or ulcers of the mouth. Increased dietary protein results in an increased requirement for vitamin B$_6$, probably because PLP is a coenzyme for many enzymes involved in amino acid metabolism. The current RDA for vitamin B$_6$ is presented in Table 33.1.

Elderly subjects (>65 years): Metabolic studies have indicated that the requirement for vitamin B$_6$ in older adults is approximately 2.0 mg per day. Several studies have found that over half of individuals >60 years consume less than the current RDA (men: 1.7 mg per day and women: 1.5 mg per day). The requirement for B$_6$ might increase with age.

33.6.2.1 Vitamin B$_6$ Deficiency, Hyperhomocysteinemia, and Cardiovascular Diseases

Vitamin B$_6$ is a cofactor for two enzymes (CBS and cystathionase) responsible for Hcy catabolism via the transsulfuration pathway. Through the transsulfuration pathways, vitamin B$_6$ is responsible for producing more cysteine and therefore glutathione, the major endogenous antioxidant [37, 38].

Even moderately elevated levels of Hcy in the blood are associated with increased risk for CVD, including heart disease and stroke. Healthy individuals utilize two different pathways to metabolize Hcy. Several large observational studies have

Tab. 33.1 Recommended dietary intake in age groups

Group	Age	mg/day
Infants	0–6 months	0.1
Infants	7–12 months	0.3
Children	1–3 years	0.5
Children	4–8 years	0.6
Children	9–13 years	1.0
Adolescents	14–18 years	1.3
Adults	19–50 years	1.3
Adults	51 years and older	1.7
Pregnancy	–	1.9
Breast feeding	–	2.0

demonstrated an association between low vitamin B_6 intake or status and increased blood Hcy levels and increased risk of CVDs. A large prospective study found that the risk of heart disease in women who consumed on average 4.6 mg of vitamin B_6 daily was only 67% of the risk in women who consumed on average 1.1 mg daily [39]. In another large prospective study, it was found that higher plasma levels of PLP were associated with a decreased risk of CVD independent of Hcy levels [40]. Further, several studies have reported that low plasma PLP status is a risk factor for coronary artery disease [41–43]. In contrast to folic acid supplementation, studies supplementing individuals with only vitamin B_6 have not resulted in significant decreases in fasting levels of Hcy. Nevertheless, post-methionine Hcy levels were effectively lowered using vitamin B_6 suggesting that the transsulfuration pathway is more active after a meal [44].

33.6.2.2 Vitamin B_6 Deficiency and Cognitive as well as Immune Dysfunction

A few clinical trials have associated cognitive decline in elderly subjects or AD with inadequate nutritional status of folic acid, vitamin B_{12}, and vitamin B_6 and thus, increased levels of Hcy [45, 46]. It has been reported that higher plasma levels of vitamin B_6 were associated with better performance on two measures of memory, but were unrelated to performance on 18 other cognitive tests [47]. Similarly, a double-blind, placebo-controlled study in 38 healthy elderly men found that vitamin B_6 supplementation improved memory but had no effect on mood or mental performance [48]. Further, a placebo-controlled trial in 211 healthy younger, middle-aged, and older women found that vitamin B_6 supplementation (75 mg per day) for five weeks improved memory performance in some age groups, but had no effect on mood [49]. Recently, a systematic review of randomized trials concluded that there is inadequate evidence that supplementation with vitamin B_6, vitamin B_{12}, or folic acid improves cognition in those with normal or impaired cognitive function [50]. Because of mixed findings, it is presently unclear whether supplementation with B vitamins might delay cognitive decline in elderly subjects. Additionally, it is not known if marginal B vitamin deficiencies, which are relatively common in the elderly, even contribute to age-associated declines in cognitive function or whether both result from processes associated with aging and/or disease.

Furthermore, low vitamin B_6 intake and nutritional status have been associated with impaired immune function, especially in the elderly. Decreased production of immune-responsive cells (lymphocytes)and interleukin-2 (an important immune reaction signaling protein) have been reported in vitamin B_6-deficient individuals [51]. Restoration of vitamin B_6 status has resulted in normalization of lymphocyte proliferation and interleukin-2 production, suggesting that adequate vitamin B_6 intake is important for optimal immune system function in older individuals. However, the amount of vitamin B_6 substitution required to restore the immune system function in elderly subjects seems to be higher than the current RDA recommendations (2.9 mg per day for men and 1.9 mg per day for women).

33.6.2.3 Disease Treatment

Because a key enzyme in the synthesis of the neurotransmitters serotonin and norepinephrine is vitamin B_6 dependent, it has been suggested that vitamin B_6 deficiency may lead to depression. However, clinical trials have not provided convincing evidence that vitamin B_6 supplementation is an effective treatment for depression [46, 52], though vitamin B_6 may have therapeutic efficacy in premenopausal women.

Vitamin B_6 has been used since the 1940s to treat nausea during pregnancy. The results of double-blind, placebo-controlled trials that used 25 mg of PN every 8 hours for three days [53] suggest that vitamin B_6 may be beneficial in alleviating morning sickness. Each study found a slight but significant reduction in nausea or vomiting in pregnant women. A recent systematic review of placebo-controlled trials on nausea during early pregnancy found vitamin B_6 to be somewhat effective [54].

Because adverse effects have been documented only from vitamin B_6 supplements and never from food sources, safety concerning only the supplemental form of vitamin B_6 (PN) is discussed. Long-term supplementation with very high doses of PN may result in painful neurological symptoms known as *sensory neuropathy*. Sensory neuropathy typically develops at doses of PN excess of 1000 mg per day. Symptoms include pain and numbness of the extremities and in severe cases, difficulty walking. However, none of the published studies in which an objective neurological examination was performed reported evidence of sensory nerve damage at intakes below 200 mg PN daily [55]. To prevent sensory neuropathy in virtually all individuals, the Food and Nutrition Board of the Institute of Medicine set the tolerable upper intake level for PN at 100 mg per day for adults. Because placebo-controlled studies have generally failed to show therapeutic benefits of high doses of PN, there is little reason to exceed the UL of 100 mg per day. Certain medications interfere with the metabolism of vitamin B_6; therefore, some individuals may be vulnerable to a vitamin B_6 deficiency if supplemental vitamin B_6 is not taken. Antituberculosis medications, including isoniazid and cycloserine, the metal chelator penicillamine, and antiparkinsonian drugs, including L-DOPA, form complexes with vitamin B_6 and thus create a functional deficiency.

33.7
Homocysteine (Hcy) as an Endogenous Toxin – Human and Experimental Studies

33.7.1
Homocysteine Toxicity in the CNS

Neurodegenerative diseases include a wide group of disorders of various etiologies and clinical features. These diseases share one important feature; that is, a disorder in protein structure, which can cause deterioration of certain nerve cells or neurons. Dementia, AD, Parkinson's disease (PD), and stroke are examples of neurodegenerative diseases that are common in elderly people. Mild cognitive impairment is a major health problem in elderly people, which will progress into dementia in about 50% of the cases [56]. AD is a multifactorial disease that is related to both genetic and acquired factors. The pathological hallmark of AD is the accumulation of neurofibrillary tangles (NFTs), neuritic plaques, and the insoluble amyloid beta (Aβ) [57, 58]. This process occurs 10–20 years before the cognitive decline [59]. Pathological mechanisms involved in neurodegenerative diseases include apoptosis, neuronal death, oxidative stress, overactivation of glutamate receptors, mitochondrial dysfunctions, and activation of caspases [60–62]. Brains of patients with AD showed abnormal redox balance and oxidative damage to proteins and nucleic acids [63, 64].

Changes in brain volume, intensity, or the presence of small infarcts are early signs of dementia. These conditions can be detected by magnetic resonance imagining (MRI) of the brain. An association between plasma concentrations of Hcy and qualitative or quantitative MRI analyses has been reported [65, 66]. Current evidence suggests that increasing the intake of the B vitamins might have protective effect on the central nervous system. This effect can be related either to lowering Hcy or to a direct effect of the vitamins.

33.7.1.1 Mechanisms of Homocysteine Toxicity in the Central Nervous System

The mechanisms by which Hcy may damage the brain are not fully understood. Hcy is toxic to neuronal cells [67–69]. Neurological damage has been reported in mice deficient in CBS enzyme (*Cbs−/+* or *Cbs−/−*), where Hcy increased by approximately 2–50-fold in comparison to wild type mice, depending on the genotype and the type of diet [70–72]. These animals show alterations in neuronal plasticity, suffer from severe retardation, and die early [73]. Animals exposed to Hcy accumulate this compound in the brain [74], suffer from restricted growth, neural or cognitive dysfunction [74, 75], and impaired brain energy metabolism [76]. Moreover, HHcy has been implicated in neural plasticity and neurodegenerative disorders in human studies [77].

Homocysteine and Hypomethylation Hcy metabolism in the brain is an important source of SAM [78]. This methyl donor plays important roles in forming and catabolism of neurotransmitters, phospholipids (phosphatidylcholine is the methylated product of phosphatidylethanolamine), DNA methylation, and in the activation of several enzymes that have essential role in the brain (i.e., protein phosphatase 2A (PP2A)) [77]. Elevated levels of Hcy can, therefore, cause damage to several key pathways in the central nervous system, either directly or by changing the methylation potential (SAM/SAH).

Experimental hyperhomocysteinemia led to increased concentrations of tHcy and SAH in the brain [79]. A lower ratio of SAM/SAH causes DNA damage and thereby apoptosis, which is one important explanation for Hcy neurotoxicity [69]. In accordance with this, SAM reduced apoptosis (by 50%) that is caused by Hcy in cortical neuronal cells [80]. Supplementing SAM after ischemia improves the blood–brain barrier and neuronal survival [81], and protects against disturbances in brain phospholipids [82].

Hcy can increase neuronal death and DNA damage [69]. Presenilin 1 (*PS1*) gene expression is one important example that has been tested in relation to DNA methylation [7, 83, 84]. PS1 is a γ-secretase that mediates the formation of Aβ from amyloid precursor protein (APP) (Figure 33.6) [83]. DNA hypomethylation causes accelerated APP processing and β-amyloid production through the upregulation of the *PS1* gene. Moreover, exogenous SAM can silence the *PS1* gene, thereby reducing Aβ production.

Extensive Aβ deposition in the brain increases as a part of the normal aging process. A prominent feature of the AD brain is the widespread cerebral deposition of Aβ within senile plaques and in cerebral and meningeal blood vessel [85, 86]. A positive association between plasma concentrations of tHcy and that of Aβ1–40 has been recently reported in aging and neurodegenerative disease [87]. A homocysteine-responsive endoplasmic reticulum protein (Herp) has been recently identified and found to enhance the γ-secretase activity and thereby Aβ1–40 accumulation in the brain [88]. On the other hand, Hcy can be toxic to neurons and can increase their vulnerability to being damaged by Aβ [89, 90]. Homocysteic acid induces the accumulation of intracellular and extracellular Aβ 42 in neuronal cells [91]. In vascular smooth muscle cells, tHcy increased Aβ toxicity and caspase-3

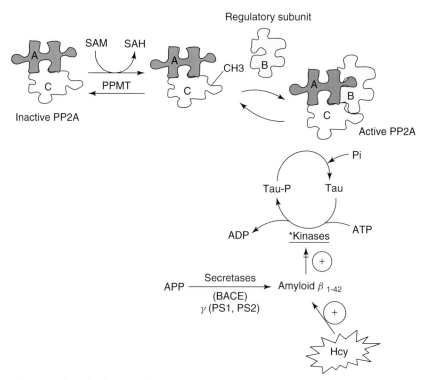

Fig. 33.6 The role of SAM and Hcy in P-tau and β-amyloid metabolism. PP2A, protein phosphatase 2A; Tau-p, phosphorylated tau; PS, presenilin; PPMT, PP2A methyl transferase; APP, amyloid precursor protein. *kinases are GSK3-beta, phosphatidylinositol 3-kinase (PI3K), and MAP kinase.

activation in a dose-dependent manner [92]. Therefore, current results suggest that tHcy accelerates dementia by stimulating Aβ deposition in the brain.

Another important biological reaction where hypomethylation can increase brain damage in AD is the formation and clearance of P-tau. Tau is an intracellular microtubule-associated protein that participates in forming the NFTs. NFTs correlate with cognitive deficits, neurodegenerative disorders, and dementia. The physiological function of tau is to facilitate tubulin assembly and to stabilize microtubules. Tau phosphorylation seems to influence its distribution and can modulate its association with plasma membrane. Increased P-tau may be related to a lower phosphatase activity or to an increased kinases activity (Figure 33.6). The enzyme protein phosphatase methyl transferase (PPMT) is involved in the regulation of PP2A, the enzyme that dephosphorylate tau protein (Figure 33.7) [93]. PPMT is SAM-dependent; hence reduced methylation capacity can increase P-tau (discussed below) by decreasing the dephosphorylation of the protein.

Increased Aβ may in turn accelerate P-tau accumulation by activating the kinases that phosphorylate tau (GSK3beta, phosphatidylinositol 3-kinase (PI3K), and MAP kinase) [94].

Several elegant experiments have been conducted aiming at testing the effect of folate deprivation on the expression of P-tau protein, PP2A enzyme, and the methyl transferase required for activation of PP2A [95]. Folate deficiency caused disturbed methylation potential in brain tissues. Furthermore, enhanced

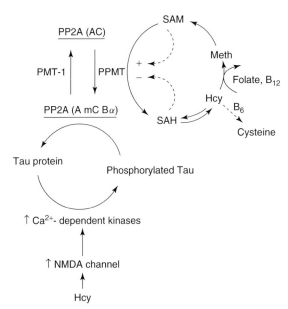

Fig. 33.7 PP2A links SAM/SAH metabolism to P-tau. SAM enhances PP2A activity or protein level and thereby dephosphorylation of tau. SAH shifts the tau to the phosphorylated form by inhibiting PP2A activity. PP2A, protein phosphatase 2A; PPMT, PP2A methyl transferase = Leucine carboxy methyl transferase; PMT-1, PP2A methyl esterase; AC, subunits A and C, mC, methylated subunit C; Bα, regulatory subunit Bα.

tau phosphorylation and cell death were related to downregulation of leucine carboxyl methyl transferase-1 and subsequent loss of ABαC complex [95]. Folate deficiency caused an enhanced expression of the nonmethylated form of PP2A enzyme. This condition did not change the existing PP2A, suggesting that binding of the ABαC complex is stable against demethylation by PME-1 [95]. Moreover, studies on AD brain have shown that the decrease in PP2A activity was related to a lower PP2A methylation and the expression of the regulatory subunit B of the enzyme [96]. In accordance with this, AD neurons showed an increased immunoreactivity of hyperphosphorylated tau (PHF-1) that is inversely correlated with downregulation of PPMT and B subunit of PP2A [96].

These results suggest that lowering tHcy might improve the methylation potential and protect against P-tau and β-amyloid accumulation in the brain.

Homocysteine and Glutamate Receptors Hcy is an endogenous glutamate receptor agonist [67,97–99] that is prone to act on N-methyl-D-aspartate (NMDA) receptor subtype [67, 100]. Homocysteic acid, an oxidative product of Hcy, is produced by brain cells and released in response to excitatory stimulation [101]. Homocysteic acid functions as an excitatory neurotransmitter by activating NMDA receptor [102]. The neurotoxicity of homocysteic acid in the brain can be blocked by using a selective NMDA antagonist [99, 103].

By binding to NMDA receptor [104], Hcy indirectly enhances calcium influx. Interestingly, in the presence of low (i.e., normal; 10 µmol l^{-1}) concentrations of glycine, Hcy acts as a partial antagonist of the glycine site of the NMDA receptor, and inhibits receptor-mediated activity (Hcy role as antagonist or neuroprotective) [67]. In contrast, when glycine levels increase in the nervous system (after stroke, head trauma, or migraine) [105], a relatively low concentration of Hcy (i.e., tHcy = 10 µmol l^{-1}) can be excitotoxic (agonist) by binding and activating NMDA receptors [67, 106, 107]. These results have suggested that Hcy may contribute to cerebral damage in patients with migraine, after stroke or after ischemia [67]. Therefore, depending on glycine concentrations, Hcy may either block the glycine site of the NMDA receptor or may act as antagonist at the glutamate site of this receptor [67].

Hcy has been shown to induce an extracellular signal-regulated kinase in the hippocampus [108]. By activating ionotrophic and metabotrophic receptors, Hcy may indirectly increase intracellular calcium levels and activate several kinases [108]. An interesting mechanism explaining the effect of folate deficiency on P-tau protein is related to enhancing the activity of one or more of the kinases responsible for tau phosphorylation in the brain [109]. This effect on kinase is thought to be related to activating NMDA channels and thereby Ca^{2+}-dependent kinase pathways by Hcy.

Homocysteine and Oxidative Stress Hcy metabolism is regulated by the redox potential in the cell [110, 111]. The activities of several enzymes (i.e., methionine synthase, CBS) that mediate the clearance of Hcy are regulated by this oxidative status [8, 110, 112]. For example, the activity of CBS is increased under conditions of oxidative stress, thus converting more Hcy into cysteine and glutathione.

Disruption of the transsulfuration pathway (*Cbs* +/−) disturbs redox homeostasis and reduces cysteine levels [70], thus contributing to neuronal damage. In contrast, the activity of methionine synthase is lower in the case of increased reactive oxygen species (ROS). Rats fed a methionine-rich diet showed elevated concentrations of tHcy in blood (20 vs. 7 µM in the control rats) and a lowered activity of glutathione peroxidase [113]. Hcy-induced oxidative stress may be worsened in the case of a reduced glutathione production.

Hcy itself can undergo autoxidation, thereby causing the disruption of redox homeostasis and affecting the redox signaling pathways in vascular and neuronal cells [110, 114, 115]. Hcy has been found to induce neurological dysfunction via oxidative stress [89, 116]. This effect can be explained by enhancing the production of ROS and oxidative deactivation of nitric oxide. Moreover, Hcy causes brain lipid peroxidation by blocking NMDA receptors [117]. Antioxidant treatment restores several toxic effects of Hcy [117].

The role of oxidative stress in neurodegeneration has been intensively studied. Oxidative stress was one important mechanism for Hcy toxicity in neuronal cells [118]. The cytotoxicity of Hcy was mitigated by antioxidants such as *N*-acetyl cysteine and vitamin E or C [89, 119, 120]. Antioxidants (vitamin E or C) also prevented memory dysfunction [119] and ATPase activity caused by Hcy in rats [120]. Folate deprivation induced a marked increase in Hcy and ROS and Aβ-induced apoptosis, while folate supplementation prevented the generation of ROS by Aβ [118]. Treatment with the SAH hydrolase inhibitor, 3-deazaadenosine (DZA), provides neuroprotection in normal and apolipoprotein E-deficient mice and in cultured neuronal cells deprived of folate and vitamin E and subjected to oxidative challenge [121]. It is, however, not known whether the effect of folate deficiency on the brain is independent of or mediated by tHcy elevation.

The transsulfuration pathway represents the metabolic link between antioxidant and methylation metabolism [122]. There is evidence suggesting an antioxidant role for SAM. SAM caused increased glutathione production, lowered lipid peroxidation by almost 65% [123], and prevented neuronal death in an experimental model of ischemia (oxidative stress) [124]. *In vivo* studies were able to demonstrate an improvement in the blood–brain barrier after transient cerebral ischemia in the presence of SAM [81]. Moreover, chronic SAM treatment (22 months) increased concentrations of glutathione and lowered lipid peroxidation in rat brain [125]. More evidence has been provided by clinical studies where B vitamins mitigate oxidative damage when administered immediately after acute ischemic stroke [126]. At least some of the neuroprotective effects of SAM can be related to its important role in enhancing the transsulfuration pathway and increasing the production of glutathione.

33.7.1.2 Homocysteine in Dementia and Cognitive Decline

An association between Hcy and dementia has been shown in several longitudinal studies. In the *MacArthur Studies of Successful Aging* [127], 370 subjects aged 70–79 were followed for a mean of seven years. The mean change in total cognitive scores over seven years was −4.3 points. Patients who were in the highest quartile of

Hcy (14.4–40 µmol l^{-1}) lost ≥9 points. Low folate also had significant effect on cognitive delay in this follow-up study. Other follow-up studies documented that Hcy predicted cognitive decline [128] and AD [129], after adjustment for low serum folate. In blood samples collected before death, higher concentrations of Hcy were related to confirmed AD compared to age-matched control subjects [130]. Moreover, in a three-year follow up, progression of the disease was worse in AD patients who had higher Hcy at entry [130]. In the Rotterdam Scan Study, the association between Hcy and neuropsychological test scores was assessed. Lower scores for psychomotor speed, memory function, and global cognitive function were found in subjects with Hcy >14 µmol l^{-1} compared to those with Hcy <8.5 µmol l^{-1} [131].

Several studies observed an association between cognitive decline with age and vitamin B_{12} status. In the Banbury B_{12} study, cognitive decline was related to low holoTC and elevated MMA and Hcy in elderly people [132]. In the Medical Research Council Cognitive Function and Aging Study, elevated MMA (indicating vitamin B_{12} deficiency) was related to bad cognitive scores in elderly nondemented people [133]. In a subgroup from the OPTIMA Study, baseline vitamin B_{12} status, indicated by total B_{12} (<308 pmol l^{-1}) and holoTC (<54 pmol l^{-1}), was a significant predictor of decreased brain volume on MRI in healthy, nondemented elderly people after five years of follow-up [134].

Thus, available studies suggested a causal relationship between HHcy and cognitive decline with age. Low vitamin B_{12} status might be one important risk factor for dementia, and the role of vitamin B_{12} seems to be only partly related to lowering Hcy. To confirm this causal link, treatment studies have been initiated. Vitamin supplementation studies in patients with dementia or those at risk for dementia, because of their high age, were, however, inconclusive.

33.7.1.3 Treatment with B Vitamin Supplements

Because of the consistent, dose-dependent relationship between tHcy and cognitive decline, lowering plasma concentrations of tHcy may have desirable effects on the progression of the disease. There is interest in whether folate and/or vitamin B_{12} may improve cognitive function or delay the progression of cognitive decline or whether this effect is related to lowering tHcy or to an independent effect of the vitamins. Several intervention studies aimed at improving symptoms or disease progression in patients with dementia.

Several, but not all, vitamin intervention studies document improvements in some measures of cognitive function. In a recent review including data from two vitamin trials, there was no evidence that folate and/or B_{12} had significant effects on cognitive function [135]. The effect of folate (2.5 mg), B_{12} (0.5 mg), and B_6 (25 mg) on cognitive function was tested in a three-months randomized, double-blind, placebo-controlled study using various combinations of the vitamins [136]. Elderly people with vascular disease were the target population and each treatment arm included 23–24 participants [136]. In this study, Stott *et al.* failed to detect any improvement in cognitive function [136]. Negative results were reported by Eussen et al. [137], where elderly people with vitamin B_{12} deficiency were treated with vitamin B_{12} and folic acid or placebo for 24 weeks. In another study testing the effect of vitamins in elderly people at risk of dementia, cognitive function was not improved during 12 weeks of the trial [138]. Trials that used vitamin B_6 for prevention or treatment of cognitive decline with age are limited and the evidence is insufficient to make any firm conclusions [46].

In an observational study in elderly people (mean age 76), Lewerin et al. found a significant association between cognitive and movement performance on the one hand and tHcy and MMA on the other hand [139]. Nevertheless, in a placebo-controlled, double-blind study, a daily dose of 0.8 mg folic acid, 0.5 mg B_{12}, and 3 mg B_6 for four months failed to improve movement or cognitive function despite lowering tHcy and MMA [139]. Despite that, the study by Lewerin et al. included almost three to five times more participants than that by Stott et al. The short duration of both studies might be a limiting factor.

The hypothesis that lowering tHcy concentrations will improve cognitive function in healthy elderly people has been tested for a longer duration. In a two-year, double-blind, placebo-controlled randomized study, subjects treated with 1 mg folic acid, 0.5 mg B_{12} and 10 mg B_6 showed no improvement in their cognitive scores compared to the placebo group, despite lowering tHcy by an average of 4.4 μmol l^{-1} [140]. In a three-year randomized, double-blind study, 818 participants were treated either with folic acid (0.8 mg) or with placebo [141]. Concentrations of tHcy were lowered by a mean of 26% and a significant improvement in some cognitive domains that decline with age has been reported. Memory scores were improved after folate supplementation and the improvement was positively related to the severity of folate deficiency [142].

There are several confounding factors that might override the effect of the B vitamins on cognition and memory. These factors are critically important for judging available results or for future studies. Hcy-lowering trials should aim at prevention and not treatment of dementia. Similarly trials using Omega-3 supplementation in patients with dementia documented improvement in patients with less severe dementia, but not in severely demented patients.

Most studies included only a small number of subjects. Additionally, their duration is too short to achieve a marked improvement. It seems important that the supplementation should start at earlier stage in the disease development and should be continued for >3 years to achieve a clinical improvement. Moreover, because baseline tHcy concentrations are related to a decline in cognitive function with age, a halting of such decline would mean a protective effect. One fact that should also be recognized is that the turnover of cells in the nervous system is negligible, whereas blood cells divide very rapidly. Therefore, vitamin treatment is known to improve hematological symptoms while neurological symptoms may take longer to improve, and may be only partially reversible. This is unsurprising given that the progression of disease can extend over several decades. Therefore, it is currently believed that ensuring sufficient B vitamin intake might be more effective in disease prevention rather than in disease treatment.

33.7.1.4 Homocysteine in Patients with Parkinson Disease

PD is the second most common neurodegenerative disease in elderly people after dementia. PD is characterized by depletion of dopamine and dysfunction and death of dopaminergic neurons. Acquired factors that enhance the risk of PD are of great interest. The association between PD and HHcy has been reported in numerous clinical studies [143, 144]. The increase in plasma Hcy in PD patients depends on folate and B_{12} status, genetic polymorphisms that influence Hcy [144]. HHcy can also arise secondary to PD itself, or to drugs commonly used for treating the disease.

A significant increase in plasma concentration of Hcy occurs in PD patients after starting L-DOPA treatment [145]. L-DOPA is methylated by catechol-*O*-methyl transferase (COMT), a SAM-dependent enzyme. The effect of combining COMT inhibitors with L-DOPA to avoid HHcy is inconclusive [146–148]. Available evidence suggests that folate, B_6, and B_{12}, but not COMT inhibitors, are the most important modifiers of the effect of L-DOPA on Hcy levels [149–151].

Higher concentrations of Hcy in PD patients are generally associated with poor progression of the disease; depression, cognitive impairment [152], vascular disease [153], and hip fracture [154–156]. Other studies failed to detect an association between vitamin intake or plasma Hcy and the incidence or clinical course of PD [157, 158]. However, these studies were limited by a low number of participants or the short term of follow-up.

Together, reducing plasma Hcy may be an important measure for secondary prevention in PD patients. Plasma concentrations of Hcy should be maintained at low levels in PD patients, especially in those patients receiving L-DOPA. A causal role for Hcy in the onset of PD has not yet been tested in large studies, despite the consistent association between PD and HHcy. There are several vitamin intervention studies in patients with PD [150, 151], but there is no information currently available about clinical outcome in treated patients.

33.7.1.5 Homocysteine in Cerebrovascular Diseases

Stroke is a clinical syndrome resulting from different underlying vascular pathologies. This heterogeneous disease includes different types. Ischemic stroke constitutes 80% of all stroke types, intracerebral hemorrhage 15%, and subarachnoid hemorrhage 5%. Transient ischemic attack (TIA) is a disorder that is similar to stroke, but is not fatal and lasts less than 24 hours. Ischemic stroke is caused by atherothromboembolism in 50% of the cases and intracranial small vessel disease (lacunar infarction) in 25% of the cases. A cardiac embolism explains approximately 20% of ischemic stroke cases.

HHcy is very common in patients with stroke and is suggested to be an independent risk factor for the disease. Systematic reviews of epidemiological and prospective studies revealed a consistent, positive, dose-dependent relationship between Hcy and stroke risk. A recent study confirmed the association between HHcy and stroke in black population and a strong association was detected with lacunar infarction in subjects with leukoaraiosis compared to those without leukoaraiosis [159]. The risk of stroke increased by 6–7% in the Rotterdam Scan

Study for each 1 µmol l^{-1} increase in plasma Hcy concentration [160]. Moreover, a graded positive association was also shown between concentrations of Hcy and the risk of stroke in the British Regional Heart Study [161]. In the Physicians Health Study, an increase (1.4-fold) in the risk of stroke was observed in subjects with Hcy >12.7 µmol l^{-1} compared with subjects with Hcy lower than this limit [162]. Similar results were reported by the Northern Manhattan Study, where the adjusted hazard ratio for a Hcy level ≥15 µmol l^{-1} compared with <10 µmol l^{-1} was 2.01 (95% CI, 1.00–4.05) for ischemic stroke [163]. An odds ratio (OR) for stroke of 2.3 was reported in subjects from the NHANES Epidemiologic Follow-up Study III in the highest quartile of Hcy compared to the lowest quartile [164]. In the Framingham Offspring Study, elevated concentration of Hcy was a risk factor for silent cerebral infarcts in 2040 people free of clinical stroke who were tested by MRI [165]. The OR for silent cerebral infarcts was 2.23 ($p < 0.001$), which is higher than that of hypertension, cholesterol, diabetes, and other common risk factors [165].

33.7.1.6 Reduction of Stroke Risk – Primary and Secondary Prevention Studies

Because the relation between Hcy elevation and stroke suggests a causal relationship, several trials have been started to test the effect of Hcy lowering on stroke prevention. Several studies have shown that risk of stroke may be significantly reduced by B vitamins. In a population-based study, Yang *et al.* [166] reported a decline in stroke-related mortality after folic acid fortification in the United States and Canada, which was stronger than that found in England and Wales during the same period [166]. Moreover, folic acid and vitamin B_6 lowered Hcy and caused slight improvement in cerebrovascular and cerebral indices [167]. In a two-year trial, high doses of B vitamins (25 mg vitamin B_6, 0.4 mg vitamin B_{12}, and 2.5 mg folic acid) were compared with low doses (200 µg B_6, 6 µg B_{12}, and 20 µg folic acid) in secondary prevention of stroke [169]. The effect of treatment on stroke risk was not significant during this trial. Nevertheless, a 21% reduction in the combined risk of stroke, coronary disease, and death was observed in a subgroup of this study population that included patients who were likely to benefit from treatment [170]. In the Heart Outcomes Prevention Evaluation (HOPE) 2 study, 5522 patients with vascular disease or diabetes were randomized to receive placebo or folic acid (2.5 mg) plus 50 mg vitamin B_6 and 1 mg B_{12} for five years; one of the primary outcomes was stroke [168]. Approximately 25% reduction in the risk of stroke was found in the supplement group compared to the placebo group (relative risk = 0.75 (0.59–0.97); $p = 0.03$) [168].

In a recent meta-analysis of secondary prevention studies [171] – a cooperation of American and Chinese universities – eight randomized treatment studies with B vitamins (folic acid, vitamin B_6, and/or B_{12}) were analyzed in regard to an alteration of risk of stroke. A significant reduction (18%) of stroke risk due to treatment with folic acid was reported. However, a reduction of stroke risk was obtained only by a treatment lasting over three years, the stroke risk being lowered by 29%. The risk reduction was also markedly greater than the average (18%) if the Hcy lowering was more than 20% (−23%), if the patients had no history of stroke (−25%), or if the patients consumed no folic acid–enriched grain products (−25%). It is concluded that folic acid supplementation can reduce the risk of stroke in primary prevention.

CVD is a heterogeneous entity and various cardiovascular endpoints can also react differently to a therapy with B vitamins. As the recent meta-analysis showed [171], an insufficient observation period of studies proved to be a major restriction of the statistical power, even though earlier studies conducted on cholesterol medication showed that longer observation periods are crucial in order to obtain risk-reduction. Furthermore, results from retrospective studies suggested that B vitamin therapy might be a safe and cost-efficient method to lower stroke risk. The effectivity, however, seems greater for stroke than for CVDs [166]. Although the usefulness of Hcy lowering by folic acid for stroke prevention has been confirmed, no definite conclusions can be drawn concerning CVDs until large meta-analyses of intervention studies with B vitamins are published. Current evidence suggests that the effect of tHcy lowering on CVD risk is smaller than that on the risk of stroke. Because HHcy is particularly common in elderly patients, B vitamin supplementation might be protective against age-related diseases. For primary and secondary prevention of degenerative diseases, a sufficient vitamin intake through food and B vitamin supplementation is of practical relevance. Periodical checkup (every two to three years) of the Hcy level and the B vitamin status is recommended for elderly people and those at risk for stroke or CVD. However, there is currently no consensus on using B vitamins for primary or secondary prevention of stroke. Nevertheless, the American Stroke Association Stroke Council reiterate the importance of meeting current recommended daily intakes of the vitamins [172].

33.7.2
Homocysteine in the Endothelial System

CVD is the leading cause of morbidity and mortality in developed countries. Traditional risk factors (i.e., cholesterol, obesity, diabetes, and smoking) do not account for the total risk. Concentration of total plasma Hcy has emerged as a nontraditional, independent and a potential modifiable risk factor for CVD. The majority of case–control studies and many prospective studies found an association between HHcy and cardiovascular risk. Experimental studies showed that Hcy can cause oxidative stress, endothelial cell dysfunction, and promote thrombogenesis. A protective role for lowering Hcy in CVD diseases has not been confirmed by several recent supplementation studies. However, all available studies have limitations and the causal role for Hcy in CVD remains unanswered.

There is considerable evidence suggesting that Hcy causes vascular dysfunction. *In vitro* or *in vivo* hyperhomocysteinemic conditions (also after a methionine-loading test) can adversely affect the endothelium [173–175]. Endothelial cell dysfunction caused by HHcy results in impaired relaxation of blood vessels. The bioavailability of the vasodilator, endothelial-derived nitric oxide, is presumed to be impaired in HHcy conditions [176]. Hcy has been found to stimulate the expression of monocyte chemoattractant protein 1 (MCP-1) in the endothelium of the developing atherosclerotic lesions, causing the movement of monocytes into the neo-intima [177]. Hcy also has a mitogenic effect on the smooth muscle cells that invade the neo-intima of the atherosclerotic lesion [178].

Hcy is known to form covalent disulfide bound with cysteine residues in cellular proteins. Examples for proteins that have been found to be homocysteinylated are albumin [179–181], prealbumin, factor Va, and matrix-associated proteins such as fibronectin [182]. Hcy has also been reported to enhance collagen production in cultured aortic smooth muscle cells [183]. Moreover, hyperhomocysteinemic patients with acute coronary syndrome showed increased collagen deposition [184] suggesting a causal relationship with HHcy.

Studies have shown that Hcy by targeting endothelial cell surface molecules can also exhibit atherothrombotic effect. *In vitro*, Hcy can inhibit the thromboresistance properties of the endothelial cell by induction of procoagulant factors, inactivation of natural anticoagulant systems, and suppression of vasodilatory and platelet-modulating factors. Hcy also inhibits the fibrinolytic system by impairing the ability of the endothelial cell to bind tissue plasminogen activator (t-PA) by interacting directly with its endothelial cell receptor, annexin II. This results in a marked decrease in the ability of annexin II to bind t-PA [185–187]. Other mechanisms by which Hcy can be atherogenic are listed in Table 33.2.

33.7.2.1 Homocysteine-Lowering Treatment in Reduction of Cardiovascular Risk

Retrospective and prospective studies also support a consistent graded and independent association between plasma concentrations of Hcy and the risk of CVD. However, data from recent large randomized controlled trials have shown almost no benefit to lowering plasma tHcy concentrations with folic acid and other B vitamins in the secondary prevention of CVD. Early investigations on patients with CBS deficiency have suggested that severe HHcy (plasma Hcy >100 µmol l^{-1}) is a risk factor for atherogenic and atherothrombotic diseases. Treating homocystinuria patients with vitamin B$_6$ and or folate proved meaningful for improving the cardiovascular outcome. A lowering of the tHcy level by a B vitamin treatment thus seems to present a promising and simple preventive measure. Numerous trials have started with the aim to test the hypothesis that lowering tHcy will decrease the risk for CVD. Several trials have been published during the last five years, but the hypothesis remains obscure and the new results added rather more complexity to the main hypothesis. Recent data of a meta-analysis of treatment studies on the lowering of stroke risk [171], however, confirmed the efficacy of tHcy lowering.

33.7.2.2 Can Treating HHcy Provide Protection against Cardiovascular Diseases?

Large meta-analyses [188, 189] of retrospective and prospective studies stress the causal correlation of HHcy and degenerative (vascular) diseases. The results suggest that each 5 µmol l^{-1} increase in plasma tHcy is associated with an OR of 1.32 for ischemic heart disease, 1.60 for thrombosis, and 1.59 for stroke. Furthermore, these meta-analyses suggested that a lowering of tHcy concentration by 3 µmol l^{-1} might reduce the risk of ischemic heart disease by 16%, thrombosis risk by 25%, or stroke risk by 24% [189].

Tab. 33.2 Atherogenic effects of homocysteine

Target	Effect
Endothelial damage	↑
VSMC proliferation	↑
Collagen synthesis	↑
MMP-9, MMP-2	↑
Enhance oxidative stress via producing H_2O_2	↑
Stimulate SOD	↑
Lipid peroxidation	↑
Leukocytes adhesion molecule	↑
sICAM-1, VCAM-1	↑
Chemotaxis (IL-8, MCP-1), vWF	↑
Proliferative fibrous plaques	↑
HSP 70 expression	↓
Mitochondrial damage	↑
ER stress	↑
NO system	↑↓
NO bioavailability	↓
ADMA	↑
Heparin sulfate	↓
Annexin II	↓
Thrombomodulin	↓
PAI-1, t-PA antigen	↑
Prothrombin fragment F1+2	↑
Inactivate factor Va, V	↓
Tissue factor	↑
Inactivate protein C	↑
Thrombin (thrombin–antithrombin complex)	↑
D-Dimer	↑
Fibronectin	↓
COX production of TXA_2 and TXB_2	↑
Activate NF-κB, SREBP, PKC	↑
HMG-CoA reductase	↑
Lipid synthesis	↑
Inactivation of PPARα and γ	↑

VSMC = vascular smooth muscle cell, ER = endoplasmic reticulum, HSP = heat shock protein, NO = nitric oxide, ADMA = asymmetric dimethylarginine, SREBP = sterol regulatory element binding protein, PKC = protein kinase C, PPAR = peroxisome proliferator activated receptor, SOD = superoxide dismutase, GPx = glutathione peroxidase, sICAM = soluble intracellular adhesion molecule, VCAM = vascular cell adhesion molecule, IL = interleukin, MCP = monocyte chemotactic protein, vWF = von Willebrand factor, PAI-1 = plasminogen activator inhibitor-1, t-PA = tissue plasminogen activator, COX = cyclooxygenase, TXA_2 = thromboxan A_2.

On the one hand, one prospective study of longer than four years observed a significant reduction of plaque within the carotid artery [190] by B vitamin supplementation. Moreover, a significant reduction of the carotid intima-media thickness in risk patients for stroke has recently been reported after a treatment with B vitamins lasting one year [191]. On the other hand, several recent treatment studies could not detect a benefit of B vitamin supplementation thus leaving open questions about the repeatedly observed association between HHcy and vascular risk in retrospective and prospective studies. Notably, available trials have inherent errors and are not optimized to address the role of tHcy lowering on primary or secondary prevention of CVD.

Worldwide more than 50 000 patients are currently included in intervention studies to determine the possible benefits of vitamin therapy (secondary prevention). By now, first intervention studies, such as VISP [169], NORVIT [192], HOPE 2 [168], WAFACS [193], and WENBIT [194], have been completed and published. One serious problem with the available studies is that instead of comparing nontreated with treated patients, the results of conventional treatment are compared with the results of conventional treatment to which vitamins are added. Most of these studies included patients with several medications that interact with tHcy metabolism or levels. Therefore, it is critically important to review the limitations of the available studies before giving a firm conclusion about the protective effect of the B vitamins against CVD and stroke.

33.7.3
Homocysteine Toxicity in Alcoholic Liver Disease

33.7.3.1 Fatty Liver and HHcy Are Mutual Findings in Chronic Alcoholism

Among the many macro- and micronutrients that can be affected by chronic alcohol ingestion are choline, betaine, and methionine, which are all essential for the generation, transport and transfer of one-carbon units to target molecules such as phospholipids, SAM, DNA, and neurotransmitters [195]. Chronic ingestion of methanol causes disturbances in methionine metabolism in the liver, leading to increased Hcy and SAH and lower SAM [196]. The activities of methionine synthase and methionine adenosine transferase are inhibited by alcohol. In contrast, the activity of betaine-homocysteine methyl transferase (BHMT) is maintained and this leads to a higher consumption and thus requirements of betaine for Hcy methylation. Additionally, alcohol consumption causes fatty liver and hepatic steatosis. Furthermore, low concentrations of folate, vitamin B_6, and betaine are common in alcoholism and have been shown to enhance the side effects of alcohol in the liver [16].

Alcohol induces disorders in Hcy catabolism and decreases the concentration of SAM/SAH that may account for lipid accumulation in the liver, and SAM can enhance liver function in chronic alcoholism [197, 198]. In addition, hypomethylation and a reduced phosphatidylcholine methyl transferase (PEMT) activity in alcoholism can be responsible for liver steatosis [199, 200]. The beneficial effect of betaine in alcoholic liver disease supports the role of BHMT in maintaining

Hcy catabolism in chronic alcoholism. Betaine has been often used in studies on alcoholic liver disease. Betaine administration (0.5% for 2–4 weeks) doubled the hepatic levels of SAM in control animals and increased by fourfold the levels of hepatic SAM in the ethanol-fed rats [201–203]. The ethanol-induced infiltration of triglycerides in the liver was also reduced by the feeding of betaine to the ethanol-fed animals. Unlike SAM and betaine, folate treatment was not able to restore the diseased liver in alcoholism [204]. These findings support the notion that betaine can enhance the activity of liver BHMT and provide more SAM for liver methyl transferases such as PEMT. Betaine administration has the capacity to elevate hepatic SAM and to prevent the ethanol-induced fatty liver [202]; probably by normalizing phosphatidylcholine synthesis and thereby very low density lipoprotein (VLDL) homeostasis.

33.7.4
Homocysteine and Bone Health

Hcy was first linked to bone diseases in patients with homocystinuria [205, 206]. These patients exhibited accelerated bone growth, skeletal deformities (genu valgus, kyphoscoliosis, pectus carinatum, etc.), flattened vertebral bodies, and a decreased bone density [207]. It was only recently that two prospective, population-based studies demonstrated a strong link between moderate HHcy and the frequency of osteoporotic fractures in elderly persons [155, 156]. A 1.4-fold increase in total fracture risk was found per 5.5 μmol l^{-1} increase in Hcy [155]. In a subgroup of the Framingham study, McLean *et al.* obtained similar results [156]. The fracture risk for men in the fourth quartile of Hcy concentration (mean Hcy 20.8 μmol l^{-1}) was 3.84 times higher than that of men in the first quartile. Patients with pernicious anemia had between 1.8- and 2.8-fold higher fracture risk compared to control subjects [208]. Therefore, there is convincing clinical data showing that HHcy and a low B vitamin status increase fracture risk.

Bone mineral density (BMD) measurement is still the gold standard to diagnose osteoporosis [209]. BMD depicts mainly bone mineralization and provides only an integral measure of bone metabolism over time and thus it contributes only 60–80% to the prediction of fracture risk [210, 211]. In addition, BMD gives only poor information about the microarchitecture of bone, which is essential for the biomechanical properties of the bone. Therefore, it is not surprising that the relation between Hcy and BMD was not impressive [212–214].

In contrast to BMD, measurement of biochemical bone turnover markers permits a kind of real-time monitoring of bone metabolism [212, 215]. Bode *et al.* observed a correlation between Hcy and circulating concentrations of the bone resorption marker carboxyterminal telopeptide of human collagen I (ICTP) [216]. A significant weak correlation between Hcy level and serum level of a bone resorption marker, ß-crosslaps, was observed in a study on 996 postmenopausal women [217]. Other bone resorption markers, deoxypyridinoline (DPD) cross-links and tartrate resistant acid phosphatase 5b (TRAP5b), showed only weak correlations with Hcy concentration [215].

Higher urinary DPD concentrations were found in women with elevated Hcy concentrations, but not in men. Similar results have been reported in other studies [212, 218]. In a study on post- and perimenopausal women, the Hcy concentration correlated positively with DPD, but not with the bone formation marker osteocalcin in serum, suggesting a shift of bone metabolism toward bone resorption [212]. Other studies, however, could not find significant results [219, 220]. In addition, most of the positive studies found significant relations only in women, but not in men. Beside an increased bone resorption, a reduced bone formation could also contribute to the relation between Hcy and fracture risk. In some studies, the association between bone formation markers, such as bone alkaline phosphatase (BAP) or osteocalcin, and Hcy has been investigated [212, 217, 220, 221].

Early clinical and experimental data suggest a reduction in osteoblasts activity by low vitamin B_{12} levels [222, 223]. Moreover, a dose–response relationship was observed between tHcy concentrations in cell cultures and osteoclast activity suggesting an enhanced bone resorption [224]. Osteoclasts remove bone matrix by acidification and proteolytic digestion, resulting in the formation of resorption pits. The activity of osteoclast *in vitro* can be quantified by the measurement of the resorption activity on standard dentine discs, the TRAP and cathepsin K (C-K) activity. A concentration-dependent increase of osteoclast activity by up to 50% at Hcy concentrations between 10 and 100 $\mu mol\, l^{-1}$ was also reported [224, 225]. In one study, the increased bone resorption was accompanied by a stimulation of p38 MAPK activity as well as generation of intracellular ROS [225]. All Hcy-induced effects could be blocked by pretreatment with the antioxidant *N*-acetyl cysteine. *In vitro* studies on primary human osteoclast showed that HHcy caused stimulation of the resorption activity, TRAP and C-K activity [226].

Together, existing studies regarding Hcy and bone turnover indicate significant correlation between Hcy and bone resorption markers, but no association between Hcy and bone formation markers. Existing cell culture data suggest a significant stimulation of osteoclast by Hcy. Considering osteoblasts, a moderate stimulation of osteoblast activity by exogenous Hcy was found, but not by B vitamin deficiency. All other studies were carried out with preosteoblastic cells or cell lines. These studies suggest an increased apoptosis and a reduced activity of preosteoblastic cells in the presence of elevated Hcy concentrations.

33.7.4.1 Bone Health and Intervention Studies with B Vitamins

At present there are only a very few trials testing the effect of B vitamins on bone health. Two studies found no effect of various Hcy-lowering treatment regimens on blood concentrations of bone turnover markers in healthy subjects [227, 228]. However, the importance of these studies is limited by the small number of subjects included and the fact that study participants were healthy. In a small intervention trial, osteoporotic subjects were treated for one year with a combination of folic acid, vitamin B_{12} and B_6, and placebo. Hcy was reduced significantly by a mean of 4.7 $\mu mol\, l^{-1}$ and the bone resorption marker DPD dropped by 13% ($p < 0.01$) after 8 and 12 months of treatment [229]. Furthermore, in subjects with Hcy above 15 $\mu mol\, l^{-1}$, B vitamin treatment tended to increase BMD at the lumbar spine from T-score −2.7 to −1.7. A double-blind, placebo-controlled intervention trial by Sato et al. [230] reported a reduction of fracture incidence by approximately 75% after supplementation with folate plus vitamin B_{12} over two years. However, the remarkable high incidence of hip fractures in the controls and the high mean concentration of plasma Hcy (19.9 µmol l^{-1}) in this study suggest that the results cannot be generalized [230]. In conclusion, the preliminary data indicative of a Hcy-lowering treatment with B vitamins show a significant reduction on fracture incidence, but the data on circulating bone turnover markers have to be substantiated. More and larger studies are needed.

33.7.5
Homocysteine Toxicity in Patients with Renal Diseases

33.7.5.1 Role of Hyperhomocysteinemia as Vascular Risk Factor in Patients with ESRD

Patients with chronic renal failure show a considerably higher rate of cardiovascular morbidity and mortality as compared to age- and sex-matched subjects with normal renal function [231]. Furthermore, mild to moderate HHcy (20–80 μmol l^{-1}) is very common in renal patients [232] and it may confer a heightened risk for atherothrombotic vessel disease in these patients [233, 234]. The hazard ratio for fatal and nonfatal atherothrombotic events was about eightfold higher in dialysis patients who had Hcy concentration ≥ 37.8 μmol l^{-1} compared to those with levels below 22.9 μmol l^{-1} [235].

In contrast, a few recent studies have reported that a decreased rather than an increased concentration of Hcy is related to a higher prevalence of CVD and poor outcome in patients with end stage renal disease (ESRD) [236, 237]. This paradoxical association between Hcy and cardiovascular outcome in dialysis patients has been referred to as a possible constituent of the so-called "reversal of the cardiovascular risks" [238]. Concentrations of Hcy correlated significantly with some nutritional markers (serum albumin, prealbumin, creatinine, and urea nitrogen) in dialysis patients, but no correlation was seen with inflammatory markers (C-reactive protein, interleukin 6, or TNF-α) [239]. It is concluded that Hcy may be a nutritional marker in dialysis patients with no association with inflammatory measures. Protein-energy malnutrition is *per se* a known risk factor for poor clinical outcome in dialysis patients. Therefore, the role of Hcy as a cardiovascular risk marker in dialysis patients may be counterbalanced by the effect of malnutrition leading to the inverse association between Hcy and cardiovascular risk. The role of Hcy as a vascular risk indicator and its role as a nutritional marker and as a marker for metabolic dysbalance should be distinguished.

33.7.5.2 Renal Function and Hyperhomocysteinemia

The kidney plays an important role in Hcy metabolism and an impaired kidney function may partly explain the metabolic abnormalities commonly found in renal patients [15]. Variation in renal function, even within the physiological range, is an important determinant for the interindividual differences in plasma concentrations of Hcy. The pathogenesis of HHcy and the dysbalanced methylation potential in renal disease are, however, not fully understood [232, 240].

The glomerular filtration rate (GFR) is a rate-limiting factor for the renal clearance of Hcy [241]. Normally, 20–30% of plasma Hcy is freely filtered, but less than 1% of total Hcy is excreted into the urine. Therefore, renal excretion of Hcy is not a major contributor of HHcy in renal disease [242–245]. The renal uptake and metabolism could account for approximately 70% of the daily elimination of Hcy from plasma [244]. However, there is no significant net renal uptake of Hcy in fasting humans with normal renal function [246]. A lower rate of plasma Hcy clearance was observed in renal patients with HHcy compared to healthy persons, which supports the role of the kidney in the elimination of this aminothiol [247]. The fractional extraction of Hcy across the human kidney varies according to renal blood flow [248]. This finding implies major alterations in the metabolism of Hcy in patients with ESRD. Furthermore, HHcy in animals produced a remarkable glomerular damage and sclerosis [249], renal tubulointerstitial injury, increased urinary albumin excretion [249], and a decrease in renal blood flow, GFR and sodium and water excretion [250]. In line with these data, treatment with folic acid and vitamin B_6 lowered not only plasma Hcy but also urinary albumin excretion in human [251].

Diabetic nephropathy is a common complication in patients with type 2 diabetes that may increase the atherothrombotic risk. Concentrations of Hcy increased significantly in patients with diabetes nephropathy with worsening of renal function, and these were not totally explained by low concentrations of serum folate [240]. Conversely, concentrations of folate, vitamins B_{12}, and B_6 were not higher in patients with better renal function. Vitamin B_{12} and/or folate are significant predictors of Hcy, CYS, MMA, and SAM in patients with type 2 diabetes who developed diabetic nephropathy [240]. The findings support the notion that increased plasma concentrations of the Hcy and methionine cycle intermediates, SAM and SAH, are related to disturbed renal function in patients with type 2 diabetes.

33.7.5.3 Reduced Remethylation Pathway in ESRD

Recent data suggest that the remethylation of Hcy to methionine in the kidney plays an important role in Hcy clearance [243, 244, 247, 252, 253]. Disturbance in the remethylation pathway may partly explain Hcy elevation in renal patients [247]. Utilizing stable isotope technique, it was proven that remethylation of Hcy is markedly decreased in dialysis patients, but Hcy transsulfuration was largely unaffected [254]. In addition, Hcy remethylation was shown to be impaired in dialysis patients [234]. In keeping with this, administration of the cofactors in the remethylation pathway (folate and B_{12}) markedly lowered serum concentrations of Hcy in renal patients [255, 256]. At least part of this effect can be attributed to improved Hcy remethylation. Metabolic signs of vitamin B_{12} deficiency are common in renal patients (elevated concentrations of MMA and Hcy). This is in contrast to normal concentrations of the vitamins usually found in those patients [255]. Lowering concentrations of MMA and Hcy in those patients after B_{12} administration may indicate the presence of a pretreatment deficiency [255]. This is also supported by the fact that plasma concentration of SAM (i.e., improved Hcy remethylation) increased in patients who were likely to show improvement in their B_{12} status.

Alterations in folate or vitamin B_{12} bioavailability, transport, or metabolism might participate to low folate status in ESRD. An impaired transmembrane transport of folate and a decreased plasma folate conjugase activity have been reported [254].

33.7.5.4 Cellular Uptake of Vitamin B_{12} in Patients with Chronic Renal Failure

HoloTC, the TC-bound vitamin B_{12}, constitutes 6–20% of the total vitamin B_{12} in serum. Only holoTC can enter the cells through a receptor-mediated process [257]. The kidney is one of the major organs that produces TC and is also rich in holoTC receptors [28, 257]. HoloTC is filtered in the glomerulus and reabsorbed in the tubule cells to enter the blood stream bound to a newly synthesized TC [28]. Concentration of holoTC may increase in patients with renal involvement, but this may not exclude cellular B_{12} deficiency [258]. The reason for holoTC accumulation in renal patients remains unclear, but it is probably related to a generalized peripheral resistance to the vitamin. Because normal serum concentrations of B vitamins in renal patients did not explain Hcy elevation, a reduced cellular metabolism of Hcy was hypothesized. We demonstrated that the uptake of vitamin B_{12} by mononuclear cells isolated from renal patients was lower than that by cells isolated from controls [259]. Nonetheless, the receptor-binding capacity was comparable between patients and controls. Renal patients showed a reduction of B_{12} uptake by 18% in average compared to the controls. Therefore, serum concentrations of vitamin B_{12} within the reference range are not likely to ensure vitamin delivery into the cells. Suprahysiological doses of vitamin B_{12} may be necessary to deliver a sufficient amount of the vitamins to the cells via mechanisms largely independent of holoTC receptor.

33.7.5.5 Response of Homocysteine to Vitamin Treatment in Dialysis Patients

Hcy-lowering treatment could be an attractive and inexpensive therapeutic strategy for reducing the burden of vascular diseases in uremic patients [260]. Folic acid is a potent Hcy-lowering agent that has been shown to induce a 16–40% decrease in Hcy in dialysis patients [261]. The addition of vitamin B_{12} (subcutaneous or intravenous) induced a marked decrease in Hcy in folate-replete patients [262]. In contrast, Hcy remained unchanged when vitamin B_{12} was administered orally or as injections given at infrequent intervals [263]. Intravenous administration of vitamin B_{12} was more effective in lowering Hcy than oral treatment. Considering the dose–response relationship, much higher doses of the vitamin are required in ESRD patients than in individuals with normal renal function to significantly lower Hcy. Although Hcy was lowered in most treatment studies, normal concentrations were not achieved in the majority of ESRD patients [261–263].

In one study, folic acid (5 mg) plus vitamin B_6 (50 mg) and B_{12} (0.7 mg) were administered three times per week intravenously to dialysis patients who had Hcy above 18 µmol l^{-1} at baseline [255]. Median Hcy reduction after four weeks was 13.4 µmol l^{-1}, which represents an average reduction of 51%. Twenty-four weeks after vitamin withdrawal, Hcy concentrations returned to values comparable to baseline (median 24.8 μmol l^{-1}). The combination of folic acid, vitamin B_{12}, and vitamin B_6 normalized serum concentrations of Hcy in almost all of the dialysis patients in this study. A marked reduction of Hcy can be achieved using a much lower doses of folic acid if this is combined with vitamin B_{12} and B_6 [255]. Lowering CYS contributed much more to the decline of Hcy level after vitamin treatment than lowering MMA.

33.7.5.6 Effect of Homocysteine-Lowering Treatment on Transmethylation

SAM is the universal methyl donor in numerous transmethylation reactions in humans. SAH hydrolase catalyzes the reversible conversion of SAH to Hcy and adenosine. Hydrolysis of SAH prevails under physiological conditions as long as the product, Hcy, is efficiently removed. However, under conditions of HHcy, the buildup of SAH is favored. Adenosine is converted into adenine nucleotides (AMP) or inosine by adenosine kinase and adenosine deaminase, respectively. A disorder in the metabolism of Hcy, SAM, or SAH may affect the other metabolites. Renal failure is associated with severe abnormalities in plasma concentrations of SAH and SAM and their ratio [240]. The ratio of SAM/SAH is an indicator for the methyl groups transferred from SAM to methyl acceptors. SAH is an inhibitor for the transmethylation reactions [264]. Moreover, increased SAH or a low ratio of SAM/SAH showed a stronger association with vascular damage than Hcy itself [265].

Therapeutic doses of B vitamins administered for one month resulted in only a limited improvement in the biomarkers of methylation (SAM and SAH) in uremic patients [255]. Concentrations of SAM increased compared to levels at start and changes of SAM were predicted by baseline vitamin B_{12} status. Therefore, enhanced Hcy remethylation after treatment may account for increase in SAM concentration in B_{12} deficient patients. Concentrations of SAH and adenosine remained unchanged after B vitamin treatment, which indicates that hydrolysis of SAH to Hcy was not increased after the treatment. SAM/SAH ratio significantly increased only in nonanemic patients. Thus, the presence of vitamin B_{12} deficiency and anemia are two important variables that affected changes of the metabolites by treatment. These findings suggest that the association between elevated Hcy and a low SAM/SAH ratio may not be monocausal. Together, there is only a little evidence that a short-term Hcy-lowering treatment could correct methylation dysbalance in uremic patients.

References

1 Grieco, A.J. (1977) Homocystinuria: pathogenetic mechanisms. *Am J Med Sci*, **273** (2), 120–132.

2 Higginbottom, M.C., Sweetman, L., and Nyhan, W.L. (1978) A syndrome of methylmalonic aciduria, homocystinuria, megaloblastic anemia and neurologic abnormalities in a vitamin B12-deficient breast-fed infant of a strict vegetarian. *N Engl J Med*, **299** (7), 317–323.

3 Liem, A., Reynierse-Buitenwerf, G.H., Zwinderman, A.H., Jukema, J.W., and Van Veldhuisen, D.J.

(2003) Secondary prevention with folic acid: effects on clinical outcomes. *J Am Coll Cardiol*, **41** (12), 2105–2113.

4 Willems, F.F., Aengevaeren, W.R., Boers, G.H., Blom, H.J., and Verheugt, F.W. (2002) Coronary endothelial function in hyperhomocysteinemia: improvement after treatment with folic acid and cobalamin in patients with coronary artery disease 2. *J Am Coll Cardiol*, **40** (4), 766–772.

5 Castro, R., Rivera, I., Martins, C., Struys, E.A., Jansen, E.E., Clode, N., Graca, L.M., Blom, H.J., Jakobs, C. and de Almeida, I.T. (2005) Intracellular S-adenosylhomocysteine increased levels are associated with DNA hypomethylation in HUVEC. *J Mol Med*, **83** (10), 831–836.

6 Guerra-Shinohara, E.M., Morita, O.E., Peres, S., Pagliusi, R.A., Sampaio Neto, L.F., D'Almeida, V., Irazusta, S.P., Allen, R.H., and Stabler, S.P. (2004) Low ratio of S-adenosylmethionine to S-adenosylhomocysteine is associated with vitamin deficiency in Brazilian pregnant women and newborns. *Am J Clin Nutr*, **80** (5), 1312–1321.

7 Scarpa, S., Fuso, A., D'Anselmi, F., and Cavallaro, R.A. (2003) Presenilin 1 gene silencing by S-adenosylmethionine: a treatment for Alzheimer disease? *FEBS Lett*, **541** (1-3), 145–148.

8 Maclean, K.N., Janosik, M., Kraus, E., Kozich, V., Allen, R.H., Raab, B.K., and Kraus, J.P. (2002) Cystathionine beta-synthase is coordinately regulated with proliferation through a redox-sensitive mechanism in cultured human cells and Saccharomyces cerevisiae. *J Cell Physiol*, **192** (1), 81–92.

9 van Beynum, I., den Heijer, M., Thomas, C.M., Afman, L., Oppenraay-van, E.D., and Blom, H.J. (2005) Total homocysteine and its predictors in Dutch children. *Am J Clin Nutr*, **81** (5), 1110–1116.

10 Refsum, H., Grindflek, A.W., Ueland, P.M., Fredriksen, A., Meyer, K., Ulvik, A., Guttormsen, A.B., Iversen, O.E., Schneede, J., and Kase, B.F. (2004) Screening for serum total homocysteine in newborn children. *Clin Chem*, **50** (10), 1769–1784.

11 Herrmann, W., Quast, S., Ullrich, M., Schultheis, H., Bodis, M., and Geisel, J. (1999) Hyperhomocysteinemia in high-aged subjects: relation of B-vitamins, folic acid, renal function and the methylenetetrahydrofolate reductase mutation. *Atherosclerosis*, **144** (1), 91–101.

12 Bjorkegren, K. and Svardsudd, K. (2001) Serum cobalamin, folate, methylmalonic acid and total homocysteine as vitamin B12 and folate tissue deficiency markers amongst elderly Swedes- -a population-based study. *J Intern Med*, **249** (5), 423–432.

13 Ventura, P., Panini, R., Verlato, C., Scarpetta, G., and Salvioli, G. (2001) Hyperhomocysteinemia and related factors in 600 hospitalized elderly subjects. *Metabolism*, **50** (12), 1466–1471.

14 Pinto, X., Vilaseca, M.A., Garcia-Giralt, N., Ferrer, I., Pala, M., Meco, J.F., Mainou, C., Ordovas, J.M., Grinberg, D., and Balcells, S. (2001) Homocysteine and the MTHFR 677C-->T allele in premature coronary artery disease. Case control and family studies. *Eur J Clin Invest*, **31** (1), 24–30.

15 Friedman, A.N., Bostom, A.G., Selhub, J., Levey, A.S., and Rosenberg, I.H. (2001) The kidney and homocysteine metabolism. *J Am Soc Nephrol*, **12** (10), 2181–2189.

16 Halsted, C.H., Villanueva, J.A., Devlin, A.M., and Chandler, C.J. (2002) Metabolic interactions of alcohol and folate. *J Nutr*, **132** (8 Suppl), 2367S–2372S.

17 de Bree, A., Verschuren, W.M., Blom, H.J., and Kromhout, D. (2001) Lifestyle factors and plasma homocysteine concentrations in a general population sample. *Am J Epidemiol*, **154** (2), 150–154.

18 Herrmann, W., Schorr, H., Obeid, R., and Geisel, J. (2003) Vitamin B-12 status, particularly holo-transcobalamin II and methylmalonic acid concentrations, and hyperhomocysteinemia in vegetarians. *Am J Clin Nutr*, **78** (1), 131–136.

19 Herrmann, W., Schorr, H., Purschwitz, K., Rassoul, F., and Richter, V. (2001) Total homocysteine, vitamin B-12, and total antioxidant status in vegetarians. *Clin Chem*, **47** (6), 1094–1101.

20 Herzlich, B. and Herbert, V. (1988) Depletion of serum holotranscobalamin II. An early sign of negative vitamin B12 balance. *Lab Invest*, **58** (3), 332–337.

21 Stroinsky A. and Schneider Z. (1987) Cobamide dependant enzymes, in *Comprehensive B-12* (eds Z. Schneider and A. Stroinsky), Printing house de Gruyter, Berlin, pp. 225–266.

22 Scott, J.M. (1997) Bioavailability of vitamin B12. *Eur J Clin Nutr*, **51** (Suppl 1), S49–S53.

23 Herbert, V. and Drivas, G. (1982) Spirulina and vitamin B 12. *J Am Med Assoc*, **248** (23), 3096–3097.

24 Carmel, R. (1985) The distribution of endogenous cobalamin among cobalamin-binding proteins in the blood in normal and abnormal states. *Am J Clin Nutr*, **41** (4), 713–719.

25 Hall, C.A. (1977) The carriers of native vitamin B12 in normal human serum. *Clin Sci Mol Med*, **53** (5), 453–457.

26 England, J.M., Down, M.C., Wise, I.J., and Linnell, J.C. (1976) The transport of endogenous vitamin B12 in normal human serum. *Clin Sci Mol Med*, **51** (1), 47–52.

27 Markle, H.V. (1996) Cobalamin. *Crit Rev Clin Lab Sci*, **33** (4), 247–356.

28 Moestrup, S.K., Birn, H., Fischer, P.B., Petersen, C.M., Verroust, P.J., Sim, R.B., Christensen, E.I., and Nexo, E. (1996) Megalin-mediated endocytosis of transcobalamin-vitamin-B12 complexes suggests a role of the receptor in vitamin-B12 homeostasis 1. *Proc Natl Acad Sci U S A*, **93** (16), 8612–8617.

29 Morris, M.S., Jacques, P.F., Rosenberg, I.H., and Selhub, J. (2002) Elevated serum methylmalonic acid concentrations are common among elderly Americans 1. *J Nutr*, **132** (9), 2799–2803.

30 Obeid, R., Jouma, M., and Herrmann, W. (2002) Cobalamin status (holo-transcobalamin, methylmalonic acid) and folate as determinants of homocysteine concentration. *Clin Chem*, **48** (11), 2064–2065.

31 Carmel, R. (1997) Cobalamin, the stomach, and aging. *Am J Clin Nutr*, **66** (4), 750–759.

32 de Bree, A., van Dusseldorp, M., Brouwer, I.A., het Hof, K.H., and Steegers-Theunissen, R.P. (1997) Folate intake in Europe: recommended, actual and desired intake. *Eur J Clin Nutr*, **51** (10), 643–660.

33 Kato, I., Dnistrian, A.M., Schwartz, M., Toniolo, P., Koenig, K., Shore, R.E., Zeleniuch-Jacquotte, A., Akhmedkhanov, A., and Riboli, E. (1999) Epidemiologic correlates of serum folate and homocysteine levels among users and non-users of vitamin supplement. *Int J Vitam Nutr Res*, **69** (5), 322–329.

34 Giovannucci, E., Stampfer, M.J., Colditz, G.A., Hunter, D.J., Fuchs, C., Rosner, B.A., Speizer, F.E., and Willett, W.C. (1998) Multivitamin use, folate, and colon cancer in women in the Nurses' Health Study. *Ann Intern Med*, **129** (7), 517–524.

35 Hamm, M.W., Mehansho, H., and Henderson, L.M. (1979) Transport and metabolism of pyridoxamine and pyridoxamine phosphate in the small intestine of the rat. *J Nutr*, **109** (9), 1552–1559.

36 Morre, D.M., Kirksey, A., and Das, G.D. (1978) Effects of vitamin B-6 deficiency on the developing central nervous system of the rat. Myelination. *J Nutr*, **108** (8), 1260–1265.

37 Vitvitsky, V., Thomas, M., Ghorpade, A., Gendelman, H.E., and Banerjee,

R. (2006) A functional transsulfuration pathway in the brain links to glutathione homeostasis. *J Biol Chem*, **281** (47), 35785–35793.

38 Martinez, M., Cuskelly, G.J., Williamson, J., Toth, J.P., and Gregory, J.F. III (2000) Vitamin B-6 deficiency in rats reduces hepatic serine hydroxymethyltransferase and cystathionine beta-synthase activities and rates of in vivo protein turnover, homocysteine remethylation and transsulfuration. *J Nutr*, **130** (5), 1115–1123.

39 Rimm, E.B., Willett, W.C., Hu, F.B., Sampson, L., Colditz, G.A., Manson, J.E., Hennekens, C., and Stampfer, M.J. (1998) Folate and vitamin B6 from diet and supplements in relation to risk of coronary heart disease among women. *J Am Med Assoc*, **279** (5), 359–364.

40 Folsom, A.R., Nieto, F.J., McGovern, P.G., Tsai, M.Y., Malinow, M.R., Eckfeldt, J.H., Hess, D.L., and Davis, C.E. (1998) Prospective study of coronary heart disease incidence in relation to fasting total homocysteine, related genetic polymorphisms, and B vitamins: the Atherosclerosis Risk in Communities (ARIC) study. *Circulation*, **98** (3), 204–210.

41 Robinson, K., Mayer, E.L., Miller, D.P., Green, R., vanLente, F., Gupta, A., Kottke, M.K., Savon, S.R., Selhub, J., Nissen, S.E. et al. (1995) Hyperhomocysteinemia and low pyridoxal phosphate. Common and independent reversible risk factors for coronary artery disease. *Circulation*, **92** (10), 2825–2830.

42 Robinson, K., Arheart, K., Refsum, H., Brattstrom, L., Boers, G., Ueland, P., Rubba, P., Palma-Reis, R., Meleady, R., Daly, L., Witteman, J., and Graham, I. (1998) Low circulating folate and vitamin B6 concentrations: risk factors for stroke, peripheral vascular disease, and coronary artery disease. European COMAC Group [see comments]. *Circulation*, **97** (5), 437–443.

43 Lin, P.T., Cheng, C.H., Liaw, Y.P., Lee, B.J., Lee, T.W., and Huang, Y.C. (2006) Low pyridoxal 5′-phosphate is associated with increased risk of coronary artery disease. *Nutrition*, **22** (11-12), 1146–1151.

44 Ubbink, J.B., Vermaak, W.J., van der Merwe, M.A., Becker, P.J., Delport, R., and Potgieter, H.C. (1994) Vitamin requirements for the treatment of hyperhomocysteinemia in humans. *J Nutr*, **124** (10), 1927–1933.

45 Miller, J.W., Green, R., Mungas, D.M., Reed, B.R., and Jagust, W.J. (2002) Homocysteine, vitamin B6, and vascular disease in AD patients. *Neurology*, **58** (10), 1471–1475.

46 Malouf, R. and Grimley, E.J. (2003) The effect of vitamin B6 on cognition. *Cochrane Database Syst Rev* 4 (Art. No.: CD004393).

47 Riggs, K.M., Spiro, A. III, Tucker, K., and Rush, D. (1996) Relations of vitamin B-12, vitamin B-6, folate, and homocysteine to cognitive performance in the Normative Aging Study. *Am J Clin Nutr*, **63** (3), 306–314.

48 Deijen, J.B., van der Beek, E.J., Orlebeke, J.F., and van den Berg, H. (1992) Vitamin B-6 supplementation in elderly men: effects on mood, memory, performance and mental effort. *Psychopharmacology (Berl)*, **109** (4), 489–496.

49 Bryan, J., Calvaresi, E., and Hughes, D. (2002) Short-term folate, vitamin B-12 or vitamin B-6 supplementation slightly affects memory performance but not mood in women of various ages. *J Nutr*, **132** (6), 1345–1356.

50 Balk, E.M., Raman, G., Tatsioni, A., Chung, M., Lau, J., and Rosenberg, I.H. (2007) Vitamin B6, B12, and folic acid supplementation and cognitive function: a systematic review of randomized trials. *Arch Intern Med*, **167** (1), 21–30.

51 Meydani, S.N., Ribaya-Mercado, J.D., Russell, R.M., Sahyoun, N., Morrow, F.D. and Gershoff, S.N. (1991) Vitamin B-6 deficiency impairs interleukin 2 production and lymphocyte proliferation in elderly adults. *Am J Clin Nutr*, **53** (5), 1275–1280.

52 Hvas, A.M., Juul, S., Bech, P., and Nexo, E. (2004) Vitamin B6 level is associated with symptoms of depression. *Psychother Psychosom*, **73** (6), 340–343.

53 Sahakian, V., Rouse, D., Sipes, S., Rose, N., and Niebyl, J. (1991) Vitamin B6 is effective therapy for nausea and vomiting of pregnancy: a randomized, double-blind placebo-controlled study. *Obstet Gynecol*, **78** (1), 33–36.

54 Jewell, D. and Young, G. (2002) Interventions for nausea and vomiting in early pregnancy. *Cochrane Database Syst Rev* 1 (Art. No.: CD000145).

55 Bender, D.A. (1999) Non-nutritional uses of vitamin B6. *Br J Nutr*, **81** (1), 7–20.

56 Wimo, A., Winblad, B., Aguero-Torres, H., and von Strauss, S.E. (2003) The magnitude of dementia occurrence in the world. *Alzheimer Dis Assoc Disord*, **17** (2), 63–67.

57 Seubert, P., Vigo-Pelfrey, C., Esch, F., Lee, M., Dovey, H., Davis, D., Sinha, S., Schlossmacher, M., Whaley, J., and Swindlehurst, C. (1992) Isolation and quantification of soluble Alzheimer's beta-peptide from biological fluids. *Nature*, **359** (6393), 325–327.

58 Hardy, J. and Selkoe, D.J. (2002) The amyloid hypothesis of Alzheimer's disease: progress and problems on the road to therapeutics. *Science*, **297** (5580), 353–356.

59 Tanemura, K., Akagi, T., Murayama, M., Kikuchi, N., Murayama, O., Hashikawa, T., Yoshiike, Y., Park, J.M., Matsuda, K., Nakao, S., Sun, X., Sato, S., Yamaguchi, H., and Takashima, A. (2001) Formation of filamentous tau aggregations in transgenic mice expressing V337M human tau. *Neurobiol Dis*, **8** (6), 1036–1045.

60 Mattson, M.P. and Duan, W. (1999) "Apoptotic" biochemical cascades in synaptic compartments: roles in adaptive plasticity and neurodegenerative disorders. *J Neurosci Res*, **58** (1), 152–166.

61 Mattson, M.P., Pedersen, W.A., Duan, W., Culmsee, C., and Camandola, S. (1999) Cellular and molecular mechanisms underlying perturbed energy metabolism and neuronal degeneration in Alzheimer's and Parkinson's diseases. *Ann N Y Acad Sci*, **893**, 154–175.

62 Su, J.H., Anderson, A.J., Cummings, B.J., and Cotman, C.W. (1994) Immunohistochemical evidence for apoptosis in Alzheimer's disease. *Neuroreport*, **5** (18), 2529–2533.

63 Pappolla, M.A., Omar, R.A., Kim, K.S., and Robakis, N.K. (1992) Immunohistochemical evidence of oxidative [corrected] stress in Alzheimer's disease. *Am J Pathol*, **140** (3), 621–628.

64 Lyras, L., Cairns, N.J., Jenner, A., Jenner, P., and Halliwell, B. (1997) An assessment of oxidative damage to proteins, lipids, and DNA in brain from patients with Alzheimer's disease. *J Neurochem*, **68** (5), 2061–2069.

65 Sachdev, P.S. (2005) Homocysteine and brain atrophy. *Prog Neuropsychopharmacol Biol Psychiatry*, **29** (7), 1152–1161.

66 Wright, C.B., Paik, M.C., Brown, T.R., Stabler, S.P., Allen, R.H., Sacco, R.L., and DeCarli, C. (2005) Total homocysteine is associated with white matter hyperintensity volume: the Northern Manhattan Study. *Stroke*, **36** (6), 1207–1211.

67 Lipton, S.A., Kim, W.K., Choi, Y.B., Kumar, S., D'Emilia, D.M., Rayudu, P.V., Arnelle, D.R. and Stamler, J.S. (1997) Neurotoxicity associated with dual actions of homocysteine at the N-methyl-D-aspartate receptor. *Proc Natl Acad Sci U S A*, **94** (11), 5923–5928.

68 Parsons, R.B., Waring, R.H., Ramsden, D.B. and Williams, A.C. (1998) In vitro effect of the cysteine metabolites homocysteic acid, homocysteine and cysteic acid

upon human neuronal cell lines. *Neurotoxicology*, **19** (4-5), 599–603.

69 Kruman, I.I., Culmsee, C., Chan, S.L., Kruman, Y., Guo, Z., Penix, L., and Mattson, M.P. (2000) Homocysteine elicits a DNA damage response in neurons that promotes apoptosis and hypersensitivity to excitotoxicity. *J Neurosci*, **20** (18), 6920–6926.

70 Vitvitsky, V., Dayal, S., Stabler, S., Zhou, Y., Wang, H., Lentz, S.R., and Banerjee, R. (2004) Perturbations in homocysteine-linked redox homeostasis in a murine model for hyperhomocysteinemia. *Am J Physiol Regul Integr Comp Physiol*, **287** (1), R39–R46.

71 Kamath, A.F., Chauhan, A.K., Kisucka, J., Dole, V.S., Loscalzo, J., Handy, D.E. and Wagner, D.D. (2006) Elevated levels of homocysteine compromise blood-brain barrier integrity in mice. *Blood*, **107** (2), 591–593.

72 Troen, A.M. (2005) The central nervous system in animal models of hyperhomocysteinemia. *Prog Neuropsychopharmacol Biol Psychiatry*, **29** (7), 1140–1151.

73 Watanabe, M., Osada, J., Aratani, Y., Kluckman, K., Reddick, R., Malinow, M.R., and Maeda, N. (1995) Mice deficient in cystathionine beta-synthase: animal models for mild and severe homocyst(e)inemia. *Proc Natl Acad Sci U S A*, **92** (5), 1585–1589.

74 Algaidi, S.A., Christie, L.A., Jenkinson, A.M., Whalley, L., Riedel, G. and Platt, B. (2005) Long-term homocysteine exposure induces alterations in spatial learning, hippocampal signalling and synaptic plasticity. *Exp Neurol*, **197** (1), 8–21.

75 Streck, E.L., Bavaresco, C.S., Netto, C.A., and Wyse, A.T. (2004) Chronic hyperhomocysteinemia provokes a memory deficit in rats in the Morris water maze task. *Behav Brain Res*, **153** (2), 377–381.

76 Streck, E.L., Delwing, D., Tagliari, B., Matte, C., Wannmacher, C.M., Wajner, M., and Wyse, A.T. (2003) Brain energy metabolism is compromised by the metabolites accumulating in homocystinuria. *Neurochem Int*, **43** (6), 597–602.

77 Mattson, M.P. and Shea, T.B. (2003) Folate and homocysteine metabolism in neural plasticity and neurodegenerative disorders. *Trends Neurosci*, **26** (3), 137–146.

78 Scott, J.M., Molloy, A.M., Kennedy, D.G., Kennedy, S., and Weir, D.G. (1994) Effects of the disruption of transmethylation in the central nervous system: an animal model. *Acta Neurol Scand Suppl*, **154**, 27–31.

79 Gharib, A., Chabannes, B., Sarda, N., and Pacheco, H. (1983) In vivo elevation of mouse brain S-adenosyl-L-homocysteine after treatment with L-homocysteine. *J Neurochem*, **40** (4), 1110–1112.

80 Ho, P.I., Ortiz, D., Rogers, E., and Shea, T.B. (2002) Multiple aspects of homocysteine neurotoxicity: glutamate excitotoxicity, kinase hyperactivation and DNA damage. *J Neurosci Res*, **70** (5), 694–702.

81 Rao, A.M., Baskaya, M.K., Maley, M.E., Kindy, M.S., and Dempsey, R.J. (1997) Beneficial effects of S-adenosyl-L-methionine on blood-brain barrier breakdown and neuronal survival after transient cerebral ischemia in gerbils. *Brain Res Mol Brain Res*, **44** (1), 134–138.

82 Trovarelli, G., De Medio, G.E., Porcellati, S., Stramentinoli, G., and Porcellati, G. (1983) The effect of S-Adenosyl-L-methionine on ischemia-induced disturbances of brain phospholipid in the gerbil. *Neurochem Res*, **8** (12), 1597–1609.

83 Fuso, A., Seminara, L., Cavallaro, R.A., D'Anselmi, F., and Scarpa, S. (2005) S-adenosylmethionine/homocysteine cycle alterations modify DNA methylation status with consequent deregulation of PS1 and BACE and beta-amyloid production. *Mol Cell Neurosci*, **28** (1), 195–204.

84 Selkoe, D.J. (2001) Presenilin, Notch, and the genesis and treatment of Alzheimer's disease. *Proc Natl Acad Sci U S A*, **98** (20), 11039–11041.

85 Masters, C.L., Multhaup, G., Simms, G., Pottgiesser, J., Martins, R.N., and Beyreuther, K. (1985) Neuronal origin of a cerebral amyloid: neurofibrillary tangles of Alzheimer's disease contain the same protein as the amyloid of plaque cores and blood vessels. *EMBO J*, **4** (11), 2757–2763.

86 Glenner, G.G., Wong, C.W., Quaranta, V., and Eanes, E.D. (1984) The amyloid deposits in Alzheimer's disease: their nature and pathogenesis. *Appl Pathol*, **2** (6), 357–369.

87 Irizarry, M.C., Gurol, M.E., Raju, S., az-Arrastia, R., Locascio, J.J., Tennis, M., Hyman, B.T., Growdon, J.H., Greenberg, S.M., and Bottiglieri, T. (2005) Association of homocysteine with plasma amyloid beta protein in aging and neurodegenerative disease. *Neurology*, **65** (9), 1402–1408.

88 Sai, X., Kawamura, Y., Kokame, K., Yamaguchi, H., Shiraishi, H., Suzuki, R., Suzuki, T., Kawaichi, M., Miyata, T., Kitamura, T., De Strooper, B., Yanagisawa, K., and Komano, H. (2002) Endoplasmic reticulum stress-inducible protein, Herp, enhances presenilin-mediated generation of amyloid beta-protein. *J Biol Chem*, **277** (15), 12915–12920.

89 Ho, P.I., Collins, S.C., Dhitavat, S., Ortiz, D., Ashline, D., Rogers, E., and Shea, T.B. (2001) Homocysteine potentiates beta-amyloid neurotoxicity: role of oxidative stress. *J Neurochem*, **78** (2), 249–253.

90 Kruman, I.I., Kumaravel, T.S., Lohani, A., Pedersen, W.A., Cutler, R.G., Kruman, Y., Haughey, N., Lee, J., Evans, M., and Mattson, M.P. (2002) Folic acid deficiency and homocysteine impair DNA repair in hippocampal neurons and sensitize them to amyloid toxicity in experimental models of Alzheimer's disease. *J Neurosci*, **22** (5), 1752–1762.

91 Hasegawa, T., Ukai, W., Jo, D.G., Xu, X., Mattson, M.P., Nakagawa, M., Araki, W., Saito, T., and Yamada, T. (2005) Homocysteic acid induces intraneuronal accumulation of neurotoxic Abeta42: implications for the pathogenesis of Alzheimer's disease. *J Neurosci Res*, **80** (6), 869–876.

92 Mok, S.S., Turner, B.J., Beyreuther, K., Masters, C.L., Barrow, C.J., and Small, D.H. (2002) Toxicity of substrate-bound amyloid peptides on vascular smooth muscle cells is enhanced by homocysteine. *Eur J Biochem*, **269** (12), 3014–3022.

93 Leulliot, N., Quevillon-Cheruel, S., Sorel, I., de La Sierra-Gallay, I.L., Collinet, B., Graille, M., Blondeau, K., Bettache, N., Poupon, A., Janin, J. and van Tilbeurgh, H. (2004) Structure of protein phosphatase methyltransferase 1 (PPM1), a leucine carboxyl methyltransferase involved in the regulation of protein phosphatase 2A activity. *J Biol Chem*, **279** (9), 8351–8358.

94 Ferreira, A., Lu, Q., Orecchio, L., and Kosik, K.S. (1997) Selective phosphorylation of adult tau isoforms in mature hippocampal neurons exposed to fibrillar A beta. *Mol Cell Neurosci*, **9** (3), 220–234.

95 Sontag, J.M., Nunbhakdi-Craig, V., Montgomery, L., Arning, E., Bottiglieri, T., and Sontag, E. (2008) Folate deficiency induces in vitro and mouse brain region-specific downregulation of leucine carboxyl methyltransferase-1 and protein phosphatase 2A B(alpha) subunit expression that correlate with enhanced tau phosphorylation. *J Neurosci*, **28** (45), 11477–11487.

96 Sontag, E., Hladik, C., Montgomery, L., Luangpirom, A., Mudrak, I., Ogris, E., and White, C.L. III (2004) Downregulation of protein phosphatase 2A carboxyl methylation and methyltransferase may contribute to Alzheimer disease pathogenesis. *J Neuropathol Exp Neurol*, **63** (10), 1080–1091.

97 Do, K.Q., Herrling, P.L., Streit, P., Turski, W.A., and Cuenod, M. (1986) In vitro release and electrophysiological effects in situ of homocysteic acid, an endogenous N-methyl-(D)-aspartic acid agonist, in

the mammalian striatum. *J Neurosci*, **6** (8), 2226–2234.
98 Ito, S., Provini, L., and Cherubini, E. (1991) L-homocysteic acid mediates synaptic excitation at NMDA receptors in the hippocampus. *Neurosci Lett*, **124** (2), 157–161.
99 Olney, J.W., Price, M.T., Salles, K.S., Labruyere, J., Ryerson, R., Mahan, K., Frierdich, G., and Samson, L. (1987) L-homocysteic acid: an endogenous excitotoxic ligand of the NMDA receptor. *Brain Res Bull*, **19** (5), 597–602.
100 Zhang, D. and Lipton, S.A. (1992) L-homocysteic acid selectively activates N-methyl-D-aspartate receptors of rat retinal ganglion cells. *Neurosci Lett*, **139** (2), 173–177.
101 Klancnik, J.M., Cuenod, M., Gahwiler, B.H., Jiang, Z.P., and Do, K.Q. (1992) Release of endogenous amino acids, including homocysteic acid and cysteine sulphinic acid, from rat hippocampal slices evoked by electrical stimulation of Schaffer collateral-commissural fibres. *Neuroscience*, **49** (3), 557–570.
102 Cuenod, M., Do, K.Q., and Streit, P. (1990) Homocysteic acid as an endogenous excitatory amino acid. *Trends Pharmacol Sci*, **11** (12), 477–478.
103 Kim, J.P., Koh, J.Y., and Choi, D.W. (1987) L-homocysteate is a potent neurotoxin on cultured cortical neurons. *Brain Res*, **437** (1), 103–110.
104 Zeise, M.L., Knopfel, T., and Zieglgansberger, W. (1988) (+/-)-beta-Parachlorophenylglutamate selectively enhances the depolarizing response to L-homocysteic acid in neocortical neurons of the rat: evidence for a specific uptake system. *Brain Res*, **443** (1-2), 373–376.
105 Alam, Z., Coombes, N., Waring, R.H., Williams, A.C., and Steventon, G.B. (1998) Plasma levels of neuroexcitatory amino acids in patients with migraine or tension headache. *J Neurol Sci*, **156** (1), 102–106.
106 Zieminska, E., Stafiej, A., and Lazarewicz, J.W. (2003) Role of group I metabotropic glutamate receptors and NMDA receptors in homocysteine-evoked acute neurodegeneration of cultured cerebellar granule neurones. *Neurochem Int*, **43** (4-5), 481–492.
107 Shi, Q., Savage, J.E., Hufeisen, S.J., Rauser, L., Grajkowska, E., Ernsberger, P., Wroblewski, J.T., Nadeau, J.H., and Roth, B.L. (2003) L-homocysteine sulfinic acid and other acidic homocysteine derivatives are potent and selective metabotropic glutamate receptor agonists. *J Pharmacol Exp Ther*, **305** (1), 131–142.
108 Robert, K., Pages, C., Ledru, A., Delabar, J., Caboche, J., and Janel, N. (2005) Regulation of extracellular signal-regulated kinase by homocysteine in hippocampus. *Neuroscience*, **133** (4), 925–935.
109 Chan, A.Y., Alsaraby, A., and Shea, T.B. (2008) Folate deprivation increases tau phosphorylation by homocysteine-induced calcium influx and by inhibition of phosphatase activity: alleviation by S-adenosyl methionine. *Brain Res*, **1199**, 133–137.
110 Zou, C.G. and Banerjee, R. (2005) Homocysteine and redox signaling. *Antioxid Redox Signal*, **7** (5-6), 547–559.
111 Banerjee, R. and Zou, C.G. (2005) Redox regulation and reaction mechanism of human cystathionine-beta-synthase: a PLP-dependent hemesensor protein. *Arch Biochem Biophys*, **433** (1), 144–156.
112 Mosharov, E., Cranford, M.R., and Banerjee, R. (2000) The quantitatively important relationship between homocysteine metabolism and glutathione synthesis by the transsulfuration pathway and its regulation by redox changes. *Biochemistry*, **39** (42), 13005–13011.
113 Baydas, G., Ozer, M., Yasar, A., Tuzcu, M., and Koz, S.T. (2005) Melatonin improves learning and memory performances impaired by hyperhomocysteinemia in rats. *Brain Res*, **1046** (1-2), 187–194.
114 Perna, A.F., Ingrosso, D., and De Santo, N.G. (2003) Homocysteine

and oxidative stress. *Amino Acids*, **25** (3-4), 409–417.

115 Weiss, N., Heydrick, S.J., Postea, O., Keller, C., Keaney, J.F. Jr, and Loscalzo, J. (2003) Influence of hyperhomocysteinemia on the cellular redox state–impact on homocysteine-induced endothelial dysfunction. *Clin Chem Lab Med*, **41** (11), 1455–1461.

116 James, S.J., Cutler, P., Melnyk, S., Jernigan, S., Janak, L., Gaylor, D.W., and Neubrander, J.A. (2004) Metabolic biomarkers of increased oxidative stress and impaired methylation capacity in children with autism. *Am J Clin Nutr*, **80** (6), 1611–1617.

117 Jara-Prado, A., Ortega-Vazquez, A., Martinez-Ruano, L., Rios, C., and Santamaria, A. (2003) Homocysteine-induced brain lipid peroxidation: effects of NMDA receptor blockade, antioxidant treatment, and nitric oxide synthase inhibition. *Neurotox Res*, **5** (4), 237–243.

118 Ho, P.I., Ashline, D., Dhitavat, S., Ortiz, D., Collins, S.C., Shea, T.B., and Rogers, E. (2003) Folate deprivation induces neurodegeneration: roles of oxidative stress and increased homocysteine. *Neurobiol Dis*, **14** (1), 32–42.

119 Reis, E.A., Zugno, A.I., Franzon, R., Tagliari, B., Matte, C., Lammers, M.L., Netto, C.A., and Wyse, A.T. (2002) Pretreatment with vitamins E and C prevent the impairment of memory caused by homocysteine administration in rats. *Metab Brain Dis*, **17** (3), 211–217.

120 Wyse, A.T., Zugno, A.I., Streck, E.L., Matte, C., Calcagnotto, T., Wannmacher, C.M., and Wajner, M. (2002) Inhibition of Na(+),K(+)-ATPase activity in hippocampus of rats subjected to acute administration of homocysteine is prevented by vitamins E and C treatment. *Neurochem Res*, **27** (12), 1685–1689.

121 Tchantchou, F., Graves, M., Ortiz, D., Rogers, E., and Shea, T.B. (2004) Dietary supplementation with 3-deaza adenosine, N-acetyl cysteine, and S-adenosyl methionine provide neuroprotection against multiple consequences of vitamin deficiency and oxidative challenge: relevance to age-related neurodegeneration. *Neuromolecular Med*, **6** (2-3), 93–103.

122 Prudova, A., Bauman, Z., Braun, A., Vitvitsky, V., Lu, S.C., and Banerjee, R. (2006) S-adenosylmethionine stabilizes cystathionine {beta}-synthase and modulates redox capacity. *Proc Natl Acad Sci USA*, **103** (17), 6489–6494.

123 Villalobos, M.A., De La Cruz, J.P., Cuerda, M.A., Ortiz, P., Smith-Agreda, J.M., and Sanchez, D.L.C. (2000) Effect of S-adenosyl-L-methionine on rat brain oxidative stress damage in a combined model of permanent focal ischemia and global ischemia-reperfusion. *Brain Res*, **883** (1), 31–40.

124 Matsui, Y., Kubo, Y., and Iwata, N. (1987) S-adenosyl-L-methionine prevents ischemic neuronal death. *Eur J Pharmacol*, **144** (2), 211–216.

125 De La Cruz, J.P., Pavia, J., Gonzalez-Correa, J.A., Ortiz, P., and Sanchez De La, C.F. (2000) Effects of chronic administration of S-adenosyl-L-methionine on brain oxidative stress in rats. *Naunyn Schmiedebergs Arch Pharmacol*, **361** (1), 47–52.

126 Ullegaddi, R., Powers, H.J., and Gariballa, S.E. (2004) B-group vitamin supplementation mitigates oxidative damage after acute ischaemic stroke. *Clin Sci (Lond)*, **107** (5), 477–484.

127 Kado, D.M., Karlamangla, A.S., Huang, M.H., Troen, A., Rowe, J.W., Selhub, J., and Seeman, T.E. (2005) Homocysteine versus the vitamins folate, B6, and B12 as predictors of cognitive function and decline in older high-functioning adults: MacArthur Studies of Successful Aging. *Am J Med*, **118** (2), 161–167.

128 Dufouil, C., Alperovitch, A., Ducros, V., and Tzourio, C. (2003) Homocysteine, white matter hyperintensities,

and cognition in healthy elderly people. *Ann Neurol*, **53** (2), 214–221.

129 Seshadri, S., Beiser, A., Selhub, J., Jacques, P.F., Rosenberg, I.H., D'Agostino, R.B., Wilson, P.W., and Wolf, P.A. (2002) Plasma homocysteine as a risk factor for dementia and Alzheimer's disease. *N Engl J Med*, **346** (7), 476–483.

130 Clarke, R., Smith, A.D., Jobst, K.A., Refsum, H., Sutton, L., and Ueland, P.M. (1998) Folate, vitamin B12, and serum total homocysteine levels in confirmed Alzheimer disease. *Arch Neurol*, **55** (11), 1449–1455.

131 Prins, N.D., den Heijer, H.T., Hofman, A., Koudstaal, P.J., Jolles, J., Clarke, R., and Breteler, M.M. (2002) Homocysteine and cognitive function in the elderly: the Rotterdam Scan Study. *Neurology*, **59** (9), 1375–1380.

132 Hin, H., Clarke, R., Sherliker, P., Atoyebi, W., Emmens, K., Birks, J., Schneede, J., Ueland, P.M., Nexo, E., Scott, J., Molloy, A., Donaghy, M., Frost, C., and Evans, J.G. (2006) Clinical relevance of low serum vitamin B12 concentrations in older people: the Banbury B12 study. *Age Ageing*, **35** (4), 416–422.

133 McCracken, C., Hudson, P., Ellis, R., and McCaddon, A. (2006) Methylmalonic acid and cognitive function in the Medical Research Council Cognitive Function and Ageing Study. *Am J Clin Nutr*, **84** (6), 1406–1411.

134 Smith, A.D., Kim, Y.I., and Refsum, H. (2008) Is folic acid good for everyone? *Am J Clin Nutr*, **87** (3), 517–533.

135 Malouf, M., Grimley, E.J., and Areosa, S.A. (2003) Folic acid with or without vitamin B12 for cognition and dementia. *Cochrane Database Syst Rev* 4 (Art. No.: CD004514).

136 Stott, D.J., MacIntosh, G., Lowe, G.D., Rumley, A., McMahon, A.D., Langhorne, P., Tait, R.C., O'Reilly, D.S., Spilg, E.G., MacDonald, J.B., MacFarlane, P.W., and Westendorp, R.G. (2005) Randomized controlled trial of homocysteine-lowering vitamin treatment in elderly patients with vascular disease. *Am J Clin Nutr*, **82** (6), 1320–1326.

137 Eussen, S.J., de Groot, L.C., Joosten, L.W., Bloo, R.J., Clarke, R., Ueland, P.M., Schneede, J., Blom, H.J., Hoefnagels, W.H., and van Staveren, W.A. (2006) Effect of oral vitamin B-12 with or without folic acid on cognitive function in older people with mild vitamin B-12 deficiency: a randomized, placebo-controlled trial. *Am J Clin Nutr*, **84** (2), 361–370.

138 Clarke, R., Harrison, G., and Richards, S. (2003) Effect of vitamins and aspirin on markers of platelet activation, oxidative stress and homocysteine in people at high risk of dementia. *J Intern Med*, **254** (1), 67–75.

139 Lewerin, C., Matousek, M., Steen, G., Johansson, B., Steen, B., and Nilsson-Ehle, H. (2005) Significant correlations of plasma homocysteine and serum methylmalonic acid with movement and cognitive performance in elderly subjects but no improvement from short-term vitamin therapy: a placebo-controlled randomized study. *Am J Clin Nutr*, **81** (5), 1155–1162.

140 McMahon, J.A., Green, T.J., Skeaff, C.M., Knight, R.G., Mann, J.I., and Williams, S.M. (2006) A controlled trial of homocysteine lowering and cognitive performance. *N Engl J Med*, **354** (26), 2764–2772.

141 Durga, J., van Boxtel, M.P., Schouten, E.G., Kok, F.J., Jolles, J., Katan, M.B., and Verhoef, P. (2007) Effect of 3-year folic acid supplementation on cognitive function in older adults in the FACIT trial: a randomised, double blind, controlled trial. *Lancet*, **369** (9557), 208–216.

142 Fioravanti, M., Ferrario, E., Massaia, M., Cappa, G., Rivolta, G., Grossi, E., and Buckley, A.E. (1998) Low folate levels in the cognitive decline of elderly patients and the efficacy of folate as a treatment for improving memory deficits. *Arch Gerontol Geriatr*, **26** (1), 1–13.

143 Blandini, F., Fancellu, R., Martignoni, E., Mangiagalli, A., Pacchetti, C., Samuele, A., and Nappi, G. (2001) Plasma homocysteine and l-dopa metabolism in patients with Parkinson disease. *Clin Chem*, **47** (6), 1102–1104.

144 Yasui, K., Kowa, H., Nakaso, K., Takeshima, T., and Nakashima, K. (2000) Plasma homocysteine and MTHFR C677T genotype in levodopa-treated patients with PD. *Neurology*, **55** (3), 437–440.

145 Yasui, K., Nakaso, K., Kowa, H., Takeshima, T., and Nakashima, K. (2003) Levodopa-induced hyperhomocysteinemia in Parkinson's disease. *Acta Neurol Scand*, **108** (1), 66–67.

146 Lamberti, P., Zoccolella, S., Iliceto, G., Armenise, E., Fraddosio, A., de Mari, M., and Livrea, P. (2005) Effects of levodopa and COMT inhibitors on plasma homocysteine in Parkinson's disease patients. *Mov Disord*, **20** (1), 69–72.

147 Valkovic, P., Benetin, J., Blazicek, P., Valkovicova, L., Gmitterova, K., and Kukumberg, P. (2005) Reduced plasma homocysteine levels in levodopa/entacapone treated Parkinson patients. *Parkinsonism Relat Disord*, **11** (4), 253–256.

148 Ostrem, J.L., Kang, G.A., Subramanian, I., Guarnieri, M., Hubble, J., Rabinowicz, A.L., and Bronstein, J. (2005) The effect of entacapone on homocysteine levels in Parkinson disease. *Neurology*, **64** (8), 1482.

149 Zesiewicz, T.A., Wecker, L., Sullivan, K.L., Merlin, L.R., and Hauser, R.A. (2006) The controversy concerning plasma homocysteine in Parkinson disease patients treated with levodopa alone or with entacapone: effects of vitamin status. *Clin Neuropharmacol*, **29** (3), 106–111.

150 Lamberti, P., Zoccolella, S., Armenise, E., Lamberti, S.V., Fraddosio, A., de Mari, M., Iliceto, G., and Livrea, P. (2005) Hyperhomocysteinemia in L-dopa treated Parkinson's disease patients: effect of cobalamin and folate administration. *Eur J Neurol*, **12** (5), 365–368.

151 Postuma, R.B., Espay, A.J., Zadikoff, C., Suchowersky, O., Martin, W.R., Lafontaine, A.L., Ranawaya, R., Camicioli, R., and Lang, A.E. (2006) Vitamins and entacapone in levodopa-induced hyperhomocysteinemia: a randomized controlled study. *Neurology*, **66** (12), 1941–1943.

152 O'Suilleabhain, P.E., Sung, V., Hernandez, C., Lacritz, L., Dewey, R.B. Jr, Bottiglieri, T., and az-Arrastia, R. (2004) Elevated plasma homocysteine level in patients with Parkinson disease: motor, affective, and cognitive associations. *Arch Neurol*, **61** (6), 865–868.

153 Rogers, J.D., Sanchez-Saffon, A., Frol, A.B., and az-Arrastia, R. (2003) Elevated plasma homocysteine levels in patients treated with levodopa: association with vascular disease. *Arch Neurol*, **60** (1), 59–64.

154 Sato, Y., Iwamoto, J., Kanoko, T., and Satoh, K. (2005) Homocysteine as a predictive factor for hip fracture in elderly women with Parkinson's disease. *Am J Med*, **118** (11), 1250–1255.

155 van Meurs, J.B., Dhonukshe-Rutten, R.A., Pluijm, S.M., van der Klift, M., de Jonge, R., Lindemans, J., de Groot, L.C., Hofman, A., Witteman, J.C., van Leeuwen, J.P., Breteler, M.M., Lips, P., Pols, H.A., and Uitterlinden, A.G. (2004) Homocysteine levels and the risk of osteoporotic fracture. *N Engl J Med*, **350** (20), 2033–2041.

156 McLean, R.R., Jacques, P.F., Selhub, J., Tucker, K.L., Samelson, E.J., Broe, K.E., Hannan, M.T., Cupples, L.A., and Kiel, D.P. (2004) Homocysteine as a predictive factor for hip fracture in older persons. *N Engl J Med*, **350** (20), 2042–2049.

157 O'Suilleabhain, P.E., Oberle, R., Bartis, C., Dewey, R.B. Jr, Bottiglieri, T., and Diaz-Arrastia, R. (2006) Clinical course in Parkinson's disease with elevated homocysteine.

Parkinsonism Relat Disord, **12** (2), 103–107.

158 Chen, H., Zhang, S.M., Schwarzschild, M.A., Hernan, M.A., Logroscino, G., Willett, W.C., and Ascherio, A. (2004) Folate intake and risk of Parkinson's disease. *Am J Epidemiol*, **160** (4), 368–375.

159 Khan, U., Crossley, C., Kalra, L., Rudd, A., Wolfe, C.D., Collinson, P., and Markus, H.S. The South London Ethnicity and Stroke Study (2008) Homocysteine and its relationship to stroke subtypes in a UK black population. *Stroke*, **39** (11), 2943–2949.

160 Bots, M.L., Launer, L.J., Lindemans, J., Hoes, A.W., Hofman, A., Witteman, J.C., Koudstaal, P.J., and Grobbee, D.E. (1999) Homocysteine and short-term risk of myocardial infarction and stroke in the elderly: the Rotterdam Study. *Arch Intern Med*, **159** (1), 38–44.

161 Malinow, M.R. (1994) Homocyst(e)ine and arterial occlusive diseases 1. *J Intern Med*, **236** (6), 603–617.

162 Verhoef, P., Hennekens, C.H., Malinow, M.R., Kok, F.J., Willett, W.C., and Stampfer, M.J. (1994) A prospective study of plasma homocyst(e)ine and risk of ischemic stroke. *Stroke*, **25** (10), 1924–1930.

163 Sacco, R.L., Anand, K., Lee, H.S., Boden-Albala, B., Stabler, S., Allen, R., and Paik, M.C. (2004) Homocysteine and the risk of ischemic stroke in a triethnic cohort: the NOrthern MAnhattan Study. *Stroke*, **35** (10), 2263–2269.

164 Giles, W.H., Croft, J.B., Greenlund, K.J., Ford, E.S., and Kittner, S.J. (1998) Total homocyst(e)ine concentration and the likelihood of nonfatal stroke: results from the Third National Health and Nutrition Examination Survey, 1988–1994. *Stroke*, **29** (12), 2473–2477.

165 Das, R.R., Seshadri, S., Beiser, A.S., Kelly-Hayes, M., Au, R., Himali, J.J., Kase, C.S., Benjamin, E.J., Polak, J.F., O'Donnell, C.J., Yoshita, M., D'Agostino, R.B. Sr, DeCarli, C., and Wolf, P.A. (2008) Prevalence and correlates of silent cerebral infarcts in the framingham offspring study. *Stroke*, **39** (11), 2929–2935.

166 Yang, Q., Botto, L.D., Erickson, J.D., Berry, R.J., Sambell, C., Johansen, H., and Friedman, J.M. (2006) Improvement in stroke mortality in Canada and the United States, 1990 to 2002. *Circulation*, **113** (10), 1335–1343.

167 Vermeulen, E.G., Stehouwer, C.D., Valk, J., van der Knapp, M., van den Berg, M., Twisk, J.W., Prevoo, W., and Rauwerda, J.A. (2004) Effect of homocysteine-lowering treatment with folic acid plus vitamin B on cerebrovascular atherosclerosis and white matter abnormalities as determined by MRA and MRI: a placebo-controlled, randomized trial. *Eur J Clin Invest*, **34** (4), 256–261.

168 Lonn, E., Yusuf, S., Arnold, M.J., Sheridan, P., Pogue, J., Micks, M., McQueen, M.J., Probstfield, J., Fodor, G., Held, C., and Genest, J. Jr (2006) Homocysteine lowering with folic acid and B vitamins in vascular disease. *N Engl J Med*, **354** (15), 1567–1577.

169 Toole, J.F., Malinow, M.R., Chambless, L.E., Spence, J.D., Pettigrew, L.C., Howard, V.J., Sides, E.G., Wang, C.H., and Stampfer, M. (2004) Lowering homocysteine in patients with ischemic stroke to prevent recurrent stroke, myocardial infarction, and death: the Vitamin Intervention for Stroke Prevention (VISP) randomized controlled trial. *J Am Med Assoc*, **291** (5), 565–575.

170 Spence, J.D., Bang, H., Chambless, L.E., and Stampfer, M.J. (2005) Vitamin Intervention For Stroke Prevention trial: an efficacy analysis. *Stroke*, **36** (11), 2404–2409.

171 Wang, X., Qin, X., Demirtas, H., Li, J., Mao, G., Huo, Y., Sun, N., Liu, L., and Xu, X. (2007) Efficacy of folic acid supplementation in stroke prevention: a meta-analysis. *Lancet*, **369** (9576), 1876–1882.

172 Goldstein, L.B., Adams, R., Alberts, M.J., Appel, L.J., Brass, L.M., Bushnell, C.D., Culebras, A., DeGraba, T.J., Gorelick, P.B., Guyton, J.R., Hart, R.G., Howard, G., Kelly-Hayes, M., Nixon, J.V., and Sacco, R.L. (2006) Primary prevention of ischemic stroke: a guideline from the American Heart Association/American Stroke Association Stroke Council: cosponsored by the Atherosclerotic Peripheral Vascular Disease Interdisciplinary Working Group; Cardiovascular Nursing Council; Clinical Cardiology Council; Nutrition, Physical Activity, and Metabolism Council; and the Quality of Care and Outcomes Research Interdisciplinary Working Group. *Circulation*, **113** (24), e873–e923.

173 Tyagi, N., Ovechkin, A.V., Lominadze, D., Moshal, K.S., and Tyagi, S.C. (2006) Mitochondrial mechanism of microvascular endothelial cells apoptosis in hyperhomocysteinemia. *J Cell Biochem*, **98** (5), 1150–1162.

174 Lee, S.J., Kim, K.M., Namkoong, S., Kim, C.K., Kang, Y.C., Lee, H., Ha, K.S., Han, J.A., Chung, H.T., Kwon, Y.G., and Kim, Y.M. (2005) Nitric oxide inhibition of homocysteine-induced human endothelial cell apoptosis by down-regulation of p53-dependent Noxa expression through the formation of S-nitrosohomocysteine. *J Biol Chem*, **280** (7), 5781–5788.

175 Jacobsen, D.W., Catanescu, O., DiBello, P.M., and Barbato, J.C. (2005) Molecular targeting by homocysteine: a mechanism for vascular pathogenesis. *Clin Chem Lab Med*, **43** (10), 1076–1083.

176 Kang, E.S., Cates, T.B., Harper, D.N., Chiang, T.M., Myers, L.K., Acchiardo, S.R., and Kimoto, M. (2001) An enzyme hydrolyzing methylated inhibitors of nitric oxide synthase is present in circulating human red blood cells. *Free Radic Res*, **35** (6), 693–707.

177 Silverman, M.D., Tumuluri, R.J., Davis, M., Lopez, G., Rosenbaum, J.T., and Lelkes, P.I. (2002) Homocysteine upregulates vascular cell adhesion molecule-1 expression in cultured human aortic endothelial cells and enhances monocyte adhesion. *Arterioscler Thromb Vasc Biol*, **22** (4), 587–592.

178 Tsai, J.C., Perrella, M.A., Yoshizumi, M., Hsieh, C.M., Haber, E., Schlegel, R., and Lee, M.E. (1994) Promotion of vascular smooth muscle cell growth by homocysteine: a link to atherosclerosis. *Proc Natl Acad Sci U S A*, **91** (14), 6369–6373.

179 Sengupta, S., Chen, H., Togawa, T., DiBello, P.M., Majors, A.K., Budy, B., Ketterer, M.E., and Jacobsen, D.W. (2001) Albumin thiolate anion is an intermediate in the formation of albumin-S-S-homocysteine. *J Biol Chem*, **276** (32), 30111–30117.

180 Sengupta, S., Wehbe, C., Majors, A.K., Ketterer, M.E., DiBello, P.M., and Jacobsen, D.W. (2001) Relative roles of albumin and ceruloplasmin in the formation of homocystine, homocysteine-cysteine-mixed disulfide, and cystine in circulation. *J Biol Chem*, **276** (50), 46896–46904.

181 Budy, B., Sengupta, S., DiBello, P.M., Kinter, M.T., and Jacobsen, D.W. (2001) A facile synthesis of homocysteine-cysteine mixed disulfide. *Anal Biochem*, **291** (2), 303–305.

182 Majors, A.K., Sengupta, S., Willard, B., Kinter, M.T., Pyeritz, R.E., and Jacobsen, D.W. (2002) Homocysteine binds to human plasma fibronectin and inhibits its interaction with fibrin. *Arterioscler Thromb Vasc Biol*, **22** (8), 1354–1359.

183 Majors, A., Ehrhart, L.A., and Pezacka, E.H. (1997) Homocysteine as a risk factor for vascular disease. Enhanced collagen production and accumulation by smooth muscle cells. *Arterioscler Thromb Vasc Biol*, **17** (10), 2074–2081.

184 Burke, A.P., Fonseca, V., Kolodgie, F., Zieske, A., Fink, L., and Virmani, R. (2002) Increased serum homocysteine and sudden death resulting from coronary atherosclerosis with

fibrous plaques. *Arterioscler Thromb Vasc Biol*, **22** (11), 1936–1941.

185 Ling, Q. and Hajjar, K.A. (2000) Inhibition of endothelial cell thromboresistance by homocysteine. *J Nutr*, **130** (2S Suppl), 373S–376S.

186 Hajjar, K.A. and Jacovina, A.T. (1998) Modulation of annexin II by homocysteine: implications for atherothrombosis. *J Investig Med*, **46** (8), 364–369.

187 Hajjar, K.A., Mauri, L., Jacovina, A.T., Zhong, F., Mirza, U.A., Padovan, J.C., and Chait, B.T. (1998) Tissue plasminogen activator binding to the annexin II tail domain. Direct modulation by homocysteine. *J Biol Chem*, **273** (16), 9987–9993.

188 Homocysteine Studies Collaboration (2002) Homocysteine and risk of ischemic heart disease and stroke: a meta- analysis. *J Am Med Assoc*, **288** (16), 2015–2022.

189 Wald, D.S., Law, M., and Morris, J.K. (2002) Homocysteine and cardiovascular disease: evidence on causality from a meta-analysis. *Br Med J*, **325** (7374), 1202.

190 Peterson, J.C. and Spence, D.J. (1998) Vitamins and progression of atherosclerosis in hyper-homocysteinaemia. *Lancet*, **351**, 263.

191 Till, U., Rohl, P., Jentsch, A., Till, H., Muller, A., Bellstedt, K., Plonne, D., Fink, H.S., Vollandt, R., Sliwka, U., Herrmann, F.H., Petermann, H., and Riezler, R. (2005) Decrease of carotid intima-media thickness in patients at risk to cerebral ischemia after supplementation with folic acid, Vitamins B6 and B12. *Atherosclerosis*, **181** (1), 131–135.

192 Bonaa, K.H., Njolstad, I., Ueland, P.M., Schirmer, H., Tverdal, A., Steigen, T., Wang, H., Nordrehaug, J.E., Arnesen, E., and Rasmussen, K. (2006) Homocysteine lowering and cardiovascular events after acute myocardial infarction. *N Engl J Med*, **354** (15), 1578–1588.

193 Albert, C.M., Cook, N.R., Gaziano, J.M., Zaharris, E., MacFadyen, J., Danielson, E., Buring, J.E., and Manson, J.E. (2008) Effect of folic acid and B vitamins on risk of cardiovascular events and total mortality among women at high risk for cardiovascular disease: a randomized trial. *J Am Med Assoc*, **299** (17), 2027–2036.

194 Ebbing, M., Bleie, O., Ueland, P.M., Nordrehaug, J.E., Nilsen, D.W., Vollset, S.E., Refsum, H., Pedersen, E.K., and Nygard, O. (2008) Mortality and cardiovascular events in patients treated with homocysteine-lowering B vitamins after coronary angiography: a randomized controlled trial. *J Am Med Assoc*, **300** (7), 795–804.

195 Barak, A.J. and Beckenhauer, H.C. (1988) The influence of ethanol on hepatic transmethylation. *Alcohol Alcohol*, **23** (1), 73–77.

196 Purohit, V. and Russo, D. (2002) Role of S-adenosyl-L-methionine in the treatment of alcoholic liver disease: introduction and summary of the symposium. *Alcohol*, **27** (3), 151–154.

197 Martinez-Chantar, M.L., Garcia-Trevijano, E.R., Latasa, M.U., Perez-Mato, I., Sanchez del Pino, M.M., Corrales, F.J., Avila, M.A., and Mato, J.M. (2002) Importance of a deficiency in S-adenosyl-L-methionine synthesis in the pathogenesis of liver injury. *Am J Clin Nutr*, **76** (5), 1177S–1182S.

198 Avila, M.A., Garcia-Trevijano, E.R., Martinez-Chantar, M.L., Latasa, M.U., Perez-Mato, I., Martinez-Cruz, L.A., del Pino, M.M., Corrales, F.J., and Mato, J.M. (2002) S-Adenosylmethionine revisited: its essential role in the regulation of liver function. *Alcohol*, **27** (3), 163–167.

199 Lieber, C.S., Robins, S.J., and Leo, M.A. (1994) Hepatic phosphatidylethanolamine methyltransferase activity is decreased by ethanol and increased by phosphatidylcholine. *Alcohol Clin Exp Res*, **18** (3), 592–595.

200 Duce, A.M., Ortiz, P., Cabrero, C., and Mato, J.M. (1988) S-adenosyl-L-methionine synthetase

and phospholipid methyltransferase are inhibited in human cirrhosis. *Hepatology*, **8** (1), 65–68.

201 Barak, A.J., Beckenhauer, H.C., and Tuma, D.J. (1996) Betaine, ethanol, and the liver: a review. *Alcohol*, **13** (4), 395–398.

202 Barak, A.J., Beckenhauer, H.C., Junnila, M., and Tuma, D.J. (1993) Dietary betaine promotes generation of hepatic S-adenosylmethionine and protects the liver from ethanol-induced fatty infiltration. *Alcohol Clin Exp Res*, **17** (3), 552–555.

203 Barak, A.J., Beckenhauer, H.C., and Tuma, D.J. (1994) S-adenosylmethionine generation and prevention of alcoholic fatty liver by betaine. *Alcohol*, **11** (6), 501–503.

204 Purohit, V., Abdelmalek, M.F., Barve, S., Benevenga, N.J., Halsted, C.H., Kaplowitz, N., Kharbanda, K.K., Liu, Q.Y., Lu, S.C., McClain, C.J., Swanson, C., and Zakhari, S. (2007) Role of S-adenosylmethionine, folate, and betaine in the treatment of alcoholic liver disease: summary of a symposium. *Am J Clin Nutr*, **86** (1), 14–24.

205 McKusick, V.A. (1966) *Heritable Disorders of Connective Tissue*, C.V. Mosby, St. Louis.

206 Carson, N.A.J. and Neill, D.W. (1962) Metabolic abnormalities detected in a survey of mentally backward individuals in Northern Ireland. *Arch Dis Child*, **37**, 505–513.

207 Parrot, F., Redonnet-Vernhet, I., Lacombe, D., and Gin, H. (2000) Osteoporosis in late-diagnosed adult homocystinuric patients. *J Inherit Metab Dis*, (4), 338–340.

208 Goerss, J.B., Kim, C.H., Atkinson, E.J., Eastell, R., O'Fallon, W.M., and Melton, L.J. III (1992) Risk of fractures in patients with pernicious anemia. *J Bone Miner Res*, **7** (5), 573–579.

209 Brown, J.P. and Josse, R.G. (2002) 2002 clinical practice guidelines for the diagnosis and management of osteoporosis in Canada. *Can Med Assoc J*, **167** (Suppl 10), S1–34.

210 Small, R.E. (2005) Uses and limitations of bone mineral density measurements in the management of osteoporosis. *MedGenMed*, **7** (2), 3.

211 Briot, K. and Roux, C. (2005) What is the role of DXA, QUS and bone markers in fracture prediction, treatment allocation and monitoring? *Best Pract Res Clin Rheumatol*, **19** (6), 951–964.

212 Herrmann, M., Kraenzlin, M., Pape, G., Sand-Hill, M., and Herrmann, W. (2005) Relation between homocysteine and biochemical bone turnover markers and bone mineral density in peri-and post-menopausal women. *Clin Chem Lab Med*, **43** (10), 1118–1123.

213 Cagnacci, A., Baldassari, F., Rivolta, G., Arangino, S., and Volpe, A. (2003) Relation of homocysteine, folate, and vitamin B12 to bone mineral density of postmenopausal women. *Bone*, **33** (6), 956–959.

214 Golbahar, J., Hamidi, A., Aminzadeh, M.A., and Omrani, G.R. (2004) Association of plasma folate, plasma total homocysteine, but not methylenetetrahydrofolate reductase C667T polymorphism, with bone mineral density in postmenopausal Iranian women: a cross-sectional study. *Bone*, **35** (3), 760–765.

215 Dhonukshe-Rutten, R.A., Pluijm, S.M., de Groot, L.C., Lips, P., Smit, J.H., and van Staveren, W.A. (2005) Homocysteine and vitamin B12 status relate to bone turnover markers, broadband ultrasound attenuation, and fractures in healthy elderly people. *J Bone Miner Res*, **20** (6), 921–929.

216 Bode, M.K., Laitinen, P., Risteli, J., Uusimaa, P., and Juvonen, T. (2000) Atherosclerosis, type 1 collagen cross-linking and homocysteine. *Atherosclerosis*, **152** (2), 531–532.

217 Gerdhem, P., Ivaska, K.K., Isaksson, A., Pettersson, K., Vaananen, H.K., Obrant, K.J., and Akesson, K. (2007) Associations between homocysteine, bone turnover, BMD, mortality, and

fracture risk in elderly women. *J Bone Miner Res*, **22** (1), 127–134.

218 Nilsson, K., Gustafson, L., Isaksson, A., and Hultberg, B. (2005) Plasma homocysteine and markers of bone metabolism in psychogeriatric patients. *Scand J Clin Lab Invest*, **65** (8), 671–680.

219 Abrahamsen, B., Jorgensen, H.L., Nielsen, T.L., Andersen, M., Haug, E., Schwarz, P., Hagen, C., and Brixen, K. (2006) MTHFR c.677C>T polymorphism as an independent predictor of peak bone mass in Danish men--results from the Odense Androgen Study. *Bone*, **38** (2), 215–219.

220 Perier, M.A., Gineyts, E., Munoz, F., Sornay-Rendu, E., and Delmas, P.D. (2007) Homocysteine and fracture risk in postmenopausal women: the OFELY study. *Osteoporos Int*, **18** (10), 1329–1336.

221 Kuriyama, M., Ueno, K., Uno, H., Kawada, Y., Akimoto, S., Noda, M., Nasu, Y., Tsushima, T., Ohmori, H., Sakai, H., Saito, Y., Meguro, N., Usami, M., Kotake, T., Suzuki, Y., Arai, Y., and Shimazaki, J. (1998) Clinical evaluation of serum prostate-specific antigen-alpha1-antichymotrypsin complex values in diagnosis of prostate cancer: a cooperative study. *Int J Urol*, **5** (1), 48–54.

222 Carmel, R., Lau, K.H., Baylink, D.J., Saxena, S., and Singer, F.R. (1988) Cobalamin and osteoblast-specific proteins. *N Engl J Med*, **319** (2), 70–75.

223 Kim, G.S., Kim, C.H., Park, J.Y., Lee, K.U., and Park, C.S. (1996) Effects of vitamin B12 on cell proliferation and cellular alkaline phosphatase activity in human bone marrow stromal osteoprogenitor cells and UMR106 osteoblastic cells. *Metabolism*, **45** (12), 1443–1446.

224 Herrmann, M., Widmann, T., Colaianni, G., Colucci, S., Zallone, A., and Herrmann, W. (2005) Increased osteoclast activity in the presence of increased homocysteine concentrations. *Clin Chem*, **51** (12), 2348–2353.

225 Koh, J.M., Lee, Y.S., Kim, Y.S., Kim, D.J., Kim, H.H., Park, J.Y., Lee, K.U., and Kim, G.S. (2006) Homocysteine enhances bone resorption by stimulation of osteoclast formation and activity through increased intracellular ROS generation. *J Bone Miner Res*, **21** (7), 1003–1011.

226 Herrmann, M., Schmidt, J., Umanskaya, N., Colaianni, G., Al-Marrawi, F., Widmann, T., Zallone, A., Wildemann, B., and Herrmann, W. (2007) Stimulation of osteoclast activity by low B-vitamin concentrations. *Bone*, **41**, 584–591.

227 Green, T.J., McMahon, J.A., Skeaff, C.M., Williams, S.M., and Whiting, S.J. (2007) Lowering homocysteine with B vitamins has no effect on biomarkers of bone turnover in older persons: a 2-y randomized controlled trial. *Am J Clin Nutr*, **85** (2), 460–464.

228 Herrmann, M., Stanger, O., Paulweber, B., Hufnagl, C., and Herrmann, W. (2006) Folate supplementation does not affect biochemical markers of bone turnover. *Clin Lab*, **52** (3-4), 131–136.

229 Herrmann, M., Umanskaya, N., Traber, L., Schmidt-Gayk, H., Menke, W., Lanzer, G., Lenhart, M., Schmidt, J., and Herrmann, W. (2007) The effect of B-vitamins on biochemical bone turnover markers and bone mineral density in osteoporotic patients – an one year double blind placebo controlled trial. *Clin Chem Lab Med*, **45** (12), 1785–1792.

230 Sato, Y., Honda, Y., Iwamoto, J., Kanoko, T., and Satoh, K. (2005) Effect of folate and Mecobalamin on hip fractures in patients with stroke: a randomized controlled trial. *J Am Med Assoc*, **293** (9), 1082–1088.

231 Baigent, C., Burbury, K., and Wheeler, D. (2000) Premature cardiovascular disease in chronic renal failure. *Lancet*, **356** (9224), 147–152.

232 Herrmann, W., Schorr, H., Geisel, J., and Riegel, W. (2001) Homocysteine,

cystathionine, methylmalonic acid and B-vitamins in patients with renal disease. *Clin Chem Lab Med*, **39** (8), 739–746.

233 Moustapha, A., Naso, A., Nahlawi, M., Gupta, A., Arheart, K.L., Jacobsen, D.W., Robinson, K., and Dennis, V.W. (1998) Prospective study of hyperhomocysteinemia as an adverse cardiovascular risk factor in end-stage renal disease. *Circulation*, **97** (2), 138–141 [published erratum appears in *Circulation* (1998) Feb 24; **97** (7), 711].

234 Robinson, K., Gupta, A., Dennis, V., Arheart, K., Chaudhary, D., Green, R., Vigo, P., Mayer, E.L., Selhub, J., Kutner, M., and Jacobsen, D.W. (1996) Hyperhomocysteinemia confers an independent increased risk of atherosclerosis in end-stage renal disease and is closely linked to plasma folate and pyridoxine concentrations. *Circulation*, **94** (11), 2743–2748.

235 Mallamaci, F., Zoccali, C., Tripepi, G., Fermo, I., Benedetto, F.A., Cataliotti, A., Bellanuova, I., Malatino, L.S., and Soldarini, A. (2002) Hyperhomocysteinemia predicts cardiovascular outcomes in hemodialysis patients. *Kidney Int*, **61** (2), 609–614.

236 Suliman, M.E., Qureshi, A.R., Barany, P., Stenvinkel, P., Filho, J.C., Anderstam, B., Heimburger, O., Lindholm, B., and Bergstrom, J. (2000) Hyperhomocysteinemia, nutritional status, and cardiovascular disease in hemodialysis patients. *Kidney Int*, **57** (4), 1727–1735.

237 Suliman, M.E., Stenvinkel, P., Qureshi, A.R., Barany, P., Heimburger, O., Anderstam, B., Alvestrand, A., and Lindholm, B. (2004) Hyperhomocysteinemia in relation to plasma free amino acids, biomarkers of inflammation and mortality in patients with chronic kidney disease starting dialysis therapy. *Am J Kidney Dis*, **44** (3), 455–465.

238 Kalantar-Zadeh, K., Ikizler, T.A., Block, G., Avram, M.M., and Kopple, J.D. (2003) Malnutrition-inflammation complex syndrome in dialysis patients: causes and consequences. *Am J Kidney Dis*, **42** (5), 864–881.

239 Kalantar-Zadeh, K., Block, G., Humphreys, M.H., McAllister, C.J., and Kopple, J.D. (2004) A low, rather than a high, total plasma homocysteine is an indicator of poor outcome in hemodialysis patients. *J Am Soc Nephrol*, **15** (2), 442–453.

240 Herrmann, W., Schorr, H., Obeid, R., Makowski, J., Fowler, B., and Kuhlmann, M.K. (2005) Disturbed homocysteine and methionine cycle intermediates s-adenosylhomocysteine and s-adenosylmethionine are related to degree of renal insufficiency in type 2 diabetes. *Clin Chem*, **51** (5), 891–897.

241 Wollesen, F., Brattstrom, L., Refsum, H., Ueland, P.M., Berglund, L., and Berne, C. (1999) Plasma total homocysteine and cysteine in relation to glomerular filtration rate in diabetes mellitus 2. *Kidney Int*, **55** (3), 1028–1035.

242 Stabler, S.P., Marcell, P.D., Podell, E.R., and Allen, R.H. (1987) Quantitation of total homocysteine, total cysteine, and methionine in normal serum and urine using capillary gas chromatography-mass spectrometry. *Anal Biochem*, **162** (1), 185–196.

243 Yeun, J.Y. (1998) The role of homocysteine in end stage renal disease. *Semin Dial*, **11**, 95–101.

244 Bostom, A.G., Shemin, D., Gohh, R.Y., Beaulieu, A.J., Jacques, P.F., Dworkin, L., and Selhub, J. (2000) Treatment of mild hyperhomocysteinemia in renal transplant recipients versus hemodialysis patients. *Transplantation*, **69** (10), 2128–2131.

245 Janssen, M.J., van den Berg, M., Stehouwer, C.D., and Boers, G.H. (1995) Hyperhomocysteinemia: a role in the accelerated atherogenesis of chronic renal failure? *Neth J Med*, **46** (5), 244–251.

246 van Guldener, C., Janssen, M.J., Lambert, J., ter Wee, P.M., Donker,

A.J., and Stehouwer, C.D. (1998) Folic acid treatment of hyperhomocysteinemia in peritoneal dialysis patients: no change in endothelial function after long-term therapy. *Perit Dial Int*, **18** (3), 282–289.
247 Guttormsen, A.B., Ueland, P.M., Svarstad, E., and Refsum, H. (1997) Kinetic basis of hyperhomocysteinemia in patients with chronic renal failure. *Kidney Int*, **52** (2), 495–502.
248 Garibotto, G., Saffioti, S., Russo, R., Verzola, D., Cappelli, V., Aloisi, F., and Sofia, A. (2003) Malnutrition in peritoneal dialysis patients: causes and diagnosis. *Contrib Nephrol*, (140), 112–121.
249 Kumagai, H., Katoh, S., Hirosawa, K., Kimura, M., Hishida, A., and Ikegaya, N. (2002) Renal tubulointerstitial injury in weanling rats with hyperhomocysteinemia. *Kidney Int*, **62** (4), 1219–1228.
250 Chen, Y.F., Li, P.L., and Zou, A.P. (2002) Effect of hyperhomocysteinemia on plasma or tissue adenosine levels and renal function. *Circulation*, **106** (10), 1275–1281.
251 Vermeulen, E.G., Rauwerda, J.A., van den Berg, M., de Jong, S.C., Schalkwijk, C., Twisk, J.W., and Stehouwer, C.D. (2003) Homocysteine-lowering treatment with folic acid plus vitamin B6 lowers urinary albumin excretion but not plasma markers of endothelial function or C-reactive protein: further analysis of secondary end-points of a randomized clinical trial. *Eur J Clin Invest*, **33** (3), 209–215.
252 Arnadottir, M., Hultberg, B., Nilsson, E.P., and Thysell, H. (1996) The effect of reduced glomerular filtration rate on plasma total homocysteine concentration. *Scand J Clin Lab Invest*, **56** (1), 41–46.
253 Guttormsen, A.B., Schneede, J., Ueland, P.M., and Refsum, H. (1996) Kinetics of total plasma homocysteine in subjects with hyperhomocysteinemia due to folate or cobalamin deficiency. *Am J Clin Nutr*, **63** (2), 194–202.
254 van Guldener, C., Kulik, W., Berger, R., Dijkstra, D.A., Jakobs, C., Reijngoud, D.J., Donker, A.J., Stehouwer, C.D., and De Meer, K. (1999) Homocysteine and methionine metabolism in ESRD: A stable isotope study. *Kidney Int*, **56** (3), 1064–1071.
255 Obeid, R., Kuhlmann, M.K., Kohler, H., and Herrmann, W. (2005) Response of homocysteine, cystathionine, and methylmalonic acid to vitamin treatment in dialysis patients. *Clin Chem*, **51** (1), 196–201.
256 Henning, B.F., Zidek, W., Riezler, R., Graefe, U., and Tepel, M. (2001) Homocyst(e)ine metabolism in hemodialysis patients treated with vitamins B6, B12 and folate 1. *Res Exp Med (Berl)*, **200** (3), 155–168.
257 Seetharam, B. and Li, N. (2000) Transcobalamin II and its cell surface receptor. *Vitam Horm*, **59**, 337–366.
258 Snow, C.F. (1999) Laboratory diagnosis of vitamin B12 and folate deficiency: a guide for the primary care physician [see comments]. *Arch Intern Med*, **159** (12), 1289–1298.
259 Obeid, R., Kuhlmann, M., Kirsch, C.M., and Herrmann, W. (2003) Evidence that cellular uptake of vitamin B12 is impaired in patients with chronic renal failure. *Nephron Clin Pract*, **99** (2), c42–c48.
260 Righetti, M., Ferrario, G.M., Milani, S., Serbelloni, P., La Rosa, L., Uccellini, M., and Sessa, A. (2003) Effects of folic acid treatment on homocysteine levels and vascular disease in hemodialysis patients. *Med Sci Monit*, **9** (4), I19–I24.
261 Sunder-Plassmann, G., Fodinger, M., Buchmayer, H., Papagiannopoulos, M., Wojcik, J., Kletzmayr, J., Enzenberger, B., Janata, O., Winkelmayer, W.C., Paul, G., Auinger, M., Barnas, U., and Horl, W.H. (2000) Effect of high dose folic acid therapy on hyperhomocysteinemia in hemodialysis patients: results of the Vienna multicenter study. *J Am Soc Nephrol*, **11** (6), 1106–1116.

262 Baragetti, I., Furiani, S., Dorighet, V., Corghi, E., Bamonti, C.F., and Buccianti, G. (2004) Effect of vitamin B12 on homocysteine plasma concentration in hemodialysis patients. *Clin Nephrol*, **61** (2), 161–162.

263 Arnadottir, M. and Hultberg, B. (2003) The effect of vitamin B12 on total plasma homocysteine concentration in folate-replete hemodialysis patients. *Clin Nephrol*, **59** (3), 186–189.

264 Schatz, R.A., Wilens, T.E., and Sellinger, O.Z. (1981) Decreased transmethylation of biogenic amines after in vivo elevation of brain S-adenosyl-l-homocysteine. *J Neurochem*, **36** (5), 1739–1748.

265 Kerins, D.M., Koury, M.J., Capdevila, A., Rana, S., and Wagner, C. (2001) Plasma S-adenosylhomocysteine is a more sensitive indicator of cardiovascular disease than plasma homocysteine. *Am J Clin Nutr*, **74** (6), 723–729.

34
Prevention of Oxidative Stress–Induced Diseases by Natural Dietary Compounds: The Mechanism of Actions

Tin Oo Khor, Ka-Lung Cheung, Avantika Barve, Harold L. Newmark, and Ah-Ng Tony Kong

Excessive and prolonged oxidative stresses induced by endogenous and exogenous toxins have been implicated in many diseases such as cancers and neurodegenerative disorders. Natural dietary compounds are believed to be able to prevent and treat these oxidative stress–induced diseases by protecting the cells or tissues from the deleterious effects of these insults. The cytoprotective effects can be achieved through direct scavenging of reactive nitrogen/oxygen species (RNOS) or indirectly by the induction of phase II detoxifying/antioxidative enzymes through the leucine zipper transcription factor nuclear factor erythroid 2-related factor 1 (Nrf2). This chapter focuses on the mechanism of actions by which some of these natural dietary compounds exert their cytoprotective effects.

34.1
Introduction

Mammalian cells are constantly undergoing respiration to acquire the energy they need. The by-products generated, namely, reactive oxygen species (ROS), free radicals, and reactive nitrogen species (RNS) are known to have adverse effects on cells and are therefore considered as endogenous toxins.

One of the sources of endogenous toxin is ROS. A number of enzymes and biochemical reactions are identified to be responsible for the production of ROS in mammalian cells. These include reactions of mitochondrial respiratory chain (oxidative phosphorylation), metabolism of arachidonic acid, enzymes of the cytochrome 450-system, xanthine oxidoreductase (XO), and peroxidases [1]. The sources of ROS in mammalian cells depend on the tissue and environmental factors. For example, the major source of ROS in endothelial cells is nicotinamide adenine dinucleotide/NADPH oxidase (NADPH) system, which is activated by growth factors, cytokines, shear stress and hypoxia, and so on [2, 3]. NADPH oxidase catalyzes the transfer of an electron to oxygen in the reaction NAD(P)H

Endogenous Toxins. Diet, Genetics, Disease and Treatment.
Edited by Peter J. O'Brien and W. Robert Bruce
Copyright © 2010 WILEY-VCH Verlag GmbH & Co. KGaA, Weinheim
ISBN: 978-3-527-32363-0

+ 2 O_2 → $NAD(P)^+$ + H^+ + 2 O_2^-. However, XO catalyzes the oxidation of hypoxanthine to xanthine and xanthine to uric acid and converts molecular oxygen to O_2^- and H_2O_2. Indeed, the main product of the xanthine oxidase reaction is H_2O_2 rather than O_2^-, which can be utilized to defend against infectious pathogens [4].

Another important source of ROS is the metabolism of arachidonic acid (AA) catalyzed by cyclooxygenase (COX) and lipoxygenase (LOX) [5]. LOX are nonheme-containing dioxygenases that oxidize polyunsaturated fatty acids (AA) to hydroxyl fatty acid derivatives, during which O_2^- is produced as by-product. In addition to the oxidative metabolic processes of AA by COX and LOX, AA itself has also been reported to activate NADPH oxidase directly, thereby inducing ROS generation. Mitochondria are another source of ROS. Oxidative phosphorylation takes place by which electrons are transferred from NADH or FADH2 to molecular oxygen. The electrons are transferred via a series of chemical reactions (complex I, II, III, and IV) and more than 98% of electrons in this transport chain is coupled with ATP production. Less than 2% of electrons leaks out for the formation of O_2^- [1]. Excessive ROS attack DNA readily, generating a variety of DNA lesions such as oxidized bases and strand breaks, which are potentially devastating to normal cell physiology, leading to mutagenesis or cell death.

Not only ROS but RNS are also important sources of endogenous toxins. RNS include various nitric oxide–derived compounds (nitric oxide (NO), nitroxyl (HNO), nitrosonium cation (NO^+)) and higher oxides of nitrogen (S-nitrosothiols (RSNOs) and $ONOO^-$) [6]. Nitric oxide (NO) is a ubiquitous intracellular messenger regulating various physiological processes and is especially important in cardiovascular functions. Although NO at low concentration is nontoxic, it is considered to be deleterious under pathological conditions because of its high reactivity. NO and O_2^- react by an enzyme-independent reaction to form peroxynitrite ($ONOO^-$), which can further decompose to yield NO_2^- and NO_3^-. RNS play an important role in the physiologic regulation of many living cells, such as smooth muscle cells, cardiomyocytes, platelets, nerve cells, and juxtaglomerular cells. They possess pleiotropic properties on cellular targets after both posttranslational modifications and interactions with ROS [6]. Elevated levels of RNS have been implicated in cellular injury and death by inducing nitrosative stress and have been linked to Alzheimer's disease [7].

Naturally occurring compounds with potent cytoprotective effects against oxidative stress have been noted as a plausible approach for clinical interventions of diseases such as cancers and neurodegenerative illnesses [8]. Some of these agents are believed to protect the cells or tissues from the malicious attack of exogenous carcinogens and/or endogenous reactive oxygen/nitrogen species (RONS) by direct scavenging of these harmful RONS. In addition, many natural dietary compounds were found to be able to protect the cell or tissue from oxidative damages by induction of several detoxifying/antioxidant enzymes through the leucine zipper-type transcription factor nuclear factor erythroid 2-related factor 1 (Nrf2) [9].

34.2
Cytoprotective Effect of Dietary Compounds: Antioxidant Effects

Protection of cellular elements against excessive and prolonged oxidative stress induced by exogenous as well as endogenous toxins is a pivotal step in the prevention of cancer, neurological disorders, and other diseases. Accumulating evidence suggests that many phytochemicals can effectively prevent the deleterious effect of oxidative stress and confer cytoprotection through the induction of a coordinated battery of enzymes that are intrinsic to the cell's ROS defenses [10, 11]. It is believed that the cytoprotective effect of these phytochemicals are elicited through the activation of antioxidant/detoxification signaling pathway regulated by Nrf2. Nrf2 plays an essential role in the regulation of many genes encoding proteins such as phase II detoxifying and antioxidant enzymes. Phase II enzymes have been shown to be able to directly scavenge ROS and/or metabolize potentially harmful electrophiles to maintain the homeostatic redox tone of normal cells [10, 11]. Under homeostatic conditions, Nrf2 is mainly sequestered in the cytoplasm by a cytoskeleton-binding protein called Keap1. Upon oxidative or electrophilic insults, Nrf2 will translocate into the nucleus where it binds with the antioxidant response elements (AREs) located in the 5'-flanking region of many genes, including phase II/antioxidant and others [12], and transcriptionally activates these genes. Among the antioxidant genes that are regulated by Nrf2 are NAD(P)H:quinone oxidoreductase (NQO1), heme oxygenase-1 (HO-1), thioredoxin reductase 1, glutamate-cysteine ligase (GCL) modifier subunit, and GCL catalytic subunit [9]. The modulation of phase II detoxifying/antioxidant enzymes by dietary phytochemicals can be achieved through different collaborative signaling pathways depending on the cellular and tissue context as depicted in Figure 34.1. As demonstrated in Figure 34.1, dietary phytochemicals can act by directly disrupting the association between Keap-1 and Nrf2, inhibiting proteosomal degradation of Nrf2, and increasing transcription and/or translation of Nrf2, ultimately leading to increased nuclear translocation of Nrf2, which results in enhanced gene expression of the detoxifying/antioxidant enzymes [13]. Additionally, several phytochemicals have been shown to activate upstream signaling pathways such as MAPK, PI3K, PKC, and PERK leading to increased nuclear translocation of Nrf2 and enhanced gene transactivation activity [14–16]. Phenolic and sulfur-containing compounds are two major classes of dietary phytochemicals functioning as inducers of detoxifying enzymes. The mechanism of action of few of these phytochemicals is discussed below.

34.2.1
Phenolic Dietary Compounds

Phenolic compounds are widely distributed in the plant kingdom. Polyphenols and flavanoids are two most studied phenolic dietary compounds. Examples of polyphenols include epigallocatechin-3-gallate (EGCG) from green tea, curcumin from turmeric, and resveratrol from grapes, whereas flavonoids are exemplified by genistein from soy.

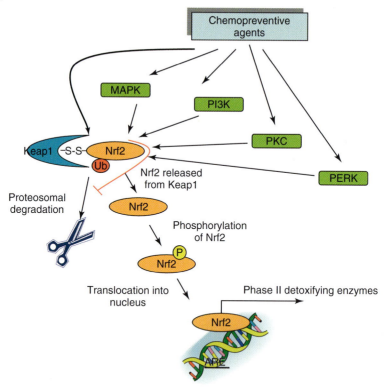

Fig. 34.1 The modulation of phase II detoxifying/antioxidant enzymes by dietary phytochemicals through different collaborative signaling pathways.

34.2.1.1 EGCG

Consumption of tea has been reported to prevent cancer, heart disease, and many other diseases [17]. Fresh tea leaves (*Camellia sinensis*) contain a high amount of catechins, a group of flavonoids, which is known to constitute 30–45% of the solid green tea extract [18]. Among the tea catechins, EGCG is the major constituent, accounting for more than 10% of the extract dry weight, followed by (−)-epigallocatechin (EGC) > (−)-epicatechin (EC) ≥ (−)-epicatechin-3-gallate (ECG). As most active catechins, EGCG possess strong antioxidant activity. EGCG has been reported to inhibit UVB-induced hydrogen peroxide (H_2O_2) production and H_2O_2-mediated phosphorylation of MAPK signaling pathways in normal human epidermal keratinocytes (NHEKs). Treatment of EGCG (20 μg ml^{-1} of media) before UVB (30 mJ cm^{-2}) exposure inhibited UVB-induced H_2O_2 production concomitant with the inhibition of UVB-induced phosphorylation of extracellular signal-regulated protein kinase (ERK)1/2, JNK, and p38 proteins [19]. In another study, Erba *et al.* demonstrated that supplementation of the Jurkat T cell line with green tea extract/EGCG decreases oxidative damage due to iron treatment through inhibition of lipid peroxidation [20]. EGCG and other catechins protect erythrocytes against oxidative stress induced by *tert*-butyl hydroperoxide (*t*-BHP) [21]. In addition, EGCG has been shown to protect against oxidative stress–induced mitochondria-dependent apoptosis in human lens epithelial cells. EGCG blocks the accumulation of intracellular ROS and the loss of mitochondrial membrane potential induced by H_2O_2 through the modulation of caspases, the Bcl-2 family, and the MAPK and Akt pathways in human lens epithelial cells [22]. On the other hand, EGCG prevents molecular degradation in oxidative stress conditions by directly altering the subcellular ROS production, glutathione (GSH) metabolism, and cytochrome P450 2E1 activity [23]. Green tea extract and EGCG have also been reported to exert potent neuroprotection in both *in vivo* and *in vitro* models of neurodegeneration. For example, EGCG is reported to protect neuronal cells against lead-induced oxidative damage. Using an *in vivo* rodent model, Yin *et al.* showed that EGCG supplementation following lead intoxication resulted in increases in the GSH and superoxide dismutase (SOD) levels and increases in the long-term potentiation (LTP) amplitude. Malondialdehyde (MDA) level, a major lipid peroxidation by-product, which increased following exposure to lead, was also significantly suppressed by EGCG treatment. In hippocampal neuron culture model, exposure to lead (20 μM) significantly inhibited the viability of neurons, which was followed by an accumulation of ROS and a decrease of mitochondrial membrane potential ($\Delta\psi m$). Treatment by EGCG (10–50 μM) effectively increased cell viability, decreased ROS formation, and improved "$\Delta\psi m$" in hippocampal neurons exposed to lead [24]. Likewise, Levites *et al.* found that EGCG can prevent striatal dopamine depletion and substantia nigra dopamine-containing neuronal loss when given chronically to mice treated with the parkinsonism-inducing neurotoxin, *N*-methyl-4-phenyl-1,2,3,6-tetrahydropyridine (MPTP) [25].

Although the precise mechanisms by which tea extracts or EGCG exert their cytoprotective/neuroprotective effects are still unknown, it is believed that a wide spectrum of cellular signaling events may be accounting for their biological actions. In fact, direct scavenging of ROS induced by endogenous toxins or neurotoxins is proposed to be one of the major mechanisms underlying the cytoprotection by EGCG. In addition, EGCG is a potent inducer of phase II detoxifying enzymes through activation of Nrf2 signaling pathway. *In vitro* and *in vivo* studies demonstrated that EGCG can induce a distinct set of antioxidant enzymes in different organs or cultured cells. For example, EGCG induces expression of glutathione peroxidase (GPx), catalase (CAT), and quinone reductase in small bowel, liver, and lung of SKH-1 hairless mice [26]. EGCG is also reported to induce Nrf2-mediated detoxifying enzymes such as GCL, manganese superoxide dismutase (MnSOD), and HO-1 [27]. In our laboratory, we have shown that EGCG induces the expression of catalytic subunit of GCL, γ-glutamyltransferase 1 (GGT1), and HO-1 in an Nrf2-dependent manner [28], and treatment of human hepatoma (HepG2) cells with the green tea extract induced expression of phase II detoxifying enzymes through ARE activation [29].

It has been proposed that EGCG activates Nrf2 signaling pathway through (i) activation of upstream MAPK cascades and (ii) direct interaction with cysteine residue of Keap1, leading to the release and nuclear translocation of Nrf2 [30].

It has been shown that EGCG can be autoxidized at the polyphenolic groups to form biologically active dimer with concomitant generation of ROS [31, 32]. The generation of ROS and reactive form of EGCG will facilitate the nuclear translocation of Nrf2 by disrupting the Nrf2–Keap1 complex. In a recent microarray study [28], we have reported that CREB-binding protein (CBP) and other coactivators were upregulated in mice liver treated with EGCG in an Nrf2-dependent manner. Katoh et al. [33] showed that two domains of Nrf2 (Neh4 and Neh5) cooperatively bind CBP and synergistically activate transcription. In addition, we have also demonstrated [34] previously, using a Gal4-Luc reporter cotransfection assay system in HepG2 cells, that the nuclear transcriptional coactivator CBP that can bind to Nrf2 transactivation domain and can be activated by ERK cascade showed synergistic stimulation with Raf on the transactivation activities of both the chimera Gal4-Nrf2 (1–370) and the full-length Nrf2. These findings suggest that CBP and other nuclear coactivators may serve as putative EGCG-regulated nuclear interacting partners of Nrf2 in eliciting the cytoprotection effects of EGCG.

34.2.1.2 Curcumin

Curcumin (diferuloylmethane) is a polyphenol derived from the plant *Curcuma longa*, commonly called *turmeric*. Curcumin is a well known cancer chemopreventive agent with strong anti-inflammatory and antioxidant activities [35]. Long-term dietary curcumin retarded adenoma development in $APC^{min/+}$ mice by suppressing the expression of Cox-2 protein and decreased levels of two oxidative DNA adducts, the pyrimidopurinone adduct of deoxyguanosine (M1dG) and 8-oxo-7,8-dihydro-2'-deoxyguanosine (8-oxo-dG) [36]. Curcumin was also reported to inhibit AOM-induced rat colon carcinogenesis by suppression of prostaglandin (PG) and thromboxane (Tx) formation [37]. In addition, curcumin has been shown to exhibit hepatoprotective and nephroprotective activities, suppress thrombosis, protect against myocardial infarction, and to have hypoglycemic and antirheumatic properties [38]. The antioxidant activity of curcumin has been reported almost 30 years ago by Sharma [39]. Several recent studies demonstrated that curcumin is an effective scavenger of ROS [40, 41]. The Michael acceptor functionalities and phenolic hydroxyl groups of curcumin can scavenge oxygen- and nitrogen-centered reactive intermediates directly and potently. Therefore, exposure of micro range of curcumin has been shown to scavenge ROS and RNS *in vitro* and *in vivo* [42]. Additionally, curcumin is found to be able to prevent lipid peroxidation and protein oxidation caused by 2,2-azobis(2-amindinopropane) hydrochloride or Fe/ascorbate in isolated, deenergized mitochondria [43].

Besides its direct scavenging effect, curcumin can also elicit its indirect antioxidant effect through induction of phase II detoxifying enzymes. For example, treatment of astrocytes with curcumin protected glucose oxidase–mediated oxidative stress, probably through induction of HO-1, NQO-1, and glutathione-*S*-transferase (GST) [44]. We have previously shown that curcumin can modulate the expression of some genes encoding phase II detoxifying enzymes in an Nrf2-dependent manner in the liver and small intestine of mouse [45] as well as in the prostate of TRAMP mice [46]. Thus, curcumin could act as a bifunctional antioxidant, protecting normal cells through direct oxidants scavenging activity or indirectly through induction of phase II detoxifying/antioxidant enzymes.

34.2.1.3 Dibenzoylmethane

Dibenzoylmethane (DBM), a minor constituent of licorice and a β-ketone analog of curcumin, has been shown to be a promising anticancer, antimutagenic, and cancer chemopreventive compound. We have recently shown that DBM when given alone or in combination with sulforaphane (SFN) significantly inhibited the development of familial adenomatous polyposis in $Apc^{-/+}$ mouse [47]. Likewise, DBM has been reported to inhibit 7,12-dimethylbenz[a]anthracene (DMBA)-induced mammary tumors and lymphomas/leukemias in Sencar mice [48]. DBM is also a very strong antimutagenic agent that could effectively inhibit mutagenicity induced by several heterocyclic amines [49]. Recent study indicates that DBM can mediate the induction of phase II enzymes by Nrf2 activation and inhibit benzo[a]pyrene-induced DNA adducts by enhancing its detoxification in lungs. In addition, Takano et al. demonstrated that 14–26(2,2'-dimethoxydibenzoylmethane), a DBM derivative, can protect dopaminergic neurons against both oxidative stress and endoplasmic reticulum stress [50].

34.2.1.4 Resveratrol

Resveratrol (3,4',5-trihydroxystilbene) is a phytoalexin found in grapes, plums, and red wines. The interest for research of resveratrol was initiated by the French paradox, a paradoxical observation that a low incidence of cardiovascular diseases may coexist with a high-fat diet and moderate consumption of red wine. Resveratrol exists as cis- and trans isomers, and the trans form is relatively stable if it is protected from high pH and light. The mechanism by which resveratrol exerts its cardiovascular protection include inhibition of platelet aggregation, arterial vasodilation mediated by NO, LDL (low-density lipoprotein)–cholesterol oxidation, antioxidant effects, induction of cardioprotective protein, and insulin sensitization [51]. The antioxidant effects of resveratrol have been widely investigated and have been implicated in preventing inflammation, prolonging life span, preventing oxidative DNA damage and cancer. Different mechanisms have been proposed to explain resveratrol's antioxidant potential. These include inhibition of ROS production, direct scavenging of free radicals, and/or induction of antioxidant genes, which in turn produce more endogenous antioxidants. It is suggested that resveratrol competes with coenzyme Q to decrease the oxidative chain complex III activity, the site of ROS generation in mitochondria [52]. On the other hand, resveratrol inhibits lipid peroxidation induced by Fenton reaction products, which again decreases the production of free radicals. Using an electron paramagnetic resonance spectroscopy in combination with 5-(diethoxyphosphoryl)-5-methylpyrroline-N-oxide (DEPMPO)-spin trapping technique, resveratrol was found to directly scavenge superoxide anions generated from both potassium superoxide and the xanthine oxidase/xanthine system [53]. In addition, resveratrol can also activate Nrf2–EpRE pathway, leading to the induction of various antioxidant genes, including HO-1, GCL, and mitochondrial superoxide dismutase (MnSOD) [54, 55]. Resveratrol has been shown to be effective in elevating antioxidant capacity in brain as a result of an increase in both MnSOD protein level and activity [56]. Apart from that, it was also shown that resveratrol pretreatment effectively protected hepatocytes in culture exposed to oxidative stress by increasing the activities of CAT, SOD, GPx, NQO, and GST, all of which are common Nrf2-regulating enzymes protecting cells from oxidative stress [57].

34.2.1.5 Genistein

Genistein, also known as *phytoestrogen*, is a naturally occurring isoflavone that has been identified predominantly in soybean. Genistein has been widely reported to be responsible for the beneficial effects of soybean foods against cancer, cardiovascular disease, and osteoporosis [58]. Antioxidant activity and induction of detoxification enzymes have been proposed to be among the major molecular mechanisms by which genistein exert its cytoprotective effect. For example, genistein has been shown to protect cells against ROS by scavenging free radicals and reducing the expression of stress-response-related genes [59, 60]. Recently, Siow *et al.* proposed that isoflavones may protect against cardiovascular disease by virtue of their ability to activate intracellular signaling pathways, leading to increased NO bioavailability and an upregulation of antioxidant gene expression via the key transcription factors NFkappaB and Nrf2 [61]. In addition, genistein and daidizein have been shown to offer protection against oxidative stress–induced endothelial injury by evoking increased expression of γ-glutamyl cysteine synthetase and nuclear expression of Nrf2 [62]. In our laboratory, we have investigated the pharmacogenomics of soy isoflavones using animal model. By comparing the gene expression profiles elicited by soy isoflavones in Nrf2-deficient- and wild-type mice, we have identified numerous genes performing a variety of biological and physiological functions and exhibit Nrf2 dependency [63].

34.2.2
Isothiocyanates

Consumption of vegetables of *Cruciferae* family (known as *cruciferous vegetables*) has been shown to be highly protective against the risks of different types of cancer. Cruciferous vegetables are a group of vegetables named by their cross-shaped flowers, which include broccoli, Brussel sprouts, watercress, cabbage, kale, cauliflower, kohlrabi, and turnip. Compared with other family of vegetables, unique chemical property of these vegetables is that they contain a large amount of glucosinolates (known by the trivial names sinigrin and sinalbin). Owing to this fact, the chemopreventive effects of cruciferous vegetables were ascribed to the pharmacological activities of the glucosinolates. Supporting this notion, numerous studies have consistently demonstrated that animals fed with glucosinolate products together with a carcinogen developed fewer tumors in comparison with control animals not receiving the glucosinolate products [64]. In addition, analysis of epidemiological evidence of many cohort studies and case-controlled studies has clearly illustrated that consumption of cruciferous vegetables is inversely related to the cancer risks in humans [65].

Glucosinolates in cruciferous vegetables exist as inactive N-hydroxysulfate with the sulfur linked to the β-glucose and variable side chains. During the process of food preparation, glucosinolates will be degraded by myrosinase by hydrolysis. Myrosinase hydrolyzes all the carbon groups in the glucosinolate, which results in the production of a number of biologically active chemopreventive isothiocyanates and indole 3-carbinols (I3Cs). Characterized by the chemical structure of $-N=C=S$ group, the central carbon of ITCs' moiety is highly electrophilic and can easily undergo conjugation reactions with intracellular reduced glutathiones (GSH) and/or thiol groups of proteins. These reactions reduce GSH levels within the cells leading to the direct interaction with the thiol groups of some signaling proteins, thereby triggering the intracellular oxidative stress and subsequent activation of various cellular signal transduction pathways. The ITC-GSH conjugates such as dithiocarbamates (DTCs) formed are then serially metabolized into cysteinylglycine, cysteine, and N-acetylcysteine (NAC) through enzymatic modifications and excreted in the urine through a metabolic pathway called the *mercapturic acid pathway* [66]. The cancer chemopreventive effect of ITCs and their metabolic products has been linked to their strong phase II detoxifying enzymes inducer property.

To date, more than 100 structurally distinct glucosinolates have been identified predominantly, but not exclusively, from a variety of cruciferous vegetables [67]. Among them are ally isothiocyanate (AITC) from cabbage, mustard, and horseradish, benzyl isothiocyanate (BITC), phenethyl isothiocyanate (PEITC) from watercress, SFN from broccoli, and erucin from daikon.

SFN is the predominant ITC found in broccoli, and it has been found to inhibit carcinogen-induced mammary gland tumorigenesis [68], colonic aberrant crypt foci [69, 70], stomach tumors [71], and lung cancer [72] in rats/mice. In our laboratory, we have demonstrated that SFN and PEITC can inhibit the development of adenomatous polyposis in $APC^{min/+}$ mice [47, 73, 74], and PEITC can inhibit prostate carcinogenesis in TRAMP mice [46] as well as human prostate cancer PC-3 xenografts in nude mice [75]. It appears that significant portion of the chemopreventive effects of isothiocyanates may be associated with the inhibition of the metabolic activation of carcinogens by cytochrome P450s (phase I), coupled with strong induction of phase II detoxifying and cellular defensive enzymes as well as proapoptotic signaling pathways. In fact, *in vitro* and *in vivo* studies indicate that ITCs are potent inducer of ARE-regulated enzymes. SFN has been reported to differentially regulate the activation of MAPK, Nrf2, and phase II enzymes. SFN is also found to interact with Keap-1 by covalent binding to thiol groups of this inhibitory protein leading to release of Nrf2 [76]. We have previously shown that PEITC can induce phase II enzyme HO-1 through Nrf2 signaling pathway [77]. We found that ARE activity and HO-1 expression strongly increased after treatment with PEITC. PEITC also increased the phosphorylation of ERK1/2 and JNK1/2 and caused release of Nrf2 from sequestration by Keap1, and its subsequent translocation into the nucleus. Importantly, PEITC-induced ARE activity was attenuated by inhibition of ERK and JNK signaling. These results suggest that PEITC can activate Nrf2 signaling pathway coupling with activation of upstream MAPK signaling cascades [77] leading to a coordinated protection against carcinogenesis in the intestine of $APC^{min/+}$ mice [47, 73, 74], in the prostate of the TRAMP mice [46], as well as in advance prostate cancer PC-3 xenografts in nude mice [75].

34.2.3
Garlic Organosulfur Compounds (OSCs)

The bioactive components of garlic are mostly garlic organosulfur compounds (OSCs), which can be classified as either water soluble, such as S-allyl cysteine, or oil soluble, such as diallyl sulfide (DAS), diallyl disulfide (DADS), and diallyl trisulfide (DATS). Epidemiological studies have shown that the consumption of garlic and its preparations could decrease the incidence of prostate, colon, laryngeal, gastric, and stomach cancer [78–81]. It has been suggested that their chemopreventive activities are related to their influence on drug metabolizing enzymes (DMEs) and that the subtle structural differences among them have major impacts on the phase I and phase II DME system responses [82]. Likewise, the positive effects of OSCs, especially DADS and DATS, on GST, GR, NQO1, as well as H- and L-ferritin were consistently observed in animal and cellular models [83–86]. In our laboratory, we have shown that three major garlic OSCs – DAS, DADS, and DATS – differentially mediate the transcriptional levels of NQO1 and HO1 [87]. The third sulfur in the structure of OSCs appears to have a major contribution to this bioactivity, with the allyl-containing OSCs being more potent than the propyl-containing OSCs. Upregulation of detoxifying enzymes by garlic OSCs occurs through Nrf2 protein accumulation and ARE activation, which might due in part to the stress signals originating from the oxidative stress and/or calcium-dependent signaling pathways [87].

34.2.4
Vitamin E

Vitamin E is a group of eight structurally related compounds: α-, β-, γ-, and δ-tocopherols (α-, β-, γ-, and δ-T) and α-, β-, γ-, and δ-tocotrienols (α-, β-, γ-, and δ-TTEs). Known as a strong antioxidant, the efficacy of vitamin E in preventing chronic diseases such as arthrosclerosis, cardiovascular diseases, and cancer has been extensively studied [88]. Vitamin E, a potent peroxyl radical scavenger, is a chain breaking antioxidant that prevents the propagation of free radical damage in biological membranes [89]. The biological activity of the various forms of vitamin E appears to correlate with their antioxidant activities. The order of relative peroxyl radical scavenging reactivities of α-, β-, γ-, and δ-TTE (100, 60, 25, and 27, respectively) is the same as the relative order of their biological activities (1.5, 0.75, 0.15, and 0.05 mg $1U^{-1}$, respectively), determined by the classic fetal resorption assay in the rats [90]. However, the protective effects of γ-T having only recently been recognized are now the subject of active investigation. Owing to its strong nucleophilic properties, γ-T exhibits greater efficiency than α-T in trapping ROS and RNS [91]. In addition, Cooney et al. reported that γ-T was superior to α-T in suppressing the transformation of murine fibroblasts incubated with chemical carcinogen 3-methylcholanthrene [92]. In addition to the antioxidant activity, induction of phase II detoxification enzymes through Nrf2 signaling pathway has been linked with the cytoprotective effect of vitamin E. Ogawa et al. reported that gamma-tocopheryl quinone (γ-TQ), not alpha-tocopheryl quinone (α-TQ), induces adaptive response through upregulation of cellular GSH and cysteine availability via activation of ATF4, which is a coactivator of Nrf2 [93]. Additionally, we have also recently reported that γ-T-enriched tocopherols suppressed the development of prostate intraepithelial neoplasia (PIN) and prostate tumorigenesis in the TRAMP mice possibly by induction of Nrf2 and its related detoxifying and antioxidant enzymes [94].

34.3
Conclusion Remarks

Maintaining an optimum redox tone of the cell plays a pivotal role in the homeostasis of tissues. Excessive ROS or RNS production will lead to a state of oxidative stress that contributes to many pathological conditions, including cancer and neurological diseases. Natural dietary compounds have been identified as promising agents to protect cells against the detrimental effects of oxidative stress. These compounds can either directly scavenge the reactive ROS or RNS or indirectly protect the cells from oxidative damages through activation of Nrf2 signaling pathway. As most of these compounds are an integral part of our diets, they are relatively safe for long-term consumption. Therefore, natural dietary compounds could be used as chemopreventive agents to prevent diseases that are initiated or promoted by altered cellular redox tone, including cancer and neurological illnesses.

34.4
Tribute to Professor Harold L. Newmark

Professor Dr Harold L. Newmark is an Adjunct Professor, Department of Chemical Biology, Ernest Mario School of Pharmacy, Rutgers University, Piscataway, NJ, and Member, The Cancer Institute of New Jersey. Harold was asked to write a Chapter for this book, but passed on the job when he reached his 90th birthday.

In 1980, at the age of 62, Harold "retired" from a highly successful industrial career at Roche to devote his energies full-time to a new career in cancer research. Since then, he made pioneering contributions in several areas related to nutrition and cancer prevention. These include (i) the inhibitory effects of vitamin C, vitamin E, and plant phenolics on nitrosamine formation in foods and in vivo in animals, (ii) the protective effect of plant phenolics as inhibitors of mutagenesis and carcinogenesis by polycyclic aromatic hydrocarbon diol epoxides, (iii) the protective effect of plant phenolics on intestinal carcinogenesis in the min+ mouse, (iv) the potential importance of squalene and olive oil as inhibitors of carcinogenesis, (v) development of the high risk Western style diet (high fat, low calcium, and low vitamin D) that causes "spontaneous" colon cancer in mice, and (vi) the inhibitory effects of high calcium and vitamin D on colon and breast cancer in animal models. During the past few years, epidemiology studies by others indicated that individuals with a high calcium intake have decreased risk of colon cancer, verifying the significance of Dr Newmark's earlier animal studies on the importance of calcium for the prevention of colon cancer.

Dr Newmark is a very humble man who has contributed enormously to the pharmaceutical sciences and to nutrition and cancer prevention. He is a wonderful role model for people all over the world.

Acknowledgments

We thank all the members of Ah-Ng Tony Kong's laboratory for their helpful discussions. This study was supported in part by Institutional Funds and by RO1-CA094828, RO1-CA073674, and R01-CA118947 to Dr Ah-Ng Tony Kong from the National Institutes of Health (NIH).

References

1 Siekmeier, R., Grammer, T., and Marz, W. (2008) Role of oxidants, nitric oxide and asymmetric dimethylarginine in endothelial function. *J Cardiovasc Pharmacol Ther*.

2 Frey, R.S., Ushio-Fukai, M., and Malik, A. (2008) NADPH oxidase-dependent signaling in endothelial cells: role in physiology and pathophysiology. *Antioxid Redox Signal*.

3 Griendling, K.K., Sorescu, D., and Ushio-Fukai, M. (2000) NAD(P)H oxidase: role in cardiovascular biology and disease. *Circ Res*, **86**, 494–501.

4 Nishino, T., Okamoto, K., Eger, B.T., Pai, E.F., and Nishino, T. (2008) Mammalian xanthine oxidoreductase – mechanism of transition from xanthine dehydrogenase to xanthine oxidase. *FEBS J*, **275**, 3278–3289.

5 Edderkaoui, M., Hong, P., Vaquero, E.C. et al. (2005) Extracellular matrix stimulates reactive oxygen species production and increases pancreatic cancer cell survival through 5-lipoxygenase and NADPH oxidase. *Am J Physiol Gastrointest Liver Physiol*, **289**, G1137–G1147.

6 Martinez, M.C. and Andriantsitohaina, R. (2008) Reactive nitrogen species: molecular mechanisms and potential significance in health and disease. *Antioxid Redox Signal*.

7 Allen, B.W., Demchenko, I.T., and Piantadosi, C.A. (2008) Two faces of nitric oxide: implications for cellular mechanisms of oxygen toxicity. *J Appl Physiol*.

8 Khor, T.O., Yu, S., and Kong, A.N. (2008) Dietary cancer chemopreventive agents – targeting inflammation and Nrf2 signaling pathway. *Planta Med*, **74**, 1540–1547.

9 Gopalakrishnan, A. and Tony Kong, A.N. (2008) Anticarcinogenesis by dietary phytochemicals: cytoprotection

by Nrf2 in normal cells and cytotoxicity by modulation of transcription factors NF-kappa B and AP-1 in abnormal cancer cells. *Food Chem Toxicol*, **46**, 1257–1270.
10 Sporn, M.B. and Liby, K.T. (2005) Cancer chemoprevention: scientific promise, clinical uncertainty. *Nat Clin Pract Oncol*, **2**, 518–525.
11 Nair, S., Li, W., and Kong, A.N. (2007) Natural dietary anti-cancer chemopreventive compounds: redox-mediated differential signaling mechanisms in cytoprotection of normal cells versus cytotoxicity in tumor cells. *Acta Pharmacol Sin*, **28**, 459–472.
12 Prawan, A., Khor, T.O., Li, W., and Kong, A.N. (2007) Application of pharmacogenomics to dietary cancer chemoprevention. *Curr Pharmacogenomics*, **5**, 190–200.
13 Li, W. and Kong, A.N. (2008) Molecular mechanisms of Nrf2-mediated antioxidant response. *Mol Carcinog*.
14 Chen, C. and Kong, A.N. (2004) Dietary chemopreventive compounds and ARE/EpRE signaling. *Free Radic Biol Med*, **36**, 1505–1516.
15 Lee, J.S. and Surh, Y.J. (2005) Nrf2 as a novel molecular target for chemoprevention. *Cancer Lett*, **224**, 171–184.
16 Kwak, M.K., Wakabayashi, N., and Kensler, T.W. (2004) Chemoprevention through the Keap1-Nrf2 signaling pathway by phase 2 enzyme inducers. *Mutat Res*, **555**, 133–148.
17 Yang, C.S., Maliakal, P., and Meng, X. (2002) Inhibition of carcinogenesis by tea. *Annu Rev Pharmacol Toxicol*, **42**, 25–54.
18 Yang, C.S. and Wang, Z.Y. (1993) Tea and cancer. *J Natl Cancer Inst*, **85**, 1038–1049.
19 Katiyar, S.K., Afaq, F., Azizuddin, K., and Mukhtar, H. (2001) Inhibition of UVB-induced oxidative stress-mediated phosphorylation of mitogen-activated protein kinase signaling pathways in cultured human epidermal keratinocytes by green tea polyphenol (-)-epigallocatechin-3-gallate. *Toxicol Appl Pharmacol*, **176**, 110–117.
20 Erba, D., Riso, P., Colombo, A., and Testolin, G. (1999) Supplementation of Jurkat T cells with green tea extract decreases oxidative damage due to iron treatment. *J Nutr*, **129**, 2130–2134.
21 Maurya, P.K. and Rizvi, S.I. (2008) Protective role of tea catechins on erythrocytes subjected to oxidative stress during human aging0. *Nat Prod Res*, **14**, 1–8.
22 Yao, K., Ye, P., Zhang, L. *et al.* (2008) Epigallocatechin gallate protects against oxidative stress-induced mitochondria-dependent apoptosis in human lens epithelial cells. *Mol Vis*, **14**, 217–223.
23 Raza, H. and John, A. (2007) In vitro protection of reactive oxygen species-induced degradation of lipids, proteins and 2-deoxyribose by tea catechins. *Food Chem Toxicol*, **45**, 1814–1820.
24 Yin, S.T., Tang, M.L., Su, L. *et al.* (2008) Effects of Epigallocatechin-3-gallate on lead-induced oxidative damage. *Toxicology*, **249**, 45–54.
25 Levites, Y., Weinreb, O., Maor, G., Youdim, M.B., and Mandel, S. (2001) Green tea polyphenol (-)-epigallocatechin-3-gallate prevents N-methyl-4-phenyl-1,2,3,6-tetrahydropyridine-induced dopaminergic neurodegeneration. *J Neurochem*, **78**, 1073–1082.
26 Khan, S.G., Katiyar, S.K., Agarwal, R., and Mukhtar, H. (1992) Enhancement of antioxidant and phase II enzymes by oral feeding of green tea polyphenols in drinking water to SKH-1 hairless mice: possible role in cancer chemoprevention. *Cancer Res*, **52**, 4050–4052.
27 Na, H.K., Kim, E.H., Jung, J.H. *et al.* (2008) (-)-Epigallocatechin gallate induces Nrf2-mediated antioxidant enzyme expression via activation of PI3K and ERK in human mammary epithelial cells. *Arch Biochem Biophys*, **476**, 171–177.

28 Shen, G., Xu, C., Hu, R. et al. (2005) Comparison of (-)-epigallocatechin-3-gallate elicited liver and small intestine gene expression profiles between C57BL/6J mice and C57BL/6J/Nrf2 (-/-) mice. *Pharm Res*, **22**, 1805–1820.

29 Chen, C., Yu, R., Owuor, E.D., and Kong, A.N. (2000) Activation of antioxidant-response element (ARE), mitogen-activated protein kinases (MAPKs) and caspases by major green tea polyphenol components during cell survival and death. *Arch Pharm Res*, **23**, 605–612.

30 Na, H.K. and Surh, Y.J. (2008) Modulation of Nrf2-mediated antioxidant and detoxifying enzyme induction by the green tea polyphenol EGCG. *Food Chem Toxicol*, **46**, 1271–1278.

31 Nakagawa, H., Hasumi, K., Woo, J.T., Nagai, K., and Wachi, M. (2004) Generation of hydrogen peroxide primarily contributes to the induction of Fe(II)-dependent apoptosis in Jurkat cells by (-)-epigallocatechin gallate. *Carcinogenesis*, **25**, 1567–1574.

32 Yang, G.Y., Liao, J., Li, C. et al. (2000) Effect of black and green tea polyphenols on c-jun phosphorylation and H(2)O(2) production in transformed and non-transformed human bronchial cell lines: possible mechanisms of cell growth inhibition and apoptosis induction. *Carcinogenesis*, **21**, 2035–2039.

33 Katoh, Y., Itoh, K., Yoshida, E. et al. (2001) Two domains of Nrf2 cooperatively bind CBP, a CREB binding protein, and synergistically activate transcription. *Genes Cells*, **6**, 857–868.

34 Shen, G., Hebbar, V., Nair, S. et al. (2004) Regulation of Nrf2 transactivation domain activity. The differential effects of mitogen-activated protein kinase cascades and synergistic stimulatory effect of Raf and CREB-binding protein. *J Biol Chem*, **279**, 23052–23060.

35 Aggarwal, B.B., Sundaram, C., Malani, N., and Ichikawa, H. (2007) Curcumin: the Indian solid gold. *Adv Exp Med Biol*, **595**, 1–75.

36 Tunstall, R.G., Sharma, R.A., Perkins, S. et al. (2006) Cyclooxygenase-2 expression and oxidative DNA adducts in murine intestinal adenomas: modification by dietary curcumin and implications for clinical trials. *Eur J Cancer*, **42**, 415–421.

37 Rao, C.V. and Reddy, B.S. (1993) Modulating effect of amount and types of dietary fat on ornithine decarboxylase, tyrosine protein kinase and prostaglandins production during colon carcinogenesis in male F344 rats. *Carcinogenesis*, **14**, 1327–1333.

38 Anand, P., Thomas, S.G., Kunnumakkara, A.B. et al. (2008) Biological activities of curcumin and its analogues (Congeners) made by man and Mother Nature. *Biochem Pharmacol*, **76**, 1590–1611.

39 Sharma, O.P. (1976) Antioxidant activity of curcumin and related compounds. *Biochem Pharmacol*, **25**, 1811–1812.

40 Rahman, I., Biswas, S.K., and Kirkham, P.A. (2006) Regulation of inflammation and redox signaling by dietary polyphenols. *Biochem Pharmacol*, **72**, 1439–1452.

41 Leu, T.H. and Maa, M.C. (2002) The molecular mechanisms for the antitumorigenic effect of curcumin. *Curr Med Chem Anticancer Agents*, **2**, 357–370.

42 Joe, B. and Lokesh, B.R. (1994) Role of capsaicin, curcumin and dietary n-3 fatty acids in lowering the generation of reactive oxygen species in rat peritoneal macrophages. *Biochim Biophys Acta*, **1224**, 255–263.

43 Wei, Q.Y., Chen, W.F., Zhou, B., Yang, L., and Liu, Z.L. (2006) Inhibition of lipid peroxidation and protein oxidation in rat liver mitochondria by curcumin and its analogues. *Biochim Biophys Acta*, **1760**, 70–77.

44 Scapagnini, G., Colombrita, C., Amadio, M. et al. (2006) Curcumin activates defensive genes and protects

neurons against oxidative stress. *Antioxid Redox Signal*, **8**, 395–403.

45 Shen, G., Xu, C., Hu, R. et al. (2006) Modulation of nuclear factor E2-related factor 2-mediated gene expression in mice liver and small intestine by cancer chemopreventive agent curcumin. *Mol Cancer Ther*, **5**, 39–51.

46 Barve, A., Khor, T.O., Hao, X. et al. (2008) Murine prostate cancer inhibition by dietary phytochemicals–curcumin and phenyethylisothiocyanate. *Pharm Res*, **25**, 2181–2189.

47 Shen, G., Khor, T.O., Hu, R. et al. (2007) Chemoprevention of familial adenomatous polyposis by natural dietary compounds sulforaphane and dibenzoylmethane alone and in combination in ApcMin/+ mouse. *Cancer Res*, **67**, 9937–9944.

48 Huang, M.T., Lou, Y.R., Xie, J.G. et al. (1998) Effect of dietary curcumin and dibenzoylmethane on formation of 7,12-dimethylbenz[a]anthracene-induced mammary tumors and lymphomas/leukemias in Sencar mice. *Carcinogenesis*, **19**, 1697–1700.

49 Shishu Singla, A.K., and Kaur I.P. (2003) Inhibitory effect of dibenzoylmethane on mutagenicity of food-derived heterocyclic amine mutagens. *Phytomedicine*, **10**, 575–582.

50 Takano, K., Kitao, Y., Tabata, Y. et al. (2007) A dibenzoylmethane derivative protects dopaminergic neurons against both oxidative stress and endoplasmic reticulum stress. *Am J Physiol Cell Physiol*, **293**, C1884–C1894.

51 de la Lastra, C.A. and Villegas, I. (2007) Resveratrol as an antioxidant and pro-oxidant agent: mechanisms and clinical implications. *Biochem Soc Trans*, **35**, 1156–1160.

52 Zini, R., Morin, C., Bertelli, A., Bertelli, A.A., and Tillement, J.P. (1999) Effects of resveratrol on the rat brain respiratory chain. *Drugs Exp Clin Res*, **25**, 87–97.

53 Jia, Z., Zhu, H., Misra, B.R. et al. (2008) EPR studies on the superoxide-scavenging capacity of the nutraceutical resveratrol. *Mol Cell Biochem*, **313**, 187–194.

54 Calabrese, V., Cornelius, C., Mancuso, C. et al. (2008) Cellular stress response: a novel target for chemoprevention and nutritional neuroprotection in aging, neurodegenerative disorders and longevity. *Neurochem Res*, **33**, 2444–2471.

55 Zhang, H., Shih, A., Rinna, A., and Forman, H.J. (2008) Resveratrol and 4-hydroxynonenal act in concert to increase glutamate cysteine ligase expression and glutathione in human bronchial epithelial cells. *Arch Biochem Biophys*, **481**, 110–115.

56 Robb, E.L., Winkelmolen, L., Visanji, N., Brotchie, J., and Stuart, J.A. (2008) Dietary resveratrol administration increases MnSOD expression and activity in mouse brain. *Biochem Biophys Res Commun*, **372**, 254–259.

57 Rubiolo, J.A., Mithieux, G., and Vega, F.V. (2008) Resveratrol protects primary rat hepatocytes against oxidative stress damage: activation of the Nrf2 transcription factor and augmented activities of antioxidant enzymes. *Eur J Pharmacol*, **591**, 66–72.

58 Bektic, J., Guggenberger, R., Eder, I.E. et al. (2005) Molecular effects of the isoflavonoid genistein in prostate cancer. *Clin Prostate Cancer*, **4**, 124–129.

59 Ruiz-Larrea, M.B., Mohan, A.R., Paganga, G. et al. (1997) Antioxidant activity of phytoestrogenic isoflavones. *Free Radic Res*, **26**, 63–70.

60 Zhou, Y. and Lee, A.S. (1998) Mechanism for the suppression of the mammalian stress response by genistein, an anticancer phytoestrogen from soy. *J Natl Cancer Inst*, **90**, 381–388.

61 Siow, R.C., Li, F.Y., Rowlands, D.J., de Winter, P., and Mann, G.E. (2007) Cardiovascular targets for estrogens and phytoestrogens: transcriptional regulation of nitric oxide synthase and antioxidant defense genes. *Free Radic Biol Med*, **42**, 909–925.

62 Mahn, K., Borras, C., Knock, G.A. et al. (2005) Dietary soy isoflavone induced increases in antioxidant and eNOS gene expression lead to improved endothelial function and reduced blood pressure in vivo. *FASEB J*, **19**, 1755–1757.

63 Barve, A., Khor, T.O., Nair, S. et al. (2008) Pharmacogenomic profile of soy isoflavone concentrate in the prostate of Nrf2 deficient and wild-type mice. *J Pharm Sci*, **97**, 4528–4545.

64 Conaway, C.C., Yang, Y.M., and Chung, F.L. (2002) Isothiocyanates as cancer chemopreventive agents: their biological activities and metabolism in rodents and humans. *Curr Drug Metab*, **3**, 233–255.

65 Verhoeven, D.T., Goldbohm, R.A., van Poppel, G., Verhagen, H., and van den Brandt, P.A. (1996) Epidemiological studies on brassica vegetables and cancer risk. *Cancer Epidemiol Biomarkers Prev*, **5**, 733–748.

66 Zhang, Y. (2004) Cancer-preventive isothiocyanates: measurement of human exposure and mechanism of action. *Mutat Res*, **555**, 173–190.

67 Fahey, J.W., Zalcmann, A.T., and Talalay, P. (2001) The chemical diversity and distribution of glucosinolates and isothiocyanates among plants. *Phytochemistry*, **56**, 5–51.

68 Zhang, Y., Kensler, T.W., Cho, C.G., Posner, G.H., and Talalay, P. (1994) Anticarcinogenic activities of sulforaphane and structurally related synthetic norbornyl isothiocyanates. *Proc Natl Acad Sci U S A*, **91**, 3147–3150.

69 Chung, F.L., Conaway, C.C., Rao, C.V., and Reddy, B.S. (2000) Chemoprevention of colonic aberrant crypt foci in Fischer rats by sulforaphane and phenethyl isothiocyanate. *Carcinogenesis*, **21**, 2287–2291.

70 Kassie, F., Uhl, M., Rabot, S. et al. (2003) Chemoprevention of 2-amino-3-methylimidazo[4,5-f]quinoline (IQ)-induced colonic and hepatic preneoplastic lesions in the F344 rat by cruciferous vegetables administered simultaneously with the carcinogen. *Carcinogenesis*, **24**, 255–261.

71 Fahey, J.W., Haristoy, X., Dolan, P.M. et al. (2002) Sulforaphane inhibits extracellular, intracellular, and antibiotic-resistant strains of Helicobacter pylori and prevents benzo[a]pyrene-induced stomach tumors. *Proc Natl Acad Sci U S A*, **99**, 7610–7615.

72 Hecht, S.S., Kenney, P.M., Wang, M., and Upadhyaya, P. (2002) Benzyl isothiocyanate: an effective inhibitor of polycyclic aromatic hydrocarbon tumorigenesis in A/J mouse lung. *Cancer Lett*, **187**, 87–94.

73 Hu, R., Khor, T.O., Shen, G. et al. (2006) Cancer chemoprevention of intestinal polyposis in ApcMin/+ mice by sulforaphane, a natural product derived from cruciferous vegetable. *Carcinogenesis*, **27**, 2038–2046.

74 Khor, T.O., Cheung, W.K., Prawan, A., Reddy, B.S., and Kong, A.N. (2008) Chemoprevention of familial adenomatous polyposis in Apc(Min/+) mice by phenethyl isothiocyanate (PEITC). *Mol Carcinog*, **47**, 321–325.

75 Khor, T.O., Keum, Y.S., Lin, W. et al. (2006) Combined inhibitory effects of curcumin and phenethyl isothiocyanate on the growth of human PC-3 prostate xenografts in immunodeficient mice. *Cancer Res*, **66**, 613–621.

76 Dinkova-Kostova, A.T., Holtzclaw, W.D., Cole, R.N. et al. (2002) Direct evidence that sulfhydryl groups of Keap1 are the sensors regulating induction of phase 2 enzymes that protect against carcinogens and oxidants. *Proc Natl Acad Sci U S A*, **99**, 11908–11913.

77 Xu, C., Yuan, X., Pan, Z. et al. (2006) Mechanism of action of isothiocyanates: the induction of ARE-regulated genes is associated with activation of ERK and JNK and the phosphorylation and nuclear

translocation of Nrf2. *Mol Cancer Ther*, **5**, 1918–1926.
78. Hsing, A.W., Chokkalingam, A.P., Gao, Y.T. et al. (2002) Allium vegetables and risk of prostate cancer: a population-based study. *J Natl Cancer Inst*, **94**, 1648–1651.
79. Steinmetz, K.A., Kushi, L.H., Bostick, R.M., Folsom, A.R., and Potter, J.D. (1994) Vegetables, fruit, and colon cancer in the Iowa Women's Health Study. *Am J Epidemiol*, **139**, 1–15.
80. Zheng, W., Blot, W.J., Shu, X.O. et al. (1992) Diet and other risk factors for laryngeal cancer in Shanghai, China. *Am J Epidemiol*, **136**, 178–191.
81. You, W.C., Blot, W.J., Chang, Y.S. et al. (1989) Allium vegetables and reduced risk of stomach cancer. *J Natl Cancer Inst*, **81**, 162–164.
82. Yang, C.S., Chhabra, S.K., Hong, J.Y., and Smith, T.J. (2001) Mechanisms of inhibition of chemical toxicity and carcinogenesis by diallyl sulfide (DAS) and related compounds from garlic. *J Nutr*, **131**, 1041S–1045S.
83. Hatono, S., Jimenez, A., and Wargovich, M.J. (1996) Chemopreventive effect of S-allylcysteine and its relationship to the detoxification enzyme glutathione S-transferase. *Carcinogenesis*, **17**, 1041–1044.
84. Wu, C.C., Sheen, L.Y., Chen, H.W., Tsai, S.J., and Lii, C.K. (2001) Effects of organosulfur compounds from garlic oil on the antioxidation system in rat liver and red blood cells. *Food Chem Toxicol*, **39**, 563–569.
85. Singh, S.V., Pan, S.S., Srivastava, S.K. et al. (1998) Differential induction of NAD(P)H:quinone oxidoreductase by anti-carcinogenic organosulfides from garlic. *Biochem Biophys Res Commun*, **244**, 917–920.
86. Thomas, M., Zhang, P., Noordine, M.L. et al. (2002) Diallyl disulfide increases rat h-ferritin, L-ferritin and transferrin receptor genes in vitro in hepatic cells and in vivo in liver. *J Nutr*, **132**, 3638–3641.
87. Chen, C., Pung, D., Leong, V. et al. (2004) Induction of detoxifying enzymes by garlic organosulfur compounds through transcription factor Nrf2: effect of chemical structure and stress signals. *Free Radic Biol Med*, **37**, 1578–1590.
88. Clarke, M.W., Burnett, J.R., and Croft, K.D. (2008) Vitamin E in human health and disease. *Crit Rev Clin Lab Sci*, **45**, 417–450.
89. Burton, G.W., Joyce, A., and Ingold, K.U. (1983) Is vitamin E the only lipid-soluble, chain-breaking antioxidant in human blood plasma and erythrocyte membranes? *Arch Biochem Biophys*, **221**, 281–290.
90. Bunyan, J., McHale, D., Green, J., and Marcinkiewicz, S. (1961) Biological potencies of epsilon- and zeta-1-tocopherol and 5-methyltocol. *Br J Nutr*, **15**, 253–257.
91. Campbell, S., Stone, W., Whaley, S., and Krishnan, K. (2003) Development of gamma (gamma)-tocopherol as a colorectal cancer chemopreventive agent. *Crit Rev Oncol Hematol*, **47**, 249–259.
92. Cooney, R.V., Franke, A.A., Harwood, P.J. et al. (1993) Gamma-tocopherol detoxification of nitrogen dioxide: superiority to alpha-tocopherol. *Proc Natl Acad Sci U S A*, **90**, 1771–1775.
93. Ogawa, Y., Saito, Y., Nishio, K. et al. (2008) Gamma-tocopheryl quinone, not alpha-tocopheryl quinone, induces adaptive response through up-regulation of cellular glutathione and cysteine availability via activation of ATF4. *Free Radic Res*, **42**, 674–687.
94. Barve, A., Khor, T.O., Nair, S. et al. (2009) Gamma-tocopherol-enriched mixed tocopherol diet inhibits prostate carcinogenesis in TRAMP mice. *Int J Cancer*, **124** (7), 1693–1699.

35
Genotoxicity of Endogenous Estrogens
James S. Wright

This chapter describes the basic chemistry and biochemistry of two of the human female sex hormones estrone (E1) and estradiol (E2), their mode of formation, and their metabolites. Some attention is also paid to the equine estrogens equilin and equilenin. Genotoxicity of the estrogens is discussed from several vantage points: the first is the work of Gustafsson and coworkers on the estrogen receptor ERβ and its relationship to the ERα receptor. These authors have shown that ligand-activated ERα causes proliferation of cell growth and hence acts as a stimulus to hormone-dependent cancers, whereas ERβ is antiproliferative and acts so as to modulate the effects of ERα. This has led to an extensive search for ERβ-selective ligands. The second viewpoint has been championed by Liehr, Cavalieri, and Bolton and coworkers, and postulates that E2, for example, can be hydroxylated on the A-ring by P450 enzymes to give the two catechols 2-OH E2 and 4-OH E2. Both metabolites are readily oxidized to their corresponding quinones, which are strongly electrophilic, but the 3-4-quinone has been developed to be carcinogenic in animal models and present in tissues of breast cancer patients. Mechanisms of carcinogenesis are discussed, including redox cycling, depletion of protein thiols, and most importantly, formation of DNA adducts. So a coherent picture of estrogen-related carcinogenesis has ben developed, where quinones act as tumor initiators and the presence of ERα-receptors fuels the cell growth. This chapter also discusses factors affecting the concentration of E2, such as the formation of estrogen sulfates. It analyzes in some detail the enzymatic reactions that control oxidation and reduction, and the reaction of protective enzymes such as COMT, which act to detoxify the catechol. Finally, some experiments are described whereby new therapeutic approaches to breast cancer are based on intervention in the genotoxic pathways, for example, by the use of antioxidants.

35.1
Introduction

Estrogen is a generic term referring to female sex hormones. The three endogenous human estrogens are 17β-estradiol (or simply "estradiol," E2), estrone (E1), and

Endogenous Toxins. Diet, Genetics, Disease and Treatment.
Edited by Peter J. O'Brien and W. Robert Bruce
Copyright © 2010 WILEY-VCH Verlag GmbH & Co. KGaA, Weinheim
ISBN: 978-3-527-32363-0

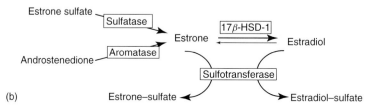

Fig. 35.1 (a) Structures for the three endogenous estrogens estrone (E1), estradiol (E2), and estriol (E3). The steroid ring numbering notation is shown for E2. (b) Formation of estradiol from estrone, and its conversion into the sulfate ester.

estriol (E3). Estrogens are derived in the body from androgens (the male sex hormones), for example, androstenedione in the presence of the aromatase enzyme forms estrone (for reviews of basic estrogen chemistry, see Refs [1–3]). Estrone is converted into the most potent natural estrogen, E2, by an enzymatic reduction, so that an equilibrium is set up between E1 and E2. In fact, estrogens exist mostly as sulfate esters, and the cell uses sulfatase enzymes to create the free estrogen as required [4]. Figure 35.1a shows the structures for E1, E2, and E3, along with an atom numbering scheme for E2. Figure 35.1b shows the enzymatic formation and loss of E1 and E2.

Estrogens belong to the classic steroid family containing the ABCD ring structure. In all three estrogens, the A-ring, from an electronic point of view, acts as a (*meta, para*)-dimethylphenol. The hydroxyl group occupies position 3, leaving positions 1, 2, and 4 vacant. As will be seen, positions 2 and 4 are important in catechol metabolite formation, whereas position 1 is important in DNA adduct formation.

The three estrogens and their derivatives act as ligands and occupy ligand-binding domains (LBDs) in the protein receptors estrogen receptor alpha (ERα) and estrogen receptor beta (ERβ). The LBDs for the two receptors are very similar and differ in only two residues. Figure 35.2a shows the 3D structure of E2, and Figure 35.2b shows E2 in the LBD where ERα and ERβ are shown together. Writing an equilibrium constant for dissociation of the ligand–receptor complex (denoted LR) as $K_d = [L][R]/[LR]$, ligand E2 has $K_d = 0.2$ nM for ERα and 0.5 nM for ERβ, that is, the ligand E2 is more tightly bound to ERα by a factor of 2.5 [5]. This is a relatively small difference, and E2 is said to be nonselective with respect to

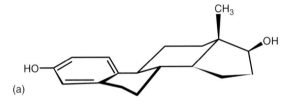

(a)

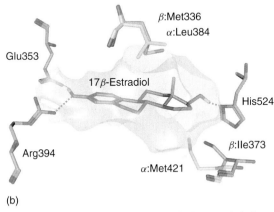

(b)

Fig. 35.2 (a) 3D structure for estradiol. (b) Estradiol shown in the binding cavity of both ERα and ERβ.

its two receptors. Similar measurements on E1 and E3 show that E2 is the most strongly bound ligand by a factor of 10–20, depending on how the binding assay is performed.

Ligand binding initiates a chain of events that eventually result in gene transcription and the generation of proteins relevant not only to female sexual characteristics but also to a host of other physiological functions affecting bone density, lipid profiles, and so on. [6]. Briefly, following ligand binding, for example, to ERα, an α–α homodimer is formed, which then binds to the DNA at an estrogen response element (ERE), and gene transcription is initiated. If the ligand can bind to ERβ a similar process occurs, creating a β–β homodimer. Because the α and β-receptors are very similar and have a common interface, it is also possible to form heterodimers α–β; this adds to the level of complexity possible in gene transcription. One generalization, which usually holds, is that the stronger the degree of ligand binding, the greater the degree of hormonal potency. Here we are assuming that the ligand is acting as an "agonist," that is, a promoter of gene transcription, and that potency is related to the amount of transcription. There are no known "endogenous antagonists," which can bind tightly, but suppress gene transcription; but there are many exogenous drugs such as tamoxifen designed to do exactly that [7].

35.2
Estrogens and Women's Health

The discussion of the risks and benefits of estrogens becomes sharply focused for postmenopausal women: to supplement or not to supplement? Hormone replacement therapy (HRT) using estradiol effectively relieves menopausal symptoms such as hot flashes and mood swings, and is protective against osteoporosis and cardiovascular disease. However, by 2003, major epidemiological studies became available showing that HRT usage definitely increased the risk of breast and uterine cancers, and the longer the exposure the greater the risk (for references to the Women's Health Initiative and related data, see Refs [4, 8–11]). HRT usage dropped precipitously, followed by a drop in cancer rates. Of course, the menopausal symptoms returned and there was much discussion of the overall risk–benefit analysis. The literature on this subject is vast and easily accessible and will not be reviewed here. We will consider the HRT-related increase in breast cancer rates to be a documented fact, and will focus on the causes of estrogen-related genotoxicity.

35.3
Causes of Estrogen Genotoxicity

35.3.1
Significance of ERα and ERβ: the Estrogen Receptors

The discovery of a second subtype of the estrogen receptor in 1996 by Kuiper, Gustafsson, and coworkers [12], now termed *ERβ* (the subtype formerly known as *ER* is now *ERα*) has proven to be important to understanding how estradiol can act as a carcinogen. Helpful reviews of the status of ERβ have been provided by Gustafsson and coworkers periodically, see Refs [2, 3, 8, 13]. Even within a subtype, there can be variants with different ligand-binding affinities and transcriptional behavior; this has been discussed by Peng *et al.* [14]. A current review, which related receptor subtypes to clinical practise in breast cancer, was given by Fox *et al.* [15].

Within two years of the discovery of ERβ, Barkhem *et al.* [16] recognized that the ligands lining the binding cavity in the two subtypes must be different enough to enable the selectivity differences they recorded in ethinyl estradiol, genistein, tamoxifen, and raloxifene, as well as several other common ligands. Already by the year 2000, Gustafsson and Warner [17] suggested that (ligand-activated) ERβ may play a protective role in breast tissue, based on their work with ERβ-knockout mice. They also drew attention to the signaling differences between ER homodimers and heterodimers [18]. In the next few years, research on properties of ERβ intensified. More data from the Gustafsson group and others solidified the central idea that ERα receptors stimulated proliferation in breast cancer cells, whereas ERβ was inhibitory. The obvious relevance to breast cancer therapy then led to a hunt for naturally occurring or synthetic ERβ-selective ligands, which continues to the present day. For example, Leitman and coworkers [18] have been investigating the components of (naturally occurring) soy products such as genistein and daidzen, which are strongly β-selective. Others, for example, Katzenellenbogen and coworkers, have been searching for, and have found synthetic ligands, which are selective for either ERα or ERβ [19–21]. The present author is also engaged in this exciting effort [22] which, however, lies beyond the scope of this review.

In the decade following the discovery of ERβ, Gustafsson and coworkers continued to elucidate its biological function. The list of relevant diseases continued to expand: by 2003, they listed depression, prostate dysfunction, leukemia, inflammatory bowel disease, and colon cancer, as well as the previously studied breast cancer [23]. The new understanding of ERs culminated in an important paper in 2003 describing a "yin-yang" interplay between subtypes [24]. The central idea here is that α- and β-receptors act in opposition to each other, thus allowing a more subtle control of physiological functions than would be possible if their functions were uncoupled. With respect to cancer, this has been broadly interpreted to mean that normal cellular growth occurs when the proper balance of ERα and ERβ occurs; outside this zone, proliferation occurs when excess ERα is present, whereas growth inhibition occurs when excess ERβ is present [25]. In later work, Gustafsson and coworkers showed that expression of ERα increased in mammary carcinogenesis [26]. They further emphasized the potential significance of discovering ERβ-selective ligands and included autoimmune and neurodegenerative diseases among potential applications [27].

Other studies supplied proof of the proliferative/inhibitory concept in HC11 mouse mammary cells [28] and in T47D human breast cancer cells [29]. In recent work, this group has expanded its focus to consider the roles of coregulatory proteins associated with the different receptors [30], and investigated the nuclear mobility of both receptors as a function of ligand type [31]. Their current view of the origin of hormone-dependent cancers could be stated as follows: certain tissue types in the body (e.g., breast epithelial cells) exist in a state of rather delicate balance determined by the concentrations of the two estrogen receptors and the ligands needed to activate them. Cell growth can become uncontrolled when ERβ is not sufficiently activated and the cancerous state begins. This theory of carcinogenesis lends itself to experimental validation, and has so far stood up well. Clearly, the contributions of Gustafsson and coworkers to understanding the link between estrogen receptors and carcinogenesis has been profound, first by their discovery of the ERβ receptor, second by studying their interaction with ERα, and third by providing simple generalizations on the proliferation/inhibition status of the combination of both receptors.

35.3.2
Quinone Formation Is Tumorigenic

In the 1990s, even before the discovery of ERβ, an alternative theory of estrogen carcinogenesis began to appear in the literature. According to this theory, the phenolic A-ring (see Figure 35.1 for steroid ring notation) could undergo enzymatic

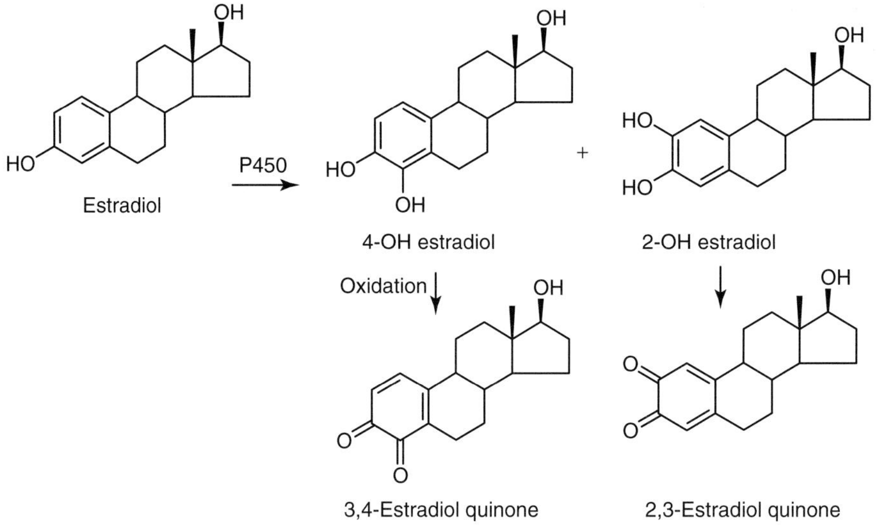

Fig. 35.3 Conversion of E2 into the catechols 2-OH E2 and 4-OH E2, followed by their conversion into the corresponding quinones.

hydroxylation to produce the catechol estrogen, followed by oxidation (either enzymatic or simply by autoxidation in the presence of oxygen) to form the estrogen quinone. Figure 35.3 shows the mechanism for formation of the principal estradiol metabolites. There are two catechols possible, 2-OH E2 and 4-OH E2, where the original phenolic OH occupies position 3. Oxidation then leads first to the semiquinone and then to the two quinones 2,3-E2 quinone and 3,4-E2 quinone. Authoritative reviews on basic quinone chemistry have been provided by Brunmark and Cadenas [32], O'Brien [33], and Bolton [34]. Brunmark and Cadenas are particularly strong in delineating basic quinone chemistry, including a description of relevant enzymes. O'Brien takes a more toxicological approach and hence is helpful in delineating cytotoxic mechanisms. Bolton relates quinone and phenol chemistry to reactions of estrogens and antiestrogens, and hence is of direct relevance to this review.

As discussed by these authors and many others, general features of steroid genotoxicity arising from quinone chemistry have slowly been clarified, and the following generalizations have emerged. The quinones are highly electrophilic and act as Michael acceptors at the β-carbon (Figure 35.4a). The quinones react most rapidly in cells with strong nucleophiles, for example, one target is the amino groups in DNA bases. Subsequent replication errors propagated during cell growth provide a mechanism for tumorigenesis. Quinones also react with sulfur atoms in protein thiols, resulting in loss of protein function. When these are critical proteins, for example, those that act as tumor suppressors, carcinogenesis can begin. Under the same category is reaction with glutathione, a tripeptide, which provides much of the reducing power in the cell; when glutathione stores are depleted, the cell enters into a state of oxidative stress. A third mechanism involving redox cycling

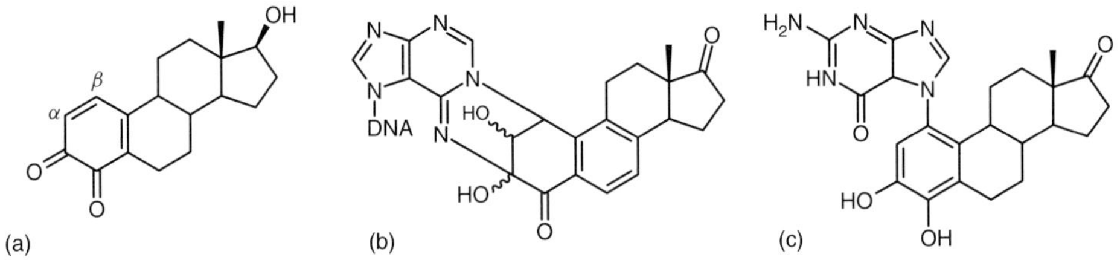

Fig. 35.4 (a) Site of Michael addition on 3,4-E2 quinone. (b) A stable DNA adduct. (c) A depurinating DNA adduct.

between the quinone and the semiquinone with production of superoxide ion is also possible. Both thiol depletion and redox cycling generate oxidative stress within the cell. The coupling of oxidative stress with DNA lesions can, in principle, accelerate the loss of regulation in the cell, leading to uncontrolled growth, that is, cancer.

Next we shall consider the contributions of several key research groups in this area, to see how their ideas developed and eventually led to the generalizations stated above. Others contributed, but the studies of Liehr and Cavalieri and coworkers were truly pioneering as far as endogenous estrogen toxicity.

35.3.3
Experimental Support for the Quinone Hypothesis

The mechanism of estrogen mutagenesis and carcinogenesis involving quinone formation was formulated in the mid-1980s by Liehr and coworkers, although the concept appeared as a hypothesis earlier in the literature [35, 36]. Probably the first definitive study of the carcinogenicity of estrogens and their catecholic derivatives was done in 1986 [37] using Syrian hamsters as the animal model. The authors measured renal carcinoma in hamsters fed estradiol (E2) or its catechol metabolites 2-OH E2, 4-OH E2 and 2-methoxy E2 (2-OMe E2). They found that E2 and 4-OH E2 stimulated tumor growth, whereas 2-OH E2 and 2-OMe E2 did not. These studies indicated the importance of the position of functional groups on the A-ring, which later became a theme of many papers in the literature. Liehr argued that the 2-OH E2 was rapidly converted by catechol O-methyl transferase (COMT) into 2-OMe E2, which lacked binding affinity to the estrogen receptor, thereby explaining the lack of carcinogenicity of 2-OH E2 (i.e., a beneficial metabolic effect). However, no study of the conversion rate of 4-OH E2 to 4-OMe E2 was performed.

In the late 1990s, Liehr and coworkers [38] drew attention to the role of cytochrome P450 enzymes in causing estrogen hydroxylation. They showed that CYP1A and CYP3A families contributed to hydroxylation at both 2- and 4-positions, but that CYP1B was mostly associated with hydroxylation at the 4-position. On the basis of hydroxylation patterns in estradiol and the equine estrogen equilenin, they proposed a free radical mechanism involving formation of the phenoxyl radical, rearrangement to a carbon-centered radical, and substitution at the radical center with Fe(IV)-OH. In a review paper titled "Is estradiol a genotoxic mutagenic carcinogen?" [39] Liehr gives the accumulating evidence for catechol estrogens, generating quinones and causing DNA lesions, as the tumor initiator, followed by hormone-receptor-mediated proliferation of the damaged cells. In other words the answer is "Yes," although obviously E2 must be a weak carcinogen.

A somewhat different perspective in 2001 was given by Liehr in "Genotoxicity of the steroidal estrogens estrone and estradiol: possible mechanism of uterine and mammary cancer development" [40], which concentrates on clarifying the difference between the initiation step in estrogen-related cancers and the growth step. The latter, he concedes may simply be receptor-mediated cell proliferation. Analysis of the mechanism of tumor induction gives renewed emphasis to the redox cycling mechanism and subsequent formation of DNA adducts with the Michael acceptor (i.e., the quinone). The redox cycling mechanism consists of a cycle where the semiquinone (an anion) transfers the extra electron to oxygen forming the quinone and superoxide anion; P450 then reduces the quinone back to the semiquinone and the cycle is repeated. This reaction is a catalytic superoxide generator, where the quinone is the catalyst and in the propagation step, oxygen is reduced to superoxide anion. Hydrogen peroxide generated from superoxide anion is capable of damaging DNA via single-strand breaks, hydroxylation of guanine bases, and by aldehyde decomposition products of lipid hydroperoxides (e.g., malondialdehyde). The latter two oxidation products had been observed in breast cancer patients [40].

An important focus for Liehr was the role of 2-OHE2 and 4-OHE2 in carcinogenesis. Accordingly, the 2-isomer is inert because it generates stable DNA adducts, whereas the 4-isomer forms unstable adducts, which decompose, eliminating a DNA base and a sugar molecule. The resulting apurinic adduct was found (by the Cavalieri group) to be much more carcinogenic than the stable adduct. Liehr often warns the reader of the complexity of the carcinogenic process. However, his bottom line is "The genotoxic and mutagenic activities of oestrogens described above are consistent with and delineate a dual role of oestrogens as carcinogens (tumour initiators) and as hormones (stimulators of cell proliferation)... the genotoxicity and mutagenicity studies of estrogens point to the catechol estrogen metabolites as the precursors of DNA damaging semiquinone/quinone reactive intermediates" [38].

Meanwhile, Cavalieri and coworkers had been carrying out studies of carcinogenesis in polycyclic aromatic hydrocarbons in the early 1990s, and recognized the important role of electrophilicity in the metabolites. They first identified the presence of a DNA adduct in 1995 [41], and by 1997 had integrated these observations into a theory on the origin of carcinogenesis. This included the surprising statement "we think that the major carcinogenic risk to humans is represented by endogenous carcinogens," in particular, the electrophilic metabolites of estrogen [42]. They were careful to differentiate between stable DNA adducts, which are formed by nucleophilic attack of a DNA base on a Michael acceptor, and depurinating adducts, which are formed with cleavage of the bond between the sugar and the DNA base. In their own words "The resultant apurinic sites in critical genes can generate mutations that initiate cancer genotoxic effects of specific electrophilic metabolites of estrogens may be at the origin of many human cancers ... " [42]. In an important paper in 2002 they identified DNA adducts involving the DNA bases guanine and adenine [43]. Figure 35.4b shows a stable DNA adduct formed with the 3,4-quinone, whereas Figure 35.4c shows a depurinating adduct. They proposed that the 3,4-quinone was responsible for the tumor initiation via its (depurinating) reaction with DNA, followed by normal hormone-mediated processes, which led to cancer growth. This provided an elegant link between the proliferation mechanism described by Gustafsson, which is hormone-receptor dependent, and the electrophilic initiation mechanism involving quinone formation. Thus, Cavalieri and Liehr were in agreement about the carcinogenic mechanism.

The quinone hypothesis continued to gain experimental support, at the same time as the experiments grew more sophisticated. In the same paper in 2002 [43], Cavalieri and coworkers described more general aspects of the initiation mechanism, and applied it to the oxidation of benzene and dopamine, where the latter is postulated to be oxidized to dopa-quinone. This, they related to the reaction with DNA to generate depurinating adducts, which could lead to neurodegeneration and potentially, Parkinson's disease. Later work by his group linked metabolites of estradiol to breast cancer [9, 44] and proposed the measurement of catechol estrogen quinones as biomarkers for susceptibility and cancer prevention [9]. Most recently, this group has attempted to intervene in the carcinogenic process by use of antioxidants such as *N*-acetylcysteine, which inhibit the oxidation of the catechol estrogen 4-OH E2 to its corresponding quinone [45], thereby reducing the formation of DNA adducts in mammary epithelial cells.

Another group that has made significant advances in understanding estrogen genotoxicity, including how to counter it, is that of Bolton. Her work on formation of quinones and quinone methides in phenols in the early 1990s provided a natural entry into the area of estrogen metabolites. In the late 1990s, she began to study the formation of metabolites from the drug Premarin, which is derived from the urine of pregnant mares. At that time Premarin was the best-selling drug in the United States, being used as an antifertility drug by premenopausal women and as a hormone supplement by (millions of) postmenopausal women. In addition to estradiol, Premarin contains equilin and equilenin, which have unsaturation in the B-ring. Figure 35.5a shows structures for equilin and equilenin. Equilenin had also been studied by Liehr and coworkers, who showed that it readily formed the 1,2-naphthoquinone structure, but for energetic reasons (loss of aromaticity in B-ring), was unable to form the 2,3-naphthoquinone [38]. Figure 35.5b shows the difference in reactivity for the two naphthoquinone structures, as applied to the equilenin framework.

In 1998, Bolton wrote an influential review on "Role of quinoids in estrogen carcinogenesis" [46]. This review covered not only quinone formation, but was particularly explicit on the role of DNA adduct formation, GSH depletion, and redox cycling, and their relevance to estrogen replacement therapy and breast cancer. In 2002, she discussed the antiestrogens such as tamoxifen, used in treatment of breast cancer [34]. Although tamoxifen is not endogenous, it does demonstrate the universality of the quinone formation mechanism. Bolton's later work, much

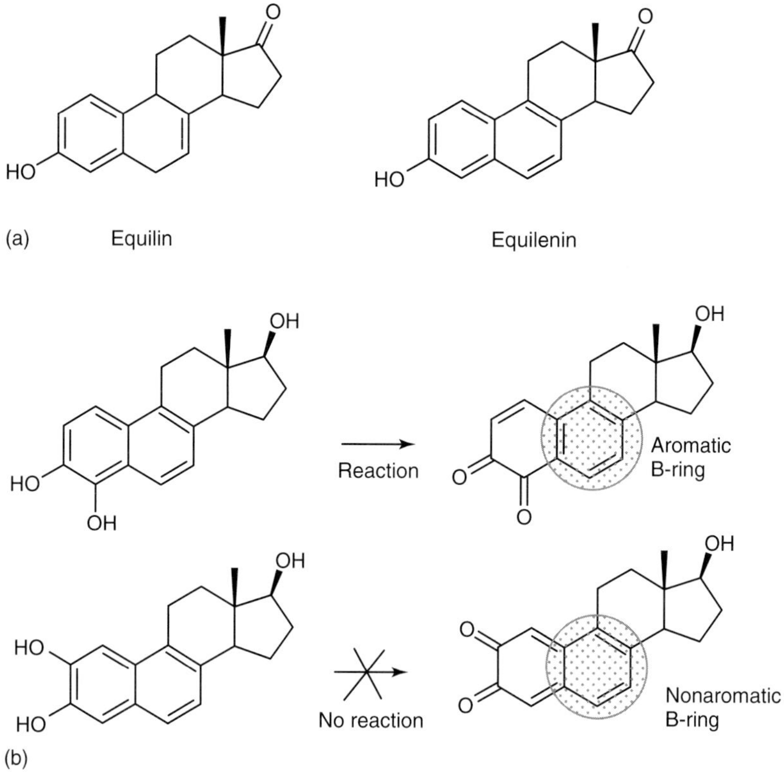

Fig. 35.5 (a) Structure of the conjugated equine estrogens equilin and equilenin. (b) Difference in reactivity of 4-OH Equilenin (reactive) and 2-OH Equilin (unreactive). Aromaticity (or lack of it) in B-ring is highlighted.

of it done with Thatcher, continued to be concerned with the equine estrogens. Finally, an interesting review article by Bolton and Thatcher [10] returned to an examination of endogenous estrogen toxicity. In addition to the recurrent themes discussed above, this review article considered the effect of estrogen metabolites on redox-sensitive enzymes, such as glutathione transferase and quinone reductase. As will be seen, enzyme modification is the focus of increased interest, because it contains the possibility of disease prevention and/or intervention.

We could argue from the work described so far that we now have a self-consistent view of the mechanism of estrogen genotoxicity, involving catechol estrogen formation, generation of mutations through DNA adducts, and oxidative stress (thiol depletion, redox cycling) leading to tumorigenesis, followed by hormone-receptor-dependent cell proliferation. This conclusion seems reasonable and was reiterated recently in an excellent article by Cavalieri and a host of distinguished coworkers, entitled "Catechol estrogen quinones as initiators of breast and other human cancers: Implications for biomarkers of susceptibility and cancer prevention" [9]. This article describes in detail the measurement of DNA adducts and experiments involving 4-OH E2 and both quinones. It introduces the concept of "estrogen homeostasis", which, when disrupted, allows the estrogen quinones to form. Homeostasis exists when the cellular environment is sufficiently reducing (good supply of GSH, ascorbate, and protein thiols); the semiquinone will then be reduced back to the nontoxic catechol instead of oxidized to the toxic quinone.

However, as far as understanding genotoxicity of estrogens, more recent investigations have shown that the description above leaves many important questions unanswered. For example, what are the concentrations of the estrogens and their receptors and what factors affect them? What are the key enzymes that affect metabolite formation, and are there endogenous inhibitors? What are the roles of coactivators and corepressors? What about the process of transcription activation or its converse, transcription suppression? Already, by 2001, these questions were beginning to be addressed, some by the same groups, but with many others contributing to the process. A continuing motivation was the search for new drugs for treatment of breast and prostate cancer, with their baffling transformation from hormone-dependent to hormone-independent phases.

35.3.4
Factors Affecting Estrogen Concentrations

Pasqualini and coworkers were already making significant contributions to the field of steroid chemistry in the 1960s. In the early 1980s, they began to study the esterification of estrone in its sulfate form, and the resulting bioavailability of estrone [47]. They soon began testing the local concentrations of the estrogen sulfate esters versus free estrogens, with surprising results: "Estrogen sulfates are quantitatively the most important form of circulating estrogens during the menstrual cycle and in the postmenopausal period. Huge quantities of estrone sulfate and estradiol sulfate are found in the breast tissues of patients with mammary carcinoma" [48]. By the early 1990s, this group realized that the interconversion of the estrogens between their esterified (sulfate) form and the free estrogen was a more important determinant of estrogen concentration than the formation pathway of estrone via androgens and the aromatase enzyme [49]. An equilibrium was established between each estrogen and its sulfate via the sulfatase enzymes (see Figure 35.5b). Denoting estradiol sulfate as E2S and estrone sulfate as E1S, an estradiol sulfatase converts E2S to E2, whereas an estradiol sulfotransferase converts E2 back to E2S, and similarly for the other two estrogens estrone and estriol. Thus, the sulfatase and sulfotransferase enzymes ensure that the thermodynamic ratios are maintained. Another key equilibrium is that between E1 and E2, because as stated previously E2 is a much more potent hormone. This equilibrium is also maintained by forward and reverse enzymes; in the conversion from E1 to E2, estrone hydrogenase is the catalyst, whereas in the reverse direction it is estradiol dehydrogenase [49]. Further, they showed that the presence or absence of these enzymes in various breast cancer cell lines were relevant to their capacity to undergo uncontrolled cell proliferation. Inhibition of the various enzymes, thus provided new intervention targets in breast cancer.

In a 1999 paper by Pasqualini and Chetrite entitled "Estrone sulfatase versus estrone sulfotransferase in human breast cancer: potential clinical applications" [50], they showed how the production of E2, with its tumor-stimulating properties, could be modulated by inhibition of the appropriate enzyme. Another paper then showed how E2 could act as an inhibitor in its own biotransformation via the formation pathway E1S → E1 → E2; this would effectively reduce the concentration of E2, which would otherwise form [51]. A comprehensive view of this area of enzymatic conversions affecting E2 concentration, appeared in 2005 under the title "Recent insight on the control of enzymes involved in estrogen formation and transformation in human breast cancer" [4]. Here the authors point out that the pathway that converts E2S into E2 is about 100 times more important than the production of E2 from the aromatase pathway. This remarkable fact has led some authors to question whether the therapeutic possibilities, for example, sulfatatase inhibitors, have been sufficiently explored [52]. Considering the intensive effort to create aromatase inhibitors for breast cancer, and their current clinical use as an alternative to the antiestrogen tamoxifen, the investigation of sulfatase inhibitors would seem to be deserving of more attention.

35.4
Enzymatic Reactions that Affect Estrogen Genotoxicity

Most of these topics have already been discussed, but several have not. Because of their importance to estrogen genotoxicity, a brief summary of the relevant enzymatic reactions, and entry points into the literature describing them, are given in this section. A recent review of many of these reactions was given by Cavalieri et al. [9].

35.4.1
Reactions that Increase Toxicity

35.4.1.1 Catechol Formation from E2

The significance of the cytochrome P450 enzyme family in converting estradiol to its genotoxic metabolites was already well studied by Liehr et al. in the 1990s [38]. They showed that the P450 1A family converted E2 into the noncarcinogenic 2-OH E2, whereas the P450 1B family converted E2 into 4-OH E2, which was easily converted into the genotoxic 3,4-quinone. On the basis of a mechanistic study, they proposed that P450 catalysis involved formation of a phenoxyl radical by H-atom abstraction from the phenolic A-ring, followed by rearrangement to a carbon-centered radical *ortho* to the initial hydroxyl group [38]. Subsequent attack by a hydroxyl radical bound to the catalyst placed an OH group at the 2- or 4-position, depending on electronic and steric constraints.

In their 2006 review, Cavalieri and coworkers discussed the role of CYP1B1, namely, that it and other 4-hydroxylases could increase the amount of the genotoxic catechol 4-OH E2 versus the much less toxic 2-OH E2. From the viewpoint of estrogen homeostasis, they postulated that overexpression of CYP1B1 could lead to large amounts of the catechols and then formation of E1- or E2-3,4-Q. If this happens, then "We postulate that unbalanced estrogen homeostasis is a condition that precedes the initiation of breast and prostate cancer" [9]. What would cause an overexpression of CYP1B1 was not discussed.

35.4.1.2 P450 and the Redox Cycle

As described previously, a redox cycle occurs as follows: the catechol estrogens, for example, 2-OH E2 are easily oxidized to their semiquinones. Denoting the catechol, which has two exchangeable H atoms as QH_2, the corresponding semiquinone is $QH^{\bullet}$. Typically, it is formed when the catechol, which has a weak OH bond, donates an H atom to another radical. Semiquinones are acidic, most with pKs below 5, so that a proton is lost and the radical anion $Q^{\bullet-}$ is formed. This transfers an electron to molecular oxygen forming superoxide, $O_2^{\bullet-}$ and the quinone Q. Q is then reduced by P450 reductase back to $Q^{\bullet-}$. The process repeats itself and a redox cycle is born, the net result of which is that Q is a catalyst for the conversion of oxygen into a superoxide anion. This redox cycle is shown in Figure 35.6. Then, another ubiquitous catalyst superoxide dismutase (SOD) rapidly converts superoxide into hydrogen peroxide. If not deactivated by catalase (CAT), hydrogen peroxide is an oxidative stressor. In the worst case, it reacts with unbound iron (Fe^{2+}) to form a hydroxyl radical, which is believed to cause DNA lesions. Thus P450 reductase, SOD, and CAT are enzymes of relevance to the redox cycle. From a toxicity perspective, only CAT could be considered protective, whereas P450 reductase leads to redox cycling and hence, causes toxicity.

35.4.1.3 Sulfatase Creates More Free Estrogen

This enzymatic reaction has already been described in sufficient detail. For more detail see Ref. [4].

35.4.1.4 Aromatase Creates More E2 via E1

Estrogens are synthesized in the cell from androgen precursors using the aromatase enzyme. Aromatase is actually an enzyme complex, and its catalytic activity has been discussed in detail by Brueggemeier et al. [11]. Both testosterone and androstenedione are substrates for this enzyme. Thus, androstenedione → E1 and testosterone → E2, when aromatase is present. Because androstenedione is the dominant form of androgen, this biosynthetic pathway yields principally, E1. E1 is converted into the more potent E2 by the action of 17β-hydroxysteroid dehydrogenase type 1 (17β-HSD-1) (Figure 35.1). In addition to the extensive development of synthetic aromatase inhibitors to reduce estrone concentration for treating breast cancer, there are a number of natural sources that show inhibitory effects. These include members of the flavone family, for example, chrysin, apigenin, and quercetin, with IC_{50} values in the low micromolar range [11].

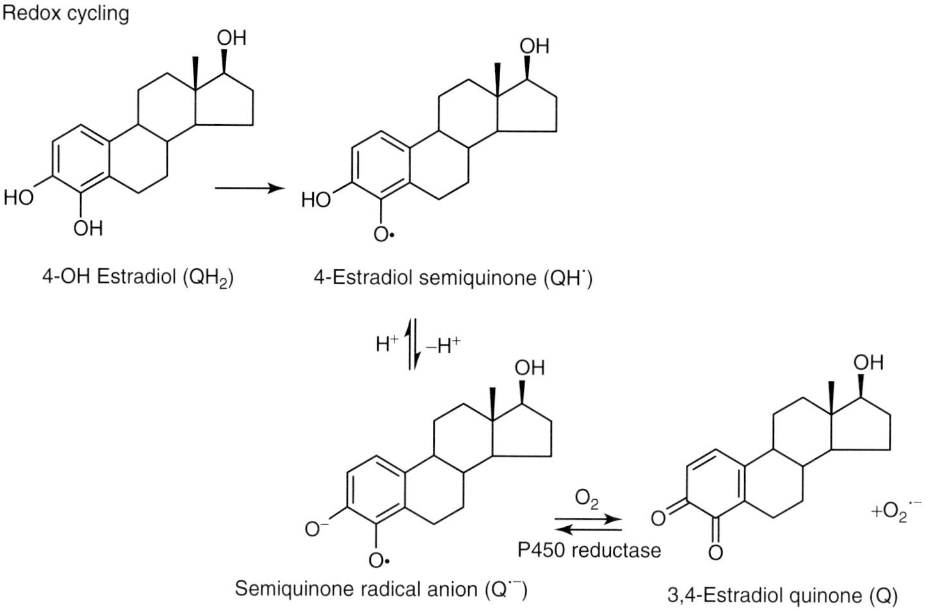

Redox cycling

Fig. 35.6 Redox cycling of 4-OH E2. Once the semiquinone is formed it ionizes, then semiquinone radical anion transfers electron to oxygen. P450 reductase reduces quinone to semiquinone radical anion, generating a redox cycle whose product is superoxide (radical) anion $O_2^{\bullet -}$.

A less explored pathway to reducing concentrations of E2 would be via inhibition of the estradiol dehydrogenase, preventing conversion of E1 into E2. This enzyme and its inhibitors have been discussed by Pasqualini and Chetrite [4].

35.4.2 Reactions that Reduce Toxicity

35.4.2.1 COMT Deactivates Catechol

The catechol estrogens derived from E2 are known to maintain significant binding affinity to ERs. Hence, in addition to the reactions leading to quinone formation, they are agonists for the ER(s) and hence, could be growth-stimulatory, for example, in breast cancer. The same is true, of course, for E2 itself, in the presence of ERα. However, in the region of the A-ring, there is a tight fit of the receptor around the ligand. Thus, methylation of the hydroxyl group causes a steric clash with the receptor, and drastically reduces the binding energy. The COMT enzyme accomplishes this O-methylation, converting catechols into ortho-methoxy alcohols, that is, 2-OH E2 into 2-OMe E2 and 4-OH E2 into 4OMe E2 [53]. None of these methoxy derivatives have significant binding affinity to ER; therefore, in a situation where catechol derivatives are problematic, COMT is a protective enzyme. On the other hand, a deliberate attempt to increase the concentration of COMT, for example, by upregulation of the enzyme, could be problematic if alcohols other than estrogens were converted indiscriminately into methoxides. For that reason there is an effort under way to develop selective COMT inhibitors [53].

A compound of current interest in breast and prostate cancer treatment is 2-OMe E2. This compound shows very little binding affinity for ER, as would be expected for a compound containing a methoxy group on the A-ring. Therefore, it must have been regarded with some amazement when it was reported that a simple endogenous metabolite could shrink solid tumors in mice [54]. Recently, the compound has entered clinical trials and shows antiangiogenic properties, which starve the tumor of its blood supply, as well as antiproliferative properties, unlike its precursor E2 [55]. The mechanism of action is under scrutiny, and may involve disruption of microtubules in the cell, causing an apoptotic response [56, 57].

35.4.2.2 NQO1 Deactivates the Quinone

NQO1 is an enzyme, which acts on quinones and causes a two-electron reduction, converting the quinone back into a catechol. This prevents the one-electron reduction of the quinone by P450 reductase, which results in redox cycling and oxidative stress; hence, NQO1 is cytoprotective. The enzyme has been well studied; for more information see the O'Brien review [33].

35.4.2.3 GSH, Ascorbate, and Antioxidant Intervention Experiments

It is well known that the endogenous reducing agent glutathione and the exogenous but ever-present ascorbate ion provide much of the reducing power in the cell. They may not inhibit the phenoxyl radical mechanism for catechol formation, but once the catechol is formed and oxidized to the semiquinone, when the reducing power is sufficient, the semiquinone will be reduced back to the catechol. Therefore, less (or no) quinone will be formed. Thus, glutathione and ascorbate molecules help to minimize the toxicity of the estrogens by generating a sort of "redox balance" or "redox homeostasis" where the dominant estrogen species are shifted away from the genotoxic quinones. Of course, if the cell were to enter a state of oxidative stress, the estrogen balance will shift toward the more oxidized state. As a consequence, the quinone concentration will increase, mutations may result. and susceptible cells will begin to proliferate.

There are two logical ways to detoxify the oxidized species (i.e., the quinones). The first is an exogenous approach: dump in as many reducing agents in the form of small-molecule antioxidants as the cell can tolerate. In 2003, Samuni and coworkers [58] wrote a paper with the title "Semiquinone radical intermediate in catecholic estrogen-mediated cytotoxicity and mutagenesis: Chemoprevention strategies with antioxidants." They showed that depletion of GSH increased the cytotoxicity to MCF-7 cells, whereas addition of ascorbate or cysteine decreased the mutation rate and the cytotoxicity. In another experiment Venugopal et al. [45] showed that the glutathione precursor N-acetylcysteine was effective at reducing the estrogen-induced transformation of mouse mammary epithelial cells.

These experiments provided a strong validation of the Liehr–Cavalieri hypothesis of quinone-induced tumorigenesis. This approach requires dietary modification, which, however, is feasible.

The second approach is to attempt to increase the concentration of antioxidant enzymes, such as SOD, CAT, GST, NQO1; in other words, to upregulate chosen reducing enzymes. Although this may sound implausible, it is not. For example, after repeated "insults" consisting of exposure to high levels of hydrogen peroxide, cells have been shown to upregulate catalase as a defense mechanism; this can improve their survival rate when exposed to the next insult. For example, Lai *et al.* [59] exposed rat cardiac myocytes to chronic H_2O_2 levels above normal and showed that CAT activity increased by about 50%. They also proved that CAT mRNA expression levels increased by up to 100% during peroxide exposure. Applied to estrogen homeostasis, the question, of course, is whether the oxidative insult which, when countered by antioxidant enzyme upregulation, provides a net benefit.

35.4.2.4 Sulfotransferase Creates Bound Estrogen

If E2 or its metabolites stimulate receptor-mediated cell proliferation, then anything that sequesters E2 into an inactive form would be cancer-protective. Antiestrogens compete with E2 for receptor sites; once bound they prevent transcription. A different way to reduce the concentration of active E2 is to convert it into an inactive form, for example, an ester. Then ester hydrolases present in the cell can release the ester back into its original (bioactive) form, but the net effect is still to decrease the effective concentration of the free estrogen. This is exactly the role of sulfotransferase; de-esterification of the ester is accomplished by sulfatase (described previously). Thus, E2 → E2S in the presence of sulfotransferase and E2S → E2 in the presence of sulfatase; an equilibrium is established, which strongly favors E2S [4]. Inhibitors that block either enzyme can affect the actual amount of E2 present. Paradoxically, E2 is an effective (endogenous) inhibitor of sulfotransferase, the very enzyme, which leads to its formation via de-esterification [4].

35.5
Conclusions

This review on genotoxicity of endogenous estrogens has covered a considerable amount of territory, even though its scope is limited to human estrogens and their metabolites. Estrogens, particularly estradiol, have been shown to be weak carcinogens and the incidence of breast cancer is related to the duration of exposure to E2. This is directly related to the question of whether postmenopausal women should take hormone supplements. HRT usage has dropped significantly since 2003, due to a demonstrated increased risk of breast and endometrial cancer by HRT users. On the other hand, the protective effects of estrogen are lost in that case; this is the other side of the risk–benefit equation.

Two major theories of carcinogenesis have been discussed, along with others that are less well developed. Gustafsson and coworkers have introduced the idea that the two estrogen receptors ERα and ERβ are intimately connected, for example, where ERβ can modulate the proliferative effects of ERα. Thus, ERβ is antiproliferative, and one consequence is the search for β-selective agonists; another is the search for α-selective antagonists. The idea is that when the cell enters a state of receptor-mediated proliferation then mutations will be amplified and carcinogenesis will begin.

The second theory, advanced by Liehr, Cavalieri, and coworkers, is that metabolites of E2, in particular, the catechol 4-OH E2, can lead to the E2-3,4-quinone. This is an electrophilic Michael acceptor and is subject to nucleophilic attack. This can lead to DNA adducts, which are mutagenic, or to thiol depletion, which places the cell in a state of oxidative stress, which has been shown to be mutagenic. A third pathway is redox cycling, another generator of oxidative stress. This theory has been supported by a growing body of evidence.

Taken together these two theories give an overview of estrogen-induced carcinogenesis. The genotoxic metabolite component acts as the tumor initiator. The hormone-dependent receptor-mediated stage provides the estrogen stimulus to cause cellular proliferation. This picture is satisfying from a conceptual point of view, and the evidence for it is strong.

Another factor that has to be considered, however, is the nature of the enzymatic reactions, which contribute to determining the effective concentration of E2. Pasqualini, Chetrite, and coworkers have drawn attention to the important sulfatase–sulfotransferase esterification reaction, which they claim is more important at determining effective E2 concentrations than the aromatase pathway, which creates estrogens from their androgen precursors. Others have focused on the role of enzymes, which affect metabolite concentrations, such as COMT and NQO1. The concept of redox homeostasis has been discussed, and factors that can disturb it.

The discussion has naturally led into some mention of intervention points, so important in attempting to modulate estrogen genotoxicity. We considered various enzyme inhibitors that act to reduce E2 availability. It is also possible in principle to upregulate protective enzymes; an example was given for catalase. This generally requires exogenous sources, although the interesting case of E2 inhibiting its own formation was given.

Finally, it is expected that this review has given the reader an idea of both the simplicity and the complexity of estrogen genotoxicity. This area of research is highly active, and new findings of significance to clinical practice are occurring constantly. A related area of research only touched upon in this review, but also very active, is the search for enzyme-selective or receptor-selective ligands to modulate a given pathway. The basic understanding of estrogen genotoxicity described here has given an excellent foundation for therapeutic intervention of clinical significance, particularly in the area of hormone-dependent disease such as breast, prostate, and endometrial cancer.

Acknowledgment

I would like to thank Dr. Hooman Shadnia (Carleton University) for a careful reading of the manuscript and many helpful comments.

References

1 Dahlman-Wright, K., Cavailles, V., Fuqua, S.A., Jordan, V.C., Katzenellenbogen, J.A., Korach, K.S., Maggi, A., Muramatsu, M., Parker, M.G., and Gustafsson, J.-A. (2006) International union of pharmacology. LXIV. Estrogen receptors. *Pharmacol Rev*, **58**, 773–781.

2 Gustafsson, J.-A. (2000) Estrogens and women's health - benefit or threat? Nobel Symposium no 113, June 29 – July 1, 1999-Karlskoga, Sweden. *J Steroid Biochem Mol Biol*, **74** (5), 243–248.

3 Nilsson, S. and Gustafsson, J.-A. (2002) Biological role of estrogen and estrogen receptors. *Crit Rev Biochem Mol Biol*, **37** (1), 1–28.

4 Pasqualini, J.R. and Chetrite, G.S. (2005) Recent insight on the control of enzymes involved in estrogen formation and transformation in human breast cancer. *J Steroid Biochem Mol Biol*, **93**, 221–236.

5 De Angelis, M., Stossi, F., Carlson, K.A., Katzenellenbogen, B.S., and Katzenellenbogen, J.A. (2005) Indazole estrogens: highly selective ligands for the estrogen receptor β. *J Med Chem*, **48**, 1132–1144.

6 Katzenellenbogen, B.S., Montano, M.M., Ediger, T.R., and Sun, J. (2000) Estrogen receptors: selective ligands, partners, and distinctive pharmacology. *Recent Prog Horm Res*, **55**, 163–195.

7 MacGregor, J.I. and Jordan, V.C. (1998) Basic guide to the mechanism of antiestrogen action. *Pharmacol Rev*, **50**, 151–196.

8 Imamov, O., Shim, G.-J., Warner, M., and Gustafsson, J.-A. (2005) Estrogen receptor beta in health and disease. *Biol Reprod*, **73**, 866–871.

9 Cavalieri, E., Chakravarti, D., Guttenplan, J., Hart, E., Ingle, J., Jankowiak, R., Muti, P., Rogan, E., Russo, J., Santen, R., and Sutter, T. (2006) Catechol estrogen quinones as initiators of breast and other human cancers: implications for biomarkers of susceptibility and cancer prevention. *Biochim Biophys Acta*, **1766**, 63–78.

10 Bolton, J.L. and Thatcher, G.R.J. (2008) Potential mechanisms of estrogen quinone carcinogenesis. *Chem Res Toxicol*, **21** (1), 93–101.

11 Brueggemeier, R.W., Hackett, J.C., and Diaz-Cruz, E.S. (2005) Aromatase inhibitors in the treatment of breast cancer. *Endocr Rev*, **26**, 331–345.

12 Kuiper, G.G., Enmark, E., Pelto-Huikko, M., Nilsson, S. and Gustafsson, J.A. (1996) Cloning of a novel estrogen receptor expressed in rat prostate and ovary. *Proc Natl Acad Sci U S A*, **93**, 5925–5930.

13 Heldring, N., Pike, A., Andersson, S., Matthews, J., Cheng, G., Hartman, J., Tujague, M., Stroem, A., Treuter, E., Warner, M., and Gustafsson, J.-A. (2007), Estrogen receptors: how do they signal and what are their targets. *Physiol Rev*, **87** (3), 905–931.

14 Peng, B., Lu, B., Leygue, E., and Murphy, L.C. (2003) Putative functional characteristics of human estrogen receptor-beta isoforms. *J Mol Endocrinol*, **30**, 13–29.

15 Fox, E.M., Davis, R.J., and Shupnik, M.A. (2008) ERβ in breast cancer – Onlooker, passive player, or active protector? *Steroids*, **73**, 1039–1051.

16 Barkhem, T., Carlsson, B., Nilsson, Y., Enmark, E., Gustafsson, J.A., and Nilsson, S. (1998) Differential response of estrogen receptor α and estrogen receptor β to partial

estrogen agonists/antagonists. *Mol Pharmacol*, **54** (1), 105–112.
17 Gustafsson, J.-A. and Warner, M. (2000) Estrogen receptor β in the breast: role in estrogen responsiveness and development of breast cancer. *J Steroid Biochem Mol Biol*, **74**, 245–248.
18 Cvoro, A., Paruthiyil, S., Jones, J.O., Tzgarakis-Foster, C., Clegg, N.J., Tatomer, D., Medina, R.T., Tagliaferri, M., Schaufele, F., Scanlan, T.S., Diamond, M.I., Cohen, I., and Leitman, D.C. (2008) Selective activation of estrogen receptor-β transcriptional pathways by an herbal extract. *Endocrinol*, **148**, 538–547.
19 Stauffer, S.R., Coletta, C.J., Tedesco, R., Sun, J., Katzenellenbogen, B.S., and Katzenellenbogen, J.A. (2000) Pyrazole ligands: structure-affinity/activity relationships of estrogen receptor α-selective agonists. *J Med Chem*, **43**, 4934–4947.
20 Harris, H.A., Katzenellenbogen, J.A., and Katzenellenbogen, B.S. (2002) Characterization of the biological roles of the estrogen receptors ER α and ERβ in estrogen target tissues in vivo through the use of an ER α-selective ligand. *Endocrinology*, **143**, 4172–4177.
21 Katzenellenbogen, J.A., Muthyala, R., and Katzenellenbogen, B.S. (2003) Nature of the ligand-binding pocket of estrogen receptor α and β: the search for subtype-selective ligands and implications for the prediction of estrogenic activity. *Pure Appl Chem*, **75**, 2397–2403.
22 Asim, M., El-Salfiti, M., Qian, Y., Choueiri, C., Salari, S., Cheng, J., Shadnia, H., Bal, M., Pratt, M.A.C., Carlson, K.E., Katzenellenbogen, J.A., Wright, J.S., and Durst, T. (2009) Deconstructing estradiol: removal of B-ring generates compounds which are potent and subtype-selective estrogen receptor agonists. *Bioorg Med Chem Lett*, **19** (4), 1250–1253.
23 Gustafsson, J.-A. (2006) ERβ scientific visions translate to clinical uses. *Climacteric*, **9**, 156–160.
24 Lindberg, M.K., Moverare, S., Skrtic, S., Gao, H., Dahlman-Wright, K., Gustafsson, J.A., and Ohlsson, C. (2003) Estrogen receptor (ER)-β reduces ER α -regulated gene transcription, supporting a ''yin yang'' relationship between ER α and ERβ in mice. *Mol Endocrinol*, **17** (2), 203–208.
25 Matthews, J. and Gustafsson, J.A. (2003) Estrogen signaling: a subtle balance between ER α and ERβ. *Mol Interv*, **3** (5), 281–292.
26 Esslimani-Sahla, M., Kramar, A., Simony-Lafontaine, J., Warner, M., Gustafsson, J.A., and Rochefort, H. (2005) Increased estrogen receptor β cx expression during mammary carcinogenesis. *Clin Cancer Res*, **11** (9), 3170–3174.
27 Koehler, K.F., Helguero, L.A., Haldosen, L.A., Warner, M., and Gustafsson, J.A. (2005) Reflections on the discovery and significance of estrogen receptor β. *Endocr Rev*, **26** (3), 465–478.
28 Helguero, L.A., Faulds, M.H., Gustafsson, J.-A., and Haldosen, L.-A. (2005) Estrogen receptors alfa (ERα) and beta (ERβ) differentially regulate proliferation and apoptosis of the normal murine mammary epithelial cell line HC11. *Oncogene*, **24** (44), 6605–6616.
29 Hartman, J., Lindberg, K., Morani, A., Inzunza, J., Stroem, A., and Gustafsson, J.-A. (2006) Estrogen receptor β inhibits angiogenesis and Growth of T47D breast cancer xenografts. *Cancer Res*, **66** (23), 11207–11213.
30 Heldring, N., Pike, A., Andersson, S., Matthews, J., Cheng, G., Hartman, J., Tujague, M., Stroem, A., Treuter, E., Warner, M., and Gustafsson, J.A. (2007) Estrogen receptors: how do they signal and what are their targets. *Physiol Rev*, **87** (3), 905–931.
31 Damdimopoulos, A.E., Spyrou, G., and Gustafsson, J.-A. (2008) Ligands differentially modify the nuclear mobility of estrogen receptors α and β. *Endocrinol*, **149** (1), 339–345.

32 Brunmark, A. and Cadenas, E. (1989) Redox and addition chemistry of quinoid compounds and its biological implications. *Free Radic Biol Med*, **7**, 435–477.

33 O'Brien, P.J. (1991) Mechanisms of quinone toxicity. *Chem Biol Interact*, **80**, 1–41.

34 Bolton, J.L. (2002) Quinoids, quinoid radicals, and phenoxyl radicals formed from estrogens and antiestrogens. *Toxicology*, **177**, 55–65.

35 Ball, P. and Knuppen, R. (1980) Catechol estrogens: chemistry, biogenesis, metabolism, occurrence and physiological significance. *Acta Endocrinol (Copenh)*, **232**, 1–127.

36 Metzler, M. (1984) Metabolism of stilbene and steroidal estrogens in elation to carcinogenicity. *Arch Toxicol*, **55**, 104–109.

37 Liehr, J.G., Fang, W.F., Sirbascu, D.A., and Ari-Ulubelen, A. (1986) Carcinogenicity of catechol estrogens in Syrian hamsters. *J Steroid Biochem*, **24**, 353–356.

38 Sarabia, S.F., Zhu, B.T., Kurosawa, T., Tohma, M., and Liehr, J.G. (1997) Mechanism of cytochrome P450-catalysed aromatic hydroxylation of estrogens. *Chem Res Toxicol*, **10**, 767–771.

39 Liehr, J.G. (2000) Is estradiol a genotoxic mutagenic carcinogen? *Endocr Rev*, **21**, 40–54.

40 Liehr, J.G. (2001) Genotoxicity of the steroidal estrogens oestrone and oestradiol: Possible mechanism of uterine and mammary cancer development. *Hum Reprod Update*, **7**, 273–281.

41 Chakravarti, D., Pelling, J.C., Cavalieri, E.L., and Rogan, E.G. (1995) Relating aromatic hydrocarbon-induced DNA adducts and c-H-ras mutations in mouse skin papillomas: the role of apurinic sites. *Proc Natl Acad Sci U S A*, **92**, 10422–10426.

42 Cavalieri, E.L., Stack, D.E., Devanesan, P.D., Todorovic, R., Dwivedy, I., Higginbotham, S., Johansson, S.L., Patil, K.D., Gross, M.L., Gooden, J.K., Ramanathan, R., Cerny, R.L., and Rogan, E.G. (1997) Molecular origin of cancer: Catechol estrogen-3-4-quinones as endogenous tumor initiators. *Proc Natl Acad Sci U S A*, **94**, 10937–10942.

43 Cavalieri, E.L., Li, K.-M., Balu, N., Saeed, M., Devanesan, P., Higginbotham, S., Zhao, J., Gross, M.L., and Rogan, E.G. (2002) Catechol ortho-quinones: the electrophilic compounds that for m depurinating DNA adducts and could initiate cancer and other diseases. *Carcinogenesis*, **23**, 1071–1077.

44 Yue, W., Santen, R.J., Wang, J.-P., Li, Y., Verderame, M.F., Bocchinfuso, W.P., Korach, K.S., Devanesan, P., Todorovic, R., Rogan, E.G., and Cavalieri, E.L. (2003) Genotoxic metabolites of estradiol in breast: potential mechanism of estradiol induced carcinogenesis. *J Steroid Biochem Mol Biol*, **86**, 477–486.

45 Venugopal, D., Zahid, M., Mailander, P.C., Meza, J.L., Rogan, E.G., Cavalieri, E.L., and Chakravarti, D. (2008) Reduction of estrogen-induced transformation of mouse mammary epithelial cells by N-acetylcysteine. *J Steroid Biochem Mol Biol*, **109**, 22–30.

46 Bolton, J.L., Pisha, E., Zhang, F., and Qiu, S. (1998) Role of quinoids in estrogen carcinogenesis. *Chem Res Toxicol*, **11**, 1113–1127.

47 Lanzone, A., Nguyen, B.L., and Pasqualini, J.R. (1983) Uptake, receptor and biological response of estrone in the fetal uterus of guinea pig. *Horm Res*, **17** (3), 168–1180.

48 Pasqualini, J.R., Gelly, C., Nguyen, B.L., and Vella, C. (1989) Importance of estrogen sulfates in breast cancer. *J Steroid Biochem*, **34**, 155–163.

49 Pasqualini, J.R., Schatz, B., Varin, C., and Nguyen, B.L. (1992) Recent data on estrogen sulfatases and sulfotransferases activities in human breast cancer. *J Steroid Biochem Mol Biol*, **41**, 323–329.

50 Pasqualini, J.R. and Chetrite, G.S. (1999) Estrone sulfatase versus estrone sulfotransferase in human

breast cancer: potential clinical applications. *J Steroid Biochem Mol Biol*, **69**, 287–292.

51 Pasqualini, J.R. and Chetrite, G. (2001) Paradoxical effect of estradiol: it can block its own bioformation in human breast cancer cells. *J Steroid Biochem Mol Biol*, **78** (1), 21–24.

52 Duncan, L., Purohit, A., Howarth, N.M., Potter, B.V.L. and Reed, M.J. (1993) Inhibition of estrone sulfatase activity by estrone-3-methylthiophosponate: a potential therapeutic agent in breast cancer. *Cancer Res*, **53**, 298–303.

53 Mannisto, P.T. and Kakkola, S. (1999) Catechol-O-methyltransferase (COMT): biochemistry, molecular biology, pharmacology and clinical efficacy of the new selective COMT inhibitors. *Pharmacol Rev*, **51**, 593–628.

54 Fotsis, T., Zhang, Y., Pepper, M.S., Adlercreutz, H., Montesano, R., Nawroth, P.P., and Schweigerer, T. (1994) The endogenous oestrogen metabolite 2-methoxyestradiol inhibits angiogenesis and suppresses tumour growth. *Nature*, **368**, 237–239.

55 Sutherland, T.E., Anderson, R.L., Hughes, R.A., Altmann, E., Schuliga, M., Ziogas, J., and Stewart, A.G. (2007) 2-Methoxyestradiol – a unique blend of activities generating a new class of anti-tumor/anti-inflammatory agents. *Drug Discov Today*, **12**, 577–584.

56 Seegers, J.C., Aveling, M.-L., Van Aswegen, C.H., Cross, M., Koch, F., and Joubert, W.S. (1989) The cytotoxic effects of estradiol-17β, catecholestradiols and methoxyestradiols on dividing MCF-7 and HeLa cells. *J Steroid Biochem*, **32**, 797–809.

57 Klauber, N., Parangi, S., Flynn, E., Hamel, E., and D'Amato, R.J. (1997) Inhibition of angiogenesis and breast cancer in mice by the microtubule inhibitors 2-methoxyestradiol and taxol. *Cancer Res*, **57**, 81–86.

58 Samuni, A.M., Chuang, E.Y., Krishna, M.C., Stein, W., DeGraff, W., Russo, A., and Mitchell, J.B. (2003) Semiquinone radical intermediate in catecholic estrogen-mediated cytotoxicity and mutagenesis: chemoprevention strategies with antioxidants. *Proc Natl Acad Sci U S A*, **100**, 5390–5395.

59 Lai, C.-C., Peng, M., Huang, W.H., and Chiui, T.H. (2006) Chronic exposure of neonatal cardiac myocytes to hydrogen peroxide enhances the expression of catalase. *J Mol Cell Cardiol*, **28**, 1157–1163.

36
Design of Nutritional Interventions for the Control of Cellular Oxidation
Elizabeth P. Ryan and Henry J. Thompson

Dietary antioxidants have received considerable attention for their potential to mitigate the pathogenesis of chronic diseases such as cancer, cardiovascular disease, and diabetes. However, accumulating evidence suggests that current approaches to evaluating the hypothesis that antioxidants reduce risk for these diseases will yield largely null results. The objective of this review and analysis is to provide a framework for moving the field beyond the current antioxidant–chronic disease conundrum through the identification of critical issues that are often overlooked in developing nutritional interventions that are intended to reduce disease risk. Because the topic of oxidants, antioxidants, and health in aggregate is unwieldy to address in a single review, this chapter will focus on issues related to reactive oxygen species, DNA oxidation, antioxidants, and cancer with the goal of distilling concepts about the design of nutritional interventions that have applicability to the field in general and that will lead to critical tests of hypotheses about various endogenous toxins and health.

36.1
Overview

Of the various endogenous toxins that have been discussed in this monograph, a good case can be made for a causal role of the oxidation of cellular macromolecules in the pathogenesis of chronic diseases. Furthermore, the use of dietary antioxidants as a method to reduce disease-inducing oxidative damage has been an area of extensive investigation that has also captured the interest of health care providers and the general public. Yet, large, prospective clinical investigations conducted over the last decade have failed to find support for many of the underlying hypotheses about disease risk reduction [1–6] and in some cases the use of antioxidant supplements has actually been reported to increase mortality [7]. On the other hand, data from oxidative endpoint biomarker studies of antioxidant activity have provided a range of positive, negative, and null findings [8, 9]. This situation suggests that a deeper level of thought and reflection is needed to provide new perspective about the emerging antioxidant–chronic disease conundrum [10].

Endogenous Toxins. Diet, Genetics, Disease and Treatment.
Edited by Peter J. O'Brien and W. Robert Bruce
Copyright © 2010 WILEY-VCH Verlag GmbH & Co. KGaA, Weinheim
ISBN: 978-3-527-32363-0

Irrespective of the reactive chemical species, the damaged macromolecule, the chronic disease, or the antioxidant that is selected for discussion, a critical review of the literature reveals a surprising array of gaps in knowledge and/or methodological limitations that constitute barriers to progress [11–15]. Nonetheless, the importance of these barriers is not given adequate consideration when large prospective trials are designed and implemented. This situation has led to speculation that many trials have been fatally flawed from their outset and that null findings are therefore predictable [12]. Given this critique, we decided to focus this chapter on a single oxidative damage–disease hypothesis with which our laboratory has had considerable experience using preclinical and clinical approaches. Our objective is to identify principles that can guide future work in the field irrespective of the oxidant, antioxidant, or disease that is investigated. Herein, we discuss issues related to efforts to test the hypothesis that oxidative damage of genomic DNA is a promutagenic event that increases the risk for the occurrence of mutations that are the basis for initiating and promoting the carcinogenic process. Thus, in the tissue in which the oxidative damage occurs, risk for cancer is increased. The corollary is that antioxidants that decrease oxidative damage to DNA reduce cancer risk. This hypothesis is illustrated in Figure 36.1 and was used to construct and sequence the sections into which this chapter is divided.

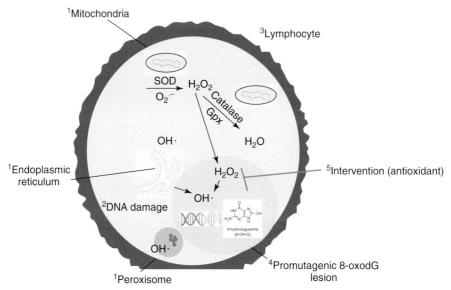

Fig. 36.1 The integration of five key elements is needed to design nutritional interventions that test the hypothesis that antioxidants reduce cancer risk via decreased oxidative DNA damage. These include (1) endogenous sources of ROS (mitochondria, peroxisomes, endoplasmic reticulum), (2) macromolecule damage (nuclear DNA 8-oxodG), (3) target tissue type (e.g., lymphocytes), (4) disease pathogenesis (ROS-induced DNA damage to oncogenes and tumor suppressor genes is promutagenic), and (5) exogenous antioxidant intervention (restore redox balance induce phase II gene transcription or endogenous antioxidant capacity).

36.2
Antioxidant–Chronic Disease Conundrum Terms

Communication in this field can sometimes be misleading because the terms used to describe oxidative status and antioxidant activities can be imprecisely applied. The language used to describe nutritional interventions in the control of oxidative stress has a powerful role in influencing the science and the interpretation of outcomes.

- **Antioxidant**: Originally was used to refer specifically to a chemical that prevented the consumption of oxygen, but can also be a molecule capable of slowing or preventing the oxidation of other molecules either enzymatically or nonenzymatically. Antioxidants terminate chain reactions by removing free radical intermediates, and inhibit other oxidation reactions by being oxidized themselves. As a result, antioxidants such as ascorbic acid or thiol compounds are often reducing agents [16, 17].
- **Hormesis**: A dose–response phenomenon characterized by low dose stimulation, high dose inhibition, resulting in either a J-shaped or an inverted U-shaped dose–response curve. Biochemical mechanisms responsible for hormetic responses are not well understood [18].
- **Nutrient**: Food or chemicals that an organism needs to live and grow or a substance used in an organism's metabolism, which must be taken in from its environment. Organic nutrients include carbohydrates, fats, proteins (or their building blocks, amino acids), and vitamins. Inorganic chemical compounds such as minerals; water, and oxygen may also be considered as nutrients. A nutrient is essential to an organism if it cannot be synthesized by the organism in sufficient quantities and must be obtained from an external source [19].
- **Oxidant/oxidizing agent**: The element or compound that gains electrons in a reduction–oxidation (redox) reaction, frequently involving the transfer of oxygen [20].
- **Reactive oxygen species ROS)**: Ions or very small molecules that include oxygen ions, free radicals, and peroxides – both inorganic and organic. They are highly reactive due to the presence of unpaired valence shell electrons. ROS form as a natural by-product of the normal metabolism of oxygen and have important roles in cell signaling. However, during times of environmental stress (e.g., ionizing radiation or heat exposure) ROS levels can increase dramatically, which can result in significant damage to cell structures. This cumulates into a situation known as *oxidative stress* [21].
- **Redox state**: A term that has historically been used to describe the ratio of the interconvertible oxidized and reduced form of a specific redox couple, for example, the balance of $NAD^+/NADH$ and $NADP^+/NADPH$ in a biological system such as a cell or organ [20].
- **Reductant/reducing agent**: The element or a compound that donates electrons in a reduction–oxidation reaction [20].

36.3
The Design of an Intervention

There are five elements illustrated in Figure 36.1 that need to be knit together to design an antioxidant–cancer intervention: the oxidizing species (endogenous toxin: the causal agent), the macromolecular target (causal link to cellular damage), the target tissue (location: the stage on which the players act), the pathogenesis of the disease (specifics of causality), and the antioxidant (the intervention). We submit that any nutritional intervention designed to mitigate the effects of an endogenous toxin will have parallel elements. Each is discussed in turn.

36.3.1
The Oxidizing Species

Life on earth has evolved in an aerobic environment. Oxygen is an obligatory component of the metabolism of the vast majority of existing life forms. Within the cells of mammalian organisms, oxygen is required for the generation of high energy phosphate compounds. During its metabolism, oxygen also participates in reduction–oxidation (redox) reactions that generate ROS. In some circumstances, the ROS serve as second messengers that convey information in cell signaling pathways constituting an essential physiological function of ROS [22]. Enzymatic reactions, occurring outside the mitochondrion, in which short-lived ROS are produced as signaling molecules include NADPH oxidase, monoamine oxidases, tyrosine hydroxylase, L-amino oxidase, lipoxygenases, cyclooxygenases, and xanthine oxidase [14, 23, 24]. Approximately 1–2% of consumed oxygen that participates in oxidative phosphorylation in mitochondria is metabolized to superoxide nonenzymatically and this constitutes an ROS species that is considered a by-product of aerobic metabolism. The concentration of superoxide is tightly regulated by enzymatic reactions whose end products are water and molecular oxygen. Superoxide radical production in the respiratory chain is primarily produced by complex I and III; however, other mitochondrial enzymes, such as complex II, glycerol-1-phosphate dehydrogenase, and dihydroorotate dehydrogenase, are also involved in the production of ROS [25, 26].

In the process of inactivating superoxide (which has a half-life of $1\,\mu s$) [27, 28], two intermediates can be formed. The first of these is hydrogen peroxide (H_2O_2), which is the product of the dismutation of superoxide by the enzyme superoxide dismutase. H_2O_2 is more stable (with a half-life of 1 ms) [27], although its stability is influenced by the pH and the redox equilibrium inside the cell. More importantly, H_2O_2 is electrically neutral and is one of the few ROS molecules that diffuses freely through cellular membranes [29]. Hydrogen peroxide is subsequently metabolized to water and molecular oxygen by glutathione peroxidase or catalase depending on the location within the cell where it is formed. These enzymes ensure that ROS products such as H_2O_2 only have a short half-life inside the cell and thus can act only at a limited distance from their site of production. A side reaction during this process is the nonenzymatic formation of hydroxyl radical as the product of the reaction between superoxide and hydrogen peroxide in the presence of a divalent cation such as iron (the Haber–Weiss Reaction). Of the ROS, hydroxyl radical has the greatest reactivity, the shortest half-life (1 ns) [27], and is directly associated with the oxidation of cellular macromolecules, including nucleic acids such as guanine.

As physiologically important signaling molecules, well regulated by-products of oxidative phosphorylation, and potential toxins/toxicants, it is not surprising that emerging evidence supports the existence of a hormetic, J- or U-shaped response to ROS in vivo [30]. In the absence of ROS for specific signaling processes, cellular dysfunction occurs [22, 31]. At low levels, normal physiological functions proceed, cellular homeostasis is maintained, and the intracellular environment is described to be in a state of redox balance. Above the levels of ROS associated with balance, a state of oxidative stress ensues; this state is associated with an increasing degree of disrupted, abnormal cellular function that has been divided into three zones of an adaptive response [18]. As intracellular levels of ROS rise above an as of yet undefined threshold level that is supraphysiological, an adaptive response is induced via the Nrf-2 transcription factor in an effort to increase endogenous antioxidant activity [32, 33]. If levels of ROS are not reduced to restore redox balance, a second tier of the response involves induction of signaling via nuclear factor kappa-light-chain-enhancer of activated B cells (NF-κB), which creates a proinflammatory response profile. A well-characterized stress response cascade initiated by forkhead box O (FoxO) transcription factors may also commence during this adaptive response zone [34, 35]. These responses would presumably then position cellular sensing machinery into a phase of monitoring cellular damage, subsequently making decisions for that particular cell's fate of death or survival.

The preceding discussion reflects the complexity of ROS regulation and illustrates a long-standing principle of toxicology, that is, the dose makes the poison. The processes described are highly dynamic and are affected not only by the type and dose of ROS present in the cell, but also by whether the ROS stimulus is transient or constitutive and whether it is induced by endogenous conditions or exogenous agents. Because of this, it is imperative to ask whether there is merit to assessment of ROS status *per se* as a dosimeter of either disease risk or antioxidant effect. Rather, in an effort to apply Ockham's razor to this dynamic intracellular landscape, we argue that a more appropriate dosimeter of disease risk and antioxidant effect is required, and therefore attention is directed to the issue of causality, specifically to the question of what target of ROS is specifically associated with the development of cancer.

36.3.2
The Macromolecular Target

Of the three ROS discussed in this chapter, it can be stated that hydroxyl radical is the most reactive species produced and has the greatest concern in terms of the oxidation of proteins, lipids, and nucleic acids. ROS generate a variety of DNA lesions, including oxidized DNA bases, abasic sites, single strand breaks, and double strand breaks. There are over 100 types of oxidative base modifications in mammalian DNA [36–41], and studies have suggested that mitochondrial DNA (mtDNA) may accumulate more oxidative DNA damage than nuclear DNA (nDNA) [42]. Increased damage to mtDNA inevitably leads to compromised mitochondrial function and integrity, and so given that damaged mitochondria are presumed to release more ROS, they could set in motion a vicious cycle of oxidative events that indirectly increase nDNA damage. Regardless of which cellular compartment suffers first, of these targets, the most compelling case can be made that the oxidation of guanine in nDNA has direct causality in the development of cancer. When hydroxyl radical is generated adjacent to nDNA, it attacks both the deoxyribose sugar and the purine and pyrimidine bases, forming multiple products. DNA bases are sensitive to ROS oxidation, particularly guanine due to its low redox potential [43]. Therefore, not surprisingly, 8-hydroxy-2-deoxyguanosine (8-oxodG) is one of the most abundant and well-characterized DNA adducts generated by ROS [40, 44]. Hydroxyl radical–induced lipid peroxidation has also been implicated in the formation of 8-oxodG [44]. It has been estimated that ∼180 guanines are oxidized to 8-oxodG per mammalian cell per day [44]. 8-oxodG is a highly mutagenic miscoding lesion that can lead to G : C to T : A transversion mutations. Although 8-oxodG is the most common product of guanine oxidation by hydroxyl radical, the various methods used to quantify the levels in experimental tumor models and human cancers have made it difficult to distinguish between normal steady-state levels that can be repaired and abnormal levels that pose an increased risk for developing cancer [45–50]. Methods used to quantify 8-oxodG levels in nDNA are listed in Table 36.1. Currently, a validated, reproducible, and standardized method to quantify the level of nDNA 8-oxodG that is considered abnormal does not exist, but will be necessary for evaluating modulation by a nutritional intervention. The issues that need to be considered in assessing 8-oxodG status in clinical studies are as follows:

- Is a snapshot of *in vivo* equilibrium levels of 8-oxodG in urine or lymphocytes representative of DNA oxidation in other tissues?
- What method of sample processing will yield reproducibly low adventitious oxidation of guanine?
- How can we uniformly validate and calibrate assays for 8-oxodG detection across labs for single site and multicenter investigations?
- What is the minimum time frame from sample collection to initiating sample processing?
- What is the allowable total sample processing time until sample is frozen for lymphocytes, tissue biopsy, or biofluids?
- How does the amount of time for sample processing and 8-oxodG methods of analysis compare with the reported half-life of 8-oxodG in nuclear DNA?

Transgenic mouse models deficient in one or more repair enzymes for 8-oxodG have served in understanding the likely causality between 8-oxodG accumulation and tumor induction. Elevated, promutagenic 8-oxodG levels were evident in some

Tab. 36.1 Detection methods used for 8-hydroxy-2-deoxyguanonsine (8-oxodG)

Methods	Considerations	References
Comet assay	Subjective analysis lends to underestimation Low spurious base oxidation and lower levels of DNA required	[21, 36, 51]
HPLC–electrochemical detection (ECD)	Guanine oxidation during DNA preparation Large amount of DNA (>30 μg) needed to minimize base oxidation Reproducible method of analysis	[36, 47, 52–54]
HPLC–MS/MS	Inaccurate detection of low levels	[55]
LC–MS/MS	Biological significance Gold standard for urine Quantitative	[56]
GC–MS	Inaccurate detection of dose response and low endogenous levels Acidic hydrolytic decomposition of DNA can produce oxidation artifact	[36, 57]
Immunological-antibody mediated	Overestimation Semiquantitative Low specificity	[56]
Radiolabeling ^{3}H, ^{32}P	High sensitivity with low amounts of DNA	[58, 59]
Ultraviolet resonance Raman (UVRR) spectroscopy	Distinct UVRR spectrum of 8-oxodG determined.	[60]

HPLC, high performance liquid chromatography; MS, mass spectrometry; GC, gas chromatography.

tissues of transgenic mouse models engineered with repair enzyme–deficient oxidative DNA damage repair genes [61, 62]. This accumulation of 8-oxodG levels in genomic DNA was associated with higher frequencies of malignancies in some tissues of transgenic mice compared to wild type mice. Specifically, G to T mutations at codon 12 of the *K-ras* gene were reported in DNA of 75% of the lung tumors of the $Ogg1^{-/-}Myh^{-/-}$ mice, establishing a causal link between deficiencies in base excision repair, DNA glycosylases correcting oxidative DNA damage, and tumorigenesis in mice. The potential for 8-oxodG adducts to affect specific cellular pathways is also supported by recent observations that the *in vivo* accumulation of 8-oxodG is not randomly distributed throughout the genome, and that there are susceptible genomic sites that are likely to be cell-type and stimulus specific [63]. An emerging concept of "chromosome territory" describes how genomic DNA can be peripherally or centrally located in the nucleus during interphase, suggesting that a particular gene location along with the chromatin structure play a role in susceptibility to oxidative base damage. Little is known regarding which parts of the genome are susceptible to oxidative damage *in vivo* at any given moment, and therefore the location of commonly mutated oncogenes and tumor suppressors at various cell cycle phases and during oxidative insult warrant further examination. Xie *et al.* revealed a pathway of lung tumorigenesis through *K-ras* activation resulting from endogenous oxidative DNA damage in a *Myh*- and *Ogg1*-deficient background, supporting the need for further investigation of these genomic "hot spots" [62]. Thus, available evidence documents that one specific oxidation product (8-oxodG), induced by a specific ROS (OH), can induce a specific mutation (G to T transversion) in a specific gene (*K-ras*) in the tumor (lung), when the cellular oxidation target was not repaired. Although not airtight, the evidence in support of a specific causal link is strong.

While recognizing the importance of simplification as a tool to focus on critical issues and to permit *in vivo* hypothesis testing, it is also important to note several caveats to the model of causality just described. First, ROS induce other forms of promutagenic nucleic acid adducts; because 8-oxodG is most prevalent, it does not mean that it is the only oncogenic oxidation product or the most important, although arguments have been made about why it merits considerable attention [49, 51, 64–67]. Secondly, as noted above, ROS regulate signaling pathways associated with cell proliferation and cell death, both of which are misregulated during carcinogenesis [22], and thus ROS may have significant effects on carcinogenesis apart from their promutagenic activity. 8-oxodG lesions may also affect the expression of specific genes via their ability to regulate the binding efficiency of transcription factors such as NF-κB to specific promoter elements [68]. Finally, the magnitude of DNA repair capacity and antioxidant defense varies across tissue types, and therefore a spectrum for the level of DNA damage that confers risk will depend on the tissue examined.

The identification of a probable causal link between cellular oxidation and cancer permits a new level of questions to be identified, answers to which are likely to play a pivotal role in designing an effective nutritional intervention. They include (i) how likely is it that ROS, specifically hydroxyl radical, will measurably drive the development of cancer if cells maintain ROS in the physiological range for the majority of the life span of a cell; (ii) what level of 8-oxodG should be considered abnormal; (iii) what component of the cellular defense system has failed and is responsible for the ability to detect abnormal levels of 8-oxodG; and (iv) how might a nutritional intervention be designed that reregulates abnormal levels of 8-oxodG?

36.3.3
The Target Tissue

Lessons that can be learned from several preclinical models of oxidative DNA damage and cancer provide important insights about what appears to be tissue-specific differences in cancer susceptibility to oxidative damage, and demonstrate the type of information needed to understand if an antioxidant is likely to have an impact on cancer in a specific tissue.

As noted above, accumulation of 8-oxodG in DNA of repair-deficient mice has been reported [69]. The finding that $myh^{-/-}/ogg1^{-/-}$ mice that are defective in base excision repair (but not singly defective animals) develop tumors of the lung, small intestine, and ovary, but not of other organs, provides the first example of a connection between accumulation of 8-oxodG in DNA and tumorigenesis. In contrast, significant accumulation of 8-oxodG in liver was not associated with malignancy. *In vivo* studies utilizing other model systems have shown that antioxidant vitamin supplementation failed to reduce liver 8-oxodG DNA levels [70, 71]. These findings suggest that multiple compensatory mechanisms predominate in liver to mitigate elevated endogenous oxidative insults, and that liver may not be an ideal target tissue for design and evaluation of nutritional interventions to reduce 8-oxodG DNA accumulation. No significant age-dependent nDNA 8-oxodG accumulation was detected in spleen, kidney, or brain of the base excision repair–deficient mice compared to wild type. Of particular importance is that differences among tissues in 8-oxodG accumulation were noted and that tumors were induced, but not in all tissues. Role of combination of repair pathways to mitigate 8-oxodG nDNA levels was demonstrated using a mouse model deficient in a mismatch repair protein ($msh2^{-/-}$) [72]. These mice exhibited increased 8-oxodG accumulation over wild type as this mismatch repair protein recognizes DNA lesions for correction during replication. Given that the oxidized dNTP pool (initially oxidized by ROS) is an important source for 8-oxodG accumulation in $msh2^{-/-}$ cell DNA, the increased steady-state levels of nDNA 8-oxodG in spleen, heart, liver, lung, and kidney indicate that mismatch repair affords a particularly important protection in these organs. Tissue accumulation of nDNA 8-oxodG is, however, inconsistent with the tumor spectrum in $msh2^{-/-}$ mice as these animals are susceptible to T cell lymphomas and intestinal cancer [72]. It can be postulated that base excision repair was sufficient to prevent tumor development; however, the absence of tumor susceptibility in organs in which 8-oxodG accumulates indicates that persistence of the oxidized purine in DNA is not, in itself, sufficient for tumor development. Consequently, we propose that evaluation of antioxidant interventions to reduce 8-oxodG accumulation and tumor development in a specific tissue should utilize preclinical models that demonstrate a causal link between a particular ROS and macromolecular target. This approach will be invaluable for future translation of nutritional interventions for the control of oxidative DNA damage by endogenous toxins in humans with identified single nucleotide polymorphisms (SNPs) in DNA repair genes [73, 74].

Interestingly, peroxiredoxin (Prdx)-$1^{-/-}$ mice on a B6129SvEv background also showed increased 8-oxodG accumulation in embryonic fibroblasts [75]. Prdx are antioxidant enzymes that have peroxidase functions, are found in both the cytoplasm and the nucleus, and work with thioredoxin-1 to effectively detoxify hydrogen peroxide. The spectrum of cancers that developed in Prdx1 mutant mice that may be associated with increased 8-oxodG levels included epithelial and mesenchymal tumors (hepatocellular carcinoma, fibrosarcoma, osteosarcoma, islet cell adenomas, and adenocarcinomas of lung and breast), the latter of which are less common in aging B6, 129SvEv mice [75]. During redox signaling, some Prdxs have also been implicated in the regulation of NF-κB through the initial activation in the cytoplasm by controlling the components affecting I-κB phosphorylation and subsequent dissociation. In principle, Prdxs could have a different function in the nucleus because NF-κB interactions with DNA are governed by a redox-sensitive cysteine (Cys^{62}) on the p50 subunit of the NF-κB dimer [76]. Together, these data implicate the Prdx family of antioxidant proteins with not only 8-oxodG accumulation and the development of cancer but also in the critical second tiered adaptive response of the redox signaling cascade. Given that redox sensors and signaling pathways for proliferation cooperate, and that selected mutations of oncogenes generate new signaling through pathways that increase cellular proliferation, the reciprocal condition must not be overlooked whereby increased proliferation can also further enhance the rate of mutation.

In total, the data reviewed in this section indicate that levels of 8-oxodG are likely to have a greater impact on risk for cancer in some tissues than others. This has important implications relative to the design of an intervention study because it underscores the difficulty of identifying susceptible tissues and the problematic nature of using relatively noninvasive procedures to assess risk in a particular organ site. For example, although oxidative DNA damage was shown to be greater in the mammary gland from breast cancer patients compared to noncancer controls [77], it is not clear if these differences are similarly reflected in concentrations of 8-oxodG in lymphocyte DNA or urine. One related factor to highlight in the clinical and biomarker studies that have reported null effects is that interventions have been studied in relatively healthy populations of individuals. As reported by us and others, using a noninvasive indicator of oxidative damage (e.g., lymphocyte 8-oxodG) in individuals at high risk for breast cancer, showed that most individuals, perhaps as high as 75%, have levels of oxidative biomarkers that are relatively low and nonresponsive to an intervention [78]. Few intervention trials have screened for oxidative abnormalities as an entry criterion, and even when that is done, there is currently no way to know if what is measured peripherally actually reflects abnormalities in redox metabolism or DNA damage in the target tissue of interest. Thus, most nutritional intervention studies are likely to be intervening in individuals who are unlikely to benefit and hence null effects should be expected as they are found in drug treatment studies when individuals without specific metabolic defects are treated with defect-specific, that is, target-specific drugs [79–82].

36.3.4
The Pathogenesis of the Disease

An important consideration for the design of an intervention study is to predict when during the carcinogenic process that oxidation is likely to contribute the pathogenesis of the disease. An equally important corollary to this issue is how an antioxidant intervention would be expected to disrupt the oxidant-mediated effects on the disease process. For this discussion, we operationally divide effects of oxidation on carcinogenesis into primary and secondary events. Primary events are defined as those related to the mutation of oncogenes and/or tumor suppressor genes. Thus, the benefit of reducing the occurrence of those events involved in the initial steps of transformation or tumor progression has to be considered in the framework of multihit models such as those established for most solid tumors, notably the colon [83]. If, in fact, the prevention of these events is the goal of an intervention, it should be expected that benefit would be observed after 10–15 years of intervention; the time frame that is predicted to be required from an initial transforming mutation until clinically detectable disease is manifest [84], yet the prospective clinical trials in which null effects have been reported [3, 5–7, 11, 37, 65, 85–87] were conducted for time frames of less than 10 years, and as noted earlier, in relatively healthy populations. Hence, it should not be a surprise if null findings are reported. If these predictions are correct, they raise an important question of whether it is possible to design nutritional intervention studies that are of sufficient duration due to issues of cost, retention of participants in a study, and maintaining adherence to an intervention protocol.

Secondary events that ROS may impact are noted in [84] and involve three major cellular processes misregulated during carcinogenesis: cell proliferation, cell death (and therefore survival), and blood vessel formation. If one attempts to simplify the complex array of signaling events involved in these processes relative to those that ROS have been shown to affect, it becomes clear that ROS-induced signaling and signaling misregulated during carcinogenesis intersect at multiple levels. The intersections related to Nrf-2 may occur early in the response to ROS to boost endogenous antioxidant capacity and then cross talk with other transcription factors, namely, NF-κB and FoxO. The relationships and specific regulation of these pathways by ROS has been the subject of a number of excellent reviews and the goal is not to review those details here [34, 35, 88–91]. Rather, the objective is to call attention to the need for predictions about how regulation of altered signaling by an antioxidant intervention would be expected to manifest in a clinical intervention study. Specifically, what quantifiable aspect of the disease would be affected, and how many years of intervention would be necessary before a clinically detectable effect on signaling should be anticipated? We expect that the most reasonable answers to these questions can be gleaned from the drug therapy arena as we are unaware of any data from nutrition interventions on which predictions can be based. In the absence of such information, two questions should be considered in planning future work: (i) how best to detect misregulation of signaling pathways in an individual and (ii) how best to measure responsiveness to an intervention? Mechanistically, of course, one must be able to identify the offending ROS and have knowledge of what nutritional intervention will restore normal signaling in the target tissue in which the misregulated function is known to be induced.

36.3.5
The Antioxidants

As noted earlier, the vast majority of life as we know it has evolved in an environment in which oxygen metabolism is an essential component of existence. Because intermediates and products of oxygen metabolism have essential functions, yet are chemically reactive, living organisms have developed systems to control the concentration and location of ROS and minimize potentially harmful side reactions. Those mechanisms can be broadly described as antioxidation. In many ways, the current approach to designing nutritional intervention studies using antioxidant supplements parallels the time when ROS were viewed as detrimental chemicals and that a primary cause of their harmful effects were environmental exposures with limited recognition of the importance of endogenous production of ROS. Only recently has the appreciation of endogenous antioxidant defense mechanisms begun to receive more attention, but in our judgment, regulation of endogenous antioxidant activity as a primary intervention mechanism has not been factored into the design or interpretation of prospective clinical trials of antioxidant–chronic disease hypotheses.

Molecules with antioxidant activity are of two sources: endogenous and exogenous and range in size from less than 1000 kDa to multisubunit proteins; together, they exert dose-dependent effects in cellular systems. However, what is less understood is that the antioxidant dose response in biological systems is not linear. Rather, it is U-shaped, that is hormetic [18]. At low and high concentrations of antioxidant "activity" or "capacity," cell survival is compromised minimally because on either side of the beneficial range, pro-oxidative, tissue damaging events are favored and the physiological functions of ROS are impaired. Moreover, essentially no consideration has been given to the need to superimpose two hormetic dose–response curves, that for ROS-mediated biological effects and that for antioxidant capacity, to fully understand how best to maintain or restore redox balance within a cell, tissue, organ system, and organism (Figure 36.2).

As outlined in preceding sections, our thesis is that individuals who are most likely to benefit from an antioxidant intervention are those experiencing a constitutively elevated level of nDNA 8-oxodG in a particular tissue. In the ideal situation, the antioxidant defense system would be interrogated to identify the deficient component(s) that accounts for aberrant 8-oxodG levels. Components of the antioxidant system to be evaluated include (i) enzymes involved in ROS production, phase I metabolism genes and superoxide dismutase, (ii) ROS detoxification enzymes (e.g., catalase, glutathione peroxidase, and Prdxs), and (iii) concentrations of low molecular weight reductants such as reduced glutathione, which is synthesized by the cells, or vitamins C and E, which are essential dietary nutrients that play a dominant role in antioxidant activity (Table 36.2). For the deficient component(s), a suitable intervention would be developed. For example, if levels of decomposition enzymes were suboptimal, a dietary phytochemical might be given to induce activity via KEAP-1: Nrf-2. If a particular radical was being produced in excess despite normal decomposition activity, a specific chemical could be delivered orally to neutralize that radical in the target tissue of interest, for example, vitamins C or E, other phytochemicals, or even a nontoxic spin trapping agent. Similarly, if repair activity is found to be low, as a result of SNPs in DNA repair genes, an intervention could be designed to induce activity and/or circumvent the consequences of the deficit. These are important opportunities to consider, and are becoming feasible to implement with the introduction of high throughput omics technologies that can be used in the clinical setting [92].

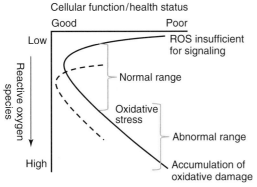

Fig. 36.2 Hypothetical approach for determination of an abnormal range of oxidative damage in a target tissue. Evaluate potential of intervention to reregulate the hormetic dose–response curves into a normal range for ROS to prevent further oxidative damage. An equation to determine the abnormal range of DNA damage is put forth as a potential tool with which to visualize what information is needed to quantify the level of oxidative damage that is abnormal and that would qualify as an entry criterion for a nutritional antioxidant intervention. The abnormal level would also be a determinant factor in the dose of antioxidant required to reregulate the oxidative stress microenvironment. With such a tool it would be possible to calculate the proportional concentration of oxidative damage that an intervention can modulate. BER, base excision repair; MMR, mismatch repair; SOD, superoxide dismutase; CAT; catalase; Gpx; glutathione peroxidase.

Tab. 36.2 Components of the antioxidant system influencing DNA damage

Genetic factors that may cause deficits in enzymatic activity	Low molecular weight reductants: nonenzymatic
ROS production Superoxide dismutase Phase I metabolism: cytochrome p450	Endogenous Glutathione (GSH/GSSG) Uric acid
	α-Lipoic acid
ROS detoxification Catalase Glutathione peroxidases Thioredoxins Peroxiredoxins	Exogenous Ascorbate (vitamin C) α-Tocopherol (vitamin E) Carotenoids Polyphenols Other redox-active phytochemicals

36.4
Designing the Next Generation of Nutritional Interventions

The five elements presented in Section 36.3 have been integrated into a template for the design of antioxidant intervention experiments to reduce cancer risk. This section lays out objectives and identifies gaps in our knowledge relative to this objective. Key points are as follows:

1. Measurement of endogenous ROS regulation
   a. difficult to measure and interpret the dynamics of various ROS types, doses, and location;
   b. ROS stimulus can be transient or constitutive as induced by endogenous conditions or exogenous agents;
   c. no current assessment of ROS status is used as a dosimeter of disease risk or antioxidant effect
   d. an appropriate dosimeter of disease risk and antioxidant effect is required to assess intervention efficacy.
2. Evidence for causal link between oxidant and macromolecular target
   a. DNA damage by ROS is linked to the development of cancer in preclinical models with insufficient repair mechanisms, but other cellular targets of ROS and tissue specificity of tumor response need further investigation;
   b. low likelihood for endogenous ROS, specifically hydroxyl radical, to measurably drive the development of cancer if cells maintain the physiological range for the majority of the cell's life span.
3. Abnormal level of oxidative damage that confers risk (e.g., 8-oxodG)
   a. preclinical and clinical validation of abnormal levels is needed for each disease endpoint;
   b. multiple combinations of genetic susceptibility factors may contribute to abnormal levels of 8-oxodG accumulation;
   c. nutritional interventions designed to reregulate abnormal levels of 8-oxodG require preclinical validation prior to clinical translation.
4. Clinical relevance of peripheral vs. target tissue levels of damage
   a. data is insufficient to support that peripheral abnormalities in redox metabolism or DNA damage (e.g., 8-oxodG in lymphocytes or urine) reflect abnormal levels in a target tissue (e.g., breast, colon, lung);
   b. the time frame must be considered from sample collection to DNA damage analysis for all tissue and biofluid types to reduce artifact DNA oxidation;
   c. it is unknown whether or not the nutritional interventions that lower peripheral oxidative damage levels also lower tissue levels.

36.4.1
Define Individuals Who Are at Increased Risk for Cancer

Although many approaches are in use, which attempt to characterize antioxidant capacity or the state of redox balance, we argue that the most information is gained if the causality factor(s) can be used as a dosimeter of disease risk.

36.4.1.1 Risk for Oxidative Stress–Induced Mutations

If the goal of the intervention is to decrease the risk for the underlying mutagenic basis of the carcinogenic process, what we have herein referred to as a *primary mechanism* by which oxidation affects carcinogenesis, then oxidized products of nucleic acids are a good candidate biomarker for cancer risk assessment. Given that a compelling case can be made for measuring 8-oxodG, there is a need for a thorough understanding of the challenges that exist in the collection of clinical samples, the measurement of this analyte in clinical materials, and in interpreting the results. Presuming that only minimally invasive procedures will be used, lymphocytes isolated from peripheral blood samples or urine are the clinical specimens of choice. Buccal mucosal cells have also been evaluated and were judged to have significant limitations due to the heterogeneity of the cells [52, 93]. Presuming that the clinical samples can be appropriately collected and evaluated, there are two critical questions that remain to be answered: (i) what level of 8-oxodG in lymphocytes or urine should be considered abnormal and (ii) how does an abnormal level of 8-oxodG in lymphocytes or urine relate to the level of oxidized base in the target tissue that is the focus of the intervention? See list given in Section 36.3.2 for additional questions regarding the use of 8-oxodG levels in clinical studies. We judge that the answers to these two questions relative to mutagenic mechanisms of carcinogenesis are unknown and represent critical gaps of knowledge that initially can and should be addressed using preclinical models such as those reported in [61, 62, 75]. To translate and confirm what is learned in preclinical models to the clinic, it is tempting to suggest that emerging imaging technologies in conjunction with spin trapping agents can be used to supply the missing information and put clinical intervention studies on a solid mechanistic footing [75, 94].

36.4.1.2 Risk for Altered Signal Transduction

As noted in Section 36.3.2, abnormal levels of ROS have been reported to regulate, at the transcriptional and posttranscriptional levels, the activity of signaling pathways misregulated during carcinogenesis, for example, via the inactivation of phosphatase activity, enhancement of signaling transduced by NF-κB, or production of cytotoxic lipid peroxides [22, 44, 78]. We argue that abnormally high levels of an oxidation damage product specifically linked to procarcinogenic cell signaling would provide a useful dosimeter to identify individuals who would be likely to respond to an antioxidant intervention. However, three gaps in knowledge need to be addressed to proceed with evaluation of the effects of an intervention on altered cell signaling attributed to oxidative damage: (i) what prominent cellular oxidation product is relevant to measure and how does it relate to altered cell signaling, (ii) in what clinical specimen obtained via relatively noninvasive techniques should the oxidation product be measured, and (iii) how does assessment of this "dosimeter" peripherally relate to what is occurring in the target tissue? In our judgment, the critical first step required to address these questions would involve experiments using appropriate preclinical models with validation in human populations prior to initiating an intervention.

36.4.2
Identify Candidate Mechanisms that Accounts for the Abnormal Levels of Oxidative Products

If an individual can be shown to produce an abnormal level of a DNA oxidation product, we judge that it is critical to determine: (i) what accounts for our ability to detect damage, (ii) how and where in a cell ROS was produced, (iii) where within the cell the induced damage is likely to be occurring, and (iv) how this information relates to the status of endogenous and exogenous antioxidant defense mechanisms. A key question is whether ROS are being produced in excess, under decomposed/neutralized, and/or if damage is not adequately repaired. The reasons for needing to know are apparent from the discussion in Section 36.3.3, but to summarize, one needs to identify the defect to determine how to intervene. We are reaching the point that initial clinical evaluation of this problem is possible via the use of SNP analyses of three sets of genes: those involved in ROS production, those involved in ROS detoxification, and those involved in the repair of ROS-mediated damage (refer to Table 36.2) [95–99]. An advantage of these are germline mutations and consequently they will be present in all cells within the body, although the extent of altered function can be inferred only in a target tissue. Emerging evidence suggests that metabolic profiling could also be used to detect altered oxidative metabolism, although this approach too suffers of a lack of specificity [100]. Again, preclinical models offer an underutilized approach by which to gain critical information essential to design effective clinical interventions.

36.4.3
Define a Nutritional Intervention that is Tailored to Correct the Defect

Available evidence indicates that many of the prospective clinical trials of antioxidant amelioration of chronic disease risk have assumed that exogenous antioxidants, whether they are essential or nonessential nutrients, in the form of supplements or whole foods, should reduce oxidative cellular events. Because the null results of clinical interventions speak for themselves, we propose that an alternative line of reasoning is required and can be grounded on two principles: establish that ROS metabolism is abnormal and identify the factors that account for abnormality. We submit, that having done this, the primary focus of the nutritional intervention should be to boost endogenous antioxidant defense mechanisms to normalize ROS production if abnormal, to increase detoxification activity if it appears low and inducible, and to enhance the activity of oxidative damage repair pathways if evidence indicates suboptimal activity. The merit of and potential role for exogenous nutrients as radical scavengers *per se* should be judged based on an understanding of hormetic dose–response curves for both ROS and antioxidant capacity (Figure 36.2). Recognition of the electrochemical potential of various reductants relative to ascorbic acid and tocopherol is also needed when these are provided to the host at nutritionally recommended levels as discussed by Rice-Evans [101, 102]. Thus, we are suggesting the identification of categorized cohorts that are characterized by a quantifiable abnormal level of oxidation, for which interventions are tailored to ameliorate the amount of defects that are classified as abnormal.

36.5
Overall Principles

From this review and analysis, we have identified a number of principles to guide the development of nutritional interventions to ameliorate the health consequences of endogenous toxins. They are as follows:

1. Identify and quantify toxin(s) that contribute to damage.
2. Establish abnormal level of the macromolecular damage product.
3. Identify the causal link between abnormal level of toxin and macromolecular target (e.g., toxin overproduction, insufficient detoxification, inadequate repair, age-related decline in endogenous antioxidant defenses).
4. Identify an intervention strategy that specifically targets the defective aspects of metabolism.
5. Conduct the intervention in defined cohorts for an extended period to assure that a measurable effect on the disease process is predictable if the problem is corrected.
6. Monitor efficacy using validated biomarkers.

36.6
The Broader Perspective

From many perspectives, the chapters in this monograph have dissected the role that endogenously produced toxins play in modulating risk of four seemingly unrelated chronic diseases. The relationships among these diseases have become better understood as data resulting from omics investigations have provided evidence of a common pathogenic basis for their occurrence [92, 103–106]. Specifically, cardiovascular disease, cancer, type II diabetes, and obesity are metabolic disorders with shared impairments in both cellular processes and metabolism, although each disease does retain unique characteristics. At the cellular level, the pathologies associated with each disease display alterations in cell proliferation, blood vessel formation, and cell death, that is, necrosis, apoptosis, and autophagy. Also common to these diseases are alterations in glucose metabolism, chronic inflammation, cellular oxidation, and chronic endotoxemia, which are attributed to a common network of cell signaling events that are misregulated in each of these disease states [103]. Emerging evidence indicates that modulation of gut microflora predisposes an individual to each of the disease processes [105, 106]. Microflora appear to be able to exert effects through either biosynthesis of new compounds or chemical transformations of ingested ones, and as a consequence, influence exposure of the host to gut microflora-associated endotoxins such as lipopolysaccharides [105–107]. Utilizing the principles distilled in this chapter, nutritional interventions may be designed as supplements and/or foods to interfere with any of these distinct cellular and molecular processes. The interrelationships discussed in this broader context are illustrated in Figure 36.3. Although it should be recognized that an intervention-related (positive) change at one site may result in no overall effect due to compensation by a (negative) change in a counterpathway, the possibility also exists for an intervention to exhibit high efficacy as a result of the multidimensions, such that targeted restoration at any one site may trigger or support the reregulation to normal function in a parallel process.

The majority of this chapter focused on the antioxidant–cancer conundrum using the results of supplemental antioxidant intervention studies to drive the discussion. However, a broad range of nutritional interventions, many of which are food based [53, 78, 108], have the capacity to modulate cellular oxidation albeit as either a primary or secondary outcome. Although the complexity of the interactions shown in Figure 36.3 is daunting, the overall principles outlined in Section 36.5 provide a framework to guide the systematic investigation of nutritional interventions intended to reduce chronic disease risk via the inhibition of pathogenic mechanisms associated with exposure to endogenously produced toxins.

36.7
Summary

This chapter was written to stimulate thought and discussion about factors that may account for emerging clinical findings of null effects of antioxidant interventions on chronic disease endpoints. The goal was to maximize the ability to view those investigations as productive failures and to minimize the tendency to judge that nutritional interventions that target oxidative mechanisms have no discernable value in chronic disease prevention. Although certain aspects of this review in fact are consistent with the notion that antioxidant interventions are unlikely to benefit many individuals, the evidence also argues that definable populations are at risk to excessive oxidative damage to cellular macromolecules and will benefit from tailored nutritional interventions [78, 108, 109]. The challenge under these circumstances is to develop feasible approaches to identify these individuals and to examine the requisite mechanistic information needed to reestablish an individual's redox balance in a highly specific manner via appropriate lifestyle interventions.

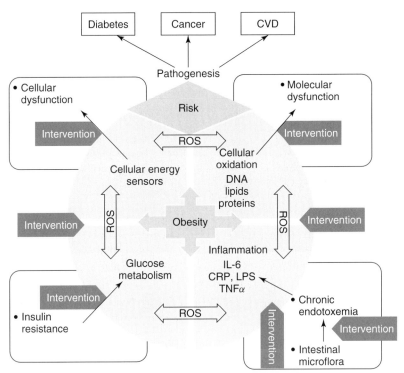

Fig. 36.3 Major chronic diseases, cancer, cardiovascular disease (CVD), type II diabetes, and obesity account for 60% of all deaths worldwide. Common to the pathogenesis of these diseases are altered glucose metabolism, chronic inflammation, increased cellular oxidation, and chronic endotoxemia. The pathogenesis of these chronic diseases has substantial cellular and molecular interrelationships that may include gram-negative bacteria in the gut that flourish and die, leaving cell wall components (e.g., lipopolysaccharide, LPS) behind to enter the systemic circulation and result in a state of chronic inflammation. Glycation products are formed at increased rates due to elevated blood glucose and may also lead to elevated levels of circulating inflammatory factors. Inflammatory processes are one stimulus for increased production of free radical species that can lead to oxidation of cellular components and result in DNA mutation and altered molecular function. Along with increases in available glucose, ROS can also perturb cellular energy sensing mechanisms with dysfunction occurring as cells lose fine control of proliferation and death pathways. Obesity results from a large number of known and unknown factors that elevate disease risk. The cumulative and interdependent effects of many processes increase risk for the development of chronic diseases. Nutritional interventions can be designed to interfere at multiple points in the pathogenesis of disease. Although the primary effect of the intervention will be to reduce abnormal levels of damage in a defined endpoint, this effect may then initiate a cascade of reprogramming other interrelated aberrant processes to normal levels.

Acknowledgments

This work was supported in part by PHS grants R01-CA125243 and U54-CA116847 from the National Cancer Institute. The authors thank John McGinley and Matthew Thompson for assistance with the materials used in the preparation of this manuscript.

References

1 Gaziano, J.M., Glynn, R.J., Christen, W.G., Kurth, T., Belanger, C., MacFadyen, J., Bubes, V., Manson, J.E., Sesso, H.D., and Buring, J.E. (2009) Vitamins E and C in the prevention of prostate and total cancer in men: the Physicians' Health Study II randomized controlled trial. *J Am Med Assoc*, **301**, 52–62.

2 Lin, J., Cook, N.R., Albert, C., Zaharris, E., Gaziano, J.M., Van, D.M., Buring, J.E., and Manson, J.E. (2009) Vitamins C and E and beta carotene supplementation and cancer risk: a randomized controlled trial. *J Natl Cancer Inst*, **101**, 14–23.

3 Lippman, S.M., Klein, E.A., Goodman, P.J., Lucia, M.S., Thompson, I.M., Ford, L.G., Parnes, H.L., Minasian, L.M., Gaziano, J.M., Hartline, J.A., Parsons, J.K., Bearden, J.D. III, Crawford, E.D., Goodman, G.E., Claudio, J., Winquist, E., Cook, E.D., Karp, D.D., Walther, P., Lieber, M.M., Kristal, A.R., Darke, A.K., Arnold, K.B., Ganz, P.A., Santella, R.M., Albanes, D., Taylor, P.R., Probstfield, J.L., Jagpal, T.J., Crowley, J.J., Meyskens, F.L. Jr, Baker, L.H., and Coltman, C.A. Jr (2009) Effect of selenium and vitamin E on risk of prostate cancer and other cancers: the Selenium and Vitamin E Cancer Prevention Trial (SELECT). *J Am Med Assoc*, **301**, 39–51.

4 Lonn, E., Bosch, J., Yusuf, S., Sheridan, P., Pogue, J., Arnold, J.M., Ross, C., Arnold, A., Sleight, P., Probstfield, J., and Dagenais, G.R. (2005) Effects of long-term vitamin E supplementation on cardiovascular events and cancer: a randomized controlled trial. *J Am Med Assoc*, **293**, 1338–1347.

5 Sesso, H.D., Buring, J.E., Christen, W.G., Kurth, T., Belanger, C., MacFadyen, J., Bubes, V., Manson, J.E., Glynn, R.J., and Gaziano, J.M. (2008) Vitamins E and C in the prevention of cardiovascular disease in men: the Physicians' Health Study II randomized controlled trial. *J Am Med Assoc*, **300**, 2123–2133.

6 Vivekananthan, D.P., Penn, M.S., Sapp, S.K., Hsu, A., and Topol, E.J. (2003) Use of antioxidant vitamins for the prevention of cardiovascular disease: meta-analysis of randomised trials. *Lancet*, **361**, 2017–2023.

7 Bjelakovic, G., Nikolova, D., Gluud, L.L., Simonetti, R.G., and Gluud, C. (2007) Mortality in randomized trials of antioxidant supplements for primary and secondary prevention: systematic review and meta-analysis. *J Am Med Assoc*, **297**, 842–857.

8 Knasmuller, S., Nersesyan, A., Misik, M., Gerner, C., Mikulits, W., Ehrlich, V., Hoelzl, C., Szakmary, A., and Wagner, K.H. (2008) Use of conventional and -omics based methods for health claims of dietary antioxidants: a critical overview. *Br J Nutr*, **99E** (Suppl 1), ES3–E52.

9 Mayne, S.T. (2003) Antioxidant nutrients and chronic disease: use of biomarkers of exposure and oxidative stress status in epidemiologic research. *J Nutr*, **133**, 933S–994S.

10 Seifried, H.E., McDonald, S.S., Anderson, D.E., Greenwald, P., and Milner, J.A. (2003) The antioxidant conundrum in cancer. *Cancer Res*, **63**, 4295–4298.

11 Frei, B. (2004) Efficacy of dietary antioxidants to prevent oxidative damage and inhibit chronic disease. *J Nutr*, **134**, 3196S–3319S.
12 Blumberg, J.B. and Frei, B. (2007) Why clinical trials of vitamin E and cardiovascular diseases may be fatally flawed. Commentary on "The relationship between dose of vitamin E and suppression of oxidative stress in humans". *Free Radic Biol Med*, **43**, 1374–1376.
13 Thompson, H.J. (2004) DNA oxidation products, antioxidant status, and cancer prevention. *J Nutr*, **134**, 3186S–3187S.
14 Valko, M., Leibfritz, D., Moncol, J., Cronin, M.T., Mazur, M., and Telser, J. (2007) Free radicals and antioxidants in normal physiological functions and human disease. *Int J Biochem Cell Biol*, **39**, 44–84.
15 Dalle-Donne, I., Rossi, R., Colombo, R., Giustarini, D., and Milzani, A. (2006) Biomarkers of oxidative damage in human disease. *Clin Chim*, **52**, 601–623.
16 Halliwell, B. (1995) How to characterize an antioxidant: an update. *Biochem Soc Symp*, **61**, 73–101.
17 Halliwell, B. (1990) How to characterize a biological antioxidant. *Free Radic Res Commun*, **9**, 1–32.
18 Calabrese, E.J. (2008) Converging concepts: adaptive response, preconditioning, and the Yerkes-Dodson Law are manifestations of hormesis. *Ageing Res Rev*, **7**, 8–20.
19 Whitney, E.A.S.R. (2007) *Understanding Nutrition*, Wadsworth Publishing.
20 Saran, M. and Bors, W. (1990) Radical reactions in vivo – an overview. *Radiat Environ Biophys*, **29**, 249–262.
21 Halliwell, B. and Gutteridge, J.M.C. (1999) *Free Radical in Biology and Medicine*, Oxford Science Publications, New York.
22 Martindale, J.L. and Holbrook, N.J. (2002) Cellular response to oxidative stress: signaling for suicide and survival. *J Cell Physiol*, **192**, 1–15.
23 Casado, J.A., Merino, J., Cid, J., Subira, M.L., and Sanchez-Ibarrola, A. (1996) Oxidizing agents and free radicals in biomedicine. *Rev Med Univ Navarra*, **40**, 31–40.
24 Coyle, J.T. and Puttfarcken, P. (1993) Oxidative stress, glutamate, and neurodegenerative disorders. *Science*, **262**, 689–695.
25 Lenaz, G. (2001) The mitochondrial production of reactive oxygen species: mechanisms and implications in human pathology. *Iubmb Life*, **52**, 159–164.
26 Murphy, M.P. (2009) How mitochondria produce reactive oxygen species. *Biochem J*, **417**, 1–13.
27 Bergendi, L., Benes, L., Durackova, Z., and Ferencik, M. (1999) Chemistry, physiology and pathology of free radicals. *Life Sci*, **65**, 1865–1874.
28 Buettner G.R. and Mason R.P. (2003) Spin-trapping methods for detecting superoxide and hydroxyl free radicals in vitro and in vivo, in *Critical Reviews of Oxidative Stress and Aging: Advances in Basic Science, Diagnostics and Intervention*, vol. **1** (eds R.G. Cutler and H. Rodriguez), World Scientific, New Jersey, London, Singapore, Hong Kong, pp. 27–38.
29 Reth, M. (2002) Hydrogen peroxide as second messenger in lymphocyte activation. *Nat Immunol*, **3**, 1129–1134.
30 Zhang, Q., Pi, J., Woods, C.G., Jarabek, A.M., Clewell, H.J., and Andersen, M.E. (2008) Hormesis and adaptive cellular control systems. *Dose Responce*, **6**, 196–208.
31 Finkel, T. and Holbrook, N.J. (2000) Oxidants, oxidative stress and the biology of ageing. *Nature*, **408**, 239–247.
32 Itoh, K., Tong, K.I., and Yamamoto, M. (2004) Molecular mechanism activating Nrf2-Keap1 pathway in regulation of adaptive response to electrophiles. *Free Radic Biol Med*, **36**, 1208–1213.
33 Thimmulappa, R.K., Mai, K.H., Srisuma, S., Kensler, T.W., Yamamato, M., and Biswal, S. (2002) Identification of Nrf2-regulated genes induced by the chemopreventive

34 Arden, K.C. (2006) Multiple roles of FOXO transcription factors in mammalian cells point to multiple roles in cancer. *Exp Gerontol*, **41**, 709–717.

35 Hu, M.C.T., Lee, D.F., Xia, W.Y., Golfman, L.S., Fu, O.Y., Yang, J.Y., Zou, Y.Y., Bao, S.L., Hanada, N., Saso, H., Kobayashi, R., and Hung, M.C. (2004) I kappa B kinase promotes tumorigenesis through inhibition of forkhead FOXO3a. *Cell*, **117**, 225–237.

36 Collins, A.R., Cadet, J., Moller, L., Poulsen, H.E., and Vina, J. (2004) Are we sure we know how to measure 8-oxo-7,8-dihydroguanine in DNA from human cells? *Arch Biochem Biophys*, **423**, 57–65.

37 Djuric, Z. and Kritschevsky, D. (1993) Modulation of oxidative DNA damage levels by dietary fat and calories. *Mutat Res*, **295**, 181–190.

38 Douki, T., Riviere, J., and Cadet, J. (2002) DNA tandem lesions containing 8-oxo-7,8-dihydroguanine and formamido residues arise from intramolecular addition of thymine peroxyl radical to guanine. *Chem Res Toxicol*, **15**, 445–454.

39 Loft, S., Astrup, A., Buemann, B., and Poulsen, H.E. (1994) Oxidative DNA damage correlates with oxygen consumption in humans. *FASEB J*, **8**, 534–537.

40 Poulsen, H.E. (2005) Oxidative DNA modifications. *Exp Toxicol Pathol*, **57** (Suppl 1), 161–169.

41 Termini, J. (2000) Hydroperoxide-induced DNA damage and mutations. *Mutat Res*, **450**, 107–124.

42 Yakes, F.M. and Van, H.B. (1997) Mitochondrial DNA damage is more extensive and persists longer than nuclear DNA damage in human cells following oxidative stress. *Proc Natl Acad Sci U S A*, **94**, 514–519.

43 Crespo-Hernandez, C.E., Close, D.M., Gorb, L., and Leszczynski, J. (2007) Determination of redox potentials for the Watson-Crick base pairs, DNA nucleosides, and relevant nucleoside analogues. *J Phys Chem B*, **111**, 5386–5395.

44 Goto, M., Ueda, K., Hashimoto, T., Fujiwara, S., Matsuyama, K., Kometani, T., and Kanazawa, K. (2008) A formation mechanism for 8-hydroxy-2′-deoxyguanosine mediated by peroxidized 2′-deoxythymidine. *Free Radic Biol Med*, **45**, 1318–1325.

45 Cadet, J., Douki, T., Frelon, S., Sauvaigo, S., Pouget, J.P., and Ravanat, J.L. (2002) Assessment of oxidative base damage to isolated and cellular DNA by HPLC-MS/MS measurement. *Free Radic Biol Med*, **33**, 441–449.

46 Cadet, J., Bellon, S., Berger, M., Bourdat, A.G., Douki, T., Duarte, V., Frelon, S., Gasparutto, D., Muller, E., Ravanat, J.L., and Sauvaigo, S. (2002) Recent aspects of oxidative DNA damage: guanine lesions, measurement and substrate specificity of DNA repair glycosylases. *Biol Chem*, **383**, 933–943.

47 Collins, A., Cadet, J., Epe, B., and Gedik, C. (1997) Problems in the measurement of 8-oxoguanine in human DNA. Report of a workshop, DNA oxidation, held in Aberdeen, UK, 19-21 January, 1997. *Carcinogenesis*, **18**, 1833–1836.

48 Djuric, Z., Heilbrun, L.K., Reading, B.A., Boomer, A., Valeriote, F.A., and Martino, S. (1991) Effects of a low-fat diet on levels of oxidative damage to DNA to human peripheral nucleated blood cells. *J Natl Cancer Inst*, **83**, 766–769.

49 Helbock, H.J., Beckman, K.B., Shigenaga, M.K., Walter, P.B., Woodall, A.A., Yeo, H.C., and Ames, B.N. (1998) DNA oxidation matters: the HPLC-electrochemical detection assay of 8-oxo-deoxyguanosine and 8-oxo-guanine. *Proc Natl Acad Sci U S A*, **95**, 288–293.

50 Ravanat, J.L., Douki, T., Duez, P., Gremaud, E., Herbert, K., Hofer, T., Lasserre, L., Saint-Pierre, C., Favier, A., and Cadet, J. (2002) Cellular background level of

8-oxo-7,8-dihydro-2′-deoxyguanosine: an isotope based method to evaluate artefactual oxidation of DNA during its extraction and subsequent work-up. *Carcinogenesis*, **23**, 1911–1918.

51 Griffiths, H.R., Moller, L., Bartosz, G., Bast, A., Bertoni-Freddari, C., Collins, A., Cooke, M., Coolen, S., Haenen, G., and Hoberg, A.M. (2002) Biomarkers. *Mol Aspects Med*, **23**, 101–208.

52 Borthakur, G., Butryee, C., Stacewicz-Sapuntzakis, M., and Bowen, P.E. (2008) Exfoliated buccal mucosa cells as a source of DNA to study oxidative stress. *Cancer Epidemiol Biomarkers Prev*, **17**, 212–219.

53 Thompson, H.J., Heimendinger, J., Gillette, C., Sedlacek, S.M., Haegele, A., O'Neill, C., and Wolfe, P. (2005) In vivo investigation of changes in biomarkers of oxidative stress induced by plant food rich diets. *J Agric Food Chem*, **53**, 6126–6132.

54 Hamilton, M.L., Guo, Z., Fuller, C.D., Van Remmen, H., Ward, W.F., Austad, S.N., Troyer, D.A., Thompson, I., and Richardson, A. (2001) A reliable assessment of 8-oxo-2-deoxyguanosine levels in nuclear and mitochondrial DNA using the sodium iodide method to isolate DNA. *Nucleic Acids Res*, **29**, 2117–2126.

55 Ohtsubo, T., Ohya, Y., Nakamura, Y., Kansui, Y., Furuichi, M., Matsumura, K., Fujii, K., Iida, M., and Nakabeppu, Y. (2007) Accumulation of 8-oxo-deoxyguanosine in cardiovascular tissues with the development of hypertension. *DNA Repair (Amst)*, **6**, 760–769.

56 Evans, M.D., Singh, R., Mistry, V., Sandhu, K., Farmer, P.B., and Cooke, M.S. (2008) Analysis of urinary 8-oxo-7,8-dihydro-purine-2′-deoxyribonucleosides by LC-MS/MS and improved ELISA. *Free Radic Res*, **42**, 831–840.

57 Lunec, J., Herbert, K.E., Jones, G.D.D., Dickinson, L., Evans, M., Mistry, N., Chauhan, D., Capper, G., and Zheng, Q. (2000) Development of a quality control material for the measurement of 8-oxo-7,8-dihydro-2′-deoxyguanosine, an in vivo marker of oxidative stress, and comparison of results from different laboratories. *Free Radic Res*, **33**, S27–S31.

58 Nagy, E., Johansson, C., Zeisig, M., and Moller, L. (2005) Oxidative stress and DNA damage caused by the urban air pollutant 3-NBA and its isomer 2-NBA in human lung cells analyzed with three independent methods. *J Chromatogr B Analyt Technol Biomed Life Sci*, **827**, 94–103.

59 Maatouk, I., Bouaicha, N., Plessis, M.J., and Perin, F. (2006) Optimization of the P-32-postlabeling/thin layer chromatography assay (P-32-TLC) for in vitro detection of 8-oxo-deoxyguanosine as a biomarker of oxidative DNA damage. *Toxicol Mech Methods*, **16**, 313–322.

60 Kundu, L.M. and Loppnow, G.R. (2007) Direct detection of 8-oxo-deoxyguanosine using UV resonance Raman spectroscopy. *Photochem Photobiol*, **83**, 600–602.

61 Russo, M.T., De, L.G., Degan, P., Parlanti, E., Dogliotti, E., Barnes, D.E., Lindahl, T., Yang, H., Miller, J.H., and Bignami, M. (2004) Accumulation of the oxidative base lesion 8-hydroxyguanine in DNA of tumor-prone mice defective in both the Myh and Ogg1 DNA glycosylases. *Cancer Res*, **64**, 4411–4414.

62 Xie, Y., Yang, H., Cunanan, C., Okamoto, K., Shibata, D., Pan, J., Barnes, D.E., Lindahl, T., McIlhatton, M., Fishel, R., and Miller, J.H. (2004) Deficiencies in mouse Myh and Ogg1 result in tumor predisposition and G to T mutations in codon 12 of the K-ras oncogene in lung tumors. *Cancer Res*, **64**, 3096–3102.

63 Toyokuni, S. (2008) Molecular mechanisms of oxidative stress-induced carcinogenesis: from epidemiology to oxygenomics. *Iubmb Life*, **60**, 441–447.

64 Haegele A.D. and Thompson H.J. (1999) Increased fruit and vegetable consumption reduces indices of oxidative DNA damage in lymphocytes and urine, in *Natural Antioxidants and Anticarcinogens in Nutrition, Health and Disease* (eds S.K. Kundu and J. Sambrook), Royal Society of Chemistry, Cambridge, pp. 423–432.

65 Halliwell, B. (1998) Can oxidative DNA damage be used as a biomarker of cancer risk in humans? Problems, resolutions and preliminary results from nutritional supplementation studies. *Free Radic Res*, **29**, 469–486.

66 Halliwell, B. (2000) Why and how should we measure oxidative DNA damage in nutritional studies? How far have we come? *Am J Clin Nutr*, **72**, 1082–1087.

67 Jackson, A.L. and Loeb, L.A. (2001) The contribution of endogenous sources of DNA damage to the multiple mutations in cancer. *Mutat Res*, **477**, 7–21.

68 Hailer-Morrison, M.K., Kotler, J.M., Martin, B.D., and Sugden, K.D. (2003) Oxidized guanine lesions as modulators of gene transcription. Altered p50 binding affinity and repair shielding by 7,8-dihydro-8-oxo-2′-deoxyguanosine lesions in the NF-kappa B promoter element. *Biochemistry*, **42**, 9761–9770.

69 Russo, M.T., De, L.G., Degan, P., and Bignami, M. (2007) Different DNA repair strategies to combat the threat from 8-oxoguanine. *Mutat Res*, **614**, 69–76.

70 Cadenas, S., Barja, G., Poulsen, H.E., and Loft, S. (1997) Oxidative DNA damage estimated by oxo8dG in the liver of guinea-pigs supplemented with graded dietary doses of ascorbic acid and alpha- tocopherol. *Carcinogenesis*, **18**, 2373–2377.

71 Umegaki, K., Ikegami, S., and Ichikawa, T. (1993) Influence of dietary vitamin-e on the 8-hydroxydeoxyguanosine levels in rat-liver DNA. *J Nutr Sci Vitaminol*, **39**, 303–310.

72 de Wind, N., Dekker, M., Berns, A., Radman, M., and te Riele, R.H. (1995) Inactivation of the mouse Msh2 gene results in mismatch repair deficiency, methylation tolerance, hyperrecombination, and predisposition to cancer. *Cell*, **82**, 321–330.

73 Mitra, A.K., Singh, N., Singh, A., Garg, V.K., Agarwal, A., Sharma, M., Chaturvedi, R., and Rath, S.K. (2008) Association of polymorphisms in base excision repair genes with the risk of breast cancer: a case-control study in North Indian women. *Oncol Res*, **17**, 127–135.

74 Sangrajrang, S., Schmezer, P., Burkholder, I., Waas, P., Boffetta, P., Brennan, P., Bartsch, H., Wiangnon, S., and Popanda, O. (2008) Polymorphisms in three base excision repair genes and breast cancer risk in Thai women. *Breast Cancer Res Treat*, **111**, 279–288.

75 Neumann, C.A., Krause, D.S., Carman, C.V., Das, S., Dubey, D.P., Abraham, J.L., Bronson, R.T., Fujiwara, Y., Orkin, S.H., and Van Etten, R.A. (2003) Essential role for the peroxiredoxin Prdx1 in erythrocyte antioxidant defence and tumour suppression. *Nature*, **424**, 561–565.

76 Neumann, C.A. and Fang, Q. (2007) Are peroxiredoxins tumor suppressors? *Curr Opin Pharmacol*, **7**, 375–380.

77 Malins, D.C., Holmes, E.H., Polissar, N.L., and Gunselman, S.J. (1993) The etiology of breast cancer. Characteristic alteration in hydroxyl radical-induced DNA base lesions during oncogenesis with potential for evaluating incidence risk. *Cancer*, **71**, 3036–3043.

78 Thompson, H.J., Heimendinger, J., Diker, A., O'Neill, C., Haegele, A., Meinecke, B., Wolfe, P., Sedlacek, S., Zhu, Z., and Jiang, W. (2006) Dietary botanical diversity affects the reduction of oxidative biomarkers in women due to high vegetable and fruit intake. *J Nutr*, **136**, 2207–2212.

79 Bertagnolli, M.M. (2007) Chemoprevention of colorectal cancer with

cyclooxygenase-2 inhibitors: two steps forward, one step back. *Lancet Oncol*, **8**, 439–443.
80 Desir, G.V. (2005) Kv1-3 potassium channel blockade as an approach to insulin resistance. *Expert Opin Ther Targets*, **9**, 571–579.
81 Johnson, J.A. (2001) Drug target pharmacogenomics: an overview. *Am J Pharmacogenomics*, **1**, 271–281.
82 Reilly, S.M. and Lee, C.H. (2008) PPAR delta as a therapeutic target in metabolic disease. *FEBS Lett*, **582**, 26–31.
83 Fearon, E.R. and Vogelstein, B. (1990) A genetic model for colorectal tumorigenesis. *Cell*, **61**, 759–767.
84 Hanahan, D. and Weinberg, R.A. (2000) The hallmarks of cancer. *Cell*, **100**, 57–70.
85 Bianchini, F., Elmstahl, S., Martinez-Garcia, C., van Kappel, A.L., Douki, T., Cadet, J., Ohshima, H., Riboli, E., and Kaaks, R. (2000) Oxidative DNA damage in human lymphocytes: correlations with plasma levels of alpha-tocopherol and carotenoids. *Carcinogenesis*, **21**, 321–324.
86 Peng, Y.M., Peng, Y.S., Lin, Y., Moon, T., Roe, D.J., and Ritenbaugh, C. (1995) Concentrations and plasma-tissue-diet relationships of carotenoids, retinoids, and tocopherols in humans. *Nutr Cancer*, **23**, 233–246.
87 Prior, R.L. (2003) Fruits and vegetables in the prevention of cellular oxidative damage. *Am J Clin Nutr*, **78**, 570S–578S.
88 Burgering, B. (2008) A brief introduction to FOXOlogy. *Oncogene*, **27**, 2258–2262.
89 Li, X.X. and Stark, G.R. (2002) NF kappa B-dependent signaling pathways. *Exp Hematol*, **30**, 285–296.
90 Li, W.G., Khor, T.O., Xu, C.J., Shen, G.X., Jeong, W.S., Yu, S., and Kong, A.N. (2008) Activation of Nrf2-antioxidant signaling attenuates NF kappa B-inflammatory response and elicits apoptosis. *Biochem Pharmacol*, **76**, 1485–1489.
91 Reddy, S.P. (2008) The antioxidant response element and oxidative stress modifiers in airway diseases. *Curr Mol Med*, **8**, 376–383.
92 Holmes, E. and Nicholson, J.K. (2007) Human metabolic phenotyping and metabolome wide association studies. *Ernst Schering Found Symp Proc*, 227–249.
93 Peng, Y.S. and Peng, Y.M. (1992) Simultaneous liquid chromatographic determination of carotenoids, retinoids, and tocopherols in human buccal mucosal cells. *Cancer Epidemiol Biomarkers Prev*, **1**, 375–382.
94 Britigan, B.E., Roeder, T.L., and Buettner, G.R. (1991) Spin traps inhibit formation of hydrogen-peroxide via the dismutation of superoxide: implications for spin trapping the hydroxyl free-radical. *Biochim Biophys Acta*, **1075**, 213–222.
95 Bai, R.K., Leal, S.M., Covarrubias, D., Liu, A., and Wong, L.J.C. (2007) Mitochondrial genetic background modifies breast cancer risk. *Cancer Res*, **67**, 4687–4694.
96 Li, D.H., Suzuki, H., Liu, B.R., Morris, J., Liu, J., Okazaki, T., Li, Y.N., Chang, P., and Abbruzzese, J.L. (2009) DNA repair gene polymorphisms and risk of pancreatic cancer. *Clin Cancer Res*, **15**, 740–746.
97 Mathers, J.C. and Hesketh, J.E. (2007) The biological revolution: understanding the impact of SNPs on diet-cancer interrelationships. *J Nutr*, **137**, 253S–258S.
98 Morgenstern, R. (2004) Oxidative stress and human genetic variation. *J Nutr*, **134**, 3173S–3174S.
99 Wang, S.S., Davis, S., Cerhan, J.R., Hartge, P., Severson, R.K., Cozen, W., Lan, Q., Welch, R., Chanock, S.J., and Rothman, N. (2006) Polymorphisms in oxidative stress genes and risk for non-Hodgkin lymphoma. *Carcinogenesis*, **27**, 1828–1834.
100 Barton, R.H., Nicholson, J.K., Elliott, P., and Holmes, E. (2008) High-throughput H-1 NMR-based metabolic analysis of human serum

and urine for large-scale epidemiological studies: validation study. *Int J Epidemiol*, **37**, 31–40.

101 Rice-Evans, C. and Miller, N. (1997) Measurement of the antioxidant status of dietary constituents, low density lipoproteins and plasma. *Prostaglandins Leukot Essent Fatty Acids*, **57**, 499–505.

102 Rice-Evans, C.A. (2000) Measurement of total antioxidant activity as a marker of antioxidant status in vivo: procedures and limitations. *Free Radic Res*, **33** (Suppl), S59–S66.

103 Marshall, S. (2006) Role of insulin, adipocyte hormones, and nutrient-sensing pathways in regulating fuel metabolism and energy homeostasis: a nutritional perspective of diabetes, obesity, and cancer. *Sci STKE*, **2006**, re7.

104 Holmes, E., Wilson, I.D., and Nicholson, J.K. (2008) Metabolic phenotyping in health and disease. *Cell*, **134**, 714–717.

105 Li, M., Wang, B., Zhang, M., Rantalainen, M., Wang, S., Zhou, H., Zhang, Y., Shen, J., Pang, X., Zhang, M., Wei, H., Chen, Y., Lu, H., Zuo, J., Su, M., Qiu, Y., Jia, W., Xiao, C., Smith, L.M., Yang, S., Holmes, E., Tang, H., Zhao, G., Nicholson, J.K., Li, L., and Zhao, L. (2008) Symbiotic gut microbes modulate human metabolic phenotypes. *Proc Natl Acad Sci U S A*, **105**, 2117–2122.

106 Martin, F.P., Dumas, M.E., Wang, Y., Legido-Quigley, C., Yap, I.K., Tang, H., Zirah, S., Murphy, G.M., Cloarec, O., Lindon, J.C., Sprenger, N., Fay, L.B., Kochhar, S., van Bladeren, B.P., Holmes, E., and Nicholson, J.K. (2007) A top-down systems biology view of microbiome-mammalian metabolic interactions in a mouse model. *Mol Syst Biol*, **3**, 112.

107 Nicholson, J.K., Holmes, E., and Elliott, P. (2008) The metabolome-wide association study: a new look at human disease risk factors. *J Proteome Res*, **7**, 3637–3638.

108 Thompson, H.J., Heimendinger, J., Sedlacek, S., Haegele, A., Diker, A., O'Neill, C., Meinecke, B., Wolfe, P., Zhu, Z., and Jiang, W. (2005) 8-Isoprostane F2alpha excretion is reduced in women by increased vegetable and fruit intake. *Am J Clin Nutr*, **82**, 768–776.

109 Ambrosone, C.B., Freudenheim, J.L., Thompson, P.A., Bowman, E., Vena, J.E., Marshall, J.R., Graham, S., Laughlin, R., Nemoto, T., and Shields, P.G. (1999) Manganese superoxide dismutase (MnSOD) genetic polymorphisms, dietary antioxidants, and risk of breast cancer. *Cancer Res*, **59**, 602–606.

Appendix
Questions for Discussion

1
What are "endogenous toxins?"

"*Endogenous toxins*" could be defined as toxins formed within the body, which can result in disease. However, as Prestwick and Dedon (Chapter 1) note, there is a "blur" between "endogenous" and "exogenous" toxins and where the toxin is formed. There is also a "blur" with relation to the body and where it begins. Can lipopolysaccharide (LPS) endotoxins formed by common colonic bacteria be considered endogenous toxins? Are compounds formed endogenously in plants that are toxic to the animal consuming them, for example, solanine, psoralins, and the cyanogenic glycosides, "endogenous toxins"? And there is the "blur" of disease. The demonstration of a clear association between suspected toxins resulting from diet and a disease state is not an easy exercise. The problem may be somewhat easier with toxins resulting from metabolic diseases, but even then, demonstration of causation can often be established only with successful intervention. Perhaps, it is easiest to consider the array of endogenous toxins considered in this book and ask what other term might be more useful? This is clearly a huge area of toxicology and a likely cause of many human diseases. The formation and toxicity of "endogenous toxins," with their association with diet, lifestyle, and genetic inborn errors of metabolism on the one hand, and disease on the other, is complex. We can all benefit from the help of others to clarify the role of endogenous toxins however we define them.

2
Why consider "alcohol drinking" in a review of endogenous toxins?

Methanol and formaldehyde, formed *in vivo* as a by-product of their methylation function [1–3], can certainly be considered endogenous toxins, but ethanol would seem to be an exogenous toxin, not an endogenous one. We have considered ethanol in this discussion both because the endogenous toxins formed in the body with the consumption of ethyl alcohol describes the complexity of the relationship between

Endogenous Toxins. Diet, Genetics, Disease and Treatment.
Edited by Peter J. O'Brien and W. Robert Bruce
Copyright © 2010 WILEY-VCH Verlag GmbH & Co. KGaA, Weinheim
ISBN: 978-3-527-32363-0

exogenous and endogenous toxins and because the studies of the toxicity of this simple molecule provide a paradigm that may help to explain other diet-associated endogenous toxins.

Ethanol is oxidized to acetaldehyde and it has been assumed that this compound is responsible for alcoholic liver disease (ALD), including alcoholic steatosis, alcoholic steatohepatitis (ASH), cirrhosis, and hepatic cancer. However, as Thiele and Klassen, and Albano point out (Chapters 4 and 18), the process is much more complex. Ethanol is also oxidized to a reactive hydroxyethyl radical (HER) by various ethanol oxidizing processes such as NADPH-dependent microsomes with iron and oxygen, a Fenton system with Fe^{2+} and H_2O_2, or a peroxynitrite oxidizing system. Another important understanding of the pathogenesis of ALD came from the demonstration that alcohol intake promotes hepatic inflammation by increasing the translocation of gut-derived endotoxins to the portal circulation where they stimulate intrahepatic Kupffer cells (see Rao, Chapter 19). "Endotoxins" are LPSs derived from the cell walls of gram-negative bacteria such as *Escherichia coli*. "Translocation" is a passive process in which LPS penetrates through the gut epithelium. Rao describes experiments that model the effect of ethanol on epithelial membrane permeability and suggests that endotoxin not "captured" by Kupffer cells results in a low level endotoxemia that could be responsible for colonic and perhaps other inflammatory processes.

3
What endogenous toxins are involved in the development of type 2 diabetes?

The three chapters on diabetes, Chapters 20–22, are each concerned with quite different endogenous toxins. What is the reason for this apparent inconsistency? A possible explanation is that different phases of the disease are considered in the three chapters. Cani and Delzenne (Chapter 20) review the interesting studies of the Brussels and Toulouse groups relating to the early development of the disease. They review the historic studies relating high-fat (and high sugar) diet(s) with obesity, insulin resistance, and the eventual development of type 2 diabetes mellitus (T2DM). The extensive studies by J.I. Gordon's group showed that high-fat diets typically induce obesity, hyperinsulinemia, and diabetes in normal experimental animals, but do not in animals that are germ free [e.g. 4]. This observation suggested the possibility that endotoxemia was involved in the development of insulin resistance and T2DM, a supposition supported by animal and clinical studies (Chapter 20). The latter showed that like ethanol, dietary fat could increase circulating LPS, and that circulating LPS was associated (in normal males) with increased dietary calories. Certainly the association of circulating LPS, obesity and diabetes risk deserves much further study, particularly as it offers a clear rationale for the use of pre- and probiotics in the endotoxemia, and thus diabetes risk.

The development and complications of T2DM is considered further by Tang, Oprescu and Giacca, and by Beisswenger (Chapters 21 and 22). The approaches taken are in a sense complementary. Giacca's group focus on the cytotoxic effects of chronic elevation of glucose and free fatty acids (FFAs), noting specifically the toxicity to the β-cells, effects on β-cell mass, and the eventual development of disease. Toxicity is attributed to an increase of oxidative stress with a decrease of glutathione and intracellular antioxidants in cells that are intrinsically sensitive to oxidative stress. They cite *in vitro*, animal, and clinical studies that demonstrate the protective effects of agents that reduce oxidative stress on the preservation of the ability to produce insulin. Beisswenger's interest is in the later stages of diabetes complications, those resulting in kidney failure, blindness and amputations, and increasing the rates of heart attack and stroke. He reviews the mechanisms resulting in protein glycation as early glycation products (EGPs), advanced glycation end products (AGEs), and deglycation reactions. Methylglyoxal, an endogenous toxic dicarbonyl that glycates proteins is formed mostly by glycolytic metabolism from glucose. These observations gain importance when one notes that insulin resistance and diabetes comprise a part of the "metabolic syndrome" including obesity, nonalcoholic fatty liver disease (NAFLD), hypertension, cardiovascular disease, all common public health problems in Western countries and more recently in Japan.

4
What endogenous sugar/fatty acid toxins are involved in the development of nonalcoholic steatohepatitis (NASH)?

There is a similar problem in considering endogenous toxins in the development of NASH and hepatitis associated with hepatitis C virus. Several different mechanisms are presented but it is not clear whether all the mechanisms could be operative or whether they are all different ways of looking at a similar problem. Jaeschke (Chapter 10) describe the role of inflammatory cells and NADPH oxidases in the formation of hypochlorite and peroxynitrite and hepatotoxicity. Hyogo, Yamagishi and Tazuma (Chapter 23) note that AGEs can be toxic to RAGE receptor–containing hepatocyte cultures *in vitro*, and that a significant elevation of the concentration of plasma glyceraldehyde–derived AGE products is found in the plasma of individuals with NASH. These observations suggest that the development of NASH from the very prevalent NAFLD depends on the oxidative stress elicited by the response of hepatic cells to circulating AGEs, perhaps through the interaction of AGEs with receptors (RAGE) in the liver.

The systematic study of the formation of AGEs from glucose described by Usui, Watanabe and Hayase (Chapter 8) describes a large number of compounds that are formed through metabolic and oxidative processes and the formation and characterization of adducts on proteins. The toxicity of the short chain sugars is also considered by Benov (Chapter 6) and the formation of adducts is further considered in Frizzell and Baynes' chapter on modification of cysteine residues in protein (Chapter 2). O'Brien *et al.* (Chapter 7) also consider the development of NASH and the 2-hit model of toxicity based on their evidence for the synergistic toxicity of fructose metabolites (e.g., glyoxal) and oxidative stress.

It is difficult at this stage to see how these various phenomena fit together. The evidence of Chapter 8 is consistent with an important role of the pathway to AGEs, but are the AGEs the ultimate endogenous toxins, or is the problem the earlier toxic carbonyls of the short chain sugars, or the even earlier oxidative stress giving rise to them? AGEs provide a robust marker of risk, but are they the toxins responsible for NASH? The measures that are available with structural studies as described by Prestwick and Dedon (Chapter 1), Frizzell and Baynes (Chapter 2), Siraki and Ehrenshaft (Chapter 3), and Usui, Watanabe and Hayase (Chapter 8) could provide the answer if combined with epidemiological studies.

Chronic infection with hepatitis virus C certainly results in an extreme oxidative stress as assessed by Chan and Bilodeau (Chapter 24). In this situation, is the 2-hit model at all applicable? Does the degree of damage depend to any degree on the formation of reactive short chain sugars or AGE?

5
Inflammation and oxidative stress are involved in many disease processes. Are these chronic diseases a direct result of oxidative stress or are they a result of even more reactive endogenous toxins formed with the interaction of oxidative stress with labile metabolites?

Martin and Boyd (Chapter 26) make a strong case for the importance of oxidative stress in the development of breast cancer, but it leaves the question of why breast tissue is particularly at risk. The reviews of Mathews and Wright (Chapters 9 and 35) suggest that estrogen metabolites could be such labile compounds. They could act synergistically with oxidative stress on breast epithelial tissue to produce genotoxic endogenous toxins specifically in estrogen-sensitive tissues. Such a mechanism could perhaps explain the high incidence of premenopausal breast cancer associated with components of the Western diet during the period of rapid mammary tissue growth.

Eyssen et al. (Chapter 27) suggest a rather different mechanism for the origin of colorectal cancer (CRC), a model more closely related to that of insulin resistance and NASH as noted above. They point out that the known risk factors for CRC expose the colon to both oxidative stress and energy excess, and suggest that risk could be associated with the interaction of these factors with the formation of reactive carbonyls.

Gieseg, Crone and Amit (Chapter 25) show that oxidation of low density lipoprotein to form oxLPL is an important event in the development of atherosclerotic plaques. This process seems to involve the oxidation of sterols and not the potentially more concentrated polyunsaturated fatty acids that are typically protective against risk of cardiovascular, diabetes, and colon and breast cancer.

The dopamine-derived neurotoxicity seen in Parkinson's disease is also associated with oxidative stress. Labandeira-Garcia (Chapter 28) suggests that the source of oxidative stress is elevated in the presence of inadequate regulation of dopamine levels. Doorn (Chapter 29) suggests that a reactive intermediate in dopamine metabolism, 3,4-dihydroxyphenylacetaldehyde, may be the source of neurotoxicity in the presence of oxidative stress. Suhr and Kim (Chapter 30) argue for the importance of tetrahydropaveroline, and its oxidation to the o-quinone, as the important, oxidative stress-sensitive step in the development of Parkinson's disease.

Oxidative stress also seems an important factor in life expectancy and aging as observed by Pamplona's group's studies (Chapter 32) with protein restriction, and by Kong's group's studies (Chapter 34) with the use of natural dietary compounds in the prevention of oxidative stress–induced diseases.

All these studies point to the likely importance of the control of oxidative stress in disease prevention as Ryan and Thompson emphasize (Chapter 36). Reduction of oxidative stress through the identification and reduction of its source from circulating lipopolysaccharide (Chapters 19, 20), from dietary heme in low calcium diets (see Corpet, Chapter 5), or from exogenous xenobiotics, and the optimal use of antioxidants, is one approach. Another could be the control and reduction of labile metabolites (energy substrates, estrogen-derived catechols, or dopamine metabolites).

Indeed the concept of an enhanced toxicity associated with the combined effects of oxidative stress and labile metabolites may have wide applicability. For instance, M. Fields and her colleagues found that diets with copper deficient diets and a fructose carbohydrate source had unexpected toxicities in rats as they resulted in hepatic toxicity, glucose intolerance, hypertriglyceridemia and hypercholesterolemia, pancreatic atrophy, arterial lesions and cardiac hypertrophy, inflammation and fibrosis, in addition to anemia [5–7]. The copper deficiency resulted in increased oxidative stress as a result of decreased superoxide dismutase activity and increased hepatic iron. Fructose is rapidly absorbed and forms reactive carbonyl metabolites [Chapters 6–8]. The resulting endogenous carbonyl toxins have many deleterious effects.

6
Endogenous toxins resulting from inborn errors of metabolism appear to be involved in an increasing number of diseases. Can dietary modification (supplements or deletions) reduce the levels of these toxins?

Potentially toxic intermediate metabolites can accumulate and behave as endogenous toxins in some micronutrient deficiencies. This could be the case in deficiencies of folate and B-vitamins, which result in increased cellular and plasma levels of potentially toxic homocysteine (Hcy) (see Basseri, Caldwell, Sengupta, and Austin, Chapter 14), hyperhomocysteinemia, and increased risk of cardiovascular disease (see Herrmann and Obeid, Chapter 33). Intermediate metabolites can also become elevated through increased flows of energy substrates, perhaps as a consequence of high glycemic index diets, or through the reduced function of a metabolizing enzyme, perhaps as a consequence of oxidative damage. For example, the potentially toxic glucose and fructose metabolites, dihydroxyacetone and glyceraldehyde, are normally rapidly oxidized by glyceraldehyde phosphate dehydrogenase (GAPDH), but GAPDH activity can be inhibited by oxidative stress. These processes are described by Benov (Chapter 6), O'Brien et al. (Chapter 7), and Hayase (Chapter 8). Diets that decrease the sudden high levels of hexose phosphates and oxidative stress, with low glycemic index, antioxidant containing diets, might be expected to reduce these risks. Such dietary interventions could also reduce oxalate levels as described by Danpure (Chapter 11).

Similar arguments could well apply to other diseases resulting from disorders of metabolism. The identification of the endogenous toxin and its source makes it possible to conceive and test for appropriate interventions. The paradigm is perhaps best illustrated by the use of dietary interventions for inborn errors of metabolism such as phenylketonuria as described by Subrahmanyam (Chapter 12). The recent demonstration of the importance of dietary sucrose on the development of gout provides another example (Strazzullo, Chapter 15). Disorders of iron and copper metabolism described by Templeton (Chapter 16) are approached this way and mitochondrial fatty acid oxidation disorders described by Sharer (Chapter 13) could be similarly approached. Interventions for chemically induced autoimmunity (Montestier, Chapter 31) and the polyglutamine neuropathies (Henderson, Chapter 17) pose more complex problems as the presumed endogenous toxins interact with immunological and neuronal cell death/cell replacement/stem cell proliferation mechanisms, respectively. Still even in these situations, measurements of endogenous toxin concentrations could provide guidance for the development of dietary interventions.

Bruce Ames with his usual perspicacity has suggested that micronutrient requirements sufficient to supply cellular energy may be insufficient to assure accurate cellular reproduction, thus assuring the short-term survival of the individual though compromising the prospects for long life [8]. In terms of our discussion, this is a suggestion that in the short term, the genotoxic effects of endogenous toxins are less important than the cytotoxic. That is, micronutrient deficiencies might result in increased genotoxic endogenous toxins under conditions in which populations still thrive. Individuals with borderline deficiencies and elevated levels of endogenous toxins would nonetheless be eliminated with chronic disease as the population ages.

7
Do differences in endogenous toxins explain the differences in chronic disease incidence with time and across populations in the world?

The incidence of chronic diseases such as ischemic heart disease, cerebral infarction, and CRC can differ greatly between countries and with time [9–11]. This could be a consequence of the widespread adoption of refined diets – diets characterized by excessive consumption of energy-rich, micronutrient-poor foods – and of a sedentary lifestyle. On the basis of the references cited above, we might expect that such diets and lifestyles would be associated with significant and possibly increasing levels of endogenous toxins and that the endogenous toxins could increase the risk of these chronic diseases within industrialized countries. But, at present we have only hypotheses. We need clear targets, endogenous toxin concentrations in different organs and cell populations, in the liver, bone marrow, colon, breast, brain etc. assessed with solid laboratory methods. We also need to know that these levels can be assessed from available body fluids for evaluation of population endogenous toxin levels. We have hardly started work on these problems.

We posed these questions, with what we hope was provocative initial discussion, to begin a wider, more thorough discussion in the future. We urge you all to add your thoughts as to the origin and importance of endogenous toxins.

>Peter J O'Brien and W. Robert Bruce
>mailto:peter.obrien@utoronto.ca and wr.bruce@utoronto.ca

References

1. Lee, E.-S., Chen, H., Hardman, C., Simm, A., and Charlton, C. (2008) Excessive S-adenosyl-L-methionine-dependent methylation increases levels of methanol, formaldehyde and formic acid in rat brain striatal homogenates: Possible role in S-adenosyl-L-methionine-induced Parkinson's disease-like disorders. *Life Sci*, **83**, 821–827.
2. Haffner, H.T., Graw, M., Besserer, K., and Stränger, J. (1998) Curvilinear increase in methanol concentration after inhibition of oxidation by ethanol: significance for the investigation of endogenous methanol concentration and formation. *Int J Legal Med*, **111**, 27–31.
3. Conaway, C.C., Whysner, J., Verna, L.K., and Williams, G.M. (1996) Formaldehyde mechanistic data and risk assessment: endogenous protection from DNA adduct formation. *Pharmacol Ther*, **71**, 29–55.
4. Backhed, F., Manchester, J.K., Semenkovich, C.F., and Gordon, J.I. (2007) Mechanisms underlying the resistance to diet-induced obesity in germ-free mice. *Proc Natl Acad Sci USA*, **104**, 979–984.
5. Fields, M., Ferretti, R.J., Reiser, S., and Smith, J.C. (1984) The severity of copper deficiency in rats is determined by the type of dietary carbohydrate. *Proc Soc Exp Biol Med*, **175**, 530–537.
6. Redman, R.S., Fields, M., Reiser, S., and Smith Jr., J.C. (1988) Dietary fructose exacerbates the cardiac abnormalities of copper deficiency in rats. *Atherosclerosis*, **74**, 202–214.
7. Fields, M. (1998) Nutritional factors adversely influencing the glucose/insulin system. *J Am Coll Nutr*, **17**, 317–321.
8. Ames, B.N. (2006) Low micronutrient intake may accelerate the degenerative diseases of aging through allocation of scarce micronutrients by triage. *Proc Natl Acad Sci*, **103**, 17589–17954.
9. Kida K., Ito T., Yang S.W., and Tanphaichitr V (1996) Effects of Western diet on risk factors of chronic disease in Asia, in *Preventive Nutrition: The Comprehensive Guide for Health Professionals* (eds A. Bendich and R.J. Deckelbaum), Humana Press, Totowa, pp. 523–535.
10. Parkin, D.M., Bray, F., Ferlay, J., and Pisani, P. (2005) Global cancer statistics, 2002. *CA Cancer J Clin*, **55**, 74–108.
11. Kono, S. (2004) The secular trend of colon cancer incidence and mortality in relation to fat and meat intake in Japan. *Eur J Cancer Prev*, **13**, 127–132.

Index

a

Abdominal fatness 673–680
Aceruloplasminemia 405–406
Acetaldehyde detoxification in liver 455
Acetaminophen hepatotoxicity 254
- mechanism 255
Acetylcholine 753
Acyl hydrolases 303
Adenosine diphosphate 773
Adiponectin 462
Adipose tissue development, gut microbiota and 513
Adiposity 660
Advanced glycation end products (AGEs), 577–589, *See also* C-7 serum AGEs associated with NASH
- association in liver 579–581
- characteristics 578–579
- circulating 581–584
- – in liver cirrhosis 581
- – in NASH 581–584
- formation 136–137, 178, 560
- formation pathways of 578–579
- – *in vitro* 579
- – *in vivo* 579, 580
- in hepatic sinusoidal endothelial cells 584
- in hepatic stellate cells 584–585
- in hepatocytes 585–588
- – AGEs-elicited insulin resistance in, molecular mechanisms 587
- – CRP gene induction in AGEs-exposed Hep3B cells 585
- – GW9662 in 586
- – IRS-1 phosphorylation 586
- – telmisartan in 586–587
- as NASH biomarker 578
- removal 562–563
Advanced lipoxidation end products (ALEs) 138
African siderosis 402

Aggregation, polyglutamine 430–432
Aging 769
Agonist 861
AhR, *See* Aryl hydrocarbon receptor (AhR)
Alanine 191
Alcohol dehydrogenase (ADH) 453
Alcohol hepatotoxicity, mechanisms 460–474
- alcoholic steatosis, 460–464, *See also* Steatohepatitis, alcoholic; Steatosis, alcoholic
Alcoholic drinks 679
- effects on tetrahydropapaveroline 734–735
Alcoholic endotoxemia, 486–490, *See also* Ethanol-induced endotoxemia
- in ALD patients 486–488
- – in human subjects 487
- bacterial translocation 489–490
- in experimental models of ALD 488–489
- gender differences in 489
- influence on organs 493–495
- – liver 493–494
- – lung 495
- – pancreas 495
Alcoholic liver disease (ALD), 451–452, 485, 908, *See also* Alcohol hepatotoxicity; Ethanol metabolism and toxicity
- binge drinking 455
- dose–effect relationship 452
- endotoxemia in 486–488
- – in experimental models of ALD 488–489
- – in human subjects 487
- in females 452–453
- genetic factors influencing 453
- histological features 452
- homocysteine toxicity 815–816
- inflammatory reactions in, mechanism promoting 464–466
- in males 452

Aldehyde dehydrogenase 189
Aldehyde reductase 188
Aldolase 179
Aldose reductase 188
ALEs, *See* Advanced lipoxidation end products (ALEs)
α1-Antitrypsin 756
α-Dicarbonyls 562
– overproduction, as diabetic complication 566–569
α-Naphthylisothiocyanate 256
α-Synuclein 700–702, 722
AMA, *See* Antimitochondrial autoantibodies (AMA)
Amadori products 558
– deglycation systems for 561
– impaired deglycation of 564–566
Aminoacetone 193
Aminochromes 701
5-Aminoimidazole-4-carboxamide ribonucleotide (AICAR) 462
Aminolevulinic acid (ALA) 403
Aminophospholipids 777
2-Aminopropanolamine 195
Amiodarone 299
Amitriptyline 299
Ammonia 717
Amodiaquine 750
AMP-dependent protein kinase (AMPK) 461–462
Ampicillin 299
Amyotrophic lateral sclerosis (ALS) 409
Anabolic steroids 299
Analogs 794
Androstenedione 871
Animal models of polyglutamine diseases 425–429
– *C. elegans* 426–427
– *Drosophila* 426–427
– fly 426–427
– manipulation strategies 426
– model dependence, critical issues in 427–429
– – protein expression 428
– – unbiased screening methods 428
– murine models 427
– worm 426–427
Antichromatin autoantibodies 751
Antiestrogens 874
Antifibrillarin 757
Antigen presenting cells (APCs) 751
Antilaminin autoantibodies 753
Anti-lipoic acid antibodies 752

Antimitochondrial autoantibodies (AMA) 752
Antimycin C 600
Antioxidant 773, 883, 891
– defense line 774
– enzymes protective upregulation, as diabetic complication 570
APCs, *See* Antigen presenting cells (APCs)
Apolipoprotein B 623
Apoptosis 759
Arginine vasopressin 294–295
Aromatase 869
Aromatic acid decarboxylase (AADC) 696
Aryl hydrocarbon receptor (AhR) 754
Aspartate 136
Ataxin-1 432
Ataxin-3 435
Ataxin-7 432
Atherosclerotic plaques 620–621
– cell-mediated oxidation 628–630
ATP7A 408
ATP7B 407
Atrophin-1 432–433
Autocrine 662
Autoimmune thyroiditis 754
Autoimmunity, chemically induced 747–761
– animal models of 756–760
– – heavy metals 756–758
– – D-penicillamine 760
– – phthalate 759–760
– – pristane 758–759
– human disease 749–756
– – D-penicillamine-induced myasthenia gravis 752–753
– – drug-induced immune hemolytic anemia 751–752
– – drug-induced lupus 749–751
– – eosinophilic pneumonia 755–756
– – goodpasture's syndrome 754–755
– – hydrocarbon exposure 754
– – mercury 753–754
– – pesticide exposure and 755
– – primary biliary cirrhosis 752
– – systemic 753–754

b

Bacterial overgrowth role in ethanol-induced endotoxemia 491
Bacterial translocation, in alcoholic endotoxemia 489–490
Baicalein 499
Bak 634
Basementmembrane proteins 757
Bax 634

β-Cell exhaustion   531
β-Cell glucotoxicity   528–532
- ATP production   531
- β-cell function impairment by chronic glucose exposure   529–532
- - insulin biosynthesis   530–531
- - insulin gene transcription   529–530
- β-cell mass   532
- chronic effect of glucose on β-cell function and mass   528–529
- late stages of insulin secretion   531–532
β-Cell insulin resistance   542
β-Cell level, type 2 diabetes at, 525–543, *See also under* Type 2 diabetes
β-Cell lipotoxicity   532–535
- β-cell function impairment by FFAs exposure   533–535
- - ATP production   534
- - insulin biosynthesis   533–534
- - insulin gene transcription   533
- β-cell mass   535
- chronic effect of FFAs on β-cell function and mass   532–533
- late stages of insulin secretion   534–535
- oxidative stress role in   538–540
β-Thalassemia   143
Benign recurrent intrahepatic cholestasis (BRIC)   295–296
Betaine aldehyde dehydrogenase   195
Bicyclol   499
Bifidobacterium spp.   517
Bile   294
- acid/salt transporters   300–301
- flow   294–295
- salt export pump   301
Bile salt transporters   294
Biliary canaliculi   293
Binge drinking and ALD   455
Biogenic aldehyde hypothesis   718
Biomarkers   868
BMD, *See* Bone mineral density (BMD)
BN, *See* Brown Norway (BN) rats
Body fatness   673–680
Bone mineral density (BMD)   816
Bosentan hepatotoxicity   300–301
Breast cancer carcinogenesis, oxidative stress in   647–663
- association of hormones, mitogens, and mutagens with   655–665
- - autocrine mechanisms   655
- - hypotheses   657
- - incidence and   656
- - levels of blood hormones   658–659
- - levels of growth factors   658–659

- - markers of lipid peroxidation, comparison of MDA and isoprostanes as   661–662
- - urinary isoprostane and MD   660
- - urinary MDA and mammographic density   659–660
- association of mitogens and MDA with   662–663
- mammographic density   652–655
- - breast cancer risk and   653
- - breast tissue composition   654
- - genetic factors   655
- - as a marker of susceptibility   654–655
- markers of   648–650
- - biological sampling   650
- - isoprostanes   649–650
- - MDA   649
- risk factors   651–652
Breast tissue age   654
6-Bromohexanoate   752
Brown Norway (BN) rats   757
Browning reaction   182, 775
Bsep   302
Buccal mucosal cells   895

C
C-7 serum AGEs associated with NASH, 577–589, *See also* Advanced glycation end products (AGEs); Nonalcoholic steatohepatitis (NASH)
- AGEs–RAGE system in liver disease, molecular mechanisms   584–588
CAG repeats   436
Caloric restriction   778
Calpain   635
Camellia sinensis   844
Captopril   299
Carbamazepine   299, 750
Carcinogen   862
Carcinogenesis   867
Cardiovascular disease   650
Carnitine palmitoyltransferase-1   184
Catalase   563, 871
- localization in liver   455
Catechol 4-OH E2   871
- redox cycling of   872
Catechol estrogen 4-OH E2   867
Catechol estrogens   662
Catechol formation   870
Catecholamines   663
Catechol-O-methyl transferase (COMT)   662, 700, 865
CD81   597
Cell death, ROS-mediated   253
Cell proliferation   655

Cellular oxidation control 881–898
– antioxidant 883
– – components influencing DNA damage 893
– broader perspective 897–898
– hormesis 883
– interventions, design of 884–889
– – antioxidants 891–893
– – disease pathogenesis 890–891
– – macromolecular target 885–888
– – oxidizing species 884–885
– – target tissue 888–890
– nutrient 883
– nutritional interventions, design of 894–897
– – abnormal levels of oxidative products, identify candidate mechanisms for 896
– – definition of 896–897
– – individuals who are at increased risk for cancer 894
– – induced mutations 895
– – risk for altered signal transduction 895
– – risk for oxidative stress 895
– overall principles 897
– oxidant/oxidizing agent 883
– redox state 883
– reductant/reducing agent 883
– ROS 883
Cellular redox signaling 140
Cellular targets, of alcoholic endotoxemia 496–497
Centenarians 770
Cephalosporins 748
Ceruloplasmin 405, 756
Chemically induced autoimmunity (CIA) 748
Chlormethiazole 471
Chlorpromazine 299, 750
Cholangiocytes 294
Cholesterol 619
Chronic alcoholism 815–816
Chronic glucose exposure, β-cell function impairment by 529–532
– insulin gene transcription 529–530
CIA, See Chemically induced autoimmunity (CIA)
Cirrhosis 452
– alcoholic cirrhosis 469–472
– circulating AGEs in 581
c-Jun N-terminal kinase (JNK) 178, 187
Claudin-1 597
Clavulanic acid 299
Clozapine 309, 750

Cobalamin 792–795
– absorption 794
– deficiency 795
– excretion 794
– folate trap, causes of 798
– homeostasis 794
– metabolic pathways 793
– sources of 794
Codeine 739
Collagen 654
Colorectal cancer 144, 673–686
– endogenous toxins, possible measures of 686
– future research 684–686
– interaction of toxicity 683–684
– lifestyle risk factors 674–684
– – associated with energy excess 675, 681–683
– – evidence 677, 682
– – oxidative stress 675–681
COMT, See Catechol-O-methyl transferase (COMT)
Concavalin A 256, 758
Coombs' test 751
Copper
– Amyotrophic lateral sclerosis (ALS) 409
– ATP7A 408
– ATP7B 407
– copper-related genes 409–410
– – COPT1 410
– – COPT2 410
– – COX17 410
– – cytochrome c oxidase 409
– – SCO1 gene 410
– – SCO2 gene 410
– physiology 406–407
– trafficking 395–411
– – in the hepatocyte 400
– – hereditary hemochromatoses 401–402
Cordarone 299
C-reactive protein 756
Cruciferous vegetables 848
Curcuma longa 846
Curcumin 846
Cyclic hemiketal fructose isomers 180
Cyclooxygenase 649, 842
Cycloserine 802
– cyclosporine-A 299, 633
Cytochrome c oxidase 409
Cytochrome P450 870
Cytochrome P4502E1 isoenzyme (CYP2E1), in ethanol oxidation 454
Cytotoxic T lymphocyte associated antigen-4 (CTLA-4) 454

d

Daidizein 848
DAT, See Dopamine transporter (DAT)
Deferoxamine 143
Dehydrogenase 700
Delayed endotoxin clearance 490–491
δ-Aminolevulinic acid synthase 2 (ALAS2) 403
Dementia 807–808
Dental amalgam 754
Dentatorubral pallidoluysian atrophy (DRPLA) 419
3-Deoxyglucosone (3DG) detoxification 559
– diabetic complications and 568–569
– pathways for 562
Deoxyguanosine 189
DIA, See D-Penicillamine-induced autoimmunity (DIA)
Diabetes 309
Diabetic atheromatoses 182
Diabetic complications, susceptibility to 557–571
– 3DG detoxification 568–569
– α-dicarbonyls overproduction 566–569
– antioxidant enzymes, protective upregulation 570
– endogenous toxins and 557–571
– glucose level elevation 557–558
– glycation/oxidation pathways, 557–558, See also Glycation products
– individuals, variations among 563–566
– – impaired deglycation of Amadori products 564–566
– – slow and rapid nonenzymatic glycation 564
– methylglyoxal, impaired degradation 567–568
– nitrosative stress and 570
– OS and 558
– – and propensity to diabetic nephropathy 569–570
Diabetic nephropathy, propensity to 569–570
Diallylsulfide 678
Dibenzoylmethane 847
Dieldrin 721
Diet 511–520
Dietary components
– alleviating tissue damage in ethanol-induced endotoxemia 499
– – glycine 499
Dietary iron 133
Dietary restriction 778–783
Dietary sugar alcohols 176

Dihydrofolate reductase 796
7,8-Dihydroneopterin 623, 631
1,4-Dihydroxynonane mercapturic acid 142
Dihydroxyacetone 190
– enzymic formation 190
– enzymic metabolism or oxidation 190
– inflammation model 190
– nonenzymic formation 190
– oxidative stress enhanced toxicity 190
– toxicity mechanisms 190
Dihydroxyacetone phosphate (DHAP) 173, 179
3,4-Dihydroxyphenylacetaldehyde (DOPAL) 715
– mechanisms for elevation in levels 720–722
– metabolism 719–720
– as a metabolite of dopamine 718–719
– reactivity with proteins 723–725
– toxicity 722
3,4-Dihydroxyphenylacetic acid (DOPAC) 700
DIIHA, See Drug-induced immune hemolytic anemia
DIL, See Drug-induced lupus (DIL)
Dimethylphenol 860
Dipalmitoylphosphatidylcholine 304
Dityrosine 136
DNA 757
DNA methyltransferase 750
DOPA quinines 701
DOPAC, See 3,4-Dihydroxyphenylacetic acid (DOPAC)
DOPAL, See 3,4-Dihydroxyphenylacetaldehyde (DOPAL)
Dopamine 663, 695, 867
– biosynthesis 715–717
– catabolism 715–726
– transporter 696, 737
Dopamine-derived neurotoxicity 695–705
– aging and 698
– animal models of parkinsonism and 698–700
– dopaminergic alterations 696–697
– dopaminergic neuron degeneration, progress of 697–698
– excitotoxicity 703–704
– Huntington's disease 703–704
– levodopa therapy for PD 697–698
– mechanisms of 700–701
– – cytosolic dopamine 700–701
– – vesicular dopamine 700–701
– neuromelanin formation 701–702
– neuroprotection and 704–705

Dopamine-derived neurotoxicity (contd.)
- neurotransmitter dopamine   695–696
- protein aggregation, interaction between   702–703
- receptor-dependent   703–704
Dopaminergic innervation   696
Dopaminergic neurons   699
- photomicrographs of   699
Dopa-quinone   867
Dorsal vagal nucleus   697
D-Penicillamine   753
D-Penicillamine-induced autoimmunity (DIA)   760
DRPLA/Smith's disease   422
Drug-induced Immune hemolytic anemia (DIIHA)   747
- proposed mechanisms of   752
Drug-induced lupus (DIL)   747, 750–751
- naturally occurring versus   750
- proposed mechanisms of   750
Dubin-Johnson Syndrome   299
Dyskinesia   697
Dystrophia myotonica (DM1)   436

e

EAAs, *See* Excitatory amino acids (EAAs)
EAE, *See* Experimental autoimmune encephalomyelitis (EAE)
EGCG, *See* Epigallocatechin-3-gallate (EGCG)
Ehlers–Danlos syndrome   407
Ehrlich adducts   307
Electron transport chain (ETC)   600
Electrophilic Michael acceptor   875
End stage renal disease (ESRD)   818
Endogenous antagonists   861
Endogenous antioxidants   630–632
Endogenous estrogens, genotoxicity of   859–875
- causes of   862
- enzymatic reactions that affect   870–872
- - catechol formation   870–871
- - creations of aromatase   871–872
- - creations of sulfatase   871
- - P450   871
- - redox cycle   871
- quinone   863–869
- - estrogen concentrations, factors affecting   869–870
- - formation is tumorigenic   863–865
- - hypothesis, experimental support for   865–869
- reactions that reduce   872–874
- - bound estrogen, sulfotransferase creates   874
- - COMT deactivates catecol   872–873
- - GSH, ascorbate, and antioxidant intervention experiments   873–874
- - NQO1 deactivates the quinone   873–874
- women's health   862
Endometrial cancer   875
Endoplasmic reticulum (ER) stress   542
Endoplasmic reticulum–associated degradation (ERAD)   433
Endotoxemia, 485–501, 511–520, *See also* Ethanol-induced endotoxemia
- definition   487
Energy excess   675, 681
Energy harvest, obesity, and metabolic disorders   514–515
Energy homeostasis, gut microbiota and   512–514
Enzymatic deglycation pathways   560–563
- AGEs, removal   562–563
- - RAGE   563
- 3-Deoxyglucosone (3DG), detoxification pathways for   562
- α-dicarbonyls as protective mechanisms   562
- for Amadori products   561
- - FN3K activity   561
- - FN3KRP in   561
- methylglyoxal (MG) detoxification   561–562
- - glutathionine (GSH)   562
- - Glyoxalase I   561–562
- - Glyoxalase II   562
- oxidative products removal, pathways to   563
Enzyme deficiency   293
Eosinophilia-myalgia   756
Eosinophilic pneumonia(EP)   755–756
EP, *See* Eosinophilic pneumonia(EP)
Epigallocatechin-3-gallate (EGCG)   844
Epilepsy   695
Epinephrine   494, 663
Equine estrogens   868
- structure   868
ER stress induction   434–436
ERE, *See* Estrogen response element (ERE)
Erythromycin   299
Escherichia coli   192
ESRD, *See* End stage renal disease (ESRD)
Estradiol   299, 302, 662, 862
Estradiol dehydrogenase   872
Estradiol sulfatase   869
Estradiol sulfate   869
Estradiol sulfotransferase   869
Estradiol-17 β-D glucuronide   302

Estrogen 859
- homeostasis 869
- women's health and 862
Estrogen receptors 862
- significance of 862–863
Estrogen response element (ERE) 861
Estrogen sulfates 869
Estrone 662, 869
Estrone hydrogenase 869
Estrone sulfatase 870
Estrone sulfate 869
ETC, See Electron transport chain (ETC)
Ethanol-induced endotoxemia, 485–501, See also Alcoholic endotoxemia; Tissue damage/injury
- causes 490–493
- - bacterial overgrowth 491
- - delayed endotoxin clearance 490–491
- - increased gut permeability to endotoxins 491–493
- - Kupffer cells role in 490
- - paracellular permeability 492
Ethanol metabolism and toxicity 453–460, 908
- acetaldehyde detoxification 455
- catalase localization 455
- CYP2E1 in ethanol oxidation 454
- cytotoxic T lymphocyte associated antigen-4 (CTLA-4) 454
- ethanol-induced oxidative stress 456–459
- - CYP2E1 expression 456
- - enzymes in 458
- - mitochondrial DNA (mtDNA) 457
- - transition metals in 457
- - vitamin E 458
- liver, metabolism in 453–455
- on methionine metabolism 459–460
- mitochondrial matrix (mtCYP2E1s) 454
Ethinyl estradiol 302, 862
Ethylene glycol 191
Eukaryotes 772
Excitatory amino acids (EAAs) 704
Experimental autoimmune encephalomyelitis (EAE) 75
Extracellular matrix 663

f

Familial intrahepatic cholestasis 1 (FIC1) 297
Fatty liver 815–816
Fenton chemistry 135
Ferric nitrilotriacetate 138
Ferric porphyrin 141
Ferritin 250

Ferroportin 398
Ferrotoxic disease 143
Fibrillarin 754
Fibrinogen 756
Fibroblasts 663
Fibrogenesis, alcohol-induced 470
Fibronectin 663
Fibrosis 756
FIC1, See Familial intrahepatic cholestasis 1 (FIC1)
FICZ, See 6-Formoindolo[3,2-b]carbazole (FICZ)
Flavanoids 843
Flaviviridae 597
Flush syndrome 455
Fly as polyglutamine diseases animal model 426–427
- C. elegans 426
- Drosophila 426
Foam cells 620
Folate 792
- biological roles 796
- deficiency 797
- metabolism 795–796
- trap, causes of 798
Folic acid 795
- derivatives 796
Formaldehyde 907
6-Formoindolo[3,2-b]carbazole (FICZ) 755
Fragile X tremor/ataxia syndrome (FXTAS) 436
Free fatty acids (FFAs) 525
- acute effect on insulin secretion 526–527
Friedreich ataxia 402
Fructans 176
Fructokinase 137
Fructosamine-3-kinase–related protein (FN3KRP) 561, 564
Fructose 173–203
- cancer risk 202
- disease prevention
- - versus fruits and vegetables 203
- fructose-oxalate path 180
- genotoxicity 202
- health concerns associated with 176–178
- - obesity 177–178
- - sugar and sugar intolerances 176–177
- increased consumption 174–176
- rat hepatocyte studies 188–201
- - dihydroxyacetone 190
- - glyceraldehyde 188–190
- - glycolaldehyde 190–192

Fructose (contd.)
- – glycolate as a peroxisomal ROS source 198
- – glyoxal 193–197
- – glyoxylate as a carbonyl toxin 198–200
- – methylglyoxal 193–197
- – oxalic acid 200–201
- sugars 178–188
- – *In Vivo* liver toxicity studies in rats 187–188
- – NASH 183–187
- – oxidation to endogenous genotoxic carbonyl products 179–183
- – oxidation to protein AGEs 179–183
- – rodent models for fatty liver 183–184
- – versus glucose metabolism 178–179

Fruit sugar 179
Furanose 179

g

γ–Interferon 619
γ–Radiolysis 136
Garlic organosulfur compounds 850
Garlic 678
Gasoline 754
Gastrointestinal permeability in ethanol-induced endotoxemia 492–493
Gender differences in endotoxemia 489
Genetic damage 655
Genetic polymorphisms 650
Genistein 843, 848
Genotoxicity 202
Glomerulonephritis 754
Glomerulosclerosis 759
Glucolipotoxicity 535–536
- fatty acid esterification products 536
- oxidative stress role in 538–540

Glucose
- acute effect on insulin secretion 526–527
- food content 175
- level elevation, diabetes and 557–558
- versus fructose 178–179

Glucose oxidation 540
Glucose-pyruvate glycolysis 180
Glucosinolates 848
Glucotoxicity 525–543
Glucuronide 302
Glutamate 695, 806
Glutathione (GSH) 700, 758
Glutathione peroxidase 1 254
Glutathione transferase 868
Glutathionine (GSH) 562
Glybenclamide 299

Glycation products, 558–560, *See also* Enzymatic deglycation pathways
- AGEs, formation 560
- Amadori products 558
- 3-Deoxyglucosone (3DG) 559
- free radical formation 559
- glyoxal 559
- Methylglyoxal (MG) 558
- slow and rapid nonenzymatic glycation 564
- synthetic pathways for 558–560
- – for oxidation products 559–560

Glycation/oxidation pathways 557–558
Glycemia 513
Glyceraldehyde 137
- detoxication metabolizing enzymes 188–189
- enzymic formation 188
- inflammation model 189–190
- oxidative stress enhanced toxicity 189–190
- toxicity mechanisms 189

Glyceraldehyde3-phosphate dehydrogenase (GAPDH) 558

Glycine
- alleviating tissue damage
- – in ethanol-induced endotoxemia 499

Glycolaldehyde 190–192, 684
- detoxication metabolizing enzymes 191
- enzymic formation 190–191
- inflammation model 192
- oxidative stress enhanced toxicity 192
- toxicity mechanisms 192

Glycolate 198
- detoxication metabolizing enzymes 198
- enzymic formation 198
- toxicity 198

Glycoprotein 597
Glyoxal 190, 193–197, 559
- detoxication metabolizing enzymes 193
- enzymic formation 193
- glyoxalase I and II 195–196
- – aldehyde/aldose reductases 195
- – betaine aldehyde dehydrogenase 195
- – 2-oxoaldehyde dehydrogenase 195
- inflammation model 197
- nonenzymic formation 193
- oxidative stress enhanced toxicity 197
- toxicity mechanisms 196–197

Glyoxalase 195
Glyoxylate 173, 198–200
- detoxication metabolizing enzymes 199
- enzymic formation 198–199
- inflammation model 200

- oxidative stress enhanced toxicity   200
- toxicity mechanisms   199
Goodpasture's syndrome (GS)   754
Granulocyte/macrophage colony stimulating factor (GM-CSF)   495
Grape sugar   179
GS, See Goodpasture's syndrome (GS)
GSH, See Glutathione (GSH)
Guanine   774, 778
Guanosine-5'-triphosphate (GTP)   630
Gut lumen   141–142
Gut microbiota   511–520
- adipose tissue development   513
- energy harvest, obesity, and metabolic disorders   512–515
- - paradoxes, question of   514–515
- energy homeostasis   512–514
- glycemia   513
- high-fat feeding   519
- insulinemia   513
- lipogenesis   513–514
- metabolic disorders   516–517
- metabolic endotoxemia   516–517
- role in inflammatory associated with obesity   515–519
- SCFA receptors   514
- selective changes in   518–519

h

Haber Weiss reaction   135
Haloperidol   299
Halothane   256
Hams-F10 media   629
HCB, See Hexachlorobenzene (HCB)
Heart outcomes prevention evaluation (HOPE)   811
Heme iron   134, 141
- experimental carcinogenesis   142
Hemin   141
Hemin effect   141–142
Hemochromatosis
- hereditary hemochromatoses   401–402
- juvenile   401
- type II (HFE2)   401
- type IIB (HFE2B)   401
- type III (HFE3)   402
- type IV (HFE4)   402
Hemojuvelin (HJV)   401
Hemolysis   752
Hemoptysis   754
Hepacivirus   597
Hepatic fibrosis   469–470
Hepatic inflammation, alcohol-induced, immune reactions role in   466–468

Hepatic injury, alcohol-induced, 451–474, See also Alcohol hepatotoxicity; Alcoholic liver disease (ALD); Ethanol metabolism and toxicity
Hepatic porphyria   755
Hepatic sinusoidal endothelial cells, AGEs in   584
Hepatic stellate cells, AGEs in   584–585
Hepatitis C, pathogenesis of   595–609
- conventional treatment   598–599
- - emerging role of antioxidants   608–609
- economical burden   596
- incidence   596
- life cycle in hepatocytes   598
- molecular basis of oxidative stress   599–609
- - adaptive immunity   601
- - innate immunity   601
- - nonphagocyte NOX   603–604
- - phagocyte NADPH oxidase (NOX2)   602–603
- - ROS induced by nonparenchymal cells of the liver   601
- pro-oxidant components   604–608
- - core protein   605–606
- - NS5A protein   606–608
- protein components   597
- transmission   596
- virus   597
Hepatobiliary System   293
Hepatocellular carcinoma (HCC)   452
Hepatocyte inflammation oxidative stress model   185
Hepatocytes, AGEs in, 585–588, See also under Advanced glycation end products (AGEs)
Hepatolenticular degeneration   407
Hepcidin   140, 398–399, 457
Hephaestin   398
Hereditary hemochromatoses   401–402
Hereditary hyperferritinemia cataract syndrome   404–405
Heterogenous nuclear ribonucleoprotein (hnRNP)-A2   759
Hexachlorobenzene (HCB)   755
HFCS, See High-fructose corn syrup (HFCS)
HgIA, See Mercury-induced autoimmunity (HgIA)
High-fat feeding and metabolic endotoxemia   519
High-fructose corn syrup (HFCS)   174–175
Histidine   136
Histone acetyl transferases (HATs)   434
Histones   756

HnRNP, *See* Heterogenous nuclear ribonucleoprotein (hnRNP)-A2
Holohaptocorrin 794
Holotranscobalamin 794
Homeostasis 869
Homocysteine, therapeutic potential for decreasing 789–821
– atherogenic effects 814
– bone health 816–818
– – intervention studies with B vitamins 817–818
– in the endothelial system 812–815
– – protection against cardiovascular diseases 813–815
– – reduction of cardiovascular risk, treatment in 813
– glutamate receptors 806
– hyperhomocysteinemia causes 791–792
– hypomethylation 803–806
– metabolism 789–791
– oxidative stress 806–807
– plasma level, physiological determinants of 791
– toxicity in alcoholic liver disease 815–816
– – fatty liver 815–816
– toxicity in patients with renal diseases 818–821
– – cellular uptake of vitamin B12 820
– – reduced remethylation pathway 819–820
– – treatment on transmethylation 821
– – vascular risk factor 818–819
– – vitamin treatment in dialysis patients 820–821
– toxicity in the CNS 802–812
– – B vitamin supplements, treatment with 808–809
– – in cerebrovascular diseases 810–811
– – dementia and cognitive decline 807–808
– – mechanisms of 803–807
– – Parkinson disease 810
– – stroke risk 811–812
– vitamin B12, role of 792–798
– vitamin B6 798–802
Homocystinuria 816
Homovanilic acid (HVA) 700, 717
HOPE, *See* Heart outcomes prevention evaluation (HOPE)
Hormesis 883
Hormone replacement therapy (HRT) 862
Huntington's disease 420–421
– clinical manifestations 421

Huntington's disease 696, 703–704
HVA, *See* Homovanilic acid (HVA)
Hydralazine 750
Hydrocarbon exposure 754
Hydrogen peroxide 250, 600, 700
Hydroimidazolones (HIs) 560
3-Hydroxyanthranilic acid 630
8-Hydroxy-2-deoxyguanonsine (8-oxodG) 886–888
– detection methods 887
8-Hydroxy-2-deoxyguanosine (8-OHdG) 648
6-Hydroxydopamine (6-OHDA) 698
Hydroxyl radical 700
Hydroxyproline 198
Hydroxypyruvate 189–190
– detoxifying enzymes 191
17β-Hydroxysteroid dehydrogenase type 1 (17β-HSD-1) :871
Hydroxytryptophans 136
Hyperglycemia 528, 559, 570
– antipsychotic-induced 309
– drugs-induced 308
Hyperhomocysteinemia 459, 462, 800, 803
– causes 791–792
– renal function and 818
Hyperphenylalaninemia 305
Hypochlorite 249, 256
– host defense functions 256
– in liver pathophysiology 256
Hypochlorous acid 256, 628
– mediated injury 257
– neutrophil-mediated formation of 258
Hypochromic microcytic anemia with iron overload 405
Hypogonadism 401
Hypohalites 256–259
– hypochlorite 256–259
– sources of hypochlorous acid 256
Hypomethylation 750, 803–806
Hypotransferrinemia 404

i

IEM, *See* Inborn errors of metabolism (IEM)
IHAs, *See* Immune hemolytic anemias (IHAs)
IKK2 183
IL-17, *See* Interleukin-17 (IL-17)
Imipramine 299
Immune hemolytic anemias (IHAs) 751
Immunoglobulins 754
Immunostimulation 754
Inborn errors of metabolism (IEM) 291–292
– of bile acid/salt transporter deficiencies 295–299

- major categories 292
- mechanisms of toxicity 305
Indoleamine 2,3-dioxygenase (IDO) 631
Induced pluriporent stem (iPS) cells 419–439
Inducible nitric oxide synthase (iNOS) 600
Inflammation 910
Inflammation model 189
Inflammatory associated with obesity, gut microbiota role in 515–519
INOS, See Inducible nitric oxide synthase (iNOS)
Insulin biosynthesis 530–531, 533–534, 540
Insulin gene transcription 529–530, 533, 540
Insulin secretory abnormalities in type 2 diabetic patients 527–528
Insulinemia 513
Interferon 596
Interferon-γ 758
Interleukin-17 (IL-17) 755
Intestinal endotoxemia 495
Intestinal fatty acid binding protein (IFABP) 495
Inulin 176
Iron 133–144
- aceruloplasminemia 405–406
- African siderosis 402
- in ethanol-induced oxidative stress 457
- Fe-/Cu-Catalyzed
- - AGEs formation 136–137
- - cellular redox signaling, role in 140
- - Fenton chemistry 135
- - lipid peroxidation and ALE 138
- - oxidation of nucleic acids 138–139
- - oxidation of proteins 135–136
- - oxidative stress cytotoxicity 139
- - protein carbonylation 136
- hereditary hyperferritinemia cataract syndrome 404–405
- hypochromic microcytic anemia with iron overload 405
- hypotransferrinemia 404
- iron fructose complexes 186
- iron overload 143–144
- molecular mechanisms 134–140
- physiology 396–401
- - in humans 396
- - mammalian cells 398
- - uptake in gut 398
- procarcinogenic effects 140–141
- - in cell culture 141
- - in food 140–141
- procarcinogenic effects in the gut 141–142
- - carcinogenesis prevention 142–143
- - free radicals in gut lumen 141
- - hemin effect in the colon of rats 141–142
- - lipid peroxidation products 142
- procarcinogenic effects inside the body 143–144
- - colorectal cancer 144
- - concentration in body fluids 143
- - overload 143
- toxic effects 134–140
- trafficking 395–411
- - diferric transferrin role in 399
- - hereditary hemochromatoses 401–402
- X-linked sideroblastic anemia 403–404
Irritable bowel syndrome (IBS) 177
Isoniazid 750, 802
Isoprostanes 647, 649
- correlation of 661
Ito cells 470

j

JNK/AP1, See c-Jun N-terminal kinase/activator protein-1 (JNK/AP1)
Juvenile hemochromatosis 401

k

Kainate 704
Keap1 843
Kennedy's disease (SBMA) 421–422
Keratinocytes 744
Kidney glomeruli 754
K-ras 887–888
Kupffer cells 252
- role in ethanol-induced endotoxemia 490
Kynurenine 136

l

L-3,4-Dihydroxyphenylalanine (L-DOPA) 696, 716
Laminin-1 757
Large neutral amino acid transporter (LNAAT) 306
LBDs, See Ligand-binding domains (LBDs)
L-buthionine sulfoximine 608
LDL, seeding of 630
L-DOPA paradox 738
L-DOPA therapy 697
Lens proteins 136, 181
Levodopa 695, 697–698
LFA-1, See Lymphocyte function antigen-1 (LFA-1)

Licorice 847
Liehr–Cavalieri hypothesis 874
Life expectancy 769–783
– dietary restriction 778–783
– – caloric restriction 778–779
– – methionine restriction 781–783
– – protein restriction 779–781
– first antioxidant defense line 774–775
– highly resistant structural components 774
– metabolism 771–772
– rate of generation of damage induced by mitochondrial free radicals 772–774
– resistance to oxidative damage 775–778
Ligand-binding domains (LBDs) 860
Lipid peroxidation 142, 650, 697
Lipogenesis 513–514
Lipopolysaccharide endotoxins 907
Lipopolysaccharides (LPS) 485–501, 516
– effect on liver 493–494
– – epinephrine in 494
Lipotoxicity 525–543
Lipoxidation 138
Lipoxygenase 701, 842
Liver
– AGEs association in 579–581
– alcoholic endotoxemia influence on 493–494
– ethanol metabolism in 453–455
– liver cancer and alcohol 472–474
Living organisms 772
LNAAT, See Large neutral amino acid transporter (LNAAT)
Locus coeruleus 697
L-tryptophan 756
Lung, alcoholic endotoxemia influence on 495
Lung alveoli 754
Lymphocyte function antigen-1 (LFA-1) 750
Lymphoproliferation 758
Lysosomal phospholipaseA2 deficiency 303
Lysosomes 139

m

Machado-Joseph disease 423–424
Macrovesicular steatosis 460
Maillard reaction 577–589
Major histocompatibility complex (MHC)-II haplotype 753
Male Wistar rats 183
Mallory bodies formation, alcohol and 469
Malonate 600
Malondialdehyde (MDA) 647, 855
– correlation of 661
– excretion by quintile of percent mammographic density 660
Malonyl-CoA/LC-CoA hypothesis 536
Malonyldialdehyde-acetaldehyde adducts (MAAs) 467
Mammalian alkaloids 733
Mammary carcinoma 869
Mammographic density (MD) 648
– association of mitogens and MDA with 662
– breast cancer risk and 653
– variations in 653
Mandelic acid 195
MAO, See Monoamine oxidase (MAO)
Mass spectrometry (MS) 650
Matrix metalloproteases (MMPs) 470
Mazindol 717
MD, See Mammographic density (MD)
MDA, See Malondialdehyde (MDA)
Mean life span, See Life expectancy
Meat 133–144
Men, ALD in 452–453
Menadione 252
Mercapturic acid pathway 849
Mercury-induced autoimmunity (HgIA) 757–758
– putative mechanisms of 758
– in rodents 757
Mesocortex 697
Metabolic disorders 514–517
Metabolic endotoxemia 516–518
– metabolic disorders 516–517
– nutritional intervention 517–518
Metabolic syndrome 681
Metabonomics 291
'Metagenome' 512
Metallothionein 139
Methamphetamine 698, 721
Methanol 907
Methionine 774, 778, 789–790
– nutritional considerations 783
– restriction 781–783
Methionine metabolism, alcohol effects on 459–460
Methyl trap 798
1-Methyl-4-phenyl-1,2,3,6-tetrahydropyridine (MPTP) 699, 736
1-Methyl-4-phenyl-2,3-dihydropyridinium (MPDP+) 699
1-Methyl-4-phenylpyridinium (MPP+) 699
Methylglyoxal (MG) 179, 137, 173, 558
– detoxication metabolizing enzymes 193–194

- detoxification 561–562
- enzymic formation 193
- glyoxalase I and II 195–196
- – aldehyde/aldose reductases 195
- – betaine aldehyde dehydrogenase 195
- – 2-oxoaldehyde dehydrogenase 195
- impaired degradation, as diabetic complication 567–568
- inflammation model 197
- nonenzymic formation 193
- oxidative stress enhanced toxicity 197
- toxicity mechanisms 196–197

MG, See Myasthenia gravis (MG)
MHC, See Major histocompatibility complex (MHC)-II haplotype
Michael acceptor 846
Microvascularization 698
Microvesicular steatosis 460
Mitochondria 633, 772
Mitochondrial matrix (mtCYP2E1s) 454
Mitogens 655
Monoamine oxidase (MAO) 700, 716
Monounsaturated fatty acyl (MUFA) 774
Morphine 739
MPDP+, See, 1-Methyl-4-phenyl-2,3-dihydropyridinium (MPDP+)
MPO, See Myeloperoxidase (MPO)
MPP+, See 1-Methyl-4-phenylpyridinium (MPP+)
MPTP, See 1-Methyl-4-phenyl-1,2,3,6-tetrahydropyridine (MPTP)
MS, See Mass spectrometry (MS)
MUFA, See Monounsaturated fatty acyl (MUFA)
Murine animal models, polyglutamine diseases 427
Mutagens 655
Myalgias 756
Myasthenia gravis (MG) 748
Myeloperoxidase (MPO) 249, 256, 628, 749
Myxathiazole 600

n

N-acetylcysteine 867, 873
NADH 191
NAFLD, See Nonalcoholic fatty liver disease (NAFLD)
Naproxen 299
1,2-Naphthoquinone 867
NASH, See Nonalcoholic steatohepatitis (NASH)
Natural killer T (NKT) cells 759
Neocortex 697

Nephrotoxicity 304
Neurodegenerative diseases 697
Neuromelanin 701–702
Neuropathological toxicity 697
Neuroprotection 704–705
Neurotoxicity 697
Neurotransmitter dopamine 695
Neurotransmitters 695
N-Formylkynurenine 136
Nicotinamide adenine dinucleotide phosphate oxidase (NOX) 600
- activation of 604
Nigrostriatal pathway 700
Nitric oxide 700
Nitric oxide synthase 249, 253
Nitroblue tetrazolium 723
Nitrosative stress and diabetic complications 570
NK, See Natural killer T (NKT) cells
NMDA, See N-methyl-D-aspartate (NMDA)
N-methyl-D-aspartate (NMDA) 806
Nonalcoholic fatty liver disease (NAFLD) 173
Nonalcoholic steatohepatitis (NASH) 174, 909
- C-7 serum AGEs associated with, 577–589, See also individual entry
- first hit 184
- hepatocyte inflammation oxidative stress model 185–186
- insulin resistance and 578
- pathological mechanisms of 578
- rodent models 183–184
- ROS formation mechanisms 187
- second hit 185–187
- simple steatosis versus 581–583
- – glyceraldehyde-derived AGE serum levels 582–584
- steatosis mechanisms 184
- two-hit hypothesis 184–187
Noncollagenous domain 754
Nonphagocyte NOX 603
Nonsteroidal anti-inflammatory drugs (NSAIDs) 747
Norepinephrine 663, 715
NOX, See Nicotinamide adenine dinucleotide phosphate oxidase (NOX)
Nrf2 843
NS5A protein 606–608
NSAIDs, See Nonsteroidal anti-inflammatory drugs (NSAIDs)
Nutrient 883
Nutritional intervention, metabolic endotoxemia and 517–518

o

Obesity 511–512, 695
- gut microbiota role in inflammatory associated with 515–519
- metabolic disorders 514–515

Olanzapine 309
Old-old 770
oMLT, *See* Oral methionine-loading test (oMLT)
Online Mendelian Inheritance in Man (OMIM) 395
o-Quinones 740
Oral methionine-loading test (oMLT) 792
Organochlorine 721
Orotic acid 183, 200–201
- detoxication metabolizing enzymes 200
- hepatic endogenous metabolite 201
- inflammation model 201
- kidney toxin 201
- oxidative stress enhanced toxicity 201
- toxicogenetic basis 200–201

Osteopontin 465
Oval cells 473
Overgrowth of bacteria in ethanol-induced endotoxemia 491
Oxidative cell death 740–742
Oxidative damage 773
Oxidative stress (OS) as an endogenous toxin 536–538, 696, 910
- breast cancer carcinogenesis 647–663
- definition 599–609
- diabetes and 558
- dihydroxyacetone 190
- ethanol-induced, 456–459, *See also under* Ethanol metabolism and toxicity
- glyceraldehyde 189–190
- glycolaldehyde 192
- glyoxal 197
- glyoxylate 200
- homocysteine 806–807
- induced by HCV core protein 605
- life expectancy 778
- methylglyoxal 197
- oxidative stress, type 2 diabetes, and β-cell dysfunction 537–538
- oxidative stress, β-cell glucotoxicity, and lipotoxicity 538–542
- oxidative stress–induced impairment of β-cell function
- – β-cell insulin resistance 542
- – downstream signaling mechanisms of 541–542
- – endoplasmic reticulum (ER) stress 542
- – glucose oxidation 540
- – by glucotoxicity and lipotoxicity, sites of 540–542
- – insulin biosynthesis 540
- – insulin gene transcription 540
- – JNK 541
- – NFκB 541
- – uncoupling protein 2, 540
- pathogenesis of hepatitis C 595
- related to CRC risk 676–681
- related to lifestyle 676–681
- ROSs
- – as a derived toxin from chronic glucose and fat exposure 536–537
- – sites of generation 537

Oxidative stress, prevention of 841–851
- cytoprotective effect of dietary compounds 843–851
- – garlic organosulfur compounds 850
- – isothiocyanates 848–849
- – phenolic dietary compounds 843–848
- – vitamin E 850–851

Oxidized low density lipoprotein (oxLDL) 137, 619–637
- apoptosis versus necrosis 635–636
- atherosclerotic plaques 628–630
- cell-mediated oxidation 628–630
- cytotoxicity mechanism 632–633
- endogenous antioxidants 630–632
- formation 622–624
- future 636–637
- mitochondria 633
- NADPH oxidase versus necrosis 635–636
- oxysterol 633–635
- – calcium 634–635
- toxicity of 625–626
- types 626–628
- – aggregated 626
- – heavily 626
- – mildly 628
- – minimally 626

oxLDL, *See* Oxidized low density lipoprotein (oxLDL)
2-Oxoaldehyde dehydrogenase 195
Oxo-histidine 136
Oxysterol 138, 628, 632–634
- calcium 634
Oxythiamine 197
Ozone layer 772

p

Pancreas, alcoholic endotoxemia influence on 495
Papaver somniferum 739

Paracellular permeability in ethanol-induced endotoxemia 492
Paracrine 662
Parkin protein 703
Parkinson's disease 695, 715, 867
- effectiveness of MAO 717
- homocysteine 810
- rat model of 699
- relevance of DOPAL 725–726
Parkinsonism 736–738
- inhibition of dopamine biosynthesis 736
- inhibition of dopamine uptake 736–737
- inhibition of mitochondrial respiration 737
- inhibition of serotonin production 738
Pathophysiology 291–309
- accumulated metabolites
- - mechanisms of toxicity associated with 305
- antipsychotic-induced hyperglycemia 309
- drugs and cholestatic liver disease 299–302
- - bosentan hepatotoxicity 300–301
- - cholestatic hepatotoxicity 302
- - troglitazone hepatotoxicity 301
- drugs-induced hyperglycemia 308
- glucose-induced oxidative stress mediated toxicity 307–308
- of hepatobiliary system 293–295
- - arginine vasopressin 294–295
- - bile flow 294–295
- hyperphenylalaninemia 305
- inborn errors of bile acid/salt transporter deficiencies 295–299
- - benign recurrent intrahepatic cholestasis (BRIC) 295–296
- - Dubin-Johnson syndrome 299
- - progressive familial cholestasis 295–298
- - sitosterolemia 299
- phenylketonuria 305
- phospholipases 302–305
- phospholipidosis 302–305
PBC, See Primary biliary cirrhosis (PBC)
Penicillamine 748
Penicillin 748
Peripheral eosinophilia 756
Pernicious anemia 792
Peroxiredoxin 254, 889
Peroxisome proliferator-activated factor-α (PPAR-α) 462
Peroxisome proliferator-activated receptor-γ (PPARγ) 498, 586

Peroxynitrite 249, 700
- acetaminophen hepatotoxicity 254
- sources of 253–254
Phagocyte NADPH Oxidase (NOX2) 602–603
Pharmacotherapy 750
Phenacetin 751
Phenolic dietary compounds 843–848
- curcumin 846
- dibenzoylmethane 847
- EGCG 844–846
- genistein 848
- resveratrol 847–848
Phenoxyl radical 870
Phenylalanine 136
Phenylketonuria 305
- antioxidant status 306
- mitochondrial damage 306
- oxidative stress 306
Phenytoin 299
Phosphatidylserine 625
Phosphodiesterases 303
Phosphofructokinase (PFK) 534
Phospholipases 302–305
- nephrotoxicity 304
Phospholipidosis 302–305
- aminoglycoside antibiotic-induced 304
Phospholipids 757
Phthalates 759
Physical activity 675
Phytoalexin 847
Phytoestrogen 848
Pictet–Spengler condensation 734
Pioglitamide 498
Plant extracts
- alleviating tissue damage in ethanol-induced endotoxemia 499–500
- - cocoa extracts 499
- - tea extracts 499
Polyglutamine (polyQ) neuropathies 419–439
- animal models of, 425–429, See also Animal models
- DRPLA/Smith's disease 422
- future perspectives 437–439
- Huntington's disease 420–421
- - clinical manifestations 421
- Kennedy's disease (SBMA) 421–422
- neurobiology 419–425
- polyglutamine-induced neural injury, mechanisms 429–437
- - CAG repeats 436
- - deubiquitination activity 433

Polyglutamine (polyQ) neuropathies (*contd.*)
– – endoplasmic reticulum–associated degradation (ERAD)   433
– – ER stress induction   434–436
– – polyglutamine aggregation   430–432
– – polyglutamine sequences, conformation   429–430
– – polyglutamine-based disruption of protein function   432–433
– – RNA-mediated toxicity   436
– – toxic gain of pathogenic polyQ proteins function   433–434
– – ubiquitin-mediated proteasomal degradation, disruption   434–436
– questions remaining   436–437
– spinocerebellar ataxia (SCA)
– – SCA1   422–423
– – SCA17   425
– – SCA2   423
– – SCA3/Machado-Joseph disease   423–424
– – SCA6   424
– – SCA7   424–425
Polyglutamine sequences, conformation   429–430
Polyphenols   843
Polyunsaturated fatty acids (PUFAs)   622, 774
Postsynaptic damage   700
Preadipocytes   662
Premarin   867
Primary biliary cirrhosis (PBC)   752
Primary hyperoxaluria type 1   200
Primary mechanism   895
Probiotics, alleviating tissue damage in ethanol-induced endotoxemia   500
Procainamide   750
Progressive familial cholestasis
– type 1   295–296
– type 2   298
– type 3   298
Prolactin   662
Propylthiouracil   750
Prostaglandins   649
Protein restriction   779
Protein tryptophan   136
Protein tyrosine   136
Proteoglycans   663
Pteroylglutamic acid   795
Pteroylmonoglutamic acid   795
PUFAs, *See* Polyunsaturated fatty acids (PUFAs)
Pulmonary hypertension   756
Pyranose   179

Pyridoxal   798
Pyridoxamine   799
Pyridoxine   798
Pyruvate dehydrogenase (PDH)   534

q
Quercetin   676
Quinidine   750
Quinines   701
Quinone   724
– conversion   864
– formation   863–865
– quinone reductase   868

r
RA, *See* Rheumatoid arthritis (RA)
Radical scavengers   896
Randle's cycle   534
RBCs, *See* Red blood cells (RBCs)
RCS, *See* Reactive carbonyl species (RCS)
Reactive carbonyl species (RCS)   173
Reactive nitrogen species (RNS)   253–255, 696
– sources of peroxynitrite   253–255
Reactive oxygen species (ROS)   632, 647, 696
– cellular redox signaling, role in   140
– mechanisms of ROS-mediated cell death   253
– nature   250
– sources   250–252, 842
Receptor for advanced glycation end products (RAGEs)   588
– in liver diseases   588
Red blood cells (RBCs)   751
– formation   799
Redox-cycling agents   251
Redox cycling mechanism   866
Redox homeostasis   873
Redox state   883
Regulatory T cells (Tregs)   755
Remethylation   789
Remethylation pathway   819
Resveratrol   847
Rheumatoid arthritis (RA)   752
Ribavirin   596
Rifampin   299
Rifamycinsv   299
RNA-mediated toxicity   436
RNS, *See* Reactive nitrogen species (RNS)
ROS, *See* Reactive oxygen species (ROS)
Rosiglitazone   308
Rotenone   600

S

S-adenosylhomocysteine 789
S-adenosylmethionine 789
Salicylic acid 722
Salsolinol 733
Saturated fatty acyl (SFA) 774
Scavenger receptors 632
Schiff base 723
Schizophrenia 695–696
SCO1 gene 410
SCO2 gene 410
Selenium 676
Sensory neuropathy 802
Serine 191
Sex hormone binding globulin (SHBG) 658
SFA, See Saturated fatty acyl (SFA)
SHBG, See Sex hormone binding globulin (SHBG)
Sitosterolemia 299
SNc, See Substantia nigra compacta (SNc)
Snowball effect 601
SOD, See Superoxide dismutase (SOD)
Sodium cyanoborohydride 723
Sorbinil 196
Spinal bulbar muscular atrophy (SBMA/Kennedy's disease) 420–421
Spinocerebellar ataxias (SCAs) 420
– SCA1 422–423
– SCA17 425
– SCA2 423
– SCA3 423–424
– SCA6 424
– SCA7 424–425
Splenocytes 750
Splenomegaly 755
Steatohepatitis, alcoholic 184, 464–469
– alcohol-induced hepatic inflammation, immune reactions role in 466–468
– – malonyldialdehyde-acetaldehyde adducts (MAAs) 467
– alcohol-mediated hepatocyte killing mechanisms 468–469
– inflammatory reactions in ALD, mechanism promoting 464–466
– – Kupffer cells in 464–465
– – osteopontin 465
– – STAT3 465
– mallory bodies formation 469
Steatosis, alcoholic 183, 452, 460–464
– hepatic triglyceride synthesis regulation 461
– in humans 463
– macrovesicular 460
– microvesicular 460
– mitochondrial functions, impairment 463
Sterol regulatory element binding protein-1 (SREBP-1) 461
Stroke 811–812
Structural elucidation of polyQ sequences 429–430
Substantia nigra compacta (SNc) 695
Sucrose 175
Sugars 178–188
Sulfamethoxazole 299
Sulfatase 869
Sulfonamides 750
Sulfotransferase 869
Superoxide dismutase (SOD) 250, 701, 871
Synthetic aromatase inhibitors 871
Synthetic drugs
– alleviating tissue damage
– – in ethanol-induced endotoxemia 498–499
Synthetic hydrocarbon ligands 755

t

T helper (Th17) cells 755
TAGEs, See Toxic advanced glycation end products (TAGEs)
Tamoxifen 658, 867
Taurocholate 300
TBA, See Thiobarbituric acid (TBA)
TBARS, See Thiobarbituric acid reactive substance (TBARS)
TCDD, See 2,3,7,8-Tetrachlorodibenzo-p-dioxin (TCDD)
T-cell lymphoproliferation 753
Tea catechins 844
Telmisartan in hepatocytes 586–587
Testosterone 871
2,3,7,8-Tetrachlorodibenzo-p-dioxin (TCDD) 755
Tetracycline 299
Tetrahydropapaveroline 733–744
– biosynthesis of 734–735
– – effects of alcohol 734–735
– – nonenzymatic versus enantioselective formation 734
– cellular protection against injuries 743
– neurotoxic potential 735–743
– – alcohol dependence 739
– – DNA damage 743
– – implications for L-DOPA paradox 738

Tetrahydropapaveroline (*contd*.)
– – neurotoxicity, biochemical mechanisms for 740
– – oxidative cell death 740–742
– – Parkinsonism 736–738
– – redox cycling 740
TGF-β, *See* Transforming growth factor beta (TGF-β)
TH, *See* Tyrosine hydroxylase (TH)
Th17, *See* T helper (Th17) cells
The other side of metabolism 772
Thiobarbituric acid (TBA) 306, 649
Thiobarbituric acid reactive substance (TBARS) 649
Thromboxane 846
Thymidylate synthase 796
Thyroglobulin 757
Tissue damage/injury
– by alcoholic endotoxemia 485–501
– – cellular targets 496–497
– – mechanism of 496–498
– – receptors and signaling 497–498
– – TLR-4 role in 497
– alleviating factors 498–500
– – bicyclol 499
– – dietary components 499
– – glycine 499
– – pioglitamide 498
– – plant extracts 499–500
– – PPARγ 498
– – probiotics 500
– – synthetic drugs 498–499
TLR-4-mediated signal transduction in alcoholic endotoxemia 497–498
TLRs, *See* Toll-like receptors (TLRs)
TNF-α, *See* Tumor necrosis factor alpha (TNF-α)
Tocopherol 624
Toll-like receptors (TLRs) 598
– TLR-7 759
TOS, *See* Toxic oil syndrome (TOS)
Toxic advanced glycation end products (TAGEs) 137
Toxic oil/Toxic oil syndrome (TOS) 756
Transcuprein 407
Transforming growth factor beta (TGF-β) 663
Transporter deficiency 293
– genetic defects 298
Transsulfuration 789
Tregs, *See* Regulatory T cells (Tregs)
Trimethoprim 299
Troglitazone hepatotoxicity 301
Troglitazone sulfate 301

Tuberoinfundibular pathway 696
Tumor necrosis factor alpha (TNF-α) 759
Tumorigenesis 864
Turmeric 846
Two-hit hypothesis 184
– of NASH 578
Type 2 diabetes 908
Type 2 diabetes at β-cell level, 525–543, *See also* β-cell glucotoxicity; β-cell lipotoxicity; Glucolipotoxicity
– insulin secretion
– – acute effect of FFA on 526–527
– – acute effect of glucose on 526–527
– insulin secretory abnormalities in 527–528
– nutrient-derived endogenous toxins in 525–543
Tyrolean infantile cirrhosis 409
Tyrosinase 701
Tyrosine 696
Tyrosine hydroxylase (TH) 696

u

Ubiquinone 250
Ubiquitin-mediated proteasomal degradation (UPS), disruption 434–436
Ubiquitin–proteosome system 696
Ultraviolet light 772
Uncoupling proteins (UCPs) 531, 540

v

Vascular disease 620–622
Vascular endothelial growth factor (VEGF) 698
Vascular thrombosis 756
VEGF, *See* vascular endothelial growth factor (VEGF)
Ventral tegmental area (VTA) 695
Very low density proteins (VLDLs) 463
Vesicular dopamine transporter 2 (VMAT2) 696
Viral hepatitis 595
Vitamin B6 798–802
– deficiency 800–802
– – cardiovascular diseases 800
– – cognitive dysfunction 801
– – hyperhomocysteinemia 800
– – immune dysfunction 801
– – treatment 802
– function 799
– – hormone function 799
– – niacin formation 799
– – nucleic acid synthesis 799
– – red blood cell formation 799

Vitamin B9 795
Vitamin B12 792–798
– cobalamin metabolism 792–795
– folate metabolism 795–798
Vitamin E 676, 850
– in ethanol-induced oxidative stress 458
VMAT2, *See* Vesicular dopamine transporter 2 (VMAT2)
VTA, *See* Ventral tegmental area (VTA)
Vulnerability 698

w

Walcott–Ralliston syndrome 530
Wegener's granulomatosis 753
Western lifestyle 673

Wilson's disease 143, 752, 407–408
Women, ALD in 452–453
Worm as polyglutamine diseases animal model 426–427
– *C. elegans* 426
– *Drosophila* 426

x

Xanthine oxidase 251, 701
X-linked cutis laxa 407
X-linked sideroblastic anemia 403–404